Connecting facets of everyday life to statistics, McClave and Sincich offer a trusted, comprehensive text that stresses the development of statistical thinking, the assessment of credibility, and the value of inferences made from data.

USING TECHNOLOGY

MINITAB: Generating a Random Sample

Step 1 Click on the "Calc" button on the MINITAB menu bar, then click on "Random Data," and finally, click on "Sample From Columns," as shown in Figure 3.M.1. The resulting dialog box appears as shown in Figure 3.M.2.

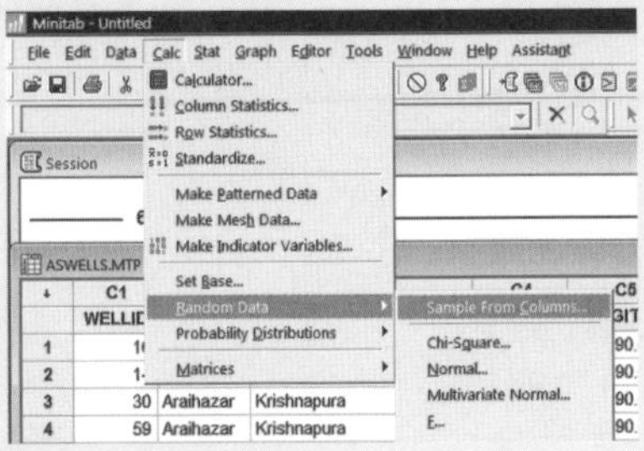

Figure 3.M.1
MINITAB menu options for sampling from a data set

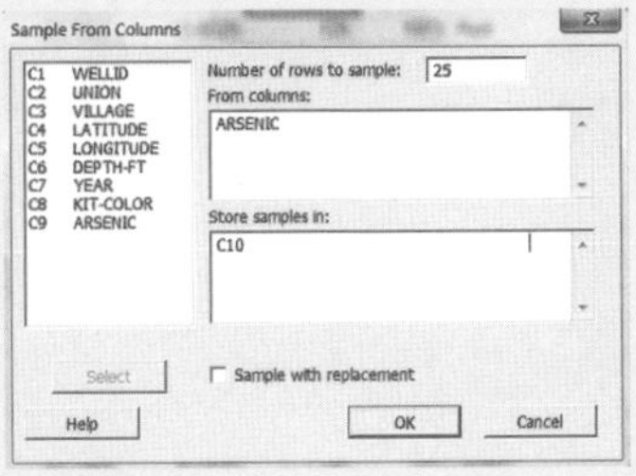

Figure 3.M.2
MINITAB options for selecting a random sample from worksheet columns

Step 2 Specify the sample size (i.e., number of rows), the variable(s) to be sampled, and the column(s) where you want to save the sample.

Step 3 Click "OK" and the MINITAB, worksheet will reappear with the values of the variable for the selected (sampled) cases in the column specified.

In MINITAB, you can also generate a sample of case numbers.

Step 1 From the MINITAB menu, click on the "Calc" button, and then click on "Random Data," and finally, click on the "Uniform" option (see Figure 3.M.1).

Step 2 In the resulting dialog box (shown in Figure 3.M.3), specify the number of cases (rows, i.e., the sample size), and the column where the case numbers selected will be stored.

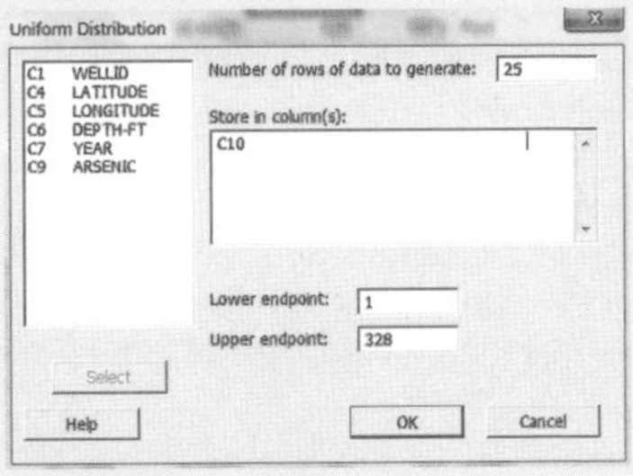

Figure 3.M.3
MINITAB options for selecting a random sample of cases

Step 3 Click "OK" and the MINITAB worksheet will reappear with the case numbers for the selected (sampled) cases in the column specified.

[*Note:* If you want the option of generating the same (identical) sample multiple times from the data set, then first click on the "Set Base" option shown in Figure 3.M.1. Specify an integer in

Page 177

USING TECHNOLOGY sections, at the end of each chapter, offer statistical software tutorials with step-by-step instructions and screen shots for MINITAB® and, where appropriate, the TI-83/84 Plus Graphing Calculator.

MORE THAN 1,800 EXERCISES are available, covering a wide range of applications that underscore the relevance of statistics to everyday life. Most of the exercises incorporate real data, and a 💿 indicates that the data set is available on the companion CD-ROM.

Ethics IN Statistics
Purposeful reporting of numerical descriptive statistics in order to mislead the target audience is considered *unethical statistical practice*.

Page 95

NEW! ETHICS BOXES appear where appropriate to highlight the importance of ethical behavior when collecting, analyzing, and interpreting statistical data.

Applying the Concepts—Intermediate

5.14 Social network densities. Social networks involve interactions (connections) between members of the network. Sociologists define network density as the ratio of actual network connections to the number of possible one-to-one connections. For example, a network with 10 members has $\binom{10}{2} = 45$ total possible connections. If that network has only 5 connections, the network density is $5/45 = .111$. Sociologists at the University of Michigan assumed that the density x of a social network would follow a uniform distribution between 0 and 1 (*Social Networks*, 2010).
   a. On average, what is the density of a randomly selected social network?
   b. What is the probability that the randomly selected network has a density higher than .7?
   c. Consider a social network with only 2 members. Explain why the uniform model would not be a good approximation for the distribution of network density.

5.15 Cycle availability of a system. In the jargon of system maintenance, "cycle availability" is defined as the probability that a system is functioning at any point in time. The U.S. Department of Defense developed a series of performance measures for assessing system cycle availability (*START*, Vol. 11, 2004). Under certain assumptions about the failure time and maintenance time of a system, cycle availability is shown to be uniformly distributed between 0 and 1. Find the following parameters for cycle availability: mean, standard deviation, 10th percentile, lower quartile, and upper quartile. Interpret the results.

5.16 Time delays at a bus stop. A bus is scheduled to stop at a certain bus stop every half hour on the hour and the half hour. At the end of the day, buses still stop after every 30 minutes, but because delays often occur earlier in the day, the bus is never early and is likely to be late. The director of the bus line claims that the length of time a bus is late is uniformly distributed and the maximum time that a bus is late is 20 minutes.
   a. If the director's claim is true, what is the expected number of minutes a bus will be late?
   b. If the director's claim is true, what is the probability that the last bus on a given day will be more than 19 minutes late?
   c. If you arrive at the bus stop at the end of a day at exactly half-past the hour and must wait more than 19 minutes for the bus, what would you conclude about the director's claim? Why?

Page 230

APPLET CORRELATION

Applet	Concept Illustrated	Description	Applet Activity
Sample from a population	Assesses how well a sample represents the population and the role that sample size plays in the process.	Produces random sample from population from specified sample size and population distribution shape. Reports mean, median, and standard deviation; applet creates plot of sample.	**4.4**, 205; **5.1**, 229; **5.3**, 242
Sampling distributions	Compares means and standard deviations of distributions; assesses effect of sample size; illustrates undbiasedness.	Simulates repeatedly choosing samples of a fixed size n from a population with specified sample size, number of samples, and shape of population distribution. Applet reports means, medians, and standard deviations; creates plots for both.	**6.1**, 290; **6.2**, 290
Random numbers	Uses a random number generator to determine the experimental units to be included in a sample.	Generates random numbers from a range of integers specified by the user.	**1.1**, 18; **1.2**, 18; **3.6**, 170; **4.1**, 188; **5.2**, 229

Long-run probability demonstrations illustrate the concept that theoretical probabilities are long-run experimental probabilities.

Simulating probability of rolling a 6	Investigates relationship between theoretical and experimental probabilities of rolling 6 as number of die rolls increases.	Reports and creates frequency histogram for each outcome of each simulated roll of a fair die. Students specify number of rolls; applet calculates and plots proportion of 6s.	**3.1**, 120; **3.2**, 120; **3.3**, 131; **3.4**, 132; **3.5**, 146
Simulating probability of rolling a 3 or 4	Investigates relationship between theoretical and experimental probabilities of rolling 3 or 4 as number of die rolls increases.	Reports outcome of each simulated roll of a fair die; creates frequency histogram for outcomes. Students specify number of rolls; applet calculates and plots proportion of 3s and 4s.	**3.3**, 131; **3.4**, 132
Simulating the probability of heads: fair coin	Investigates relationship between theoretical and experimental probabilities of getting heads as number of fair coin flips increases.	Reports outcome of each fair coin flip and creates a bar graph for outcomes. Students specify number of flips; applet calculates and plots proportion of heads.	**4.2**, 188
Simulating probability of heads: unfair coin ($P(H) = .2$)	Investigates relationship between theoretical and experimental probabilities of getting heads as number of unfair coin flips increases.	Reports outcome of each flip for a coin where heads is less likely to occur than tails and creates a bar graph for outcomes. Students specify number of flips; applet calculates and plots the proportion of heads.	**4.3**, 205
Simulating probability of heads: unfair coin ($P(H) = .8$)	Investigates relationship between theoretical and experimental probabilities of getting heads as number of unfair coin flips increases.	Reports outcome of each flip for a coin where heads is more likely to occur than tails and creates a bar graph for outcomes. Students specify number of flips; applet calculates and plots the proportion of heads.	**4.3**, 205
Simulating the stock market	Theoretical probabilities are long run experimental probabilities.	Simulates stock market fluctuation. Students specify number of days; applet reports whether stock market goes up or down daily and creates a bar graph for outcomes. Calculates and plots proportion of simulated days stock market goes up.	**4.5**, 205
Mean versus median	Investigates how skewedness and outliers affect measures of central tendency.	Students visualize relationship between mean and median by adding and deleting data points; applet automatically updates mean and median.	**2.1**, 57; **2.2**, 57; **2.3**, 57

(Continued)

Applet	Concept Illustrated	Description	Applet Activity
Standard deviation	Investigates how distribution shape and spread affect standard deviation.	Students visualize relationship between mean and standard deviation by adding and deleting data points; applet updates mean and standard deviation.	**2.4**, 64; **2.5**, 64; **2.6**, 65; **2.7**, 85
Confidence intervals for a proportion	Not all confidence intervals contain the population proportion. Investigates the meaning of 95% and 99% confidence.	Simulates selecting 100 random samples from the population and finds the 95% and 99% confidence intervals for each. Students specify population proportion and sample size; applet plots confidence intervals and reports number and proportion containing true proportion.	**7.5**, 325; **7.6**, 325
Confidence intervals for a mean (the impact of confidence level)	Not all confidence intervals contain the population mean. Investigates the meaning of 95% and 99% confidence.	Simulates selecting 100 random samples from population; finds 95% and 99% confidence intervals for each. Students specify sample size, distribution shape, and population mean and standard deviation; applet plots confidence intervals and reports number and proportion containing true mean.	**7.1**, 306; **7.2**, 307
Confidence intervals for a mean (not knowing standard deviation)	Confidence intervals obtained using the sample standard deviation are different from those obtained using the population standard deviation. Investigates effect of not knowing the population standard deviation.	Simulates selecting 100 random samples from the population and finds the 95% z-interval and 95% t-interval for each. Students specify sample size, distribution shape, and population mean and standard deviation; applet plots confidence intervals and reports number and proportion containing true mean.	**7.3**, 316; **7.4**, 316
Hypothesis tests for a proportion	Not all tests of hypotheses lead correctly to either rejecting or failing to reject the null hypothesis. Investigates the relationship between the level of confidence and the probabilities of making Type I and Type II errors.	Simulates selecting 100 random samples from population; calculates and plots z-statistic and P-value for each. Students specify population proportion, sample size, and null and alternative hypotheses; applet reports number and proportion of times null hypothesis is rejected at 0.05 and 0.01 levels.	**8.5**, 385; **8.6**, 385
Hypothesis tests for a mean	Not all tests of hypotheses lead correctly to either rejecting or failing to reject the null hypothesis. Investigates the relationship between the level of confidence and the probabilities of making Type I and Type II errors.	Simulates selecting 100 random samples from population; calculates and plots t statistic and P-value for each. Students specify population distribution shape, mean, and standard deviation; sample size, and null and alternative hypotheses; applet reports number and proportion of times null hypothesis is rejected at both 0.05 and 0.01 levels.	**8.1**, 360; **8.2**, 364; **8.3**, 364; **8.4**, 364
Correlation by eye	Correlation coefficient measures strength of linear relationship between two variables. Teaches user how to assess strength of a linear relationship from a scattergram.	Computes correlation coefficient r for a set of bivariate data plotted on a scattergram. Students add or delete points and guess value of r; applet compares guess to calculated value.	**11.2**, 585
Regression by eye	The least squares regression line has a smaller SSE than any other line that might approximate a set of bivariate data. Teaches students how to approximate the location of a regression line on a scattergram.	Computes least squares regression line for a set of bivariate data plotted on a scattergram. Students add or delete points and guess location of regression line by manipulating a line provided on the scattergram; applet plots least squares line and displays the equations and the SSEs for both lines.	**11.1**, 561

STATISTICS

STATISTICS

TWELFTH EDITION

James T. McClave

Info Tech, Inc.

University of Florida

Terry Sincich

University of South Florida

Boston   Columbus   Indianapolis   New York   San Francisco   Upper Saddle River
Amsterdam   Cape Town   Dubai   London   Madrid   Milan   Munich   Paris   Montréal   Toronto
Delhi   Mexico City   São Paulo   Sydney   Hong Kong   Seoul   Singapore   Taipei   Tokyo

Editor in Chief: Deirdre Lynch
Acquisitions Editor: Marianne Stepanian
Associate Content Editor: Dana Bettez
Editorial Assistant: Sonia Ashraf
Senior Managing Editor: Karen Wernholm
Senior Production Project Manager: Tracy Patruno
Associate Director of Design, USHE North and West: Andrea Nix
Text and Cover Designer: Barbara T. Atkinson
Digital Assets Manager: Marianne Groth
Production Coordinator: Katherine Roz
Media Producer: Jean Choe
Software Developers: Mary Durnwald and Bob Carroll
Marketing Manager: Erin Lane
Marketing Assistant: Kathleen DeChavez
Senior Author Support/Technology Specialist: Joe Vetere
Rights and Permissions Advisor: Michael Joyce
Image Manager: Rachel Youdelman
Procurement Manager: Evelyn Beaton
Procurement Specialist: Linda Cox
Senior Media Procurement Sepcialist: Ginny Michaud
Production Coordination, Composition, Illustrations: Integra
Cover Image: Crowd of small symbolic 3d figures linked by lines, ©Higyou/Shutterstock

Credits appear on page P-1, which constitutes a continuation of the copyright page.

Library of Congress Cataloging-in-Publication Data

McClave, James T.
  Statistics/ James T. McClave, Terry Sincich. —12th ed.
     p. cm
  ISBN 0-321-75593-6
1. Statistics I. Sincich, Terry. II. Title.
  QA276.M4 2013
  519.5—dc23

2011031868

1 2 3   4 5 6 7 8 9 10—CKV—15 14 13 12 11

www.pearsonhighered.com

ISBN 10: 0-321-75593-6
ISBN 13: 978-0-321-75593-3

Contents

Chapter 4   Discrete Random Variables   179

Chapter 5   Continuous Random Variables   224

Chapter 6   Sampling Distributions   271

Chapter 7   Inferences Based on a Single Sample: Estimation with Confidence Intervals   298

Chapter 8

Inferences Based on a Single Sample:
Tests of Hypothesis   349

Chapter 9

Inferences Based on a Two Samples:
Confidence Intervals and Tests of Hypotheses   409

Chapter 10

Analysis of Variance:
Comparing More than Two Means   474

Chapter 14   Nonparametric Statistics (available on CD)   14-1

Appendices

Preface

Statistics is an introductory text that emphasizes inference and sound decision-making through extensive coverage of data collection and analysis. As in earlier editions, the twelfth edition text stresses the development of statistical thinking, the assessment of credibility, and value of the inferences made from data, both by those who consume and those who produce them. It assumes a mathematical background of basic algebra.

The text incorporates the following strategies, developed from the American Statistical Association's (ASA) Guidelines for Assessment and Instruction in Statistics Education (GAISE) Project:

- Emphasize statistical literacy and develop statistical thinking
- Use real data in applications
- Use technology for developing conceptual understanding and analyzing data
- Foster active learning in the classroom
- Stress conceptual understanding rather than mere knowledge of procedures

A briefer version of the book, *A First Course in Statistics*, is available for single-semester courses that include minimal coverage of regression analysis, analysis of variance, and categorical data analysis.

New in the Twelfth Edition

- **More than 1,800 exercises, with revisions and updates to 20%.** Many new and updated exercises, based on contemporary studies and real data, have been added. Most of these exercises foster and promote critical thinking skills.
- **Updated technology.** All printouts from statistical software (SAS®, IBM® SPSS®, MINITAB®, and the TI-83/84 Plus Graphing Calculator) and corresponding instructions for use have been revised to reflect the latest versions of the software.
- **New and Revised Statistics in Action Cases.** More than one-third of the *Statistics in Action* cases are new or revised, each based on real data from a recent study.
- **Redesigned end-of-chapter summaries.** Summaries at the end of each chapter have been redesigned to make them better study aids for students. Important points are reinforced through flow graphs (which aid in selecting the appropriate statistical method) and boxed notes with key ideas, terms, symbols/notation, and formulas.
- **Emphasis on ethics.** Where appropriate, boxes have been added emphasizing the importance of ethical behavior when collecting, analyzing, and interpreting data.
- **Learning objectives.** Chapter opening *Where We're Going* bullet points include section numbers that correspond to where that concept is discussed in the chapter.

Content-Specific Changes to This Edition

- **Chapter 7 (Confidence Intervals).** The methodology for finding a confidence interval for a population mean is developed based on using either a normal (z) statistic (Section 7.2) or a Student's t-statistic (Section 7.3). Also, we added an optional section (Section 7.6) on estimating a population variance.
- **Chapter 8 (Tests of Hypothesis).** A new section emphasizing the formulation of the null and alternative hypotheses (Section 8.2) has been added.

- **Chapter 12 (Multiple Regression and Model Building).** For pedagogical purposes, the chapter is divided into three parts: First Order Models with Quantitative Independent Variables, Model Building, and Multiple Regression Diagnostics.
- **Chapter 13 (Categorical Data Analysis).** A subsection on contingency tables with fixed marginals as been added to Section 13.3.

Hallmark Strengths

We have maintained or strengthened the pedagogical features that make *Statistics* unique among introductory statistics texts. These features, which assist the student in achieving an overview of statistics and an understanding of its relevance in the world and everyday life, are as follows:

- **Use of examples as a teaching device.** Almost all new ideas are introduced and illustrated by data-based applications and examples. We believe that students better understand definitions, generalizations, and theoretical concepts *after* seeing an application. All examples have three components: (1) Problem, (2) Solution, and (3) Look Back (or Look Ahead). This step-by-step process provides students with a defined structure by which to approach problems and enhances their problem-solving skills. The Look Back feature often gives helpful hints to solving the problem and/or provides a further reflection or insight into the concept or procedure that is covered.
- **Now Work.** A *Now Work* exercise suggestion follows each example. The *Now Work* exercise (marked with the icon [NW] in exercise sets) is similar in style and concept to the text example. This provides students with an opportunity to immediately test and confirm their understanding.
- **Statistics in Action.** Each chapter begins with a case study based on an actual contemporary, controversial or high-profile issue. Relevant research questions and data from the study are presented and the proper analysis demonstrated in short *Statistics in Action Revisited* sections throughout the chapter. These motivate students to critically evaluate the findings and think through the statistical issues involved.
- **Applet Exercises.** The text is accompanied by a resource CD containing applets (short JAVA computer programs). These point-and-click applets allow students to easily run simulations that visually demonstrate some of the more difficult statistical concepts (e.g., sampling distributions and confidence intervals.) Each chapter contains several optional applet exercises in the exercise sets. They are denoted with the following icon: ⚪
- **Real data-driven exercises.** The text includes more than 1,800 exercises based on a wide variety of applications in various disciplines and research areas. Nearly all of the applied exercises use current, real data extracted from newspapers, magazines, current journals, and the Internet. Some students have difficulty learning the mechanics of statistical techniques when all problems are couched in terms of realistic applications. For this reason, all exercise sections are divided into four parts:

  Learning the Mechanics. Designed as straightforward applications of new concepts, these exercises allow students to test their ability to comprehend a mathematical concept or a definition.

  Applying the Concepts—Basic. Based on applications taken from a wide variety of journals, newspapers, and other sources, these short exercises help students begin developing the skills necessary to diagnose and analyze real-world problems.

  Applying the Concepts—Intermediate. Based on more detailed real-world applications, these exercises require students to apply their knowledge of the technique presented in the section.

  Applying the Concepts—Advanced. These more difficult real-data exercises require students to utilize their critical thinking skills.

- **Critical Thinking Challenges.**   Placed at the end of the Supplementary Exercises section only, students are asked to apply their critical thinking skills to solve one or two challenging real-life problems. These exercises expose students to real-world problems with solutions that are derived from careful, logical thought and selection of the appropriate statistical analysis tool.

- **Exploring data with statistical computer software and the graphing calculator.** We demonstrate each statistical analysis method presented using output from three leading statistical software packages: SAS, SPSS, and MINITAB. These outputs appear throughout the text in examples and exercises, exposing students to the output they will encounter in today's high-tech world. In addition, we provide output and keystroke instructions for the TI-83/84 Plus Graphing Calculator in the *Using Technology* section at the end of appropriate chapters.

- **Using Technology tutorials.**   At the end of each chapter we've included statistical software tutorials with instructions and screen shots for MINITAB and, where appropriate, the TI-83/84 Plus Graphing Calculator. These step-by-step tutorials are easily located and show students how to best use statistical software.

- **Biographies.**   Brief descriptions of famous statisticians and his/her achievements are presented in-text and in marginal boxes. With these profiles, students will develop an appreciation for the statistician's efforts and the discipline of statistics as a whole.

- **CD-ROM.**   New copies of the text are accompanied by a resource CD that contains files for all of the text examples, exercises, Statistics in Action, and Real-World case data sets marked with a ⊙. Data sets are provided in multiple formats. The CD also contains Chapter 14, Nonparametric Statistics, and a set of applets that illustrate statistical concepts.

Flexibility in Coverage

The text is written to allow the instructor flexibility in coverage of topics. Suggestions for two topics, probability and regression, are given below.

- **Probability and counting rules.**   One of the most troublesome aspects of an introductory statistics course is the study of probability. Probability poses a challenge for instructors because they must decide on the level of presentation, and students find it a difficult subject to comprehend. We believe that one cause for these problems is the mixture of probability and counting rules that occurs in most introductory texts. Consequently, we have included the counting rules (with examples) in an appendix (Appendix A) rather than in the body of Chapter 3. Thus, the instructor can control the level of coverage of probability covered.

- **Multiple regression and model building.**   This topic represents one of the most useful statistical tools for the solution of applied problems. Although an entire text could be devoted to regression modeling, we feel that we have presented coverage that is understandable, usable, and much more comprehensive that the presentations in other introductory statistics texts. We devote two full chapters to discussing the major types of inferences that can be derived from a regression analysis, showing how these results appear in the output from statistical software, and, most important, selecting multiple regression models to be used in an analysis. Thus, the instructor has the choice of a one-chapter coverage of simple linear regression (Chapter 11), a two-chapter treatment of simple and multiple regression (excluding the sections on model building in Chapter 12), or complete coverage of regression analysis, including model building and regression diagnostics. This extensive coverage of such useful statistical tools will provide added evidence to the student of the relevance of statistics to real-world problems.

- **Role of calculus in footnotes.**   Although the text is designed for students without a calculus background, footnotes explain the role of calculus in various derivations. Footnotes are also used to inform the student about some of the theory underlying certain methods of analysis. These footnotes allow additional flexibility in the mathematical and theoretical level at which the material is presented.

Supplements

Student Resources

Student's Solutions Manual, by Nancy Boudreau (Bowling Green State University), includes complete worked out solutions to all odd-numbered text exercises (ISBN-13: 978-0-321-75597-1; ISBN-10: 0-321-75597-9).

Excel® Manual (download only), by Mark Dummeldinger (University of South Florida). Available for download from www.pearsonhighered.com/mathstatsresources.

MINITAB® Manual (download only), by Keith Bower, available for download from www.pearsonhighered.com/mathstatsresources.

Graphing Calculator Manual (download only), by Susan Herring (Sonoma State University), available for download from www.pearsonhighered.com/mathstatsresources.

Study Cards for Statistics Software. This series of study cards, available for Excel, MINITAB, JMP®, SPSS, R, StatCrunch®, and TI-83/84 Plus Graphing Calculators provides students with easy step-by-step guides to the most common statistics software. Visit myPearsonstore.com for more information.

Instructor Resources

Annotated Instructor's Edition contains answers to text exercises. Annotated marginal notes include Teaching Tips, suggested exercises to reinforce the statistical concepts discussed in the text, and short answers to exercises and examples (ISBN-13: 978-0-321-75694-7; ISBN-10: 0-321-75694-0).

Instructor's Solutions Manual, by Nancy Boudreau (Bowling Green State University), includes complete worked-out solutions to all even-numbered text exercises Careful attention has been paid to ensure that all methods of solution and notation are consistent with those used in the core text (ISBN-13: 978-0-321-78340-0; ISBN-10: 0-321-78340-9).

PowerPoint® Lecture Slides include figures and tables from the textbook. Available for download from Pearson's online catalog at www.pearsonhighered.com/irc.

TestGen® (www.pearsoned.com/testgen) enables instructors to build, edit, print, and administer tests using a computerized bank of questions developed to cover all the objectives of the text. TestGen is algorithmically based, allowing instructors to create multiple but equivalent versions of the same question or test with the click of a button. Instructors can also modify test bank questions or add new questions. The software and test bank are available for download from Pearson Education's online catalog.

Online Test Bank, a test bank derived from TestGen®, is available for download from Pearson's online catalog at www.pearsonhighered.com/irc.

The Pearson Math Adjunct Support Center (http://www.pearsontutorservices.com/math-adjunct.html) is staffed by qualified instructors with more than 100 years of combined experience at both the community college and university levels. Assistance is provided for faculty in the following areas:

- Suggested syllabus consultation
- Tips on using materials packed with your book
- Book-specific content assistance
- Teaching suggestions, including advice on classroom strategies

Technology Resources

A companion **CD-ROM** is bound in new copies of *Statistics*. The CD holds a number of support materials, including:

- **Data sets** formatted as .csv, .txt, and TI files
- **Applets** (short JAVA computer programs) that allow students to run simulations that visually demonstrate statistical concepts
- **Chapter 14:** Nonparametric Statistics

Data sets are also available for download from www.pearsonhighered.com/mathstatsresources.

MathXL® for Statistics Online Course (access code required) MathXL® is the homework and assessment engine that runs MyStatLab. (MyStatLab is MathXL plus a learning management system.) With MathXL for Statistics, instructors can:

- Create, edit, and assign online homework and tests using algorithmically generated exercises correlated at the objective level to the textbook.
- Create and assign their own online exercises and import TestGen tests for added flexibility.
- Maintain records of all student work, tracked in MathXL's online gradebook.

With MathXL for Statistics, students can:

- Take chapter tests in MathXL and receive personalized study plans and/or personalized homework assignments based on their test results.

- Use the study plan and/or the homework to link directly to tutorial exercises for the objectives they need to study.
- Students can also access supplemental animations and video clips directly from selected exercises.
- Knowing that students often use external statistical software, we make it easy to copy our data sets, both from the ebook and the MyStatLab questions, into software like StatCrunch, MINITAB, Excel, and more.

MathXL for Statistics is available to qualified adopters. For more information, visit our website at www.mathxl.com, or contact your Pearson representative.

MyStatLab™ Online Course (access code required)

MyStatLab is a course management system that delivers **proven results** in helping individual students succeed.

- MyStatLab can be successfully implemented in any environment—lab-based, hybrid, fully online, traditional—and demonstrates the quantifiable difference that integrated usage has on student retention, subsequent success, and overall achievement.
- MyStatLab's comprehensive online gradebook automatically tracks students' results on tests, quizzes, homework, and in the study plan. Instructors can use the gradebook to intervene if students have trouble or to provide positive feedback. Data can be easily exported to a variety of spreadsheet programs, such as Microsoft Excel.

MyStatLab provides **engaging experiences** that personalize, stimulate, and measure learning for each student.

- **Tutorial Exercises with Multimedia Learning Aids.** The homework and practice exercises in MyStatLab align with the exercises in the textbook, and they regenerate algorithmically to give students unlimited opportunity for practice and mastery. Exercises offer immediate helpful feedback, guided solutions, sample problems, animations, videos, and eText clips for extra help at point-of-use.
- **Getting Ready for Statistics.** A library of questions now appears within the MyStatLab assessment manager to offer the developmental math topics students need for the course. These can be assigned as a prerequisite to other assignments, if desired.
- **Conceptual Question Library.** In addition to algorithmically regenerated questions that are aligned with your textbook, there is a library of 1,000 Conceptual Questions available in the assessment managers that require students to apply their statistical understanding.
- **StatCrunch.** MyStatLab includes a web-based statistical software, StatCrunch, within the online assessment platform so that students can easily analyze data sets from exercises and the text. In addition, MyStatLab includes access to www.statcrunch.com, a web site where users can access more than 13,000 shared data sets, conduct online surveys, perform complex analyses

using the powerful statistical software, and generate compelling reports.

- **Integration of Statistical Software.** Knowing that students often use external statistical software, we make it easy to copy our data sets, both from the ebook and MyStatLab questions, into software like StatCrunch, MINITAB, Excel and more. Students have access to a variety of support—Technology Instruction Videos, Technology Study Cards, and Manuals—to learn how to effectively use statistical software.
- **Expert Tutoring.** Although many students describe the whole of MyStatLab as "like having your own personal tutor," students also have access to live tutoring from Pearson. Qualified statistics instructors provide tutoring sessions for students via MyStatLab.

And, MyStatLab comes from a **trusted partner** with educational expertise and an eye on the future.

Knowing that you are using a Pearson product means knowing that you are using quality content. That means that our eTexts are accurate, that our assessment tools work, and that our questions are error-free. And whether you are just getting started with MyStatLab, or have a question along the way, we're here to help you learn about our technologies and how to incorporate them into your course.

To learn more about how MyStatLab combines proven learning applications with powerful assessment, visit www.mystatlab.com or contact your Pearson representative.

StatCrunch

StatCrunch is powerful web-based statistical software that allows users to perform complex analyses, share data sets, and generate compelling reports of their data. The vibrant online community offers more than 13,000 data sets for students to analyze.

- **Collect.** Users can upload their own data to StatCrunch or search a large library of publicly shared data sets, spanning almost any topic of interest. Also, an online survey tool allows users to quickly collect data via web-based surveys.
- **Crunch.** A full range of numerical and graphical methods allow users to analyze and gain insights from any data set. Interactive graphics help users understand statistical concepts, and are available for export to enrich reports with visual representations of data.
- **Communicate.** Reporting options help users create a wide variety of visually-appealing representations of their data.

Full access to StatCrunch is available with a MyStatLab kit, and StatCrunch is available by itself to qualified adopters. For more information, visit our website at www.statcrunch.com, or contact your Pearson representative.

The Student Edition of MINITAB is a condensed edition of the professional release of MINITAB statistical software.

It offers the full range of statistical methods and graphical capabilities, along with worksheets that can include up to 10,000 data points. Individual copies of the software can be bundled with the text (ISBN-13: 978-0-321-11313-9; ISBN-10: 0-321-11313-6).

JMP Student Edition is an easy-to-use, streamlined version of JMP desktop statistical discovery software from SAS Institute Inc. and is available for bundling with the text (ISBN-13: 978-0-321-67212-4 ISBN-10: 0-321-67212-7).

Acknowledgments

This book reflects the efforts of a great many people over a number of years. First, we would like to thank the following professors, whose reviews and comments on this and prior editions have contributed to the 12th edition:

Reviewers Involved with the Twelfth Edition of *Statistics*

Ali Arab, *Georgetown University*

Jen Case, *Jacksonville State University*

Maggie McBride, *Montana State University — Billings*

Surajit Ray, *Boston University*

JR Schott, *University of Central Florida*

Susan Schott, *University of Central Florida*

Lewis Shoemaker, *Millersville University*

Engin Sungur, *University of Minnesota — Morris*

Sherwin Toribio, *Universitiy of Wisconsin — La Crosse*

Michael Zwilling, *Mt. Union College*

Reviewers of Previous Editions

Bill Adamson, South Dakota State; Ibrahim Ahmad, Northern Illinois University; Roddy Akbari, Guilford Technical Community College; David Atkinson, Olivet Nazarene University; Mary Sue Beersman, Northeast Missouri State University; William H. Beyer, University of Akron; Marvin Bishop, Manhattan College; Patricia M. Buchanan, Pennsylvania State University; Dean S. Burbank, Gulf Coast Community College; Ann Cascarelle, St. Petersburg College; Kathryn Chaloner, University of Minnesota; Hanfeng Chen, Bowling Green State University; Gerardo Chin-Leo, The Everygreen State College; Linda Brant Collins, Iowa State University; Brant Deppa, Winona State University; John Dirkse, California State University — Bakersfield; N. B. Ebrahimi, Northern Illinois University; John Egenolf, University of Alaska — Anchorage; Dale Everson, University of Idaho; Christine Franklin, University of Georgia; Khadiga Gamgoum, Northern Virginia CC; Rudy Gideon, University of Montana; Victoria Marie Gribshaw, Seton Hill College; Larry Griffey, Florida Community College; David Groggel, Miami University at Oxford; Sneh Gulati, Florida International University; John E. Groves, California Polytechnic State University — San Luis Obispo; Dale K. Hathaway, Olivet Nazarene University; Shu-ping Hodgson, Central Michigan University; Jean L. Holton, Virginia Commonwealth University; Soon Hong, Grand Valley; Ina Parks S. Howell, Florida International University; Gary Itzkowitz, Rowan College of New Jersey; John H. Kellermeier, State University College at Plattsburgh; Golan Kibria, Florida International University; Timothy J. Killeen, University of Connecticut; William G. Koellner, Montclair State University; James R. Lackritz, San Diego State University; Diane Lambert, AT&T/Bell Laboratories; Edwin G. Landauer, Clackamas Community College; James Lang, Valencia Junior College; Glenn Larson, University of Regina; John J. Lefante, Jr., University of South Alabama; Pi-Erh Lin, Florida State University; R. Bruce Lind, University of Puget Sound; Rhonda Magel, North Dakota State University; Linda C.

Malone, University of Central Florida; Allen E. Martin, California State University—Los Angeles; Rick Martinez,Foothill College; Brenda Masters, Oklahoma State University; Leslie Matekaitis, Cal Genetics; E. Donice McCune, Stephen F. Austin State University; Mark M. Meerschaert, University of Nevada—Reno; Greg Miller, Steven F. Austin State University; Satya Narayan Mishra, University of South Alabama; Kazemi Mohammed, UNC–Charlotte; Christopher Morrell, Loyola College in Maryland; Mir Mortazavi, Eastern New Mexico University;A. Mukherjea, University of South Florida; Steve Nimmo, Morningside College (Iowa); Susan Nolan, Seton Hall University;Thomas O'Gorman, Northern Illinois University; Bernard Ostle, University of Central Florida; William B. Owen, Central Washington University;Won J. Park, Wright State University; John J. Peterson, Smith Kline & French Laboratories; Ronald Pierce, Eastern Kentucky University; Betty Rehfuss, North Dakota State University—Bottineau; Andrew Rosalsky, University of Florida; C. Bradley Russell, Clemson University; Rita Schillaber,University of Alberta; James R. Schott, University of Central Florida; Susan C. Schott, University of Central Florida; George Schultz, St. Petersburg Junior College; Carl James Schwarz, University of Manitoba; Mike Seyfried, Shippensburg University; Arvind K. Shah, University of South Alabama; Lewis Shoemaker, Millersville University; Sean Simpson, Westchester CC; Charles W. Sinclair, Portland State University; Robert K. Smidt, California Polytechnic State University—San Luis Obispo; Vasanth B. Solomon, Drake University; W. Robert Stephenson, Iowa State University;Thaddeus Tarpey, Wright State University; Kathy Taylor, Clackamas Community College; Barbara Treadwell, Western Michigan University; Dan Voss, Wright State University; Augustin Vukov, University of Toronto; Dennis D. Wackerly, University of Florida; Barbara Wainwright, Salisbury University; Matthew Wood, University of Missouri—Columbia.

Other Contributors

Special thanks are due to our ancillary authors, Nancy Boudreau, Mark Dummeldinger, Keith Bower, and Susan Herring. Thank you to our accuracy checkers, Engin Sungur and Cathleen Zucco-Teveloff, who helped to insure a highly accurate, clean text. Finally, the Pearson staff of Deirdre Lynch, Marianne Stepanian, Tracy Patruno, Dana Bettez, Sonia Ashraf, Barbara Atkinson, Jean Choe, Kathleen DeChavez, Roxanne McCarley, and Erin Lane, and Integra-Chicago's Amanda Zagnoli all helped greatly with various stages of the book and media.

Applications Index

1 Statistics, Data, and Statistical Thinking

CONTENTS

Where We're Going

- Introduce the field of statistics (1.1)
- Demonstrate how statistics applies to real-world problems (1.2)
- Introduce the language of statistics and the key elements to any statistical problem. (1.3)
- Differentiate between population and sample data (1.3)
- Differentiate between descriptive and inferential statistics (1.3)
- Identify the different types of data and data collection methods (1.4–1.5)
- Discover how critical thinking through statistics can help improve our quantitative literacy (1.6)

Statistics IN Action — Social Media Networks and the Millennial Generation

The Pew Research Center, a nonpartisan organization funded by a Philadelphia-based charity, has conducted over 100 surveys on Internet usage in the United States as part of the Pew Internet & American Life Project (PIALP). The PIALP has recently published a series of reports on teens and adults from ages 18 to 29 years—called the "Millennial Generation." In a 2010 report titled "Social Media & Mobile Internet Use," the PIALP examined the Millennial Generation's attitudes and behavior towards online social networks (e.g., Facebook, MySpace, Twitter). According to Wikipedia (the free online encyclopedia), *social media* are "media for social interaction, using highly accessible and scalable publishing techniques" such as Weblogs, Internet forums, Twitter, and social networking sites such as Facebook and MySpace.

Results from several of the survey questions asked of the teens are provided here:

- Internet Use
  When asked how often they use the Internet, teens responded:

Several times a day	36%
About once a day	27%
3–5 days a week	14%
1–2 days a week	12%
Every few weeks	7%
Less often	4%

- Social Networking
  When asked if they use social network sites like Facebook or MySpace, teens responded:

Yes	73%
No	27%

- Twitter
  When asked if they use Twitter, teens responded:

Yes	8%
No	91%

- Text Messaging
  When asked how often they send text messages on their cell phones, teens responded:

Every day	54%
Several times per week	10%
At least once a week	5%
Less than once a week	3%
Never	28%

- Average Number of Phone Calls
  On an average day, teens make and receive 10.7 phone calls on their cell phones.

- Average Number of Text Messages
  On an average day, teens send and receive 112.4 text messages on their cell phones.

In the following Statistics in Action Revisited sections, we discuss several key statistical concepts covered in this chapter that are relevant to the Pew Internet & American Life Project survey.

- Identifying the Population, Sample, and Inference (p. 9)
- Identifying the Data Collection Method and Data Type (p. 14)
- Critically Assessing the Ethics of a Statistical Study (p. 17)

Based on Pew Research Center for the People & the Press, "Social Media & Mobile Internet Use" report. © 2010 PIALF.

1.1 The Science of Statistics

What does statistics mean to you? Does it bring to mind batting averages, Gallup polls, unemployment figures, or numerical distortions of facts (lying with statistics!)? Or is it simply a college requirement you have to complete? We hope to persuade you that statistics is a meaningful, useful science whose broad scope of applications to business, government, and the physical and social sciences is almost limitless. We also want to show that statistics can lie only when they are misapplied. Finally, we wish to demonstrate the key role statistics plays in critical thinking—whether in the classroom, on the job, or in everyday life. Our objective is to leave you with the impression that the time you spend studying this subject will repay you in many ways.

The *Random House College Dictionary* defines **statistics** as "the science that deals with the collection, classification, analysis, and interpretation of information or data." Thus, a statistician isn't just someone who calculates batting averages at baseball games or tabulates the results of a Gallup poll. Professional statisticians are trained in *statistical science*. That is, they are trained in collecting information in the form of **data**, evaluating the information, and drawing conclusions from it. Furthermore, statisticians determine what information is relevant in a given problem and whether the conclusions drawn from a study are to be trusted.

> **Statistics** is the science of data. This involves collecting, classifying, summarizing, organizing, analyzing, presenting, and interpreting numerical information.

In the next section, you'll see several real-life examples of statistical applications that involve making decisions and drawing conclusions.

1.2 Types of Statistical Applications

"Statistics" means "numerical descriptions" to most people. Monthly housing starts, the failure rate of liver transplants, and the proportion of African-Americans who feel brutalized by local police all represent statistical descriptions of large sets of data collected on some phenomenon. (Later, in Section 1.4, we learn that not all data is numerical in nature.) Often the data are selected from some larger set of data whose characteristics we wish to estimate. We call this selection process *sampling*. For example, you might collect the ages of a sample of customers who shop for a particular product online to estimate the average age of *all* customers who shop online for the product. Then you could use your estimate to target the Web site's advertisements to the appropriate age group. Notice that statistics involves two different processes: (1) describing sets of data and (2) drawing conclusions (making estimates, decisions, predictions, etc.) about the sets of data on the basis of sampling. So, the applications of statistics can be divided into two broad areas: **descriptive statistics** and **inferential statistics.**

> **Descriptive statistics** utilizes numerical and graphical methods to look for patterns in a data set, to summarize the information revealed in a data set, and to present that information in a convenient form.

> **Inferential statistics** utilizes sample data to make estimates, decisions, predictions, or other generalizations about a larger set of data.

BIOGRAPHY FLORENCE NIGHTINGALE (1820–1910)

The Passionate Statistician

In Victorian England, the "Lady of the Lamp" had a mission to improve the squalid field hospital conditions of the British army during the Crimean War. Today, most historians consider Florence Nightingale to be the founder of the nursing profession. To convince members of the British Parliament of the need for supplying nursing and medical care to soldiers in the field, Nightingale compiled massive amounts of data from army files. Through a remarkable series of graphs (which included the first pie chart), she demonstrated that most of the deaths in the war either were due to illnesses contracted outside the battlefield or occurred long after battle action from wounds that went untreated. Florence Nightingale's compassion and self-sacrificing nature, coupled with her ability to collect, arrange, and present large amounts of data, led some to call her the Passionate Statistician. ■

Although we'll discuss both descriptive and inferential Statistics in the chapters that follow, the primary theme of the text is **inference.**

Let's begin by examining some studies that illustrate applications of statistics.

Study 1.1 "Best-Selling Girl Scout Cookies" (*www.girlscouts.org*)

Since 1917, the Girl Scouts of America have been selling boxes of cookies. Currently, there are eight varieties for sale: Thin Mints, Samoas, Caramel DeLites, Tagalongs,

Peanut Butter Patties, Do-si-dos, Peanut Butter Sandwiches, and Trefoils. Each of the approximately 150 million boxes of Girl Scout cookies sold each year is classified by variety. The results are summarized in Figure 1.1. From the graph, you can clearly see that the best-selling variety is Thin Mints (25%), followed by Samoas (19%) and Tagalongs (13%). Since the figure describes the various categories of boxes of Girl Scout cookies sold, the graphic is an example of descriptive statistics.

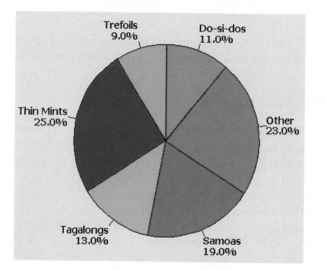

Figure 1.1

MINITAB graph of best-selling Girl Scout cookies

Based on www.girscouts.org.

Study 1.2   "Does Playing Video Games Make for Better Visual Attention Skills?" (*Journal of Articles in Support of the Null Hypothesis*, Vol. 6, No. 1, 2009)

Researchers at Griffin University (Australia) conducted a study to determine whether video game players have superior visual attention skills compared to non–video game players. Each in a sample of 65 male psychology students was classified as a video game player or a nonplayer. The two groups were then subjected to a series of visual attention tasks that included the "attentional blink" test, the "field of view" test, and the "repetition blindness" test. Except for attentional blink, no differences in the performance of the two groups were found. From this analysis, the researchers inferred "a limited role for video game playing in the modification of visual attention." Thus, inferential statistics was applied to arrive at this conclusion.

Study 1.3   "Animal Assisted Therapy…[for] Hospitalized Heart Failure Patients" (*American Heart Association Conference*, November 2005)

A team from the UCLA Medical Center and School of Nursing, led by RN Kathie Cole, conducted a study to gauge whether animal-assisted therapy can improve the physiological responses of heart failure patients. Cole and her colleagues studied 76 heart patients, randomly divided into three groups. Each person in one group of patients was visited by a human volunteer accompanied by a trained dog, each person in another group was visited by a volunteer only, and the third group was not visited at all. The researchers measured patients' physiological responses (levels of anxiety, stress, and blood pressure) before and after the visits. An analysis of the data revealed that those patients with animal-assisted therapy had significantly greater drops in levels of anxiety, stress, and blood pressure. Thus, the researchers concluded that "pet therapy has the potential to be an effective treatment…for patients hospitalized with heart failure." Like Study 1.2, this study is an example of the use of inferential statistics. The medical researchers used data from 76 patients to make inferences about the effectiveness of animal-assisted therapy for all heart failure patients.

These studies provide three real-life examples of the uses of statistics. Notice that each involves an analysis of data, either for the purpose of describing the data set (Study 1.1) or for making inferences about a data set (Studies 1.2 and 1.3).

1.3 Fundamental Elements of Statistics

Statistical methods are particularly useful for studying, analyzing, and learning about **populations** of **experimental units.**

> An **experimental** (or **observational**) **unit** is an object (e.g., person, thing, transaction, or event) about which we collect data.

> A **population** is a set of units (usually people, objects, transactions, or events) that we are interested in studying.

For example, populations may include (1) *all* employed workers in the United States, (2) *all* registered voters in California, (3) *everyone* who is afflicted with AIDS, (4) *all* the cars produced last year by a particular assembly line, (5) the *entire* stock of spare parts available at United Airlines' maintenance facility, (6) *all* sales made at the drive-in window of a McDonald's restaurant during a given year, or (7) the set of *all* accidents occurring on a particular stretch of interstate highway during a holiday period. Notice that the first three population examples (1–3) are sets (groups) of people, the next two (4–5) are sets of objects, the next (6) is a set of transactions, and the last (7) is a set of events. Notice also that *each set includes all the units in the population.*

In studying a population, we focus on one or more characteristics or properties of the units in the population. We call such characteristics **variables.** For example, we may be interested in the variables age, gender, and number of years of education of the people currently unemployed in the United States.

> A **variable** is a characteristic or property of an individual experimental (or observational) unit in the population.

The name *variable* is derived from the fact that any particular characteristic may vary among the units in a population.

In studying a particular variable, it is helpful to be able to obtain a numerical representation for it. Often, however, numerical representations are not readily available, so measurement plays an important supporting role in statistical studies. **Measurement** is the process we use to assign numbers to variables of individual population units. We might, for instance, measure the performance of the president by asking a registered voter to rate it on a scale from 1 to 10. Or we might measure the age of the U.S. workforce simply by asking each worker, "How old are you?" In other cases, measurement involves the use of instruments such as stopwatches, scales, and calipers.

If the population you wish to study is small, it is possible to measure a variable for every unit in the population. For example, if you are measuring the GPA for all incoming first-year students at your university, it is at least feasible to obtain every GPA. When we measure a variable for every unit of a population, it is called a **census** of the population. Typically, however, the populations of interest in most applications are much larger, involving perhaps many thousands, or even an infinite number, of units. Examples of large populations are those following the definition of population above, as well as all graduates of your university or college, all potential buyers of a new iPhone, and all pieces of first-class mail handled by the U.S. Post Office. For such populations, conducting a census would be prohibitively time consuming or costly. A reasonable alternative would be to select and study a *subset* (or portion) of the units in the population.

> A **sample** is a subset of the units of a population.

For example, instead of polling all 145 million registered voters in the United States during a presidential election year, a pollster might select and question a sample of just 1,500 voters. (See Figure 1.2.) If he is interested in the variable "presidential preference," he would record (measure) the preference of each vote sampled.

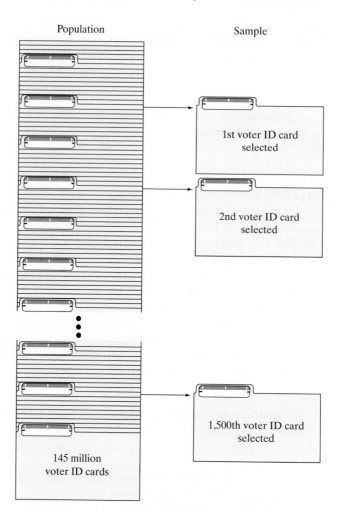

Figure 1.2

A sample of voter registration cards for all registered voters

After the variables of interest for every unit in the sample (or population) are measured, the data are analyzed, either by descriptive or inferential statistical methods. The pollster, for example, may be interested only in *describing* the voting patterns of the sample of 1,500 voters. More likely, however, he will want to use the information in the sample to make inferences about the population of all 145 million voters.

> A **statistical inference** is an estimate, prediction, or some other generalization about a population based on information contained in a sample.

That is, *we use the information contained in the smaller sample to learn about the larger population.** Thus, from the sample of 1,500 voters, the pollster may estimate the percentage of all the voters who would vote for each presidential candidate if the election were held on the day the poll was conducted, or he might use the results to predict the outcome on election day.

---

*The terms *population* and *sample* are often used to refer to the sets of measurements themselves, as well as to the units on which the measurements are made. When a single variable of interest is being measured, this usage causes little confusion. But when the terminology is ambiguous, we'll refer to the measurements as *population data sets* and *sample data sets*, respectively.

Example 1.1

Key Elements of a Statistical Problem— Ages of TV Viewers

Problem According to *Variety* (Aug. 27, 2009), the average age of viewers of live television programs broadcast on CBS, NBC, and ABC is 51 years. Suppose a rival network (e.g., Fox) executive hypothesizes that the average age of Fox viewers is less than 51. To test her hypothesis, she samples 200 Fox viewers and determines the age of each.

a. Describe the population.
b. Describe the variable of interest.
c. Describe the sample.
d. Describe the inference.

Solution

a. The population is the set of units of interest to the TV executive, which is the set of all Fox viewers.
b. The age (in years) of each viewer is the variable of interest.
c. The sample must be a subset of the population. In this case, it is the 200 Fox viewers selected by the executive.
d. The inference of interest involves the *generalization* of the information contained in the sample of 200 viewers to the population of all Fox viewers. In particular, the executive wants to *estimate* the average age of the viewers in order to determine whether it is less than 51 years. She might accomplish this by calculating the average age of the sample and using that average to estimate the average age of the population.

Look Back A key to diagnosing a statistical problem is to identify the data set collected (in this example, the ages of the 200 Fox viewers) as a population or a sample.

Example 1.2

Key Elements of a Statistical Problem— Pepsi vs. Coca-Cola

Problem "Cola wars" is the popular term for the intense competition between Coca-Cola and Pepsi displayed in their marketing campaigns, which have featured movie and television stars, rock videos, athletic endorsements, and claims of consumer preference based on taste tests. Suppose, as part of a Pepsi marketing campaign, 1,000 cola consumers are given a blind taste test (i.e., a taste test in which the two brand names are disguised). Each consumer is asked to state a preference for brand A or brand B.

a. Describe the population.
b. Describe the variable of interest.
c. Describe the sample.
d. Describe the inference.

Solution

a. Since we are interested in the responses of cola consumers in a taste test, a cola consumer is the experimental unit. Thus, the population of interest is the collection or set of all cola consumers.
b. The characteristic that Pepsi wants to measure is the consumer's cola preference, as revealed under the conditions of a blind taste test, so *cola preference* is the variable of interest.
c. The sample is the 1,000 cola consumers selected from the population of all cola consumers.
d. The inference of interest is the *generalization* of the cola preferences of the 1,000 sampled consumers to the population of all cola consumers. In particular, the preferences of the consumers in the sample can be used to *estimate* the percentages of cola consumers who prefer each brand.

Look Back In determining whether the study is inferential or descriptive, we assess whether Pepsi is interested in the responses of only the 1,000 sampled customers (descriptive statistics) or in the responses of the entire population of consumers (inferential statistics).

Now Work Exercise 1.14b

The preceding definitions and examples identify four of the five elements of an inferential statistical problem: a population, one or more variables of interest, a sample, and an inference. But making the inference is only part of the story; we also need to know its **reliability**—that is, how good the inference is. The only way we can be certain that an inference about a population is correct is to include the entire population in our sample. However, because of *resource constraints* (i.e., insufficient time or money), we usually can't work with whole populations, so we base our inferences on just a portion of the population (a sample). Thus, we introduce an element of *uncertainty* into our inferences. Consequently, whenever possible, it is important to determine and report the reliability of each inference made. Reliability, then, is the fifth element of inferential statistical problems.

The **measure of reliability** that accompanies an inference separates the science of statistics from the art of fortune-telling. A palm reader, like a statistician, may examine a sample (your hand) and make inferences about the population (your life). However, unlike statistical inferences, the palm reader's inferences include no measure of reliability.

Suppose, like the TV executive in Example 1.1, we are interested in the *error of estimation* (i.e., the difference between the average age of a population of TV viewers and the average age of a sample of viewers). Using statistical methods, we can determine a *bound on the estimation error*. This bound is simply a number that our estimation error (the difference between the average age of the sample and the average age of the population) is not likely to exceed. We'll see in later chapters that this bound is a measure of the uncertainty of our inference. The reliability of statistical inferences is discussed throughout this text. For now, we simply want you to realize that an inference is incomplete without a measure of its reliability.

> A **measure of reliability** is a statement (usually quantitative) about the degree of uncertainty associated with a statistical inference.

Let's conclude this section with a summary of the elements of descriptive and of inferential statistical problems and an example to illustrate a measure of reliability.

Four Elements of Descriptive Statistical Problems

1. The population or sample of interest
2. One or more variables (characteristics of the population or sample units) that are to be investigated
3. Tables, graphs, or numerical summary tools
4. Identification of patterns in the data

Five Elements of Inferential Statistical Problems

1. The population of interest
2. One or more variables (characteristics of the population units) that are to be investigated
3. The sample of population units
4. The inference about the population based on information contained in the sample
5. A measure of the reliability of the inference

Example 1.3
Reliability of an Inference— Pepsi vs. Coca-Cola

Problem Refer to Example 1.2, in which the preferences of 1,000 cola consumers were indicated in a taste test. Describe how the reliability of an inference concerning the preferences of all cola consumers in the Pepsi bottler's marketing region could be measured.

Solution When the preferences of 1,000 consumers are used to estimate those of all consumers in a region, the estimate will not exactly mirror the preferences of the population. For example, if the taste test shows that 56% of the 1,000 cola consumers preferred Pepsi, it does not follow (nor is it likely) that exactly 56% of all cola drinkers in the region prefer Pepsi. Nevertheless, we can use sound statistical reasoning (which we'll explore later in the text) to ensure that the sampling procedure will generate estimates that are almost certainly within a specified limit of the true percentage of all cola consumers who prefer Pepsi. For example, such reasoning might assure us that the estimate of the preference for Pepsi is almost certainly within 5% of the preference of the population. The implication is that the actual preference for Pepsi is between 51% [i.e., $(56 - 5)\%$] and 61% [i.e., $(56 + 5)\%$]—that is, $(56 \pm 5)\%$. This interval represents a measure of the reliability of the inference.

Look Ahead The interval 56 ± 5 is called a *confidence interval*, since we are confident that the true percentage of cola consumers who prefer Pepsi in a taste test falls into the range $(51, 61)$. In Chapter 7, we learn how to assess the degree of confidence (e.g., a 90% or 95% level of confidence) in the interval.

Statistics IN Action   Revisited   Identifying the Population, Sample, and Inference

Consider the 2010 Pew Internet & American Life Project survey on social networking and cell phone use by teenagers. In particular, consider the survey results on social networking sites like Facebook or MySpace. The experimental unit for the study is a teenager (the person answering the question), and the variable measured is the response ("yes" or "no") to the question.

The Pew Research Center reported that approximately 800 teens participated in the study. Obviously, that number is not all of the teenagers in the United States. Consequently, the 800 responses represent a sample selected from the much larger population of all American teenagers.

Earlier surveys found that 55% of American teenagers used an online social networking site in 2006, and 65%

in 2008. These are descriptive statistics that provide information on the popularity of social networking in past years. Since 73% of the surveyed teens in 2010 used an online social networking site, the Pew Research Center inferred that teens' usage of social networking cites continues its upward trend, with more and more teens getting online each year. That is, the researchers used the descriptive statistics from the sample to make an inference about the current population of American teenagers' use of social networking.

1.4 Types of Data

You have learned that statistics is the science of data and that data are obtained by measuring the values of one or more variables on the units in the sample (or population). All data (and hence the variables we measure) can be classified as one of two general types: **quantitative data** and **qualitative data.**

Quantitative data are data that are measured on a naturally occurring numerical scale.* The following are examples of quantitative data:

1. The temperature (in degrees Celsius) at which each piece in a sample of 20 pieces of heat-resistant plastic begins to melt

---

*Quantitative data can be subclassified as either *interval data* or *ratio data*. For ratio data, the origin (i.e., the value 0) is a meaningful number. But the origin has no meaning with interval data. Consequently, we can add and subtract interval data, but we can't multiply and divide them. Of the four quantitative data sets listed as examples, (1) and (3) are interval data while (2) and (4) are ratio data.

2. The current unemployment rate (measured as a percentage) in each of the 50 states

3. The scores of a sample of 150 law school applicants on the LSAT, a standardized law school entrance exam administered nationwide

4. The number of convicted murderers who receive the death penalty each year over a 10-year period

> **Quantitative data** are measurements that are recorded on a naturally occurring numerical scale.

In contrast, qualitative data cannot be measured on a natural numerical scale; they can only be classified into categories.* (For this reason, this type of data is also called *categorical data*.) Examples of qualitative data include the following:

1. The political party affiliation (Democrat, Republican, or Independent) in a sample of 50 voters

2. The defective status (defective or not) of each of 100 computer chips manufactured by Intel

3. The size of a car (subcompact, compact, midsize, or full size) rented by each of a sample of 30 business travelers

4. A taste tester's ranking (best, worst, etc.) of four brands of barbecue sauce for a panel of 10 testers

Often, we assign arbitrary numerical values to qualitative data for ease of computer entry and analysis. But these assigned numerical values are simply codes: They cannot be meaningfully added, subtracted, multiplied, or divided. For example, we might code Democrat = 1, Republican = 2, and Independent = 3. Similarly, a taste tester might rank the barbecue sauces from 1 (best) to 4 (worst). These are simply arbitrarily selected numerical codes for the categories and have no utility beyond that.

> **Qualitative** (or **categorical**) **data** are measurements that cannot be measured on a natural numerical scale; they can only be classified into one of a group of categories.

Example 1.4
Data Types—Army Corps of Engineers Study of a Contaminated River

Problem Chemical and manufacturing plants often discharge toxic-waste materials such as DDT into nearby rivers and streams. These toxins can adversely affect the plants and animals inhabiting the river and the riverbank. The U.S. Army Corps of Engineers conducted a study of fish in the Tennessee River (in Alabama) and its three tributary creeks: Flint Creek, Limestone Creek, and Spring Creek. A total of 144 fish were captured, and the following variables were measured for each:

1. River/creek where each fish was captured

2. Species (channel catfish, largemouth bass, or smallmouth buffalo fish)

3. Length (centimeters)

4. Weight (grams)

5. DDT concentration (parts per million)

---

*Qualitative data can be subclassified as either *nominal data* or *ordinal data*. The categories of an ordinal data set can be ranked or meaningfully ordered, but the categories of a nominal data set can't be ordered. Of the four qualitative data sets listed as examples, (1) and (2) are nominal and (3) and (4) are ordinal.

(For future analyses, these data are saved in the **FISHDDT** file.) Classify each of the five variables measured as quantitative or qualitative.

Solution The variables length, weight, and DDT concentration are quantitative because each is measured on a numerical scale: length in centimeters, weight in grams, and DDT in parts per million. In contrast, river/creek and species cannot be measured quantitatively: They can only be classified into categories (e.g., channel catfish, largemouth bass, and smallmouth buffalo fish for species). Consequently, data on river/creek and species are qualitative.

Look Ahead It is essential that you understand whether the data you are interested in are quantitative or qualitative, since the statistical method appropriate for describing, reporting, and analyzing the data depends on the data type (quantitative or qualitative).

Now Work Exercise 1.12

We demonstrate many useful methods for analyzing quantitative and qualitative data in the remaining chapters of the text. But first, we discuss some important ideas on data collection in the next section.

1.5 Collecting Data

Once you decide on the type of data—quantitative or qualitative—appropriate for the problem at hand, you'll need to collect the data. Generally, you can obtain data in three different ways:

1. From a *published source*

2. From a *designed experiment*

3. From an *observational study* (e.g., a *survey*)

Sometimes, the data set of interest has already been collected for you and is available in a **published source**, such as a book, journal, or newspaper. For example, you may want to examine and summarize the divorce rates (i.e., number of divorces per 1,000 population) in the 50 states of the United States. You can find this data set (as well as numerous other data sets) at your library in the *Statistical Abstract of the United States*, published annually by the U.S. government. Similarly, someone who is interested in monthly mortgage applications for new home construction would find this data set in the *Survey of Current Business*, another government publication. Other examples of published data sources include *The Wall Street Journal* (financial data) and *The Sporting News* (sports information). The Internet (World Wide Web) now provides a medium by which data from published sources are readily obtained.*

A second method of collecting data involves conducting a **designed experiment**, in which the researcher exerts strict control over the units (people, objects, or things) in the study. For example, an often-cited medical study investigated the potential of aspirin in preventing heart attacks. Volunteer physicians were divided into two groups: the *treatment* group and the *control* group. Each physician in the treatment group took one aspirin tablet a day for one year, while each physician in the control group took an aspirin-free placebo made to look like an aspirin tablet. The researchers—not the physicians under study—controlled who received the aspirin (the treatment) and who received the placebo. As you'll learn in Chapter 10, a properly designed experiment allows you to extract more information from the data than is possible with an uncontrolled study.

Finally, observational studies can be employed to collect data. In an **observational study**, the researcher observes the experimental units in their natural setting and records the variable(s) of interest. For example, a child psychologist might observe and record the

---

*With published data, we often make a distinction between the *primary source* and a *secondary source*. If the publisher is the original collector of the data, the source is primary. Otherwise, the data is secondary-source data.

level of aggressive behavior of a sample of fifth graders playing on a school playground. Similarly, a zoologist may observe and measure the weights of newborn elephants born in captivity. Unlike a designed experiment, an observational study is a study in which the researcher makes no attempt to control any aspect of the experimental units.

The most common type of observational study is a **survey**, where the researcher samples a group of people, asks one or more questions, and records the responses. Probably the most familiar type of survey is the political poll, conducted by any one of a number of organizations (e.g., Harris, Gallup, Roper, and CNN) and designed to predict the outcome of a political election. Another familiar survey is the Nielsen survey, which provides the major networks with information on the most-watched programs on television. Surveys can be conducted through the mail, with telephone interviews, or with in-person interviews. Although in-person surveys are more expensive than mail or telephone surveys, they may be necessary when complex information is to be collected.

> A **designed experiment** is a data collection method where the researcher exerts full control over the characteristics of the experimental units sampled. These experiments typically involve a group of experimental units that are assigned the *treatment* and an untreated (or *control*) group.

> An **observational study** is a data collection method where the experimental units sampled are observed in their natural setting. No attempt is made to control the characteristics of the experimental units sampled. (Examples include *opinion polls* and *surveys*.)

Regardless of which data collection method is employed, it is likely that the data will be a sample from some population. And if we wish to apply inferential statistics, we must obtain a **representative sample.**

> A **representative sample** exhibits characteristics typical of those possessed by the target population.

For example, consider a political poll conducted during a presidential election year. Assume that the pollster wants to estimate the percentage of all 145 million registered voters in the United States who favor the incumbent president. The pollster would be unwise to base the estimate on survey data collected for a sample of voters from the incumbent's own state. Such an estimate would almost certainly be *biased* high; consequently, it would not be very reliable.

The most common way to satisfy the representative sample requirement is to select a random sample. A **random sample** ensures that every subset of fixed size in the population has the same chance of being included in the sample. If the pollster samples 1,500 of the 145 million voters in the population so that every subset of 1,500 voters has an equal chance of being selected, she has devised a random sample. The procedure for selecting a random sample is discussed in Chapter 3. Here, we look at two examples involving actual sampling studies.

> A **random sample** of n experimental units is a sample selected from the population in such a way that every different sample of size n has an equal chance of selection. (See Section 3.7 for how to generate a random sample.)*

---

*In statistical sampling theory, this sample is better known as a *simple random sample*.

Example 1.5

Method of Data Collection—Internet Addiction Study

Problem What percentage of Web users are addicted to the Internet? To find out, a psychologist designed a series of 10 questions based on a widely used set of criteria for gambling addiction and distributed them through the Web site *ABCNews.com*. (A sample question: "Do you use the Internet to escape problems?") A total of 17,251 Web users responded to the questionnaire. If participants answered "yes" to at least half of the questions, they were viewed as addicted. The findings, released at an annual meeting of the American Psychological Association, revealed that 990 respondents, or 5.7%, are addicted to the Internet.

a. Identify the data collection method.

b. Identify the target population.

c. Are the sample data representative of the population?

Solution

a. The data collection method is a survey: 17,251 Internet users responded to the questions posed at the *ABCNews.com* Web site.

b. Since the Web site can be accessed by anyone surfing the Internet, presumably the target population is *all* Internet users.

c. Because the 17,251 respondents clearly make up a subset of the target population, they do form a sample. Whether or not the sample is representative is unclear, since we are given no information on the 17,251 respondents. However, a survey like this one in which the respondents are *self-selected* (i.e., each Internet user who saw the survey chose whether or not to respond to it) often suffers from *nonresponse bias*. It is possible that many Internet users who chose not to respond (or who never saw the survey) would have answered the questions differently, leading to a higher (or lower) percentage of affirmative answers.

Look Back Any inferences based on survey samples that employ self-selection are suspect due to potential nonresponse bias.

Example 1.6

Method of Data Collection—Study of Susceptibility to Hypnosis

Problem Psychologists at the University of Tennessee carried out a study of the susceptibility of people to hypnosis (*Psychological Assessment*, Mar. 1995). In a random sample of 130 undergraduate psychology students at the university, each experienced both traditional hypnosis and computer-assisted hypnosis. Approximately half were randomly assigned to undergo the traditional procedure first, followed by the computer-assisted procedure. The other half were randomly assigned to experience computer-assisted hypnosis first, then traditional hypnosis. Following the hypnosis episodes, all students filled out questionnaires designed to measure a student's susceptibility to hypnosis. The susceptibility scores of the two groups of students were compared.

a. Identify the data collection method.

b. Is the sample data representative of the target population?

Solution

a. Here, the experimental units are the psychology students. Since the researchers controlled which type of hypnosis—traditional or computer assisted—the students experienced first (through random assignment), a designed experiment was used to collect the data.

b. The sample of 130 psychology students was randomly selected from all psychology students at the University of Tennessee. If the target population is *all University of Tennessee psychology students*, it is likely that the sample is representative. However, the researchers warn that the sample data should not be used to make inferences about other, more general, populations.

Look Ahead By using randomization in a designed experiment, the researcher is attempting to eliminate different types of bias, including self-selection bias. (We discuss this type of bias in the next section.)

Now Work Exercise 1.15

Statistics IN Action **Revisited** Identifying the Data Collection Method and Data Type

In the Pew Internet & American Life Project report, American teenagers are asked to respond to a variety of questions about Internet, cell phone, and social networking site usage. According to the report, the data were obtained through phone interviews in the continental United States of 800 teenagers who were 12 to 17 years old at the time. Consequently, the data collection method is a survey (observational study).

Both quantitative and qualitative data were collected in the survey. For example, the survey question asking teens if they use social networking sites is phrased to elicit a "yes" or "no" response. Since the responses produced for this question are categorical in nature, these data are qualitative. However, the question asking teens for the number of text messages they send and receive per day will give meaningful numerical responses, such as 5, 10, 12, etc. Thus, these data are quantitative.

1.6 The Role of Statistics in Critical Thinking and Ethics

According to H. G. Wells, author of such science-fiction classics as *The War of the Worlds* and *The Time Machine,* "Statistical thinking will one day be as necessary for efficient citizenship as the ability to read and write." Written more than a hundred years ago, Wells's prediction is proving true today.

The growth in data collection associated with scientific phenomena, business operations, and government activities (quality control, statistical auditing, forecasting, etc.) has been remarkable in the past several decades. Every day the media present us with published results of political, economic, and social surveys. In increasing government emphasis on drug and product testing, for example, we see vivid evidence of the need for *quantitative literacy* (i.e., the ability to evaluate data intelligently). Consequently, each of us has to develop a discerning sense—an ability to use rational thought to interpret and understand the meaning of data. Quantitative literacy can help you make intelligent decisions, inferences, and generalizations; that is, it helps you *think critically* using statistics.

BIOGRAPHY H. G. WELLS (1866–1946)

Writer and Novelist

English-born Herbert George Wells published his first novel, *The Time Machine,* in 1895 as a parody of the English class division and as a satirical warning that human progress is inevitable. Although most famous as a science-fiction novelist, Wells was a prolific writer as a journalist, sociologist, historian, and philosopher. Wells's prediction about statistical thinking is just one of a plethora of observations he made about life on this world. Here are a few more of H. G. Wells's more famous quotes:

"Advertising is legalized lying."

"Crude classification and false generalizations are the curse of organized life."

"The crisis of today is the joke of tomorrow."

"Fools make researchers and wise men exploit them."

"The only true measure of success is the ratio between what we might have done and what we might have been on the one hand, and the thing we have made and the things we have made of ourselves on the other." ∎

> **Statistical thinking** involves applying rational thought and the science of statistics to critically assess data and inferences. Fundamental to the thought process is that variation exists in populations of data.

To gain some insight into the role statistics plays in critical thinking, we present two examples of some misleading or faulty surveys.

Example 1.7
Biased Sample— Motorcyclists and Helmets

Problem An article in the *New York Times* considered the question of whether motorcyclists should be required by law to wear helmets. In supporting his argument for no helmets, the editor of a magazine for Harley-Davidson bikers presented the results of one study that claimed "nine states without helmet laws had a lower fatality rate (3.05 deaths per 10,000 motorcycles) than those that mandated helmets (3.38)" and

a survey that found "of 2,500 bikers at a rally, 98% of the respondents opposed such laws." Based on this information, do you think it is safer to ride a motorcycle without a helmet? What further statistical information would you like?

Solution You can use "statistical thinking" to help you critically evaluate the study. For example, before you can evaluate the validity of the 98% estimate, you would want to know how the data were collected. If a survey was, in fact, conducted, it's possible that the 2,500 bikers in the sample were not selected at random from the target population of all bikers, but rather were "self-selected." (Remember, they were all attending a rally—a rally, likely, for bikers who oppose the law.) If the respondents were likely to have strong opinions regarding the helmet law (e.g., to strongly oppose the law), the resulting estimate is probably biased high. Also, if the biased sample was intentional, with the sole purpose to mislead the public, the researchers would be guilty of **unethical statistical practice**.

You would also want more information about the study comparing the motorcycle fatality rate of the nine states without a helmet law to those states that mandate helmets. Were the data obtained from a published source? Were all 50 states included in the study, or were only certain states selected? That is, are you seeing sample data or population data? Furthermore, do the helmet laws vary among states? If so, can you really compare the fatality rates?

Look Back Questions such as these led a group of mathematics and statistics teachers attending an American Statistical Association course to discover a scientific and statistically sound study on helmets. The study reported a dramatic *decline* in motorcycle crash deaths after California passed its helmet law.

Now Work Exercise 1.31c

Example 1.8
Manipulative or Ambiguous Survey Questions— Howard Stern on Sirius Radio

Problem Talk-show host, Howard Stern moved his controversial radio program from free, over-the-air (AM/FM) radio to Sirius satellite radio. The move was perceived in the industry to boost satellite radio subscriptions. This led American Media Services, a developer of AM/FM radio properties, to solicit a January 2006 nationwide random-digit dialing phone survey of 1,008 people. The purpose of the survey was to determine how much interest Americans really have

in buying satellite radio service. After providing some background on Howard Stern's controversial radio program, one of the questions asked, "How likely are you to purchase a subscription to satellite radio after Howard Stern's move to Sirius?" The result: Eighty-six percent of the respondents stated that they weren't likely to buy satellite radio because of Stern's move. Consequently, American Media Services concluded that "the Howard Stern Factor is overrated" and that "few Americans expect to purchase satellite radio"—claims that made the headlines of news reports and Weblogs. Do you agree?

Solution First, we need to recognize that American Media Services has a vested interest in the outcome of the survey—the company makes its money from over-the-air broadcast radio stations. Second, although the phone survey was conducted using random-digit dialing, there is no information provided on the response rate. It's possible that nonrespondents (people who were not home or refused to answer the survey questions) tend to be people who use cell phones more than their landline phone, and, consequently, are more likely to use the latest in electronic technology, including satellite radio. Finally, the survey question itself is ambiguous. Do the respondents have negative feelings about satellite radio, Howard Stern, or both? If not for Howard Stern's program, would the respondents be more likely to buy satellite radio? To the critical thinker, it's unclear what the results of the survey imply.

Look Back Examining the survey results from the perspective of satellite radio providers, 14% of the respondents indicated that they would be likely to purchase satellite radio. Projecting the 14% back to the population of all American adults, this figure represents about 50 million people; what is interpreted as "few Americans" by American Media Services could be music to the ears of satellite radio providers.

Now Work Exercise 1.29b

Ethics IN Statistics

Intentionally selecting a biased sample in order to produce misleading statistics is considered *unethical statistical practice*.

As with many statistical studies, both the motorcycle helmet study and the satellite radio study are based on survey data. Most of the problems with these surveys result from the use of *nonrandom* samples. These samples are subject to errors such as **selection bias, nonresponse bias** (recall Example 1.5), and **measurement error.** Researchers who are aware of these problems yet continue to use the sample data to make inferences are practicing unethical statistics.

> **Selection bias** results when a subset of the experimental units in the population is excluded so that these units have no chance of being selected in the sample.

> **Nonresponse bias** results when the researchers conducting a survey or study are unable to obtain data on all experimental units selected for the sample.

> **Measurement error** refers to inaccuracies in the values of the data recorded. In surveys, this kind of error may be due to ambiguous or leading questions and the inter-viewer's effect on the respondent.

Statistics IN Action Revisited | Critically Assessing the Ethics of a Statistical Study

The results from the Pew Internet & American Life Project (PIALP) survey on social networking and cell phone use by teenagers led the research center to make conclusions such as "teens continue to be avid users of social networking Web sites" and that the percentage "continues to climb upwards." The survey included a sample of 800 teens. In order to assess the validity of this inference, a critical thinker would consider several issues.

First, is the sample representative of all American teenagers? In the PIALP report, the claim is made that the survey results are based on "telephone interviews with a nationally representative sample of teenagers." Two samples of teenagers (and their parents) were selected. The first sample consisted of landline telephone numbers randomly selected (using a random number generator) from all landline numbers listed in the continental United States. The second sample consisted of cellular telephone numbers randomly selected from all U.S. cell numbers that were not associated with directory-listed landline numbers. As many as seven attempts were made to contact and interview the teenager at each number. These contact calls were staggered over times of day and days of the week in an attempt to reduce nonresponse bias. Clearly, there was no attemtp to bias the results of the study by intentionally selecting a sample with a certain point of view. In fact, just the opposite occurred. Consequently, the sample of 800 teenagers appears to be a truly random sample of all teenagers in the United States;

therefore, the sample is very likely to be representative of the population of interest.

Second, is the sample of 800 teenagers large enough to draw a reliable inference? Although there are over 25 million teenagers living in the continental United States, we will learn in later chapters that a random sample of size 800 will allow the researchers to draw a reliable inference. In fact, the PIALP report states that the margin of sampling error for the study is about 4%. This means that anywhere from 69% to 77% ($73\% \pm 4\%$) of all the teens in the United States regularly visit a social networking site.

Third, is the wording of the survey questions ambiguous or leading? Such questions are often crafted to elicit an answer that supports a specific point of view. For example, consider a hypothetical question about Twitter: "Twitter is becoming more and more popular with affluent teenagers. Have you ever used or desired to use Twitter in order to keep up with your friends?" The question is meant to elicit a "yes" answer and bias the results of the study. If you examine the six questions in the actual study (see p. 2), you will conclude that all are very straightforward and clear. Consequently, the survey results appear to be both ethical and valid.

In the remaining chapters of the text, you'll become familiar with the tools essential for building a firm foundation in statistics and statistical thinking.

CHAPTER NOTES

Key Terms

Census 5
Data 2
Descriptive statistics 3
Designed experiment 11
Experimental (or observational) unit 5
Inference 3
Inferential statistics 3
Measure of reliability 8
Measurement 5
Measurement error 16
Nonresponse bias 16
Observational study 12
Population 5
Published source 11

Qualitative (or categorical) data 10
Quantitative data 10
Random sample 12
Reliability (of an inference) 8
Representative sample 12
Sample 5
Selection bias 16
Statistical inference 6
Statistical thinking 15
Statistics 3
Survey 12
Unethical statistical practice 15
Variable 5

Key Ideas

Types of Statistical Applications

Descriptive
1. Identify **population** or **sample** (collection of **experimental units**)
2. Identify **variable(s)**
3. Collect **data**
4. **Describe** data

Inferential
1. Identify **population** (collection of *all* **experimental units**)
2. Identify **variable(s)**

3. Collect **sample** data (*subset* of population)
4. **Inference** about population based on sample
5. **Measure of reliability** of inference

Types of Data

1. **Quantitative** (numerical in nature)
2. **Qualitative** (categorical in nature)

Data Collection Methods

1. **Published source**
2. **Observational** (e.g., survey)
3. **Designed experiment**

Problems with Nonrandom Samples

1. **Selection bias**
2. **Nonresponse bias**
3. **Measurement error**

Exercises 1.1–1.33

Understanding the Principles

1.1 What is statistics?

1.2 Explain the difference between descriptive and inferential statistics.

1.3 List and define the five elements of an inferential statistical analysis.

1.4 List the four major methods of collecting data, and explain their differences.

1.5 Explain the difference between quantitative and qualitative data.

1.6 Explain how populations and variables differ.

1.7 Explain how populations and samples differ.

1.8 What is a representative sample? What is its value?

1.9 Why would a statistician consider an inference incomplete without an accompanying measure of its reliability?

1.10 Define statistical thinking.

1.11 Suppose you're given a data set that classifies each sample unit into one of four categories: A, B, C, or D. You plan to create a computer database consisting of these data, and you decide to code the data as A = 1, B = 2, C = 3, and D = 4. Are the data consisting of the classifications A, B, C, and D qualitative or quantitative? After the data are input as 1, 2, 3, or 4, are they qualitative or quantitative? Explain your answers.

Applet Exercise 1.1

The *Random Numbers* applet generates a list of *n* random numbers from 1 to *N*, where *n* is the size of the sample and *N* is the size of the population. The list generated often contains repetitions of one or more numbers.

   a. Using the applet *Random Numbers*, enter 1 for the minimum value, 10 for the maximum value, and 10 for the number of samples. Then click on *Sample*. Look at the results, and list any numbers that are repeated and the number of times each of these numbers occurs.

   b. Repeat part (a), changing the maximum value to 20 and keeping the size of the sample fixed at 10. If you still have repetitions, repeat the process, increasing the maximum value by 10 each time but keeping the size of the sample fixed. What is the smallest maximum value for which you had no repetitions?

   c. Describe the relationship between the population size (maximum value) and the number of repetitions in the list of random numbers as the population size increases and the sample size remains the same. What

can you conclude about using a random number generator to choose a relatively small sample from a large population?

Applet Exercise 1.2

The *Random Numbers* applet can be used to select a random sample from a population, but can it be used to simulate data? In parts (a) and (b), you will use the applet to create data sets. Then you will explore whether those data sets are realistic.

   a. Consider the number of customers waiting in line to order at a fast-food outlet. Use the *Random Numbers* applet to simulate this data set by setting the minimum value equal to 0, the maximum value equal to 99, and the sample size equal to 30. Explain what the numbers in the list produced by the applet represent in the context of the problem. Do the numbers produced by the applet seem reasonable? Explain.

   b. Use the *Random Numbers* applet to simulate grades on a statistics test by setting the minimum value equal to 0, the maximum value equal to 100, and the sample size equal to 30. Explain what the numbers in the list produced by the applet represent in this context. Do the numbers produced by the applet seem reasonable? Explain.

   c. Referring to parts (a) and (b), why do the randomly generated data seem more reasonable in one situation than in the other? Comment on the usefulness of using a random-number generator to produce data.

Applying the Concepts—Basic

1.12 **College application.** Colleges and universities are requiring an increasing amount of information about applicants before making acceptance and financial aid decisions. Classify each of the following types of data required on a college application as quantitative or qualitative.

   a. High school GPA

   b. High school class rank

   c. Applicant's score on the SAT or ACT

   d. Gender of applicant

   e. Parents' income

   f. Age of applicant

1.13 **Ground motion of earthquakes.** In the *Journal of Earthquake Engineering* (Nov. 2004), a team of civil and environmental engineers studied the ground motion characteristics of 15 earthquakes that occurred around the world since 1940. Three (of many) variables measured on each earthquake were the type of ground motion (short, long, or forward directive), the magnitude of the earthquake (on the Richter scale), and peak ground acceleration (feet per second). One of the goals of the study was to estimate the inelastic spectra of any ground motion cycle.

a. Identify the experimental units for this study.

b. Do the data for the 15 earthquakes represent a population or a sample? Explain.

c. Define the variables measured and classify them as quantitative or qualitative.

1.14 **Sprint speed training.** *The Sport Journal* (Winter 2004) reported on a study of a speed-training program for high school football players. Each participant was timed in a 40-yard sprint both before and after training. The researchers measured two variables: (1) the difference between the before and after sprint times (in seconds), and (2) the category of improvement ("improved," "no change," and "worse") for each player.

a. Identify the type (quantitative or qualitative) of each variable measured.

NW **b.** A total of 14 high school football players participated in the speed-training program. Does the data set collected represent a population or a sample? Explain.

1.15 **Going online for health information.** *A cyberchondriac* is defined as a person who regularly searches the Web for health care information. Each year a Harris Poll is conducted to determine the number of cyberchondriacs in the United States. In 2008, the Harris Poll surveyed 1,010 U.S. adults by telephone and asked the following questions:

1. Have you ever gone online to look for health care information?

2. How many times per month do you look for health care information online?

3. In the past year, have you ever discussed with your doctor the information you found online?

a. For each question, determine whether the type of data collected is quantitative or qualitative.

b. Do the data collected for the 1,010 adults represent a sample or a population? Explain.

1.16 **Student GPAs.** Consider the set of all students enrolled in your statistics course this term. Suppose you're interested in learning about the current grade point averages (GPAs) of this group.

a. Define the population and variable of interest.

b. Is the variable qualitative or quantitative?

c. Suppose you determine the GPA of every member of the class. Would this determination represent a census or a sample?

d. Suppose you determine the GPA of 10 members of the class. Would this determination represent a census or a sample?

e. If you determine the GPA of every member of the class and then calculate the average, how much reliability does your calculation have as an "estimate" of the class average GPA?

f. If you determine the GPA of 10 members of the class and then calculate the average, will the number you get necessarily be the same as the average GPA for the whole class? On what factors would you expect the reliability of the estimate to depend?

g. What must be true in order for the sample of 10 students you select from your class to be considered a random sample?

1.17 **Treasury deficit prior to the Civil War.** In *Civil War History* (June 2009), historian Jane Flaherty researched the condition of the U.S. Treasury on the eve of the Civil War in 1861. Between 1854 and 1857 (under President Franklin Pierce), the annual surplus/deficit was +18.8, +6.7, +5.3, and +1.3 million dollars, respectively. In contrast, between 1858 and 1861 (under President James Buchanan), the annual surplus/deficit was −27.3, −16.2, −7.2, and −25.2 million dollars, respectively. Flaherty used these data to aid in portraying the exhausted condition of the U.S. Treasury when Abraham Lincoln took office in 1861. Does this study represent a descriptive or inferential statistical study? Explain.

1.18 **Extinct birds.** Biologists at the University of California (Riverside) are studying the patterns of extinction in the New Zealand bird population. (*Evolutionary Ecology Research*, July 2003.) At the time of the Maori colonization of New Zealand (prior to European contact), the following variables were measured for each bird species:

a. Flight capability (volant or flightless)

b. Type of habitat (aquatic, ground terrestrial, or aerial terrestrial)

c. Nesting site (ground, cavity within ground, tree, cavity above ground)

d. Nest density (high or low)

e. Diet (fish, vertebrates, vegetables, or invertebrates)

f. Body mass (grams)

g. Egg length (millimeters)

h. Extinct status (extinct, absent from island, present)

Identify each variable as quantitative or qualitative.

1.19 **Study of quality of drinking water.** *Disasters* (Vol. 28, 2004) published a study of the effects of a tropical cyclone on the quality of drinking water on a remote Pacific island. Water samples (size 500 milliliters) were collected approximately four weeks after Cyclone Ami hit the island. The following variables were recorded for each water sample:

a. Town where sample was collected

b. Type of water supply (river intake, stream, or borehole)

c. Acidic level (pH scale, 1 to 14)

d. Turbidity level (nephalometric turbidity units = NTUs)

e. Temperature (degrees centigrade)

f. Number of fecal coliforms per 100 milliliters

g. Free-chlorine residual (milligrams per liter)

h. Presence of hydrogen sulphide (yes or no)

Identify each variable as quantitative or qualitative.

Applying the Concepts—Intermediate

1.20 **Herbal medicines.** *The American Association of Nurse Anesthetists Journal* (Feb. 2000) published the results of a study on the use of herbal medicines before surgery. Of 500 surgical patients randomly selected for the study, 51% used herbal or alternative medicines (e.g., garlic, ginkgo, kava, fish oil) against their doctor's advice prior to surgery.

a. Do the 500 surgical patients represent a population or a sample? Explain.

b. If your answer was "sample" in part **a**, is the sample representative of the population? If you answered "population" in part **a**, explain how to obtain a representative sample from the population.

c. For each patient, what variable is measured? Are the data collected quantitative or qualitative?

1.21 **National Bridge Inventory.** All highway bridges in the United States are inspected periodically for structural deficiency by the Federal Highway Administration (FHWA). Data from the FHWA inspections are compiled into the National Bridge Inventory (NBI). Several of the nearly 100 variables maintained by the NBI are listed next. Classify each variable as quantitative or qualitative.
   a. Length of maximum span (feet)
   b. Number of vehicle lanes
   c. Toll bridge (yes or no)
   d. Average daily traffic
   e. Condition of deck (good, fair, or poor)
   f. Bypass or detour length (miles)
   g. Type of route (interstate, U.S., state, county, or city)

1.22 **Annual survey of computer crimes.** The Computer Security Institute (CSI) conducts an annual survey of computer crime committed at U.S. businesses. CSI sends survey questionnaires to computer security personnel at all U.S. corporations and government agencies. In 2006, 616 organizations responded to the CSI survey. Fifty-two percent of the respondents admitted unauthorized use of computer systems at their firms during the year. (*Computer Security Issues & Trends*, Spring 2006.)
   a. Identify the population of interest to CSI.
   b. Identify the data collection method used by CSI. Are there any potential biases in the method used?
   c. Describe the variable measured in the CSI survey. Is it quantitative or qualitative?
   d. What inference can be made from the result of the study?

1.23 **CT scanning for lung cancer.** According to the American Lung Association, lung cancer accounts for 28% of all cancer deaths in the United States. A new type of screening for lung cancer, computed tomography (CT), has been developed. Medical researchers believe that CT scans are more sensitive than regular X-rays in pinpointing small tumors. The H. Lee Moffitt Cancer Center at the University of South Florida is currently conducting a clinical trial of 50,000 smokers nationwide to compare the effectiveness of CT scans with X-rays for detecting lung cancer. (*Todays' Tomorrows*, Fall 2002.) Each participating smoker is randomly assigned to one of two screening methods, CT or chest X-ray, and his or her progress tracked over time. The age at which the scanning method first detects a tumor is the variable of interest.
   a. Identify the data collection method used by the cancer researchers.
   b. Identify the experimental units of the study.
   c. Identify the type (quantitative or qualitative) of the variable measured.
   d. Identify the population and sample.
   e. What is the inference that will ultimately be drawn from the clinical trial?

1.24 **Satellite radio in cars.** A recent survey conducted for the National Association of Broadcasters investigated satellite radio subscriber service and usage. The June 2007 survey, conducted by Wilson Research Strategies, consisted of a random sample of 501 satellite radio subscribers. One of the questions of interest was, "Do you have a satellite radio receiver in your car?" The survey found that 396 subscribers did, in fact, have a satellite receiver in their car.
   a. Identify the population of interest to the National Association of Broadcasters.
   b. Based on the survey question, what is the variable of interest?
   c. Does the variable produce quantitative or qualitative data?
   d. Describe that sample of interest.
   e. What inference can be made from the survey results?

1.25 **Massage vs. rest in boxing.** Does a massage enable the muscles of tired athletes to recover from exertion faster than usual? To answer this question, researchers recruited eight amateur boxers to participate in an experiment. (*British Journal of Sports Medicine*, April 2000.) After a 10-minute workout in which each boxer threw 400 punches, half the boxers were given a 20-minute massage and half just rested for 20 minutes. Before they returned to the ring for a second workout, the heart rate (beats per minute) and blood lactate level (micromoles) were recorded for each boxer. The researchers found no difference in the means of the two groups of boxers for either variable.
   a. Identify the data collection method used by the researchers.
   b. Identify the experimental units of the study.
   c. Identify the variables measured and their type (quantitative or qualitative).
   d. What is the inference drawn from the analysis?
   e. Comment on whether this inference can be made about all athletes.

1.26 **Insomnia and education.** Is insomnia related to education status? Researchers at the Universities of Memphis, Alabama at Birmingham, and Tennessee investigated this question in the *Journal of Abnormal Psychology* (Feb. 2005). Adults living in Tennessee were selected to participate in the study, which used a random-digit telephone dialing procedure. Two of the many variables measured for each of the 575 study participants were number of years of education and insomnia status (normal sleeper or chronic insomniac). The researchers discovered that the fewer the years of education, the more likely the person was to have chronic insomnia.
   a. Identify the population and sample of interest to the researchers.
   b. Identify the data collection method. Are there any potential biases in the method used?
   c. Describe the variables measured in the study as quantitative or qualitative.
   d. What inference did the researchers make?

1.27 **Guilt in decision making.** The effect of the emotion of guilt on how a decision maker focuses on a problem was investigated in the *Journal of Behavioral Decision Making* (January 2007). A total of 155 volunteer students participated in the experiement, where each was randomly assigned to one of three emotional states (guilt, anger, or neutral) through a reading/writing task. Immediately after the task, the students were presented with a decision problem (e.g., whether or not to spend money on repairing a very old car). The researchers found that a higher proportion of students in the guilty-state group chose not to repair the car than those in the neutral-state and anger-state groups.
   a. Identify the population, sample, and variables measured for this study.
   b. Identify the data collection method used.
   c. What inference was made by the researcher?

d. In later chapters you will learn that the reliability of an inference is related to the size of the sample used. In addition to sample size, what factors might affect the reliability of the inference drawn in this study?

Applying the Concepts—Advanced

1.28 **Object recall study.** Are men or women more adept at remembering where they leave misplaced items (such as car keys)? According to University of Florida psychology professor Robin West, women show greater competence in actually finding these objects (*Explore*, Fall 1998). Approximately 300 men and women from Gainesville, Florida, participated in a study in which each person placed 20 common objects in a 12-room "virtual" house represented on a computer screen. Thirty minutes later, the subjects were asked to recall where they put each of the objects. For each object, a recall variable was measured as "yes" or "no."

a. Identify the population of interest to the psychology professor.

b. Identify the sample.

c. Does the study involve descriptive or inferential statistics? Explain.

d. Are the variables measured in the study quantitative or qualitative?

1.29 **Dating and disclosure.** As an adolescent, did you voluntarily disclose information about dating and romantic relationships to your parents? This was the research question of interest in the *Journal of Adolescence* (Apr. 2010). A sample of 222 high school students was recruited to participate in the study. Some of the many variables measured on each student were age (years), gender, dating experience (number of dates), and the extent to which the student was willing to tell his/her parent (without being asked) about a dating issue (e.g., how late the daters stayed out). The responses for the last variable were categorized as "never tell," "rarely tell," "sometimes tell," "almost always tell," and "always tell."

a. Identify the data type for each variable.

NW **b.** The study was unclear on exactly how the sample of students was selected, stating only that "participants were recruited from health or government classes in a primarily European American middle-class school district." Based on this information, what are the potential caveats to using the sample to make inferences on dating and disclosure to parents for all high school students?

1.30 **Success/failure of software reuse.** The PROMISE Software Engineering Repository, hosted by the University of Ottawa, is a collection of publicly available data sets to serve researchers in building prediction software models. A PROMISE data set on software reuse, saved in the **SWREUSE** file, provides information on the success or failure of reusing previously developed software for each in a sample of 24 new software development projects. (*Data source: IEEE Transactions on Software Engineering*, Vol. 28, 2002.) Of the 24 projects, 9 were judged failures and 15 were successfully implemented.

a. Identify the experimental units for this study.

b. Describe the population from which the sample is selected.

c. What is the variable of interest in the study? Is it quantitative or qualitative?

d. Critically evaluate the statement "Since $15/24 = .625$, it follows that 62.5% of all new software development projects will be successfully implemented."

Critical Thinking Challenges

1.31 **Your choice for a mom.** After running stories on current First Lady Michelle Obama and former vice presidential candidate Sarah Palin on consecutive weeks, *USA Weekend* magazine asked its readers on Mother's Day, "Who would you rather have as your mom, Sarah Palin or Michelle Obama?" Readers were asked to vote online at www.usaweekend.com. Based on over 34,000 votes cast, the results were: Obama—71%, Palin—29%. (*USA Weekend Magazine* press release, May 11, 2010.)

a. What type of data collection method is used in this study?

b. Is the data collected quantitative or qualitative? Explain.

NW **c.** Discuss the validity of the study results. What are the potential problems with running a poll where voting is done online?

1.32 **Poll on alien spacecraft.** "Have you ever seen anything that you believe was a spacecraft from another planet?" This was the question put to 1,500 American adults in a national poll conducted by ABC News and *The Washington Post*. The pollsters used random-digit telephone dialing to contact adult Americans until 1,500 responded. Ten percent (i.e., 150) of the respondents answered that they had, in fact, seen an alien spacecraft. (*Chance*, Summer 1997.) No information was provided on how many adults were called and, for one reason or another, did not answer the question.

a. Identify the data collection method.

b. Identify the target population.

c. Comment on the validity of the survey results.

1.33 ***20/20 survey exposés.*** The popular prime-time ABC television program *20/20* presented several misleading (and possibly unethical) surveys in a segment titled "Facts or Fiction? Exposés of So-Called Surveys" (March 31, 1995). The information reported from four of these surveys, conducted by businesses or special-interest groups with specific objectives in mind, are given. (Actual survey facts are provided in parentheses.)

Quaker Oats Study: Eating oat bran is a cheap and easy way to reduce your cholesterol count. (*Fact:* Diet must consist of nothing but oat bran to achieve a slightly lower cholesterol count.)

March of Dimes Report: Domestic violence causes more birth defects than all medical issues combined. (*Fact:* No study—false report.)

American Association of University Women (AAUW) study: Only 29% of high school girls are happy with themselves, compared with 66% of elementary school girls. (*Fact:* Of 3,000 high school girls, 29% responded "Always true" to the statement, "I am happy the way I am." Most answered, "Sort of true" and "Sometimes true.")

Food Research and Action Center study: One in four American children under age 12 is hungry or at risk of hunger. (*Fact:* Survey results are based on responses to the questions "Do you ever cut the size of meals?," "Do you ever eat less than you feel you should?" and "Did you ever rely on limited numbers of foods to feed your children because you were running out of money to buy food for a meal?")

a. Refer to the Quaker Oats study relating oat bran to cholesterol levels. Discuss why it is unethical to report the results as stated.

b. Consider the false March of Dimes report on domestic violence and birth defects. Discuss the type of data required to investigate the impact of domestic violence

on birth defects. What data collection method would you recommend?

c. Refer to the AAUW study of the self-esteem of high school girls. Explain why the results of the study are likely to be misleading. What data might be appropriate for assessing the self-esteem of high school girls?

d. Refer to the Food Research and Action Center study of hunger in America. Explain why the results of the study are likely to be misleading. What data would provide insight into the proportion of hungry American children?

Activity   Data in the News

Scan your daily newspaper or weekly news magazine, or search the Internet for articles that contain data. The data might be a summary of the results of a public opinion poll, the results of a vote by the U.S. Senate, or a list of crime rates, birth or death rates, etc. For each article you find, answer the following questions:

a. Do the data constitute a sample or an entire population? If a sample has been taken, clearly identify both the sample and the population; otherwise, identify the population.

b. What type of data (quantitative or qualitative) has been collected?

c. What is the source of the data?

d. If a sample has been observed, is it likely to be representative of the population?

e. If a sample has been observed, does the article present an explicit (or implied) inference about the population of interest? If so, state the inference made in the article.

f. If an inference has been made, has a measure of reliability been included? What is it?

g. Use your answers to questions d–f to critically evaluate the article. Comment on the ethics of the study.

References

What Is a Survey? American Statistical Association (F. Scheuren, editor), 2nd. ed. 2005. (*www.amstat.org*)

Careers in Statistics, American Statistical Association, Biometric Society, Institute of Mathematical Statistics and Statistical Society of Canada, 2008. (*www.amstat.org*)

Cochran, W. G. *Sampling Techniques*, 3rd ed. New York: Wiley, 1977.

Deming, W. E. *Sample Design in Business Research.* New York: Wiley, 1963.

Ethical Guidelines for Statistical Practice. American Statistical Association, 1999. (*www.amstat.org*)

Hansen, M. H., Hurwitz, W. N., and Madow, W. G. *Sample Survey Methods and Theory*, Vol. 1. New York: Wiley, 1953.

Hoerl, R., and Snee, R. *Statistical Thinking: Improving Business Performance.* Boston: Duxbury, 2002.

Kirk, R. E., ed. *Statistical Issues: A Reader for the Behavioral Sciences.* Monterey, CA: Brooks/Cole, 1972.

Kish, L. *Survey Sampling.* New York: Wiley, 1965 (paperback: 1995).

Peck, R., Casella, G., Cobb, G., Hoerl, R., Nolan, D., Starbuck, R., and Stern, H. *Statisics: A Guide to the Unknown*, 4th ed. Monterey, CA: Brooks/Cole, 2006.

Scheaffer, R., Mendenhall, W., and Ott, R. L. *Elementary Survey Sampling*, 6th ed. Boston: Duxbury, 2005.

USING TECHNOLOGY

MINITAB: Accessing and Listing Data

When you start a MINITAB session, you will see a screen similar to Figure 1.M.1. The bottom portion of the screen is an empty spreadsheet—called a MINITAB worksheet—with columns representing variables and rows representing observations (or cases). The very top of the screen is the MINITAB main menu bar, with buttons for the different functions and procedures available in MINITAB. Once you have entered

data into the spreadsheet, you can analyze the data by clicking the appropriate menu buttons. The results will appear in the Session window.

Entering Data

Create a MINITAB data file by entering data directly into the worksheet. Figure 1.M.2 shows data entered for a variable called "RATIO." Name the variables (columns) by typing in the name of each variable in the box below the column number.

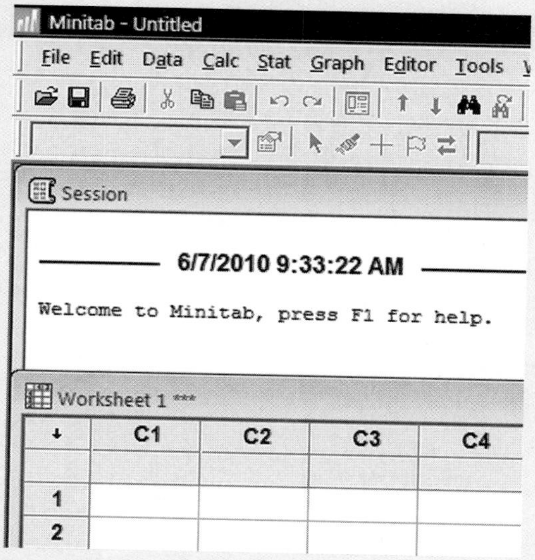

Figure 1.M.1 Initial screen viewed by the MINITAB user

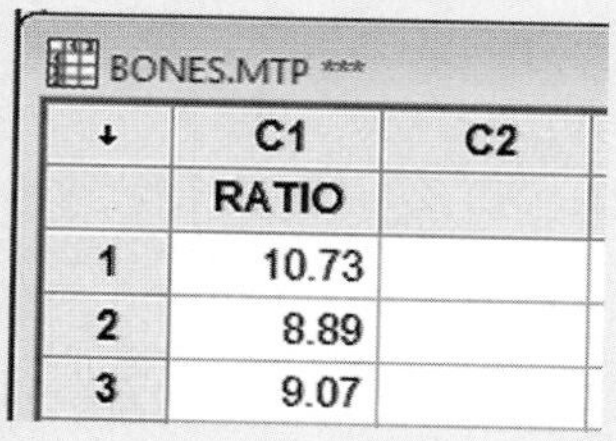

Figure 1.M.2 Data entered into the MINITAB worksheet

Accessing External Data from a File

Step 1 Click the "File" button on the menu bar, and then click "Open Worksheet" as shown in Figure 1.M.3. A dialog box similar to Figure 1.M.4 will appear.

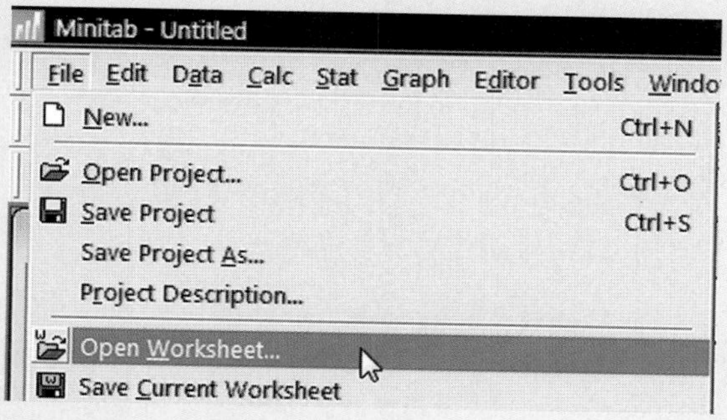

Figure 1.M.3 MINITAB options for reading data from an external file

Step 2 Specify the disk drive and folder that contains the external data file and the file type, and then click on the file name, as shown in Figure 1.M.4.

Step 3 If the data set contains qualitative data or data with special characters, click on the "Options" button as shown in Figure 1.M.4. The Options dialog box, shown in Figure 1.M.5, will appear.

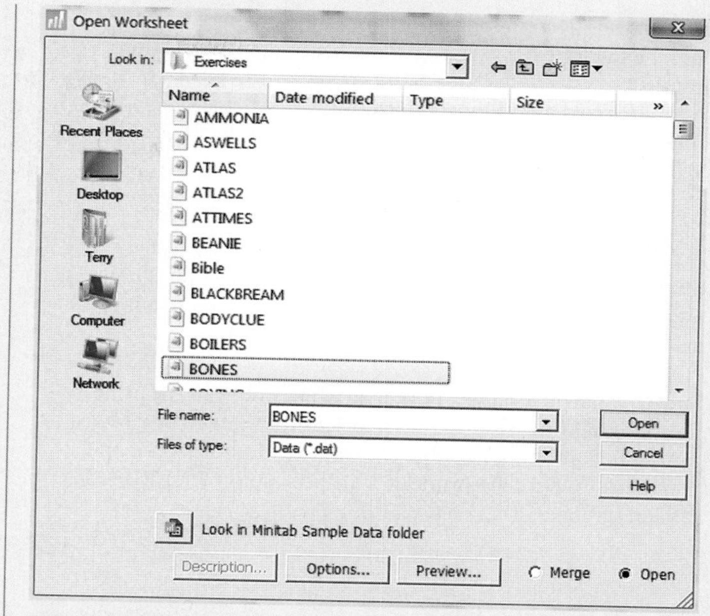

Figure 1.M.4 Selecting the external data file in MINITAB

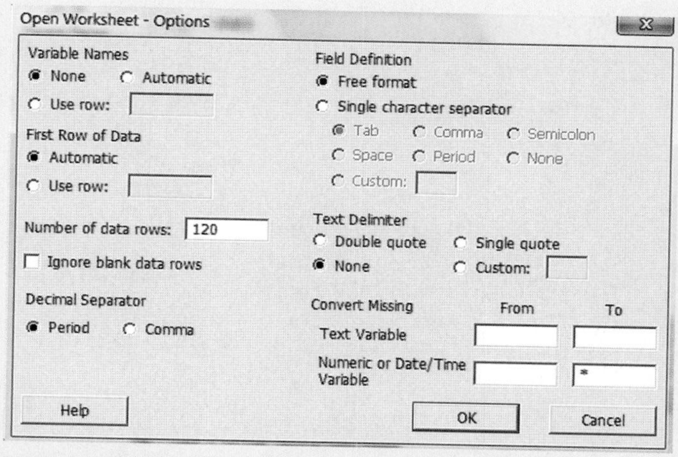

Figure 1.M.5 Selecting the MINITAB data input options

Step 4 Specify the appropriate options for the data set, and then click "OK" to return to the "Open Worksheet" dialog box (Figure 1.M.4).

Step 5 Click "Open" and the MINITAB worksheet will appear with the data from the external data file, as shown in Figure 1.M.6.

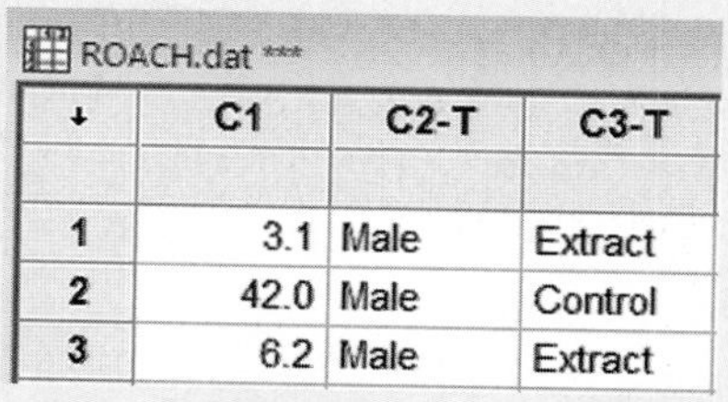

Figure 1.M.6 MINITAB worksheet with the imported data

Reminder: The variables (columns) can be named by typing in the name of each variable in the box under the column number.

To access a previously saved MINITAB worksheet, click "File," then "Open Worksheet," then select the MINITAB file.

Listing (Printing) Data

Step 1 Click on the "Data" button on the MINITAB main menu bar, and then click on "Display Data." The resulting menu, or dialog box, appears as in Figure 1.M.7.

Step 2 Enter the names of the variables you want to print in the "Columns, constants, and matrices to display" box (you can do this by simply double clicking on the variables), and then click "OK." The printout will show up on your MINITAB session screen.

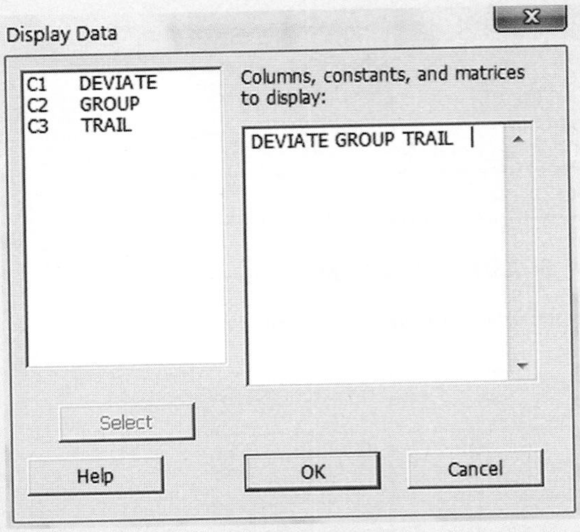

Figure 1.M.7 MINITAB Display Data dialog box

2 Methods for Describing Sets of Data

CONTENTS

Where We've Been

- Examined the difference between inferential and descriptive statistics
- Described the key elements of a statistical problem
- Learned about the two types of data: quantitative and qualitative
- Discussed the role of statistical thinking in managerial decision making

Where We're Going

- Describe qualitative data with graphs (2.1)
- Describe quantitative data with graphs (2.2)
- Describe quantitative data with numerical measures (2.3–2.8)
- Describe the relationship between two quantitative variables with a graph (2.9)
- Detecting descriptive methods which distort the truth (2.10)

Statistics IN Action   Body Image Dissatisfaction: Real or Imagined?

"Everything has beauty, but not everyone sees it" — ancient Chinese sage Confucius

"Body dissatisfaction can produce extreme body-shaping behaviors, such as eating disorders. Women and girls can't help being exposed to ultra-thin models in advertising whose body size is unrealistic and unhealthy. There is good evidence already that exposure to these unhealthy models leads a large proportion of women to feel dissatisfied with their own bodies." — Helga Dittmar, University of Sussex researcher

"Action figures (like G.I. Joe) present subtle messages of unrealistic role models of well-sculpted, heavily muscled, 'perfect' bodies that little boys see as their role models." — Sondra Kronberg, director of Eating Disorder Associates Treatment & Referral Centers

"By age 13, 53% of American girls are unhappy with their bodies; that figure grows to 78% by the time girls reach 17. In another study on fifth graders, 10-year-old girls and boys told researchers they were dissatisfied with their own bodies after watching a music video by Britney Spears or a clip from the TV show Friends. *And adolescent girls who viewed commercials depicting unrealistically thin models felt 'less confident, more angry, and more dissatisfied with their weight and appearance.'"* — statistics posted by the National Institute on Media and the Family

Are you dissatisfied with your physical appearance? Do you have a negative image of your own body? In today's media-driven society, many of us would answer yes to these questions (as the statistics in the previous quote show). Much research has been conducted on the body images of normal adolescents and adults. However, there is a lack of information on how patients with a body image disorder evaluate their own appearance, health, and fitness. To fill this gap, researchers from the Department of Psychiatry and Human Behavior at Brown University conducted and published a study in *Body Image: An International Journal of Research*, January 2010.

Data were collected on 92 patients diagnosed with *body dysmorphic disorder* (*BDD*). This disorder "is characterized by a distressing or impairing preoccupation with an imagined or slight defect in appearance that causes clinically significant distress or functional impairment." (Patients were also diagnosed with additional mental disorders, such as major depression or social phobia, called a comorbid disorder.) Each patient completed the Multidimensional Body-Self Relations

Questionnaire (MBSRQ). The questionnaire elicits responses that assess how satisfied one is with his/her appearance, health, fitness, and weight. In this Statistics in Action, our focus is on the appearance evaluations (e.g., "How satisfied are you with your physical attractiveness and looks?") of the BDD patients. The scores for each of seven appearance items (questions) were recorded on 5-point Likert scales, where the possible responses are 1 = definitely dissatisfied, 2 = somewhat dissatisfied, 3 = neutral, 4 = somewhat satisfied, and 5 = definitely satisfied. These scores were summed and a total appearance score (ranging from 7 to 35 points) was recorded for each patient.

The data for the study (simulated on the basis of summary statistics presented in the journal article) are provided in the **BDD** file. For each of the 92 patients in the experiment, the following variables were measured:

1. *Total Appearance Score* (points)
2. *Gender* (M or F)
3. *Comorbid Disorder* (Major Depression, Social Phobia, Obsessive Compulsive, or Anorexia/Bulimia Nervosa)
4. *Dissatisfied with Looks* (Yes or No)

For this application, we are interested in evaluating the data collected on the BDD patients. Specifically, we want to know if (1) BDD females tend to be more dissatisfied with their looks than BDD males, (2) certain comorbid disorders lead to a higher level of dissatisfaction with body appearance, and (3) BDD patients have lower appearance scores than normal people. We apply the graphical and numerical descriptive techniques of this chapter to the BDD data to answer these questions in the following Statistics in Action Revisited sections:

Statistics IN Action Revisited

- Interpreting Pie Charts for the Body Image Data (p. 31)
- Interpreting Histograms for the Body Image Data (p. 44)
- Interpreting Descriptive Statistics for the Body Image Data (p. 70)
- Detecting Outliers in the Body Image Data (p. 84)

Data Set: BDD

Suppose you wish to evaluate the mathematical capabilities of a class of 1,000 first-year college students, based on their quantitative Scholastic Aptitude Test (SAT) scores. How would you describe these 1,000 measurements? Characteristics of interest include the typical, or most frequent, SAT score; the variability in the scores; the highest and lowest scores; the "shape" of the data; and whether the data set contains any unusual scores. Extracting this information isn't easy. The 1,000 scores provide too many bits of information for our minds to comprehend. Clearly, we need some method for summarizing and characterizing the information in such a data set. Methods for describing data sets are also essential for statistical inference. Most populations make for large data sets. Consequently, we need methods for describing a data set that let us make inferences about a population on the basis of information contained in a sample.

Two methods for describing data are presented in this chapter, one *graphical* and the other *numerical*. Both play an important role in statistics. Section 2.1 presents both graphical and numerical methods for describing qualitative data. Graphical methods for describing quantitative data are illustrated in Sections 2.2, 2.8, and 2.9; numerical descriptive methods for quantitative data are presented in Sections 2.3–2.7. We end the chapter with a section on the *misuse* of descriptive techniques.

2.1 Describing Qualitative Data

Consider a study of aphasia published in the *Journal of Communication Disorders* (Mar. 1995). Aphasia is the "impairment or loss of the faculty of using or understanding spoken or written language." Three types of aphasia have been identified by researchers: Broca's, conduction, and anomic. The researchers wanted to determine whether one type of aphasia occurs more often than any other and, if so, how often. Consequently, they measured the type of aphasia for a sample of 22 adult aphasics. Table 2.1 gives the type of aphasia diagnosed for each aphasic in the sample.

For this study, the variable of interest, type of aphasia, is qualitative in nature. Qualitative data are nonnumerical in nature; thus, the value of a qualitative variable can only be classified into categories called *classes*. The possible types of aphasia—Broca's, conduction, and anomic—represent the classes for this qualitative variable. We can summarize such data numerically in two ways: (1) by computing the *class frequency*—the number of observations in the data set that fall into each class—or (2) by computing the *class relative frequency*—the proportion of the total number of observations falling into each class.

A **class** is one of the categories into which qualitative data can be classified.

The **class frequency** is the number of observations in the data set that fall into a particular class.

The **class relative frequency** is the class frequency divided by the total number of observations in the data set; that is,

$$\text{class relative frequency} = \frac{\text{class frequency}}{n}$$

Table 2.1	**Data on 22 Adult Aphasias**		
Subject	Type of Aphasia	Subject	Type of Aphasia
1	Broca's	12	Broca's
2	Anomic	13	Anomic
3	Anomic	14	Broca's
4	Conduction	15	Anomic
5	Broca's	16	Anomic
6	Conduction	17	Anomic
7	Conduction	18	Conduction
8	Anomic	19	Broca's
9	Conduction	20	Anomic
10	Anomic	21	Conduction
11	Conduction	22	Anomic

Based on Li, E. C., Williams, S. E., and Volpe, R. D. "The effects of topic and listener familiarity of discourse variables in procedural and narrative discourse tasks." *The Journal of Communication Disorders*, Vol. 28, No. 1, Mar. 1995, p. 44 (Table 1).

Data Set: APHASIA

> The **class percentage** is the class relative frequency multiplied by 100; that is,
>
> class percentage = (class relative frequency) × 100

Examining Table 2.1, we observe that 10 aphasics in the study were diagnosed as suffering from anomic aphasia, 5 from Broca's aphasia, and 7 from conduction aphasia. These numbers—10, 5, and 7—represent the class frequencies for the three classes and are shown in the summary table, Figure 2.1, produced with SPSS.

TYPE

		Frequency	Percent	Valid Percent	Cumulative Percent
Valid	Anomic	10	45.5	45.5	45.5
	Brocas	5	22.7	22.7	68.2
	Conduction	7	31.8	31.8	100.0
	Total	22	100.0	100.0	

Figure 2.1
SPSS summary table for types of aphasia

Figure 2.1 also gives the relative frequency of each of the three aphasia classes. From the class relative frequency definition, we calculate the relative frequency by dividing the class frequency by the total number of observations in the data set. Thus, the relative frequencies for the three types of aphasia are

$$\text{Anomic:} \frac{10}{22} = .455$$

$$\text{Broca's:} \frac{5}{22} = .227$$

$$\text{Conduction:} \frac{7}{22} = .318$$

These values, expressed as a percent, are shown in the SPSS summary table of Figure 2.1. From these relative frequencies, we observe that nearly half (45.5%) of the 22 subjects in the study are suffering from anomic aphasia.

Although the summary table of Figure 2.1 adequately describes the data of Table 2.1, we often want a graphical presentation as well. Figures 2.2 and 2.3 show two of the most widely used graphical methods for describing qualitative data: bar graphs and pie charts. Figure 2.2 shows the frequencies of the three types of aphasia in a **bar graph** produced with SAS. Note that the height of the rectangle, or "bar," over each class is equal to the class frequency. (Optionally, the bar heights can be proportional to class relative frequencies.)

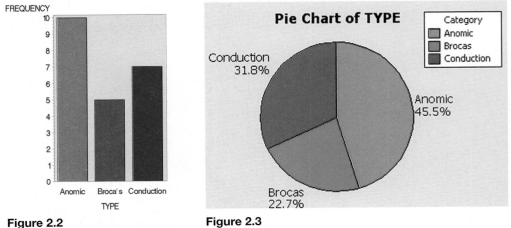

Figure 2.2
SAS bar graph for type of aphasia

Figure 2.3
MINITAB pie chart for type of aphasia

In contrast, Figure 2.3 shows the relative frequencies of the three types of aphasia in a **pie chart** generated with MINITAB. Note that the pie is a circle (spanning 360°) and the size (angle) of the "pie slice" assigned to each class is proportional to the class relative frequency. For example, the slice assigned to anomic aphasia is 45.5% of 360°, or $(.455)(360°) = 163.8°$.

Before leaving the data set in Table 2.1, consider the bar graph shown in Figure 2.4, produced with SPSS. Note that the bars for the types of aphasia are arranged in descending order of height, from left to right across the horizontal axis. That is, the tallest bar (Anomic) is positioned at the far left and the shortest bar (Broca's) is at the far right. This rearrangement of the bars in a bar graph is called a **Pareto diagram**. One goal of a Pareto diagram (named for the Italian economist Vilfredo Pareto) is to make it easy to locate the "most important" categories—those with the largest frequencies.

BIOGRAPHY

VILFREDO PARETO

(1843–1923)

The Pareto Principle

Born in Paris to an Italian aristo-cratic family, Vilfredo Pareto was educated at the University of Turin, where he studied engineering and mathematics. After the death of his parents, Pareto quit his job as an engineer and began writing and lecturing on the evils of the economic policies of the Italian government. While at the University of Lausanne in Switzerland in 1896, he published his first paper, *Cours d'économie politique*. In the paper, Pareto derived a com-plicated mathematical formula to prove that the distribution of income and wealth in society is not random, but that a consistent pattern appears throughout history in all societies. Essentially, Pareto showed that approximately 80% of the total wealth in a society lies with only 20% of the families. This famous law about the "vital few and the trivial many" is widely known as the Pareto principle in economics. ∎

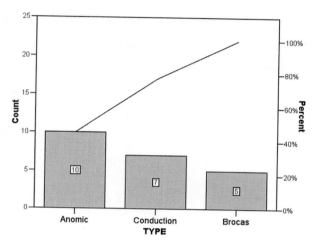

Figure 2.4

SPSS Pareto diagram for type of aphasia

Summary of Graphical Descriptive Methods for Qualitative Data

Bar Graph: The categories (classes) of the qualitative variable are represented by bars, where the height of each bar is either the class frequency, class relative frequency, or class percentage.

Pie Chart: The categories (classes) of the qualitative variable are represented by slices of a pie (circle). The size of each slice is proportional to the class relative frequency.

Pareto Diagram: A bar graph with the categories (classes) of the qualitative variable (i.e., the bars) arranged by height in descending order from left to right.

Now Work Exercise 2.6

Let's look at a practical example that requires interpretation of the graphical results.

Example 2.1

Graphing and Summarizing Qualitative Data— Drug Designed to Reduce Blood Loss

Problem A group of cardiac physicians in southwest Florida has been studying a new drug designed to reduce blood loss in coronary bypass operations. Blood loss data for 114 cor-onary bypass patients (some who received a dosage of the drug and others who did not) are saved in the **BLOODLOSS** file. Although the drug shows promise in reducing blood loss, the physicians are concerned about possible side effects and complications. So their data set includes not only the qualitative variable DRUG, which indicates whether or not the patient received the drug, but also the qualitative variable COMP, which specifies the type (if any) of complication experienced by the patient. The four values of COMP are (1) redo surgery, (2) post-op infection, (3) both, or (4) none.

a. Figure 2.5, generated by SAS, shows summary tables for the two qualitative variables, DRUG and COMP. Interpret the results.

b. Interpret the MINITAB and SPSS printouts shown in Figures 2.6 and 2.7, respectively.

The FREQ Procedure

DRUG	Frequency	Percent	Cumulative Frequency	Cumulative Percent
NO	57	50.00	57	50.00
YES	57	50.00	114	100.00

COMP	Frequency	Percent	Cumulative Frequency	Cumulative Percent
BOTH	6	5.26	6	5.26
INFECT	15	13.16	21	18.42
NONE	79	69.30	100	87.72
REDO	14	12.28	114	100.00

Figure 2.5
SAS summary tables for DRUG and COMP

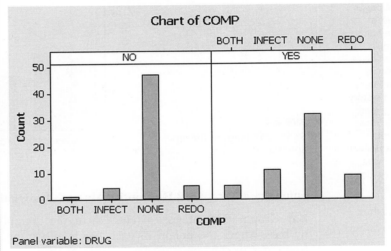

Figure 2.6
MINITAB side-by-side bar graphs for COMP by value of DRUG

COMP

DRUG			Frequency	Percent	Valid Percent	Cumulative Percent
NO	Valid	BOTH	1	1.8	1.8	1.8
		INFECT	4	7.0	7.0	8.8
		NONE	47	82.5	82.5	91.2
		REDO	5	8.8	8.8	100.0
		Total	57	100.0	100.0	
YES	Valid	BOTH	5	8.8	8.8	8.8
		INFECT	11	19.3	19.3	28.1
		NONE	32	56.1	56.1	84.2
		REDO	9	15.8	15.8	100.0
		Total	57	100.0	100.0	

Figure 2.7
SPSS summary tables for COMP by value of drug

Solution

a. The top table in Figure 2.5 is a summary frequency table for DRUG. Note that exactly half (57) of the 114 coronary bypass patients received the drug and half did not. The bottom table in Figure 2.5 is a summary frequency table for COMP. The class percentages are given in the Percent column. We see that about 69% of the 114 patients had no complications, leaving about 31% who experienced either a redo surgery, a post-op infection, or both.

b. Figure 2.6 is a MINITAB side-by-side bar graph of the data. The four bars in the left-side graph represent the frequencies of COMP for the 57 patients who did not receive the drug; the four bars in the right-side graph represent the frequencies of COMP for the 57 patients who did receive a dosage of the drug. The graph clearly shows that patients who did not receive the drug suffered fewer complications. The exact percentages are displayed in the SPSS summary tables of Figure 2.7. Over 56% of the patients who got the drug had no complications, compared with about 83% for the patients who got no drug.

Look Back Although these results show that the drug may be effective in reducing blood loss, Figures 2.6 and 2.7 imply that patients on the drug may have a higher risk of incurring complications. But before using this information to make a decision about the drug, the physicians will need to provide a measure of reliability for the inference. That is, the physicians will want to know whether the difference between the percentages of patients with complications observed in this sample of 114 patients is generalizable to the population of all coronary bypass patients.

Now Work Exercise 2.15

Statistics IN Action | Revisited | Interpreting Pie Charts for the Body Image Data

In the *Body Image: An International Journal of Research* (Jan. 2010) study, Brown University researchers measured several qualitative (categorical) variables for each of 92 body dysmorphic disorder (BDD) patients: *Gender* (M or F), *Comorbid Disorder* (Major Depression, Social Phobia, Obsessive Compulsive Disorder—OCD, or Anorexia/Bulimia Nervosa), and *Dissatisfied with Looks* (Yes or No). [*Note:* "Yes" values for Dissatisfied with Looks correspond to total appearance evaluation responses of 20 points (out of 35) or less. "No" values correspond to totals of 21 points or more.] Pie charts and bar graphs can be used to summarize and describe the responses for these variables. Recall that the data are saved in the **BDD** file. We used MINITAB to create pie charts for these variables.

Figure SIA2.1 shows individual pie charts for each of the three qualitative variables, Gender, Disorder, and Dissatisfied with Looks. First, notice that of the 92 BDD patients, 65% are females and 35% are males. Of these patients, the most common comorbid disorder is major depression (42%), followed by social phobia (35%) and OCD (20%). The third pie chart shows that 77% of the BDD patients are dissatisfied in some way with their body.

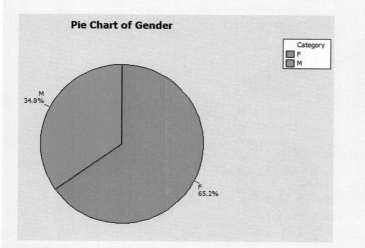

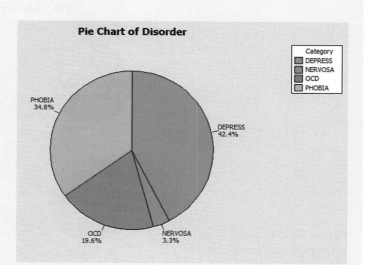

Figure SIA2.1

MINITAB pie charts for Gender, Disorder, and Dissatisfied with Looks

(continued)

Statistics IN Action
(continued)

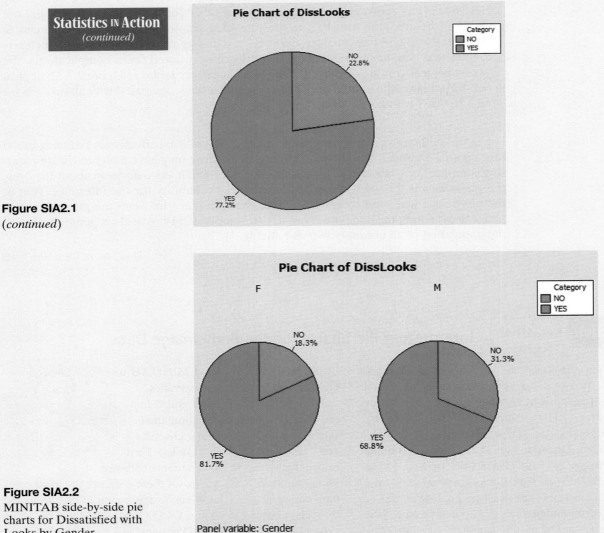

Figure SIA2.1
(*continued*)

Figure SIA2.2
MINITAB side-by-side pie charts for Dissatisfied with Looks by Gender

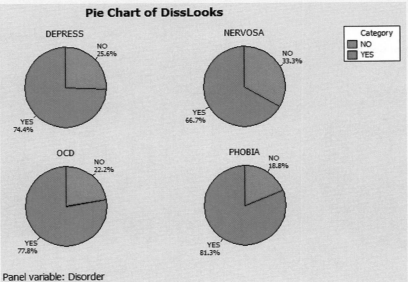

Figure SIA2.3
MINITAB side-by-side Pie charts for Dissatisfied with Looks by Comorbid Disorder

Statistics IN Action
(continued)

Of interest in the study is whether BDD females tend to be more dissatisfied with their looks than BDD males. We can gain insight into this question by forming side-by-side pie charts of the Dissatisfied with Looks variable, one chart for females and one for males. These pie charts are shown in Figure SIA2.2. You can see that about 82% of the females are dissatisfied in some way with their body as compared to about 69% of the males. Thus, it does appear that BDD females tend to be more dissatisfied with their looks than males, at least for this sample of patients.

Also of interest is whether certain comorbid disorders lead to a higher level of dissatisfaction with body appearance. Figure SIA2.3 is a set of side-by-side pie charts for Dissatisfied with Looks, one chart for each of four comorbid disorders. The graphs show that the percentage of BDD patients who are dissatisfied in some way with their body image range from about 67% for those with a nervosa disorder to about 81% for those diagnosed with a social phobia.

Data Set: BBD

⚠ **CAUTION** The information produced in these pie charts should be limited to describing the sample of 92 BDD patients. If one is interested in making inferences about the population of all BDD patients (as were the Brown University researchers), inferential statistical methods need to be applied to the data. These methods are the topics of later chapters. ▲

Exercises 2.1–2.21

Understanding the Principles

2.1 Explain the difference between class frequency, class relative frequency, and class percentage for a qualitative variable.

2.2 Explain the difference between a bar graph and a pie chart.

2.3 Explain the difference between a bar graph and a Pareto diagram.

Learning the Mechanics

2.4 Complete the following table:

Grade on Statistics Exam	Frequency	Relative Frequency
A: 90–100		.08
B: 80–89	36	
C: 65–79	90	
D: 50–64	30	
F: Below 50	28	
Total	200	1.00

2.5 A qualitative variable with three classes (X, Y, and Z) is measured for each of 20 units randomly sampled from a target population. The data (observed class for each unit) are as follows:

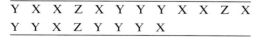

Y X X Z X Y Y Y X X Z X
Y Y X Z Y Y Y X

a. Compute the frequency for each of the three classes.
b. Compute the relative frequency for each of the three classes.
c. Display the results from part **a** in a frequency bar graph.
d. Display the results from part **b** in a pie chart.

Applying the Concepts—Basic

2.6 **Study of importance of libraries.** In *The Canadian Journal of Information and Library Science* (Vol. 33, 2009), researchers from the University of Western Ontario reported on mall shoppers' opinions about libraries and their importance to today's society. Each in a sample of over 200 mall shoppers was asked the following question: "In today's world, with Internet access and online and large book sellers like Amazon, do you think libraries have become more, less, or the same in importance to their community?" The accompanying graphic summarizes the mall shoppers' responses.

a. What type of graph is shown?
b. Identify the qualitative variable described in the graph.
c. From the graph, identify the most common response.
d. Convert the graph into a Pareto diagram. Interpret the results.

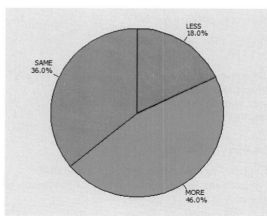

2.7 Cheek teeth of extinct primates. The characteristics of cheek teeth (e.g., molars) can provide anthropologists with information on the dietary habits of extinct mammals. The cheek teeth of an extinct primate species was the subject of research reported in the *American Journal of Physical Anthropology* (Vol. 142, 2010). These data are saved in the **CHEEKTEETH** file. A total of 18 cheek teeth extracted from skulls discovered in western Wyoming were analyzed. Each tooth was classified according to degree of wear (unworn, slight, light-moderate, moderate, moderate-heavy, or heavy). The 18 measurements are listed here.

Data on Degree of Wear	
Unknown	Slight
Unknown	Slight
Unknown	Heavy
Moderate	Unworn
Slight	Light-moderate
Unknown	Light-moderate
Moderate-heavy	Moderate
Moderate	Unworn
Slight	Unknown

a. Identify the variable measured in the study and its type (quantitative or qualitative).
b. Count the number of cheek teeth in each wear category.
c. Calculate the relative frequency for each wear category.
d. Construct a relative frequency bar graph for the data.
e. Construct a Pareto diagram for the data.
f. Identify the degree of wear category that occurred most often in the sample of 18 cheek teeth.

2.8 Humane education and classroom pets. In grade school, did your teacher have a pet in the classroom? Many teachers do so in order to promote humane education. The *Journal of Moral Education* (Mar. 2010) published a study designed to examine elementary school teachers' experiences with humane education and animals in the classroom. Based on a survey of 75 elementary school teachers, the following results were reported:

Survey Question	Response	Number of Teachers
Q1: Do you keep classroom pets?	Yes	61
	No	14
Q2: Do you allow visits by pets?	Yes	35
	No	40
Q3: What formal humane education program does your school employ?	Character counts	21
	Roots of empathy	9
	Teacher designed	3
	Other	14
	None	28

Based on *Journal of Moral Education*, March 2010.

a. Explain why the data collected is qualitative.
b. For each survey question, summarize the results with a graph.
c. For this sample of 75 teachers, write two or three sentences describing their experience with humane education and classroom pets.

2.9 Japanese reading levels. University of Hawaii language professors C. Hitosugi and R. Day incorporated a 10-week extensive reading program into a second-semester Japanese language course in an effort to improve students' Japanese reading comprehension. (*Reading in a Foreign Language*, Apr. 2004.) The professors collected 266 books originally written for Japanese children and required their students to read at least 40 of them as part of the grade in the course. The books were categorized into reading levels (color coded for easy selection) according to length and complexity. The reading levels for the 266 books are summarized in the following table:

Reading Level	Number
Level 1 (Red)	39
Level 2 (Blue)	76
Level 3 (Yellow)	50
Level 4 (Pink)	87
Level 5 (Orange)	11
Level 6 (Green)	3
Total	266

Source: Hitosugi, C. I., and Day, R. R. "Extensive reading in Japanese," *Reading in a Foreign Language*, Vol. 16, No. 1. Apr. 2004 (Table 2). Reprinted with permission from the National Foreign Language Resource Center, University of Hawaii.

a. Calculate the proportion of books at reading level 1 (red).
b. Repeat part a for each of the remaining reading levels.
c. Verify that the proportions in parts **a** and **b** sum to 1.
d. Use the previous results to form a bar graph for the reading levels.
e. Construct a Pareto diagram for the data. Use the diagram to identify the reading level that occurs most often.

2.10 PIN pad shipments. Personal identification number (PIN) pads are devices that connect to point-of-sale electronic cash registers for debit and credit card purchases. The PIN pad allows the customer card to be accessed and the PIN encrypted before it is sent to the transaction manager. *The Nilson Report* (Oct. 2008) listed the volume of PIN pad shipments by manufacturers worldwide in 2007. For the 12 manufacturers listed in the table, a total of 334,039 PIN pads were shipped in 2007 (These data are saved in the **PINPADS** file.)

Manufacturer	Number Shipped (units)
Bitel	13,500
CyberNet	16,200
Fujian Landi	119,000
Glintt (ParaRede)	5,990
Intelligent	4,562
KwangWoo	42,000
Omron	20,000
Pax Tech.	10,072
ProvencoCadmus	20,000
SZZT Electronics	67,300
Toshiba TEC	12,415
Urmet	3,000

Based on *The Nilson Report*, No. 913, Oct. 2008 (p. 9).

a. One of the 334,039 PIN pads is selected and the manufacturer of the pad is determined. What type of data (quantitative or qualitative) is measured?

MINITAB Output for Exercise 2.12

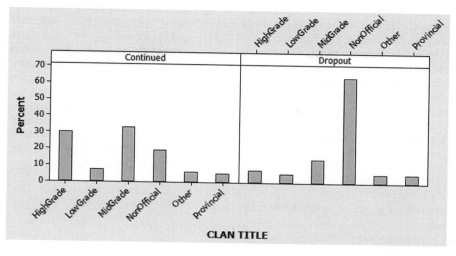

b. Construct a frequency bar chart for the data summarized in the table.

c. Convert the frequency bar chart, part **b,** into a Pareto diagram. Interpret the results.

2.11 Estimating the rhino population. The International Rhino Federation estimates that there are 17,800 rhinoceroses living in the wild in Africa and Asia. A breakdown of the number of rhinos of each species is reported in the accompanying table:

Rhino Species	Population Estimate
African Black	3,610
African White	11,330
(Asian) Sumatran	300
(Asian) Javan	60
(Asian) Indian	2,500
Total	17,800

Based on International Rhino Federation, Mar. 2007.

a. Construct a relative frequency table for the data.
b. Display the relative frequencies in a bar graph.

c. What proportion of the 17,800 rhinos are African rhinos? Asian?

Applying the Concepts—Intermediate

2.12 Genealogy research. Writing in the *Journal of Family History* (Jan. 2010), Hong Kong University historian Sangkuk Lee investigated the genealogy of a certain Korean clan. Of interest to Lee was whether or not a family name continued into the next generation in the clan genealogy (called a *continued line*) or dropped out (called a *dropout line*). Lee used side-by-side pie charts to describe the rate at which certain occupational titles of clan individuals occurred within each line type. Similarly constructed MINITAB bar charts are shown above.
a. Identify the two qualitative variables graphed in the side-by-side pie charts.
b. Give a full interpretation of the charts. Identify the major differences (if any) between the two line groups.

Based on Sangkuk Lee, *Journal of Family History* (Jan 2010).

2.13 Characteristics of ice melt ponds. The National Snow and Ice Data Center (NSIDC) collects data on the albedo, depth, and physical characteristics of ice melt ponds in

SAS output for Exercise 2.13

ICETYPE	Frequency	Percent	Cumulative Frequency	Cumulative Percent
First-year	88	17.46	88	17.46
Landfast	196	38.89	284	56.35
Multi-year	220	43.65	504	100.00

The FREQ Procedure

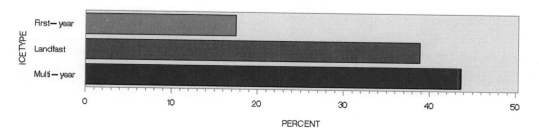

the Canadian Arctic. Environmental engineers at the University of Colorado are using these data to study how climate affects the sea ice. Data on 504 ice melt ponds located in the Barrow Strait in the Canadian Arctic are saved in the **PONDICE** file. One variable of interest is the type of ice observed for each pond, classified as first-year ice, multiyear ice, or landfast ice. An SAS summary table and a horizontal bar graph that describe the types of ice of the 504 melt ponds are shown at the bottom of page 35.

a. Of the 504 melt ponds, what proportion had landfast ice?

b. The University of Colorado researchers estimated that about 17% of melt ponds in the Canadian Arctic have first-year ice. Do you agree?

c. Convert the horizontal bar graph into a Pareto diagram. Interpret the graph.

2.14 **Excavating ancient pottery.** Archaeologists excavating the ancient Greek settlement at Phylakopi classified the pottery found in trenches. (*Chance*, Fall 2000.) The accompanying table describes the collection of 837 pottery pieces uncovered in a particular layer at the excavation site. Construct and interpret a graph that will aid the archaeologists in understanding the distribution of the types of pottery found at the site.

Pot Category	Number Found
Burnished	133
Monochrome	460
Slipped	55
Painted in curvilinear decoration	14
Painted in geometric decoration	165
Painted in naturalistic decoration	4
Cycladic white clay	4
Conical cup clay	2
Total	837

Based on Berg, I., and Bliedon, S. "The pots of Phylakopi: Applying statistical techniques to archaeology." *Chance*, Vol. 13, No. 4, Fall 2000.

2.15 **Satellites in orbit.** According to the Union of Concerned Scientists (www.ucsusa.org), as of April 1, 2009, there were 442 low earth orbit (LEO) and 366 geosynchronous orbit (GEO) satellites in space. Each satellite is owned by an entity in either the government, military, commercial, or civil sector. A breakdown of the number of satellites in orbit for each sector is displayed in the accompanying table. Use this information to construct a pair of graphs that compare the ownership sectors of LEO and GEO satellites in orbit. What observations do you have about the data?

	LEO Satellites	GEO Satellites
Government	193	45
Military	88	67
Commercial	130	253
Civil	31	1
Total	442	366

2.16 **Sociology fieldwork methods.** University of New Mexico professor Jane Hood investigated the fieldwork methods used by qualitative sociologists. (*Teaching Sociology*, July 2006.) Searching for all published journal articles, dissertations, and conference proceedings over the previous seven years in the *Sociological Abstracts* database, she discovered that fieldwork methods could be categorized as follows: Interview, Observation plus Participation, Observation Only, and Grounded Theory. The accompanying table shows the number of papers in each category. (These data are saved in the **FIELDWORK** file.) Use an appropriate graph to portray the results of the study. Interpret the graph.

Fieldwork Method	Number of Papers
Interview	5,079
Observation + Participation	1,042
Observation Only	848
Grounded Theory	537

Based on Hood, J. C. "Teaching against the text: The case of qualitative methods." *Teaching Sociology*, Vol. 34, Issue 3, p. 207 (Exhibit 2).

2.17 **Best-paid CEOs.** *Forbes* magazine conducts an annual survey of the salaries of chief executive officers. In addition to salary information, *Forbes* collects and reports personal data on the CEOs, including their level of education and age. Do most CEOs have advanced degrees, such as master's degrees or doctorates? The data in the table below (saved in the **FORBES40** file) represent the highest degree earned by each of the top 40 best-paid CEOs of 2010. Use a graphical method to summarize the highest degree earned by these CEOs. What is your opinion about whether most CEOs have advanced degrees?

MBA	PhD	Bachelor's	Bachelor's
PhD	Bachelor's	Bachelor's	MBA
None	Law	Bachelor's	Bachelor's
Bachelor's	MBA	None	Master's
Bachelor's	MBA	MBA	MBA
PhD	Bachelor's	None	None
Bachelor's	MBA	MBA	MBA
Masters	Bachelor's	Law	Bachelor's
MBA	MBA	None	Bachelor's
Bachelor's	None	Law	Master's

Based on "40 highest-paid CEOs of 2010" from "What the boss makes: CEO compensation." *Forbes*, April 28, 2010.

2.18 **Do you believe in the Bible?** In its annual General Social Survey (GSS), the National Opinion Research Center (NORC) elicits the opinions of Americans on a wide variety of social topics. One question in the survey asked about a person's belief in the Bible. Data for the approximately 2,800 Americans who participated in the 2004 GSS are saved in the **BIBLE** file. Each respondent selected from one of the following answers: (1) The Bible is the actual word of God and is to be taken literally; (2) the Bible is the inspired word of God, but not everything is to be taken literally; (3) the Bible is an ancient book of fables; and (4) the Bible has some other origin, but is recorded by men. The variable "Bible1" contains the responses.

(*Note:* A response value of 8 represents "Don't Know"; a value of 9 represents a missing value.)

a. Summarize the responses to the Bible question in the form of a relative frequency table.

b. Summarize the responses to the Bible question in a pie chart.

c. Write a few sentences that give a practical interpretation of the results shown in the summary table and graph.

Applying the Concepts—Advanced

2.19 Museum management. What criteria do museums use to evaluate their performance? In a worldwide survey reported in *Museum Management and Curatorship* (June 2010), managers of 30 leading museums of contemporary art were asked to provide the performance measure used most often. A summary of the results is provided in the table. The researcher concluded that "there is a large amount of variation within the museum community with regard to… performance measurement and evaluation." Do you agree? Use a graph to support your conclusion.

Performance Measure	Number of Museums
Total visitors	8
Paying visitors	5
Big shows	6
Funds raised	7
Members	4

2.20 Extinct New Zealand birds. Refer to the *Evolutionary Ecology Research* (July 2003) study of the patterns of extinction in the New Zealand bird population, Exercise 1.18 (p. 19). Data on flight capability (volant or flightless), habitat (aquatic, ground terrestrial, or aerial terrestrial), nesting site (ground, cavity within ground, tree, or cavity above ground), nest density (high or low), diet (fish, vertebrates, vegetables, or invertebrates), body mass (grams), egg length (millimeters), and extinct status (extinct, absent from island, or present) for 132 bird species that existed at the time of the Maori colonization of New Zealand are saved in the **NZBIRDS** file. Use a graphical method to investigate the theory that extinct status is related to flight capability, habitat, and nest density.

2.21 Groundwater contamination in wells. In New Hampshire, about half the counties mandate the use of reformulated gasoline. This has led to an increase in the contamination of groundwater with methyl *tert*-butyl ether (MTBE). *Environmental Science & Technology* (Jan. 2005) reported on the factors related to MTBE contamination in public and private New Hamsphire wells. Data were collected on a sample of 223 wells. These data are saved in the **MTBE** file. Three of the variables are qualitative in nature: well class (public or private), aquifer (bedrock or unconsolidated), and detectable level of MTBE (below limit or detect). [*Note:* A detectable level of MTBE occurs if the MTBE value exceeds .2 microgram per liter.] The data on 11 selected wells are shown in the accompanying table.

(11 selected observations from 223)

Well Class	Aquifer	Detect MTBE?
Private	Bedrock	Below Limit
Private	Bedrock	Below Limit
Public	Unconsolidated	Detect
Public	Unconsolidated	Below Limit
Public	Unconsolidated	Below Limit
Public	Unconsolidated	Below Limit
Public	Unconsolidated	Detect
Public	Unconsolidated	Below Limit
Public	Unconsolidated	Below Limit
Public	Bedrock	Detect
Public	Bedrock	Detect

Based on Ayotte, J. D., Argue, D. M., and McGarry, F. J. "Methyl *tert*-Butyl ether occurrence and related factors in public and private wells in southeast New Hampshire." *Environmental Science & Technology*, Vol. 39, No. 1, Jan. 2005, pp. 9-16.

a. Use graphical methods to describe each of the three qualitative variables for all 223 wells.

b. Use side-by-side bar charts to compare the proportions of contaminated wells for private and public well classes.

c. Use side-by-side bar charts to compare the proportions of contaminated wells for bedrock and unconsolidated aquifers.

d. What inferences can be made from the bar charts of parts **a–c**?

2.2 Graphical Methods for Describing Quantitative Data

Recall from Section 1.4 that quantitative data sets consist of data that are recorded on a meaningful numerical scale. To describe, summarize, and detect patterns in such data, we can use three graphical methods: *dot plots, stem-and-leaf displays,* and *histograms*. Since most statistical software packages can be used to construct these displays, we'll focus here on their interpretation rather than their construction.

For example, the Environmental Protection Agency (EPA) performs extensive tests on all new car models to determine their mileage ratings. Suppose that the 100 measurements in Table 2.2 represent the results of such tests on a certain new car model. How can we summarize the information in this rather large sample?

Table 2.2	EPA Mileage Ratings on 100 Cars			
36.3	41.0	36.9	37.1	44.9
32.7	37.3	41.2	36.6	32.9
40.5	36.5	37.6	33.9	40.2
36.2	37.9	36.0	37.9	35.9
38.5	39.0	35.5	34.8	38.6
36.3	36.8	32.5	36.4	40.5
41.0	31.8	37.3	33.1	37.0
37.0	37.2	40.7	37.4	37.1
37.1	40.3	36.7	37.0	33.9
39.9	36.9	32.9	33.8	39.8
36.8	30.0	37.2	42.1	36.7
36.5	33.2	37.4	37.5	33.6
36.4	37.7	37.7	40.0	34.2
38.2	38.3	35.7	35.6	35.1
39.4	35.3	34.4	38.8	39.7
36.6	36.1	38.2	38.4	39.3
37.6	37.0	38.7	39.0	35.8
37.8	35.9	35.6	36.7	34.5
40.1	38.0	35.2	34.8	39.5
34.0	36.8	35.0	38.1	36.9

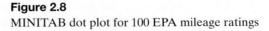 *Data Set:* EPAGAS

A visual inspection of the data indicates some obvious facts. For example, most of the mileages are in the 30s, with a smaller fraction in the 40s. But it is difficult to provide much additional information on the 100 mileage ratings without resorting to some method of summarizing the data. One such method is a dot plot.

Dot Plots

A MINITAB **dot plot** for the 100 EPA mileage ratings is shown in Figure 2.8. The horizontal axis of the figure is a scale for the quantitative variable in miles per gallon (mpg). The rounded (to the nearest half gallon) numerical value of each measurement in the data set is located on the horizontal scale by a dot. When data values repeat, the dots are placed above one another, forming a pile at that particular numerical location. As you can see, this dot plot verifies that almost all of the mileage ratings are in the 30s, with most falling between 35 and 40 miles per gallon.

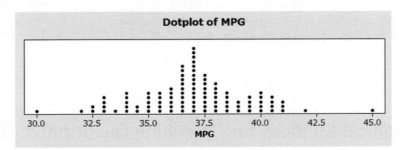

Figure 2.8
MINITAB dot plot for 100 EPA mileage ratings

Stem-and-Leaf Display

Another graphical representation of these same data, a MINITAB **stem-and-leaf display**, is shown in Figure 2.9. In this display, the *stem* is the portion of the measurement (mpg) to the left of the decimal point, while the remaining portion, to the right of the decimal point, is the *leaf*.

In Figure 2.9, the stems for the data set are listed in the second column, from the smallest (30) to the largest (44). Then the leaf for each observation is listed to the right, in the row of the display corresponding to the observation's stem.* For example, the leaf 3 of the first observation (36.3) in Table 2.2 appears in the row corresponding to the stem 36. Similarly, the leaf 7 for the second observation (32.7) in Table 2.2 appears in the row corresponding to the stem 32, while the leaf 5 for the third observation (40.5) appears in the row corresponding to the stem 40. (The stems and leaves for these first three observations are highlighted in Figure 2.9.) Typically, the leaves in each row are ordered as shown in the MINITAB stem-and-leaf display.

Stem-and-Leaf Display: MPG

```
Stem-and-leaf of MPG   N = 100
Leaf Unit = 0.10


    1      30   0
    2      31   8
    6      32   5799
   12      33   126899
   18      34   024588
   29      35   01235667899
   49      36   01233445566777888999
  (21)     37   000011122334456677899
   30      38   0122345678
   20      39   00345789
   12      40   0123557
    5      41   002
    2      42   1
    1      43
    1      44   9
```

Figure 2.9

MINITAB stem-and-leaf display for 100 mileage ratings

The stem-and-leaf display presents another compact picture of the data set. You can see at a glance that the 100 mileage readings were distributed between 30.0 and 44.9, with most of them falling in stem rows 35 to 39. The 6 leaves in stem row 34 indicate that 6 of the 100 readings were at least 34.0, but less than 35.0. Similarly, the 11 leaves in stem row 35 indicate that 11 of the 100 readings were at least 35.0, but less than 36.0. Only five cars had readings equal to 41 or larger, and only one was as low as 30.

The definitions of the stem and leaf for a data set can be modified to alter the graphical description. For example, suppose we had defined the stem as the tens digit for

JOHN TUKEY (1915–2000)

The Picasso of Statistics

Like the legendary artist Pablo Picasso, who mastered and revolutionized a variety of art forms during his lifetime, John Tukey is recognized for his contributions to many subfields of statistics. Born in Massachusetts, Tukey was home schooled, graduated with his bachelor's and master's degrees in chemistry from Brown University, and received his Ph.D. in mathematics from Princeton University. While at Bell Telephone Laboratories in the 1960s and early 1970s, Tukey developed exploratory data analysis, a set of graphical descriptive methods for summarizing and presenting huge amounts of data. Many of these tools, including the stem-and-leaf display and the box plot, are now standard features of modern statistical software packages. (In fact, it was Tukey himself who coined the term *software* for computer programs.) ∎

---

*The first column of the MINITAB stem-and-leaf display represents the cumulative number of measurements from the class interval to the nearest extreme class interval.

the gas mileage data, rather than the ones and tens digits. With this definition, the stems and leaves corresponding to the measurements 36.3 and 32.7 would be as follows:

Stem	Leaf		Stem	Leaf
3	6		3	2

Note that the decimal portion of the numbers has been dropped. Generally, only one digit is displayed in the leaf.

If you look at the data, you'll see why we didn't define the stem this way. All the mileage measurements fall into the 30s and 40s, so all the leaves would fall into just two stem rows in this display. The resulting picture would not be nearly as informative as Figure 2.9.

Now Work Exercise 2.27

Histograms

An SPSS **histogram** for the 100 EPA mileage readings given in Table 2.2 is shown in Figure 2.10. The horizontal axis of the figure, which gives the miles per gallon for a given automobile, is divided into **class intervals**, commencing with the interval from 30–31 and proceeding in intervals of equal size to 44–45 mpg. The vertical axis gives the number (or *frequency*) of the 100 readings that fall into each interval. It appears that about 21 of the 100 cars, or 21%, attained a mileage between 37 and 38 mpg. This class interval contains the highest frequency, and the intervals tend to contain a smaller number of the measurements as the mileages get smaller or larger.

Histograms can be used to display either the frequency or relative frequency of the measurements falling into the class intervals. The class intervals, frequencies, and relative frequencies for the EPA car mileage data are shown in the summary table, Table 2.3.*

By summing the relative frequencies in the intervals 35–36, 36–37, 37–38, and 38–39, you find that 65% of the mileages are between 35 and 39. Similarly, only 2% of

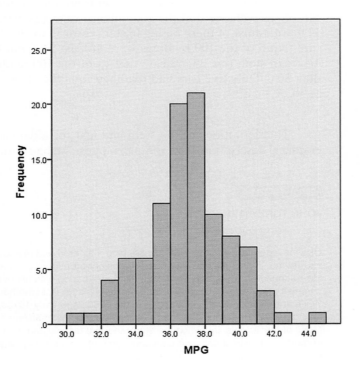

Figure 2.10
SPSS histogram for 100 EPA gas mileage ratings

*SPSS, like many software packages, will classify an observation that falls on the borderline of a class interval into the next-highest interval. For example, the gas mileage of 37.0, which falls on the border between the class intervals 36–37 and 37–38, is classified into the 37–38 class. The frequencies in Table 2.3 reflect this convention.

Table 2.3	Class Intervals, Frequencies, and Relative Frequencies for the Gas Mileage Data	
Class Interval	Frequency	Relative Frequency
30–31	1	0.01
31–32	1	0.01
32–33	4	0.04
33–34	6	0.06
34–35	6	0.06
35–36	11	0.11
36–37	20	0.20
37–38	21	0.21
38–39	10	0.10
39–40	8	0.08
40–41	7	0.07
41–42	3	0.03
42–43	1	0.01
43–44	0	0.00
44–45	1	0.01
Totals	100	1.00

the cars obtained a mileage rating over 42.0. Many other summary statements can be made by further examining the histogram and accompanying summary table. Note that the sum of all class frequencies will always equal the sample size n.

In interpreting a histogram, consider two important facts. First, the proportion of the total area under the histogram that falls above a particular interval on the x-axis is equal to the relative frequency of measurements falling into that interval. For example, the relative frequency for the class interval 37–38 in Figure 2.10 is .20. Consequently, the rectangle above the interval contains .20 of the total area under the histogram.

Second, imagine the appearance of the relative frequency histogram for a very large set of data (representing, say, a population). As the number of measurements in a data set is increased, you can obtain a better description of the data by decreasing the width of the class intervals. When the class intervals become small enough, a relative frequency histogram will (for all practical purposes) appear as a smooth curve. (See Figure 2.11.)

Figure 2.11

The effect of the size of a data set on the outline of a histogram

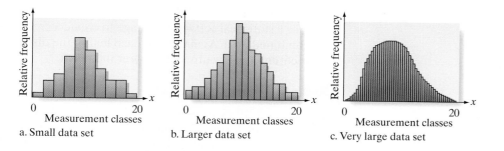

a. Small data set   b. Larger data set   c. Very large data set

Some recommendations for selecting the number of intervals in a histogram for smaller data sets are given in the following box:

Determining the Number of Classes in a Histogram

Number of Observations in Data Set	Number of Classes
Fewer than 25	5–6
25–50	7–14
More than 50	15–20

While histograms provide good visual descriptions of data sets—particularly very large ones—they do not let us identify individual measurements. In contrast, each of the original measurements is visible to some extent in a dot plot and is clearly visible in a stem-and-leaf display. The stem-and-leaf display arranges the data in ascending order, so it's easy to locate the individual measurements. For example, in Figure 2.9 we can easily see that two of the gas mileage measurements are equal to 36.3, but we can't see that fact by inspecting the histogram in Figure 2.10. However, stem-and-leaf displays can become unwieldy for very large data sets. A very large number of stems and leaves causes the vertical and horizontal dimensions of the display to become cumbersome, diminishing the usefulness of the visual display.

Example 2.2

Graphing a Quantitative Variable—The "Water-Level Task"

Problem Over 60 years ago, famous child psychologist Jean Piaget devised a test of basic perceptual and conceptual skills dubbed the "water-level task." Subjects were shown a drawing of a glass being held at a 45° angle and asked to draw a line representing the true surface of the water. Today, research psychologists continue to use the task to test the perception of both adults and children. In one study, the water-level task was given to several groups that included 20 male bartenders and 20 female waitresses (*Psychological Science*, March 1995). For each participant, the researchers measured the deviation (in angle degrees) of the judged line from the true line. These deviations (simulated on the basis of summary results presented in the journal article) are shown in Table 2.4. [*Note:* Deviations can be negative if the judged angle is smaller than the angle of the true line.]

Table 2.4	Water-Level Task Deviations (angle degrees)
Bartenders:	−9 6 10 6 10 −3 7 8 6 14 7 8 −5 2 −1 0 2 3 0 2
Waitresses:	7 10 25 8 10 8 12 9 35 10 12 11 7 10 21 −1 4 0 16 −1

Data Set: WLTASK

a. Use a statistical software package to create a frequency histogram for the combined data in Table 2.4. Then, shade the area under the histogram that corresponds to deviations recorded for waitresses. Interpret the result.

b. Use a statistical software package to create a create a stem-and-leaf display for these combined data. Again, shade each leaf of the display that corresponds to a deviation recorded for a waitress. Interpret the result.

Solution:

a. We used SPSS to generate the frequency histogram, shown in Figure 2.12. Note that SPSS formed 20 classes, with class intervals −20 to −10, −10 to −5, ..., 30 to 35, and 35 to 40. This histogram clearly shows the clustering of the deviation angles between 0° and 15°, with a few deviations in the upper end of the distribution (greater than 20°). SPSS used green bars to shade the areas of the histogram that correspond to the measurements for waitresses. The graph clearly shows that waitresses tend to have greater (positive) deviations than do bartenders, and fewer deviations near 0° relative to bartenders.

b. We used MINITAB to produce the stem-and-leaf display in Figure 2.13. Note that the stem (the second column on the printout) represents the first digit (including 0) in the deviation angle measurement while the leaf (the third column on the printout) represents the second digit. Thus, the leaf 5 in the stem 2 row represents the deviation angle of 25°. The shaded leaves represent deviations recorded for waitresses. As with the histogram, the stem-and-leaf display shows that deviations for waitresses tend to appear in the upper tail of the distribution. Together, the graphs imply that waitresses tend to overestimate the angle of the true line relative to bartenders.

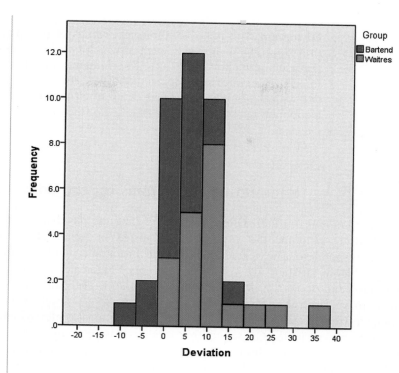

Figure 2.12

SPSS histogram for task deviations

Look Back As is usually the case with data sets that are not too large (say, fewer than 100 measurements), the stem-and-leaf display provides more detail than the histogram without being unwieldy. For instance, the stem-and-leaf display in Figure 2.13 clearly indicates the values of the individual measurements in the data set. For example, the largest deviation angle (representing the measurement 35°) is shown in the last stem row. By contrast, histograms are most useful for displaying very large data sets when the overall shape of the distribution of measurements is more important than the identification of individual measurements.

Stem-and-Leaf Display: Deviation

```
Stem-and-leaf of Deviation  N  = 40
Leaf Unit = 1.0


  2     -0  95
  7     -0  31110
 14      0  0022234
(12)     0  666777788889
 14      1  0000001224
  4      1  6
  3      2  1
  2      2  5
  1      3
  1      3  5
```

Figure 2.13

MINITAB stem-and-leaf display for task deviations

Now Work Exercise 2.31

Most statistical software packages can be used to generate histograms, stem-and-leaf displays, and dot plots. All three are useful tools for graphically describing data sets. We recommend that you generate and compare the displays whenever you can.

Summary of Graphical Descriptive Methods for Quantitative Data

Dot Plot: The numerical value of each quantitative measurement in the data set is represented by a dot on a horizontal scale. When data values repeat, the dots are placed above one another vertically.

Stem-and-Leaf Display: The numerical value of the quantitative variable is partitioned into a "stem" and a "leaf." The possible stems are listed in order in a column. The leaf for each quantitative measurement in the data set is placed in the corresponding stem row. Leaves for observations with the same stem value are listed in increasing order horizontally.

(continued)

> **Histogram:** The possible numerical values of the quantitative variable are partitioned into class intervals, each of which has the same width. These intervals form the scale of the horizontal axis. The frequency or relative frequency of observations in each class interval is determined. A vertical bar is placed over each class interval, with the height of the bar equal to either the class frequency or class relative frequency.

Statistics IN Action Revisited   Interpreting Histograms for the Body Image Data

In the *Body Image: An International Journal of Research* (Jan. 2010) study of 92 BDD patients, the researchers asked each patient to respond to a series of questions on body image (e.g., "How satisfied are you with your physical attractiveness and looks?"). Recall that the scores were summed to yield an Appearance Evaluation score which ranged from 7 to 35 points. This score represents a quantitative variable. Consequently, to graphically investigate whether BDD females tend to be more dissatisfied with their looks than BDD males, we can form side-by-side histograms for the total score, one histogram for females and one for males. These histograms are shown in Figure SIA2.4.

Like the pie charts in the previous Statistics in Action Revisited section, the histograms tend to support the theory. For females, the histogram for appearance evaluation score is centered at about 17 points, while for males the histogram is centered higher, at about 20 points. Also from the histograms you can see that about 55% of the female patients had a score of less than 20, compared to only about 45% of the males. Again, the histograms seem to indicate that BDD females tend to be more dissatisfied with their looks than males. In later chapters, we'll learn how to attach a measure of reliability to such an inference.

Data Set: BBD

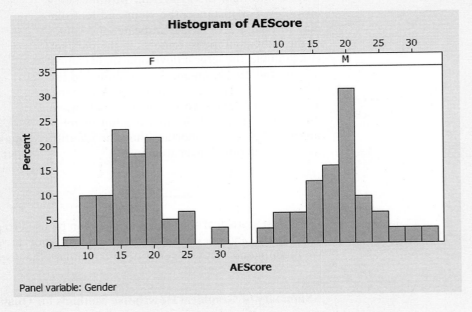

Figure SIA2.4
MINITAB side-by-side histograms for Appearance Evaluation by Gender

Exercises 2.22–2.43

Understanding the Principles

2.22 Explain the difference between a bar graph and a histogram.

2.23 Explain the difference between a dot plot and a stem-and-leaf display.

2.24 Explain the difference between the stem and the leaf in a stem-and-leaf display.

2.25 In a histogram, what are the class intervals?

2.26 How many classes are recommended in a histogram of a data set with more than 50 observations?

Learning the Mechanics

2.27 Consider the stem-and-leaf display shown here:

Stem	Leaf
5	1
4	457
3	00036
2	1134599
1	2248
0	012

a. How many observations were in the original data set?

b. In the bottom row of the stem-and-leaf display, identify the stem, the leaves, and the numbers in the original data set represented by this stem and its leaves.

c. Re-create all the numbers in the data set, and construct a dot plot.

2.28 Graph the relative frequency histogram for the 500 measurements summarized in the accompanying relative frequency table.

Class Interval	Relative Frequency
.5–2.5	.10
2.5–4.5	.15
4.5–6.5	.25
6.5–8.5	.20
8.5–10.5	.05
10.5–12.5	.10
12.5–14.5	.10
14.5–16.5	.05

2.29 Refer to Exercise 2.28. Calculate the number of the 500 measurements falling into each of the measurement classes. Then graph a frequency histogram of these data.

2.30 Consider the MINITAB histogram shown below.

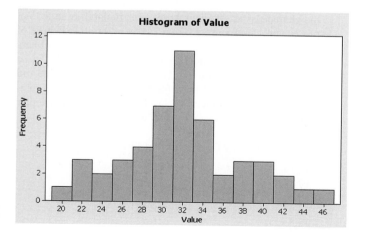

a. Is this a frequency histogram or a relative frequency histogram? Explain.

b. How many class intervals were used in the construction of this histogram?

c. How many measurements are there in the data set described by this histogram?

Applying the Concepts—Basic

2.31 **Body length of armadillos.** A group of environmentalists reported the results of a study of giant armadillos inhabiting the southeastern region of Venezuela. A sample of 80 armadillos was captured and the body length (not including the tail) of each was measured (in centimeters). A MINITAB graph summarizing the data is shown below.

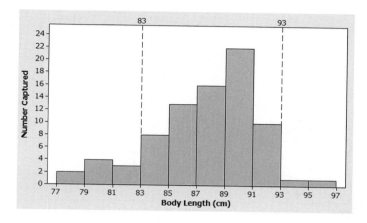

a. What type of graph is employed?

b. How many armadillos have body lengths between 87 and 91 centimeters?

c. What proportion of the armadillos have body lengths between 87 and 91 centimeters?

d. The dotted vertical lines on the graph show the minimum and maximum sizes of giant armadillos that can be legally captured for commercial purposes. What proportion of the captured armadillos are illegal?

2.32 **Is honey a cough remedy?** Does a teaspoon of honey before bed really calm a child's cough? To test the folk remedy, pediatric researchers at Pennsylvania State University carried out a designed study involving a sample of 105 children who were ill with an upper respiratory tract infection (*Archives of Pediatrics and Adolescent Medicine*, Dec. 2007). On the first night, parents rated their children's cough symptoms on a scale from 0 (no problems at all) to 30 (extremely severe). On the second night, the parents were instructed to give their sick child a dosage of liquid "medicine" prior to bedtime. Unknown to the parents, some were given a dosage of dextromethorphan (DM)—an over-the-counter cough medicine—while others were given a similar dose of honey. Also, a third group of parents (the control group) gave their sick children no dosage at all. Again, the parents rated their children's cough symptoms, and the improvement in total cough symptoms score was determined for each child. The data (improvement scores) for the study are shown in the table (p. 46) and saved in the **HONEYCOUGH** file.

a. Construct a dot plot for the coughing improvement scores for the 35 children in the honey dosage group.

b. Refer to part **a.** What coughing improvement score occurred most often in the honey dosage group?

MINITAB dot plot for Exercise 2.32

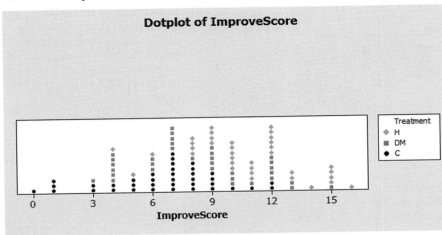

Data for Exercise 2.32

Honey Dosage:	12	11	15	11	10	13	10	4	15	16	9
	14	10	6	10	8	11	12	12	8	12	9
	11	15	10	15	9	13	8	12	10	8	9
	5	12									

DM Dosage:	4	6	9	4	7	7	7	9	12	10	11	6
	3	4	9	12	7	6	8	12	12	4	12	13
	7	10	13	9	4	4	10	15	9			

No Dosage (Control):	5	8	6	1	0	8	12	8	7	7	1	6	7	7
	12	7	9	7	9	5	11	9	5	6	8	8	6	7
	10	9	4	8	7	3	1	4	3					

Based on Paul, I. M., et al. "Effect of honey, dextromethorphan, and no treatment on nocturnal cough and sleep quality for coughing children and their parents." *Archives of Pediatrics and Adolescent Medicine*, Vol. 161, No. 12, Dec. 2007 (data simulated).

c. A MINITAB dot plot for the improvement scores of all three groups is shown above. Note that the green dots represent the children who received a dose of honey, the red dots represent those who got the DM dosage, and the black dots represent the children in the control group. What conclusions can pediatric researchers draw from the graph? Do you agree with the statement (extracted from the article), "Honey may be a preferable treatment for the cough and sleep difficulty associated with childhood upper respiratory tract infection"?

2.33 **Reading Japanese books.** Refer to the *Reading in a Foreign Language* (Apr. 2004) experiment to improve the Japanese reading comprehension levels of University of Hawaii students, Exercise 2.9 (p. 34). Fourteen students participated in a 10-week extensive reading program in a second-semester Japanese language course. The number of books read by each student and the student's grade in the course are listed in the next table and saved in the **JAPANESE** file.

a. Construct a stem-and-leaf display for the number of books read by the students.

b. Highlight (or circle) the leaves in the display that correspond to students who earned an A grade in the course. What inference can you make about these students?

Number of Books	Course Grade
53	A
42	A
40	A
40	B
39	A
34	A
34	A
30	A
28	B
24	A
22	C
21	B
20	B
16	B

Source: Hitosugi, C. I., and Day, R. R. "Extensive reading in Japanese." *Reading in a Foreign Language,* Vol. 16, No. 1, Apr. 2004 (Table 4). Reprinted with permission from the National Foreign Language Resource Center, University of Hawaii.

2.34 **Cheek teeth of extinct primates.** Refer to the *American Journal of Physical Anthropology* (Vol. 142, 2010) study of the characteristics of cheek teeth (e.g., molars) in an extinct primate species, Exercise 2.7 (p. 34). In addition to degree of wear, the researchers recorded the dentary depth of molars (in millimeters) for 18 cheek teeth extracted from skulls. These depth measurements are listed in the accompanying table and saved in the **CHEEKTEETH** file. Summarize the data graphically with a stem-and-leaf display. Is there a particular molar depth that occurs more frequently in the sample?

Data on Dentary Depth (mm) of Molars	
18.12	16.55
19.48	15.70
19.36	17.83
15.94	13.25
15.83	16.12
19.70	18.13
15.76	14.02
17.00	14.04
13.96	16.20

Based on Boyer, D. M., Evans, A. R., and Jernvall, J. "Evidence of dietary differentiation among late Paleocene–Early Eocene Plesiadapids (Mammalia, Primates)." *American Journal of Physical Anthropology*, Vol. 142, © 2010 (Table A3).

2.35 **Sanitation inspection of cruise ships.** To minimize the potential for gastrointestinal disease outbreaks, all passenger cruise ships arriving at U.S. ports are subject to unannounced sanitation inspections. Ships are rated on a 100-point scale by the Centers for Disease Control and Prevention. A score of 86 or higher indicates that the ship is providing an accepted standard of sanitation. The latest (as of Jan. 2010) sanitation scores for 186 cruise ships are saved in the **SHIPSANIT** file. The first five and last five observations in the data set are listed in the following table:

Ship Name	Sanitation Score
Adventure of the Seas	98
AID Aaura	100
Albatross	69
Amadea	84
Amsterdam	99
⋮	⋮
Westerdam	99
Wind Spirit	97
Wind Surf	95
Zaandam	99
Zuiderdam	95

Based on National Center for Environmental Health, Centers for Disease Control and Prevention, Jan. 6, 2010.

a. Generate a stem-and-leaf display of the data. Identify the stems and leaves of the graph.
b. Use the stem-and-leaf display to estimate the proportion of ships that have an accepted sanitation standard.
c. Locate the inspection score of 69 (*Albatrass*) on the stem-and-leaf display.

Applying the Concepts—Intermediate

2.36 **Radioactive lichen.** Lichen has a high absorbance capacity for radiation fallout from nuclear accidents. Since lichen is a major food source for Alaskan caribou, and caribou are, in turn, a major food source for many Alaskan villagers, it is important to monitor the level of radioactivity in lichen. Researchers at the University of Alaska, Fairbanks, collected data on nine lichen specimens at various locations for this purpose. The amount of the radioactive element cesium-137 was measured (in microcuries per milliliter) for each specimen. These data are saved in the **LICHEN** file. The data values, converted to logarithms, are given in the following table (note that the closer the value is to zero, the greater is the amount of cesium in the specimen).

Location			
Bethel	−5.50	−5.00	
Eagle Summit	−4.15	−4.85	
Moose Pass	−6.05		
Turnagain Pass	−5.00		
Wickersham Dome	−4.10	−4.50	−4.60

Based on Lichen Radionuclide Baseline Research Project, 2003, p. 25. *Orion*, University of Alaska–Fairbanks.

a. Construct a dot plot for the nine measurements.
b. Construct a stem-and-leaf display for the nine measurements.
c. Construct a histogram plot of the nine measurements.
d. Which of the three graphs in parts **a–c,** respectively, is most informative?

e. What proportion of the measurements has a radioactivity level of −5.00 or lower?

2.37 **Sound waves from a basketball.** An experiment was conducted to characterize sound waves in a spherical cavity (*American Journal of Physics*, June 2010). A fully inflated basketball, hanging from rubber bands, was struck with a metal rod, producing a series of metallic-sounding pings. Of particular interest were the frequencies of sound waves resulting from the first 24 resonances (echoes). A mathematical formula, well known in physics, was used to compute the theoretical frequencies. The data are saved in the **BBALL** file. These frequencies (measured in hertz) are listed in the table. Use a graphical method to describe the distribution of sound frequencies for the first 24 resonances.

Resonance	Frequency
1	979
2	1572
3	2113
4	2122
5	2659
6	2795
7	3181
8	3431
9	3638
10	3694
11	4038
12	4203
13	4334
14	4631
15	4711
16	4993
17	5130
18	5210
19	5214
20	5633
21	5779
22	5836
23	6259
24	6339

Based on Russell, D. A. "Basketballs as spherical acoustic cavities." *American Journal of Physics*, Vol. 78, No. 6, June 2010 (Table I).

2.38 **Crab spiders hiding on flowers.** Crab spiders use camouflage to hide on flowers while lying in wait to prey on other insects. Ecologists theorize that this natural camouflage also enables the spiders to hide from their own predators, such as birds and lizards. Researchers at the French Museum of Natural History conducted a field test of this theory and published the results in *Behavioral Ecology* (Jan. 2005). They collected a sample of 10 adult female crab spiders, each sitting on the yellow central part of a daisy. The chromatic contrast between each spider and the flower it was sitting on was measured numerically with a spectroradiometer, on which higher values indicate a greater contrast (and, presumably, easier detection by predators). The data for the 10 crab spiders are shown in the following table and saved in the **SPIDER** file.

57	75	116	37	96	61	56	2	43	32

Based on Thery, M., et al. "Specific color sensitivities of prey and predator explain camouflage in different visual systems." *Behavioral Ecology*, Vol. 16, No. 1, Jan. 2005 (Table 1).

a. Summarize the chromatic contrast measurements for the 10 spiders with a stem-and-leaf display.

b. For birds, the detection threshold is 70. (A contrast of 70 or greater allows the bird to see the spider.) Locate the spiders that can be seen by bird predators by circling their respective contrast values on the stem-and-leaf display.

c. Use the result of part **b** to make an inference about the likelihood of a bird detecting a crab spider sitting on the yellow central part of a daisy.

2.39 **Research on brain specimens.** The *postmortem interval* (PMI) is defined as the time elapsed (in days) between death and an autopsy. Knowledge of the PMI is considered essential to conducting medical research on human cadavers. The data in the accompanying table are the PMIs of 22 human brain specimens obtained at autopsy in a recent study (*Brain and Language*, June 1995). Describe the PMI data graphically with a dot plot. On the basis of the plot, make a summary statement about the PMIs of the 22 human brain specimens. (The data are saved in the **BRAINPMI** file.)

Postmortem Intervals for 22 Human Brain Specimens							
5.5	14.5	6.0	5.5	5.3	5.8	11.0	6.1
7.0	14.5	10.4	4.6	4.3	7.2	10.5	6.5
3.3	7.0	4.1	6.2	10.4	4.9		

Based on Hayes, T. L., and Lewis, D. A. "Anatomical specialization of the anterior motor speech area: Hemispheric differences in magnopyramidal neurons." *Brain and Language*, Vol. 49, No. 3, June 1995, p. 292 (Table 1).

2.40 **Research on eating disorders.** Data from a psychology experiment were reported and analyzed in *The American Statistician* (May 2001). Two samples of female students participated in the experiment. One sample consisted of 11 students known to suffer from the eating disorder bulimia; the other sample consisted of 14 students with normal eating habits. Each student completed a questionnaire from which a "fear of negative evaluation" (FNE) score was produced. (The higher the score, the greater was the fear of negative evaluation.) The data are saved in the **BULIMIA** file and displayed in the following table:

Bulimic students: 21 13 10 20 25 19 16 21 24 13 14

Normal students: 13 6 16 13 8 19 23 18 11 19 7 10 15 20

Based on Randles, R. H. "On neutral responses (zeros) in the sign test and ties in the Wilcoxon–Mann–Whitney test." *The American Statistician*, Vol. 55, No. 2, May 2001 (Figure 3).

a. Construct a dot plot or stem-and-leaf display for the FNE scores of all 25 female students.

b. Highlight the bulimic students on the graph you made in part **a**. Does it appear that bulimics tend to have a greater fear of negative evaluation? Explain.

c. Why is it important to attach a measure of reliability to the inference made in part **b**?

Applying the Concepts—Advanced

2.41 **College protests of labor exploitation.** The United Students Against Sweatshops (USAS) was formed by students on U.S. and Canadian college campuses in 1999 to protest labor exploitation in the apparel industry. Clark University sociologist Robert Ross analyzed the USAS movement in the *Journal of World-Systems Research* (Winter 2004). Between 1999 and 2000, there were 18 student "sit-ins" for a "sweat-free campus" organized at several universities. These data are saved in the **SITIN** file. The following table gives the duration (in days) of each sit-in, as well as the number of student arrests. Do the data support the theory that sit-ins of longer duratian are more likely to lead to arrests? Support your answer with a graph.

Sit-in	Year	University	Duration (days)	Number of Arrests	Tier Ranking
1	1999	Duke	1	0	1st
2	1999	Georgetown	4	0	1st
3	1999	Wisconsin	1	0	1st
4	1999	Michigan	1	0	1st
5	1999	Fairfield	1	0	1st
6	1999	North Carolina	1	0	1st
7	1999	Arizona	10	0	1st
8	2000	Toronto	11	0	1st
9	2000	Pennsylvania	9	0	1st
10	2000	Macalester	2	0	1st
11	2000	Michigan	3	0	1st
12	2000	Wisconsin	4	54	1st
13	2000	Tulane	12	0	1st
14	2000	SUNY Albany	1	11	2nd
15	2000	Oregon	3	14	2nd
16	2000	Purdue	12	0	2nd
17	2000	Iowa	4	16	2nd
18	2000	Kentucky	1	12	2nd

Based on Ross, R. J. S. "From antisweatshop to global justice to antiwar: How the new new left is the same and different from the old new left." *Journal of World-Systems Research*, Vol. X, No. 1, Winter 2004 (Table 1 and 3).

2.42 **Comparing SAT scores.** Educators are constantly evaluating the efficacy of public schools in the education and training of U.S. students. One quantitative assessment of change over time is the difference in scores on the SAT, which has been used for decades by colleges and universities as one criterion for admission. The **SATSCORES** file contains the average SAT scores for each of the 50 states and the District of Columbia for 2009 and 2005. Selected observations are shown in the following table:

State	2009	2005
Alabama	1,109	1,126
Alaska	1,036	1,042
Arizona	1,037	1,056
Arkansas	1,144	1,115
California	1,013	1,026
⋮	⋮	⋮
Wisconsin	1,202	1,191
Wyoming	1,135	1,087

Based on College Entrance Examination Board, 2010.

a. Use graphs to display the two SAT score distributions. How have the distributions of state scores changed from 1990 to 2005?

b. As another method of comparing the 1990 and 2005 average SAT scores, compute the **paired difference** by subtracting the 1990 score from the 2005 score for each state. Summarize these differences with a graph.

c. Interpret the graph you made in part **b.** How do your conclusions compare with those of part **a?**

d. Identify the state with the largest improvement in the SAT score between 2005 and 2009.

2.43 **Phishing attacks to e-mail accounts.** *Phishing* is the term used to describe an attempt to extract personal/financial information (e.g., PIN numbers), credit card information, bank account numbers) from unsuspecting people through fraudulent e-mail. An article in *Chance* (Summer 2007) demonstrates how statistics can help identify phishing attempts and make e-commerce safer. Data from an actual phishing attack against an organization were used to determine whether the attack may have been an "inside job" that originated within the company. The company set up a publicized e-mail account—called a "fraud box"—which enabled employees to notify it if they suspected an e-mail phishing attack. The interarrival times, i.e., the time differences (in seconds), for 267 fraud box e-mail notifications were recorded. Researchers showed that if there is minimal or no collaboration or collusion from within the company, the interarrival times would have a frequency distribution similar to the one shown in the accompanying figure. The 267 interarrival times are saved in the **PHISHING** file. Construct a frequency histogram for the interarrival times. Give your opinion on whether the phishing attack against the organization was an "inside job."

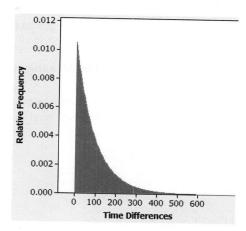

2.3 Summation Notation

Now that we've examined some graphical techniques for summarizing and describing quantitative data sets, we turn to numerical methods for accomplishing that objective. Before giving the formulas for calculating numerical descriptive measures, let's look at some shorthand notation that will simplify our calculation instructions. Remember that such notation is used for one reason only: to avoid repeating the same verbal descriptions over and over. If you mentally substitute the verbal definition of a symbol each time you read it, you'll soon get used to the symbol.

We denote the measurements of a quantitative data set as

$$x_1, x_2, x_3, \ldots, x_n$$

where x_1 is the first measurement in the data set, x_2 is the second measurement in the data set, x_3 is the third measurement in the data set, $\ldots$, and x_n is the nth (and last) measurement in the data set. Thus, if we have five measurements in a set of data, we will write x_1, x_2, x_3, x_4, x_5 to represent the measurements. If the actual numbers are 5, 3, 8, 5, and 4, we have $x_1 = 5, x_2 = 3, x_3 = 8, x_4 = 5,$ and $x_5 = 4$.

Most of the formulas we use require a summation of numbers. For example, one sum we'll need to obtain is the sum of all the measurements in the data set, or $x_1 + x_2 + x_3 + \cdots + x_n$. To shorten the notation, we use the symbol $\sum$ for the summation. That is, $x_1 + x_2 + x_3 + \cdots + x_n = \sum_{i=1}^{n} x_i$. Verbally translate $\sum_{i=1}^{n} x_i$ as follows:

"The sum of the measurements whose typical member is x_i, beginning with the member x_1 and ending with the member x_n."

Suppose, as in our earlier example, $x_1 = 5, x_2 = 3, x_3 = 8, x_4 = 5,$ *and* $x_5 = 4$.

Then the sum of the five measurements, denoted $\sum_{i=1}^{5} x_i$, is obtained as follows:

$$\sum_{i=1}^{5} x_i = x_1 + x_2 + x_3 + x_4 + x_5$$

$$= 5 + 3 + 8 + 5 + 4 = 25$$

Another important calculation requires that we square each measurement and then sum the squares. The notation for this sum is $\sum_{i=1}^{n} x_i^2$. For the five measurements, we have

$$\sum_{i=1}^{5} x_i^2 = x_1^2 + x_2^2 + x_3^2 + x_4^2 + x_5^2$$
$$= 5^2 + 3^2 + 8^2 + 5^2 + 4^2$$
$$= 25 + 9 + 64 + 25 + 16 = 139$$

In general, the symbol following the summation sign $\sum$ represents the variable (or function of the variable) that is to be summed.

The Meaning of Summation Notation $\sum_{i=1}^{n} x_i$

Sum the measurements of the variable that appears to the right of the summation symbol, beginning with the first measurement and ending with the nth measurement.

Exercises 2.44–2.47

Learning the Mechanics

Note: In all exercises, $\sum$ represents $\sum_{i=1}^{n}$.

2.44 A data set contains the observations 5, 1, 3, 2, 1. Find
 a. $\sum x$ **b.** $\sum x^2$ **c.** $\sum(x-1)$
 d. $(\sum x - 1)^2$ **e.** $(\sum x)^2$

2.45 Suppose a data set contains the observations 3, 8, 4, 5, 3, 4, 6. Find

 a. $\sum x$ **b.** $\sum x^2$ **c.** $\sum(x-5)^2$
 d. $\sum(x-2)^2$ **e.** $(\sum x)^2$

2.46 Refer to Exercise 2.44. Find
 a. $\sum x^2 - \dfrac{(\sum x)^2}{5}$ **b.** $\sum(x-2)^2$ **c.** $\sum x^2 - 10$

2.47 A data set contains the observations 6, 0, −2, −1, 3. Find
 a. $\sum x$ **b.** $\sum x^2$ **c.** $\sum x^2 - \dfrac{(\sum x)^2}{5}$

2.4 Numerical Measures of Central Tendency

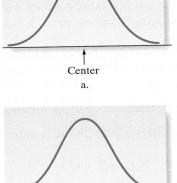

When we speak of a data set, we refer to either a sample or a population. If statistical inference is our goal, we'll ultimately wish to use sample **numerical descriptive measures** to make inferences about the corresponding measures for a population.

As you'll see, a large number of numerical methods are available to describe quantitative data sets. Most of these methods measure one of two data characteristics:

1. The **central tendency** of the set of measurements—that is, the tendency of the data to cluster, or center, about certain numerical values. (See Figure 2.14a.)

2. The **variability** of the set of measurements—that is, the spread of the data. (See Figure 2.14b.)

In this section, we concentrate on **measures of central tendency.** In the next section, we discuss measures of variability.

The most popular and best understood measure of central tendency for a quantitative data set is the *arithmetic mean* (or simply the *mean*) of the data set.

> The **mean** of a set of quantitative data is the sum of the measurements, divided by the number of measurements contained in the data set.

Center
a.

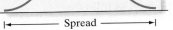

Spread
b.

Figure 2.14
Numerical descriptive measures

In everyday terms, the mean is the average value of the data set and is often used to represent a "typical" value. We denote the **mean of a sample** of measurements by $\bar{x}$ (read "x-bar"), and represent the formula for its calculation as shown in the following box:

Formula for a Sample Mean

$$\bar{x} = \frac{\sum_{i=1}^{n} x_i}{n}$$

Example 2.3
Computing the Sample Mean

Problem Calculate the mean of the following five sample measurements: 5, 3, 8, 5, 6.

Solution Using the definition of sample mean and the summation notation, we find that

$$\bar{x} = \frac{\sum_{i=1}^{5} x_i}{5} = \frac{5 + 3 + 8 + 5 + 6}{5} = \frac{27}{5} = 5.4$$

Thus, the mean of this sample is 5.4.

Look Back There is no specific rule for rounding when calculating $\bar{x}$, because $\bar{x}$ is specifically defined to be the sum of all measurements, divided by n; that is, it is a specific fraction. When $\bar{x}$ is used for descriptive purposes, it is often convenient to round the calculated value of $\bar{x}$ to the number of significant figures used for the original measurements. When $\bar{x}$ is to be used in other calculations, however, it may be necessary to retain more significant figures.

Now Work Exercise 2.57

Example 2.4
Finding the Mean on a Printout— Mean Gas Mileage

Problem Calculate the sample mean for the 100 EPA mileages given in Table 2.2.

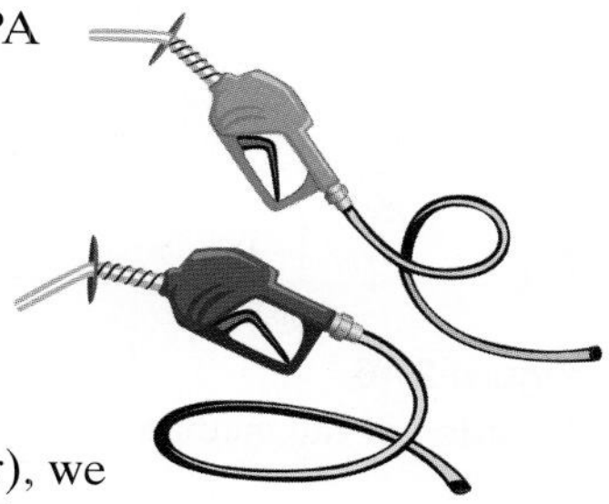

Solution The mean gas mileage for the 100 cars is denoted

$$\bar{x} = \frac{\sum_{i=1}^{100} x_i}{100}$$

Rather than compute $\bar{x}$ by hand (or even with a calculator), we employed SAS to compute the mean. The SAS printout is shown in Figure 2.15. The sample mean, highlighted on the printout, is $\bar{x} = 36.9940$.

Figure 2.15
SAS numerical descriptive measures for 100 EPA gas mileages

The MEANS Procedure

Analysis Variable : MPG

Mean	Std Dev	Variance	N	Minimum	Maximum	Median
36.9940000	2.4178971	5.8462263	100	30.0000000	44.9000000	37.0000000

Look Back Given this information, you can visualize a distribution of gas mileage readings centered in the vicinity of $\bar{x} \approx 37$. An examination of the relative frequency histogram (Figure 2.10) confirms that $\bar{x}$ does in fact fall near the center of the distribution.

The sample mean $\bar{x}$ will play an important role in accomplishing our objective of making inferences about populations on the basis of information about the sample. For this reason, we need to use a different symbol for the *mean of a population*—the mean of the set of measurements on every unit in the population. We use the Greek letter μ (mu) for the population mean.

> **Symbols for the Sample Mean and the Population Mean**
>
> In this text, we adopt a general policy of using Greek letters to represent numerical descriptive measures of the population and Roman letters to represent corresponding descriptive measures of the sample. The symbols for the mean are
>
> $$\bar{x} = \text{Sample mean} \qquad \mu = \text{Population mean}$$

We'll often use the sample mean $\bar{x}$ to estimate (make an inference about) the population mean μ. For example, the EPA mileages for the population consisting of *all* cars has a mean equal to some value μ. Our sample of 100 cars yielded mileages with a mean of $\bar{x} = 36.9940$. If, as is usually the case, we don't have access to the measurements for the entire population, we could use $\bar{x}$ as an estimator or approximator for μ. Then we'd need to know something about the reliability of our inference. That is, we'd need to know how accurately we might expect $\bar{x}$ to estimate μ. In Chapter 7, we'll find that this accuracy depends on two factors:

1. The *size of the sample.* The larger the sample, the more accurate the estimate will tend to be.

2. The *variability, or spread, of the data.* All other factors remaining constant, the more variable the data, the less accurate is the estimate.

Another important measure of central tendency is the **median.**

> The **median** of a quantitative data set is the middle number when the measurements are arranged in ascending (or descending) order.

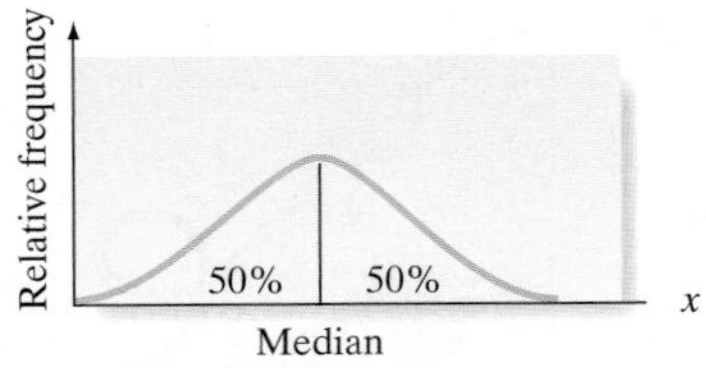

The median is of most value in describing large data sets. If a data set is characterized by a relative frequency histogram (Figure 2.16), the median is the point on the x-axis such that half the area under the histogram lies above the median and half lies below. [*Note*: In Section 2.2, we observed that the relative frequency associated with a particular interval on the x-axis is proportional to the amount of area under the histogram that lies above the interval.] We denote the *median* of a *sample* by M.

Figure 2.16
Location of the median

> **Calculating a Sample Median M**
> Arrange the n measurements from the smallest to the largest.
> 1. If n is odd, M is the middle number.
> 2. If n is even, M is the mean of the middle two numbers.

Example 2.5
Computing the Median

Problem Consider the following sample of $n = 7$ measurements: 5, 7, 4, 5, 20, 6, 2.

a. Calculate the median M of this sample.

b. Eliminate the last measurement (the 2), and calculate the median of the remaining $n = 6$ measurements.

Solution

a. The seven measurements in the sample are ranked in ascending order: 2, 4, 5, 5, 6, 7, 20. Because the number of measurements is odd, the median is the middle measurement. Thus, the median of this sample is $M = 5$.

b. After removing the 2 from the set of measurements, we rank the sample measurements in ascending order as follows: 4, 5, 5, 6, 7, 20. Now the number of measurements is even, so we average the middle two measurements. The median is $M = (5 + 6)/2 = 5.5$.

Look Back When the sample size n is even (as in part **b**), exactly half of the measurements will fall below the calculated median M. However, when n is odd (as in part **a**), the percentage of measurements that fall below M is approximately 50%. The approximation improves as n increases.

Now Work Exercise 2.54

In certain situations, the median may be a better measure of central tendency than the mean. In particular, the median is less sensitive than the mean to extremely large or small measurements. Note, for instance, that all but one of the measurements in part **a** of Example 2.5 are close to $x = 5$. The single relatively large measurement, $x = 20$, does not affect the value of the median, 5, but it causes the mean, $\bar{x} = 7$, to lie to the right of most of the measurements.

As another example of data for which the central tendency is better described by the median than the mean, consider the household incomes of a community being studied by a sociologist. The presence of just a few households with very high incomes will affect the mean more than the median. Thus, the median will provide a more accurate picture of the typical income for the community. The mean could exceed the vast majority of the sample measurements (household incomes), making it a misleading measure of central tendency.

Example 2.6

Finding the Median on a Printout—Median Gas Mileage

Problem Calculate the median for the 100 EPA mileages given in Table 2.2. Compare the median with the mean computed in Example 2.4.

Solution For this large data set, we again resort to a computer analysis. The median is highlighted on the SAS printout displayed in Figure 2.15 (p. 51). You can see that the median is 37.0. Thus, half of the 100 mileages in the data set fall below 37.0 and half lie above 37.0. Note that the median, 37.0, and the mean, 36.9940, are almost equal, a relationship that indicates a lack of **skewness** in the data. In other words, the data exhibit a tendency to have as many measurements in the left tail of the distribution as in the right tail. (Recall the histogram of Figure 2.10.)

Look Back In general, extreme values (large or small) affect the mean more than the median, since these values are used explicitly in the calculation of the mean. The median is not affected directly by extreme measurements, since only the middle measurement (or two middle measurements) is explicitly used to calculate the median. Consequently, if measurements are pulled toward one end of the distribution, the mean will shift toward that tail more than the median will.

A data set is said to be **skewed** if one tail of the distribution has more extreme observations than the other tail.

A comparison of the mean and the median gives us a general method for detecting skewness in data sets, as shown in the next box. With *rightward skewed* data, the right tail (high end) of the distribution has more extreme observations. These few, but large, measurements pull the mean away from the median towards the right; that is, rightward skewness indicates that the mean is greater than the median. Conversely, with *leftward skewed* data, the left tail (low end) of the distribution has more extreme observations. These few, but small, measurements also pull the mean away from the median, but towards the left; consequently, leftward skewness implies that the mean is smaller than the median.

Detecting Skewness by Comparing the Mean and the Median

If the data set is skewed to the right, then the median is less than the mean.

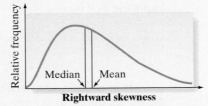

If the data set is symmetric, then the mean equals the median.

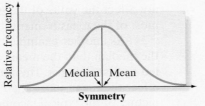

If the data set is skewed to the left, then the mean is less than the median.

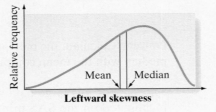

Now Work Exercise 2.53

A third measure of central tendency is the **mode** of a set of measurements.

The **mode** is the measurement that occurs most frequently in the data set.

Therefore, the mode shows where the data tend to concentrate.

Example 2.7	
Finding the Mode	

Problem Each of 10 taste testers rated a new brand of barbecue sauce on a 10-point scale, where 1 = awful and 10 = excellent. Find the mode for the following 10 ratings:

8 7 9 6 8 10 9 9 5 7

Solution Since 9 occurs most often (three times), the mode of the ten taste ratings is 9.

Look Back Note that the data are actually qualitative in nature (e.g., "awful," "excellent"). The mode is particularly useful for describing qualitative data. The modal category is simply the category (or class) that occurs most often.

Now Work Exercise 2.55

Because it emphasizes data concentration, the mode is also used with quantitative data sets to locate the region in which much of the data is concentrated. A retailer of men's clothing would be interested in the modal neck size and sleeve length of potential customers. The modal income class of the laborers in the United States is of interest to the U.S. Department of Labor.

For some quantitative data sets, the mode may r
ple, consider the EPA mileage ratings in Table 2.2. '
that the gas mileage of 37.0 occurs most often (fou
not particularly useful as a measure of central te

A more meaningful measure can be obt·
for quantitative data. The measurement class c
is called the **modal class.** Several definitions exist tc.
within a modal class, but the simplest is to define the mo~
class. For example, examine the frequency histogram for ι.
Figure 2.10 (p. 40). You can see that the modal class is the interva
midpoint) is 37.5 This modal class (and the mode itself) identifies th~
data are most concentrated and, in that sense, is a measure of central tenu~
for most applications involving quantitative data, the mean and median prо
descriptive information than the mode.

Example 2.8

Comparing the Mean, Median, and Mode—Earthquake Aftershocks

Problem Seismologists use the term "aftershock" to describe the smaller earthquakes that follow a main earthquake. Following the Northridge earthquake in 1994, the Los Angeles area experienced a record 2,929 aftershocks in a three-week period. The magnitudes (measured on the Richter scale) of these aftershocks as well as their interarrival times (in minutes) were recorded by the U.S. Geological Survey. (The data are saved in the **EARTHQUAKE** file.) Today seismologists continue to use these data to model future earthquake characteristics. Find and interpret the mean, median, and mode for both of these variables. Which measure of central tendency is better for describing the magnitude distribution? The distribution of interarrival times?

Solution Measures of central tendency for the two variables, magnitude and interarrival time, were produced using MINITAB. The means, medians, and modes are displayed in Figure 2.17.

Figure 2.17
MINITAB descriptive statistics for earthquake data

Descriptive Statistics: MAGNITUDE, INT-TIME

Variable	N	Mean	Median	Mode	N for Mode
MAGNITUDE	2929	2.1197	2.0000	1.8	298
INT-TIME	2928	9.771	6.000	2	354

For magnitude, the mean, median, and mode are 2.12, 2.00, and 1.8, respectively, on the Richter scale. The average magnitude is 2.12; half the magnitudes fall below 2.0; and the most commonly occurring magnitude is 1.8. These values are nearly identical, with the mean slightly larger than the median. This implies a slight rightward skewness in the data, which is shown graphically in the MINITAB histogram for magnitude displayed in Figure 2.18a. Because the distribution is nearly symmetric, any of the three measures would be adequate for describing the "center" of the earthquake aftershock magnitude distribution.

The mean, median, and mode of the interarrival times of the aftershocks are 9.77, 6.0, and 2.0 minutes, respectively. On average, the aftershocks arrive 9.77 minutes apart; half the aftershocks have interarrival times below 6.0 minutes; and the most commonly occurring interarrival time is 2.0 minutes. Note that the mean is much larger than the median, implying that the distribution of interarrival times is highly skewed to the right. This extreme rightward skewness is shown graphically in the histogram, in Figure 2.18b. The skewness is due to several exceptionally large interarrival times. Consequently, we would probably want to use the median of 6.0 minutes as the "typical" interarrival time for the aftershocks. You can see that the mode of 2.0 minutes is not very descriptive of the "center" of the interarrival time distribution.

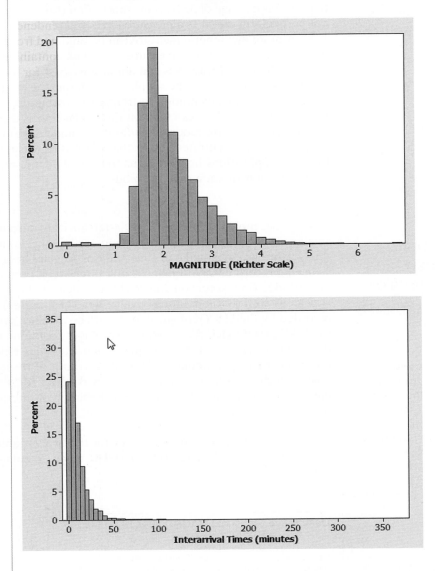

Figure 2.18a
MINITAB Histogram for
Magnitudes of Aftershocks

Figure 2.18b
MINITAB Histogram for
Inter-Arrival Times of
Aftershocks

Look Back The choice of which measure of central tendency to use will depend on the properties of the data set analyzed and the application of interest. Consequently, it is vital that you understand how the mean, median, and mode are computed.

Now Work Exercise 2.68

Exercises 2.48–2.70

Understanding the Principles

2.48 Explain the difference between a measure of central tendency and a measure of variability.

2.49 Give three different measures of central tendency.

2.50 What is the symbol used to represent the sample mean? The population mean?

2.51 What two factors affect the accuracy of the sample mean as an estimate of the population mean?

2.52 Explain the concept of a skewed distribution.

2.53 Describe how the mean compares with the median for a
[NW] distribution as follows:
    a. Skewed to the left
    b. Skewed to the right
    c. Symmetric

Learning the Mechanics

2.54 Calculate the mean and median of the following grade
[NW] point averages:

        3.2   2.5   2.1   3.7   2.8   2.0

2.55 Calculate the mode, mean, and median of the following data:

18 10 15 13 17 15 12 15 18 16 11

2.56 Construct one data set consisting of five measurements, and another consisting of six measurements, for which the medians are equal.

2.57 Calculate the mean for samples for which

NW **a.** $n = 10$, $\sum x = 85$

b. $n = 16$, $\sum x = 400$

c. $n = 45$, $\sum x = 35$

d. $n = 18$, $\sum x = 242$

2.58 Calculate the mean, median, and mode for each of the following samples:
a. 7, –2, 3, 3, 0, 4
b. 2, 3, 5, 3, 2, 3, 4, 3, 5, 1, 2, 3, 4
c. 51, 50, 47, 50, 48, 41, 59, 68, 45, 37

Applet Exercise 2.1

Use the applet entitled *Mean versus Median* to find the mean and median of each of the three data sets presented in Exercise 2.58. For each data set, set the lower limit to a number less than all of the data, set the upper limit to a number greater than all of the data, and then click on *Update*. Click on the approximate location of each data item on the number line. You can get rid of a point by dragging it to the trash can. To clear the graph between data sets, simply click on the trash can.

a. Compare the means and medians generated by the applet with those you calculated by hand in Exercise 2.58. If there are differences, explain why the applet might give values slightly different from the hand-calculated values.

b. Despite providing only approximate values of the mean and median of a data set, describe some advantages of using the applet to find those values.

Applet Exercise 2.2

Use the applet *Mean versus Median* to illustrate your descriptions in Exercise 2.53. For each part **a, b,** and **c,** create a data set with 10 items that has the given property. Using the applet, verify that the mean and median have the relationship you described in Exercise 2.53.

Applet Exercise 2.3

Use the applet *Mean versus Median* to study the effect that an extreme value has on the difference between the mean and median. Begin by setting appropriate limits and plotting the following data on the number line provided in the applet:

0 6 7 7 8 8 8 9 9 10

a. Describe the shape of the distribution and record the value of the mean and median. On the basis of the shape of the distribution, do the mean and median have the relationship that you would expect?

b. Replace the extreme value of 0 with 2, then 4, and then 6. Record the mean and median each time. Describe what is happening to the mean as 0 is replaced, in turn, by the higher numbers stated. What is happening to the

median? How is the difference between the mean and the median changing?

c. Now replace 0 with 8. What values does the applet give you for the mean and the median? Explain why the mean and the median should now be the same.

Applying the Concepts—Basic

2.59 **Characteristics of a rock fall.** In *Environmental Geology* (Vol. 58, 2009) computer simulation was employed to estimate how far a block from a collapsing rock wall will bounce—called *rebound length*—down a soil slope. Based on the depth, location, and angle of block-soil impact marks left on the slope from an actual rockfall, the following 13 rebound lengths (in meters) were estimated. The data are saved in the **ROCKFALL** file. Compute the mean and median of the rebound lengths and interpret these values.

10.94 13.71 11.38 7.26 17.83 11.92 11.87 5.44 13.35 4.90 5.85 5.10 6.77

Based on Paronuzzi, P. "Rockfall-induced block propagation on a soil slope, northern Italy," *Environmental Geology*, Vol. 58, 2009 (Table 2).

2.60 **Most powerful businesswomen in America.** *Fortune* (Oct. 16, 2008) published a list of the 50 most powerful women in business in the United States. The data on age (in years) and title of each of these 50 women are stored in the **WPOWER50** file. The first five and last five observations of the data set are listed in the table below. Descriptive statistics for the ages are displayed on the MINITAB printouts (p. 58).

Rank	Name	Age	Company	Title
1	Indra Nooyi	52	PepsiCo	CEO/Chairman
2	Irene Rosenfeld	55	Kraft Foods	CEO/Chairman
3	Pat Woertz	55	Archer Daniels Midland	CEO/Chairman
4	Anne Mulcahy	55	Xerox	CEO/Chairman
5	Angela Braley	47	Wellpoint	CEO/President
⋮	⋮	⋮	⋮	⋮
46	Lorrie Norrington	48	eBay	CEO
47	Terri Dial	58	Citigroup	CEO
48	Lynn Elsenhans	52	Sunoco	CEO/President
49	Cathie Black	64	Hearst Magazines	President
50	Marissa Mayer	33	Google	Vice president

Based on "50 most powerful businesswomen," *Fortune*, October 16, 2008.

a. Find the mean, median, and modal age of these 50 women on the printout. Interpret these values.

b. What do the mean and median indicate about the skewness of the age distribution?

c. Find the modal age class on the histogram

2.61 **Reading Japanese books.** Refer to the *Reading in a Foreign Language* (Apr. 2004) experiment to improve the Japanese reading comprehension levels of 14 University of Hawaii students, Exercise 2.33 (p. 46). The data are saved in the **JAPANESE** file. The number of books read by each student and the student's course grade are repeated in the table on the next page.

MINITAB Output for Exercise 2.60

Descriptive Statistics: AGE

Variable	N	Mean	StDev	Minimum	Median	Maximum	Mode	N for Mode
AGE	50	50.020	6.444	28.000	51.000	64.000	54	7

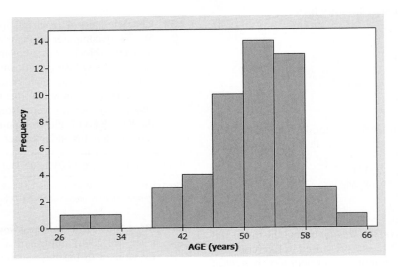

Data for Exercise 2.61

Number of Books	Course Grade	Number of Books	Course Grade
53	A	30	A
42	A	28	B
40	A	24	A
40	B	22	C
39	A	21	B
34	A	20	B
34	A	16	B

Reprinted with permission from the National Foreign Language Resource Center, University of Hawaii.

a. Find the mean, median, and mode of the number of books read. Interpret these values.

b. What do the mean and median indicate about the skewness of the distribution of the data?

2.62 **Is honey a cough remedy?** Refer to the *Archives of Pediatrics and Adolescent Medicine* (Dec. 2007) study of honey as a remedy for coughing, Exercise 2.32 (p. 45). Recall that the 105 ill children in the sample were randomly divided into three groups: those who received a dosage of an over-the-counter cough medicine (DM), those who received a dosage of honey (H), and those who received no dosage (control group). The coughing improvement scores (saved in the **HONEYCOUGH** file) for the patients are reproduced in the next table.

a. Find the median improvement score for the honey dosage group.

b. Find the median improvement score for the DM dosage group.

c. Find the median improvement score for the control group.

d. Based on the results, parts **a–c,** what conclusions can pediatric researchers draw? (We show how to support these conclusions with a measure of reliability in subsequent chapters.)

Honey Dosage:	12	11	15	11	10	13	10	4	15	16	9
	14	10	6	10	8	11	12	12	8	12	9
	11	15	10	15	9	13	8	12	10	8	9
	5	12									

DM Dosage:	4	6	9	4	7	7	7	9	12	10	11
	6	3	4	9	12	7	6	8	12	12	4
	12	13	7	10	13	9	4	4	10	15	9

No Dosage (Control):	5	8	6	1	0	8	12	8	7	7	1
	6	7	7	12	7	9	7	9	5	11	9
	5	6	8	8	6	7	10	9	4	8	7
	3	1	4	3							

Based on Paul, I. M., et al. "Effect of honey, dextromethorphan, and no treatment on nocturnal cough and sleep quality for coughing children and their parents." *Archives of Pediatrics and Adolescent Medicine*, Vol. 161, No. 12, Dec. 2007 (data simulated).

2.63 **Radioactive lichen.** Refer to the University of Alaska study to monitor the level of radioactivity in lichen, presented in Exercise 2.36 (p. 47). The amount of the radioactive element cesium-137 (measured in microcuries per milliliter) for each of nine lichen specimens is repeated in the following table and saved in the **LICHEN** file:

Location			
Bethel	− 5.50	− 5.00	
Eagle Summit	− 4.15	− 4.85	
Moose Pass	− 6.05		
Turnagain Pass	− 5.00		
Wickersham Dome	− 4.10	− 4.50	− 4.60

Based on Lichen Radionuclide Baseline Research Project, 2003, p. 25. *Orion*, University of Alaska-Fairbanks.

a. Find the mean, median, and mode of the radioactivity levels.

b. Interpret the value of each measure of central tendency calculated in part **a**.

Applying the Concepts—Intermediate

2.64 **Ranking driving performance of professional golfers.** A group of Northeastern University researchers developed a new method for ranking the total driving performance of golfers on the Professional Golf Association (PGA) tour (*The Sport Journal*, Winter 2007). The method requires knowing a golfer's average driving distance (yards) and driving accuracy (percent of drives that land in the fairway). The values of these two variables are used to compute a driving performance index. Data for the top 40 PGA golfers (ranked by the new method) are saved in the **PGADRIVER** file. The first five and last five observations are listed in the accompanying table.

Rank	Player	Driving Distance (yards)	Driving Accuracy (%)	Driving Performance Index
1	Woods	316.1	54.6	3.58
2	Perry	304.7	63.4	3.48
3	Gutschewski	310.5	57.9	3.27
4	Wetterich	311.7	56.6	3.18
5	Hearn	295.2	68.5	2.82
⋮	⋮	⋮	⋮	⋮
36	Senden	291	66	1.31
37	Mickelson	300	58.7	1.30
38	Watney	298.9	59.4	1.26
39	Trahan	295.8	61.8	1.23
40	Pappas	309.4	50.6	1.17

Based on Wiseman, F., et al. "A new method for ranking total driving performance on the PGA Tour," *Sports Journal*, Vol. 10, No. 1, Winter 2007 (Table 2).

a. Find the mean, median, and mode for the 40 driving performance index values.

b. Interpret each of the measures of central tendency calculated in part **a.**

c. Use the results from part **a** to make a statement about the type of skewness in the distribution of driving performance indexes. Support your statement with a graph.

2.65 **Training zoo animals.** "The Training Game" is an activity used in psychology in which one person shapes an arbitrary behavior by selectively reinforcing the movements of another person. A group of 15 psychology students at the Georgia Institute of Technology played the game at Zoo Atlanta while participating in an experimental psychology laboratory in which they assisted in the training of animals (*Teaching of Psychology*, May 1998). At the end of the session, each student was asked to rate the statement: "'The Training Game is a great way for students to understand the animal's perspective during training." Responses were recorded on a 7-point scale ranging from 1 (strongly disagree) to 7 (strongly agree). The 15 responses were summarized as follows:

$$\text{mean} = 5.87, \text{mode} = 6.$$

a. Interpret the measures of central tendency in the words of the problem.

b. What type of skewness (if any) is likely to be present in the distribution of student responses? Explain.

2.66 **Symmetric or skewed?** Would you expect the data sets that follow to possess relative frequency distributions that are symmetric, skewed to the right, or skewed to the left? Explain.

a. The salaries of all persons employed by a large university

b. The grades on an easy test

c. The grades on a difficult test

d. The amounts of time students in your class studied last week

e. The ages of automobiles on a used-car lot

f. The amounts of time spent by students on a difficult examination (maximum time is 50 minutes)

2.67 **Cheek teeth of extinct primates.** Refer to the *American Journal of Physical Anthropology* (Vol. 142, 2010) study of the characteristics of cheek teeth (e.g., molars) in an extinct primate species, Exercise 2.34 (p. 46). The data on dentary depth of molars (in millimeters) for 18 cheek teeth extracted from skulls are reproduced below and saved in the **CHEEKTEETH** file.

Data on Dentary Depth (mm) of Molars	
18.12	16.55
19.48	15.70
19.36	17.83
15.94	13.25
15.83	16.12
19.70	18.13
15.76	14.02
17.00	14.04
13.96	16.20

Based on Boyer, D. M., Evans, A. R., and Jernvall, J. "Evidence of dietary differentiation among Late Paleocene–Early Eocene Plesiadapids (Mammalia, Primates)." *American Journal of Physical Anthropology*, Vol. 142, © 2010 (Table A3).

a. Find and interpret the mean of the data set. If the largest depth measurement in the sample were doubled, how would the mean change? Would it increase or decrease?

b. Find and interpret the median of the data set. If the largest depth measurement in the sample were doubled, how would the median change? Would it increase or decrease?

c. Note that there is no single measurement that occurs more than once. How does this fact impact the mode?

2.68 **Mongolian desert ants.** The *Journal of Biogeography* (Dec. 2003) published an article on the first comprehensive study of ants in Mongolia (Central Asia). Botanists placed seed baits at 11 study sites and observed the ant species attracted to each site. Some of the data recorded at each study site are provided in the table on the next page and saved in the **GOBIANTS** file.

a. Find the mean, median, and mode for the number of ant species discovered at the 11 sites. Interpret each of these values.

b. Which measure of central tendency would you recommend to describe the center of the number-of-ant-species distribution? Explain.

c. Find the mean, median, and mode for the percentage of total plant cover at the five Dry Steppe sites only.

d. Find the mean, median, and mode for the percentage of total plant cover at the six Gobi Desert sites only.

Data for Exercise 2.68

Site	Region	Annual Rainfall (mm)	Max. Daily Temp. (°C)	Total Plant Cover (%)	Number of Ant Species	Species Diversity Index
1	Dry Steppe	196	5.7	40	3	.89
2	Dry Steppe	196	5.7	52	3	.83
3	Dry Steppe	179	7.0	40	52	1.31
4	Dry Steppe	197	8.0	43	7	1.48
5	Dry Steppe	149	8.5	27	5	.97
6	Gobi Desert	112	10.7	30	49	.46
7	Gobi Desert	125	11.4	16	5	1.23
8	Gobi Desert	99	10.9	30	4	
9	Gobi Desert	125	11.4	56	4	.76
10	Gobi Desert	84	11.4	22	5	1.26
11	Gobi Desert	115	11.4	14	4	.69

Based on Pfeiffer, M., et al. "Community organization and species richness of ants in Mongolia along an ecological gradient from steppe to Gobi desert." *Journal of Biogeography,* Vol. 30, No. 12, Dec. 2003 (Table 1 and 2).

e. On the basis of the results of parts **c** and **d,** does the center of the distribution for total plant cover percentage appear to be different at the two regions?

Applying the Concepts—Advanced

2.69 **Eye refractive study.** The conventional method of measuring the refractive status of an eye involves three quantities: (1) sphere power, (2) cylinder power, and (3) axis. Optometric researchers studied the variation in these three measures of refraction (*Optometry and Vision Science,* June 1995). Twenty-five successive refractive measurements were obtained on the eyes of over 100 university students. The cylinder power measurements for the left eye of one particular student (ID #11) are listed in the table and saved in the **LEFTEYE** file. [*Note:* All measurements are negative values.]

.08	.08	1.07	.09	.16	.04	.07	.17	.11
.06	.12	.17	.20	.12	.17	.09	.07	.16
.15	.16	.09	.06	.10	.21	.06		

Based on Rubin, A., and Harris, W. F. "Refractive variation during autorefraction: Multivariate distribution of refractive status." *Optometry and Vision Science: Official Publication of the American Academy of Optometry,* Vol. 72, No. 6, June 1995, p. 409 (Table 4).

a. Find measures of central tendency for the data and interpret their values.

b. Note that the data contain one unusually large (negative) cylinder power measurement relative to the other measurements in the data set. Find this measurement. (In Section 2.8, we call this value an **outlier**).

c. Delete the outlier of part **b** from the data set and recalculate the measures of central tendency. Which measure is most affected by the deletion of the outlier?

2.70 **Active nuclear power plants.** The U.S. Energy Information Administration monitors all nuclear power plants operating in the United States. The table below lists the number of active nuclear power plants operating in each of a sample of 20 states. All of the data are saved in the **NUCLEAR** file.

a. Find the mean, median, and mode of this data set.

b. Eliminate the largest value from the data set and repeat part **a.** What effect does dropping this measurement have on the measures of central tendency found in part **a?**

c. Arrange the 20 values in the table from lowest to highest. Next, eliminate the lowest two values and the highest two values from the data set, and find the mean of the remaining data values. The result is called a *10% trimmed mean,* since it is calculated after removing the highest 10% and the lowest 10% of the data values. What advantages does a trimmed mean have over the regular arithmetic mean?

Table for Exercise 2.70

State	Number of Power Plants	State	Number of Power Plants
Alabama	5	New Hampshire	1
Arizona	2	New York	6
California	4	North Carolina	5
Florida	5	Ohio	3
Georgia	4	Pennsylvania	9
Illinois	11	South Carolina	7
Kansas	1	Tennessee	3
Louisiana	2	Texas	4
Massachusetts	1	Vermont	1
Mississippi	1	Wisconsin	3

Based on *Statistical Abstract of the United States,* 2010 (Table 907). U.S. Energy Information Administration, *Electric Power Annual.*

2.5 Numerical Measures of Variability

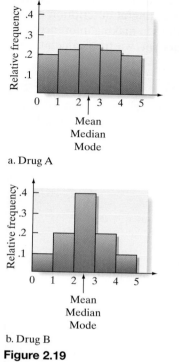

a. Drug A

b. Drug B

Figure 2.19

Response time histograms for two drugs

Measures of central tendency provide only a partial description of a quantitative data set. The description is incomplete without a **measure of the variability,** or **spread,** of the data set. Knowledge of the data set's variability, along with knowledge of its center, can help us visualize the shape of the data set as well as its extreme values.

For example, suppose we want to compare response time to a stimulus for subjects treated with two different drugs, A and B. The histograms for the response times (in seconds) for each of the two drugs are shown in Figure 2.19. If you examine the two histograms, you'll notice that both data sets are symmetric, with equal modes, medians, and means. However, Drug A (Figure 2.19a) has response times spread with almost equal relative frequency over the measurement classes, while Drug B (Figure 2.19b) has most of its response times clustered about its center. Thus, the response times for Drug B are *less variable* than those for Drug A. Consequently, you can see that we need a measure of variability as well as a measure of central tendency to describe a data set.

Perhaps the simplest measure of the variability of a quantitative data set is its *range*.

> The **range** of a quantitative data set is equal to the largest measurement minus the smallest measurement.

The range is easy to compute and easy to understand, but it is a rather insensitive measure of data variation when the data sets are large. This is because two data sets can have the same range and be vastly different with respect to data variation. The phenomenon is demonstrated in Figure 2.19: Both distributions of data shown in the figure have the same range, but we already know that the response times for Drug B are much less variable than those for Drug A. Thus, you can see that the range does not always detect differences in data variation for large data sets.

Let's see if we can find a measure of data variation that is more sensitive than the range. Consider the two samples in Table 2.5: Each has five measurements. (We have ordered the numbers for convenience.) Note that both samples have a mean of 3 and

Table 2.5 Two Hypothetical Data Sets

	Sample 1	Sample 2
Measurements	1, 2, 3, 4, 5	2, 3, 3, 3, 4
Mean	$\bar{x} = \dfrac{1 + 2 + 3 + 4 + 5}{5} = \dfrac{15}{5} = 3$	$\bar{x} = \dfrac{2 + 3 + 3 + 3 + 4}{5} = \dfrac{15}{5} = 3$
Deviations of measurement values from $\bar{x}$	$(1 - 3), (2 - 3), (3 - 3), (4 - 3),$ $(5 - 3)$ or $-2, -1, 0, 1, 2$	$(2 - 3), (3 - 3), (3 - 3), (3 - 3),$ $(4 - 3)$ or $-1, 0, 0, 0, 1$

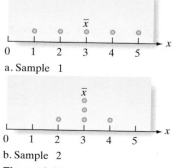

a. Sample 1

b. Sample 2

Figure 2.20

Dot plots for deviations in Table 2.5

that we have calculated the distance and direction, or *deviation,* between each measurement and the mean. What information do these deviations contain? If they tend to be large in magnitude, as in sample 1, the data are spread out, or highly variable. If the deviations are mostly small, as in sample 2, the data are clustered around the mean, $\bar{x}$, and therefore do not exhibit much variability. You can see that these deviations, displayed graphically in Figure 2.20, provide information about the variability of the sample measurements.

The next step is to condense the information in these distances into a single numerical measure of variability. Averaging the deviations from $\bar{x}$ won't help, because the negative and positive deviations cancel; that is, the sum of the deviations (and thus the average deviation) is always equal to zero.

Two methods come to mind for dealing with the fact that positive and negative deviations from the mean cancel. The first is to treat all the deviations as though they were positive, ignoring the sign of the negative values. We won't pursue this line of thought because the resulting measure of variability (the mean of the absolute values of the deviations) presents analytical difficulties beyond the scope of this text. A second method of eliminating the minus signs associated with the deviations is to square them. The quantity we can calculate from the squared deviations provides a meaningful description of the variability of a data set and presents fewer analytical difficulties in making inferences.

To use the squared deviations calculated from a data set, we first calculate the *sample variance.*

> The **sample variance** for a sample of n measurements is equal to the sum of the squared deviations from the mean, divided by $(n - 1)$. The symbol s^2 is used to represent the sample variance.

> **Formula for the Sample Variance:**
>
> $$s^2 = \frac{\sum_{i=1}^{n} (x_i - \bar{x})^2}{n - 1}$$
>
> *Note:* A shortcut formula for calculating s^2 is
>
> $$s^2 = \frac{\sum_{i=1}^{n} x_i^2 - \frac{\left(\sum_{i=1}^{n} x_i\right)^2}{n}}{n - 1}$$

Referring to the two samples in Table 2.5, you can calculate the variance for sample 1 as follows:

$$s^2 = \frac{(1 - 3)^2 + (2 - 3)^2 + (3 - 3)^2 + (4 - 3)^2 + (5 - 3)^2}{5 - 1}$$

$$= \frac{4 + 1 + 0 + 1 + 4}{4} = 2.5$$

The second step in finding a meaningful measure of data variability is to calculate the *standard deviation* of the data set.

> The **sample standard deviation**, s, is defined as the positive square root of the sample variance, s^2, or, mathematically,
>
> $$s = \sqrt{s^2}$$

The population variance, denoted by the symbol σ^2 (sigma squared), is the average of the squared deviations from the mean, μ, of the measurements on *all* units in the population, and σ (sigma) is the square root of this quantity.

> **Symbols for Variance and Standard Deviation**
>
> s^2 = Sample variance
> s = Sample standard deviation
> σ^2 = Population variance
> σ = Population standard deviation

Notice that, unlike the variance, the standard deviation is expressed in the original units of measurement. For example, if the original measurements are in dollars, the variance is expressed in the peculiar units "dollars squared," but the standard deviation is expressed in dollars.

You may wonder why we use the divisor $(n - 1)$ instead of n when calculating the sample variance. Wouldn't using n seem more logical, so that the sample variance would be the average squared distance from the mean? The trouble is, using n tends to produce an underestimate of the population variance σ^2. So we use $(n - 1)$ in the denominator to provide the appropriate correction for this tendency.* Since sample statistics such as s^2 are used primarily to estimate population parameters such as σ^2, $(n - 1)$ is preferred to n in defining the sample variance.

Example 2.9

Computing Measures of Variation

Problem Calculate the variance and standard deviation of the following sample: 2, 3, 3, 3, 4.

Solution As the number of measurements increases, calculating s^2 and s becomes very tedious. Fortunately, as we show in Example 2.10, we can use a statistical software package (or a calculator) to find these values. If you must calculate these quantities by hand, it is advantageous to use the shortcut formula provided on page 62.

To do this, we need two summations: $\sum x$ and $\sum x^2$. These can easily be obtained from the following type of tabulation:

x	x^2
2	4
3	9
3	9
3	9
4	16
$\sum x = 15$	$\sum x^2 = 47$

Then we use[†]

$$s^2 = \frac{\sum_{i=1}^{n} x_i^2 - \frac{\left(\sum_{i=1}^{n} x_i\right)^2}{n}}{n - 1} = \frac{47 - \frac{(15)^2}{5}}{5 - 1} = \frac{2}{4} = .5$$

$$s = \sqrt{.5} = .71$$

Look Back As the sample size n increases, these calculations can become very tedious. As the next example shows, we can use the computer to find s^2 and s.

Now Work Exercise 2.78a

Example 2.10

Finding Measures of Variation on a Printout

Problem Use the computer to find the sample variance s^2 and the sample standard deviation s for the 100 gas mileage readings given in Table 2.2.

Solution The SAS printout describing the gas mileage data is reproduced in Figure 2.21. The variance and standard deviation, highlighted on the printout, are $s^2 = 5.85$ and $s = 2.42$ (rounded to two decimal places).

You now know that the standard deviation measures the variability of a set of data, and you know how to calculate the standard deviation. The larger the standard deviation,

---

Appropriate here means that s^2, with a divisor of $(n - 1)$, is an *unbiased estimator* of σ^2. We define and discuss unbiasedness of estimators in Chapter 6.

[†]In calculating s^2, how many decimal places should you carry? Although there are no rules for the rounding procedure, it is reasonable to retain twice as many decimal places in s^2 as you ultimately wish to have in s. If, for example, you wish to calculate s to the nearest hundredth (two decimal places), you should calculate s^2 to the nearest ten-thousandth (four decimal places).

the more variable the data are. The smaller the standard deviation, the less variation there is in the data. But how can we practically interpret the standard deviation and use it to make inferences? This is the topic of Section 2.6.

Figure 2.21

Reproduction of SAS numerical descriptive measures for 100 EPA mileages

The MEANS Procedure						
Analysis Variable : MPG						
Mean	Std Dev	Variance	N	Minimum	Maximum	Median
36.9940000	2.4178971	5.8462263	100	30.0000000	44.9000000	37.0000000

Exercises 2.71–2.89

Understanding the Principles

2.71 What is the range of a data set?

2.72 What is the primary disadvantage of using the range to compare the variability of data sets?

2.73 Describe the sample variance in words rather than with a formula. Do the same with the population variance.

2.74 Can the variance of a data set ever be negative? Explain. Can the variance ever be smaller than the standard deviation? Explain.

2.75 If the standard deviation increases, does this imply that the data are more variable or less variable?

Learning the Mechanics

2.76 Calculate the variance and standard deviation for samples for which

    a. $n = 10$, $\sum x^2 = 84$, $\sum x = 20$

    b. $n = 40$, $\sum x^2 = 380$, $\sum x = 100$

    c. $n = 20$, $\sum x^2 = 18$, $\sum x = 17$

2.77 Calculate the range, variance, and standard deviation for the following samples:
    a. 39, 42, 40, 37, 41
    b. 100, 4, 7, 96, 80, 3, 1, 10, 2
    c. 100, 4, 7, 30, 80, 30, 42, 2

2.78 Calculate the range, variance, and standard deviation for
[NW] the following samples:
    a. 4, 2, 1, 0, 1
    b. 1, 6, 2, 2, 3, 0, 3
    c. 8, −2, 1, 3, 5, 4, 4, 1, 3
    d. 0, 2, 0, 0, −1, 1, −2, 1, 0, −1, 1, −1, 0, −3, −2, −1, 0, 1

2.79 Using only integers between 0 and 10, construct two data sets with at least 10 observations each such that the two sets have the same mean, but different variances. Construct dot plots for each of your data sets, and mark the mean of each data set on its dot plot.

2.80 Using only integers between 0 and 10, construct two data sets with at least 10 observations each such that the two sets have the same range, but different means. Construct a dot plot for each of your data sets, and mark the mean of each data set on its dot plot.

2.81 Consider the following sample of five measurements: 2, 1, 1, 0, 3.
    a. Calculate the range, s^2, and s.
    b. Add 3 to each measurement and repeat part **a**.

    c. Subtract 4 from each measurement and repeat part **a**.

    d. Considering your answers to parts **a**, **b**, and **c**, what seems to be the effect on the variability of a data set of adding the same number to or subtracting the same number from each measurement?

2.82 Compute s^2, and s for each of the data sets listed. Where appropriate, specify the units in which your answer is expressed.
    a. 3, 1, 10, 10, 4
    b. 8 feet, 10 feet, 32 feet, 5 feet
    c. −1, −4, −3, 1, −4, −4
    d. 1/5 ounce, 1/5 ounce, 1/5 ounce, 2/5 ounce, 1/5 ounce, 4/5 ounce

Applet Exercise 2.4

Use the applet entitled *Standard Deviation* to find the standard deviation of each of the four data sets listed in Exercise 2.78. For each data set, set the lower limit to a number less than all of the data, set the upper limit to a number greater than all of the data, and then click on *Update*. Click on the approximate location of each data item on the number line. You can get rid of a point by dragging it to the trash can. To clear the graph between data sets, simply click on the trash can.
    a. Compare the standard deviations generated by the applet with those you calculated by hand in Exercise 2.78. If there are differences, explain why the applet might give values slightly different from the hand-calculated values.
    b. Despite the fact that it provides a slightly different value of the standard deviation of a data set, describe some advantages of using the applet.

Applet Exercise 2.5

Use the applet *Standard Deviation* to study the effect that multiplying or dividing each number in a data set by the same number has on the standard deviation. Begin by setting appropriate limits and plotting the given data on the number line provided in the applet.

          0 1 1 1 2 2 3 4

    a. Record the standard deviation. Then multiply each data item by 2, plot the new data items, and record the standard deviation. Repeat the process, first multiplying each of the original data items by 3 and then by 4. Describe what happens to the standard deviation as the data items are multiplied by ever higher numbers. Divide each standard deviation by the standard deviation of the original data set. Do you see a pattern? Explain.

b. Divide each of the original data items by 2, plot the new data, and record the standard deviation. Repeat the process, first dividing each of the original data items by 3 and then by 4. Describe what happens to the standard deviation as the data items are divided by ever higher numbers. Divide each standard deviation by the standard deviation of the original data set. Do you see a pattern? Explain.

c. Using your results from parts **a** and **b**, describe what happens to the standard deviation of a data set when each of the data items in the set is multiplied or divided by a fixed number n. Experiment by repeating parts **a** and **b** for other data sets if you need to.

🔘 Applet Exercise 2.6

Use the applet *Standard Deviation* to study the effect that an extreme value has on the standard deviation. Begin by setting appropriate limits and plotting the following data on the number line provided in the applet:

0 6 7 7 8 8 8 9 9 10

a. Record the standard deviation. Replace the extreme value of 0 with 2, then 4, and then 6. Record the standard deviation each time. Describe what happens to the standard deviation as 0 is replaced by ever higher numbers.

b. How would the standard deviation of the data set compare with the original standard deviation if the 0 were replaced by 16? Explain.

Applying the Concepts—Basic

2.83 **Shell lengths of sea turtles.** *Aquatic Biology* (Vol. 9, 2010) reported on a study of green sea turtles inhabiting the Grand Cayman South Sound lagoon. The data on curved carapace (shell) length (in centimeters) for 76 captured turtles are saved in the **TURTLES** file. Descriptive statistics for the data are shown on the accompanying MINITAB printout.

Descriptive Statistics: Length

Variable	N	Mean	StDev	Variance	Minimum	Maximum	Range
Length	76	55.47	11.34	128.57	30.37	81.63	51.26

---

a. Locate the range of the shell lengths on the printout.
b. Locate the variance of the shell lengths on the printout.
c. Locate the standard deviation of the shell lengths on the printout.
d. If the target of your interest is these specific 76 captured turtles, what symbols would you use to represent the variance and standard deviation?

2.84 **Reading Japanese books.** Refer to the *Reading in a Foreign Language* (Apr. 2004) experiment to improve the Japanese reading comprehension levels of 14 University of Hawaii students, presented in Exercise 2.61 (p. 57). The data on number of books read and grade for each student are saved in the **JAPANESE** file.
a. Find the range, variance, and standard deviation of the number of books read by students who earned an A grade.
b. Find the range, variance, and standard deviation of the number of books read by students who earned either a B or C grade.

c. Refer to parts **a** and **b**. Which of the two groups of students has a more variable distribution for number of books read?

2.85 **Is honey a cough remedy?** Refer to the *Archives of Pediatrics and Adolescent Medicine* (Dec. 2007) study of honey as a remedy for coughing, Exercise 2.62 (p. 58). The coughing improvement scores (saved in the **HONEYCOUGH** file) for the patients in the over-the-counter cough medicine dosage (DM) group, honey dosage group, and control group are reproduced in the accompanying table.

Honey Dosage:	12	11	15	11	10	13	10	4	15	16	9
	14	10	6	10	8	11	12	12	8	12	9
	11	15	10	15	9	13	8	12	10	8	9
	5	12									
DM Dosage:	4	6	9	4	7	7	7	9	12	10	11
	6	3	4	9	12	7	6	8	12	12	4
	12	13	7	10	13	9	4	4	10	15	9
No Dosage (Control):	5	8	6	1	0	8	12	8	7	7	1
	6	7	7	12	7	9	7	9	5	11	9
	5	6	8	86	7	10	9	4	8	7	3
	1	4	3								

Based on Paul, I. M., et al. "Effect of honey, dextromethorphan, and no treatment on nocturnal cough and sleep quality for coughing children and their parents." *Archives of Pediatrics and Adolescent Medicine*, Vol. 161, No. 12, Dec. 2007 (data simulated).

a. Find the standard deviation of the improvement scores for the honey dosage group.
b. Find the standard deviation of the improvement scores for the DM dosage group.
c. Find the standard deviation of the improvement scores for the control group.
d. Based on the results, parts **a–c**, which group appears to have the most variability in coughing improvement scores? The least variability?

Applying the Concepts—Intermediate

2.86 **Characteristics of a rockfall.** Refer to the *Environmental Geology* (Vol. 58, 2009) study of how far a block from a collapsing rock wall will bounce, Exercise 2.59 (p. 57). The rebound lengths (meters) for a sample of 13 rock bounces are reproduced in the table below and saved in the **ROCKFALL** file.

10.94	13.71	11.38	7.26	17.83	11.92	11.87
5.44	13.35	4.90	5.85	5.10	6.77	

Based on Paronuzzi, P. "Rockfall-induced block propagation on a soil slope, northern Italy." *Environmental Geology*, Vol. 58, 2009 (Table 2).

a. Compute the range of the 13 rebound lengths. Give the units of measurement of the range.
b. Compute the variance of the 13 rebound lengths. Give the units of measurement of the variance.
c. Compute the standard deviation of the 13 rebound lengths. Give the units of measurement of the standard deviation.

2.87 **Cheek teeth of extinct primates.** Refer to the *American Journal of Physical Anthropology* (Vol. 142, 2010) study of the characteristics of cheek teeth (e.g., molars) in an extinct primate species, Exercise 2.67 (p. 59). The data on dentary depth of molars (in millimeters) for 18 cheek teeth extracted from skulls are reproduced in the next table (p. 66) and saved in the **CHEECKTEETH** file.

Data for Exercise 2.87

Data on Dentary Depth (mm) of Molars		
18.12	15.76	13.25
19.48	17.00	16.12
19.36	13.96	18.13
15.94	16.55	14.02
15.83	15.70	14.04
19.70	17.83	16.20

Based on Boyer, D. M., Evans, A. R., and Jernvall, J. "Evidence of Dietary Differentiation Among Late Paleocene–Early Eocene Plesiadapids (Mammalia, Primates)," *American Journal of Physical Anthropology*, Vol. 142, 2010 (Table A3).

a. Find the range of the data set. If the largest depth measurement in the sample were doubled, how would the range change? Would it increase or decrease?

b. Find the variance of the data set. If the largest depth measurement in the sample were doubled, how would the variance change? Would it increase or decrease?

c. Find the standard deviation of the data set. If the largest depth measurement in the sample were doubled, how would the standard deviation change? Would it increase or decrease?

2.88 **Most powerful women in America.** Refer to Exercise 2.60 (p. 57) and *Fortune's* list of the 50 most powerful women in America. The data are stored in the **WPOWER50** file.

a. Find the range of the ages of these 50 women.

b. Find the variance of the ages of these 50 women.

c. Find the standard deviation of the ages for these 50 women.

d. Suppose the standard deviation of the ages of the most powerful women in Europe is 10 years. For which location, the United States or Europe, is the age data more variable?

e. If the largest age in the data set is omitted, would the standard deviation increase or decrease? Verify your answer.

2.89 **Active nuclear power plants.** Refer to Exercise 2.70 (p. 60) and the U.S. Energy Information Administration's data on the number of nuclear power plants operating in each of 20 states. The data are saved in the **NUCLEAR** file.

a. Find the range, variance, and standard deviation of this data set.

b. Eliminate the largest value from the data set and repeat part **a**. What effect does dropping this measurement have on the measures of variation found in part **a**?

c. Eliminate the smallest and largest value from the data set and repeat part **a**. What effect does dropping both of these measurements have on the measures of variation found in part **a**?

2.6 Interpreting the Standard Deviation

We've seen that if we are comparing the variability of two samples selected from a population, the sample with the larger standard deviation is the more variable of the two. Thus, we know how to interpret the standard deviation on a relative or comparative basis, but we haven't explained how it provides a measure of variability for a single sample.

To understand how the standard deviation provides a measure of variability of a data set, consider the following questions: How many measurements are within one standard deviation of the mean? How many measurements are within two standard deviations? For any particular data set, we can answer these questions by counting the number of measurement in each of the intervals. However, finding an answer that applies to *any* set of data, whether a population or a sample, is more problematic.

Tables 2.6 and 2.7 give two sets of answers to the questions of how many measurements fall within one, two, and three standard deviations of the mean. The first, which applies to *any* set of data, is derived from a theorem proved by the Russian mathematician P. L. Chebyshev (1821–1894). The second, which applies to **mound-shaped, symmetric distributions** of data (for which the mean, median, and mode are all about the same), is based upon empirical evidence that has accumulated over the years. However, the percentages given for the intervals in Table 2.7 provide remarkably good approximations even when the distribution of the data is slightly skewed or asymmetric. Note that the rules apply to either population or sample data sets.

BIOGRAPHY **PAFNUTY L. CHEBYSHEV (1821–1894)**

The Splendid Russian Mathematician

P. L. Chebyshev was educated in mathematical science at Moscow University, eventually earning his master's degree. Following his graduation, Chebyshev joined St. Petersburg (Russia) University as a professor, becoming part of the well-known "Petersburg mathematical school." It was here that he proved his famous theorem about the probability of a measurement being within k standard deviations of the mean (Table 2.6). His fluency in French allowed him to gain international recognition in probability theory. In fact, Chebyshev once objected to being described as a "splendid Russian mathematician," saying he surely was a "worldwide mathematician." One student remembered Chebyshev as "a wonderful lecturer" who "was always prompt for class. As soon as the bell sounded, he immediately dropped the chalk and, limping, left the auditorium." ■

Table 2.6   Interpreting the Standard Deviation: Chebyshev's Rule

Chebyshev's rule applies to any data set, regardless of the shape of the frequency distribution of the data.

  a. It is possible that very few of the measurements will fall within one standard deviation of the mean [i.e., within the interval $(\bar{x} - s, \bar{x} + s)$ for samples and $(\mu - \sigma, \mu + \sigma)$ for populations].

  b. At least $\frac{3}{4}$ of the measurements will fall within two standard deviations of the mean [i.e., within the interval $(\bar{x} - 2s, \bar{x} + 2s)$ for samples and $(\mu - 2\sigma, \mu + 2\sigma)$ for populations].

  c. At least $\frac{8}{9}$ of the measurements will fall within three standard deviations of the mean [i.e., within the interval $(\bar{x} - 3s, \bar{x} + 3s)$ for samples and $(\mu - 3\sigma, \mu + 3\sigma)$ for populations].

  d. Generally, for any number k greater than 1, at least $(1 - 1/k^2)$ of the measurements will fall within k standard deviations of the mean [i.e., within the interval $(\bar{x} - ks, \bar{x} + ks)$ for samples and $(\mu - k\sigma, \mu + k\sigma)$ for populations].

Table 2.7   Interpreting the Standard Deviation: The Empirical Rule

The **empirical rule** is a rule of thumb that applies to data sets with frequency distributions that are mound shaped and symmetric, as follows:

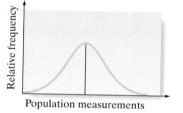

  a. Approximately 68% of the measurements will fall within one standard deviation of the mean [i.e., within the interval $(\bar{x} - s, \bar{x} + s)$ for samples and $(\mu - \sigma, \mu + \sigma)$ for populations].

  b. Approximately 95% of the measurements will fall within two standard deviations of the mean [i.e., within the interval $(\bar{x} - 2s, \bar{x} + 2s)$ for samples and $(\mu - 2\sigma, \mu + 2\sigma)$ for populations].

  c. Approximately 99.7% (essentially all) of the measurements will fall within three standard deviations of the mean [i.e., within the interval $(\bar{x} - 3s, \bar{x} + 3s)$ for samples and $(\mu - 3\sigma, \mu + 3\sigma)$ for populations].

Example 2.11
Interpreting the Standard Deviation— Rat-in-Maze Experiment

Problem Thirty students in an experimental psychology class use various techniques to train a rat to move through a maze. At the end of the course, each student's rat is timed as it negotiates the maze. The results (in minutes) are listed in Table 2.8. Determine the fraction of the 30 measurements in the intervals $\bar{x} \pm s$, $\bar{x} \pm 2s$, and $\bar{x} \pm 3s$, and compare the results with those predicted in Tables 2.6 and 2.7.

Table 2.8 Times (in Minutes) of 30 Rats Running through a Maze									
1.97	.60	4.02	3.20	1.15	6.06	4.44	2.02	3.37	3.65
1.74	2.75	3.81	9.70	8.29	5.63	5.21	4.55	7.60	3.16
3.77	5.36	1.06	1.71	2.47	4.25	1.93	5.15	2.06	1.65

Data Set: RATMAZE

Solution First, we entered the data into the computer and used MINITAB to produce summary statistics. The mean and standard deviation of the sample data, highlighted on the printout shown in Figure 2.22, are

$$\bar{x} = 3.74 \text{ minutes} \qquad s = 2.20 \text{ minutes}$$

rounded to two decimal places.

Figure 2.22
MINITAB descriptive statistics for rat maze times

Descriptive Statistics: RUNTIME

Variable	N	Mean	StDev	Minimum	Median	Maximum
RUNTIME	30	3.744	2.198	0.600	3.510	9.700

Now we form the interval

$$(\bar{x} - s, \bar{x} + s) = (3.74 - 2.20, 3.74 + 2.20) = (1.54, 5.94)$$

A check of the measurements shows that 23 of the times are within this one standard-deviation interval around the mean. This number represents 23/30, or $\approx$ 77%, of the sample measurements.

The next interval of interest is

$$(\bar{x} - 2s, \bar{x} + 2s) = (3.74 - 4.40, 3.74 + 4.40) = (-.66, 8.14)$$

All but two of the times are within this interval, so 28/30, or approximately 93%, are within two standard deviations of $\bar{x}$.

Finally, the three-standard-deviation interval around $\bar{x}$ is

$$(\bar{x} - 3s, \bar{x} + 3s) = (3.74 - 6.60, 3.74 + 6.60) = (-2.86, 10.34)$$

All of the times fall within three standard deviations of the mean.

These one-, two-, and three-standard-deviation percentages (77%, 93%, and 100%) agree fairly well with the approximations of 68%, 95%, and 100% given by the empirical rule (Table 2.7).

Look Back If you look at the MINITAB frequency histogram for this data set (Figure 2.23), you'll note that the distribution is not really mound shaped, nor is it extremely skewed. Thus, we get reasonably good results from the mound-shaped approximations. Of course, we know from Chebyshev's rule (Table 2.6) that no matter what the shape of the distribution, we would expect at least 75% and at least 89% of the measurements to lie within two and three standard deviations of $\bar{x}$, respectively.

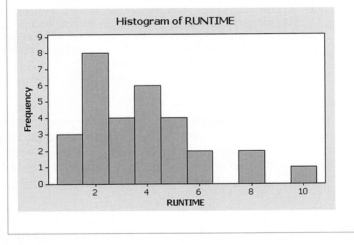

Figure 2.23
MINITAB histogram of rat maze times

Now Work Exercise 2.94

Example 2.12

Checking the Calculation of the Sample Standard Deviation

Problem Chebyshev's rule and the empirical rule are useful as a check on the calculation of the standard deviation. For example, suppose we calculated the standard deviation for the gas mileage data (Table 2.2) to be 5.85. Are there any "clues" in the data that enable us to judge whether this number is reasonable?

Solution The range of the mileage data in Table 2.2 is $44.9 - 30.0 = 14.9$. From Chebyshev's rule and the empirical rule, we know that most of the measurements (approximately 95% if the distribution is mound shaped) will be within two standard deviations of the mean. And regardless of the shape of the distribution and the number of measurements, almost all of them will fall within three standard deviations of the mean. Consequently, we would expect the range of the measurements to be between 4 (i.e., $\pm 2s$) and 6 (i.e., $\pm 3s$)

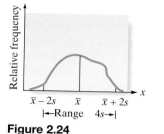

Figure 2.24

The relation between the range and the standard deviation

standard deviations in length. (See Figure 2.24.) For the car mileage data, this means that s should fall between

$$\frac{\text{Range}}{6} = \frac{14.9}{6} = 2.48 \text{ and } \frac{\text{Range}}{4} = \frac{14.9}{4} = 3.73$$

Hence, the standard deviation should not be much larger than 1/4 of the range, particularly for the data set with 100 measurements. Thus, we have reason to believe that the calculation of 5.85 is too large. A check of our work reveals that 5.85 is the variance s^2, not the standard deviation s. (See Example 2.10.) We "forgot" to take the square root (a common error); the correct value is $s = 2.42$. Note that this value is slightly smaller than the range divided by 6 (2.48). The larger the data set, the greater is the tendency for very large or very small measurements (extreme values) to appear, and when they do, the range may exceed six standard deviations.

Look Back In examples and exercises, we'll sometimes use $s \approx \text{range}/4$ to obtain a crude, and usually conservatively large, approximation for s. However, we stress that this is no substitute for calculating the exact value of s when possible.

Now Work Exercise 2.95

In the next example, we use the concepts in Chebyshev's rule and the empirical rule to build the foundation for making statistical inferences.

Example 2.13

Making a Statistical Inference—Car Battery Guarantee

Problem A manufacturer of automobile batteries claims that the average length of life for its grade A battery is 60 months. However, the guarantee on this brand is for just 36 months. Suppose the standard deviation of the life length is known to be 10 months and the frequency distribution of the life-length data is known to be mound shaped.

a. Approximately what percentage of the manufacturer's grade A batteries will last more than 50 months, assuming that the manufacturer's claim is true?

b. Approximately what percentage of the manufacturer's batteries will last less than 40 months, assuming that the manufacturer's claim is true?

c. Suppose your battery lasts 37 months. What could you infer about the manufacturer's claim?

Solution If the distribution of life length is assumed to be mound shaped with a mean of 60 months and a standard deviation of 10 months, it would appear as shown in Figure 2.25. Note that we can take advantage of the fact that mound-shaped distributions are (approximately) symmetric about the mean, so that the percentages given by the empirical rule can be split equally between the halves of the distribution on each side of the mean.

For example, since approximately 68% of the measurements will fall within one standard deviation of the mean, the distribution's symmetry implies that approximately

Figure 2.25

Battery life-length distribution: manufacturer's claim assumed true

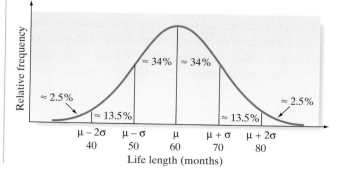

(1/2)(68%) = 34% of the measurements will fall between the mean and one standard deviation on each side. This concept is illustrated in Figure 2.25. The figure also shows that 2.5% of the measurements lie beyond two standard deviations in each direction from the mean. This result follows from the fact that if approximately 95% of the measurements fall within two standard deviations of the mean, then about 5% fall outside two standard deviations; if the distribution is approximately symmetric, then about 2.5% of the measurements fall beyond two standard deviations on each side of the mean.

a. It is easy to see in Figure 2.25 that the percentage of batteries lasting more than 50 months is approximately 34% (between 50 and 60 months) plus 50% (greater than 60 months). Thus, approximately 84% of the batteries should have a life exceeding 50 months.

b. The percentage of batteries that last less than 40 months can also be easily determined from Figure 2.25: Approximately 2.5% of the batteries should fail prior to 40 months, assuming that the manufacturer's claim is true.

c. If you are so unfortunate that your grade A battery fails at 37 months, you can make one of two inferences: Either your battery was one of the approximately 2.5% that fail prior to 40 months, or something about the manufacturer's claim is not true. Because the chances are so small that a battery fails before 40 months, you would have good reason to have serious doubts about the manufacturer's claim. A mean smaller than 60 months or a standard deviation longer than 10 months would each increase the likelihood of failure prior to 40 months.*

Look Back The approximations given in Figure 2.25 are more dependent on the assumption of a **mound-shaped distribution** than are the assumptions stated in the empirical rule (Table 2.7), because the approximations in Figure 2.25 depend on the (approximate) symmetry of the mound-shaped distribution. We saw in Example 2.11 that the empirical rule can yield good approximations even for skewed distributions. This will *not* be true of the approximations in Figure 2.25; the distribution *must* be mound shaped and (approximately) symmetric.

Example 2.13 is our initial demonstration of the statistical inference-making process. At this point, you should realize that we'll use sample information (in Example 2.13, your battery's failure at 37 months) to make inferences about the population (in Example 2.13, the manufacturer's claim about the length of life for the population of all batteries). We'll build on this foundation as we proceed.

Statistics IN Action   Revisited   Interpreting Descriptive Statistics for the Body Image Data

We return to the analysis of the data from the *Body Image: An International Journal of Research* (Jan. 2010) study of 92 BDD patients. Recall that the quantitative variable of interest in the study is Total Appearance Evaluation score (ranging from 7 to 35 points). One of the questions of interest to the Brown University researchers was whether female BDD patients were less satisfied in their body image than normal females (i.e., females with no disorders), and whether male BDD patients were less satisfied in their body image than normal males. The analysis involved comparing the mean score for the BDD patients to the mean

score for a normal group (called a "norm"). The appearance evaluation "norms" for females and males were 23.5 and 24.4, respectively.

MINITAB descriptive statistics for the data in the **BDD** file are displayed in Figure SIA2.5, with means and standard deviations highlighted. The sample mean score for females is 17.05 while the sample mean for males is 19.0. Note that both of these values fall well below the "norm" for each respective gender. Consequently, *for this sample of BDD*

---

*The assumption that the distribution is mound shaped and symmetric may also be incorrect. However, if the distribution were skewed to the right, as life-length distributions often tend to be, the percentage of measurements more than two standard deviations below the mean would be even less than 2.5%.

Statistics IN Action
(continued)

Figure SIA2.5
MINITAB descriptive statistics for appearance evaluation by gender

Descriptive Statistics: AEScore

Variable	Gender	N	Mean	StDev	Variance	Minimum	Maximum
AEScore	F	60	17.050	4.757	22.625	7.000	30.000
	M	32	19.000	5.424	29.419	7.000	32.000

patients, both females and males have a lower average opinion about their body image than do normal females and males.

To interpret the standard deviations, we substitute into the formula, $\bar{x} \pm 2s$, to obtain the intervals:

Females: $\bar{x} \pm 2s = 17.05 \pm 2(4.76) = 17.05 \pm 9.52$
  $= (7.53, 26.57)$

Males: $\bar{x} \pm 2s = 19.00 \pm 2(5.42) = 19.00 \pm 10.84$
  $= (8.16, 29.84)$

From Chebyshev's rule (Table 2.6), we know that at least 75% of the females in the sample of BDD patients will have an Appearance Evaluation score anywhere between 7.53 and 26.57 points. Similarly, we know that at least 75% of the males in the sample of BDD patients will have an Appearance Evaluation score anywhere between 8.16 and 29.84 points. You can see that the interval for the females lies slightly below the interval for males. How does this information help in determining whether the female BDD patients and male BDD patients have mean scores below the "norm" for their gender? In Chapters 7 and 8 we will learn about an inferential method that uses both the sample mean and sample standard deviation to determine whether the population mean of the BDD patients is really smaller than the "norm" value.

Data Set: BBD

Exercises 2.90–2.107

Understanding the Principles

2.90 To what kind of data sets can Chebyshev's rule be applied? How about the empirical rule?

2.91 The output from a statistical computer program indicates that the mean and standard deviation of a data set consisting of 200 measurements are $1,500 and $300, respectively.
   a. What are the units of measurement of the variable of interest? On the basis of the units, what type of data is this, quantitative or qualitative?
   b. What can be said about the number of measurements between $900 and $2,100? between $600 and $2,400? between $1,200 and $1,800? between $1,500 and $2,100?

2.92 For any set of data, what can be said about the percentage of the measurements contained in each of the following intervals?
   a. $\bar{x} - s$ to $\bar{x} + s$
   b. $\bar{x} - 2s$ to $\bar{x} + 2s$
   c. $\bar{x} - 3s$ to $\bar{x} + 3s$

2.93 For a set of data with a mound-shaped relative frequency distribution, what can be said about the percentage of the measurements contained in each of the intervals specified in Exercise 2.92?

Learning the Mechanics

2.94 The following is a sample of 25 measurements. The data are saved in the **LM2_94** file.

7	6	6	11	8	9	11	9	10	8	7	7	5
9	10	7	7	7	7	9	12	10	10	8	6	

NW **a.** Compute $\bar{x}$, s^2, and s for this sample.
   b. Count the number of measurements in the intervals $\bar{x} \pm s$, $\bar{x} \pm 2s$, and $\bar{x} \pm 3s$. Express each count as a percentage of the total number of measurements.

   c. Compare the percentages found in part **b** with the percentages given by the empirical rule and Chebyshev's rule.
   d. Calculate the range and use it to obtain a rough approximation for s. Does the result compare favorably with the actual value for s found in part **a**?

2.95 Given a data set with a largest value of 760 and a smallest value of 135, what would you estimate the standard deviation to be? Explain the logic behind the procedure you used to estimate the standard deviation. Suppose the standard deviation is reported to be 25. Is this number reasonable? Explain.

Applying the Concepts—Basic

2.96 **Most powerful women in America.** Refer to the *Fortune* list of the 50 most powerful women in America, saved in the **WPOWER50** file. In Exercise 2.60 (p. 57) you found the mean age of the 50 women in the data set, and in Exercise 2.88 (p. 66) you found the standard deviation. Use these values to form an interval that will contain at least 75% of the ages in the data set.

2.97 **Sanitation inspection of cruise ships.** Refer to Exercise 2.35 (p. 47) and the Centers for Disease Control and Prevention listing of the sanitation scores for 186 cruise ships. The data are saved in the **SHIPSANIT** file.
   a. Find the mean and standard deviation of the sanitation scores.
   b. Calculate the intervals $\bar{x} \pm s$, $\bar{x} \pm 2s$, and $\bar{x} \pm 3s$.
   c. Find the percentage of measurements in the data set that fall within each of the intervals in part **b**. Do these percentages agree with either Chebyshev's rule or the empirical rule?

2.98 **Motivation of drug dealers.** Researchers at Georgia State University investigated the personality characteristics of drug dealers in order to shed light on their motivation for participating in the illegal drug market (*Applied Psychology in Criminal Justice*, Sept. 2009). The sample consisted of 100 convicted drug dealers who attended a court-mandated counseling program. Each dealer was scored on the Wanting Recognition (WR) Scale, which provides a quantitative measure of a person's level of need for approval and sensitivity to social situations. (Higher scores indicate a greater need for approval.) The sample of drug dealers had a mean WR score of 39, with a standard deviation of 6. Assume the distribution of WR scores for drug dealers is mound shaped and symmetric.

a. Give a range of WR scores that will contain about 95% of the scores in the drug dealer sample.

b. What proportion of the drug dealers will have WR scores above 51?

c. Give a range of WR sores that will contain nearly all the scores in the drug dealer sample.

2.99 Dentists' use of anesthetics. A study published in *Current Allergy & Clinical Immunology* (Mar. 2004) investigated allergic reactions of dental patients to local anesthetics. Based on a survey of dental practitioners, the study reported that the mean number of units (ampoules) of local anesthetics used per week by dentists was 79, with a standard deviation of 23. Suppose we want to determine the percentage of dentists who use less than 102 units of local anesthetics per week.

a. Assuming that nothing is known about the shape of the distribution for the data, what percentage of dentists use less than 102 units of local anesthetics per week?

b. Assuming that the data has a mound-shaped distribution, what percentage of dentists use less than 102 units of local anesthetics per week?

Applying the Concepts—Intermediate

2.100 Laptop use in middle school. Many middle schools have initiated a program that provides every student with a free laptop (notebook) computer. Student usage of laptops at a middle school that participates in the initiative was investigated in *American Secondary Education* (Fall 2009). In a sample of 106 students, the researchers reported the following statistics on how many minutes per day each student used his or her laptop for taking notes: $\bar{x} = 13.2, s = 19.5$

a. Compute the interval, $\bar{x} \pm 2s$.

b. Explain why the distribution of laptop usage for taking notes for these 106 students cannot be symmetric.

c. Given your answer to part **b,** what percentage of the 106 students have laptop usages that fall within the $\bar{x} \pm 2s$ interval?

2.101 Hand washing versus hand rubbing. In hospitals, washing the hands with soap is emphasized as the single most important measure in the prevention of infections. As an alternative to hand washing, some hospitals allow health workers to rub their hands with an alcohol-based antiseptic. The *British Medical Journal* (Aug. 17, 2002) reported on a study to compare the effectiveness of washing the hands with soap and rubbing the hands with alcohol. One group of health care workers used hand rubbing, while a second group used hand washing to clean their hands. The bacterial count (number of colony-forming units) on the hand of each worker was recorded. The table gives descriptive statistics on bacteria counts for the two groups of health care workers.

	Mean	Standard Deviation
Hand rubbing	35	59
Hand washing	69	106

a. For hand rubbers, form an interval that contains at least 75% of the bacterial counts. (*Note:* The bacterial count cannot be less than 0.)

b. Repeat part **a** for hand washers.

c. On the basis of your results in parts **a** and **b,** make an inference about the effectiveness of the two hand-cleaning methods.

2.102 Extinct New Zealand birds. Refer to the *Evolutionary Ecology Research* (July 2003) study of the patterns of extinction in the New Zealand bird population, presented in Exercise 2.20 (p. 37). Consider the data on the egg length (measured in millimeters) for the 132 bird species saved in the **NZBIRDS** file.

a. Find the mean and standard deviation of the egg lengths.

b. Form an interval that can be used to predict the egg length of a bird species found in New Zealand.

2.103 Velocity of Winchester bullets. The *American Rifleman* (June 1993) reported on the velocity of ammunition fired from the FEG P9R pistol, a 9-mm gun manufactured in Hungary. Field tests revealed that Winchester bullets fired from the pistol had a mean velocity (at 15 feet) of 936 feet per second and a standard deviation of 10 feet per second. Tests were also conducted with Uzi and Black Hills ammunition.

a. Describe the velocity distribution of Winchester bullets fired from the FEG P9R pistol.

b. A bullet whose brand is unknown is fired from the FEG P9R pistol. Suppose the velocity (at 15 feet) of the bullet is 1,000 feet per second. Is the bullet likely to be manufactured by Winchester? Explain.

2.104 Sentence complexity study. A study published in *Applied Psycholinguistics* (June 1998) compared the language skills of young children (16–30 months old) from low-income and middle-income families. A total of 260 children—65 in the low-income and 195 in the middle-income group—completed the Communicative Development Inventory (CDI) exam. One of the variables measured on each child was sentence complexity score. Summary statistics for the scores of the two groups are reproduced in the accompanying table. Use this information to sketch a graph of the sentence complexity score distribution for each income group. (Assume that the distributions are mound shaped and symmetric.) Compare the distributions. What can you infer?

	Low Income	Middle Income
Sample Size	65	195
Mean	7.62	15.55
Standard Deviation	8.91	12.24
Minimum	0	0
Maximum	36	37

Based on Arriaga, R. I., et al. "Scores on the MacArthur Communicative Development Inventory of children from low-income and middle-income families." *Applied Psycholinguistics,* Vol. 19, No. 2, June 1998, p. 217 (Table 7).

Applying the Concepts—Advanced

2.105 Improving SAT scores. The National Education Longitudinal Survey (NELS) tracks a nationally representative sample of U.S. students from eighth grade through high school and college. Research published in *Chance* (Winter 2001) examined the SAT scores of 265 NELS students who paid a private tutor to help them improve their scores. The table summarizes the changes in both the SAT-Mathematics and SAT-Verbal scores for these students.

	SAT-Math	SAT-Verbal
Mean change in score	19	7
Standard deviation of score changes	65	49

a. Suppose one of the 265 students who paid a private tutor is selected at random. Give an interval that is likely to contain the change in this student's SAT-Math score.

b. Repeat part **a** for the SAT-Verbal score.

c. Suppose the selected student's score increased on one of the SAT tests by 140 points. Which test, the SAT-Math or SAT-Verbal, is the one most likely to have had the 140-point increase? Explain.

2.106 Animal-assisted therapy for heart patients. Recall the *American Heart Association Conference* (Nov. 2005) study to gauge whether animal-assisted therapy can improve the physiological responses of heart failure patients. (See Chapter 1, Study 1.3, p. 4.) A team of nurses from the UCLA Medical Center randomly divided 76 heart patients into three groups. Each patient in group T was visited by a human volunteer accompanied by a trained dog, each patient in group V was visited by a volunteer only, and the patients in group C were not visited at all. The anxiety level of each patient was measured (in points) both before and after the visits. The accompanying table gives summary statistics for the drop in anxiety level for patients in the three groups. Suppose the anxiety level of a patient selected from the study had a drop of 22.5 points. From which group is the patient more likely to have come? Explain.

	Sample Size	Mean Drop	Std. Dev.
Group T: Volunteer + trained dog	26	10.5	7.6
Group V: Volunteer only	25	3.9	7.5
Group C: Control group (no visit)	25	1.4	7.5

Based on Cole, K., et al. "Animal assisted therapy decreases hemodynamics, plasma epinephrine and state anxiety in hospitalized heart failure patients." *American Journal of Critical Care*, 2007, 16: 575–585.

2.107 Land purchase decision. A buyer for a lumber company must decide whether to buy a piece of land containing 5,000 pine trees. If 1,000 of the trees are at least 40 feet tall, the buyer will purchase the land; otherwise, he won't. The owner of the land reports that the height of the trees has a mean of 30 feet and a standard deviation of 3 feet. On the basis of this information, what is the buyer's decision?

2.7 Numerical Measures of Relative Standing

We've seen that numerical measures of central tendency and variability describe the general nature of a quantitative data set (either a sample or a population). In addition, we may be interested in describing the *relative* quantitative location of a particular measurement within a data set. Descriptive measures of the relationship of a measurement to the rest of the data are called **measures of relative standing.**

One measure of the relative standing of a measurement is its **percentile ranking.** For example, suppose you scored an 80 on a test and you want to know how you fared in comparison with others in your class. If the instructor tells you that you scored at the 90th percentile, it means that 90% of the grades were lower than yours and 10% were higher. Thus, if the scores were described by the relative frequency histogram in Figure 2.26, the 90th percentile would be located at a point such that 90% of the total area under the relative frequency histogram lies below the 90th percentile and 10% lies above. If the instructor tells you that you scored in the 50th percentile (the median of the data set), 50% of the test grades would be lower than yours and 50% would be higher.

Percentile rankings are of practical value only with large data sets. Finding them involves a process similar to the one used in finding a median. The measurements are ranked in order, and a rule is selected to define the location of each percentile. Since we are interested primarily in interpreting the percentile rankings of measurements (rather than in finding particular percentiles for a data set), we define the *p*th *percentile* of a data set as follows.

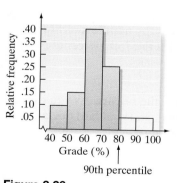

Figure 2.26

Location of 90th percentile for test grades

For any set of *n* measurements (arranged in ascending or descending order), the ***p*th percentile** is a number such that *p*% of the measurements fall below that number and $(100 - p)$% fall above it.

Example 2.14

Finding and Interpreting Percentiles—The "Water Level Task"

Problem Refer to the water level task deviations for the 40 subjects in Table 2.4. An SPSS printout describing the data is shown in Figure 2.27. Locate the 25th percentile and 95th percentile on the printout, and interpret the associated values.

Solution Both the 25th percentile and 95th percentile are highlighted on the SPSS printout. The values in question are 2.0 and 24.8, respectively. Our interpretations are as follows: 25% of the 40 deviations fall below 2.0 and 95% of the deviations fall below 24.8.

Case Processing Summary

	Cases					
	Valid		Missing		Total	
	N	Percent	N	Percent	N	Percent
Deviation	40	100.0%	0	.0%	40	100.0%

Percentiles

		Percentiles						
		5	10	25	50	75	90	95
Weighted Average (Definition 1)	Deviation	-4.90	-1.00	2.00	7.00	10.00	15.80	24.80
Tukey's Hinges	Deviation			2.00	7.00	10.00		

Figure 2.27

SPSS percentile statistics for water-level task deviations

Look Back The method for computing percentiles with small data sets varies according to the software used. As the sample size increases, the percentiles from the different software packages will converge to a single number.

Now Work Exercise 2.109

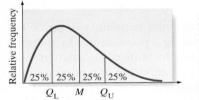

Figure 2.28

The quartiles for a data set

Percentiles that partition a data set into four categories, each category containing exactly 25% of the measurements, are called **quartiles.** The *lower quartile (Q_L)* is the 25th percentile, the *middle quartile (M)* is the median or 50th percentile, and the *upper quartile (Q_U)* is the 75th percentile, as shown in Figure 2.28. Therefore, in Example 2.14, we have (from the SPSS printout, Figure 2.27), $Q_L = 2.0$, $M = 7.0$, and $Q_U = 10.0$. Quartiles will prove useful in finding unusual observations in a data set (Section 2.8).

> The **lower quartile (Q_L)** is the 25th percentile of a data set. The **middle quartile (M)** is the median or 50th percentile. The **upper quartile (Q_U)** is the 75th percentile.

Another measure of relative standing in popular use is the **z-score.** As you can see in the following definition, the z-score makes use of the mean and standard deviation of a data set in order to specify the relative location of the measurement.

> The **sample z-score** for a measurement x is
>
> $$z = \frac{x - \bar{x}}{s}$$
>
> The **population z-score** for a measurement x is
>
> $$z = \frac{x - \mu}{\sigma}$$

Note that the z-score is calculated by subtracting $\bar{x}$ (or μ) from the measurement x and then dividing the result by s (or σ). The final result, the **z-score**, represents the distance between a given measurement x and the mean, expressed in standard deviations.

Example 2.15

Finding a z-Score

Problem Suppose a sample of 2,000 high school seniors' verbal SAT scores is selected. The mean and standard deviation are

$$\bar{x} = 550 \quad s = 75$$

Suppose Joe Smith's score is 475. What is his sample z-score?

Solution Joe Smith's verbal SAT score lies below the mean score of the 2,000 seniors, as shown in Figure 2.29.

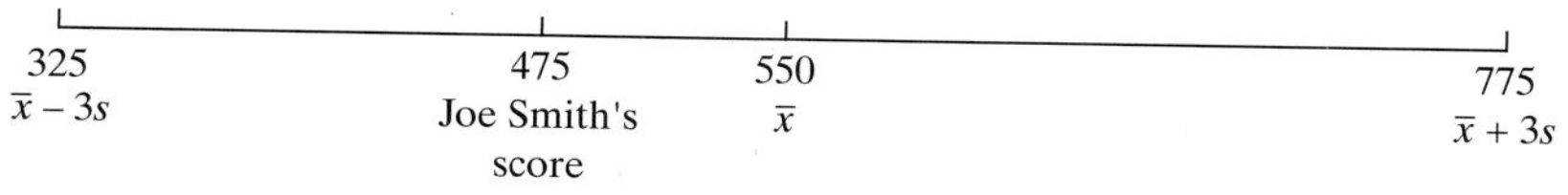

Figure 2.29
Verbal SAT scores of high school seniors

325	475	550	775
$\bar{x} - 3s$	Joe Smith's score	$\bar{x}$	$\bar{x} + 3s$

We compute

$$z = \frac{x - \bar{x}}{s} = \frac{475 - 550}{75} = -1.0$$

which tells us that Joe Smith's score is 1.0 standard deviation *below* the sample mean; in short, his sample z-score is -1.0.

Look Back The numerical value of the z-score reflects the relative standing of the measurement. A large positive z-score implies that the measurement is larger than almost all other measurements, whereas a large negative z-score indicates that the measurement is smaller than almost every other measurement. If a z-score is 0 or near 0, the measurement is located at or near the mean of the sample or population.

Now Work Exercise 2.111

We can be more specific if we know that the frequency distribution of the measurements is mound shaped. In this case, the following interpretation of the z-score can be given:

Interpretation of z-Scores for Mound-Shaped Distributions of Data

1. Approximately 68% of the measurements will have a z-score between -1 and 1.

2. Approximately 95% of the measurements will have a z-score between -2 and 2.

3. Approximately 99.7% (almost all) of the measurements will have a z-score between -3 and 3.

Note that this interpretation of z-scores is identical to that given by the empirical rule for mound-shaped distributions (Table 2.7). The statement that a measurement falls into the interval from $(\mu - \sigma)$ to $(\mu + \sigma)$ is equivalent to the statement that a measurement has a population z-score between -1 and 1, since all measurements between $(\mu - \sigma)$ and $(\mu + \sigma)$ are within one standard deviation of μ. These z-scores are displayed in Figure 2.30.

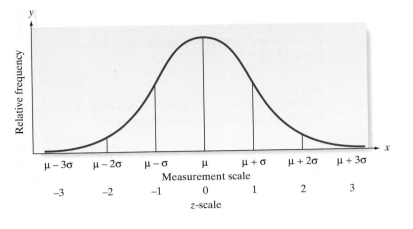

Figure 2.30
Population z-scores for a
mound-shaped distribution

Exercises 2.108–2.124

Understanding the Principles

2.108 For a quantitative data set
    a. What is the 50th percentile called?
    b. Define Q_L.
    c. Define Qu.

2.109 Give the percentage of measurements in a data set that are
    above and below each of the following percentiles:
    a. 75th percentile
    b. 50th percentile
    c. 20th percentile
    d. 84th percentile

2.110 For mound-shaped data, what percentage of measurements have a z-score between -2 and 2?

Learning the Mechanics

2.111 Compute the z-score corresponding to each of the following values of x:
    a. $x = 40, s = 5, \bar{x} = 30$
    b. $x = 90, \mu = 89, \sigma = 2$
    c. $\mu = 50, \sigma = 5, x = 50$
    d. $s = 4, x = 20, \bar{x} = 30$
    e. In parts **a–d**, state whether the z-score locates x within a sample or within a population.
    f. In parts **a–d**, state whether each value of x lies above or below the mean and by how many standard deviations.

2.112 Compare the z-scores to decide which of the following x values lie the greatest distance above the mean and the greatest distance below the mean:
    a. $x = 100, \mu = 50, \sigma = 25$
    b. $x = 1, \mu = 4, \sigma = 1$
    c. $x = 0, \mu = 200, \sigma = 100$
    d. $x = 10, \mu = 5, \sigma = 3$

2.113 Suppose that 40 and 90 are two elements of a population data set and that their z-scores are -2 and 3, respectively. Using only this information, is it possible to determine the population's mean and standard deviation? If so, find them. If not, explain why it's not possible.

Applying the Concepts—Basic

2.114 **Math scores of eighth graders.** According to the National Center for Education Statistics (2009), scores on a mathematics assessment test for United States eighth graders have a mean of 283, a 25th percentile of 259, a 75th percentile of 308, and a 90th percentile of 329. Interpret each of these numerical descriptive measures.

2.115 **Drivers stopped by police.** According to the Bureau of Justice Statistics (June 2006), 74% of all licensed drivers stopped by police are 25 years or older. Give a percentile ranking for the age of 25 years in the distribution of all ages of licensed drivers stopped by police.

2.116 **Sanitation inspection of cruise ships.** Refer to the sanitation levels of cruise ships presented in Exercise 2.35 (p. 47) and saved in the **SHIPSANIT** file.
    a. Give a measure of relative standing for the *Nautilus Explorer's* score of 74. Interpret the result.
    b. Give a measure of relative standing for the *Rotterdam's* score of 92. Interpret the result.

2.117 **Motivation of drug dealers.** Refer to the *Applied Psychology in Criminal Justice* (Sept. 2009) study of convicted drug dealers' motivations, Exercise 2.98 (p. 71). Recall that the sample of drug dealers had a mean Wanting Recognition (WR) score of 39 points, with a standard deviation of 6 points.
    a. Find and interpret the z-score for a drug dealer with a WR score of 30 points.
    b. What proportion of the sampled drug dealers had WR scores below 39 points? (Assume the distribution of WR scores is mound shaped and symmetric.)

Applying the Concepts—Intermediate

2.118 **Reading Japanese books.** Refer to the *Reading in a Foreign Language* (Apr. 2004) experiment to improve the Japanese reading comprehension levels of 14 University of Hawaii students, presented in Exercise 2.61 (p. 57). The data on number of books read and grade for each student are saved in the **JAPANESE** file.

  a. Find the mean and standard deviation of the number of books read by students who earned an A grade. Find the z-score for an A student who read 40 books. Interpret the result.

  b. Find the mean and standard deviation of the number of books read by students who earned either a B or a C grade. Find the z-score for a B or C student who read 40 books. Interpret the result.

  c. Refer to parts **a** and **b**. Which of the two groups of students is more likely to have read 40 books? Explain.

2.119 Extinct New Zealand birds. Refer to the *Evolutionary Ecology Research* (July 2003) study of the patterns of extinction in the New Zealand bird population, first presented in Exercise 2.20 (p. 37). Again, consider the data on the egg length (measured in millimeters) for the 132 bird species saved in the **NZBIRDS** file.

  a. Find the 10th percentile for the egg length distribution and interpret its value.

  b. The *moas (P. australis)* is a bird species with an egg length of 205 millimeters. Find the z-score for this species of bird and interpret its value.

2.120 Lead in drinking water. The U.S. Environmental Protection Agency (EPA) sets a limit on the amount of lead permitted in drinking water. The EPA *Action Level* for lead is .015 milligram per liter (mg/L) of water. Under EPA guidelines, if 90% of a water system's study samples have a lead concentration less than .015 mg/L, the water is considered safe for drinking. I (coauthor Sincich) received a report on a study of lead levels in the drinking water of homes in my subdivision. The 90th percentile of the study sample had a lead concentration of .00372 mg/L. Are water customers in my subdivision at risk of drinking water with unhealthy lead levels? Explain.

2.121 Ranking Ph.D. programs in economics. Thousands of students apply for admission to graduate schools in economics each year with the intention of obtaining a Ph.D. The *Southern Economic Journal* (Apr. 2008) published a guide to graduate study in economics by ranking the Ph.D. programs at 129 colleges and universities. Each program was evaluated according to the number of publications published by faculty teaching in the Ph.D. program and by the quality of the publications. Data obtained from the Social Science Citation Index (SSCI) were used to calculate an overall productivity score for each Ph.D. program. The mean and standard deviation of these 129 productivity scores were then used to compute a z-score for each economics program. Harvard University had the highest z-score ($z = 5.08$) and, hence, was the top-ranked school; Howard University was ranked last since it had the lowest z-score ($z = -0.81$). The data (z-scores) for all 129 economic programs are saved in the **ECOPHD** file.

  a. Interpret the z-score for Harvard University.

  b. Interpret the z-score for Howard University.

  c. The authors of the *Southern Economic Journal* article note that "only 44 of the 129 schools have positive z-scores, indicating that the distribution of overall productivity is skewed to the right." Do you agree? (Check your answer by constructing a histogram for the z-scores in the **ECOPHD** file.)

2.122 Blue versus red exam study. In a study of how external clues influence performance, psychology professors at the University of Alberta and Pennsylvania State University gave two different forms of a midterm examination to a large group of introductory psychology students. The questions on the exam were identical and in the same order, but one exam was printed on blue paper and the other on red paper (*Teaching Psychology*, May 1998). Grading only the difficult questions on the exam, the researchers found that scores on the blue exam had a distribution with a mean of 53% and a standard deviation of 15%, while scores on the red exam had a distribution with a mean of 39% and a standard deviation of 12%. (Assume that both distributions are approximately mound shaped and symmetric.)

  a. Give an interpretation of the standard deviation for the students who took the blue exam.

  b. Give an interpretation of the standard deviation for the students who took the red exam.

  c. Suppose a student is selected at random from the group of students who participated in the study and the student's score on the difficult questions is 20%. Which exam form is the student more likely to have taken, the blue or the red exam? Explain.

Applying the Concepts—Advanced

2.123 GPAs of students. At one university, the students are given z-scores at the end of each semester, rather than the traditional GPAs. The mean and standard deviation of all students' cumulative GPAs, on which the z-scores are based, are 2.7 and .5, respectively.

  a. Translate each of the following z-scores to corresponding GPA scores: $z = 2.0, z = -1.0, z = .5, z = -2.5$.

  b. Students with z-scores below -1.6 are put on probation. What is the corresponding probationary GPA?

  c. The president of the university wishes to graduate the top 16% of the students with *cum laude* honors and the top 2.5% with *summa cum laude* honors. Where (approximately) should the limits be set in terms of z-scores? In terms of GPAs? What assumption, if any, did you make about the distribution of the GPAs at the university?

2.124 Ranking Ph.D. programs in economics (cont'd). Refer to the *Southern Economic Journal* (Apr. 2008) study of Ph.D. programs in economics, Exercise 2.121. The authors also made the following observation: "A noticeable feature of this skewness is that distinction between schools diminishes as the rank declines. For example, the top-ranked school, Harvard, has a z-score of 5.08, and the fifth-ranked school, Yale, has a z-score of 2.18, a substantial difference. However,...the 70th-ranked school, the University of Massachusetts, has a z-score of -0.43, and the 80th-ranked school, the University of Delaware, has a z-score of -0.50, a very small difference. [Consequently] the ordinal rankings presented in much of the literature that ranks economics departments miss the fact that below a relatively small group of top programs, the differences in [overall] productivity become fairly small." Do you agree?

2.8 Methods for Detecting Outliers: Box Plots and z-Scores

Sometimes it is important to identify inconsistent or unusual measurements in a data set. An observation that is unusually large or small relative to the data values we want to describe is called an **outlier**.

Outliers are often attributable to one of several causes. First, the measurement associated with the outlier may be invalid. For example, the experimental procedure used to generate the measurement may have malfunctioned, the experimenter may have misrecorded the measurement, or the data might have been coded incorrectly in the computer. Second, the outlier may be the result of a misclassified measurement. That is, the measurement belongs to a population different from that from which the rest of the sample was drawn. Finally, the measurement associated with the outlier may be recorded correctly and from the same population as the rest of the sample, but represent a rare (chance) event. Such outliers occur most often when the relative frequency distribution of the sample data is extremely skewed, because a skewed distribution has a tendency to include extremely large or small observations relative to the others in the data set.

> An observation (or measurement) that is unusually large or small relative to the other values in a data set is called an **outlier**. Outliers typically are attributable to one of the following causes:
>
> 1. The measurement is observed, recorded, or entered into the computer incorrectly.
> 2. The measurement comes from a different population.
> 3. The measurement is correct, but represents a rare (chance) event.

Two useful methods for detecting outliers, one graphical and one numerical, are **box plots** and z-scores. The box plot is based on the quartiles (defined in Section 2.7) of a data set. Specifically, a box plot is based on the *interquartile range (IQR)*—the distance between the lower and upper quartiles:

$$IQR = Q_U - Q_L$$

> The **interquartile range (IQR)** is the distance between the lower and upper quartiles:
> $$IQR = Q_U - Q_L$$

An annotated MINITAB box plot for the gas mileage data (see Table 2.2) is shown in Figure 2.31.[*] Note that a rectangle (the *box*) is drawn, with the bottom and top of the rectangle (the **hinges**) drawn at the quartiles Q_L and Q_U, respectively. Recall that QL represents the 25th percentile and Qu represents the 75th percentile. By definition, then, the "middle" 50% of the observations—those between Q_L and Q_U—fall inside the box. For the gas mileage data, these quartiles are at 35.625 and 38.375. Thus,

$$IQR = 38.375 - 35.625 = 2.75$$

The median is shown at 37 by a horizontal line within the box.

To guide the construction of the "tails" of the box plot, two sets of limits, called **inner fences** and **outer fences,** are used. Neither set of fences actually appears on the plot. Inner fences are located at a distance of 1.5(IQR) from the hinges. Emanating from the hinges of the box are vertical lines called the **whiskers.** The two whiskers extend to the most extreme observation inside the inner fences. For example, the inner fence on the lower side of the gas mileage box plot is

$$\text{Lower inner fence} = \text{Lower hinge} - 1.5(IQR)$$
$$= 35.625 - 1.5(2.75)$$
$$= 35.625 - 4.125 = 31.5$$

---

[*]Although box plots can be generated by hand, the amount of detail required makes them particularly well suited for computer generation. We use computer software to generate the box plots in this section.

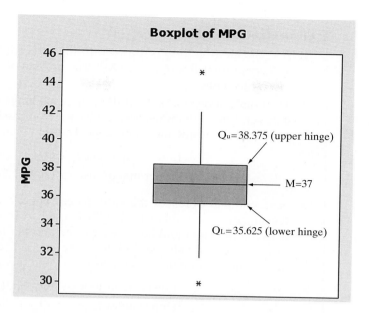

Figure 2.31

Annotated MINITAB box plot for EPA gas mileages

The smallest measurement *inside* this fence is the second-smallest measurement, 31.8. Thus, the lower whisker extends to 31.8. Similarly, the upper whisker extends to 42.1, the largest measurement inside the upper inner fence:

$$
\begin{aligned}
\text{Upper inner fence} &= \text{Upper hinge} + 1.5(\text{IQR}) \\
&= 38.375 + 1.5(2.75) \\
&= 38.375 + 4.125 = 42.5
\end{aligned}
$$

Values that are beyond the inner fences are deemed *potential outliers* because they are extreme values that represent relatively rare occurrences. In fact, for mound-shaped distributions, less than 1% of the observations are expected to fall outside the inner fences. Two of the 100 gas mileage measurements, 30.0 and 44.9, fall beyond the inner fences, one on each end of the distribution. Each of these potential outliers is represented by a common symbol (an asterisk in MINITAB).

The other two imaginary fences, the outer fences, are defined at a distance 3(IQR) from each end of the box. Measurements that fall beyond the outer fences (also represented by an asterisk in MINITAB) are very extreme measurements that require special analysis. Since less than one-hundredth of 1% (.01% or .0001) of the measurements from mound-shaped distributions are expected to fall beyond the outer fences, these measurements are considered to be *outliers*. Because there are no measurements of gas mileage beyond the outer fences, there are no outliers.

Recall that outliers may be incorrectly recorded observations, members of a population different from the rest of the sample, or, at the least, very unusual measurements from the same population. The box plot of Figure 2.31 detected two potential outliers: the two gas mileage measurements beyond the inner fences. When we analyze these measurements, we find that they are correctly recorded. Perhaps they represent mileages that correspond to exceptional models of the car being tested or to unusual gas mixtures. Outlier analysis often reveals useful information of this kind and therefore plays an important role in the statistical inference-making process.

In addition to detecting outliers, box plots provide useful information on the variation in a data set. The elements (and nomenclature) of box plots are summarized in the box. Some aids to the interpretation of box plots are also given.

Elements of a Box Plot

1. A rectangle (the **box**) is drawn with the ends (the **hinges**) drawn at the lower and upper quartiles (Q_L and Q_U). The median of the data is shown in the box, usually by a line.

2. The points at distances 1.5(IQR) from each hinge mark the **inner fences** of the data set. Lines (the **whiskers**) are drawn from each hinge to the most extreme measurement inside the inner fence. Thus,

$$\text{Lower inner fence} = Q_L - 1.5(\text{IQR})$$
$$\text{Upper inner fence} = Q_U + 1.5(\text{IQR})$$

3. A second pair of fences, the **outer fences**, appears at a distance of 3(IQR) from the hinges. One symbol (e.g., "*") is used to represent measurements falling between the inner and outer fences, and another (e.g., "0") is used to represent measurements that lie beyond the outer fences. Thus, outer fences are not shown unless one or more measurements lie beyond them. We have

$$\text{Lower outer fence} = Q_L - 3(\text{IQR})$$
$$\text{Upper outer fence} = Q_U + 3(\text{IQR})$$

4. The symbols used to represent the median and the extreme data points (those beyond the fences) will vary with the software you use to construct the box plot. (You may use your own symbols if you are constructing a box plot by hand.) You should consult the program's documentation to determine exactly which symbols are used.

Aids to the Interpretation of Box Plots

1. The line (median) inside the box represents the "center" of the distribution of data.

2. Examine the length of the box. The IQR is a measure of the sample's variability and is especially useful for the comparison of two samples. (See Example 2.16.)

3. Visually compare the lengths of the whiskers. If one is clearly longer, the distribution of the data is probably skewed in the direction of the longer whisker.

4. Analyze any measurements that lie beyond the fences. Less than 5% should fall beyond the inner fences, even for very skewed distributions. Measurements beyond the outer fences are probably outliers, with one of the following explanations:

   a. The measurement is incorrect. It may have been observed, recorded, or entered into the computer incorrectly.

   b. The measurement belongs to a population different from the population that the rest of the sample was drawn from. (See Example 2.17.)

   c. The measurement is correct *and* from the same population as the rest of the sample. Generally, we accept this explanation only after carefully ruling out all others.

Example 2.16

Computer-Generated Box Plot—The "Water-Level Task"

Problem Refer to the *Psychological Science* experiment, called the "water-level task," Examples 2.2 and 2.14. Use statistical software to produce a box plot for the 40 water-level task deviation angles shown in Table 2.4. Identify any outliers in the data set.

Solution The 40 deviations were entered into MINITAB, and a box plot produced in Figure 2.32. Recall (from Example 2.14) that $Q_L = 2$, $M = 7$, and $Q_U = 10$. The middle line in the box represents the median deviation of $M = 7$. The lower hinge of the box represents $Q_L = 2$, while the upper hinge represents $Q_U = 10$. Note that there are two extreme observations (represented by asterisks) that are beyond the upper inner fence, but inside the upper outer fence. These two values are outliers. Examination of the data reveals that these measurements correspond to deviation angles of 25° and 35°.

Look Ahead Before removing the outliers from the data set, a good analyst will make a concerted effort to find the cause of the outliers. For example, both the outliers are deviation measurements for waitresses. An investigation may discover that both waitresses were new on the job the day the experiment was conducted. If the study target population is experienced bartenders and waitresses, these two observations would be dropped from the analysis.

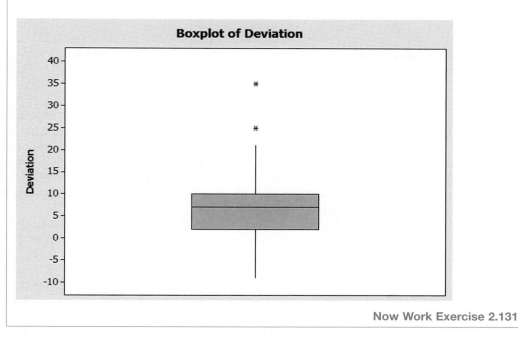

Boxplot of Deviation

Figure 2.32
MINITAB box plot for water-level task deviations

Now Work Exercise 2.131

Example 2.17
Comparing Box Plots—Stimulus Reaction Study

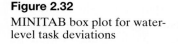

Problem A Ph.D. student in psychology conducted a stimulus reaction experiment as a part of her dissertation research. She subjected 50 subjects to a threatening stimulus and 50 to a nonthreatening stimulus. The reaction times of all 100 students, recorded to the nearest tenth of a second, are listed in Table 2.9. Box plots of the two resulting samples of reaction times, generated with SAS, are shown in Figure 2.33. Interpret the box plots.

Table 2.9   Reaction Times of Students

Nonthreatening Stimulus

2.0	1.8	2.3	2.1	2.0	2.2	2.1	2.2	2.1	2.1
2.0	2.0	1.8	1.9	2.2	2.0	2.2	2.4	2.1	2.0
2.2	2.1	2.2	1.9	1.7	2.0	2.0	2.3	2.1	1.9
2.0	2.2	1.6	2.1	2.3	2.0	2.0	2.0	2.2	2.6
2.0	2.0	1.9	1.9	2.2	2.3	1.8	1.7	1.7	1.8

Threatening Stimulus

1.8	1.7	1.4	2.1	1.3	1.5	1.6	1.8	1.5	1.4
1.4	2.0	1.5	1.8	1.4	1.7	1.7	1.7	1.4	1.9
1.9	1.7	1.6	2.5	1.6	1.6	1.8	1.7	1.9	1.9
1.5	1.8	1.6	1.9	1.3	1.5	1.6	1.5	1.6	1.5
1.3	1.7	1.3	1.7	1.7	1.8	1.6	1.7	1.7	1.7

Data Set: REACTION

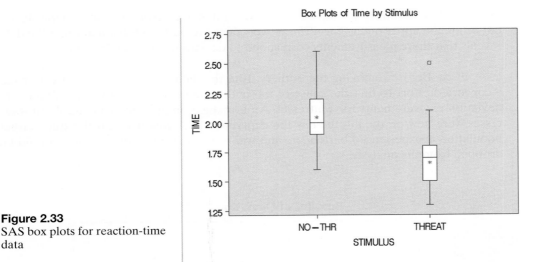

Figure 2.33
SAS box plots for reaction-time data

Solution In SAS, the median is represented by the horizontal line through the box, while the asterisk (*) represents the mean. Analysis of the box plots on the same numerical scale reveals that the distribution of times corresponding to the threatening stimulus lies below that of the nonthreatening stimulus. The implication is that the reaction times tend to be faster to the threatening stimulus. Note, too, that the upper whiskers of both samples are longer than the lower whiskers, indicating that the reaction times are positively skewed.

No observations in the two samples fall between the inner and outer fences. However, there is one outlier: the observation of 2.5 seconds corresponding to the threatening stimulus that is beyond the outer fence (denoted by the square symbol in SAS). When the researcher examined her notes from the experiments, she found that the subject whose time was beyond the outer fence had mistakenly been given the nonthreatening stimulus. You can see in Figure 2.33 that his time would have been within the upper whisker if moved to the box plot corresponding to the nonthreatening stimulus. The box plots should be reconstructed, since they will both change slightly when this misclassified reaction time is moved from one sample to the other.

Look Ahead The researcher concluded that the reactions to the threatening stimulus were faster than those to the nonthreatening stimulus. However, she was asked by her Ph.D. committee whether the results were *statistically significant*. Their question addresses the issue of whether the observed difference between the samples might be attributable to chance or sampling variation rather than to real differences between the populations. To answer this question, the researcher must use inferential statistics rather than graphical descriptions. We discuss how to compare two samples by means of inferential statistics in Chapter 9.

The next example illustrates how z-scores can be used to detect outliers and make inferences.

**Example 2.18
Inference Using
z-Scores**

Problem Suppose a female bank employee believes that her salary is low as a result of sex discrimination. To substantiate her belief, she collects information on the salaries of her male counterparts in the banking business. She finds that their salaries have a mean of $64,000 and a standard deviation of $2,000. Her salary is $57,000. Does this information support her claim of sex discrimination?

Solution The analysis might proceed as follows: First, we calculate the z-score for the woman's salary with respect to those of her male counterparts. Thus,

$$z = \frac{\$57{,}000 - \$64{,}000}{\$2{,}000} = -3.5$$

The implication is that the woman's salary is 3.5 standard deviations *below* the mean of the male salary distribution. Furthermore, if a check of the male salary data shows that the frequency distribution is mound shaped, we can infer that very few salaries in this distribution should have a z-score less than −3, as shown in Figure 2.34. Clearly, a z-score of −3.5 represents an outlier. Either her salary is from a distribution different from the male salary distribution, or it is a very unusual (highly improbable) measurement from a salary distribution no different from the male distribution.

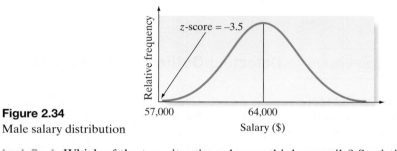

Figure 2.34
Male salary distribution

Look Back Which of the two situations do you think prevails? Statistical thinking would lead us to conclude that her salary does not come from the male salary distribution, lending support to the female bank employee's claim of sex discrimination. However, the careful investigator should require more information before inferring that sex discrimination is the cause. We would want to know more about the data collection technique the woman used and more about her competence at her job. Also, perhaps other factors, such as length of employment, should be considered in the analysis.

Now Work Exercise 2.129

Examples 2.17 and 2.18 exemplify an approach to statistical inference that might be called the **rare-event approach.** An experimenter hypothesizes a specific frequency distribution to describe a population of measurements. Then a sample of measurements is drawn from the population. If the experimenter finds it unlikely that the sample came from the hypothesized distribution, the hypothesis is judged to be false. Thus, in Example 2.18, the woman believes that her salary reflects discrimination. She hypothesizes that her salary should be just another measurement in the distribution of her male counterparts' salaries if no discrimination exists. However, it is so unlikely that the sample (in this case, her salary) came from the male frequency distribution that she rejects that hypothesis, concluding that the distribution from which her salary was drawn is different from the distribution for the men.

This rare-event approach to inference making is discussed further in later chapters. Proper application of the approach requires a knowledge of probability, the subject of our next chapter.

We conclude this section with some rules of thumb for detecting outliers.

Rules of Thumb for Detecting Outliers[*]

Box Plots: Observations falling between the inner and outer fences are deemed *suspect outliers*. Observations falling beyond the outer fence are deemed *highly suspect outliers*.

(continued)

---

[*]The z-score and box plot methods both establish rule-of-thumb limits outside of which a measurement is deemed to be an outlier. Usually, the two methods produce similar results. However, the presence of one or more outliers in a data set can inflate the computed value of s. Consequently, it will be less likely that an errant observation would have a z-score larger than |3|. In contrast, the values of the quartiles used to calculate the intervals for a box plot are not affected by the presence of outliers.

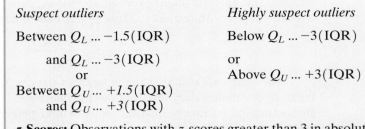

Suspect outliers	Highly suspect outliers
Between $Q_L \ldots -1.5(\text{IQR})$ and $Q_L \ldots -3(\text{IQR})$ or Between $Q_U \ldots +1.5(\text{IQR})$ and $Q_U \ldots +3(\text{IQR})$	Below $Q_L \ldots -3(\text{IQR})$ or Above $Q_U \ldots +3(\text{IQR})$

z-Scores: Observations with z-scores greater than 3 in absolute value are considered *outliers*. For some highly skewed data sets, observations with z-scores greater than 2 in absolute value *may be outliers*.

Possible Outliers	Outliers
$|z| > 2$	$|z| > 3$

Statistics IN Action | Revisited | Detecting Outliers in the Body Image Data

In the *Body Image: An International Journal of Research* (Jan. 2010) study of 92 BDD patients, the quantitative variable of interest is Total Appearance Evaluation score (ranging from 7 to 35 points). Are there any unusual scores in the BDD data set? We will apply both the box plot and z-score method to aid in identifying any outliers in the data. Since from previous analyses, there appears to be a difference in the distribution of appearance evaluation scores for males and females, we will analyze the data by gender.

To employ the z-score method, we require the mean and standard deviation of the data for each gender. These values were already computed in the previous Statistics in Action Revisited section. For females, $\bar{x} = 17.05$ and $s = 4.76$; for males, $\bar{x} = 19.0$ and $s = 5.42$ (see Figure SIA2.5). Then, the 3-standard-deviation interval for each gender is:

Females: $\bar{x} \pm 3s = 17.05 \pm 3(4.76) = 17.05 \pm 14.28 = (2.77, 31.33)$
Males: $\bar{x} \pm 3s = 19.00 \pm 3(5.42) = 19.00 \pm 16.26 = (2.74, 35.26)$

If you examine the appearance evaluation scores in the **BDD** file, you will find that none of the scores fall beyond the 3-standard-deviation interval for each group. Consequently, if we use the z-score approach, there are no highly suspect outliers in the data.

Box plots for the data are shown in Figure SIA2.6. Although several suspect outliers (asterisks) are shown on the box plot for each gender, there are no highly suspect outliers (zeros) shown. That is, no data points fall beyond the outer fences of the box plots. As with the z-score approach, the box plot method does not detect any highly suspect outliers.

[*Note:* If we were to detect one or more highly suspect outliers, we should investigate whether or not to include the observation in any analysis which leads to an inference about the population of BDD patients. Is the outlier a legitimate value (in which case it will remain in the data set for analysis) or is the outlier associated with a subject that is not a member of the population of interest—say, a person who is misdiagnosed with BDD (in which case it will be removed from the data set prior to analysis).]

Data Set: BBD

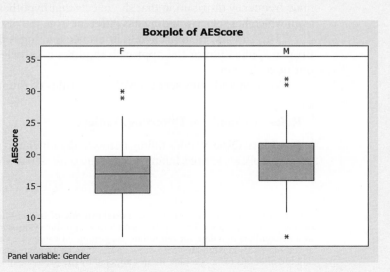

Figure SIA2.6
MINITAB box plots for Appearance Evaluation score by Gender

Exercises 2.125–2.142

Understanding the Principles

2.125 Define an outlier.

2.126 What is the interquartile range?

2.127 What are the hinges of a box plot?

2.128 With mound-shaped data, what proportion of the measurements have *z*-scores between −3 and 3?

Learning the Mechanics

2.129 A sample data set has a mean of 57 and a standard
NW deviation of 11. Determine whether each of the following sample measurements is an outlier.
   a. 65
   b. 21
   c. 72
   d. 98

2.130 Suppose a data set consisting of exam scores has a lower quartile $Q_L = 60$, a median $M = 75$, and an upper quartile $Q_U = 85$. The scores on the exam range from 18 to 100. Without having the actual scores available to you, construct as much of the box plot as possible.

2.131 Consider the following horizontal box plot:
NW

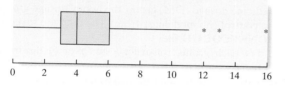

   a. What is the median of the data set (approximately)?
   b. What are the upper and lower quartiles of the data set (approximately)?
   c. What is the interquartile range of the data set (approximately)?
   d. Is the data set skewed to the left, skewed to the right, or symmetric?
   e. What percentage of the measurements in the data set lie to the right of the median? To the left of the upper quartile?
   f. Identify any outliers in the data.

2.132 Consider the following two sample data sets, saved in the
   LM2_132 file:

Sample A			Sample B		
121	171	158	171	152	170
173	184	163	168	169	171
157	85	145	190	183	185
165	172	196	140	173	206
170	159	172	172	174	169
161	187	100	199	151	180
142	166	171	167	170	188

   a. Construct a box plot for each data set.
   b. Identify any outliers that may exist in the two data sets.

Applet Exercise 2.7

Use the applet *Standard Deviation* to determine whether an item in a data set may be an outlier. Begin by setting appropriate limits and plotting the given data on the number line provided in the applet. Here is the data set:

10 80 80 85 85 85 85 90 90 90 90 90 95 95 95 95 100 100

a. The green arrow shows the approximate location of the mean. Multiply the standard deviation given by the applet by 3. Is the data item 10 more than three standard deviations away from the green arrow (the mean)? Can you conclude that the 10 is an outlier?

b. Using the mean and standard deviation from part **a**, move the point at 10 on your plot to a point that appears to be about three standard deviations from the mean. Repeat the process in part **a** for the new plot and the new suspected outlier.

c. When you replaced the extreme value in part **a** with a number that appeared to be within three standard deviations of the mean, the standard deviation got smaller and the mean moved to the right, yielding a new data set whose extreme value was *not* within three standard deviations of the mean. Continue to replace the extreme value with higher numbers until the new value is within three standard deviations of the mean in the new data set. Use trial and error to estimate the smallest number that can replace the 10 in the original data set so that the replacement is not considered to be an outlier.

Applying the Concepts—Basic

2.133 **Treating psoriasis with the "Doctorfish of Kangal."** Psoriasis is a skin disorder with no known cure and no proven effective pharmacological treatment. An alternative treatment for psoriasis is ichthyotherapy also known as therapy with the "Doctorfish of Kangal." Fish from the hot pools of Kangal, Turkey, feed on the skin scales of bathers, reportedly reducing the symptoms of psoriasis. In one study, 67 patients diagnosed with psoriasis underwent three weeks of ichthyotherapy. (*Evidence-Based Research in Complementary and Alternative Medicine*, Dec. 2006). The Psoriasis Area Severity Index (PASI) of each patient was measured both before and after treatment. (The lower the PASI score, the better is the skin condition.) Box plots of the PASI scores, both before (baseline) and after three weeks of ichthyotherapy treatment, are shown in the accompanying diagram.
   a. Find the approximate 25th percentile, the median, and the 75th percentile for the PASI scores before treatment.
   b. Find the approximate 25th percentile, the median, and the 75th percentile for the PASI scores after treatment.
   c. Comment on the effectiveness of ichthyotherapy in treating psoriasis.

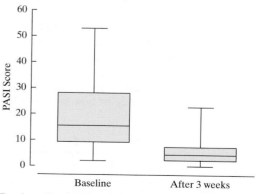

Based on Grassberger, M., and Hoch, W. "Ichthyotherapy as alternative treatment for patients with psoriasis: A pilot study." *Evidence-Based Research in Complementary and Alternative Medicine*, Vol. 3, No. 4, Dec. 2004 (Figure 3). National Center for Complementary and Alternative Medicine.

2.134 Dentists' use of anesthetics. Refer to the *Current Allergy & Clinical Immunology* study of the use of local anesthetics in dentistry, presented in Exercise 2.99 (p. 72). Recall that the mean number of units (ampoules) of local anesthetics used per week by dentists was 79, with a standard deviation of 23. Consider a dentist who used 175 units of local anesthetics in a week.
  a. Find the *z*-score for this measurement.
  b. Would you consider the measurement to be an outlier? Explain.
  c. Give several reasons the outlier may have occurred.

2.135 Research on brain specimens. Refer to the *Brain and Language* data on postmortem intervals (PMIs) of 22 human brain specimens, presented in Exercise 2.39 (p. 48). The data are saved in the **BRAINPMI** file. The mean and standard deviation of the PMI values are 7.3 and 3.18, respectively.
  a. Find the *z*-score for the PMI value of 3.3.
  b. Is the PMI value of 3.3 considered an outlier? Explain.

2.136 Estimating the age of glacial drifts. Tills are glacial drifts consisting of a mixture of clay, sand, gravel, and boulders. Engineers from the University of Washington's Department of Earth and Space Sciences studied the chemical makeup of buried tills in order to estimate the age of the glacial drifts in Wisconsin (*American Journal of Science*, Jan. 2005). The ratio of the elements aluminum (Al) and beryllium (Be) in sediment is related to the duration of burial. The Al–Be ratios for a sample of 26 buried till specimens are given in the accompanying table and saved in the **TILLRATIO** file. An SPSS printout with descriptive statistics for the Al–Be ratio is shown at the bottom of the page.

3.75	4.05	3.81	3.23	3.13	3.30	3.21	3.32	4.09	3.90	5.06	3.85	3.88
4.06	4.56	3.60	3.27	4.09	3.38	3.37	2.73	2.95	2.25	2.73	2.55	3.06

Based on *American Journal of Science*, Vol. 305, No. 1, Jan. 2005, p. 16 (Table 2).

  a. Find and interpret the *z*-score associated with the largest ratio, the smallest ratio, and the mean ratio.
  b. Would you consider the largest ratio to be unusually large? Why or why not?
  c. Construct a box plot for the data and identify any outliers.

Applying the Concepts—Intermediate

2.137 Sanitation inspection of cruise ships. Refer to the data on sanitation levels of cruise ships, presented in Exercise 2.35 (p. 47). These data are saved in the **SHIPSANIT** file.
  a. Use the box plot method to detect any outliers in the data.
  b. Use the *z*-score method to detect any outliers in the data.
  c. Do the two methods agree? If not, explain why.

2.138 Characteristics of a rockfall. Refer to the *Environmental Geology* (Vol. 58, 2009) study of how far a block from a collapsing rockwall will bounce, Exercise 2.59 (p. 57). The computer-simulated rebound lengths (in meters) for 13 block-soil impact marks left on a slope from an actual rockfall (saved in the **ROCKWELL** file) are reproduced in the table. Do you detect any outliers in the data? Explain.

10.94	13.71	11.38	7.26	17.83	11.92	11.87	5.44	13.35	4.90	5.85	5.10	6.77

Based on Paronuzzi, P. "Rockfall-induced block propagation on a soil slope, northern Italy." *Environmental Geology*, Vol. 58, 2009 (Table 2).

2.139 Most powerful women in America. Refer to the *Fortune* ranking of the 50 most powerful women in America, presented in Exercise 2.60 (p. 57). (These data are saved in the **WPOWER50** file.) Use side-by-side box plots to compare the ages of the women in three groups, based on their position within the firm: Group 1 (CEO, CEO/ Chairman, or CEO/president); Group 2 (COO/President, EVP/Pres., chairman, COO, EVP/COO, or president); Group 3 (EVP, executive, SVP, SVP, VP, or direct). Do you detect outliers?

2.140 Is honey a cough remedy? Refer to the *Archives of Pediatrics and Adolescent Medicine* (Dec. 2007) study of honey as a remedy for coughing, Exercise 2.62 (p. 58). Recall that coughing improvement scores were recorded for children in three groups: the over-the-counter cough medicine dosage (DM) group, the honey dosage group, and the control (no medicine) group. The data are saved in the **HONEYCOUGH** file.
  a. For each group, construct a box plot of the improvement scores.
  b. How do the median improvement scores compare for the three groups?
  c. How does the variability in improvement scores compare for the three groups?
  d. Do you detect any outliers in any of the three coughing improvement score distributions?

2.141 Comparing SAT Scores. Refer to Exercise 2.42 (p. 48), in which we compared state average SAT scores in 2009 and 2005. The data are saved in the **SATSCORES** file.
  a. Construct side-by-side box plots of the SAT scores for the two years.
  b. Compare the variability of the SAT scores for the two years.
  c. Are any state SAT scores outliers in either year? If so, identify them.

Applying the Concepts—Advanced

2.142 Library book checkouts. A city librarian claims that books have been checked out an average of seven (or more) times in the last year. You suspect he has exaggerated the checkout rate (book usage) and that the mean number of checkouts per book per year is, in fact, less than seven. Using the computerized card catalog, you randomly select one book and find that it has been checked out four times

Descriptive Statistics

	N	Minimum	Maximum	Mean	Std. Deviation
RATIO	26	2.25	5.06	3.5069	.63439
Valid N (listwise)	26				

SPSS Output for Exercise 2.136

in the last year. Assume that the standard deviation of the number of checkouts per book per year is approximately 1.

a. If the mean number of checkouts per book per year really is 7, what is the z-score corresponding to four?

b. Considering your answer to part **a**, do you have reason to believe that the librarian's claim is incorrect?

c. If you knew that the distribution of the number of checkouts was mound shaped, would your answer to part **b** change? Explain.

d. If the standard deviation of the number of checkouts per book per year were 2 (instead of 1), would your answers to parts **b** and **c** change? Explain.

2.9 Graphing Bivariate Relationships (Optional)

The claim is often made that the crime rate and the unemployment rate are "highly correlated." Another popular belief is that smoking and lung cancer are "related." Some people even believe that the Dow Jones Industrial Average and the lengths of fashionable skirts are "associated." The words *correlated*, *related*, and *associated* imply a relationship between two variables—in the examples just mentioned, two *quantitative* variables.

One way to describe the relationship between two quantitative variables—called a **bivariate relationship**—is to construct a **scatterplot.** A scatterplot is a two-dimensional plot, with one variable's values plotted along the vertical axis and the others along the horizontal axis. For example, Figure 2.35 is a scatterplot relating (1) the cost of mechanical work (heating, ventilating, and plumbing) to (2) the floor area of the building, for a sample of 26 factory and warehouse buildings. Note that the scatterplot suggests a general tendency for mechanical cost to increase as building floor area increases.

When an increase in one variable is generally associated with an increase in the second variable, we say that the two variables are "positively related" or "positively correlated."* Figure 2.35 implies that mechanical cost and floor area are positively correlated. Alternatively, if one variable has a tendency to decrease as the other increases, we say the variables are "negatively correlated." Figure 2.36 shows several hypothetical scatterplots that portray a positive bivariate relationship (Figure 2.36a), a negative bivariate relationship (Figure 2.36b), and a situation in which the two variables are unrelated (Figure 2.36c).

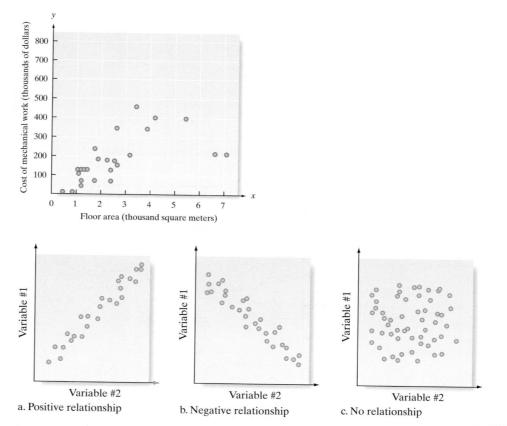

Figure 2.35
Scatterplot of cost vs. floor area

Figure 2.36
Hypothetical bivariate relationships

a. Positive relationship

b. Negative relationship

c. No relationship

*A formal definition of correlation is given in Chapter 11. There, we will learn that correlation measures the strength of the linear (or straight-line) relationship between two quantitative variables.

Example 2.19
Graphing Bivariate Data—Hospital Application

Problem A medical item used to administer to a hospital patient is called a *factor*. For example, factors can be intravenous (IV) tubing, IV fluid, needles, shave kits, bedpans, diapers, dressings, medications, and even code carts. The coronary care unit at Bayonet Point Hospital (St. Petersburg, Florida) recently investigated the relationship between the number of factors administered per patient and the patient's length of stay (in days). Data on these two variables for a sample of 50 coronary care patients are given in Table 2.10. Use a scatterplot to describe the relationship between the two variables of interest: number of factors and length of stay.

Table 2.10 Medfactors Data on Patients' Factors and Length of Stay

Number of Factors	Length of Stay (days)	Number of Factors	Length of Stay (days)
231	9	354	11
323	7	142	7
113	8	286	9
208	5	341	10
162	4	201	5
117	4	158	11
159	6	243	6
169	9	156	6
55	6	184	7
77	3	115	4
103	4	202	6
147	6	206	5
230	6	360	6
78	3	84	3
525	9	331	9
121	7	302	7
248	5	60	2
233	8	110	2
260	4	131	5
224	7	364	4
472	12	180	7
220	8	134	6
383	6	401	15
301	9	155	4
262	7	338	8

Based on Bayonet Point Hospital, Coronary Care Unit.

Data Set: MEDFACTORS

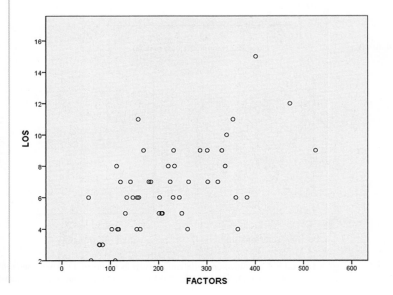

Figure 2.37
SPSS scatterplot for medical factors data in Table 2.10

Solution Rather than construct the plot by hand, we resort to a statistical software package. The SPSS plot of the data in Table 2.10, with length of stay (LOS) on the vertical axis and number of factors (FACTORS) on the horizontal axis, is shown in Figure 2.37.

Although the plotted points exhibit a fair amount of variation, the scatterplot clearly shows an increasing trend. It appears that a patient's length of stay is positively correlated with the number of factors administered to the patient.

Look Back If hospital administrators can be confident that the sample trend shown in Figure 2.37 accurately describes the trend in the population, then they may use this information to improve their forecasts of lengths of stay for future patients.

Now Work Exercise 2.147

The scatterplot is a simple, but powerful, tool for describing a bivariate relationship. However, keep in mind that it is only a graph. No measure of reliability can be attached to inferences made about bivariate populations based on scatterplots of sample data. The statistical tools that enable us to make inferences about bivariate relationships are presented in Chapter 11.

Exercises 2.143–2.158

Understanding the Principles

2.143 Define a bivariate relationship.

2.144 For what types of variables, quantitative or qualitative, are scatterplots useful?

2.145 What is the difference between positive association and negative association as it pertains to the relationship between two variables?

Learning the Mechanics

2.146 Construct a scatterplot for the data in the table that follows. Do you detect a trend?

Variable #1:	.5	1	1.5	2	2.5	3	3.5	4	4.5	5
Variable #2:	2	1	3	4	6	10	9	12	17	17

2.147 Construct a scatterplot for the data in the table that follows.
NW Do you detect a trend?

Variable #1:	5	3	−1	2	7	6	4	0	8
Variable #2:	14	3	10	1	8	5	3	2	12

Applying the Concepts—Basic

2.148 Does elevation impact hitting performance in baseball? The Colorado Rockies play their major league home baseball games in Coors Field, Denver. Each year, the Rockies are among the leaders in team batting statistics (e.g., home runs, batting average, and slugging percentage). Many baseball experts attribute this phenomenon to the "thin air" of Denver—called the "mile-high city" due to its elevation. *Chance* (Winter 2006) investigated the effects of elevation on slugging percentage in Major League Baseball. Data were compiled on players' composite slugging percentage at each of 29 cities for the 2003 season, as well as each city's elevation (feet above sea level). The data are saved in the **MLBPARKS** file. (Selected observations are shown in

the next table.) Construct a scatterplot for the data. Do you detect a trend?

City	Slug Pct	Elevation
Anaheim	.480	160
Arlington	.605	616
Atlanta	.530	1050
Baltimore	.505	130
Boston	.505	20
⋮	⋮	⋮
Denver	.625	5277
⋮	⋮	⋮
Seattle	.550	350
San Francisco	.510	63
St. Louis	.570	465
Tampa	.500	10
Toronto	.535	566

Based on Schaffer, J., and Heiny, E. L. "The effects of elevation on slugging percentage in Major League Baseball." *Chance*, Vol. 19, No. 1, Winter 2006 (Figure 2).

2.149 Comparing SAT scores. Refer to Exercise 2.42 (p. 48) and the data on state SAT scores saved in the **SATSCORES** file. Consider a scatterplot for the data, with 2009 SAT score on the vertical axis and 2005 SAT score on the horizontal axis. What type of trend would you expect to observe? Why? Create the scatterplot and check your answer.

2.150 Feeding behavior of fish. Zoologists at the University of Western Australia conducted a study of the feeding behavior of black bream (a type of fish) spawned in aquariums. (*Brain, Behavior and Evolution*, April 2000). In one experiment, the zoologists recorded the number of aggressive strikes of two black bream feeding at the bottom of the aquarium in the 10-minute period following the addition of food. The number of strikes and age of the fish (in days) were recorded approximately each week for nine weeks,

Data for Exercise 2.150

Week	Number of Strikes	Age of Fish (days)
1	85	120
2	63	136
3	34	150
4	39	155
5	58	162
6	35	169
7	57	178
8	12	184
9	15	190

Based on Shand, J., et al. "Variability in the location of the retinal gan-glion cell area centralis is correlated with ontogenetic changes in feeding behavior in the Blackbream, *Acanthopagrus 'butcher'*." *Brain and Behavior*, Vol. 55, No. 4, April 2000 (Figure 10H).

as shown in the table above. The data are saved in the **BLACKBREAM** file.

a. Construct a scatterplot of the data, with number of strikes on the *y*-axis and age of the fish on the *x*-axis.

b. Examine the scatterplot of part **a.** Do you detect a trend?

2.151 **New method for blood typing.** In *Analytical Chemistry* (May 2010), medical researchers tested a new method of typing blood using low cost paper. Blood drops were applied to the paper and the rate of absorption (called *blood wicking*) was measured. The table gives the wicking lengths (in millimeters) for six blood drops, each at a different antibody concentration. (The data are saved in the **BLOODTYPE** file.) Construct a plot to investigate the relationship between wicking length and antibody concentration. What do you observe?

Droplet	Length (mm)	Concentration
1	22.50	0.0
2	16.00	0.2
3	13.50	0.4
4	14.00	0.6
5	13.75	0.8
6	12.50	1.0

Based on Khan, M. S. "Paper diagnostic for instant blood typing." *Analytical Chemistry*, Vol. 82, No. 10, May 2010 (Figure 4b).

2.152 **Sound waves from a basketball.** Refer to the *American Journal of Physics* (June 2010) study of sound waves in a spherical cavity, Exercise 2.37 (p. 47). The frequencies of sound waves (estimated using a mathematical formula) resulting from the first 24 resonances (echoes) after striking a basketball with a metal rod are reproduced in the table in the next column. (The data are saved in the **BBALL** file.) Graph the data in a scatterplot, with frequency on the vertical axis and resonance number on the horizontal axis. Since a mathematical formula was used to estimate frequency, the researcher expects an increasing trend with very little variation. Does the graph support the researcher's theory?

Based on Russell, D. A. "Basketballs as spherical acoustic cavities." *American Journal of Physics*, Vol. 78, No. 6, June 2010 (Table I).

Applying the Concepts—Intermediate

2.153 **College protests of labor exploitation.** Refer to the *Journal of World-Systems Research* (Winter 2004) study of 14 student sit-ins for a "sweat-free campus," presented in

Exercise 2.41 (p. 48). The **SITIN** file contains data on the duration (in days) of each sit-in, as well as data on the number of student arrests.

a. Use a scatterplot to graph the relationship between duration and number of arrests. Do you detect a trend?

b. Repeat part **a,** but graph only the data for sit-ins in which there was at least one arrest. Do you detect a trend?

c. Comment on the reliability of the trend you detected in part **b.**

Resonance	Frequency
1	979
2	1572
3	2113
4	2122
5	2659
6	2795
7	3181
8	3431
9	3638
10	3694
11	4038
12	4203
13	4334
14	4631
15	4711
16	4993
17	5130
18	5210
19	5214
20	5633
21	5779
22	5836
23	6259
24	6339

Source: Russell, D. A. "Basketballs as spherical acoustic cavities." *American Journal of Physics*, Vol. 48, No. 6, June 2010 (Table I).

2.154 **Are geography journals worth their cost?** In *Geoforum* (Vol. 37, Nov. 2006), Simon Fraser University professor Nicholas Blomley assessed whether the price of a geography journal is correlated with quality. He collected pricing data (cost for a 1-year subscription, in U.S. dollars) for a sample of 28 geography journals. In addition to cost, three other variables were measured: journal impact factor (JIF), defined as the average number of times articles from the journal have been cited; number of citations for a journal over the past five years; and relative price index (RPI), a measure developed by economists. [*Note:* A journal with an RPI less than 1.25 is considered a "good value."] The data for the 28 geography journals are saved in the **GEOJRNL** file. (Selected observations are shown in the table on page 91.)

a. Construct a scatterplot for the variables JIF and cost. Do you detect a trend?

b. Construct a scatterplot for the variables number of cites and cost. Do you detect a trend?

c. Construct a scatterplot for the variables RPI and cost. Do you detect a trend?

Data for Exercise 2.154

Journal	Cost ($)	JIF	Cites	RPI
J.Econ.Geogr.	468	3.139	207	1.16
Prog. Hum. Geog.	624	2.943	544	0.77
T.I. Brit.Geogr.	499	2.388	249	1.11
Econ.Geogr.	90	2.325	173	0.30
A.A.A. Geogr.	698	2.115	377	0.93
⋮	⋮	⋮	⋮	⋮
Geogr.Anal.	213	0.902	106	0.88
Geogr.J.	223	0.857	81	0.94
Appl.Geogr.	646	0.853	74	3.38

Based on Blomley, N. "Is this journal worth US$118?" *Geoforum*, Vol. 37, No. 6, Nov. 2006 (Tables 1, 2, and 3).

2.155 Mongolian desert ants. Refer to the *Journal of Biogeography* (Dec. 2003) study of ants in Mongolia, presented in Exercise 2.68 (p. 59). Data on annual rainfall, maximum daily temperature, percentage of plant cover, number of ant species, and species diversity index recorded at each of 11 study sites are saved in the **GOBIANTS** file.

   a. Construct a scatterplot to investigate the relationship between annual rainfall and maximum daily temperature. What type of trend (if any) do you detect?

   b. Use scatterplots to investigate the relationship that annual rainfall has with each of the other four variables in the data set. Are the other variables positively or negatively related to rainfall?

2.156 Forest fragmentation study. Ecologists classify the cause of forest fragmentation as either anthropogenic (i.e., due to human development activities, such as road construction or logging) or natural in origin (e.g., due to wetlands or wildfire). *Conservation Ecology* (Dec. 2003) published an article on the causes of fragmentation for 54 South American forests. Using advanced high-resolution satellite imagery, the researchers developed two fragmentation indexes for each forest—one for anthropogenic fragmentation and one for fragmentation from natural causes. The values of these two indexes (where higher values indicate more fragmentation) for five of the forests in the sample are shown in the table below. The data for all 54 forests are saved in the **FORFRAG** file.

Ecoregion (forest)	Anthropogenic Index	Natural Origin Index
Araucaria moist forests	34.09	30.08
Atlantic Coast *restingas*	40.87	27.60
Bahia coastal forests	44.75	28.16
Bahia interior forests	37.58	27.44
Bolivian *Yungas*	12.40	16.75

Based on Wade, T. G., et al. "Distribution and causes of global forest fragmentation." *Conservation Ecology*, Vol. 72, No. 2, Dec. 2003 (Table 6).

   a. Ecologists theorize that an approximately linear (straight-line) relationship exists between the two fragmentation indexes. Graph the data for all 54 forests. Does the graph support the theory?

   b. Delete the data for the three forests with the largest anthropogenic indexes, and reconstruct the graph of part **a.** Comment on the ecologists' theory.

Applying the Concepts—Advanced

2.157 Ranking driving performance of professional golfers. Refer to *The Sport Journal* (Winter 2007) analysis of a new method for ranking the total driving performance of golfers on the PGA tour, Exercise 2.64 (p. 59). Recall that the method uses both the average driving distance (in yards) and the driving accuracy (percent of drives that land in the fairway). Data on these two variables for the top 40 PGA golfers are saved in the **PGADRIVER** file. A professional golfer is practicing a new swing to increase his average driving distance. However, he is concerned that his driving accuracy will be lower. Is his concern reasonable? Explain.

2.158 Spreading rate of spilled liquid. A contract engineer at DuPont Corp. studied the rate at which a spilled volatile liquid will spread across a surface (*Chemical Engineering Progress*, Jan. 2005). Suppose that 50 gallons of methanol spills onto a level surface outdoors. The engineer uses derived empirical formulas (assuming a state of turbulence-free convection) to calculate the mass (in pounds) of the spill after a period ranging from 0 to 60 minutes. The calculated mass values are given in the table below and saved in the **LIQUIDSPILL** file. Is there evidence to indicate that the mass of the spill tends to diminish as time increases?

Time (minutes)	Mass (pounds)
0	6.64
1	6.34
2	6.04
4	5.47
6	4.94
8	4.44
10	3.98
12	3.55
14	3.15
16	2.79
18	2.45
20	2.14
22	1.86
24	1.60
26	1.37
28	1.17
30	0.98
35	0.60
40	0.34
45	0.17
50	0.06
55	0.02
60	0.00

Based on Barry, J. "Estimating rates of spreading and evaporation of volatile liquids." *Chemical Engineering Progress*, Vol. 101, No. 1. Jan. 2005.

2.10 Distorting the Truth with Descriptive Statistics

A picture may be "worth a thousand words," but pictures can also color messages or distort them. In fact, the pictures displayed in statistics—histograms, bar charts, and other graphical images—are susceptible to distortion, so we have to examine each of them with care. Accordingly, we begin this section by mentioning a few of the pitfalls to watch for in interpreting a chart or a graph. Then we discuss how numerical descriptive statistics can be used to distort the truth.

Graphical Distortions

One common way to change the impression conveyed by a graph is to alter the scale on the vertical axis, the horizontal axis, or both. For example, consider the data on collisions of large marine vessels operating in European waters over a certain five-year period, summarized in Table 2.11. Figure 2.38 is a MINITAB bar graph showing the frequency of collisions for each of the three locations listed in the table. The graph shows that in-port collisions occur more often than collisions at sea or collisions in restricted waters.

Table 2.11	Collisions of Marine Vessels by Location
Location	Number of Ships
At Sea	376
In Restricted Waters	273
In Port	478
Total	1,127

Based on *The Dock and Harbour Authority*.  *Data Set:* COLLISION

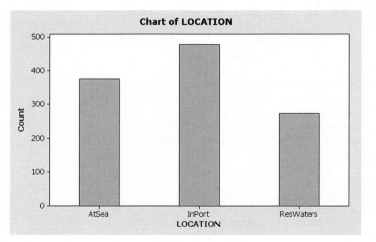

Figure 2.38
MINITAB bar graph of vessel collisions by location

Now, suppose you want to use the same data to exaggerate the difference between the number of in-port collisions and the number of collisions in restricted waters. One way to do this is to increase the distance between successive units on the vertical axis—that is, *stretch* the vertical axis by graphing only a few units per inch. A telltale sign of stretching is a long vertical axis, but this indication is often hidden by starting the vertical axis at some point above the origin, 0. Such a graph is shown in the SPSS printout in Figure 2.39. By starting the bar chart at 250 collisions (instead of 0), it appears that the frequency of in-port collisions is many times greater than the frequency of collisions in restricted waters.

Ethics IN Statistics

Intentionally distorting a graph to portray a particular viewpoint is considered unethical statistical practice.

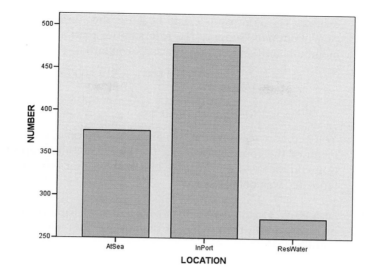

Figure 2.39

SPSS bar graph of vessel collisions by location, with adjusted vertical axis

Another method of achieving visual distortion with bar graphs is by making the width of the bars proportional to their height. For example, look at the bar chart in Figure 2.40a, which depicts the percentage of the total number of motor vehicle deaths in a year that occurred on each of four major highways. Now suppose we make both the width and the height grow as the percentage of fatal accidents grows. This change is shown in Figure 2.40b. The distortion is that the reader may tend to equate the *area* of the bars with the percentage of deaths occurring at each highway when, in fact, the true relative frequency of fatal accidents is proportional only to the *height* of the bars.

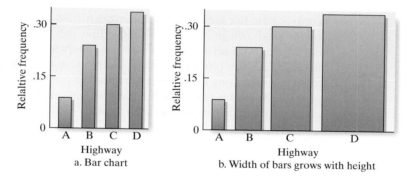

Figure 2.40

Relative frequency of fatal motor vehicle accidents on each of four major highways

Although we've discussed only a few of the ways that graphs can be used to convey misleading pictures of phenomena, the lesson is clear: Look at all graphical descriptions of data with a critical eye. In particular, check the axes and the size of the units on each axis. Ignore the visual changes, and concentrate on the actual numerical changes indicated by the graph or chart.

Misleading Numerical Descriptive Statistics

The information in a data set can also be distorted by using numerical descriptive measures, as Example 2.20 shows.

Example 2.20

Misleading Descriptive Statistics

Problem Suppose you're considering working for a small law firm—one that currently has a senior member and three junior members. You inquire about the salary you could expect to earn if you join the firm. Unfortunately, you receive two answers:

Answer A: The senior member tells you that an "average employee" earns $107,500.

Answer B: One of the junior members later tells you that an "average employee" earns $95,000.

Which answer can you believe?

Solution The confusion exists because the phrase "average employee" has not been clearly defined. Suppose the four salaries paid are $95,000 for each of the three junior members and $145,000 for the senior member. Then,

$$Mean = \frac{3(\$95,000) + \$145,000}{4} = \frac{\$430,000}{4} = \$107,500$$

$$Median = \$95,000$$

You can now see how the two answers were obtained: The senior member reported the mean of the four salaries, and the junior member reported the median. The information you received was distorted because neither person stated which measure of central tendency was being used.

Look Back On the basis of our earlier discussion of the mean and median, we would probably prefer the median as the number that best describes the salary of the "average employee."

Another distortion of information in a sample occurs when *only* a measure of central tendency is reported. Both a measure of central tendency and a measure of variability are needed to obtain an accurate mental image of a data set.

Suppose, for instance, that you want to buy a new car and are trying to decide which of two models to purchase. Since energy and economy are both important issues, you decide to purchase model A because its EPA mileage rating is 32 miles per gallon in the city, whereas the mileage rating for model B is only 30 miles per gallon in the city.

However, you may have acted too quickly. How much variability is associated with the ratings? As an extreme example, suppose that further investigation reveals that the standard deviation for model A mileages is 5 miles per gallon, whereas that for model B is only 1 mile per gallon. If the mileages form a mound-shaped distribution, they might appear as shown in Figure 2.41. Note that the larger amount of variability associated with model A implies that more risk is involved in purchasing that model. That is, the particular car you purchase is more likely to have a mileage rating that will differ greatly from the EPA rating of 32 miles per gallon if you purchase model A, while a model B car is not likely to vary from the 30-miles-per-gallon rating by more than 2 miles per gallon.

We conclude this section with another example on distorting the truth with numerical descriptive measures.

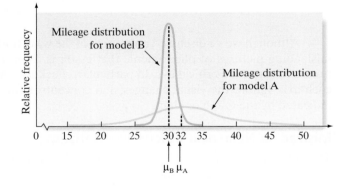

Figure 2.41

Mileage distributions for two car models

Example 2.21
More Misleading Descriptive Statistics— Delinquent Children

Problem *Children Out of School in America* is a report on the delinquency of school-age children written by the Children's Defense Fund (CDF). Consider the following three reported results of the CDF survey.

- Reported result: Twenty-five percent of the 16- and 17-year-olds in the Portland, Maine, Bayside East Housing Project were out of school. Actual data: *Only eight children were surveyed; two were found to be out of school.*

- Reported result: Of all the secondary school students who had been suspended more than once in census tract 22 in Columbia, South Carolina, 33% had been suspended two times and 67% had been suspended three or more times. Actual data: *CDF found only three children in that entire census tract who had been suspended; one child was suspended twice and the other two children three or more times.*

- Reported result: In the Portland Bayside East Housing Project, 50% of all the secondary school children who had been suspended more than once had been suspended three or more times. Actual data: *The survey found just two secondary school children who had been suspended in that area; one of them had been suspended three or more times.*

Identify the potential distortions in the results reported by the CDF.

Solution In each of these examples, the reporting of percentages (i.e., relative frequencies) instead of the numbers themselves is misleading. No inference we might draw from the examples cited would be reliable. (We'll see how to measure the reliability of estimated percentages in Chapter 7.) In short, either the report should state the numbers alone instead of percentages, or, better yet, it should state that the numbers were too small to report by region.

Look Back If several regions were combined, the numbers (and percentages) would be more meaningful.

Ethics IN Statistics

Purposeful reporting of numerical descriptive statistics in order to mislead the target audience is considered *unethical statistical practice.*

CHAPTER NOTES

Key Terms

Note: Starred (*) terms are from the optional sections in this chapter.

Bar graph 29
Bivariate relationship* 87
Box plots* 78
Central tendency 50
Chebyshev's rule 67
Class 27
Class percentage 28
Class interval 40
Class frequency 27
Class relative frequency 27
Dot plot 38
Empirical rule 67
Hinge* 78
Histogram 40
Inner fence* 78
Interquartile range (IQR)* 78
Lower quartile (Q_L)* 74
Mean 50
Measures of central tendency 50
Measures of relative standing 73
Measures of variability
　(or spread) 61
Median 52
Middle quartile (M)* 74

Modal class 55
Mode 54
Mound-shaped distribution 66
Numerical descriptive
　measures 50
Outer fence* 78
Outlier* 78
Pareto diagram 29
pth percentile 73
Percentile ranking 73
Pie chart 29
Quartiles* 74
Range 61
Rare-event approach* 83
Scatterplot* 87
Skewness/skewed 53
Spread 61
Standard deviation 62
Stem-and-leaf display 38
Symmetric distribution 66
Summation notation 50
Upper quartile (Q_U)* 74
Variability 50
Variance 62
Whiskers* 78
z-score 74

Key Symbols

	Sample	Population
Mean:	$\bar{x}$	μ
Variance:	s^2	σ^2
Std. Dev.	s	σ
Median:	M	
Lower Quartile:	Q_L	
Upper Quartile:	Q_U	
Interquartile Range:	IQR	

Key Ideas

Describing QUALITATIVE Data

1. Identify **category** classes
2. Determine **class frequencies**
3. **Class relative frequency** = (class frequency)/n
4. Graph relative frequencies

Pie Chart:

Bar Graph:

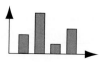

Pareto Diagram:

Graphing QUANTITATIVE Data

One Variable:

1. Identify class intervals
2. Determine class interval frequencies
3. Class interval relative frequency = (class interval frequency)/n
4. Graph class interval relative frequencies

Dot Plot:

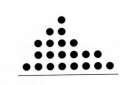

Box plot:

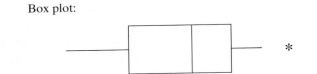

Stem-and-Leaf Display:

1	3
2	2489
3	126678
4	37
5	2

Two Variables:

Scatterplot:

Histogram:

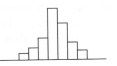

Numerical Description of QUANTITATIVE Data

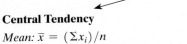

Central Tendency

Mean: $\bar{x} = (\Sigma x_i)/n$

Median: Middle value (or mean of two middle values) when data ranked in order

Mode: Value that occurs most often

Variation

Range: Difference between largest and smallest value

Variance:

$$s^2 = \frac{\Sigma(x_i - \bar{x})^2}{n-1} = \frac{\Sigma x_i^2 - \frac{(\Sigma x_i)^2}{n}}{n-1}$$

Std Dev.: $s = \sqrt{s^2}$

Interquartile Range: $\text{IQR} = Q_U - Q_L$

Relative Standing

Percentile Score: Percentage of values that fall below x-score

z-score: $z = (x - \bar{x})/s$

Rules for Describing Quantitative Data

Interval	Chebyshev's Rule	Empirical Rule
$\bar{x} \pm s$	At least 0%	$\approx 68\%$
$\bar{x} \pm 2s$	At least 75%	$\approx 95\%$
$\bar{x} \pm 3s$	At least 89%	$\approx$ All

Rules for Detecting Quantitative Outliers

Method	Suspect	Highly Suspect
Box plot:	Values between inner and outer fences	Values beyond outer fences
z-score:	$2 < |z| < 3$	$|z| > 3$

Guide to Selecting the Data Description Method

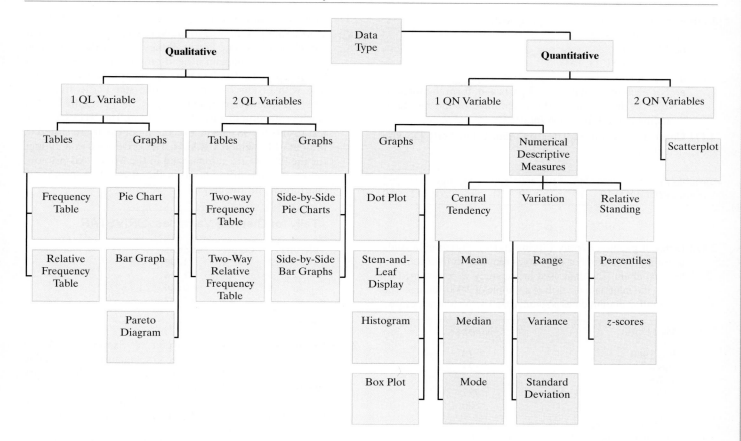

Supplementary Exercises 2.159–2.198

Note: Starred (*) exercises refer to the optional section in this chapter.

Understanding the Principles

2.159 Discuss conditions under which the median is preferred to the mean as a measure of central tendency.

2.160 Explain why we generally prefer the standard deviation to the range as a measure of variation.

2.161 Give a situation in which we would prefer using a stem-and-leaf display over a histogram in describing quantitative data graphically.

2.162 Give a situation in which we would prefer using a box plot over z-scores to detect an outlier.

2.163 Give a technique that is used to distort information shown on a graph.

Learning the Mechanics

2.164 Construct a relative frequency histogram for the data summarized in the following table:

Measurement Class	Relative Frequency	Measurement Class	Relative Frequency
.00–.75	.02	5.25–6.00	.15
.75–1.50	.01	6.00–6.75	.12
1.50–2.25	.03	6.75–7.50	.09
2.25–3.00	.05	7.50–8.25	.05
3.00–3.75	.10	8.25–9.00	.04
3.75–4.50	.14	9.00–9.75	.01
4.50–5.25	.19		

2.165 Consider the following three measurements: 50, 70, 80. Find the z-score for each measurement if they are from a population with a mean and standard deviation equal to
a. $\mu = 60, \sigma = 10$
b. $\mu = 60, \sigma = 5$
c. $\mu = 40, \sigma = 10$
d. $\mu = 40, \sigma = 100$

2.166 Refer to Exercise 2.165. For parts **a–d,** determine whether the three measurements 50, 70, and 80 are outliers.

2.167 Compute s^2 for data sets with the following characteristics:
a. $\sum_{i=1}^{n} x_i^2 = 246, \sum_{i=1}^{n} x_i = 63, n = 22$
b. $\sum_{i=1}^{n} x_i^2 = 666, \sum_{i=1}^{n} x_i = 106, n = 25$
c. $\sum_{i=1}^{n} x_i^2 = 76, \sum_{i=1}^{n} x_i = 11, n = 7$

2.168 If the range of a set of data is 20, find a rough approximation to the standard deviation of the data set.

2.169 For each of the data sets in parts **a–c,** compute $\bar{x}$, s^2, and s. If appropriate, specify the units in which your answers are expressed.
a. 4, 6, 6, 5, 6, 7
b. –$1, $4, –$3, $0, –$3, –$6
c. 3/5%, 4/5%, 2/5%, 1/5%, 1/16%
d. Calculate the range of each data set in parts **a–c.**

97

2.170 For each of the data sets in parts **a–d,** compute $\bar{x}$, s^2, and s.
   a. 13, 1, 10, 3, 3
   b. 13, 6, 6, 0
   c. 1, 0, 1, 10, 11, 11, 15
   d. 3, 3, 3, 3
   e. For each of the data sets in parts **a–d,** form the interval $\bar{x} \pm 2s$ and calculate the percentage of the measurements that fall into that interval.

*2.171 Construct a scatterplot for the data listed here. Do you detect any trends?

| Variable #1: | 174 | 268 | 345 | 119 | 400 | 520 | 190 | 448 | 307 | 252 |
| Variable #2: | 8 | 10 | 15 | 7 | 22 | 31 | 15 | 20 | 11 | 9 |

Applying the Concepts—Basic

2.172 Dates of pennies. *Chance* (Spring 2000) reported on a study to estimate the number of pennies required to fill a coin collector's album. The data used in the study were obtained by noting the mint date on each of a sample of 2,000 pennies. The distribution of mint dates is summarized in the following table:

Mint Date	Number
Pre-1960	18
1960s	125
1970s	330
1980s	727
1990s	800

Source: Lu, S., and Skiena, S. "Filling a penny album." *Chance,* Vol. 13, No. 2, Spring 2000, p. 26 (Table 1). Copyright 2000 by the American Statistical Association. All rights reserved.

   a. Identify the experimental unit for the study.
   b. Identify the variable measured.
   c. What proportion of pennies in the sample have mint dates in the 1960s?
   d. Construct a pie chart to describe the distribution of mint dates for the 2,000 sampled pennies.

2.173 Al Qaeda attacks on the United States. An article in *Studies in Conflict & Terrorism* (Vol. 29, 2006) conducted an empirical analysis of incidents involving suicide terrorist attacks. The data in the table that follows are the number of individual suicide bombings or attacks for each in a sample of 21 incidents involving an attack against the United States by the Al Qaeda terrorist group. (For example, the infamous September 11, 2001, Al Qaeda attack involved four separate attacks with hijacked airplanes: Two planes crashed into the World Trade Center towers, one battered into the Pentagon, and a fourth plane shattered in a field in Pennsylvania.) Summarize the data (saved in the **ALQAEDA** file) with a dot plot. Interpret the results.

1	4	2	1	2	1	1
2	1	1	3	1	5	1
1	4	2	1	2	1	2

Based on Moghadam, A. "Suicide terrorism, occupation, and the globalization of martyrdom: A critique of *Dying to Win,*" *Studies in Conflict & Terrorism,* Vol. 29, No. 8, 2006 (Table 3).

2.174 Crash tests on new cars. Each year, the National Highway Traffic Safety Administration (NHTSA) crash tests new car models to determine how well they protect the driver and front-seat passenger in a head-on collision. The NHTSA has developed a "star" scoring system for the frontal crash test, with results ranging from one star (*) to five stars (*****). The more stars in the rating, the better the level of crash protection in a head-on collision. The NHTSA crash test results for 98 cars in a recent model year are stored in the data file named **CRASH**. The driver-side star ratings for the 98 cars are summarized in the MINITAB printout below. Use the information in the printout to form a pie chart. Interpret the graph.

Tally for Discrete Variables: DRIVSTAR

DRIVSTAR	Count	Percent
2	4	4.08
3	17	17.35
4	59	60.20
5	18	18.37
N=	98	

2.175 Crash tests on new cars. Refer to Exercise 2.174 and the NHTSA crash test data. One quantitative variable recorded by the NHTSA is driver's severity of head injury (measured on a scale from 0 to 1,500). Numerical descriptive statistics for the 98 driver head-injury ratings in the **CRASH** file are displayed in the MINITAB printout at the bottom of the page.
   a. Interpret each of the statistics shown on the printout.
   b. Find the z-score for a driver head-injury rating of 408. Interpret the result.

2.176 Competition for bird nest holes. The *Condor* (May 1995) published a study of competition for nest holes among collared flycatchers, a bird species. The authors collected the data for the study by periodically inspecting nest boxes located on the island of Gotland in Sweden. The nest boxes were grouped into 14 discrete locations (called plots). The accompanying table gives the number of flycatchers killed and the number of flycatchers breeding at each plot. The data are saved in the **CONDOR** file.

	Number Killed					
	0	1	2	3	4	5
	30	124	34	38	28	30
Number of	46	68	26			
Breeders	132	86				
	100	32				
	6					

Reprinted with permission of the author and the Cooper Ornithological Club, University of California Press, from *The Condor,* "Interspecific competition for nest holes causes adult mortality in the collared flycatcher" by J. Merila and D. A. Wiggins. *The Condor,* Vol. 97, No. 2, May 1995, p. 447. Permission conveyed through Copyright Clearance Center, Inc.

Descriptive Statistics: DRIVHEAD

Variable	N	Mean	StDev	Minimum	Q1	Median	Q3	Maximum
DRIVHEAD	98	603.7	185.4	216.0	475.0	605.0	724.3	1240.0

a. Calculate the mean, median, and mode for the number of flycatchers killed at the 14 study plots.

b. Interpret the measures of central tendency you found in part **a.**

*c. Graphically examine the relationship between number killed and number of breeders. Do you detect a trend?

2.177 Ratings of chess players. The United States Chess Federation (USCF) establishes a numerical rating for each competitive chess player. The USCF rating is a number between 0 and 4,000 that changes over time, depending on the outcome of tournament games. The higher the rating, the better (more successful) the player is. The following table describes the rating distribution of 65,455 active USCF members. The data are saved in the **USCF** file.

Classification	Rating Range	Number of Players
Senior Master	2,800 to 2,899	1
Senior Master	2,700 to 2,799	13
Senior Master	2,600 to 2,699	66
Senior Master	2,500 to 2,599	87
Senior Master	2,400 to 2,499	133
Master	2,300 to 2,399	231
Master	2,200 to 2,299	691
Expert	2,100 to 2,199	783
Expert	2,000 to 2,099	1,516
Class A	1,900 to 1,999	1,907
Class A	1,800 to 1,899	2,682
Class B	1,700 to 1,799	3,026
Class B	1,600 to 1,699	3,437
Class C	1,500 to 1,599	3,582
Class C	1,400 to 1,499	3,386
Class D	1,300 to 1,399	3,139
Class D	1,200 to 1,299	3,153
Class E	1,100 to 1,199	2,973
Class E	1,000 to 1,099	3,021
Class F	900 to 999	3,338
Class F	800 to 899	3,520
Class G	700 to 799	3,829
Class G	600 to 699	3,946
Class H	500 to 599	3,783
Class H	400 to 499	3,544
Class I	300 to 399	3,142
Class I	200 to 299	2,547
Class J	0 to 199	3,979
	Total Members	65,455
	Average Rating	1,068

Based on United States Chess Federation; www.uschess.org.

a. Convert the information in the table into a graph that portrays the distribution of USCF ratings.

b. What percentage of players have a USCF rating of 2,000 or higher? (These players are "experts," "masters," or "senior masters.")

c. The mean USCF rating of the 65,455 players is 1,068. Give a practical interpretation of this value.

2.178 Groundwater contamination in wells. Refer to the *Environmental Science & Technology* (Jan. 2005) study of the factors related to MTBE contamination in 223 New Hampshire wells, presented in Exercise 2.21 (p. 37). The data are saved in the **MTBE** file. Two of the many quantitative variables measured for each well are the pH level (standard units) and the MTBE level (micrograms per liter).

a. Construct a histogram for the pH levels of the sampled wells. From the histogram, estimate the proportion of wells with pH values less than 7.0.

b. For those wells with detectible levels of MTBE, construct a histogram for the MTBE values. From the histogram, estimate the proportion of contaminated wells with MTBE values that exceed 5 micrograms per liter.

c. Find the mean and standard deviation for the pH levels of the sampled wells, and construct the interval $\bar{x} \pm 2s$. Estimate the percentage of wells with pH levels that fall within the interval. What rule did you apply to obtain the estimate? Explain.

d. Find the mean and standard deviation for the MTBE levels of the sampled wells, and construct the interval $\bar{x} \pm 2s$. Estimate the percentage of wells with MTBE levels that fall within the interval. What rule did you apply to obtain the estimate? Explain.

2.179 "Made in the USA" survey. "Made in the USA" is a claim stated in many product advertisements or on product labels. Advertisers want consumers to believe that their product is manufactured with 100% U.S. labor and materials—which is often not the case. What does "Made in the USA" mean to the typical consumer? To answer this question, a group of marketing professors conducted an experiment at a shopping mall in Muncie, Indiana. (*Journal of Global Business*, Spring 2002.) They asked every fourth adult entrant to the mall to participate in the study. A total of 106 shoppers agreed to answer the question, "'Made in the USA' means what percentage of U.S. labor and materials?" The responses of the 106 shoppers are summarized as follows: 64 shoppers responded 100%; 20 shoppers stated 75 to 99%; 18 shoppers stated 50 to 74%; and 4 shoppers said less than 50%.

a. What type of data collection method was used?

b. What type of variable, quantitative or qualitative, was measured?

c. Present the data in graphical form. Use the graph to make a statement about the percentage of consumers who believe that "Made in the USA" means 100% U.S. labor and materials.

Based on " 'Made in the USA': Consumer perceptions, deception and policy alternatives." *Journal of Global Business*, Vol. 13, No. 24, Spring 2002 (Table 3).

2.180 Achievement test scores. The distribution of scores on a nationally administered college achievement test has a median of 520 and a mean of 540.

a. Explain why it is possible for the mean to exceed the median for this distribution of measurements.

b. Suppose you are told that the 90th percentile is 660. What does this mean?

c. Suppose you are told that you scored at the 94th percentile. Interpret this statement.

Applying the Concepts—Intermediate

2.181 New-book reviews. *Choice* magazine provides new-book reviews in each issue. A random sample of 375 *Choice* book reviews in American history, geography, and area studies was selected, and the "overall opinion" of the book stated in each review was ascertained (*Library*

Acquisitions: Practice and Theory, Vol. 19, 1995). Overall opinion was coded as follows: 1 = would not recommend, 2 = cautions or very little recommendation, 3 = little or no preference, 4 = favorable/recommended, 5 = outstanding/significant contribution. The data are summarized in the following table:

Opinion Code	Number of Reviews
1	19
2	37
3	35
4	238
5	46
Total	375

Based on Carlo, P. W., and Natowitz, A. "Choice book reviews in American history, geography, and area studies: An analysis for 1988–1993." *Library Acquisitions: Practice & Theory*, Vol. 19, No. 2, p. 159 (Figure 1).

a. Use a Pareto diagram to find the opinion that occurred most often. What proportion of the books reviewed received this opinion?
b. Do you agree with the following statement extracted from the study: "A majority (more than 75%) of books reviewed are evaluated favorably and recommended for purchase"?

2.182 Archaeologists' study of ring diagrams. Archaeologists gain insight into the social life of ancient tribes by measuring the distance (in centimeters) of each artifact found at a site from the "central hearth" (i.e., the middle of an artifact scatter). A graphical summary of these distances, in the form of a histogram, is called a *ring diagram*. (*World Archaeology*, Oct. 1997). Ring diagrams for two archaeological sites (A and G) in Europe are shown below and in the next column.
a. Identify the type of skewness (if any) present in the data collected at the two sites.
b. Archaeologists have associated unimodal ring diagrams with open-air hearths and multimodal ring diagrams with hearths inside dwellings. Can you identify the type of hearth (*open air or inside dwelling*) that was most likely present at each site?

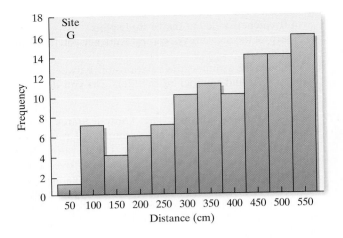

2.183 Best-paid CEOs. Refer to the *Forbes* list of the 40 best-paid chief executive officers, Exercise 2.17 (p. 36). The salary and age of each CEO are saved in the **FORBES40** file.
*a. Use a graph to portray the relationship between a CEO's salary and age. Do you detect a trend?
b. According to Chebyshev's rule, what percentage of the age measurements would you expect to find in the intervals $\bar{x} \pm .75s$, $\bar{x} \pm 2.5s$, and $\bar{x} \pm 4s$?
c. What percentage of the age measurements actually fall into the intervals of part b? Compare your results with those of part **b**.
d. Repeat parts **b** and **c** for salary.

2.184 Ammonia in car exhaust. Three-way catalytic converters have been installed in new vehicles in order to reduce pollutants from motor vehicle exhaust emissions. However, these converters unintentionally increase the level of ammonia in the air. *Environmental Science & Technology* (Sept. 1, 2000) published a study on the ammonia levels near the exit ramp of a San Francisco highway tunnel. The data in the table represent daily ammonia concentrations (parts per million) on eight randomly selected days during the afternoon drive time in the summer of a recent year. The data are saved in the **AMMONIA** file. Find an interval that is likely to contain the ammonia level for a randamly selected day during afternoon drive time.

1.53	1.50	1.37	1.51	1.55	1.42	1.41	1.48

2.185 Whistling dolphins. Marine scientists who study dolphin communication have discovered that bottlenose dolphins exhibit an individualized whistle contour known as their *signature whistle*. A study was conducted to categorize the signature whistles of 10 captive adult bottlenose dolphins in socially interactive contexts (*Ethology*, July 1995). A total of 185 whistles was recorded during the study period; each whistle contour was analyzed and assigned to a category by a contour similarity (CS) technique. The results are reported in the table on the next page and saved in the **DOLPHIN** file. Use a graphical method to summarize the results. Do you detect any patterns in the data that might be helpful to marine scientists?

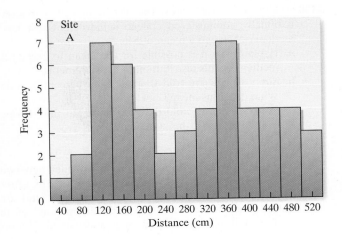

Data for Exercise 2.185

Whistle Category	Number of Whistles
Type a	97
Type b	15
Type c	9
Type d	7
Type e	7
Type f	2
Type g	2
Type h	2
Type i	2
Type j	4
Type k	13
Other types	25

Based on McCowan, B., and Reiss, D. "Quantitative comparison of whistle repertoires from captive adult bottlenose dolphins (Delphiniae, *Tursiops truncates*): A re-evaluation of the signature whistle hypothesis." *Ethology*, Vol. 100, No. 3, July 1995, p. 200 (Table 2).

2.186 Speed of light from galaxies. Astronomers theorize that cold dark matter caused the formation of galaxies and clusters of galaxies in the universe. The theoretical model for cold dark matter requires an estimate of the velocities of light emitted from galaxy clusters. *The Astronomical Journal* (July 1995) published a study of observed velocities of galaxies in four different clusters. Galaxy velocity was measured in kilometers per second (km/s), using a spectrograph and high-power telescope.

a. The observed velocities of 103 galaxies located in the cluster named A2142 are summarized in the accompanying histogram. Comment on whether the empirical rule is applicable to describing the velocity distribution for this cluster.

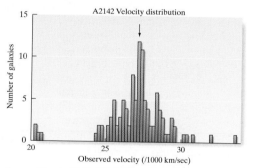

Source: Oegerle, W. R., Hill, J. M., and Fitchett, M. J. "Observations of high dispersion clusters of galaxies: Constraints on cold dark matter." *The Astronomical Journal*, Vol. 110, No. 1, July 1995, p. 37. Reproduced by permission of the AAS.

b. The mean and standard deviation of the 103 velocities observed in galaxy cluster A2142 were reported as $\bar{x} = 27,117$ km/s and $s = 1,280$ km/s, respectively. Use this information to construct an interval that captures approximately 95% of the velocities of the galaxies in the cluster.

c. Recommend a single velocity value to be used in the CDM model for galaxy cluster A2142. Explain your reasoning.

*2.187 Freckling of superalloy ingots. Freckles are defects that sometimes form during the solidification of alloy ingots. A freckle index has been developed to measure the level of freckling on the ingot. A team of engineers conducted several experiments to measure the freckle index of a certain type of superalloy (*Journal of Metallurgy*, Sept. 2004). The data for $n = 18$ alloy tests are shown in the table and saved in the **FRECKLE** file.

12.6	22.0	4.1	16.4	1.4	2.4
16.8	10.0	3.2	30.1	6.8	14.6
2.5	12.0	33.4	22.2	8.1	15.1

Based on Yang, W. H. "A freckle criterion for the solidification of superalloys with a tilted solidification front." *JOM: Journal of the Minerals, Metals and Materials Society*, Vol. 56, No. 9, Sept. 2004.

a. Construct a box plot for the data and use it to find any outliers.

b. Find and interpret the z-scores associated with the alloys you identified in part **a.**

2.188 Children's use of pronouns. Clinical observations suggest that specifically language-impaired (SLI) children have great difficulty with the proper use of pronouns. This phenomenon was investigated and reported in the *Journal of Communication Disorders* (Mar. 1995). Thirty children, all from low-income families, participated in the study. Ten were five-year-old SLI children, 10 were younger (three-year-old) normally developing (YND) children, and 10 were older (five-year-old) normally developing (OND) children. The table contains the group, deviation intelligence quotient (DIQ), and percentage of pronoun errors observed for each of the 30 subjects. (The data for these variables, as well as gender, are saved in the **SLI** file.)

a. Identify the variables in the data set as quantitative or qualitative.

b. Why is it nonsensical to compute numerical descriptive measures for qualitative variables?

c. Compute measures of central tendency for DIQ for the 10 SLI children.

Table for Exercise 2.188

YND		SLI		OND	
DIQ	Pronoun Errors (%)	DIQ	Pronoun Errors (%)	DIQ	Pronoun Errors (%)
110	94.40	86	60.00	110	0
92	19.05	86	40.00	113	0
92	62.50	94	31.58	113	0
100	18.75	98	66.67	109	0
86	0.00	89	42.86	92	0
105	55.00	84	27.27	108	0
90	100.00	110	33.33	95	0
96	86.67	107	0.00	87	0
90	32.43	87	0.00	94	0
92	0.00	95	0.00	98	0

Based on M. E. Moore, "Error analysis of pronouns by normal and language-impaired children." *Journal of Communication Disorders*, Vol. 28, No. 1, Moore, M.E., p. 62 (Table 2), p. 67 (Table 5).

d. Compute measures of central tendency for DIQ for the 10 YND children.

e. Compute measures of central tendency for DIQ for the 10 OND children.

f. Use the results from parts **c–e** to compare the DIQ central tendencies of the three groups of children. Is it reasonable to use a single number (e.g., mean or median) to describe the center of the DIQ distribution? Or should three "centers" be calculated, one for each of the three groups of children? Explain.

g. Repeat parts **c–f** for the percentage of pronoun errors.

*h. Plot all the data to investigate a possible trend between DIQ and proper use of pronouns. What do you observe?

*i. Plot the data for the 10 SLI children only. Is there a trend between DIQ and proper use of pronouns?

2.189 Oil spill impact on seabirds. The *Journal of Agricultural, Biological, and Environmental Statistics* (Sept. 2000) published a study on the impact of the *Exxon Valdez* tanker oil spill on the seabird population in Prince William Sound, Alaska. A subset of the data analyzed is stored in the **EVOS** file. Data were collected on 96 shoreline locations (called *transects*) of constant width, but variable length. For each transect, the number of seabirds found is recorded, as are the length (in kilometers) of the transect and whether or not the transect was in an oiled area. (The first five and last five observations in the **EVOS** file are listed in the accompanying table.)

Transect	Seabirds	Length	Oil
1	0	4.06	No
2	0	6.51	No
3	54	6.76	No
4	0	4.26	No
5	14	3.59	No
⋮	⋮	⋮	⋮
92	7	3.40	Yes
93	4	6.67	Yes
94	0	3.29	Yes
95	0	6.22	Yes
96	27	8.94	Yes

Based on McDonald, T. L., Erickson, W. P., and McDonald, L. L. "Analysis of count data from before–after control-impact studies." *Journal of Agricultural, Biological, and Environmental Statistics,* Vol. 5, No. 3, Sept. 2000, pp. 277–8 (Table A.1).

a. Identify the variables measured as quantitative or qualitative.

b. Identify the experimental unit.

c. Use a pie chart to describe the percentage of transects in oiled and unoiled areas.

*d. Use a graphical method to examine the relationship between observed number of seabirds and transect length.

e. Observed seabird density is defined as observed count divided by length of transect. MINITAB descriptive statistics for seabird densities in unoiled and oiled transects are displayed in the printout at the bottom of the page. Assess whether the distribution of seabird densities differs for transects in oiled and unoiled areas.

f. For unoiled transects, give an interval of values that is likely to contain at least 75% of the seabird densities.

g. For oiled transects, give an interval of values that is likely to contain at least 75% of the seabird densities.

h. Which type of transect, an oiled or unoiled one, is more likely to have a seabird density of 16? Explain.

2.190 Benford's Law of Numbers. According to *Benford's law*, certain digits $(1, 2, 3, \ldots, 9)$ are more likely to occur as the first significant digit in a randomly selected number than are other digits. For example, the law predicts that the number "1" is the most likely to occur (30% of the time) as the first digit. In a study reported in the *American Scientist* (July–Aug. 1998) to test Benford's law, 743 first-year college students were asked to write down a six-digit number at random. The first significant digit of each number was recorded and its distribution summarized in the following table. These data are saved in the **DIGITS** file. Describe the first digit of the "random guess" data with an appropriate graph. Does the graph support Benford's law? Explain.

First Digit	Number of Occurrences
1	109
2	75
3	77
4	99
5	72
6	117
7	89
8	62
9	43
Total	743

Based on Hill, T. P. "The first digit phenomenon." *American Scientist,* Vol. 86, No. 4, July–Aug. 1998, p. 363 (Figure 5).

Applying the Concepts—Advanced

2.191 Speed of light from galaxies. Refer to *The Astronomical Journal* study of galaxy velocities, presented in Exercise 2.186 (p. 101). A second cluster of galaxies, named A1775, is thought to be a *double cluster*—that is, two clusters of galaxies in close proximity. Fifty-one velocity observations (in kilometers per second, km/s) from cluster A1775 are listed in the table on the next page. The data are saved in the **GALAXY2** file.

MINITAB Output for Exercise 2.189

Descriptive Statistics: Density

Variable	Oil	N	Mean	StDev	Minimum	Q1	Median	Q3	Maximum
Density	no	36	3.27	6.70	0.000	0.000	0.890	3.87	36.23
	yes	60	3.495	5.968	0.0000	0.000	0.700	5.233	32.836

Data for Exercise 2.191

22,922	20,210	21,911	19,225	18,792	21,993	23,059
20,785	22,781	23,303	22,192	19,462	19,057	23,017
20,186	23,292	19,408	24,909	19,866	22,891	23,121
19,673	23,261	22,796	22,355	19,807	23,432	22,625
22,744	22,426	19,111	18,933	22,417	19,595	23,408
22,809	19,619	22,738	18,499	19,130	23,220	22,647
22,718	22,779	19,026	22,513	19,740	22,682	19,179
19,404	22,193					

Based on Oegerle, W. R., Hill, J. M., and Fitchett, M. J. "Observations of high dispersion clusters of galaxies: Constraints on cold dark matter." *The Astronomical Journal*, Vol. 110, No. 1, July 1995, (Table 1 and Figure 1).

a. Use a graphical method to describe the velocity distribution of galaxy cluster A1775.

b. Examine the graph you created in part **a.** Is there evidence to support the double-cluster theory? Explain.

c. Calculate numerical descriptive measures (e.g., mean and standard deviation) for galaxy velocities in cluster A1775. Depending on your answer to part **b,** you may need to calculate two sets of numerical descriptive measures, one for each of the clusters (say, A1775A and A1775B) within the double cluster.

d. Suppose you observe a galaxy velocity of 20,000 km/s. Is this galaxy likely to belong to cluster A1775A or A1775B? Explain.

2.192 Standardized test "average." *US News & World Report* reported on many factors contributing to the breakdown of public education. One study mentioned in the article found that over 90% of the nation's school districts reported that their students were scoring "above the national average" on standardized tests. Using your knowledge of measures of central tendency, explain why the schools' reports are incorrect. Does your analysis change if the term "average" refers to the mean? To the median? Explain what effect this misinformation might have on the perception of the nation's schools.

2.193 Zinc phosphide in pest control. A chemical company produces a substance composed of 98% cracked corn particles and 2% zinc phosphide for use in controlling rat populations in sugarcane fields. Production must be carefully controlled to maintain the 2% zinc phosphide, because too much zinc phosphide will cause damage to the sugarcane and too little will be ineffective in controlling the rat population. Records from past production indicate that the distribution of the actual percentage of zinc phosphide present in the substance is approximately mound shaped, with a mean of 2.0% and a standard deviation of .08%. Suppose one batch chosen randomly actually contains 1.80% zinc phosphide. Does this indicate that there is too little zinc phosphide in this production? Explain your reasoning.

2.194 Risk of jail suicides. Suicide is the leading cause of death of Americans incarcerated in correctional facilities. To determine what factors increase the risk of suicide in urban jails, a group of researchers collected data on all 37 suicides that occurred over a 15-year period in the Wayne County Jail in Detroit, Michigan (*American Journal of Psychiatry*, July 1995). The data on each suicide victim are saved in the **SUICIDE** file. Selected observations are shown in the table at the top of page 104.

a. Identify the type (quantitative or qualitative) of each variable measured.

b. Are suicides at the jail more likely to be committed by inmates charged with murder/manslaughter or with lesser crimes? Illustrate with a graph.

c. Are suicides at the jail more likely to be committed at night? Illustrate with a graph.

d. What is the mean length of time an inmate is in jail before committing suicide? What is the median? Interpret these two numbers.

e. Is it likely that a future suicide at the jail will occur after 200 days? Explain.

f. Have suicides at the jail declined over the years? Support your answer with a graph.

2.195 Salaries of professional athletes. The salaries of superstar professional athletes receive much attention in the media. The multimillion-dollar long-term contract is now commonplace among this elite group. Nevertheless, rarely does a season pass without negotiations between one or more of the players' associations and team owners for additional salary and fringe benefits for *all* players in their particular sports.

a. If a players' association wanted to support its argument for higher "average" salaries, which measure of central tendency do you think it should use? Why?

b. To refute the argument, which measure of central tendency should the owners apply to the players' salaries? Why?

2.196 Grades in statistics. The final grades given by two professors in introductory statistics courses have been carefully examined. The students in the first professor's class had a grade point average of 3.0 and a standard deviation of .2. Those in the second professor's class had grade points with an average of 3.0 and a standard deviation of 1.0. If you had a choice, which professor would you take for this course?

Critical Thinking Challenges

2.197 The Hite Report. Researcher Shere Hite shocked conservative America with her famous Hite Report on the permissive sexual attitudes of American men and women. In her book *Women and Love: A Cultural Revolution in Progress* (Knopf Press, 1988), Hite reveals some startling statistics describing how women feel about contemporary relationships:

- Eighty-four percent are not emotionally satisfied with their relationship.
- Ninety-five percent report "emotional and psychological harassment" from their men.
- Seventy percent of those married five years or more are having extramarital affairs.
- Only 13% of those married more than two years are "in love."

Hite conducted the survey by mailing out 100,000 questionnaires to women across the country over a seven-year period. The questionnaires were mailed to a wide variety of organizations, including church groups, women's voting and political groups, women's rights organizations, and counseling and walk-in centers for women. Organizational leaders were asked to circulate the questionnaires to their members. Hite also relied on volunteer respondents who wrote in for copies of the questionnaire. Each questionnaire consisted of 127 open-ended questions, many with numerous subquestions and follow-ups. Hite's instructions read, "It is not necessary to answer

Data for Exercise 2.194

Victim	Days in Jail before Suicide	Marital Status	Race	Murder/Manslaughter Charge	Time of Suicide	Year
1	3	Married	W	Yes	Night	1972
2	4	Single	W	Yes	Night	1987
3	5	Single	NW	Yes	Afternoon	1975
4	7	Widowed	NW	Yes	Night	1981
5	10	Single	NW	Yes	Afternoon	1982
⋮	⋮	⋮	⋮	⋮	⋮	⋮
36	41	Single	NW	No	Night	1985
37	86	Married	W	No	Night	1968

Based on DuRand, C. J., et al. "A quarter century of suicide in a major urban jail: Implications for community psychiatry." *American Journal of Psychiatry,* Vol. 152, No. 7, July 1995, p. 1078 (Table 1).

every question! Feel free to skip around and answer those questions you choose." Approximately 4,500 completed questionnaires were returned, a response rate of 4.5%. These questionnaires form the data set from which the preceding percentages were determined. Hite claims that the 4,500 women respondents are a representative sample of all women in the United States and, therefore, that the survey results imply that vast numbers of women are "suffering a lot of pain in their love relationships with men." Critically assess the survey results. Do you believe they are reliable?

2.198 No Child Left Behind Act. According to the government, federal spending on K–12 education has increased dramatically over the past 20 years, but student performance has stayed essentially the same. Hence, former President George Bush signed into law the No Child Left Behind Act, a bill that promised improved student achievement for all U.S. children. *Chance* (Fall 2003) reported on a graphic that was designed to support the legislation. The graphic, obtained from the U.S. Department of Education Web site (www.ed.gov), is reproduced here. The bars in the graph represent annual federal spending on education, in billions of dollars (left-side vertical axis). The horizontal line represents the annual average 4th-grade children's reading ability score (right-side vertical axis). Critically assess the information portrayed in the graph. Does it, in fact, support the government's position that our children are not making classroom improvements despite federal spending on education? Use the following facts (divulged in the *Chance* article) to help you frame your answer: (1) The U.S. student population has also increased dramatically over the past

20 years, (2) 4th-grade reading test scores are designed to have an average of 250 with a standard deviation of 50, and (3) the reading test scores of 7th and 12th graders and the mathematics scores of 4th graders did improve substantially over the past 20 years.

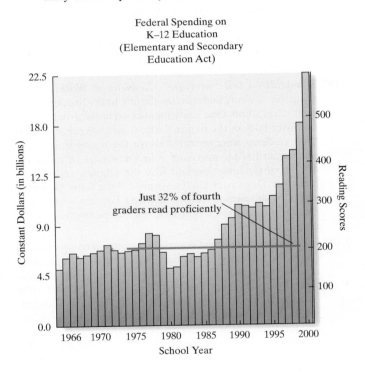

Federal Spending on K–12 Education (Elementary and Secondary Education Act)

Just 32% of fourth graders read proficiently

Activity   Describing Data from Popular Sources

We list here several sources of real-life data sets. (Many in the list have been obtained from Wasserman and Bernero's *Statistics Sources,* and many can be accessed via the Internet.) This index of data sources is far from complete, but is a useful reference for anyone interested in finding almost any type of data.

First we list some almanacs:
Baseball Almanac (www.baseball-almanac.com)
Information Please Almanac (www.infoplease.com)
World Almanac and Book of Facts (www.worldalmanac.com)

United States Government publications are also rich sources of data:

Agricultural Statistics (www.nass.usda.gov)
Digest of Educational Statistics (nces.ed.gov)
Handbook of Labor Statistics (www.bls.gov)
Housing and Urban Development Statistics (www.huduser.org)
Social Indicators (www.unstats.un.org)
Uniform Crime Reports (www.fbi.gov)
Vital Statistics of the United States (www.cdc.gov)
Economic Indicators (www.economicindicators.gov)
Monthly Labor Review (www.bls.gov)
Survey of Current Business (www.bea.gov)
Statistical Abstract of the United States (www.census.gov)

Many data sources are published on an annual basis:
Commodity Yearbook (www.crbtrader.com)
The World Bank Data Catalog (data.worldbank.org)
Yearbook of American and Canadian Churches (www.thearda.com)
Standard and Poor's Corporation,Trade and Securities: Statistics (published monthly)

Some sources contain data that are international in scope:
Compendium of Social Statistics (www.esewa.un.org)
Stokholm International Peace Research Institute Yearbook (www.sipri.org)
United Nations Statistical Yearbook (www.unstats.un.org)
World Handbook of Political and Social Indicators (www.lepsr.umich.edu)

Here are a few miscellaneous data sources:
Federal Drug Data Sources (www.whitehousedrugpolicy.gov)
Internet Data Sources for Social Scientists (www.ciser.cornell.edu)

Climate Data Sources (www.realclimate.org)
World Health Organization Statistics (www.who.int)

Utilizing the data sources listed, sources suggested by your instructor, or your own resourcefulness, find one real-life quantitative data set that stems from an area of particular interest to you.

a. Describe the data set by using a relative frequency histogram, stem-and-leaf display, or dot plot.

b. Find the mean, median, variance, standard deviation, and range of the data set.

c. Use Chebyshev's rule and the empirical rule to describe the distribution of this data set. Count the actual number of observations that fall within one, two, and three standard deviations of the mean of the data set, and compare these counts with the description of the data set you developed in part **b.**

References

Huff, D. *How to Lie with Statistics*. New York: Norton, 1954.

Mendenhall, W., Beaver, R. J., and Beaver, B. M. *Introduction to Probability and Statistics*, 12th ed. North Scituate, MA: Duxbury, 2006.

Tufte, E. R. *Beautiful Evidence.* Cheshire, CT: Graphics Press, 2006.

Tufte, E. R. *Envisioning Information*. Cheshire, CT: Graphics Press, 1990.

Tufte, E. R. *Visual Display of Quantitative Information*. Cheshire, CT: Graphics Press, 1983.

Tufte, E. R. *Visual Explanations*. Cheshire, CT: Graphics Press, 1997.

Tukey, J. *Exploratory Data Analysis*. Reading, MA: Addison-Wesley, 1977.

USING TECHNOLOGY

MINITAB: Describing Data

Graphing Data

Step 1 Click on the "Graph" button on the MINITAB menu bar.

Step 2 Click on the graph of your choice (bar, pie, scatterplot, histogram, dot plot, or stem-and-leaf) to view the appropriate dialog box. The dialog box for a histogram is shown in Figure 2.M.1.

Step 3 Make the appropriate variable selections and click "OK" to view the graph.

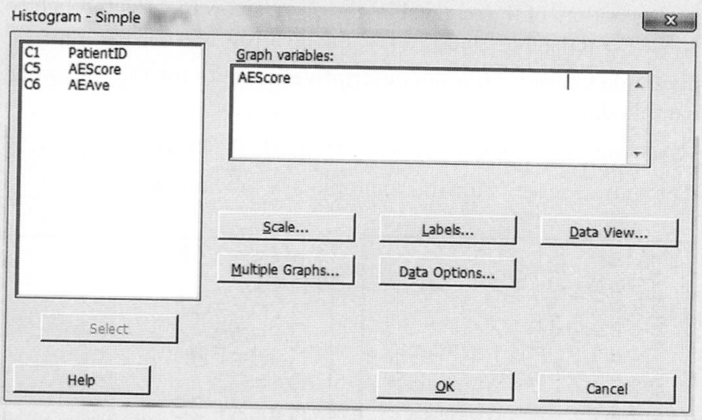

Figure 2.M.1 MINITAB histogram dialog box

Numerical Descriptive Statistics

Step 1 Click on the "Stat" button on the main menu bar, click on "Basic Statistics," and then click on "Display Descriptive Statistics." The resulting dialog box appears in Figure 2.M.2.

Step 2 Select the quantitative variables you want to analyze and place them in the "Variables" box. You can control which

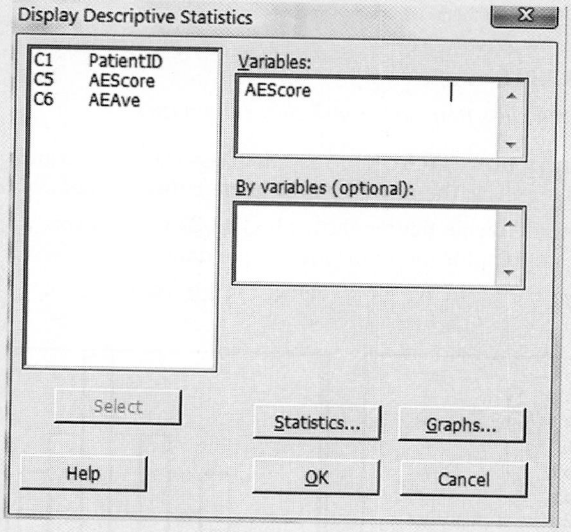

Figure 2.M.2 MINITAB descriptive statistics dialog box

descriptive statistics appear by clicking the "Statistics" button on the dialog box and making your selections. (As an option, you can create histograms and dot plots for the data by clicking the "Graphs" button and making the appropriate selections.)

Step 3 Click "OK" to view the descriptive statistics printout.

TI-83/TI-84 Plus Graphing Calculator: Describing Data

Histogram from Raw Data

Step 1 *Enter the data*

• Press **STAT** and select **1:Edit**

Note: If the list already contains data, clear the old data. Use the up arrow to highlight "**L1**." Press **CLEAR ENTER.** Use the arrow and **ENTER** keys to enter the data set into **L1.**

Step 2 *Set up the histogram plot*

• Press **2nd** and press **Y =** for **STAT PLOT**

• Press **1** for **Plot 1**

• Set the cursor so that **ON** is flashing

• For **Type,** use the arrow and Enter keys to highlight and select the histogram

• For **Xlist,** choose the column containing the data (in most cases, **L1**)

Note: Press **2nd 1** for **L1 Freq** should be set to 1.

Step 3 *Select your window settings*

• Press **WINDOW** and adjust the settings as follows:

$$X \text{ min} = \text{lowest class boundary}$$
$$X \text{ max} = \text{highest class boundary}$$
$$X \text{ sel} = \text{class width}$$
$$Y \text{ min} = 0$$
$$Y \text{ max} \geq \text{greatest class frequency}$$
$$Y \text{ scl} = 1$$
$$X \text{ res} = 1$$

Step 4 *View the graph*

• Press **GRAPH**

Optional *Read class frequencies and class boundaries*

Step 5 You can press **TRACE** to read the class frequencies and class boundaries. Use the arrow keys to move between bars.

Example The following figures show TI-83/TI-84 window settings and histogram for the following sample data:

86, 70, 62, 98, 73, 56, 53, 92, 86, 37, 62, 83, 78, 49, 78, 37, 67, 79, 57

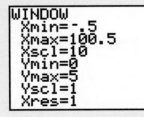
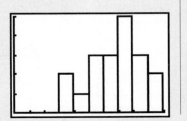

Histogram from a Frequency Table

Step 1 *Enter the data*

• Press **STAT** and select **1:Edit**

Note: If a list already contains data, clear the old data. Use the up arrow to highlight the list name, "**L1**" or "**L2**."

• Press **CLEAR ENTER**

• Enter the midpoint of each class into **L1**

• Enter the class frequencies or relative frequencies into **L2**

Step 2 *Set up the histogram plot*

• Press **2nd** and **Y =** for **STATPLOT**

• Press **1** for **Plot 1**

• Set the cursor so that **ON** is flashing

• For **Type,** use the arrow and Enter keys to highlight and select the histogram

• For **Xlist,** choose the column containing the midpoints

• For **Freq,** choose the column containing the frequencies or relative frequencies

Step 3–4 *Follow steps 3–4 given above.*

Note: To set up the Window for relative frequencies, be sure to set **Y max** to a value that is greater than or equal to the largest relative frequency.

One-Variable Descriptive Statistics

Step 1 *Enter the data*

• Press **STAT** and select **1:Edit**

Note: If the list already contains data, clear the old data. Use the up arrow to highlight "**L1**." Press **CLEAR ENTER.**

• Use the arrow and ENTER keys to enter the data set into L1

Step 2 *Calculate descriptive statistics*

• Press **STAT**

• Press the right arrow key to highlight **CALC**

• Press **ENTER** for **1-Var Stats**

• Enter the name of the list containing your data

• Press **2nd 1** for **L1** (or **2nd 2** for **L2,** etc.)

• Press **ENTER**

You should see the statistics on your screen. Some of the statistics are off the bottom of the screen. Use the down arrow to scroll through to see the remaining statistics. Use the up arrow to scroll back up.

Example Consider finding descriptive statistics for the following sample data set:

86, 70, 62, 98, 73, 56, 53, 92, 86, 37, 62, 83, 78, 49, 78, 37, 67, 79, 57

The output screens for this example are shown below:

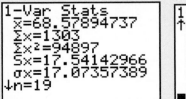
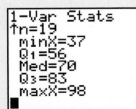

Sorting Data (to find the Mode)

The descriptive statistics do not include the mode. To find the mode, sort your data as follows:

- Press **STAT**
- Press **2** for **SORTA(**
- Enter the name of the list your data are in. If your data are in **L1,** press **2nd 1**
- Press **ENTER**
- The screen will say: **DONE**
- To see the sorted data, press **STAT** and select **1:Edit**
- Scroll down through the list and locate the data value that occurs most frequently

Box Plot

Step 1 *Enter the data*
- Press **STAT** and select **1:Edit**
  Note: If the list already contains data, clear the old data. Use the up arrow to highlight "**L1.**"
- Press **CLEAR ENTER**
- Use the arrow and **ENTER** keys to enter the data set into **L1**

Step 2 *Set up the box plot*
- Press **2nd Y =** for **STAT PLOT**
- Press **1** for **Plot 1**
- Set the cursor so that "**ON**" is flashing
- For **TYPE,** use the right arrow to scroll through the plot icons and select the box plot in the middle of the second row
- For **XLIST**, choose **L1**
- Set **FREQ** to **1**

Step 3 *View the graph*
- Press **ZOOM** and select **9:ZoomStat**

Optional *Read the five number summary*
- Press **TRACE**
- Use the left and right arrow keys to move between **minX, Q1, Med, Q3,** and **maxX**

Example Make a box plot for the given data:

86, 70, 62, 98, 73, 56, 53, 92, 86, 37, 62, 83, 78, 49, 78, 37, 67, 79, 57

The output screen for this example is shown below.

Scatterplots

Step 1 *Enter the data*
- Press **STAT** and select **1:Edit**

Note: If a list already contains data, clear the old data. Use the up arrow to highlight the list name, "**L1**" or "**L2.**"
- Press **CLEAR ENTER**
- Enter your *x*-data in **L1** and your *y*-data in **L2**

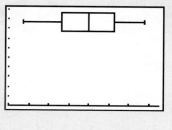

Step 2 *Set up the scatterplot*
- Press **2nd Y =** for **STAT PLOT**
- Press **1** for **Plot 1**
- Set the cursor so that **ON** is flashing
- For **Type,** use the arrow and Enter keys to highlight and select the scatterplot (first icon in the first row)
- For **Xlist,** choose the column containing the *x*-data
- For **Ylist,** choose the column containing the *y*-data

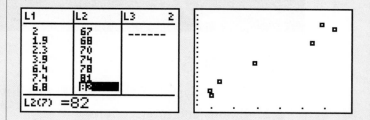

Step 3 *View the scatterplot*
- Press **ZOOM 9** for **ZoomStat**

Example The figures below show a table of data entered on the T1-84 and the scatterplot of the data obtained using the steps given above.

3 Probability

CONTENTS

Where We've Been

- Identified the objective of inferential statistics: to make inferences about a population on the basis of information in a sample
- Introduced graphical and numerical descriptive measures for both quantitative and qualitative data

Where We're Going

- Develop probability as a measure of uncertainty (3.1)
- Introduce basic rules for finding probabilities (3.2–3.6)
- Use a probability as a measure of reliability for an inference (3.2–3.6)
- Use of probability in random sampling (3.7)
- Provide more advanced rules for finding probabilities (3.8–3.9)

Statistics IN Action   Lotto Buster! Can You Improve Your Chance of Winning?

"Welcome to the Wonderful World of Lottery Busters." So *began the premier issue of* Lottery Buster, *a monthly publication for players of the state lottery games.* Lottery Buster *provides interesting facts and figures on the 42 state lotteries and 2 multistate lotteries currently operating in the United States and, more importantly, tips on how to increase a player's odds of winning the lottery.*

In 1963, New Hampshire became the first state in modern times to authorize a state lottery as an alternative to increasing taxes. (Prior to that time, beginning in 1895, lotteries were banned in America because they were thought to be corrupt.) Since then, lotteries have become immensely popular, for two reasons. First, they lure you with the opportunity to win millions of dollars with a $1 investment, and second, when you lose, at least you believe that your money is going to a good cause. Many state lotteries, like Florida's, earmark a high percentage of lottery revenues to fund state education.

The popularity of the state lottery has brought with it an avalanche of "experts" and "mathematical wizards" (such as the editors of *Lottery Buster*) who provide advice on how to win the lottery—for a fee, of course! Many offer guaranteed "systems" of winning through computer software products with catchy names such as Lotto Wizard, Lottorobics, Win4D, and Lotto-Luck.

For example, most knowledgeable lottery players would agree that the "golden rule" or "first rule" in winning lotteries is *game selection*. State lotteries generally offer three types of games: Instant (scratch-off tickets or online) games, Daily Numbers (Pick-3 or Pick-4), and the weekly Pick-6 Lotto game.

One version of the Instant game involves scratching off the thin opaque covering on a ticket with the edge of a coin to determine whether you have won or lost. The cost of the ticket ranges from 50¢ to $1, and the amount won ranges from $1 to $100,000 in most states to as much as $1 million in others. *Lottery Buster* advises against playing the Instant game because it is "a pure chance play, and you can win only by dumb luck. No skill can be applied to this game."

The Daily Numbers game permits you to choose either a three-digit (Pick-3) or four-digit (Pick-4) number at a cost of $1 per ticket. Each night, the winning number is drawn. If your number matches the winning number, you win a large sum of money, usually $100,000. You do have some control over the Daily Numbers game (since you pick the numbers that you play); consequently, there are strategies available to increase your chances of winning. However, the Daily Numbers game, like the Instant game, is not available for out-of-state play.

To play Pick-6 Lotto, you select six numbers of your choice from a field of numbers ranging from 1 to N, where N depends on which state's game you are playing. For example, Florida's current Lotto game involves picking six numbers ranging from 1 to 53. The cost of a ticket is $1, and the payoff, if your six numbers match the winning numbers drawn, is $7 million or more, depending on the number of tickets purchased. (To date, Florida has had the largest state weekly payoff, over $200 million.) In addition to capturing the grand prize, you can win second-, third-, and fourth-prize payoffs by matching five, four, and three of the six numbers drawn, respectively. And you don't have to be a resident of the state to play the state's Lotto game.

In this chapter, several Statistics in Action Revisited examples demonstrate how to use the basic concepts of probability to compute the odds of winning a state lottery game and to assess the validity of the strategies suggested by lottery "experts."

Statistics IN Action Revisited

- The Probability of Winning Lotto (p. 119)
- The Probability of Winning Lotto with a Wheeling System (p. 130)
- The Probability of Winning Cash 3 or Play 4 (p. 144)

Recall that one branch of statistics is concerned with decisions made about a population on the basis of information learned about a sample. You can see how this is accomplished more easily if you understand the relationship between population and sample—a relationship that becomes clearer if we reverse the statistical procedure of making inferences from sample to population. In this chapter, we assume that the population is *known* and calculate the chances of obtaining various samples from the population. Thus, we show that probability is the reverse of statistics: *In probability, we use information about the population to infer the probable nature of the sample.*

Probability plays an important role in making inferences. Suppose, for example, you have an opportunity to invest in an oil exploration company. Past records show that, out of 10 previous oil drillings (a sample of the company's experiences), all 10 came up dry. What do you conclude? Do you think the chances are better than 50:50 that the company will hit a gusher? Should you invest in this company? Chances are, your answer to these questions will be an emphatic "No!" However, if the company's exploratory prowess is sufficient to hit a producing well 50% of the time, a record of 10 dry wells out of 10 drilled is an event that is just too improbable.

Or suppose you're playing poker with what your opponents assure you is a well-shuffled deck of cards. In three consecutive five-card hands, the person on your right is dealt four aces. On the basis of this sample of three deals, do you think the cards are being adequately shuffled? Again, your answer is likely to be "No," because dealing three hands of four aces is just too improbable if the cards were properly shuffled.

Note that the decisions concerning the potential success of the oil-drilling company and the adequacy of card shuffling both involve knowing the chance—or probability—of a certain sample result. Both situations were contrived so that you could easily conclude that the probabilities of the sample results were small. Unfortunately, the probabilities of many observed sample results aren't so easy to evaluate intuitively. In these cases, we need the assistance of a theory of probability.

3.1 Events, Sample Spaces, and Probability

Let's begin our treatment of probability with straightforward examples that are easily described. With the aid of these simple examples, we can introduce important definitions that will help us develop the notion of probability more easily.

Suppose a coin is tossed once and the up face is recorded. The result we see is called an *observation*, or *measurement*, and the process of making an observation is called an *experiment*. Notice that our definition of experiment is broader than the one used in the physical sciences, which brings to mind test tubes, microscopes, and other laboratory equipment. Statistical experiments may include, in addition to these things, recording an Internet user's preference for a Web browser, recording a voter's opinion on an important political issue, measuring the amount of dissolved oxygen in a polluted river, observing the level of anxiety of a test taker, counting the number of errors in an inventory, and observing the fraction of insects killed by a new insecticide. The point is that a statistical experiment can be almost any act of observation, as long as the outcome is uncertain.

> An **experiment** is an act or process of observation that leads to a single outcome that cannot be predicted with certainty.

Consider another simple experiment consisting of tossing a die and observing the number on the up face. The six possible outcomes of this experiment are as follows:

1. Observe a 1.

2. Observe a 2.

3. Observe a 3.

4. Observe a 4.

5. Observe a 5.

6. Observe a 6.

Note that if this experiment is conducted once, *you can observe one and only one of these six basic outcomes, and the outcome cannot be predicted with certainty*. Also, these outcomes cannot be decomposed into more basic ones. Because observing the outcome of an experiment is similar to selecting a sample from a population, the basic possible outcomes of an experiment are called **sample points**.*

> A **sample point** is the most basic outcome of an experiment.

*Alternatively, the term *simple event* can be used.

Example 3.1

Listing Sample Points for a Coin-Tossing Experiment

Problem Two coins are tossed, and their up faces are recorded. List all the sample points for this experiment.

Solution Even for a seemingly trivial experiment, we must be careful when listing the sample points. At first glance, we might expect one of three basic outcomes: Observe two heads; Observe two tails; or Observe one head and one tail. However, further reflection reveals that the last of these, Observe one head and one tail, can be decomposed into two outcomes: Head on coin 1, Tail on coin 2; and Tail on coin 1, Head on coin 2.

A useful tool for illustrating this notion is a **tree diagram.** Figure 3.1 shows a tree diagram for this experiment. At the top of the "tree" there are two branches, representing

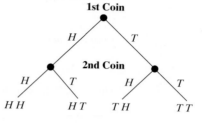

Figure 3.1
Tree diagram for the coin-tossing experiment

the two outcomes (H or T) for the first tossed coin. Each of these outcomes results in two more branches, representing the two outcomes (H or T) for the second tossed coin. Consequently, after tossing both coins, you can see that we have four sample points:

1. Observe HH

2. Observe HT

3. Observe TH

4. Observe TT

where H in the first position means "Head on coin 1," H in the second position means "Head on coin 2," and so on.

Look Back Even if the coins are identical in appearance, they are, in fact, two distinct coins. Thus, the sample points must account for this distinction.

Now Work Exercise 3.15a

We often wish to refer to the collection of all the sample points of an experiment. This collection is called the *sample space* of the experiment. For example, there are six sample points in the sample space associated with the die-toss experiment. The sample spaces for the experiments discussed thus far are shown in Table 3.1.

> The **sample space** of an experiment is the collection of all its sample points.

Just as graphs are useful in describing sets of data, a pictorial method for presenting the sample space will often be useful. Figure 3.2 shows such a representation for each of the experiments in Table 3.1. In each case, the sample space is shown as a closed figure, labeled S, containing all possible sample points. Each sample point is represented by a solid dot (i.e., a "point") and labeled accordingly. Such graphical representations are called **Venn diagrams.**

○ H ○ T

S

a. Experiment: Observe the up face on a coin

○ 1   ○ 2   ○ 3
○ 4   ○ 5   ○ 6

S

b. Experiment: Observe the up face on a die

○ HH ○ HT
○ TH ○ TT

S

c. Experiment: Observe the up faces on two coins

Figure 3.2

Venn diagrams for the three experiments from Table 3.1

Table 3.1   Experiments and Their Sample Spaces

Experiment:	Observe the up face on a coin.
Sample Space:	1. Observe a head.
	2. Observe a tail.

This sample space can be represented in set notation as a set containing two sample points:

$$S: \{H, T\}$$

Here, H represents the sample point Observe a head and T represents the sample point Observe a tail.

Experiment:	Observe the up face on a die.
Sample Space:	1. Observe a 1.
	2. Observe a 2.
	3. Observe a 3.
	4. Observe a 4.
	5. Observe a 5.
	6. Observe a 6.

This sample space can be represented in set notation as a set of six sample points:

$$S: \{1, 2, 3, 4, 5, 6\}$$

Experiment:	Observe the up faces on two coins.
Sample Space:	1. Observe HH.
	2. Observe HT.
	3. Observe TH.
	4. Observe TT.

This sample space can be represented in set notation as a set of four sample points:

$$S: \{HH, HT, TH, TT\}$$

Now that we know that an experiment will result in *only one* basic outcome—called a sample point—and that the sample space is the collection of all possible sample points, we're ready to discuss the probabilities of the sample points. You have undoubtedly used the term *probability* and have some intuitive idea about its meaning. Probability is generally used synonymously with "chance," "odds," and similar concepts. For example, if a fair coin is tossed, we might reason that both the sample points Observe a head and Observe a tail have the same *chance* of occurring. Thus, we might state, "The probability of observing a head is 50%" or "The odds of seeing a head are 50:50." Both of these statements are based on an informal knowledge of probability. We'll begin our treatment of probability by using such informal concepts and then solidify what we mean later.

The probability of a sample point is a number between 0 and 1 which measures the likelihood that the outcome will occur when the experiment is performed. This number is usually taken to be the relative frequency of the occurrence of a sample point in a very long series of repetitions of an experiment.* For example, if we are assigning probabilities to the two sample points in the coin-toss experiment (Observe a head and Observe a tail), we might reason that if we toss a balanced coin a very large number of times, the sample points Observe a head and Observe a tail will occur with the same relative frequency of .5.

Our reasoning is supported by Figure 3.3. The figure plots the relative frequency of the number of times that a head occurs in simulations (by computer) of the toss of a coin N times, where N ranges from as few as 25 tosses to as many as 1,500 tosses. You can see that when N is large (e.g., $N = 1,500$), the relative frequency is converging to .5. Thus, the probability of each sample point in the coin-tossing experiment is .5.

For some experiments, we may have little or no information on the relative frequency of occurrence of the sample points; consequently, we must assign probabilities to the sample points on the basis of general information about the experiment. For example,

---

*The result derives from an axiom in probability theory called the **law of large numbers.** Phrased informally, the law states that the relative frequency of the number of times that an outcome occurs when an experiment is replicated over and over again (i.e., a large number of times) approaches the true (or theoretical) probability of the outcome.

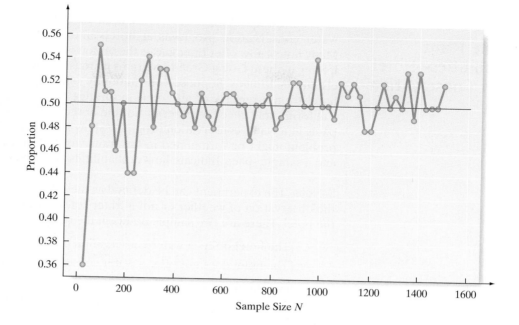

Figure 3.3

Proportion of heads in N tosses of a coin

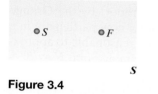

Figure 3.4

Experiment: invest in a business venture and observe whether it succeeds (S) or fails (F)

if the experiment is about investing in a business venture and observing whether it succeeds or fails, the sample space would appear as in Figure 3.4.

We are unlikely to be able to assign probabilities to the sample points of this experiment based on a long series of repetitions, since unique factors govern each performance of this kind of experiment. Instead, we may consider factors such as the personnel managing the venture, the general state of the economy at the time, the rate of success of similar ventures, and any other information deemed pertinent. If we finally decide that the venture has an 80% chance of succeeding, we assign a probability of .8 to the sample point Success. This probability can be interpreted as a measure of our degree of belief in the outcome of the business venture; that is, it is a *subjective prob-ability.* Notice, however, that such probabilities should be based on expert information that is carefully assessed. If not, we may be misled on any decisions based on these probabilities or based on any calculations in which they appear. [*Note:* For a text that deals in detail with the subjective evaluation of probabilities, see Winkler (1972) or Lindley (1985).]

No matter how you assign the probabilities to sample points, the probabilities assigned must obey two rules:

Probability Rules for Sample Points

Let p_i represent the probability of sample point i. Then

1. All sample point probabilities *must* lie between 0 and 1 (i.e., $0 \leq p_i \leq 1$).

2. The probabilities of all the sample points within a sample space *must* sum to 1 (i.e., $\Sigma p_i = 1$).

Assigning probabilities to sample points is easy for some experiments. For exam-ple, if the experiment is to toss a fair coin and observe the face, we would probably all agree to assign a probability of $\frac{1}{2}$ to the two sample points, Observe a head and Observe a tail. However, many experiments have sample points whose probabilities are more difficult to assign.

Example 3.2
Sample Point Probabilities—Hotel Water Conservation

Problem *Going Green* is the term used to describe water conservation programs at hotels and motels. Many hotels now offer their guests the option of participating in these Going Green programs by reusing towels and bed linens. Suppose you randomly select one hotel from a registry of all hotels in Orange County, California, and check whether or not the hotel participates in a water conservation program. Show how this problem might be formulated in the framework of an experiment with sample points and a sample space. Indicate how probabilities might be assigned to the sample points.

Solution The experiment can be defined as the selection of an Orange County hotel and the observation of whether or not a water conservation program is offered to guests at the hotel. There are two sample points in the sample space for this experiment:

C: {The hotel offers a water conservation program.}

N: {The hotel does not offer a water conservation program.}

The difference between this and the coin-toss experiment becomes apparent when we attempt to assign probabilities to the two sample points. What probability should we assign to sample point *C*? If you answer .5, you are assuming that the events *C* and *N* occur with equal likelihood, just like the sample points Observe a head and Observe a tail in the coin-toss experiment. But assigning sample point probabilities for the hotel water conservation experiment is not so easy. In fact, a recent report by the Orange County Water District stated that 70% of the hotels in the county now participate in a water conservation program of some type. In that case, it might be reasonable to approximate the probability of the sample point *C* as .7 and that of the sample point *N* as .3.

Look Back Here we see that the sample points are not always equally likely, so assigning probabilities to them can be complicated, particularly for experiments that represent real applications (as opposed to coin- and die-toss experiments).

Now Work Exercise 3.20

Although the probabilities of sample points are often of interest in their own right, it is usually probabilities of *collections* of sample points that are important. Example 3.3 demonstrates this point.

Example 3.3
Probability of a Collection of Sample Points— Die-Tossing Experiment

Problem A fair die is tossed, and the up face is observed. If the face is even, you win $1. Otherwise, you lose $1. What is the probability that you win?

Solution Recall that the sample space for this experiment contains six sample points:

$$S: \{1,2,3,4,5,6\}$$

Since the die is balanced, we assign a probability of $1/6$ to each of the sample points in this sample space. An even number will occur if one of the sample points Observe a 2, Observe a 4, or Observe a 6 occurs. A collection of sample points such as this is called an *event*, which we denote by the letter *A*. Since the event *A* contains three sample points—all with probability $1/6$—and since no sample points can occur simultaneously, we reason that the probability of *A* is the sum of the probabilities of the sample points in *A*. Thus, the probability of *A* is $1/6 + 1/6 + 1/6 = 1/2$.

Look Back On the basis of this notion of probability, *in the long run*, you will win $1 half the time and lose $1 half the time.

Now Work Exercise 3.14

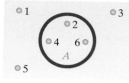

Figure 3.5

Die-toss experiment with event A, observe an even number

Figure 3.5 is a Venn diagram depicting the sample space associated with a die-toss experiment and the event A, Observe an even number. The event A is represented by the closed figure inside the sample space **S.** This closed figure A comprises all of the sample points that make up event A.

To decide which sample points belong to the set associated with an event A, test each sample point in the sample space **S.** If event A occurs, then that sample point is in the event A. For example, the event A, Observe an even number, in the die-toss experiment will occur if the sample point Observe a 2 occurs. By the same reasoning, the sample points Observe a 4 and Observe a 6 are also in event A.

To summarize, we have demonstrated that an event can be defined in words, or it can be defined as a specific set of sample points. This leads us to the following general definition of an *event*:

> An **event** is a specific collection of sample points.

Example 3.4

Probability of an Event—Coin-Tossing Experiment

Problem Consider the experiment of tossing two *unbalanced* coins. Because the coins are *not* balanced, their outcomes (H or T) are not equiprobable. Suppose the correct probabilities associated with the sample points are given in the accompanying table. [*Note:* The necessary properties for assigning probabilities to sample points are satisfied.] Consider the events

$$A: \{\text{Observe exactly one head.}\}$$
$$B: \{\text{Observe at least one head.}\}$$

Calculate the probability of A and the probability of B.

Sample Point	Probability
HH	$\frac{4}{9}$
HT	$\frac{2}{9}$
TH	$\frac{2}{9}$
TT	$\frac{1}{9}$

Solution Event A contains the sample points HT and TH. Since two or more sample points cannot occur at the same time, we can easily calculate the probability of event A by summing the probabilities of the two sample points. Thus, the probability of observing exactly one head (event A), denoted by the symbol $P(A)$, is

$$P(A) = P(\text{Observe } HT.) + P(\text{Observe } TH.) = \frac{2}{9} + \frac{2}{9} = \frac{4}{9}$$

Similarly, since B contains the sample points HH, HT, and TH, it follows that

$$P(B) = \frac{4}{9} + \frac{2}{9} + \frac{2}{9} = \frac{8}{9}$$

Look Back Again, these probabilities should be interpreted *in the long run*. For example, $P(B) = {}^{8}\!/_{9} \approx .89$ implies that if we were to toss two coins an infinite number of times, we would observe at least two heads on about 89% of the tosses.

Now Work Exercise 3.11

The preceding example leads us to a general procedure for finding the probability of an event A:

Probability of an Event

The probability of an event A is calculated by summing the probabilities of the sample points in the sample space for A.

Thus, we can summarize the steps for calculating the probability of any event, as indicated in the next box.

Steps for Calculating Probabilities of Events

1. Define the experiment; that is, describe the process used to make an observation and the type of observation that will be recorded.
2. List the sample points.
3. Assign probabilities to the sample points.
4. Determine the collection of sample points contained in the event of interest.
5. Sum the sample point probabilities to get the probability of the event.

Example 3.5

Applying the Five Steps to Find a Probability—Study of Divorced Couples

Problem The American Association for Marriage and Family Therapy (AAMFT) is a group of professional therapists and family practitioners that treats many of the nation's couples and families. The AAMFT released the findings of a study that tracked the post-divorce history of 100 pairs of former spouses with children. Each divorced couple was classified into one of four groups, nicknamed "perfect pals (PP)," "cooperative colleagues (CC)," "angry associates (AA)," and "fiery foes (FF)." The proportions classified into each group are shown in Table 3.2.

Table 3.2	Results of AAMFT Study of Divorced Couples
Group	Proportion
Perfect Pals (PP) (Joint-custody parents who get along well)	.12
Cooperative Colleagues (CC) (Occasional conflict, likely to be remarried)	.38
Angry Associates (AA) (Cooperate on children-related issues only, conflicting otherwise)	.25
Fiery Foes (FF) (Communicate only through children, hostile toward each other)	.25

Suppose one of the 100 couples is selected at random.

a. Define the experiment that generated the data in Table 3.2, and list the sample points.
b. Assign probabilities to the sample points.
c. What is the probability that the former spouses are "fiery foes?"
d. What is the probability that the former spouses have at least some conflict in their relationship?

Solution

a. The experiment is the act of classifying the randomly selected couple. The sample points—the simplest outcomes of the experiment—are the four groups (categories) listed in Table 3.2. They are shown in the Venn diagram in Figure 3.6.

b. If, as in Example 3.1, we were to assign equal probabilities in this case, each of the response categories would have a probability of one-fourth (1/4), or .25. But by examining Table 3.2, you can see that equal probabilities are not reasonable in this

Figure 3.6
Venn diagram for AAMFT survey

case, because the response percentages are not all the same in the four categories. It is more reasonable to assign a probability equal to the response proportion in each class, as shown in Table 3.3.*

Table 3.3	Sample Point Probabilities for AAMFT Survey
Sample Point	Probability
PP	.12
CC	.38
AA	.25
FF	.25

c. The event that the former spouses are "fiery foes" corresponds to the sample point FF. Consequently, the probability of the event is the probability of the sample point. From Table 3.3, we find that $P(FF) = .25$. Therefore, there is a .25 probability (or one-fourth chance) that the couple we select are "fiery foes."

d. The event that the former spouses have at least some conflict in their relationship, call it event C, is not a sample point, because it consists of more than one of the response classifications (the sample points). In fact, as shown in Figure 3.6, the event C consists of three sample points: CC, AA, and FF. The probability of C is defined to be the sum of the probabilities of the sample points in C:

$$P(C) = P(CC) + P(AA) + P(FF) = .38 + .25 + .25 = .88$$

Thus, the chance that we observe a divorced couple with some degree of conflict in their relationship is .88—a fairly high probability.

Look Back The key to solving this problem is to follow the steps outlined in the box. We defined the experiment (Step 1) and listed the sample points (Step 2) in part **a.** The assignment of probabilities to the sample points (Step 3) was done in part **b.** For each probability in parts **c** and **d,** we identified the collection of sample points in the event (Step 4) and summed their probabilities (Step 5).

Now Work Exercise 3.23

The preceding examples have one thing in common: The number of sample points in each of the sample spaces was small; hence, the sample points were easy to identify and list. How can we manage this when the sample points run into the thousands or millions? For example, suppose you wish to select 5 marines for a dangerous mission from a division of 1,000. Then each different group of 5 marines would represent a sample point. How can you determine the number of sample points associated with this experiment?

One method of determining the number of sample points for a complex experiment is to develop a counting system. Start by examining a simple version of the experiment. For example, see if you can develop a system for counting the number of ways to select 2 marines from a total of 4. If the marines are represented by the symbols M_1, M_2, M_3, and M_4, the sample points could be listed in the following pattern:

(M_1, M_2) (M_2, M_3) (M_3, M_4)
(M_1, M_3) (M_2, M_4)
(M_1, M_4)

Note the pattern and now try a more complex situation—say, sampling 3 marines out of 5. List the sample points and observe the pattern. Finally, see if you can deduce the

---

*Since the response percentages were based on a sample of divorced couples, these assigned probabilities are estimates of the true population response percentages. You will learn how to measure the reliability of probability estimates in Chapter 7.

pattern for the general case. Perhaps you can program a computer to produce the matching and counting for the number of samples of 5 selected from a total of 1,000.

A second method of determining the number of sample points for an experiment is to use **combinatorial mathematics.** This branch of mathematics is concerned with developing counting rules for given situations. For example, there is a simple rule for finding the number of different samples of 5 marines selected from 1,000. This rule, called the **combinations rule,** is given in the box.

Combinations Rule

Suppose a sample of n elements is to be drawn without replacement from a set of N elements. Then the number of different samples possible is denoted by $\binom{N}{n}$ and is equal to

$$\binom{N}{n} = \frac{N!}{n!(N-n)!}$$

where

$$n! = n(n-1)(n-2)\cdots(3)(2)(1)$$

and similarly for $N!$ and $(N-n)!$ For example, $5! = 5\cdot4\cdot3\cdot2\cdot1$. [*Note:* The quantity 0! is defined to be equal to 1.]

Example 3.6

Using the Combinations Rule— Selecting 2 Marines from 4

Problem Consider the task of choosing 2 marines from a platoon of 4 to send on a dangerous mission. Use the combinations counting rule to determine how many different selections can be made.

Solution For this example, $N = 4, n = 2$, and

$$\binom{4}{2} = \frac{4!}{2!2!} = \frac{4\cdot3\cdot2\cdot1}{(2\cdot1)(2\cdot1)} = 6$$

Look Back You can see that this answer agrees with the number of sample points listed at the top of the page.

Now Work Exercise 3.13

Example 3.7

Using The Combinations Rule— Selecting 5 movies from 20

Problem Suppose a movie reviewer for a newspaper reviews 5 movies each month. This month, the reviewer has 20 new movies from which to make the selection. How many different samples of 5 movies can be selected from the 20?

Solution For this example, $N = 20$ and $n = 5$. Then the number of different samples of 5 that can be selected from the 20 movies is

$$\binom{20}{5} = \frac{20!}{5!(20-5)!} = \frac{20!}{5!15!}$$

$$= \frac{20\cdot19\cdot18\cdots\cdots3\cdot2\cdot1}{(5\cdot4\cdot3\cdot2\cdot1)(15\cdot14\cdot13\cdots\cdots3\cdot2\cdot1)} = 15{,}504$$

Look Back You can see that attempting to list all the sample points for this experiment would be an extremely tedious and time-consuming, if not practically impossible, task.

The combinations rule is just one of a large number of counting rules that have been developed by combinatorial mathematicians. This counting rule applies to situations in which the experiment calls for selecting n elements from a total of N elements, without replacing each element before the next is selected. Several other basic counting rules are presented in optional Section 3.8.

Statistics IN Action | **Revisited** | ### The Probability of Winning Lotto

In Florida's state lottery game, called Pick-6 Lotto, you select six numbers of your choice from a set of numbers ranging from 1 to 53. We can apply the combinations rule to determine the total number of combinations of 6 numbers selected from 53 (i.e., the total number of sample points [or possible winning tickets]). Here, $N = 53$ and $n = 6$; therefore, we have

$$\binom{N}{n} = \frac{N!}{n!(N-n)!} = \frac{53!}{6!47!}$$

$$= \frac{(53)(52)(51)(50)(49)(48)(47!)}{(6)(5)(4)(3)(2)(1)(47!)}$$

$$= 22,957,480$$

Now, since the Lotto balls are selected at random, each of these 22,957,480 combinations is equally likely to occur. Therefore, the probability of winning Lotto is

$$P(\text{Win } 6/53 \text{ Lotto}) = 1/(22,957,480) = .00000004356$$

This probability is often stated as follows: The odds of winning the game with a single ticket are 1 in 22,957,480, or 1 in approximately 23 million. For all practical purposes, this probability is 0, implying that you have almost no chance of winning the lottery with a single ticket. Yet each week there is almost always a winner in the Florida Lotto. This apparent contradiction can be explained with the following analogy:

Suppose there is a line of minivans, front to back, from New York City to Los Angeles, California. Based on the distance between the two cities and the length of a standard minivan, there would be approximately 23 million minivans in line. Lottery officials will select, at random, one of the minivans and put a check for $10 million in the glove compartment. For a cost of $1, you may roam the country and select one (and only one) minivan and check the glove compartment. Do you think you will find $10 million in the minivan you choose? You can be almost certain that you won't. But now permit anyone to enter the lottery for $1 and suppose that 50 million people do so. With such a large number of participants, it is very likely that someone will find the minivan with the $10 million—but it almost certainly won't be you! (This example illustrates an axiom in statistics called the law of large numbers. See the footnote at the bottom of p. 112.)

Exercises 3.1–3.34

Understanding the Principles

3.1 What is an experiment?

3.2 What are the most basic outcomes of an experiment called?

3.3 Define the sample space.

3.4 What is a Venn diagram?

3.5 Give two probability rules for sample points.

3.6 What is an event?

3.7 How do you find the probability of an event made up of several sample points?

3.8 Give a scenario where the combinations rule is appropriate for counting the number of sample points.

Learning the Mechanics

3.9 An experiment results in one of the following sample points: E_1, E_2, E_3, E_4, and E_5.

a. Find $P(E_3)$ if $P(E_1) = .1, P(E_2) = .2, P(E_4) = .1$, and $P(E_5) = .1$

b. Find $P(E_3)$ if $P(E_1) = P(E_3), P(E_2) = .1$, $P(E_4) = .2$, and $P(E_5) = .1$

c. Find $P(E_3)$ if $P(E_1) = P(E_2) = P(E_4) = P(E_5) = .1$

3.10 The following Venn diagram describes the sample space of a particular experiment and events A and B:

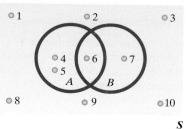

a. Suppose the sample points are equally likely. Find $P(A)$ and $P(B)$.

b. Suppose $P(1) = P(2) = P(3) = P(4) = P(5) = \frac{1}{20}$ and $P(6) = P(7) = P(8) = P(9) = P(10) = \frac{3}{20}$. Find $P(A)$ and $P(B)$.

3.11 The sample space for an experiment contains five sample [NW] points with probabilities as shown in the table. Find the probability of each of the following events:

A: {Either 1, 2, or 3 occurs.}

B: {Either 1, 3, or 5 occurs.}

C: {4 does not occur.}

Sample Points	Probabilities
1	.05
2	.20
3	.30
4	.30
5	.15

3.12 Compute each of the following:

a. $\binom{9}{4}$ b. $\binom{7}{2}$ c. $\binom{4}{4}$

d. $\binom{5}{0}$ e. $\binom{6}{5}$

3.13 Compute the number of ways you can select n elements [NW] from N elements for each of the following:

a. $n = 2, N = 5$

b. $n = 3, N = 6$

c. $n = 5, N = 20$

3.14 Two fair dice are tossed, and the up face on each die is [NW] recorded.

a. List the 36 sample points contained in the sample space.

b. Assign probabilities to the sample points.

c. Find the probability of observing each of the following events:

A: {A 3 appears on each of the two dice.}

B: {The sum of the numbers is even.}

C: {The sum of the numbers is equal to 7.}

D: {A 5 appears on at least one of the dice.}

E: {The sum of the numbers is 10 or more.}

3.15 Two marbles are drawn at random and without replacement from a box containing two blue marbles and three red marbles.

[NW] a. List the sample points.

b. Assign probabilities to the sample points.

c. Determine the probability of observing each of the following events:

A: {Two blue marbles are drawn.}

B: {A red and a blue marble are drawn.}

C: {Two red marbles are drawn.}

3.16 Simulate the experiment described in Exercise 3.15, using any five identically shaped objects, two of which are one color and three another. Mix the objects, draw two, record the results, and then replace the objects. Repeat the experiment a large number of times (at least 100). Calculate the proportion of times events A, B, and C occur. How do these proportions compare with the probabilities you calculated in Exercise 3.15? Should these proportions equal the probabilities? Explain.

Applet Exercise 3.1

Use the applet entitled *Simulating the Probability of Rolling a 6* to explore the relationship between the proportion of sixes rolled on several rolls of a die and the theoretical probability of rolling a 6 on a fair die.

a. To simulate rolling a die one time, click on the *Roll* button on the screen while $n = 1$. The outcome of the roll appears in the list at the right, and the cumulative proportion of sixes for one roll is shown above the graph and as a point on the graph corresponding to 1. Click *Reset* and repeat the process with $n = 1$ several times. What are the possible values of the cumulative proportion of sixes for one roll of a die? Can the cumulative proportion of sixes for one roll of a die equal the theoretical probability of rolling a 6 on a fair die? Explain.

b. Set $n = 10$ and click the *Roll* button. Repeat this several times, resetting after each time. Record the cumulative proportion of sixes for each roll. Compare the cumulative proportions for $n = 10$ with those for $n = 1$ in part **a.** Which tend to be closer to the theoretical probability of rolling a 6 on a fair die?

c. Repeat part **b** for $n = 1,000$, comparing the cumulative proportions for $n = 1,000$ with those for $n = 1$ in part **a** and for $n = 10$ in part **b.**

d. On the basis of your results for parts **a, b,** and **c,** do you believe that we could justifiably conclude that a die is unfair because we rolled it 10 times and didn't roll any sixes? Explain.

Applet Exercise 3.2

Use the applet entitled *Simulating the Probability of a Head with a Fair Coin* to explore the relationship between the proportion of heads on several flips of a coin and the theoretical probability of getting heads on one flip of a fair coin.

a. Repeat parts **a–c** of Applet Exercise 3.1 for the experiment of flipping a coin and the event of getting heads.

b. On the basis of your results for part **a,** do you believe that we could justifiably conclude that a coin is unfair because we flipped it 10 times and didn't roll any heads? Explain.

Applying the Concepts—Basic

3.17 **Colors of M&Ms candies.** In 1940, Forrest E. Mars, Sr., formed the Mars Corporation in order to produce chocolate candies with a sugar shell that could be sold throughout the year and wouldn't melt during the summer. Originally, M&Ms Plain Chocolate Candies came in only a brown color. Today, M&Ms in standard bags come in six colors: brown, yellow, red, blue, orange, and green. According to Mars Corporation, today 24% of all M&Ms produced are blue, 20% are orange, 16% are green, 14% are yellow, 13% are brown, and 13% are red. Suppose you purchase a randomly selected bag of M&Ms Plain Chocolate Candies and randomly select one of the M&Ms from the bag. The color of the selected M&M is of interest.

a. Identify the outcomes (sample points) of this experiment.

b. Assign reasonable probabilities to the outcomes, part **a.**

c. What is the probability that the selected M&M is brown (the original color)?

d. In 1960, the colors red, green, and yellow were added to brown M&Ms. What is the probability that the selected M&M is either red, green, or yellow?

e. In 1995, based on voting by American consumers, the color blue was added to the M&M mix. What is the probability that the selected M&M is not blue?

3.18 **Rare underwater sounds.** A study of underwater sounds in a specific region of the Pacific Ocean focused on scarce sounds, such as humpback whale screams, dolphin whistles, and sounds from passing ships (*Acoustical Physics,* Vol. 56, 2010). During the month of September (non–rainy season), research revealed the following probabilities of rare sounds: $P(\text{whale scream}) = .03$, $P(\text{ship sound}) = .14$, and $P(\text{rain}) = 0$. If a sound is picked up by the acoustical equipment placed in this region of the Pacific Ocean, is it more likely to be a whale scream or a sound from a passing ship? Explain.

3.19 **USDA chicken inspection.** The United States Department of Agriculture (USDA) reports that, under its standard inspection system, one in every 100 slaughtered chickens passes inspection for fecal contamination.

a. If a slaughtered chicken is selected at random, what is the probability that it passes inspection for fecal contamination?

b. The probability of part **a** was based on a USDA study which found that 306 of 32,075 chicken carcasses passed inspection for fecal contamination. Do you agree with the USDA's statement about the likelihood of a slaughtered chicken passing inspection for fecal contamination?

3.20 **African rhinos.** Two species of rhinoceros native to Africa
NW are black rhinos and white rhinos. The International Rhino Federation estimates that the African rhinoceros population consists of 3,610 white rhinos and 11,330 black rhinos. Suppose one rhino is selected at random from the African rhino population and its species (black or white) is observed.

a. List the sample points for this experiment.

b. Assign probabilities to the sample points on the basis of the estimates made by the International Rhino Federation.

3.21 **Going online for health information.** A *cyberchondriac* is defined as a person who regularly searches the Web for health care information. A recent Harris Poll surveyed 1,010 U.S. adults by telephone and asked each respondent how often (in the past month) he or she looked for health care information online. The results are summarized in the following table. Consider the response category of a randomly selected person who participated in the Harris poll.

Response (# per month)	Percentage of Respondents
None	25
1 or 2	31
3–5	25
6–9	5
10 or more	14
Total	100%

Based on "Cyberchondriac" poll. *The Harris Poll,* July 29, 2008 (Table 3).

a. List the sample points for the experiment.

b. Assign reasonable probabilities to the sample points.

c. Find the probability that the respondent looks for health care information online more than two times per month.

3.22 **Health risks to beachgoers.** According to a University of Florida veterinary researcher, the longer a beachgoer sits in wet sand or stays in the water, the higher the health risk (*University of Florida News*, Jan. 29, 2008). Using data collected at Fort Lauderdale, Hollywood, and Hobie Beach, the researcher discovered the following: (1) 6 out of 1,000 people exposed to wet sand for a 10-minute period will acquire gastroenteritis; (2) 12 out of 100 people exposed to wet sand for two consecutive hours will acquire gastroenteritis; (3) 7 out of 1,000 people exposed to ocean water for a 10-minute period will acquire gastroenteritis; and, (4) 7 out of 100 people exposed to ocean water for a 70-minute period will acquire gastroenteritis.

a. If a beachgoer spends 10 minutes in the wet sand, what is the probability that he or she will acquire gastroenteritis?

b. If a beachgoer spends two hours in the wet sand, what is the probability that he or she will acquire gastroenteritis?

c. If a beachgoer spends 10 minutes in the ocean water, what is the probability that he or she will acquire gastroenteritis?

d. If a beachgoer spends 70 minutes in the ocean water, what is the probability that he or she will acquire gastroenteritis?

Applying the Concepts—Intermediate

3.23 **Sociology fieldwork methods.** Refer to University of New
NW Mexico professor Jane Hood's study of the fieldwork methods used by qualitative sociologists, presented in Exercise 2.16 (p. 36). Recall that she discovered that fieldwork methods could be classified into four distinct categories: Interview, Observation plus Participation, Observation Only, and Grounded Theory. The table that follows, reproduced from *Teaching Sociology* (July 2006), and saved in the **FIELDWORK** file, gives the number of sociology field research papers in each category. Suppose we randomly select one of these research papers and determine the method used. Find the probability that the method used is either Interview or Grounded Theory.

Fieldwork Method	Number of Papers
Interview	5,079
Observation plus Participation	1,042
Observation Only	848
Grounded Theory	537

Based on Hood, J. C. "Teaching against the text: The case of qualitative methods." *Teaching Sociology*, Vol. 34, Issue 3 (Exhibit 2).

3.24 **Chance of rain.** Answer the following question posed in the *Atlanta Journal-Constitution*: When a meteorologist says, "The probability of rain this afternoon is .4," does it mean that it will be raining 40% of the time during the afternoon?

3.25 **Cheek teeth of extinct primates.** Refer to the *American Journal of Physical Anthropology* (Vol. 142, 2010) study of the dietary habits of extinct mammals, Exercise 2.7

(p. 34). Recall that 18 cheek teeth extracted from skulls of an extinct primate species discovered in western Wyoming were analyzed. Each tooth was classified according to degree of wear (unworn, slight, light-moderate, moderate, moderate-heavy, or heavy). The 18 measurements are reproduced in the accompanying table and saved in the **CHEEKTEETH** file. One tooth is randomly selected from the 18 cheek teeth. What is the probability that the tooth shows a slight or moderate amount of wear?

Data on Degree of Wear	
Unknown	Slight
Unknown	Slight
Unknown	Heavy
Moderate	Unworn
Slight	Light-moderate
Unknown	Light-moderate
Moderate-heavy	Moderate
Moderate	Unworn
Slight	Unknown

3.26 **Museum management.** Refer to the *Museum Management and Curatorship* (June 2010) study of the criteria used to evaluate museum performance, Exercise 2.19 (p. 37). Recall that the managers of 30 leading museums of contemporary art were asked to provide the performance measure used most often. A summary of the results is reproduced in the table. One of the 30 museums is selected at random. Find the probability that the museum uses big shows most often as a performance measure.

Performance Measure	Number of Museums
Total visitors	8
Paying visitors	5
Big shows	6
Funds raised	7
Members	4

3.27 **Choosing portable grill displays.** University of Maryland marketing professor R. W. Hamilton studied how people attempt to influence the choices of others by offering undesirable alternatives (*Journal of Consumer Research*, Mar. 2003). Such a phenomenon typically occurs when family members propose a vacation spot, friends recommend a restaurant for dinner, and realtors show the buyer potential homesites. In one phase of the study, the researcher had each of 124 college students select showroom displays for portable grills. Five different displays (representing five different-sized grills) were available, but only three would be selected. The students were instructed to select the displays to maximize purchases of Grill #2 (a smaller grill).

a. In how many possible ways can the three-grill displays be selected from the 5 displays? List the possibilities.

b. The next table shows the grill display combinations and number of each selected by the 124 students. Use this information to assign reasonable probabilities to the different display combinations.

c. Find the probability that a student who participated in the study selected a display combination involving Grill #1.

Grill Display Combination	Number of Students
1–2–3	35
1–2–4	8
1–2–5	42
2–3–4	4
2–3–5	1
2–4–5	34

Based on Hamilton, R. W. "Why do people suggest what they do not want? Using context effects to influence others' choices." *Journal of Consumer Research*, Vol. 29, Mar. 2003, Table 1.

3.28 **ESL students and plagiarism.** The *Journal of Education and Human Development* (Vol. 3, 2009) investigated the causes of plagiarism among six English-as-a-second language (ESL) students taking a master's course in linguistics. All students in the class wrote two essays, one in the middle and one at the end of the semester. After the first essay, the students were instructed on how to avoid plagiarism in the second essay. Of the six ESL students, three admitted to plagiarizing on the first essay. Only one ESL student admitted to plagiarizing on the second essay. (This student, who also plagiarized on the first essay, claimed she misplaced her notes on plagiarism.)

a. If one of the six ESL students is randomly selected, what is the probability that he or she plagiarized on the first essay?

b. Consider the results (plagiarism or not) for the six ESL students on the second essay. List the possible outcomes (e.g., Students 1 and 3 plagiarize, the others do not).

c. Refer to part **b.** Assume that, despite the instruction on plagiarism, the ESL students are just as likely as not to plagiarize on the second essay. What is the probability that no more than one of the ESL students plagiarizes on the second essay?

3.29 **Jai alai Quinella bet.** The Quinella bet at the parimutuel game of jai alai consists of picking the jai alai players that will place first and second in a game, *irrespective of order*. In jai alai, eight players (numbered 1, 2, 3,..., 8) compete in every game.

a. How many different Quinella bets are possible?

b. Suppose you bet the Quinella combination 2–7. If the players are of equal ability, what is the probability that you win the bet?

3.30 **Groundwater contamination in wells.** Refer to the *Environmental Science & Technology* (Jan. 2005) study of methyl *tert*-butyl ether (MTBE) contamination in New Hampshire wells, presented in Exercise 2.21 (p. 37). Data collected for a sample of 223 wells are saved in the **MTBE** file. Recall that each well was classified according to its class (public or private), aquifer (bedrock or unconsolidated), and detectable level of MTBE (below limit or detectable).

a. Consider an experiment in which the class, aquifer, and detectable MTBE level of a well are observed. List the sample points for this experiment. [*Hint:* One sample point is Private/Bedrock/BelowLimit.]

b. Use statistical software to find the number of the 223 wells in each sample point outcome. Then use this information to compute probabilities for the sample points.

c. Find and interpret the probability that a well has a detectable level of MTBE.

Applying the Concepts—Advanced

3.31 **Lead bullets as forensic evidence.** *Chance* (Summer 2004) published an article on the use of lead bullets as forensic evidence in a federal criminal case. Typically, the Federal Bureau of Investigation (FBI) will use a laboratory method to match the lead in a bullet found at a crime scene with unexpended lead cartridges found in the possession of a suspect. The value of this evidence depends on the chance of a *false positive*—that is, the probability that the FBI finds a match, given that the lead at the crime scene and the lead in the possession of the suspect are actually from two different "melts," or sources. To estimate the false positive rate, the FBI collected 1,837 bullets that the agency was confident all came from different melts. Then, using its established criteria, the FBI examined every possible pair of bullets and counted the number of matches. According to *Chance*, the FBI found 693 matches. Use this information to compute the chance of a false positive. Is this probability small enough for you to have confidence in the FBI's forensic evidence?

3.32 **Matching socks.** Consider the following question posed to Marilyn vos Savant in her weekly newspaper column, "Ask Marilyn":

I have two pairs of argyle socks, and they look nearly identical—one navy blue and the other black. [When doing the laundry] my wife matches the socks incorrectly much more often than she does correctly…. If all four socks are in front of her, it seems to me that her chances are 50% for a wrong match and 50% for a right match. What do you think?

Source: *Parade Magazine*, Feb. 27, 1994.

Use your knowledge of probability to answer this question. [*Hint:* List the sample points in the experiment.]

3.33 **Post-op nausea study.** Nausea and vomiting after surgery are common side effects of anesthesia and painkillers. Six different drugs, varying in cost, were compared for their effectiveness in preventing nausea and vomiting (*New*

England Journal of Medicine, June 10, 2004). The medical researchers looked at all possible combinations of the drugs as treatments, including a single drug, as well as two-drug, three-drug, four-drug, five-drug, and six-drug combinations.
   a. How many two-drug combinations of the six drugs are possible?
   b. How many three-drug combinations of the six drugs are possible?
   c. How many four-drug combinations of the six drugs are possible?
   d. How many five-drug combinations of the six drugs are possible?
   e. The researchers stated that a total of 64 drug combinations were tested as treatments for nausea. Verify that there are 64 ways that the six drugs can be combined. (Remember to include the one-drug and six-drug combinations, as well as the control treatment of no drugs.)

3.34 **Dominant versus recessive traits.** An individual's genetic makeup is determined by the genes obtained from each parent. For every genetic trait, each parent possesses a gene pair, and each parent contributes one-half of this gene pair, with equal probability, to his or her offspring, forming a new gene pair. The offspring's traits (eye color, baldness, etc.) come from this new gene pair, each gene of which possesses some characteristic.

For the gene pair that determines eye color, each gene trait may be one of two types: dominant brown (B) or recessive blue (b). A person possessing the gene pair BB or Bb has brown eyes, whereas the gene pair bb produces blue eyes.
   a. Suppose both parents of an individual are brown eyed, each with a gene pair of type Bb. What is the probability that a randomly selected child of this couple will have blue eyes?
   b. If one parent has brown eyes, type Bb, and the other has blue eyes, what is the probability that a randomly selected child of this couple will have blue eyes?
   c. Suppose one parent is brown eyed with a gene pair of type BB. What is the probability that a child has blue eyes?

3.2 Unions and Intersections

An event can often be viewed as a composition of two or more other events. Such events, which are called **compound events,** can be formed (composed) in two ways.

> The **union** of two events A and B is the event that occurs if either A or B (or both) occurs on a single performance of the experiment. We denote the union of events A and B by the symbol $A \cup B$. $A \cup B$ consists of all the sample points that belong to A or B or both. (See Figure 3.7a.)

> The **intersection** of two events A and B is the event that occurs if both A and B occur on a single performance of the experiment. We write $A \cap B$ for the intersection of A and B. $A \cap B$ consists of all the sample points belonging to *both A and B.* (See Figure 3.7b.)

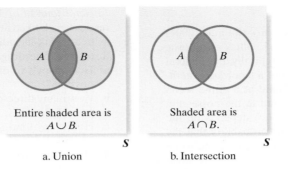

Figure 3.7
Venn diagrams for union and intersection

Entire shaded area is $A \cup B$.

Shaded area is $A \cap B$.

a. Union

b. Intersection

Example 3.8

Probabilities of Unions and Intersections— Die-Toss Experiment

Problem Consider a die-toss experiment in which the following events are defined:

> A: {Toss an even number.}
> B: {Toss a number less than or equal to 3.}

a. Describe $A \cup B$ for this experiment.
b. Describe $A \cap B$ for this experiment.
c. Calculate $P(A \cup B)$ and $P(A \cap B)$, assuming that the die is fair.

Solution Draw the Venn diagram as shown in Figure 3.8.

a. The union of A and B is the event that occurs if we observe either an even number, a number less than or equal to 3, or both on a single throw of the die. Consequently, the sample points in the event $A \cup B$ are those for which A occurs, B occurs, or both A and B occur. Checking the sample points in the entire sample space, we find that the collection of sample points in the union of A and B is

$$A \cup B = \{1, 2, 3, 4, 6\}$$

b. The intersection of A and B is the event that occurs if we observe *both* an even number and a number less than or equal to 3 on a single throw of the die. Checking the sample points to see which imply the occurrence of *both* events A and B, we see that the intersection contains only one sample point:

$$A \cap B = \{2\}$$

In other words, the intersection of A and B is the sample point Observe a 2.

c. Recalling that the probability of an event is the sum of the probabilities of the sample points of which the event is composed, we have

$$P(A \cup B) = P(1) + P(2) + P(3) + P(4) + P(6)$$
$$= \frac{1}{6} + \frac{1}{6} + \frac{1}{6} + \frac{1}{6} + \frac{1}{6} = \frac{5}{6}$$

and

$$P(A \cap B) = P(2) = \frac{1}{6}$$

Look Back Since the six sample points are equally likely, the probabilities in part **c** are simply the number of sample points in the event of interest, divided by 6.

Now Work Exercise 3.45a–d

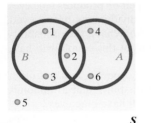

Figure 3.8
Venn diagram for die toss

Unions and intersections can be defined for more than two events. For example, the event $A \cup B \cup C$ represents the union of three events: A, B, and C. This event, which includes the set of sample points in A, B, or C, will occur if any one (or more) of the events A, B, and C occurs. Similarly, the intersection $A \cap B \cap C$ is the event that all three of the events A, B, and C occur. Therefore, $A \cap B \cap C$ is the set of sample points that are in all three of the events A, B, and C.

Example 3.9

Finding Probabilities From a Two-Way Table—Mother's Race versus Maternal Age

Problem *Family Planning Perspectives* reported on a study of over 200,000 births in New Jersey over a recent two-year period. The study investigated the link between the mother's race and the age at which she gave birth (called maternal age). The percentages of the total number of births in New Jersey, by the maternal age and race classifications, are given in Table 3.4.

Table 3.4	Percentage of New Jersey Birth Mothers, by Age and Race	
	Race	
Maternal Age (years)	White	Black
≤17	2%	2%
18–19	3%	2%
20–29	41%	12%
≥30	33%	5%

Based on Reichman, N. E., and Pagnini, D. L. "Maternal age and birth outcomes: Data from New Jersey." *Family Planning Perspectives*, Vol. 29, No. 6, Nov./Dec. 1997, p. 269 (adapted from Table 1).

This table is called a **two-way table,** since responses are classified according to two variables: maternal age (rows) and race (columns).

Define the following event:

> *A*: {A New Jersey birth mother is white.}
>
> *B*: {A New Jersey mother was a teenager when giving birth.}

a. Find $P(A)$ and $P(B)$.

b. Find $P(A \cup B)$.

c. Find $P(A \cap B)$.

Solution Following the steps for calculating probabilities of events, we first note that the objective is to characterize the race and maternal age distribution of New Jersey birth mothers. To accomplish this objective, we define the experiment to consist of selecting a birth mother from the collection of all New Jersey birth mothers during the two-year period of the study and observing her race and maternal age class. The sample points are the eight different age–race classifications:

E_1: {≤ 17 yrs., white}	E_5: {≤ 17 yrs., black}
E_2: {18–19 yrs., white}	E_6: {18–19 yrs., black}
E_3: {20–29 yrs., white}	E_7: {20–29 yrs., black}
E_4: {≥ 30 yrs., white}	E_8: {≥ 30 yrs., black}

Next, we assign probabilities to the sample points. If we blindly select one of the birth mothers, the probability that she will occupy a particular age–race classification is just the proportion, or relative frequency, of birth mothers in that classification. These proportions (as percentages) are given in Table 3.4. Thus,

$P(E_1)$ = Relative frequency of birth mothers in age–race class {≤ 17 yrs., white} = .02

$P(E_2)$ = .03

$P(E_3)$ = .41

$P(E_4)$ = .33

$P(E_5)$ = .02

$P(E_6)$ = .02

$P(E_7)$ = .12

$P(E_8)$ = .05

You may verify that the sample point probabilities sum to 1.

a. To find $P(A)$, we first determine the collection of sample points contained in event A. Since A is defined as {white}, we see from Table 3.4 that A contains the four sample points represented by the first column of the table. In other words, the event A consists of the race classification {white} in all four age classifications. The probability of A is the sum of the probabilities of the sample points in A:

$$P(A) = P(E_1) + P(E_2) + P(E_3) + P(E_4) = .02 + .03 + .41 + .33 = .79$$

Similarly, B = {teenage mother, age ≤ 19 years} consists of the four sample points in the first and second rows of Table 3.4:

$$P(B) = P(E_1) + P(E_2) + P(E_5) + P(E_6) = .02 + .03 + .02 + .02 = .09$$

b. The union of events A and B, $A \cup B$, consists of all sample points in *either A or B (or both)*. That is, the union of A and B consists of all birth mothers who are white *or* who gave birth as a teenager. In Table 3.4, this is any sample point found in the first column *or* the first two rows. Thus,

$$P(A \cup B) = .02 + .03 + .41 + .33 + .02 + .02 = .83$$

c. The intersection of events A and B, $A \cap B$, consists of all sample points in *both A and B*. That is, the intersection of A and B consists of all birth mothers who are white *and* who gave birth as a teenager. In Table 3.4, this is any sample point found in the first column *and* the first two rows. Thus,

$$P(A \cap B) = .02 + .03 = .05.$$

Look Back As in previous problems, the key to finding the probabilities of parts **b** and **c** is to identify the sample points that make up the event of interest. In a two-way table such as Table 3.4, the total number of sample points will be equal to the number of rows times the number of columns.

Now Work Exercise 3.47f–g

3.3 Complementary Events

A very useful concept in the calculation of event probabilities is the notion of **complementary events**:

> The **complement** of an event A is the event that A does *not* occur—that is, the event consisting of all sample points that are not in event A. We denote the complement of A by A^c.

An event A is a collection of sample points, and the sample points included in A^c are those not in A. Figure 3.9 demonstrates this idea. Note from the figure that all sample points in S are included in *either* A or A^c and that *no* sample point is in both A and A^c. This leads us to conclude that the probabilities of an event and its complement *must sum to 1*:

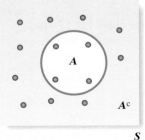

Figure 3.9
Venn diagram of complementary events

> **Rule of Complements**
> The sum of the probabilities of complementary events equals 1; that is,
> $$P(A) + P(A^c) = 1.$$

In many probability problems, calculating the probability of the complement of the event of interest is easier than calculating the event itself. Then, because

$$P(A) + P(A^c) = 1$$

we can calculate $P(A)$ by using the relationship

$$P(A) = 1 - P(A^c).$$

Example 3.10
Probability of a Complementary Event—Coin-Toss Experiment

Problem Consider the experiment of tossing fair coins. Define the following event: A: {Observing at least one head}.

a. Find $P(A)$ if 2 coins are tossed.

b. Find $P(A)$ if 10 coins are tossed.

Solution

a. When 2 coins are tossed, we know that the event A: {Observe at least one head.} consists of the sample points

$$A: \{HH, HT, TH\}$$

The complement of A is defined as the event that occurs when A does not occur. Therefore,

$$A^c: \{\text{Observe no heads.}\} = \{TT\}$$

This complementary relationship is shown in Figure 3.10. Since the coins are balanced, we have

$$P(A^c) = P(TT) = \frac{1}{4}$$

and

$$P(A) = 1 - P(A^c) = 1 - \frac{1}{4} = \frac{3}{4}.$$

b. We solve this problem by following the five steps for calculating probabilities of events. (See Section 3.1.)

Step 1 Define the experiment. The experiment is to record the results of the 10 tosses of the coin.

Step 2 List the sample points. A sample point consists of a particular sequence of 10 heads and tails. Thus, one sample point is $HHTTTHTHTT$, which denotes head on first toss, head on second toss, tail on third toss, etc. Others are $HTHHHTTTTT$ and $THHTHTHTTH$. Obviously, the number of sample points is very large—too many to list. It can be shown (see Section 3.8) that there are $2^{10} = 1,024$ sample points for this experiment.

Step 3 Assign probabilities. Since the coin is fair, each sequence of heads and tails has the same chance of occurring; therefore, all the sample points are equally likely. Then

$$P(\text{Each sample point}) = \frac{1}{1,024}$$

Step 4 Determine the sample points in event A. A sample point is in A if at least one H appears in the sequence of 10 tosses. However, if we consider the complement of A, we find that

$$A^c = \{\text{No heads are observed in 10 tosses.}\}$$

Thus, A^c contains only one sample point:

$$A^c: \{TTTTTTTTTT\}$$

and $P(A^c) = \dfrac{1}{1,024}$

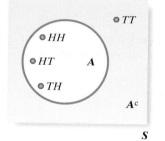

Figure 3.10
Complementary events in the toss of two coins

Step 5   Now we use the relationship of complementary events to find $P(A)$:

$$P(A) = 1 - P(A^c) = 1 - \frac{1}{1{,}024} = \frac{1{,}023}{1{,}024} = .999$$

Look Back In part **a,** we can find $P(A)$ by summing the probabilities of the sample points $HH, HT,$ and TH in A. Many times, however, it is easier to find $P(A^c)$ by using the rule of complements.

Look Forward Since $P(A) = .999$ in part **b,** we are virtually certain of observing at least one head in 10 tosses of the coin.

Now Work Exercise 3.45e–f

3.4 The Additive Rule and Mutually Exclusive Events

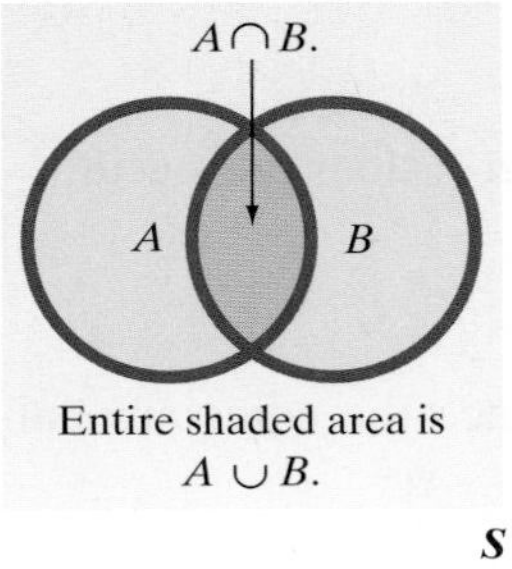

Entire shaded area is $A \cup B$.

Figure 3.11
Venn diagram of union

In Section 3.2, we saw how to determine which sample points are contained in a union of two sets and how to calculate the probability of the union by summing the separate probabilities of the sample points in the union. It is also possible to obtain the probability of the union of two events by using the **additive rule of probability.**

By studying the Venn diagram in Figure 3.11, you can see that the probability of the union of two events A and B can be obtained by summing $P(A)$ and $P(B)$ and subtracting $P(A \cap B)$. We must subtract $P(A \cap B)$ because the sample point probabilities in $A \cap B$ have been included twice—once in $P(A)$ and once in $P(B)$.

The formula for calculating the probability of the union of two events is given in the next box.

> **Additive Rule of Probability**
>
> The probability of the union of events A and B is the sum of the probability of event A and the probability of event B, minus the probability of the intersection of events A and B; that is
>
> $$P(A \cup B) = P(A) + P(B) - P(A \cap B)$$

Example 3.11

Applying the Additive Rule— Hospital Admissions Study

Problem Hospital records show that 12% of all patients are admitted for surgical treatment, 16% are admitted for obstetrics, and 2% receive both obstetrics and surgical treatment. If a new patient is admitted to the hospital, what is the probability that the patient will be admitted for surgery, for obstetrics, or for both?

Solution Consider the following events:

A: {A patient admitted to the hospital receives surgical treatment.}
B: {A patient admitted to the hospital receives obstetrics treatment.}

Then, from the given information,

$$P(A) = .12$$
$$P(B) = .16$$

and the probability of the event that a patient receives both obstetrics and surgical treatment is

$$P(A \cap B) = .02$$

The event that a patient admitted to the hospital receives either surgical treatment, obstetrics treatment, or both is the union $A \cup B$, the probability of which is given by the additive rule of probability:

$$P(A \cup B) = P(A) + P(B) - P(A \cap B)$$
$$= .12 + .16 - .02 = .26$$

Thus, 26% of all patients admitted to the hospital receive either surgical treatment, obstetrics treatment, or both.

Look Back From the information given, it is not possible to list and assign probabilities to all the sample points. Consequently, we cannot proceed through the five-step process (p. 116) for finding $P(A \cup B)$, and we must use the additive rule.

Now Work Exercise 3.43

A very special relationship exists between events A and B when $A \cap B$ contains no sample points. In this case, we call the events A and B *mutually exclusive events*.

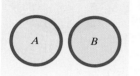

Figure 3.12
Venn diagram of mutually exclusive events

> Events A and B are **mutually exclusive** if $A \cap B$ contains no sample points—that is, if A and B have no sample points in common. For mutually exclusive events,
>
> $$P(A \cap B) = 0$$

Figure 3.12 shows a Venn diagram of two mutually exclusive events. The events A and B have no sample points in common; that is, A and B cannot occur simultaneously, and $P(A \cap B) = 0$. Thus, we have the important relationship given in the next box.

> **Probability of Union of Two Mutually Exclusive Events**
> If two events A and B are *mutually exclusive*, the probability of the union of A and B equals the sum of the probability of A and the probability of B; that is, $P(A \cup B) = P(A) + P(B)$.

⚠ **CAUTION** The preceding formula is *false* if the events are *not* mutually exclusive. In that case (i.e., two non-mutually exclusive events), you must apply the general additive rule of probability. ▲

Example 3.12
Union of Two Mutually Exclusive Events—Coin-Tossing Experiment

Problem Consider the experiment of tossing two balanced coins. Find the probability of observing *at least* one head.

Solution Define the events

$$A: \{\text{Observe at least one head.}\}$$
$$B: \{\text{Observe exactly one head.}\}$$
$$C: \{\text{Observe exactly two heads.}\}$$

Note that

$$A = B \cup C$$

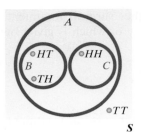

Figure 3.13
Venn diagram for coin-toss experiment

and that $B \cap C$ contains no sample points. (See Figure 3.13.) Thus, B and C are mutually exclusive, and it follows that

$$P(A) = P(B \cup C) = P(B) + P(C) = \frac{1}{2} + \frac{1}{4} = \frac{3}{4}$$

Look Back Although this example is quite simple, it shows us that writing events with verbal descriptions that include the phrases "at least" or "at most" as unions of mutually exclusive events is very useful. This practice enables us to find the probability of the event by adding the probabilities of the mutually exclusive events.

Statistics IN Action | **Revisited** | The Probability of Winning Lotto with a Wheeling System

Refer to Florida's Pick-6 Lotto game, in which you select six numbers of your choice from a field of numbers ranging from 1 to 53. In Section 3.1, we learned that the probability of winning Lotto on a single ticket is only 1 in approximately 23 million. The "experts" at Lotto Buster recommend many strategies for increasing the odds of winning the lottery. One strategy is to employ a *wheeling system*. In a complete wheeling system, you select more than six numbers—say, seven—and play every combination of six of those seven numbers.

For example, suppose you choose to "wheel" the following seven numbers: 2, 7, 18, 23, 30, 32, and 51. Every combination of six of these seven numbers is listed in Table SIA3.1. You can see that there are seven different possibilities. (Use the combinations rule with $N = 7$ and $n = 6$ to verify this.) Thus, we would purchase seven tickets (at a cost of $7) corresponding to these different combinations in a complete wheeling system.

#1 or Ticket #2 or Ticket #3 or Ticket #4 or Ticket #5 or Ticket #6 or Ticket #7 is the winning combination. Note that this probability is stated with the use of the word *or*, implying a *union* of seven events. Letting T_1 represent the event that Ticket #1 wins, and defining $T_2, T_3, \dots, T_7$ in a similar fashion, we want to find

$$P(T_1 \text{ or } T_2 \text{ or } T_3 \text{ or } T_4 \text{ or } T_5 \text{ or } T_6 \text{ or } T_7)$$

Recall (Section 3.2) that the 22,957,480 possible combinations in Pick-6 Lotto are mutually exclusive and equally likely to occur. Consequently, the probability of the union of the seven events is simply the sum of the probabilities of the individual events, where each event has probability $1/(22,957,480)$:

$$P(\text{win Lotto with 7 Wheeled Numbers})$$
$$= P(T_1 \text{ or } T_2 \text{ or } T_3 \text{ or } T_4 \text{ or } T_5 \text{ or } T_6 \text{ or } T_7)$$
$$= 7/(22,957,480) = .0000003$$

Table SIA3.1	**Wheeling the Seven Numbers 2, 7, 18, 23, 30, 32, and 51**					
Ticket #1	2	7	18	23	30	32
Ticket #2	2	7	18	23	30	51
Ticket #3	2	7	18	23	32	51
Ticket #4	2	7	18	30	32	51
Ticket #5	2	7	23	30	32	51
Ticket #6	2	18	23	30	32	51
Ticket #7	7	18	23	30	32	51

To determine whether this strategy does, in fact, increase our odds of winning, we need to find the probability that one of these seven combinations occurs during the 6/53 Lotto draw. That is, we need to find the probability that either Ticket

In terms of odds, we now have 3 chances in 10 million of winning the Lotto with the complete wheeling system. Technically, the "experts" are correct: Our odds of winning Lotto have increased (from 1 in 23 million). However, the probability of winning is so close to 0 that we question whether the $7 spent on lottery tickets is worth the negligible increase in odds. In fact, it can be shown that to increase your chance of winning the 6/53 Lotto to 1 chance in 100 (i.e., .01) by means of a complete wheeling system, you would have to wheel 26 of your favorite numbers—a total of 230,230 combinations at a cost of $230,230!

Exercises 3.35–3.62

Understanding the Principles

3.35 Define in words the union of two events.

3.36 Define in words the intersection of two events.

3.37 Define in words the complement of an event.

3.38 State the rule of complements.

3.39 State the additive rule of probability for any two events.

3.40 Define in words mutually exclusive events.

3.41 State the additive rule of probability for mutually exclusive events.

Learning the Mechanics

3.42 Suppose $P(A) = .4$, $P(B) = .7$ and $P(A \cap B) = .3$. Find the following probabilities:
- **a.** $P(B^c)$
- **b.** $P(A^c)$
- **c.** $P(A \cup B)$

3.43 A fair coin is tossed three times, and the events A and B
NW are defined as follows:

> A: {At least one head is observed.}
> B: {The number of heads observed is odd.}

- **a.** Identify the sample points in the events A, B, $A \cup B$, A^c, and $A \cap B$.
- **b.** Find $P(A)$, $P(B)$, $P(A \cup B)$, $P(A^c)$, and $P(A \cap B)$ by summing the probabilities of the appropriate sample points.
- **c.** Use the additive rule to find $P(A \cup B)$. Compare your answer with the one you obtained in part **b.**
- **d.** Are the events A and B mutually exclusive? Why?

3.44 A pair of fair dice is tossed. Define the following events:

> A: {You will roll a 7 (i.e., the sum of the numbers of dots on the upper faces of the two dice is equal to 7).}
> B: {At least one of the two dice is showing a 4.}

- **a.** Identify the sample points in the events A, B, $A \cap B$, $A \cup B$, and A^c.
- **b.** Find $P(A)$, $P(B)$, $P(A \cap B)$, $P(A \cup B)$, and $P(A^c)$ and by summing the probabilities of the appropriate sample points.
- **c.** Use the additive rule to find $P(A \cup B)$. Compare your answer with that for the same event in part **b.**
- **d.** Are A and B mutually exclusive? Why?

3.45 Consider the following Venn diagram, where
NW $P(E_2) = P(E_3) = {}^1/_5$, $P(E_4) = P(E_5) = {}^1/_{20}$, $P(E_6) = {}^1/_{10}$, $P(E_7) = {}^1/_5$:

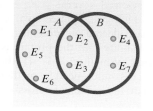

Find each of the following probabilities:
- **a.** $P(A)$
- **b.** $P(B)$
- **c.** $P(A \cup B)$
- **d.** $P(A \cap B)$
- **e.** $P(A^c)$
- **f.** $P(B^c)$
- **g.** $P(A \cup A^c)$
- **h.** $P(A^c \cap B)$

3.46 Consider the following Venn diagram, where

$P(E_1) = .10$, $P(E_2) = .05$, $P(E_3) = P(E_4) = .2$, $P(E_5) = .06$, $P(E_6) = .3$, $P(E_7) = .06$, and $P(E_8) = .03$:

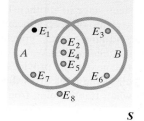

Find the following probabilities:
- **a.** $P(A^c)$
- **b.** $P(B^c)$
- **c.** $P(A^c \cap B)$
- **d.** $P(A \cup B)$
- **e.** $P(A \cap B)$
- **f.** $P(A^c \cup B^c)$
- **g.** Are events A and B mutually exclusive? Why?

3.47 The outcomes of two variables are (Low, Medium, High)
NW and (On, Off), respectively. An experiment is conducted in which the outcomes of each of the two variables are observed. The probabilities associated with each of the six possible outcome pairs are given in the following table:

	Low	Medium	High
On	.50	.10	.05
Off	.25	.07	.03

Consider the following events:

A: {On}
B: {Medium or On}
C: {Off and Low}
D: {High}

- **a.** Find $P(A)$.
- **b.** Find $P(B)$.
- **c.** Find $P(C)$.
- **d.** Find $P(D)$.
- **e.** Find $P(A^c)$.
- **f.** Find $P(A \cup B)$.
- **g.** Find $P(A \cap B)$.
- **h.** Consider each possible pair of events taken from the events A, B, C, and D. List the pairs of events that are mutually exclusive. Justify your choices.

3.48 Three fair coins are tossed. We wish to find the probability of the event A: {Observe at least one head.}
- **a.** Express A as the union of three mutually exclusive events. Using the expression you wrote, find the probability of A.
- **b.** Express A as the complement of an event. Using the expression you wrote, find the probability of A.

⬤ Applet Exercise 3.3

Use the applets entitled *Simulating the Probability of Rolling a 6* and *Simulating the Probability of Rolling a 3 or 4* to explore the additive rule of probability.
- **a.** Explain why the applet *Simulating the Probability of Rolling a 6* can also be used to simulate the probability of rolling a 3. Then use the applet with $n = 1000$ to simulate the probability of rolling a 3. Record the cumulative proportion. Repeat the process to simulate the probability of rolling a 4.
- **b.** Use the applet *Simulating the Probability of Rolling a 3 or 4* with $n = 1000$ to simulate the probability of rolling a 3 or 4. Record the cumulative proportion.
- **c.** Add the two cumulative proportions from part **a.** How does this sum compare with the cumulative proportion in part **b**? How does your answer illustrate the additive rule for probability?

⊙ Applet Exercise 3.4

Use the applets entitled *Simulating the Probability of Rolling a 6* and *Simulating the Probability of Rolling a 3 or 4* to simulate the probability of the complement of an event.

a. Explain how the applet *Simulating the Probability of Rolling a 6* can also be used to simulate the probability of the event *rolling a 1, 2, 3, 4, or 5.* Then use the applet with $n = 1000$ to simulate this probability.

b. Explain how the applet *Simulating the Probability of Rolling a 3 or 4* can also be used to simulate the probability of the event *rolling a 1, 2, 5, or 6.* Then use the applet with $n = 1000$ to simulate this probability.

c. Which applet could be used to simulate the probability of the event *rolling a 1, 2, 3, or 4?* Explain.

Applying the Concepts—Basic

3.49 **Social networking Web sites in the United Kingdom.** In the United States, MySpace and FaceBook are considered the two most popular social networking Web sites. In the United Kingdom (UK), the competition for social networking is between MySpace and Bebo. According to Nielsen/NetRatings (April 2006), 4% of UK citizens visit MySpace, 3% visit Bebo, and 1% visit both MySpace and Bebo.

a. Draw a Venn diagram to illustrate the use of social networking sites in the United Kingdom.

b. Find the probability that a UK citizen visits either the MySpace or Bebo social networking sites.

c. Use your answer to part **b** to find the probability that a UK citizen does not visit either social networking site.

3.50 **Study of analysts' forecasts.** The *Journal of Accounting Research* (March 2008) published a study on relationship incentives and degree of optimism among analysts' forecasts. Participants were analysts at either a large or small brokerage firm who made their forecasts either early or late in the quarter. Also, some analysts were only concerned with making an accurate forecast, while others were also interested in their relationship with management. Suppose one of these analysts is randomly selected. Consider the following events:

$A = \{$The analyst is only concerned with making an accurate forecast.$\}$
$B = \{$The analyst makes the forecast early in the quarter.$\}$
$C = \{$the analyst is from a small brokerage firm.$\}$

Describe each of the following events in terms of unions, intersections, and complements (e.g., $A \cup B, A \cap B, A^C$, etc.).

a. The analyst makes an early forecast and is only concerned with accuracy.

b. The analyst is not only concerned with accuracy.

c. The analyst is from a small brokerage firm or makes an early forecast.

d. The analyst makes a late forecast and is not only concerned with accuracy.

3.51 **Toxic chemical incidents.** *Process Safety Progress* (Sept. 2004) reported on an emergency response system for incidents involving toxic chemicals in Taiwan. The system has logged over 250 incidents since being implemented. The next table gives a breakdown of the locations where these incidents occurred. Consider the location of a toxic chemical incident in Taiwan.

a. List the sample points for this experiment.

b. Assign reasonable probabilities to the sample points.

c. What is the probability that the incident occurs in a school laboratory?

d. What is the probability that the incident occurs in either a chemical or a nonchemical plant?

e. What is the probability that the incident does not occur in transit?

Location	Percent of Incidents
School laboratory	6%
In transit	26%
Chemical plant	21%
Nonchemical plant	35%
Other	12%
Total	100%

Based on Chen, J. R., et al. "Emergency response of toxic chemicals in Taiwan: The system and case studies." *Process Safety Progress*, Vol. 23, No. 3, Sept. 2004 (Figure 5a).

3.52 **Scanning errors at Wal-Mart.** The National Institute of Standards and Technology (NIST) mandates that for every 100 items scanned through the electronic checkout scanner at a retail store, no more than 2 should have an inaccurate price. A recent study of the accuracy of checkout scanners at Wal-Mart stores in California (*Tampa Tribune*, Nov. 22, 2005) showed that, of the 60 Wal-Mart stores investigated, 52 violated the NIST scanner accuracy standard. If 1 of the 60 Wal-Mart stores is randomly selected, what is the probability that that store does not violate the NIST scanner accuracy standard?

3.53 **Gene expression profiling.** Gene expression profiling is a state-of-the-art method for determining the biology of cells. In *Briefings in Functional Genomics and Proteomics* (Dec. 2006), biologists at Pacific Northwest National Laboratory reviewed several gene expression profiling methods. The biologists applied two of the methods (A and B) to data collected on proteins in human mammary cells. The probability that the protein is cross-referenced (i.e., identified) by method A is .41, the probability that the protein is cross-referenced by method B is .42, and the probability that the protein is cross-referenced by both methods is .40.

a. Draw a Venn diagram to illustrate the results of the gene-profiling analysis.

b. Find the probability that the protein is cross-referenced by either method A or method B.

c. On the basis of your answer to part **b**, find the probability that the protein is not cross-referenced by either method.

3.54 **Sleep apnea and sleep stage transitioning.** Sleep apnea is a common sleep disorder characterized by collapses of the upper airway during sleep. *Chance* (Winter 2009) investigated the role of sleep apnea in how people transition from one sleep stage to another. The various stages of sleep for a large group of sleep apnea patients were monitored in 30-second intervals, or "epochs." For each epoch, sleep stage was categorized as Wake, REM, or Non-REM. The table on the next page provides a summary of the results. Each cell of the table gives the number of epochs that occurred when transitioning from the previous sleep stage to the current sleep stage. Consider the previous and current sleep stage of a randomly selected epoch from the study.

	Previous Sleep Stage		
Current Stage	Non-REM	REM	Wake
Non-REM	31,880	160	1,733
REM	346	7,609	175
Wake	1,588	358	6,079
Totals	33,814	8,127	7,987

Based on Caffo, B. S., et al. "An overview of observational sleep research with application to sleep stage transitioning." *Chance*, Vol. 22, No. 1, Winter 2009 (Table 2).

a. List the sample points for this experiment.
b. Assign reasonable probabilities to the sample points.
c. What is the probability that the current sleep stage of the epoch is REM?
d. What is the probability that the previous sleep stage of the epoch is Wake?
e. What is the probability that the epoch transitioned from the REM sleep stage to the Non-REM sleep stage?

Applying the Concepts—Intermediate

3.55 **Guilt in decision making.** The effect of guilt emotion on how a decision maker focuses on the problem was investigated in the Jan. 2007 issue of the *Journal of Behavioral Decision Making* (see Exercise 1.27, p. 20). A total of 171 volunteer students participated in the experiment, where each was randomly assigned to one of three emotional states (guilt, anger, or neutral) through a reading/writing task. Immediately after the task, the students were presented with a decision problem where the stated option has predominantly negative features (e.g., spending money on repairing a very old car). The results (number responding in each category) are summarized in the accompanying table. Suppose one of the 171 participants is selected at random.

Emotional State	Choose Stated Option	Do Not Choose Stated Option	Totals
Guilt	45	12	57
Anger	8	50	58
Neutral	7	49	56
Totals	60	111	171

Based on Gangemi, A., and Mancini, F. "Guilt and focusing in decision-making." *Journal of Behavioral Decision Making*, Vol. 20, Jan. 2007 (Table 2).

a. Find the probability that the respondent is assigned to the guilty state.
b. Find the probability that the respondent chooses the stated option (repair the car).
c. Find the probability that the respondent is assigned to the guilty state and chooses the stated option.
d. Find the probability that the respondent is assigned to the guilty state or chooses the stated option.

3.56 **Abortion provider survey.** The Alan Guttmacher Institute Abortion Provider Survey is a survey of all 238 known nonhospital abortion providers in the United States (*Perspectives on Sexual and Reproductive Health*, Jan./Feb. 2003). For one part of the survey, the 358 providers were classified according to case load (number of abortions performed per year) and whether they permit their patients to take the abortion drug misoprostol at home or require the patients to return to the abortion facility to receive the drug. The responses are summarized in the accompanying table. Suppose we select, at random, one of the 358 providers and observe the provider's case load (fewer than 50, or 50 or more) and home use of the drug (yes or no).

Permit Drug at Home	Number of Abortions		
	Fewer than 50	50 or More	Totals
Yes	170	130	300
No	48	10	58
Totals	218	140	358

Based on Henshaw, S. K., and Finer, L. B. "The accessibility of abortion services in the United States, 2001." *Perspectives on Sexual and Reproductive Health*, Vol. 35, No. 1, Jan./Feb. 2003 (Table 4).

a. Find the probability that the provider permits home use of the abortion drug.
b. Find the probability that the provider permits home use of the drug or has a case load of fewer than 50 abortions.
c. Find the probability that the provider permits home use of the drug and has a case load of fewer than 50 abortions.

3.57 **Fighting probability of fallow deer bucks.** In *Aggressive Behavior* (Jan./Feb. 2007), zoologists investigated the likelihood of fallow deer bucks fighting during the mating season. During a 270-hour observation period, the researchers recorded 205 encounters between two bucks. Of these, 167 involved one buck clearly initiating the encounter with the other. In these 167 initiated encounters, the zoologists kept track of whether or not a physical contact fight occurred and whether the initiator ultimately won or lost the encounter. (The buck that is driven away by the other is considered the loser.) A summary of the 167 initiated encounters is provided in the accompanying table. Suppose we select one of these 167 encounters and note the outcome (fight status and winner).

a. What is the probability that a fight occurs and the initiator wins?
b. What is the probability that no fight occurs?
c. What is the probability that there is no clear winner?
d. What is the probability that a fight occurs or the initiator loses?
e. Are the events "No clear winner" and "initiator loses" mutually exclusive?

	Initiator Wins	No Clear Winner	Initiator Loses	Totals
Fight	26	23	15	64
No Fight	80	12	11	103
Totals	106	35	26	167

Based on Bartos, L. et al. "Estimation of the probability of fighting in fallow deer (*Dama dama*) during the rut." *Aggressive Behavior*, Vol. 33, Jan./ Feb. 2007, pp. 7–13.

3.58 **Cell phone handoff behavior.** A "handoff" is a term used in wireless communications to describe the process of a cell phone moving from a coverage area of one base station to another. Each base station has multiple channels (called color codes) that allow it to communicate with the cell phone. The *Journal of Engineering, Computing and Architecture* (Vol. 3., 2009) published a study of cell phone handoff behavior. During a sample driving trip which involved crossing from one base station to another, the different color codes accessed by the cell phone were monitored and recorded. The table below shows the number of times each color code was accessed for two identical driving trips, each using a different cell phone model. (*Note:* The table is similar to the one published in the article.) Suppose you randomly select one point during the combined driving trips.

	Color Code				
	0	5	b	c	Total
Model 1	20	35	40	0	85
Model 2	15	50	6	4	75
Total	35	85	46	4	160

a. What is the probability that the cell phone is using color code 5?

b. What is the probability that the cell phone is using color code 5 or color code 0?

c. What is the probability that the cell phone used is Model 2 and the color code is 0?

3.59 **Federal civil trial appeals.** The *Journal of the American Law and Economics Association* (Vol. 3, 2001) published the results of a study of appeals of federal civil trials. The accompanying table, extracted from the article, gives a breakdown of 2,143 civil cases that were appealed by either the plaintiff or the defendant. The outcome of the appeal, as well as the type of trial (judge or jury), was determined for each civil case. Suppose one of the 2,143 cases is selected at random and both the outcome of the appeal and the type of trial are observed.

	Jury	Judge	Totals
Plaintiff trial win—reversed	194	71	265
Plaintiff trial win—affirmed/dismissed	429	240	669
Defendant trial win—reversed	111	68	179
Defendant trial win—affirmed/ dismissed	731	299	1,030
Totals	1,465	678	2,143

a. Find $P(A)$, where $A = \{\text{jury trial}\}$.

b. Find $P(B)$, where $B = \{\text{plaintiff trial win is reversed}\}$.

c. Are A and B mutually exclusive events?

d. Find $P(A^c)$.

e. Find $P(A \cup B)$.

f. Find $P(A \cap B)$.

3.60 **Employee behavior problems.** The *Organizational Development Journal* (Summer 2006) reported on the results of a survey of human resource officers (HROs) at major firms. The focus of the study was employee behavior, namely absenteeism, promptness to work, and turnover. The study found that 55% of the HROs had problems with absenteeism. Also, 41% of the HROs had problems with turnover. Suppose that 22% of the HROs had problems with both absenteeism and turnover.

a. Find the probability that a human resource officer selected from the group surveyed had problems with employee absenteeism or employee turnover.

b. Find the probability that a human resource officer selected from the group surveyed did not have problems with employee absenteeism.

c. Find the probability that a human resource officer selected from the group surveyed did not have problems with employee absenteeism nor with employee turnover.

3.61 **Chemical signals of mice.** The ability of a mouse to recognize the odor of a potential predator (e.g., a cat) is essential to the mouse's survival. The chemical makeup of these odors—called kairomones—was the subject of a study published in *Cell* (May 14, 2010). Typically, the source of these odors are major urinary proteins (Mups). Cells collected from lab mice were exposed to Mups from rodent species A, Mups from rodent species B, and kairomones (from a cat). The accompanying Venn diagram shows the proportion of cells that chemically responded to each of the three odors. (*Note*: A cell may respond to more than a single odor.)

a. What is the probability that a lab mouse responds to all three source odors?

b. What is the probability that a lab mouse responds to the kairomone?

c. What is the probability that a lab mouse responds to Mups A and Mups B, but not the kairomone?

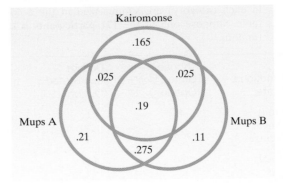

Applying the Concepts—Advanced

3.62 **Galileo's passe-dix game.** Passe-dix is a game of chance played with three fair dice. Players bet whether the sum of the faces shown on the dice will be above or below 10. During the late 16th century, the astronomer and mathematician Galileo Galilei was asked by the Grand Duke of Tuscany to explain why "the chance of throwing a total of 9 with three fair dice was less than that of throwing a total of 10." (*Interstat*, Jan. 2004). The Grand Duke believed that the chance should be the same, since "there are an equal number of partitions of the numbers 9 and 10." Find the flaw in the Grand Duke's reasoning and answer the question posed to Galileo.

3.5 Conditional Probability

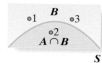

Figure 3.14
Reduced sample space for the die-toss experiment: given that event B has occurred

The event probabilities we've been discussing give the relative frequencies of the occurrences of the events when the experiment is repeated a very large number of times. Such probabilities are often called **unconditional probabilities**, because no special conditions are assumed other than those which define the experiment.

Often, however, we have additional knowledge that might affect the outcome of an experiment, so we may need to alter the probability of an event of interest. A probability that reflects such additional knowledge is called the **conditional probability** of the event. For example, we've seen that the probability of observing an even number (event A) on a toss of a fair die is $1/2$. But suppose we're given the information that on a particular throw of the die the result was a number less than or equal to 3 (event B). Would the probability of observing an even number on that throw of the die still be equal to $1/2$? It can't be, because making the assumption that B has occurred reduces the sample space from six sample points to three sample points (namely, those contained in event B). This reduced sample space is as shown in Figure 3.14.

Because the sample points for the die-toss experiment are equally likely, each of the three sample points in the reduced sample space is assigned an equal *conditional probability* of $1/3$. Since the only even number of the three in the reduced sample space B is the number 2 and the die is fair, we conclude that the probability that A occurs *given that B occurs* is $1/3$. We use the symbol $P(A\,|\,B)$ to represent the probability of event A given that event B occurs. For the die-toss example,

$$P(A\,|\,B) = \frac{1}{3}$$

To get the probability of event A given that event B occurs, we proceed as follows: We divide the probability of the part of A that falls within the reduced sample space B, namely, $P(A \cap B)$, by the total probability of the reduced sample space, namely, $P(B)$. Thus, for the die-toss example with event A: {Observe an even number.} and event B: {Observe a number less than or equal to 3.} we find that

$$P(A\,|\,B) = \frac{P(A \cap B)}{P(B)} = \frac{P(2)}{P(1) + P(2) + P(3)} = \frac{1/6}{3/6} = \frac{1}{3}$$

The formula for $P(A\,|\,B)$ is true in general:

Conditional Probability Formula

To find the *conditional probability that event A occurs given that event B occurs*, divide the probability that *both* A and B occur by the probability that B occurs; that is,

$$P(A\,|\,B) = \frac{P(A \cap B)}{P(B)} \quad [\text{We assume that } P(B) \neq 0.]$$

This formula adjusts the probability of $A \cap B$ from its original value in the complete sample space **S** to a conditional probability in the reduced sample space B. On the one hand, if the sample points in the complete sample space are equally likely, then the formula will assign equal probabilities to the sample points in the reduced sample space, as in the die-toss experiment. On the other hand, if the sample points have unequal probabilities, the formula will assign conditional probabilities proportional to the probabilities in the complete sample space. The latter situation is illustrated in the next three examples.

Example 3.13

The Conditional Probability Formula—Executives Who Cheat at Golf

Problem To develop programs for business travelers staying at convention hotels, Hyatt Hotels Corp. commissioned a study of executives who play golf. The study revealed that 55% of the executives admitted that they had cheated at golf. Also, 20% of the executives admitted that they had cheated at golf and had lied in business. Given that an executive had cheated at golf, what is the probability that the executive also had lied in business?

Solution Let's define events A and B as follows:

$$A = \{\text{Executive had cheated at golf.}\}$$
$$B = \{\text{Executive had lied in business.}\}$$

From the study, we know that 55% of executives had cheated at golf, so $P(A) = .55$. Now, executives who *both* cheat at golf (event A) *and* lie in business (event B) represent the compound event $A \cap B$. From the study, $P(A \cap B) = .20$. We want to know the probability that an executive lied in business (event B) given that he or she cheated at golf (event A); that is, we want to know the conditional probability $P(B|A)$. Applying the conditional probability formula, we have

$$P(B|A) = \frac{P(A \cap B)}{P(A)} = \frac{.20}{.55} = .364$$

Thus, given that a certain executive had cheated at golf, the probability that the executive also had lied in business is .364.

Look Back One of the keys to applying the formula correctly is to write the information in the study in the form of probability statements involving the events of interest. The word "and" in the clause "cheat at golf *and* lie in business" implies an intersection of the two events A and B. The word "given" in the phrase "*given* that an executive cheats at golf" implies that event A is the given event.

Now Work Exercise 3.79

Example 3.14

Applying the Conditional Probability Formula in a Two-Way Table—Smoking and Cancer

Problem Many medical researchers have conducted experiments to examine the relationship between cigarette smoking and cancer. Consider an individual randomly selected from the adult male population. Let A represent the event that the individual smokes, and let A^c denote the complement of A (the event that the individual does not smoke). Similarly, let B represent the event that the individual develops cancer, and let B^c be the complement of that event. Then the four sample points associated with the experiment are shown in Figure 3.15, and their probabilities for a certain section of the United States are given in Table 3.5. Use these sample point probabilities to examine the relationship between smoking and cancer.

$A \cap B$
(Smoker, cancer)

$A \cap B^c$
(Smoker, no cancer)

$A^c \cap B$
(Nonsmoker, cancer)

$A^c \cap B^c$
(Nonsmoker, no cancer)

S

Figure 3.15

Sample space for Example 3.15

Table 3.5	Probabilities of Smoking and Developing Cancer	
	Develops Cancer	
Smoker	Yes, B	No, B^c
Yes, A	.05	.20
No, A^c	.03	.72

Solution One method of determining whether the given probabilities indicate that smoking and cancer are related is to compare the *conditional probability* that an adult male acquires cancer given that he smokes with the conditional probability that an adult male acquires cancer given that he does not smoke [i.e., compare $P(B|A)$ with $P(B|A^c)$].

First, we consider the reduced sample space A corresponding to adult male smokers. This reduced sample space is highlighted in Figure 3.15. The two sample points $A \cap B$ and $A \cap B^c$ are contained in this reduced sample space, and the adjusted probabilities of these two sample points are the two conditional probabilities

$$P(B|A) = \frac{P(A \cap B)}{P(A)} \quad \text{and} \quad P(B^c|A) = \frac{P(A \cap B^c)}{P(A)}$$

The probability of event A is the sum of the probabilities of the sample points in A:

$$P(A) = P(A \cap B) + P(A \cap B^c) = .05 + .20 = .25$$

Then the values of the two conditional probabilities in the reduced sample space A are

$$P(B|A) = \frac{.05}{.25} = .20 \quad \text{and} \quad P(B^c|A) = \frac{.20}{.25} = .80$$

These two numbers represent the probabilities that an adult male smoker develops cancer and does not develop cancer, respectively.

In a like manner, the conditional probabilities of an adult male nonsmoker developing cancer and not developing cancer are

$$P(B|A^c) = \frac{P(A^c \cap B)}{P(A^c)} = \frac{.03}{.75} = .04$$

$$P(B^c|A^c) = \frac{P(A^c \cap B^c)}{P(A^c)} = \frac{.72}{.75} = .96$$

Two of the conditional probabilities give some insight into the relationship between cancer and smoking: the probability of developing cancer given that the adult male is a smoker, and the probability of developing cancer given that the adult male is not a smoker. The conditional probability that an adult male smoker develops cancer (.20) is five times the probability that a nonsmoker develops cancer (.04). This relationship does not imply that smoking *causes* cancer, but it does suggest a pronounced link between smoking and cancer.

Look Back Notice that the conditional probabilities $P(B^c|A) = .80$ and $P(B|A) = .20$ are in the same 4-to-1 ratio as the original unconditional probabilities, .20 and .05. The conditional probability formula simply adjusts the unconditional probabilities so that they add to 1 in the reduced sample space A of adult male smokers. Similarly, the conditional probabilities $P(B^c|A^c) = .96$ and $P(B|A^c) = .04$ are in the same 24-to-1 ratio as the unconditional probabilities .72 and .03.

Now Work Exercise 3.67a–b

Example 3.15
Conditional Probability in a Two-Way Table—Consumer Complaints

Problem The Federal Trade Commission's investigation of consumer product complaints has generated much interest on the part of manufacturers in the quality of their products. A manufacturer of an electromechanical kitchen utensil conducted an analysis of a large number of consumer complaints and found that they fell into the six categories shown in Table 3.6. If a consumer complaint is received, what is the probability that the cause of the complaint was the appearance of the product given that the complaint originated during the guarantee period?

Table 3.6   Distribution of Product Complaints

	Reason for Complaint			
	Electrical	Mechanical	Appearance	Totals
During Guarantee Period	18%	13%	32%	63%
After Guarantee Period	12%	22%	3%	37%
Totals	30%	35%	35%	100%

Solution Let A represent the event that the cause of a particular complaint is the appearance of the product, and let B represent the event that the complaint occurred during the guarantee period. Checking Table 3.6, you can see that $(18 + 13 + 32)\% = 63\%$ of the complaints occur during the guarantee period. Hence, $P(B) = .63$. The percentage of complaints that were caused by appearance and occurred during the guarantee period (the event $A \cap B$) is 32%. Therefore, $P(A \cap B) = .32$.

Using these probability values, we can calculate the conditional probability $P(A|B)$ that the cause of a complaint is appearance given that the complaint occurred during the guarantee time:

$$P(A|B) = \frac{P(A \cap B)}{P(B)} = \frac{.32}{.63} = .51$$

Thus, slightly more than half of the complaints that occurred during the guarantee period were due to scratches, dents, or other imperfections in the surface of the kitchen devices.

Look Back Note that the answer $\frac{.32}{.63}$ is the proportion for the event of interest A (.32), divided by the row total proportion for the given event B (.63). That is, it is the proportion of the time that A occurs together within the given event B.

Now Work Exercise 3.78a

3.6 The Multiplicative Rule and Independent Events

The probability of an intersection of two events can be calculated with the *multiplicative rule*, which employs the conditional probabilities we defined in the previous section. Actually, we've already developed the formula in another context. Recall that the conditional probability of B given A is

$$P(B|A) = \frac{P(A \cap B)}{P(A)}$$

Multiplying both sides of this equation by $P(A)$, we obtain a formula for the probability of the intersection of events A and B. This formula is often called the **multiplicative rule of probability.**

> **Multiplicative Rule of Probability**
>
> $P(A \cap B) = P(A)P(B|A)$ or, equivalently, $P(A \cap B) = P(B)P(A|B)$

Example 3.16
The Multiplicative Rule—a Famous Psychological Experiment

Problem In a classic psychology study conducted in the early 1960s, Stanley Milgram performed a series of experiments in which a teacher is asked to shock a learner who is attempting to memorize word pairs whenever the learner gives the wrong answer. The shock levels increase with each successive wrong answer. (Unknown to the teacher, the shocks are not real.) Two events of interest are

A: {The teacher "applies" a severe shock (450 volts).}
B: {The learner protests verbally prior to receiving the shock.}

A recent application of Milgram's shock study revealed that $P(B) = .5$ and $P(A|B) = .7$. On the basis of this information, what is the probability that a learner will protest verbally *and* a teacher will apply a severe shock? That is, find $P(A \cap B)$.

Solution We want to calculate $P(A \cap B)$. Using the formula for the multiplicative rule, we obtain

$$P(A \cap B) = P(B) \, P(A|B) = (.5)(.7) = .35$$

Thus, about 35% of the time the learner will give verbal protest and the teacher will apply a severe shock.

Look Back The multiplicative rule can be expressed in two ways: $P(A \cap B) = P(A) \, P(B|A)$, or $P(A \cap B) = P(B) \, P(A|B)$. Select the formula that involves a given event for which you know the probability (e.g., event B in the example).

Now Work Exercise 3.80

Intersections often contain only a few sample points. In this case, the probability of an intersection is easy to calculate by summing the appropriate sample point probabilities. However, the formula for calculating intersection probabilities is invaluable when the intersection contains numerous sample points, as the next example illustrates.

Example 3.17
Applying The Multiplicative Rule—Study of Welfare Workers

Problem A county welfare agency employs 10 welfare workers who interview prospective food stamp recipients. Periodically, the supervisor selects, at random, the forms completed by two workers and subsequently audits them for illegal deductions. Unknown to the supervisor, three of the workers have regularly been giving illegal deductions to applicants. What is the probability that both of the workers chosen have been giving illegal deductions?

Solution Define the following two events:

A: {First worker selected gives illegal deductions.}
B: {Second worker selected gives illegal deductions.}

We want to find the probability that both workers selected have been giving illegal deductions. This event can be restated as {First worker gives illegal deductions *and* second worker gives illegal deductions.}. Thus, we want to find the probability of the intersection $A \cap B$. Applying the multiplicative rule, we have

$$P(A \cap B) = P(A)P(B|A)$$

To find $P(A)$, it is helpful to consider the experiment as selecting 1 worker from the 10. Then the sample space for the experiment contains 10 sample points (representing the 10 welfare workers), in which the 3 workers giving illegal deductions are denoted by the

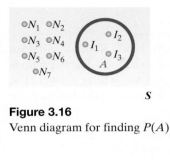

Figure 3.16
Venn diagram for finding $P(A)$

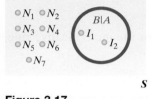

Figure 3.17
Venn diagram for finding $P(B|A)$

symbol I (I_1, I_2, I_3) and the 7 workers not giving illegal deductions are denoted by the symbol N ($N_1, \ldots, N_7$). The resulting Venn diagram is illustrated in Figure 3.16.

Since the first worker is selected at random from the 10, it is reasonable to assign equal probabilities to the 10 sample points. Thus, each sample point has a probability of $1/10$. The sample points in event A are $\{I_1, I_2, I_3\}$—the three workers who are giving illegal deductions. Thus,

$$P(A) = P(I_1) + P(I_2) + P(I_3) = \frac{1}{10} + \frac{1}{10} + \frac{1}{10} = \frac{3}{10}$$

To find the conditional probability $P(B|A)$, we need to alter the sample space **S**. Since we know that A has occurred (i.e., the first worker selected is giving illegal deductions), only 2 of the 9 remaining workers in the sample space are giving illegal deductions. The Venn diagram for this new sample space is shown in Figure 3.17.

Each of these nine sample points is equally likely, so each is assigned a probability of $1/9$. Since the event $B|A$ contains the sample points $\{I_1, I_2\}$, we have

$$P(B|A) = P(I_1) + P(I_2) = \frac{1}{9} + \frac{1}{9} = \frac{2}{9}$$

Substituting $P(A) = 3/10$ and $P(B|A) = 2/9$ into the formula for the multiplicative rule, we find that

$$P(A \cap B) = P(A)P(B|A) = \left(\frac{3}{10}\right)\left(\frac{2}{9}\right) = \frac{6}{90} = \frac{1}{15}$$

Thus, there is a 1-in-15 chance that both workers chosen by the supervisor have been giving illegal deductions to food stamp recipients.

Look Back The key words *both* and *and* in the statement "both A and B occur" imply an intersection of two events. This, in turn, implies that we should *multiply* probabilities to obtain the answer.

Now Work Exercise 3.71

The sample-space approach is only one way to solve the problem posed in Example 3.17. An alternative method employs the **tree diagram** (introduced in Example 3.1). Tree diagrams are helpful for calculating the probability of an intersection.

To illustrate, a tree diagram for Example 3.17 is displayed in Figure 3.18. The tree begins at the far left with two branches. These branches represent the two possible outcomes N (no illegal deductions) and I (illegal deductions) for the first worker selected. The unconditional probability of each outcome is given (in parentheses) on the appropriate branch. That is, for the first worker selected, $P(N) = 7/10$ and $P(I) = 3/10$. (These unconditional probabilities can be obtained by summing sample point probabilities as in Example 3.17.)

The next level of the tree diagram (moving to the right) represents the outcomes for the second worker selected. The probabilities shown here are conditional probabilities, since the outcome for the first worker is assumed to be known. For example, if the first worker is giving illegal deductions (I), the probability that the second worker is also giving illegal deductions (I) is $2/9$, because, of the 9 workers left to be selected, only 2 remain who are giving illegal deductions. This conditional probability, $2/9$, is shown in parentheses on the bottom branch of Figure 3.18.

Finally, the four possible outcomes of the experiment are shown at the end of each of the four tree branches. These events are intersections of two events (outcome of first worker *and* outcome of second worker). Consequently, the multiplicative rule is applied to calculate each probability, as shown in Figure 3.18. You can see that the intersection $\{I \cap I\}$—the event that both workers selected are giving illegal deductions—has probability $6/90 = 1/15$, which is the same as the value obtained in Example 3.17.

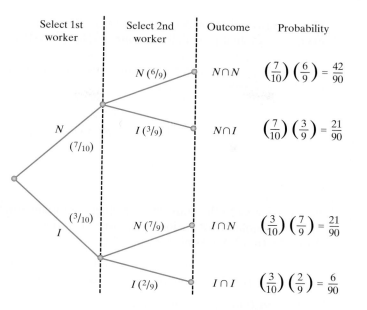

Figure 3.18
Tree diagram for Example 3.17

In Section 3.5, we showed that the probability of event A may be substantially altered by the knowledge that an event B has occurred. However, this will not always be the case; in some instances, the assumption that event B has occurred will *not* alter the probability of event A at all. When this occurs, we say that the two events A and B are *independent events*.

Events A and B are **independent events** if the occurrence of B does not alter the probability that A has occurred; that is, events A and B are independent if

$$P(A|B) = P(A)$$

When events A and B are independent, it is also true that

$$P(B|A) = P(B)$$

Events that are not independent are said to be **dependent.**

Example 3.18
Checking for Independence— Die-Tossing Experiment

Problem Consider the experiment of tossing a fair die, and let

$$A = \{\text{Observe an even number.}\}$$
$$B = \{\text{Observe a number less than or equal to 4.}\}$$

Are A and B independent events?

Solution The Venn diagram for this experiment is shown in Figure 3.19. We first calculate

$$P(A) = P(2) + P(4) + P(6) = \frac{1}{2}$$

$$P(B) = P(1) + P(2) + P(3) + P(4) = \frac{4}{6} = \frac{2}{3}$$

$$P(A \cap B) = P(2) + P(4) = \frac{2}{6} = \frac{1}{3}$$

Now, assuming that B has occurred, we see that the conditional probability of A given B is

$$P(A|B) = \frac{P(A \cap B)}{P(B)} = \frac{1/3}{2/3} = \frac{1}{2} = P(A)$$

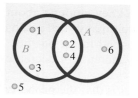

Figure 3.19
Venn diagram for die-toss experiment

Thus, assuming that the occurence of event B does not alter the probability of observing an even number, that probability remains $\frac{1}{2}$. Therefore, the events A and B are independent.

Look Back Note that if we calculate the conditional probability of B given A, our conclusion is the same:

$$P(B|A) = \frac{P(A \cap B)}{P(A)} = \frac{1/3}{1/2} = \frac{2}{3} = P(B)$$

**Example 3.19
Checking for Independence—Consumer Complaint Study**

Problem Refer to the consumer product complaint study in Example 3.15. The percentages of complaints of various types during and after the guarantee period are shown in Table 3.6 (p. 138). Define the following events:

A: {The cause of the complaint is the appearance of the product.}
B: {The complaint occurred during the guarantee term.}

Are A and B independent events?

Solution Events A and B are independent if $P(A|B) = P(A)$. In Example 3.15, we calculated $P(A|B)$ to be .51, and from Table 3.6, we see that

$$P(A) = .32 + .03 = .35$$

Therefore, $P(A|B)$ is not equal to $P(A)$, and A and B are dependent events.

Now Work Exercise 3.67c

BLAISE PASCAL (1623–1662)

Solver of the Chevalier's Dilemma

As a precocious child growing up in France, Blaise Pascal showed an early inclination toward mathematics. Although his father would not permit Pascal to study mathematics before the age of 15 (removing all math texts from his house), at age 12 Blaise discovered on his own that the sum of the angles of a triangle are two right angles.

Pascal went on to become a distinguished mathematician, as well as a physicist, a theologian, and the inventor of the first digital calculator. Most historians attribute the beginning of the study of probability to the correspondence between Pascal and Pierre de Fermat in 1654. The two solved the Chevalier's Dilemma, a gambling problem related to Pascal by his friend and Paris gambler the Chevalier de Mere. The problem involved determining the expected number of times one could roll two dice without throwing a double 6. (Pascal proved that the "break-even" point was 25 rolls.) ∎

To gain an intuitive understanding of independence, think of situations in which the occurrence of one event does not alter the probability that a second event will occur. For example, new medical procedures are often tested on laboratory animals. The scientists conducting the tests generally try to perform the procedures on the animals so that the results for one animal do not affect the results for the others. That is, the event that the procedure is successful on one animal is *independent* of the result for another. In this way, the scientists can get a more accurate idea of the efficacy of the procedure than if the results were dependent, with the success or failure for one animal affecting the results for other animals.

As a second example, consider an election poll in which 1,000 registered voters are asked their preference between two candidates. Pollsters try to use procedures for selecting a sample of voters so that the responses are independent. That is, the objective of the pollster is to select the sample so that one polled voter's preference for candidate A does not alter the probability that a second polled voter prefers candidate A.

Now consider the world of sports. Do you think the results of a batter's successive trips to the plate in baseball, or of a basketball player's successive shots at the basket, are independent? For example, if a basketball player makes two successive shots, is the probability of making the next shot altered from its value if the result of the first shot is not known? If a player makes two shots in a row, the probability of a third successful shot is likely to be different from what we would assign if we knew nothing about the first two shots. Why should this be so? Research has shown that many such results in sports tend to be *dependent* because players (and even teams) tend to get on "hot" and "cold" streaks, during which their probabilities of success may increase or decrease significantly.

We will make three final points about independence. The first is that the property of independence, unlike the property of mutual exclusivity generally cannot be shown on, or gleaned from, a Venn diagram. This means that *you can't trust your intuition.* In general, the only way to check for independence is by performing the calculations of the probabilities in the definition.

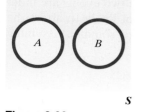

Figure 3.20

Mutually exclusive events are dependent events

The second point concerns the relationship between the properties of mutual exclusivity and independence. Suppose that events A and B are mutually exclusive, as shown in Figure 3.20, and that both events have nonzero probabilities. Are these events independent or dependent? That is, does the assumption that B occurs alter the probability of the occurrence of A? It certainly does, because if we assume that B has occurred, it is impossible for A to have occurred simultaneously. That is, $P(A|B) = 0$. Thus, *mutually exclusive events are dependent events*, since $P(A) \neq P(A|B)$.

The third point is that the probability of the intersection of independent events is very easy to calculate. Referring to the formula for calculating the probability of an intersection, we find that

$$P(A \cap B) = P(A)P(B|A)$$

Thus, since $P(B|A) = P(B)$ when A and B are independent, we have the following useful rule:

Probability of Intersection of Two Independent Events

If events A and B are independent, then the probability of the intersection of A and B equals the product of the probabilities of A and B; that is,

$$P(A \cap B) = P(A)P(B)$$

The converse is also true: If $P(A \cap B) = P(A)P(B)$, then events A and B are independent.

In the die-toss experiment, we showed in Example 3.18 that the two events A: {Observe an even number.} and B: {Observe a number less than or equal to 4.} are independent if the die is fair. Thus,

$$P(A \cap B) = P(A)P(B) = \left(\frac{1}{2}\right)\left(\frac{2}{3}\right) = \frac{1}{3}$$

This result agrees with the one we obtained in the example:

$$P(A \cap B) = P(2) + P(4) = \frac{2}{6} = \frac{1}{3}$$

Example 3.20

Probability of Independent Events Occurring Simultaneously— Divorced Couples Study

Problem Recall from Example 3.5 (p. 116) that the American Association for Marriage and Family Therapy (AAMFT) found that 25% of divorced couples are classified as "fiery foes" (i.e., they communicate through their children and are hostile toward each other).

a. What is the probability that in a sample of 2 divorced couples, both are classified as "fiery foes?"

b. What is the probability that in a sample of 10 divorced couples, all 10 are classified as "fiery foes?"

Solution

a. Let F_1 represent the event that divorced couple 1 is classified as a pair of "fiery foes" and F_2 represent the event that divorced couple 2 is also classified as a pair of "fiery foes." Then the event that *both* couples are "fiery foes" is the intersection of the two events, $F_1 \cap F_2$. On the basis of the AAMFT survey which found that 25% of divorced couples are "fiery foes," we could reasonably conclude that $P(F_1) = .25$ and $P(F_2) = .25$. However, in order to compute the probability of $F_1 \cap F_2$ from the multiplicative rule, we must make the assumption that the two events are independent. Since the classification of any divorced couple is not likely to affect the classification of another divorced couple, this assumption is reasonable. Assuming independence, we have

$$P(F_1 \cap F_2) = P(F_1)P(F_2) = (.25)(.25) = .0625$$

b. To see how to compute the probability that 10 of 10 divorced couples will all be classified as "fiery foes," first consider the event that 3 of 3 couples are "fiery foes." If F_3 represents the event that the third divorced couple are "fiery foes," then we want to compute the probability of the intersection $F_1 \cap F_2 \cap F_3$. Again assuming independence of the classifications, we have

$$P(F_1 \cap F_2 \cap F_3) = P(F_1)P(F_2)P(F_3) = (.25)(.25)(.25) = .015625$$

Similar reasoning leads us to the conclusion that the intersection of 10 such events can be calculated as follows:

$$P(F_1 \cap F_2 \cap F_3 \cap \ldots \cap F_{10}) = P(F_1)P(F_2)\ldots P(F_{10}) = (.25)^{10} = .000001$$

Thus, the probability that 10 of 10 divorced couples sampled are all classified as "fiery foes" is 1 in 1 million, assuming that the probability of each couple being classified as "fiery foes" is .25 and that the classification decisions are independent.

Look Back The very small probability in part **b** makes it extremely unlikely that 10 of 10 divorced couples are "fiery foes." If this event should actually occur, we should question the probability of .25 provided by the AAMFT and used in the calculation—it is likely to be much higher. (This conclusion is another application of the rare-event approach to statistical inference.)

Now Work Exercise 3.83

Statistics IN Action | **Revisited**   **The Probability of Winning Cash 3 or Play 4**

In addition to the biweekly Lotto 6/53, the Florida Lottery runs several other games. Two popular daily games are Cash 3 and Play 4. In Cash 3, players pay $1 to select three numbers in order, where each number ranges from 0 to 9. If the three numbers selected (e.g., 2–8–4) match exactly the order of the three numbers drawn, the player wins $500. Play 4 is similar to Cash 3, but players must match four numbers (each number ranging from 0 to 9). For a $1 Play 4 ticket (e.g., 3–8–3–0), the player will win $5,000 if the numbers match the order of the four numbers drawn.

During the official drawing for Cash 3, 10 table tennis balls numbered 0, 1, 2, 3, 4, 5, 6, 7, 8, and 9 are placed into each of three chambers. The balls in the first chamber are colored pink, the balls in the second chamber are blue, and the balls in the third chamber are yellow. One ball of each color is randomly drawn, with the official order as pink–blue–yellow. In Play 4, a fourth chamber with orange balls is added, and the official order is pink–blue–yellow–orange. Since the draws of the colored balls are random

and independent, we can apply an extension of the probability rule for the intersection of two independent events to find the odds of winning Cash 3 and Play 4. The probability of matching a numbered ball being drawn from a chamber is 1/10; therefore,

$$
\begin{aligned}
P(\text{Win Cash 3}) &= P(\text{match pink } and \text{ match blue} \\
&\quad and \text{ match yellow}) \\
&= P(\text{match pink}) \times P(\text{match blue}) \times \\
&\quad P(\text{match yellow}) \\
&= (1/10)(1/10)(1/10) = 1/1000 = .001
\end{aligned}
$$

$$
\begin{aligned}
P(\text{Win Play 4}) &= P(\text{match pink } and \text{ match blue} \\
&\quad \text{and match yellow } and \text{ match orange}) \\
&= P(\text{match pink}) \times P(\text{match blue}) \times \\
&\quad P(\text{match yellow}) \times P(\text{match orange}) \\
&= (1/10)(1/10)(1/10) \\
&= 1/10,000 = .0001
\end{aligned}
$$

Statistics IN Action
(continued)

Although the odds of winning one of these daily games is much better than the odds of winning Lotto 6/53, there is still only a 1 in 1,000 chance (for Cash 3) or 1 in 10,000 chance (for Play 4) of winning the daily game. And the payoffs ($500 or $5,000) are much smaller. In fact, it can be shown that you will lose an average of 50¢ every time you play either Cash 3 or Play 4!

Exercises 3.63–3.97

Understanding the Principles

3.63 Explain the difference between an unconditional probability and a conditional probability.

3.64 Give the multiplicative rule of probability for
a. two independent events.
b. any two events.

3.65 Give the formula for finding $P(B|A)$.

3.66 Defend or refute each of the following statements:
a. Dependent events are always mutually exclusive.
b. Mutually exclusive events are always dependent.
c. Independent events are always mutually exclusive.

Learning the Mechanics

3.67 For two events A and B, $P(A) = .4, P(B) = .2$, and
[NW] $P(A \cap B) = .1$.
a. Find $P(A|B)$.
b. Find $P(B|A)$.
c. Are A and B independent events?

3.68 For two events A and B, $P(A) = .4, P(B) = .2$, and $P(A|B) = .6$.
a. Find $P(A \cap B)$.
b. Find $P(B|A)$.

3.69 For two independent events A and B, $P(A) = .4$ and $P(B) = .2$.
a. Find $P(A \cap B)$.
b. Find $P(A|B)$.
c. Find $P(A \cup B)$.

3.70 An experiment results in one of three mutually exclusive events A, B, and C. It is known that $P(A) = .30, P(B) = .55$, and $P(C) = .15$, Find each of the following probabilities:
a. $P(A \cup B)$
b. $P(A \cap B))$
c. $P(A|B)$
d. $P(B \cup C))$
e. Are B and C independent events? Explain.

3.71 Consider the experiment defined by the accompanying
[NW] Venn diagram, with the sample space S containing five sample points. The sample points are assigned the following probabilities: $P(E_1) = .1, P(E_2) = .1, P(E_3) = .2, P(E_4) = .5, P(E_5) = .1$.

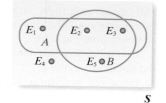

S

a. Calculate $P(A), P(B)$, and $P(A \cap B)$.
b. Suppose we know that event A has occurred, so the reduced sample space consists of the three sample points in A: E_1, E_2, and E_3. Use the formula for conditional probability to determine the probabilities of these three sample points given that A has occurred. Verify that the conditional probabilities are in the same ratio to one another as the original sample point probabilities and that they sum to 1.
c. Calculate the conditional probability $P(B|A)$ in two ways: First, sum $P(E_2|A)$ and $P(E_3|A)$, since these sample points represent the event that B occurs given that A has occurred. Second, use the formula for conditional probability:

$$P(B|A) = \frac{P(A \cap B)}{P(A)}$$

Verify that the two methods yield the same result.

3.72 An experiment results in one of five sample points with the following probabilities: $P(E_1) = .22, P(E_2) = .31, P(E_3) = .15, P(E_4) = .22$, and $P(E_5) = .1$. The following events have been defined:

$$A: \{E_1, E_3\}$$
$$B: \{E_2, E_3, E_4\}$$
$$C: \{E_1, E_5\}$$

Find each of the following probabilities:
a. $P(A)$
b. $P(B)$
c. $P(A \cap B)$
d. $P(A|B)$
e. $P(B \cap C)$
f. $P(C|B)$
g. Consider each pair of events A and B, A and C, and B and C. Are any of the pairs of events independent? Why?

3.73 Two fair dice are tossed, and the following events are defined:

$A: \{$The sum of the numbers showing is odd.$\}$
$B: \{$The sum of the numbers showing is 9, 11, or 12.$\}$

Are events A and B independent? Why?

3.74 A sample space contains six sample points and events A, B, and C, as shown in the Venn diagram on page 146. The probabilities of the sample points are $P(1) = .20, P(2) = .05, P(3) = .30, P(4) = .10, P(5) = .10$, and $P(6) = .25$.

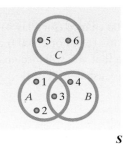

S

a. Which pairs of events, if any, are mutually exclusive? Why?
b. Which pairs of events, if any, are independent? Why?
c. Find $P(A \cup B)$ by adding the probabilities of the sample points and then by using the additive rule. Verify that the answers agree. Repeat for $P(A \cup C)$.

3.75 A box contains two white, two red, and two blue poker chips. Two chips are randomly chosen without replacement, and their colors are noted. Define the following events:

> *A*: {Both chips are of the same color.}
> *B*: {Both chips are red.}
> *C*: {At least one chip is red or white.}

Find $P(B|A)$, $P(B|A^c)$, $P(B|C)$, $P(A|C)$, and $P(C|A^c)$.

💿 Applet Exercise 3.5

Use the applet entitled *Simulating the Probability of Rolling a 6* to simulate conditional probabilities. Begin by running the applet twice with $n = 10$, without resetting between runs. The data on your screen represent 20 rolls of a die. The diagram above the *Roll* button shows the frequency of each of the six possible outcomes. Use this information to find each of the following probabilities:
a. The probability of 6 given that the outcome is 5 or 6
b. The probability of 6 given that the outcome is even
c. The probability of 4 or 6 given that the outcome is even
d. The probability of 4 or 6 given that the outcome is odd

Applying the Concepts—Basic

3.76 **Do you have a library card?** According to a September 2008 Harris poll, 68% of all American adults have a public library card. The percentages do differ by gender—62% of males have library cards compared to 73% of females.
a. Consider the three probabilities: $P(A) = .68$, $P(A|B) = .62$, and $P(A|C) = .73$. Define events *A*, *B*, and *C*.
b. Assuming that half of all American adults are males and half are females, what is the probability that an American adult is a female who owns a library card?

3.77 **National firearms survey.** The Harvard School of Public Health conducted a study of privately held firearm stock in the United States. In a representative household telephone survey of 2,770 adults, 26% reported that they own at least one gun (*Injury Prevention,* Jan. 2007). Of those that own a gun, 5% own a handgun. Suppose one of the 2,770 adults surveyed is randomly selected.
a. What is the probability that the adult owns at least one gun?

b. What is the probability that the adult owns at least one gun and the gun is a handgun?

3.78 **Guilt in decision making.** Refer to the *Journal of Behavioral Decision Making* (Jan. 2007) study of the effect of guilt emotion on how a decision maker focuses on the problem, Exercise 3.55 (p. 133). The results (number responding in each category) for the 171 study participants are reproduced in the accompanying table. Suppose one of the 171 participants is selected at random.

Emotional State	Choose Stated Option	Do Not Choose Stated Option	Totals
Guilt	45	12	57
Anger	8	50	58
Neutral	7	49	56
Totals	60	111	171

Based on Gangemi, A., and Mancini, F. "Guilt and focusing in decision-making." *Journal of Behavioral Decision Making,* Vol. 20, Jan. 2007 (Table 2).

a. Given the respondent is assigned to the guilty state, what is the probability that the respondent chooses the stated option?
b. If the respondent does not choose to repair the car, what is the probability of the respondent being in the anger state?
c. Are the events {repair the car} and {guilty state} independent?

3.79 **Speeding linked to fatal car crashes.** According to the National Highway Traffic and Safety Administration's National Center for Statistics and Analysis (NCSA), "Speeding is one of the most prevalent factors contributing to fatal traffic crashes" (*NHTSA Technical Report,* Aug. 2005). The probability that speeding is a cause of a fatal crash is .3. Furthermore, the probability that speeding and missing a curve are causes of a fatal crash is .12. Given that speeding is a cause of a fatal crash, what is the probability that the crash occurred on a curve?

3.80 **Twitter and wireless Internet.** According to the *Pew Internet & American Life Project* (Oct. 2009), 54% of Internet users have a wireless connection to the Internet via a laptop, cell phone, game console, or other mobile device. Of these wireless users, 25% use Twitter to share updates about themselves. What is the probability that an Internet user has a wireless connection and uses Twitter?

3.81 **Appraisals and negative emotions.** According to psychological theory, each negative emotion one has is linked to some thought (appraisal) about the environment. This theory was tested in the journal *Cognition and Emotion* (Vol. 24, 2010). Over a period of two days, undergraduate students were asked to respond to events that occurred within the past 15 minutes. Each student rated the negative emotions he or she felt (e.g., anger, sadness, fear, or guilt) and gave an appraisal of the event (e.g., unpleasant or unfair). A total of 10,797 responses was evaluated in the study, of which 594 indicated a perceived unfairness appraisal. Of these unfairness appraisals, the students reported feeling angry in 127 of them. Consider a response randomly selected from the study.
a. What is the probability that the response indicates a perceived unfairness appraisal?
b. Given a perceived unfairness appraisal, what is the probability of an angry emotion?

3.82 **Study of ancient pottery.** Refer to the *Chance* (Fall 2000) study of ancient pottery found at the Greek settlement of Phylakopi, presented in Exercise 2.14 (p. 36). Of the 837 pottery pieces uncovered at the excavation site, 183 were painted. These painted pieces included 14 painted in a curvilinear decoration, 165 painted in a geometric decoration, and 4 painted in a naturalistic decoration. Suppose 1 of the 837 pottery pieces is selected and examined.

   a. What is the probability that the pottery piece is painted?

   b. Given that the pottery piece is painted, what is the probability that it is painted in a curvilinear decoration?

Applying the Concepts—Intermediate

3.83 **Health risks to beachgoers.** Refer to the University of Florida study of health risks to beachgoers, Exercise 3.22 (p. 121). According to the study, 6 out of 1,000 people exposed to wet sand for a 10-minute period will acquire gastroenteritis and 7 out of 1,000 people exposed to ocean water for a 10-minute period will acquire gastroenteritis. Suppose you and a friend go to the beach. Over the next 10 minutes, you play in the wet sand while your friend swims in the ocean. What is the probability that at least one of you acquires gastroenteritis? (What assumption did you need to make in order to find the probability?)

3.84 **Sleep apnea and sleep stage transitioning.** Refer to the sleep apnea study published in *Chance* (Winter 2009), Exercise 3.54 (p. 132). Recall that the various stages of sleep for a large group of sleep apnea patients were monitored in 30-second "epochs," and the sleep stage for the previous and current epoch was determined. A summary table is reproduced below. One of the epochs in the study is selected.

	Previous Sleep Stage		
Current Stage	**Non-REM**	**REM**	**Wake**
Non-REM	31,880	160	1,733
REM	346	7,609	175
Wake	1,588	358	6,079
Totals	33,814	8,127	7,987

Based on Caffo, B. S., et al. "An overview of observational sleep research with application to sleep stage transitioning." *Chance*, Vol. 22, No. 1, Winter 2009 (Table 2).

   a. Given that the previous sleep stage for the epoch was the Wake state, what is the probability that the current sleep stage is REM?

   b. Given that the current sleep stage for the epoch is REM, what is the probability that the previous sleep stage was not the Wake state?

   c. Are the events {previous stage is REM} and {current stage is REM} mutually exclusive?

   d. Are the events {previous stage is REM} and {current stage is REM} independent?

   e. Are the events {previous stage is Wake} and {current stage is Wake} independent?

3.85 **Are you really being served red snapper?** Red snapper is a rare and expensive reef fish served at upscale restaurants. Federal law prohibits restaurants from serving a cheaper, look-alike variety of fish (e.g., vermillion snapper or lane snapper) to customers who order red snapper. Researchers at the University of North Carolina used DNA analysis to examine fish specimens labeled "red snapper" that were purchased form vendors across the country (*Nature*, July 15, 2004). The DNA tests revealed that 77% of the specimens were not red snapper, but the cheaper, look-alike variety of fish.

   a. Assuming that the results of the DNA analysis are valid, what is the probability that you are actually served red snapper the next time you order it at a restaurant?

   b. If there are five customers at a restaurant, all who have ordered red snapper, what is the probability that at least one customer is actually served red snapper?

3.86 **Fighting probability of fallow deer bucks.** Refer to the *Aggressive Behavior* (Jan./Feb. 2007) study of fallow deer bucks fighting during the mating season, presented in Exercise 3.57 (p. 133). Recall that researchers recorded 167 encounters between two bucks, one of which clearly initiated the encounter with the other. A summary of the fight status of the initiated encounters is provided in the accompanying table. Suppose we select 1 of these 167 encounters and note the outcome (fight status and winner).

   a. Given that a fight occurs, what is the probability that the initiator wins?

   b. Given no fight, what is the probability that the initiator wins?

   c. Are the events "no fight" and "initiator wins" independent?

	Initiator Wins	No Clear Winner	Initiator Loses	Totals
Fight	26	23	15	64
No Fight	80	12	11	103
Totals	106	35	26	167

Based on Bartos, L. et al. "Estimation of the probability of fighting in fallow deer (*Dama dama*) during the rut." *Aggressive Behavior*, Vol. 33, Jan./Feb., 2007, pp. 7–13.

3.87 **Extinct New Zealand birds.** Refer to the *Evolutionary Ecology Research* (July 2003) study of the patterns of extinction in the New Zealand bird population, presented in Exercise 2.20 (p. 37). Consider the data on extinction status (extinct, absent from island, present) for the 132 bird species. The data are saved in the **NZBIRDS** file and are summarized in the accompanying MINITAB printout. Suppose you randomly select 10 of the 132 bird species (without replacement) and record the extinction status of each.

   a. What is the probability that the first species you select is extinct? (*Note:* Extinct = Yes on the MINITAB printout.)

   b. Suppose the first 9 species you select are all extinct. What is the probability that the 10th species you select is extinct?

Tally for Discrete Variables: Extinct

Extinct	Count	Percent
Absent	16	12.12
No	78	59.09
Yes	38	28.79
N=	132	

3.88 Probability of winning a war. Before a country enters into a war, a prudent government will assess the cost, utility, and probability of a victory. University of Georgia professor P. L. Sullivan has developed a statistical model for determining the probability of winning a war based on a government's capabilities and resources (*Journal of Conflict Resolution,* Vol. 51, 2007). Now consider the current U.S.–Iraq conflict. One researcher used the model to estimate that the probability of a successful regime change in Iraq was .70 prior to the start of the war. Of course, we now know that the successful regime change was achieved. However, the model also estimates that given the mission is extended to support a weak Iraq government, the probability of ultimate success is only .26. Assume these probabilities are accurate.

  a. Prior to the start of the U.S.–Iraq war, what is the probability that a successful regime change is not achieved?

  b. Given that the mission is extended to support a weak Iraq government, what is the probability that a successful regime change is ultimately achieved?

  c. Suppose the probability of the United States extending the mission to support a weak Iraq government was .55. Find the probability that the mission is extended and results in a successful regime change.

3.89 Ambulance response time. *Geographical Analysis* (Jan. 2010) presented a study of Emergency Medical Services (EMS) ability to meet the demand for an ambulance. In one example, the researchers presented the following scenario. An ambulance station has one vehicle and two demand locations, A and B. The probability that the ambulance can travel to a location in under eight minutes is .58 for location A and .42 for location B. The probability that the ambulance is busy at any point in time is .3.

  a. Find the probability that EMS can meet demand for an ambulance at location A.

  b. Find the probability that EMS can meet demand for an ambulance at location B.

3.90 Intrusion detection systems. A computer intrusion detection system (IDS) is designed to provide an alarm whenever someone intrudes (e.g., through unauthorized access) into a computer system. A probabilistic evaluation of a system with two independently operating intrusion detection systems (a double IDS) was published in the *Journal of Research of the National Institute of Standards and Technology* (Nov.–Dec. 2003). Consider a double IDS with system A and system B. If there is an intruder, system A sounds an alarm with probability .9 and system B sounds an alarm with probability .95. If there is no intruder, the probability that system A sounds an alarm (i.e., a false alarm) is .2 and the probability that system B sounds an alarm is .1.

  a. Use symbols to express the four probabilities just given.

  b. If there is an intruder, what is the probability that both systems sound an alarm?

  c. If there is no intruder, what is the probability that both systems sound an alarm?

  d. Given that there is an intruder, what is the probability that at least one of the systems sounds an alarm?

3.91 Detecting traces of TNT. University of Florida researchers in the Department of Materials Science and Engineering have invented a technique that rapidly detects traces of TNT (*Today*, Spring 2005). The method, which involves shining a laser on a potentially contaminated object, provides instantaneous results and gives no false positives. In this application, a false positive would occur if the laser detected traces of TNT when, in fact, no TNT were actually present on the object. Let A be the event that the laser light detects traces of TNT. Let B be the event that the object contains no traces of TNT. The probability of a false positive is 0. Write this probability in terms of A and B, using symbols such as "∪", "∩", and " | ".

3.92 Forest fragmentation study. Refer to the *Conservation Ecology* (Dec. 2003) study of the causes of forest fragmentation, presented in Exercise 2.156 (p. 91). Recall that the researchers used advanced high-resolution satellite imagery to develop fragmentation indexes for each forest. A 3×3 grid was superimposed over an aerial photo of the forest, and each square (pixel) of the grid was classified as forest (F), as earmarked for anthropogenic land use (A), or as natural land cover (N). An example of one such grid is shown here. The edges of the grid (where an "edge" is an imaginary line that separates any two adjacent pixels) are classified as F–A, F–N, A–A, A–N, N–N, or F–F edges.

A	A	N
N	F	F
N	F	F

  a. Note that there are 12 edges inside the grid. Classify each edge as F–A, F–N, A–A, A–N, N–N or F–F.

  b. The researchers calculated the fragmentation index by considering only the F-edges in the grid. Count the number of F-edges. (These edges represent the sample space for the experiment.)

  c. If an F-edge is selected at random, find the probability that it is an F–A edge. (This probability is proportional to the anthropogenic fragmentation index calculated by the researchers.)

  d. If an F edge is selected at random, find the probability that it is an F–N edge. (This probability is proportional to the natural fragmentation index calculated by the researchers.)

3.93 Seeded players at a shuffleboard tournament. Shuffleboard is an outdoor game popular with senior citizens. In single-elimination shuffleboard tournaments, players are paired and play a match against each other, with the winners moving on to play another match. The tournament concludes in a final that matches two undefeated players. The winner of the final match is the overall tournament champion. *Chance* (Winter 2006) investigated the probability of winning a shuffleboard tournament with 64 players. As an example, the authors considered a small tournament with only 4 players, named A, B, C, and D. This tournament involves a total of three matches. In the first round, A plays B in one match and C plays D in the other match. Then the winners play in the final round.

a. List the different outcomes of the tournament. (For example, one outcome is "*A* and *C* win first-round matches and *A* wins the final.")

b. If the players are of equal ability, what is the probability that *A* wins the tournament?

c. Suppose the players are not of equal ability. The accompanying table gives the likelihood of one player defeating the other. (*Example:* The probability that *A* defeats *B* is .9.) What is the probability that *A* wins the tournament?

Outcome	Probability
A defeats *B*	.9
A defeats *C*	.7
A defeats *D*	.6
B defeats *C*	.1
B defeats *D*	.2
C defeats *D*	.4

Applying the Concepts—Advanced

3.94 **Random mutation of cells.** *Chance* (Spring 2010) presented an article on the random mutation hypothesis developed by microbiologists. Under this hypothesis, when a wild-type organic cell (e.g., a bacteria cell) divides, there is a chance that at least one of the two "daughter" cells is a mutant. When a mutant cell divides, both offspring will be mutant. The schematic below shows a possible pedigree from a single cell that has divided. Note that one "daughter" cell is mutant (●) and one is a normal cell (○).

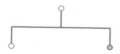

a. Consider a single, normal cell that divides into two offspring. List the different possible pedigrees.

b. Assume that a "daughter" cell is equally likely to be mutant or normal. What is the probability that a single, normal cell that divides into two offspring will result in at least one mutant cell?

c. Now assume that the probability of a mutant "daughter" cell is .2. What is the probability that a single, normal cell that divides into two offspring will result in at least one mutant cell?

d. The schematic below shows a possible second-generation pedigree from a single cell that has divided. Note that the first generation mutant cell automatically produces two mutant cells in the second generation. List the different possible second generation pedigrees. (*Hint:* Use your answer to part **a.**)

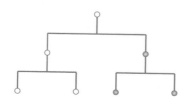

e. Assume that a "daughter" cell is equally likely to be mutant or normal. What is the probability that a single, normal cell that divides into two offspring will result in at least one mutant cell after the second generation?

3.95 **Testing a psychic's ability.** Consider an experiment in which 10 identical small boxes are placed side by side on a table. A crystal is placed at random inside one of the boxes. A self-professed "psychic" is asked to pick the box that contains the crystal.

a. If the "psychic" simply guesses, what is the probability that she picks the box with the crystal?

b. If the experiment is repeated seven times, what is the probability that the "psychic" guesses correctly at least once?

c. A group called the Tampa Bay Skeptics recently tested a self-proclaimed "psychic" by administering the preceding experiment seven times. The "psychic" failed to pick the correct box all seven times (*Tampa Tribune*, Sept. 20, 1998). What would you infer about this person's psychic ability?

3.96 **Risk of a natural-gas pipeline accident.** *Process Safety Progress* (Dec. 2004) published a risk analysis for a natural-gas pipeline between Bolivia and Brazil. The most likely scenario for an accident would be natural-gas leakage from a hole in the pipeline. The probability that the leak ignites immediately (causing a jet fire) is .01. If the leak does not immediately ignite, it may result in the delayed ignition of a gas cloud. Given no immediate ignition, the probability of delayed ignition (causing a flash fire) is .01. If there is no delayed ignition, the gas cloud will disperse harmlessly. Suppose a leak occurs in the natural-gas pipeline. Find the probability that either a jet fire or a flash fire will occur. Illustrate with a tree diagram.

3.97 **NBA draft lottery.** The National Basketball Association (NBA) utilizes a lottery to determine the order in which teams draft amateur (high school and college) players. Teams with the poorest won–lost records have the best chance of obtaining the first selection (and, presumably, best player) in the draft. Under the current lottery system, 14 table tennis balls numbered 1 through 14 are placed in a drum, and 4 balls are drawn. There are 1,001 possible combinations when 4 balls are drawn out of 14 without regard to their order of selection. Prior to the drawing, 1,000 combinations are assigned to the 13 teams with the worst records—the lottery teams—based on their order of finish during the regular season. (The team with the worst record is assigned 250 of these combinations, the team with the second-worst record is assigned 200 of the combinations, the team with the third-worst record is assigned 157 of the combinations, etc. The table on page 150 gives the number of combinations assigned to each lottery team.) Once the 4 balls are drawn to the top of the drum to determine a four-digit combination, the team that has been assigned that combination will receive the first pick. The 4 balls are then placed back in the drum, and the process is repeated to determine the second, third, and subsequent draft picks. (*Note:* If the one unassigned combination is drawn, the balls are drawn to the top again.)

NBA Lottery Team	Number of Combinations
Worst record	250
2nd-worst record	200
3rd-worst record	157
4th-worst record	120
5th-worst record	89
6th-worst record	64
7th-worst record	44
8th-worst record	29
9th-worst record	18
10th-worst record	11
11th-worst record	7
12th-worst record	6
13th-worst record	5

Based on McCann, M. A. "Illegal defense: The irrational economics of banning high school players from the NBA draft." *Virginia Sports and Entertainment Law Journal,* Vol. 3, No. 2, Spring 2004, (Table 5).

a. Demonstrate that there are 1,001 four-number combinations of the numbers 1 through 14.
b. For each of the 13 Lottery teams, find the probability that the team obtains the first pick in the draft.
c. Given that the team with the second-worst record obtains the number-1 pick, find the probability that the team with the worst record obtains the second pick of the draft.
d. Given that the team with the third-worst record obtains the number-1 pick, find the probability that the team with the worst record obtains the second pick of the draft.
e. Find the probability that the team with the worst record obtains the second pick of the draft, assuming that it did not obtain the first pick.

3.7 Random Sampling

How a sample is selected from a population is of vital importance in statistical inference because the probability of an observed sample will be used to infer the characteristics of the sampled population. To illustrate, suppose you deal yourself four cards from a deck of 52 cards, and all four cards are aces. Do you conclude that your deck is an ordinary bridge deck, containing only four aces, or do you conclude that the deck is stacked with more than four aces? It depends on how the cards were drawn. If the four aces are always placed at the top of a standard bridge deck, drawing four aces is not unusual—it is certain. If, however, the cards are thoroughly mixed, drawing four aces in a sample of four cards is highly improbable. The point, of course, is that in order to use the observed sample of four cards to draw inferences about the population (the deck of 52 cards), you need to know how the sample was selected from the deck.

One of the simplest and most frequently employed sampling procedures is implied in the previous examples and exercises. It produces what is known as a *random sample*. We learned in Section 1.5 (p. 11) that a random sample is likely to be *representative* of the population that it is selected from.

> If *n* elements are selected from a population in such a way that every set of *n* elements in the population has an equal probability of being selected, then the *n* elements are said to be a **random sample.***

If a population is not too large and the elements can be numbered on slips of paper, poker chips, etc., you can physically mix the slips of paper or chips and remove *n* elements from the total. The numbers that appear on the chips selected would indicate the population elements to be included in the sample. Since it is often difficult to achieve a thorough mix, such a procedure only provides an approximation to random sampling. Most researchers rely on **random-number generators** to automatically generate a random sample. Random-number generators are available in tabular form, are built into most statistical software packages, and are available on the Internet (e.g., at www.random.org).

---

*Strictly speaking, this is a *simple random sample*. There are many different types of random samples. The simple random sample is the most common.

Example 3.21

Selecting a Random Sample—5 Households from 100,000

Problem Suppose you wish to randomly sample five households from a population of 100,000 households.

a. How many different samples can be selected?
b. Use a random-number generator to select a random sample.

Solution

a. Since we want to select $n = 5$ objects (households) from $N = 100,000$, we apply the combinations rule of Section 3.1:

$$\binom{N}{n} = \binom{100,000}{5} = \frac{100,000!}{5!\,99,995!}$$
$$= \frac{100,000 \cdot 99,999 \cdot 99,998 \cdot 99,997 \cdot 99,996}{5 \cdot 4 \cdot 3 \cdot 2 \cdot 1}$$
$$= 8.33 \times 10^{22}$$

Thus, there are 83.3 billion trillion different samples of five households that can be selected from 100,000.

b. To ensure that each of the possible samples has an equal chance of being selected, as is required for random sampling, we can employ a **random-number table**, such as Table I of Appendix A. Random-number tables are constructed in such a way that every number occurs with (approximately) equal probability. Furthermore, the occurrence of any one number in a position is independent of any of the other numbers that appear in the table. To use a table of random numbers, number the N elements in the population from 1 to N. Then turn to Table I and select a starting number in the table. Proceeding from this number either across the row or down the column, remove and record n numbers from the table.

To illustrate, first we number the households in the population from 1 to 100,000. Then we turn to a page of Table I—say, the first page. (A partial reproduction of the first page of Table I is shown in Table 3.7.) Now we arbitrarily select a starting number—say, the random number appearing in the third row, second column. This number is 48,360. Then we proceed down the second column to obtain the remaining four random numbers. In this case, we have selected five random numbers, which are highlighted in Table 3.7. Using the first five digits to represent households from 1 to 99,999 and the number 00000 to represent household 100,000, we can see that the households numbered

48,360   93,093   39,975   6,907   72,905

should be included in our sample.

Table 3.7 Partial Reproduction of Table I in Appendix A

Row/Column	1	2	3	4	5	6
1	10480	15011	01536	02011	81647	91646
2	22368	46573	25595	85393	30995	89198
3	24130	48360	22527	97265	76393	64809
4	42167	93093	06243	61680	07856	16376
5	37670	39975	81837	16656	06121	91782
6	77921	06907	11008	42751	27756	53498
7	99562	72905	56420	69994	98872	31016
8	96301	91977	05463	07972	18876	20922
9	89579	14342	63661	10281	17453	18103
10	85475	36857	53342	53988	53060	59533
11	28918	69578	88231	33276	70997	79936
12	63553	40961	48235	03427	49626	69445
13	09429	93969	52636	92737	88974	33488

Note: Use only the necessary number of digits in each random number to identify the element to be included in the sample. If, in the course of recording the *n* numbers from the table, you select a number that has already been selected, simply discard the duplicate and select a replacement at the end of the sequence. Thus, you may have to record more than *n* numbers from the table to obtain a sample of *n* unique numbers.

Look Back Can we be sure that all 83.3 billion trillion samples have an equal chance of being selected? The fact is, we can't; but to the extent that the random number table contains truly random sequences of digits, the sample should be very close to random.

Now Work Exercise 3.102

Table I in Appendix A is just one example of a random-number generator. For most scientific studies that require a large random sample, computers are used to generate the random sample. The SAS, MINITAB, and SPSS statistical software packages all have easy-to-use random-number generators.

For example, suppose we required a random sample of *n* = 50 households from the population of 100,000 households in Example 3.21. Here, we might employ the MINITAB random-number generator. Figure 3.21 shows a MINITAB printout listing 50 random numbers (from a population of 100,000). The households with these identification numbers would be included in the random sample.

Figure 3.21

MINITAB worksheet with random sample of 50 households

HOUSE50.MTW ***

	C1	C2	C3	C4	C5
	HouseID1	HouseID2	HouseID3	HouseID4	HouseID5
1	2036	22037	39884	62123	75750
2	3161	22055	44687	62421	80044
3	3930	22226	44689	63749	81857
4	9838	22773	45422	69079	81918
5	9961	22820	48104	69802	91680
6	11623	32989	50226	70895	94987
7	14114	33119	50443	71029	96110
8	17244	35045	50806	72879	97679
9	17598	35922	50813	73962	99232
10	18769	39272	61059	75517	99789
11					

Recall our discussion of designed experiments in Section 1.5 (p. 11). The twin notions of random selection and randomization are key to conducting good research with a designed experiment. The next example illustrates a basic application.

Example 3.22

Randomization In a Designed Experiment— Clinical Trial

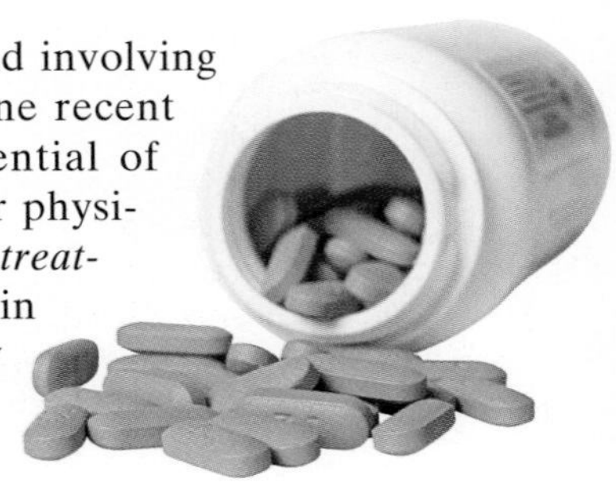

Problem A designed experiment in the medical field involving human subjects is referred to as a *clinical trial*. One recent clinical trial was designed to determine the potential of using aspirin in preventing heart attacks. Volunteer physicians were randomly divided into two groups: the *treatment* group and the *control* group. Each physician in the treatment group took one aspirin tablet a day for one year, while the physicians in the control group took an aspirin-free placebo made to look

⧉ ASPIRIN-STUDY.MTW **

↓	C1	C2	C3
	Physician	Treatment	
1	1	3	
2	2	11	
3	3	10	
4	4	14	
5	5	2	
6	6	7	
7	7	6	
8	8	16	
9	9	17	
10	10	9	
11	11		
12	12		
13	13		
14	14		
15	15		
16	16		
17	17		
18	18		
19	19		
20	20		
21			

Figure 3.22

MINITAB worksheet with random assignment of physicians

identical to an aspirin tablet. Since the physicians did not know to which group—treatment or control—they were assigned, the clinical trial is called a *blind study*. Assume that 20 physicians volunteered for the study. Use a random-number generator to randomly assign half of the physicians to the treatment group and half to the control group.

Solution   Essentially, we want to select a random sample of 10 physicians from the 20. The first 10 selected will be assigned to the treatment group; the remaining 10 will be assigned to the control group. (Alternatively, we could randomly assign each physician, one by one, to either the treatment or control group. However, this would not guarantee exactly 10 physicians in each group.)

The MINITAB random-sample procedure was employed, producing the printout shown in Figure 3.22. Numbering the physicians from 1 to 20, we see that physicians 3, 11, 10, 14, 2, 7, 6, 16, 17, and 9 are assigned to receive the aspirin (the treatment). The remaining physicians are assigned the placebo (the control).

Now Work Exercise 3.107

Ethics IN Statistics   Intentionally selecting a nonrandom sample in an effort to support a particular viewpoint is considered *unethical statistical practice*.

Exercises 3.98–3.111

Understanding the Principles

3.98 Define a random sample.

3.99 How are random samples related to representative samples?

3.100 What is a random-number generator?

3.101 Define a blind study.

Learning the Mechanics

3.102 Suppose you wish to sample $n = 2$ elements from a total of $N = 10$ elements.
[NW]
   a. Count the number of different samples that can be drawn, first by listing them and then by using combinatorial mathematics.
   b. If random sampling is to be employed, what is the probability that any particular sample will be selected?
   c. Show how to use the random-number table, Table I in Appendix A, to select a random sample of 2 elements from a population of 10 elements. Perform the sampling procedure 20 times. Do any two of the samples contain the same 2 elements? Given your answer to part **b,** did you expect repeated samples?

3.103 Suppose you wish to sample $n = 3$ elements from a total of $N = 600$ elements.
   a. Count the number of different samples by using combinatorial mathematics.
   b. If random sampling is to be employed, what is the probability that any particular sample will be selected?
   c. Show how to use the random-number table, Table I in Appendix A, to select a random sample of 3 elements

from a population of 600 elements. Perform the sampling procedure 20 times. Do any two of the samples contain the same 3 elements? Given your answer to part **b,** did you expect repeated samples?
   d. Use the computer to generate a random sample of 3 from the population of 600 elements.

3.104 Suppose that a population contains $N = 200,000$ elements. Use a computer or Table I of Appendix A to select a random sample of $n = 10$ elements from the population. Explain how you selected your sample.

Applying the Concepts—Basic

3.105 **Random-digit dialing.** To ascertain the effectiveness of their advertising campaigns, firms frequently conduct telephone interviews with consumers by using *random-digit dialing*. With this method, a random-number generator mechanically creates the sample of phone numbers to be called.
   a. Explain how the random-number table (Table I of Appendix A) or a computer could be used to generate a sample of seven-digit telephone numbers.
   b. Use the procedure you described in part **a** to generate a sample of 10 seven-digit telephone numbers.
   c. Use the procedure you described in part **a** to generate 5 seven-digit telephone numbers whose first three digits are 373.

3.106 **Census sampling.** In addition to its decennial enumeration of the population, the U.S. Census Bureau regularly samples the population for demographic information such as income, family size, employment, and marital status. Suppose the Bureau plans to sample 1,000 households in a city that has a total of 534,322 households. Show how the Bureau could use the random-number table in Appendix A or a computer to generate the sample. Select the first 10 households to be included in the sample.

Applying the Concepts—Intermediate

3.107 **Kiwifruit as an iron supplement.** Massey University (New [NW] Zealand) researchers have designed a 16-week study to determine the effectiveness of gold kiwifruit as an iron supplement in women with an iron deficiency (*BMC Public Health*, Vol. 10, 2010). Half the women in the study will receive an iron-fortified breakfast cereal with a banana, and half will receive an iron-fortified breakfast cereal with gold kiwifruit. The researchers are planning for 100 women to participate in the study. Number the women from 1 to 100. Then use a random-number generator to select 50 of these women. These 50 will receive the kiwi-fruit breakfast; the remaining 50 will receive the banana breakfast.

3.108 **Stress and diet study.** Researchers at the University of South Florida investigated the effects of diet on the physiology and behavior of rats (*USF News*, Oct. 17, 2009). One group of rats was fed a typical American diet (40% fat, 40% protein, 20% carbohydrates); a second group of rats was fed an Atkins-type diet (70% fat, 20% protein, 10% carbohydrates); and, a third (control) group was fed rat chow. Assume the experiment utilized 50 rats. Use a random-number generator to randomly assign the rats to the three diet groups.

3.109 **Study of the effect of songs with violent lyrics.** A series of designed experiments was conducted to examine the effects of songs with violent lyrics on aggressive thoughts and hostile feelings (*Journal of Personality & Social Psychology*, May 2003). In one of the experiments, college students were randomly assigned to listen to one of four types of songs: (1) a humorous song with violent lyrics, (2) a humorous song with nonviolent lyrics, (3) a nonhumorous song with violent lyrics, and (4) a nonhumorous song with nonviolent lyrics. Assuming that 120 students participated in the study, use a random-number generator to randomly assign 30 students to each of the four experimental conditions.

3.110 **Auditing an accounting system.** In auditing a firm's financial statements, an auditor is required to assess the operational effectiveness of the accounting system. In performing the assessment, the auditor frequently relies on a random sample of actual transactions (Stickney and Weil, *Financial Accounting: An Introduction to Concepts, Methods, and Uses*, 2002). A particular firm has 5,382 customer accounts that are numbered from 0001 to 5382.

a. One account is to be selected at random for audit. What is the probability that account number 3,241 is selected?

b. Draw a random sample of 10 accounts and explain in detail the procedure you used.

c. Refer to part **b**. The following are two possible random samples of size 10:

Sample Number 1				
5011	0082	0963	0772	3415
2663	1126	0008	0026	4189

Sample Number 2				
0001	0003	0005	0007	0009
0002	0004	0006	0008	0010

Is one more likely to be selected than the other? Explain.

Applying the Concepts—Advanced

3.111 **Selecting archaeological dig sites.** Archaeologists plan to perform test digs at a location they believe was inhabited several thousand years ago. The site is approximately 10,000 meters long and 5,000 meters wide. They first draw rectangular grids over the area, consisting of lines every 100 meters, creating a total of $100 \cdot 50 = 5,000$ intersections (not counting one of the outer boundaries). The plan is to randomly sample 50 intersection points and dig at the sampled intersections. Explain how you could use a random-number generator to obtain a random sample of 50 intersections. Develop at least two plans: one that numbers the intersections from 1 to 5,000 prior to selection and another that selects the row and column of each sampled intersection (from the total of 100 rows and 50 columns).

3.8 Some Additional Counting Rules (Optional)

In Section 3.1, we pointed out that experiments sometimes have so many sample points that it is impractical to list them all. However, many of these experiments possess sample points with identical characteristics. If you can develop a **counting rule** to count the number of sample points for such an experiment, it can be used to aid in the solution of the problems. The combinations rule for selecting n elements from N elements without regard to order was presented in Section 3.1 (p. 118). In this optional section, we give three additional counting rules: the *multiplicative rule*, the *permutations rule*, and the *partitions rule*.

Example 3.23

Applying the Multiplicative Rule— Airline Shipping

Problem A product can be shipped by four airlines, and each airline can ship via three different routes. How many distinct ways exist to ship the product?

Solution A pictorial representation of the different ways to ship the product will aid in counting them. The tree diagram (see Section 3.6) is useful for this purpose. Consider the tree diagram shown in Figure 3.23. At the starting point (stage 1), there are four choices—the different airlines—to begin the journey. Once we have chosen an airline (stage 2), there are three choices—the different routes—to complete the shipment and reach the final destination. Thus, the tree diagram clearly shows that there are $(4)(3) = 12$ distinct ways to ship the product.

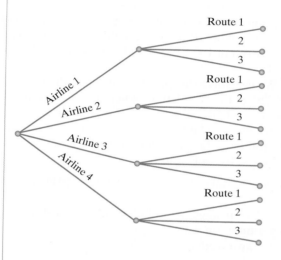

Figure 3.23

Tree diagram for shipping problem

Look Back The method of solving this example can be generalized to any number of stages with sets of different elements. This method is given by the *multiplicative rule*.

Now Work Exercise 3.122c

The Multiplicative Rule

You have k sets of elements, n_1 in the first set, n_2 in the second set, ... , and n_k in the kth set. Suppose you wish to form a sample of k elements by *taking one element from each* of the k sets. Then the number of different samples that can be formed is the product

$$n_1 n_2 n_3 \ \ldots \ n_k$$

Example 3.24

Applying the Multiplicative Rule— Hiring Executives

Problem There are 20 candidates for three different executive positions: $E_1, E_2,$ and E_3. How many different ways could you fill the positions?

Solution For this example, there are $k = 3$ sets of elements, corresponding to

Set 1: Candidates available to fill position E_1

Set 2: Candidates remaining (after filling E_1) that are available to fill E_2

Set 3: Candidates remaining (after filling E_1 and E_2) that are available to fill E_3

The numbers of elements in the sets are $n_1 = 20, n_2 = 19,$ and $n_3 = 18$. Therefore, the number of different ways of filling the three positions is given by the multiplicative rule as $n_1 n_2 n_3 = (20)(19)(18) = 6,840$.

Example 3.25

Applying the Multiplicative Rule— 10 Coin Tosses

Problem Consider the experiment discussed in Example 3.106 of tossing a coin 10 times. Show how we found that there were $2^{10} = 1{,}024$ sample points for this experiment.

Solution There are $k = 10$ sets of elements for this experiment. Each set contains two elements: a head and a tail. Thus, there are

$$(2)(2)(2)(2)(2)(2)(2)(2)(2)(2) = 2^{10} = 1{,}024$$

different outcomes (sample points) of this experiment.

Example 3.26

Applying the Permutations Rule— Selecting Soldiers for a Mission

Problem Suppose there are five dangerous military missions, each requiring one soldier. In how many different ways can 5 soldiers from a squadron of 100 be assigned to these five missions?

Solution We can solve this problem by using the multiplicative rule. The entire set of 100 soldiers is available for the first mission, and after the selection of 1 soldier for that mission, 99 are available for the second mission, etc. Thus, the total number of different ways of choosing 5 soldiers for the five missions is

$$n_1 n_2 n_3 n_4 n_5 = (100)(99)(98)(97)(96) = 9{,}034{,}502{,}400$$

Look Back The arrangement of elements in a distinct order is called a **permutation**. There are more than 9 billion different *permutations* of 5 elements (soldiers) drawn from a set of 100 elements!

Permutations Rule

Given a *single set* of N different elements, you wish to select n elements from the N and *arrange* them within n positions. The number of different **permutations** of the N elements taken n at a time is denoted by P_n^N and is equal to

$$P_n^N = N(N-1)(N-2)\cdot\cdots\cdot(N-n+1) = \frac{N!}{(N-n)!}$$

where $n! = n(n-1)(n-2)\ldots(3)(2)(1)$ and is called n *factorial*. (For example, $5! = 5\cdot4\cdot3\cdot2\cdot1 = 120$.) The quantity $0!$ is defined to be 1.

Example 3.27

Applying the Permutations Rule— Driving Routes

Problem Suppose you wish to drive, in sequence, from a starting point to each of five cities, and you wish to compare the distances—and, ultimately, the costs—of the different routes. How many different routes would you have to compare?

Solution Denote the cities as $C_1, C_2, \ldots, C_5$. Then a route moving from the starting point to C_2 to C_1 to C_3 to C_4 to C_5 would be represented as $C_2 C_1 C_3 C_4 C_5$. The total number of routes would equal the number of ways you could rearrange the $N = 5$ cities into $n = 5$ positions. This number is

$$P_n^N = P_5^5 = \frac{5!}{(5-5)!} = \frac{5!}{0!} = \frac{5\cdot4\cdot3\cdot2\cdot1}{1} = 120$$

(Recall that $0! = 1$.)

Now Work Exercise 3.123c

Example 3.28
Applying the Partitions Rule— Assigning Workers

Problem Suppose you are supervising four construction workers. You must assign three to job 1 and one to job 2. In how many different ways can you make this assignment?

Solution To begin, suppose each worker is to be assigned to a distinct job. Then, using the multiplicative rule, you obtain $(4)(3)(2)(1) = 24$ ways of assigning the workers to four distinct jobs. The 24 ways are listed in four groups in Table 3.8 (where $ABCD$ signifies that worker A was assigned the first distinct job, worker B the second, etc.).

Table 3.8	Workers Assigned to Four Jobs, Example 3.28		
Group 1	Group 2	Group 3	Group 4
$ABCD$	$ABDC$	$ACDB$	$BCDA$
$ACBD$	$ADBC$	$ADCB$	$BDCA$
$BACD$	$BADC$	$CADB$	$CBDA$
$BCAD$	$BDAC$	$CDAB$	$CDBA$
$CABD$	$DABC$	$DACB$	$DBCA$
$CBAD$	$DBAC$	$DCAB$	$DCBA$

Now suppose the first three positions represent job 1 and the last position represents job 2. You can see that all the listings in group 1 represent the same outcome of the experiment of interest. That is, workers A, B, and C are assigned to job 1 and worker D to job 2. Similarly, group 2 listings are equivalent, as are group 3 and group 4 listings. Thus, there are only four different assignments of four workers to the two jobs. These assignments are shown in Table 3.9.

Table 3.9

Workers Assigned to Two Jobs, Example 3.28

Job 1	Job 2
ABC	D
ABD	C
ACD	B
BCD	A

Look Back To generalize this result, we point out that the final result can be found by

$$\frac{(4)(3)(2)(1)}{(3)(2)(1)(1)} = 4$$

The numerator, $(4)(3)(2)(1)$, is the number of different ways (*permutations*) the workers could be assigned four distinct jobs. In the denominator, the division by $(3)(2)(1)$ is to remove the duplicated permutations resulting from the fact that three workers are assigned the same job, and the division by 1 is associated with the worker assigned to job 2.

Now Work Exercise 3.124

Partitions Rule

Suppose you wish to partition a *single* set of N different elements into k sets, with the first set containing n_1 elements, the second containing n_2 elements, ..., and the kth set containing n_k elements. Then the number of different partitions is

$$\frac{N!}{n_1!n_2! \ldots n_k!}, \text{ where } n_1 + n_2 + n_3 + \cdots + n_k = N$$

Example 3.29

Applying the Partitions Rule— Assigning Workers

Problem Suppose you have 12 construction workers and you wish to assign 3 to job-site 1, 4 to jobsite 2, and 5 to jobsite 3. In how many different ways can you make this assignment?

Solution For this example, $k = 3$ (corresponding to the $k = 3$ different jobsites), $N = 12, n_1 = 3, n_2 = 4$, and $n_3 = 5$. Then the number of different ways to assign the workers to the jobsites is

$$\frac{N!}{n_1!n_2!n_3!} = \frac{12!}{3!4!5!} = \frac{12 \cdot 11 \cdot 10 \cdot \cdots \cdot 3 \cdot 2 \cdot 1}{(3 \cdot 2 \cdot 1)(4 \cdot 3 \cdot 2 \cdot 1)(5 \cdot 4 \cdot 3 \cdot 2 \cdot 1)} = 27{,}720$$

Example 3.30

Applying the Combinations Rule— Selecting Soldiers for a Mission

Problem Five soldiers from a squadron of 100 are to be chosen for a dangerous mission. In how many ways can groups of 5 be formed?

Solution This problem is equivalent to sampling $n = 5$ elements from a set of $N = 100$ elements. Here, we apply the combinations rule from Section 3.1. The number of ways is the number of possible combinations of 5 soldiers selected from 100, or

$$\binom{100}{5} = \frac{100!}{(5!)(95!)} = \frac{100 \cdot 99 \cdot 98 \cdot 97 \cdot 96 \cdot 95 \cdot 94 \cdot \cdots \cdot 2 \cdot 1}{(5 \cdot 4 \cdot 3 \cdot 2 \cdot 1)(95 \cdot 94 \cdot \cdots \cdot 2 \cdot 1)}$$

$$= \frac{100 \cdot 99 \cdot 98 \cdot 97 \cdot 96}{5 \cdot 4 \cdot 3 \cdot 2 \cdot 1} = 75{,}287{,}520$$

Look Back Compare this result with that of Example 3.26, where we found that the number of permutations of 5 elements drawn from 100 was more than 9 billion. Because the order of the elements does not affect combinations, there are *fewer* combinations than permutations.

Now Work Exercise 3.123a

When working a probability problem, you should carefully examine the experiment to determine whether you can use one or more of the rules we have discussed in this section. Let's see how these rules (summarized in the next box) can help solve a probability problem.

Summary of Counting Rules

1. *Multiplicative rule.* If you are drawing *one element from each of k sets of elements*, where the sizes of the sets are $n_1, n_2, \cdot \ldots \cdot, n_k$, then the number of different results is

$$n_1 n_2 n_3 \ldots n_k$$

2. *Permutations rule.* If you are drawing *n elements from a set of N elements and arranging the n elements in a distinct order*, then the number of different results is

$$P_n^N = \frac{N!}{(N - n)!}$$

3. *Partitions rule.* If you are partitioning the *elements of a set of N elements into k groups consisting* of $n_1, n_2, \ldots, n_k$ elements ($n_1 + \cdots + n_k = N$), then the number of different results is

$$\frac{N!}{n_1!n_2! \cdot \ldots \cdot n_k!}$$

> **4.** *Combinations rule.* If you are drawing *n elements from a set of N elements without regard to the order of the n elements*, then the number of different results is
>
> $$\binom{N}{n} = \frac{N!}{n!(N-n)!}$$
>
> [*Note:* The combinations rule is a special case of the partitions rule when $k = 2$.]

Example 3.31

Applying Several Counting Rules— Ranking Detergents

Problem A consumer testing service is commissioned to rank the top 3 brands of laundry detergent. Ten brands are to be included in the study.

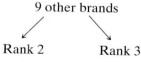

a. In how many different ways can the consumer testing service arrive at the final ranking?

b. If the testing service can distinguish no difference among the brands and therefore arrives at the final ranking by random selection, what is the probability that company Z's brand is ranked first? in the top three?

Solution

a. Since the testing service is drawing 3 elements (brands) from a set of 10 elements and arranging the 3 elements into a distinct order, we use the permutations rule to find the number of different results:

$$P_3^{10} = \frac{10!}{(10-3)!} = 10 \cdot 9 \cdot 8 = 720$$

b. The steps for calculating the probability of interest are as follows:

Step 1 The experiment is to select and rank 3 brands of detergent from 10 brands.

Step 2 There are too many sample points to list. However, we know from part **a** that there are 720 different outcomes (i.e., sample points) of this experiment.

Step 3 If we assume that the testing service determines the rankings at random, each of the 720 sample points should have an equal probability of occurrence. Thus,

$$P(\text{Each sample point}) = \frac{1}{720}$$

Step 4 One event of interest to company Z is that its brand receive top ranking. We will call this event A. The list of sample points that result in the occurrence of event A is long, but the *number* of sample points contained in event A is determined by breaking event A into two parts:

Company Z brand → Rank 1

9 other brands → Rank 2, Rank 3

Only 1 possibility

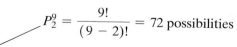

$$P_2^9 = \frac{9!}{(9-2)!} = 72 \text{ possibilities}$$

Event A

$$1 \cdot 72 = 72 \text{ sample points}$$

Thus, event A can occur in 72 different ways.

Now define B as the event that company Z's brand is ranked in the top three. Since event B specifies only that brand Z appear in the top three, we repeat the previous calculations, fixing brand Z in position 2 and then in position 3. We conclude that the number of sample points contained in event B is $3(72) = 216$.

Step 5 The final step is to calculate the probabilities of events A and B. Since the 720 sample points are equally likely to occur, we find that

$$P(A) = \frac{\text{Number of sample points in } A}{\text{Total number of sample points}} = \frac{72}{720} = \frac{1}{10}$$

Similarly,

$$P(B) = \frac{216}{720} = \frac{3}{10}$$

Example 3.32

Applying Several Counting Rules— Top 3 Detergents

Problem Refer to Example 3.31. Suppose the consumer testing service is to choose the top 3 laundry detergents from the group of 10, *but is not to rank them.*

a. In how many different ways can the testing service choose the 3 brands to be designated as top detergents?

b. Assuming that the testing service makes its choice by a random selection and that company X has 2 brands in the group of 10, what is the probability that exactly 1 of company X's brands is selected in the top 3? At least 1?

Solution

a. The testing service is selecting 3 elements (brands) from a set of 10 elements *without regard to order*, so we can apply the combinations rule to determine the number of different results:

$$\binom{10}{3} = \frac{10!}{3!(10-3)!} = \frac{10 \cdot 9 \cdot 8}{3 \cdot 2 \cdot 1} = 120$$

b. The steps for calculating the probabilities of interest are as follows:

Step 1 The experiment is to select (*but not rank*) three brands from 10.

Step 2 There are 120 sample points for this experiment.

Step 3 Since the selection is made at random,

$$P(\text{Each sample point}) = \frac{1}{120}$$

Step 4 Now, define events A and B as follows:

$$A: \{\text{Exactly one company X brand is selected.}\}$$
$$B: \{\text{At least one company X brand is selected.}\}$$

Since each of the sample points is equally likely to occur, we need only know the number of sample points in A and B to determine their probabilities.

For event A to occur, exactly 1 brand from company X must be selected, along with 2 of the remaining 8 brands from other companies. We thus break A into two parts:

2 company X brands   8 other brands
↓   ↓
Select 1   Select 2

$$\binom{2}{1} = 2 \text{ different possibilities} \qquad \binom{8}{2} = \frac{8!}{2!(8-2)!} = 28 \text{ different possibilities}$$

Event A

$$2 \cdot 28 = 56 \text{ sample points}$$

Note that the 1 company X brand can be selected in 2 ways, whereas the 2 other brands can be selected in 28 ways. (We use the combinations rule because the order of selection is not important.) Now we use the multiplicative rule to combine 1 of the 2 ways to select a company X brand with 1 of the 28 ways to select two other brands, yielding a total of 56 sample points for event A.

To count the sample points contained in B, we first note that

$$B = A \cup C$$

where A is defined as before and we have

$$C: \{\text{Both company X brands are selected.}\}$$

We have determined that A contains 56 sample points. Using the same method for event C, we find that

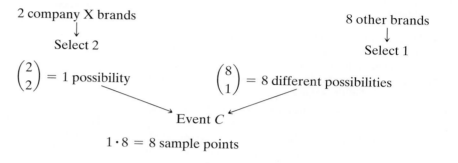

2 company X brands → Select 2
$\binom{2}{2} = 1$ possibility

8 other brands → Select 1
$\binom{8}{1} = 8$ different possibilities

Event C

$1 \cdot 8 = 8$ sample points

Thus, event C contains eight sample points, and $P(C) = \dfrac{8}{120}$.

Step 5 Since all the sample points are equally likely,

$$P(A) = \frac{\text{Number of sample ponts in } A}{\text{Total number of sample points}} = \frac{56}{120} = \frac{7}{15}$$

$$P(B) = P(A \cup C) = P(A) + P(C) - P(A \cap C)$$

where $P(A \cap C) = 0$ because the testing service cannot select exactly one *and* exactly two of company X's brands; that is, A and C are mutually exclusive events. Thus,

$$P(B) = \frac{56}{120} + \frac{8}{120} = \frac{64}{120} = \frac{8}{15}$$

Exercises 3.112–3.137

Understanding the Principles

3.112 Give a scenario where the multiplicative rule applies.

3.113 Give a scenario where the permutations rule applies.

3.114 Give a scenario where the partitions rule applies.

3.115 What is the difference between the permutations rule and the combinations rule?

Learning the Mechanics

3.116 Find the numerical values of

a. $\binom{6}{3}$ **b.** P_2^5 **c.** P_2^4 **d.** $\binom{100}{98}$

e. $\binom{50}{50}$ **f.** $\binom{50}{0}$ **g.** P_3^5 **h.** P_0^{10}

3.117 Use the multiplicative rule to determine the number of sample points in the sample space corresponding to the experiment of tossing a coin the following number of times:

a. 2 times   **b.** 3 times

c. 5 times   **d.** n times

3.118 Determine the number of sample points contained in the sample space when you toss the following:

a. 1 die   **b.** 2 dice

c. 4 dice   **d.** n dice

3.119 An experiment consists of choosing objects without regard to order. How many sample points are there if you choose the following?

a. 3 objects from 7   **b.** 2 objects from 6

c. 2 objects from 30   **d.** 8 objects from 10

e. r objects from

Applying the Concepts—Basic

3.120 Cheek teeth of extinct primates. Refer to the *American Journal of Physical Anthropology* (Vol. 142, 2010) study of the dietary habits of extinct mammals, Exercise 3.25 (p. 121). Recall that 18 cheek teeth extracted from skulls of an extinct primate species discovered in western Wyoming were analyzed. Each tooth was classified according to degree of wear (unworn, slight, light-moderate, moderate,

moderate-heavy, or heavy), with the 18 measurements shown in the accompanying table. (These data are saved in the **CHEEKTEETH** file)

Data on Degree of Wear	
Unknown	Slight
Unknown	Slight
Unknown	Heavy
Moderate	Unworn
Slight	Light-moderate
Unknown	Light-moderate
Moderate-heavy	Moderate
Moderate	Unworn
Slight	Unknown

a. Suppose the researcher will randomly select one tooth from each wear category for a more detailed analysis. How many different samples are possible?

b. Repeat part **a**, but do not include any teeth classified as "unknown" in the sample.

3.121 Choosing portable grill displays. Refer to the *Journal of Consumer Research* (March 2003) study of how people attempt to influence the choices of others, Exercise 3.27 (p. 122). Recall that students selected three portable grill displays to be compared from an offering of five different grill displays.

a. Use a counting rule to count the number of ways the three displays can be selected from the five available displays to form a three-grill-display combination.

b. The researchers informed students to select the three displays in order to convince people to choose Grill #2. Consequently, Grill #2 was a required selection. Use a counting rule to count the number of different ways the three grill displays can be selected from the five displays if Grill #2 must be selected. (Your answer should agree with the answer in 3.27a.)

c. Now suppose the three selected grills will be set up in a specific order for viewing by a customer. (The customer views one grill first, then the second, and finally the third grill.) Again, Grill #2 must be one of the three selected. How many different ways can the three grill displays be selected if customers view the grills in order?

3.122 Monitoring impedance to leg movements. In an experiment designed to monitor impedance to leg movement, Korean engineers attached electrodes to the ankles and knees of volunteers. Of interest were the voltage readings between pairs of electrodes (*IEICE Transactions on Information & Systems*, Jan. 2005). These readings were used to determine the signal-to-noise ratio (SNR) of impedance changes such as knee flexes and hip extensions.

a. Six voltage electrodes were attached to key parts of the ankle. How many electrode pairs on the ankle are possible?

b. Ten voltage electrodes were attached to key parts of the knee. How many electrode pairs on the knee are possible?

[NW] **c.** Determine the number of possible electrode pairs such that one electrode is attached to the knee and one is attached to the ankle.

3.123 Picking a basketball team. Suppose you are to choose a [NW] basketball team (five players) from eight available athletes.

a. How many ways can you choose a team (ignoring positions)?

b. How many ways can you choose a team composed of two guards, two forwards, and a center?

c. How many ways can you choose a team composed of one each of a point guard, shooting guard, power forward, small forward, and center?

3.124 Selecting project teams. Suppose you are managing 10 [NW] employees and you need to form three teams to work on different projects. Assume that each employee may serve on any team. In how many different ways can the teams be formed if the number of members on each project team are as follows:

a. 3, 3, 4   **b.** 2, 3, 5
c. 1, 4, 5   **d.** 2, 4, 4

3.125 Traveling between cities. A salesperson living in city A wishes to visit four cities B, C, D, and E.

a. If the cities are all connected by airlines, how many different travel plans could be constructed to visit each city exactly once and then return home?

b. Suppose all cities are connected, except that B and C are not directly connected. How many different flight plans would be available to the salesperson?

3.126 U.S. Zip codes. The nine-digit Zip code has become an integral part of the U.S. postal system.

a. How many different Zip codes are available for use by the postal service? (Assume that any nine-digit number can be used as a Zip code.)

b. The first three digits of a Chicago Zip code are 606. If no other city in the United States has these first three digits as part of its Zip, how many different Zip codes can exist in Chicago?

Applying the Concepts—Intermediate

3.127 Kiwifruit as an iron supplement. Refer to the *BMC Public Health* (Vol. 10, 2010) study of the effectiveness of gold kiwifruit as an iron supplement in women with an iron deficiency, Exercise 3.107 (p. 154). Recall that 50 women were assigned to receive an iron-fortified breakfast cereal with a banana, and 50 women were assigned to receive an iron-fortified breakfast cereal with gold kiwifruit. At the conclusion of the 16-week study, the medical researchers will select two women from each group and perform a full physical examination on each. How many possible ways can the selections be made?

3.128 Randomization in a study of TV commercials. Gonzaga University professors conducted a study of over 1,500 television commercials and published their results in the *Journal of Sociology, Social Work and Social Welfare* (Vol. 2, 2008). Commercials from eight networks—ABC, FAM, FOX, MTV, ESPN, CBS, CNN, and NBC—were sampled during an eight-day period, with one network randomly selected each day. The table below shows the order determined by random draw:

ABC—July 6 (Wed)
FAM—July 7 (Thr)
FOX—July 9 (Sat)
MTV—July 10 (Sun)
ESPN—July 11 (Mon)
CBS—July 12 (Tue)
CNN—July 16 (Sat)
NBC—July 17 (Sun)

a. Determine the number of possible orderings of the networks over the eight days.

b. What is the probability that ESPN is selected on Monday, July 11th?

c. What is the probability that MTV is selected on a Sunday?

3.129 Multilevel marketing schemes. Successful companies such as Amway, Herbalife, and NuSkin employ multilevel marketing (MLM) schemes to distribute their products. In MLM, each distributor recruits several additional distributors to work at the first level. Each of these first-level distributors, in turn, recruits distributors to work at the second level. Distributors at any level continue to recruit distributors, forming additional levels. A distributor at any particular level typically makes a 5% commission on the sales of all distributors at lower levels in his or her "group."

a. Elaine is a distributor for an MLM company. Elaine recruits six distributors to work for her at the first level. Each of these distributors recruits an additional five distributors to work at the second level. How many distributors are under Elaine?

b. Refer to part **a.** Suppose each distributor at Elaine's second level recruits seven third-level distributors, and each of these third-level distributors recruits five level-four distributors. Now how many distributors are under Elaine? (Your answer should give you an insight into why some MLM distributors can earn $100,000 in commissions per month!)

3.130 Mathematical theory of partitions. Mathematicians at the University of Florida solved a 30-year-old mathematics problem with the use of the theory of partitions (*Explore,* Fall 2000). In mathematical terminology, a partition is a representation of an integer as a sum of positive integers. (For example, the number 3 has three possible partitions: $3, 2 + 1,$ and $1 + 1 + 1$.) The researchers solved the problem by using "colored partitions" of a number, where the colors correspond to the four suits—red hearts, red diamonds, black spades, and black clubs—in a standard 52-card bridge deck. Consider forming colored partitions of an integer.

a. How many colored partitions of the number 3 are possible? [*Hint:* One partition is 3♥; another is 2♦ + 1♣.]

b. How many colored partitions of the number 5 are possible?

3.131 Matching medical students with residencies. The National Resident Matching Program (NRMP) is a service provided by the Association of American Medical Colleges to match graduating medical students with residency appointments at hospitals. After students and hospital officials have evaluated each other, they submit rank-order lists of their preferences to the NRMP. Using a matching algorithm, the NRMP then generates final, nonnegotiable assignments of students to the residency programs of hospitals (www.nrmp.org). Assume that three graduating medical students (#1, #2, and #3) have applied for positions at three different hospitals (*A, B,* and *C*), each of which has one and only one resident opening.

a. How many different assignments of medical students to hospitals are possible? List them.

b. Suppose student #1 prefers hospital B. If the NRMP algorithm is entirely random, what is the probability that the student is assigned to hospital B?

3.132 Florida license plates. In the mid-1980s, the state of Florida ran out of combinations of letters and numbers for its license plates. Then, each license plate contained three letters of the alphabet, followed by three digits selected from the 10 digits 0, 1, 2, … , 9.

a. How many different license plates did this system allow?

b. New Florida tags were obtained by reversing the procedure, starting with three digits followed by three letters. How many new tags did the new system provide?

c. Since the new tag numbers were added to the old numbers, what is the total number of licenses available for registration in Florida?

3.133 Selecting a maintenance support system. ARTHUR is the Norwegian Army's high-tech radar system designed to identify and track "unfriendly" artillery grenades, calculate where enemy positions are, and direct counterattacks on the enemy. In the *Journal of Quality in Maintenance Engineering* (Vol. 9, 2003), researchers used an analytic hierarchy process to help build a preferred maintenance organization for ARTHUR. The process requires the builder to select alternatives in three different stages (called echelons). In the first echelon, the builder must choose one of two mobile units (regular soldiers or soldiers with engineering training). In the second echelon, the builder chooses one of three heavy mobile units (units in the Norwegian Army, units from Supplier 2, or shared units). Finally, in the third echelon, the builder chooses one of three maintenance workshops (Norwegian Army, Supplier 1, or Supplier 2).

a. How many maintenance organization alternatives exist when choices are made in the three echelons?

b. The researchers determined that only four of the alternatives in part **a** are feasible alternatives for ARTHUR. If one of the alternatives is randomly selected, what is the probability that it is a feasible alternative?

3.134 Selecting new-car options. A company sells midlevel models of automobiles in five different styles. A buyer can get an automobile in one of eight colors and with either standard or automatic transmission. Would it be reasonable to expect a dealer to stock at least one automobile in every combination of style, color, and transmission? At a minimum, how many automobiles would the dealer have to stock?

3.135 Studying exam questions. A college professor hands out a list of 10 questions, 5 of which will appear on the final examination for the course. One of the students taking the course is pressed for time and can prepare for only 7 of the 10 questions on the list. Suppose the professor chooses the 5 questions at random from the 10.

a. What is the probability that the student will be prepared for all 5 questions that appear on the final examination?

b. What is the probability that the student will be prepared for fewer than 3 questions?

c. What is the probability that the student will be prepared for exactly 4 questions?

3.136 Volleyball positions. Intercollegiate volleyball rules require that after the opposing team has lost its serve, each of the six members of the serving team must rotate into new positions on the court. Hence, each player must be able to play all six different positions. How many different team combinations, by player and position, are possible during a volleyball game? If players are initially assigned to the positions in a random manner, find the probability that the best server on the team is in the serving position.

Applying the Concepts—Advanced

3.137 A straight flush in poker. Consider 5-card poker hands dealt from a standard 52-card bridge deck. Two important events are

$$A: \{\text{You draw a flush.}\}$$
$$B: \{\text{You draw a straight.}\}$$

[*Note:* A *flush* consists of any 5 cards of the same suit. A *straight* consists of any 5 cards with values in sequence. In a straight, the cards may be of any suit, and an ace may be considered as having a value of 1 or, alternatively, a value higher than a king.]

a. How many different 5-card hands can be dealt from a 52-card bridge deck?

b. Find $P(A)$.

c. Find $P(B)$.

d. The event that both A and B occur—that is $A \cap B$—is called a *straight flush*. Find $P(A \cap B)$.

3.9 Bayes's Rule (Optional)

An early attempt to employ probability in making inferences is the basis for a branch of statistical methodology known as **Bayesian statistical methods.** The logic employed by the English philosopher Thomas Bayes in the mid-1700s involves converting an unknown conditional probability, say, $P(B|A)$, to one involving a known conditional probability, say, $P(A|B)$. The method is illustrated in the next example.

Example 3.33

Applying Bayes's Logic—Intruder Detection System

Problem An unmanned monitoring system uses high-tech video equipment and microprocessors to detect intruders. A prototype system has been developed and is in use outdoors at a weapons munitions plant. The system is designed to detect intruders with a probability of .90. However, the design engineers expect this probability to vary with the weather. The system automatically records the weather condition each time an intruder is detected. On the basis of a series of controlled tests in which an intruder was released at the plant under various weather conditions, the following information is available: Given that the intruder was, in fact, detected by the system, the weather was clear 75% of the time, cloudy 20% of the time, and raining 5% of the time. When the system failed to detect the intruder, 60% of the days were clear, 30% cloudy, and 10% rainy. Use this information to find the probability of detecting an intruder, given rainy weather. (Assume that an intruder has been released at the plant.)

Solution Define D to be the event that the intruder is detected by the system. Then D^c is the event that the system failed to detect the intruder. Our goal is to calculate the conditional probability $P(D|\text{Rainy})$. From the statement of the problem, the following information is available:

$$P(D) = .90 \qquad\qquad PD^c = .10$$
$$P(\text{Clear}|D) = .75 \qquad P(\text{Clear}|D^c) = .60$$
$$P(\text{Cloudy}|D) = .20 \qquad P(\text{Cloudy}|D^c) = .30$$
$$P(\text{Rainy}|D) = .05 \qquad P(\text{Rainy}|D^c) = .10$$

Note that $P(D|\text{Rainy})$ is not one of the conditional probabilities that is known. However, we can find

$$P(\text{Rainy} \cap D) = P(D)P(\text{Rainy}|D) = (.90)(.05) = .045$$

and

$$P(\text{Rainy} \cap D^c) = P(D^c)P(\text{Rainy}|D^c) = (.10)(.10) = .01$$

by the multiplicative probability rule. These two probabilities are highlighted on the tree diagram for the problem in Figure 3.24.

Now, the event Rainy is the union of two mutually exclusive events: (Rainy $\cap D$) and (Rainy $\cap D^c$). Thus, applying the additive probability rule, we have

$$P(\text{Rainy}) = P(\text{Rainy} \cap D) + P(\text{Rainy} \cap D^c) = .045 + .01 = .055$$

We now apply the formula for conditional probability to obtain

$$P(D \mid \text{Rainy}) = \frac{P(\text{Rainy} \cap D)}{P(\text{Rainy})} = \frac{P(\text{Rainy} \cap D)}{P(\text{Rainy} \cap D) + P(\text{Rainy} \cap D^c)} = \frac{.045}{.055} = .818$$

Therefore, under rainy weather conditions, the prototype system can detect the intruder with a probability of .818, a value lower than the designed probability of .90.

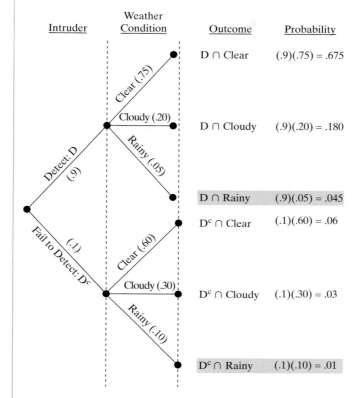

Figure 3.24
Tree Diagram for Example 3.33

The technique utilized in Example 3.33, called **Bayes's rule**, can be applied when an observed event A occurs with any one of several mutually exclusive and exhaustive events, $B_1, B_2, \ldots, B_k$. The formula for finding the appropriate conditional probabilities is given in the next box.

THOMAS BAYES (1702–1761)

The Inverse Probabilist

The Reverend Thomas Bayes was an ordained English Presbyterian minister who became a Fellow of the Royal Statistical Society without benefit of any formal training in mathematics or any published papers in science during his lifetime. His manipulation of the formula for conditional probability in 1761 is now known as Bayes's theorem. At the time and for 200 years afterward, the use of Bayes's theorem (or inverse probability, as it was called) was controversial and, to some, considered an inappropriate practice. It was not until the 1960s that the power of the Bayesian approach to decision making began to be tapped. ■

Bayes's Rule

Given k mutually exclusive and exhaustive events, $B_1, B_2, \ldots, B_k$ such that $P(B_1) + P(B_2) + \cdots + P(B_k) = 1$, and given an observed event A, it follows that

$$P(B_i \mid A) = P(B_i \cap A) / P(A)$$

$$= \frac{P(B_i)P(A \mid B_i)}{P(B_1)P(A \mid B_1) + P(B_2)P(A \mid B_2) + \cdots + P(B_k)P(A \mid B_k)}$$

In applying Bayes's rule to Example 3.33, the observed event $A = \{\text{Rainy}\}$ and the $k = 2$ mutually exclusive and exhaustive events are the complementary events $D = \{\text{intruder detected}\}$ and $D^c = \{\text{intruder not detected}\}$. Hence, we obtain the formula

$$P(D \mid \text{Rainy}) = \frac{P(D)P(\text{Rainy} \mid D)}{P(D)P(\text{Rainy} \mid D) + P(D^c)P(\text{Rainy} \mid D^c)}$$

$$= \frac{(.90)(.05)}{(.90)(.05) + (.10)(.10)} = .818$$

Example 3.34
Bayes's Rule Application— Wheelchair Control

Problem Electric wheelchairs are difficult to maneuver for many disabled people. In a paper presented at the *1st International Workshop on Advances in Service Robotics* (March 2003), researchers applied Bayes's rule to evaluate an "intelligent" robotic controller that aims to capture the intent of a wheelchair user and aid in navigation. Consider the following scenario. From a certain location in a room, a wheelchair user will either (1) turn sharply to the left and navigate through a door, (2) proceed straight to the other side of the room, or (3) turn slightly right and stop at a table. Denote these three events as D (for door), S (Straight), and T (for table). Based on previous trips, $P(D) = .5$, $P(S) = .2$, and $P(T) = .3$. The wheelchair is installed with a robot-controlled joystick. When the user intends to go through the door, she points the joystick straight 30% of the time; when the user intends to go straight, she points the joystick straight 40% of the time; and, when the user intends to go to the table, she points the joystick straight 5% of the time. If the wheelchair user points the joystick straight, what is her most likely destination?

Solution Let $J = \{$joystick is pointed straight$\}$. The user intention percentages can be restated as the following conditional probabilities: $P(J|D) = .3$, $P(J|S) = .4$, and $P(J \cdot T) = .05$. Since the user has pointed the joystick straight, we want to find the following probabilities: $P(D|J)$, $P(S|J)$, and $P(T|J)$. Now, the three events, D, S, and T, represent mutually exclusive and exhaustive events, where $P(D) = .5$, $P(S) = .2$, and $P(T) = .3$. Consequently, we can apply Bayes's rule as follows:

$$P(D|J) = P(J|D) \cdot P(D)/[P(J|D) \cdot P(D) + P(J|S) \cdot P(S) + P(J|T) \cdot P(T)]$$
$$= (.3)(.5)/[(.3)(.5) + (.4)(.2) + (.05)(.3)] = .15/.245 = .612$$
$$P(S|J) = P(J|S) \cdot P(S)/[P(J|D) \cdot P(D) + P(J|S) \cdot P(S) + P(J|T) \cdot P(T)]$$
$$= (.4)(.2)/[(.3)(.5) + (.4)(.2) + (.05)(.3)] = .08/.245 = .327$$
$$P(T|J) = P(J|T) \cdot P(T)/[P(J|D) \cdot P(D) + P(J|S) \cdot P(S) + P(J|T) \cdot P(T)]$$
$$= (.05)(.3)/[(.3)(.5) + (.4)(.2) + (.05)(.3)] = .015/.245 = .061$$

Note that the largest conditional probability is $P(D|J) = .612$. Thus, if the joystick is pointed straight, the wheelchair user is most likely headed through the door.

Now Work: Exercise 3.142

Exercises 3.138–3.153

Understanding the Principles

3.138 Explain the difference between the two probabilities $P(A|B)$ and $P(B|A)$

3.139 Why is Bayes's rule unnecessary for finding $P(B|A)$ if events A and B are independent?

3.140 Why is Bayes's rule unnecessary for finding $P(B|A)$ if events A and B are mutually exclusive?

Learning the Mechanics

3.141 Suppose the events B_1 and B_2 are mutually exclusive and complementary events such that $P(B_1) = .75$ and $P(B_2) = .25$. Consider another event A such that $P(A|B_1) = .3$ and $P(A|B_2) = .5$.
  a. Find $P(B_1 \cap A)$.
  b. Find $P(B_2 \cap A)$.
  c. Find $P(A)$, using the results in part **a** and **b**.
  d. Find $P(B_1|A)$.
  e. Find $P(B_2|A)$.

3.142 Suppose the events B_1, B_2, and B_3 are mutually exclusive and complementary events such that $P(B_1) = .2$, $P(B_2) = .15$, and $P(B_3) = .65$. Consider another event A such that $P(A|B_1) = .4$, $P(A|B_2) = .25$, and $P(A|B_3) = .6$. Use Bayes's rule to find
  a. $P(B_1|A)$
  b. $P(B_2|A)$
  c. $P(B_3|A)$

3.143 Suppose the events B_1, B_2, and B_3 are mutually exclusive and complementary events such that $P(B_1) = .2$, $P(B_2) = .15$, and $P(B_3) = .65$. Consider another event A such that $P(A) = .4$. If A is independent of B_1, B_2, and B_3, use Bayes's rule to show that $P(B_1|A) = P(B_1) = .2$.

Applying the Concepts—Basic

3.144 Reverse-engineering gene identification. In *Molecular Systems Biology* (Vol. 3, 2007), geneticists at the University of Naples (Italy) used reverse engineering to identify genes. They calculated $P(G|D)$, where D is a gene expression data set of interest and G is a graphical identifier. Several graphical identifiers were investigated. Suppose that, for two different graphical identifiers G_1 and G_2, $P(D|G_1) = .5$ and $P(D|G_2) = .3$. Also, $P(D) = .34$, $P(G_1) = .2$, and $P(G_2) = .8$.

a. Use Bayes's rule to find $P(G_1|D)$.

b. Use Bayes's rule to find $P(G_2|D)$.

3.145 Drug testing in athletes. Due to inaccuracies in drug-testing procedures (e.g., false positives and false negatives), in the medical field the results of a drug test represent only one factor in a physician's diagnosis. Yet when Olympic athletes are tested for illegal drug use (i.e., doping), the results of a single test are used to ban the athlete from competition. In *Chance* (Spring 2004), University of Texas biostatisticians D. A. Berry and L. Chastain demonstrated the application of Bayes's rule to making inferences about testosterone abuse among Olympic athletes. They used the following example: In a population of 1,000 athletes, suppose 100 are illegally using testosterone. Of the users, suppose 50 would test positive for testosterone. Of the nonusers, suppose 9 would test positive.

a. Given that the athlete is a user, find the probability that a drug test for testosterone will yield a positive result. (This probability represents the sensitivity of the drug test.)

b. Given that the athlete is a nonuser, find the probability that a drug test for testosterone will yield a negative result. (This probability represents the specificity of the drug test.)

c. If an athlete tests positive for testosterone, use Bayes's rule to find the probability that the athlete is really doping. (This probability represents the *positive predictive value* of the drug test.)

3.146 Contaminated fish. Refer to the U.S. Army Corps of Engineers' study on the DDT contamination of fish in the Tennessee River (in Alabama), presented in Example 1.4 (p. 10). Part of the investigation focused on how far upstream the contaminated fish have migrated. (A fish is considered to be contaminated if its measured DDT concentration is greater than 5.0 parts per million.)

a. Considering only the contaminated fish captured from the Tennessee River, the data reveal that 52% of the fish are found between 275 and 300 miles upstream, 39% are found from 305 to 325 miles upstream, and 9% are found from 330 to 350 miles upstream. Use these percentages to determine the probabilities $P(275 - 300)$, $P(305 - 325)$, and $P(330 - 350)$.

b. Given that a contaminated fish is found a certain distance upstream, the probability that it is a channel catfish (CC) is determined from the data as $P(CC|275 - 300) = .775$, $P(CC|305 - 325) = .77$, and $P(CC|330 - 350) = .86$. If a contaminated channel catfish is captured from the Tennessee River, what is the probability that it was captured $275 - 300$ miles upstream?

3.147 Tests for Down syndrome. Currently, there are three diagnostic tests available for chromosome abnormalities in a developing fetus: triple serum marker screening, ultrasound, and amniocentesis. The safest (to both the mother and fetus) and least expensive of the three is the ultrasound test. Two San Diego State University statisticians investigated the accuracy of using ultrasound to test for Down syndrome (*Chance*, Summer 2007). Let D denote that the

fetus has a genetic marker for Down syndrome and N denote that the ultrasound test is normal (i.e., no indication of chromosome abnormalities). Then, the statisticians desire the probability $P(D|N)$. Use Bayes's rule and the following probabilities (provided in the article) to find the desired probability: $P(D) = 1/80$, $P(D^C) = 79/80$, $P(N|D) = 1/2$, $P(N^C|D) = 1/2$, $P(N|D^C) = 1$ and, $P(N^C|D^C) = 0$.

Applying the Concepts—Intermediate

3.148 HIV testing and false positives. Bayes's rule was applied to the problem of HIV testing in *The American Statistician* (Aug. 2008). In North America, the probability of a person having HIV is .008. A test for HIV yields either a positive or negative result. Given that a person has HIV, the probability of a positive test result is .99. (This probability is called the *sensitivity* of the test.) Given that a person does not have HIV, the probability of a negative test result is also .99. (This probability is called the *specificity* of the test.) The authors of the article are interested in the probability that a person actually has HIV given that the test is positive.

a. Find the probability of interest for a North American by using Bayes's rule.

b. In East Asia, the probability of a person having HIV is only .001. Find the probability of interest for an East Asian by using Bayes's rule. (Assume that both the sensitivity and specificity of the test are .99.)

c. Typically, if one tests positive for HIV, a follow-up test is administered. What is the probability that a North American has HIV given that both tests are positive? (Assume that the tests are independent.)

d. Repeat part **c** for an East Asian.

3.149 Mining for dolomite. Dolomite is a valuable mineral that is found in sedimentary rock. During mining operations, dolomite is often confused with shale. The radioactivity features of rock can aid miners in distinguishing between dolomite and shale rock zones. For example, if the gamma ray reading of a rock zone exceeds 60 API units, the area is considered to be mostly shale (and is not mined); if the gamma ray reading of a rock zone is less than 60 API units, the area is considered to be abundant in dolomite (and is mined). Data on 771 core samples in a rock quarry collected by the Kansas Geological Survey revealed that 476 of the samples are dolomite and 295 of the samples are shale. Of the 476 dolomite core samples, 34 had a gamma ray reading greater than 60. Of the 295 shale core samples, 280 had a gamma ray reading greater than 60. Suppose you obtain a gamma ray reading greater than 60 at a certain depth of the rock quarry. Should this area be mined?

3.150 Nondestructive evaluation. Nondestructive evaluation (NDE) describes methods that quantitatively characterize materials, tissues, and structures by noninvasive means, such as X-ray computed tomography, ultrasonics, and acoustic emission. Recently, NDE was used to detect defects in steel castings (*JOM*, May 2005). Assume that the probability that NDE detects a "hit" (i.e., predicts a defect in a steel casting) when, in fact, a defect exists is .97. (This is often called the probability of detection.) Assume also that the probability that NDE detects a "hit" when, in fact, no defect exists is .005. (This is called the probability of a false call.) Past experience has shown that a defect occurs once in every 100 steel castings. If NDE detects a "hit" for a particular steel casting, what is the probability that an actual defect exists?

3.151 Intrusion detection systems. Refer to the *Journal of Research of the National Institute of Standards and Technology* (Nov.–Dec. 2003) study of a double intrusion detection system with independent systems, presented in Exercise 3.90 (p. 148). Recall that if there is an intruder, system A sounds an alarm with probability .9 and system B sounds an alarm with probability .95. If there is no intruder, system A sounds an alarm with probability .2 and system B sounds an alarm with probability .1. Now, assume that the probability of an intruder is .4. If both systems sound an alarm, what is the probability that there is an intruder?

3.152 Repairing a computer system. The local area network (LAN) for the College of Business computing system at a large university is temporarily shut down for repairs. Previous shutdowns have been due to hardware failure, software failure, or power failure. Maintenance engineers have determined that the probabilities of hardware, software, and power problems are .01, .05, and .02, respectively. They have also determined that if the system experiences hardware problems, it shuts down 73% of the time. Similarly, if software problems occur, the system shuts down 12% of the time; and, if power failure occurs, the system shuts down 88% of the time. What is the probability that the current shutdown of the LAN is due to hardware failure? software failure? power failure?

Applying the Concepts—Advanced

3.153 Forensic analysis of JFK assassination bullets. Following the assassination of President John F. Kennedy (JFK) in 1963, the House Select Committee on Assassinations (HSCA) conducted an official government investigation. The HSCA concluded that although there was a probable conspiracy, involving at least one additional shooter other than Lee Harvey Oswald, the additional shooter missed all limousine occupants. A recent analysis of assassination bullet fragments, reported in the *Annals of Applied Statistics* (Vol. 1, 2007), contradicted these findings, concluding that evidence used to rule out a second assassin by the HSCA is fundamentally flawed. It is well documented that at least two different bullets were the source of bullet fragments used in the assassination. Let E = {bullet evidence used by the HSCA}, T = {two bullets used in the assassination}, and T^C = {more than two bullets used in the assassination}. Given the evidence (E), which is more likely to have occurred—two bullets used (T) or more than two bullets used (T^C)?

a. The researchers demonstrated that the ratio, $P(T \mid E)/P(T^C \mid E)$, is less than 1. Explain why this result supports the theory of more than two bullets used in the assassination of JFK.

b. To obtain the result in part **a**, the researchers first showed that

$$P(T \mid E)/P(T^C \mid E) = [P(E \mid T) \cdot P(T)]/[P(E \mid T^C) \cdot P(T^C)]$$

Demonstrate this equality using Bayes's theorem.

CHAPTER NOTES

Key Terms

Note: Items marked with an asterisk (*) are from the optional sections in this chapter.

Additive rule of probability 128
*Bayes's rule 165
*Bayesian statistical methods 164
Combinations rule 118
Combinatorial mathematics 118
Complement 126
Complementary events 126
Compound event 123
Conditional probability 135
*Counting rule 154
Dependent events 141
Event 115
Experiment 110
Independent events 141
Intersection 123
Law of large numbers 112

*Multiplicative (counting) rule 155
Multiplicative rule of
   probability 138
Mutually exclusive events 129
Odds 175
*Partitions rule 157
*Permutation 156
*Permutations rule 156
Probability rules for sample
   points 113
Random-number generator 150
Random-number table 151
Random sample 150
Sample point 110
Sample space 111
Tree diagram 111
Two-way table 125
Unconditional probabilities 135
Union 123
Venn diagram 111

Key Symbols

S	**Sample Space** (collection of all sample points)
$A: \{1,2\}$	Set of **sample points in event** A
$P(A)$	**Probability** of event A
$A \cup B$	**Union** of events A and B (either A or B occurs)
$A \cap B$	**Intersection** of events A and B (both A and B occur)
A^c	**Complement** of A (event A does not occur)
$A \mid B$	Event A occurs, **given that** event B occurs
$\binom{N}{n}$	Number of **combinations** of N elements taken n at a time
$N!$	N **factorial** $= N(N-1)(N-2)\ldots(2)(1)$

Key Ideas

Probability Rules for k Sample Points, $S_1, S_2, S_3, \ldots, S_k$
1. $0 \le P(E_i) \le 1$
2. $\Sigma P(E_i) = 1$

Random Sample

All possible such samples have equal probability of being selected

Bayes's Rule

For mutually exclusive events, $B_1, B_2, B_3, \ldots, B_k$, such that $P(B_1) + P(B_2) + \cdots + P(B_k) = 1$,

$$P(B_i \mid A) = \frac{P(B_i)P(A \mid B_i)}{P(B_1)P(A \mid B_1) + P(B_2)P(A \mid B_2) + \cdots + P(B_k)P(A \mid B_k)}$$

Combinations Rule

Counting number of samples of n elements selected from N elements

$$\binom{N}{n} = \frac{N!}{n!(N-n)!}$$

$$= \frac{N(N-1)(N-2)\ldots(N-n+1)}{n(n-1)(n-2)\ldots(2)(1)}$$

Guide to Selecting Probability Rules

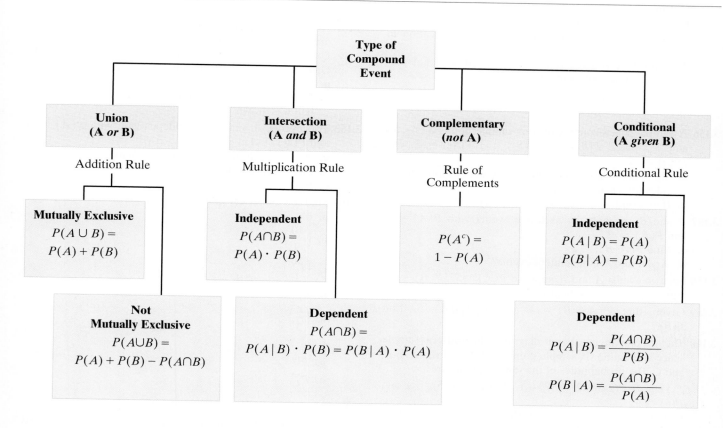

Supplementary Exercises 3.154–3.199

Note: Starred (*) exercises refer to the optional sections in this chapter.

Understanding the Principles

3.154 Which of the following pairs of events are mutually exclusive?
  a. $A = \{$The Tampa Bay Rays win the World Series next year.$\}$
   $B = \{$Evan Longoria, Rays infielder, hits 75 home runs runs next year.$\}$
  b. $A = \{$Psychiatric patient Tony responds to a stimulus within 5 seconds.$\}$
   $B = \{$Psychiatric patient Tony has the fastest stimulus response time of 2.3 seconds.$\}$
  c. $A = \{$High school graduate Cindy enrolls at the University of South Florida next year.$\}$
   $B = \{$High school graduate Cindy does not enroll in college next year.$\}$

3.155 Use the symbols $\cap$, $\cup$, $|$, and ^{c}, to convert the following statements into compound events involving events A and B, where $A = \{$You purchase a notebook computer$\}$ and $B = \{$You vacation in Europe$\}$:
  a. You purchase a notebook computer or vacation in Europe.
  b. You will not vacation in Europe.
  c. You purchase a notebook computer and vacation in Europe.
  d. Given that you vacation in Europe, you will not purchase a notebook computer.

Learning the Mechanics

3.156 A sample space consists of four sample points S_1, S_2, S_3, and S_4, where $P(S_1) = .2, P(S_2) = .1, P(S_3) = .3$, and $P(S_4) = .4$.
  a. Show that the sample points obey the two probability rules for a sample space.
  b. If an event $A = \{S_1, S_4,\}$ find P(A).

3.157 A and B are mutually exclusive events, with $P(A) = .2$ and $P(B) = .3$.
  a. Find $P(A|B)$.
  b. Are A and B independent events?

3.158 For two events A and B, suppose $P(A) = .7, P(B) = .5$, and $P(A \cap B) = .4$. Find $P(A \cup B)$.

3.159 Given that $P(A \cap B) = .4$ and $P(A|B) = .8$, find $P(B)$.

3.160 The accompanying Venn diagram illustrates a sample space containing six sample points and three events, A, B, and C. The probabilities of the sample points are $P(1) = .3, P(2) = .2, P(3) = .1, P(4) = .1, P(5) = .1,$ and $P(6) = .2$.
  a. Find $P(A \cap B), P(B \cap C), P(A \cup C), P(A \cup B \cup C)$, $P(B^c), P(A^c \cap B), P(B|C)$, and $P(B|A)$.
  b. Are A and B independent? Mutually exclusive? Why?
  c. Are B and C independent? Mutually exclusive? Why?

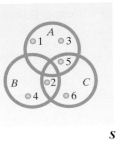

S

3.161 A fair die is tossed, and the up face is noted. If the number is even, the die is tossed again; if the number is odd, a fair coin is tossed. Consider the following events:

$$A: \{\text{A head appears on the coin.}\}$$
$$B: \{\text{The die is tossed only one time.}\}$$

  a. List the sample points in the sample space.
  b. Give the probability for each of the sample points.
  c. Find $P(A)$ and $P(B)$.
  d. Identify the sample points in $A^c, B^c, A \cap B$, and $A \cup B$.
  e. Find $P(A^c), P(B^c), P(A \cap B), P(A \cup B), P(A|B)$, and $P(B|A)$.
  f. Are A and B mutually exclusive events? Independent events? Why?

3.162 A balanced die is thrown once. If a 4 appears, a ball is drawn from urn 1; otherwise, a ball is drawn from urn 2. Urn 1 contains four red, three white, and three black balls. Urn 2 contains six red and four white balls.
  a. Find the probability that a red ball is drawn.
  b. Find the probability that urn 1 was used given that a red ball was drawn.

3.163 Two events, A and B, are independent, with $P(A) = .3$ and $P(B) = .1$.
  a. Are A and B mutually exclusive? Why?
  b. Find $P(A|B)$ and $P(B|A)$.
  c. Find $P(A \cup B)$.

3.164 Find the numerical value of
  a. 6!
  b. $\binom{10}{9}$
  c. $\binom{10}{1}$
  d. P_2^6
  e. $\binom{6}{3}$
  f. 0!
  g. P_4^{10}
  h. P_2^{50}

🔵 Applet Exercise 3.6

Use the applet entitled *Random Numbers* to generate a list of 50 numbers between 1 and 100, inclusive. Use the list to find each of the following probabilities:
  a. The probability that a number chosen from the list is less than or equal to 50

b. The probability that a number chosen from the list is even

c. The probability that a number chosen from the list is both less than or equal to 50 and even

d. The probability that a number chosen from the list is less than or equal to 50, given that the number is even

e. Do your results from parts **a–d** support the conclusion that the events *less than or equal to 50* and *even* are independent? Explain.

Applying the Concepts—Basic

3.165 Post office violence. *The Wall Street Journal* (Sept. 1, 2000) reported on an independent study of postal workers and violence at post offices. In a sample of 12,000 postal workers, 600 were physically assaulted on the job in a recent year. Use this information to estimate the probability that a randomly selected postal worker will be physically assaulted on the job during the year.

3.166 Sterile couples in Jordan. A sterile family is a couple that has no children by their deliberate choice or because they are biologically infertile. Couples who are childless by chance are not considered to be sterile. Researchers at Yarmouk University (in Jordan) estimated the proportion of sterile couples in that country to be .06 (*Journal of Data Science*, July 2003). Also, 64% of the sterile couples in Jordan are infertile. Find the probability that a Jordanian couple is both sterile and infertile.

3.167 NHTSA new car crash testing. Refer to the National Highway Traffic Safety Administration (NHTSA) crash tests of new car models, presented in Exercise 2.174 (p. 98). Recall that the NHTSA has developed a "star" scoring system, with results ranging from one star (*) to five stars (*****). The more stars in the rating, the better the level of crash protection in a head-on collision. A summary of the driver-side star ratings for 98 cars (saved in the **CRASH** file) is reproduced in the accompanying MINITAB printout. Assume that one of the 98 cars is selected at random. State whether each of the following is true or false:

a. The probability that the car has a rating of two stars is 4.

b. The probability that the car has a rating of four or five stars is .7857.

c. The probability that the car has a rating of one star is 0.

d. The car has a better chance of having a two-star rating than of having a five-star rating.

Tally for Discrete Variables: DRIVSTAR

DRIVSTAR	Count	Percent
2	4	4.08
3	17	17.35
4	59	60.20
5	18	18.37
N=	98	

3.168 Selecting a sample. A random sample of five students is to be selected from 50 sociology majors for participation in a special program.

a. In how many different ways can the sample be drawn?

b. Show how the random-number table, Table I of Appendix A, can be used to select the sample of students.

3.169 Mathematics achievement test. According to the National Center for Education Statistics, only 5% of U. S. eighth graders score above 655 on a mathematics assessment test. Find the probability that a randomly selected eighth grader has a score of 655 or below on the mathematics assessment test.

3.170 Oil spill impact on seabirds. Refer to the *Journal of Agricultural, Biological, and Environmental Statistics* (Sept. 2000) study of the impact of the *Exxon Valdez* tanker oil spill on the seabird population in Prince William Sound, Alaska, presented in Exercise 2.189 (p. 102). Recall that data were collected on 96 shoreline locations (called transects), and it was determined whether or not the transect was in an oiled area. The data are stored in the file called **EVOS**. From the data, estimate the probability that a randomly selected transect in Prince William Sound is contaminated with oil.

3.171 Fungi in beech forest trees. Beechwood forests in East Central Europe are being threatened by dynamic changes in land ownership and economic upheaval. The current status of the beech tree species in this area was evaluated by Hungarian university professors in *Applied Ecology and Environmental Research* (Vol. 1, 2003). Of 188 beech trees surveyed, 49 had been damaged by fungi. Depending on the species of fungus, damage will occur on either the trunk, branches, or leaves of the tree. In the damaged trees, the trunk was affected 85% of the time, the leaves 10% of the time, and the branches 5% of the time.

a. Give a reasonable estimate of the probability of a beech tree in East Central Europe being damaged by fungi.

b. A fungus-damaged beech tree is selected at random, and the area (trunk, leaf, or branch) affected is observed. List the sample points for this experiment, and assign a reasonable probability to each one.

3.172 Herbal medicines survey. Refer to *The American Association of Nurse Anesthetists Journal* (Feb. 2000) study on the use of herbal medicines before surgery, presented in Exercise 1.20 (p. 19). Recall that 51% of surgical patients use herbal medicines against their doctor's advice prior to surgery.

a. What is the probability that a randomly selected surgical patient will use herbal medicines against his or her doctor's advice?

b. What is the probability that in a sample of two independently selected surgical patients, both will use herbal medicines against their doctor's advice?

c. What is the probability that in a sample of five independently selected surgical patients, all five will use herbal medicines against their doctor's advice?

d. Would you expect the event in part **c** to occur? Explain.

3.173 Beach erosional hot spots. Beaches that exhibit high erosion rates relative to the surrounding beach are defined as *erosional hot spots.* The U.S. Army Corps of Engineers is conducting a study of beach hot spots. Through an online questionnaire, data are collected on six beach hot spots. The data are listed in the next table.

Beach Hot Spot	Beach Condition	Nearshore Bar Condition	Long-Term Erosion Rate (miles/year)
Miami Beach, FL	No dunes/flat	Single, shore parallel	4
Coney Island, NY	No dunes/flat	Other	13
Surfside, CA	Bluff/scarp	Single, shore parallel	35
Monmouth Beach, NJ	Single dune	Planar	Not estimated
Ocean City, NJ	Single dune	Other	Not estimated
Spring Lake, NJ	Not observed	Planar	14

Based on *Identification and Characterization of Erosional Hotspots*, William & Mary Virginia Institute of Marine Science, U.S. Army Corps of Engineers Project Report, Mar. 18, 2002.

a. Suppose you record the nearshore bar condition of each beach hot spot. Give the sample space for this experiment.

b. Find the probabilities of the sample points in the sample space you defined in part **a**.

c. What is the probability that a beach hot spot has either a planar or single, shore-parallel nearshore bar condition?

d. Now suppose you record the beach condition of each beach hot spot. Give the sample space for this experiment.

e. Find the probabilities of the sample points in the sample space you defined in part **d**.

f. What is the probability that the condition of the beach at a particular beach hot spot is not flat?

3.174 Chemical insect attractant. An entomologist is studying the effect of a chemical sex attractant (pheromone) on insects. Several insects are released at a site equidistant from the pheromone under study and a control substance. If the pheromone has an effect, more insects will travel toward it rather than toward the control. Otherwise, the insects are equally likely to travel in either direction. Suppose the pheromone under study has no effect, so that it is equally likely that an insect will move toward the pheromone or toward the control. Suppose five insects are released.

a. List or count the number of different ways the insects can travel.

b. What is the chance that all five travel toward the pheromone?

c. What is the chance that exactly four travel toward the pheromone?

d. What inference would you make if the event in part **c** actually occurs? Explain.

Applying the Concepts—Intermediate

3.175 Winning at roulette. *Roulette* is a very popular game in many American casinos. In Roulette, a ball spins on a circular wheel that is divided into 38 arcs of equal length, bearing the numbers 00, 0, 1, 2, ... , 35, 36. The number of the arc on which the ball stops is the outcome of one play of the game. The numbers are also colored in the manner shown in the following table.

Red: 1, 3, 5, 7, 9, 12, 14, 16, 18, 19, 21, 23, 25, 27, 30, 32, 34, 36
Black: 2, 4, 6, 8, 10, 11, 13, 15, 17, 20, 22, 24, 26, 28, 29, 31, 33, 35
Green: 00, 0

Players may place bets on the table in a variety of ways, including bets on odd, even, red, black, high, low, etc. Consider the following events:

A: {The outcome is an odd number (00 and 0 are considered neither odd nor even.)}
B: {The outcome is a black number.}
C: {The outcome is a low number (1–18).}

a. Define the event $A \cap B$ as a specific set of sample points.

b. Define the event $A \cup B$ as a specific set of sample points.

c. Find $P(A)$, $P(B)$, $P(A \cap B)$, $P(A \cup B)$, and $P(C)$ by summing the probabilities of the appropriate sample points.

d. Define the event $A \cap B \cap C$ as a specific set of sample points.

e. Use the additive rule to find $P(A \cup B)$. Are events A and B mutually exclusive? Why?

f. Find $P(A \cap B \cap C)$ by summing the probabilities of the sample points given in part **d**.

g. Define the event $(A \cup B \cup C)$ as a specific set of sample points.

h. Find $P(A \cup B \cup C)$ by summing the probabilities of the sample points given in part **g**.

3.176 Stacking in the NBA. In professional sports, *stacking* is a term used to describe the practice of African-American players being excluded from certain positions because of race. To illustrate the stacking phenomenon, the *Sociology of Sport Journal* (Vol. 14, 1997) presented the table shown here. The table summarizes the race and positions of 368 National Basketball Association (NBA) players in 1993. Suppose an NBA player is selected at random from that year's player pool.

	Position			
	Guard	Forward	Center	Totals
White	26	30	28	84
Black	128	122	34	284
Totals	154	152	62	368

a. What is the probability that the player is white?

b. What is the probability that the player is a center?

c. What is the probability that the player is African-American and plays guard?

d. What is the probability that the player is not a guard?

e. What is the probability that the player is white or a center?

f. Given that the player is white, what is the probability that he is a center?

g. Given that the player is African-American, what is the probability that he is a center?

h. Are the events {White player} and {Center} independent?

i. Use your answers to parts **f–h** to make an inference about stacking in the NBA.

*3.177 Errors in estimating job costs.** A construction company employs three sales engineers. Engineers 1, 2, and 3 estimate the costs of 30%, 20%, and 50%, respectively, of all jobs bid on by the company. For $i = 1, 2, 3$, define E_i to be the event that a job is estimated by engineer i. The following probabilities describe the rates at which the engineers make serious errors in estimating costs:

$P(\text{error}|E_1) = .01$, $P(\text{error}|E_2) = .03$, and $P(\text{error}|E_3) = .02$

a. If a particular bid results in a serious error in estimating job cost, what is the probability that the error was made by engineer 1?

b. If a particular bid results in a serious error in estimating job cost, what is the probability that the error was made by engineer 2?

c. If a particular bid results in a serious error in estimating job cost, what is the probability that the error was made by engineer 3?

d. Based on the probabilities given in parts **a–c**, which engineer is most likely responsible for making the serious error?

3.178 Elderly wheelchair user study. The *American Journal of Public Health* (Jan. 2002) reported on a study of elderly wheelchair users who live at home. A sample of 306 wheelchair users, age 65 or older, were surveyed about whether they had an injurious fall during the year and whether their home featured any one of five structural modifications: bathroom modifications, widened doorways/hallways, kitchen modifications, installed railings, and easy-open doors. The responses are summarized in the accompanying table. Suppose we select, at random, one of the 306 wheelchair users surveyed.

Home Features	Injurious Fall(s)	No Falls	Totals
All 5	2	7	9
At least 1, but not all	26	162	188
None	20	89	109
Totals	48	258	306

Based on Berg, K., Hines, M., and Allen, S. "Wheelchair users at home: Few home modifications and many injurious falls." *American Journal of Public Health,* Vol. 92, No. 1, Jan. 2002 (Table 1).

a. Find the probability that the wheelchair user had an injurious fall.

b. Find the probability that the wheelchair user had all five features installed in the home.

c. Find the probability that the wheelchair user had no falls and none of the features installed in the home.

d. Given no features installed in the home, find the probability of an injurious fall.

3.179 Shooting free throws. In college basketball games, a player may be afforded the opportunity to shoot two consecutive foul shots (free throws).

a. Suppose a player who makes (i.e., scores on) 80% of his foul shots has been awarded two free throws. If the two throws are considered independent, what is the probability that the player makes both shots? exactly one? neither shot?

b. Suppose a player who makes 80% of his first attempted foul shots has been awarded two free throws and the outcome on the second shot is dependent on the outcome of the first shot. In fact, if this player makes the first shot, he makes 90% of the second shots; and if he misses the first shot, he makes 70% of the second shots. In this case, what is the probability that the player makes both shots? exactly one? neither shot?

c. In parts **a** and **b**, we considered two ways of *modeling* the probability that a basketball player makes two consecutive foul shots. Which model do you think gives a more realistic explanation of the outcome of shooting foul shots; that is, do you think two consecutive foul shots are independent or dependent? Explain.

3.180 Lie detector test. Consider a lie detector called the Computerized Voice Stress Analyzer (CVSA). The manufacturer claims that the CVSA is 98% accurate and, unlike a polygraph machine, will not be thrown off by drugs and medical factors. However, laboratory studies by the U.S. Defense Department found that the CVSA had an accuracy rate of 49.8%, slightly less than pure chance. Suppose the CVSA is used to test the veracity of four suspects. Assume that the suspects' responses are independent.

a. If the manufacturer's claim is true, what is the probability that the CVSA will correctly determine the veracity of all four suspects?

b. If the manufacturer's claim is true, what is the probability that the CVSA will yield an incorrect result for at least one of the four suspects?

c. Suppose that in a laboratory experiment conducted by the U.S. Defense Department on four suspects, the CVSA yielded incorrect results for two of the suspects. Make an inference about the true accuracy rate of the new lie detector.

3.181 Maize seeds. The genetic origin and properties of maize (modern-day corn) were investigated in *Economic Botany*. Seeds from maize ears carry either single spikelets or paired spikelets, but not both. Progeny tests on approximately 600 maize ears revealed the following information: Forty percent of all seeds carry single spikelets, while 60% carry paired spikelets. A seed with single spikelets will produce maize ears with single spikelets 29% of the time and paired spikelets 71% of the time. A seed with paired spikelets will produce maize ears with single spikelets 26% of the time and paired spikelets 74% of the time.

a. Find the probability that a randomly selected maize ear seed carries a single spikelet and produces ears with single spikelets.

b. Find the probability that a randomly selected maize ear seed produces ears with paired spikelets.

*3.182 Selecting committee members. The Republican governor of a state with no income tax is appointing a committee of five members to consider changes in the income tax law. There are 15 state representatives—seven Democrats and eight Republicans—available for appointment to the committee. Assume that the governor selects the committee of five members randomly from the 15 representatives.

a. In how many different ways can the committee members be selected?

b. What is the probability that no Democrat is appointed to the committee? If this were to occur, what would you conclude about the assumption that the appointments were made at random? Why?

c. What is the probability that the majority of the committee members are Republican? If this were to occur, would you have reason to doubt that the governor made the selections randomly? Why?

3.183 Series and parallel systems. Consider the two systems shown in the schematic on page 174. System A operates properly only if all three components operate properly. (The three components are said to operate *in series*.) The probability of failure for system A components 1, 2, and 3 is .12, .09, and .11, respectively. Assume that the components operate independently of each other.

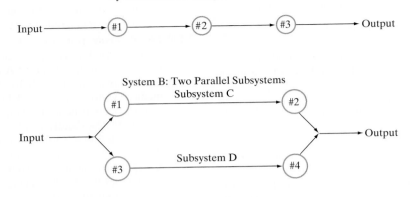

System B comprises two subsystems said to operate *in parallel.* Each subsystem has two components that operate in series. System B will operate properly as long as at least one of the subsystems functions properly. The probability of failure for each component in the system is .1. Assume that the components operate independently of each other.

a. Find the probability that System A operates properly.

b. What is the probability that at least one of the components in System A will fail and therefore that the system will fail?

c. Find the probability that System B operates properly.

d. Find the probability that exactly one subsystem in System B fails.

e. Find the probability that System B fails to operate properly.

f. How many parallel subsystems like the two shown here would be required to guarantee that the system would operate properly at least 99% of the time?

3.184 Monitoring the quality of power equipment. *Mechanical Engineering* (Feb. 2005) reported on the need for wireless networks to monitor the quality of industrial equipment. For example, consider Eaton Corp., a company that develops distribution products. Eaton estimates that 90% of the electrical switching devices it sells can monitor the quality of the power running through the device. Eaton further estimates that, of the buyers of electrical switching devices capable of monitoring quality, 90% do not wire the equipment up for that purpose. Use this information to estimate the probability that an Eaton electrical switching device is capable of monitoring power quality and is wired up for that purpose.

*3.185 Purchasing microchips. An important component of your desktop or laptop personal computer (PC) is a microchip. The next table gives the proportions of microchips that a certain PC manufacturer purchases from seven suppliers:

Supplier	Proportion
S_1	.15
S_2	.05
S_3	.10
S_4	.20
S_5	.12
S_6	.20
S_7	.18

a. It is known that the proportions of defective microchips produced by the seven suppliers are .001, .0003, .0007, .006, .0002, .0002, and .001, respectively. If a single PC microchip failure is observed, which supplier is most likely responsible?

b. Suppose the seven suppliers produce defective microchips at the same rate, .0005. If a single PC microchip failure is observed, which supplier is most likely responsible?

*3.186 Dream experiment in psychology. A clinical psychologist is given tapes of 20 subjects discussing their recent dreams. The psychologist is told that 10 are high-anxiety individuals and 10 are low-anxiety individuals. The psychologist's task is to select the 10 high-anxiety subjects.

a. Count the number of sample points for this experiment.

b. Assuming that the psychologist is guessing, assign probabilities to each of the sample points.

c. Find the probability that the psychologist guesses all classifications correctly.

d. Find the probability that the psychologist guesses at least 9 of the 10 high-anxiety subjects correctly.

3.187 Cigar smoking and cancer. The *Journal of the National Cancer Institute* (Feb. 16, 2000) published the results of a study that investigated the association between cigar smoking and death from tobacco-related cancers. Data were obtained for a national sample of 137,243 American men. The results are summarized in the table on page 175. Each male in the study was classified according to his cigar-smoking status and whether or not he died from a tobacco-related cancer.

a. Find the probability that a randomly selected man never smoked cigars and died from cancer.

b. Find the probability that a randomly selected man was a former cigar smoker and died from cancer.

c. Find the probability that a randomly selected man was a current cigar smoker and died from cancer.

d. Given that a male was a current cigar smoker, find the probability that he died from cancer.

e. Given that a male never smoked cigars, find the probability that he died from cancer.

Cigars	Died from Cancer		
	Yes	No	Totals
Never Smoked	782	120,747	121,529
Former Smoker	91	7,757	7,848
Current Smoker	141	7,725	7,866
Totals	1,014	136,229	137,243

Based on Shapiro, J. A., Jacobs, E. J., and Thun, M. J. "Cigar smoking in men and risk of death from tobacco-related cancers." *Journal of the National Cancer Institute,* Vol. 92, No. 4, Feb. 16, 2000 (Table 2).

Applying the Concepts—Advanced

3.188 Sex composition patterns of children in families. In having children, is there a genetic factor that causes some families to favor one sex over the other? That is, does having boys or girls "run in the family"? This was the question of interest in *Chance* (Fall 2001). Using data collected on children's sex for over 4,000 American families that had at least two children, the researchers compiled the accompanying table. Make an inference about whether having boys or girls "runs in the family."

Sex Composition of First Two Children	Frequency
Boy–Boy	1,085
Boy–Girl	1,086
Girl–Boy	1,111
Girl–Girl	926
Total	4,208

Based on Rodgers, J. L., and Doughty, D. "Does having boys or girls run in the family?" *Chance*, Vol. 14, No. 4, Fall 2001 (Table 3).

3.189 Odds of winning a horse race. Handicappers for horse races express their beliefs about the probability of each horse winning a race in terms of **odds.** If the probability of event E is $P(E)$, then the *odds in favor of E* are $P(E)$ to $1 - P(E)$. Thus, if a handicapper assesses a probability of .25 that Smarty Jones will win the Belmont Stakes, the odds in favor of Smarty Jones are $^{25}/_{100}$ to $^{75}/_{100}$, or 1 to 3. It follows that the *odds against E* are $1 - P(E)$ to $P(E)$, or 3 to 1 against a win by Smarty Jones. In general, if the odds in favor of event E are a to b, then $P(E) = a/(a + b)$.

a. A second handicapper assesses the probability of a win by Smarty Jones to be $^1/_3$. According to the second handicapper, what are the odds in favor of a Smarty Jones win?

b. A third handicapper assesses the odds in favor of Smarty Jones to be 1 to 1. According to the third handicapper, what is the probability of a Smarty Jones win?

c. A fourth handicapper assesses the odds against Smarty Jones winning to be 3 to 2. Find this handicapper's assessment of the probability that Smarty Jones will win.

3.190 Chance of winning at blackjack. Blackjack, a favorite game of gamblers, is played by a dealer and at least one opponent. At the outset of the game, 2 cards of a 52-card bridge deck are dealt to the player and 2 cards to the dealer. Drawing an ace and a face card is called *blackjack*. If the dealer does not draw a blackjack and the player does, the player wins. If both the dealer and player draw blackjack, a "push" (i.e., a tie) occurs.

a. What is the probability that the dealer will draw a blackjack?

b. What is the probability that the player wins with a blackjack?

3.191 Finding an organ transplant match. One of the problems encountered with organ transplants is the body's rejection of the transplanted tissue. If the antigens attached to the tissue cells of the donor and receiver match, the body will accept the transplanted tissue. Although the antigens in identical twins always match, the probability of a match in other siblings is .25, and that of a match in two people from the population at large is .001. Suppose you need a kidney and you have two brothers and a sister.

a. If one of your three siblings offers a kidney, what is the probability that the antigens will match?

b. If all three siblings offer a kidney, what is the probability that all three antigens will match?

c. If all three siblings offer a kidney, what is the probability that none of the antigens will match?

d. Repeat parts **b** and **c,** this time assuming that the three donors were obtained from the population at large.

3.192 Language impairment in children. Children who develop unexpected difficulties with the spoken language are often diagnosed as *specifically language impaired* (SLI). A study published in the *Journal of Speech, Language, and Hearing Research* (Dec. 1997) investigated the incidence of SLI in kindergarten children. As an initial screen, each in a national sample of over 7,000 children was given a test for language performance. The percentages of children who passed and failed the screen were 73.8% and 26.2%, respectively. All children who failed the screen were tested clinically for SLI. About one-third of those who passed the screen were randomly selected and also tested for SLI. The percentage of children diagnosed with SLI in the "failed screen" group was 20.5%; the percentage diagnosed with SLI in the "pass screen" group was 2.8%.

a. For this problem, let "pass" represent a child who passed the language performance screen, "fail" represent a child who failed the screen, and "SLI" represent a child diagnosed with SLI. Now find each of the following probabilities: $P(\text{Pass})$, $P(\text{Fail})$, $P(\text{SLI}|\text{Pass})$, and $P(\text{SLI}|\text{Fail})$.

b. Use the probabilities from part **a** to find $P(\text{Pass} \cap \text{SLI})$ and $P(\text{Fail} \cap \text{SLI})$. What probability law did you use to calculate these probabilities?

c. Use the probabilities from part **b** to find $P(\text{SLI})$. What probability law did you use to calculate this probability?

3.193 Accuracy of pregnancy tests. Seventy-five percent of all women who submit to pregnancy tests are really pregnant. A certain pregnancy test gives a *false positive* result with probability .02 and a *valid positive result* with probability .99. If a particular woman's test is positive, what is the probability that she really is pregnant? [*Hint:* If A is the event that a woman is pregnant and B is the event that the pregnancy test is positive, then B is the union of the two mutually exclusive events $A \cap B$ and $A^c \cap B$. Also, the probability of a false positive result may be written as $P(B|A^c) = .02$.]

3.194 Chance of winning at "craps." A version of the dice game "craps" is played in the following manner. A player starts by rolling two balanced dice. If the roll (the sum of the two numbers showing on the dice) results in a 7 or 11, the player wins. If the roll results in a 2 or a 3 (called *craps*), the player loses. For any other roll outcome, the player continues to throw the dice until the original roll outcome recurs (in which case the player wins) or until a 7 occurs (in which case the player loses).

a. What is the probability that a player wins the game on the first roll of the dice?

b. What is the probability that a player loses the game on the first roll of the dice?

c. If the player throws a total of 4 on the first roll, what is the probability that the game ends (win or lose) on the next roll?

*3.195 **The perfect bridge hand.** According to a morning news program, a very rare event recently occurred in Dubuque, Iowa. Each of four women playing bridge was astounded to note that she had been dealt a perfect bridge hand. That is, one woman was dealt all 13 spades, another all 13 hearts, another all the diamonds, and another all the clubs. What is the probability of this rare event?

3.196 **Odd Man Out.** Three people play a game called "Odd Man Out." In this game, each player flips a fair coin until the outcome (heads or tails) for one of the players is not the same as that for the other two players. This player is then "the odd man out" and loses the game. Find the probability that the game ends (i.e., either exactly one of the coins will fall heads or exactly one of the coins will fall tails) after only one toss by each player. Suppose one of the players, hoping to reduce his chances of being the odd man out, uses a two-headed coin. Will this ploy be successful? Solve by listing the sample points in the sample space.

Critical Thinking Challenges

3.197 **"Let's Make a Deal."** Marilyn vos Savant, who is listed in the *Guinness Book of World Records Hall of Fame* as having the "Highest IQ," writes a weekly column in the Sunday newspaper supplement *Parade Magazine*. Her column, "Ask Marilyn," is devoted to games of skill, puzzles, and mind-bending riddles. In one issue (*Parade Magazine*, Feb. 24, 1991), vos Savant posed the following question:

Suppose you're on a game show and you're given a choice of three doors. Behind one door is a car; behind the others, goats. You pick a door—say, #1—and the host, who knows what's behind the doors, opens another door—say #3—which has a goat. He then says to you, "Do you want to pick door #2?" Is it to your advantage to switch your choice?

Marilyn's answer: "Yes, you should switch. The first door has a $1/3$ chance of winning [the car], but the second has a $2/3$ chance [of winning the car]." Predictably, vos Savant's surprising answer elicited thousands of critical letters, many of them from Ph.D. mathematicians, that disagreed with her. Who is correct, the Ph.D.s or Marilyn?

3.198 **Flawed Pentium computer chip.** In October 1994, a flaw was discovered in the Pentium microchip installed in personal computers. The chip produced an incorrect result when dividing two numbers. Intel, the manufacturer of the Pentium chip, initially announced that such an error would occur once in 9 billion division operations, or "once every 27,000 years," for a typical user; consequently, Intel did not immediately offer to replace the chip.

Depending on the procedure, statistical software packages (e.g., SAS) may perform an extremely large number of divisions to produce required output. For heavy users of the software, 1 billion divisions over a short time frame is not unusual. Will the flawed chip be a problem for a heavy SAS user? [*Note:* Two months after the flaw was discovered, Intel agreed to replace all Pentium chips free of charge.]

3.199 **Most likely coin-toss sequence.** In *Parade Magazine*'s (Nov. 26, 2000) column "Ask Marilyn," the following question was posed: "I have just tossed a [balanced] coin 10 times, and I ask you to guess which of the following three sequences was the result. One (and only one) of the sequences is genuine."

1. H H H H H H H H H H
2. H H T T H T T H H H
3. T T T T T T T T T T

Marilyn's answer to the question posed was "Though the chances of the three specific sequences occurring randomly are equal … it's reasonable for us to choose sequence (2) as the most likely genuine result." Do you agree?

Activity | Exit Polls

Exit polls are conducted in selected locations as voters leave their polling places after voting. In addition to being used to predict the outcome of elections before the votes are counted, these polls seek to gauge tendencies among voters. The results are usually stated in terms of conditional probabilities.

The accompanying table shows the results of exit polling which suggest that men were almost evenly split on voting for John McCain or Barack Obama, while women were more likely to vote for Obama in the 2008 presidential election. In addition, the table suggests that more women than men voted in the election. The six percentages in the last three columns represent conditional probabilities, where the given event is gender.

2008 Presidential Election, Vote by Gender

	Obama	McCain	Other
Male (47%)	49%	48%	3%
Female (53%)	56%	43%	1%

Based on CNN.com

1. Find similar exit poll results in which voters are categorized by race, income, education, or some other criterion for a recent national, state, or local election. Choose two different examples, and interpret the percentages given as probabilities, or conditional probabilities where appropriate.

2. Use the multiplicative rule of probability to find the probabilities related to the percentages given. [For example, in the accompanying table, find *P*(Obama and Male).] Then interpret each of these probabilities and use them to determine the total percentage of the electorate that voted for each candidate.

3. Describe a situation in which a political group might use a form of exit polling to gauge voters' opinions on a "hot- button" topic (e.g., global warming). Identify the political group, the "hot-button" topic, the criterion used to categorize voters, and how the voters' opinions will be determined. Then describe how the results will be summarized as conditional probabilities. How might the results of the poll be used to support a particular agenda?

References

Bennett, D. J. *Randomness*. Cambridge, MA: Harvard University Press, 1998.

Epstein, R. A. *The Theory of Gambling and Statistical Logic*, rev. ed. New York: Academic Press, 1977.

Feller, W. *An Introduction to Probability Theory and Its Applications*, 3rd ed., Vol. 1. New York: Wiley, 1968.

Lindley, D. V. *Making Decisions*, 2nd ed. London: Wiley, 1985.

Parzen, E. *Modern Probability Theory and Its Applications*. New York: Wiley, 1960.

Wackerly, D., Mendenhall, W., and Scheaffer, R. L. *Mathematical Statistics with Applications*, 7th ed. Boston: Duxbury, 2008.

Williams, B. *A Sampler on Sampling*. New York: Wiley, 1978.

Winkler, R. L. *An Introduction to Bayesian Inference and Decision*. New York: Holt, Rinehart and Winston, 1972.

Wright, G., and Ayton, P., eds. *Subjective Probability*. New York: Wiley, 1994.

USING TECHNOLOGY

MINITAB: Generating a Random Sample

Step 1 Click on the "Calc" button on the MINITAB menu bar, then click on "Random Data," and finally, click on "Sample From Columns," as shown in Figure 3.M.1. The resulting dialog box appears as shown in Figure 3.M.2.

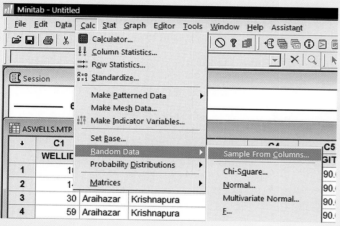

Figure 3.M.1
MINITAB menu options for sampling from a data set

Figure 3.M.2
MINITAB options for selecting a random sample from worksheet columns

Step 2 Specify the sample size (i.e., number of rows), the variable(s) to be sampled, and the column(s) where you want to save the sample.

Step 3 Click "OK" and the MINITAB, worksheet will reappear with the values of the variable for the selected (sampled) cases in the column specified.

In MINITAB, you can also generate a sample of case numbers.

Step 1 From the MINITAB menu, click on the "Calc" button, and then click on "Random Data," and finally, click on the "Uniform" option (see Figure 3.M.1).

Step 2 In the resulting dialog box (shown in Figure 3.M.3), specify the number of cases (rows, i.e., the sample size), and the column where the case numbers selected will be stored.

Figure 3.M.3
MINITAB options for selecting a random sample of cases

Step 3 Click "OK" and the MINITAB worksheet will reappear with the case numbers for the selected (sampled) cases in the column specified.

[*Note:* If you want the option of generating the same (identical) sample multiple times from the data set, then first click on the "Set Base" option shown in Figure 3.M.1. Specify an integer in the resulting dialog box. If you always select the same integer, MINITAB will select the same sample when you choose the random sampling options.]

TI-83/TI-84 Plus Graphing Calculator: Combinations and Permutations

Combinations—Choosing *n* Elements from *N* Elements

Step 1 Enter the value of *N*.

Step 2 Press **MATH**, then select **PRB.**

Step 3 Press **nCr** for combinations.

Step 4 Enter the value of *n*, then press **ENTER.**

Permutations—Choosing *n* Elements from *N* Elements

Step 1 Enter the value of *N*.

Step 2 Press **MATH.** Then select **PRB.**

Step 3 Press **nPr** for permutations.

Step 4 Enter the value of *n*, then press **ENTER.**

4 Discrete Random Variables

CONTENTS

Where We've Been

- Used probability to make an inference about a population from data in an observed sample
- Used probability to measure the reliability of the inference

Where We're Going

- Develop the notion of a random variable (4.1)
- Learn that many types of numerical data are observed values of discrete random variables (4.1)
- Learn the probabilistic properties of discrete random variables (4.2–4.3)
- Study several important types of discrete random variables and their probability models (4.4–4.6)

The American Statistician (May 1991) described an interesting application of a discrete probability distribution in a case involving illegal drugs. It all started with a "bust" in a midsized Florida city. During the bust, police seized approximately 500 foil packets of a white, powdery substance, presumably cocaine. Since it is not a crime to buy or sell nonnarcotic cocaine look-alikes (e.g., inert powders), detectives had to prove that the packets actually contained cocaine in order to convict their suspects of drug trafficking. When the police laboratory randomly selected and chemically tested 4 of the packets, all 4 tested positive for cocaine. This finding led to the conviction of the traffickers.

After the conviction, the police decided to use the remaining foil packets (i.e., those not tested) in reverse sting operations. Two of these packets were randomly selected and sold by undercover officers to a buyer. Between the sale and the arrest, however, the buyer disposed of the evidence. The key question is, Beyond a reasonable doubt, did the defendant really purchase cocaine?

In court, the defendant's attorney argued that his client should not be convicted, because the police could not prove that the missing foil packets contained cocaine. The police contended, however, that since 4 of the original packets tested positive for cocaine, the 2 packets sold in the reverse sting were also highly likely to contain cocaine. In this chapter, two Statistics in Action Revisited examples demonstrate how to use probability models to solve the dilemma posed by the police's reverse sting. (The case represented Florida's first cocaine-possession conviction without the actual physical evidence.)

Statistics IN Action Revisited

- Using the Binomial Model to Solve the Cocaine Sting Case (p. 204)
- Using the Hypergeometric Model to Solve the Cocaine Sting Case (p. 214)

You may have noticed that many of the examples of experiments in Chapter 3 generated quantitative (numerical) observations. The unemployment rate, the percentage of voters favoring a particular candidate, the cost of textbooks for a school term, and the amount of pesticide in the discharge waters of a chemical plant are all examples of numerical measurements of some phenomenon. Thus, most experiments have sample points that correspond to values of some numerical variable.

To illustrate, consider the coin-tossing experiment of Chapter 3. Figure 4.1 is a Venn diagram showing the sample points when two coins are tossed and the up faces (heads or tails) of the coins are observed. One possible numerical outcome is the total number of heads observed. This value (0, 1, or 2) is shown in parentheses on the Venn diagram, with one numerical value associated with each sample point. In the jargon of probability, the variable "total number of heads observed in two tosses of a coin" is called a *random variable*.

HH
(2)

HT
(1)

TT
(0)

TH
(1)

S

Figure 4.1

Venn diagram for coin-tossing experiment

> A **random variable** is a variable that assumes numerical values associated with the random outcomes of an experiment, where one (and only one) numerical value is assigned to each sample point.

The term *random variable* is more meaningful than the term *variable* alone, because the adjective *random* indicates that the coin-tossing experiment may result in one of the several possible values of the variable—0, 1, and 2—according to the *random* outcomes of the experiment: *HH, HT, TH*, and *TT*. Similarly, if the experiment is to count the number of customers who use the drive-up window of a bank each day, the random variable (the number of customers) will vary from day to day, partly because of the random phenomena that influence whether customers use the window. Thus, the possible values of this random variable range from 0 to the maximum number of customers the window can serve in a day.

We define two different types of random variables, *discrete* and *continuous*, in Section 4.1. Then we spend the remainder of the chapter discussing specific types of discrete random variables and the aspects that make them important to the statistician. We discuss continuous random variables in Chapter 5.

4.1 Two Types of Random Variables

Recall that the sample-point probabilities corresponding to an experiment must sum to 1. Dividing one unit of probability among the sample points in a sample space and, consequently, assigning probabilities to the values of a random variable is not always as easy as the examples in Chapter 3 might lead you to believe. If the number of sample points can be completely listed, the job is straightforward. But if the experiment results in an infinite number of sample points that are impossible to list, the task of assigning probabilities to the sample points is impossible without the aid of a probability model. The next three examples demonstrate the need for different probability models, depending on the number of values that a random variable can assume.

Example 4.1

Values of a Discrete Random Variable— Wine Ratings

Problem A panel of 10 experts for the *Wine Spectator* (a national publication) is asked to taste a new white wine and assign it a rating of 0, 1, 2, or 3. A score is then obtained by adding together the ratings of the 10 experts. How many values can this random variable assume?

Solution A sample point is a sequence of 10 numbers associated with the rating of each expert. For example, one sample point is

$$\{1, 0, 0, 1, 2, 0, 0, 3, 1, 0\}$$

The random variable assigns a score to each one of these sample points by adding the 10 numbers together. Thus, the smallest score is 0 (if all 10 ratings are 0), and the largest score is 30 (if all 10 ratings are 3). Since every integer between 0 and 30 is a possible score, the random variable denoted by the symbol x can assume 31 values. Note that the value of the random variable for the sample point shown here is $x = 8$.*

Look Back This is an example of a *discrete random variable*, since there is a finite number of distinct possible values. Whenever all the possible values a random variable can assume can be listed (or *counted*), the random variable is *discrete*.

Example 4.2

Values of a Discrete Random Variable— EPA Application

Problem Suppose the Environmental Protection Agency (EPA) takes readings once a month on the amount of pesticide in the discharge water of a chemical company. If the amount of pesticide exceeds the maximum level set by the EPA, the company is forced to take corrective action and may be subject to penalty. Consider the random variable number x of months before the company's discharge exceeds the EPA's maximum level. What values can x assume?

Solution The company's discharge of pesticide may exceed the maximum allowable level on the first month of testing, the second month of testing, etc. It is possible that the company's discharge will *never* exceed the maximum level. Thus, the set of possible values for

---

*The standard mathematical convention is to use a capital letter (e.g., X) to denote the theoretical random variable. The possible values (or realizations) of the random variable are typically denoted with a lowercase letter (e.g., x). Thus, in Example 4.1, the random variable X can take on the values $x = 0, 1, 2, \ldots, 30$. Since this notation can be confusing for introductory statistics students, we simplify the notation by using the lowercase x to represent the random variable throughout.

the number of months until the level is first exceeded is the set of all positive integers $1, 2, 3, 4, \ldots$.

Look Back If we can list the values of a random variable x, even though the list is never ending, we call the list **countable** and the corresponding random variable **discrete**. Thus, the number of months until the company's discharge first exceeds the limit is a **discrete random variable.**

Now Work Exercise 4.7

Example 4.3

Values of a Continuous Random Variable—Another EPA Application

Problem Refer to Example 4.2. A second random variable of interest is the amount x of pesticide (in milligrams per liter) found in the monthly sample of discharge waters from the same chemical company. What values can this random variable assume?

Solution Some possible values of x are 1.7, 28.42, and 100.987 milligrams per liter. Unlike the *number* of months before the company's discharge exceeds the EPA's maximum level, the set of all possible values for the *amount* of discharge *cannot* be listed (i.e., is not countable). The possible values for the amount x of pesticide would correspond to the points on the interval between 0 and the largest possible value the amount of the discharge could attain, the maximum number of milligrams that could occupy 1 liter of volume. (Practically, the interval would be much smaller, say, between 0 and 500 milligrams per liter.)

Look Ahead When the values of a random variable are not countable, but instead correspond to the points on some interval, we call the variable a *continuous random variable*. Thus, the *amount* of pesticide in the chemical plant's discharge waters is a *continuous random variable*.

Now Work Exercise 4.9

> Random variables that can assume a *countable* number of values are called **discrete.**

> Random variables that can assume values corresponding to any of the points contained in an interval are called **continuous.**

The following are examples of discrete random variables:

1. The number of seizures an epileptic patient has in a given week: $x = 0, 1, 2, \ldots$

2. The number of voters in a sample of 500 who favor impeachment of the president: $x = 0, 1, 2, \ldots, 500$

3. The shoe size of a tennis player: $x = \ldots\ 5, 5\frac{1}{2}, 6, 6\frac{1}{2}, 7, 7\frac{1}{2} \ldots$

4. The change received for paying a bill: $x = 1\cent, 2\cent, 3\cent, \ldots, \$1, \$1.01, \$1.02, \ldots$

5. The number of customers waiting to be served in a restaurant at a particular time: $x = 0, 1, 2, \ldots$

Note that several of the examples of discrete random variables begin with the words *The number of.* ... This wording is very common, since the discrete random variables most frequently observed are counts. The following are examples of continuous random variables:

1. The length of time (in seconds) between arrivals at a hospital clinic: $0 \leq x \leq \infty$ (infinity)

2. The length of time (in minutes) it takes a student to complete a one-hour exam: $0 \leq x \leq 60$

3. The amount (in ounces) of carbonated beverage loaded into a 12-ounce can in a can-filling operation: $0 \leq x \leq 12$

4. The depth (in feet) at which a successful oil-drilling venture first strikes oil: $0 \leq x \leq c$, where c is the maximum depth obtainable

5. The weight (in pounds) of a food item bought in a supermarket: $0 \leq x \leq 500$ [*Note:* Theoretically, there is no upper limit on x, but it is unlikely that it would exceed 500 pounds.]

Discrete random variables and their probability distributions are discussed in this chapter. Continuous random variables and their probability distributions are the topic of Chapter 5.

Exercises 4.1–4.14

Understanding the Principles

4.1 What is a random variable?

4.2 How do discrete and continuous random variables differ?

Applying the Concepts—Basic

4.3 **Type of Random Variable.** Classify the following random variables according to whether they are discrete or continuous:
 a. The number of words spelled correctly by a student on a spelling test
 b. The amount of water flowing through the Hoover Dam in a day
 c. The length of time an employee is late for work
 d. The number of bacteria in a particular cubic centimeter of drinking water
 e. The amount of carbon monoxide produced per gallon of unleaded gas
 f. Your weight

4.4 **Type of Random Variable.** Identify the following random variables as discrete or continuous:
 a. The amount of flu vaccine in a syringe
 b. The heart rate (number of beats per minute) of an American male
 c. The time it takes a student to complete an examination
 d. The barometric pressure at a given location
 e. The number of registered voters who vote in a national election
 f. Your score on the either the SAT or ACT

4.5 **Type of Random Variable.** Identify the following variables as discrete or continuous:
 a. The difference in reaction time to the same stimulus before and after training
 b. The number of violent crimes committed per month in your community
 c. The number of commercial aircraft near-misses per month
 d. The number of winners each week in a state lottery
 e. The number of free throws made per game by a basketball team
 f. The distance traveled by a school bus each day

4.6 **NHTSA crash tests.** The National Highway Traffic Safety Administration (NHTSA) has developed a driver-side "star" scoring system for crash-testing new cars.

Each crash-tested car is given a rating ranging from one star (*) to five stars (*****): the more stars in the rating, the better is the level of crash protection in a head-on collision. Suppose that a car is selected and its driver-side star rating is determined. Let x equal the number of stars in the rating. Is x a discrete or continuous random variable?

4.7 **Customers in line at a Subway shop.** The number of customers, x, waiting in line to order sandwiches at a Subway shop at noon is of interest to the store manager. What values can x assume? Is x a discrete or continuous random variable?

4.8 **Sound waves from a basketball.** Refer to the *American Journal of Physics* (June 2010) experiment on sound waves produced from striking a basketball, Exercise 2.37 (p. 47). Recall that the frequencies of sound wave echoes resulting from striking a hanging basketball with a metal rod were recorded. Classify the random variable, frequency (measured in hertz) of an echo, as discrete or continuous.

4.9 **Mongolian desert ants.** Refer to the *Journal of Biogeography* (Dec. 2003) study of ants in Mongolia, presented in Exercise 2.68 (p. 59). Two of the several variables recorded at each of 11 study sites were annual rainfall (in millimeters) and number of ant species. Identify these variables as discrete or continuous.

4.10 **Motivation of drug dealers.** Refer to the *Applied Psychology in Criminal Justice* (Sept. 2009) study of the personality characteristics of drug dealers, Exercise 2.98 (p. 71). For each of 100 convicted drug dealers, the researchers measured several variables, including the number of prior felony arrests x. Is x a discrete or continuous random variable? Explain.

Applying the Concepts—Intermediate

4.11 **Psychology.** Give an example of a discrete random variable of interest to a psychologist.

4.12 **Sociology.** Give an example of a discrete random variable of interest to a sociologist.

4.13 **Nursing.** Give an example of a discrete random variable of interest to a hospital nurse.

4.14 **Art history.** Give an example of a discrete random variable of interest to an art historian.

4.2 Probability Distributions for Discrete Random Variables

A complete description of a discrete random variable requires that we *specify all the values the random variable can assume* and *the probability associated with each value*. To illustrate, consider Example 4.4.

Example 4.4

Finding a Probability Distribution— Coin-Tossing Experiment

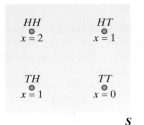

Figure 4.2
Venn diagram for the two-coin-toss experiment

Problem Recall the experiment of tossing two coins (p. 180), and let x be the number of heads observed. Find the probability associated with each value of the random variable x, assuming that the two coins are fair.

Solution The sample space and sample points for this experiment are reproduced in Figure 4.2. Note that the random variable x can assume values 0, 1, 2. Recall (from Chapter 3) that the probability associated with each of the four sample points is $1/4$. Then, identifying the probabilities of the sample points associated with each of these values of x, we have

$$P(x = 0) = P(TT) = \frac{1}{4}$$

$$P(x = 1) = P(TH) + P(HT) = \frac{1}{4} + \frac{1}{4} = \frac{1}{2}$$

$$P(x = 2) = P(HH) = \frac{1}{4}$$

Thus, we now know the values the random variable can assume (0, 1, 2) and how the probability is *distributed over* those values ($1/4$, $1/2$, $1/4$). This dual specification completely describes the random variable and is referred to as the *probability distribution*, denoted by the symbol $p(x)$.* The probability distribution for the coin-toss example is shown in tabular form in Table 4.1 and in graphic form in Figure 4.3 Since the probability distribution for a discrete random variable is concentrated at specific points (values of x), the graph in Figure 4.3a represents the probabilities as the heights of vertical lines over the corresponding values of x. Although the representation of the probability distribution as a histogram, as in Figure 4.3b, is less precise (since the probability is spread over a unit interval), the histogram representation will prove useful when we approximate probabilities of certain discrete random variables in Section 4.4.

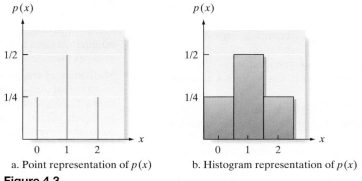

a. Point representation of $p(x)$ b. Histogram representation of $p(x)$

Figure 4.3
Probability distribution for coin-toss experiment: graphical form

Table 4.1

Probability Distribution for Coin-Toss Experiment: Tabular Form

x	$p(x)$
0	$1/4$
1	$1/2$
2	$1/4$

Look Ahead We could also present the probability distribution for x as a formula, but this would unnecessarily complicate a very simple example. We give the formulas for the probability distributions of some common discrete random variables later in the chapter.

Now Work Exercise 4.21

---

*In standard mathematical notation, the probability that a random variable X takes on a value x is denoted $P(X = x) = p(x)$. Thus, $P(X = 0) = p(0)$, $P(X = 1) = p(1)$, etc. In this text, we adopt the simpler $p(x)$ notation.

The **probability distribution of a discrete random variable** is a graph, table, or formula that specifies the probability associated with each possible value that the random variable can assume.

Two requirements must be satisfied by all probability distributions for discrete random variables:

Requirements for the Probability Distribution of a Discrete Random Variable x

1. $p(x) \geq 0$ for all values of x.

2. $\sum p(x) = 1$

where the summation of $p(x)$ is over all possible values of x.*

Example 4.5

Probability Distribution from a Graph—Playing Craps

Problem Craps is a popular casino game in which a player throws two dice and bets on the outcome (the sum total of the dots showing on the upper faces of the two dice). Consider a $5 wager. On the first toss (called the *come-out* roll), if the total is 7 or 11 the roller wins $5. If the outcome is a 2, 3, or 12, the roller loses $5 (i.e., the roller wins −$5). For any other outcome (4, 5, 6, 8, 9, or 10) a *point* is established and no money is lost or won on that roll (i.e., the roller wins $0). In a computer simulation of repeated tosses of two dice, the outcome x of the come-out roll wager (−$5, $0, or +$5) was recorded. A relative frequency histogram summarizing the results is shown in Figure 4.4. Use the histogram to find the approximate probability distribution of x.

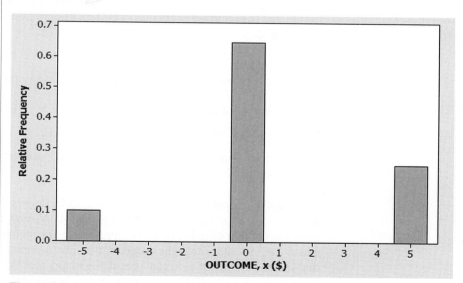

Figure 4.4

MINITAB Histogram for $5 Wager on Come-Out Roll in Craps

Solution The histogram shows that the relative frequencies of the outcomes $x = -\$5$, $x = \$0$, and $x = \$5$ are .1, .65, and .25, respectively. For example, in repeated tosses of two dice, 25% of the outcomes resulted in a sum of 7 or 11 (a $5 win for

*Unless otherwise indicated, summations will always be over all possible values of x.

the roller). Based on our long-run definition of probability given in Chapter 3, these relative frequencies estimate the probabilities of the three outcomes. Consequently, the approximate probability distribution of x, the outcome of the come-out wager in craps, is $p(-\$5) = .1, p(\$0) = .65$, and $p(\$5) = .25$. Note that these probabilities sum to 1.

Look Back When two dice are tossed, there is a total of 36 possible outcomes. (Can you list these outcomes, or sample points?) Of these, 4 result in a sum of 2, 3, or 12; 24 result in a sum of 4, 5, 6, 8, 9, or 10; and 8 result in a sum of 7 or 11. Using the rules of probability established in Chapter 3, you can show that the actual probability distribution for x is $p(-\$5) = 4/36 = .1111, p(\$0) = 24/36 = .6667$, and $p(\$5) = 8/36 = .2222$.

Now Work Exercise 4.17

Examples 4.4 and 4.5 illustrate how the probability distribution for a discrete random variable can be derived, but for many practical situations the task is much more difficult. Fortunately, numerous experiments and associated discrete random variables observed in nature possess identical characteristics. Thus, you might observe a random variable in a psychology experiment that would possess the same probability distribution as a random variable observed in an engineering experiment or a social sample survey. We classify random variables according to type of experiment, derive the probability distribution for each of the different types, and then use the appropriate probability distribution when a particular type of random variable is observed in a practical situation. The probability distributions for most commonly occurring discrete random variables have already been derived. This fact simplifies the problem of finding the probability distributions for random variables, as the next example illustrates.

Example 4.6
Probability Distribution Using a Formula—Texas Droughts

Problem A drought is a period of abnormal dry weather that causes serious problems in the farming industry of the region. University of Arizona researchers used historical annual data to study the severity of droughts in Texas (*Journal of Hydrologic Engineering*, Sept./Oct. 2003). The researchers showed that the distribution of x, the number of consecutive years that must be sampled until a dry (drought) year is observed, can be modeled using the formula

$$p(x) = (.3)(.7)^{x-1}, x = 1, 2, 3, \ldots$$

Find the probability that exactly 3 years must be sampled before a drought year occurs.

Solution We want to find the probability that $x = 3$. Using the formula, we have

$$p(3) = (.3)(.7)^{3-1} = (.3)(.7)^2 = (.3)(.49) = .147$$

Thus, there is about a 15% chance that exactly 3 years must be sampled before a drought year occurs in Texas.

Look Back The probability of interest can also be derived using the principles of probability developed in Chapter 3. The event of interest is $N_1 N_2 D_3$, where N_1 represents no drought occurs in the first sampled year, N_2 represents no drought occurs in the second sampled year, and D_3 represents a drought occurs in the third sampled year. The researchers discovered that the probability of a drought occurring in any sampled year is .3 (and, consequently, the probability of no drought occurring in any sampled year is .7). Using the multiplicative rule of probability for independent events, the probability of interest is $(.7)(.7)(.3) = .147$.

In Sections 4.4–4.6, we describe three important types of discrete random variables, give their probability distributions, and explain where and how they can be applied in practice. (Mathematical derivations of the probability distributions will be omitted, but these details can be found in the chapter references.)

But first, in Section 4.3, we discuss some descriptive measures of these sometimes complex probability distributions. Since probability distributions are analogous to the relative frequency distributions of Chapter 2, it should be no surprise that the mean and standard deviation are useful descriptive measures.

Exercises 4.15–4.33

Understanding the Principles

4.15 Give three different ways of representing the probability distribution of a discrete random variable.

Learning the Mechanics

4.16 Consider the following probability distribution:

x	−4	0	1	3
$p(x)$	.1	.2	.4	.3

  a. List the values that x may assume.
  b. What value of x is most probable?
  c. What is the probability that x is greater than 0?
  d. What is the probability that $x = -2$?

4.17 A discrete random variable x can assume five possible values: 20, 21, 22, 23, and 24. The MINITAB histogram at the bottom of the page shows the likelihood of each value.
  a. What is $p(22)$?
  b. What is the probability that x equals 20 or 24?
  c. What is $P(x \le 23)$?

4.18 Explain why each of the following is or is not a valid probability distribution for a discrete random variable x:

a.
x	0	1	2	3
$p(x)$	.2	.3	.3	.2

b.
x	−2	−1	0
$p(x)$	.25	.50	.20

c.
x	4	9	20
$p(x)$	−.3	1.0	.3

d.
x	2	3	5	6
$p(x)$	.15	.20	.40	.35

4.19 The random variable x has the following discrete probability distribution:

x	10	11	12	13	14
$p(x)$	.2	.3	.2	.1	.2

Since the values that x can assume are mutually exclusive events, the event $\{x \le 12\}$ is the union of three mutually exclusive events:

$$\{x = 10\} \cup \{x = 11\} \cup \{x = 12\}$$

  a. Find $P(x \le 12)$.
  b. Find $P(x > 12)$.
  c. Find $P(x \le 14)$.
  d. Find $P(x = 14)$.
  e. Find $P(x \le 11$ or $x > 12)$.

4.20 The random variable x has the discrete probability distribution shown here:

x	−2	−1	0	1	2
$p(x)$	.10	.15	.40	.30	.05

  a. Find $P(x \le 0)$.
  b. Find $P(x > -1)$.
  c. Find $P(-1 \le x \le 1)$.
  d. Find $P(x < 2)$.
  e. Find $P(-1 < x < 2)$.
  f. Find $P(x < 1)$.

4.21 Toss three fair coins, and let x equal the number of heads observed.
  a. Identify the sample points associated with this experiment, and assign a value of x to each sample point.
  b. Calculate $p(x)$ for each value of x.

Histogram for Exercise 4.17

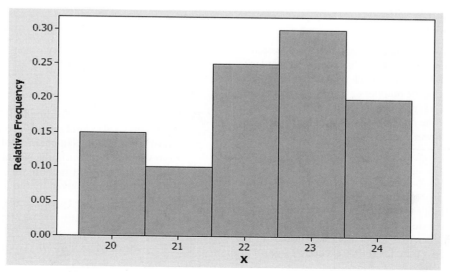

c. Construct a probability histogram for $p(x)$.

d. What is $P(x = 2 \text{ or } x = 3)$? $^1/_2$

Applet Exercise 4.1

Use the applet entitled *Random Numbers* to generate a list of 25 numbers between 1 and 3, inclusive. Let x represent a number chosen from this list.

a. What are the possible values of x?

b. Give the probability distribution for x in table form.

c. Let y be a number randomly chosen from the set $\{1, 2, 3\}$. Give the probability distribution for y in table form.

d. Compare the probability distributions of x and y in parts **b** and **c**. Why should these distributions be approximately the same?

Applet Exercise 4.2

Run the applet entitled *Simulating the Probability of a Head with a Fair Coin* 10 times with $n = 2$, resetting between runs, to simulate flipping two coins 10 times. Count and record the number of heads each time. Let x represent the number of heads on a single flip of the coins.

a. What are the possible values of x?

b. Use the results of the simulation to find the probability distribution for x in table form.

c. Explain why the probability of exactly two heads should be close to .25.

Applying the Concepts—Basic

4.22 **NHTSA crash tests.** Refer to the National Highway Traffic Safety Administration (NHTSA) crash tests of new car models, presented in Exercise 4.6 (p. 183). A summary of the driver-side star ratings for the 98 cars in the **CRASH** file is reproduced in the accompanying MINITAB printout. Assume that 1 of the 98 cars is selected at random, and let x equal the number of stars in the car's driver-side star rating.

a. Use the information in the printout to find the probability distribution for x.

b. Find $P(x = 5)$.

c. Find $P(x \leq 2)$.

Tally for Discrete Variables: DRIVSTAR

DRIVSTAR	Count	Percent
2	4	4.08
3	17	17.35
4	59	60.20
5	18	18.37
N=	98	

4.23 **Dust mite allergies.** A dust mite allergen level that exceeds 2 micrograms per gram ($\mu g/g$) of dust has been associated with the development of allergies. Consider a random sample of four homes, and let x be the number of homes with a dust mite level that exceeds 2 $\mu g/g$. The probability distribution for x, based on a study by the National Institute of Environmental Health Sciences, is shown in the following table:

x	0	1	2	3	4
$p(x)$	.09	.30	.37	.20	.04

a. Verify that the probabilities for x in the table sum to 1.

b. Find the probability that three or four of the homes in the sample have a dust mite level that exceeds 2 $\mu g/g$.

c. Find the probability that fewer than two homes in the sample have a dust mite level that exceeds 2 $\mu g/g$.

4.24 **Controlling the water hyacinth.** Entomologists are continually searching for new biological agents to control the water hyacinth, one of the world's worst aquatic weeds. An insect that naturally feeds on the water hyacinth is the delphacid. Female delphacids lay anywhere from one to four eggs onto a water hyacinth blade. The *Annals of the Entomological Society of America* (Jan. 2005) published a study of the life cycle of a South American delphacid species. The following table gives the percentages of water hyacinth blades that have one, two, three, and four delphacid eggs:

	One Egg	Two Eggs	Three Eggs	Four Eggs
Percentage of Blades	40	54	2	4

Source: Sosa, A. J., et al. "Life history of *Megamelus scutellaris* with description of immature stages," *Annals of the Entomological Society of America*, Vol. 98, No. 1, Jan. 2005 (adapted from Table 1).

a. One of the water hyacinth blades in the study is randomly selected, and x, the number of delphacid eggs on the blade, is observed. Give the probability distribution of x.

b. What is the probability that the blade has at least three delphacid eggs?

4.25 **Gender in two-child families.** *Human Biology* (Feb. 2009) published a study on the gender of children in two-child families. In populations where it is just as likely to have a boy as a girl, the probabilities of having two girls, two boys, or a boy and a girl are well known. Let x represent the number of boys in a two-child family.

a. List the possible ways (sample points) in which a two-child family can be gender-configured. (For example, BG represents the event that the first child is a boy and the second is a girl.)

b. Assuming boys are just as likely as girls to be born, assign probabilities to the sample points in part **a**.

c. Use the probabilities, part **a**, to find the probability distribution for x.

d. The article reported on the results of the National Health Interview Survey (NHIS) of almost 43,000 two-child families. The table gives the proportion of families with each gender configuration. Use this information to revise the probability distribution for x.

Gender Configuration	Proportion
Girl-girl (GG)	.222
Boy-girl (BG)	.259
Girl-boy (GB)	.254
Boy-boy (BB)	.265

Applying the Concepts—Intermediate

4.26 **Solar energy cells.** According to *Wired* (June 2008), 35% of the world's solar energy cells are manufactured in China. Consider a random sample of 5 solar energy cells, and let x represent the number in the sample that are manufactured in China. In the next section, we show that the probability distribution for x is given by the formula,

$$p(x) = \frac{(5!)(.35)^x(.65)^{5-x}}{(x!)(5-x)!},$$

where $n! = (n)(n-1)(n-2) \ldots (2)(1)$

a. Explain why x is a discrete random variable.

b. Find $p(x)$ for $x = 0, 1, 2, 3, 4,$ and 5.

c. Show that the properties for a discrete probability distribution are satisfied.

d. Find the probability that at least 4 of the 5 solar energy cells in the sample are manufactured in China.

4.27 Contaminated gun cartridges. A weapons manufacturer uses a liquid propellant to produce gun cartridges. During the manufacturing process, the propellant can get mixed with another liquid to produce a contaminated cartridge. A University of South Florida statistician hired by the company to investigate the level of contamination in the stored cartridges found that 23% of the cartridges in a particular lot were contaminated. Suppose you randomly sample (without replacement) gun cartridges from this lot until you find a contaminated one. Let x be the number of cartridges sampled until a contaminated one is found. It is known that the probability distribution for x is given by the formula

$$p(x) = (.23)(.77)^{x-1}, x = 1, 2, 3, \ldots$$

a. Find $p(1)$. Interpret this result.

b. Find $p(5)$. Interpret this result.

c. Find $P(x \geq 2)$. Interpret this result.

4.28 Ryder Cup miracle in golf. The Ryder Cup is a three-day golf tournament played between a team of golf professionals from the United States and a team from Europe. A total of 28 matches are played between the teams; one point is awarded to the team winning a match and half a point is awarded to each team if the match ends in a tie (draw). The team with the most points wins the tournament. In 1999, the United States was losing 10 points to 6 when it miraculously won 8.5 of a possible 12 points on the last day of the tournament to seal the win. On the last day, 12 single matches are played. A total of 8.5 points can be won in a variety of ways, as shown in the accompanying table. Given one team scores at least 8.5 points on the last day of the tournament, *Chance* (Fall 2009) determined the probabilities of each of these outcomes assuming each team is equally likely to win a match. Let x be the points scored by the winning team on the last day of the tournament when the team scores at least 8.5 points. Find the probability distribution of x.

Wins	Ties	Points	Probability
5	7	8.5	.000123
6	5	8.5	.008823
6	6	9.0	.000456
7	3	8.5	.128030
7	4	9.0	.020086
7	5	9.5	.001257
8	1	8.5	.325213
8	2	9.0	.153044
8	3	9.5	.032014
8	4	10.0	.002514
9	0	9.0	.115178
9	1	9.5	.108400
9	2	10.0	.032901
9	3	10.5	.003561
10	0	10.0	.034552
10	1	10.5	.021675
10	2	11.0	.003401
11	0	11.0	.006284
11	1	11.0	.001972
12	0	12.0	.000518

4.29 Choosing portable grill displays. Refer to the *Journal of Consumer Research* (Mar. 2003) marketing study of influencing consumer choices by offering undesirable alternatives, presented in Exercise 3.27 (p. 122). Recall that each of 124 college students selected showroom displays for portable grills. Five different displays (representing five different-sized grills) were available, but the students were instructed to select only three displays in order to maximize purchases of Grill #2 (a smaller grill). The table that follows shows the grill display combinations and number of each selected by the 124 students. Suppose 1 of the 124 students is selected at random. Let x represent the sum of the grill numbers selected by that student. (This sum is an indicator of the size of the grills selected.)

a. Find the probability distribution for x.

b. What is the probability that x exceeds 10?

Grill Display Combination	Number of Students
1–2–3	35
1–2–4	8
1–2–5	42
2–3–4	4
2–3–5	1
2–4–5	34

Based on Hamilton, R. W. "Why do people suggest what they do not want? Using context effects to influence others' choices." *Journal of Consumer Research,* Vol. 29, Mar. 2003 (Table 1).

4.30 Federal civil trial appeals. Refer to the *Journal of the American Law and Economics Association* (Vol. 3, 2001) study of appeals of federal civil trials, presented in Exercise 3.59 (p. 134). A breakdown of the 678 civil cases that were originally tried in front of a judge (rather than a jury) and appealed by either the plaintiff or defendant is reproduced in the accompanying table. Suppose each civil case is awarded points (positive or negative) on the basis of the outcome of the appeal for the purpose of evaluating federal judges. If the appeal is affirmed or dismissed, +5 points are awarded. If the appeal of a plaintiff trial win is reversed, −1 point is awarded. If the appeal of a defendant trial win is reversed, −3 points are awarded. Suppose 1 of the 678 cases is selected at random, and the number x of points awarded is determined. Find and graph the probability distribution for x.

Outcome of Appeal	Number of Cases
Plaintiff trial win — reversed	71
Plaintiff trial win — affirmed/ dismissed	240
Defendant trial win — reversed	68
Defendant trial win — affirmed/ dismissed	299
Total	678

Applying the Concepts — Advanced

4.31 X and Y chromosomes. Every human possesses two sex chromosomes. A copy of one or the other (equally likely) is contributed to an offspring. Males have one X chromosome and one Y chromosome. Females have two X chromosomes. If a couple has three children, what is the probability that they have at least one boy?

4.32 Punnett square for earlobes. Geneticists use a grid—called a *Punnett square*—to display all possible gene combinations in genetic crosses. (The grid is named for Reginald Punnett, a British geneticist who developed the method in the early 1900s.) The accompanying figure is a Punnett square for a cross involving human earlobes. In humans, free earlobes (E) are dominant over attached earlobes (e). Consequently, the gene pairs EE and Ee will result in free earlobes, while the gene pair ee results in attached earlobes. Consider a couple with genes as shown in the accompanying Punnett square. Suppose the couple has seven children. Let *x* represent the number of children with attached earlobes (i.e., with the gene pair ee). Find the probability distribution of *x*.

4.33 Robot-sensor system configuration. Engineers at Broadcom Corp. and Simon Fraser University collaborated on research involving a robot-sensor system in an unknown environment (*The International Journal of Robotics Research*, Dec. 2004). As an example, the engineers presented the three-point, single-link robotic system shown in the accompanying figure. Each point (A, B, or C) in the physical space of the system has either an "obstacle" status or a "free" status. There are two single links in the system: $A \leftrightarrow B$ and $B \leftrightarrow C$. A link has a "free" status if and only if both points in the link are "free"; otherwise the link has an "obstacle" status. Of interest is the random variable *x*: the total number of links in the system that are "free."

a. List the possible values of *x* for the system.

b. The researchers stated that the probability of any point in the system having a "free" status is .5. Assuming that the three points in the system operate independently, find the probability distribution for *x*.

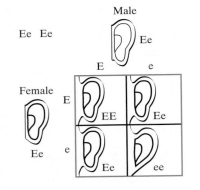

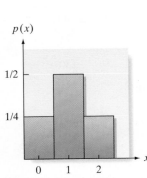

4.3 Expected Values of Discrete Random Variables

If a discrete random variable *x* were observed a very large number of times and the data generated were arranged in a relative frequency distribution, the relative frequency distribution would be indistinguishable from the probability distribution for the random variable. Thus, the probability distribution for a random variable is a theoretical model for the relative frequency distribution of a population. To the extent that the two distributions are equivalent (and we will assume that they are), the probability distribution for *x* possesses a mean μ and a variance σ^2 that are identical to the corresponding descriptive measures for the population. This section explains how you can find the mean value for a random variable. We illustrate the procedure with an example.

Examine the probability distribution for *x* (the number of heads observed in the toss of two fair coins) in Figure 4.5. Try to locate the mean of the distribution intuitively. We may reason that the mean μ of this distribution is equal to 1 as follows: In a large number of experiments—say, 100,000—$1/4$ (or 25,000) should result in $x = 0$ heads, $1/2$ (or 50,000) in $x = 1$ head, and $1/4$ (or 25,000) in $x = 2$ heads. Therefore, the average number of heads is

$$\mu = \frac{0(25,000) + 1(50,000) + 2(25,000)}{100,000} = 0(^1/_4) + 1(^1/_2) + 2(^1/_4)$$

$$= 0 + ^1/_2 + ^1/_2 = 1$$

Note that to get the population mean of the random variable *x*, we multiply each possible value of *x* by its probability $p(x)$, and then we sum this product over all possible values of *x*. The *mean of x* is also referred to as the *expected value of x*, denoted $E(x)$.

The **mean,** or **expected value,** of a discrete random variable *x* is

$$\mu = E(x) = \Sigma x p(x)$$

Expected is a mathematical term and should not be interpreted as it is typically used. Specifically, *a random variable might never be equal to its "expected value."* Rather,

Figure 4.5

Probability distribution for a two-coin toss

the expected value is the mean of the probability distribution, or a measure of its central tendency. You can think of μ as the mean value of x in a *very large* (actually, *infinite*) number of repetitions of the experiment in which the values of x occur in proportions equivalent to the probabilities of x.

Example 4.7
Finding an Expected Value—An Insurance Application

Problem Suppose you work for an insurance company and you sell a $10,000 one-year term insurance policy at an annual premium of $290. Actuarial tables show that the probability of death during the next year for a person of your customer's age, sex, health, etc., is .001. What is the expected gain (amount of money made by the company) for a policy of this type?

Solution The experiment is to observe whether the customer survives the upcoming year. The probabilities associated with the two sample points, Live and Die, are .999 and .001, respectively. The random variable you are interested in is the gain x, which can assume the values shown in the following table:

Gain x	Sample Point	Probability
$290	Customer lives	.999
−$9,710	Customer dies	.001

If the customer lives, the company gains the $290 premium as profit. If the customer dies, the gain is negative because the company must pay $10,000, for a net "gain" of $(290 − 10,000) = −$9,710$. The expected gain is therefore

$$\mu = E(x) = \Sigma x p(x) = (290)(.999) + (-9{,}710)(.001) = \$280$$

In other words, if the company were to sell a very large number of $10,000 one-year policies to customers possessing the characteristics described, it would (on the average) net $280 per sale in the next year.

Look Back Note that $E(x)$ need not equal a possible value of x. That is, the expected value is $280, but x will equal either $290 or −$9,710 each time the experiment is performed (a policy is sold and a year elapses). The expected value is a measure of central tendency—and in this case represents the average over a very large number of one-year policies—but is not a possible value of x.

Now Work Exercise 4.42

We learned in Chapter 2 that the mean and other measures of central tendency tell only part of the story about a set of data. The same is true about probability distributions: We need to measure variability as well. Since a probability distribution can be viewed as a representation of a population, we will use the population variance to measure its variability.

The *population variance* σ^2 is defined as the average of the squared distance of x from the population mean μ. Since x is a random variable, the squared distance, $(x - \mu)^2$, is also a random variable. Using the same logic we employed to find the mean value of x, we calculate the mean value of $(x - \mu)^2$ by multiplying all possible values of $(x - \mu)^2$ by $p(x)$ and then summing over all possible x values.* This quantity,

$$E\left[(x - \mu)^2\right] = \sum_{\text{all } x}(x - \mu)^2 p(x)$$

is also called the *expected value of the squared distance from the mean;* that is, $\sigma^2 = E\left[(x - \mu)^2\right]$. The standard deviation of x is defined as the square root of the variance σ^2.

> The **variance** of a random variable x is
> $$\sigma^2 = E\left[(x - \mu)^2\right] = \Sigma(x - \mu)^2 p(x) = \Sigma x^2 p(x) - \mu^2$$

*It can be shown that $E\left[(x - \mu)^2\right] = E(x^2) - \mu^2$, where $E(x^2) = \Sigma x^2 p(x)$. Note the similarity between this expression and the shortcut formula $\Sigma(x - \bar{x})^2 = \Sigma x^2 - (\Sigma x)^2/n$ given in Chapter 2.

> The **standard deviation** of a discrete random variable is equal to the square root of the variance, or $\sigma = \sqrt{\sigma^2}$.

Knowing the mean μ and standard deviation σ of the probability distribution of x, in conjunction with Chebyshev's rule (Table 2.6) and the Empirical rule (Table 2.7), we can make statements about the likelihood of values of x falling within the intervals $\mu \pm \sigma, \mu \pm 2\sigma$, and $\mu \pm 3\sigma$. These probabilities are given in the following box:

Chebyshev's Rule and Empirical Rule for a Discrete Random Variable

Let x be a discrete random variable with probability distribution $p(x)$, mean μ, and standard deviation σ. Then, depending on the shape of $p(x)$, the following probability statements can be made:

	Chebyshev's Rule	Empirical Rule
	Applies to any probability distribution (see Figure 4.6a)	Applies to probability distributions that are mound shaped and symmetric (see Figure 4.6b)
$P(\mu - \sigma < x < \mu + \sigma)$	≥ 0	$\approx .68$
$P(\mu - 2\sigma < x < \mu + 2\sigma)$	$\geq \frac{3}{4}$	$\approx .95$
$P(\mu - 3\sigma < x < \mu + 3\sigma)$	$\geq \frac{8}{9}$	≈ 1.00

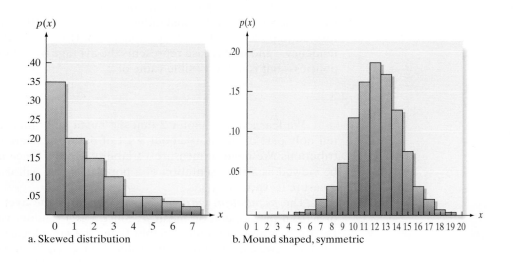

Figure 4.6

Shapes of two probability distributions for a discrete random variable x

a. Skewed distribution

b. Mound shaped, symmetric

Example 4.8

Finding μ and σ—Skin Cancer Treatment

Problem Medical research has shown that a certain type of chemotherapy is successful 70% of the time when used to treat skin cancer. Suppose five skin cancer patients are treated with this type of chemotherapy, and let x equal the number of successful cures out of the five. The probability distribution for the number x of successful cures out of five is given in the following table:

x	0	1	2	3	4	5
$p(x)$	.002	.029	.132	.309	.360	.168

a. Find $\mu = E(x)$. Interpret the result.

b. Find $\sigma = \sqrt{E[(x - \mu)^2]}$. Interpret the result.

c. Graph $p(x)$. Locate μ and the interval $\mu \pm 2\sigma$ on the graph. Use either Chebyshev's rule or the empirical rule to approximate the probability that x falls into this interval. Compare your result with the actual probability.

d. Would you expect to observe fewer than two successful cures out of five?

Solution

a. Applying the formula for μ, we obtain

$$\mu = E(x) = \Sigma xp(x)$$
$$= 0(.002) + 1(.029) + 2(.132) + 3(.309) + 4(.360) + 5(.168) = 3.50$$

On average, the number of successful cures out of five skin cancer patients treated with chemotherapy will equal 3.5. Remember that this expected value has meaning only when the experiment—treating five skin cancer patients with chemotherapy—is repeated a large number of times.

b. Now we calculate the variance of x:

$$\sigma^2 = E[(x - \mu)^2] = \Sigma(x - \mu)^2 p(x)$$
$$= (0 - 3.5)^2(.002) + (1 - 3.5)^2(.029) + (2 - 3.5)^2(.132)$$
$$+ (3 - 3.5)^2(.309) + (4 - 3.5)^2(.360) + (5 - 3.5)^2(.168)$$
$$= 1.05$$

Thus, the standard deviation is

$$\sigma = \sqrt{\sigma^2} = \sqrt{1.05} = 1.02$$

This value measures the spread of the probability distribution of x, the number of successful cures out of five. A more useful interpretation is obtained by answering parts **c** and **d**.

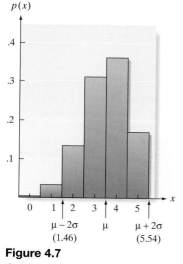

$p(x)$

Figure 4.7
Graph of $p(x)$ for Example 4.8

c. The graph of $p(x)$ is shown in Figure 4.7, with the mean μ and the interval $\mu \pm 2\sigma = 3.50 \pm 2(1.02) = 3.50 \pm 2.04 = (1.46, 5.54)$ also indicated. Note particularly that $\mu = 3.5$ locates the center of the probability distribution. Since this distribution is a theoretical relative frequency distribution that is moderately mound shaped (see Figure 4.7), we expect (from Chebyshev's rule) at least 75% and, more likely (from the empirical rule), approximately 95%, of observed x values to fall between 1.46 and 5.54. You can see from the figure that the actual probability that x falls in the interval $\mu \pm 2\sigma$ includes the sum of $p(x)$ for the values $x = 2, x = 3, x = 4$, and $x = 5$. This probability is $p(2) + p(3) + p(4) + p(5) = .132 + .309 + .360 + .168 = .969$. Therefore, 96.9% of the probability distribution lies within two standard deviations of the mean. This percentage is consistent with both Chebyshev's rule and the empirical rule.

d. Fewer than two successful cures out of five implies that $x = 0$ or $x = 1$. Both of these values of x lie outside the interval $\mu \pm 2\sigma$, and the empirical rule tells us that such a result is unlikely (approximate probability of .05). The exact probability, $P(x \le 1)$, is $p(0) + p(1) = .002 + .029 = .031$. Consequently, in a single experiment in which five skin cancer patients are treated with chemotherapy, we would not expect to observe fewer than two successful cures.

Now Work Exercise 4.40

Exercises 4.34–4.49

Understanding the Principles

4.34 What does the expected value of a random variable represent?

4.35 Will $E(x)$ always be equal to a specific value of the random variable x?

4.36 For a mound-shaped, symmetric distribution, what is the probability that x falls into the interval $\mu \pm 2\sigma$?

Learning the Mechanics

4.37 Consider the probability distribution shown for the random variable x here:

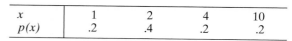

x	1	2	4	10
$p(x)$	.2	.4	.2	.2

a. Find $\mu = E(x)$.

b. Find $\sigma^2 = E[(x - \mu)^2]$.

c. Find σ.

d. Interpret the value you obtained for μ.

e. In this case, can the random variable x ever assume the value μ? Explain.

f. In general, can a random variable ever assume a value equal to its expected value? Explain.

4.38 Consider the probability distribution for the random variable x shown here:

x	10	20	30	40	50	60
$p(x)$	.05	.20	.30	.25	.10	.10

a. Find μ, σ^2, and σ.

b. Graph $p(x)$.

c. Locate μ and the interval $\mu \pm 2\sigma$ on your graph. What is the probability that x will fall within the interval $\mu \pm 2\sigma$?

4.39 Consider the probability distributions shown here:

x	0	1	2		y	0	1	2
$p(x)$	.3	.4	.3		$p(y)$	.1	.8	.1

a. Use your intuition to find the mean for each distribution. How did you arrive at your choice?

b. Which distribution appears to be more variable? Why?

c. Calculate μ and σ^2 for each distribution. Compare these answers with your answers in parts **a** and **b**.

4.40 Consider the probability distribution shown here:

NW

x	−4	−3	−2	−1	0	1	2	3	4
$p(x)$	.02	.07	.10	.15	.30	.18	.10	.06	.02

a. Calculate μ, σ^2, and σ.

b. Graph $p(x)$. Locate $\mu, \mu - 2\sigma$, and $\mu + 2\sigma$ on the graph.

c. What is the probability that x will fall into the interval $\mu \pm 2\sigma$?

Applying the Concepts—Basic

4.41 **Gender in two-child families.** Refer to the *Human Biology* (Feb. 2009) study on the gender of children in two-child families, Exercise 4.25 (p. 188). Recall that the National Health Interview Survey (NHIS) of almost 43,000 two-child families yielded the following table. You used this information in Exercise 4.25d to find the probability distribution for x, the number of boys in a two-child family. Find $E(x)$ and give a practical interpretation of its value.

Gender Configuration	Proportion
Girl-girl (GG)	.222
Boy-girl (BG)	.259
Girl-boy (GB)	.254
Boy-boy (BB)	.265

4.42 **NHTSA car crash tests.** Refer to Exercise 4.22 (p. 188), in which you found the probability distribution for x, the number of stars in a randomly selected car's driver-side

NW

crash rating. Find $\mu = E(x)$ for this distribution and interpret the result practically.

4.43 **Dust mite allergies.** Exercise 4.23 (p. 188) gives the probability distribution for the number x of homes with high dust mite levels. The probability distribution is reproduced here:

Number of Homes, x	$p(x)$
0	.09
1	.30
2	.37
3	.20
4	.04

a. Find $E(x)$. Give a meaningful interpretation of the result.

b. Find σ.

c. Find the exact probability that x is in the interval $\mu \pm 2\sigma$. Compare your answer with Chebyshev's rule and the empirical rule.

4.44 **Controlling the water hyacinth.** Refer to the *Annals of the Entomological Society of America* (Jan. 2005) study of the number of insect eggs, x, on a blade of water hyacinth, presented in Exercise 4.24 (p. 188). The probability distribution for x is reproduced in the accompanying table. Find the mean of the probability distribution for x, and interpret its value.

Number of eggs, x	1	2	3	4
$p(x)$	.40	.54	.02	.04

Source: Sosa, A. J., et al. "Life history of *Megamelus scutellaris* with description of immature stages," *Annals of the Entomological Society of America*, Vol. 98, No. 1, Jan. 2005 (adapted from Table 1).

Applying the Concepts—Intermediate

4.45 **Beach erosional hot spots.** Refer to the U.S. Army Corps of Engineers' study of beach erosional hot spots, presented in Exercise 3.173 (p. 171). The data on the nearshore bar condition for six beach hot spots are reproduced in the accompanying table. Suppose you randomly select two of these six beaches and count x, the total number in the sample with a planar nearshore bar condition.

Beach Hot Spot	Nearshore Bar Condition
Miami Beach, FL	Single, shore parallel
Coney Island, NY	Other
Surfside, CA	Single, shore parallel
Monmouth Beach, NJ	Planar
Ocean City, NJ	Other
Spring Lake, NJ	Planar

Based on *Identification and Characterization of Erosional Hotspots*, William & Mary Virginia Institute of Marine Science, U.S. Army Corps of Engineers Project Report, March 18, 2002.

a. List all possible pairs of beach hot spots that can be selected from the six.

b. Assign probabilities to the outcomes in part **a.**

c. For each outcome in part **a,** determine the value of x.

d. Form a probability distribution table for x.

e. Find the expected value of x. Interpret the result.

4.46 **Expected Lotto winnings.** The chance of winning Florida's Pick-6 Lotto game is 1 in approximately 23 million. Suppose you buy a $1 Lotto ticket in anticipation of winning the $7 million grand prize. Calculate your expected net winnings for this single ticket. Interpret the result.

4.47 **Expected winnings in roulette.** In the popular casino game of roulette, you can bet on whether the ball will fall in an arc on the wheel colored red, black, or green. You showed (Exercise 3.175, p. 172) that the probability of a red outcome is 18/38, that of a black outcome is 18/38, and that of a green outcome is 2/38. Suppose you make a $5 bet on red. Find your expected net winnings for this single bet. Interpret the result.

4.48 **The Showcase Showdown.** On the popular television game show *The Price is Right*, contestants can play "The Showcase Showdown." The game involves a large wheel with 20 nickel values, 5, 10, 15, 20,…, 95, 100, marked on it. Contestants spin the wheel once or twice, with the objective of obtaining the highest total score *without going over a dollar (100)*. [According to the *American Statistician* (Aug. 1995), the optimal strategy for the first spinner in a three-player game is to spin a second time only if the value of the initial spin is 65 or less.] Let x represent the score of a single contestant playing "The Showcase Showdown." Assume a "fair" wheel (i.e., a wheel with equally likely outcomes). If the total of the player's spins exceeds 100, the total score is set to 0.
a. If the player is permitted only one spin of the wheel, find the probability distrubtion for x.
b. Refer to part **a.** Find $E(x)$ and interpret this value.
c. Refer to part **a.** Give a range of values within which x is likely to fall.
d. Suppose the player will spin the wheel twice, no matter what the outcome of the first spin. Find the probability distribution for x.
e. What assumption did you make to obtain the probability distribution in part **d**? Is it a reasonable assumption?
f. Find μ and σ for the probability distribution of part **d,** and interpret the results.
g. Refer to part **d.** What is the probability that in two spins the player's total score exceeds a dollar (i.e., is set to 0)?
h. Suppose the player obtains a 20 on the first spin and decides to spin again. Find the probability distribution for x.
i. Refer to part **h.** What is the probability that the player's total score exceeds a dollar?
j. Given that the player obtains a 65 on the first spin and decides to spin again, find the probability that the player's total score exceeds a dollar.
k. Repeat part **j** for different first-spin outcomes. Use this information to suggest a strategy for the one-player game.

Applying the Concepts—Advanced

4.49 **Parlay card betting.** Odds makers try to predict which professional and college football teams will win and by how much (the *spread*). If the odds makers do this accurately, adding the spread to the underdog's score should make the final score a tie. Suppose a bookie will give you $6 for every $1 you risk if you pick the winners in three different football games (adjusted by the spread) on a "parlay" card. What is the bookie's expected earnings per dollar wagered? Interpret this value.

4.4 The Binomial Random Variable

Many experiments result in *dichotomous* responses (i.e., responses for which there exist two possible alternatives, such as Yes–No, Pass–Fail, Defective–Nondefective, or Male–Female). A simple example of such an experiment is the coin-toss experiment. A coin is tossed a number of times, say, 10. Each toss results in one of two outcomes, Head or Tail, and the probability of observing each of these two outcomes remains the same for each of the 10 tosses. Ultimately, we are interested in the probability distribution of x, the number of heads observed. Many other experiments are equivalent to tossing a coin (either balanced or unbalanced) a fixed number n of times and observing the number x of times that one of the two possible outcomes occurs. Random variables that possess these characteristics are called **binomial random variables.**

Public opinion and consumer preference polls (e.g., the CNN, Gallup, and Harris polls) frequently yield observations on binomial random variables. For example, suppose a sample of 100 students is selected from a large student body and each person is asked whether he or she favors (a Head) or opposes (a Tail) a certain campus issue. Suppose we are interested in x, the number of students in the sample who favor the issue. Sampling 100 students is analogous to tossing the coin 100 times. Thus, you can see that opinion polls which record the number of people who favor a certain issue are real-life equivalents of coin-toss experiments. We have been describing a **binomial experiment,** identified by the following characteristics:

> **Characteristics of a Binomial Random Variable**
>
> 1. The experiment consists of n identical trials.
> 2. There are only two possible outcomes on each trial. We will denote one outcome by S (for Success) and the other by F (for Failure).
> 3. The probability of S remains the same from trial to trial. This probability is denoted by p, and the probability of F is denoted by $q = 1 - p$.
> 4. The trials are independent.
> 5. The binomial random variable x is the number of S's in n trials.

BIOGRAPHY **JACOB BERNOULLI (1654–1705)**

The Bernoulli Distribution

Son of a magistrate and spice maker in Basel, Switzerland, Jacob Bernoulli completed a degree in theology at the University of Basel. While at the university, however, he studied mathematics secretly and against the will of his father. Jacob taught mathematics to his younger brother Johan, and they both went on to become distinguished European mathematicians. At first the brothers collaborated on the problems of the time (e.g., calculus); unfortunately, they later became bitter mathematical rivals. Jacob applied his philosophical training and mathematical intuition to probability and the theory of games of chance, where he developed the law of large numbers. In his book *Ars Conjectandi*, published in 1713 (eight years after his death), the binomial distribution was first proposed. Jacob showed that the binomial distribution was a sum of independent 0–1 variables, now known as Bernoulli random variables. ∎

Example 4.9
Assessing Whether x Is a Binomial

Problem For the following examples, decide whether x is a binomial random variable:

a. A university scholarship committee must select two students to receive a scholarship for the next academic year. The committee receives 10 applications for the scholarships—6 from male students and 4 from female students. Suppose the applicants are all equally qualified, so that the selections are randomly made. Let x be the number of female students who receive a scholarship.

b. Before marketing a new product on a large scale, many companies conduct a consumer-preference survey to determine whether the product is likely to be successful. Suppose a company develops a new diet soda and then conducts a taste-preference survey in which 100 randomly chosen consumers state their preferences from among the new soda and the two leading sellers. Let x be the number of the 100 who choose the new brand over the two others.

c. Some surveys are conducted by using a method of sampling other than simple random sampling (defined in Chapter 3). For example, suppose a television cable company plans to conduct a survey to determine the fraction of households in a certain city that would use its new fiber-optic service (FiOS). The sampling method is to choose a city block at random and then survey every household on that block. This sampling technique is called *cluster sampling*. Suppose 10 blocks are so sampled, producing a total of 124 household responses. Let x be the number of the 124 households that would use FiOS.

Solution

a. In checking the binomial characteristics, a problem arises with both characteristic 3 (probabilities the same across trials) and characteristic 4 (independence). On the one hand, given that the first student selected is female, the probability that the second chosen is female is $^3/_9$. On the other hand, given that the first selection is a male student, the probability that the second is female is $^4/_9$. Thus, the conditional probability of a Success (choosing a female student to receive a scholarship) on the second trial (selection) depends on the outcome of the first trial, and the trials are therefore dependent. Since the trials are *not independent*, this variable is not a binomial random variable. (This variable is actually a *hypergeometric* random variable, the topic of optional Section 4.6.)

b. Surveys that produce dichotomous responses and use random-sampling techniques are classic examples of binomial experiments. In this example, each randomly selected consumer either states a preference for the new diet soda or does not. The sample of 100 consumers is a very small proportion of the totality of potential consumers, so the response of one would be, for all practical purposes, independent of another.* Thus, x is a binomial random variable.

c. This example is a survey with dichotomous responses (Yes or No to using FiOS), but the sampling method is not simple random sampling. Again, the binomial characteristic of independent trials would probably not be satisfied. The responses of households within a particular block would be dependent, since households within a block tend to be similar with respect to income, level of education, and general interests. Thus, the binomial model would not be satisfactory for x if the cluster sampling technique were employed.

Look Back Nonbinomial variables with two outcomes on every trial typically occur because they do not satisfy characteristic 3 or characteristic 4 of a binomial distribution listed in the previous box.

Now Work Exercise 4.64a

Example 4.10

Deriving the Binomial Probability Distribution— Passing a Physical Fitness Exam

Problem The Heart Association claims that only 10% of U.S. adults over 30 years of age meet the minimum requirements established by the President's Council on Fitness, Sports, and Nutrition. Suppose four adults are randomly selected and each is given the fitness test.

a. Use the steps given in Chapter 3 (box on p. 116) to find the probability that none of the four adults passes the test.

b. Find the probability that three of the four adults pass the test.

c. Let x represent the number of the four adults who pass the fitness test. Explain why x is a binomial random variable.

d. Use the answers to parts **a** and **b** to derive a formula for $p(x)$, the probability distribution of the binomial random variable x.

Solution

a. **1.** The first step is to define the experiment. Here we are interested in observing the fitness test results of each of the four adults: pass (S) or fail (F).

   2. Next, we list the sample points associated with the experiment. Each sample point consists of the test results of the four adults. For example, $SSSS$ represents the sample point denoting that all four adults pass, while $FSSS$ represents the sample point denoting that adult 1 fails, while adults 2, 3, and 4 pass the test. The 16 sample points are listed in Table 4.2.

---

*In most real-life applications of the binomial distribution, the population of interest has a finite number of elements (trials), denoted N. When N is large and the sample size n is small relative to N, say, $n/N \leq .05$, the sampling procedure, for all practical purposes, satisfies the conditions of a binomial experiment.

Table 4.2		Sample Points for Fitness Test of Example 4.10		
SSSS	*FSSS*	*FFSS*	*SFFF*	*FFFF*
	SFSS	*FSFS*	*FSFF*	
	SSFS	*FSSF*	*FFSF*	
	SSSF	*SFFS*	*FFFS*	
		SFSF		
		SSFF		

3. We now assign probabilities to the sample points. Note that each sample point can be viewed as the intersection of four adults' test results, and assuming that the results are independent, the probability of each sample point can be obtained by the multiplicative rule as follows:

$$P(SSSS) = P[(\text{adult 1 passes}) \cap (\text{adult 2 passes})$$
$$\cap\ (\text{adult 3 passes}) \cap (\text{adult 4 passes})]$$
$$= P(\text{adult 1 passes}) \times P(\text{adult 2 passes})$$
$$\times\ P(\text{adult 3 passes}) \times P(\text{adult 4 passes})$$
$$= (.1)(.1)(.1)(.1) = (.1)^4 = .0001$$

All other sample-point probabilities are calculated by similar reasoning. For example,

$$P(FSSS) = (.9)(.1)(.1)(.1) = .0009$$

You can check that this reasoning results in sample-point probabilities that add to 1 over the 16 points in the sample space.

4. Finally, we add the appropriate sample-point probabilities to obtain the desired event probability. The event of interest is that all four adults fail the fitness test. In Table 4.2, we find only one sample point, *FFFF*, contained in this event. All other sample points imply that at least one adult passes. Thus,

$$P(\text{All four adults fail}) = P(FFFF) = (.9)^4 = .6561$$

b. The event that three of the four adults pass the fitness test consists of the four sample points in the second column of Table 4.2: *FSSS, SFSS, SSFS,* and *SSSF*. To obtain the event probability, we add the sample-point probabilities:

$$P(3 \text{ of } 4 \text{ adults pass}) = P(FSSS) + P(SFSS) + P(SSFS) + P(SSSF)$$
$$= (.1)^3(.9) + (.1)^3(.9) + (.1)^3(.9) + (.1)^3(.9)$$
$$= 4(.1)^3(.9) = .0036$$

Note that each of the four sample-point probabilities is the same, because each sample point consists of three *S*'s and one *F*; the order does not affect the probability because the adults' test results are (assumed) independent.

c. We can characterize this experiment as consisting of four identical trials: the four test results. There are two possible outcomes to each trial, *S* or *F*, and the probability of passing, $p = .1$, is the same for each trial. Finally, we are assuming that each adult's test result is independent of all others, so that the four trials are independent. Then it follows that x, the number of the four adults who pass the fitness test, is a binomial random variable.

d. The event probabilities in parts **a** and **b** provide insight into the formula for the probability distribution $p(x)$. First, consider the event that three adults pass (part **b**). We found that

$P(x = 3) =$
(Number of sample points for which $x = 3$) $\times (.1)^{\text{Number of successes}} \times (.9)^{\text{Number of failures}}$
$= 4(.1)^3(.9)^1$

In general, we can use combinatorial mathematics to count the number of sample points. For example,

Number of sample points for which $x = 3$

= Number of different ways of selecting 3 successes in the 4 trials

$$= \binom{4}{3} = \frac{4!}{3!(4-3)!} = \frac{4 \cdot 3 \cdot 2 \cdot 1}{(3 \cdot 2 \cdot 1) \cdot 1} = 4$$

The formula that works for any value of x can be deduced as follows:

$$P(x = 3) = \binom{4}{3}(.1)^3(.9)^1 = \binom{4}{x}(.1)^x(.9)^{4-x}$$

The component $\binom{4}{x}$ counts the number of sample points with x successes, and the component $(.1)^x(.9)^{4-x}$ is the probability associated with each sample point having x successes.

For the general binomial experiment, with n trials and probability p of Success on each trial, the probability of x successes is

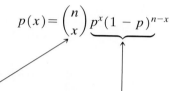

$$p(x) = \binom{n}{x} \underbrace{p^x(1-p)^{n-x}}$$

Number of simple   Probability of x S's
events with   and $(n-x)$ F's in
x S's   any simple event

Look Ahead In theory, you could always resort to the principles developed in this example to calculate binomial probabilities; just list the sample points and sum their probabilities. However, as the number of trials (n) increases, the number of sample points grows very rapidly. (The number of sample points is 2^n.) Thus, we prefer the formula for calculating binomial probabilities, since its use avoids listing sample points.

The **binomial distribution*** is summarized in the following box:

The Binomial Probability Distribution

$$p(x) = \binom{n}{x}p^x q^{n-x} \qquad (x = 0, 1, 2, \ldots, n)$$

where

p = Probability of a success on a single trial
$q = 1 - p$
n = Number of trials
x = Number of successes in n trials

$$\binom{n}{x} = \frac{n!}{x!(n-x)!}$$

As noted in Chapter 3, the symbol 5! Means $5 \cdot 4 \cdot 3 \cdot 2 \cdot 1 = 120$. Similarly, $n! = n(n-1)(n-2) \ldots 3 \cdot 2 \cdot 1$. (Remember, $0! = 1$.)

---

*The binomial distribution is so named because the probabilities, $p(x), x = 0, 1, \ldots, n$, are terms of the binomial expansion, $(q + p)^n$.

Example 4.11

Applying the Binomial Distribution— Physical Fitness Problem

Problem Refer to Example 4.10. Use the formula for a binomial random variable to find the probability distribution of x, where x is the number of adults who pass the fitness test. Graph the distribution.

Solution For this application, we have $n = 4$ trials. Since a success S is defined as an adult who passes the test, $p = P(S) = .1$ and $q = 1 - p = .9$. Substituting $n = 4, p = .1$, and $q = .9$ into the formula for $p(x)$, we obtain

$$p(0) = \frac{4!}{0!(4-0)!}(.1)^0(.9)^{4-0} = \frac{4\cdot 3\cdot 2\cdot 1}{(1)(4\cdot 3\cdot 2\cdot 1)}(.1)^0(.9)^4 = 1(.1)^0(.9)^4 = .6561$$

$$p(1) = \frac{4!}{1!(4-1)!}(.1)^1(.9)^{4-1} = \frac{4\cdot 3\cdot 2\cdot 1}{(1)(3\cdot 2\cdot 1)}(.1)^1(.9)^3 = 4(.1)(.9)^3 = .2916$$

$$p(2) = \frac{4!}{2!(4-2)!}(.1)^2(.9)^{4-2} = \frac{4\cdot 3\cdot 2\cdot 1}{(2\cdot 1)(2\cdot 1)}(.1)^2(.9)^2 = 6(.1)^2(.9)^2 = .0486$$

$$p(3) = \frac{4!}{3!(4-3)!}(.1)^3(.9)^{4-3} = \frac{4\cdot 3\cdot 2\cdot 1}{(3\cdot 2\cdot 1)(1)}(.1)^3(.9)^1 = 4(.1)^3(.9) = .0036$$

$$p(4) = \frac{4!}{4!(4-4)!}(.1)^4(.9)^{4-4} = \frac{4\cdot 3\cdot 2\cdot 1}{(4\cdot 3\cdot 2\cdot 1)(1)}(.1)^4(.9)^0 = 1(.1)^4(.9) = .0001$$

Look Back Note that these probabilities, listed in Table 4.3, sum to 1. A graph of this probability distribution is shown in Figure 4.8.

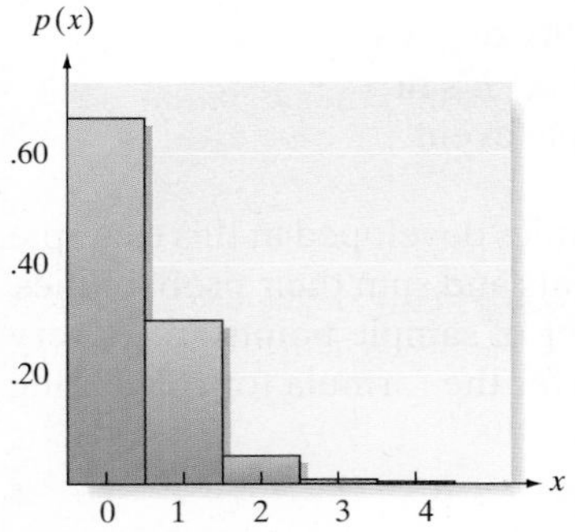

Figure 4.8
Probability distribution for physical fitness example: graphical form

Table 4.3	Probability Distribution for Physical Fitness Example: Tabular Form
x	$p(x)$
0	.6561
1	.2916
2	.0486
3	.0036
4	.0001

Now Work Exercise 4.55

Example 4.12

Finding μ and σ—Physical Fitness Problem

Problem Refer to Examples 4.10 and 4.11. Calculate μ and σ, the mean and standard deviation, respectively, of the number of the four adults who pass the test. Interpret the results.

Solution From Section 4.3, we know that the mean of a discrete probability distribution is

$$\mu = \sum xp(x)$$

Referring to Table 4.3, the probability distribution for the number x who pass the fitness test, we find that

$$\mu = 0(.6561) + 1(.2916) + 2(.0486) + 3(.0036) + 4(.0001) = .4$$
$$= 4(.1) = np$$

Thus, in the long run, the average number of adults (out of four) who pass the test is only .4. [*Note: The relationship $\mu = np$ holds in general for a binomial random variable.*]

The variance is

$$\sigma^2 = \Sigma(x - \mu)^2 p(x) = \Sigma(x - .4)^2 p(x)$$
$$= (0 - .4)^2(.6561) + (1 - .4)^2(.2916) + (2 - .4)^2(.0486)$$
$$+ (3 - .4)^2(.0036) + (4 - .4)^2(.0001)$$
$$= .104976 + .104976 + .124416 + .024336 + .001296$$
$$= .36 = 4(.1)(.9) = npq$$

[*Note: The relationship* $\sigma^2 = npq$ *holds in general for a binomial random variable.*]
Finally, the standard deviation of the number who pass the fitness test is

$$\sigma = \sqrt{\sigma^2} = \sqrt{.36} = .6$$

Since the distribution shown in Figure 4.8 is skewed right, we should apply Chebyshev's rule to describe where most of the x-values fall. According to the rule, at least 75% of the x values will fall into the interval $\mu \pm 2\sigma = .4 \pm 2(.6) = (-.8, 1.6)$. Since x cannot be negative, we expect (i.e., in the long run) the number of adults out of four who pass the fitness test to be less than 1.6.

Look Back Examining Figure 4.8, you can see that all observations equal to 0 or 1 will fall within the interval $(-.8, 1.6)$. The probabilities corresponding to these values (from Table 4.3) are .6561 and .2916, respectively. Summing them, we obtain $.6561 + .2916 = .9477 \approx .95$. This result is closer to the value stated by the empirical rule. In practice, researchers have found that the proportion of observations that fall within two standard deviations of the mean for many skewed distributions will be close to .95.

We emphasize that you need not use the expectation summation rules to calculate μ and σ^2 for a binomial random variable. You can find them easily from the formulas $\mu = np$ and $\sigma^2 = npq$.

Mean, Variance, and Standard Deviation for a Binomial Random Variable

Mean: $\mu = np$

Variance: $\sigma^2 = npq$

Standard deviation: $\sigma = \sqrt{npq}$

Using Binomial Tables

Calculating binomial probabilities becomes tedious when n is large. For some values of n and p, the binomial probabilities have been tabulated in Table II of Appendix A. Part of that table is shown in Table 4.4; a graph of the binomial probability distribution for $n = 10$ and $p = .10$ is shown in Figure 4.9.

Table 4.4   Reproduction of Part of Table II of Appendix A: Binomial Probabilities for $n = 10$

k \ p	.01	.05	.10	.20	.30	.40	.50	.60	.70	.80	.90	.95	.99
0	.904	.599	.349	.107	.028	.006	.001	.000	.000	.000	.000	.000	.000
1	.996	.914	.736	.376	.149	.046	.011	.002	.000	.000	.000	.000	.000
2	1.000	.988	.930	.678	.383	.167	.055	.012	.002	.000	.000	.000	.000
3	1.000	.999	.987	.879	.650	.382	.172	.055	.011	.001	.000	.000	.000
4	1.000	1.000	.998	.967	.850	.633	.377	.166	.047	.006	.000	.000	.000
5	1.000	1.000	1.000	.994	.953	.834	.623	.367	.150	.033	.002	.000	.000
6	1.000	1.000	1.000	.999	.989	.945	.828	.618	.350	.121	.013	.001	.000
7	1.000	1.000	1.000	1.000	.998	.988	.945	.833	.617	.322	.070	.012	.000
8	1.000	1.000	1.000	1.000	1.000	.988	.989	.954	.851	.624	.264	.086	.004
9	1.000	1.000	1.000	1.000	1.000	1.000	.999	.994	.972	.893	.651	.401	.096

Table II actually contains a total of nine tables, labeled **(a)** through **(i)**, one each corresponding to $n = 5, 6, 7, 8, 9, 10, 15, 20$, and 25, respectively. In each of these tables, the columns correspond to values of p and the rows correspond to values of the random variable x. The entries in the table represent **cumulative binomial probabilities.** For example, the entry in the column corresponding to $p = .10$ and the row corresponding to $x = 2$ is .930 (highlighted), and its interpretation is

$$P(x \leq 2) = P(x = 0) + P(x = 1) + P(x = 2) = .930$$

This probability is also highlighted in the graphical representation of the binomial distribution with $n = 10$ and $p = .10$ in Figure 4.9

You can also use Table II to find the probability that x equals a specific value. For example, suppose you want to find the probability that $x = 2$ in the binomial distribution with $n = 10$ and $p = .10$ This probability is found by subtraction as follows:

$$P(x = 2) = \left[P(x = 0) + P(x = 1) + P(x = 2) \right] - \left[P(x = 0) + P(x = 1) \right]$$
$$= P(x \leq 2) - P(x \leq 1) = .930 - .736 = .194$$

The probability that a binomial random variable exceeds a specified value can be found from Table II together with the notion of complementary events. For example, to find the probability that x exceeds 2 when $n = 10$ and $p = .10$, we use

$$P(x > 2) = 1 - P(x \leq 2) = 1 - .930 = .070$$

Note that this probability is represented by the *un*highlighted portion of the graph in Figure 4.9

All probabilities in Table II are rounded to three decimal places. Thus, although none of the binomial probabilities in the table is exactly zero, some are small enough (less than .0005) to round to .000. For example, using the formula to find $P(x = 0)$ when $n = 10$ and $p = .6$, we obtain

$$P(x = 0) = \binom{10}{0}(.6)^0(.4)^{10-0} = .4^{10} = .00010486$$

but this is rounded to .000 in Table II of Appendix A. (See Table 4.4.)

Similarly, none of the table entries is exactly 1.0, but when the cumulative probabilities exceed .9995, they are rounded to 1.000. The row corresponding to the largest possible value for x, $x = n$, is omitted, because all the cumulative probabilities in that row are equal to 1.0 (exactly). For example, in Table 4.4 with $n = 10$, $P(x \leq 10) = 1.0$, no matter what the value of p.

The next example further illustrates the use of Table II.

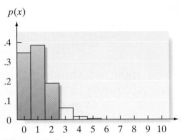

$p(x)$

Figure 4.9

Binomial probability distribution for $n = 10$ and $p = .10$, with $P(x \leq 2)$ highlighted

---

Example 4.13

Using the Binomial Table—Voting for Mayor

Problem Suppose a poll of 20 voters is taken in a large city. The purpose is to determine x, the number who favor a certain candidate for mayor. Suppose that 60% of all the city's voters favor the candidate.

a. Find the mean and standard deviation of x.

b. Use Table II of Appendix A to find the probability that $x \leq 10$.

c. Use Table II to find the probability that $x > 12$.

d. Use Table II to find the probability that $x = 11$.

e. Graph the probability distribution of x, and locate the interval $\mu \pm 2\sigma$ on the graph.

Solution

a. The number of voters polled is presumably small compared with the total number of eligible voters in the city. Thus, we may treat x, the number of the 20 who favor the mayoral candidate, as a binomial random variable. The value of p is the fraction of the total number of voters who favor the candidate (i.e., $p = .6$). Therefore, we calculate the mean and variance:

$$\mu = np = 20(.6) = 12$$
$$\sigma^2 = npq = 20(.6)(.4) = 4.8$$
$$\sigma = \sqrt{4.8} = 2.19$$

b. Looking in the row for $k = 10$ and the column for $p = .6$ of Table II (Appendix A) for $n = 20$, we find the value .245. Thus,

$$P(x \le 10) = .245$$

c. To find the probability

$$P(x > 12) = \sum_{x=13}^{20} p(x)$$

we use the fact that for all probability distributions,

$$\sum_{all\ x} p(x) = 1.$$

Therefore,

$$P(x > 12) = 1 - P(x \le 12) = 1 - \sum_{x=0}^{12} p(x)$$

Consulting Table II of Appendix A, we find the entry in row $k = 12$, column $p = .6$ to be .584. Thus,

$$P(x > 12) = 1 - .584 = .416$$

d. To find the probability that exactly 11 voters favor the candidate, recall that the entries in Table II are cumulative probabilities and use the relationship

$$P(x = 11) = [p(0) + p(1) + \cdots + p(11)] - [p(0) + p(1) + \cdots + p(10)]$$
$$= P(x \le 11) - P(x \le 10)$$

Then

$$P(x = 11) = .404 - .245 = .159$$

e. The probability distribution for x is shown in Figure 4.10. Note that

$$\mu - 2\sigma = 12 - 2(2.2) = 7.6 \qquad \mu + 2\sigma = 12 + 2(2.2) = 16.4$$

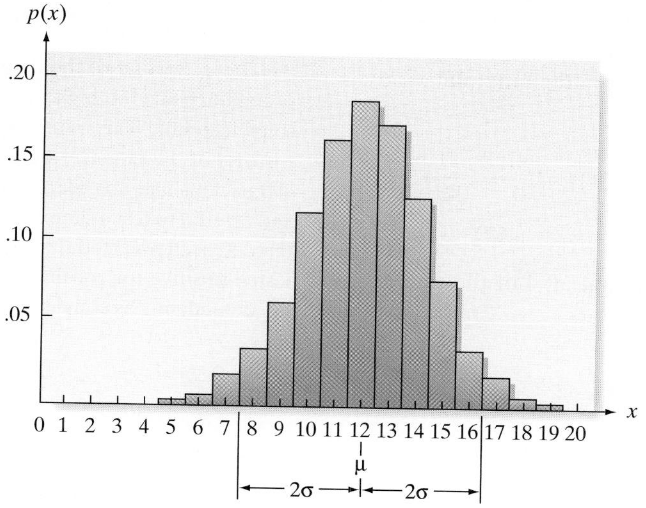

Figure 4.10

The binomial probability distribution for x in Example 4.13; $n = 20$ and $p = .6$

The interval $\mu - 2\sigma$ to $\mu + 2\sigma$ also is shown in Figure 4.10. The probability that x falls into the interval $\mu + 2\sigma$ is $P(x = 8, 9, 10, \ldots, 16) = P(x \le 16) - P(x \le 7)$ .984 − .021 = .963. Note that this probability is very close to the .95 given by the empirical rule. Thus, we expect the number of voters in the sample of 20 who favor the mayoral candidate to be between 8 and 16.

Now Work Exercise 4.58

Statistics IN Action Revisited — Using the Binomial Model to Solve the Cocaine Sting Case

Refer to the reverse cocaine sting case described on p. 180. During a drug bust, police seized 496 foil packets of a white, powdery substance that appeared to be cocaine. The police laboratory randomly selected 4 packets and found that all 4 tested positive for cocaine. This finding led to the conviction of the drug traffickers. Following the conviction, the police used 2 of the remaining 492 foil packets (i.e., those not tested) in a reverse sting operation. The 2 randomly selected packets were sold by undercover officers to a buyer who disposed of the evidence before being arrested. Is there evidence beyond a reasonable doubt that the 2 packets contained cocaine?

To solve the dilemma, we will assume that, of the 496 original packets confiscated, 331 contained cocaine and 165 contained an inert (legal) powder. (A statistician hired as an expert witness on the case showed that the chance of the defendant being found not guilty is maximized when 331 packets contain cocaine and 165 do not.) First, we'll find the probability that 4 packets randomly selected from the original 496 will test positive for cocaine. Then we'll find the probability that the 2 packets sold in the reverse sting did not contain cocaine. Finally, we'll find the probability that both events occur (i.e., that the first 4 packets selected test positive for cocaine, but that the next 2 packets selected do not). In each of these probability calculations, we will apply the binomial probability distribution to approximate the probabilities.

Let x be the number of packets that contain cocaine in a sample of n selected from the 496 packets. Here, we are defining a success as a packet that contains cocaine. If n is small, say, $n = 2$ or $n = 4$, then x has an approximate binomial distribution with probability of success $p = 331/496 \approx .67$.

The probability that the first 4 packets selected contain cocaine [i.e., $P(x = 4)$] is obtained from the binomial formula with $n = 4$ and $p = .67$:

$$P(x = 4) = p(4) = \binom{4}{4}p^4(1 - p)^0 = \frac{4!(.67)^4(.33)^0}{4!0!}$$
$$= (.67)^4 = .202$$

Thus, there is about a 20% chance that all 4 of the randomly selected packets contain cocaine.

Given that 4 of the original packets tested positive for cocaine, the probability that the 2 packets randomly selected and sold in the reverse sting *do not* contain cocaine is approximated by a binomial distribution with $n = 2$ and $p = .67$. Since a success is a packet with cocaine, we find $P(x = 0)$:

$$P(x = 0) = p(0) = \binom{2}{0}p^0(1 - p)^2 = \frac{2!(.67)^0(.33)^2}{0!2!}$$
$$= (.33)^2 = .109$$

Finally, to compute the probability that, of the original 496 foil packets, the first 4 selected (at random) test positive for cocaine and the next two selected (at random) test negative, we employ the multiplicative law of probability. Let A be the event that the first 4 packets test positive. Let B be the event that the next 2 packets test negative. We want to find the probability of both events occurring [i.e., $P(A \text{ and } B) = P(A \cap B) = P(B|A)P(A)$]. Note that this probability is the product of the two previously calculated probabilities:

$$P(A \text{ and } B) = (.109)(.201) = .022$$

Consequently, there is only a .022 probability (i.e., about 2 chances in a hundred) that the first 4 packets will test positive for cocaine and the next 2 packets will test negative for cocaine. A reasonable jury would likely believe that an event with such a small probability is unlikely to occur and conclude that the 2 "lost" packets contained cocaine. In other words, most of us would infer that the defendant in the reverse cocaine sting was guilty of drug trafficking.

[*Epilogue*: Several of the defendant's lawyers believed that the .022 probability was too high for jurors to conclude guilt "beyond a reasonable doubt." The argument was made moot, however, when, to the surprise of the defense, the prosecution revealed that the remaining 490 packets had not been used in any other reverse sting operations and offered to test a sample of them. On the advice of the statistician, the defense requested that an additional 20 packets be tested. All 20 tested positive for cocaine! As a consequence of this new evidence, the defendant was convicted by the jury.]

Exercises 4.50–4.76

Understanding the Principles

4.50 Give the five characteristics of a binomial random variable.

4.51 Give the formula for $p(x)$ for a binomial random variable with $n = 7$ and $p = .2$.

4.52 Consider the following binomial probability distribution:

$$p(x) = \binom{5}{x}(.7)^x(.3)^{5-x} \quad (x = 0, 1, 2, \ldots, 5)$$

a. How many trials (n) are in the experiment?

b. What is the value of p, the probability of success?

Learning the Mechanics

4.53 Compute the following:

a. $\dfrac{6!}{2!(6 - 2)!}$ **b.** $\dbinom{5}{2}$ **c.** $\dbinom{7}{0}$

d. $\binom{6}{6}$ **e.** $\binom{4}{3}$

4.54 Refer to Exercise 4.52.
a. Graph the probability distribution.
b. Find the mean and standard deviation of x.
c. Show the mean and the two-standard-deviation interval on each side of the mean on the graph you drew in part **a**.

4.55 If x is a binomial random variable, compute $p(x)$ for each
[NW] of the following cases:
a. $n = 5, x = 1, p = .2$
b. $n = 4, x = 2, q = .4$
c. $n = 3, x = 0, p = .7$
d. $n = 5, x = 3, p = .1$
e. $n = 4, x = 2, q = .6$
f. $n = 3, x = 1, p = .9$

4.56 Suppose x is a binomial random variable with $n = 3$ and $p = .3$.
a. Calculate the value of $p(x), x = 0, 1, 2, 3$, using the formula for a binomial probability distribution.
b. Using your answers to part **a**, give the probability distribution for x in tabular form.

4.57 If x is a binomial random variable, calculate μ, σ^2, and σ for each of the following:
a. $n = 25, p = .5$
b. $n = 80, p = .2$
c. $n = 100, p = .6$
d. $n = 70, p = .9$
e. $n = 60, p = .8$
f. $n = 1,000, p = .04$

4.58 If x is a binomial random variable, use Table II in Appendix
[NW] A to find the following probabilities:
a. $P(x = 2)$ for $n = 10, p = .4$
b. $P(x \le 5)$ for $n = 15, p = .6$
c. $P(x > 1)$ for $n = 5, p = .1$

4.59 If x is a binomial random variable, use Table II in Appendix A to find the following probabilities:
a. $P(x < 10)$ for $n = 25, p = .7$
b. $P(x \ge 10)$ for $n = 15, p = .9$
c. $P(x = 2)$ for $n = 20, p = .2$

4.60 Suppose x is a binomial random variable with $n = 5$ and $p = .5$. Compute $p(x)$ for $x = 0, 1, 2, 3, 4$, and 5, using the following two methods:
a. List the sample points (take S for Success and F for Failure on each trial) corresponding to each value of x, assign probabilities to each sample point, and obtain $p(x)$ by adding sample–point probabilities.
b. Use the formula for the binomial probability distribution to obtain $p(x)$.

4.61 The binomial probability distribution is a family of probability distributions with each single distribution depending on the values of n and p. Assume that x is a binomial random variable with $n = 4$.
a. Determine a value of p such that the probability distribution of x is symmetric.
b. Determine a value of p such that the probability distribution of x is skewed to the right.
c. Determine a value of p such that the probability distribution of x is skewed to the left.
d. Graph each of the binomial distributions you obtained in parts **a, b**, and **c**. Locate the mean for each distribution on its graph.

e. In general, for what values of p will a binomial distribution be symmetric? skewed to the right? skewed to the left?

Applet Exercise 4.3

Use the applets entitled *Simulating the Probability of a Head with an Unfair Coin* ($P(H) = 0.2$) and *Simulating the Probability of a Head with an Unfair Coin* ($P(H) = 0.8$) to study the mean μ of a binomial distribution.
a. Run each applet once with $n = 1,000$ and record the cumulative proportions. How does the cumulative proportion for each applet compare with the value of $P(H)$ given for the applet?
b. Using the cumulative proportion from each applet as p, compute $\mu = np$ for each applet, where $n = 1,000$. What does the value of μ represent in terms of the results obtained from running each applet in part **a**?
c. In your own words, describe what the mean μ of a binomial distribution represents.

Applet Exercise 4.4

Open the applet entitled *Sample from a Population*. On the pull-down menu to the right of the top graph, select *Binary*. Set $n = 10$ as the sample size and repeatedly choose samples from the population. For each sample, record the number of 1's in the sample. Let x be the number of 1's in a sample of size 10. Explain why x is a binomial random variable.

Applet Exercise 4.5

Use the applet entitled *Simulating the Stock Market* to estimate the probability that the stock market will go up each of the next two days. Repeatedly run the applet for $n = 2$, recording the number of ups each time. Use the proportion of 2's among your results as the estimate of the probability. Compare your answer with the binomial probability where $x = 2, n = 2$, and $p = 0.5$.

Applying the Concepts—Basic

4.62 Where will you get your next pet? According to an Associated Press/Petside.com poll, half of all pet owners would get their next dog or cat from a shelter (*USA Today*, May 12, 2010). Consider a random sample of 10 pet owners and define x as the number of pet owners who would acquire their next dog or cat from a shelter. Assume that x is a binomial random variable.
a. For this binomial experiment, define a success.
b. For this binomial experiment, what is n?
c. For this binomial experiment, what is p?
d. Find $P(x = 7)$.
e. Find $P(x \ge 3)$.
f. Find $P(x > 8)$.

4.63 Chemical signals of mice. Refer to the *Cell* (May 14, 2010) study of the ability of a mouse to recognize the odor of a potential predator, Exercise 3.61 (p. 134). Recall that the sources of these odors are typically major urinary proteins (Mups). In an experiment, 40% of lab mice cells exposed to chemically produced cat Mups responded positively (i.e., recognized the danger of the lurking predator). Consider a sample of 100 lab mice cells, each exposed to chemically produced cat Mups. Let x represent the number of cells that respond positively.
a. Explain why the probability distribution of x can be approximated by the binomial distribution.

b. Find $E(x)$ and interpret its value, practically.

c. Find the variance of x.

d. Give an interval that is likely to contain the value of x.

4.64 Analysis of bottled water. Is the bottled water you're drinking really purified water? A study of various brands of bottled water conducted by the Natural Resources Defense Council found that 25% of bottled water is just tap water packaged in a bottle (*Scientific American*, July 2003). Consider a sample of five bottled-water brands, and let x equal the number of these brands that use tap water.

[NW]

a. Explain why x is (approximately) a binomial random variable.

b. Give the probability distribution for x as a formula.

c. Find $P(x = 2)$.

d. Find $P(x \leq 1)$.

4.65 Tracking missiles with satellite imagery. The U.S. government has devoted considerable funding to missile defense research over the past 20 years. The latest development is the Space-Based Infrared System (SBIRS), which uses satellite imagery to detect and track missiles (*Chance*, Summer 2005). The probability that an intruding object (e.g., a missile) will be detected on a flight track by SBIRS is .8. Consider a sample of 20 simulated tracks, each with an intruding object. Let x equal the number of these tracks on which SBIRS detects the object.

a. Demonstrate that x is (approximately) a binomial random variable.

b. Give the values of p and n for the binomial distribution.

c. Find $P(x = 15)$, the probability that SBIRS will detect the object on exactly 15 tracks.

d. Find $P(x \geq 15)$, the probability that SBIRS will detect the object on at least 15 tracks.

e. Find $E(x)$ and interpret the result.

4.66 Caesarian births. The American College of Obstetricians and Gynecologists reports that 32% of all births in the United States take place by Caesarian section each year. (*National Vital Statistics Reports*, Mar. 2010).

a. In a random sample of 1,000 births, how many, on average, will take place by Caesarian section?

b. What is the standard deviation of the number of Caesarian section births in a sample of 1,000 births?

c. Use your answers to parts **a** and **b** to form an interval that is likely to contain the number of Caesarian section births in a sample of 1,000 births.

Apply the Concepts—Intermediate

4.67 Hotel guest satisfaction. Each year, J. D. Power and Associates publishes the results of its North American Hotel Guest Satisfaction Index Study. For 2009, the study revealed that 66% of hotel guests were aware of the hotel's "green" conservation program. Among these guests, 72% actually participate in the program by reusing towels and bed linens. In a random sample of 15 hotel guests, consider the number (x) of guests who are aware and participate in the hotel's conservation efforts.

a. Explain why x is (approximately) a binomial random variable.

b. Use the rules of probability to determine the value of p for this binomial experiment.

c. Assume $p = .45$. Find the probability that at least 10 of the 15 hotel guests are aware of and participate in the hotel's conservation efforts.

4.68 Immediate feedback to incorrect exam answers. Researchers from the Educational Testing Service (ETS) found that providing immediate feedback to students answering open-ended questions can dramatically improve students' future performance on exams (*Educational and Psychological Measurement,* Feb. 2010). The ETS researchers used questions from the Graduate Record Examination (GRE) in the experiment. After obtaining feedback, students could revise their answers. Consider one of these questions. Initially, 50% of the students answered the question correctly. After providing immediate feedback to students who answered incorrectly, 70% answered correctly. Consider a bank of 100 open-ended questions similar to those on the GRE.

a. In a random sample of 20 students, what is the probability that more than half initially answer the question correctly?

b. Refer to part **a**. After providing immediate feedback, what is the probability that more than half of the students answer the question correctly?

4.69 Making your vote count. The recent Democratic and Republican presidential state primary elections were highlighted by the difference in the way winning candidates were awarded delegates. In Republican states, the winner is awarded all the state's delegates; conversely, the Democratic state winner is awarded delegates in proportion to the percentage of votes. This led to a *Chance* (Fall 2007) article on making your vote count. Consider a scenario where you are one of five county commissioners voting on an issue, where each commissioner is equally likely to vote for or against.

a. Your vote counts (i.e., is the decisive vote) only if the other four voters split, 2 in favor and 2 against. Use the binomial distribution to find the probability that your vote counts.

b. If you convince two other commissioners to "vote in bloc" (i.e., you all agree to vote among yourselves first, and whatever the majority decides is the way all three will vote, guaranteeing that the issue is decided by the bloc), your vote counts only if these 2 commissioners split their bloc votes, 1 in favor and 1 against. Again, use the binomial distribution to find the probability that your vote counts.

4.70 Student gambling on sports. A study of gambling activity at the University of West Georgia (UWG) discovered that 60% of the male students wagered on sports the previous year (*The Sport Journal*, Fall 2006). Consider a random sample of 50 UWG male students. How many of these students would you expect to have gambled on sports the previous year? Give a range that is likely to include the number of male students who have gambled on sports.

4.71 Victims of domestic abuse. According to researchers at Dan Jones & Associates, 1 in every 3 women has been a victim of domestic abuse (Domestic Violence: Incidence and Prevalence Study, Sept.–Dec., 2005). This probability was obtained from a survey of 1,000 adult women residing in Utah. Suppose we randomly sample 15 women and find that 4 have been abused.

a. What is the probability of observing 4 or more abused women in a sample of 15 if the proportion p of women who are victims of domestic abuse is really $p = {}^1/_3$?

b. Many experts on domestic violence believe that the proportion of women who are domestically abused is closer to $p = .10$. Calculate the probability of observing 4 or more abused women in a sample of 15 if $p = .10$.

c. Why might your answers to parts **a** and **b** lead you to believe that $p = \frac{1}{3}$?

4.72 Chickens with fecal contamination. The United States Department of Agriculture (USDA) reports that, under its standard inspection system, one in every 100 slaughtered chickens passes inspection with fecal contamination (*Tampa Tribune*, Mar. 31, 2000). In Exercise 3.19 (p. 121), you found the probability that a randomly selected slaughtered chicken passes inspection with fecal contamination. Now find the probability that, in a random sample of 5 slaughtered chickens, at least one passes inspection with fecal contamination.

4.73 Testing a psychic's ESP. Refer to Exercise 3.95 (p. 149) and the experiment conducted by the Tampa Bay Skeptics to see whether an acclaimed psychic has extrasensory perception (ESP). Recall that a crystal was placed, at random, inside 1 of 10 identical boxes lying side by side on a table. The experiment was repeated seven times, and x, the number of correct decisions, was recorded. (Assume that the seven trials are independent.)

a. If the psychic is guessing (i.e., if the psychic does *not* possess ESP), what is the value of p, the probability of a correct decision on each trial?

b. If the psychic is guessing, what is the expected number of correct decisions in seven trials?

c. If the psychic is guessing, what is the probability of no correct decisions in seven trials?

d. Now suppose the psychic has ESP and $p = .5$. What is the probability that the psychic guesses incorrectly in all seven trials?

e. Refer to part **d**. Recall that the psychic failed to select the box with the crystal on all seven trials. Is this evidence against the psychic having ESP? Explain.

Applying the Concepts—Advanced

4.74 Assigning a passing grade. A literature professor decides to give a 20-question true–false quiz to determine who

has read an assigned novel. She wants to choose the passing grade such that the probability of passing a student who guesses on every question is less than .05. What score should she set as the lowest passing grade?

4.75 USGA golf ball specifications. According to the U.S. Golf Association (USGA), "The weight of the [golf] ball shall not be greater than 1.620 ounces avoirdupois (45.93 grams)…. The diameter of the ball shall not be less than 1.680 inches…. The velocity of the ball shall not be greater than 250 feet per second" (USGA, 2006). The USGA periodically checks the specifications of golf balls sold in the United States by randomly sampling balls from pro shops around the country. Two dozen of each kind are sampled, and if more than three do not meet size or velocity requirements, that kind of ball is removed from the USGA's approved-ball list.

a. What assumptions must be made and what information must be known in order to use the binomial probability distribution to calculate the probability that the USGA will remove a particular kind of golf ball from its approved-ball list?

b. Suppose 10% of all balls produced by a particular manufacturer are less than 1.680 inches in diameter, and assume that the number of such balls, x, in a sample of two dozen balls can be adequately characterized by a binomial probability distribution. Find the mean and standard deviation of the binomial distribution.

c. Refer to part **b**. If x has a binomial distribution, then so does the number, y, of balls in the sample that meet the USGA's minimum diameter. [*Note: $x + y = 24$.*] Describe the distribution of y. In particular, what are p, q, and n? Also, find $E(y)$ and the standard deviation of y.

4.76 Does having boys run in the family? *Chance* (Fall 2001) reported that the eight men in the Rodgers family produced 24 biological children over four generations. Of these 24 children, 21 were boys and 3 were girls. How likely is it for a family of 24 children to have 21 boys? Use the binomial distribution and the fact that 50% of the babies born in the United States are male to answer the question. Do you agree with the statement, "Rodgers men produce boys"?

4.5 The Poisson Random Variable (Optional)

A type of probability distribution that is often useful in describing the number of rare events that will occur during a specific period or in a specific area or volume is the **Poisson distribution** (named after the 18th-century physicist and mathematician Siméon Poisson). Typical examples of random variables for which the Poisson probability distribution provides a good model are

1. The number of traffic accidents per month at a busy intersection

2. The number of noticeable surface defects (scratches, dents, etc.) found by quality inspectors on a new automobile

3. The number of parts per million (ppm) of some toxin found in the water or air emissions from a manufacturing plant

4. The number of diseased trees per acre of a certain woodland

5. The number of death claims received per day by an insurance company

6. The number of unscheduled admissions per day to a hospital

---

Characteristics of a Poisson Random Variable

1. The experiment consists of counting the number of times a certain event occurs during a given unit of time or in a given area or volume (or weight, distance, or any other unit of measurement).

2. The probability that an event occurs in a given unit of time, area, or volume is the same for all the units.

3. The number of events that occur in one unit of time, area, or volume is independent of the number that occur in other units.

4. The mean (or expected) number of events in each unit is denoted by the Greek letter lambda (λ).

The characteristics of the **Poisson random variable** are usually difficult to verify for practical examples. The examples given here satisfy them well enough that the Poisson distribution provides a good model in many instances.* As with all probability models, the real test of the adequacy of the Poisson model is whether it provides a reasonable approximation to reality—that is, whether empirical data support it.

The probability distribution, mean, and variance for a Poisson random variable are shown in the next box.

Probability Distribution, Mean, and Variance for a Poisson Random Variable

$$p(x) = \frac{\lambda^x e^{-\lambda}}{x!} \quad (x = 0, 1, 2, \dots) \quad \mu = \lambda \quad \sigma^2 = \lambda$$

where

λ = Mean number of events during a given unit of time, area, volume, etc.

$e = 2.71828\dots$

BIOGRAPHY **SIMÉON D. POISSON (1781–1840)**

A Lifetime Mathematician

Growing up in France during the French Revolution, Siméon-Denis Poisson was sent away by his father to become an apprentice surgeon, but he lacked the manual dexterity to perform the delicate procedures required and returned home. He eventually enrolled in the École Polytechnique to study mathematics. In his final year of study, Poisson wrote a paper on the theory of equations that was of such quality that he was allowed to graduate without taking the final examination. Two years later, he was named a professor at the university. During his illustrious career, Poisson published between 300 and 400 mathematics papers. He is most known for his 1837 paper in which he first set forth the distribution of a rare event—the Poisson distribution (although the distribution was actually described years earlier by one of the Bernoulli brothers). Poisson dedicated his life to mathematics, once stating that "Life is good for only two things: to study mathematics and to teach it." ∎

The calculation of Poisson probabilities is made easier by the use of Table III in Appendix A, which gives the cumulative probabilities $P(x \leq k)$ for various values of λ. The use of Table III is illustrated in Example 4.14.

---

*The Poisson probability distribution also provides a good approximation to a binomial probability distribution with mean $\lambda = np$ when n is large and p is small (say, $np \leq 7$).

Example 4.14

Finding Poisson Probabilities—Whale Sightings

Problem Ecologists often use the number of reported sightings of a rare species of animal to estimate the remaining population size. For example, suppose the number x of reported sightings per week of blue whales is recorded. Assume that x has (approximately) a Poisson probability distribution. Furthermore, assume that the average number of weekly sightings is 2.6.

a. Find the mean and standard deviation of x, the number of blue-whale sightings per week. Interpret the results.

b. Use Table III to find the probability that fewer than two sightings are made during a given week.

c. Use Table III to find the probability that more than five sightings are made during a given week.

d. Use Table III to find the probability that exactly five sightings are made during a given week.

Solution

a. The mean and variance of a Poisson random variable are both equal to λ. Thus, for this example,

$$\mu = \lambda = 2.6$$
$$\sigma^2 = \lambda = 2.6$$

Then the standard deviation of x is

$$\sigma = \sqrt{2.6} = 1.61$$

Remember that the mean measures the central tendency of the distribution and does not necessarily equal a possible value of x. In this example, the mean is 2.6 sightings, and although there cannot be 2.6 sightings during a given week, the average number of weekly sightings is 2.6. Similarly, the standard deviation of 1.61 measures the variability of the number of sightings per week. Perhaps a more helpful measure is the interval $\mu \pm 2\sigma$, which in this case stretches from $-.62$ to 5.82. We expect the number of sightings to fall into this interval most of the time—with at least 75% relative frequency (according to Chebyshev's rule) and probably closer to 95% relative frequency (the empirical rule). The mean and the two-standard-deviation interval around it are shown in Figure 4.11.

b. A partial reproduction of Table III is shown in Table 4.5. The rows of the table correspond to different values of λ, and the columns correspond to different values k of the Poisson random variable x. The entries in the table (like the binomial probabilities in Table II) give the cumulative probability $P(x \le k)$. To find the probability that fewer than two sightings are made during a given week, we first note that

$$P(x < 2) = P(x \le 1)$$

This probability is a cumulative probability and therefore is the entry in Table III in the row corresponding to $\lambda = 2.6$ and the column corresponding to $k = 1$. The entry is .267, shown highlighted in Table 4.5. This probability corresponds to the highlighted area in Figure 4.11 and may be interpreted as meaning that there is a 26.7% chance that fewer than two sightings will be made during a given week.

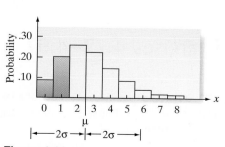

Figure 4.11

Probability distribution for number of blue-whale sightings

c. To find the probability that more than five sightings are made during a given week, we consider the complementary event

Table 4.5		Reproduction of Part of Table III in Appendix A								
λ \ k	0	1	2	3	4	5	6	7	8	9
2.2	.111	.355	.623	.819	.928	.975	.993	.998	1.000	1.000
2.4	.091	.308	.570	.779	.904	.964	.988	.997	.999	1.000
2.6	.074	.267	.518	.736	.877	.951	.983	.995	.999	1.000
2.8	.061	.231	.469	.692	.848	.935	.976	.992	.998	.999
3.0	.050	.199	.423	.647	.815	.916	.966	.988	.996	.999
3.2	.041	.171	.380	.603	.781	.895	.955	.983	.994	.998
3.4	.033	.147	.340	.558	.744	.871	.942	.977	.992	.997
3.6	.027	.126	.303	.515	.706	.844	.927	.969	.988	.996
3.8	.022	.107	.269	.473	.668	.816	.909	.960	.984	.994
4.0	.018	.092	.238	.433	.629	.785	.889	.949	.979	.992
4.2	.015	.078	.210	.395	.590	.753	.867	.936	.972	.989
4.4	.012	.066	.185	.359	.551	.720	.844	.921	.964	.985
4.6	.010	.056	.163	.326	.513	.686	.818	.905	.955	.980
4.8	.008	.048	.143	.294	.476	.651	.791	.887	.944	.975
5.0	.007	.040	.125	.265	.440	.616	.762	.867	.932	.968
5.2	.006	.034	.109	.238	.406	.581	.732	.845	.918	.960
5.4	.005	.029	.095	.213	.373	.546	.702	.822	.903	.951
5.6	.004	.024	.082	.191	.342	.512	.670	.797	.886	.941
5.8	.003	.021	.072	.170	.313	.478	.638	.771	.867	.929
6.0	.002	.017	.062	.151	.285	.446	.606	.744	.847	.916

$$P(x > 5) = 1 - P(x \le 5) = 1 - .951 = .049$$

where .951 is the entry in Table III corresponding to $\lambda = 2.6$ and $k = 5$. (See Table 4.5.) Note from Figure 4.11 that this probability is represented by the area in the interval $\mu \pm 2\sigma$, or $-.62$ to 5.82. Then the number of sightings should exceed 5—or, equivalently, should be more than two standard deviations from the mean—during only about 4.9% of all weeks. Note that this percentage agrees remarkably well with that given by the empirical rule for mound-shaped distributions, which tells us to expect approximately 5% of the measurements (values of the random variable) to lie farther than two standard deviations from the mean.

d. To use Table III to find the probability that *exactly* five sightings are made during a given week, we must write the probability as the difference between two cumulative probabilities:

$$P(x = 5) = P(x \le 5) - P(x \le 4) = .951 - .877 = .074$$

Now Work Exercise 4.83

Note that the probabilities in Table III are all rounded to three decimal places. Thus, although, in theory, a Poisson random variable can assume infinitely large values, the values of k in Table III are extended only until the cumulative probability is 1.000. This does not mean that x *cannot* assume larger values, but only that the likelihood is less than .001 (in fact, less than .0005) that it will do so.

Finally, you may need to calculate Poisson probabilities for values of λ not found in Table III. You may be able to obtain an adequate approximation by interpolation, but if not, consult more extensive tables for the Poisson distribution.

Exercises 4.77–4.97

Understanding the Principles

4.77 Give the four characteristics of a Poisson random variable.

4.78 Consider a Poisson random variable with probability distribution

$$p(x) = \frac{10^x e^{-10}}{x!} \quad (x = 0, 1, 2, \dots)$$

What is the value of λ?

4.79 Consider the Poisson probability distribution shown here:

$$p(x) = \frac{3^x e^{-3}}{x!} \quad (x = 0, 1, 2, \dots)$$

What is the value of λ?

Learning the Mechanics

4.80 Refer to Exercise 4.78.
a. Graph the probability distribution.

b. Find the mean and standard deviation of x.

4.81 Refer to Exercise 4.79.
a. Graph the probability distribution.
b. Find the mean and standard deviation of x.

4.82 Given that x is a random variable for which a Poisson probability distribution provides a good approximation, use Table III in Appendix A to compute the following:
a. $P(x \le 2)$ when $\lambda = 1$
b. $P(x \le 2)$ when $\lambda = 2$
c. $P(x \le 2)$ when $\lambda = 3$
d. What happens to the probability of the event $\{x \le 2\}$ as λ increases from 1 to 3? Is this intuitively reasonable?

4.83 Assume that x is a random variable having a Poisson probability distribution with a mean of 1.5. Use Table III in Appendix A to find the following probabilities:
a. $P(x \le 3)$
b. $P(x \ge 3)$
c. $P(x = 3)$
d. $P(x = 0)$
e. $P(x > 0)$
f. $P(x > 6)$

4.84 Suppose x is a random variable for which a Poisson probability distribution with $\lambda = 1$ provides a good characterization.
a. Graph $p(x)$ for $x = 0, 1, 2, \dots, 9$.
b. Find μ and σ for x, and locate μ and the interval $\mu \pm 2\sigma$ on the graph.
c. What is the probability that x will fall within the interval $\mu \pm 2\sigma$?

4.85 Suppose x is a random variable for which a Poisson probability distribution with $\lambda = 3$ provides a good characterization.
a. Graph $p(x)$ for $x = 0, 1, 2, \dots, 9$.
b. Find μ and σ for x, and locate μ and the interval $\mu \pm 2\sigma$ on the graph.
c. What is the probability that x will fall within the interval $\mu \pm 2\sigma$?

4.86 As mentioned in Section 4.5, when n is large, p is small, and $np \le 7$, the Poisson probability distribution provides a good approximation to the binomial probability distribution. Since we provide exact binomial probabilities (Table II in Appendix A) for relatively small values of n, you can investigate the adequacy of the approximation for $n = 25$. Use Table II to find $p(0), p(1)$, and $p(2)$ for $n = 25$ and $p = .05$. Calculate the corresponding Poisson approximations, using $\lambda = \mu = np$. [*Note:* These approximations are reasonably good for n as small as 25, but to use such an approximation in a practical situation we would prefer to have $n \ge 100$.]

Applying the Concepts—Basic

4.87 **Eye fixation experiment.** Cognitive scientists at the University of Massachusetts designed an experiment to measure x, the number of times a reader's eye fixated on a single word before moving past that word (*Memory and Cognition*, Sept. 1997). For this experiment, x was found to have a mean of 1. Suppose one of the readers in the experiment is randomly selected, and assume that x has a Poisson distribution.
a. Find $P(x = 0)$.
b. Find $P(x > 1)$.
c. Find $P(x \le 2)$.

4.88 **Noise in laser imaging.** Penumbrol imaging is a technique used by nuclear engineers for imaging objects (e.g., X-rays and lasers) that emit high-energy photons. In *IEICE Transactions on Information & Systems* (Apr. 2005), researchers demonstrated that penumbrol images are always degraded by noise, where the number x of noise events occurring in a unit of time follows a Poisson process with mean λ. Suppose that $\lambda = 9$ for a particular image.
a. Find and interpret the mean of x.
b. Find the standard deviation of x.
c. The signal-to-noise ratio (SNR) for a penumbrol image is defined as SNR $= \mu/\sigma$, where μ and σ are the mean and standard deviation, respectively, of the noise process. Find the SNR for x.

4.89 **Spare line replacement units.** The U.S. Department of Defense Reliability Analysis Center publishes Selected Topics in Assurance Related Technologies (START) sheets to help improve the quality of manufactured components and systems. One START sheet, titled "Application of the Poisson Distribution" (Vol. 9, No. 1, 2002), focuses on a spare line replacement unit (LRU). The number of LRUs that fail in any 10,000-hour period is assumed to follow a Poisson distribution with a mean of 1.2.
a. Find the probability that there are no LRU failures during the next 10,000 hours of operation.
b. Find the probability that there are at least two LRU failures during the next 10,000 hours of operation.

4.90 **Rare planet transits.** A "planet transit" is a rare celestial event in which a planet appears to cross in front of its star as seen from Earth. The planet transit causes a noticeable dip in the star's brightness, allowing scientists to detect a new planet even though it is not directly visible. The National Aeronautics and Space Administration (NASA) recently launched its Kepler mission, designed to discover new planets in the Milky Way by detecting extrasolar planet transits. After one year of the mission in which 3,000 stars were monitored, NASA announced that 5 planet transits were detected (NASA, American Astronomical Society, Jan. 4, 2010). Assume that the number of planet transits discovered for every 3,000 stars follows a Poisson distribution with $\lambda = 5$. What is the probability that, in the next 3,000 stars monitored by the Kepler mission, more than 10 planet transits will be seen?

4.91 **Airline fatalities.** U.S. airlines average about 1.6 fatalities per month. (*Statistical Abstract of the United States: 2010*). Assume that the probability distribution for x, the number of fatalities per month, can be approximated by a Poisson probability distribution.
a. What is the probability that no fatalities will occur during any given month?
b. What is the probability that one fatality will occur during any given month?
c. Find $E(x)$ and the standard deviation of x.

Applying the Concepts—Intermediate

4.92 **Making high-stakes insurance decisions.** The *Journal of Economic Psychology* (Sept. 2008) published the results of a high-stakes experiment where subjects were asked how much they would pay for insuring a valuable painting. The painting was threatened by fire and theft, hence the need for insurance. To make the risk realistic, the subjects were informed that if it rained on exactly 24 days in July, the painting was considered to be stolen; and, if it rained on exactly 23 days in August, the painting was considered to be destroyed by fire. Although the probability of these two events, "fire" and "theft," was ambiguous for the subjects, the researchers estimated their probabilities of occurrence at .0001. Rain frequencies for the months of July and August were shown to follow a Poisson distribution with a mean of 10 days per month.

a. Find the probability that it will rain on exactly 24 days in July.

b. Find the probability that it will rain on exactly 23 days in August.

c. Are the probabilities, parts **a** and **b**, good approximations to the probabilities of "fire" and "theft"?

4.93 **LAN videoconferencing.** A network administrator is installing a videoconferencing module in a local area network (LAN) computer system. Of interest is the capacity of the LAN to handle users who attempt to call in for videoconferencing during peak hours. Calls are blocked if the user finds that all LAN lines are "busy." The capacity is directly related to the rate at which calls are blocked ("Traffic Engineering Model for LAN Video Conferencing," Intel, 2005). Let x equal the number of calls blocked during the peak hour (busy) videoconferencing call time. The network administrator believes that x has a Poisson distribution with mean $\lambda = 5$.

a. Find the probability that fewer than 3 calls are blocked during the peak hour.

b. Find $E(x)$ and interpret its value.

c. Is it likely that no calls will be blocked during the peak hour?

d. If, in fact, $x = 0$ during a randomly selected peak hour, what would you infer about the value of λ? Explain.

4.94 **Davy Crockett's use of words.** Davy Crockett, a U.S. congressman during the 1830s, published *A Narrative of the Life of David Crockett, Written by Himself.* In a *Chance* (Spring 1999) article, researchers determined that x, the number of times (per 1,000 words) that the word *though* appears in the *Life* narrative, has a mean of .25. Assume that x has an approximate Poisson distribution.

a. Find the variance of x.

b. Calculate the interval $\mu \pm 2\sigma$. What proportion of the time will x fall into this interval?

c. In 1836, Davy Crockett died at the Alamo defending Texans from an attack by the Mexican general Santa Anna. The following year, the narrative *Colonel Crockett's Exploits and Adventures in Texas* was published. Some historians have doubts whether Crockett was the actual author of this narrative. In the *Texas* narrative, "though" appears twice in the first 1,000 words. Use this information and your answer to part **b** to make an inference about whether Crockett actually wrote the *Texas* narrative. (Assume that the *Life* narrative was really written by Davy Crockett.)

4.95 **Customer arrivals at a bakery.** As part of a project targeted at improving the services of a local bakery, a management consultant (L. Lei of Rutgers University) monitored customer arrivals for several Saturdays and Sundays. Using the arrival data, she estimated the average number of customer arrivals per 10-minute period on Saturdays to be 6.2. She assumed that arrivals per 10-minute interval followed the Poisson distribution in the table below, some of whose values are missing.

a. Compute the missing probabilities.

b. Graph the distribution.

c. Find μ and σ, and show the intervals $\mu \pm \sigma, \mu \pm 2\sigma$, and $\mu \pm 3\sigma$ on your graph of part **b.**

d. The owner of the bakery claims that more than 75 customers per hour enter the store on Saturdays. On the basis of the consultant's data, is this likely? Explain.

4.96 **Flaws in plastic-coated wire.** The British Columbia Institute of Technology provides on its Web site (www.math.bcit.ca) practical applications of statistics to mechanical engineering. The following is a Poisson application. A roll of plastic-coated wire has an average of .8 flaws per 4-meter length of wire. Suppose a quality control engineer will sample a 4-meter length of wire from a roll of wire 220 meters in length. If no flaws are found in the sample, the engineer will accept the entire roll of wire. What is the probability that the roll will be rejected? What assumption did you make to find this probability?

Applying the Concepts—Advanced

4.97 **Waiting for a car wash.** A certain automatic car wash takes exactly 5 minutes to wash a car. On the average, 10 cars per hour arrive at the car wash. Suppose that, 30 minutes before closing time, five cars are in line. If the car wash is in continuous use until closing time, is it likely anyone will be in line at closing time?

Table for Exercise 4.95

x	0	1	2	3	4	5	6	7	8	9	10	11	12	13
$p(x)$	.002	.013	—	.081	.125	.155	—	.142	.110	.076	—	.026	.014	.007

Based on Lei, L. *Dorsi's Bakery: Modeling Service Operations.* Graduate School of Management, Rutgers University, 1993.

4.6 The Hypergeometric Random Variable (Optional)

The **hypergeometric probability distribution** provides a realistic model for some types of enumerative (countable) data. The characteristics of the hypergeometric distribution are listed in the following box:

> **Characteristics of a Hypergeometric Random Variable**
>
> 1. The experiment consists of randomly drawing n elements without replacement from a set of N elements, r of which are S's (for Success) and $(N - r)$ of which are F's (for Failure).
> 2. The hypergeometric random variable x is the number of S's in the draw of n elements.

Note that both the hypergeometric and binomial characteristics stipulate that each draw, or trial, results in one of two outcomes. The basic difference between these random variables is that the hypergeometric trials are dependent, while the binomial trials are independent. The draws are dependent because the probability of drawing an S (or an F) is dependent on what occurred on preceding draws.

To illustrate the dependence between trials, we note that the probability of drawing an S on the first draw is r/N. Then the probability of drawing an S on the second draw depends on the outcome of the first. It will be either $(r - 1)/(N - 1)$ or $r/(N - 1)$, depending on whether the first draw was an S or an F. Consequently, the results of the draws represent dependent events.

For example, suppose we define x as the number of women hired in a random selection of three applicants from a total of six men and four women. This random variable satisfies the characteristics of a **hypergeometric random variable** with $N = 10$ and $n = 3$. The possible outcomes on each trial are either the selection of a female (S) or the selection of a male (F). Another example of a hypergeometric random variable is the number x of defective large-screen plasma televisions in a random selection of $n = 4$ from a shipment of $N = 8$ TVs. Finally, as a third example, suppose $n = 5$ stocks are randomly selected from a list of $N = 15$ stocks. Then the number x of the five companies selected that pay regular dividends to stockholders is a hypergeometric random variable.

The hypergeometric probability distribution is summarized in the following box:

> **Probability Distribution, Mean, and Variance of the Hypergeometric Random Variable**
>
> $$p(x) = \frac{\binom{r}{x}\binom{N - r}{n - x}}{\binom{N}{n}} \quad [x = \text{Maximum}\,[0, n - (N - r)], \ldots, \text{Minimum}(r, n)]$$
>
> $$\mu = \frac{nr}{N} \qquad \sigma^2 = \frac{r(N - r)n(N - n)}{N^2(N - 1)}$$
>
> where
>
> N = Total number of elements
>
> r = Number of S's in the N elements
>
> n = Number of elements drawn
>
> x = Number of S's drawn in the n elements

Example 4.15

Applying the Hypergeometric Distribution— Selecting Teaching Assistants

Problem Suppose a professor randomly selects three new teaching assistants from a total of 10 applicants—six male and four female students. Let x be the number of females who are hired.

a. Find the mean and standard deviation of x.

b. Find the probability that no females are hired.

Solution

a. Since x is a hypergeometric random variable with $N = 10, n = 3$, and $r = 4$, the mean and variance are

$$\mu = \frac{nr}{N} = \frac{(3)(4)}{10} = 1.2$$

$$\sigma^2 = \frac{r(N-r)n(N-n)}{N^2(N-1)} = \frac{4(10-4)3(10-3)}{(10)^2(10-1)}$$

$$= \frac{(4)(6)(3)(7)}{(100)(9)} = .56$$

The standard deviation is

$$\sigma = \sqrt{.56} = .75$$

b. The probability that no female students are hired by the professor, assuming that the selection is truly random, is

$$P(x = 0) = p(0) = \frac{\binom{4}{0}\binom{10-4}{3-0}}{\binom{10}{3}}$$

$$= \frac{\dfrac{4!}{0!(4-0)!}\dfrac{6!}{3!(6-3)!}}{\dfrac{10!}{3!(10-3)!}} = \frac{(1)(20)}{120} = \frac{1}{6}$$

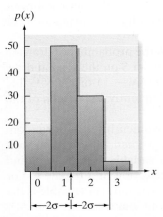

$p(x)$

.50
.40
.30
.20
.10

0   1   2   3 → x
         μ
$\leftarrow 2\sigma \rightarrow \leftarrow 2\sigma \rightarrow$

Figure 4.12
Probability distribution for x in Example 4.15

Look Back The entire probability distribution for x is shown in Figure 4.12. The mean $\mu = 1.2$ and the interval $\mu \pm 2\sigma = (-.3, 2.7)$ are indicated. You can see that if this random variable were to be observed over and over again a large number of times, most of the values of x would fall within the interval $\mu \pm 2\sigma$.

Now Work Exercise 4.115

Statistics IN Action | **Revisited** | ## Using the Hypergeometric Model to Solve the Cocaine Sting Case

The reverse cocaine sting case described on p. 180 and solved in Section 4.4 can also be solved by applying the hypergeometric distribution. In fact, the probabilities obtained with the hypergeometric distribution are *exact* probabilities, compared with the approximate probabilities obtained with the binomial distribution.

Our objective, you will recall, is to find the probability that 4 packets (randomly selected from 496 packets confiscated in a drug bust) will contain cocaine and 2 packets (randomly selected from the remaining 492) will not contain cocaine. We assumed that, of the 496 original packets, 331 contained cocaine and 165 contained an inert (legal) powder. Since we are sampling *without replacement* from the 496 packets, the probability of a success (i.e., the probability of a packet containing cocaine) does not remain *exactly* the same from trial to trial. For example, for the first randomly selected packet, the probability of a success is 331/496 = .66734. If we find that the first 3 packets selected contain cocaine, the probability of success for the 4th packet selected is now 328/493 = .66531. You can see that these probabilities are not exactly the same. Hence, the binomial distribution will only approximate

the distribution of x, the number of packets that contain cocaine in a sample of size n.

To find the probability that 4 packets randomly selected from the original 496 will test positive for cocaine under the hypergeometric distribution, we first identify the parameters of the distribution:

$N = 496$ is the total number of packets in the population
$S = 331$ is the number of successes (cocaine packets) in the population
$n = 4$ is the sample size
$x = 4$ is the number of successes (cocaine packets) in the sample

Substituting into the formula for $p(x)$ (above), we obtain

$$P(x = 4) = p(4) = \frac{\binom{331}{4}\binom{165}{0}}{\binom{496}{4}} = \frac{\left(\dfrac{331!}{4!327!}\right)\left(\dfrac{165!}{0!165!}\right)}{\left(\dfrac{496!}{4!492!}\right)}$$

$$= .197$$

Statistics IN Action
(continued)

To find the probability that 2 packets randomly selected from the remaining 492 will test negative for cocaine, *assuming that the first 4 packets tested positive*, we identify the parameters of the relevant hypergeometric distribution:

$N = 492$ is the total number of packets in the population
$S = 327$ is the number of successes (cocaine packets) in the population
$n = 2$ is the sample size
$x = 0$ is the number of successes (cocaine packets) in the sample

Again, we substitute into the formula for $p(x)$ (page 214) to obtain

$$P(x = 0) = p(0) = \frac{\binom{327}{0}\binom{165}{2}}{\binom{492}{2}} = \frac{\left(\frac{327!}{0!327!}\right)\left(\frac{165!}{2!163!}\right)}{\left(\frac{492!}{2!490!}\right)}$$

$$= .112$$

By the multiplicative law of probability, the probability that the first 4 packets test positive for cocaine and the next 2 packets test negative is the product of the two probabilities just given:

$$P\,(\text{first 4 positive and next 2 negative}) = (.197)(.112)$$
$$= .0221$$

Note that this exact probability is almost identical to the approximate probability computed with the binomial distribution in Section 4.4.

Exercises 4.98–4.117

Understanding the Principles

4.98 Give the characteristics of a hypergeometric distribution.

4.99 How do binomial and hypergeometric random variables differ? In what respects are they similar?

4.100 Explain the difference between sampling with replacement and sampling without replacement.

Learning the Mechanics

4.101 Given that x is a hypergeometric random variable, compute $p(x)$ for each of the following cases:
 a. $N = 5, n = 3, r = 3, x = 1$
 b. $N = 9, n = 5, r = 3, x = 3$
 c. $N = 4, n = 2, r = 2, x = 2$
 d. $N = 4, n = 2, r = 2, x = 0$

4.102 Given that x is a hypergeometric random variable with $N = 8, n = 3$, and $r = 5$, compute the following:
 a. $P(x = 1)$
 b. $P(x = 0)$
 c. $P(x = 3)$
 d. $P(x \geq 4)$

4.103 Given that x is a hypergeometric random variable with $N = 12, n = 8$, and $r = 6$:
 a. Display the probability distribution for x in tabular form.
 b. Compute μ and σ for x.
 c. Graph $p(x)$, and locate μ and the interval $\mu \pm 2\sigma$ on the graph.
 d. What is the probability that x will fall within the interval $\mu \pm 2\sigma$?

4.104 Use the results of Exercise 4.103 to find the following probabilities:
 a. $P(x = 1)$
 b. $P(x = 4)$
 c. $P(x \leq 4)$
 d. $P(x \geq 5)$
 e. $P(x < 3)$
 f. $P(x \geq 8)$

4.105 Suppose you plan to sample 10 items from a population of 100 items and would like to determine the probability of observing 4 defective items in the sample. Which probability distribution should you use to compute this probability under the conditions listed here? Justify your answers.
 a. The sample is drawn without replacement.
 b. The sample is drawn with replacement.

4.106 Given that x is a hypergeometric random variable with $N = 10, n = 5$, and $r = 7$:
 a. Display the probability distribution for x in tabular form.
 b. Compute the mean and variance of x.
 c. Graph $p(x)$, and locate μ and the interval $\mu \pm 2\sigma$ on the graph.
 d. What is the probability that x will fall within the interval $\mu \pm 2\sigma$?

Applying the Concepts — Basic

4.107 Mail rooms contaminated with anthrax. In the fall of 2001, there was a highly publicized outbreak of anthrax cases among U.S. Postal Service workers. In *Chance* (Spring 2002), research statisticians discussed the problem of sampling mail rooms for the presence of anthrax spores. Let x equal the number of mail rooms contaminated with anthrax spores in a random sample of 3 mail rooms selected from a population of 100 mail rooms. If 20 of the 100 mail rooms are contaminated with anthrax, researchers showed that the probability distribution for x is given by the formula

$$p(x) = \frac{\binom{20}{x}\binom{80}{3 - x}}{\binom{100}{3}}$$

 a. What is the name given to this probability distribution?
 b. Specify r, N, and n for this distribution.
 c. Find $p(0)$.

d. Find $p(1)$.

e. Find $p(2)$.

f. Find $p(3)$.

4.108 **ESL students and plagiarism.** Refer to the *Journal of Education and Human Development* (Vol. 3, 2009) investigation of plagiarism among English-as-a-second language (ESL) students, Exercise 3.28 (p. 122). Recall that of six ESL students taking a master's course in linguistics, three admitted to plagiarizing on an essay. Two of these six students will be selected to write an original article for the college newspaper. If the two students are selected at random, what is the probability that at least one plagiarized on the essay?

4.109 **On-site treatment of hazardous waste.** The Resource Conservation and Recovery Act mandates the tracking and disposal of hazardous waste produced at U.S. facilities. *Professional Geographer* (Feb. 2000) reported the hazardous-waste generation and disposal characteristics of 209 facilities. Only 8 of these facilities treated hazardous waste on-site.

a. In a random sample of 10 of the 209 facilities, what is the expected number in the sample that treat hazardous waste on-site? Interpret this result.

b. Find the probability that 4 of the 10 selected facilities treat hazardous waste on-site.

4.110 **Guilt in decision making.** Refer to the *Journal of Behavioral Decision Making* (Jan., 2007) study of how guilty feelings impact decisions, Exercise 3.55 (p. 133). Recall that 57 students were assigned to a guilty state through a reading/writing task. Immediately after the task, the students were presented with a decision problem where the stated option has predominantly negative features (e.g., spending money on repairing a very old car). Of these 57 students, 45 chose the stated option. Suppose 10 of the 57 guilty-state students are selected at random. Define x as the number in the sample of 10 who chose the stated option.

a. Find $P(x = 5)$.

b. Find $P(x = 8)$.

c. What is the expected value (mean) of x?

4.111 **Contaminated gun cartridges.** Refer to the investigation of contaminated gun cartridges at a weapons manufacturer, presented in Exercise 4.27 (p. 189). In a sample of 158 cartridges from a certain lot, 36 were found to be contaminated and 122 were "clean." If you randomly select 5 of these 158 cartridges, what is the probability that all 5 will be "clean"?

Applying the Concepts—Intermediate

4.112 **Lot inspection sampling.** Imagine that you are purchasing small lots of a manufactured product. If it is very costly to test a single item, it may be desirable to test a sample of items from the lot instead of testing every item in the lot. Suppose each lot contains 10 items. You decide to sample 4 items per lot and reject the lot if you observe 1 or more defectives.

a. If the lot contains 1 defective item, what is the probability that you will accept the lot?

b. What is the probability that you will accept the lot if it contains 2 defective items?

4.113 **Extinct New Zealand birds.** Refer to the *Evolutionary Ecology Research* (July 2003) study of the patterns of extinction in the New Zealand bird population, presented in Exercise 3.87 (p. 147). Of the 132 bird species listed in the **NZBIRDS** file, 38 are extinct. Suppose you randomly select 10 of the 132 bird species (without replacement) and record the extinction status of each.

a. What is the probability that exactly 5 of the 10 species you select are extinct?

b. What is the probability that at most 1 species is extinct?

4.114 **Cell phone handoff behavior.** Refer to the *Journal of Engineering, Computing and Architecture* (Vol. 3., 2009) study of cell phone handoff behavior, Exercise 3.58 (p. 134). Recall that a "handoff" describes the process of a cell phone moving from one base channel (identified by a color code) to another. During a particular driving trip a cell phone changed channels (color codes) 85 times. Color code "b" was accessed 40 times on the trip. You randomly select 7 of the 85 handoffs. How likely is it that the cell phone accesses color code "b" only twice for these 7 handoffs?

4.115 **Testing for spoiled wine.** Suppose that you are purchasing cases of wine (12 bottles per case) and that, periodically, you select a test case to determine the adequacy of the bottles' seals. To do this, you randomly select and test 3 bottles in the case. If a case contains 1 spoiled bottle of wine, what is the probability that this bottle will turn up in your sample?

Applying the Concepts—Advanced

4.116 **Gender discrimination suit.** The *Journal of Business & Economic Statistics* (July 2000) presented a case in which a charge of gender discrimination was filed against the U.S. Postal Service. At the time, there were 302 U.S. Postal Service employees (229 men and 73 women) who applied for promotion. Of the 72 employees who were awarded promotion, 5 were female. Make an inference about whether or not females at the U.S. Postal Service were promoted fairly.

4.117 **Awarding of home improvement grants.** A curious event was described in the *Minneapolis Star and Tribune*. The Minneapolis Community Development Agency (MCDA) makes home improvement grants each year to homeowners in depressed city neighborhoods. Of the $708,000 granted one year, $233,000 was awarded by the city council via a "random selection" of 140 homeowners' applications from among a total of 743 applications: 601 from the north side, and 142 from the south side, of Minneapolis. Oddly, all 140 grants awarded were from the north side—clearly a highly improbable outcome if, in fact, the 140 winners were randomly selected from among the 743 applicants.

a. Suppose the 140 winning applications were randomly selected from among the total of 743, and let x equal the number in the sample from the north side. Find the mean and standard deviation of x.

b. Use the results of part **a** to support a contention that the grant winners were not randomly selected.

CHAPTER NOTES

Key Terms

Note: Starred (*) terms are from the optional sections in this chapter.

Binomial distribution 199
Binomial experiment 195
Binomial random variable 195
Continuous random variable 182
Countable 182
Cumulative binomial
  probabilities 202
Discrete random variable 182
Expected value 190
*Hypergeometric probability
  distribution 212

*Hypergeometric random
  variable 213
Mean value of a discrete random
  variable 190
*Poisson distribution 207
*Poisson random variable 208
Probability distribution
  of a discrete random
  variable 185
Random variable 180
Standard deviation of a discrete
  random variable 192
Variance of a random
  variable 191

Key Symbols

$p(x)$ Probability distribution for discrete random variable x

$f(x)$ Probability distribution for continuous random variable x

S Outcome of binomial trial denoted "success"
F Outcome of binomial trial denoted "failure"
P $P(S)$ in binomial trial
q $P(F)$ in binomial trial $= 1 - p$
e Constant used in Poisson probability distribution;
     $e = 2.71828\ldots$

Key Ideas

Properties of Discrete Probability Distributions

1. $p(x) \geq 0$

2. $\displaystyle\sum_{\text{all } x} p(x) = 1$

Guide to Selecting a Discrete Probability Distribution

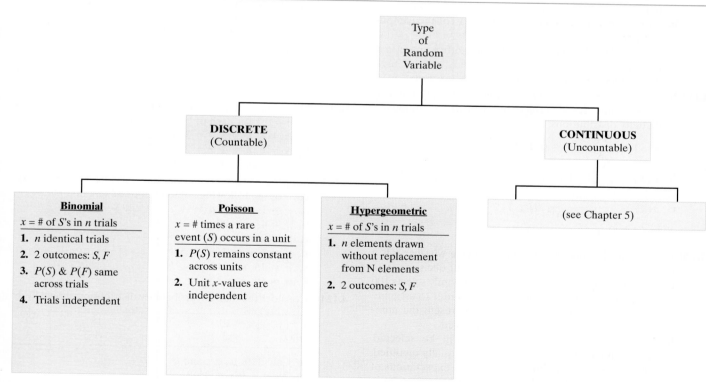

Key Formulas

Random Variable	Probability Distribution	Mean	Variance
General Discrete:	Table, formula, or graph for $p(x)$	$\sum_{\text{all } x} x \cdot p(x)$	$\sum_{\text{all } x} (x - \mu)^2 \cdot p(x)$
Binomial:	$p(x) = \binom{n}{x} p^x q^{n-x}$	np	npq
Poisson:	$p(x) = \dfrac{\lambda^x e^{-\lambda}}{x!}$	λ	λ
Hypergeometric:	$p(x) = \dfrac{\binom{r}{x}\binom{N-r}{n-x}}{\binom{N}{x}}$	$\dfrac{nr}{N}$	$\dfrac{r(N-r)\,n(N-n)}{N^2\,(N-1)}$

Supplementary Exercises 4.118–4.148

Note: Starred (*) exercises refer to the optional sections in this chapter.

Understanding the Principles

4.118 Which of the following describe discrete random variables, and which describe continuous random variables?
  a. The length of time that an exercise physiologist's program takes to elevate her client's heart rate to 140 beats per minute
  b. The number of crimes committed on a college campus per year
  c. The number of square feet of vacant office space in a large city
  d. The number of voters who favor a new tax proposal

4.119 Identify the type of random variable—binomial, Poisson, or hypergeometric—described by each of the following probability distributions:
  *a. $p(x) = \dfrac{.5^x e^{-.5}}{x!}$ $(x = 0, 1, 2, \ldots)$
  b. $p(x) = \binom{6}{x}(.2)^x(.8)^{6-x}$ $(x = 0, 1, 2, \ldots, 6)$
  c. $p(x) = \dfrac{10!}{x!(10-x)!}(.9)^x(.1)^{10-x}$
  $(x = 0, 1, 2, \ldots, 10)$

4.120 For each of the following examples, decide whether x is a binomial random variable and explain your decision:
  a. A manufacturer of computer chips randomly selects 100 chips from each hour's production in order to estimate the proportion of defectives. Let x represent the number of defectives in the 100 chips sampled.
  b. Of five applicants for a job, two will be selected. Although all applicants appear to be equally qualified, only three have the ability to fulfill the expectations of the company. Suppose that the two selections are made at random from the five applicants, and let x be the number of qualified applicants selected.
  c. A software developer establishes a support hot line for customers to call in with questions regarding use of the software. Let x represent the number of calls received on the hot line during a specified workday.
  d. Florida is one of a minority of states with no state income tax. A poll of 1,000 registered voters is conducted to determine how many would favor a state income tax in light of the state's current fiscal condition. Let x be the number in the sample who would favor the tax.

Learning the Mechanics

4.121 Suppose x is a binomial random variable with $n = 20$ and $p = .7$.
  a. Find $P(x = 14)$.
  b. Find $P(x \le 12)$.
  c. Find $P(x > 12)$.
  d. Find $P(9 \le x \le 18)$.
  e. Find $P(8 < x < 18)$.
  f. Find μ, σ^2, and σ.
  g. What is the probability that x is in the interval $\mu \pm 2\sigma$?

*4.122 Given that x is a hypergeometric random variable, compute $p(x)$ for each of the following cases:
  a. $N = 8, n = 5, r = 3, x = 2$
  b. $N = 6, n = 2, r = 2, x = 2$
  c. $N = 5, n = 4, r = 4, x = 3$

4.123 Suppose x is a binomial random variable. Find $p(x)$ for each of the following combinations of $x, n,$ and p:
  a. $x = 1, n = 3, p = .1$
  b. $x = 4, n = 20, p = .3$
  c. $x = 0, n = 2, p = .4$
  d. $x = 4, n = 5, p = .5$
  e. $n = 15, x = 12, p = .9$
  f. $n = 10, x = 8, p = .6$

4.124 Consider the discrete probability distribution shown here:

x	10	12	18	20
$p(x)$	.2	.3	.1	.4

  a. Calculate μ, σ^2, and σ.
  b. What is $P(x < 15)$?
  c. Calculate $\mu \pm 2\sigma$.
  d. What is the probability that x is in the interval $\mu \pm 2\sigma$?

*4.125 Suppose x is a Poisson random variable. Compute $p(x)$ for each of the following cases:
  a. $\lambda = 2, x = 3$
  b. $\lambda = 1, x = 4$
  c. $\lambda = .5, x = 2$

Applying the Concepts—Basic

4.126 Use of laughing gas. According to the American Dental Association, 60% of all dentists use nitrous oxide ("laughing gas") in their practice. If x equals the number of dentists in a random sample of five dentists who use laughing gas in practice, then the probability distribution of x is

x	0	1	2	3	4	5
$p(x)$	.0102	.0768	.2304	.3456	.2592	.0778

a. Find the probability that the number of dentists using laughing gas in the sample of five is less than 2.
b. Find $E(x)$ and interpret the result.
c. Show that the distribution of x is binomial with $n = 5$ and $p = .6$.

4.127 Benford's Law of Numbers. Refer to the *American Scientist* (July–Aug. 1998) study of which integer is most likely to occur as the first significant digit in a randomly selected number (Benford's law), presented in Exercise 2.190 (p. 102). The table giving the frequency of each integer selected as the first digit in a six-digit random number is reproduced here:

First Digit	Frequency of Occurrence
1	109
2	75
3	77
4	99
5	72
6	117
7	89
8	62
9	43
Total	743

a. Construct a probability distribution for the first significant digit x.
b. Find $E(x)$.
c. If possible, give a practical interpretation of $E(x)$.

4.128 Parents who condone spanking. According to a nationwide survey, 60% of parents with young children condone spanking their child as a regular form of punishment (*Tampa Tribune*, Oct. 5, 2000). Consider a random sample of three people, each of whom is a parent with young children. Assume that x, the number in the sample who condone spanking, is a binomial random variable.
a. What is the probability that none of the three parents condones spanking as a regular form of punishment for their children?
b. What is the probability that at least one condones spanking as a regular form of punishment?
c. Give the mean and standard deviation of x. Interpret the results.

4.129 Quit-smoking program. According to the University of South Florida's Tobacco Research and Intervention Program, only 5% of the nation's cigarette smokers ever enter into a treatment program to help them quit smoking (*USF Magazine*, Spring 2000). In a random sample of 200 smokers, let x be the number who enter into a treatment program.
a. Explain why x is a binomial random variable (to a reasonable degree of approximation).
b. What is the value of p? Interpret this value.
c. What is the expected value of x? Interpret this value.

*4.130 Emergency rescue vehicle use. An emergency rescue vehicle is used an average of 1.3 times daily. Use the Poisson distribution to find
a. The probability that the vehicle will be used exactly twice tomorrow
b. The probability that the vehicle will be used more than twice
c. The probability that the vehicle will be used exactly three times

4.131 Married-women study. According to *Women's Day* magazine (Jan. 2007), only 50% of married women would marry their current husbands again if given the chance. Consider the number x in a random sample of 10 married women who would marry their husbands again. Identify the discrete probability distribution that best models the distribution of x. Explain.

*4.132 Dutch elm disease. A nursery advertises that it has 10 elm trees for sale. Unknown to the nursery, 3 of the trees have already been infected with Dutch elm disease and will die within a year.
a. If a buyer purchases two trees, what is the probability that both trees will be healthy?
b. Refer to part **a**. What is the probability that at least 1 of the trees is infected?

4.133 New book reviews. In Exercise 2.181 (p. 99), you read about a study of book reviews in American history, geography, and area studies published in *Choice* magazine (*Library Acquisitions: Practice and Theory*, Vol. 19, 1995). The overall rating stated in each review was ascertained and recorded as follows: 1 = would not recommend, 2 = cautious or very little recommendation, 3 = little or no preference, 4 = favorable/recommended, 5 = outstanding/significant contribution. Based on a sample of 375 book reviews, the probability distribution of rating, x, follows.

Booking Rating x	$p(x)$
1	.051
2	.099
3	.093
4	.635
5	.122

a. Is this a valid probability distribution?
b. What is the probability that a book reviewed in *Choice* has a rating of 1?
c. What is the probability that a book reviewed in *Choice* has a rating of at least 4?
d. What is the probability that a book reviewed in *Choice* has a rating of 2 or 3?
e. Find $E(x)$.
f. Interpret the result, part **e**.

*4.134 Deep-draft vessel casualties. Economists at the University of New Mexico modeled the number of casualties (deaths or missing persons) experienced by a deep-draft U.S. flag vessel over a three-year period as a Poisson random variable x. The researchers estimated $E(x)$ to be .03 (*Management Science*, Jan. 1999).
a. Find the variance of x.
b. Discuss the conditions that would make the researchers' Poisson assumption plausible.
c. What is the probability that a deep-draft U.S. flag vessel will have no casualties over a three-year period?

4.135 **Belief in an afterlife.** A national poll conducted by *The New York Times* (May 7, 2000) revealed that 80% of Americans believe that after you die, some part of you lives on, either in a next life on earth or in heaven. Consider a random sample of 10 Americans and count x, the number who believe in life after death.
   a. Find $P(x = 3)$.
   b. Find $P(x \le 7)$.
   c. Find $P(x > 4)$.

Applying the Concepts—Intermediate

4.136 **Fungi in beech forest trees.** Refer to the *Applied Ecology and Environmental Research* (Vol. 1, 2003) study of beech trees damaged by fungi, presented in Exercise 3.171 (p. 171). The researchers found that 25% of the beech trees in east central Europe had been damaged by fungi. Consider a sample of 20 beech trees from this area.
   a. What is the probability that fewer than half are damaged by fungi?
   b. What is the probability that more than 15 are damaged by fungi?
   c. How many of the sampled trees would you expect to be damaged by fungi?

4.137 **Pesticides on food samples.** The Food and Drug Administration (FDA) produces a quarterly report called *Total Diet Study*. The FDA's report covers more than 200 food items, each of which is analyzed for dangerous chemical compounds. One *Total Diet Study* reported that no pesticides at all were found in 65% of the domestically produced food samples. Consider a random sample of 800 food items analyzed for the presence of pesticides.
   a. Compute μ and σ for the random variable x, the number of food items found without any trace of pesticide.
   b. On the basis of a sample of 800 food items, is it likely that you would observe fewer than half without any traces of pesticide? Explain.

*4.138 **Accepting or rejecting a shipment.** By mistake, a manufacturer of DVD recording systems includes 3 defective systems in a shipment of 10 going out to a small retailer. The retailer has decided to accept the shipment of DVD recorders only if none is found to be defective. Upon receipt of the shipment, the retailer examines only 5 of the systems.
   a. What is the probability that the shipment will be rejected?
   b. If the retailer inspects 6 of the systems, what is the probability that the shipment will be accepted?

4.139 **Parents' behavior at a gym meet.** *Pediatric Exercise Science* (Feb. 1995) published an article on the behavior of parents at competitive youth gymnastic meets. On the basis of a survey of the gymnasts, the researchers estimated the probability of a parent "yelling" at his or her child before, during, or after the meet as .05. In a random sample of 20 parents attending a gymnastic meet, find the probability that at least 1 parent yells at his or her child before, during, or after the meet.

*4.140 **Birds in butterfly hot spots.** "Hot spots" are species-rich geographical areas. A *Nature* (Sept. 1993) study estimated the probability of a bird species in Great Britain inhabiting a butterfly hot spot at .70. Consider a random sample of 4 British bird species selected from a total of 10 tagged species. Assume that 7 of the 10 tagged species inhabit a butterfly hot spot.
   a. What is the probability that exactly half of the 4 bird species sampled inhabit a butterfly hot spot?
   b. What is the probability that at least 1 of the 4 bird species sampled inhabits a butterfly hot spot?

*4.141 **Vinyl chloride emissions.** The Environmental Protection Agency (EPA) limits the amount of vinyl chloride in plant air emissions to no more than 10 parts per million. Suppose the mean emission of vinyl chloride for a particular plant is 4 parts per million. Assume that the number x of parts per million of vinyl chloride in air samples follows a Poisson probability distribution.
   a. What is the standard deviation of x for the plant?
   b. Is it likely that a sample of air from the plant would yield a value of x that would exceed the EPA limit? Explain.
   c. Discuss conditions that would make the Poisson assumption plausible.

4.142 **Countries that allow a free press.** The degree to which democratic and nondemocratic countries attempt to control the news media was examined in the *Journal of Peace Research* (Nov. 1997). The article reported that 80% of all democratic regimes allow a free press. In contrast, 10% of all nondemocratic regimes allow a free press.
   a. In a random sample of 50 democratic regimes, how many would you expect to allow a free press? Give a range that is highly likely to include the number of democratic regimes with a free press.
   b. In a random sample of 50 nondemocratic regimes, how many would you expect to allow a free press? Give a range that is highly likely to include the number of nondemocratic regimes with a free press.

4.143 **Efficacy of an insecticide.** The efficacy of insecticides is often measured by the dose necessary to kill a certain percentage of insects. Suppose a certain dose of a new insecticide is supposed to kill 80% of the insects that receive it. To test the claim, 25 insects are exposed to the insecticide.
   a. If the insecticide really kills 80% of the exposed insects, what is the probability that fewer than 15 die?
   b. If you observed such a result, what would you conclude about the new insecticide? Explain your logic.

*4.144 **Crime Watch neighborhood.** In many cities, neighborhood Crime Watch groups are formed in an attempt to reduce the amount of criminal activity. Suppose one neighborhood that has experienced an average of 10 crimes per year organizes such a group. During the first year following the creation of the group, 3 crimes are committed in the neighborhood.
   a. Use the Poisson distribution to calculate the probability that 3 or fewer crimes are committed in a year, assuming that the average number is still 10 crimes per year.
   b. Do you think this event provides some evidence that the Crime Watch group has been effective in this neighborhood?

Applying the Concepts—Advanced

4.145 **How many questionnaires to mail?** The probability that a person responds to a mailed questionnaire is .4. How many questionnaires should be mailed if you want to be reasonably certain that at least 100 will be returned?

4.146 **Reliability of a "one-shot" device.** A "one-shot" device can be used only once; after use, the device (e.g., a nuclear weapon, space shuttle, automobile air bag) either is destroyed or must be rebuilt. The destructive nature of a one-shot device makes repeated testing either impractical or too costly. Hence, the reliability of such a device must be determined with minimal testing. Consider a one-shot device that has some probability p of failure. Of course, the true value of p is unknown, so designers will specify a value of p which is the largest defective rate that they are willing to accept. Designers will conduct n tests of the device and determine the success or failure of each test. If the number of observed failures, x, is less than or equal to some specified value k, then the device is considered to have the desired failure rate. Consequently, the designers want to know the minimum sample size n needed so that observing K or fewer defectives in the sample will demonstrate that the true probability of failure for the one-shot device is no greater than p.

a. Suppose the desired failure rate for a one-shot device is $p = .10$. Suppose also that designers will conduct $n = 20$ tests of the device and conclude that the device is performing to specifications if $K = 1$ (i.e., if 1 or no failure is observed in the sample). Find $P(x \le pbc)$.

b. In reliability analysis, $1 - P(x \le K)$ is often called the *level of confidence* for concluding that the true failure rate is less than or equal to p. Find the level of confidence for the one-shot device described in part **a**. In your opinion, is this an acceptable level? Explain.

c. Demonstrate that the confidence level can be increased by either (1) increasing the sample size n or (2) decreasing the number K of failures allowed in the sample.

d. Typically, designers want a confidence level of .90, .95, or .99. Find the values of n and K to use so that the designers can conclude with at least 95% confidence that the failure rate for the one-shot device of part **a** is no greater than $p = .10$.

Note: The U.S. Department of Defense Reliability Analysis Center (DoD RAC) provides designers with free access to tables and toolboxes that give the minimum sample size n required to obtain a desired confidence level for a specified number of observed failures in the sample.

***4.147** **Emergency room bed availability.** The mean number of patients admitted per day to the emergency room of a small hospital is 2.5. If, on a given day, there are only four beds available for new patients, what is the probability that the hospital will not have enough beds to accommodate its newly admitted patients?

Critical Thinking Challenge

***4.148** **Space shuttle disaster.** On January 28, 1986, the space shuttle *Challenger* exploded, killing all seven astronauts aboard. An investigation concluded that the explosion was caused by the failure of the O ring seal in the joint between the two lower segments of the right solid rocket booster. In a report made one year prior to the catastrophe, the National Aeronautics and Space Administration (NASA) claimed that the probability of such a failure was about $1/60,000$, or about once in every 60,000 flights. But a risk-assessment study conducted for the Air Force at about the same time assessed the probability to be $1/35$, or about once in every 35 missions. (*Note:* The shuttle had flown 24 successful missions prior to the disaster.) Given the events of January 28, 1986, which risk assessment—NASA's or the Air Force's—appears to be more appropriate?

Activity	Simulating Binomial Probabilities Using Poker Chips

Consider the following random variables:

1. The number x of people who recover from a certain disease out of a sample of five patients

2. The number x of voters in a sample of five who favor a method of tax reform

3. The number x of hits a baseball player gets in five official times at bat

In each case, x is a binomial random variable (or approximately so) with $n = 5$ trials. Assume that in each case the probability of success is .3. (What is a success in each of the examples?) If this were true, the probability distribution for x would be the same for each of the three examples. To obtain a relative frequency histogram for x, conduct the following experiment: Place 10 poker chips (pennies, marbles, or any 10 *identical* items) in a bowl, and mark 3 of the 10 "Success." The remaining 7 will be marked "Failure." Randomly select a chip from the 10, observing whether it was a success or failure. Then return the chip, randomly select a second chip from the 10 available chips, and record this outcome. Repeat the process until a total of five trials has been conducted. Count the number x of successes observed in the five trials. Repeat the entire process 100 times to obtain 100 observed values of x.

a. Use the 100 values of x obtained from the simulation to construct a relative frequency histogram for x. Note that this histogram is an approximation to $p(x)$.

b. Calculate the exact values of $p(x)$ for $n = 5$ and $p = .3$, and compare these values with the approximations found in part **a**.

c. If you were to repeat the simulation an extremely large number of times (say, 100,000), how do you think the relative frequency histogram and true probability distribution would compare?

References

Hogg, R. V., McKean, J. W., and Craig, A. *Introduction to Mathematical Statistics*, 6th ed. Upper Saddle River, NJ: Prentice Hall, 2005.

Larsen, R. J., and Marx, M. L. An *Introduction to Mathematical Statistics and Its Applications*, 4th ed. Upper Saddle River, N.J.: Prentice Hall, 2005.

Parzen, E. *Modern Probability Theory and Its Applications*. New York: Wiley, 1960.

Wackerly, D., Mendenhall, W., and Scheaffer, R. L. *Mathematical Statistics with Applications*, 7th ed. North Scituate, MA: Duxbury, 2008.

USING TECHNOLOGY

MINITAB: Discrete Probabilities

Step 1 Select the "Calc" button on the MINITAB menu bar, click on "Probability Distributions," and then finally select the discrete distribution of your choice (e.g., "Binomial"), as shown in Figure 4.M.1.

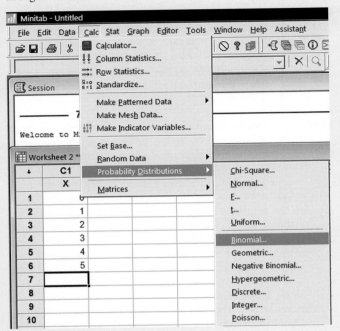

Figure 4.M.1
MINITAB menu options for discrete probabilities

Step 2 Select either "Probability" or "Cumulative probability" on the resulting dialog box.

Step 3 Specify the parameters of the distribution (e.g., sample size n, probability of success p).

[*Note*: For the Poisson option, enter the mean λ. For the hypergeometric option, enter N, r, and n.]

Step 4 Specify the value x of interest in the "Input constant" box.

Step 5 Click "OK." The probability for the value of x will appear on the MINITAB session window.

[*Note*: Figure 4.M.2 gives the specifications for finding $p(x = 2)$ in a binomial distribution with $n = 5$ and $p = .3$.]

TI-83/TI-84 Plus Graphing Calculator: Discrete Random Variables and Probabilities

Calculating the Mean and Standard Deviation of a Discrete Random Variable

Step 1 *Enter the data*
- Press **STAT** and select **1:Edit**

(*Note*: If the lists already contain data, clear the old data. Use the up **ARROW** to highlight 'L1.')

Figure 4.M.2
MINITAB binomial distribution dialog box

- Press **CLEAR ENTER**.
- Use the up **ARROW** to highlight 'L2'
- Press **CLEAR ENTER**
- Use the **ARROW** and **ENTER** keys to enter the x-values of the variable into **L1**
- Use the **ARROW** and **ENTER** keys to enter the probabilities, P(x), into **L2**

Step 2 *Access the Calc Menu*
- Press **STAT**
- Arrow right to **CALC**
- Select **1-Var Stats**
- Press **ENTER**
- Press **2nd 1** for **L1**
- Press **COMMA**
- Select **2nd 2** for **L2**
- Press **ENTER**

The mean and standard deviation will be displayed on the screen, as will the quartiles, min, and max.

Calculating Binomial Probabilities

I. $P(x = k)$

To compute the probability of k successes in n trials where p is the probability of success for each trial, use the **binompdf(** command. *Binompdf* stands for "binomial probability density function." This command is under the **DISTR**ibution menu and has the format **binompdf(n, p, k)**.

Example Compute the probability of 5 successes in 8 trials where the probability of success for a single trial is 40%. In this example, $n = 8, p = .4$, and $k = 5$.

Step 1 *Enter the binomial parameter*
- Press **2nd VARS** for **DISTR**
- Press the down **ARROW** key until **binompdf** is highlighted

- Press **ENTER**
- After **binompdf(**, type **8, .4, 5** (*Note*: Be sure to use the **COMMA** key between each parameter)
- Press **ENTER**
- You should see

```
binompdf(8,.4,5)
            .12386304
```

Thus, $P(x = k)$ is about 12.4%.

II. $P(x \leq k)$

To compute the probability of k or fewer successes in n trials where the p is probability of success for each trial, use the **binomcdf(** command. *Binomcdf* stands for "binomial **cumulative** probability density function." This command is under the **DISTR**ibution menu and has the format **binomcdf(***n, p, k***)**.

Example Compute the probability of 5 or fewer successes in 8 trials where the probability of success for a single trial is 40%. In this example, $n = 8$, $p = .4$, and $k = 5$.

Step 2 *Enter the binomial parameters*
- Press **2nd VARS** for **DISTR**
- Press down the **ARROW** key until **binomcdf** is highlighted
- Press **ENTER**
- After **binomcdf(**, type **8, .4, 5**
- Press **ENTER**
- You should see

```
binomcdf(8,.4,5)
            .95019264
```

Thus, $P(x \leq 5)$ is about 95%.

III. $P(x < k)$, $P(x > k)$, $P(x \geq k)$

To find the probability of less than k successes $P(x < k)$, more than k successes $P(x > k)$, or at least k successes $P(x \geq k)$, variations of the **binomcdf(** command must be used as shown below.

- $P(x < k)$ use **binomcdf(***n, p, k − 1***)**
- $P(x > k)$ use **1 − binomcdf(***n, p, k***)**
- $P(x \geq k)$ use **1 − binomcdf(***n, p, k − 1***)**

Calculating Poisson Probabilities

I. $P(x \leq k)$

To compute $P(x = k)$, the probability of exactly k successes in a specified interval where λ is the mean number of successes in the interval, use the **poissonpdf(** command. *Poissonpdf* stands for "Poisson probability density function." This command is under the **DISTR**ibution menu and has the format **poissonpdf()**.

Example Suppose that the number, x, of reported sightings per week of blue whales is recorded. Assume that x has approximately a Poisson probability distribution and that the average number of weekly sightings is 2.6. Compute the probability that exactly five sightings are made during a given week. In this example, $\lambda = 2.6$ and $k = 5$.

Step 1 *Enter the Poisson parameters*
- Press **2nd VARS** for **DISTR**
- Press the down **ARROW** key until **poissonpdf** is highlighted
- Press **ENTER**
- After **poissonpdf(**, type **2.6,5** (*Note*: Be sure to use the **COMMA** key between each parameter.)

- Press **ENTER**
- You should see

```
poissonpdf(2.6,5
)
            .0735393591
```

Thus, the $P(x = 5)$ is about 7.4%.

II. $P(x \leq k)$

To compute the probability of k or fewer successes in a specified interval where λ is the mean number of successes in the interval, use the **poissoncdf(** command. *Poissoncdf* stands for "Poisson *cumulative* probability density function." This command is under the **DISTR**ibution menu and has the format **poissoncdf(**.

Example In the preceding example, compute the probability that five or fewer sightings are made during a given week. In this example, $\lambda = 2.6$ and $k = 5$.

Step 1 *Enter the Poisson parameters*
- Press **2nd VARS** for **DISTR**
- Press the down **ARROW** key until **poissoncdf** is highlighted
- Press **ENTER**
- After **poissoncdf(**, type **2.6, 5**
- Press **ENTER**
- You should see

```
poissoncdf(2.6,5
)
            .9509628481
```

Thus, the $P(x = 5)$ is about 95.1%.

III. $P(x < k)$, $P(x > k)$, $P(x \geq k)$

To find the probability of less than k successes, more than k successes, or at least k successes, variations of the **poissoncdf(** command must be used as shown below.

- $P(x < k)$ use **poissoncdf(**$\lambda, k − 1$**)**
- $P(x > k)$ use **1− poissoncdf(**λ, k**)**
- $P(x \geq k)$ use **1− poissoncdf(**$\lambda, k − 1$**)**

Calculating Hypergeometric Probabilities

Step 1 Compute $\binom{r}{x}$ using the keystrokes for combinations (see Using Technology, Chapter 3).

Step 2 Press **STO>**, **ALPHA**, **4**, then press **ENTER** to save the result of $\binom{r}{x}$ as the variable **T**.

Step 3 Compute $\binom{N-r}{n-x}$ using the keystrokes for combinations (see Using Technology, Chapter 3).

Step 4 Press **STO>**, **ALPHA**, **5**, then press **ENTER** to save the result of $\binom{N-r}{n-x}$ as the variable **U**.

Step 5 Compute $\binom{N}{x}$ using the keystrokes for combinations (see Using Technology, Chapter 3).

Step 6 Press **STO>**, **ALPHA**, **6**, then press **ENTER** to save the result of $\binom{N}{x}$ as the variable **V**.

Step 7 Press **ALPHA**, **4**, ×, **ALPHA**, **5**, ÷, **ALPHA**, **6** then press **ENTER**.

The probability will be displayed on the window.

5 Continuous Random Variables

CONTENTS

Where We've Been

- Learned how to find probabilities of discrete events by using the probability rules of Chapter 3
- Discussed probability models (distributions) for discrete random variables in Chapter 4

Where We're Going

- Develop the notion of a probability distribution for a continuous random variable. (5.1)
- Study several important types of continuous random variables and their probability models. (5.2–5.3, 5.6)
- Introduce the normal probability distribution as one of the most useful distributions in statistics. (5.3–5.5)

Statistics IN Action Super Weapons Development—Is the Hit Ratio Optimized?

The U.S. Army is working with a major defense contractor to develop a "super" weapon. The weapon is designed to fire a large number of sharp tungsten bullets—called flechettes—with a single shot that will destroy a large number of enemy soldiers. Flechettes are about the size of an average nail, with small fins at one end to stabilize them in flight. Since World War I, when France dropped them in large quantities from aircraft on masses of ground troops, munitions experts have experimented with using flechettes in a variety of guns. The problem with using flechettes as ammunition is accuracy: Current weapons that fire large quantities of flechettes have unsatisfactory hit ratios when fired at long distances.

The defense contractor (not named here for both confidentiality and security reasons) has developed a prototype gun that fires 1,100 flechettes with a single round. In range tests, three 2-feet-wide targets were set up a distance of 500 meters (approximately 1,500 feet) from the weapon. With a number line used as a reference, the centers of the three targets were at 0, 5, and 10 feet, respectively, as shown in Figure SIA5.1. The prototype gun was aimed at the middle target (center at 5 feet) and fired once. The point where each of the 1,100 flechettes landed at the 500-meter distance was measured with the use of a horizontal and vertical grid. For the purposes of this application, only the horizontal measurements are considered. These 1,100 measurements are saved in the **MOAGUN** file. (The data are simulated for confidentiality reasons.) For example, a flechette with a value of $x = 5.5$ hit the middle target, but a flechette with a value of $x = 2.0$ did not hit any of the three targets. (See Figure SIA5.1.)

The defense contractor is interested in the likelihood of any one of the targets being hit by a flechette and, in particular, wants to set the gun specifications to maximize the number of hits. The weapon is designed to have a mean horizontal value equal to the aim point (e.g., $\mu = 5$ feet when aimed at the center target). By changing specifications, the contractor can vary the standard deviation σ. The **MOAGUN** file contains flechette measurements for three different range tests: one with a standard deviation of $\sigma = 1$ foot, one with $\sigma = 2$ feet, and one with $\sigma = 4$ feet.

In this chapter, two Statistics in Action Revisited examples demonstrate how we can use one of the probability models discussed in the chapter—the normal probability distribution—to aid the defense contractor in developing its "super" weapon.

Statistics IN Action Revisited

- Using the Normal Model to Maximize the Probability of a Hit with the Super Weapon (p. 240)

- Assessing whether the Normal Distribution Is Appropriate for Modeling the Super Weapon Hit Data (p. 247)

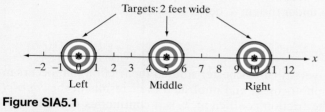

Figure SIA5.1

Target placement on gun range

Data Set: MOAGUN

In this chapter, we'll consider some continuous random variables that are commonly encountered. Recall that a **continuous random variable** is a random variable that can assume any value within some interval or intervals. For example, the length of time between a person's visits to a doctor, the thickness of sheets of steel produced in a rolling mill, and the yield of wheat per acre of farmland are all continuous random variables. The methodology we employ to describe continuous random variables will necessarily be somewhat different from that used to describe discrete random variables. First we discuss the general form of **continuous probability distributions,** and then we explore three specific types that are used in making statistical decisions. The normal probability distribution, which plays a basic and important role in both the theory and application of statistics, is essential to the study of most of the subsequent chapters in this book. The other types have practical applications, but a study of these topics is optional.

5.1 Continuous Probability Distributions

The graphical form of the probability distribution for a continuous random variable x is a smooth curve that might appear as shown in Figure 5.1. This curve, a function of x, is denoted by the symbol $f(x)$ and is variously called a **probability density function,** a **frequency function,** or a **probability distribution.**

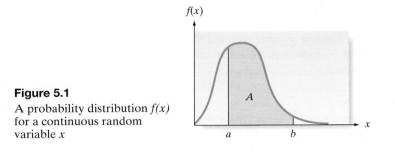

Figure 5.1

A probability distribution *f(x)* for a continuous random variable *x*

The areas under a probability distribution correspond to probabilities for *x*. For example, the area *A* beneath the curve between the two points *a* and *b*, as shown in Figure 5.1, is the probability that *x* assumes a value between *a* and *b* ($a < x < b$). Because there is no area over a single point, say, $x = a$, it follows that (according to our model) the probability associated with a particular value of *x* is equal to 0; that is, $P(x = a) = 0$, and hence $P(a < x < b) = P(a \leq x \leq b)$. In other words, the probability is the same regardless of whether or not you include the endpoints of the interval. Also, because areas over intervals represent probabilities, it follows that the total area under a probability distribution—the total probability assigned to the set of all values of *x*—should equal 1. Note that probability distributions for continuous random variables possess different shapes, depending on the relative frequency distributions of real data that the probability distributions are supposed to model.

> The **probability distribution for a continuous random variable, *x*,** can be represented by a smooth curve—a function of *x*, denoted *f(x)*. The curve is called a **density function** or **frequency function**. The probability that *x* falls between two values, *a* and *b*, i.e., $P(a < x < b)$, is the area under the curve between *a* and *b*.

Example 5.1

Finding a Continuous Probability—Paper Friction in a Photocopier

Problem Researchers at the University of Rochester studied the friction that occurs in the paper-feeding process of a photocopier and published their results in the *Journal of Engineering for Industry*. The friction coefficient, *x*, is a continuous random variable that measures the degree of friction between two adjacent sheets of paper in the feeder stack. The random variable can be modeled using the smooth curve (called a *triangular density function*) shown in Figure 5.2. Note that the friction coefficient, *x*, ranges between the values 5 and 15. Find the probability that the friction coefficient is less than 10.

Solution To find the probability that the friction coefficient is less than 10, we need to compute $P(5 < x < 10)$. This is the area between $x = 5$ and $x = 10$ shaded on the graph in Figure 5.2. Note that this area is represented by a right triangle with a base of length 5 and a height of .2. Since the area of a triangle is $(1/2)(\text{base}) \times (\text{height})$, then

$$P(5 < x < 10) = (1/2)(\text{base}) \times (\text{height}) = (.5)(5)(.2) = .5$$

Look Ahead The areas under most probability distributions are obtained by means of calculus or numerical methods.* Because these methods often involve difficult procedures, we give the areas for some of the most common probability distributions in tabular form in Appendix A. Then, to find the area between two values of *x*, say, $x = a$ and $x = b$, you simply have to consult the appropriate table.

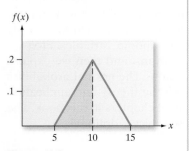

Figure 5.2

Density Function for Friction Coefficient, Example 5.1

---

*Students with a knowledge of calculus should note that the probability that *x* assumes a value in the interval $a < x < b$ is $P(a < x < b) = \int_a^b f(x)\, dx$, assuming that the integral exists. As with the requirements for a discrete probability distribution, we require that $f(x) \geq 0$ and $\int_{-\infty}^{\infty} f(x)\, dx = 1$.

For each of the continuous random variables presented in this chapter, we will give the formula for the probability distribution, along with its mean μ and standard deviation σ. These two numbers will enable you to make some approximate probability statements about a random variable even when you do not have access to a table of areas under the probability distribution.

5.2 The Uniform Distribution

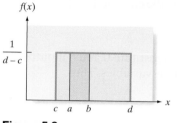

$f(x)$

$\dfrac{1}{d-c}$

c a b d x

Figure 5.3

The uniform probability distribution

All the probability problems discussed in Chapter 3 had sample spaces that contained a finite number of sample points. In many of these problems—for example, in those involving a die toss or coin toss—the sample points were assigned equal probabilities. For continuous random variables, there is an infinite number of values in the sample space, but in some cases the values may appear to be equally likely. For example, if a short exists in a 5-meter stretch of electrical wire, it may have an equal probability of being in any particular 1-centimeter segment along the line. Or if a safety inspector plans to choose a time at random during the 4 afternoon work hours to pay a surprise visit to a certain area of a plant, then each 1-minute time interval in this 4-hour-work period will have an equally likely chance of being selected for the visit.

Continuous random variables that appear to have *equally likely outcomes over their range of possible values* possess a **uniform probability distribution**, perhaps the simplest of all continuous probability distributions. Suppose the random variable x can assume values only in an interval $c \le x \le d$. Then the uniform frequency function has a rectangular shape, as shown in Figure 5.3. Note that the possible values of x consist of all points in the interval between point c and point d, inclusive. The height of $f(x)$ is constant in that interval and equals $1/(d-c)$. Therefore, the total area under $f(x)$ is given by

$$\text{Total area of rectangle} = (\text{Base})(\text{Height}) = (d-c)\left(\frac{1}{d-c}\right) = 1$$

The uniform probability distribution provides a model for continuous random variables that are *evenly distributed* over a certain interval. That is, a **uniform random variable** is just as likely to assume a value in one interval as it is to assume a value in any other interval of equal size. There is no clustering of values around any one value; instead, there is an even spread over the entire region of possible values.

The uniform distribution is sometimes referred to as the **randomness distribution**, since one way of generating a uniform random variable is to perform an experiment in which a point is *randomly selected* on the horizontal axis between the points c and d. If we were to repeat this experiment infinitely often, we would create a uniform probability distribution like that shown in Figure 5.3. The random selection of points in an interval can also be used to generate random numbers such as those in Table I of Appendix A. Recall that random numbers are selected in such a way that every number has an equal probability of selection. Therefore, random numbers are realizations of a uniform random variable. (Random numbers were used to draw random samples in Section 3.7.) The formulas for the uniform probability distribution, its mean, and its standard deviation are shown in the next box.

Probability Distribution for a Uniform Random Variable x

Probability density function: $f(x) = \dfrac{1}{d-c}$ $(c \le x \le d)$

Mean: $\mu = \dfrac{c+d}{2}$ Standard deviation: $\sigma = \dfrac{d-c}{\sqrt{12}}$

$P(a < x < b) = (b-a)/(d-c), c \le a < b \le d$

Suppose the interval $a < x < b$ lies within the domain of x; that is, it falls within the larger interval $c \le x \le d$. Then the probability that x assumes a value within the interval $a < x < b$ is equal to the area of the rectangle over the interval, namely, $(b - a)/(d - c)$.* (See the shaded area in Figure 5.3.)

Example 5.2
Applying the Uniform Distribution— Used-Car Warranties

Problem An unprincipled used-car dealer sells a car to an unsuspecting buyer, even though the dealer knows that the car will have a major breakdown within the next 6 months. The dealer provides a warranty of 45 days on all cars sold. Let x represent the length of time until the breakdown occurs. Assume that x is a uniform random variable with values between 0 and 6 months.

a. Calculate and interpret the mean and standard deviation of x.

b. Graph the probability distribution of x, and show the mean on the horizontal axis. Also show one- and two-standard-deviation intervals around the mean.

c. Calculate the probability that the breakdown occurs while the car is still under warranty.

Solution

a. To calculate the mean and standard deviation for x, we substitute 0 and 6 months for c and d, respectively, in the formulas for uniform random variables. Thus,

$$\mu = \frac{c + d}{2} = \frac{0 + 6}{2} = 3 \text{ months}$$

$$\sigma = \frac{d - c}{\sqrt{12}} = \frac{6 - 0}{\sqrt{12}} = \frac{6}{3.464} = 1.73 \text{ months}$$

Our interpretations of μ and σ follow:
The average length of time x until breakdown for all similar used cars is $\mu = 3$ months. From Chebyshev's theorem (Table 2.6, p. 67), we know that at least 75% of the values of x in the distribution will fall into the interval

$$\mu \pm 2\sigma = 3 \pm 2(1.73)$$
$$= 3 \pm 3.46$$

or between $-.46$ and 6.46 months. Consequently, we expect the length of time until breakdown to be less than 6.46 months at least 75% of the time.

b. The uniform probability distribution is

$$f(x) = \frac{1}{d - c} = \frac{1}{6 - 0} = \frac{1}{6} \quad (0 \leq x \leq 6)$$

The graph of this function is shown in Figure 5.4. The mean and the one- and two-standard-deviation intervals around the mean are shown on the horizontal axis. Note that the entire distribution of x lies within the $\mu \pm 2\sigma$ interval. (This result demonstrates, once again, the conservativeness of Chebyshev's theorem.)

Figure 5.4
Distribution for x in
Example 5.2

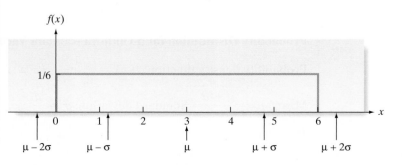

*The student with a knowledge of calculus should note that

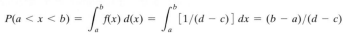

$$P(a < x < b) = \int_a^b f(x)\, d(x) = \int_a^b [1/(d - c)]\, dx = (b - a)/(d - c)$$

c. To find the probability that the car is still under warranty when it breaks down, we must find the probability that x is less than 45 days, or (about) 1.5 months. As indicated in Figure 5.5, we need to calculate the area under the frequency function $f(x)$ between the points $x = 0$ and $x = 1.5$. Therefore, in this case $a = 0$ and $b = 1.5$. Applying the formula in the box, we have

$$P(x < 1.5) = (1.5 - 0)/(6 - 0) = 1.5/6 = .25$$

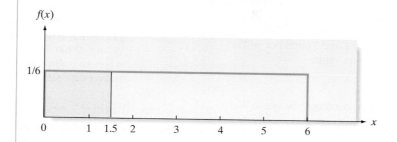

Figure 5.5

Probability that car breaks down within 1.5 months of purchase

That is, there is a 25% chance that the car will break down while under warranty.

Look Back The calculated probability in part **c** is the area of a rectangle with base $1.5 - 0 = 1.5$ and height $= 1/6$. Alternatively, the probability that the unsuspecting car buyer will be able to have the car repaired at the dealer's expense is $P(0 < x < 1.5) = (\text{Base})(\text{Height}) = (1.5)(1/6) = .25$.

Now Work Exercise 5.9

Exercises 5.1–5.19

Understanding the Principles

5.1 Give a characteristic of a uniform random variable.

5.2 The uniform distribution is sometimes referred to as the _____ distribution. (Fill in the blank.)

Learning the Mechanics

5.3 Suppose x is a random variable best described by a uniform probability distribution with $c = 10$ and $d = 30$.
   a. Find $f(x)$.
   b. Find the mean and standard deviation of x.
   c. Graph $f(x)$, and locate μ and the interval $\mu \pm 2\sigma$ on the graph. Note that the probability that x assumes a value within the interval $\mu \pm 2\sigma$ is equal to 1.

5.4 Refer to Exercise 5.3. Find the following probabilities:
   a. $P(10 \leq x \leq 25)$ **b.** $P(20 < x < 30)$
   c. $P(x \geq 25)$ **d.** $P(x \leq 10)$
   e. $P(x \leq 25)$ **f.** $P(20.5 \leq x \leq 25.5)$

5.5 Suppose x is a random variable best described by a uniform probability distribution with $c = 2$ and $d = 4$.
   a. Find $f(x)$.
   b. Find the mean and standard deviation of x.
   c. Find $P(\mu - \sigma \leq x \leq \mu + \sigma)$.
   d. Find $P(x > 2.78)$.
   e. Find $P(2.4 \leq x \leq 3.7)$.
   f. Find $P(x < 2)$.

5.6 Refer to Exercise 5.5. Find the value of a that makes each of the following probability statements true:
   a. $P(x \geq a) = .5$
   b. $P(x \leq a) = .2$

   c. $P(x \leq a) = 0$
   d. $P(2.5 \leq x \leq a) = .5$

5.7 The random variable x is best described by a uniform probability distribution with $c = 100$ and $d = 200$. Find the probability that x assumes a value
   a. More than two standard deviations from μ
   b. Less than three standard deviations from μ
   c. Within two standard deviations of μ

5.8 The random variable x is best described by a uniform probability distribution with mean 50 and standard deviation 5. Find c, d, and $f(x)$. Graph the probability distribution.

Applet Exercise 5.1

Open the applet entitled *Sample from a Population*. On the pull-down menu to the right of the top graph, select *Uniform*. The box to the left of the top graph displays the population mean, median, and standard deviation.
   a. Run the applet for each available value of n on the pull-down menu for the sample size. Go from the smallest to the largest value of n. For each value of n, observe the shape of the graph of the sample data and record the mean, median, and standard deviation of the sample.
   b. Describe what happens to the shape of the graph and the mean, median, and standard deviation of the sample as the sample size increases.

Applet Exercise 5.2

Suppose we set the *Random Numbers* applet to generate one number between 1 and 100, inclusive. We let the value of the random variable x be the number generated when the *Sample* button is clicked. Explain why the distribution of x is approximately uniform even though x is a discrete rather than continuous random variable.

Applying the Concepts—Basic

5.9 **Uranium in the Earth's crust.** The *American Mineralogist*
[NW] (October 2009) published a study of the evolution of uranium minerals in the Earth's crust. Researchers estimate that the trace amount of uranium x in reservoirs follows a uniform distribution ranging between 1 and 3 parts per million.
  a. Find $E(x)$ and interpret its value.
  b. Compute $P(2 < x < 2.5)$.
  c. Compute $P(x < 1.75)$.

5.10 **Requests to a Web server.** According to Brighton Webs LTD, a British company that specializes in data analysis, the arrival time of requests to a Web server within each hour can be modeled by a uniform distribution (www.brighton-webs.co.uk). Specifically, the number of seconds x from the start of the hour that the request is made is uniformly distributed between 0 and 3,600 seconds. Find the probability that a request is made to a Web server sometime during the last 15 minutes of the hour.

5.11 **New method for detecting anthrax.** Researchers at the University of South Florida Center for Biological Defense have developed a safe method for rapidly detecting anthrax spores in powders and on surfaces (*USF Magazine*, Summer 2002). The method has been found to work well even when there are very few anthrax spores in a powder specimen. Consider a powder specimen that has exactly 10 anthrax spores. Suppose that the number of anthrax spores in the sample detected by the new method follows an approximate uniform distribution between 0 and 10.
  a. Find the probability that 8 or fewer anthrax spores are detected in the powder specimen.
  b. Find the probability that between 2 and 5 anthrax spores are detected in the powder specimen.

5.12 **Random numbers.** The data set listed here was created in the MINITAB random-number generator. Construct a relative frequency histogram for the data (saved in the **RANUNI** file). Except for the expected variation in relative frequencies among the class intervals, does your histogram suggest that the data are observations on a uniform random variable with $c = 0$ and $d = 100$? Explain.

38.8759	98.0716	64.5788	60.8422	.8413
99.3734	31.8792	32.9847	.7434	93.3017
12.4337	11.7828	87.4506	94.1727	23.0892
47.0121	43.3629	50.7119	88.2612	69.2875
62.6626	55.6267	78.3936	28.6777	71.6829
44.0466	57.8870	71.8318	28.9622	23.0278
35.6438	38.6584	46.7404	11.2159	96.1009
95.3660	21.5478	87.7819	12.0605	75.1015

5.13 **Maintaining pipe wall temperature.** Maintaining a constant temperature in a pipe wall in some hot process applications is critical. A new technique that utilizes bolt-on trace elements to maintain the temperature was presented in the *Journal of Heat Transfer* (November 2000). Without bolt-on trace elements, the pipe wall temperature of a switch condenser used to produce plastic has a uniform distribution ranging from 260° to 290°F. When several bolt-on trace elements are attached to the piping, the wall temperature is uniform from 278° to 285°F.
  a. Ideally, the pipe wall temperature should range between 280° and 284°F. What is the probability that the temperature will fall into this ideal range when no bolt-on trace elements are used? when bolt-on trace elements are attached to the pipe?
  b. When the temperature is 268°F or lower, the hot liquid plastic hardens (or "plates"), causing a buildup in the piping. What is the probability of the plastic plating when no bolt-on trace elements are used? when bolt-on trace elements are attached to the pipe?

Applying the Concepts—Intermediate

5.14 **Social network densities.** Social networks involve interactions (connections) between members of the network. Sociologists define network density as the ratio of actual network connections to the number of possible one-to-one connections. For example, a network with 10 members has $\binom{10}{2} = 45$ total possible connections. If that network has only 5 connections, the network density is 5/45 = .111. Sociologists at the University of Michigan assumed that the density x of a social network would follow a uniform distribution between 0 and 1 (*Social Networks*, 2010).
  a. On average, what is the density of a randomly selected social network?
  b. What is the probability that the randomly selected network has a density higher than .7?
  c. Consider a social network with only 2 members. Explain why the uniform model would not be a good approximation for the distribution of network density.

5.15 **Cycle availability of a system.** In the jargon of system maintenance, "cycle availability" is defined as the probability that a system is functioning at any point in time. The U.S. Department of Defense developed a series of performance measures for assessing system cycle availability (*START*, Vol. 11, 2004). Under certain assumptions about the failure time and maintenance time of a system, cycle availability is shown to be uniformly distributed between 0 and 1. Find the following parameters for cycle availability: mean, standard deviation, 10th percentile, lower quartile, and upper quartile. Interpret the results.

5.16 **Time delays at a bus stop.** A bus is scheduled to stop at a certain bus stop every half hour on the hour and the half hour. At the end of the day, buses still stop after every 30 minutes, but because delays often occur earlier in the day, the bus is never early and is likely to be late. The director of the bus line claims that the length of time a bus is late is uniformly distributed and the maximum time that a bus is late is 20 minutes.
  a. If the director's claim is true, what is the expected number of minutes a bus will be late?
  b. If the director's claim is true, what is the probability that the last bus on a given day will be more than 19 minutes late?
  c. If you arrive at the bus stop at the end of a day at exactly half-past the hour and must wait more than 19 minutes for the bus, what would you conclude about the director's claim? Why?

5.17 Soft-drink dispenser. The manager of a local soft-drink bottling company believes that when a new beverage-dispensing machine is set to dispense 7 ounces, it in fact dispenses an amount x at random anywhere between 6.5 and 7.5 ounces, inclusive. Suppose x has a uniform probability distribution.

a. Is the amount dispensed by the beverage machine a discrete or a continuous random variable? Explain.

b. Graph the frequency function for x, the amount of beverage the manager believes is dispensed by the new machine when it is set to dispense 7 ounces.

c. Find the mean and standard deviation for the distribution graphed in part **b**, and locate the mean and the interval $\mu \pm 2\sigma$ on the graph.

d. Find $P(x \geq 7)$.

e. Find $P(x < 6)$.

f. Find $P(6.5 \leq x \leq 7.25)$.

g. What is the probability that each of the next six bottles filled by the new machine will contain more than 7.25 ounces of beverage? Assume that the amount of beverage dispensed in one bottle is independent of the amount dispensed in another bottle.

Applying the Concepts—Advanced

5.18 Reliability of a robotic device. The *reliability* of a piece of equipment is frequently defined to be the probability p that the equipment performs its intended function successfully for a given period under specific conditions (Render and Heizer, *Principles of Operations Management*, 1995).

Because p varies from one point in time to another, some reliability analysts treat p as if it were a random variable. Suppose an analyst characterizes the uncertainty about the reliability of a particular robotic device used in an automobile assembly line by means of the following distribution:

$$f(p) = \begin{cases} 1 & 0 \leq p \leq 1 \\ 0 & \text{otherwise} \end{cases}$$

a. Graph the analyst's probability distribution for p.

b. Find the mean and variance of p.

c. According to the analyst's probability distribution for p, what is the probability that p is greater than .95? Less than .95?

d. Suppose the analyst receives the additional information that p is definitely between .90 and .95, but that there is complete uncertainty about where it lies between these values. Describe the probability distribution the analyst should now use to describe p.

5.19 Gouges on a spindle. A tool-and-die machine shop produces extremely high-tolerance spindles. The spindles are 18-inch slender rods used in a variety of military equipment. A piece of equipment used in the manufacture of the spindles malfunctions on occasion and places a single gouge somewhere on the spindle. However, if the spindle can be cut so that it has 14 consecutive inches without a gouge, then the spindle can be salvaged for other purposes. Assuming that the location of the gouge along the spindle is best described by a uniform distribution, what is the probability that a defective spindle can be salvaged?

5.3 The Normal Distribution

One of the most commonly observed continuous random variables has a **bell-shaped probability distribution** (or **bell curve**), as shown in Figure 5.6. It is known as a **normal random variable** and its probability distribution is called a **normal distribution.**

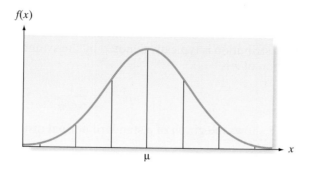

Figure 5.6

A normal probability distribution

The normal distribution plays a very important role in the science of statistical inference. Moreover, many phenomena generate random variables with probability distributions that are very well approximated by a normal distribution. For example, the error made in measuring a person's blood pressure may be a normal random variable, and the probability distribution for the yearly rainfall in a certain region might be approximated by a normal probability distribution. You can determine the adequacy of the normal approximation to an existing population of data by comparing the relative frequency distribution of a large sample of the data with the normal probability distribution. Methods for detecting disagreement between a set of data and the assumption of normality are presented in Section 5.4.

The normal distribution is perfectly symmetric about its mean μ, as can be seen in the examples in Figure 5.7. Its spread is determined by the value of its standard deviation

CARL F. GAUSS (1777–1855)

The Gaussian Distribution

The normal distribution began in the 18th century as a theoretical distribution for errors in disciplines in which fluctuations in nature were believed to behave randomly. Although he may not have been the first to discover the formula, the normal distribution was named the Gaussian distribution after Carl Friedrich Gauss. A well-known and respected German mathematician, physicist, and astronomer, Gauss applied the normal distribution while studying the motion of planets and stars. Gauss's prowess as a mathematician was exemplified by one of his most important discoveries: At the young age of 22, Gauss constructed a regular 17-gon by ruler and compasses— a feat that was the most major advance in mathematics since the time of the ancient Greeks. In addition to publishing close to 200 scientific papers, Gauss invented the heliograph as well as a primitive telegraph. ∎

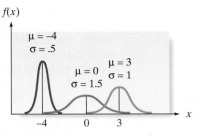

$f(x)$

Figure 5.7

Several normal distributions with different means and standard deviations

σ. The formula for the normal probability distribution is shown in the next box. When plotted, this formula yields a curve like that shown in Figure 5.6.

Probability Distribution for a Normal Random Variable x

Probability density function: $f(x) = \dfrac{1}{\sigma\sqrt{2\pi}}e^{-(1/2)[(x-\mu)/\sigma]^2}$

where

μ = Mean of the normal random variable x
σ = Standard deviation
π = 3.1416 . . .
e = 2.71828 . . .

$P(x < a)$ is obtained from a table of normal probabilities.

Note that the mean μ and standard deviation σ appear in this formula, so that no separate formulas for μ and σ are necessary. To graph the normal curve, we have to know the numerical values of μ and σ. Computing the area over intervals under the normal probability distribution is a difficult task.* Consequently, we will use the computed areas listed in Table IV of Appendix A. Although there are an infinitely large number of normal curves—one for each pair of values of μ and σ—we have formed a single table that will apply to any normal curve.

Table IV is based on a normal distribution with mean $\mu = 0$ and standard deviation $\sigma = 1$, called a *standard normal distribution*. A random variable with a standard normal distribution is typically denoted by the symbol z. The formula for the probability distribution of z is

$$f(z) = \frac{1}{\sqrt{2\pi}}e^{-(1/2)z^2}$$

Figure 5.8 shows the graph of a standard normal distribution.

$f(z)$

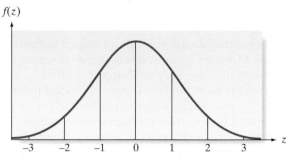

Figure 5.8

Standard normal distribution: $\mu = 0, \sigma = 1$

*The student with a knowledge of calculus should note that there is no closed-form expression for $P(a < x < b) = \int_a^b f(x)\,dx$ for the normal probability distribution. The value of this definite integral can be obtained to any desired degree of accuracy by numerical approximation. For this reason, it is tabulated for the user.

The **standard normal distribution** is a normal distribution with $\mu = 0$ and $\sigma = 1$. A random variable with a standard normal distribution, denoted by the symbol z, is called a **standard normal random variable.**

Since we will ultimately convert all normal random variables to standard normal variables in order to use Table IV to find probabilities, it is important that you learn to use Table IV well. A partial reproduction of that table is shown in Table 5.1. Note that the values of the standard normal random variable z are listed in the left-hand column. The entries in the body of the table give the area (probability) between 0 and z. Examples 5.3–5.6 illustrate the use of the table.

Table 5.1   Reproduction of Part of Table IV in Appendix A

$f(z)$

z	.00	.01	.02	.03	.04	.05	.06	.07	.08	.09
.0	.0000	.0040	.0080	.0120	.0160	.0199	.0239	.0279	.0319	.0359
.1	.0398	.0438	.0478	.0517	.0557	.0596	.0636	.0675	.0714	.0753
.2	.0793	.0832	.0871	.0910	.0948	.0987	.1026	.1064	.1103	.1141
.3	.1179	.1217	.1255	.1293	.1331	.1368	.1406	.1443	.1480	.1517
.4	.1554	.1591	.1628	.1664	.1700	.1736	.1772	.1808	.1844	.1879
.5	.1915	.1950	.1985	.2019	.2054	.2088	.2123	.2157	.2190	.2224
.6	.2257	.2291	.2324	.2357	.2389	.2422	.2454	.2486	.2517	.2549
.7	.2580	.2611	.2642	.2673	.2704	.2734	.2764	.2794	.2823	.2852
.8	.2881	.2910	.2939	.2967	.2995	.3023	.3051	.3078	.3106	.3133
.9	.3159	.3186	.3212	.3238	.3264	.3289	.3315	.3340	.3365	.3389
1.0	.3413	.3438	.3461	.3485	.3508	.3531	.3554	.3577	.3599	.3621
1.1	.3643	.3665	.3686	.3708	.3729	.3749	.3770	.3790	.3810	.3830
1.2	.3849	.3869	.3888	.3907	.3925	.3944	.3962	.3980	.3997	.4015
1.3	.4032	.4049	.4066	.4082	.4099	.4115	.4131	.4147	.4162	.4177
1.4	.4192	.4207	.4222	.4236	.4251	.4265	.4279	.4292	.4306	.4319
1.5	.4332	.4345	.4357	.4370	.4382	.4394	.4406	.4418	.4429	.4441

Example 5.3

Using the Standard Normal Table to find $P(-z_0 < z < z_0)$

Problem Find the probability that the standard normal random variable z falls between -1.33 and $+1.33$.

Solution The standard normal distribution is shown again in Figure 5.9. Since all probabilities associated with standard normal random variables can be depicted as areas under the standard normal curve, you should always draw the curve and then equate the desired probability to an area.

In this example, we want to find the probability that z falls between -1.33 and $+1.33$, which is equivalent to the area between -1.33 and $+1.33$, shown highlighted in Figure 5.9. Table IV gives the area between $z = 0$ and any positive value of z, so that if

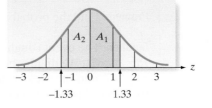

Figure 5.9
Areas under the standard normal curve for Example 5.3

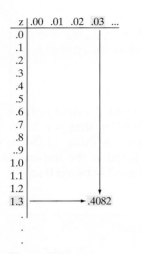

Figure 5.10

Finding $z = 1.33$ in the standard normal table, Example 5.3

we look up $z = 1.33$ (the value in the 1.3 row and .03 column, as shown in Figure 5.10), we find that the area between $z = 0$ and $z = 1.33$ is .4082. This is the area labeled A_1 in Figure 5.9. To find the area A_2 located between $z = 0$ and $z = -1.33$, we note that the symmetry of the normal distribution implies that the area between $z = 0$ and any point to the left is equal to the area between $z = 0$ and the point equidistant to the right. Thus, in this example the area between $z = 0$ and $z = -1.33$ is equal to the area between $z = 0$ and $z = +1.33$. That is,

$$A_1 = A_2 = .4082$$

The probability that z falls between -1.33 and $+1.33$ is the sum of the areas of A_1 and A_2. We summarize in probabilistic notation:

$$P(-1.33 < z < +1.33) = P(-1.33 < z < 0) + P(0 < z \leq 1.33)$$
$$= A_1 + A_2 = .4082 + .4082 = .8164$$

Look Back Remember that "<" and "≤" are equivalent in events involving z, because the inclusion (or exclusion) of a single point does not alter the probability of an event involving a continuous random variable.

Now Work Exercise 5.25 e–f

Example 5.4

Using the Standard Normal Table to find $P(z > z_0)$

Problem Find the probability that a standard normal random variable exceeds 1.64; that is, find $P(z > 1.64)$.

Solution The area under the standard normal distribution to the right of 1.64 is the highlighted area labeled A_1 in Figure 5.11. This area represents the probability that z exceeds 1.64. However, when we look up $z = 1.64$ in Table IV, we must remember that the probability given in the table corresponds to the area between $z = 0$ and $z = 1.64$ (the area labeled A_2 in Figure 5.11). From Table IV, we find that $A_2 = .4495$. To find the area A_1 to the right of 1.64, we make use of two facts:

1. The standard normal distribution is symmetric about its mean, $z = 0$.
2. The total area under the standard normal probability distribution equals 1.

Taken together, these two facts imply that the areas on either side of the mean, $z = 0$, equal .5; thus, the area to the right of $z = 0$ in Figure 5.11 is $A_1 + A_2 = .5$. Then

$$P(z > 1.64) = A_1 = .5 - A_2 = .5 - .4495 = .0505$$

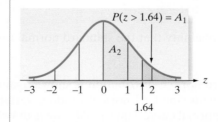

Figure 5.11

Areas under the standard normal curve for Example 5.4

Look Back To attach some practical significance to this probability, note that the implication is that the chance of a standard normal random variable exceeding 1.64 is only about .05.

Now Work Exercise 5.26a

Example 5.5

Using the Standard Normal Table to find $P(z < z_0)$

Problem Find the probability that a standard normal random variable lies to the left of .67.

Solution The event sought is shown as the highlighted area in Figure 5.12. We want to find $P(z < .67)$. We divide the highlighted area into two parts: the area A_1 between $z = 0$ and $z = .67$, and the area A_2 to the left of $z = 0$. We must always make such a division when the desired area lies on both sides of the mean ($z = 0$) because Table IV contains areas between $z = 0$ and the point you look up. We look up $z = .67$ in Table IV to find that $A_1 = .2486$. The symmetry of the standard normal distribution also implies that half the distribution lies on each side of the mean, so the area A_2 to the left of $z = 0$ is .5. Then

$$P(z < .67) = A_1 + A_2 = .2486 + .5 = .7486$$

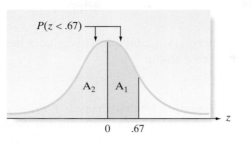

Figure 5.12

Areas under the standard normal curve for Example 5.5

Look Back Note that this probability is approximately .75. Thus, about 75% of the time, the standard normal random variable z will fall below .67. This statement implies that $z = .67$ represents the approximate 75th percentile (or upper quartile) of the standard normal distribution.

Now Work Exercise 5.26h

Example 5.6
Using the Standard Normal Table to find $P(|z| > z_0)$

Problem Find the probability that a standard normal random variable exceeds 1.96 in absolute value.

Solution The event sought is shown highlighted in Figure 5.13. We want to find

$$P(|z| > 1.96) = P(z < -1.96 \text{ or } z > 1.96)$$

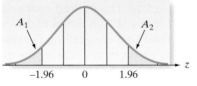

Figure 5.13

Areas under the standard normal curve for Example 5.6

Note that the total highlighted area is the sum of the two areas A_1 and A_2—areas that are equal because of the symmetry of the normal distribution.

We look up $z = 1.96$ and find the area between $z = 0$ and $z = 1.96$ to be .4750. Then A_2, the area to the right of 1.96, is $.5 - .4750 = .0250$, so that

$$P(|z| > 1.96) = A_1 + A_2 = .0250 + .0250 = .05$$

Look Back We emphasize, again, the importance of sketching the standard normal curve in finding normal probabilities.

To apply Table IV to a normal random variable x with any mean μ and any standard deviation σ, we must first convert the value of x to a z-score. The population z-score for a measurement was defined in Section 2.7 as the *distance* between the measurement and the population mean, divided by the population standard deviation. Thus, the z-score gives the distance between a measurement and the mean in units equal to the standard deviation. In symbolic form, the z-score for the measurement x is

$$z = \frac{x - \mu}{\sigma}$$

Note that when $x = \mu$, we obtain $z = 0$.

An important property of the normal distribution is that if x is normally distributed with any mean and any standard deviation, z is *always* normally distributed with mean 0 and standard deviation 1. That is, z is a standard normal random variable.

Property of Normal Distributions

If x is a normal random variable with mean μ and standard deviation σ, then the random variable z defined by the formula

$$z = \frac{x - \mu}{\sigma}$$

has a standard normal distribution. The value z describes the number of standard deviations between x and μ.

Recall from Example 5.6 that $P(|z| > 1.96) = .05$. This probability, coupled with our interpretation of z, implies that any normal random variable lies more than 1.96 standard deviations from its mean only 5% of the time. Compare this statement with the empirical rule (Chapter 2), which tells us that about 5% of the measurements in mound-shaped distributions will lie beyond two standard deviations from the mean. The normal distribution actually provides the model on which the empirical rule is based, along with much "empirical" experience with real data that often approximately obey the rule, whether drawn from a normal distribution or not.

Example 5.7

Finding a Normal Probability—Cell Phone Application

Problem Assume that the length of time, x, between charges of a cellular phone is normally distributed with a mean of 10 hours and a standard deviation of 1.5 hours. Find the probability that the cell phone will last between 8 and 12 hours between charges.

Solution The normal distribution with mean $\mu = 10$ and $\sigma = 1.5$ is shown in Figure 5.14. The desired probability that the cell phone lasts between 8 and 12 hours is highlighted. In order to find that probability, we must first convert the distribution to a standard normal distribution, which we do by calculating the z-score:

$$z = \frac{x - \mu}{\sigma}$$

The z-scores corresponding to the important values of x are shown beneath the x values on the horizontal axis in Figure 5.14. Note that $z = 0$ corresponds to the mean of $\mu = 10$ hours, whereas the x values 8 and 12 yield z-scores of -1.33 and $+1.33$, respectively. Thus, the event that the cell phone lasts between 8 and 12 hours is equivalent to the event that a standard normal random variable lies between -1.33 and $+1.33$. We found this probability in Example 5.3 (see Figure 5.9) by doubling the area corresponding to $z = 1.33$ in Table IV. That is,

$$P(8 \le x \le 12) = P(-1.33 \le z \le 1.33) = 2(.4082) = .8164$$

Now Work Exercise 5.31a

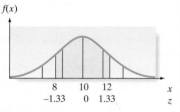

$f(x)$

| 8 | 10 | 12 | x |
| -1.33 | 0 | 1.33 | z |

Figure 5.14

Areas under the normal curve for Example 5.7

The steps to follow in calculating a probability corresponding to a normal random variable are shown in the following box:

Steps for Finding a Probability Corresponding to a Normal Random Variable

1. Sketch the normal distribution and indicate the mean of the random variable x. Then shade the area corresponding to the probability you want to find.
2. Convert the boundaries of the shaded area from x values to standard normal random variable z values by using the formula

$$z = \frac{x - \mu}{\sigma}$$

Show the z values under the corresponding x values on your sketch.

3. Use Table IV in Appendix A to find the areas corresponding to the z values. If necessary, use the symmetry of the normal distribution to find areas corresponding to negative z values and the fact that the total area on each side of the mean equals .5 to convert the areas from Table IV to the probabilities of the event you have shaded.

Example 5.8

Using Normal Probabilities to Make an Inference— Advertised Gas Mileage

Problem Suppose an automobile manufacturer introduces a new model that has an advertised mean in-city mileage of 27 miles per gallon. Although such advertisements seldom report any measure of variability, suppose you write the manufacturer for the details of the tests and you find that the standard deviation is 3 miles per gallon. This information leads you to formulate a probability model for the random variable x, the in-city mileage for this car model. You believe that the probability distribution of x can be approximated by a normal distribution with a mean of 27 and a standard deviation of 3.

a. If you were to buy this model of automobile, what is the probability that you would purchase one that averages less than 20 miles per gallon for in-city driving? In other words, find $P(x < 20)$.

b. Suppose you purchase one of these new models and it does get less than 20 miles per gallon for in-city driving. Should you conclude that your probability model is incorrect?

Solution

a. The probability model proposed for x, the in-city mileage, is shown in Figure 5.15. We are interested in finding the area A to the left of 20, since that area corresponds to the probability that a measurement chosen from this distribution falls below 20. In other words, if this model is correct, the area A represents the fraction of cars that can be expected to get less than 20 miles per gallon for in-city driving. To find A, we first calculate the z value corresponding to $x = 20$. That is,

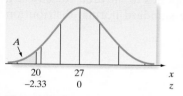

Figure 5.15
Area under the normal curve for Example 5.8

$$z = \frac{x - \mu}{\sigma} = \frac{20 - 27}{3} = -\frac{7}{3} = -2.33$$

Then

$$P(x < 20) = P(z < -2.33)$$

as indicated by the highlighted area in Figure 5.15. Since Table IV gives only areas to the right of the mean (and because the normal distribution is symmetric about its mean), we look up 2.33 in Table IV and find that the corresponding area is .4901. This is equal to the area between $z = 0$ and $z = -2.33$, so we find that

$$P(x < 20) = A = .5 - .4901 = .0099 \approx .01$$

According to this probability model, you should have only about a 1% chance of purchasing a car of this make with an in-city mileage under 20 miles per gallon.

b. Now you are asked to make an inference based on a sample: the car you purchased. You are getting less than 20 miles per gallon for in-city driving. What do you infer? We think you will agree that one of two possibilities exists:

1. The probability model is correct. You simply were unfortunate to have purchased one of the cars in the 1% that get less than 20 miles per gallon in the city.
2. The probability model is incorrect. Perhaps the assumption of a normal distribution is unwarranted, or the mean of 27 is an overestimate, or the standard

deviation of 3 is an underestimate, or some combination of these errors occurred. At any rate, the form of the actual probability model certainly merits further investigation.

You have no way of knowing with certainty which possibility is correct, but the evidence points to the second one. We are again relying on the rare-event approach to statistical inference that we introduced earlier. The sample (one measurement in this case) was so unlikely to have been drawn from the proposed probability model that it casts serious doubt on the model. We would be inclined to believe that the model is somehow in error.

Look Back In applying the rare-event approach, the calculated probability must be small (say, less than or equal to .05) in order to infer that the observed event is, indeed, unlikely.

Now Work Exercise 5.40

Occasionally you will be given a probability and will want to find the values of the normal random variable that correspond to that probability. For example, suppose the scores on a college entrance examination are known to be normally distributed and a certain prestigious university will consider for admission only those applicants whose scores exceed the 90th percentile of the test score distribution. To determine the minimum score for consideration for admission, you will need to be able to use Table IV in reverse, as demonstrated in the next example.

Example 5.9
Using the Normal Table in Reverse

Problem Find the value of z—call it z_0—in the standard normal distribution that will be exceeded only 10% of the time. That is, find z_0 such that $P(z \geq z_0) = .10$.

Solution In this case, we are given a probability, or an area, and are asked to find the value of the standard normal random variable that corresponds to the area. Specifically, we want to find the value z_0 such that only 10% of the standard normal distribution exceeds z_0. (See Figure 5.16.)

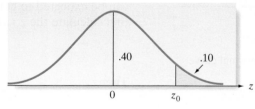

Figure 5.16
Areas under the standard normal curve for Example 5.9

We know that the total area to the right of the mean $z = 0$, is .5, which implies that z_0 must lie to the right of 0 ($z_0 > 0$). To pinpoint the value, we use the fact that the area to the right of z_0 is .10, which implies that the area between $z = 0$ and z_0 is $.5 - .1 = .4$. But areas between $z = 0$ and some other z value are exactly the types given in Table IV. Therefore, we look up the area .4000 in the body of Table IV and find that the corresponding z value is (to the closest approximation) $z_0 = 1.28$. The implication is that the point 1.28 standard deviations above the mean is the 90th percentile of a normal distribution.

Look Back As with earlier problems, it is critical to correctly draw the normal probability sought on the normal curve. The placement of z_0 to the left or right of 0 is the key. Be sure to shade the probability (area) involving z_0. If it does not agree with the probability sought (i.e., if the shaded area is greater than .5 and the probability sought is smaller than .5), then you need to place z_0 on the opposite side of 0.

Example 5.10

Using the Normal Table in Reverse

Problem Find the value of z_0 such that 95% of the standard normal z values lie between $-z_0$ and $+z_0$; that is, find $P(-z_0 \le z \le z_0) = .95$.

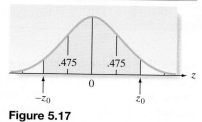

Figure 5.17
Areas under the standard normal curve for Example 5.10

Solution Here we wish to move an equal distance z_0 in the positive and negative directions from the mean $z = 0$ until 95% of the standard normal distribution is enclosed. This means that the area on each side of the mean will be equal to $\frac{1}{2}(.95) = .475$, as shown in Figure 5.17. Since the area between $z = 0$ and z_0 is .475, we look up .475 in the body of Table IV to find the value $z_0 = 1.96$. Thus, as we found in reverse order in Example 5.6, 95% of a normal distribution lies between $+1.96$ and -1.96 standard deviations of the mean.

Now Work Exercise 5.28a

Now that you have learned to use Table IV to find a standard normal z value that corresponds to a specified probability, we demonstrate a practical application in Example 5.11.

Example 5.11

The Normal Table in Reverse—College Entrance Exam Application

Problem Suppose the scores x on a college entrance examination are normally distributed with a mean of 550 and a standard deviation of 100. A certain prestigious university will consider for admission only those applicants whose scores exceed the 90th percentile of the distribution. Find the minimum score an applicant must achieve in order to receive consideration for admission to the university.

Solution In this example, we want to find a score x_0 such that 90% of the scores (x values) in the distribution fall below x_0 and only 10% fall above x_0. That is,

$$P(x \le x_0) = .90$$

Converting x to a standard normal random variable where $\mu = 550$ and $\sigma = 100$, we have

$$P(x \le x_0) = P\left(z \le \frac{x_0 - \mu}{\sigma}\right)$$

$$= P\left(z \le \frac{x_0 - 550}{100}\right) = .90$$

In Example 5.9 (see Figure 5.16), we found the 90th percentile of the standard normal distribution to be $z_0 = 1.28$. That is, we found that $P(z \le 1.28) = .90$. Consequently, we know that the minimum test score x_0 corresponds to a z-score of 1.28; in other words,

$$\frac{x_0 - 550}{100} = 1.28$$

If we solve this equation for x_0, we find that

$$x_0 = 550 + 1.28(100) = 550 + 128 = 678$$

This x value is shown in Figure 5.18. Thus, the 90th percentile of the test score distribution is 678. That is to say, an applicant must score at least 678 on the entrance exam to receive consideration for admission by the university.

Look Back As the example shows, in practical applications of the normal table in reverse, first find the value of z_0 and then use the z-score formula in reverse to convert the value to the units of x.

Now Work Exercise 5.43e

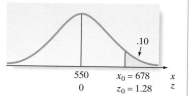

Figure 5.18
Area under the normal curve for Example 5.11

Statistics IN Action	Revisited	Using the Normal Model to Maximize the Probability of a Hit with the Super Weapon

Recall that a defense contractor has developed a prototype gun for the U.S. Army that fires 1,100 flechettes with a single round. The specifications of the weapon are set so that when the gun is aimed at a target 500 meters away, the mean horizontal grid value of the flechettes is equal to the aim point. In the range test, the weapon was aimed at the center target in Figure SIA5.1; thus, $\mu = 5$ feet. For three different tests, the standard deviation was set at $\sigma = 1$ foot, $\sigma = 2$ feet, and $\sigma = 4$ feet. From past experience, the defense contractor has found that the distribution of the horizontal flechette measurements is closely approximated by a normal distribution. Therefore, we can use the normal distribution to find the probability that a single flechette shot from the weapon will hit any one of the three targets. Recall from Figure SIA5.1 that the three targets range from -1 to 1, 4 to 6, and 9 to 11 feet on the horizontal grid.

Consider first the middle target. Letting x represent the horizontal measurement for a flechette shot from the gun, we see that the flechette will hit the target if $4 \leq x \leq 6$. Using the normal probability table (Table 4, Appendix I), we then find that the probability that this flechette will hit the target when $\mu = 5$ and $\sigma = 1$ is

Middle: $P(4 \leq x \leq 6) = P\left(\dfrac{4-5}{1} < z < \dfrac{6-5}{1}\right)$

$\sigma = 1$

$$= P(-1 < z < 1)$$
$$= 2(.3413) = .6826$$

Similarly, we find the probabilities that the flechette hits the left and right targets shown in Figure SIA5.1:

Left: $P(-1 \leq x \leq 1) = P\left(\dfrac{-1-5}{1} < z < \dfrac{1-5}{1}\right)$

$\sigma = 1$

$$= P(-6 < z < -4) \approx 0$$

Right: $P(9 \leq x \leq 11) = P\left(\dfrac{9-5}{1} < z < \dfrac{11-5}{1}\right)$

$\sigma = 1$

$$= P(4 < z < 6) \approx 0$$

You can see that there is about a 68% chance that a flechette will hit the middle target, but virtually no chance that one will hit the left or right target when the standard deviation is set at 1 foot.

To find these three probabilities for $\sigma = 2$ and $\sigma = 4$, we use the normal probability function in MINITAB. Figure SIA5.2 is a MINITAB worksheet giving the cumulative probabilities of a normal random variable falling below the x values in the first column. The cumulative probabilities for $\sigma = 2$ and $\sigma = 4$ are given in the columns named "sigma2" and "sigma4," respectively.

Using the cumulative probabilities in the figure to find the three probabilities when $\sigma = 2$, we have

Middle: $P(4 \leq x \leq 6) = P(x \leq 6) - P(x \leq 4)$

$\sigma = 2$

$$= .6915 - .3085 = .3830$$

Left: $P(-1 \leq x \leq 1) = P(x \leq 1) - P(x \leq -1)$

$\sigma = 2$

$$= .0227 - .0013 = .0214$$

Right: $P(9 \leq x \leq 11) = P(x \leq 11) - P(x \leq 9)$

$\sigma = 2$

$$= .9987 - .9773 = .0214$$

Thus, when $\sigma = 2$, there is about a 38% chance that a flechette will hit the middle target, a 2% chance that one will hit the left target, and a 2% chance that one will hit the right target. The probability that a flechette will hit either the middle or the left or the right target is simply the sum of these three probabilities (an application of the additive rule of probability). This sum is .3830 + .0214 + .0214 = .4258; consequently, there is about a 42% chance of hitting any one of the three targets when specifications are set so that $\sigma = 2$.

Figure SIA5.2

MINITAB worksheet with cumulative normal probabilities

↓	C1	C2	C3	C4	C5
	x	sigma1	sigma2	sigma4	
1	-1	0.00000	0.001350	0.066807	
2	1	0.00003	0.022750	0.158655	
3	4	0.15866	0.308538	0.401294	
4	6	0.84134	0.691462	0.598706	
5	9	0.99997	0.977250	0.841345	
6	11	1.00000	0.998650	0.933193	
7					

Statistics IN Action
(continued)

Now we use the cumulative probabilities in Figure SIA5.2 to find the three hit probabilities when $\sigma = 4$:

Middle: $P(4 \leq x \leq 6) = P(x \leq 6) - P(x \leq 4)$

$\sigma = 4$

$= .5987 - .4013 = .1974$

Left: $P(-1 \leq x \leq 1) = P(x \leq 1) - P(x \leq -1)$

$\sigma = 4$

$= .1587 - .0668 = .0919$

Right: $P(9 \leq x \leq 11) = P(x \leq 11) - P(x \leq 9)$

$\sigma = 4$

$= .9332 - .8413 = .0919$

Thus, when $\sigma = 4$, there is about a 20% chance that a flechette will hit the middle target, a 9% chance that one will hit

the left target, and a 9% chance that one will hit the right target. The probability that a flechette will hit any one of the three targets is $.1974 + .0919 + .0919 = .3812$.

These probability calculations reveal a few patterns. First, the probability of hitting the middle target (the target at which the gun is aimed) is reduced as the standard deviation is increased. Obviously, then, if the U.S. Army wants to maximize the chance of hitting the target that the prototype gun is aimed at, it will want specifications set with a small value of σ. But if the Army wants to hit multiple targets with a single shot of the weapon, σ should be increased. With a larger σ, not as many of the flechettes will hit the target aimed at, but more will hit peripheral targets. Whether σ should be set at 4 or 6 (or some other value) depends on how high of a hit rate is required for the peripheral targets.

Exercises 5.20–5.52

Understanding the Principles

5.20 Describe the shape of a normal probability distribution.

5.21 What is the name given to a normal distribution when $\mu = 0$ and $\sigma = 1$?

5.22 If x has a normal distribution with mean μ and standard deviation σ, describe the distribution of $z = (x - \mu)/\sigma$.

Learning the Mechanics

5.23 Find the area under the standard normal probability distribution between the following pairs of z-scores:
 a. $z = 0$ and $z = 2.00$
 b. $z = 0$ and $z = 1.00$
 c. $z = 0$ and $z = 3$
 d. $z = 0$ and $z = .58$

5.24 Find the area under the standard normal distribution between the following pairs of z-scores:
 a. $z = -2.00$ and $z = 0$
 b. $z = -1.00$ and $z = 0$
 c. $z = -1.69$ and $z = 0$
 d. $z = -.58$ and $z = 0$

5.25 Find each of the following probabilities for a standard normal random variable z:
 a. $P(z = 1)$
 b. $P(z \leq 1)$
 c. $P(z < 1)$
 d. $P(z > 1)$
 NW **e.** $P(-1 \leq z \leq 1)$
 NW **f.** $P(-2 \leq z \leq 2)$
 g. $P(-2.16 \leq z \leq .55)$
 NW **h.** $P(-.42 < z < 1.96)$

5.26 Find the following probabilities for the standard normal random variable z:
 NW **a.** $P(z > 1.46)$
 b. $P(z < -1.56)$
 c. $P(.67 \leq z \leq 2.41)$
 d. $P(-1.96 \leq z < -.33)$

 e. $P(z \geq 0)$
 f. $P(-2.33 < z < 1.50)$
 g. $P(z \geq -2.33)$
 NW **h.** $P(z < 2.33)$

5.27 Give the z-score for a measurement from a normal distribution for the following:
 a. 1 standard deviation above the mean
 b. 1 standard deviation below the mean
 c. Equal to the mean
 d. 2.5 standard deviations below the mean
 e. 3 standard deviations above the mean

5.28 Find a value z_0 of the standard normal random variable z such that
 NW **a.** $P(z \leq z_0) = .0401$
 b. $P(-z_0 \leq z \leq z_0) = .95$
 c. $P(-z_0 \leq z \leq z_0) = .90$
 d. $P(-z_0 \leq z \leq z_0) = .8740$
 e. $P(-z_0 \leq z \leq 0) = .2967$
 f. $P(-2 < z < z_0) = .9710$
 g. $P(z \geq z_0) = .5$
 h. $P(z \geq z_0) = .0057$

5.29 Find a value z_0 of the standard normal random variable z such that
 a. $P(z \geq z_0) = .05$
 b. $P(z \geq z_0) = .025$
 c. $P(z \leq z_0) = .025$
 d. $P(z \geq z_0) = .10$
 e. $P(z > z_0) = .10$

5.30 Suppose the random variable x is best described by a normal distribution with $\mu = 25$ and $\sigma = 5$. Find the z-score that corresponds to each of the following x values:
 a. $x = 25$
 b. $x = 30$
 c. $x = 37.5$
 d. $x = 10$
 e. $x = 50$
 f. $x = 32$

5.31 Suppose x is a normally distributed random variable with $\mu = 11$ and $\sigma = 2$. Find each of the following:

[NW] **a.** $P(10 \leq x \leq 12)$
b. $P(6 \leq x \leq 10)$
c. $P(13 \leq x \leq 16)$
d. $P(7.8 \leq x \leq 12.6)$
e. $P(x \geq 13.24)$
f. $P(x \geq 7.62)$

5.32 Suppose x is a normally distributed random variable with $\mu = 30$ and $\sigma = 8$. Find a value x_0 of the random variable x such that

a. $P(x \geq x_0) = .5$
b. $P(x < x_0) = .025$
c. $P(x > x_0) = .10$
d. $P(x > x_0) = .95$

5.33 Refer to Exercise 5.32. Find x_0 such that

a. 10% of the values of x are less than x_0.
b. 80% of the values of x are less than x_0.
c. 1% of the values of x are greater than x_0.

5.34 Suppose x is a normally distributed random variable with mean 100 and standard deviation 8. Draw a rough graph of the distribution of x. Locate μ and the interval $\mu \pm 2\sigma$ on the graph. Find the following probabilities:

a. $P(\mu - 2\sigma \leq x \leq \mu + 2\sigma)$ .9544
b. $P(x \geq \mu + 2\sigma)$
c. $P(x \leq 92)$
d. $P(92 \leq x \leq 116)$
e. $P(92 \leq x \leq 96)$
f. $P(76 \leq x \leq 124)$

5.35 The random variable x has a normal distribution with standard deviation 25. It is known that the probability that x exceeds 150 is .90. Find the mean μ of the probability distribution.

5.36 The random variable x has a normal distribution with $\mu = 300$ and $\sigma = 30$.

a. Find the probability that x assumes a value more than two standard deviations from its mean. More than three standard deviations from μ.
b. Find the probability that x assumes a value within one standard deviation of its mean. Within two standard deviations of μ.
c. Find the value of x that represents the 80th percentile of this distribution. The 10th percentile.

Applet Exercise 5.3

Open the applet entitled *Sample from a Population*. On the pull-down menu to the right of the top graph, select *Bell shaped*. The box to the left of the top graph displays the population mean, median, and standard deviation.

a. Run the applet for each available value of n on the pull-down menu for the sample size. Go from the smallest to the largest value of n. For each value of n, observe the shape of the graph of the sample data and record the mean, median, and standard deviation of the sample.
b. Describe what happens to the shape of the graph and the mean, median, and standard deviation of the sample as the sample size increases.

Applying the Concepts—Basic

5.37 **Dental anxiety study.** To gauge their fear of going to a dentist, a random sample of adults completed the Modified Dental Anxiety Scale questionnaire (*BMC Oral Health*,

Vol. 9, 2009). Scores on the scale range from zero (no anxiety) to 25 (extreme anxiety). The mean score was 11 and the standard deviation was 4. Assume that the distribution of all scores on the Modified Dental Anxiety Scale is approximately normal with $\mu = 11$ and $\sigma = 4$.

a. Suppose you score a 10 on the Modified Dental Anxiety Scale. Find the z-value for your score.
b. Find the probability that someone scores between 10 and 15 on the Modified Dental Anxiety Scale.
c. Find the probability that someone scores above 20 on the Modified Dental Anxiety Scale.

5.38 **Tomato as a taste modifier.** Miraculin—a protein naturally produced in a rare tropical fruit—can convert a sour taste into a sweet taste. Consequently, miraculin has the potential to be an alternative low-calorie sweetener. In *Plant Science* (May 2010), a group of Japanese environmental scientists investigated the ability of a hybrid tomato plant to produce miraculin. For a particular generation of the tomato plant, the amount x of miraculin produced (measured in micrograms per gram of fresh weight) had a mean of 105.3 and a standard deviation of 8.0. Assume that x is normally distributed.

a. Find $P(x > 120)$.
b. Find $P(100 < x < 110)$.
c. Find the value a for which $P(x < a) = .25$.

5.39 **Most powerful American women.** Refer to the *Fortune* (Oct. 16, 2008) list of the 50 most powerful women in America, presented in Exercise 2.60 (p. 57). Recall that the data on age (in years) of each woman is stored in the **WPOWER50** file. The ages in the data set can be shown to be approximately normally distributed with a mean of 50 years and a standard deviation of 6.4 years. A powerful woman is randomly selected from the data, and her age is observed.

a. Find the probability that her age will fall between 55 and 60 years.
b. Find the probability that her age will fall between 48 and 52 years.
c. Find the probability that her age will be less than 35 years.
d. Find the probability that her age will exceed 40 years.
e. A 25-year-old woman claims she is one of the 50 most powerful women. Do you believe her? Explain.

5.40 **Casino gaming.** Casino gaming yields over $35 billion in revenue each year in the United States. In *Chance* (Spring 2005), University of Denver statistician R. C. Hannum discussed the business of casino gaming and its reliance on the laws of probability. Casino games of pure chance (e.g., craps, roulette, baccarat, and keno) always yield a "house advantage." For example, in the game of double-zero roulette, the expected casino win percentage is 5.26% on bets made on whether the outcome will be either black or red. (This percentage implies that for every $5 bet on black or red, the casino will earn a net of about 25 cents.) It can be shown that in 100 roulette plays on black/red, the average casino win percentage is normally distributed with mean 5.26% and standard deviation 10%. Let x represent the average casino win percentage after 100 bets on black/red in double-zero roulette.

a. Find $P(x > 0)$. (This is the probability that the casino wins money.)
b. Find $P(5 < x < 15)$.

c. Find $P(x < 1)$.

d. If you observed an average casino win percentage of -25% after 100 roulette bets on black/red, what would you conclude?

5.41 Transmission delays in wireless technology. Resource Reservation Protocol (RSVP) was originally designed to establish signaling links for stationary networks. In *Mobile Networks and Applications* (Dec. 2003), RSVP was applied to mobile wireless technology (e.g., a PC notebook with wireless LAN card for Internet access). A simulation study revealed that the transmission delay (measured in milliseconds) of an RSVP–linked wireless device has an approximate normal distribution with mean $\mu = 48.5$ milliseconds and $\sigma = 8.5$ milliseconds.

a. What is the probability that the transmission delay is less than 57 milliseconds?

b. What is the probability that the transmission delay is between 40 and 60 milliseconds?

5.42 Shell lengths of sea turtles. Refer to the *Aquatic Biology* (Vol. 9, 2010) study of green sea turtles inhabiting the Grand Cayman South Sound lagoon, Exercise 2.83 (p. 65). Researchers discovered that the curved carapace (shell) length of these turtles is approximately normally distributed with mean 55.7 centimeters and standard deviation 11.5 centimeters.

a. The minimum and maximum size limits for captured turtles in the legal marine turtle fishery are 40 cm and 60 cm, respectively. How likely are you to capture a green sea turtle that is considered illegal?

b. What maximum limit, L, should be set so that only 10% of the turtles captured have shell lengths greater than L?

5.43 NHTSA crash safety tests. Refer to the National Highway Traffic Safety Administration (NHTSA) crash test data for new cars, presented in Exercise 2.174 (p. 98) and saved in the **CRASH** file. One of the variables measured is the severity of a driver's head injury when the car is in a head-on collision with a fixed barrier while traveling at 35 miles per hour. The more points assigned to the head injury rating, the more severe is the injury. The head injury ratings can be shown to be approximately normally distributed with a mean of 605 points and a standard deviation of 185 points. One of the crash-tested cars is randomly selected from the data, and the driver's head injury rating is observed.

a. Find the probability that the rating will fall between 500 and 700 points.

b. Find the probability that the rating will fall between 400 and 500 points.

c. Find the probability that the rating will be less than 850 points.

d. Find the probability that the rating will exceed 1,000 points.

NW **e.** What rating will only 10% of the crash-tested cars exceed?

Applying the Concepts—Intermediate

5.44 Optimal goal target in soccer. When attempting to score a goal in soccer, where should you aim your shot? Should you aim for a goalpost (as some soccer coaches teach), the middle of the goal, or some other target? To answer these questions, *Chance* (Fall 2009) utilized the normal probability distribution. Suppose the accuracy x of a professional soccer player's

shots follows a normal distribution with a mean of zero feet and a standard deviation of 3 feet. (For example, if the player hits his target, $x = 0$; if he misses his target by 2 feet to the right, $x = 2$; and if he misses 1 foot to the left, $x = -1$.) Now, a regulation soccer goal is 24 feet wide. Assume that a goalkeeper will stop (save) all shots within 9 feet of where he is standing; all other shots on goal will score. Consider a goalkeeper who stands in the middle of the goal.

a. If the player aims for the right goalpost, what is the probability that he will score?

b. If the player aims for the center of the goal, what is the probability that he will score?

c. If the player aims for halfway between the right goalpost and the outer limit of the goalkeeper's reach, what is the probability that he will score?

5.45 Ambulance response time. Ambulance response time is measured as the time (in minutes) between the initial call to emergency medical services (EMS) and when the patient is reached by ambulance. *Geographical Analysis* (Vol. 41, 2009) investigated the characteristics of ambulance response time for EMS calls in Edmonton, Alberta. For a particular EMS station (call it Station A), ambulance response time is known to be normally distributed with $\mu = 7.5$ minutes and $\sigma = 2.5$ minutes.

a. Regulations require that 90% of all emergency calls should be reached in 9 minutes or less. Are the regulations met at EMS Station A? Explain.

b. A randomly selected EMS call in Edmonton has an ambulance response time of 2 minutes. Is it likely that this call was serviced by Station A? Explain.

5.46 Rating employee performance. Almost all companies utilize some type of year-end performance review for their employees. Human Resources (HR) at the University of Texas Health Science Center provides guidelines for supervisors rating their subordinates. For example, raters are advised to examine their ratings for a tendency to be either too lenient or too harsh. According to HR, "if you have this tendency, consider using a normal distribution—10% of employees (rated) exemplary, 20% distinguished, 40% competent, 20% marginal, and 10% unacceptable." Suppose you are rating an employee's performance on a scale of 1 (lowest) to 100 (highest). Also, assume the ratings follow a normal distribution with a mean of 50 and a standard deviation of 15.

a. What is the lowest rating you should give to an "exemplary" employee if you follow the University of Texas HR guidelines?

b. What is the lowest rating you should give to a "competent" employee if you follow the University of Texas HR guidelines?

5.47 Range of women's heights. In *Chance* (Winter 2007), Yale Law School professor Ian Ayres published the results of a study he conducted with his son and daughter on whether college students could estimate a range for women's heights. The students were shown a graph of a normal distribution of heights and were asked, "The average height of women over 20 years old in the United States is 64 inches. Using your intuition, please give your best estimate of the range of heights that would include $C\%$ of women over 20 years old. Please make sure that the center of the range is the average height of 64 inches." The value of C was randomly selected as 50%, 75%, 90%, 95%, or 99% for each student surveyed.

a. Give your estimate of the range for $C = 50\%$ of women's heights.

b. Give your estimate of the range for $C = 75\%$ of women's heights.

c. Give your estimate of the range for $C = 90\%$ of women's heights.

d. Give your estimate of the range for $C = 95\%$ of women's heights.

e. Give your estimate of the range for $C = 99\%$ of women's heights.

f. The standard deviation of heights for women over 20 years old is known to be 2.6 inches. Use this information to revise your answers to parts **a–e**.

g. Which value of C has the most accurate estimated range? (*Note:* The researchers found that college students were most accurate for $C = 90\%$ and $C = 95\%$.)

5.48 **Alcohol, threats, and electric shocks.** A group of Florida State University psychologists examined the effects of alcohol on the reactions of people to a threat (*Journal of Abnormal Psychology*, Vol. 107, 1998). After obtaining a specified blood alcohol level, the psychologists placed experimental subjects in a room and threatened them with electric shocks. Using sophisticated equipment to monitor the subjects' eye movements, the psychologists recorded the startle response (measured in milliseconds) of each subject. The mean and standard deviation of the startle responses were 37.9 and 12.4, respectively. Assume that the startle response x for a person with the specified blood alcohol level is approximately normally distributed.

a. Find the probability that x is between 40 and 50 milliseconds.

b. Find the probability that x is less than 30 milliseconds.

c. Give an interval for x centered around 37.9 milliseconds so that the probability that x falls in the interval is .95.

d. Ten percent of the experimental subjects have startle responses above what value?

5.49 **California's electoral college votes.** During a presidential election, each state is allotted a different number of votes to the electoral college depending on population. For example, California is allotted 55 votes (the most) while several states (including the District of Columbia) are allotted 3 votes each (the least). When a presidential candidate wins the popular vote in a state, the candidate wins all the electoral college votes in that state. To become president, a candidate must win 270 of the total of 538 votes in the electoral college. *Chance* (Winter 2010) demonstrated the impact of winning California on the presidential election. Assuming a candidate wins California's 55 votes, the number of additional electoral college votes the candidate will win can be approximated by a normal distribution with $\mu = 241.5$ votes and $\sigma = 49.8$ votes. If a presidential candidate wins the popular vote in California, what are the chances that he or she becomes the next U.S. president?

Applying the Concepts—Advanced

5.50 **Industrial filling process.** The characteristics of an industrial filling process in which an expensive liquid is injected into a container were investigated in the *Journal of Quality Technology* (July 1999). The quantity injected per container is approximately normally distributed with mean 10 units and standard deviation .2 unit. Each unit of fill costs $20. If a container contains less than 10 units (i.e., is underfilled), it must be reprocessed at a cost of $10. A properly filled container sells for $230.

a. Find the probability that a container is underfilled.

b. A container is initially underfilled and must be reprocessed. Upon refilling, it contains 10.6 units. How much profit will the company make on this container?

c. The operations manager adjusts the mean of the filling process upward to 10.5 units in order to make the probability of underfilling approximately zero. Under these conditions, what is the expected profit per container?

5.51 **Box plots and the standard normal distribution.** What relationship exists between the standard normal distribution and the box-plot methodology (optional Section 2.8) for describing distributions of data by means of quartiles? The answer depends on the true underlying probability distribution of the data. Assume for the remainder of this exercise that the distribution is normal.

a. Calculate the values z_L and z_U of the standard normal random variable z that correspond, respectively, to the hinges of the box plot (i.e., the lower and upper quartiles Q_L and Q_U) of the probability distribution.

b. Calculate the z values that correspond to the inner fences of the box plot for a normal probability distribution.

c. Calculate the z values that correspond to the outer fences of the box plot for a normal probability distribution.

d. What is the probability that an observation lies beyond the inner fences of a normal probability distribution? The outer fences?

e. Can you now better understand why the inner and outer fences of a box plot are used to detect outliers in a distribution? Explain.

5.52 **Load on frame structures.** In the *Journal of the International Association for Shell and Spatial Structures* (April 2004), Japanese environmental researchers studied the performance of truss-and-frame structures subjected to uncertain loads. The load was assumed to have a normal distribution with a mean of 20 thousand pounds. Also, the probability that the load is between 10 and 30 thousand pounds is .95. On the basis of this information, find the standard deviation of the load distribution.

5.4 Descriptive Methods for Assessing Normality

In the chapters that follow, we learn how to make inferences about the population on the basis of information contained in the sample. Several of these techniques are based on the assumption that the population is approximately normally distributed. Consequently, it will be important to determine whether the sample data come from a normal population before we can apply these techniques properly.

A number of descriptive methods can be used to check for normality. In this section, we consider the four methods summarized in the following box:

Determining whether the Data Are from an Approximately Normal Distribution

1. Construct either a histogram or stem-and-leaf display for the data, and note the shape of the graph. If the data are approximately normal, the shape of the histogram or stem-and-leaf display will be similar to the normal curve shown in Figure 5.6 (i.e., the display will be mound shaped and symmetric about the mean).

2. Compute the intervals $\bar{x} \pm s$, $\bar{x} \pm 2s$, and $\bar{x} \pm 3s$, and determine the percentage of measurements falling into each. If the data are approximately normal, the percentages will be approximately equal to 68%, 95%, and 100%, respectively.

3. Find the interquartile range IQR and standard deviation s for the sample, and then calculate the ratio IQR/s. If the data are approximately normal, then IQR/$s \approx 1.3$.

4. Construct a *normal probability plot* for the data. If the data are approximately normal, the points will fall (approximately) on a straight line.

The first two methods come directly from the properties of a normal distribution established in Section 5.3. Method 3 is based on the fact that for normal distributions, the z values corresponding to the 25th and 75th percentiles are $-.67$ and $.67$, respectively. (See Example 5.5.) Since $\sigma = 1$ for a standard normal distribution,

$$\frac{\text{IQR}}{\sigma} = \frac{Q_U - Q_L}{\sigma} = \frac{.67 - (-.67)}{1} = 1.34$$

The final descriptive method for checking normality is based on a *normal probability plot*. In such a plot, the observations in a data set are ordered from smallest to largest and are then plotted against the expected z-scores of observations calculated under the assumption that the data come from a normal distribution. When the data are, in fact, normally distributed, a linear (straight-line) trend will result. A nonlinear trend in the plot suggests that the data are nonnormal.

A **normal probability plot** for a data set is a scatterplot with the ranked data values on one axis and their corresponding expected z-scores from a standard normal distribution on the other axis. [*Note:* Computation of the expected standard normal z-scores is beyond the scope of this text. Therefore, we will rely on available statistical software packages to generate a normal probability plot.]

Example 5.12

Checking for Normal Data—EPA Estimated Gas Mileages

Problem The EPA mileage ratings on 100 cars, first presented in Chapter 2 (p. 38), are reproduced in Table 5.2. Recall that these data are saved in the **EPAGAS** file. Numerical and graphical descriptive measures for the data are shown on the MINITAB and SPSS printouts presented in Figure 5.19a–c. Determine whether the EPA mileage ratings are from an approximate normal distribution.

Solution As a first check, we examine the MINITAB histogram of the data shown in Figure 5.19a. Clearly, the mileages fall into an approximately mound shaped, symmetric distribution centered around the mean of about 37 mpg. Note that a normal curve is superimposed on the figure. Therefore, check #1 in the box indicates that the data are approximately normal.

Table 5.2	EPA Gas Mileage Ratings for 100 Cars (miles per gallon)								
36.3	41.0	36.9	37.1	44.9	36.8	30.0	37.2	42.1	36.7
32.7	37.3	41.2	36.6	32.9	36.5	33.2	37.4	37.5	33.6
40.5	36.5	37.6	33.9	40.2	36.4	37.7	37.7	40.0	34.2
36.2	37.9	36.0	37.9	35.9	38.2	38.3	35.7	35.6	35.1
38.5	39.0	35.5	34.8	38.6	39.4	35.3	34.4	38.8	39.7
36.3	36.8	32.5	36.4	40.5	36.6	36.1	38.2	38.4	39.3
41.0	31.8	37.3	33.1	37.0	37.6	37.0	38.7	39.0	35.8
37.0	37.2	40.7	37.4	37.1	37.8	35.9	35.6	36.7	34.5
37.1	40.3	36.7	37.0	33.9	40.1	38.0	35.2	34.8	39.5
39.9	36.9	32.9	33.8	39.8	34.0	36.8	35.0	38.1	36.9

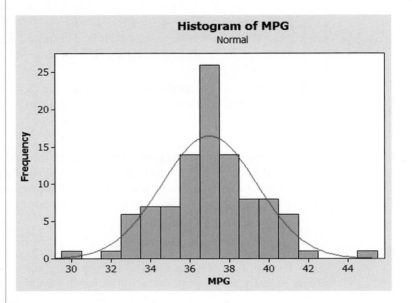

Figure 5.19a

MINITAB histogram for gas mileage data

Figure 5.19b

MINITAB descriptive statistics for gas mileage data

Descriptive Statistics: MPG

Variable	N	Mean	StDev	Minimum	Q1	Median	Q3	Maximum
MPG	100	36.994	2.418	30.000	35.625	37.000	38.375	44.900

To apply check #2, we obtain $\bar{x} = 37$ and $s = 2.4$ from the MINITAB printout of Figure 5.19b. The intervals $\bar{x} \pm s$, $\bar{x} \pm 2s$, and $\bar{x} \pm 3s$ are shown in Table 5.3, as is the percentage of mileage ratings that fall into each interval. (We obtained these results in Section 2.6, p. 66.) These percentages agree almost exactly with those from a normal distribution.

Check #3 in the box requires that we find the ratio IQR/s. From Figure 5.19b, the 25th percentile (labeled Q_1 by MINITAB) is $Q_L = 35.625$ and the 75th percentile (labeled Q_3 by MINITAB) is $Q_U = 38.375$. Then IQR $= Q_U - Q_L = 2.75$, and the ratio is

$$\frac{\text{IQR}}{s} = \frac{2.75}{2.4} = 1.15$$

Since this value is approximately equal to 1.3, we have further confirmation that the data are approximately normal.

A fourth descriptive method is to interpret a normal probability plot. An SPSS normal probability plot of the mileage data is shown in Figure 5.19c. Notice that the ordered mileage values (shown on the horizontal axis) fall reasonably close to a straight line when plotted against the expected values from a normal distribution. Thus, check #4 also suggests that the EPA mileage data are approximately normally distributed.

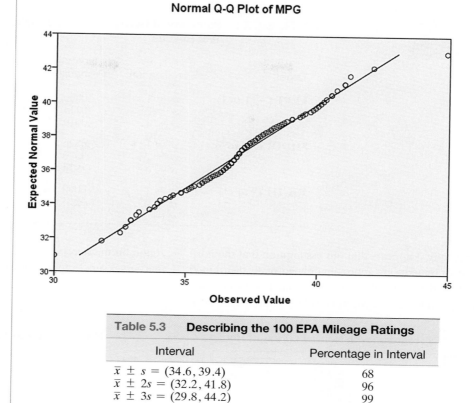

Figure 5.19c

SPSS normal probability plot for gas mileage data

Table 5.3	Describing the 100 EPA Mileage Ratings
Interval	Percentage in Interval
$\bar{x} \pm s = (34.6, 39.4)$	68
$\bar{x} \pm 2s = (32.2, 41.8)$	96
$\bar{x} \pm 3s = (29.8, 44.2)$	99

Look Back The checks for normality given in the box are simple, yet powerful, techniques to apply, but they are only descriptive in nature. It is possible (although unlikely) that the data are nonnormal even when the checks are reasonably satisfied. Thus, we should be careful not to claim that the 100 EPA mileage ratings are, in fact, normally distributed. We can only state that it is reasonable to believe that the data are from a normal distribution.*

Now Work Exercise 5.60

As we will learn in the next chapter, several inferential methods of analysis require the data to be approximately normal. If the data are clearly nonnormal, inferences derived from the method may be invalid. Therefore, it is advisable to check the normality of the data prior to conducting any analysis.

Statistics in Action Revisited

Assessing whether the Normal Distribution Is Appropriate for Modeling the Super Weapon Hit Data

In *Statistics in Action Revisited* in Section 5.3, we used the normal distribution to find the probability that a single flechette from a super weapon that shoots 1,100 flechettes at once hits one of three targets at 500 meters. Recall that for three range tests, the weapon was always aimed at the center target (i.e., the specification mean was set at $\mu = 5$ feet), but the specification standard deviation was varied at $\sigma = 1$ foot, $\sigma = 2$ feet, and $\sigma = 4$ feet. Table SIA5.1 shows the calculated normal probabilities of hitting the three targets for the different

values of σ, as well as the actual results of the three range tests. (Recall that the actual data are saved in the **MOAGUN** file.) You can see that the proportion of the 1,100 flechettes that actually hit each target—called the hit ratio—agrees very well with the estimated probability of a hit derived from the normal distribution.

(continued)

*Statistical tests of normality that provide a measure of reliability for the inference are available. However, these tests tend to be very sensitive to slight departures from normality (i.e., they tend to reject the hypothesis of normality for any distribution that is not perfectly symmetrical and mound shaped). Consult the references (see especially Ramsey & Ramsey, 1990) if you want to learn more about these tests.

Statistics IN Action
(continued)

Table SIA 5.1 **Summary of Normal Probability Calculations and Actual Range Test Results**

Target	Specification	Normal Probability	Actual Number of Hits	Hit Ratio (Hits/1,100)
LEFT (− 1 to 1)	$\sigma = 1$	.0000	0	.000
	$\sigma = 2$	.0214	30	.027
	$\sigma = 4$	.0919	73	.066
MIDDLE (4 to 6)	$\sigma = 1$	.6826	764	.695
	$\sigma = 2$	.3820	409	.372
	$\sigma = 4$	.1974	242	.220
RIGHT (9 to 11)	$\sigma = 1$	.0000	0	.000
	$\sigma = 2$	.0214	23	.021
	$\sigma = 4$	.0919	93	.085

Consequently, it appears that our assumption that the horizontal hit measurements are approximately normally distributed is reasonably satisfied. Further evidence that this assumption is satisfied is provided by the MINITAB histograms of the horizontal hit measurements shown in Figures SIA5.3a–c. The normal curves superimposed on the histograms fit the data very well.

Data Set: MOAGUN

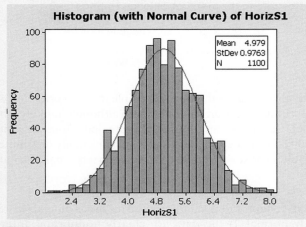

Figure SIA5.3a
MINITAB histogram for the horizontal hit measurements when $\sigma = 1$

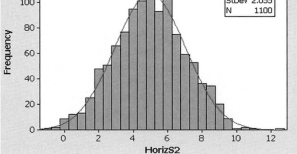

Figure SIA 5.3b
MINITAB histogram for the horizontal hit measurements when $\sigma = 2$

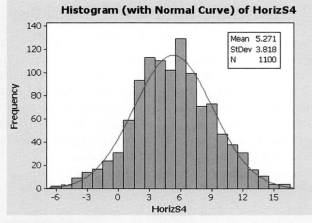

Figure SAI 5.3c
MINITAB histogram for the horizontal hit measurements when $\sigma = 4$

Exercises 5.53–5.72

Understanding the Principles

5.53 Why is it important to check whether the sample data come from a normal population?

5.54 Give four methods for determining whether the sample data come from a normal population.

5.55 If a population data set is normally distributed, what is the proportion of measurements you would expect to fall within the following intervals?
 a. $\mu \pm \sigma$ **b.** $\mu \pm 2\sigma$ **c.** $\mu \pm 3\sigma$

5.56 What is a normal probability plot and how is it used?

Learning the Mechanics

5.57 Normal probability plots for three data sets are shown below. Which plot indicates that the data are approximately normally distributed?

5.58 Consider a sample data set with the following summary statistics: $s = 95, Q_L = 72,$ and $Q_U = 195$.
 a. Calculate IQR.
 b. Calculate IQR/s.
 c. Is the value of IQR/s approximately equal to 1.3? What does this imply?

5.59 Examine the following sample data saved in the **LM5_59** file.

32	48	25	135	53	37	5	39	213	165
109	40	1	146	39	25	21	66	64	57
197	60	112	10	155	134	301	304	107	82
35	81	60	95	401	308	180	3	200	59

 a. Construct a stem-and-leaf plot to assess whether the data are from an approximately normal distribution.
 b. Find the values of $Q_L, Q_U,$ and s for the sample data.
 c. Use the results from part **b** to assess the normality of the data.
 d. Generate a normal probability plot for the data, and use it to assess whether the data are approximately normal.

5.60 Examine the following sample data saved in the **LM5_60** file.

5.9	5.3	1.6	7.4	8.6	3.2	2.1
4.0	7.3	8.4	5.9	6.7	4.5	6.3
6.0	9.7	3.5	3.1	4.3	3.3	8.4
4.6	8.2	6.5	1.1	5.0	9.4	6.4

 a. Construct a stem-and-leaf plot to assess whether the data are from an approximately normal distribution.
 b. Compute s for the sample data.
 c. Find the values of Q_L and Q_U and the value of s from part **b** to assess whether the data come from an approximately normal distribution.
 d. Generate a normal probability plot for the data, and use it to assess whether the data are approximately normal.

Applying the Concepts—Basic

5.61 **Most powerful American women.** Refer to the *Fortune* (Oct. 16, 2008) list of the 50 most powerful women in America. In Exercise 5.39 (p. 242), you assumed that the ages (in years) of these women are approximately normally distributed. A MINITAB printout with summary statistics for the age variable saved in the **WPOWER50** file is reproduced below.
 a. Use the relevant statistics on the printout to find the interquartile range IQR.
 b. Locate the value of the standard deviation s on the printout.
 c. Use the results from parts **a** and **b** to demonstrate that the age distribution is approximately normal.
 d. Construct a relative frequency histogram for the age data. Use this graph to support your assumption of normality.

Descriptive Statistics: AGE

Variable	N	Mean	StDev	Minimum	Q1	Median	Q3	Maximum
AGE	50	50.020	6.444	28.000	47.000	51.000	54.000	64.000

5.62 **Shear strength of rock fractures.** Understanding the characteristics of rock masses, especially the nature of the fractures, are essential when building dams and power plants. The shear strength of rock fractures was investigated in *Engineering Geology* (May 12, 2010). The Joint Roughness Coefficient (JRC) was used to measure shear strength. Civil engineers collected JRC data for over 750

Plots for Exercise 5.57

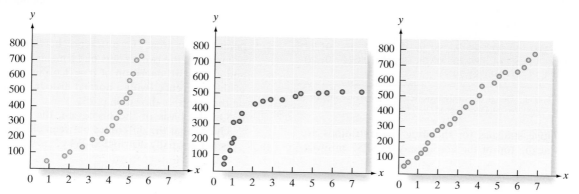

rock fractures. The results (simulated from information provided in the article) are summarized in the SPSS histogram shown below. Should the engineers use the normal probability distribution to model the behavior of shear strength for rock fractures? Explain.

SPSS Output for Exercise 5.62

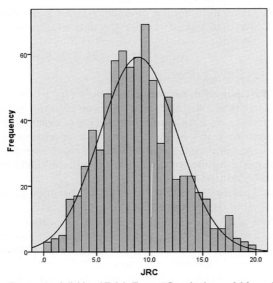

Based on Pooyan Asadollahi and Fulvio Tonon, "Constitutive model for rock fractures: Revisiting Barton's empirical model." *Engineering Geology*, Vol. 113, no. 1, pp. 11–32.

5.63 **Software file updates.** Software configuration management was used to monitor a software engineering team's performance at Motorola, Inc. (*Software Quality Professional*, Nov. 2004). One of the variables of interest was the number of updates to a file changed because of a problem report. Summary statistics for $n = 421$ files yielded the following results: $\bar{x} = 4.71$, $s = 6.09$, $Q_L = 1$, and $Q_U = 6$. Are these data approximately normally distributed? Explain.

5.64 **Drug content assessment.** Scientists at GlaxoSmithKline Medicines Research Center used high-performance liquid chromatography (HPLC) to determine the amount of drug in a tablet produced by the company (*Analytical Chemistry*, Dec. 15, 2009). Drug concentrations (measured as a percentage) for 50 randomly selected tablets are listed in the accompanying table and saved in the **DRUGCON** file.

91.28	92.83	89.35	91.90	82.85	94.83	89.83	89.00	84.62
86.96	88.32	91.17	83.86	89.74	92.24	92.59	84.21	89.36
90.96	92.85	89.39	89.82	89.91	92.16	88.67	89.35	86.51
89.04	91.82	93.02	88.32	88.76	89.26	90.36	87.16	91.74
86.12	92.10	83.33	87.61	88.20	92.78	86.35	93.84	91.20
93.44	86.77	83.77	93.19	81.79				

Based on Borman, P. J., et al., "Design and analysis of method equivalence studies." *Analytical Chemistry*, Vol. 81, No. 24, December 15, 2009 (Table 3).

a. Descriptive statistics for the drug concentrations are shown at the top of the accompanying SPSS printout.

SPSS Output for Exercise 5.64

Descriptives

			Statistic	Std. Error
Content	Mean		89.2906	.45021
	95% Confidence Interval for Mean	Lower Bound	88.3859	
		Upper Bound	90.1953	
	5% Trimmed Mean		89.3963	
	Median		89.3750	
	Variance		10.134	
	Std. Deviation		3.18344	
	Minimum		81.79	
	Maximum		94.83	
	Range		13.04	
	Interquartile Range		4.84	
	Skewness		-.544	.337
	Kurtosis		-.389	.662

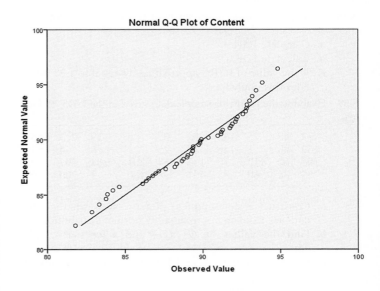

Use this information to assess whether the data are approximately normal.

b. An SPSS normal probability plot is shown above. Use this information to assess whether the data are approximately normal.

5.65 **Estimating glacier elevations.** Digital elevation models (DEMs) are now used to estimate elevations and slopes of remote regions. In *Arctic, Antarctic, and Alpine Research* (May 2004), geographers analyzed reading errors from maps produced by DEMs. Two readers of a DEM map of White Glacier (in Canada) estimated elevations at 400 points in the area. The difference between the elevation estimates of the two readers had a mean of $\mu = .28$ meter and a standard deviation of $\sigma = 1.6$ meters. A histogram of the difference (with a normal histogram superimposed on the graph) is shown on p. 251.

a. On the basis of the histogram, the researchers concluded that the difference between elevation estimates is not normally distributed. Why?

Histogram for Exercise 5.65

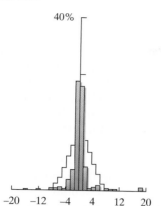

"Uncertainty in digital elevation models of Axel Heiberg Island. Arctic Canada," *Arctic, Antarctic, and Alpine Research,* Vol. 36, No. 2, May 2004 (Figure 3). © Regents of the University of Colorado. Reprinted with permission.

b. Will the interval $\mu \pm 2\sigma$ contain more than 95%, exactly 95%, or less than 95% of the 400 elevation differences? Explain.

Applying the Concepts—Intermediate

5.66 Habitats of endangered species. An evaluation of the habitats of endangered salmon species was performed in *Conservation Ecology* (December 2003). The researchers identified 734 sites (habitats) for Chinook, coho, or steelhead salmon species in Oregon and assigned a habitat quality score to each site. (Scores range from 0 to 36 points, with lower scores indicating poorly maintained or degraded habitats.) The data are saved in the **HABITAT** file. Give your opinion on whether the habitat quality score is normally distributed.

5.67 Baseball batting averages. Major League Baseball (MLB) has two leagues: the American League (AL), which utilizes the designated hitter (DH) to bat for the pitcher, and the National League (NL), which does not allow the DH. A player's batting average is computed by dividing the player's total number of hits by his official number of at bats. The batting averages for all AL and NL players with at least 100 official at bats during the 2006 season are stored in the **MLB2009-AL** and **MLB2009-NL** files, respectively. Determine whether each batting average distribution is approximately normal.

5.68 Ranking the driving performance of professional golfers. Refer to *The Sport Journal* (Winter 2007) article on a new method for ranking the driving performance of PGA golfers, presented in Exercise 2.64 (p. 59). Recall that the method incorporates a golfer's average driving distance (yards) and driving accuracy (percentage of drives that land in the fairway) into a driving performance index. Data on these three variables for the top 40 PGA golfers are saved in the **PGADRIVER** file. Determine which of the variables—driving distance, driving accuracy, and driving performance index—are approximately normally distributed.

5.69 NHTSA crash tests. Refer to the National Highway Traffic Safety Administration (NHTSA) crash test data

for new cars. In Exercise 5.43 (p. 243), you assumed that the driver's head injury rating is approximately normally distributed. Apply the methods of this chapter to the data saved in the **CRASH** file to support that assumption.

5.70 Cruise ship sanitation scores. Refer to the data on the Jan. 2010 sanitation scores for 186 cruise ships, presented in Exercise 2.35 (p. 47). The data are saved in the **SHIPSANIT** file. Assess whether the sanitation scores are approximately normally distributed.

Applying the Concepts—Advanced

5.71 Laptop use in middle school. Refer to the *American Secondary Education* (Fall 2009) study of student usage of laptops at a middle school, Exercise 2.100 (p. 72). Recall that x, the number of minutes per day each student used his or her laptop for taking notes, has a mean of 13.2 with a standard deviation of 19.5. Demonstrate why the probability distribution of x is unlikely to be normal.

5.72 Ranking Ph.D. programs in economics. Refer to the *Southern Economic Journal* (Apr. 2008) rankings of Ph.D. programs in economics at 129 colleges and universities, Exercise 2.121 (p. 77). Recall that the number of publications published by faculty teaching in the Ph.D. program and the quality of the publications were used to calculate an overall productivity score for each program. The mean and standard deviation of these 129 productivity scores were then used to compute a z-score for each economics program. The data (z-scores) for all 129 economic programs are saved in the **ECOPHD** file. A MINITAB normal probability plot for the z-scores is shown in the accompanying printout.

a. Use the graph to assess whether the data are approximately normal.

b. Based on the graph, determine the nature of the skewness of the data.

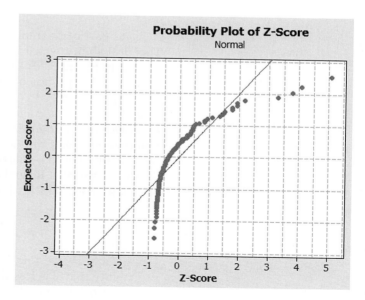

5.5 Approximating a Binomial Distribution with a Normal Distribution (Optional)

When the discrete binomial random variable (Section 4.4) can assume a large number of values, the calculation of its probabilities may become tedious. To contend with this problem, we provide tables in Appendix A that give the probabilities for some values of n and p, but these tables are by necessity incomplete. Recall that the binomial probability table (Table II) can be used only for $n = 5, 6, 7, 8, 9, 10, 15, 20$, or 25. To deal with this limitation, we seek approximation procedures for calculating the probabilities associated with a binomial probability distribution.

When n is large, a normal probability distribution may be used to provide a good approximation to the probability distribution of a binomial random variable. To show how this approximation works, we refer to Example 4.13, in which we used the binomial distribution to model the number x of 20 eligible voters who favor a candidate. We assumed that 60% of all the eligible voters favored the candidate. The mean and standard deviation of x were found to be $\mu = 12$ and $\sigma = 2.2$, respectively. The binomial distribution for $n = 20$ and $p = .6$ is shown in Figure 5.20, and the approximating normal distribution with mean $\mu = 12$ and standard deviation $\sigma = 2.2$ is superimposed.

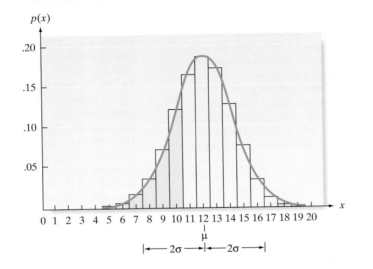

Figure 5.20

Binomial distribution for $n = 20, p = .6$ and normal distribution with $\mu = 12, \sigma = 2.2$

As part of Example 4.13, we used Table II to find the probability that $x \leq 10$. This probability, which is equal to the sum of the areas contained in the rectangles (shown in Figure 5.20) that correspond to $p(0), p(1), p(2), \ldots, p(10)$, was found to equal .245. The portion of the normal curve that would be used to approximate the area $p(0) + p(1) + p(2) + \cdots + p(10)$ is highlighted in Figure 5.20. *Note that this highlighted area lies to the left of 10.5 (not 10), so we may include all of the probability in the rectangle corresponding to $p(10)$.* Because we are approximating a discrete distribution (the binomial) with a continuous distribution (the normal), we call the use of 10.5 (instead of 10 or 11) a **correction for continuity**. That is, we are correcting the discrete distribution so that it can be approximated by the continuous one. The use of the correction for continuity leads to the calculation of the following standard normal z-value:

$$z = \frac{x - \mu}{\sigma} = \frac{10.5 - 12}{2.2} = -.68$$

Using Table IV, we find the area between $z = 0$ and $z = .68$ to be .2517. Then the probability that x is less than or equal to 10 is approximated by the area under the normal distribution to the left of 10.5, shown highlighted in Figure 5.20. That is,

$$P(x \leq 10) \approx P(z \leq -.68) = .5 - P(-.68 < z \leq 0) = .5 - .2517 = .2483$$

The approximation differs only slightly from the exact binomial probability, .245. Of course, when tables of exact binomial probabilities are available, we will use the exact value rather than a normal approximation.

The normal distribution will not always provide a good approximation to binomial probabilities. The following is a useful rule of thumb to determine when n is large enough for the approximation to be effective: *The interval $\mu \pm 3\sigma$ should lie within the range of the binomial random variable x (i.e., from 0 to n) in order for the normal approximation to be adequate.* The rule works well because almost all of the normal distribution falls within three standard deviations of the mean, so if this interval is contained within the range of x values, there is "room" for the normal approximation to work.

As shown in Figure 5.21a for the preceding example with $n = 20$ and $p = .6$, the interval $\mu \pm 3\sigma = 12 + 3(2.2) = (5.4, 18.6)$ lies within the range from 0 to 20. However, if we were to try to use the normal approximation with $n = 10$ and $p = .1$, the interval $\mu \pm 3\sigma$ becomes $1 \pm 3(.95)$, or $(-1.85, 3.85)$. As shown in Figure 5.21b, this interval is not contained within the range of x, since $x = 0$ is the lower bound for a binomial random variable. Note in Figure 5.21b that the normal distribution will not "fit" in the range of x; therefore, it will not provide a good approximation to the binomial probabilities.

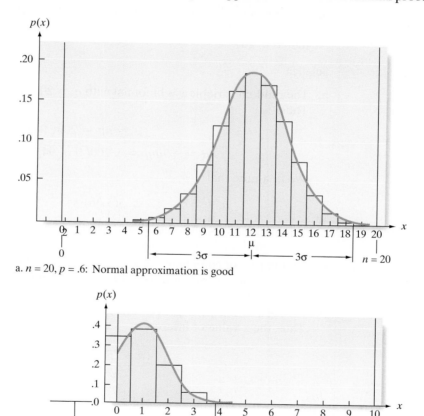

a. $n = 20, p = .6$: Normal approximation is good

b. $n = 10, p = .1$: Normal approximation is poor

Figure 5.21

Rule of thumb for normal approximation to binomial probabilities

---

BIOGRAPHY   ABRAHAM DE MOIVRE (1667–1754)

Advisor to Gamblers

French-born mathematician Abraham de Moivre moved to London when he was 21 years old to escape religious persecution. In England, he earned a living first as a traveling teacher of mathematics and then as an advisor to gamblers, underwriters, and annuity brokers. De Moivre's major contributions to probability theory are contained in two of his books: *The Doctrine of Chances* (1718) and *Miscellanea Analytica* (1730). In these works, he defines statistical independence, develops the formula for the normal probability distribution, and derives the normal curve as an approximation to the binomial distribution. Despite his eminence as a mathematician, de Moivre died in poverty. He is famous for using an arithmetic progression to predict the day of his death. ∎

Example 5.13

Approximating a Binomial Probability with the Normal Distribution—Lot Acceptance Sampling

Problem One problem with any product that is mass produced (e.g., a graphing calculator) is quality control. The process must be monitored or audited to be sure that the output of the process conforms to requirements. One monitoring method is *lot acceptance sampling*, in which items being produced are sampled at various stages of the production process and are carefully inspected. The lot of items from which the sample is drawn is then accepted or rejected on the basis of the number of defectives in the sample. Lots that are accepted may be sent forward for further processing or may be shipped to customers; lots that are rejected may be reworked or scrapped. For example, suppose a manufacturer of calculators chooses 200 stamped circuits from the day's production and determines x, the number of defective circuits in the sample. Suppose that up to a 6% rate of defectives is considered acceptable for the process.

a. Find the mean and standard deviation of x, assuming that the rate of defectives is 6%.

b. Use the normal approximation to determine the probability that 20 or more defectives are observed in the sample of 200 circuits (i.e., find the approximate probability that $x \geq 20$).

Solution

a. The random variable x is binomial with $n = 200$ and the fraction defective $p = .06$. Thus,

$$\mu = np = 200(.06) = 12$$

$$\sigma = \sqrt{npq} = \sqrt{200(.06)(.94)} = \sqrt{11.28} = 3.36$$

We first note that

$$\mu \pm 3\sigma = 12 \pm 3(3.36) = 12 \pm 10.08 = (1.92, 22.08)$$

lies completely within the range from 0 to 200. Therefore, a normal probability distribution should provide an adequate approximation to this binomial distribution.

b. By the rule of complements, $P(x \geq 20) = 1 - P(x \leq 19)$. To find the approximating area corresponding to $x \leq 19$, refer to Figure 5.22. Note that we want to include all the binomial probability histograms from 0 to 19, inclusive. Since the event is of the form $x \leq a$, the proper correction for continuity is $a + .5 = 19 + .5 = 19.5$. Thus, the z value of interest is

$$z = \frac{(a + .5) - \mu}{\sigma} = \frac{19.5 - 12}{3.36} = 2.23$$

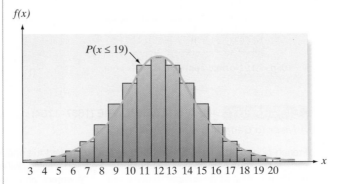

Figure 5.22

Normal approximation to the binomial distribution with $n = 200, p = .06$

Referring to Table IV in Appendix A, we find that the area to the right of the mean, 0, corresponding to $z = 2.23$ (see Figure 5.23) is .4871. So the area $A = P(z \leq 2.23)$ is

$$A = .5 + .4871 = .9871$$

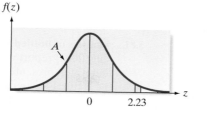

Figure 5.23
Standard normal distribution

Thus, the normal approximation to the binomial probability we seek is

$$P(x \geq 20) = 1 - P(x \leq 19) \approx 1 - .9871 = .0129$$

In other words, *if, in fact, the true fraction of defectives is .06*, then the probability that 20 or more defectives will be observed in a sample of 200 circuits is extremely small.

Look Back If the manufacturer observes $x \geq 20$, the likely reason is that the process is producing more than the acceptable 6% defectives. The lot acceptance sampling procedure is another example of using the rare-event approach to make inferences.

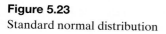

> **Using a Normal Distribution to Approximate Binomial Probabilities**
>
> 1. After you have determined n and p for the binomial distribution, calculate the interval
>
> $$\mu \pm 3\sigma = np \pm 3\sqrt{npq}$$
>
> If the interval lies in the range from 0 to n, the normal distribution will provide a reasonable approximation to the probabilities of most binomial events.
>
> 2. Express the binomial probability to be approximated in the form $P(x \leq a)$ or $P(x \leq b) - P(x \leq a)$. For example,
>
> $$P(x < 3) = P(x \leq 2)$$
> $$P(x \geq 5) = 1 - P(x \leq 4)$$
> $$P(7 \leq x \leq 10) = P(x \leq 10) - P(x \leq 6)$$
>
> 3. For each value of interest, a, the correction for continuity is $(a + .5)$ and the corresponding standard normal z value is
>
> $$z = \frac{(a + .5) - \mu}{\sigma} \text{ (see Figure 5.24)}$$
>
> 4. Sketch the approximating normal distribution and shade the area corresponding to the probability of the event of interest, as in Figure 5.24. Verify that the rectangles you have included in the shaded area correspond to the probability you wish to approximate. Use Table IV and the z value(s) you calculated in step 3 to find the shaded area. This is the approximate probability of the binomial event.

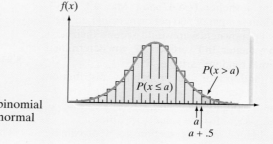

Figure 5.24
Approximating binomial probabilities by normal probabilities

The steps for approximating a binomial probability by a normal probability are given in the following box:

Exercises 5.73–5.92

Understanding the Principles

5.73 For large n (say, $n = 100$), why is it advantageous to use the normal distribution to approximate a binomial probability?

5.74 Why do we need a correction for continuity when approximating a binomial probability with the normal distribution?

Learning the Mechanics

5.75 Suppose x is a binomial random variable with $p = .4$ and
NW $n = 25$.
   a. Would it be appropriate to approximate the probability distribution of x with a normal distribution? Explain.
   b. Assuming that a normal distribution provides an adequate approximation to the distribution of x, what are the mean and variance of the approximating normal distribution?
   c. Use Table II of Appendix A to find the exact value of $P(x \geq 9)$.
   d. Use the normal approximation to find $P(x \geq 9)$.

5.76 Assume that x is a binomial random variable with n and p as specified in parts **a–f** that follow. For which cases would it be appropriate to use a normal distribution to approximate the binomial distribution?
   a. $n = 100, p = .01$ **b.** $n = 20, p = .6$
   c. $n = 10, p = .4$ **d.** $n = 1,000, p = .05$
   e. $n = 100, p = .8$ **f.** $n = 35, p = .7$

5.77 Assume that x is a binomial random variable with $n = 25$ and $p = .5$. Use Table II of Appendix A and the normal approximation to find the exact and approximate values, respectively, of the following probabilities:
   a. $P(x \leq 11)$
   b. $P(x \geq 16)$
   c. $P(8 \leq x \leq 16)$

5.78 Assume that x is a binomial random variable with $n = 1,000$ and $p = .50$. Find each of the following probabilities:
   a. $P(x > 500)$ **b.** $P(490 \leq x < 500)$
   c. $P(x > 550)$

5.79 Assume that x is a binomial random variable with $n = 100$ and $p = .40$. Use a normal approximation to find the following:
   a. $P(x \leq 35)$ **b.** $P(40 \leq x \leq 50)$
   c. $P(x \geq 38)$

Applying the Concepts — Basic

5.80 **Where will you get your next pet?** Refer to Exercise 4.62 (p. 205) and the Associated Press/Petside.com poll that revealed that half of all pet owners would get their next dog or cat from a shelter (*USAToday*, May 12, 2010). Consider a random sample of 500 pet owners and define x as the number of pet owners who would acquire their next dog or cat from a shelter. In Exercise 4.62, you determined that x is a binomial random variable.
   a. Compute μ and σ for the probability distribution of x.
   b. Compute the z-score for the value $x = 240$.
   c. Compute the z-score for the value $x = 270$.
   d. Use the technique of this section to approximate $P(240 < x < 270)$.

5.81 **Analysis of bottled water.** Refer to the *Scientific American* (July 2003) report on whether bottled water is really purified water, presented in Exercise 4.64 (p. 206). Recall that the Natural Resources Defense Council found that 25% of bottled-water brands fill their bottles with just tap water. In a random sample of 65 bottled-water brands, let x be the number that contain tap water.
   a. Find the mean of x.
   b. Find the standard deviation of x.
   c. Find the z-score for the value $x = 20$.
   d. Find the approximate probability that 20 or more of the 65 sampled bottled-water brands will contain tap water.

5.82 **LASIK surgery complications.** According to recent studies, 1% of all patients who undergo laser surgery (i.e., LASIK) to correct their vision have serious post-laser vision problems (*All About Vision*, 2006). In a random sample of 100,000 LASIK patients, let x be the number who experience serious post-laser vision problems.
   a. Find $E(x)$.
   b. Find $Var(x)$.
   c. Find the z-score for $x = 950$.
   d. Find the approximate probability that fewer than 950 patients in a sample of 100,000 will experience serious post-laser vision problems.

5.83 **Melanoma deaths.** According to the *American Cancer Society*, melanoma, a form of skin cancer, kills 15% of Americans who suffer from the disease each year. Consider a sample of 10,000 melanoma patients.
   a. What are the expected value and variance of x, the number of the 10,000 melanoma patients who die of the affliction this year?
   b. Find the probability that x will exceed 1,600 patients per year.
   c. Would you expect x, the number of patients dying of melanoma, to exceed 6,500 in any single year? Explain.

5.84 **Caesarian birth study.** In Exercise 4.66 (p. 206), you learned that 32% of all births in the United States occur by Caesarian section each year (*National Vital Statistics Reports*, Mar. 2010). In a random sample of 1,000 births this year, let x be the number that occur by Caesarian section.
   a. Find the mean of x. (This value should agree with your answer to Exercise 4.66 **a.**)
   b. Find the standard deviation of x. (This value should agree with your answer to Exercise 4.66**b.**)
   c. Find the z-score for the value $x = 200.5$.
   d. Find the approximate probability that the number of Caesarian sections in a sample of 1,000 births is less than or equal to 200.

Applying the Concepts — Intermediate

5.85 **Ecotoxicological survival study.** The *Journal of Agricultural, Biological and Environmental Statistics* (Sept. 2000) gave an evaluation of the risk posed by hazardous pollutants. In the experiment, guppies (all the same age and size) were released into a tank of natural seawater polluted with the pesticide dieldrin and the number of guppies surviving after five days was determined. The researchers estimated that the probability of any single guppy surviving is .60. If 300 guppies are released into the polluted tank, estimate the probability that fewer than 100 guppies survive after five days.

5.86 Chemical signals of mice. Refer to the *Cell* (May 14, 2010) study of the ability of a mouse to recognize the odor of a potential predator, Exercise 4.63 (p. 205). You learned that 40% of lab mice cells exposed to chemically produced major urinary proteins (Mups) from a cat responded positively (i.e., recognized the danger of the lurking predator). Again, consider a sample of 100 lab mice cells, each exposed to chemically produced cat Mups, and let *x* represent the number of cells that respond positively. How likely is it that less than half of the cells respond positively to cat Mups?

5.87 Defects in semiconductor wafers. The computer chips in notebook and laptop computers are produced from semiconductor wafers. Certain semiconductor wafers are exposed to an environment that generates up to 100 possible defects per wafer. The number of defects per wafer, *x*, was found to follow a binomial distribution if the manufacturing process is stable and generates defects that are randomly distributed on the wafers (*IEEE Transactions on Semiconductor Manufacturing*, May 1995). Let *p* represent the probability that a defect occurs at any one of the 100 points of the wafer. For each of the following cases, determine whether the normal approximation can be used to characterize *x*:
a. $p = .01$ **b.** $p = .50$ **c.** $p = .90$

5.88 Victims of domestic abuse. In Exercise 4.71 (p. 206), you learned that some researchers believe that one in every three women has been a victim of domestic abuse (Domestic Vidence: Incidence and Prevalance study, Sep–Dec., 2005).
a. For a random sample of 150 women, what is the approximate probability that more than half are victims of domestic abuse?
b. For a random sample of 150 women, what is the approximate probability that fewer than 50 are victims of domestic abuse?
c. Would you expect to observe fewer than 30 domestically abused women in a sample of 150? Explain.

5.89 Hotel guest satisfaction. Refer to the 2009 North American Hotel Guest Satisfaction Index Study, Exercise 4.67 (p. 206). You determined that the probability of a hotel guest participating in the hotel's "green" conservation program by reusing towels and bed linens is .45. Suppose a large hotel chain randomly samples 200 of its guests. The chain's national director claims that more than 110 of these guests participated in the conservation program. Do you believe this claim? Explain.

Applying the Concepts—Advanced

5.90 Body fat in men. The percentage of fat in the bodies of American men is an approximately normal random variable with mean equal to 15% and standard deviation equal to 2%.
a. If these values were used to describe the body fat of men in the U.S. Army, and if a measure of 20% or more body fat characterizes the person as obese, what is the approximate probability that a random sample of 10,000 soldiers will contain fewer than 50 who would actually be characterized as obese?
b. If the army actually were to check the percentage of body fat for a random sample of 10,000 men, and if only 30 contained 20% (or higher) body fat, would you conclude that the army was successful in reducing the percentage of obese men below the percentage in the general population? Explain your reasoning.

5.91 Luggage inspection at Newark airport. *New Jersey Business* reports that Newark International Airport's terminal handles an average of 3,000 international passengers an hour, but is capable of handling twice that number. Also, after scanning all luggage, 20% of arriving international passengers are detained for intrusive luggage inspection. The inspection facility can handle 600 passengers an hour without unreasonable delays for the travelers.
a. When international passengers arrive at the rate of 1,500 per hour, what is the expected number of passengers who will be detained for luggage inspection?
b. In the future, it is expected that as many as 4,000 international passengers will arrive per hour. When that occurs, what is the expected number of passengers who will be detained for luggage inspection?
c. Refer to part **b.** Find the approximate probability that more than 600 international passengers will be detained for luggage inspection. (This is also the probability that travelers will experience unreasonable luggage inspection delays.)

5.92 Waiting time at an emergency room. According to *Health Affairs* (Oct. 28, 2004), the median time a patient waits to see a doctor in a typical U.S. emergency room is 30 minutes. On a day when 150 patients visit the emergency room, what is the approximate probability that
a. More than half will wait more than 30 minutes?
b. More than 85 will wait more than 30 minutes?
c. More than 60, but fewer than 90, will wait more than 30 minutes?

5.6 The Exponential Distribution (Optional)

The length of time between emergency arrivals at a hospital, the length of time between breakdowns of manufacturing equipment, the length of time between catastrophic events (floods, earthquakes, etc.), and the distance traveled by a wildlife ecologist between sightings of an endangered species are all random phenomena that we might want to describe probabilistically. The length of time or the distance between occurrences of random events like these can often be described by the **exponential probability distribution.** For this reason, the exponential distribution is sometimes called the **waiting-time distribution.** The formula for the exponential probability distribution is shown in the following box, along with its mean and standard deviation.

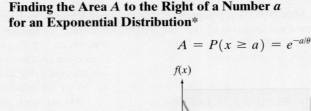

Probability Distribution for an Exponential Random Variable x

Probability density function: $f(x) = \dfrac{1}{\theta}e^{-x/\theta}$ $(x > 0)$

Mean: $\mu = \theta$ Standard deviation: $\sigma = \theta$

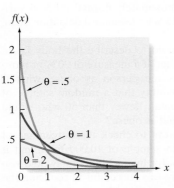

$f(x)$

$\theta = .5$

$\theta = 1$

$\theta = 2$

Figure 5.25
Exponential distributions

Unlike the normal distribution, which has a shape and location determined by the values of the two quantities μ and σ, the shape of the exponential distribution is governed by a single quantity: θ. Further, it is a probability distribution with the property that its mean equals its standard deviation. Exponential distributions corresponding to $\theta = .5, 1$, and 2 are shown in Figure 5.25.

To calculate probabilities of **exponential random variables,** we need to be able to find areas under the exponential probability distribution. Suppose we want to find the area A to the right of some number a, as shown in Figure 5.26. This area can be calculated by means of the formula shown in the box that follows. Use Table V in Appendix A or a calculator with an exponential function to find the value of $e^{-a/\theta}$ after substituting the appropriate numerical values for θ and a.

Finding the Area A to the Right of a Number a for an Exponential Distribution*

$$A = P(x \geq a) = e^{-a/\theta}$$

$f(x)$

A

Figure 5.26
The area A to the right of a number a for an exponential distribution

a

Example 5.14

Finding an Exponential Probability—Hospital Emergency Arrivals

$f(x)$

.6

.4

.2

A

0 1 2 3 4 5 6 7 8 x

Figure 5.27
Area to the right of $a = 5$ for Example 5.14

Problem Suppose the length of time (in hours) between emergency arrivals at a certain hospital is modeled as an exponential distribution with $\theta = 2$. What is the probability that more than 5 hours pass without an emergency arrival?

Solution The probability we want is the area A to the right of $a = 5$ in Figure 5.27. To find this probability, we use the area formula:

$$A = e^{-a/\theta} = e^{-(5/2)} = e^{-2.5}$$

Referring to Table V, we find that

$$A = e^{-2.5} = .082085$$

Our exponential model indicates that the probability that more than 5 hours pass between emergency arrivals is about .08 for this hospital.

Look Back The value $e^{-2.5}$ can also be found with a standard hand calculator. Or you can find the desired probability with a statistical software package.

Now Work Exercise 5.97

---

*For students with a knowledge of calculus, the highlighted area in Figure 5.26 corresponds to the integral
$$\int_a^\infty \frac{1}{\theta}e^{-x/\theta}\,dx = -e^{-10}\int_a^\infty = e^{-a/\theta}.$$

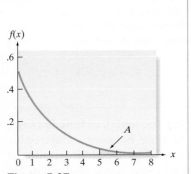

Example 5.15

The Mean and Variance of an Exponential Random Variable— Length of Life of a Microwave Oven

Problem A manufacturer of microwave ovens is trying to determine the length of warranty period it should attach to its magnetron tube, the most critical component in the oven. Preliminary testing has shown that the length of life (in years), x, of a magnetron tube has an exponential probability distribution with $\theta = 6.25$.

a. Find the mean and standard deviation of x.

b. Suppose a warranty period of five years is attached to the magnetron tube. What fraction of tubes must the manufacturer plan to replace, assuming that the exponential model with $\theta = 6.25$ is correct?

c. What should the warranty period be if the manufacturer wants to replace only 27.5% of the tubes?

d. Find the probability that the length of life of a magnetron tube will fall within the interval $\mu - 2\sigma$ to $\mu + 2\sigma$.

Solution

a. Since $\theta = \mu = \sigma$, both μ and σ equal 6.25.

b. To find the fraction of tubes that will have to be replaced before the five-year warranty period expires, we need to find the area between 0 and 5 under the distribution. This area, A, is shown in Figure 5.28. To find the required probability, we recall the formula

$$P(x > a) = e^{-a/\theta}$$

Using this formula, we find that

$$P(x > 5) = e^{-a/\theta} = e^{-5/6.25} = e^{-.80} = .449329$$

(See Table V.) To find the area A, we use the complementary relationship:

$$P(x \le 5) = 1 - P(x > 5) = 1 - .449329 = .550671$$

So approximately 55% of the magnetron tubes will have to be replaced during the five-year warranty period.

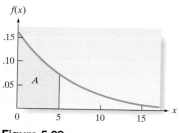
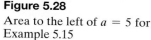

$f(x)$

.15
.10
.05
A

0   5   10   15 x

Figure 5.28
Area to the left of $a = 5$ for Example 5.15

c. Let W represent the warranty period in years. If the manufacturer only wants to replace 27.5% of the tubes, then

$$P(x < W) = .275$$

Using the rule of complements and the exponential formula, we have

$$P(x > W) = e^{-W/6.25} = (1 - .275) = .725$$

To solve for W, we take the natural logarithm of both sides of the equation:

$$ln(e^{-W/6.25}) = -W/6.25 = ln(.725)$$

Solving for W, we obtain

$$W = (-6.25)ln(.725) = (-6.25)(-.322) = 2.01$$

Consequently, if the manufacturer only wants to replace 27.5% of the tubes, it will need to specify a warranty period of $W = 2$ years.

d. We would expect the probability that the life of a magnetron tube, x, falls within the interval $\mu - 2\sigma$ to $\mu + 2\sigma$ to be quite large. A graph of the exponential distribution showing the interval from $\mu - 2\sigma$ to $\mu + 2\sigma$ is given in Figure 5.29. Since the point $\mu - 2\sigma$ lies below $x = 0$, we need to find only the area between $x = 0$ and $x = \mu + 2\sigma = 6.25 + 2(6.25) = 18.75$.

This area, P, which is highlighted in Figure 5.29, is

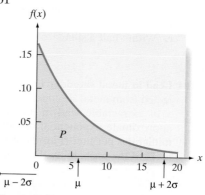

$f(x)$

.15
.10
.05
P

0   5   10   15   20 x

$\mu - 2\sigma$ μ $\mu + 2\sigma$

Figure 5.29
Area in the interval $\mu \pm 2\sigma$ for Example 5.15

$$P = 1 - P(x > 18.75) = 1 - e^{-18.75/\theta} = 1 - e^{-18.75/6.25} = 1 - e^{-3}$$

Using Table V or a calculator, we find that $e^{-3} = .049787$. Therefore, the probability that the life x of a magnetron tube will fall within the interval $\mu - 2\sigma$ to $\mu + 2\sigma$ is

$$P = 1 - e^{-3} = 1 - .049787 = .950213$$

Look Back You can see that this probability agrees well with the interpretation of a standard deviation given by the empirical rule (Table 2.7, p. 67), even though the probability distribution that we were given is not mound shaped. (It is strongly skewed to the right.)

Exercises 5.93–5.111

Understanding the Principles

5.93 What are the characteristics of an exponential random variable?

5.94 The exponential distribution is often called the _____ distribution. (Fill in the blank.)

Learning the Mechanics

5.95 The random variables x and y have exponential distributions with $\theta = 3$ and $\theta = 1$, respectively. Using Table V in Appendix A, carefully plot both distributions on the same set of axes.

5.96 Use Table V in Appendix A to determine the value of $e^{-a/\theta}$ for each of the following cases:
  a. $\theta = 1, a = 1$ **b.** $\theta = 1, a = 2.5$
  c. $\theta = .4, a = 3$ **d.** $\theta = .2, a = .3$

5.97 Suppose x has an exponential distribution with $\theta = 1$. Find
  [NW] the following probabilities:
  a. $P(x > 1)$ **b.** $P(x \le 3)$
  c. $P(x > 1.5)$ **d.** $P(x \le 5)$

5.98 Suppose x has an exponential distribution with $\theta = 2.5$. Find the following probabilities:
  a. $P(x \le 4)$ **b.** $P(x > 5)$
  c. $P(x \le 2)$ **d.** $P(x > 3)$

5.99 Suppose the random variable x has an exponential probability distribution with $\theta = 1$. Find the mean and standard deviation of x. Find the probability that x will assume a value within the interval $\mu \pm 2\sigma$.

5.100 The random variable x can be adequately approximated by an exponential probability distribution with $\theta = 2$. Find the probability that x assumes a value
  a. More than three standard deviations from μ.
  b. Less than two standard deviations from μ.
  c. Within half a standard deviation of μ.

Applying the Concepts—Basic

5.101 Lead in metal shredder residue. On the basis of data collected from metal shredders across the nation, the amount x of extractable lead in metal shredder residue has an approximate exponential distribution with mean $\theta = 2.5$ milligrams per liter (Florida Shredder's Association).
  a. Find the probability that x is greater than 2 milligrams per liter.
  b. Find the probability that x is less than 5 milligrams per liter.

5.102 Critical-part failures in NASCAR vehicles. In NASCAR races such as the Daytona 500, 43 drivers start the race;

however, about 10% of the cars do not finish due to the failure of critical parts. University of Portland professors conducted a study of critical-part failures from 36 NASCAR races (*The Sport Journal*, Winter 2007). The researchers discovered that the time (in hours) until the first critical-part failure is exponentially distributed with a mean of .10 hour.
  a. Find the probability that the time until the first critical-part failure is 1 hour or more.
  b. Find the probability that the time until the first critical-part failure is less than 30 minutes.

5.103 Phishing attacks to e-mail accounts. Refer to the *Chance* (Summer 2007) article on phishing attacks at a company, Exercise 2.43 (p. 49). Recall that *phishing* describes an attempt to extract personal/financial information through fraudulent e-mail. The company set up a publicized e-mail account—called a "fraud box"—which enabled employees to notify it if they suspected an e-mail phishing attack. If there is minimal or no collaboration or collusion from within the company, the interarrival times (i.e., the time between successive e-mail notifications, in seconds) have an approximate exponential distribution with a mean of 95 seconds.
  a. What is the probability of observing an interarrival time of at least 2 minutes?
  b. Data for a sample of 267 interarrival times are saved in the **PHISHING** file. Do the data appear to follow an exponential distribution?

5.104 Preventative maintenance tests. The optimal scheduling of preventative maintenance tests of some (but not all) of n independently operating components was developed in *Reliability Engineering and System Safety* (Jan. 2006). The time (in hours) between failures of a component was approximated by an exponentially distributed random variable with mean θ.
  a. Suppose $\theta = 1,000$ hours. Find the probability that the time between component failures ranges between 1,200 and 1,500 hours.
  b. Again, assume $\theta = 1,000$ hours. Find the probability that the time between component failures is at least 1,200 hours.
  c. Given that the time between failures is at least 1,200 hours, what is the probability that the time between failures is less than 1,500 hours?

Applying the Concepts—Intermediate

5.105 NHL overtime games. In the National Hockey League (NHL), games that are tied at the end of three periods are sent into "sudden-death" overtime. In overtime, the team to score the first goal wins. An analysis of NHL overtime games showed that the length of time elapsed before the winning goal is scored has an exponential distribution with mean 9.15 minutes (*Chance*, Winter 1995).

a. For a randomly selected overtime NHL game, find the probability that the winning goal is scored in 3 minutes or less.

b. In the NHL, each period (including overtime) lasts 20 minutes. If neither team scores a goal in overtime, the game is considered a tie. What is the probability of an NHL game ending in a tie?

5.106 Forest fragmentation study. Refer to the *Conservation Ecology* (Dec. 2003) study on the causes of fragmentation in 54 South American forests, presented in Exercise 2.156 (p. 91). Recall that the cause is classified as either anthropogenic or natural in origin. The anthropogenic fragmentation index (saved in the **FORFRAG** file) for the South American forests has an approximate exponential distribution with a mean of 23.

a. Find the probability that a South American forest has an anthropogenic fragmentation index between 20 and 40.

b. Find the probability that a South American forest has an anthropogenic fragmentation index below 50.

c. The natural fragmentation index (also saved in the **FORFRAG** file) does not have an approximate exponential distribution. Why?

5.107 Ship-to-shore transfer times. Lack of port facilities or shallow water may require cargo on a large ship to be transferred to a pier in smaller craft. The smaller craft may have to cycle back and forth from ship to shore many times. Researchers G. Horne (Center for Naval Analysis) and T. Irony (George Washington University) developed models of this transfer process that provide estimates of ship-to-shore transfer times (*Naval Research Logistics*, Vol. 41, 1994). They used an exponential distribution to model the time between arrivals of the smaller craft at the pier.

a. Assume that the mean time between arrivals at the pier is 17 minutes. Give the value of θ for this exponential distribution. Graph the distribution.

b. Suppose there is only one unloading zone at the pier available for the small craft to use. If the first craft docks at 10:00 A.M. and doesn't finish unloading until 10:15 A.M., what is the probability that the second craft will arrive at the unloading zone and have to wait before docking?

5.108 Reliability of CD-ROMs. In *Reliability Ques* (March 2004), the exponential distribution was used to model the lengths of life of CD-ROM drives in a two-drive system. The two CD-ROM drives operate independently, and at least one drive must be operating for the system to operate successfully. Both drives have a mean length of life of 25,000 hours.

a. The reliability $R(t)$ of a single CD-ROM drive is the probability that the life of the drive exceeds t hours. Give a formula for $R(t)$.

b. Use the result from part **a** to find the probability that the life of the single CD-ROM drive exceeds 8,760 hours (the number of hours of operation in a year).

c. The reliability $S(t)$ of the two-CD-ROM-drive system is the probability that the life of at least one drive exceeds t hours. Give a formula for S(t). [*Hint:* Use the rule of

complements and the fact that the two drives operate independently.]

d. Use the result from part **c** to find the probability that the two-drive CD-ROM system has a life whose length exceeds 8,760 hours.

e. Compare the probabilities you found in parts **b** and **d**.

5.109 Product failure behavior. An article in *Hotwire* (Dec. 2002) discussed the length of time till failure of a product produced at Hewlett-Packard. At the end of the product's lifetime, the time till failure is modeled using an exponential distribution with mean 500 thousand hours. In reliability jargon, this is known as the "wear-out" distribution for the product. During its normal (useful) life, assume the product's time till failure is uniformly distributed over the range 100 thousand to 1 million hours.

a. At the end of the product's lifetime, find the probability that the product fails before 700 thousand hours.

b. During its normal (useful) life, find the probability that the product fails before 700 thousand hours.

c. Show that the probability of the product failing before 830 thousand hours is approximately the same for both the normal (useful) life distribution and the wear-out distribution.

Applying the Concepts—Advanced

5.110 Acceptance sampling of a product. An essential tool in the monitoring of the quality of a manufactured product is *acceptance sampling*. An acceptance sampling plan involves knowing the distribution of the life length of the item produced and determining how many items to inspect from the manufacturing process. The *Journal of Applied Statistics* (Apr. 2010) demonstrated the use of the exponential distribution as a model for the life length x of an item (e.g., a bullet). The article also discussed the importance of using the median of the lifetime distribution as a measure of product quality, since half of the items in a manufactured lot will have life lengths exceeding the median. For an exponential distribution with mean θ, give an expression for the median of the distribution. [*Hint*: Your answer will be a function of θ.]

5.111 Length of life of a halogen bulb. For a certain type of halogen light bulb, an old bulb that has been in use for a while tends to have a longer life than a new bulb. Let x represent the life (in hours) of a new halogen light bulb, and assume that x has an exponential distribution with mean $\theta = 250$ hours. According to *Microelectronics and Reliability* (Jan. 1986), the "life" distribution of x is considered *new better than used* (NBU) if

$$P(x > a + b) \leq P(x > a)P(x > b)$$

Alternatively, a "life" distribution is considered *new worse than used* (NWU) if

$$P(x > a + b) \geq P(x > a)P(x > b)$$

a. Show that when $a = 300$ and $b = 200$, the exponential distribution is both NBU and NWU.

b. Choose any two positive numbers a and b, and repeat part **a**.

c. Show that, in general, for any positive a and b, the exponential distribution with mean θ is both NBU and NWU. Such a "life" distribution is said to be *new same as used*, or *memoryless*. Explain why.

CHAPTER NOTES

Key Terms

Note: Starred (*) terms are from the optional sections in this chapter.

Bell curve 231
Bell-shaped probability
    distribution 231
Continuous probability
    distribution 225
Continuous random variable 225
*Correction for continuity 252
*Exponential probability
    distribution 257
*Exponential random
    variables 258
Frequency function 225
Normal distribution 231

Normal probability
    plot 245
Normal random variable 231
Probability density
    function 225
Probability distribution 225
Randomness distribution 227
Standard normal
    distribution 233
Standard normal random
    variable 233
Uniform probability
    distribution 227
Uniform random
    variable 227
*Waiting-time distribution 257

Key Symbols

f(x) Probability distribution (density function) for continuous random variable x

e Constant used in normal probability distributions, $e = 2.71828\ldots$

π Constant used in normal probability distributions, $\pi = 3.1416\ldots$

θ Mean of exponential random variable

Key Ideas

Properties of Continuous Probability Distributions

1. $P(x = a) = 0$
2. $P(a < x < b)$ is area under curve between a and b

Methods for Assessing Normality

(1) *Histogram*

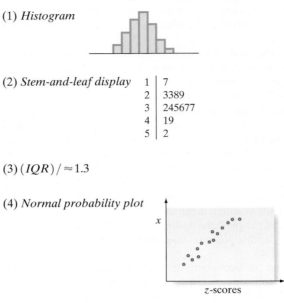

(2) *Stem-and-leaf display*

1	7
2	3389
3	245677
4	19
5	2

(3) $(IQR)/ \approx 1.3$

(4) *Normal probability plot*

x

z-scores

Normal Approximation to Binomial

x is binomial (n, p)

$$P(\le a) \approx P\{z < (a + .5) - \mu\}$$

Key Formulas

Random Variable	Prob. Dist'n	Mean	Variance
Uniform:	$f(x) = 1/(d-c)$ $(c \le x \le d)$	$(c + d)/2$	$(d-c)^2/12$
Normal:	$f(x) = \dfrac{1}{\sigma\sqrt{2\pi}}e^{-1/2[(x-\mu)/\sigma]^2}$	σ	σ^2
Standard Normal:	$f(z) = \dfrac{1}{\sqrt{2\pi}}e^{-1/2(z)^2}$ $z = (x - \mu)/\sigma$	$\mu = 0$	$\sigma^2 = 1$
*Exponential:	$f(x) = \dfrac{1}{\theta}e^{-x/\theta}$	$\mu = \theta$	$\sigma = \theta$

Guide to Selecting a Probability Distribution

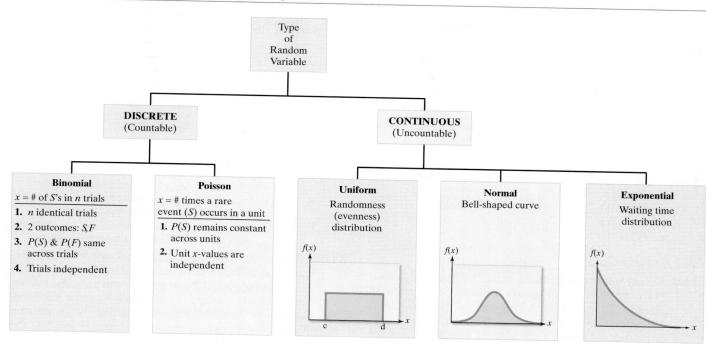

Supplementary Exercises 5.112–5.153

Note: Starred (*) exercises refer to the optional sections in this chapter.

Understanding the Principles

*5.112 Consider the continuous random variables that follow. Give the probability distribution (uniform, normal, or exponential) that is likely to best approximate the distribution of the random variable:
   a. Score on an IQ test
   b. Time (in minutes) waiting in line at a supermarket checkout counter
   c. Amount of liquid (in ounces) dispensed into a can of soda
   d. Difference between SAT scores for tests taken at two different times

*5.113 Identify the type of continuous random variable—uniform, normal, or exponential—described by each of the following probability density functions:
   a. $f(x) = (e^{-x/7})/7, x > 0$
   b. $f(x) = 1/20, 5 < x < 25$
   c. $f(x) = \dfrac{e^{-.5[(x-10)/5]^2}}{5\sqrt{2\pi}}$

Learning the Mechanics

5.114 Assume that x is a random variable best described by a uniform distribution with $c = 40$ and $d = 70$.
   a. Find $f(x)$.
   b. Find the mean and standard deviation of x.
   c. Graph the probability distribution for x, and locate its mean and the interval $\mu \pm 2\sigma$ on the graph.
   d. Find $P(x \le 45)$.

   e. Find $(P \ge 58)$.
   f. Find $P(x \le 100)$.
   g. Find $P(\mu - \sigma \le x \le \mu + \sigma)$.
   h. Find $P(x > 60)$.

5.115 Find the following probabilities for the standard normal random variable z:
   a. $P(z \le 2.1)$
   b. $P(z \ge 2.1)$
   c. $P(z \ge -1.65)$
   d. $P(-2.13 \le z \le -.41)$
   e. $P(-1.45 \le z \le 2.15)$
   f. $P(z \le -1.43)$

5.116 Find a z-score, say, z_0, such that
   a. $P(z \le z_0) = .8708$
   b. $P(z \ge z_0) = .0526$
   c. $P(z \le z_0) = .5$
   d. $P(-z_0 \le z \le z_0) = .8164$
   e. $P(z \ge z_0) = .8023$
   f. $P(z \ge z_0) = .0041$

5.117 The random variable x has a normal distribution with $\mu = 70$ and $\sigma = 10$. Find the following probabilities:
   a. $P(x \le 75)$
   b. $P(x \ge 90)$
   c. $P(60 \le x \le 75)$
   d. $P(x > 75)$
   e. $P(x = 75)$
   f. $P(x \le 95)$

5.118 The random variable x has a normal distribution with $\mu = 40$ and $\sigma^2 = 36$. Find a value of x, say, x_0, such that
   a. $P(x \ge x_0) = .5$
   b. $P(x \le x_0) = .9911$

c. $P(x \le x_0) = .0028$
d. $P(x \ge x_0) = .0228$
e. $P(x \le x_0) = .1003$
f. $P(x \ge x_0) = .7995$

*5.119 Assume that x is a binomial random variable with $n = 100$ and $p = .5$. Use the normal probability distribution to approximate the following probabilities:
a. $P(x \le 48)$
b. $P(50 \le x \le 65)$
c. $P(x \ge 70)$
d. $P(55 \le x \le 58)$
e. $P(x = 62)$
f. $P(x \le 49 \text{ or } x \ge 72)$

*5.120 Assume that x has an exponential distribution with $\theta = 3$. Find
a. $P(x \le 1)$
b. $P(x > 1)$
c. $P(x = 1)$
d. $P(x \le 6)$
e. $P(2 \le x \le 10)$

Applying the Concepts—Basic

*5.121 **Galaxy velocity study.** Refer to the *Astronomical Journal* (July 1995) study of galaxy velocities, presented in Exercise 2.186 (p. 101). A histogram of the velocities of 103 galaxies located in a particular cluster named A2142 is reproduced below. Comment on whether or not the galaxy velocities are approximately normally distributed.

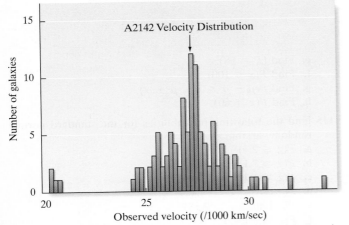

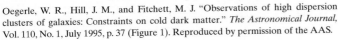

Oegerle, W. R., Hill, J. M., and Fitchett, M. J. "Observations of high dispersion clusters of galaxies: Constraints on cold dark matter." *The Astronomical Journal*, Vol. 110, No. 1, July 1995, p. 37 (Figure 1). Reproduced by permission of the AAS.

5.122 **Trajectory of an electrical circuit.** Researchers at the University of California–Berkeley have designed, built, and tested a switched-capacitor circuit for generating random signals (*International Journal of Circuit Theory and Applications*, May–June 1990). The circuit's trajectory was shown to be uniformly distributed on the interval $(0, 1)$.
a. Give the mean and variance of the circuit's trajectory.
b. Compute the probability that the trajectory falls between .2 and .4.
c. Would you expect to observe a trajectory that exceeds .995? Explain.

5.123 **Passing the FCAT math test.** All Florida high schools require their students to demonstrate competence in mathematics by scoring 70% or above on the FCAT mathematics achievement test. The FCAT math scores of those students taking the test for the first time are normally distributed with a mean of 77% and a standard deviation

of 7.3%. What percentage of students who take the test for the first time will pass it?

5.124 **Improving SAT scores.** Refer to the *Chance* (Winter 2001) study of students who paid a private tutor to help them improve their SAT scores, presented in Exercise 2.105 (p. 73). The table summarizing the changes in both the SAT-Mathematics and SAT-Verbal scores for these students is reproduced here. Assume that both distributions of SAT score changes are approximately normal.

	SAT-Math	SAT-Verbal
Mean change in score	19	7
Standard deviation of changes in score	65	49

a. What is the probability that a student increases his or her score on the SAT-Math test by at least 50 points?
b. What is the probability that a student increases his or her score on the SAT-Verbal test by at least 50 points?

5.125 **Alkalinity of river water.** The alkalinity level of water specimens collected from the Han River in Seoul, Korea, has a mean of 50 milligrams per liter and a standard deviation of 3.2 milligrams per liter (*Environmental Science & Engineering*, Sept., 1, 2000). Assume that the distribution of alkalinity levels is approximately normal, and find the probability that a water specimen collected from the river has an alkalinity level
a. exceeding 45 milligrams per liter.
b. below 55 milligrams per liter.
c. between 51 and 52 milligrams per liter.

5.126 **Weights of contaminated fish.** Refer to the U.S. Army Corps of Engineers data on contaminated fish in the Tennessee River, presented in Example 1.4 (p. 10). Recall that one of the variables measured for each captured fish is weight (in grams). The weights in the data set (saved in the **DDT** file) can be shown to be approximately normally distributed with a mean of 1,050 grams and a standard deviation of 375 grams. An observation is randomly selected from the data and the fish weight is observed.
a. Find the probability that the weight will fall between 1,000 and 1,400 grams.
b. Find the probability that the weight will fall between 800 and 1,000 grams.
c. Find the probability that the weight will be less than 1,750 grams.
d. Find the probability that the weight will exceed 500 grams. .9292
e. Ninety-five percent of the captured fish have a weight below what value? 1,666.875
f. Another variable saved in the **DDT** file is the amount of DDT (in parts per million) detected in each fish captured. Determine whether the DDT level is approximately normal.

*5.127 **Sickle-cell anemia.** Eight percent of the African-American population is known to carry the trait for sickle-cell anemia (Sickle Cell-Information Center, Atlanta, GA). If 1,000 African-Americans are sampled at random, what is the approximate probability that
a. More than 175 carry the trait?
b. Fewer than 140 carry the trait?

5.128 **Where will you get your next pet?** In Exercise 4.62 (p. 205), you learned that 50% of pet owners will get their next dog or cat from a Shelter (*USA Today*, May 12,

2010). In a random sample of 200 pet owners, let x be the number who will get their next dog or cat from a shelter.

a. Find the mean of x.

b. Find the standard deviation of x.

c. Find the z-score for the value $x = 110$.

d. Find the approximate probability that x is less than or equal to 110.

5.129 Tropical island temperatures. Records indicate that the daily high January temperatures on a tropical island tend to have a uniform distribution over the interval from 75°F to 90°F. A tourist arrives on the island on a randomly selected day in January.

a. What is the probability that the temperature will be above 80°F?

b. What is the probability that the temperature will be between 80°F and 85°F?

c. What is the expected temperature?

*5.130 Hospital patient interarrival times. The length of time between arrivals at a hospital clinic has an approximately exponential probability distribution. Suppose the mean time between arrivals for patients at a clinic is 4 minutes.

a. What is the probability that a particular interarrival time (the time between the arrival of two patients) is less than 1 minute?

b. What is the probability that the next four interarrival times are all less than 1 minute?

c. What is the probability that an interarrival time will exceed 10 minutes?

5.131 Dart-throwing errors. How accurate are you at the game of darts? Researchers at Iowa State University attempted to develop a probability model for dart throws (*Chance*, Summer 1997). For each of 590 throws made at certain targets on a dart board, the distance from the dart to the target point was measured (to the nearest millimeter). Descriptive statistics for the distances from the target for the 590 throws are as follows:

$$\bar{x} = 24.4 \text{ mm}$$
$$s = 12.8 \text{ mm}$$
$$Q_L = 14 \text{ mm}$$
$$Q_U = 34 \text{ mm}$$

Use this information to decide whether the error distribution is approximately normal.

Applying the Concepts—Intermediate

5.132 Marine losses for an oil company. The frequency distribution shown in the accompanying table depicts the property and marine losses incurred by a large oil company over a two-year period. In the insurance business, each "loss" interval is called a *layer*. *Research Review* (Summer 1998) demonstrated that analysts often treat the actual loss value within a layer as a uniform random variable.

Layer	Marine Losses ($ thousands)	Relative Frequency
1	0–10	.927
2	10–50	.053
3	50–100	.010
4	100–250	.006
5	250–500	.003
6	500–1,000	.001
7	1,000–2,500	.000

Based on Cozzolino, J. M., and Mikola, P. J. "Applications of the piecewise constant Pareto distribution." *Research Review*, Summer 1998, pp. 39-59.

a. Use the uniform distribution to find the mean loss amount in layer 2.

b. Use the uniform distribution to find the mean loss amount in layer 6.

c. If a loss occurs in layer 2, what is the probability that it exceeds $30,000?

d. If a loss occurs in layer 6, what is the probability that it is between $750,000 and $800,000?

5.133 Visually impaired students. The *Journal of Visual Impairment & Blindness* (May–June 1997) published a study of the lifestyles of visually impaired students. Using diaries, the students kept track of several variables, including number of hours of sleep obtained in a typical day. These visually impaired students had a mean of 9.06 hours and a standard deviation of 2.11 hours. Assume that the distribution of the number of hours of sleep for this group of students is approximately normal.

a. Find the probability that a visually impaired student obtains less than 6 hours of sleep on a typical day.

b. Find the probability that a visually impaired student gets between 8 and 10 hours of sleep on a typical day.

c. Twenty percent of all visually impaired students obtain less than how many hours of sleep on a typical day?

5.134 Comparison of exam scores: red versus blue exam. Refer to the *Teaching Psychology* (May 1998) study of how external clues influence performance, presented in Exercise 2.122 (p. 77). Recall that two different forms of a midterm psychology examination were given, one printed on blue paper and the other on red paper. Grading only the difficult questions, the researchers found that scores on the blue exam had a distribution with a mean of 53% and a standard deviation of 15%, while scores on the red exam had a distribution with a mean of 39% and a standard deviation of 12%. Assuming that both distributions are approximately normal, on which exam is a student more likely to score below 20% on the difficult questions, the blue one or the red one? (Compare your answer with that of Exercise 2.122**c.**)

5.135 Waiting for an elevator. The manager of a large department store with three floors reports that the time a customer on the second floor must wait for an elevator has a uniform distribution ranging from 0 to 4 minutes. If it takes the elevator 15 seconds to go from floor to floor, find the probability that a hurried customer can reach the first floor in less than 1.5 minutes after pushing the second-floor elevator button.

5.136 Length of gestation for pregnant women. On the basis of data from the National Center for Health Statistics, N. Wetzel used the normal distribution to model the length of gestation for pregnant U.S. women (*Chance*, Spring 2001). Gestation has a mean length of 280 days with a standard deviation of 20 days.

a. Find the probability that the length of gestation is between 275.5 and 276.5 days. (This estimate is the probability that a woman has her baby 4 days earlier than the "average" due date.)

b. Find the probability that the length of gestation is between 258.5 and 259.5 days. (This estimate is the probability that a woman has her baby 21 days earlier than the "average" due date.)

c. Find the probability that the length of gestation is between 254.5 and 255.5 days. (This estimate is the probability that a woman has her baby 25 days earlier than the "average" due date.)

d. The *Chance* article referenced a newspaper story about three sisters who all gave birth on the same day (March 11, 1998). Karralee had her baby 4 days early, Marrianne had her baby 21 days early, and Jennifer had her baby 25 days early. Use the results from parts **a** − **c** to estimate the probability that three women have their babies 4, 21, and 25 days early, respectively. Assume that the births are independent events.

5.137 Galaxy velocity study. Recall *The Astronomical Journal* (July 1995) study of galaxy velocity, Exercise 5.121 (p. 264). The observed velocity of a galaxy located in galaxy cluster A2142 was found to have a normal distribution with a mean of 27,117 kilometers per second (km/s) and a standard deviation of 1,280 km/s. A galaxy with a velocity of 24,350 km/s is observed. Comment on the likelihood of this galaxy being located in cluster A2142.

*5.138 Parents who condone spanking. In Exercise 4.128 (p. 219), you learned that 60% of parents with young children condone spanking their child as a regular form of punishment (*Tampa Tribune*, Oct. 5, 2000). A child psychologist with 150 parent clients claims that no more than 20 of the parents condone spanking. Do you believe this claim? Explain.

*5.139 Fungi in beech forest trees. Refer to the *Applied Ecology and Environmental Research* (Vol. 1, 2003) study of beech trees damaged by fungi, presented in Exercise 4.136 (p. 220). Recall that the researchers found that 25% of the beech trees in east central Europe have been damaged by fungi. In an east central European forest with 200 beech trees, how likely is it that more than half of the trees have been damaged by fungi?

5.140 Mile run times. A physical-fitness association is including the mile run in its secondary-school fitness test for students. The time for this event is approximately normally distributed with a mean of 450 seconds and a standard deviation of 40 seconds. If the association wants to designate the fastest 10% of secondary-school students as "excellent," what time should the association set for this criterion?

5.141 Wear-out of used display panels. Wear-out failure time of electronic components is often assumed to have a normal distribution. Can the normal distribution be applied to the wear-out of used manufactured products, such as colored display panels? A lot of 50 used display panels was purchased by an outlet store. Each panel displays 12 to 18 color characters. Prior to acquisition, the panels had been used for about one-third of their expected lifetimes. The data in the accompanying table (saved in the **PANELFAIL** file) give the failure times (in years) of the 50 used panels. Use the techniques of this section to determine whether the used panel wear-out times are approximately normally distributed.

5.142 Fitness of cardiac patients. The physical fitness of a patient is often measured by the patient's maximum oxygen

uptake (recorded in milliliters per kilogram, ml/kg). The mean maximum oxygen uptake for cardiac patients who regularly participate in sports or exercise programs was found to be 24.1, with a standard deviation of 6.30 (*Adapted Physical Activity Quarterly*, Oct. 1997). Assume that this distribution is approximately normal.

a. What is the probability that a cardiac patient who regularly participates in sports has a maximum oxygen uptake of at least 20 ml/kg?

b. What is the probability that a cardiac patient who regularly exercises has a maximum oxygen uptake of 10.5 ml/kg or lower?

c. Consider a cardiac patient with a maximum oxygen uptake of 10.5. Is it likely that this patient participates regularly in sports or exercise programs? Explain.

5.143 Forest development following wildfires. *Ecological Applications* (May 1995) published a study on the development of forests following wildfires in the Pacific Northwest. One variable of interest to the researcher was tree diameter at breast height 110 years after the fire.

a. The population of Douglas fir trees was shown to have an approximately normal diameter distribution with $\mu = 50$ centimeters (cm) and $\sigma = 12$ cm. Find the diameter d such that 30% of the Douglas fir trees in the population have diameters that exceed d.

b. Another species of tree, western hemlock, was found to have a breast height diameter distribution that resembled an exponential distribution with $\theta = 30$ centimeters. Find the probability that a western hemlock tree growing in the forest damaged by wildfire 110 years ago has a diameter that exceeds 25 centimeters.

5.144 Sedimentary deposits in reservoirs. Geologists have successfully used statistical models to evaluate the nature of sedimentary deposits (called *facies*) in reservoirs. One of the models' key parameters is the proportion P of facies bodies in a reservoir. An article in *Mathematical Geology* (Apr. 1995) demonstrated that the number of facies bodies that must be sampled to estimate P satisfactorily is approximately normally distributed with $\mu = 99$ and $\sigma = 4.3$. How many facies bodies are required to estimate P satisfactorily for 99% of the reservoirs evaluated?

5.145 Normal curve approximation. A. K. Shah published a simple approximation for areas under the normal curve in *The American Statistician* (Feb. 1985). Shah showed that the area A under the standard normal curve between 0 and z is

$$A \approx \begin{cases} z(4.4 - z)/10 & \text{for } 0 \le z \le 2.2 \\ .49 & \text{for } 2.2 < z < 2.6 \\ .50 & \text{for } z \ge 2.6 \end{cases}$$

a. Use Shah's approximation to find
  i. $P(0 < z < 1.2)$
  ii. $P(0 < z < 2.5)$
  iii. $P(z > .8)$
  iv. $P(z < 1.0)$

Data for Exercise 5.141

0.01	1.21	1.71	2.30	2.96	0.19	1.22	1.75	2.30	2.98	0.51
1.24	1.77	2.41	3.19	0.57	1.48	1.79	2.44	3.25	0.70	1.54
1.88	2.57	3.31	0.73	1.59	1.90	2.61	1.19	0.75	1.61	1.93
2.62	3.50	0.75	1.61	2.01	2.72	3.50	1.11	1.62	2.16	2.76
3.50	1.16	1.62	2.18	2.84	3.50					

Based on Irony, T. Z., Lauretto, M., Pereira, C., and Stern, J. M. "A Weibull Wearout Test: Full Bayesian Approach," paper presented at *Mathematical Sciences Colloquium*, Binghamton University, Binghamton, UK, December 2001.

b. Find the exact probabilities in part **a.**

c. Shah showed that the approximation has a maximum absolute error of .0052. Verify this for the approximations in part **a.**

*5.146 Spruce budworm infestation. An infestation of a certain species of caterpillar called the spruce budworm can cause extensive damage to the timberlands of the northern United States. It is known that an outbreak of this type of infestation occurs, on the average, every 30 years. Assuming that this phenomenon obeys an exponential probability law, what is the probability that catastrophic outbreaks of spruce budworm infestation will occur within six years of each other?

5.147 Maximum time to take a test. A professor believes that if a class is allowed to work on an examination as long as desired, the times spent by the students would be approximately normal with mean 40 minutes and standard deviation 6 minutes. About how long should be allotted for the examination if the professor wants almost all (say, 97.5%) of the class to finish?

*5.148 Machine repair times. An article in *IEEE Transactions* (Mar. 1990) gave an example of a flexible manufacturing system with four machines operating independently. The repair rates for the machines (i.e., the time, in hours, it takes to repair a failed machine) are exponentially distributed with means $\mu_1 = 1, \mu_2 = 2, \mu_3 = .5,$ and $\mu_4 = .5,$ respectively.

a. Find the probability that the repair time for machine 1 exceeds 1 hour.

b. Repeat part **a** for machine 2.

c. Repeat part **a** for machines 3 and 4.

d. If all four machines fail simultaneously, find the probability that the repair time for the entire system exceeds 1 hour.

Applying the Concepts—Advanced

5.149 Dye discharged in paint. A machine used to regulate the amount of dye dispensed for mixing shades of paint can be set so that it discharges an average of μ milliliters (mL) of dye per can of paint. The amount of dye discharged is known to have a normal distribution with a standard deviation of .4 mL. If more than 6 mL of dye are discharged when making a certain shade of blue paint, the shade is unacceptable. Determine the setting for μ so that only 1% of the cans of paint will be unacceptable.

*5.150 Accidents at a plant. The number of serious accidents in a manufacturing plant has (approximately) a Poisson probability distribution with a mean of two serious accidents per month. It can be shown that if x, the number of events per unit time, has a Poisson distribution with mean λ, then the time between two successive events has an exponential probability distribution with mean $\theta = 1/\lambda$.

a. If an accident occurs today, what is the probability that the next serious accident will not occur within the next month?

b. What is the probability that more than one serious accident will occur within the next month?

5.151 Water retention of soil cores. A team of soil scientists investigated the water retention properties of soil cores sampled from an uncropped field consisting of silt loam (*Soil Science*, Jan. 1995). At a pressure of .1 megapascal (MPa), the water content of the soil (measured in cubic meters of water per cubic meter of soil) was determined to be approximately normally distributed with $\mu = .27$ and $\sigma = .04$. In addition to water content readings at a pressure of .1 MPa, measurements were obtained at pressures 0, .005, .01, .03, and 1.5 MPa. Consider a soil core with a water content reading of .14. Is it likely that this reading was obtained at a pressure of .1 MPa? Explain.

Critical Thinking Challenges

*5.152 Weights of corn chip bags. The net weight per bag of a certain brand of corn chips is listed as 10 ounces. The weight of chips actually dispensed in each bag by an automated machine, when operating to specifications, is a normal random variable with mean 10.5 ounces and standard deviation .25 ounce. A quality control inspector will randomly select 1,500 bags from the automated process and measure the weight of each bag precisely. Ideally, all of the bags should contain at least 10 ounces of corn chips. However, if at least 97% of the bags contain 10 ounces or more, the process is considered "under control" and no adjustments to the machine are deemed necessary. Assess the likelihood that the quality control inspector will find fewer than 97% of the bags with 10 ounces or more even if the process is "under control."

5.153 IQs and *The Bell Curve*. In their controversial book *The Bell Curve* (Free Press, 1994), Professors Richard J. Herrnstein (a Harvard psychologist who died while the book was in production) and Charles Murray (a political scientist at MIT) explored, as the subtitle states, "intelligence and class structure in American life." *The Bell Curve* employs statistical analyses heavily in an attempt to support the authors' positions. Since the book's publication, many expert statisticians have raised doubts about the authors' statistical methods and the inferences drawn from them. (See, for example, "Wringing *The Bell Curve*: A cautionary tale about the relationships among race, genes, and IQ," *Chance*, Summer 1995.) One of the many controversies sparked by the book is the authors' tenet that level of intelligence (or lack thereof) is a cause of a wide range of intractable social problems, including constrained economic mobility. The measure of intelligence chosen by the authors is the well-known intelligent quotient (IQ). Numerous tests have been developed to measure IQ; Herrnstein and Murray use the Armed Forces Qualification Test (AFQT), originally designed to measure the cognitive ability of military recruits. Psychologists traditionally treat IQ as a random variable having a normal distribution with mean $\mu = 100$ and standard deviation $\sigma = 15$.

In their book, Herrnstein and Murray refer to five cognitive classes of people defined by percentiles of the normal distribution. Class I ("very bright") consists of those with IQs above the 95th percentile; Class II ("bright") are those with IQs between the 75th and 95th percentiles; Class III ("normal") includes IQs between the 25th and 75th percentiles; Class IV ("dull") are those with IQs between the 5th and 25th percentiles; and Class V ("very dull") are IQs below the 5th percentile.

a. Assuming that the distribution of IQ is accurately represented by the normal curve, determine the proportion of people with IQs in each of the five cognitive classes defined by Herrnstein and Murray.

b. Although Herrnstein and Murray define the cognitive classes in terms of percentiles, they stress that IQ scores should be compared with z-scores, not percentiles. In other words, it is more informative to give the difference in z-scores for two IQ scores than it is to give the difference in percentiles. Do you agree?

c. Researchers have found that scores on many intelligence tests are decidedly nonnormal. Some distributions are skewed toward higher scores, others toward lower scores. How would the proportions in the five cognitive classes defined in part **a** differ for an IQ distribution that is skewed right? Skewed left?

Activity   The Normal Approximation to the Binomial

For large values of n, the computational effort involved in working with the binomial probability distribution is considerable. Fortunately, in many instances the normal distribution provides a good approximation to the binomial distribution. This activity was designed to enable you to demonstrate to yourself how well the normal distribution approximates the binomial distribution.

a. Let the random variable x have a binomial probability with $n = 10$ and $p = .5$. Using the binomial distribution, find the probability that x takes on a value in each of the following intervals: $\mu \pm \sigma$, $\mu \pm 2\sigma$, and $\mu \pm 3\sigma$.

b. Find the probabilities requested in part **a** by using a normal approximation to the given binomial distribution.

c. Find the magnitude of the difference between each of the three probabilities as determined by the binomial distribution and by the normal approximation.

d. Letting x have a binomial distribution with $n = 20$ and $p = .5$, repeat parts **a**, **b**, and **c**. Notice that the probability estimates provided by the normal distribution are more accurate for $n = 20$ than for $n = 10$.

e. Letting x have a binomial distribution with $n = 20$ and $p = .01$, repeat parts **a**, **b**, and **c**. Notice that the probability estimates provided by the normal distribution are poor in this case. Explain why this occurs.

References

Hogg, R. V., McKean, J. W., and Craig, A. T. *Introduction to Mathematical Statistics*, 6th ed. Upper Saddle River, NJ: Prentice Hall, 2005.

Lindgren, B. W. *Statistical Theory*, 4th ed. New York: Chapman & Hall, 1993.

Mood, A. M., Graybill, F. A., and Boes, D. C. *Introduction to the Theory of Statistics*, 3rd ed. New York: McGraw-Hill, 1974.

Ramsey, P. P., and Ramsey, P. H. "Simple tests of normality in small samples." *Journal of Quality Technology*, Vol. 22, 1990.

Ross, S. M. *Stochastic Processes*, 2nd ed. New York: Wiley, 1996.

Wackerly, D., Mendenhall, W., and Scheaffer, R. L. *Mathematical Statistics with Applications*, 7th ed. North Scituate, MA: Duxbury, 2008.

Winkler, R. L., and Hays, W. *Statistics: Probability, Inference, and Decision*, 2nd ed. New York: Holt, Rinehart and Winston, 1975.

USING TECHNOLOGY

MINITAB: Continuous Random Variable Probabilities and Normal Probability Plots

Continuous Probabilities

Step 1 Select the "Calc" button on the MINITAB menu bar, click on "Probability Distributions," and then finally select the Continuous distribution of your choice (e.g., "Normal"), as shown in Figure 5.M.1.

Step 2 Select either "Probability" or "Cumulative probability" on the resulting dialog box.

Step 3 Specify the parameters of the distribution (e.g., the range for a uniform distributions, μ and σ for a normal distributions and the mean θ for an exponential distribution).

Step 4 Specify the value x of interest in the "Input constant" box.

Step 5 Click "OK". The probability for the value of x will appear on the MINITAB session window.

Figure 5.M.1
MINITAB menu options for continuous probabilities

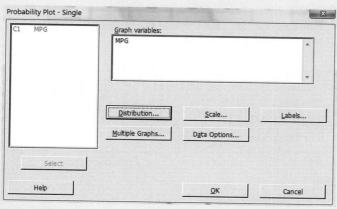

Figure 5.M.2
MINITAB normal distribution dialog box

[*Note*: Figure 5.M.2 gives the specifications for finding $P(x \leq 23)$ in a normal distribution with $\mu = 24.5$ and $\sigma = 1$.]

Normal Probability Plot

Step 1 Select "Graph" on the MINITAB menu bar, and then click on "Probability Plot," as shown in Figure 5.M.3.

Figure 5.M.3
MINITAB options for a normal probability plot

Figure 5.M.4
MINITAB normal probability plot dialog box

Step 2 Select "Single" (for one variable) on the next box, and the dialog box will appear as shown in Figure 5.M.4.

Step 3 Specify the variable of interest in the "Graph variables" box, and then click the "Distribution" button and select the "Normal" option. Click "OK" to return to the Probability Plot dialog box.

Step 4 Click "OK" to generate the normal probability plot.

TI-83/TI-84 Plus Graphing Calculator: Normal Random Variable and Normal Probability Plots

Graphing the Area under the Standard Normal Curve

Step 1 *Turn off all plots*
- Press **2nd PRGM** and select **1:ClrDraw**
- Press **ENTER,** and 'Done' will appear on the screen
- Press **2nd Y=** and select **4:PlotsOff**
- Press **ENTER,** and 'Done' will appear on the screen

Step 2 *Set the viewing window (Recall that almost all of the area under the standard normal curve falls between −5 and 5. A height of 0.5 is a good choice for Ymax.)*

Note: When entering a negative number, be sure to use the negative sign **(−)**, and not the minus sign.
- Set **Xmin** = −5
- **Xmax** = 5
- **Xscl** = 1
- **Ymin** = 0
- **Ymax** = .5
- **Yscl** = 0
- **Xres** = 1

Step 3 *View graph*
- Press **2nd VARS**
- Arrow right to **DRAW**
- Press **ENTER** to select **1:ShadeNorm(**
- Enter your lower limit (e.g., −5)
- Press **COMMA**
- Enter your upper limit (e.g., 1.5)
- Press **)**
- Press **ENTER**

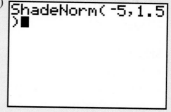

  The graph will be displayed along with the area, lower limit, and upper limit.

  Thus, $P(Z < 1.5) = .9332$

Finding Normal Probabilities without a Graph

To compute probabilities for a normal distribution, use the **normalcdf(** command. *Normalcdf* stands for "normal cumulative density function." This command is under the **DISTR**ibution menu and has the format **normalcdf(*lower limit, upper limit, mean, standard deviation*)**.

Step 1 *Find the probability*

- Press **2nd VARS** for **DISTR** and select **Normalcdf(**
- After **Normalcdf(**, type in the lower limit
- Press **COMMA**
- Enter the upper limit
- Press **COMMA**
- Enter the mean
- Press **COMMA**
- Enter the standard deviation
- Press **)**
- Press **ENTER**

  The probability will be displayed on the screen.

  Example What is $P(x < 115)$ for a normal distribution with $\mu = 100$ and $\sigma = 10$? In this example, the lower limit is $-\infty$, the upper limit is 115, the mean is 100, and the standard deviation is 10. To represent $-\infty$ on the calculator, enter **(–) 1**, press **2nd** and press the **COMMA** key for **EE**, and then press **99**. The screen appears as follows:

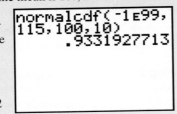

  Thus, $P(x < 115)$ *is* $= .9332$

Finding Normal Probabilities with a Graph

Step 1 *Turn off all plots*

- Press **Y =** and **CLEAR** all functions from the Y registers
- Press **2nd Y =** and select **4:PlotsOff**
- Press **ENTER ENTER**, and 'Done' will appear on the screen

Step 2 *Set the viewing window (These values depend on the mean and standard deviation of the data.)* Note: When entering a negative number, be sure to use the negative sign **(–),** not the minus sign.

- Press **WINDOW**
- Set **Xmin** $= \mu - 5\sigma$
- **Xmax** $= \mu + 5\sigma$
- **Xscl** $= \sigma$
- **Ymin** $= -.125/\sigma$
- **Ymax** $= .5/\sigma$
- **Yscl** $= 1$
- **Xres** $= 1$

Step 3 *View graph*

- Press **2nd VARS**
- **ARROW** right to **DRAW**
- Press **ENTER** to select **1:ShadeNorm(**
- Enter the lower limit
- Press **COMMA**
- Enter the upper limit
- Press **COMMA**
- Enter the mean

- Press **COMMA**
- Enter the standard deviation
- Press **)**
- Press **ENTER**

  The graph will be displayed along with the area, lower limit, and upper limit

  Example What is $P(x < 115)$ for a normal distribution with $\mu = 100$ and $\sigma = 10$? In this example, the lower limit is $-\infty$, the upper limit is 115, the mean is 100, and the standard deviation is 10. To represent $-\infty$ on the calculator, enter **(–) 1**, press **2nd** and press the **comma** key for **EE**, and then press **99**. The screens appear as follows:

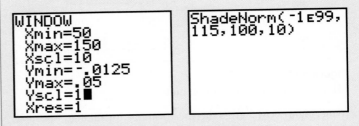

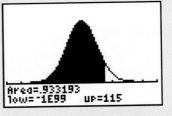

Graphing a Normal Probability Plot

Step 1 *Enter the data*

- Press **STAT** and select **1:Edit**

  Note: If the list already contains data, clear the old data. Use the up arrow to highlight **L1**

- Press **CLEAR ENTER**
- Use the **ARROW** and **ENTER** keys to enter the data set into **L1**

Step 2 *Set up the normal probability plot*

- Press **Y =** and **CLEAR** all functions from the Y registers
- Press **2nd** and press **Y =** for **STAT PLOT**
- Press **1** for **Plot 1**
- Set the cursor so that **ON** is flashing
- For **Type**, use the **ARROW** and **ENTER** keys to highlight and select the last graph in the bottom row
- For **Data List**, choose the column containing the data (in most cases, L1)

  (*Note*: Press **2nd 1** for **L1**)

- For **Data Axis**, choose **X** and press **ENTER**

Step 3 *View plot*

- Press **ZOOM 9**

  Your data will be displayed against the expected *z*-scores from a normal distribution. If you see a "generally" linear relationship, your data set is approximately normal.

6 Sampling Distributions

CONTENTS

Where We've Been

- Learned that the objective of most statistical investigations is inference—that is, making decisions about a population on the basis of information about a sample

- Discovered that sample statistics such as the sample mean and sample variance can be used to make decisions

- Found that probability distributions for random variables are used to construct theoretical models of populations

Where We're Going

- Establish that a sample statistic is a random variable with a probability distribution (6.1)

- Define a *sampling distribution* as the probability distribution of a sample statistic (6.1)

- Give two important properties of sampling distributions (6.2)

- Learn that the sampling distribution of the sample mean tends to be approximately normal (6.3)

271

Statistics IN Action The Insomnia Pill: Is It Effective?

More than 15 years ago, neuroscientists at the Massachusetts Institute of Technology (MIT) began experimenting with melatonin—a hormone secreted by the pineal gland in the brain—as a sleep-inducing hormone. Their original study, published in the *Proceedings of the National Academy of Sciences* (March 1994), brought encouraging news to insomniacs and travelers who suffer from jet lag: Melatonin was discovered to be effective in reducing sleep onset latency (the amount of time between a person lying down to sleep and the actual onset of stage one sleep). Furthermore, since the hormone is naturally produced, it is nonaddictive.

Since then, pharmaceutical companies that produce melatonin pills for treating insomnia have received negative reports on the efficacy of the drug. Many people don't think the melatonin pills work. Consequently, the MIT researchers undertook a follow-up study and published their results in *Sleep Medicine Reviews* (Feb. 2005). They reported that commercially available melatonin pills contain too high a dosage of the drug. When the melatonin receptors in the brain are exposed to too much of the hormone, they become unresponsive. The researchers' analysis of all previous studies on melatonin as a sleep inducer confirmed that, taken in small doses, melatonin is effective in reducing sleep onset latency.

In this Statistics in Action, our focus is on the original MIT study. Young male volunteers were given various doses of melatonin or a placebo (a dummy medication containing no melatonin). Then they were placed in a dark room at midday and told to close their eyes for 30 minutes. The variable of interest was sleep onset latency (in minutes).

According to the lead investigator, Professor Richard Wurtman, "Our volunteers fall asleep in 5 or 6 minutes on

melatonin, while those on placebo take about 15 minutes." Wurtman warned, however, that uncontrolled doses of melatonin could cause mood-altering side effects.

Now, consider a random sample of 40 young males, each of whom is given a dosage of the sleep-inducing hormone melatonin. The times taken (in minutes) to fall asleep for these 40 males are listed in Table SIA6.1 and saved in the **INSOMNIA** file. The researchers know that with the placebo (i.e., no hormone), the mean sleep onset latency is $\mu = 15$ minutes and the standard deviation is $\sigma = 10$ minutes. They want to use the data to make an inference about the true value of μ for those taking the melatonin. Specifically, the researchers want to know whether melatonin is an effective drug against insomnia.

In this chapter, a Statistics in Action Revisited example demonstrates how we can use one of the topics discussed in the chapter—the Central Limit Theorem—to make an inference about the effectiveness of melatonin as a sleep-inducing hormone.

Statistics IN Action Revisited

- Making an Inference about the Mean Sleep Onset Latency for Insomnia Pill Takers (p. 288)

Table SIA6.1	Times Taken (in Minutes) for 40 Male Volunteers to Fall Asleep								
7.6	2.1	1.4	1.5	3.6	17.0	14.9	4.4	4.7	20.1
7.7	2.4	8.1	1.5	10.3	1.3	3.7	2.5	3.4	10.7
2.9	1.5	15.9	3.0	1.9	8.5	6.1	4.5	2.2	2.6
7.0	6.4	2.8	2.8	22.8	1.5	4.6	2.0	6.3	3.2

[Note: These data are simulated sleep times based on summary information provided in the MIT study.]   ⊙ *Data Set:* INSOMNIA

In Chapters 4 and 5, we assumed that we knew the probability distribution of a random variable, and using this knowledge, we were able to compute the mean, variance, and probabilities associated with the random variable. However, in most practical applications, this information is not available. To illustrate, in Example 4.13 (p. 202), we calculated the probability that the binomial random variable x, the number of 20 polled voters who favored a certain mayoral candidate, assumed specific values. To do this, it was necessary to assume some value for p, the proportion of all voters who favored the candidate. Thus, for the purposes of illustration, we assumed that $p = .6$ when, in all likelihood, the exact value of p would be unknown. In fact, the probable purpose of taking the poll is to estimate p. Similarly, when we modeled the in-city gas mileage of a certain automobile model, we used the normal probability distribution with an *assumed* mean and standard deviation of 27 and 3 miles per gallon, respectively. In most situations, the true mean and standard deviation are unknown quantities that have to be estimated. Numerical quantities that describe probability distributions are called *parameters*. Thus, p, the probability of a success in a binomial experiment, and μ and σ, the mean and standard deviation, respectively, of a normal distribution, are examples of parameters.

> A **parameter** is a numerical descriptive measure of a population. Because it is based on the observations in the population, its value is almost always unknown.

We have also discussed the sample mean $\bar{x}$, sample variance s^2, sample standard deviation s, and the like, which are numerical descriptive measures calculated from the sample. (See Table 6.1 for a list of the statistics covered so far in this text.) We will often use the information contained in these *sample statistics* to make inferences about the parameters of a population.

> A **sample statistic** is a numerical descriptive measure of a sample. It is calculated from the observations in the sample.

Table 6.1	List of Population Parameters and Corresponding Sample Statistics	
	Population Parameter	Sample Statistic
Mean:	μ	$\bar{x}$
Variance:	σ^2	s^2
Standard deviation:	σ	s
Binomial proportion:	p	$\hat{p}$

Note that the term *statistic* refers to a *sample* quantity and the term *parameter* refers to a *population* quantity.

Before we can show you how to use sample statistics to make inferences about population parameters, we need to be able to evaluate their properties. Does one sample statistic contain more information than another about a population parameter? On what basis should we choose the "best" statistic for making inferences about a parameter? The purpose of this chapter is to answer these questions.

6.1 The Concept of a Sampling Distribution

If we want to estimate a parameter of a population—say, the population mean μ—we can use a number of sample statistics for our estimate. Two possibilities are the sample mean $\bar{x}$ and the sample median M. Which of these do you think will provide a better estimate of μ?

Before answering this question, consider the following example: Toss a fair die, and let x equal the number of dots showing on the up face. Suppose the die is tossed three times, producing the sample measurements 2, 2, 6. The sample mean is then $\bar{x} = 3.33$, and the sample median is $M = 2$. Since the population mean of x is $\mu = 3.5$, you can see that, for this sample of three measurements, the sample mean $\bar{x}$ provides an estimate that falls closer to μ than does the sample median (see Figure 6.1a). Now suppose we toss the die three more times and obtain the sample measurements 3, 4, 6. Then the mean and median of this sample are $\bar{x} = 4.33$ and $M = 4$, respectively. This time M is closer to μ. (See Figure 6.1b.)

This simple example illustrates an important point: Neither the sample mean nor the sample median will *always* fall closer to the population mean. Consequently, we

Figure 6.1

Comparing the sample mean ($\bar{x}$) and sample median (M) as estimators of the population mean (μ)

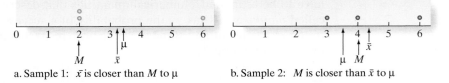

a. Sample 1: $\bar{x}$ is closer than M to μ b. Sample 2: M is closer than $\bar{x}$ to μ

cannot compare these two sample statistics or, in general, any two sample statistics on the basis of their performance with a single sample. Instead, we need to recognize that sample statistics are themselves random variables, because different samples can lead to different values for the sample statistics. As random variables, sample statistics must be judged and compared on the basis of their probability distributions (i.e., the *collection* of values and associated probabilities of each statistic that would be obtained if the sampling experiment were repeated a *very large number of times*). We will illustrate this concept with another example.

Suppose it is known that in a certain part of Canada the daily high temperature recorded for all past months of January has a mean $\mu = 10°F$ and a standard deviation $\sigma = 5°F$. Consider an experiment consisting of randomly selecting 25 daily high temperatures from the records of past months of January and calculating the sample mean $\bar{x}$. If this experiment were repeated a very large number of times, the value of $\bar{x}$ would vary from sample to sample. For example, the first sample of 25 temperature measurements might have a mean $\bar{x} = 9.8$, the second sample a mean $\bar{x} = 11.4$, the third sample a mean $\bar{x} = 10.5$, etc. If the sampling experiment were repeated a very large number of times, the resulting histogram of sample means would be approximately the probability distribution of $\bar{x}$. If $\bar{x}$ is a good estimator of μ, we would expect the values of $\bar{x}$ to cluster around μ as shown in Figure 6.2. This probability distribution is called a *sampling distribution*, because it is generated by repeating a sampling experiment a very large number of times.

> The **sampling distribution** of a sample statistic calculated from a sample of n temperature measurements is the probability distribution of the statistic.

In actual practice, the sampling distribution of a statistic is obtained mathematically or (at least approximately) by simulating the sample on a computer, using a procedure similar to that just described.

If $\bar{x}$ has been calculated from a sample of $n = 25$ measurements selected from a population with mean $\mu = 10$ and standard deviation $\sigma = 5$, the sampling distribution (Figure 6.2) provides information about the behavior of $\bar{x}$ in repeated sampling.

Figure 6.2

Sampling distribution for $\bar{x}$ based on a sample of $n = 25$ temperature measurements

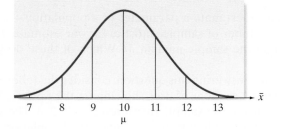

For example, the probability that you will draw a sample of 25 measurements and obtain a value of $\bar{x}$ in the interval $9 \le \bar{x} \le 10$ will be the area under the sampling distribution over that interval.

Since the properties of a statistic are typified by its sampling distribution, it follows that, to compare two sample statistics, you compare their sampling distributions.

For example, if you have two statistics, A and B, for estimating the same parameter (for purposes of illustration, suppose the parameter is the population variance σ^2), and if their sampling distributions are as shown in Figure 6.3, you would prefer statistic A over statistic B. You would do so because the sampling distribution for statistic A centers over σ^2 and has less spread (variation) than the sampling distribution for statistic B. Then, when you draw a single sample in a practical sampling situation, the probability is higher that statistic A will fall nearer σ^2.

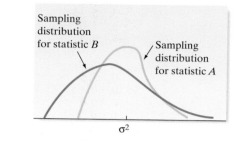

Figure 6.3

Two sampling distributions for estimating the population variance σ^2

Remember that, in practice, we will not know the numerical value of the unknown parameter σ^2, so we will not know whether statistic A or statistic B is closer to σ^2 for a particular sample. We have to rely on our knowledge of the theoretical sampling distributions to choose the best sample statistic and then use it sample after sample. The procedure for finding the sampling distribution for a statistic is demonstrated in Example 6.1.

Example 6.1

Finding a Sampling Distribution— Come-Out Roll in Craps

Problem Consider the popular casino game of craps, in which a player throws two dice and bets on the outcome (the sum total of the dots showing on the upper faces of the two dice). In Example 4.5 (p. 185), we looked at the possible outcomes of a $5 wager on the first toss (called the *come-out* roll). Recall that if the sum total of the dice is 7 or 11, the roller wins $5; if the total is a 2, 3, or 12, the roller loses $5 (i.e., the roller "wins" −$5); and, for any other total (4, 5, 6, 8, 9, or 10) no money is lost or won on that roll (i.e., the roller wins $0). Let x represent the result of the come-out roll wager (−$5, $0, or +$5). We showed in Example 4.5 that the actual probability distribution of x is:

Outcome of wager, x	−5	0	5
$p(x)$	1/9	6/9	2/9

Now, consider a random sample of $n = 3$ come-out rolls.

a. Find the sampling distribution of the sample mean, $\bar{x}$.

b. Find the sampling distribution of the sample median, M.

Solution The outcomes for every possible sample of $n = 3$ come-out rolls are listed in Table 6.2, along with the sample mean and median. The probability of each sample is obtained using the Multiplicative Rule. For example, the probability of the sample $(0, 0, 5)$ is $p(0) \cdot p(0) \cdot p(5) = (6/9)(6/9)(2/9) = 72/729 = .099$. The probability for each sample is also listed in Table 6.2. Note that the sum of these probabilities is equal to 1.

a. From Table 6.2, you can see that $\bar{x}$ can assume the values −5, −3.33, −1.67, 0, 1.67, 3.33, and 5. Because $\bar{x} = -5$ occurs only in one sample, $P(\bar{x} = -5) = 1/729$. Similarly, $\bar{x} = -3.33$ occurs in three samples, $(-5, -5, 0)$, $(-5, 0, -5)$ and, $(0, -5, -5)$. Therefore, $P(\bar{x} = -3.33) = 6/729 + 6/729 + 6/729 = 18/729$. Calculating the

Table 6.2   All Possible Samples of $n = 3$ Come-Out Rolls in Craps

Possible Samples	$\bar{x}$	M	Probability
$-5, -5, -5$	-5	-5	$(1/9)(1/9)(1/9) = 1/729$
$-5, -5, 0$	-3.33	-5	$(1/9)(1/9)(6/9) = 6/729$
$-5, -5, 5$	-1.67	-5	$(1/9)(1/9)(2/9) = 2/729$
$-5, \ 0, -5$	-3.33	-5	$(1/9)(6/9)(1/9) = 6/729$
$-5, \ 0, 0$	-1.67	0	$(1/9)(6/9)(6/9) = 36/729$
$-5, \ 0, 5$	0	0	$(1/9)(6/9)(2/9) = 12/729$
$-5, \ 5, -5$	-1.67	-5	$(1/9)(2/9)(1/9) = 2/729$
$-5, \ 5, 0$	0	0	$(1/9)(2/9)(6/9) = 12/729$
$-5, \ 5, 5$	1.67	5	$(1/9)(2/9)(2/9) = 4/729$
$0, -5, -5$	-3.33	-5	$(6/9)(1/9)(1/9) = 6/729$
$0, -5, 0$	-1.67	0	$(6/9)(1/9)(6/9) = 36/729$
$0, -5, 5$	0	0	$(6/9)(1/9)(2/9) = 12/729$
$0, \ 0, -5$	-1.67	0	$(6/9)(6/9)(1/9) = 36/729$
$0, \ 0, 0$	0	0	$(6/9)(6/9)(6/9) = 216/729$
$0, \ 0, 5$	1.67	0	$(6/9)(6/9)(2/9) = 72/729$
$0, \ 5, -5$	0	0	$(6/9)(2/9)(1/9) = 12/729$
$0, \ 5, 0$	1.67	0	$(6/9)(2/9)(6/9) = 72/729$
$0, \ 5, 5$	3.33	5	$(6/9)(2/9)(2/9) = 24/729$
$5, -5, -5$	-1.67	-5	$(2/9)(1/9)(1/9) = 2/729$
$5, -5, 0$	0	0	$(2/9)(1/9)(6/9) = 12/729$
$5, -5, 5$	1.67	5	$(2/9)(1/9)(2/9) = 4/729$
$5, \ 0, -5$	0	0	$(2/9)(6/9)(1/9) = 12/729$
$5, \ 0, 0$	1.67	0	$(2/9)(6/9)(6/9) = 72/729$
$5, \ 0, 5$	3.33	5	$(2/9)(6/9)(2/9) = 24/729$
$5, \ 5, -5$	1.67	5	$(2/9)(2/9)(1/9) = 4/729$
$5, \ 5, 0$	3.33	5	$(2/9)(2/9)(6/9) = 24/729$
$5, \ 5, 5$	5	5	$(2/9)(2/9)(2/9) = 8/729$

$$Sum = 729/729 = 1$$

probabilities of the remaining values of $\bar{x}$ and arranging them in a table, we obtain the following probability distribution:

$\bar{x}$	-5	-3.33	-1.67	0	1.67	3.33	5
$p(\bar{x})$	$1/729 =$ .0014	$18/729 =$ .0247	$114/729 =$ .1564	$288/729 =$ .3951	$228/729 =$ .3127	$72/729 =$ .0988	$8/729 =$ .0110

This is the sampling distribution for $\bar{x}$ because it specifies the probability associated with each possible value of $\bar{x}$. You can see that the most likely mean outcome after 3 randomly selected come-out rolls is $\bar{x} = \$0$; this result occurs with probability $288/729 = .3951$.

b. In Table 6.2, you can see that the median M can assume one of three values: $-5, 0,$ and 5. The value $M = -5$ occurs in 7 different samples. Therefore, $P(M = -5)$ is the sum of the probabilities associated with these 7 samples; that is, $P(M = -5) = 1/729 + 6/729 + 2/729 + 6/729 + 2/729 + 6/729 + 2/729 = 25/729$. Similarly, $M = 0$ occurs in 13 samples and $M = 5$ occurs in 7 samples. These probabilities are obtained by summing the probabilities of their respective sample points. After performing these calculations, we obtain the following probability distribution for the median M:

M	-5	0	5
$p(M)$	$25/729 = .0343$	$612/729 = .8395$	$92/729 = .1262$

Once again, the most likely median outcome after 3 randomly selected come-out rolls is $\bar{x} = \$0$—a result that occurs with probability $612/729 = .8395$.

Look Back The sampling distributions of parts **a** and **b** are found by first listing all possible distinct values of the statistic and then calculating the probability of each value. Note that if the values of x were equally likely, the 27 sample points in Table 6.2 would all have the same probability of occurring, namely, $1/27$.

Now Work Exercise 6.3

Example 6.1 demonstrates the procedure for finding the exact sampling distribution of a statistic when the number of different samples that could be selected from the population is relatively small. In the real world, populations often consist of a large number of different values, making samples difficult (or impossible) to enumerate. When this situation occurs, we may choose to obtain the approximate sampling distribution for a statistic by simulating the sampling over and over again and recording the proportion of times different values of the statistic occur. Example 6.2 illustrates this procedure.

Example 6.2

Simulating a Sampling Distribution— Thickness of Steel Sheets

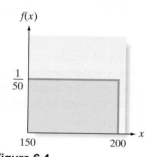

Figure 6.4

Uniform distribution for thickness of steel sheets

Problem The rolling machine of a steel manufacturer produces sheets of steel of varying thickness. The thickness of a steel sheet follows a uniform distribution with values between 150 and 200 millimeters. Suppose we perform the following experiment over and over again: Randomly sample 11 steel sheets from the production line and record the thickness x of each. Calculate the two sample statistics

$$\bar{x} = \text{Sample mean} = \frac{\Sigma x}{11}$$

$$M = \text{Median} = \text{Sixth sample measurement when the 11 thicknesses}$$
$$\text{are arranged in ascending order}$$

Obtain approximations to the sampling distributions of $\bar{x}$ and M.

Solution The population of thicknesses follows the uniform distribution shown in Figure 6.4. We used MINITAB to generate 1,000 samples from this population, each with $n = 11$ observations. Then we computed $\bar{x}$ and M for each sample. Our goal is to obtain approximations to the sampling distributions of $\bar{x}$ and M in order to find out which sample statistic ($\bar{x}$ or M) contains more information about μ. [*Note:* In this particular example, we *know* that the population mean is $\mu = 175$ mm. (See Section 5.2.)] The first 10 of the 1,000 samples generated are presented in Table 6.3. For instance, the first computer-generated sample from the uniform distribution contained the following measurements (arranged in ascending order): 151, 157, 162, 169, 171, 173, 181, 182, 187, 188, and 193 millimeters. The sample mean $\bar{x}$ and median M computed for this sample are

$$\bar{x} = \frac{151 + 157 + \cdots + 193}{11} = 174.0$$

$$M = \text{Sixth ordered measurement} = 173$$

The MINITAB relative frequency histograms for $\bar{x}$ and M for the 1,000 samples of size $n = 11$ are shown in Figure 6.5. These histograms represent approximations to the true sampling distributions of $\bar{x}$ and M.

Look Back You can see that the values of $\bar{x}$ tend to cluster around μ to a greater extent than do the values of M. Thus, on the basis of the observed sampling distributions, we

Table 6.3	First 10 Samples of $n = 11$ Thickness Measurements from Uniform Distribution												
Sample	Thickness Measurements											Mean	Median
1	173	171	187	151	188	181	182	157	162	169	193	174.00	173
2	181	190	182	171	187	177	162	172	188	200	193	182.09	182
3	192	195	187	187	172	164	164	189	179	182	173	180.36	182
4	173	157	150	154	168	174	171	182	200	181	187	172.45	173
5	169	160	167	170	197	159	174	174	161	173	160	169.46	169
6	179	170	167	174	173	178	173	170	173	198	187	176.55	173
7	166	177	162	171	154	177	154	179	175	185	193	172.09	175
8	164	199	152	153	163	156	184	151	198	167	180	169.73	164
9	181	193	151	166	180	199	180	184	182	181	175	179.27	181
10	155	199	199	171	172	157	173	187	190	185	150	176.18	173

Data Set: SIMUNI

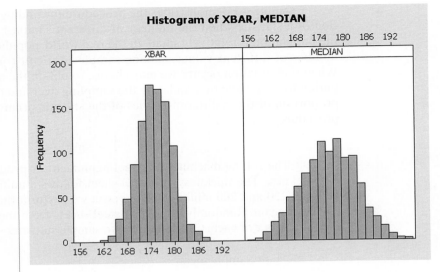

Figure 6.5
MINITAB histograms for sample mean and sample median, Example 6.2

conclude that $\bar{x}$ contains more information about μ than M does—at least for samples of $n = 11$ measurements from the uniform distribution.

Now Work Exercise 6.8

As noted earlier, many sampling distributions can be derived mathematically, but the theory necessary to do so is beyond the scope of this text. Consequently, when we need to know the properties of a statistic, we will present its sampling distribution and simply describe its properties. Several of the important properties we look for in sampling distributions are discussed in the next section.

Exercises 6.1–6.9

Understanding the Principles

6.1 What is the difference between a population parameter and a sample statistic?

6.2 What is a sampling distribution of a sample statistic?

Learning the Mechanics

6.3 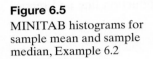 The probability distribution shown here describes a population of measurements that can assume values of 0, 2, 4, and 6, each of which occurs with the same relative frequency:

x	0	2	4	6
$p(x)$	$\frac{1}{4}$	$\frac{1}{4}$	$\frac{1}{4}$	$\frac{1}{4}$

a. List all the different samples of $n = 2$ measurements that can be selected from this population.
b. Calculate the mean of each different sample listed in part **a.**
c. If a sample of $n = 2$ measurements is randomly selected from the population, what is the probability that a specific sample will be selected?
d. Assume that a random sample of $n = 2$ measurements is selected from the population. List the different values of $\bar{x}$ found in part **b,** and find the probability of each. Then give the sampling distribution of the sample mean $\bar{x}$ in tabular form.
e. Construct a probability histogram for the sampling distribution of $\bar{x}$.

6.4 Simulate sampling from the population described in Exercise 6.3 by marking the values of x, one on each of four identical coins (or poker chips, etc.). Place the coins (marked 0, 2, 4, and 6) into a bag, randomly select one, and observe its value. Replace this coin, draw a second coin, and observe its value. Finally, calculate the mean $\bar{x}$ for this sample of $n = 2$ observations randomly selected from the population (Exercise 6.3, part **b**). Replace the coins, mix them, and, using the same procedure, select a sample of $n = 2$ observations from the population. Record the numbers and calculate $\bar{x}$ for this sample. Repeat this sampling process until you acquire 100 values of $\bar{x}$. Construct a relative frequency distribution for these 100 sample means. Compare this distribution with the exact sampling distribution of $\bar{x}$ found in part **e** of Exercise 6.3. [*Note:* The distribution obtained in this exercise is an approximation to the exact sampling distribution. However, if you were to repeat the sampling procedure, drawing two coins not 100 times, but 10,000 times, then the relative frequency distribution for the 10,000 sample means would be almost identical to the sampling distribution of $\bar{x}$ found in Exercise 6.3, part **e.**]

6.5 Consider the population described by the probability distribution shown here:

x	1	2	3	4	5
$p(x)$	.2	.3	.2	.2	.1

The random variable x is observed twice. If these observations are independent, verify that the different samples of size 2 and their probabilities are as follows:

Sample	Probability	Sample	Probability
1, 1	.04	3, 4	.04
1, 2	.06	3, 5	.02
1, 3	.04	4, 1	.04
1, 4	.04	4, 2	.06
1, 5	.02	4, 3	.04
2, 1	.06	4, 4	.04
2, 2	.09	4, 5	.02
2, 3	.06	5, 1	.02
2, 4	.06	5, 2	.03
2, 5	.03	5, 3	.02
3, 1	.04	5, 4	.02
3, 2	.06	5, 5	.01
3, 3	.04		

a. Find the sampling distribution of the sample mean $\bar{x}$.
b. Construct a probability histogram for the sampling distribution of $\bar{x}$.
c. What is the probability that $\bar{x}$ is 4.5 or larger?
d. Would you expect to observe a value of $\bar{x}$ equal to 4.5 or larger? Explain.

6.6 Refer to Exercise 6.5 and find $E(x) = \mu$. Then use the sampling distribution of $\bar{x}$ found in Exercise 6.5 to find the expected value of $\bar{x}$. Note that $E(\bar{x}) = \mu$.

6.7 Refer to Exercise 6.5. Assume that a random sample of $n = 2$ measurements is randomly selected from the population.

a. List the different values that the sample median M may assume, and find the probability of each. Then give the sampling distribution of the sample median.
b. Construct a probability histogram for the sampling distribution of the sample median, and compare it with the probability histogram for the sample mean (Exercise 6.5, part b).

6.8 In Example 6.2, we use the computer to generate 1,000 [NW] samples, each containing $n = 11$ observations, from a uniform distribution over the interval from 150 to 200. Now use the computer to generate 500 samples, each containing $n = 15$ observations, from that same population.

a. Calculate the sample mean for each sample. To approximate the sampling distribution of $\bar{x}$, construct a relative frequency histogram for the 500 values of $\bar{x}$.
b. Repeat part **a** for the sample median. Compare this approximate sampling distribution with the approximate sampling distribution of $\bar{x}$ found in part **a**.

6.9 Consider a population that contains values of x equal to 00, 01, 02, 03, … , 96, 97, 98, 99. Assume that these values occur with equal probability. Use the computer to generate 500 samples, each containing $n = 25$ measurements, from this population. Calculate the sample mean $\bar{x}$ and sample variance s^2 for each of the 500 samples.

a. To approximate the sampling distribution of $\bar{x}$, construct a relative frequency histogram for the 500 values of $\bar{x}$.
b. Repeat part **a** for the 500 values of s^2.

6.2 Properties of Sampling Distributions: Unbiasedness and Minimum Variance

The simplest type of statistic used to make inferences about a population parameter is a *point estimator*—a rule or formula which tells us how to use the sample data to calculate a single number that is intended to estimate the value of some population parameter. For example, the sample mean $\bar{x}$ is a point estimator of the population mean μ. Similarly, the sample variance s^2 is a point estimator of the population variance σ^2.

> A **point estimator** of a population parameter is a rule or formula which tells us how to use the sample data to calculate a single number that can be used as an *estimate* of the population parameter.

Often, many different point estimators can be found to estimate the same parameter. Each will have a sampling distribution that provides information about the point estimator. By examining the sampling distribution, we can determine how large the difference between an estimate and the true value of the parameter (called the **error of estimation**) is likely to be. We can also tell whether an estimator is more likely to overestimate or to underestimate a parameter.

Example 6.3

Comparing Two Statistics

Problem Suppose two statistics, A and B, exist to estimate the same population parameter θ (theta). (Note that θ could be any parameter: μ, σ^2, σ, etc.) Suppose the two statistics have sampling distributions as shown in Figure 6.6. On the basis of these sampling distributions, which statistic is more attractive as an estimator of θ?

Solution As a first consideration, we would like the sampling distribution to center over the value of the parameter we wish to estimate. One way to characterize this property is in terms of the mean of the sampling distribution. Consequently, we say that a statistic is

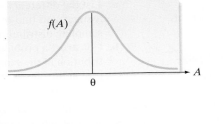 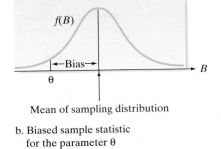

Mean of sampling distribution

a. Unbiased sample statistic
   for the parameter θ

b. Biased sample statistic
   for the parameter θ

Figure 6.6
Sampling distributions of
unbiased and biased estimators

unbiased if the mean of the sampling distribution is equal to the parameter it is intended to estimate. This situation is shown in Figure 6.6a, where the mean μ_A of statistic A is equal to θ. If the mean of a sampling distribution is *not* equal to the parameter it is intended to estimate, the statistic is said to be *biased*. The sampling distribution for a biased statistic is shown in Figure 6.6b. The mean μ_B of the sampling distribution for statistic B is not equal to θ; in fact, it is shifted to the right of θ.

Look Back You can see that biased statistics tend either to overestimate or to underestimate a parameter. Consequently, when other properties of statistics tend to be equivalent, we will choose an unbiased statistic to estimate a parameter of interest.[*]

Now Work Exercise 6.14

> If the sampling distribution of a sample statistic has a mean equal to the population parameter the statistic is intended to estimate, the statistic is said to be **an unbiased estimate** of the parameter.
>
> If the mean of the sampling distribution is not equal to the parameter, the statistic is said to be a **biased estimate** of the parameter.

The standard deviation of a sampling distribution measures another important property of statistics: the spread of these estimates generated by repeated sampling. Suppose two statistics, A and B, are both unbiased estimators of the population parameter. Since the means of the two sampling distributions are the same, we turn to their standard deviations in order to decide which will provide estimates that fall closer to the unknown population parameter we are estimating. Naturally, we will choose the sample statistic that has the smaller standard deviation. Figure 6.7 depicts sampling distributions for A and B. Note that the standard deviation

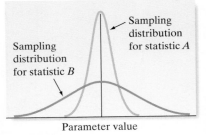

Figure 6.7
Sampling distributions for two
unbiased estimators

of the distribution for A is smaller than the standard deviation of the distribution for B, indicating that, over a large number of samples, the values of A cluster more closely around the unknown population parameter than do the values of B. Stated differently, the probability that A is close to the parameter value is higher than the probability that B is close to the parameter value.

In sum, to make an inference about a population parameter, we use the sample statistic with a sampling distribution that is unbiased and has a small standard deviation (usually smaller than the standard deviation of other unbiased sample statistics). The

---

[*]Unbiased statistics do not exist for all parameters of interest, but they do exist for all the parameters considered in this text.

derivation of this sample statistic will not concern us, because the "best" statistic for estimating specific parameters is a matter of record. We will simply present an unbiased estimator with its standard deviation for each population parameter we consider. [*Note:* The standard deviation of the sampling distribution of a statistic is also called the **standard error of the statistic.**]

Example 6.4 **Biased and Unbiased Estimators—Craps Application**	**Problem** Refer to Example 6.1 and the outcome x of a $5 wager in craps. We found the sampling distributions of the sample mean $\bar{x}$ and the sample median M for random samples of $n = 3$ dice rolls from a population defined by the following probability distribution:

x	-5	0	5
$p(x)$	$1/9$	$6/9$	$2/9$

The sampling distributions of $\bar{x}$ and M were found to be as follows:

$\bar{x}$	-5	-3.33	-1.67	0	1.67	3.33	5
$p(\bar{x})$	$1/729$	$18/729$	$114/729$	$288/729$	$228/729$	$72/729$	$8/729$

M	-5	0	5
$p(M)$	$25/729$	$612/729$	$92/729$

a. Show that $\bar{x}$ is an unbiased estimator of μ in this situation.

b. Show that M is a biased estimator of μ in this situation.

Solution

a. The expected value of a discrete random variable x (see Section 4.3) is defined as $E(x) = \Sigma x p(x)$, where the summation is over all values of x. Then

$$E(x) = \mu = \Sigma x p(x) = (-5)(1/9) + (0)(6/9) + (5)(2/9) = \frac{5}{9} = .556$$

The expected value of the discrete random variable $\bar{x}$ is

$$E(\bar{x}) = \Sigma(\bar{x}) p(\bar{x})$$

summed over all values of $\bar{x}$, or

$$E(\bar{x}) = (-5)(1/729) + (-3.33)(18/729) + (-1.67)(114/729) + \cdots + (5)(8/729) = .556$$

Since $E(\bar{x}) = \mu$, $\bar{x}$ is an unbiased estimator of μ.

b. The expected value of the sample median M is

$$E(M) = \Sigma M p(M) = (-5)(25/729) + (0)(612/729) + (5)(92/729) = \frac{335}{729} = .460$$

Since the expected value of M is not equal to μ ($\mu = .556$), the sample median M is a biased estimator of μ.

Now Work Exercise 6.16 a, b, c, d

Example 6.5 **Variance of Estimators—Craps Application**	**Problem** Refer to Example 6.4 and find the standard deviations of the sampling distributions of $\bar{x}$ and M. Which statistic would appear to be a better estimator of μ? **Solution** The variance of the sampling distribution of $\bar{x}$ (we denote it by the symbol $\sigma_{\bar{x}}^2$) is found to be

$$\sigma_{\bar{x}}^2 = E\{[\bar{x} - E(\bar{x})]^2\} = \Sigma(\bar{x} - \mu)^2 p(\bar{x})$$

where, from Example 6.4,

$$E(\bar{x}) = \mu = .556$$

Then

$$\sigma_{\bar{x}}^2 = (-5-.556)^2(^1/_{729}) + (-3.33-.556)^2(^{18}/_{729}) + \cdots (5-.556)^2(^8/_{729})$$
$$= 2.675$$

and

$$\sigma_{\bar{x}} = \sqrt{2.675} = 1.64$$

Similarly, the variance of the sampling distribution of M (we denote it by σ_M^2) is

$$\sigma_M^2 = E\{[M - E(M)]^2\}$$

where, from Example 6.4, the expected value of M is $E(M) = .460$. Then

$$\sigma_M^2 = E\{[M - E(M)]^2\} = \Sigma [M - E(M)]^2 p(M)$$
$$= (-5-.460)(^{25}/_{729}) + (0-.460)(^{612}/_{729}) + (5-.460)(^{92}/_{729}) = 3.801$$

and

$$\sigma_M = \sqrt{3.801} = 1.95$$

Which statistic appears to be the better estimator for the population mean μ, the sample mean $\bar{x}$ or the median M? To answer this question, we compare the sampling distributions of the two statistics. The sampling distribution of the sample median M is biased (i.e., it is located to the left of the mean μ), and its standard deviation $\sigma_M = 1.95$ is larger than the standard deviation of the sampling distribution of $\bar{x}$, $\sigma_{\bar{x}} = 1.64$. Consequently, for the population in question, the sample mean $\bar{x}$ would be a better estimator of the population mean μ than the sample median M would be.

Look Back Ideally, we desire an estimator that is unbiased *and* has the smallest variance among all unbiased estimators. We call this statistic the **minimum-variance unbiased estimator (MVUE)**.

Now Work Exercise 6.16 e, f

Exercises 6.10–6.20

Understanding the Principles

6.10 What is a point estimator of a population parameter?

6.11 What is the difference between a biased and unbiased estimator?

6.12 What is the MVUE for a parameter?

6.13 What are the properties of an ideal estimator?

Learning the Mechanics

6.14 Consider the following probability distribution:
[NW]

x	0	1	4
$p(x)$	$^1/_3$	$^1/_3$	$^1/_3$

a. Find μ and σ^2.

b. Find the sampling distribution of the sample mean $\bar{x}$ for a random sample of $n = 2$ measurements from this distribution.

c. Show that $\bar{x}$ is an unbiased estimator of μ. [*Hint:* Show that $E(\bar{x}) = \Sigma \bar{x} p(\bar{x}) = \mu$.]

d. Find the sampling distribution of the sample variance s^2 for a random sample of $n = 2$ measurements from this distribution.

e. Show that s^2 is an unbiased estimator for σ^2.

6.15 Consider the following probability distribution:

x	2	4	9
$p(x)$	$^1/_3$	$^1/_3$	$^1/_3$

a. Calculate μ for this distribution.

b. Find the sampling distribution of the sample mean $\bar{x}$ for a random sample of $n = 3$ measurements from this distribution, and show that $\bar{x}$ is an unbiased estimator of μ.

c. Find the sampling distribution of the sample median M for a random sample of $n = 3$ measurements from this distribution, and show that the median is a biased estimator of μ.

d. If you wanted to use a sample of three measurements from this population to estimate μ, which estimator would you use? Why?

6.16 Consider the following probability distribution:

[NW]

x	0	1	2
$p(x)$	$1/3$	$1/3$	$1/3$

a. Find μ.

b. For a random sample of $n = 3$ observations from this distribution, find the sampling distribution of the sample mean.

c. Find the sampling distribution of the median of a sample of $n = 3$ observations from this population.

d. Refer to parts **b** and **c**, and show that both the mean and median are unbiased estimators of μ for this population.

e. Find the variances of the sampling distributions of the sample mean and the sample median.

f. Which estimator would you use to estimate μ? Why?

6.17 Use the computer to generate 500 samples, each containing $n = 25$ measurements, from a population that contains values of x equal to $1, 2, \dots, 48, 49, 50$. Assume that these values of x are equally likely. Calculate the sample mean $\bar{x}$ and median M for each sample. Construct relative frequency histograms for the 500 values of $\bar{x}$ and the 500 values of M. Use these approximations to the sampling distributions of $\bar{x}$ and M to answer the following questions:

a. Does it appear that $\bar{x}$ and M are unbiased estimators of the population mean? [*Note:* $\mu = 25.5$.]

b. Which sampling distribution displays greater variation?

6.18 Refer to Exercise 6.5.

a. Show that $\bar{x}$ is an unbiased estimator of μ.

b. Find $\sigma_{\bar{x}}^2$.

c. Find the probability that $\bar{x}$ will fall within $2\sigma_{\bar{x}}$ of μ.

6.19 Refer to Exercise 6.5.

a. Find the sampling distribution of s^2.

b. Find the population variance σ^2.

c. Show that s^2 is an unbiased estimator of σ^2.

d. Find the sampling distribution of the sample standard deviation s.

e. Show that s is a biased estimator of σ.

6.20 Refer to Exercise 6.7, in which we found the sampling distribution of the sample median. Is the median an unbiased estimator of the population mean μ?

6.3 The Sampling Distribution of $\bar{x}$ and the Central Limit Theorem

Estimating the mean useful life of automobiles, the mean number of crimes per month in a large city, and the mean yield per acre of a new soybean hybrid are practical problems with something in common. In each case, we are interested in making an inference about the mean μ of some population. As we mentioned in Chapter 2, the sample mean $\bar{x}$ is, in general, a good estimator of μ. We now develop pertinent information about the sampling distribution for this useful statistic. We will show that $\bar{x}$ is the minimum-variance unbiased estimator (MVUE) of μ.

Example 6.6

Describing the Sampling Distribution of $\bar{x}$

Problem Suppose a population has the uniform probability distribution given in Figure 6.8. Then the mean and standard deviation of this probability distribution are, respectively, $\mu = 175$ and $\sigma = 14.43$. (See Section 5.2 for the formulas for μ and σ.) Now suppose a sample of 11 measurements is selected from this population. Describe the sampling distribution of the sample mean $\bar{x}$ based on the 1,000 sampling experiments discussed in Example 6.2.

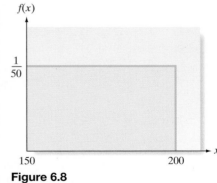

Figure 6.8

Sampled uniform population

Solution Recall that in Example 6.2 we generated 1,000 samples of $n = 11$ measurements each. The MINITAB histogram for the 1,000 sample means is shown in Figure 6.9, with a normal probability distribution superimposed. You can see that this normal probability distribution approximates the computer-generated sampling distribution very well.

To fully describe a normal probability distribution, it is necessary to know its mean and standard deviation. MINITAB gives these statistics for the 1,000 $\bar{x}$'s in the upper right corner of the histogram of Figure 6.9. You can see that the mean is 175.2 and the standard deviation is 4.383.

To summarize our findings based on 1,000 samples, each consisting of 11 measurements from a uniform population, the sampling distribution of $\bar{x}$ appears to be approximately normal with a mean of about 175 and a standard deviation of about 4.38.

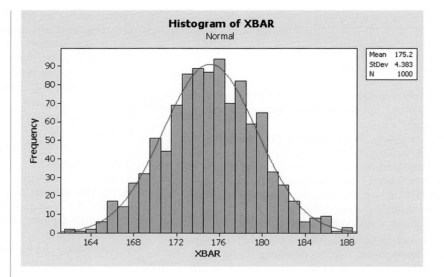

Figure 6.9
MINITAB histogram for
sample mean in 1,000 samples

Look Back Note that the simulated value $\mu_{\bar{x}} = 175.2$ is very close to $\mu = 175$ for the uniform distribution; that is, the simulated sampling distribution of $\bar{x}$ appears to provide an unbiased estimate of μ.

The true sampling distribution of $\bar{x}$ has the properties given in the next box, assuming only that a random sample of n observations has been selected from *any* population.

Properties of the Sampling Distribution of $\bar{x}$

1. The mean of the sampling distribution of $\bar{x}$ equals the mean of the sampled population. That is, $\mu_{\bar{x}} = E(\bar{x}) = \mu$.

2. The standard deviation of the sampling distribution of $\bar{x}$ equals

$$\frac{\text{Standard deviation of sampled population}}{\text{Square root of sample size}}$$

That is, $\sigma_{\bar{x}} = \sigma/\sqrt{n}$*

The standard deviation $\sigma_{\bar{x}}$ is often referred to as the **standard error of the mean**.

You can see that our approximation to $\mu_{\bar{x}}$ in Example 6.6 was precise, since property 1 assures us that the mean is the same as that of the sampled population: 175. Property 2 tells us how to calculate the standard deviation of the sampling distribution of $\bar{x}$. Substituting $\sigma = 14.43$ (the standard deviation of the sampled uniform distribution) and the sample size $n = 11$ into the formula for $\sigma_{\bar{x}}$, we find that

$$\sigma_{\bar{x}} = \frac{\sigma}{\sqrt{n}} = \frac{14.43}{\sqrt{11}} = 4.35$$

Thus, the approximation we obtained in Example 6.6, $\sigma_{\bar{x}} \approx 4.38$, is very close to the exact value, $\sigma_{\bar{x}} = 4.35$. It can be shown (proof omitted) that the value of $\sigma_{\bar{x}}^2$ is the smallest variance among all unbiased estimators of μ; thus, $\bar{x}$ is the MVUE for μ.

What about the shape of the sampling distribution? Two theorems provide this information. One is applicable whenever the original population data are normally

---

*If the sample size n is large relative to the number N of elements in the population (e.g., 5% or more), $\sigma/\sqrt{n}$ must be multiplied by the finite population correction factor $\sqrt{(N-n)/(N-1)}$. In most sampling situations, this correction factor will be close to 1 and can be ignored.

distributed. The other, applicable when the sample size n is large, represents one of the most important theoretical results in statistics: the **Central Limit Theorem**.

Theorem 6.1

If a random sample of n observations is selected from a population with a normal distribution, the sampling distribution of $\bar{x}$ will be a normal distribution.

Theorem 6.2: Central Limit Theorem

Consider a random sample of n observations selected from a population (*any* population) with mean μ and standard deviation σ. Then, when n is sufficiently large, the sampling distribution of $\bar{x}$ will be approximately a normal distribution with mean $\mu_{\bar{x}} = \mu$ and standard deviation $\sigma_{\bar{x}} = \sigma/\sqrt{n}$. The larger the sample size, the better will be the normal approximation to the sampling distribution of $\bar{x}$.[*]

Thus, for sufficiently large samples, the sampling distribution of $\bar{x}$ is approximately normal. How large must the sample size n be so that the normal distribution provides a good approximation to the sampling distribution of $\bar{x}$? The answer depends on the shape of the distribution of the sampled population, as shown by Figure 6.10. Generally speaking, the greater the skewness of the sampled population distribution, the larger the

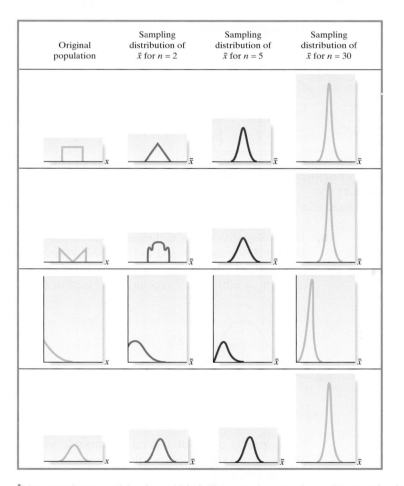

Figure 6.10

Sampling distributions of $\bar{x}$ for different populations and different sample sizes

[*]Moreover, because of the Central Limit Theorem, the sum of a random sample of n observations, Σx, will possess a sampling distribution that is approximately normal for large samples. This distribution will have a mean equal to $n\mu$ and a variance equal to $n\sigma^2$. Proof of the Central Limit Theorem is beyond the scope of this book, but it can be found in many mathematical statistics texts.

sample size must be before the normal distribution is an adequate approximation to the sampling distribution of $\bar{x}$. For most sampled populations, sample sizes of $n \geq 30$ will suffice for the normal approximation to be reasonable.

Example 6.7

Using the Central Limit Theorem to Find a Probability

Problem Suppose we have selected a random sample of $n = 36$ observations from a population with mean equal to 80 and standard deviation equal to 6. It is known that the population is not extremely skewed.

a. Sketch the relative frequency distributions for the population and for the sampling distribution of the sample mean $\bar{x}$.

b. Find the probability that $\bar{x}$ will be larger than 82.

Solution

a. We do not know the exact shape of the population relative frequency distribution, but we do know that it should be centered about $\mu = 80$, its spread should be measured by $\sigma = 6$, and it is not highly skewed. One possibility is shown in Figure 6.11a. From the Central Limit Theorem, we know that the sampling distribution of $\bar{x}$ will be approximately normal, since the sampled population distribution is not extremely skewed. We also know that the sampling distribution will have mean

and standard deviation
$$\mu_{\bar{x}} = \mu = 80$$

$$\sigma_{\bar{x}} = \frac{\sigma}{\sqrt{n}} = \frac{6}{\sqrt{36}} = 1$$

The sampling distribution of $\bar{x}$ is shown in Figure 6.11b.

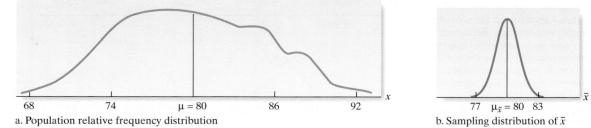

a. Population relative frequency distribution

Figure 6.11
A population relative frequency distribution and the sampling distribution for $\bar{x}$

b. Sampling distribution of $\bar{x}$

b. The probability that $\bar{x}$ will exceed 82 is equal to the highlighted area in Figure 6.12. To find this area, we need to find the z value corresponding to $\bar{x} = 82$. Recall that the standard normal random variable z is the difference of any normally distributed random variable and its mean, expressed in units of its standard deviation. Since $\bar{x}$

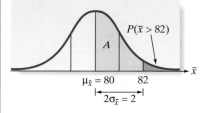

Figure 6.12
The sampling distribution of $\bar{x}$

is a normally distributed random variable with mean $\mu_{\bar{x}} = \mu$ and standard deviation $\sigma_{\bar{x}} = \sigma/\sqrt{n}$, it follows that the standard normal z value corresponding to the sample mean $\bar{x}$ is

$$z = \frac{(\text{Normal random variable}) - (\text{Mean})}{\text{Standard deviation}} = \frac{\bar{x} - \mu_{\bar{x}}}{\sigma_{\bar{x}}}$$

Therefore, for $\bar{x} = 82$, we have

$$z = \frac{\bar{x} - \mu_{\bar{x}}}{\sigma_{\bar{x}}} = \frac{82 - 80}{1} = 2$$

The area A in Figure 6.12 corresponding to $z = 2$ is given in the table of areas under the normal curve (see Table IV of Appendix A) as .4772. Therefore, the tail area corresponding to the probability that $\bar{x}$ exceeds 82 is

$$P(\bar{x} > 82) = P(z > 2) = .5 - .4772 = .0228$$

Look Back The key to finding the probability in part **b** is to recognize that the distribution of $\bar{x}$ is normal with $\mu_{\bar{x}} = \mu$ and $\sigma_{\bar{x}} = \sigma/\sqrt{n}$.

Now Work Exercise 6.30

Example 6.8

Application of the Central Limit Theorem—Testing a Manufacturer's Claim

Problem A manufacturer of automobile batteries claims that the distribution of the lengths of life of its best battery has a mean of 54 months and a standard deviation of 6 months. Suppose a consumer group decides to check the claim by purchasing a sample of 50 of the batteries and subjecting them to tests that estimate the battery's life.

a. Assuming that the manufacturer's claim is true, describe the sampling distribution of the mean lifetime of a sample of 50 batteries.

b. Assuming that the manufacturer's claim is true, what is the probability that the consumer group's sample has a mean life of 52 or fewer months?

Solution

a. Even though we have no information about the shape of the probability distribution of the lives of the batteries, we can use the Central Limit Theorem to deduce that the sampling distribution for a sample mean lifetime of 50 batteries is approximately normally distributed. Furthermore, the mean of this sampling distribution is the same as the mean of the sampled population, which is $\mu = 54$ months according to the manufacturer's claim. Finally, the standard deviation of the sampling distribution is given by

$$\sigma_{\bar{x}} = \frac{\sigma}{\sqrt{n}} = \frac{6}{\sqrt{50}} = .85 \text{ month}$$

Note that we used the claimed standard deviation of the sampled population, $\sigma = 6$ months. Thus, if we assume that the claim is true, then the sampling distribution for the mean life of the 50 batteries sampled must be as shown in Figure 6.13.

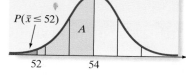

Figure 6.13
Sampling distribution of $\bar{x}$ in Example 6.8 for $n = 50$

b. If the manufacturer's claim is true, the probability that the consumer group observes a mean battery life of 52 or fewer months for its sample of 50 batteries, $P(\bar{x} \leq 52)$, is equivalent to the highlighted area in Figure 6.13. Since the sampling distribution is approximately normal, we can find this area by computing the standard normal z value:

$$z = \frac{\bar{x} - \mu_{\bar{x}}}{\sigma_{\bar{x}}} = \frac{\bar{x} - \mu}{\sigma_{\bar{x}}} = \frac{52 - 54}{.85} = -2.35$$

Here, $\mu_{\bar{x}}$, the mean of the sampling distribution of $\bar{x}$, is equal to μ, the mean of the lifetimes of the sampled population, and $\sigma_{\bar{x}}$ is the standard deviation of the sampling distribution of $\bar{x}$. Note that z is the familiar standardized distance (z-score) of Section 2.7, and since $\bar{x}$ is approximately normally distributed, it will possess the standard normal distribution of Section 5.3.

The area A shown in Figure 6.13 between $\bar{x} = 52$ and $\bar{x} = 54$ (corresponding to $z = -2.35$) is found in Table IV of Appendix A to be .4906. Therefore, the area to the left of $\bar{x} = 52$ is

$$P(\bar{x} \leq 52) = .5 - A = .5 - .4906 = .0094$$

Thus, the probability that the consumer group will observe a sample mean of 52 or less is only .0094 if the manufacturer's claim is true.

Look Back If the 50 batteries tested do exhibit a mean of 52 or fewer months, the consumer group will have strong evidence that the manufacturer's claim is untrue, because such an event is very unlikely to occur if the claim is true. (This is still another application of the *rare-event approach* to statistical inference.)

Now Work Exercise 6.37

We conclude this section with two final comments on the sampling distribution of $\bar{x}$. First, from the formula $\sigma_{\bar{x}} = \sigma/\sqrt{n}$, we see that the standard deviation of the sampling distribution of $\bar{x}$ gets smaller as the sample size n gets larger. For example, we computed $\sigma_{\bar{x}} = .85$ when $n = 50$ in Example 6.8. However, for $n = 100$, we obtain $\sigma_{\bar{x}} = \sigma/\sqrt{n} = 6/\sqrt{100} = .60$. This relationship will hold true for most of the sample statistics encountered in this text. That is, *the standard deviation of the sampling distribution decreases as the sample size increases.* Consequently, the larger the sample size, the more accurate the sample statistic (e.g., $\bar{x}$) is in estimating a population parameter (e.g., μ). We will use this result in Chapter 7 to help us determine the sample size needed to obtain a specified accuracy of estimation.

Our second comment concerns the Central Limit Theorem. In addition to providing a very useful approximation for the sampling distribution of a sample mean, the Central Limit Theorem offers an explanation for the fact that many relative frequency distributions of data possess mound-shaped distributions. Many of the measurements we take in various areas of research are really means or sums of a large number of small phenomena. For example, a year's growth of a pine seedling is the total of the numerous individual components that affect the plant's growth. Similarly, we can view the length of time a construction company takes to build a house as the total of the times taken to complete a multitude of distinct jobs, and we can regard the monthly demand for blood at a hospital as the total of the many individual patients' needs. Whether or not the observations entering into these sums satisfy the assumptions basic to the Central Limit Theorem is open to question; however, it is a fact that many distributions of data in nature are mound shaped and possess the appearance of normal distributions.

Statistics IN Action | Revisited — Making an Inference about the Mean Sleep Onset Latency for Insomnia Pill Takers

In a Massachusetts Institute of Technology (MIT) study, each member of a sample of 40 young male volunteers was given a dosage of the sleep-inducing hormone melatonin, was placed in a dark room at midday, and was told to close his eyes for 30 minutes. The researchers measured the time (in minutes) elapsed before each volunteer fell asleep—called sleep onset latency. Recall that the data (shown in Table SIA6.1) are saved in the **INSOMNIA** file.

Previous research established that with the placebo (i.e., no hormone), the mean sleep onset latency is $\mu = 15$ minutes and the standard deviation is $\sigma = 10$ minutes. If the true

value of μ for those taking the melatonin is $\mu < 15$ (i.e., if, on average, the volunteers fall asleep faster with the drug than with the placebo), then the researchers can infer that melatonin is an effective drug against insomnia.

Descriptive statistics for the 40 sleep times are displayed in the SPSS printout shown in Figure SIA6.1. You can see that the mean for the sample is $\bar{x} = 5.935$ minutes. If the drug

Statistics IN Action
(continued)

is not effective in reducing sleep times, then the distribution of sleep times will be no different from the distribution with the placebo. That is, if the drug is not effective, the mean and standard deviation of the population of sleep times are $\mu = 15$ and $\sigma = 10$. If this is true, how likely is it to observe a sample mean that is below 6 minutes?

Descriptive Statistics

	N	Minimum	Maximum	Mean	Std. Deviation
SLEEPTIME	40	1.3	22.8	5.935	5.3917
Valid N (listwise)	40				

Figure SIA6.1

SPSS descriptive statistics for sleep time data

To answer this question, we desire the probability $P(\bar{x} < 6)$. To find this probability, we invoke the Central Limit Theorem. According to the theorem, the sampling distribution of $\bar{x}$ has the following mean and standard deviation:

$$\mu_{\bar{x}} = \mu = 15$$
$$\sigma_{\bar{x}} = \sigma/\sqrt{n} = 10/\sqrt{40} = 1.58$$

The theorem also states that $\bar{x}$ is approximately normally distributed. Therefore, we find the desired probability (using the standard normal table) as follows:

$$P(\bar{x} < 6) = P\left(z < \frac{6 - \mu_{\bar{x}}}{\sigma_{\bar{x}}}\right)$$
$$= P\left(z < \frac{6 - 15}{1.58}\right) = P(z < -5.70) \approx 0$$

In other words, the probability that we observe a sample mean below $\bar{x} = 6$ minutes *if the mean and standard deviation of the sleep times are*, respectively, $\mu = 15$ and $\sigma = 10$ (*i.e., if the drug is not effective*), is almost 0. Therefore, either the drug is not effective and the researchers have observed an extremely rare event (one with almost no chance of happening), or the true value of μ for those taking the melatonin pill is much less than 15 minutes. The rare-event approach to making statistical inferences, of course, would favor the second conclusion. Melatonin appears to be an effective insomnia pill, one that lowers the average time it takes the volunteers to fall asleep.

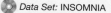 *Data Set:* INSOMNIA

Exercises 6.21–6.47

Understanding the Principles

6.21 What do the symbols $\mu_{\bar{x}}$ and $\sigma_{\bar{x}}$ represent?

6.22 How does the mean of the sampling distribution of $\bar{x}$ relate to the mean of the population from which the sample is selected?

6.23 How does the standard deviation of the sampling distribution of $\bar{x}$ relate to the standard deviation of the population from which the sample is selected?

6.24 Another name given to the standard deviation of $\bar{x}$ is the _____. (Fill in the blank.)

6.25 State the Central Limit Theorem.

6.26 Will the sampling distribution of $\bar{x}$ always be approximately normally distributed? Explain.

Learning the Mechanics

6.27 Suppose a random sample of n measurements is selected from a population with mean $\mu = 100$ and variance $\sigma^2 = 100$. For each of the following values of n, give the mean and standard deviation of the sampling distribution of the sample mean $\bar{x}$.
   a. $n = 4$ **b.** $n = 25$
   c. $n = 100$ **d.** $n = 50$
   e. $n = 500$ **f.** $n = 1,000$

6.28 Suppose a random sample of $n = 25$ measurements is selected from a population with mean μ and standard deviation σ. For each of the following values of μ and σ, give the values of $\mu_{\bar{x}}$ and $\sigma_{\bar{x}}$.
   a. $\mu = 10, \sigma = 3$
   b. $\mu = 100, \sigma = 25$
   c. $\mu = 20, \sigma = 40$
   d. $\mu = 10, \sigma = 100$

6.29 Consider the following probability distribution:

x	1	2	3	8
$p(x)$	.1	.4	.4	.1

   a. Find μ, σ^2, and σ.
   b. Find the sampling distribution of $\bar{x}$ for random samples of $n = 2$ measurements from this distribution by listing all possible values of $\bar{x}$, and find the probability associated with each.
   c. Use the results of part **b** to calculate $\mu_{\bar{x}}$ and $\sigma_{\bar{x}}$. Confirm that $\mu_{\bar{x}} = \mu$ and that $\sigma_{\bar{x}} = \sigma/\sqrt{n} = \sigma/\sqrt{2}$.

6.30 A random sample of $n = 64$ observations is drawn from a population with a mean equal to 20 and standard deviation equal to 16.
   a. Give the mean and standard deviation of the (repeated) sampling distribution of $\bar{x}$.
   b. Describe the shape of the sampling distribution of $\bar{x}$. Does your answer depend on the sample size?
   c. Calculate the standard normal z-score corresponding to a value of $\bar{x} = 16$.
   d. Calculate the standard normal z-score corresponding to $\bar{x} = 23$.
   e. Find $P(\bar{x} < 16)$.
   f. Find $P(\bar{x} > 23)$.
   g. Find $P(16 < \bar{x} < 23)$.

6.31 A random sample of $n = 100$ observations is selected from a population with $\mu = 30$ and $\sigma = 16$.
   a. Find $\mu_{\bar{x}}$ and $\sigma_{\bar{x}}$.
   b. Describe the shape of the sampling distribution of $\bar{x}$.
   c. Find $P(\bar{x} \geq 28)$.
   d. Find $P(22.1 \leq \bar{x} \leq 26.8)$.

e. Find $P(\bar{x} \leq 28.2)$.

f. Find $P(\bar{x} \geq 27.0)$.

6.32 A random sample of $n = 900$ observations is selected from a population with $\mu = 100$ and $\sigma = 10$.

a. What are the largest and smallest values of $\bar{x}$ that you would expect to see?

b. How far, at the most, would you expect $\bar{x}$ to deviate from μ?

c. Did you have to know μ to answer part **b**? Explain.

6.33 Consider a population that contains values of x equal to $0, 1, 2, \ldots, 97, 98, 99$. Assume that the values of x are equally likely. For each of the following values of n, use the computer to generate 500 random samples and calculate $\bar{x}$ for each sample. For each sample size, construct a relative frequency histogram of the 500 values of $\bar{x}$. What changes occur in the histograms as the value of n increases? What similarities exist? Use $n = 2, n = 5, n = 10, n = 30$, and $n = 50$.

Applet Exercise 6.1

Open the applet entitled *Sampling Distribution*. On the pull-down menu to the right of the top graph, select *Binary*.

a. Run the applet for the sample size $n = 10$ and the number of samples $N = 1000$. Observe the shape of the graph of the sample proportions, and record the mean, median, and standard deviation of the sample proportions.

b. How does the mean of the sample proportions compare with the mean $\mu = 0.5$ of the original distribution?

c. Use the formula $\sigma = \sqrt{np(1-p)}$, where $n = 1$ and $p = 0.5$, to compute the standard deviation of the original distribution. Divide the result by $\sqrt{10}$, the square root of the sample size used in the sampling distribution. How does this result compare with the standard deviation of the sample proportions?

d. Explain how the graph of the distribution of sample proportions suggests that the distribution may be approximately normal.

e. Explain how the results of parts **b–d** illustrate the Central Limit Theorem.

Applet Exercise 6.2

Open the applet entitled *Sampling Distributions*. On the pull-down menu to the right of the top graph, select *Uniform*. The box to the left of the top graph displays the population mean, median, and standard deviation of the original distribution.

a. Run the applet for the sample size $n = 30$ and the number of samples $N = 1000$. Observe the shape of the graph of the sample means, and record the mean, median, and standard deviation of the sample means.

b. How does the mean of the sample means compare with the mean of the original distribution?

c. Divide the standard deviation of the original distribution by $\sqrt{30}$, the square root of the sample size used in the sampling distribution. How does this result compare with the standard deviation of the sample proportions?

d. Explain how the graph of the distribution of sample means suggests that the distribution may be approximately normal.

e. Explain how the results of parts **b–d** illustrate the Central Limit Theorem.

Applying the Concepts—Basic

6.34 **Phishing attacks to e-mail accounts.** In Exercise 2.43 (p. 49), you learned that *phishing* describes an attempt to extract personal/financial information from unsuspecting people through fraudulent e-mail. Data from an actual phishing attack against an organization were presented in *Chance* (Summer 2007). The interarrival times, i.e., the time differences (in seconds), for 267 fraud box e-mail notifications were recorded and are saved in the **PHISHING** file. For this exercise, consider these interarrival times to represent the population of interest.

a. In Exercise 2.43 you constructed a histogram for the interarrival times. Describe the shape of the population of interarrival times.

b. Find the mean and standard deviation of the population of interarrival times.

c. Now consider a random sample of $n = 40$ interarrival times selected from the population. Describe the shape of the sampling distribution of $\bar{x}$, the sample mean. Theoretically, what are $\mu_{\bar{x}}$ and $\sigma_{\bar{x}}$?

d. Find $P(\bar{x} < 90)$.

e. Use a random number generator to select a random sample of $n = 40$ interarrival times from the population, and calculate the value of $\bar{x}$. (Every student in the class should do this.)

f. Refer to part **e**. Obtain the values of $\bar{x}$ computed by the students and combine them into a single data set. Form a histogram for these values of $\bar{x}$. Is the shape approximately normal?

g. Refer to part **f**. Find the mean and standard deviation of the $\bar{x}$-values. Do these values approximate $\mu_{\bar{x}}$ and $\sigma_{\bar{x}}$, respectively?

6.35 **Physical activity of obese young adults.** In a study on the physical activity of young adults, pediatric researchers measured overall physical activity as the total number of registered movements (counts) over a period of time and then computed the number of counts per minute (cpm) for each subject (*International Journal of Obesity*, January 2007). The study revealed that the overall physical activity of obese young adults has a mean of $\mu = 320$ cpm and a standard deviation of $\sigma = 100$ cpm. (In comparison, the mean for young adults of normal weight is 540 cpm.) In a random sample of $n = 100$ obese young adults, consider the sample mean counts per minute, $\bar{x}$.

a. Describe the sampling distribution of $\bar{x}$.

b. What is the probability that the mean overall physical activity level of the sample is between 300 and 310 cpm?

c. What is the probability that the mean overall physical activity level of the sample is greater than 360 cpm?

6.36 **Research on eating disorders.** Refer to *The American Statistician* (May 2001) study of female students who suffer from bulimia, presented in Exercise 2.40 (p. 48). Recall that each student completed a questionnaire from which a "fear of negative evaluation" (FNE) score was produced. (The higher the score, the greater is the fear of negative evaluation.) Suppose the FNE scores of bulimic students have a distribution with mean $\mu = 18$ and standard deviation $\sigma = 5$. Now, consider a random sample of 45 female students with bulimia.

a. What is the probability that the sample mean FNE score is greater than 17.5?

b. What is the probability that the sample mean FNE score is between 18 and 18.5?

c. What is the probability that the sample mean FNE score is less than 18.5?

6.37 **Dentists' use of anesthetics.** Refer to the *Current Allergy & Clinical Immunology* (March 2004) study of allergic reactions of dental patients to local anesthetics, presented in Exercise 2.99 (p. 72). The study reported that the mean number of units (ampoules) of local anesthetics used per week by dentists was 79, with a standard deviation of 23. Consider a random sample of 100 dentists, and let $\bar{x}$ represent the mean number of units of local anesthetic used per week for the sample.

a. What is $\mu_{\bar{x}}$?

b. What is $\sigma_{\bar{x}}$?

c. Describe the shape of the sampling distribution of $\bar{x}$.

d. Find the z-score for the value $\bar{x} = 80$ ampoules.

e. Find $P(\bar{x} > 80)$.

6.38 **Shell lengths of sea turtles.** Refer to the *Aquatic Biology* (Vol. 9, 2010) study of green sea turtles inhabiting the Grand Cayman South Sound lagoon, Exercise 2.83 (p. 65). Research shows that the curved carapace (shell) lengths of these turtles has a distribution with mean $\mu = 50$ cm and standard deviation $\sigma = 10$ cm. In the study, $n = 76$ green sea turtles were captured from the lagoon; the mean shell length for the sample was $\bar{x} = 55.5$ cm. How likely is it to observe a sample mean of 55.5 cm or larger?

6.39 **Tomato as a taste modifier.** Miraculin is a protein naturally produced in a rare tropical fruit that can convert a sour taste into a sweet taste. Refer to the *Plant Science* (May 2010) investigation of the ability of a hybrid tomato plant to produce miraculin, Exercise 5.38 (p. 242). Recall that the amount x of miraculin produced in the plant had a mean of 105.3 micro-grams per gram of fresh weight with a standard deviation of 8.0. Consider a random sample of $n = 64$ hybrid tomato plants and let $\bar{x}$ represent the sample mean amount of miraculin produced. Would you expect to observe a value of $\bar{x}$ less than 103 micrograms per gram of fresh weight? Explain.

Applying the Concepts—Intermediate

6.40 **Uranium in the Earth's crust.** Refer to the *American Mineralogist* (October 2009) study of the evolution of uranium minerals in the Earth's crust, Exercise 5.9 (p. 230). Recall that researchers estimate that the trace amount of uranium x in reservoirs follows a uniform distribution ranging between 1 and 3 parts per million. In a random sample of $n = 60$ reservoirs, let $\bar{x}$ represent the sample mean amount of uranium.

a. Find $E(\bar{x})$ and interpret its value.

b. Find $Var(\bar{x})$.

c. Describe the shape of the sampling distribution of $\bar{x}$.

d. Find the probability that $\bar{x}$ is between 1.5 ppm and 2.5 ppm.

e. Find the probability that $\bar{x}$ exceeds 2.2 ppm.

6.41 **Critical part failures in NASCAR vehicles.** Refer to *The Sport Journal* (Winter, 2007) analysis of critical part failures at NASCAR races, Exercise 5.102 (p. 260). Recall that researchers found that the time x (in hours) until the first critical part failure is exponentially distributed with $\mu = .10$ and $\sigma = .10$. Now, consider a random sample of $n = 50$ NASCAR races and let $\bar{x}$ represent the sample mean time until the first critical part failure.

a. Find $E(\bar{x})$ and $Var(\bar{x})$.

b. Although x has an exponential distribution, the sampling distribution of $\bar{x}$ is approximately normal. Why?

c. Find the probability that the sample mean time until the first critical part failure exceeds .13 hour.

6.42 **Cost of unleaded fuel.** According to the American Automobile Association (AAA), the average cost of a gallon of regular unleaded fuel at gas stations in August 2010 was \$2.78 (*AAA Fuel Gauge Report*). Assume that the standard deviation of such costs is \$.15. Suppose that a random sample of $n = 100$ gas stations is selected from the population and the cost per gallon of regular unleaded fuel is determined for each. Consider $\bar{x}$, the sample mean cost per gallon.

a. Calculate $\mu_{\bar{x}}$ and $\sigma_{\bar{x}}$.

b. What is the approximate probability that the sample has a mean fuel cost between \$2.78 and \$2.80?

c. What is the approximate probability that the sample has a mean fuel cost that exceeds \$2.80?

d. How would the sampling distribution of $\bar{x}$ change if the sample size n were doubled from 100 to 200? How do your answers to parts **b** and **c** change when the sample size is doubled?

6.43 **Levelness of concrete slabs.** Geotechnical engineers use water-level "manometer" surveys to assess the levelness of newly constructed concrete slabs. Elevations are typically measured at eight points on the slab; of interest is the maximum differential between elevations. The *Journal of Performance of Constructed Facilities* (Feb. 2005) published an article on the levelness of slabs in California residential developments. Elevation data collected on over 1,300 concrete slabs *before tensioning* revealed that the maximum differential x has a mean of $\mu = .53$ inch and a standard deviation of $\sigma = .193$ inch. Consider a sample of $n = 50$ slabs selected from those surveyed, and let $\bar{x}$ represent the mean of the sample.

a. Describe fully the sampling distribution of $\bar{x}$.

b. Find $P(\bar{x} > .58)$.

c. The study also revealed that the mean maximum differential of concrete slabs measured *after tensioning and loading* is $\mu = .58$ inch. Suppose the sample data yield $\bar{x} = .59$ inch. Comment on whether the sample measurements were obtained before tensioning or after tensioning and loading.

6.44 **Motivation of drug dealers.** Refer to the *Applied Psychology in Criminal Justice* (Sept. 2009) investigation of the personality characteristics of drug dealers, Exercise 2.98 (p. 71). Convicted drug dealers were scored on the Wanting Recognition (WR) Scale—a scale which provides a quantitative measure of a person's level of need for approval and sensitivity to social situations. (Higher scores indicate a greater need for approval.) Based on the study results, we can assume that the WR scores for the population of convicted drug dealers has a mean of 40 and a standard deviation of 5. Suppose that in a sample of $n = 100$ people, the mean WR scale score is $\bar{x} = 42$. Is this sample likely to have been selected from the population of convicted drug dealers? Explain.

6.45 Is exposure to a chemical in Teflon-coated cookware hazardous? Perfluorooctanoic acid (PFOA) is a chemical used in Teflon®-coated cookware to prevent food from sticking. The Environmental Protection Agency is investigating the potential risk of PFOA as a cancer-causing agent (*Science News Online*, August 27, 2005). It is known that the blood concentration of PFOA in the general population has a mean of $\mu = 6$ parts per billion (ppb) and a standard deviation of $\sigma = 10$ ppb. *Science News Online* reported on tests for PFOA exposure conducted on a sample of 326 people who live near DuPont's Teflon-making Washington (West Virginia) Works facility.

   a. What is the probability that the average blood concentration of PFOA in the sample is greater than 7.5 ppb?

   b. The actual study resulted in $\bar{x} = 300$ ppb. Use this information to make an inference about the true mean (μ) PFOA concentration for the population that lives near DuPont's Teflon facility.

Applying the Concepts—Advanced

6.46 Test of Knowledge about Epilepsy. The Test of Knowledge about Epilepsy (KAE), which is designed to measure attitudes toward persons with epilepsy, uses 20 multiple-choice items, all of which are incorrect. For each person, two scores (ranging from 0 to 20) are obtained: an attitude score (KAE-A) and a general-knowledge score (KAE-GK). On the basis of a large-scale study of college students, the distribution of KAE-A scores has a mean of $\mu = 11.92$ and a standard deviation of $\sigma = 2.95$ while the distribution of KAE-GK scores has a mean of $\mu = 6.35$ and a standard deviation of $\sigma = 2.12$ (*Rehabilitative Psychology*, Spring 1995). Consider a random sample of 100 college students, and suppose you observe a sample mean KAE score of 6.5. Is this result more likely to be the mean of the attitude scores (KAE-A) or the general-knowledge scores (KAE-GK)? Explain.

6.47 Hand washing versus hand rubbing. Refer to the *British Medical Journal* (August 17, 2002) study comparing the effectiveness of hand washing with soap and hand rubbing with alcohol, presented in Exercise 2.101 (p. 72). Health care workers who used hand rubbing had a mean bacterial count of 35 per hand with a standard deviation of 59. Health care workers who used hand washing had a mean bacterial count of 69 per hand with a standard deviation of 106. In a random sample of 50 health care workers, all using the same method of cleaning their hands, the mean bacterial count per hand ($\bar{x}$) is less than 30. Give your opinion on whether this sample of workers used hand rubbing with alcohol or hand washing with soap.

CHAPTER NOTES

Key Terms

Biased estimate 280
Central Limit Theorem 285
Error of estimation 279
Minimum-variance unbiased
   estimator (MVUE) 282
Parameter 273

Point estimator 279
Sample statistic 273
Sampling distribution 274
Standard error of the mean 284
Standard error of the statistic 281
Unbiased estimate 280

Key Formulas

	Mean	Standard Deviation	z-score
Sampling distribution of $\bar{x}$	$\mu_{\bar{x}} = \mu$	$\sigma_{\bar{x}} = \dfrac{\sigma}{\sqrt{n}}$	$z = \dfrac{\bar{x} - \mu_{\bar{x}}}{\sigma_{\bar{x}}} = \dfrac{\bar{x} - \mu}{\sigma/\sqrt{n}}$

Key Ideas

Sampling distribution of a statistic—the theoretical probability distribution of the statistic in repeated sampling

Unbiased estimator—a statistic with a sampling distribution mean equal to the population parameter being estimated

Central Limit Theorem—the sampling distribution of the sample mean, $\bar{x}$, is approximately normal for large n (e.g., $n \geq 30$).

$\bar{x}$ is the **minimum-variance unbiased estimator (MVUE)** of μ.

Key Symbols

θ	Population parameter (general)
$\mu_{\bar{x}}$	True mean of sampling distribution of $\bar{x}$
$\sigma_{\bar{x}}$	True standard deviation of sampling distribution of $\bar{x}$

Generating the Sampling Distribution of $\bar{X}$

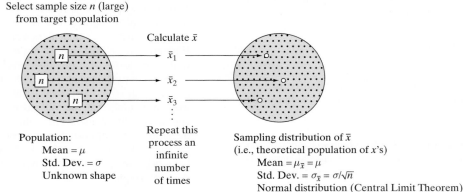

Select sample size n (large) from target population

Calculate $\bar{x}$

$\bar{x}_1$
$\bar{x}_2$
$\bar{x}_3$

Population:
  Mean $= \mu$
  Std. Dev. $= \sigma$
  Unknown shape

Repeat this process an infinite number of times

Sampling distribution of $\bar{x}$
(i.e., theoretical population of x's)
  Mean $= \mu_{\bar{x}} = \mu$
  Std. Dev. $= \sigma_{\bar{x}} = \sigma/\sqrt{n}$
  Normal distribution (Central Limit Theorem)

Supplementary Exercises 6.48–6.74

Understanding the Principles

6.48 Describe how you could obtain the simulated sampling distribution of a sample statistic.

6.49 **True or False.** The sample mean, $\bar{x}$, will always be equal to $\mu_{\bar{x}}$.

6.50 **True or False.** The sampling distribution of $\bar{x}$ is normally distributed, regardless of the size of the sample n.

6.51 **True or False.** The standard error of $\bar{x}$ will always be smaller than σ.

Learning the Mechanics

6.52 The standard deviation (or, as it is usually called, the *standard error*) of the sampling distribution for the sample mean, $\bar{x}$, is equal to the standard deviation of the population from which the sample was selected, divided by the square root of the sample size. That is,

$$\sigma_{\bar{x}} = \frac{\sigma}{\sqrt{n}}$$

a. As the sample size is increased, what happens to the standard error of $\bar{x}$? Why is this property considered important?

b. Suppose a sample statistic has a standard error that is not a function of the sample size. In other words, the standard error remains constant as n changes. What would this imply about the statistic as an estimator of a population parameter?

c. Suppose another unbiased estimator (call it A) of the population mean is a sample statistic with a standard error equal to

$$\sigma_A = \frac{\sigma}{\sqrt[3]{n}}$$

Which of the sample statistics, $\bar{x}$ or A, is preferable as an estimator of the population mean? Why?

d. Suppose that the population standard deviation σ is equal to 10 and that the sample size is 64. Calculate the standard errors of $\bar{x}$ and A. Assuming that the sampling distribution of A is approximately normal, interpret the standard errors. Why is the assumption of (approximate) normality unnecessary for the sampling distribution of $\bar{x}$?

6.53 Consider a sample statistic A. As with all sample statistics, A is computed by utilizing a specified function (formula) of the sample measurements. (For example, if A were the sample mean, the specified formula would sum the measurements and divide by the number of measurements.)

a. Describe what we mean by the phrase "the sampling distribution of the sample statistic A."

b. Suppose A is to be used to estimate a population parameter α. What is meant by the assertion that A is an unbiased estimator of α?

c. Consider another sample statistic, B. Assume that B is also an unbiased estimator of the population parameter α. How can we use the sampling distributions of A and B to decide which is the better estimator of α?

d. If the sample sizes on which A and B are based are large, can we apply the Central Limit Theorem and assert that the sampling distributions of A and B are approximately normal? Why or why not?

6.54 A random sample of 40 observations is to be drawn from a large population of measurements. It is known that 30% of the measurements in the population are 1's, 20% are 2's, 20% are 3's, and 30% are 4's.

a. Give the mean and standard deviation of the (repeated) sampling distribution of $\bar{x}$, the sample mean of the 40 observations.

b. Describe the shape of the sampling distribution of $\bar{x}$. Does your answer depend on the sample size?

6.55 A random sample of $n = 68$ observations is selected from a population with $\mu = 19.6$ and $\sigma = 3.2$. Approximate each of the following probabilities:

a. $P(\bar{x} \leq 19.6)$
b. $P(\bar{x} \leq 19)$
c. $P(\bar{x} \geq 20.1)$
d. $P(19.2 \leq \bar{x} \leq 20.6)$

6.56 Use a statistical software package to generate 100 random samples of size $n = 2$ from a population characterized by a normal probability distribution with a mean of 100 and a standard deviation of 10. Compute $\bar{x}$ for each sample, and plot a frequency distribution for the 100 values of $\bar{x}$. Repeat this process for $n = 5, 10, 30$, and 50. How does the fact that the sampled population is normal affect the sampling distribution of $\bar{x}$?

6.57 Use a statistical software package to generate 100 random samples of size $n = 2$ from a population characterized by a uniform probability distribution with $c = 0$ and $d = 10$. Compute $\bar{x}$ for each sample, and plot a frequency distribution for the 100 $\bar{x}$ values. Repeat this process for $n = 5, 10, 30$, and 50. Explain how your plots illustrate the Central Limit Theorem.

6.58 Suppose x equals the number of heads observed when a single coin is tossed; that is, $x = 0$ or $x = 1$. The population corresponding to x is the set of 0's and 1's generated when the coin is tossed repeatedly a large number of times. Suppose we select $n = 2$ observations from this population. (That is, we toss the coin twice and observe two values of x.)

a. List the three different samples (combinations of 0's and 1's) that could be obtained.

b. Calculate the value of $\bar{x}$ for each of the samples.

c. List the values that $\bar{x}$ can assume, and find the probabilities of observing these values.

d. Construct a graph of the sampling distribution of $\bar{x}$.

6.59 A random sample of size n is to be drawn from a large population with mean 100 and standard deviation 10, and the sample mean $\bar{x}$ is to be calculated. To see the effect of different sample sizes on the standard deviation of the sampling distribution of $\bar{x}$, plot $\sigma/\sqrt{n}$ against n for $n = 1, 5, 10, 20, 30, 40$, and 50.

Applying the Concepts—Basic

6.60 **Salaries of travel professionals.** According to a National Business Travel Association (NBTA) 2008 survey, the average salary of a travel management professional is $97,300. Assume that the standard deviation of such salaries is $30,000. Consider a random sample of 50 travel

management professionals, and let $\bar{x}$ represent the mean salary for the sample.
  a. What is $\mu_{\bar{x}}$?
  b. What is $\sigma_{\bar{x}}$?
  c. Describe the shape of the sampling distribution of $\bar{x}$.
  d. Find the z-score for the value $\bar{x} = \$89,500$.
  e. Find $P(\bar{x} > 89,500)$.

6.61 **Children's attitude toward reading.** In the journal *Knowledge Quest* (January/February 2002), education professors at the University of Southern California investigated children's attitudes toward reading. One study measured third through sixth graders' attitudes toward recreational reading on a 140-point scale (where higher scores indicate a more positive attitude). The mean score for this population of children was 106 points, with a standard deviation of 16.4 points. Consider a random sample of 36 children from this population, and let $\bar{x}$ represent the mean recreational reading attitude score for the sample.
  a. What is $\mu_{\bar{x}}$?
  b. What is $\sigma_{\bar{x}}$?
  c. Describe the shape of the sampling distribution of $\bar{x}$.
  d. Find the z-score for the value $\bar{x} = 100$ points.
  e. Find $P(\bar{x} < 100)$.

6.62 **Violence and stress.** Interpersonal violence (e.g., rape) generally leads to psychological stress for the victim. *Clinical Psychology Review* (Vol. 15, 1995) reported on the results of all recently published studies of the relationship between interpersonal violence and psychological stress. The distribution of the time elapsed between the violent incident and the initial sign of stress has a mean of 5.1 years and a standard deviation of 6.1 years. Consider a random sample of $n = 150$ victims of interpersonal violence. Let $\bar{x}$ represent the mean time elapsed between the violent act and the first sign of stress for the sampled victims.
  a. Give the mean and standard deviation of the sampling distribution of $\bar{x}$.
  b. Will the sampling distribution of $\bar{x}$ be approximately normal? Explain.
  c. Find $P(\bar{x} > 5.5)$.
  d. Find $P(4 < \bar{x} < 5)$.

6.63 **Improving SAT scores.** Refer to the *Chance* (Winter 2001) examination of SAT scores of students who pay a private tutor to help them improve their results, presented in Exercise 2.105 (p. 73). On the SAT-Mathematics test, these students had a mean change in score of +19 points, with a standard deviation of 65 points. In a random sample of 100 students who pay a private tutor to help them improve their results, what is the likelihood that the change in the sample mean score is less than 10 points?

Applying the Concepts—Intermediate

6.64 **Producing machine bearings.** To determine whether a metal lathe that produces machine bearings is properly adjusted, a random sample of 25 bearings is collected and the diameter of each is measured.
  a. If the standard deviation of the diameters of the bearings measured over a long period of time is .001 inch, what is the approximate probability that the mean diameter $\bar{x}$ of the sample of 25 bearings will lie within .0001 inch of the population mean diameter of the bearings?
  b. If the population of diameters has an extremely skewed distribution, how will your approximation in part **a** be affected?

6.65 **Quality control.** Refer to Exercise 6.64. The mean diameter of the bearings produced by the machine is supposed to be .5 inch. The company decides to use the sample mean from Exercise 6.64 to decide whether the process is in control (i.e., whether it is producing bearings with a mean diameter of .5 inch). The machine will be considered out of control if the mean of the sample of $n = 25$ diameters is less than .4994 inch or larger than .5006 inch. If the true mean diameter of the bearings produced by the machine is .501 inch, what is the approximate probability that the test will imply that the process is out of control?

6.66 **Length of job tenure.** Researchers at the Terry College of Business at the University of Georgia sampled 344 business students and asked them this question: "Over the course of your lifetime, what is the maximum number of years you expect to work for any one employer?" The sample resulted in $\bar{x} = 19.1$ years. Assume that the sample of students was randomly selected from the 6,000 undergraduate students at the Terry College and that $\sigma = 6$ years.
  a. Describe the sampling distribution of $\bar{x}$.
  b. If the mean for the 6,000 undergraduate students is $\mu = 18.5$ years, find $P(\bar{x} > 19.1)$
  c. If the mean for the 6,000 undergraduate students is $\mu = 19.5$ years, find $P(\bar{x} > 19.1)$
  d. If $P(\bar{x} > 19.1) = .5$, what is μ?
  e. If $P(\bar{x} > 19.1) = .2$, is μ greater than or less than 19.1 years? Explain.

6.67 **Susceptibility to hypnosis.** The Computer-Assisted Hypnosis Scale (CAHS) is designed to measure a person's susceptibility to hypnosis. In computer-assisted hypnosis, the computer serves as a facilitator of hypnosis by using digitized speech processing coupled with interactive involvement with the hypnotic subject. CAHS scores range from 0 (no susceptibility) to 12 (extremely high susceptibility). A study in *Psychological Assessment* (March 1995) reported a mean CAHS score of 4.59 and a standard deviation of 2.95 for University of Tennessee undergraduates. Assume that $\mu = 4.59$ and $\sigma = 2.95$ for this population. Suppose a psychologist uses the CAHS to test a random sample of 50 subjects.
  a. Would you expect to observe a sample mean CAHS score of $\bar{x} = 6$ or higher? Explain.
  b. Suppose the psychologist actually observes $\bar{x} = 6.2$. On the basis of your answer to part **a**, make an inference about the population from which the sample was selected.

6.68 **Supercooling temperature of frogs.** Many species of terrestrial tree frogs that hibernate at or near the ground surface can survive prolonged exposure to low winter temperatures. In freezing conditions, the frog's body temperature, called its *supercooling temperature*, remains relatively higher because of an accumulation of glycerol in its body fluids. A study in *Science* revealed that the supercooling temperature of terrestrial frogs frozen at $-6°C$ has a relative frequency distribution with a mean of $-2°C$ and a standard deviation of .3°C. Consider the mean supercooling temperature $\bar{x}$ of a random sample of $n = 42$ terrestrial frogs frozen at $-6°C$.
  a. Find the probability that $\bar{x}$ exceeds $-2.05°C$.
  b. Find the probability that $\bar{x}$ falls between $-2.20°C$ and $-2.10°C$.

6.69 **Surface roughness of pipe.** The journal *Anti-Corrosion Methods and Materials* (Vol. 50, 2003) published a study of the surface roughness of oil field pipes. A scanning probe instrument was used to measure the surface roughness x (in micrometers) of 20 sampled sections of coated interior pipe. Consider the sample mean $\bar{x}$.

a. Assume that the surface roughness distribution has a mean of $\mu = 1.8$ micrometers and a standard deviation of $\sigma = .5$ micrometer. Use this information to find the probability that $\bar{x}$ exceeds 1.85 micrometers.

b. The sample data from the study are listed in the table and saved in the **ROUGHPIPE** file. Compute $\bar{x}$.

c. On the basis of the result from part **b,** comment on the validity of the assumptions made in part **a.**

1.72	2.50	2.16	2.13	1.06	2.24	2.31	2.03	1.09	1.40
2.57	2.64	1.26	2.05	1.19	2.13	1.27	1.51	2.41	1.95

Based on Farshad, F., and Pesacreta, T. "Coated pipe interior surface roughness as measured by three scanning probe instruments." *Anti-Corrosion Methods and Materials*, Vol. 50, No. 1, 2003 (Table II).

6.70 Verbal ability of delinquents. Over the past 20 years, the *Journal of Abnormal Psychology* has published numerous studies on the relationship between juvenile delinquency and poor verbal abilities. Assume that scores on a verbal IQ test have a population mean $\mu = 107$ and a population standard deviation $\sigma = 15$.

a. What shape would you expect the sampling distribution of $\bar{x}$ for $n = 84$ juveniles to have? Does your answer depend on the shape of the distribution of verbal IQ scores for all juveniles?

b. Assuming that the population mean and standard deviation for juveniles with no record of delinquency are the same as those for all juveniles, approximate the probability that the sample mean verbal IQ for $n = 84$ juveniles will be 110 or more. State any assumptions you make.

c. In one study cited, the researchers found that a sample of $n = 84$ juveniles with no record of delinquency had a mean verbal IQ of $\bar{x} = 110$. Considering your answer to part **b,** do you think that the population mean and standard deviation for nondelinquent juveniles are the same as those for all juveniles? Explain.

Applying the Concepts—Advanced

6.71 Flaws in aluminum siding. [*Note:* This exercise refers to an optional section in Chapter 4.] A building contractor has decided to purchase a load of factory-reject aluminum siding as long as the average number of flaws per piece of siding in a sample of size 35 from the factory's reject pile is 2.1 or less. If it is known that the number of flaws per piece of siding in the factory's reject pile has a Poisson probability distribution with a mean of 2.5, find the approximate probability that the contractor will not purchase a load of siding. [*Hint:* If x is a Poisson random variable with mean λ, then $\sigma_{\bar{x}}^2$ equals λ/n.]

6.72 Machine repair time. [*Note:* This exercise refers to an optional section in Chapter 5.] An article in *Industrial Engineering* (August 1990) discussed the importance of modeling machine downtime correctly in simulation studies. As an illustration, the researcher considered a single-machine-tool system with repair times (in minutes) that can be modeled by an exponential distribution with $\theta = 60$. (See optional Section 5.6.) Of interest is the mean repair time $\bar{x}$ of a sample of 100 machine breakdowns.

a. Find $E(\bar{x})$ and the variance of $\bar{x}$.

b. What probability distribution provides the best model of the sampling distribution of $\bar{x}$? Why?

c. Calculate the probability that the mean repair time $\bar{x}$ is no longer than 30 minutes.

Critical Thinking Challenges

6.73 Soft-drink bottles. A soft-drink bottler purchases glass bottles from a vendor. The bottles are required to have an internal pressure of at least 150 pounds per square inch (psi). A prospective bottle vendor claims that its production process yields bottles with a mean internal pressure of 157 psi and a standard deviation of 3 psi. The bottler strikes an agreement with the vendor that permits the bottler to sample from the vendor's production process to verify the vendor's claim. The bottler randomly selects 40 bottles from the last 10,000 produced, measures the internal pressure of each, and finds the mean pressure for the sample to be 1.3 psi below the process mean cited by the vendor.

a. Assuming the vendor's claim to be true, what is the probability of obtaining a sample mean this far or farther below the process mean? What does your answer suggest about the validity of the vendor's claim?

b. If the process standard deviation were 3 psi as claimed by the vendor, but the mean were 156 psi, would the observed sample result be more or less likely than in part **a**? What if the mean were 158 psi?

c. If the process mean were 157 psi as claimed, but the process standard deviation were 2 psi, would the sample result be more or less likely than in part **a**? What if instead the standard deviation were 6 psi?

6.74 Fecal pollution at Huntington Beach. The state of California mandates fecal indicator bacteria monitoring at all public beaches. When the concentration of fecal bacteria in the water exceeds a certain limit (400 colony-forming units of fecal coliform per 100 milliliters), local health officials must post a sign (called surf zone posting) warning beachgoers of potential health risks upon entering the water. For fecal bacteria, the state uses a single-sample standard; that is, if the fecal limit is exceeded in a single sample of water, surf zone posting is mandatory. This single-sample standard policy has led to a recent rash of beach closures in California.

Joon Ha Kim and Stanley B. Grant, engineers at the University of California at Irvine, conducted a study of the surf water quality at Huntington Beach in California and reported the results in *Environmental Science & Technology* (September 2004). The researchers found that beach closings were occurring despite low pollution levels in some instances, while in others signs were not posted when the fecal limit was exceeded. They attributed these "surf zone posting errors" to the variable nature of water quality in the surf zone (for example, fecal bacteria concentration tends to be higher during ebb tide and at night) and the inherent time delay between when a water sample is collected and when a sign is posted or removed. In order to prevent posting errors, the researchers recommend using an averaging method, rather than a single sample, to determine unsafe water quality. (For example, one simple averaging method is to take a random sample of multiple water specimens and compare the average fecal bacteria level of the sample with the limit of 400 cfu/100 mL in order to determine whether the water is safe.)

Discuss the pros and cons of using the single-sample standard versus the averaging method. Part of your discussion should address the probability of posting a sign when in fact the water is safe and the probability of posting a sign when in fact the water is unsafe. (Assume that the fecal bacteria concentrations of water specimens at Huntington Beach follow an approximately normal distribution.)

Activity   Simulating a Sampling Distribution—Cell Phone Usage

[*Note:* This activity is designed for small groups or the entire class.] Consider the length of time a student spends making a cell phone call, sending/retrieving a text message, or accessing e-mail on his/her cell phone. Let *x* represent the length of time, in seconds, for a single cell phone activity (call, text, or e-mail). Here, we are interested in the sampling distribution of $\bar{x}$, the mean length of time for a sample of size *n* cell phone activities.

1. Keep track of the time lengths for all cell phone activities you engage in over the next week.

2. Pool your time length data with data from other class members or the entire class so that the pooled data set has at least 100 observations. Designate someone in the group to calculate the mean and standard deviation of the pooled data set.

3. Devise a convenient way to choose random samples from the pooled data set. (For example, you could assign each observation a number beginning with "1" and use a random number generator to select a sample.)

4. Choose a random sample of size $n = 30$ from the pooled data, and find the mean of the sample. Group members should repeat the process of choosing a sample of size $n = 30$ from the pooled data and finding the sample mean until the group has accumulated at least 25 sample means. (Call this data set *Sample Means.*)

5. Find the mean and standard deviation of the *Sample Means* data set. Also, form a histogram for the *Sample Means* data set. Explain how the Central Limit Theorem is illustrated in this activity.

References

Hogg, R. V., McKean, J. W., and Craig, A. T. *Introduction to Mathematical Statistics*, 6th ed. Upper Saddle River, NJ: Prentice Hall, 2005.

Larsen, R. J., and Marx, M. L. *An Introduction to Mathematical Statistics and Its Applications*, 4th ed. Upper Saddle River, NJ: Prentice Hall, 2005.

Lindgren, B. W. *Statistical Theory*, 3d ed. New York: Macmillan, 1976.

Wackerly, D., Mendenhall, W., and Scheaffer, R. L. *Mathematical Statistics with Applications*, 7th ed. North Scituate, MA: Duxbury, 2008.

USING TECHNOLOGY

MINITAB: Simulating a Sampling Distribution

Step 1 Select "Calc" on the MINITAB menu bar, and then click on "Random Data" (see Figure 6.M.1).

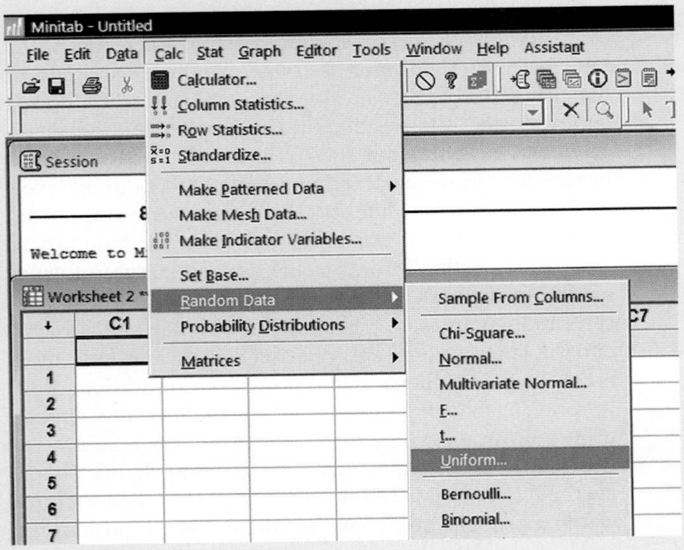

Figure 6.M.1
MINITAB options for generating random data

Step 2 On the resulting menu list, click on the distribution of your choice (e.g., "Uniform"). A dialog box similar to the one (the Uniform Distribution) shown in Figure 6.M.2 will appear.

Step 3 Specify the number of samples (e.g., 1,000) to generate in the "Number of rows to generate" box and the columns where the data will be stored in the "Store in columns" box. (The number of columns will be equal to the sample size, e.g.,

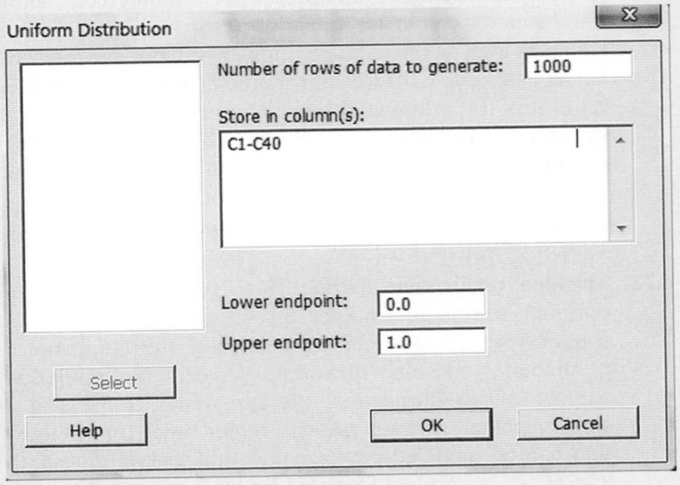

Figure 6.M.2
MINITAB dialog box for simulating the uniform distribution

40.) Finally, specify the parameters of the distribution (e.g., the lower and upper range of the uniform distribution). When you click "OK," the simulated data will appear on the MINITAB worksheet.

Step 4 Calculate the value of the sample statistic of interest for each sample. To do this, click on the "Calc" button on the MINITAB menu bar, and then click on "Row Statistics," as shown in Figure 6.M.3. The resulting dialog box appears in Figure 6.M.4.

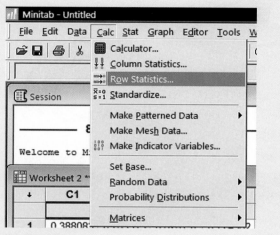

Figure 6.M.3
MINITAB selections for generating sample statistics for the simulated data

Figure 6.M.4
MINITAB row statistics dialog box

Step 5 Check the sample statistic (e.g., the mean) you want to calculate, specify the "Input variables" (or columns), and the column where you want the value of the sample statistic to be saved. Click "OK" and the value of the statistic for each sample will appear on the MINITAB worksheet.

[*Note:* Use the MINITAB menu choices provided in the Using Technology section in Chapter 2, p. 106, to generate a histogram of the sampling distribution of the statistic or to find the mean and variance of the sampling distribution.]

7 Inferences Based on a Single Sample
Estimation with Confidence Intervals

CONTENTS

Where We've Been

- Learned that populations are characterized by numerical descriptive measures—called *parameters*
- Found that decisions about population parameters are based on *statistics* computed from the sample
- Discovered that *inferences* about parameters are subject to uncertainty and that this uncertainty is reflected in the *sampling distribution* of a statistic

Where We're Going

- Estimate a population parameter (mean, proportion, or variance) on the basis of a sample selected from the population (7.1)
- Use the sampling distribution of a statistic to form a confidence interval for the population parameter (7.2–7.4, 7.6)
- Show how to select the proper sample size for estimating a population parameter (7.5)

Statistics in Action   Medicare Fraud Investigations

United States Department of Justice (USDOJ) press release (May 8, 2008): *Eleven people have been indicted in . . . a targeted criminal, civil and administrative effort against individuals and health care companies that fraudulently bill the Medicare program. The indictments in the Central District of California resulted from the creation of a multi-agency team of federal, state and local investigators designed specifically to combat Medicare fraud through the use of real-time analysis of Medicare billing data.*

USDOJ press release (Jun 27, 2008): *The owners of four Miami-based health care corporations were sentenced and remanded to prison yesterday for their roles in schemes to defraud the Medicare program. Collectively, the three defendants through their companies collected more than $14 million from the Medicare program for unnecessary medicine, durable medical equipment (DME) and home health care services.*

USDOJ press release (Oct. 7, 2008): *Eight Miami–Dade County residents have been charged in a 16-count indictment for their alleged roles in a Medicare fraud scheme involving fake HIV infusion treatments.*

As the above press releases imply, the U.S. Department of Justice (USDOJ) and its Medicare Fraud Strike Force conduct investigations into suspected fraud and abuse of the Medicare system by health care providers. According to published reports, the Strike Force was responsible for almost 25% of the Medicare fraud charges brought nationwide in 2007.

One way in which Medicare fraud occurs is through the use of "upcoding," which refers to the practice of providers coding Medicare claims at a higher level of care than was actually provided to the patient. For example, suppose a particular kind of claim can be coded at three levels, where Level 1 is a routine office visit, Level 2 is a thorough examination involving advanced diagnostic tests, and Level 3 involves performing minor surgery. The amount of Medicare payment is higher for each increased level of claim. Thus, upcoding would occur if Level 1 services were billed at Level 2 or Level 3 payments, or if Level 2 services were billed at Level 3 payment.

The USDOJ relies on sound statistical methods to help identify Medicare fraud. Once the USDOJ has determined that possible upcoding has occurred, it next seeks to further investigate whether it is the result of legitimate practice (perhaps the provider is a specialist giving higher levels of care) or the result of fraudulent action on the part of the provider. To further its investigation, the USDOJ will next ask a statistician to select a sample of the provider's claims. For example, the statistician might determine that a random sample of 52 claims from the 1,000 claims in question will provide a sufficient sample to estimate the overcharge reliably. The USDOJ then asks a health care expert to audit each of the medical files corresponding to the sampled claims and determine whether the level of care matches the level billed by the provider, and, if not, to determine what level should have been billed. Once the audit has been completed, the USDOJ will calculate the overcharge.

In this chapter, we present a recent Medicare fraud case investigated by the USDOJ. Results for the audit of 52 sampled claims, with the amount paid for each claim, the amount disallowed by the auditor, and the amount that should have been paid for each claim, are saved in the **MCFRAUD** file.* Knowing that a total of $103,500 was paid for the 1,000 claims, the USDOJ wants to use the sample results to extrapolate the overpayment amount to the entire population of 1,000 claims.

Statistics in Action Revisited

- Estimating the Mean Overpayment (p. 315)
- Estimating the Coding Error Rate (p. 324)
- Determining Sample Size (p. 331)

Data Set: MCFRAUD

7.1 Identifying and Estimating the Target Parameter

In this chapter, our goal is to estimate the value of an unknown population parameter, such as a population mean or a proportion from a binomial population. For example, we might want to know the mean gas mileage for a new car model, the average expected life of a flat-screen computer monitor, or the proportion of Iraq War veterans with post-traumatic stress syndrome.

You'll see that different techniques are used for estimating a mean or proportion, depending on whether a sample contains a large or small number of measurements. Nevertheless, our objectives remain the same: We want to use the sample information to

*Data provided (with permission) from Info Tech, Inc., Gainesville, Florida.

estimate the population parameter of interest (called the *target parameter*) and to assess the reliability of the estimate.

> The unknown population parameter (e.g., mean or proportion) that we are interested in estimating is called the **target parameter.**

Often, there are one or more key words in the statement of the problem that indicate the appropriate target parameter. Some key words associated with the parameters covered in this chapter are listed in the following box:

Determining the Target Parameter

Parameter	Key Words or Phrases	Type of Data
μ	Mean; average	Quantitative
p	Proportion; percentage; fraction; rate	Qualitative
σ^2 (optional)	Variance; variability; spread	Quantitative

For the examples given in the first paragraph of this section, the words *mean* in "mean gas mileage" and *average* in "average life expectancy" imply that the target parameter is the population mean μ. The word *proportion* in "proportion of Iraq War veterans with post-traumatic stress syndrome" indicates that the target parameter is the binomial proportion p.

In addition to key words and phrases, the type of data (quantitative or qualitative) collected is indicative of the target parameter. With quantitative data, you are likely to be estimating the mean or variance of the data. With qualitative data with two outcomes (success or failure), the binomial proportion of successes is likely to be the parameter of interest.

A single number calculated from the sample that estimates a target population parameter is called a **point estimator.** For example, we'll use the sample mean, $\bar{x}$, to estimate the population mean μ. Consequently, $\bar{x}$ is a point estimator. Similarly, we'll learn that the sample proportion of successes, denoted $\hat{p}$, is a point estimator for the binomial proportion p and that the sample variance s^2 is a point estimator for the population variance σ^2. Also, we will attach a measure of reliability to our estimate by obtaining an **interval estimator**—a range of numbers that contain the target parameter with a high degree of confidence. For this reason the interval estimate is also called a **confidence interval.**

> A **point estimator** of a population parameter is a rule or formula that tells us how to use the sample data to calculate a *single* number that can be used as an *estimate* of the target parameter.

> An **interval estimator (or confidence interval)** is a formula that tells us how to use the sample data to calculate an *interval* that *estimates* the target parameter.

We consider methods for estimating a population mean in Sections 7.2 and 7.3. Estimating a population proportion is presented in Section 7.4. We show how to determine the sample sizes necessary for reliable estimates of the target parameters in Section 7.5. Finally, in optional Section 7.6 we discuss estimation of a population variance.

7.2 Confidence Interval for a Population Mean: Normal (z) Statistic

Suppose a large hospital wants to estimate the average length of time patients remain in the hospital. Hence, the hospital's target parameter is the population mean μ. To accomplish this objective, the hospital administrators plan to randomly sample 100 of all previous patients' records and to use the sample mean $\bar{x}$ of the lengths of stay to estimate μ, the mean of *all* patients' visits. The sample mean $\bar{x}$ represents a *point estimator* of the population mean μ. How can we assess the accuracy of this large-sample point estimator?

According to the Central Limit Theorem, the sampling distribution of the sample mean is approximately normal for large samples, as shown in Figure 7.1. Let us calculate the interval estimator:

$$\bar{x} \pm 1.96\sigma_{\bar{x}} = \bar{x} \pm \frac{1.96\sigma}{\sqrt{n}}$$

That is, we form an interval from 1.96 standard deviations below the sample mean to 1.96 standard deviations above the mean. *Prior to drawing the sample*, what are the chances that this interval will enclose μ, the population mean?

To answer this question, refer to Figure 7.1. If the 100 measurements yield a value of $\bar{x}$ that falls between the two lines on either side of μ (i.e., within 1.96 standard deviations of μ), then the interval $\bar{x} \pm 1.96\sigma_{\bar{x}}$ will contain μ; if $\bar{x}$ falls outside these boundaries, the interval $\bar{x} \pm 1.96\sigma_{\bar{x}}$ will not contain μ. From Section 6.3, we know the area under the normal curve (the sampling distribution of $\bar{x}$) between these boundaries is exactly .95. Thus, the interval $\bar{x} \pm 1.96\sigma_{\bar{x}}$ will contain μ with a probability equal to .95.

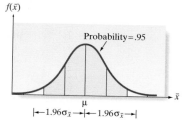

Figure 7.1

Sampling distribution of $\bar{x}$

Example 7.1

Estimating the Mean Hospital Length of Stay, σ Known

Problem Consider the large hospital that wants to estimate the average length of stay of its patients, μ. The hospital randomly samples $n = 100$ of its patients and finds that the sample mean length of stay is $\bar{x} = 4.5$ days. Also, suppose it is known that the standard deviation of the length of stay for all hospital patients is $\sigma = 4$ days. Use the interval estimator $\bar{x} \pm 1.96\sigma_{\bar{x}}$ to calculate a confidence interval for the target parameter, μ.

Solution Substituting $\bar{x} = 4.5$ and $\sigma = 4$ into the interval estimator formula, we obtain:

$$\bar{x} \pm 1.96\sigma_{\bar{x}} = \bar{x} \pm (1.96)\sigma/\sqrt{n} = 4.5 \pm (1.96)(4/\sqrt{100}) = 4.5 \pm .78$$

Or, (3.72, 5.28).

Look Back Since we know the probability that the interval, $\bar{x} \pm 1.96\sigma_{\bar{x}}$, will contain μ is .95, we call the interval estimator a *95% confidence interval* for μ.

Now Work Exercise 7.9a

The interval $\bar{x} \pm 1.96\sigma_{\bar{x}}$ in Example 7.1 is called a *large-sample* 95% confidence interval for the population mean μ. The term *large-sample* refers to the sample being of sufficiently large size that we can apply the Central Limit Theorem and the **normal (z) statistic** to determine the form of the sampling distribution of $\bar{x}$. Empirical research suggests that a sample size n exceeding a value between 20 and 30 will usually yield a sampling distribution of $\bar{x}$ that is approximately normal. This result led many practitioners to adopt the rule of thumb that a sample size of $n \geq 30$ is required to use large-sample confidence interval procedures. Keep in mind, though, that 30 is not a magical number and, in fact, is quite arbitrary.

Also, note that the large-sample interval estimator requires knowing the value of the population standard deviation, σ. In most (if not nearly all) practical applications, however, the value of σ will be unknown. For large samples, the fact that σ is unknown poses only a minor problem since the sample standard deviation s provides a very good approximation to σ. The next example illustrates the more realistic large-sample confidence interval procedure.

Example 7.2

Estimating the Mean Hospital Length of Stay, σ Unknown

Problem Refer to Example 7.1 and the problem of estimating μ, the average length of stay of a hospital's patients. The lengths of stay (in days) for the $n = 100$ sampled patients are shown in Table 7.1. Use the data to find a 95% confidence interval for μ and interpret the result.

Table 7.1   Lengths of Stay (in Days) for 100 Patients

2	3	8	6	4	4	6	4	2	5
8	10	4	4	4	2	1	3	2	10
1	3	2	3	4	3	5	2	4	1
2	9	1	7	17	9	9	9	4	4
1	1	1	3	1	6	3	3	2	5
1	3	3	14	2	3	9	6	6	3
5	1	4	6	11	22	1	9	6	5
2	2	5	4	3	6	1	5	1	6
17	1	2	4	5	4	4	3	2	3
3	5	2	3	3	2	10	2	4	2

Data Set: HOSPLOS

Solution The hospital almost surely does not know the true standard deviation, σ, of the population of lengths of stay. However, since the sample size is large, we will use the sample standard deviation, s, as an estimate for σ in the confidence interval formula. An SAS printout of summary statistics for the sample of 100 lengths of stay is shown at the top of Figure 7.2. From the shaded portion of the printout, we find $\bar{x} = 4.53$ days and $s = 3.68$ days. Substituting these values into the interval estimator formula, we obtain:

$$\bar{x} \pm (1.96)\sigma/\sqrt{n} \approx \bar{x} \pm (1.96)s/\sqrt{n} = 4.53 \pm (1.96)(3.68)/\sqrt{100} = 4.53 \pm .72$$

Or, (3.81, 5.25). That is, we estimate the mean length of stay for all hospital patients to fall within the interval 3.81 to 5.25 days.

Look Back The confidence interval is also shown at the bottom of the SAS printout, Figure 7.2. Note that the endpoints of the interval vary slightly from those computed in

```
Sample Statistics for LOS

      N          Mean        Std. Dev.        Std. Error
   ------------------------------------------------------
     100          4.53          3.68             0.37

Hypothesis Test

     Null hypothesis:    Mean of LOS  =  0
     Alternative:        Mean of LOS ^= 0

            t Statistic        Df        Prob > t
         --------------------------------------------
              12.318           99         <.0001


95 % Confidence Interval for the Mean

        Lower Limit:              3.80
        Upper Limit:              5.26
```

Figure 7.2

SAS printout with summary statistics and 95% confidence interval for data on 100 hospital stays the example. This is due to the fact that when σ is unknown and n is large, the sampling distribution of $\bar{x}$ will deviate slightly from the normal (z) distribution. In practice, these differences can be ignored.

Now Work Exercise 7.13

Can we be sure that μ, the true mean, is in the interval from 3.81 to 5.25 in Example 7.2? We cannot be certain, but we can be reasonably confident that it is. This confidence is derived from the knowledge that if we were to draw repeated random samples of 100 measurements from this population and form the interval $\bar{x} \pm 1.96\sigma_{\bar{x}}$ each time, 95% of the intervals would contain μ. We have no way of knowing (without looking at all the patients' records) whether our sample interval is one of the 95% that contain μ or one of the 5% that do not, but the odds certainly favor its containing μ. The probability, .95, that measures the confidence we can place in the interval estimate is called a *confidence coefficient*. The percentage, 95%, is called the *confidence level* for the interval estimate.

> The **confidence coefficient** is the probability that an interval estimator encloses the population parameter—that is, the relative frequency with which the interval estimator encloses the population parameter when the estimator is used repeatedly a very large number of times. The **confidence level** is the confidence coefficient expressed as a percentage.

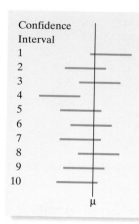

Confidence Interval

1
2
3
4
5
6
7
8
9
10

μ

Figure 7.3

Confidence intervals for μ: 10 samples

Now we have seen how an interval can be used to estimate a population mean. When we use an interval estimator, we can usually calculate the probability that the estimation *process* will result in an interval that contains the true value of the population mean. That is, the probability that the interval contains the parameter in repeated usage is usually known. Figure 7.3 shows what happens when 10 different samples are drawn from a population and a confidence interval for μ is calculated from each. The location of μ is indicated by the vertical line in the figure. Ten confidence intervals, each based on one of 10 samples, are shown as horizontal line segments. Note that the confidence intervals move from sample to sample, sometimes containing μ and other times missing μ. *If our confidence level is 95%, then in the long run, 95% of our sample confidence intervals will contain μ.*

Suppose you wish to choose a confidence coefficient other than .95. Notice in Figure 7.1 that the confidence coefficient .95 is equal to the total area under the sampling distribution, less .05 of the area, which is divided equally between the two tails. Using this idea, we can construct a confidence interval with any desired confidence coefficient by increasing or decreasing the area (call it α) assigned to the tails of the sampling distribution. (See Figure 7.4.) For example, if we place the area $\alpha/2$ in each tail and if $z_{\alpha/2}$ is the z value such that $\alpha/2$ will lie to its right, then the confidence interval with confidence coefficient $(1 - \alpha)$ is

$$\bar{x} \pm z_{\alpha/2}\sigma_{\bar{x}}$$

Figure 7.4

Locating $z_{\alpha/2}$ on the standard normal curve

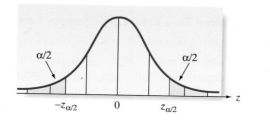

Speaking Statistics with a Polish Accent

Polish-born Jerzy Neyman was educated at the University of Kharkov (Russia) in elementary mathematics, but taught himself graduate mathematics by studying journal articles on the subject. After receiving his doctorate in 1924 from the University of Warsaw (Poland), Neyman accepted a position at University College (London). There, he developed a friendship with Egon Pearson; Neyman and Pearson together developed the theory of hypothesis testing (Chapter 8). In a 1934 talk to the Royal Statistical Society, Neyman first proposed the idea of interval estimation, which he called "confidence intervals." (It is interesting that Neyman rarely receives credit in textbooks as the originator of the confidence interval procedure.) In 1938, he emigrated to the United States and went to the University of California at Berkeley, where he built one of the strongest statistics departments in the country. Jerzy Neyman is considered one of the great founders of modern statistics. He was a superb teacher and innovative researcher who loved his students, always sharing his ideas with them. Neyman's influence on those he met is best expressed by a quote from prominent statistician David Salsburg: "We have all learned to speak statistics with a Polish accent." ∎

The value z_α is defined as the value of the standard normal random variable z such that the area α will lie to its right. In other words, $P(z > z_\alpha) = \alpha$.

To illustrate, for a confidence coefficient of .90, we have $(1 - \alpha) = .90$, $\alpha = .10$, and $\alpha/2 = .05$; $z_{.05}$ is the z value that locates area .05 in the upper tail of the sampling distribution. Recall that Table IV in Appendix A gives the areas between the mean and a specified z-value. Since the total area to the right of the mean is .5, we find that $z_{.05}$ will be the z value corresponding to an area of $.5 - .05 = .45$ to the right of the mean. (See Figure 7.5.) This z value is $z_{.05} = 1.645$.

Confidence coefficients used in practice usually range from .90 to .99. The most commonly used confidence coefficients with corresponding values of α and $z_{\alpha/2}$ are shown in Table 7.2.

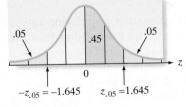

Figure 7.5
The z value ($z_{.05}$) corresponding to an area equal to .05 in the upper tail of the z-distribution

Table 7.2	**Commonly Used Values of $z_{\alpha/2}$**		
Confidence Level			
$100(1 - \alpha)\%$	α	$\alpha/2$	$z_{\alpha/2}$
90%	.10	.05	1.645
95%	.05	.025	1.96
99%	.01	.005	2.575

Now Work Exercise 7.7

Large-Sample $100(1 - \alpha)\%$ Confidence Interval for μ, Based on a Normal (z) Statistic

σ *known*: $\quad \bar{x} \pm (z_{\alpha/2})\sigma_{\bar{x}} = \bar{x} \pm (z_{\alpha/2})(\sigma/\sqrt{n})$

σ *unknown*: $\quad \bar{x} \pm (z_{\alpha/2})\sigma_{\bar{x}} \approx \bar{x} \pm (z_{\alpha/2})(s/\sqrt{n})$

where $z_{\alpha/2}$ is the z-value corresponding to an area $\alpha/2$ in the tail of a standard normal distribution (see Figure 7.4), $\sigma_{\bar{x}}$ is the standard deviation of the sampling distribution of $\bar{x}$, σ is the standard deviation of the population, and s is the standard deviation of the sample.

Conditions Required for a Valid Large-Sample Confidence Interval for μ

1. A random sample is selected from the target population.
2. The sample size n is large (i.e., $n \geq 30$). (Due to the Central Limit Theorem, this condition guarantees that the sampling distribution of $\bar{x}$ is approximately normal. Also, for large n, s will be a good estimator of σ.)

Example 7.3

A Large–Sample Confidence Interval for μ— Mean Number of Unoccupied Seats per Flight

Problem Unoccupied seats on flights cause airlines to lose revenue. Suppose a large airline wants to estimate its average number of unoccupied seats per flight over the past year. To accomplish this, the records of 225 flights are randomly selected, and the number of unoccupied seats is noted for each of the sampled flights. (The data are saved in the **AIRNOSHOWS** file.) Descriptive statistics for the data are displayed in the MINITAB printout of Figure 7.6.

Estimate μ, the mean number of unoccupied seats per flight during the past year, using a 90% confidence interval.

Solution The form of a large-sample 90% confidence interval for a population mean (based on the *z*-statistic) is:

$$\bar{x} \pm z_{\alpha/2}\sigma_{\bar{x}} = \bar{x} \pm z_{.05}\sigma_{\bar{x}} = \bar{x} \pm 1.645\left(\frac{\sigma}{\sqrt{n}}\right)$$

From Figure 7.6, we find (after rounding) that $\bar{x} = 11.6$. Since we do not know the value of σ (the standard deviation of the number of unoccupied seats per flight for all flights of the year), we use our best approximation—the sample standard deviation, $s = 4.1$, shown on the MINITAB printout. Then the 90% confidence interval is approximately

$$11.6 \pm 1.645\left(\frac{4.1}{\sqrt{225}}\right) = 11.6 \pm .45$$

or from 11.15 to 12.05. That is, at the 90% confidence level, we estimate the mean number of unoccupied seats per flight to be between 11.15 and 12.05 during the sampled year. This result is verified (except for rounding) on the right side of the MINITAB printout in Figure 7.6.

Figure 7.6

MINITAB printout with descriptive statistics and 90% confidence interval for Example 7.3

Variable	N	Mean	StDev	SE Mean	90% CI
NOSHOWS	225	11.596	4.103	0.274	(11.144, 12.047)

Look Back We stress that the confidence level for this example, 90%, refers to the procedure used. If we were to apply that procedure repeatedly to different samples, approximately 90% of the intervals would contain μ. Although we do not know for sure whether this particular interval (11.15, 12.05) is one of the 90% that contain μ or one of the 10% that do not, our knowledge of probability gives us "confidence" that the interval contains μ.

Now Work Exercise 7.14a

The interpretation of confidence intervals for a population mean is summarized in the next box.

Interpretation of a Confidence Interval for a Population Mean

When we form a $100(1 - \alpha)\%$ confidence interval for μ, we usually express our confidence in the interval with a statement such as "We can be $100(1 - \alpha)\%$ confident that μ lies between the lower and upper bounds of the confidence interval," where, for a particular application, we substitute the appropriate numerical values for the level of confidence and for the lower and upper bounds. *The statement reflects our confidence in the estimation process, rather than in the particular interval that is calculated from the sample data.* We know that repeated application of the same procedure will result in different lower and upper bounds on the interval. Furthermore, we know that $100(1 - \alpha)\%$ of the resulting intervals will contain μ. There is (usually) no way to determine whether any particular interval is one of those which contain μ or one of those which do not. However, unlike point estimators, confidence intervals have some measure of reliability—the confidence coefficient—associated with them. For that reason, they are generally preferred to point estimators.

Sometimes, the estimation procedure yields a confidence interval that is too wide for our purposes. In this case, we will want to reduce the width of the interval to obtain a more precise estimate of μ. One way to accomplish that is to decrease the confidence coefficient, $1 - \alpha$. For example, consider the problem of estimating the mean length of stay, μ for hospital patients. Recall that for a sample of 100 patients, $\bar{x} = 4.53$ days and $s = 3.68$ days. A 90% confidence interval for μ is

$$\bar{x} \pm 1.645(\sigma/\sqrt{n}) \approx 4.53 \pm (1.645)(3.68)/\sqrt{100} = 4.53 \pm .61$$

or (3.92, 5.14). You can see that this interval is narrower than the previously calculated 95% confidence interval, (3.81, 5.25). Unfortunately, we also have "less confidence" in the 90% confidence interval. An alternative method used to decrease the width of an interval without sacrificing "confidence" is to increase the sample size n. We demonstrate this method in Section 7.5.

Exercises 7.1–7.26

Understanding the Principles

7.1 Define the target parameter.

7.2 What is the confidence coefficient in a 90% confidence interval for μ?

7.3 Explain the difference between an interval estimator and a point estimator for μ.

7.4 Explain what is meant by the statement "We are 95% confident that an interval estimate contains μ."

7.5 Will a large-sample confidence interval be valid if the population from which the sample is taken is not normally distributed? Explain.

7.6 What conditions are required to form a valid large-sample confidence interval for μ?

Learning the Mechanics

7.7 Find $z_{\alpha/2}$ for each of the following:
NW **a.** $\alpha = .10$ **b.** $\alpha = .01$
     c. $\alpha = .05$ **d.** $\alpha = .20$

7.8 What is the confidence level of each of the following confidence intervals for μ?

a. $\bar{x} \pm 1.96\left(\dfrac{\sigma}{\sqrt{n}}\right)$

b. $\bar{x} \pm 1.645\left(\dfrac{\sigma}{\sqrt{n}}\right)$

c. $\bar{x} \pm 2.575\left(\dfrac{\sigma}{\sqrt{n}}\right)$

d. $\bar{x} \pm 1.28\left(\dfrac{\sigma}{\sqrt{n}}\right)$

e. $\bar{x} \pm .99\left(\dfrac{\sigma}{\sqrt{n}}\right)$

7.9 A random sample of n measurements was selected from a
NW population with unknown mean μ and standard deviation $\sigma = 20$. Calculate a 95% confidence interval for μ for each of the following situations:
  a. $n = 75, \bar{x} = 28$
  b. $n = 200, \bar{x} = 102$
  c. $n = 100, \bar{x} = 15$
  d. $n = 100, \bar{x} = 4.05$
  e. Is the assumption that the underlying population of measurements is normally distributed necessary to ensure the validity of the confidence intervals in parts **a–d**? Explain.

7.10 A random sample of 90 observations produced a mean $\bar{x} = 25.9$ and a standard deviation $s = 2.7$.
  a. Find a 95% confidence interval for the population mean μ.
  b. Find a 90% confidence interval for μ.
  c. Find a 99% confidence interval for μ.

7.11 A random sample of 100 observations from a normally distributed population possesses a mean equal to 83.2 and a standard deviation equal to 6.4.
  a. Find a 95% confidence interval for μ.
  b. What do you mean when you say that a confidence coefficient is .95?
  c. Find a 99% confidence interval for μ.
  d. What happens to the width of a confidence interval as the value of the confidence coefficient is increased while the sample size is held fixed?
  e. Would your confidence intervals of parts **a** and **c** be valid if the distribution of the original population were not normal? Explain.

7.12 The mean and standard deviation of a random sample of n measurements are equal to 33.9 and 3.3, respectively.
  a. Find a 95% confidence interval for μ if $n = 100$.
  b. Find a 95% confidence interval for μ if $n = 400$.
  c. Find the widths of the confidence intervals you calculated in parts **a** and **b**. What is the effect on the width of a confidence interval of quadrupling the sample size while holding the confidence coefficient fixed?

Applet Exercise 7.1

Use the applet entitled *Confidence Intervals for a Mean (the impact of confidence level)* to investigate the situation in Exercise 7.11 further. For this exercise, assume that $\mu = 83.2$ is the population mean and $\sigma = 6.4$ is the population standard deviation.
  a. Using $n = 100$ and the normal distribution with mean and standard deviation as just given, run the applet one time. How many of the 95% confidence intervals contain the mean? How many would you expect to contain the mean? How many of the 99% confidence intervals contain the mean? How many would you expect to contain the mean?
  b. Which confidence level has a greater frequency of intervals that contain the mean? Is this result what you would expect? Explain.
  c. Without clearing, run the applet several more times. What happens to the proportion of 95% confidence

intervals that contain the mean as you run the applet more and more? What happens to the proportion of 99% confidence intervals that contain the mean as you run the applet more and more? Interpret these results in terms of the meanings of the 95% confidence interval and the 99% confidence interval.

d. Change the distribution to *right skewed*, clear, and run the applet several more times. Do you get the same results as in part **c**? Would you change your answer to part **e** of Exercise 7.11? Explain.

● Applet Exercise 7.2

Use the applet entitled *Confidence Intervals for a Mean (the impact of confidence level)* to investigate the effect of the sample size on the proportion of confidence intervals that contain the mean when the underlying distribution is skewed. Set the distribution to *right skewed*, the mean to 10, and the standard deviation to 1.

a. Using $n = 30$, run the applet several times without clearing. What happens to the proportion of 95% confidence intervals that contain the mean as you run the applet more and more? What happens to the proportion of 99% confidence intervals that contain the mean as you run the applet more and more? Do the proportions seem to be approaching the values that you would expect?

b. Clear and run the applet several times, using $n = 100$. What happens to the proportions of 95% confidence intervals and 99% confidence intervals that contain the mean this time? How do these results compare with your results in part **a**?

c. Clear and run the applet several times, using $n = 1000$. How do the results compare with your results in parts **a** and **b**?

d. Describe the effect of sample size on the likelihood that a confidence interval contains the mean for a skewed distribution.

Applying the Concepts—Basic

7.13 **Latex allergy in health care workers.** Health care workers who use latex gloves with glove powder on a daily basis are particularly susceptible to developing a latex allergy. Symptoms of a latex allergy include conjunctivitis, hand eczema, nasal congestion, a skin rash, and shortness of breath. Each in a sample of 46 hospital employees who were diagnosed with latex allergy based on a skin-prick test reported on their exposure to latex gloves (*Current Allergy & Clinical Immunology,* March 2004). Summary statistics for the number of latex gloves used per week are $\bar{x} = 19.3$ and $s = 11.9$.

a. Give a point estimate for the average number of latex gloves used per week by all health care workers with a latex allergy.

b. Form a 95% confidence interval for the average number of latex gloves used per week by all health care workers with a latex allergy.

c. Give a practical interpretation of the interval you found in part **b**.

d. Give the conditions required for the interval in part **b** to be valid.

7.14 **Lipid profiles of hypertensive patients.** Hypertension is diagnosed if a patient's systolic blood pressure exceeds 140 mmHg and diastolic blood pressure exceeds 90 mmHg.

A study of the lipid profiles of hypertensive patients was carried out and the results published in *Biology and Medicine* (Vol. 2, 2010). Data on fasting blood sugar (milligrams/deciliter) and magnesium (milligrams/deciliter) in blood specimens collected from 50 patients diagnosed with hypertension were collected and are stored in the **HYPER** file. Biochemists used these data to establish a benchmark for fasting blood sugar (FBS) and magnesium (MAG) levels in hypertensive patients. The accompanying MINITAB printout gives 90% confidence intervals for the mean fasting blood sugar and mean magnesium level.

a. Locate and interpret the 90% confidence interval for mean fasting blood sugar on the printout.

b. Locate and interpret the 90% confidence interval for mean magnesium level on the printout.

c. If the biochemists increase the confidence level to 95%, what will happen to the width of the intervals?

d. If the biochemists increase the sample of hypertensive patients from 50 to 100, what will likely happen to the width of the intervals?

Variable	N	Mean	StDev	SE Mean	90% CI
FBS	50	101.62	34.13	4.83	(93.53, 109.71)
MAG	50	1.94100	0.05607	0.00793	(1.92771, 1.95429)

7.15 **Motivation of drug dealers.** Refer to the *Applied Psychology in Criminal Justice* (Sept. 2009) study of the personality characteristics of drug dealers, Exercise 2.98 (p. 71). Recall that each in a sample of 100 convicted drug dealers was scored on the Wanting Recognition (WR) Scale, which provides a quantitative measure of a person's level of need for approval and sensitivity to social situations. (Higher scores indicate a greater need for approval.) The sample of drug dealers had a mean WR score of 39, with a standard deviation of 6. Use this information to find an interval estimate of the mean WR score for all convicted drug dealers. Use a confidence coefficient of 99%. Interpret the result.

7.16 **Laptop use in middle school.** Refer to the *American Secondary Education* (Fall 2009) study of middle school students' usage of laptop computers, Exercise 2.100 (p. 72). Recall that for a sample of 106 students, the researchers reported the following statistics on how many minutes per day each student used a laptop for taking notes: $\bar{x} = 13.2, s = 19.5$. Now the researchers want to estimate the average amount of time per day laptops are used for taking notes for all middle school students across the country.

a. Define the target parameter for the researchers in words and symbols.

b. In Exercise 2.100**a**, you computed the interval, $\bar{x} \pm 2s$. Explain why this formula should not be used as an interval estimate for the target parameter.

c. Calculate a 90% confidence interval for the target parameter. Interpret the result.

d. Explain what the phrase "90% confidence" implies in your answer to part **c**.

e. In Exercise 2.100**b** you concluded that the distribution of laptop usage for taking notes for the sample of 106 students cannot be symmetric. Consequently, the distribution cannot be normal. Does this compromise the validity of the interval estimate, part **c**? Explain.

7.17 **Albedo of ice melt ponds.** Refer to the National Snow and Ice Data Center (NSIDC) collection of data on the albedo, depth, and physical characteristics of ice-melt ponds in the Canadian Arctic, presented in Exercise 2.13 (p. 35). Albedo is the ratio of the light reflected by the ice to that received by it. (High albedo values give a white appearance to the ice.) Visible albedo values were recorded for a sample of 504 ice-melt ponds located in the Barrow Strait in the Canadian Arctic; these data are saved in the **PONDICE** file.

   a. Find a 90% confidence interval for the true mean visible albedo value of all Canadian Arctic ice ponds.

   b. Give both a practical and a theoretical interpretation of the interval.

   c. Recall from Exercise 2.13 that the type of ice for each pond was classified as first-year ice, multiyear ice, or landfast ice. Find 90% confidence intervals for the mean visible albedo for each of the three types of ice. Interpret the intervals.

7.18 **Extinct New Zealand birds.** Refer to the *Evolutionary Ecology Research* (July 2003) study of the patterns of extinction in the New Zealand bird population, presented in Exercise 2.102 (p. 72). Suppose you are interested in estimating the mean egg length (in millimeters) for the New Zealand bird population.

   a. What is the target parameter?

   b. Recall that the egg lengths for 132 bird species are saved in the **NZBIRDS** file. Obtain a random sample of 50 egg lengths from the data set.

   c. Find the mean and standard deviation of the 50 egg lengths you obtained in part **b.**

   d. Use the information from part **c** to form a 99% confidence interval for the true mean egg length of a bird species found in New Zealand.

   e. Give a practical interpretation of the interval you found in part **d.**

Applying the Concepts—Intermediate

7.19 **Personality and aggressive behavior.** How does personality impact aggressive behavior? A team of university psychologists conducted a review of studies that examined the relationship between personality and aggressive behavior (*Psychological Bulletin*, Vol. 132, 2006). One variable of interest to the researchers was the difference between the aggressive behavior level of individuals in the study who scored high on a personality test and those who scored low on the test. This variable, standardized to be between −7 and 7, was called "effect size". (A large positive effect size indicates that those who score high on the personality test are more aggressive than those who score low.) The researchers collected the effect sizes for a sample of $n = 109$ studies published in psychology journals. This data is saved in the **PERAGGR** file. A dot plot and summary statistics for effect size are shown in the MINITAB printouts at the bottom of the page. Of interest to the researchers is the true mean effect size μ for all psychological studies of personality and aggressive behavior.

   a. Identify the parameter of interest to the researchers.

   b. Examine the dot plot. Does effect size have a normal distribution? Explain why your answer is irrelevant to the subsequent analysis.

   c. Locate a 95% confidence interval for μ on the accompanying printout. Interpret the result.

   d. If the true mean effect size exceeds 0, then the researchers will conclude that in the population, those who score high on a personality test are more aggressive than those who score low. Can the researchers draw this conclusion? Explain.

7.20 **Shell lengths of sea turtles.** Refer to the *Aquatic Biology* (Vol. 9, 2010) study of green sea turtles inhabiting the Grand Cayman South Sound lagoon, Exercise 2.83 (p. 65). The data on curved carapace (shell) length, measured in centimeters, for 76 captured turtles are displayed in the table and saved in the **TURTLES** file. Environmentalists want to estimate the true mean shell length of all green sea turtles in the lagoon.

33.96	30.37	32.57	31.50	36.46	35.54	36.16	35.32	35.99
39.55	44.33	42.73	42.15	42.43	49.96	46.04	48.76	47.78
45.81	49.05	49.65	49.71	54.29	52.01	51.15	54.42	52.62
53.27	54.07	50.40	53.69	51.30	54.29	54.58	55.11	57.65
56.35	55.68	58.40	58.06	57.79	56.54	57.03	57.64	59.27
64.79	61.96	60.08	62.34	63.84	60.61	64.91	60.35	62.63
63.33	63.00	64.55	60.03	64.75	60.24	69.01	65.07	65.77
65.30	68.24	65.28	67.54	68.49	66.98	65.67	70.26	70.94
70.52	72.01	74.34	81.63					

   a. Define the parameter of interest to the environmentalists.

   b. Use the data in the **TURTLES** file to find a point estimate of the target parameter.

   c. Compute a 95% confidence interval for the target parameter. Interpret the result.

   d. Suppose a biologist claims that the mean shell length of all green sea turtles in the lagoon is 60 cm. Make an inference about the validity of this claim.

7.21 **Colored string preferred by chickens.** Animal behaviorists have discovered that the more domestic chickens

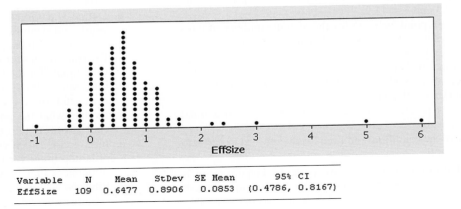

Variable	N	Mean	StDev	SE Mean	95% CI
EffSize	109	0.6477	0.8906	0.0853	(0.4786, 0.8167)

MINITAB Output for Exercise 7.19

peck at objects placed in their environment, the healthier the chickens seem to be. White string has been found to be a particularly attractive pecking stimulus. In one experiment, 72 chickens were exposed to a string stimulus. Instead of white string, blue string was used. The number of pecks each chicken took at the blue string over a specified interval of time was recorded. Summary statistics for the 72 chickens were $\bar{x} = 1.13$ pecks and $s = 2.21$ pecks (*Applied Animal Behaviour Science*, October 2000).

a. Use a 99% confidence interval to estimate the population mean number of pecks made by chickens pecking at blue string. Interpret the result.

b. Previous research has shown that $\mu = 7.5$ pecks if chickens are exposed to white string. Based on the results you found in part **a**, is there evidence that chickens are more apt to peck at white string than blue string? Explain.

7.22 **Speed training in football.** A key statistic used by football coaches to evaluate players is a player's 40-yard sprint time. Can a drill be developed for improving a player's speed in the sprint? Researchers at Northern Kentucky University designed and tested a speed-training program for junior varsity and varsity high school football players (*The Sport Journal*, Winter 2004). The training program included 50-yard sprints run at varying speeds, high knee running sprints, butt kick sprints, "crazy legs" straddle runs, quick feet drills, jumping, power skipping, and all-out sprinting. Each in a sample of 38 high school athletes was timed in a 40-yard sprint prior to the start of the training program and timed again after completing the program. The decreases in times (measured in seconds) are listed in the table and saved in the **SPRINT** file. [*Note:* A negative decrease implies that the athlete's time after completion of the program was higher than his time prior to training.] The goal of the research is to demonstrate that the training program is effective in improving 40-yard sprint times.

−.01	.1	.1	.24	.25	.05	.28	.25	.2	.14
.32	.34	.3	.09	.05	0	.04	.17	0	.21
.15	.3	.02	.12	.14	.1	.08	.5	.36	.1
.01	.9	.34	.38	.44	.08	0	0		

Based on Gray, M., & Sauerbeck, J. A. "Speed training program for high school football players." *The Sport Journal*, Vol. 7, No. 1, Winter 2004 (Table 2).

a. Find a 95% confidence interval for the true mean decrease in sprint times for the population of all football players who participate in the speed-training program.

b. Based on the confidence interval, is the training program really effective in improving the mean 40-yard sprint time of high school football players? Explain.

7.23 **Improving SAT scores.** Refer to the *Chance* (Winter 2001) and National Education Longitudinal Survey (NELS) study of 265 students who paid a private tutor to help them improve their SAT scores, presented in Exercise 2.101 (p. 72). The changes in both the SAT-Mathematics and SAT-Verbal scores for these students are reproduced in the following table:

	SAT-Math	SAT-Verbal
Mean change in score	19	7
Standard deviation of score changes	65	49

a. Construct and interpret a 95% confidence interval for the population mean change in SAT-Mathematics score for students who pay a private tutor.

b. Repeat part **a** for the population mean change in SAT-Verbal score.

c. Suppose the true population mean change in score on one of the SAT tests for all students who paid a private tutor is 15. Which of the two tests, SAT-Mathematics or SAT-Verbal, is most likely to have this mean change? Explain.

7.24 **Attention time given to twins.** Psychologists have found that twins, in their early years, tend to have lower IQs and pick up language more slowly than nontwins. (*Wisconsin Twin Research Newsletter*, Winter 2004). The slower intellectual growth of most twins may be caused by benign parental neglect. Suppose it is desired to estimate the mean attention time given to twins per week by their parents. A sample of 50 sets of $2\frac{1}{2}$-year-old twin boys is taken, and at the end of 1 week, the attention time given to each pair is recorded. The data (in hours) are listed in the following table and saved in the **ATTIMES** file. Find a 90% confidence interval for the mean attention time given to all twin boys by their parents. Interpret the confidence interval.

20.7	16.7	22.5	12.1	2.9
23.5	6.4	1.3	39.6	35.6
10.9	7.1	46.0	23.4	29.4
44.1	13.8	24.3	9.3	3.4
15.7	46.6	10.6	6.7	5.4
14.0	20.7	48.2	7.7	22.2
20.3	34.0	44.5	23.8	20.0
43.1	14.3	21.9	17.5	9.6
36.4	0.8	1.1	19.3	14.6
32.5	19.1	36.9	27.9	14.0

Applying the Concepts—Advanced

7.25 **Study of undergraduate problem drinking.** In *Alcohol & Alcoholism* (Jan/Feb. 2007), psychologists at the University of Pennsylvania compared the levels of alcohol consumption of male and female freshman students. Each student was asked to estimate the amount of alcohol (beer, wine, or liquor) they consume in a typical week. Summary statistics for 128 males and 184 females are provided in the accompanying table.

a. For each gender, find a 95% confidence interval for mean weekly alcohol consumption.

b. Prior to sampling, what is the probability that at least one of the two confidence intervals will not contain the population mean it estimates? Assume that the two intervals are independent.

c. Based on the two confidence intervals, what inference can you make about which gender consumes the most alcohol, on average, per week? (*Caution:* In Chapter 9, we will learn about a more valid method of comparing population means.)

	Males	Females
Sample size, n	128	184
Mean (ounces), $\bar{x}$	16.79	10.79
Standard deviation, s	13.57	11.53

Based on Leeman, R. F., Fenton, M., & Volpicelli, J. R. "Impaired control and undergraduate problem drinking." *Alcohol & Alcoholism*, Vol. 42, No. 1, Jan/Feb. 2007 (Table 1).

7.26 **The Raid test kitchen.** According to scientists, the cockroach has had 300 million years to develop a resistance to destruction. In a study conducted by researchers for S. C. Johnson & Son, Inc. (manufacturers of Raid® and Off®), 5,000 roaches (the expected number in a roach-infested house) were released in the Raid test kitchen. One week later, the kitchen was fumigated and 16,298 dead roaches were counted, a gain of 11,298 roaches for the 1-week period. Assume that none of the original roaches died during the 1-week period and that the standard deviation of x, the number of roaches produced per roach in a 1-week period, is 1.5. Use the number of roaches produced by the sample of 5,000 roaches to find a 95% confidence interval for the mean number of roaches produced per week for each roach in a typical roach-infested house.

7.3 Confidence Interval for a Population Mean: Student's t–Statistic

Federal legislation requires pharmaceutical companies to perform extensive tests on new drugs before they can be marketed. Initially, a new drug is tested on animals. If the drug is deemed safe after this first phase of testing, the pharmaceutical company is then permitted to begin human testing on a limited basis. During this second phase, inferences must be made about the safety of the drug on the basis of information obtained from very small samples.

Suppose a pharmaceutical company must estimate the average increase in blood pressure of patients who take a certain new drug. Assume that only six patients (randomly selected from the population of all patients) can be used in the initial phase of human testing. The use of a *small sample* in making an inference about μ presents two immediate problems when we attempt to use the standard normal z as a test statistic.

BIOGRAPHY

WILLIAM S. GOSSET (1876–1937)

Student's t-Distribution

At the age of 23, William Gosset earned a degree in chemistry and mathematics at prestigious Oxford University. He was immediately hired by the Guinness Brewing Company in Dublin, Ireland, for his expertise in chemistry. However, Gosset's mathematical skills allowed him to solve numerous practical problems associated with brewing beer. For example, Gosset applied the Poisson distribution to model the number of yeast cells per unit volume in the fermentation process. His most important discovery was the t-distribution in 1908. Since most applied researchers worked with small samples, Gosset was interested in the behavior of the mean in the case of small samples. He tediously took numerous small sets of numbers, calculated the mean and standard deviation of each, obtained their t-ratio, and plotted the results on graph paper. The shape of the distribution was always the same: the t-distribution. Under company policy, employees were forbidden to publish their research results, so Gosset used the pen name *Student* to publish a paper on the subject. Hence, the distribution has been called Student's t-distribution. ∎

Problem 1 The shape of the sampling distribution of the sample mean $\bar{x}$ (and the z-statistic) now depends on the shape of the population that is sampled. We can no longer assume that the sampling distribution of $\bar{x}$ is approximately normal, because the Central Limit Theorem ensures normality only for samples that are sufficiently large.

Solution to Problem 1 The sampling distribution of $\bar{x}$ (and z) is exactly normal even for relatively small samples *if the sampled population is normal*. It is approximately normal if the sampled population is approximately normal.

Problem 2 The population standard deviation σ is almost always unknown. Although it is still true that $\sigma_{\bar{x}} = \sigma/\sqrt{n}$, the sample standard deviation s may provide a poor approximation for σ when the sample size is small.

Solution to Problem 2 Instead of using the standard normal statistic

$$z = \frac{\bar{x} - \mu}{\sigma_{\bar{x}}} = \frac{\bar{x} - \mu}{\sigma/\sqrt{n}}$$

which requires knowledge of, or a good approximation to, σ, we define and use the statistic

$$t = \frac{\bar{x} - \mu}{s/\sqrt{n}}$$

in which the sample standard deviation s replaces the population standard deviation σ.

If we are sampling from a normal distribution, the **t-statistic** has a sampling distribution very much like that of the z-statistic: mound shaped, symmetric, and with mean 0. The primary difference between the sampling distributions of t and z is that the t-statistic is more variable than the z, a property that follows intuitively when you realize that t contains two random quantities ($\bar{x}$ and s), whereas z contains only one ($\bar{x}$).

The actual amount of variability in the sampling distribution of t depends on the sample size n. A convenient way of expressing this dependence is to say that the t statistic has $(n - 1)$ **degrees of freedom (df).*** Recall that the quantity $(n - 1)$ is the divisor that appears in the formula for s^2. This number plays a key role in the sampling distribution of

---

*Since degrees of freedom are related to the sample size n, it is helpful to think of the number of degrees of freedom as the amount of information in the sample available for estimating the target parameter.

s^2 and appears in discussions of other statistics in later chapters. In particular, the smaller the number of degrees of freedom associated with the *t*-statistic, the more variable will be its sampling distribution.

In Figure 7.7, we show both the sampling distribution of z and the sampling distribution of a *t*-statistic with both 4 and 20 df. You can see that the *t*-distribution is more variable than the *z*-distribution and that this variability increases as the degrees of freedom decrease. You can also see that the increased variability of the *t*-statistic means that the *t*-value, t_α, that locates an area α in the upper tail of the *t*-distribution is larger than the corresponding value, z_α. For any given value of α the *t*-value, t_α, increases as the number of degrees of freedom (df) decreases. Values of *t* that will be used in forming small-sample confidence intervals of μ are given in Table VI of Appendix A. A partial reproduction of this table is shown in Table 7.3. Note that t_α values are listed for various degrees of freedom, where α refers to the tail area under the *t*-distribution to the right of t_α.

For example, if we want the *t*-value with an area of .025 to its right and 4 df, we look in the table under the column $t_{.025}$ for the entry in the row corresponding to 4 df. This entry is $t_{.025} = 2.776$, shaded in Table 7.3 and shown in Figure 7.8. The corresponding standard normal *z*-score is $z_{.025} = 1.96$. Note that the last row of Table VI, where df $= \infty$ (infinity), contains the standard normal *z*-values. This follows from the fact that as the sample size n grows very large, s becomes closer to σ and thus *t* becomes closer in distribution to *z*. In fact,

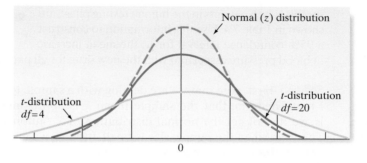

Figure 7.7

Standard normal (z) distribution and *t*-distributions

Table 7.3   Reproduction of Part of Table VI in Appendix A

Degrees of Freedom	$t_{.100}$	$t_{.050}$	$t_{.025}$	$t_{.010}$	$t_{.005}$	$t_{.001}$	$t_{.0005}$
1	3.078	6.314	12.706	31.821	63.657	318.31	636.62
2	1.886	2.920	4.303	6.965	9.925	22.326	31.598
3	1.638	2.353	3.182	4.541	5.841	10.213	12.924
4	1.533	2.132	2.776	3.747	4.604	7.173	8.610
5	1.476	2.015	2.571	3.365	4.032	5.893	6.869
6	1.440	1.943	2.447	3.143	3.707	5.208	5.959
7	1.415	1.895	2.365	2.998	3.499	4.785	5.408
8	1.397	1.860	2.306	2.896	3.355	4.501	5.041
9	1.383	1.833	2.262	2.821	3.250	4.297	4.781
10	1.372	1.812	2.228	2.764	3.169	4.144	4.587
11	1.363	1.796	2.201	2.718	3.106	4.025	4.437
12	1.356	1.782	2.179	2.681	3.055	3.930	4.318
13	1.350	1.771	2.160	2.650	3.012	3.852	4.221
14	1.345	1.761	2.145	2.624	2.977	3.787	4.140
15	1.341	1.753	2.131	2.602	2.947	3.733	4.073
⋮	⋮	⋮	⋮	⋮	⋮	⋮	⋮
∞	1.282	1.645	1.960	2.326	2.576	3.090	3.291

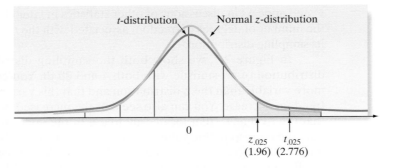

Figure 7.8

The $t_{.025}$ value in a t-distribution with 4 df, and the corresponding $z_{.025}$ value

when df $= 29$, there is little difference between corresponding tabulated values of z and t. Thus, we choose the arbitrary cutoff of $n = 30$ (df $= 29$) to distinguish between large-sample and small-sample inferential techniques when σ is unknown.

Example 7.4

Confidence Interval for Mean Blood Pressure Increase, t-Statistic

Problem Consider the pharmaceutical company that desires an estimate of the mean increase in blood pressure of patients who take a new drug. The blood pressure increases (measured in points) for the $n = 6$ patients in the human testing phase are shown in Table 7.4. Use this information to construct a 95% confidence interval for μ, the mean increase in blood pressure associated with the new drug for all patients in the population.

Solution First, note that we are dealing with a sample too small to assume, by the Central Limit Theorem, that the sample mean $\bar{x}$ is approximately normally distributed. That is, we do not get the normal distribution of $\bar{x}$ "automatically" from the Central Limit Theorem when the sample size is small. Instead, we must *assume* that the measured variable, in this case the increase in blood pressure, is normally distributed in order for the distribution of $\bar{x}$ to be normal.

Second, unless we are fortunate enough to know the population standard deviation σ, which in this case represents the standard deviation of *all* the patients' increases in blood pressure when they take the new drug, we cannot use the standard normal z-statistic to form our confidence interval for μ. Instead, we must use the t-distribution, with $(n - 1)$ degrees of freedom.

In this case, $n - 1 = 5$ df, and the t-value is found in Table 7.3 to be

$$t_{.025} = 2.571 \text{ with 5 df}$$

Recall that the large-sample confidence interval would have been of the form

$$\bar{x} \pm z_{\alpha/2}\sigma_{\bar{x}} = \bar{x} \pm z_{\alpha/2}\frac{\sigma}{\sqrt{n}} = \bar{x} \pm z_{.025}\frac{\sigma}{\sqrt{n}}$$

where 95% is the desired confidence level. To form the interval for a small sample from *a normal distribution, we simply substitute t for z and s for σ in the preceding formula, yielding*

$$\bar{x} \pm t_{\alpha/2}\frac{s}{\sqrt{n}}$$

An SPSS printout showing descriptive statistics for the six blood pressure increases is displayed in Figure 7.9. Note that $\bar{x} = 2.283$ and $s = .950$. Substituting these numerical values into the confidence interval formula, we get

$$2.283 \pm (2.571)\left(\frac{.950}{\sqrt{6}}\right) = 2.283 \pm .997$$

or 1.286 to 3.280 points. Note that this interval agrees (except for rounding) with the confidence interval highlighted on the SPSS printout in Figure 7.9.

Table 7.4   Blood Pressure Increases (Points) for Six Patients

1.7	3.0	.8	3.4	2.7	2.1

Data Set: BPINCR

Descriptives

BPINCR			Statistic	Std. Error
Mean			2.283	.3877
95% Confidence Interval for Mean	Lower Bound		1.287	
	Upper Bound		3.280	
5% Trimmed Mean			2.304	
Median			2.400	
Variance			.902	
Std. Deviation			.9496	
Minimum			.8	
Maximum			3.4	
Range			2.6	
Interquartile Range			1.625	
Skewness			-.573	.845
Kurtosis			-.389	1.741

Figure 7.9

SPSS confidence interval for mean blood pressure increase

We interpret the interval as follows: We can be 95% confident that the mean increase in blood pressure associated with taking this new drug is between 1.286 and 3.28 points. As with our large-sample interval estimates, our confidence is in the *process*, not in this particular interval. We know that if we were to repeatedly use this estimation procedure, 95% of the confidence intervals produced would contain the true mean μ, *assuming that the probability distribution of changes in blood pressure from which our sample was selected is normal.* The latter assumption is necessary for the small-sample interval to be valid.

Look Back What price did we pay for having to utilize a small sample to make the inference? First, we had to assume that the underlying population is normally distributed, and if the assumption is invalid, our interval might also be invalid.* Second, we had to form the interval by using a *t* value of 2.571 rather than a *z* value of 1.96, resulting in a wider interval to achieve the same 95% level of confidence. If the interval from 1.286 to 3.28 is too wide to be of much use, we know how to remedy the situation: Increase the number of patients sampled in order to decrease the width of the interval (on average).

Now Work Exercise 7.29

The procedure for forming a small-sample confidence interval is summarized in the accompanying boxes.

Small-Sample $100(1-\alpha)$ Confidence Interval for μ, *t*-Statistic

σ unknown: $\bar{x} \pm (t_{\alpha/2}) (s/\sqrt{n})$

where $t_{\alpha/2}$ is the *t*-value corresponding to an area $\alpha/2$ in the upper tail of the Student's *t*-distribution based on $(n-1)$ degrees of freedom.

Note: If σ is *known* and condition #2 (below) is met, then use the *z*-statistic to find the confidence interval: $\bar{x} \pm (z_{\alpha/2})(\sigma/\sqrt{n})$

Conditions Required for a Valid Small-Sample Confidence Interval for μ

1. A random sample is selected from the target population.
2. The population has a relative frequency distribution that is approximately normal.

---

*By *invalid*, we mean that the probability that the procedure will yield an interval that contains μ is not equal to $(1 - \alpha)$. Generally, if the underlying population is approximately normal, the confidence coefficient will approximate the probability that the interval contains μ.

Example 7.5

A Small–Sample Confidence Interval for μ—Destructive Sampling

Problem Some quality control experiments require *destructive sampling* (i.e., the test to determine whether the item is defective destroys the item) in order to measure a particular characteristic of the product. The cost of destructive sampling often dictates small samples. Suppose a manufacturer of printers for personal computers wishes to estimate the mean number of characters printed before the printhead fails. The printer manufacturer tests $n = 15$ printheads and records the number of characters printed until failure for each. These 15 measurements (in millions of characters) are listed in Table 7.5, followed by a MINITAB summary statistics printout in Figure 7.10.

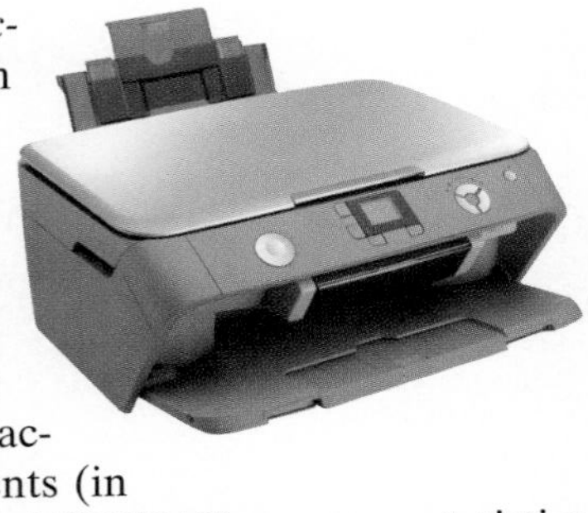

a. Form a 99% confidence interval for the mean number of characters printed before the printhead fails. Interpret the result.

b. What assumption is required for the interval you found in part **a** to be valid? Is that assumption reasonably satisfied?

Table 7.5	**Number of Characters (in Millions) for $n = 15$ Printhead Tests**

| 1.13 | 1.55 | 1.43 | .92 | 1.25 | 1.36 | 1.32 | .85 | 1.07 | 1.48 | 1.20 | 1.33 | 1.18 | 1.22 | 1.29 |

Data Set: PRINTHEAD

Figure 7.10
MINITAB printout with descriptive statistics and 99% confidence interval for Example 7.5

Variable	N	Mean	StDev	SE Mean	99% CI
NUMCHAR	15	1.23867	0.19316	0.04987	(1.09020, 1.38714)

Solution

a. For this small sample ($n = 15$), we use the *t*-statistic to form the confidence interval. We use a confidence coefficient of .99 and $n - 1 = 14$ degrees of freedom to find $t_{\alpha/2}$ in Table VI:

$$t_{\alpha/2} = t_{.005} = 2.977$$

[*Note:* The small sample forces us to extend the interval almost three standard deviations (of $\bar{x}$) on each side of the sample mean in order to form the 99% confidence interval.] From the MINITAB printout shown in Figure 7.10, we find that $\bar{x} = 1.24$ and $s = .19$. Substituting these (rounded) values into the confidence interval formula, we obtain

$$\bar{x} \pm t_{.005}\left(\frac{s}{\sqrt{n}}\right) = 1.24 \pm 2.977\left(\frac{.19}{\sqrt{15}}\right)$$
$$= 1.24 \pm .15 \text{ or } (1.09, 1.39)$$

This interval is highlighted in Figure 7.10.

Our interpretation is as follows: The manufacturer can be 99% confident that the printhead has a mean life of between 1.09 and 1.39 million characters. If the manufacturer were to advertise that the mean life of its printheads is (at least) 1 million characters, the interval would support such a claim. Our confidence is derived from the fact that 99% of the intervals formed in repeated applications of this procedure will contain μ.

b. Since n is small, we must assume that the number of characters printed before the printhead fails is a random variable from a normal distribution. That is, we assume that the population from which the sample of 15 measurements is selected is distributed normally. One way to check this assumption is to graph the distribution of data in Table 7.5. If the sample data are approximately normally distributed, then the population from which the sample is selected is very likely to be normal. A MINITAB stem-and-leaf plot for the sample data is displayed in Figure 7.11. The

---

Stem-and-Leaf Display: NUMBER

```
Stem-and-leaf of NUMBER   N = 15
Leaf Unit = 0.010


  1    8   5
  2    9   2
  3   10   7
  5   11   38
 (4)  12   0259
  6   13   236
  3   14   38
  1   15   5
```

Figure 7.11

MINITAB stem-and-leaf display of data in Table 7.5

distribution is mound shaped and nearly symmetric. Therefore, the assumption of normality appears to be reasonably satisfied.

Look Back Other checks for normality, such as a normal probability plot and the ratio IQR/*s*, may also be used to verify the normality condition.

Now Work Exercise 7.39

We have emphasized throughout this section that an assumption that the population is approximately normally distributed is necessary for making small-sample inferences about μ when the *t*-statistic is used. Although many phenomena do have approximately normal distributions, it is also true that many random phenomena have distributions that are not normal or even mound shaped. Empirical evidence acquired over the years has shown that the *t*-distribution is rather insensitive to moderate departures from normality. That is, the use of the *t*-statistic when sampling from slightly skewed mound-shaped populations generally produces credible results; however, for cases in which the distribution is distinctly nonnormal, we must either take a large sample or use a *nonparametric method* (the topic of Chapter 14).

What Do You Do When the Population Relative Frequency Distribution Departs Greatly from Normality?

Answer: Use the nonparametric statistical methods of Chapter 14 (available on the text Resource CD).

Statistics IN Action | **Revisited**   Estimating the Mean Overpayment

Refer to the Medicare fraud investigation described in the Statistics in Action (p. 299). Recall that the United States Department of Justice (USDOJ) obtained a random sample of 52 claims from a population of 1,000 Medicare claims. For each claim, the amount paid, the amount disallowed (denied) by the auditor, and the amount that should have been paid (allowed) were recorded and saved in the **MCFRAUD** file. The USDOJ wants to use these data to calculate an estimate of the overpayment for all 1,000 claims in the population.

One way to do this is to first use the sample data to estimate the mean overpayment per claim for the population, then use the estimated mean to extrapolate the overpayment amount to the population of all 1,000 claims. The difference between the amount paid and the amount allowed by the auditor represents the overpayment for each claim. This value is recorded as the amount denied in the **MCFRAUD** file. These overpayment amounts are listed in the accompanying table.

MINITAB software is used to find a 95% confidence interval for μ, the mean overpayment amount. The MINITAB printout is displayed in Figure SIA.7.1. The 95% confidence interval for μ, (highlighted on the printout) is (16.51, 27.13). Thus, the USDOJ can be 95% confident that the mean overpayment

(continued)

Statistics IN Action
(continued)

amount for the population of 1,000 claims is between $16.51 and $27.13.

Now, let x_i represent the overpayment amount for the i the claim. If the true mean μ were known, then the total overpayment amount for all 1,000 claims would be equal to

$$\sum_{i=1}^{1,000} x_i = (1,000)\left[\sum_{i=1}^{1,000} x_i/(1,000)\right] = (1,000)\mu$$

Consequently, to estimate the total overpayment amount for the 1,000 claims, the USDOJ will simply multiply the end points of the interval by 1,000. This yields the 95% confidence interval ($16,510, $27,130).* Typically, the USDOJ is willing to give the Medicare provider in question the benefit of the doubt by demanding a repayment equal to the lower 95% confidence bound—in this case, $16,510.

Table SIA7.1	Overpayment Amounts for Sample of 52 Claims						
$0.00	$31.00	$0.00	$37.20	$37.20	$0.00	$43.40	$0.00
$37.20	$43.40	$0.00	$37.20	$0.00	$24.80	$0.00	$0.00
$37.20	$0.00	$37.20	$0.00	$37.20	$37.20	$37.20	$0.00
$37.20	$37.20	$0.00	$0.00	$37.20	$0.00	$43.40	$37.20
$0.00	$37.20	$0.00	$37.20	$37.20	$0.00	$37.20	$37.20
$0.00	$37.20	$43.40	$0.00	$37.20	$37.20	$37.20	$0.00
$43.40	$0.00	$43.40	$0.00				

Data Set: MCFRAUD

Figure SIA7.1

MINITAB confidence interval for mean overpayment

```
Variable        N   Mean  StDev  SE Mean      95% CI
Denied_Amount  52  21.82  19.08     2.65  (16.51, 27.13)
```

Exercises 7.27–7.47

Understanding the Principles

7.27 State the two problems (and corresponding solutions) that arise with using a small sample to estimate μ.

7.28 Compare the shapes of the z- and t-distributions.

7.29 Explain the differences in the sampling distributions
[NW] of $\bar{x}$ for large and small samples under the following assumptions:
  a. The variable of interest, x, is normally distributed.
  b. Nothing is known about the distribution of the variable x.

Applet Exercise 7.3

Use the applet entitled *Confidence Intervals for a Mean (the impact of not knowing the standard deviation)* to compare proportions of z-intervals and t-intervals that contain the mean for a population that is normally distributed.
  a. Using $n = 5$ and the normal distribution with mean 50 and standard deviation 10, run the applet several times. How do the proportions of z-intervals and t-intervals that contain the mean compare?
  b. Repeart part **a** first for $n = 10$ and then for $n = 20$. Compare your results with those you obtained in part **a**.
  c. Describe any patterns you observe between the proportion of z-intervals that contain the mean and the

proportion of t-intervals that contain the mean as the sample size increases.

Applet Exercise 7.4

Use the applet entitled *Confidence Intervals for a Mean (the impact of not knowing the standard deviation)* to compare proportions of z-intervals and t-intervals that contain the mean for a population with a skewed distribution.
  a. Using $n = 5$ and the right-skewed distribution with mean 50 and standard deviation 10, run the applet several times. How do the proportions of z-intervals and t-intervals that contain the mean compare?
  b. Repeat part **a** first for $n = 10$ and then for $n = 20$. Compare your results with those you obtained in part **a**.
  c. Describe any patterns you observe between the proportion of z-intervals that contain the mean and the proportion of t-intervals that contain the mean as the sample size increases.
  d. How does the skewness of the underlying distribution affect the proportions of z-intervals and t-intervals that contain the mean?

Learning the Mechanics

7.30 Suppose you have selected a random sample of $n = 7$ measurements from a normal distribution. Compare the

*This interval represents an approximation to the true 95% confidence interval for the total amount of overpayment. The precise interval involves use of a continuity correction for population size. Consult the references for the use of a finite population correction factor in interval estimation.

standard normal *z*-values with the corresponding *t*-values if you were forming the following confidence intervals:
a. 80% confidence interval
b. 90% confidence interval
c. 95% confidence interval
d. 98% confidence interval
e. 99% confidence interval
f. Use the table values you obtained in parts **a–e** to sketch the *z*- and *t*-distributions. What are the similarities and differences?

7.31 Let t_0 be a specific value of *t*. Use Table VI in Appendix A to find t_0 values such that the following statements are true:
a. $P(t \geq t_0) = .025$, where df = 10
b. $P(t \geq t_0) = .01$, where df = 17
c. $P(t \leq t_0) = .005$, where df = 6
d. $P(t \leq t_0) = .05$, where df = 13

7.32 Let t_0 be a particular value of *t*. Use Table VI of Appendix A to find t_0 values such that the following statements are true:
a. $P(-t_0 < t < t_0) = .95$, where df = 16
b. $P(t \leq -t_0 \text{ or } t \geq t_0) = .05$, where df = 16
c. $P(t \leq t_0) = .05$, where df = 16
d. $P(t \leq -t_0 \text{ or } t \geq t_0) = .10$, where df = 12
e. $P(t \leq -t_0 \text{ or } t \geq t_0) = .01$, where df = 8

7.33 The following random sample was selected from a normal distribution: 4, 6, 3, 5, 9, 3.
a. Construct a 90% confidence interval for the population mean μ.
b. Construct a 95% confidence interval for the population mean μ.
c. Construct a 99% confidence interval for the population mean μ.
d. Assume that the sample mean $\bar{x}$ and sample standard deviation *s* remain exactly the same as those you just calculated, but that they are based on a sample of *n* = 25 observations rather than *n* = 6 observations. Repeat parts **a–c**. What is the effect of increasing the sample size on the width of the confidence intervals?

7.34 The following sample of 16 measurements, saved in the **LM7_34** file, was selected from a population that is approximately normally distributed:

91	80	99	110	95	106	78	121	106	100
97	82	100	83	115	104				

a. Construct an 80% confidence interval for the population mean.
b. Construct a 95% confidence interval for the population mean, and compare the width of this interval with that of part **a**.
c. Carefully interpret each of the confidence intervals, and explain why the 80% confidence interval is narrower.

Applying the Concepts—Basic

7.35 **Lobster trap placement.** Strategic placement of lobster traps is one of the keys for a successful lobster fisherman. An observational study of teams fishing for the red spiny lobster in Baja California Sur, Mexico, was conducted and the results published in *Bulletin of Marine Science* (April, 2010). One of the variables of interest was the average distance separating traps—called *trap spacing*—

deployed by the same team of fishermen. Trap spacing measurements (in meters) for a sample of seven teams of red spiny lobster fishermen are shown in the accompanying table (and saved in the **TRAPSPACE** file). Of interest is the mean trap spacing for the population of red spiny lobster fishermen fishing in Baja California Sur, Mexico.

93	99	105	94	82	70	86

Based on Shester, G. G. "Explaining catch variation among Baja California lobster fishers through spatial analysis of trap-placement decisions." *Bulletin of Marine Science*, Vol. 86, No. 2, April 2010 (Table 1), pp. 479–498.

a. Identify the target parameter for this study.
b. Compute a point estimate of the target parameter.
c. What is the problem with using the normal (*z*) statistic to find a confidence interval for the target parameter?
d. Find a 95% confidence interval for the target parameter.
e. Give a practical interpretation of the interval, part **d**.
f. What conditions must be satisfied for the interval, part **d**, to be valid?

7.36 **Shell lengths of sea turtles.** Refer to the *Aquatic Biology* (Vol. 9, 2010) study of green sea turtles inhabiting the Grand Cayman South Sound lagoon, Exercise 7.20 (p. 308). Time-depth recorders were deployed on 6 of the 76 captured turtles. The time-depth recorders allowed the environmentalists to track the movement of the sea turtles in the lagoon. These 6 turtles had a mean shell length of 52.9 cm with a standard deviation of 6.8 cm.
a. Use the information on the 6 tracked turtles to estimate, with 99% confidence, the true mean shell length of all green sea turtles in the lagoon. Interpret the result.
b. What assumption about the distribution of shell lengths must be true in order for the confidence interval, part a, to be valid? Is this assumption reasonably satisfied? (Use the data saved in the **TURTLES** file to help you answer this question.)

7.37 **Duration of daylight in western Pennsylvania.** What area of the United States has the least amount of daylight, on average? Having grown up in western Pennsylvania, co-author Sincich wonders if it is his hometown of Sharon, PA. Data on the number of minutes of daylight per day in Sharon, PA, for 12 randomly selected days (one each month) in 2009 were obtained from the *Naval Oceanography Portal* Web site (*aa.usno.navy.mil/USNO/astronomical-applications/data-services*). The data are listed in the table and saved in the **DAYLIGHT** file. Descriptive statistics and a 95% confidence interval for the mean are produced in the SPSS printout on page 318.
a. Locate the confidence interval on the printout and give the value of the confidence coefficient.
b. Use the descriptive statistics on the printout to calculate the 95% confidence interval. Be sure your answer agrees with the interval shown on the printout.
c. Practically interpret the confidence interval.
d. Comment on the method of sampling. Do you think the sample is representative of the target population?

591	618	704	831	896	909	865	839	748	672	583	565

Based on Astronomical Applications Dept., U.S. Naval Observatory, Washington, DC.

SPSS Output for Exercise 7.37

Descriptives

			Statistic	Std. Error
DAYLIGHT	Mean		735.08	37.402
	95% Confidence Interval for Mean	Lower Bound	652.76	
		Upper Bound	817.40	
	5% Trimmed Mean		734.87	
	Median		726.00	
	Variance		16786.992	
	Std. Deviation		129.565	
	Minimum		565	
	Maximum		909	
	Range		344	
	Interquartile Range		261	
	Skewness		.016	.637
	Kurtosis		-1.752	1.232

7.38 **Assessing the bending strength of a wooden roof.** The white wood material used for the roof of an ancient Japanese temple is imported from Northern Europe. The wooden roof must withstand as much as 100 centimeters of snow in the winter. Architects at Tohoku University (in Japan) conducted a study to estimate the mean bending strength of the white wood roof (*Journal of the International Association for Shell and Spatial Structures,* Aug. 2004). A sample of 25 pieces of the imported wood was tested and yielded the following statistics on breaking strength (in MPa): $\bar{x} = 75.4, s = 10.9$. Estimate the true mean breaking strength of the white wood with a 90% confidence interval. Interpret the result.

7.39 **Radioactive lichen.** Refer to the Lichen Radionuclide Baseline Research project at the University of Alaska, presented in Exercise 2.36 (p. 47). Recall that the researchers collected 9 lichen specimens and measured the amount (in microcuries per milliliter) of the radioactive element cesium-137 for each. (The natural logarithms of the data values are saved in the **LICHEN** file.) A MINITAB printout with summary statistics for the actual data is shown below.
  a. Give a point estimate for the mean amount of cesium in lichen specimens collected in Alaska.
  b. Give the *t*-value used in a small-sample 95% confidence interval for the true mean amount of cesium in Alaskan lichen specimens.
  c. Use the result you obtained in part **b** and the values of $\bar{x}$ and s shown on the MINITAB printout to form a 95% confidence interval for the true mean amount of cesium in Alaskan lichen specimens.
  d. Check the interval you found in part **c** with the 95% confidence interval shown on the MINITAB printout.
  e. Give a practical interpretation for the interval you obtained in part **c.**

Variable	N	Mean	StDev	SE Mean	95% CI
CESIUM	9	0.009027	0.004854	0.001618	(0.005296, 0.012759)

7.40 **Rainfall and desert ants.** Refer to the *Journal of Biogeography* (December 2003) study of ants and their habitat in the desert of Central Asia, presented in Exercise 2.68 (p. 59). Recall that botanists randomly selected five sites in the Dry Steppe region and six sites in the Gobi Desert where ants were observed. One of the variables of interest is the annual rainfall (in millimeters) at each site. (The data are saved in the **GOBIANTS** file.) Summary statistics for the annual rainfall at each site are provided in the SAS printout below.
  a. Give a point estimate for the average annual rainfall amount at ant sites in the Dry Steppe region of Central Asia.
  b. Give the *t*-value used in a small-sample 90% confidence interval for the true average annual rainfall amount at ant sites in the Dry Steppe region.
  c. Use the result you obtained in part **b** and the values of $\bar{x}$ and s shown on the SAS printout to form a 90% confidence interval for the target parameter.
  d. Give a practical interpretation for the interval you found in part **c.**
  e. Use the data in the **GOBIANTS** file to check the validity of the confidence interval you found in part **c.**
  f. Repeat parts **a–e** for the Gobi Desert region of Central Asia.

The MEANS Procedure

Analysis Variable : RAIN

REGION	N Obs	Mean	Std Dev
DryStepp	5	183.4000000	20.6470337
Gobi	6	110.0000000	15.9749804

Applying the Concepts—Intermediate

7.41 **Pitch memory of amusiacs.** Congenital amusia is a disorder that impacts one's perception of music. A team of psychologists and neuroscientists tested the pitch memory of individuals diagnosed with amusia and reported their results in *Advances in Cognitive Psychology* (Vol. 6, 2010). Each in a sample of 17 amusiacs listened to a series of tone pairs, where each tone pair was a standard tone followed by a comparison tone. For each tone pair, the subjects were asked to determine if the tones were the same or different. In one trial, the tones were separated by 1 second. In a second trial, the tones were separated by 5 seconds. (In theory, the longer the delay between tones, the less likely one is to detect a difference between the tones.) Scores in the two trials were compared for each amusiac. The mean score difference was .11 with a standard deviation of .19. Use this information to form a 90% confidence interval for the true mean score difference for all amusiacs. Interpret the result. What assumption about the population of score differences must hold true for the interval to be valid?

7.42 **Studies on treating Alzheimer's disease.** Alzheimer's disease is a progressive disease of the brain. Much research has been conducted on how to treat Alzheimer's. The journal *eCAM* (November 2006) published an article that critiqued the quality of the methodology used in studies on Alzheimer treatment. For each in a sample of 13 studies, the quality of the methodology was measured on the Wong scale, with scores ranging from 9 (low quality) to 27 (high quality). The data are shown in the table (p. 319) and saved in the **TREATAD** file. Estimate, with a 99% confidence interval, the mean quality μ of all studies on the treatment of Alzheimer's disease. Interpret the result.

Data for Exercise 7.42

22	21	18	19	20	15	19	20	15	20	17	20	13

Based on Chiappelli, F., et al. "Evidence-based research in complementary and alternative medicine III: Treatment of patients with Alzheimer's disease." *eCAM*, Vol. 3, No. 4, Nov. 2006 (Table 1).

7.43 **Reproduction of bacteria-infected spider mites.** Zoologists in Japan investigated the reproductive traits of spider mites with a bacterial infection (*Heredity*, Jan. 2007). Male and female pairs of infected spider mites were mated in a laboratory and the number of eggs produced by each female recorded. Summary statistics for several samples are provided in the accompanying table. Note that, in some samples, one or both infected spider mites were treated with antibiotic prior to mating.
   a. For each type of female–male pair, construct and interpret a 90% confidence interval for the population mean number of eggs produced by the female spider mite.
   b. Identify the type of female–male pair that appears to produce the highest mean number of eggs.

Female–Male Pairs	Sample Size	Mean # of Eggs	Standard Deviation
Both untreated	29	20.9	3.34
Male treated	23	20.3	3.50
Female treated	18	22.9	4.37
Both treated	21	18.6	2.11

Based on Gotoh, T., Noda, H., & Ito, S. "Cardinium symbionts cause cytoplasmic incompatibility in spider mites." *Heredity*, Vol. 98, No. 1, Jan. 2007 (Table 2).

7.44 **Minimizing tractor skidding distance.** In planning for a new forest road to be used for tree harvesting, planners must select the location that will minimize tractor skidding distance. In the *Journal of Forest Engineering* (July 1999), researchers wanted to estimate the true mean skidding distance along a new road in a European forest. The skidding distances (in meters) were measured at 20 randomly selected road sites. These values are given in the accompanying table and saved in the **SKIDDING** file.

488	350	457	199	285	409	435	574	439	546
385	295	184	261	273	400	311	312	141	425

Based on Tujek, J., & Pacola, E. "Algorithms for skidding distance modeling on a raster Digital Terrain Model." *Journal of Forest Engineering*, Vol. 10, No. 1, July 1999 (Table 1).

   a. Estimate, with a 95% confidence interval, the true mean skidding distance of the road.
   b. Give a practical interpretation of the interval you found in part **a.**
   c. What conditions are required for the inference you made in part b to be valid? Are these conditions reasonably satisfied?
   d. A logger working on the road claims that the mean skidding distance is at least 425 meters. Do you agree?

7.45 **Antigens for a parasitic roundworm in birds.** *Ascaridia galli* is a parasitic roundworm that attacks the intestines of birds, especially chickens and turkeys. Scientists are working on a synthetic vaccine (antigen) for the parasite. The success of the vaccine hinges on the characteristics of DNA in peptide (protein) produced by the antigen. In the journal *Gene Therapy and Molecular Biology* (June 2009), scientists tested alleles of antigen-produced protein for level of peptide. For a sample of 4 alleles, the mean peptide score was 1.43 and the standard deviation was .13.
   a. Use this information to construct a 90% confidence interval for the true mean peptide score in alleles of the antigen-produced protein.
   b. Interpret the interval for the scientists.
   c. What is meant by the phrase "90% confidence"?

7.46 **Eating disorders in females.** The "fear of negative evaluation" (FNE) scores for 11 bulimic female students and 14 normal female students, first presented in Exercise 2.40 and saved in the **BULIMIA** file (p. 49), are reproduced at the bottom of the page. (Recall that the higher the score, the greater is the FNE.)
   a. Construct a 95% confidence interval for the mean FNE score of the population of bulimic female students. Interpret the result.
   b. Construct a 95% confidence interval for the mean FNE score of the population of normal female students. Interpret the result.
   c. What assumptions are required for the intervals of parts **a** and **b** to be statistically valid? Are these assumptions reasonably satisfied? Explain.

Based on Randles, R. H. "On neutral responses (zeros) in the sign test and ties in the Wilcoxon-Mann-Whitney test." *The American Statistician*, May 2001, Vol. 55, No. 2 (Figure 3).

Applying the Concepts—Advanced

7.47 **Study on waking sleepers early.** Scientists have discovered increased levels of the hormone adrenocorticotropin in people just before they awake from sleeping (*Nature*, January 7, 1999). In the study described, 15 subjects were monitored during their sleep after being told that they would be woken at a particular time. One hour prior to the designated wake-up time, the adrenocorticotropin level (pg/mL) was measured in each, with the following results:

$$\bar{x} = 37.3 \qquad s = 13.9$$

   a. Use a 95% confidence interval to estimate the true mean adrenocorticotropin level of sleepers one hour prior to waking.
   b. Interpret the interval you found in part **a** in the words of the problem.
   c. The researchers also found that if the subjects were woken three hours earlier than they anticipated, the average adrenocorticotropin level was 25.5 pg/mL. Assume that $\mu = 25.5$ for all sleepers who are woken three hours earlier than expected. Use the interval from part **a** to make an inference about the mean adrenocorticotropin level of sleepers under two conditions: one hour before the anticipated wake-up time and three hours before the anticipated wake-up time.

Bulimic students:	21	13	10	20	25	19	16	21	24	13	14			
Normal students:	13	6	16	13	8	19	23	18	11	19	7	10	15	20

Source: Randles, R. H. "On neutral responses (zeros) in the sign test and ties in the Wilcoxon–Mann–Whitney Test." *The American Statistician*, Vol. 55, No. 2, May 2001 (Figure 3).

7.4 Large–Sample Confidence Interval for a Population Proportion

The number of public-opinion polls has grown at an astounding rate in recent years. Almost daily, the news media report the results of some poll. Pollsters regularly determine the percentage of people in favor of the president's welfare-reform program, the fraction of voters in favor of a certain candidate, the fraction of customers who favor a particular brand of wine, and the proportion of people who smoke cigarettes. In each case, we are interested in estimating the percentage (or proportion) of some group with a certain characteristic. In this section, we consider methods for making inferences about population proportions when the sample is large.

Example 7.6

Estimating a Population Proportion—Fraction Who Trust the President

Problem Public-opinion polls are conducted regularly to estimate the fraction of U.S. citizens who trust the president. Suppose 1,000 people are randomly chosen and 637 answer that they trust the president. How would you estimate the true fraction of *all* U.S. citizens who trust the president?

Solution What we have really asked is how you would estimate the probability p of success in a binomial experiment in which p is the probability that a person chosen trusts the president. One logical method of estimating p for the population is to use the proportion of successes in the sample. That is, we can estimate p by calculating

$$\hat{p} = \frac{\text{Number of people sampled who trust the president}}{\text{Number of people sampled}}$$

where $\hat{p}$ is read "p hat." Thus, in this case,

$$\hat{p} = \frac{637}{1,000} = .637$$

Look Back To determine the reliability of the estimator $\hat{p}$, we need to know its sampling distribution. That is, if we were to draw samples of 1,000 people over and over again, each time calculating a new estimate $\hat{p}$, what would be the frequency distribution of all the $\hat{p}$-values? The answer lies in viewing $\hat{p}$ as the average, or mean, number of successes per trial over the n trials. If each success is assigned a value equal to 1 and each failure is assigned a value of 0, then the sum of all n sample observations is x, the total number of successes, and $\hat{p} = x/n$ is the average, or mean, number of successes per trial in the n trials. The Central Limit Theorem tells us that the *relative frequency distribution of the sample mean for any population is approximately normal for sufficiently large samples.*

Now Work Exercise 7.56a

The repeated sampling distribution of $\hat{p}$ has the characteristics listed in the next box and shown in Figure 7.12.

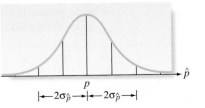

Figure 7.12
Sampling distribution of $\hat{p}$

Sampling Distribution of $\hat{p}$

1. The mean of the sampling distribution of $\hat{p}$ is p; that is, $\hat{p}$ is an unbiased estimator of p.

2. The standard deviation of the sampling distribution of $\hat{p}$ is $\sqrt{pq/n}$; that is, $\sigma_p = \sqrt{pq/n}$, where $q = 1 - p$.

3. For large samples, the sampling distribution of $\hat{p}$ is approximately normal. A sample size is considered large if both $n\hat{p} \geq 15$ and $n\hat{q} \geq 15$.

The fact that $\hat{p}$ is a "sample mean fraction of successes" allows us to form confidence intervals about p in a manner that is completely analogous to that used for large-sample estimation of μ.

Large-Sample Confidence Interval for p

$$\hat{p} \pm z_{\alpha/2}\sigma_{\hat{p}} = \hat{p} \pm z_{\alpha/2}\sqrt{\frac{pq}{n}} \approx \hat{p} \pm z_{\alpha/2}\sqrt{\frac{\hat{p}\hat{q}}{n}}$$

where $\hat{p} = \dfrac{x}{n}$ and $\hat{q} = 1 - \hat{p}$

Note: When n is large, $\hat{p}$ can approximate the value of p in the formula for $\sigma_{\hat{p}}$.

Conditions Required for a Valid Large-Sample Confidence Interval for p

1. A random sample is selected from the target population.

2. The sample size n is large. (This condition will be satisfied if both $n\hat{p} \geq 15$ and $n\hat{q} \geq 15$. Note that $n\hat{p}$ and $n\hat{q}$ are simply the number of successes and number of failures, respectively, in the sample.

Thus, if 637 of 1,000 U.S. citizens say they trust the president, a 95% confidence interval for the proportion of *all* U.S. citizens who trust the president is

$$\hat{p} \pm z_{\alpha/2}\sigma_{\hat{p}} = .637 \pm 1.96 \sqrt{\frac{pq}{1,000}}$$

where $q = 1 - p$. Just as we needed an approximation for σ in calculating a large-sample confidence interval for μ, we now need an approximation for p. As Table 7.6 shows, the approximation for p does not have to be especially accurate, because the value of $\sqrt{pq}$ needed for the confidence interval is relatively insensitive to changes in p. Therefore, we can use $\hat{p}$ to approximate p. Keeping in mind that $\hat{q} = 1 - \hat{p}$, we substitute these values into the formula for the confidence interval:

$$\begin{aligned}\hat{p} \pm 1.96\sqrt{pq/1,000} &\approx \hat{p} \pm 1.96\sqrt{\hat{p}\hat{q}/1,000} \\ &= .637 \pm 1.96\sqrt{(.637)(.363)/1,000} = .637 \pm .030 \\ &= (.607, .667)\end{aligned}$$

Table 7.6 Values of pq for Several Different Values of p

p	pq	$\sqrt{pq}$
.5	.25	.50
.6 or .4	.24	.49
.7 or .3	.21	.46
.8 or .2	.16	.40
.9 or .1	.09	.30

Then we can be 95% confident that the interval from 60.7% to 66.7% contains the true percentage of *all* U.S. citizens who trust the president. That is, in repeated constructions of confidence intervals, approximately 95% of all samples would produce confidence intervals that enclose p. Note that the guidelines for interpreting a confidence interval about μ also apply to interpreting a confidence interval for p, because p is the "population mean fraction of successes" in a binomial experiment.

Example 7.7

A Large-Sample Confidence Interval for p—Proportion Optimistic about the Economy

Problem Many public polling agencies conduct surveys to determine the current consumer sentiment concerning the state of the economy. For example, the Bureau of Economic and Business Research (BEBR) at the University of Florida conducts quarterly surveys to gauge consumer sentiment in the Sunshine State. Suppose that BEBR randomly samples 484 consumers and finds that only 157 are optimistic about the state of the economy. Use a 90% confidence interval to estimate the proportion of all consumers in Florida who are optimistic about the state of the economy. Based on the confidence interval, can BEBR infer that a minority of Florida consumers is optimistic about the economy?

Solution The number x of the 484 sampled consumers who are optimistic about the Florida economy is a binomial random variable if we can assume that the sample was randomly selected from the population of Florida consumers and that the poll was conducted identically for each consumer sampled.

The point estimate of the proportion of Florida consumers who are optimistic about the economy is

$$\hat{p} = \frac{x}{n} = \frac{157}{484} = .324$$

We first check to be sure that the sample size is sufficiently large that the normal distribution provides a reasonable approximation to the sampling distribution of $\hat{p}$. We require the number of successes in the sample, $n\hat{p}$, and the number of failures, $n\hat{q}$, both to be at least 15. Since the number of successes is $n\hat{p} = 157$ and the number of failures is $n\hat{q} = 327$, we may conclude that the normal approximation is reasonable.

We now proceed to form the 90% confidence interval for p, the true proportion of Florida consumers who are optimistic about the state of the economy:

$$\hat{p} \pm z_{\alpha/2}\sigma_{\hat{p}} = \hat{p} \pm z_{\alpha/2}\sqrt{\frac{pq}{n}} \approx \hat{p} \pm z_{\alpha/2}\sqrt{\frac{\hat{p}\hat{q}}{n}}$$

$$= .324 \pm 1.645\sqrt{\frac{(.324)(.676)}{484}} = .324 \pm .035 = (.289, .359)$$

(This interval is also shown on the MINITAB printout of Figure 7.13.) Thus, we can be 90% confident that the proportion of all Florida consumers who are confident about the economy is between .289 and .359. As always, our confidence stems from the fact that 90% of all similarly formed intervals will contain the true proportion p and not from any knowledge about whether this particular interval does.

Figure 7.13

MINITAB printout with 90% confidence interval for p

Test and CI for One Proportion

```
Sample   X    N   Sample p        90% CI
1       157  484  0.324380  (0.289379, 0.359381)

Using the normal approximation.
```

Can we conclude on the basis of this interval that a minority of Florida consumers is optimistic about the economy? If we wished to use this interval to infer that a minority is optimistic, the interval would have to support the inference that p is less than .5—that is, that less than 50% of Florida consumers are optimistic about the economy. Note that the interval contains only values below .5. Therefore, we can conclude on the basis of this 90% confidence interval that the true value of p is less than .5.

Look Ahead If the confidence interval includes .5 (e.g., an interval from .42 to .54), then we could not conclude that the true proportion of consumers who are optimistic is less than .5. (This is because it is possible that p is as high as .54.)

Now Work Exercise 7.59

We conclude this section with a warning and an illustrative example.

⚠ **CAUTION** Unless n is extremely large, the large-sample procedure presented in this section performs poorly when p is near 0 or near 1. ▲

The problem stated in the above warning can be illustrated as follows. Suppose you want to estimate the proportion of people who die from a bee sting using a sample size of $n = 100$. This proportion is likely to be near 0, say $p \approx .001$. If so, then $np \approx 100(.001) = .1$ is less than the recommended value of 15 (see conditions in the box on p. 321). Consequently, a confidence interval for p based on a sample of $p = 100$ will probably be misleading.

To overcome this potential problem, an *extremely* large sample size is required. Since the value of n required to satisfy "extremely large" is difficult to determine, statisticians (see Agresti & Coull, 1998) have proposed an alternative method, based on the Wilson (1927) point estimator of p. **Wilson's adjustment for estimating p** is outlined in the next box. Researchers have shown that this confidence interval works well for any p, even when the sample size n is very small.

Adjusted $(1 - \alpha)$ 100% Confidence Interval for a Population Proportion p

$$\tilde{p} \pm z_{\alpha/2} \sqrt{\frac{\tilde{p}(1 - \tilde{p})}{n + 4}}$$

where $\tilde{p} = \dfrac{x + 2}{n + 4}$ is the adjusted sample proportion of observations with the characteristic of interest, x is the number of successes in the sample, and n is the sample size.

Example 7.8

Using the Adjusted Confidence Interval Procedure— Proportion Who are Victims of a Violent Crime

Problem According to *True Odds: How Risk Affects Your Everyday Life* (Walsh, 1997), the probability of being the victim of a violent crime is less than .01. Suppose that, in a random sample of 200 Americans, 3 were victims of a violent crime. Use a 95% confidence interval to estimate the true proportion of Americans who were victims of a violent crime.

Solution Let p represent the true proportion of Americans who were victims of a violent crime. Since p is near 0, an "extremely large" sample is required to estimate its value by the usual large-sample method. Here, the number of "successes," 3, is less than 15. Thus, we doubt whether the sample size of 200 is large enough, to apply the large-sample method. Alternatively, we will apply Wilson's adjustment outlined in the box.

Because the number of "successes" (i.e., number of victims of a violent crime) in the sample is $x = 3$, the adjusted sample proportion is

$$\tilde{p} = \frac{x + 2}{n + 4} = \frac{3 + 2}{200 + 4} = \frac{5}{204} = .025$$

Note that this adjusted sample proportion is obtained by adding a total of four observations—two "successes" and two "failures"—to the sample data. Substituting $\tilde{p} = .025$ into the equation for a 95% confidence interval, we obtain

$$\tilde{p} \pm 1.96\sqrt{\frac{\tilde{p}(1 - \tilde{p})}{n + 4}} = .025 \pm 1.96\sqrt{\frac{(.025)(.975)}{204}}$$

$$= .025 \pm .021$$

or (.004, .046). Consequently, we are 95% confident that the true prfoportion of Americans who are victims of a violent crime falls between .004 and .046.

Look Back: If we apply the standard large-sample confidence interval formula, where $\hat{p} = 3/200 = .015$, we obtain

$$\hat{p} \pm 1.96\sqrt{\frac{\hat{p}\hat{q}}{200}} = .015 \pm 1.96\sqrt{\frac{(.015)(.985)}{200}}$$

$$= .015 \pm .017$$

or, $(-.002, .032)$. Note that the interval estimate contains negative (nonsensical) values for the true proportion. Such a result is typical when the large-sample method is misapplied.

Now Work Exercise 7.65b

Statistics IN Action | Revisited — Estimating the Coding Error Rate

In the previous Statistics in Action Revisited (p. 315), we showed how to estimate the mean overpayment amount for claims in a Medicare fraud study. In addition to estimating overcharges, the USDOJ also is interested in estimating the *coding error rate* of a Medicare provider. The coding error rate is defined as the proportion of Medicare claims that are coded incorrectly. Thus, for this inference, the USDOJ is interested in estimating a population proportion, p. Typically, the USDOJ finds that about 50% of the claims in a Medicare fraud case are incorrectly coded.

If you examine the sample data in Table SIA.7.1, you can verify that of the 52 audited claims, 31 were determined to be coded incorrectly, resulting in an overcharge. These are the claims with a disallowed amount greater than $0. Therefore, an estimate of the coding error rate, p, for this Medicare provider is

$$\hat{p} = 31/52 = .596$$

A 95% confidence interval for p can be obtained with the use of a confidence interval formula or with statistical software.

The **MCFRAUD** file includes a qualitative variable (called "Coding Error") at two levels, where "Yes" represents that the claim is coded incorrectly and "No" represents that the claim is coded correctly. Thus, the USDOJ desires an estimate of the proportion of "Yes" values for the "Coding Error" variable. A confidence interval for the proportion of incorrectly coded claims is highlighted on the accompanying MINITAB printout, Figure SIA.7.2. The interval, (.45, .73), implies that the true coding error rate for the 1,000 claims in the population falls between .45 and .73, with 95% confidence. Note that the value .5—the proportion of incorrectly coded claims expected by the USDOJ—falls within the 95% confidence interval.

Data Set: MCFRAUD

Figure SIA7.2

MINITAB confidence interval for coding error rate

```
Event = Yes

Variable       X   N  Sample p         95% CI
Coding_Error  31  52  0.596154  (0.451016, 0.729940)
```

Exercises 7.48–7.67

Understanding the Principles

7.48 Describe the sampling distribution of $\hat{p}$ on the basis of large samples of size n. That is, give the mean, the standard deviation, and the (approximate) shape of the distribution of $\hat{p}$ when large samples of size n are (repeatedly) selected from the binomial distribution with probability p of success.

7.49 Explain the meaning of the phrase "$\hat{p}$ is an unbiased estimator of p."

7.50 If p is near 0 or 1, how large a sample is needed to employ the large-sample confidence interval procedure?

Applet Exercise 7.5

Use the applet entitled *Confidence Intervals for a Proportion* to investigate the effect of the value of p on the number of confidence intervals that contain the population proportion p for a fixed sample size. For this exercise, use sample size $n = 10$.

a. Run the applet several times without clearing for $p = .1$. What proportion of the 95% confidence intervals contain p? What proportion of the 99% confidence intervals contain p? Do the results surprise you? Explain.

b. Repeat part **a** for each value of p: $p = .2, p = .3$, $p = .4$. $p = .5, p = .6, p = .7, p = .8$, and $p = .9$.

c. Which value of p yields the greatest proportion of each type of interval that contain p?

d. Based on your results, what values of p will yield more reliable confidence intervals for a fixed sample size n? Explain.

Applet Exercise 7.6

Use the applet entitled *Confidence Intervals for a Proportion* to investigate the effect of the sample size on the number of confidence intervals that contain the population proportion p for a value of p close to 0 or 1.

a. Run the applet several times without clearing for $p = .5$ and $n = 50$. Record the proportion of the 99% confidence intervals containing p.

b. Now set $p = .1$ and run the applet several times without clearing for $n = 50$. How does the proportion of the 99% confidence intervals containing p compare with that in part **a**?

c. Repeat part **b**, keeping $p = .1$ and increasing the sample size by 50 until you find a sample size that yields a similar proportion of the 99% confidence intervals containing p as that in part **a**.

d. Based on your results, describe how the value of p affects the sample size needed to guarantee a certain level of confidence.

Learning the Mechanics

7.51 A random sample of size $n = 196$ yielded $\hat{p} = .64$.

a. Is the sample size large enough to use the methods of this section to construct a confidence interval for p? Explain.

b. Construct a 95% confidence interval for p.

c. Interpret the 95% confidence interval.

d. Explain what is meant by the phrase "95% confidence interval."

7.52 A random sample of size $n = 144$ yielded $\hat{p} = .76$.

a. Is the sample size large enough to use the methods of this section to construct a confidence interval for p? Explain.

b. Construct a 90% confidence interval for p.

c. What assumption is necessary to ensure the validity of this confidence interval?

7.53 For the binomial sample information summarized in each part, indicate whether the sample size is large enough to use the methods of this chapter to construct a confidence interval for p.

a. $n = 500, \hat{p} = .05$
b. $n = 100, \hat{p} = .05$
c. $n = 10, \hat{p} = .5$
d. $n = 10, \hat{p} = .3$

7.54 A random sample of 50 consumers taste-tested a new snack food. Their responses were coded (0: do not like; 1: like; 2: indifferent), recorded and saved in the **SNACK** file as follows:

1	0	0	1	2	0	1	1	0	0
0	1	0	2	0	2	2	0	0	1
1	0	0	0	0	1	0	2	0	0
0	1	0	0	1	0	0	1	0	1
0	2	0	0	1	1	0	0	0	1

a. Use an 80% confidence interval to estimate the proportion of consumers who like the snack food.

b. Provide a statistical interpretation for the confidence interval you constructed in part **a**.

Applying the Concepts — Basic

7.55 **Is Starbucks coffee overpriced?** The *Minneapolis Star Tribune* (August 12, 2008) reported that 73% of Americans say that Starbucks coffee is overpriced. The source of this information was a national telephone survey of 1,000 American adults conducted by Rasmussen Reports.

a. Identify the population of interest in this study.

b. Identify the sample for the study.

c. Identify the parameter of interest in the study.

d. Find and interpret a 95% confidence interval for the parameter of interest.

7.56 **Satellite radio in cars.** Refer to the June 2007 survey on satellite radio subscriber service and usage conducted for the National Association of Broadcasters, Exercise 1.24 (p. 20). Recall that a random sample of 501 satellite radio subscribers were asked, "Do you have a satellite radio receiver in your car?" The survey found that 396 subscribers did, in fact, have a satellite receiver in their car.

[NW] **a.** From the sample, calculate an estimate of the true proportion of satellite radio subscribers who have a satellite radio receiver in their car.

b. Form a 90% confidence interval for the estimate, part **a**.

c. Give a practical interpretation of the interval, part **b**.

7.57 **National Firearms Survey.** Refer to the Harvard School of Public Health survey to determine the size and composition of privately held firearm stock in the United States, presented in Exercise 3.77 (p. 146). Recall that, in a representative household telephone survey of 2,770 adults, 26% reported that they own at least one gun (*Injury Prevention*, Jan. 2007). The researchers want to estimate the true percentage of adults in the United States that own at least one gun.

a. Identify the population of interest to the researchers.
b. Identify the parameter of interest to the researchers.
c. Compute an estimate of the population parameter.
d. Form a 99% confidence interval around the estimate.
e. Interpret the confidence interval practically.
f. Explain the meaning of the phrase "99% confident."

7.58 **Are you really being served red snapper?** Refer to the *Nature* (July 15, 2004) study of fish specimens labeled "red snapper," presented in Exercise 3.85 (p. 147). Recall that federal law prohibits restaurants from serving a cheaper look-alike variety of fish (e.g., vermillion snapper or lane snapper) to customers who order red snapper. In an effort to estimate the true proportion of fillets that are really red snapper, a team of University of North Carolina (UNC) researchers analyzed the meat from each in a sample of 22 "red snapper" fish fillets purchased from vendors across the United States. DNA tests revealed that 17 of the 22 fillets (or 77%) were not red snapper, but the cheaper look-alike variety of fish.

a. Identify the parameter of interest to the UNC researchers.
b. Explain why a large-sample confidence interval is inappropriate to apply in this study.
c. Use Wilson's adjustment to construct a 95% confidence interval for the parameter of interest.
d. Give a practical interpretation of the confidence interval.

7.59 **What we do when we are sick at home.** *USA Today* (Feb. 15, 2007) reported on the results of an opinion poll in which adults were asked what one thing they are most likely to do when they are home sick with a cold or the flu. In the survey, 63% said that they are most likely to sleep and 18% said that they would watch television. Although the sample size was not reported, typically opinion polls include approximately 1,000 randomly selected respondents.

a. Assuming a sample size of 1,000 for this poll, construct a 95% confidence interval for the true percentage of all adults who would choose to sleep when they are at home sick.
b. If the true percentage of adults who would choose to sleep when they are at home sick is 70%, would you be surprised? Explain.

7.60 **Speed training in football.** Refer to *The Sport Journal* (Winter 2004) study on the effectiveness of a speed-training program for football players, Exercise 7.22 (p. 309). The decreases in 40-yard sprint times (time after training minus time before training) for 38 players are saved in the **SPRINT** file. An athlete's sprint performance will be classified as "Improved" if the "after" time is less than the "before" time, and classified as "Not Improved" if otherwise.

a. Find an estimate for the true proportion of all high school athletes who attain improved sprint times after participating in the speed-training program.
b. Convert the estimate, part **a**, into a 95% confidence interval. Give a practical interpretation of the result.

Applying the Concepts—Intermediate

7.61 **Teenagers' use of emoticons in school writing.** Due to the popularity of instant messaging and social networking sites among American teenagers, informal elements such as emoticons (e.g., the symbol ":)" to represent a smile) and abbreviations (e.g., "LOL" for "laughing out loud") have worked their way into teenagers' school writing assignments. In a survey cosponsored by the National Commission on Writing at the College Board, the *Pew Internet and American Life Project* (April 2008) interviewed 700 randomly selected U.S. teenagers by telephone on their writing habits. Overall, 448 of the teenagers admitted using at least one informal element in school writing assignments. Based on the survey results, construct a 99% confidence interval for the proportion of all U.S. teenagers who have used at least one informal element in school writing assignments. Give a practical interpretation of the interval, one that is relevant to the National Commission on Writing at the College Board.

7.62 **Characteristics of ice-melt ponds.** Refer to the University of Colorado study of ice-melt ponds in the Canadian Arctic, presented in Exercise 2.13 (p. 35). Environmental engineers are using data collected by the National Snow and Ice Data Center to learn how climate affects the sea ice. Data on 504 ice melt ponds are saved in the **PONDICE** file. Of these 504 melt ponds, 88 were classified as having "first-year ice." Recall that the researchers estimated that about 17% of melt ponds in the Canadian Arctic have first-year ice. Use the methodology of this chapter to estimate, with 90% confidence, the percentage of all ice-melt ponds in the Canadian Arctic that have first-year ice. Give a practical interpretation of the results.

7.63 **Do you believe in the Bible?** Refer to the National Opinion Research Center's General Social Survey (GSS), presented in Exercise 2.18 (p. 36). Data on the approximately 2,800 Americans who participated in the 2004 GSS are saved in the **BIBLE** file. Recall that one question in the survey asked about a person's belief in the Bible. Suppose you want to estimate the proportion of all Americans who believe that the Bible is the actual word of God and is to be taken literally. (*Note:* The variable "Bible1" contains the responses to this question.)

a. Use a 95% confidence interval to estimate the proportion of interest.
b. Give a practical interpretation of the interval you used in part **a**.
c. Discuss how the survey methodology can affect the validity of the results.

7.64 **Do you think you smell?** If you falsely believe you emit an unpleasant, foul, or offensive body odor, you may suffer from olfactory reference syndrome (ORS). The disorder disables patients, who often isolate themselves and consider suicide. Psychiatrists disagree over how prevalent ORS is in the human population. *Depression and Anxiety* (June 2010) discussed one self-reported survey of 2,481 university students in Japan. The study reported that 52 of the students were "concerned with emitting a strange bodily odor."

a. Use the survey results to estimate the true proportion of all people in the world who suffer from ORS. Place 95% confidence bounds around the estimate and interpret the result.
b. Give several reasons why the inference, part **a**, may be invalid. Explain.

7.65 **Splinting in mountain-climbing accidents.** The most common injury that occurs among mountain climbers is trauma to the lower extremity (leg). Consequently, rescuers must be proficient in immobilizing and splinting of fractures. In *High Altitude Medicine & Biology* (Vol. 10, 2009), researchers provided official recommendations for mountain emergency medicine. As part of the document, the researchers examined the likelihood of needing certain types of splints. A Scottish Mountain Rescue study reported that there was 1 femoral shaft splint needed among 333 live casualties. The researchers will use this study to estimate the proportion of all mountain casualties that require a femoral shaft splint.

   a. Is the sample large enough to apply the large-sample estimation method of this section? Show why or why not.

[NW] **b.** Use Wilson's adjustment to find a 95% confidence interval for the true proportion of all mountain casualties that require a femoral shaft splint. Interpret the result.

7.66 Studies on treating Alzheimer's disease. Refer to the journal *eCAM* (November 2006) assessment of the quality of the methodology used in studies of the treatment for Alzheimer's disease, presented in Exercise 7.42 (p. 318). Data on the quality of the methodology (using the Wong scale) for each in a sample of 13 studies are reproduced in the next table and saved in the **TREATAD** file. According to the researchers, a study with a Wong score below 18 used

a methodology that "fails to support the author's conclusions" about the treatment of Alzheimer's. Use Wilson's adjustment to estimate the proportion of all studies on the treatment of Alzheimer's disease with a Wong score below 18. Construct a 99% confidence interval around the estimate and interpret the result.

| 22 | 21 | 18 | 19 | 20 | 15 | 19 | 20 | 15 | 20 | 17 | 20 | 21 |

Based on Chiappelli, F., et al. "Evidence-based research in complementary and alternative medicine III: Treatment of patients with Alzheimer's disease." *eCAM*, Vol. 3, No. 4, Nov. 2006 (Table 1).

Applying the Concepts—Advanced

7.67 Latex allergy in health care workers. Refer to the *Current Allergy & Clinical Immunology* (March 2004) study of health care workers who use latex gloves, presented in Exercise 7.13 (p. 307). In addition to the 46 hospital employees who were diagnosed with a latex allergy on the basis of a skin-prick test, another 37 health care workers were diagnosed with the allergy by means of a latex-specific serum test. Of these 83 workers with a confirmed latex allergy, only 36 suspected that they had the allergy when they were asked about it on a questionnaire. Make a statement about the likelihood that a health care worker with a latex allergy suspects that he or she actually has the allergy. Attach a measure of reliability to your inference.

7.5 Determining the Sample Size

Recall from Section 1.5 that one way to collect the relevant data for a study used to make inferences about a population is to implement a designed (planned) experiment. Perhaps the most important design decision faced by the analyst is to determine the size of the sample. We show in this section that the appropriate sample size for making an inference about a population mean or proportion depends on the desired reliability.

Estimating a Population Mean

Consider Example 7.1 (p. 301), in which we estimated the mean length of stay for patients in a large hospital. A sample of 100 patients' records produced the 95% confidence interval $\bar{x} \pm 1.96\sigma_{\bar{x}} = 4.5 \pm .78$. Consequently, our estimate $\bar{x}$ was within .78 day of the true mean length of stay, μ, for all the hospital's patients at the 95% confidence level. That is, the 95% confidence interval for μ was $2(.78) = 1.56$ days wide when 100 accounts were sampled. This is illustrated in Figure 7.14a.

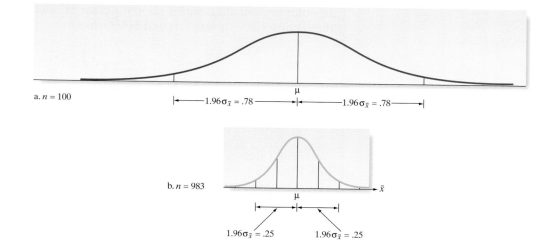

Figure 7.14

Relationship between sample size and width of confidence interval: hospital-stay example

Now suppose we want to estimate μ to within .25 day with 95% confidence. That is, we want to narrow the width of the confidence interval from 1.56 days to .50 day, as shown in Figure 7.14b. How much will the sample size have to be increased to accomplish this? If we want the estimator $\bar{x}$ to be within .25 day of μ, then we must have

$$1.96\sigma_{\bar{x}} = .25 \text{ or, equivalently, } 1.96\left(\frac{\sigma}{\sqrt{n}}\right) = .25$$

Note that we are using $\sigma_{\bar{x}} = \dfrac{\sigma}{\sqrt{n}}$ in the formula, since we are dealing with the sampling distribution of $\bar{x}$ (the estimator of μ).

The necessary sample size is obtained by solving this equation for n. First, we need the value of σ. In Example 7.1, we assumed that the standard deviation of length of stay was $\sigma = 4$ days. Thus,

$$1.96\left(\frac{\sigma}{\sqrt{n}}\right) = 1.96\left(\frac{4}{\sqrt{n}}\right) = .25$$

$$\sqrt{n} = \frac{1.96(4)}{.25} = 31.36$$

$$n = (31.36)^2 = 983.45$$

Consequently, over 983 patients' records will have to be sampled to estimate the mean length of stay, μ, to within .25 day with (approximately) 95% confidence. The confidence interval resulting from a sample of this size will be approximately .50 day wide. (See Figure 7.14b.)

In general, we express the reliability associated with a confidence interval for the population mean μ by specifying the **sampling error** within which we want to estimate μ with $100(1 - \alpha)$% confidence. The sampling error (denoted SE) is then equal to the half-width of the confidence interval, as shown in Figure 7.15.

The procedure for finding the sample size necessary to estimate μ with a specific sampling error is given in the following box. Note that if σ is unknown (as is usually the case, in practice) you will need to estimate the value of σ.

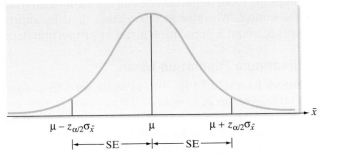

Figure 7.15

Specifying the sampling error SE as the half-width of a confidence interval

---

Determination of Sample Size for $100(1 - \alpha)$% Confidence Intervals for μ

In order to estimate μ with a sampling error SE and with $100(1 - \alpha)$% confidence, the required sample size is found as follows:

$$z_{\alpha/2}\left(\frac{\sigma}{\sqrt{n}}\right) = SE$$

The solution for n is given by the equation

$$n = \frac{(z_{\alpha/2})^2\sigma^2}{(SE)^2}$$

Note: The value of σ is usually unknown. It can be estimated by the standard deviation s from a previous sample. Alternatively, we may approximate the range R of observations in the population and (conservatively) estimate $\sigma \approx R/4$. In any case, you should round the value of n obtained *upward* to ensure that the sample size will be sufficient to achieve the specified reliability.

Example 7.9
Sample Size for Estimating μ— Mean Inflation Pressure of Footballs

Problem Suppose the manufacturer of official NFL footballs uses a machine to inflate the new balls to a pressure of 13.5 pounds. When the machine is properly calibrated, the mean inflation pressure is 13.5 pounds, but uncontrollable factors cause the pressures of individual footballs to vary randomly from about 13.3 to 13.7 pounds. For quality control purposes, the manufacturer wishes to estimate the mean inflation pressure to within .025 pound of its true value with a 99% confidence interval. What sample size should be specified for the experiment?

Solution We desire a 99% confidence interval that estimates μ with a sampling error of SE = .025 pound. For a 99% confidence interval, we have $z_{\alpha/2} = z_{.005} = 2.575$. To estimate σ, we note that the range of observations is $R = 13.7 - 13.3 = .4$ and we use $\sigma \approx R/4 = .1$. Next, we employ the formula derived in the box to find the sample size n:

$$n = \frac{(z_{\alpha/2})^2 \sigma^2}{(\text{SE})^2} \approx \frac{(2.575)^2(.1)^2}{(.025)^2} = 106.09$$

We round this up to $n = 107$. Realizing that σ was approximated by $R/4$, we might even advise that the sample size be specified as $n = 110$ to be more certain of attaining the objective of a 99% confidence interval with a sampling error of .025 pound or less.

Look Back To determine the value of the sampling error SE, look for the value that follows the key words "estimate μ to within . . . ."

Now Work Exercise 7.83

Sometimes the formula will lead to a solution that indicates a small sample size is sufficient to achieve the confidence interval goal. Unfortunately, the procedures and assumptions for small samples differ from those for large samples, as we discovered in Section 7.3. Therefore, if the formulas yield a small sample size, one simple strategy is to select a sample size $n = 30$.

Estimating a Population Proportion

The method just outlined is easily applied to a population proportion p. To illustrate, in Example 7.6 (p. 320) a pollster used a sample of 1,000 U.S. citizens to calculate a 95% confidence interval for the proportion who trust the president, obtaining the interval $.637 \pm .03$. Suppose the pollster wishes to estimate more precisely the proportion who trust the president, say, to within .015 with a 95% confidence interval.

The pollster wants a confidence interval for p with a sampling error SE = .015. The sample size required to generate such an interval is found by solving the following equation for n:

$$z_{\alpha/2}\sigma_{\hat{p}} = \text{SE} \qquad \text{or} \qquad z_{\alpha/2}\sqrt{\frac{pq}{n}} = .015 \qquad \text{(see Figure 7.16)}$$

Since a 95% confidence interval is desired, the appropriate z value is $z_{\alpha/2} = z_{.025} = 1.96$. We must approximate the value of the product pq before we can solve the equation

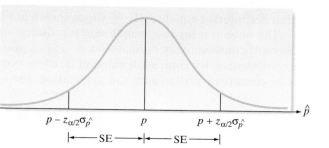

Figure 7.16

Specifying the sampling error SE of a confidence interval for a population proportion p

for n. As shown in Table 7.6 (p. 321), the closer the values of p and q to .5, the larger is the product pq. Thus, to find a conservatively large sample size that will generate a confidence interval with the specified reliability, we generally choose an approximation of p close to .5. In the case of the proportion of U.S. citizens who trust the president, how-ever, we have an initial sample estimate of $\hat{p} = .637$. A conservatively large esti-mate of pq can therefore be obtained by using, say, $p = .60$. We now substitute into the equation and solve for n:

$$1.96\sqrt{\frac{(.60)(.40)}{n}} = .015$$

$$n = \frac{(1.96)^2(.60)(.40)}{(.015)^2} = 4{,}097.7 \approx 4{,}098$$

The pollster must sample about 4,098 U.S. citizens to estimate the percentage who trust the president with a confidence interval of width .03.

The procedure for finding the sample size necessary to estimate a population proportion p with a specified sampling error SE is given in the following box:

Determination of Sample Size for $100(1-\alpha)\%$ Confidence Interval for p

In order to estimate a binomial probability p with sampling error SE and with $100(1 - \alpha)\%$ confidence, the required sample size is found by solving the following equation for n:

$$z_{\alpha/2}\sqrt{\frac{pq}{n}} = \text{SE}$$

The solution for n can be written as follows:

$$n = \frac{(z_{\alpha/2})^2(pq)}{(\text{SE})^2}$$

Note: Because the value of the product pq is unknown, it can be estimated by the sample fraction of successes, $\hat{p}$, from a previous sample. Remember (Table 7.6) that the value of pq is at its maximum when p equals .5, so you can obtain conservatively large values of n by approximating p by .5 or values close to .5. In any case, you should round the value of n obtained *upward* to ensure that the sample size will be sufficient to achieve the specified reliability.

Example 7.10

Sample Size For Estimating p — Fraction of Defective Cell Phones

Problem A cellular telephone manufacturer that entered the post-regulation market quickly has an initial problem with excessive cus-tomer complaints and consequent returns of cell phones for repair or replacement. The manufacturer wants to estimate the magni-tude of the problem in order to design a quality control program. How many cellular telephones should be sampled and checked in order to estimate the fraction defective, p, to within .01 with 90% confidence?

Solution In order to estimate p to within .01 of its true value, we set the half-width of the confidence interval equal to SE = .01, as shown in Figure 7.17.

The equation for the sample size n requires an estimate of the product pq. We could most conservatively estimate $pq = .25$ (i.e., use $p = .5$), but this estimate may be too conservative. By contrast, a value of .1, corresponding to 10% defective, will prob-ably be conservatively large for this application. The solution is therefore

$$n = \frac{(z_{\alpha/2})^2(pq)}{(SE)^2} = \frac{(1.645)^2(.1)(.9)}{(.01)^2} = 2{,}435.4 \approx 2{,}436$$

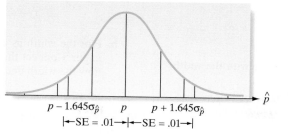

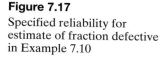

Figure 7.17

Specified reliability for estimate of fraction defective in Example 7.10

Thus, the manufacturer should sample 2,436 telephones in order to estimate the fraction defective, p, to within .01 with 90% confidence.

Look Back Remember that this answer depends on our approximation of pq, for which we used .09. If the fraction defective is closer to .05 than .10, we can use a sample of 1,286 telephones (check this) to estimate p to within .01 with 90% confidence.

Now Work Exercise 7.82

Ethics IN Statistics

In sampling, intentional omission of experimental units (e.g., respondents in a survey) in order to bias the results towards a particular view or outcome is considered *unethical statistical practice*.

The cost of sampling will also play an important role in the final determination of the sample size to be selected to estimate either μ or p. Although more complex formulas can be derived to balance the reliability and cost considerations, we will solve for the necessary sample size and note that the sampling budget may be a limiting factor. (Consult the references for a more complete treatment of this problem.) Once the sample size n is determined, be sure to devise a sampling plan that will ensure that a representative sample is selected from the target population.

Statistics IN Action   Revisited   Determining Sample Size

In the previous Statistics in Action applications in this chapter, we used confidence intervals (1) to estimate μ, the mean overpayment amount for claims in a Medicare fraud study, and (2) to estimate p, the coding error rate (i.e., proportion of claims that are incorrectly coded) of a Medicare provider. Both of these confidence intervals were based on selecting a random sample of 52 claims from the population of claims handled by the Medicare provider. How does the USDOJ determine how many claims to sample for auditing?

Consider the problem of estimating the coding error rate, p, As stated in a previous Statistics in Action Revisited, the USDOJ typically finds that about 50% of the claims in a Medicare fraud case are incorrectly coded. Suppose the USDOJ wants to estimate the true coding error rate of a Medicare provider to within .1 with 95% confidence. How many claims should be randomly sampled for audit in order to attain the desired estimate?

Here, the USDOJ desires a sampling error of SE $= .1$, a confidence level of $1 - \alpha = .95$ (for which $z_{\alpha/2} = 1.96$), and

uses an estimate $p \approx .50$. Substituting these values into the sample size formula (p. 328), we obtain.

$$
\begin{aligned}
n &= (z_{\alpha/2})^2 (pq)/(\text{SE})^2 \\
&= (1.96)^2 (.5)(.5)/(.1)^2 \\
&= 96.04
\end{aligned}
$$

Consequently, the USDOJ should audit about 97 randomly selected claims to attain a 95% confidence interval for p with a sampling error of .10.

[*Note:* You may wonder why the sample actually used in the fraud analysis included only 52 claims. The sampling strategy employed involved more than selecting a simple random sample; rather, it used a more sophisticated sampling scheme, called stratified random sampling. The 52 claims represented the sample for just one of the strata.]

Exercises 7.68–7.92

Understanding the Principles

7.68 How does the sampling error SE compare with the width of a confidence interval?

7.69 **True or false.** For a specified sampling error SE, increasing the confidence level $(1 - \alpha)$ will lead to a larger n in determining the sample size.

7.70 **True or false.** For a fixed confidence level $(1 - \alpha)$, increasing the sampling error SE will lead to a smaller n in determining the sample size.

Learning the Mechanics

7.71 If you wish to estimate a population mean to within .2 with a 95% confidence interval and you know from previous sampling that σ^2 is approximately equal to 5.4, how many observations would you have to include in your sample?

7.72 If nothing is known about p, .5 can be substituted for p in the sample-size formula for a population proportion. But when this is done, the resulting sample size may be larger than needed. Under what circumstances will using $p = .5$ in the sample-size formula yield a sample size larger than is needed to construct a confidence interval for p with a specified bound and a specified confidence level?

7.73 Suppose you wish to estimate a population mean correct to within .15 with a confidence level of .90. You do not know σ^2, but you know that the observations will range in value between 31 and 39.
  a. Find the approximate sample size that will produce the desired accuracy of the estimate. You wish to be conservative to ensure that the sample size will be ample for achieving the desired accuracy of the estimate. [*Hint:* Using your knowledge of data variation from Section 2.5, assume that the range of the observations will equal 4σ.]
  b. Calculate the approximate sample size, making the less conservative assumption that the range of the observations is equal to 6σ.

7.74 In each case, find the approximate sample size required to construct a 95% confidence interval for p that has sampling error SE = .06.
  a. Assume that p is near .3.
  b. Assume that you have no prior knowledge about p, but you wish to be certain that your sample is large enough to achieve the specified accuracy for the estimate.

7.75 The following is a 90% confidence interval for p: (.26, .54). How large was the sample used to construct this interval?

7.76 It costs you $10 to draw a sample of size $n = 1$ and measure the attribute of interest. You have a budget of $1,200.
  a. Do you have sufficient funds to estimate the population mean for the attribute of interest with a 95% confidence interval 4 units in width? Assume that $\sigma = 12$.
  b. If a 90% confidence level were used, would your answer to part **a** change? Explain.

7.77 Suppose you wish to estimate the mean of a normal population with a 95% confidence interval and you know from prior information that $\sigma^2 \approx 1$.
  a. To see the effect of the sample size on the width of the confidence interval, calculate the width of the confidence interval for $n = 16, 25, 49, 100,$ and 400.

  b. Plot the width as a function of sample size n on graph paper. Connect the points by a smooth curve, and note how the width decreases as n increases.

Applying the Concepts—Basic

7.78 **Risk of home burglary in cul-de-sacs.** Research published in the *Journal of Quantitative Criminology* (March 2010) revealed that the risk of burglaries in homes located on cul-de-sacs is lower than for homes on major roads. Suppose you want to estimate the true percentage of cul-de-sac homes in your home city that were burglarized in the past year. Devise a sampling plan so that your estimate will be accurate to within 2% of the true value using a confidence coefficient of 95%. How many cul-de-sac homes need to be sampled and what information do you need to collect for each sampled home?

7.79 **Lobster trap placement.** Refer to the *Bulletin of Marine Science* (April 2010) study of lobster trap placement, Exercise 7.35 (p. 317). Recall that you used a 95% confidence interval to estimate the mean trap spacing (in meters) for the population of red spiny lobster fishermen fishing in Baja California Sur, Mexico. How many teams of fishermen would need to be sampled in order to reduce the width of the confidence interval to 5 meters? Use the sample standard deviation from Exercise 7.35 in your calculation.

7.80 **Radioactive lichen.** Refer to the Alaskan Lichen Radionuclide Baseline Research study, presented in Exercise 7.39 (p. 318). In a sample of $n = 9$ lichen specimens, the researchers found the mean and standard deviation of the amount of the radioactive element, cesium-137, that was present to be .009 and .005 microcurie per milliliter, respectively. Suppose the researchers want to increase the sample size in order to estimate the mean μ to within .001 microcurie per milliliter of its true value, using a 95% confidence interval.
  a. What is the confidence level desired by the researchers?
  b. What is the sampling error desired by the researchers?
  c. Compute the sample size necessary to obtain the desired estimate.

7.81 **Scanning errors at Wal-Mart.** Refer to the National Institute for Standards and Technology (NIST) study of the accuracy of checkout scanners at Wal-Mart stores in California, presented in Exercise 3.52 (p. 132). NIST sets standards so that no more than 2 of every 100 items scanned through an electronic checkout scanner can have an inaccurate price. Recall that in a sample of 60 Wal-Mart stores, 52 violated the NIST scanner accuracy standard (*Tampa Tribune*, Nov. 22, 2005). Suppose you want to estimate the true proportion of Wal-Mart stores in California that violate the NIST standard.
  a. Explain why the large-sample methodology of Section 7.4 is inappropriate for this study.
  b. Determine the number of Wal-Mart stores that must be sampled in order to estimate the true proportion to within .05 with 90% confidence, using the large-sample method.

7.82 **Aluminum cans contaminated by fire.** A gigantic warehouse located in Tampa, Florida, stores approximately

60 million empty aluminum beer and soda cans. Recently, a fire occurred at the warehouse. The smoke from the fire contaminated many of the cans with blackspot, rendering them unusable. A University of South Florida statistician was hired by the insurance company to estimate p, the true proportion of cans in the warehouse that were contaminated by the fire. How many aluminum cans should be randomly sampled to estimate the true proportion to within .02 with 90% confidence?

7.83 **Assessing the bending strength of a wooden roof.** Refer
[NW] to the *Journal of the International Association for Shell and Spatial Structures* (Aug. 2004) study to estimate the mean bending strength of imported white wood used on the roof of an ancient Japanese temple, presented in Exercise 7.38 (p. 318). Suppose you want to estimate the true mean breaking strength of the white wood to within 4 MPa, using a 90% confidence interval. How many pieces of the imported wood need to be tested? Recall that the sample standard deviation of the breaking strengths found in the study was 10.9 Mpa.

Applying the Concepts—Intermediate

7.84 **Duration of daylight in western Pennsylvania.** Refer to the *Naval Oceanography Portal* data on number of minutes of daylight per day in Sharon, PA, Exercise 7.37 (p. 317). An estimate of the mean number of minutes of daylight per day was obtained using data collected for 12 randomly selected days (one each month) in 2009.
   a. Determine the number of days in 2009 that need to be sampled in order to estimate the desired mean to within 45 minutes of its true value with 95% confidence.
   b. Based on your answer, part **a,** develop a sampling plan that will likely result in a random sample that is representative of the population.
   c. Go to the Web site, http://aa.usno.navy.mil/USNO/astronomical-applications/data-services, and collect the data for Sharon, PA in year 2009 using your sampling plan.
   d. Use the data, part **c,** to construct a 95% confidence interval for the desired mean. Does your interval have the desired width?

7.85 **Pitch memory of amusiacs.** Refer to the *Advances in Cognitive Psychology* (Vol. 6, 2010) study of pitch memory of amusiacs, Exercise 7.41 (p. 318). Recall that diagnosed amusiacs listened to a series of tone pairs and were asked to determine if the tones were the same or different. In the first trial, the tones were separated by 1 second; in the second trial, the tones were separated by 5 seconds. The variable of interest was the difference between scores on the two trials. How many amusiacs would need to participate in the study in order to estimate the true mean score difference for all amusiacs to within .05 with 90% confidence?

7.86 **Speed training in football.** Refer to *The Sport Journal* (Winter 2004) study on the effectiveness of a speed training program for football players, Exercise 7.60 (p. 326). The confidence interval for the population proportion of athletes who improve their sprint times after participating in the speed-training program was fairly wide due to the relatively small number of athletes ($n = 38$) who participated in the program. In order to find a valid, narrower confidence interval for the true proportion, the researchers would need to include more athletes in the study. How many high school athletes should be sampled

to estimate the true proportion to within .03 with 95% confidence?

7.87 **Bacteria in bottled water.** Is the bottled water you drink safe? The Natural Resources Defense Council warns that the bottled water you are drinking may contain more bacteria and other potentially carcinogenic chemicals than are allowed by state and federal regulations. Of the more than 1,000 bottles studied, nearly one-third exceeded government levels. Suppose that the Natural Resources Defense Council wants an updated estimate of the population proportion of bottled water that violates at least one government standard. Determine the sample size (number of bottles) needed to estimate this proportion to within ± 0.01 with 99% confidence.

7.88 **Asthma drug study.** The chemical benzalkonium chloride (BAC) is an antibacterial agent that is added to some asthma medications to prevent contamination. Researchers at the University of Florida College of Pharmacy have discovered that adding BAC to asthma drugs can cause airway constriction in patients. In a sample of 18 asthmatic patients, each of whom received a heavy dose of BAC, 10 experienced a significant drop in breathing capacity (*Journal of Allergy and Clinical Immunology*, January 2001). Based on this information, a 95% confidence interval for the true percentage of asthmatic patients who experience breathing difficulties after taking BAC is (.326, .785).
   a. Why might the confidence interval lead to an erroneous inference?
   b. How many asthma patients must be included in the study in order to estimate the true percentage who experience a significant drop in breathing capacity to within 4% with a 95% confidence interval?

7.89 **Do you think you smell?** Refer to the *Depression and Anxiety* (June 2010) study of patients who suffer from olfactory reference syndrome (ORS), Exercise 7.64 (p. 326). Recall that psychiatrists disagree over how prevalent ORS is in the human population. Suppose you want to estimate the true proportion of U.S. adults who suffer from ORS using a 99% confidence interval. Determine the size of the sample necessary to attain a sampling error no larger than .04.

7.90 **Caffeine content of coffee.** According to a Food and Drug Administration (FDA) study, a cup of coffee contains an average of 115 milligrams (mg) of caffeine, with the amount per cup ranging from 60 to 180 mg. Suppose you want to repeat the FDA experiment in order to obtain an estimate of the mean caffeine content in a cup of coffee correct to within 5 mg with 95% confidence. How many cups of coffee would have to be included in your sample?

7.91 **USGA golf ball tests.** The United States Golf Association (USGA) tests all new brands of golf balls to ensure that they meet USGA specifications. One test conducted is intended to measure the average distance traveled when the ball is hit by a machine called "Iron Byron." Suppose the USGA wishes to estimate the mean distance for a new brand to within 1 yard with 90% confidence. Assume that past tests have indicated that the standard deviation of the distances Iron Byron hits golf balls is approximately 10 yards. How many golf balls should be hit by Iron Byron to achieve the desired accuracy in estimating the mean?

Applying the Concepts—Advanced

7.92 **Preventing production of defective items.** It costs more to produce defective items—since they must be scrapped or reworked—than it does to produce nondefective items. This simple fact suggests that manufacturers should ensure the quality of their products by perfecting their production processes instead of depending on inspection of finished products (Deming, 1986). In order to better understand a particular metal stamping process, a manufacturer wishes to estimate the mean length of items produced by the process during the past 24 hours.

a. How many parts should be sampled in order to estimate the population mean to within .1 millimeter (mm) with 90% confidence? Previous studies of this machine have indicated that the standard deviation of lengths produced by the stamping operation is about 2 mm.

b. Time permits the use of a sample size no larger than 100. If a 90% confidence interval for is constructed with $n = 100$, will it be wider or narrower than would have been obtained using the sample size determined in part **a**? Explain.

c. If management requires that μ be estimated to within .1 mm and that a sample size of no more than 100 be used, what is (approximately) the maximum confidence level that could be attained for a confidence interval that meets management's specifications?

7.6 Confidence Interval for a Population Variance (Optional)

In the previous sections, we considered interval estimation for population means or proportions. In this optional section, we discuss a confidence interval for a population variance, σ^2.

Recall Example 1.4 (p. 10) and the U.S. Army Corps of Engineers study of contaminated fish in the Tennessee River, Alabama. It is important for the Corps of Engineers to know how stable the weights of the contaminated fish are. That is, how large is the variation in the fish weights? The keyword "variation" indicates that the target population parameter is σ^2, the variance of the weights of all contaminated fish inhabiting the Tennessee River. Of course, the exact value of σ^2 will be unknown. Consequently, the Corps of Engineers wants to estimate its value with a high level of confidence.

Intuitively, it seems reasonable to use the sample variance, s^2, to estimate σ^2. However, unlike with sample means and proportions, the sampling distribution of s^2 does not follow a normal (z) distribution or a Student's t distribution. Rather, when certain assumptions are satisfied (we discuss these later), the sampling distribution of s^2 possesses approximately a **chi-square (χ^2) distribution**. The chi-square probability distribution, like the t distribution, is characterized by a quantity called the *degrees of freedom* (df) associated with the distribution. Several chi-square distributions with different df values are shown in Figure 7.18. You can see that unlike z and t distributions, the chi-square distribution is not symmetric about 0.

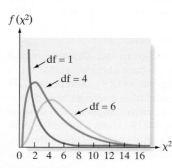

$f(\chi^2)$

df = 1
df = 4
df = 6

0 2 4 6 8 10 12 14 16 χ^2

Figure 7.18
Several χ^2 Probability Distributions

The upper-tail areas for this distribution have been tabulated and are given in Table VII in Appendix A, a portion of which is reproduced in Table 7.7. The table gives the values of χ^2, denoted as χ^2_α, that locate an area of α in the upper tail of the chi-square distribution; that is, $P(\chi^2 > \chi^2_\alpha) = \alpha$. As with the t-statistic, the degrees of freedom associated with s^2 are $(n-1)$. Thus, for $n = 10$ and an upper-tail value $\alpha = .05$, you will have $n-1 = 9$ df and $\chi^2_{.05} = 16.9190$ (highlighted in Table 7.7).

The chi-square distribution is used to find a confidence interval for σ^2, as shown in the box. An illustrative example follows.

A 100(1−α) Confidence Interval for σ^2

$$\frac{(n-1)s^2}{\chi^2_{\alpha/2}} \leq \sigma^2 \leq \frac{(n-1)s^2}{\chi^2_{(1-\alpha/2)}}$$

where $\chi^2_{\alpha/2}$ and $\chi^2_{(1-\alpha/2)}$ are values corresponding to an area of $\alpha/2$ in the right (upper) and left (lower) tails, respectively, of the chi-square distribution based on $(n-1)$ degrees of freedom.

Table 7.7   Reproduction of Part of Table VII in Appendix A: Critical Values of Chi Square

$f(\chi^2)$

α

χ_α^2

0 χ^2

Degrees of Freedom	$\chi^2_{.100}$	$\chi^2_{.050}$	$\chi^2_{.025}$	$\chi^2_{.010}$	$\chi^2_{.005}$
1	2.70554	3.84146	5.02389	6.63490	7.87944
2	4.60517	5.99147	7.37776	9.21034	10.5966
3	6.25139	7.81473	9.34840	11.3449	12.8381
4	7.77944	9.48773	11.1433	13.2767	14.8602
5	9.23635	11.0705	12.8325	15.0863	16.7496
6	10.6446	12.5916	14.4494	16.8119	18.5476
7	12.0170	14.0671	16.0128	18.4753	20.2777
8	13.3616	15.5073	17.5346	20.0902	21.9550
9	14.6837	16.9190	19.0228	21.6660	23.5893
10	15.9871	18.3070	20.4831	23.2093	25.1882
11	17.2750	19.6751	21.9200	24.7250	25.7569
12	18.5494	21.0261	23.3367	26.2170	28.2995
13	19.8119	22.3621	24.7356	27.6883	29.8194
14	21.0642	23.6848	26.1190	29.1413	31.3193
15	22.3072	24.9958	27.4884	30.5779	32.8013
16	23.5418	26.2862	28.8454	31.9999	34.2672
17	24.7690	27.5871	30.1910	33.4087	35.7185
18	25.9894	28.8693	31.5264	34.8053	37.1564
19	27.2036	30.1435	32.8523	36.1908	38.5822

Conditions Required for a Valid Confidence Interval for σ^2

1. A random sample is selected from the target population.

2. The population of interest has a relative frequency distribution that is approximately normal.

Example 7.11

Estimating the Weight Variance, σ^2, of Contaminated Fish

Problem Refer to the U.S. Army Corps of Engineers study of contaminated fish in the Tennessee River. The Corps of Engineers has collected data for a random sample of 144 fish contaminated with DDT. (The engineers made sure to capture contaminated fish in several different randomly selected streams and tributaries of the river.) The fish weights (in grams) are saved in the **FISHDDT** file. The Army Corps of Engineers wants to estimate the true variation in fish weights in order to determine whether the fish are stable enough to allow further testing for DDT contamination.

a. Use the sample data to find a 95% confidence interval for the parameter of interest.

b. Determine whether the confidence interval, part **a**, is valid.

Solution

a. Here, the target parameter is σ^2, the variance of the population of weights of contaminated fish. First, we need to find the sample variance, s^2, to compute the interval estimate. The MINITAB printout, Figure 7.19, gives descriptive statistics for

Figure 7.19

MINITAB descriptive statistics for fish weights, Example 7.11

Descriptive Statistics: WEIGHT

Variable	N	Mean	StDev	Minimum	Median	Maximum
WEIGHT	144	1049.7	376.5	173.0	1000.0	2302.0

the sample weights saved in the **FISHDDT** file. You can see that $s = 376.5$ grams. Consequently, $s^2 = (376.5)^2 = 141,752.25$.

Next, we require the critical values $\chi^2_{\alpha/2}$ and $\chi^2_{(1-\alpha/2)}$ for a chi-square distribution. For a 95% confidence interval, $\alpha = .05$, $\alpha/2 = .025$, and $(1 - \alpha/2) = .975$. Therefore, we need $\chi^2_{.025}$ and $\chi^2_{.975}$. Now, for a sample size $n = 144$, the degrees of freedom associated with the distribution is df $= (n - 1) = 143$. Looking in the df $= 150$ row of Table VII, Appendix A (the row with the df value closest to 143), we find $\chi^2_{.025} = 185.800$ and $\chi^2_{.975} = 117.985$.

Substituting the appropriate values into the formula given in the box, we obtain

$$\frac{(144 - 1)(376.5)^2}{185.500} \leq \sigma^2 \leq \frac{(144 - 1)(376.5)^2}{117.985}$$

Or,

$$109,275 \leq \sigma^2 \leq 171,806$$

Thus, the Army Corps of Engineers can be 95% confident that the variance in weights of the population of contaminated fish ranges between 109,275 and 171,806.

b. According to the box, two conditions are required for the confidence interval to be valid. First, the sample must be randomly selected from the population. The Army Corps of Engineers did, indeed, collect a random sample of contaminated fish, making sure to sample fish from different locations in the Tennessee River. Second, the population data (the fish weights) must be approximately normally distributed. A MINITAB histogram for the sampled fish weights (with a normal curve superimposed) is displayed in Figure 7.20. Clearly, the data appear to be approximately normally distributed. Thus, the confidence interval is valid.

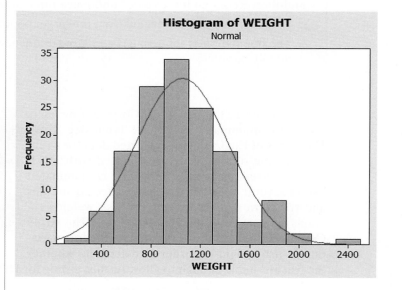

Figure 7.20

MINITAB histogram of fish weights, Example 7.11

Look Ahead Will this confidence interval be practically useful in helping the Corps of Engineers decide whether the weights of the fish are stable? Only if it is clear what a weight variance of, say, 150,000 grams^2 implies. Most likely, the Corps of Engineers will want the interval in the same units as the weight measurement—grams. Consequently, a confidence interval for σ, the standard deviation of the population of fish weights, is desired. We demonstrate how to obtain this interval estimate in the next example.

Now Work Exercise 7.103a

Example 7.12

Estimating the Weight Standard Deviation, σ, of Contaminated Fish

Problem Refer to Example 7.11. Find a 95% confidence interval for σ, the true standard deviation of the contaminated fish weights.

Solution A confidence interval for σ is obtained by taking the square roots of the lower and upper endpoints of a confidence interval for σ^2. Consequently, the 95% confidence interval for σ is:

$$\sqrt{109{,}275} \le \sigma \le \sqrt{171{,}806}$$

Or,

$$330.5 \le \sigma \le 414.5$$

Thus, the engineers can be 95% confident that the true standard deviation of fish weights is between 330.5 grams and 414.5 grams.

Look Back Suppose the Corps of Engineers' threshold is $\sigma = 500$ grams. That is, if the standard deviation in fish weights is 500 grams or higher, further DDT contamination tests will be suspended due to the unstableness of the fish weights. Since the 95% confidence interval for σ lies below 500 grams, the engineers will continue the DDT contamination tests on the fish.

Now Work Exercise 7.103b

⚠ **CAUTION** The procedure for estimating either σ^2 or σ requires an assumption regardless of whether the sample size n is large or small (see the conditions in the box). The sampled data must come from a population that has an approximate normal distribution. Unlike small sample confidence intervals for μ based on the t-distribution, *slight to moderate departures from normality will render the chi-square confidence interval for σ^2 invalid.* ▲

Exercises 7.93–7.108

Understanding the Principles

7.93 What sampling distribution is used to find an interval estimate for σ^2?

7.94 What conditions are required for a valid confidence interval for σ^2?

7.95 How many degrees of freedom are associated with a chi-square sampling distribution for a sample of size n?

Learning the Mechanics

7.96 For each of the following combinations of α and degrees of freedom (df), use Table VII in Appendix B to find the values of $\chi^2_{\alpha/2}$ and $\chi^2_{(1-\alpha/2)}$ that would be used to form a confidence interval for σ^2.
  a. $\alpha = .05, df = 7$
  b. $\alpha = .10, df = 16$
  c. $\alpha = .01, df = 20$
  d. $\alpha = .05, df = 20$

7.97 Given the following values of $\bar{x}$, s, and n, form a 90% confidence interval for σ^2.
  a. $\bar{x} = 21, s = 2.5, n = 50$
  b. $\bar{x} = 1.3, s = .02, n = 15$
  c. $\bar{x} = 167, s = 31.6, n = 22$
  d. $\bar{x} = 9.4, s = 1.5, n = 5$

7.98 Refer to Exercise 7.97. For each part, **a–d,** form a 90% confidence interval for σ.

7.99 A random sample of $n = 6$ observations from a normal distribution resulted in the data shown in the table and saved in the **LM7_99** file. Compute a 95% confidence interval for σ^2.

8	2	3	7	11	6

Applying the Concepts—Basics

7.100 Motivation of drug dealers. Refer to the *Applied Psychology in Criminal Justice* (Sept. 2009) study of the personality characteristics of convicted drug dealers, Exercise 7.15 (p. 307). A random sample of 100 drug dealers had a mean Wanting Recognition (WR) score of 39 points, with a standard deviation of 6 points. The researchers also are interested in σ^2, the variation in WR scores for all convicted drug dealers.
  a. Identify the target parameter, in symbols and words.
  b. Compute a 99% confidence interval for σ^2.
  c. What does it mean to say that the target parameter lies within the interval with "99% confidence"?
  d. What assumption about the data must be satisfied in order for the confidence interval to be valid?
  e. To obtain a practical interpretation of the interval, part **b,** explain why a confidence interval for the standard deviation, σ, is desired.

f. Use the results, part **b,** to compute a 99% confidence interval for σ. Give a practical interpretation of the interval.

7.101 Laptop use in middle school. Refer to the *American Secondary Education* (Fall 2009) study of 106 middle school students' usage of laptop computers, Exercise 7.16 (p. 307). The number of minutes per day each student used a laptop for taking notes was recorded, with the following results: $\bar{x} = 13.2, s = 19.5$.
  a. Find and interpret a 95% confidence interval for the variance in the amount of time per day laptops are used for taking notes for all middle school students across the country.
  b. Recall that in Exercise 2.100b you concluded that the distribution of laptop usage for taking notes cannot be normally distributed. Does this compromise the validity of the interval estimate, part **a**? Explain.

7.102 Antigens for a parasitic roundworm in birds. Refer to the *Gene Therapy and Molecular Biology* (June 2009) study of DNA in peptide (protein) produced by antigens for a parasitic roundworm in birds, Exercise 7.45 (p. 319). Recall that scientists tested each in a sample of 4 alleles of antigen-produced protein for level of peptide. The results were: $\bar{x} = 1.43$ and $s = .13$. Use this information to construct a 90% confidence interval for the true variation in peptide scores for alleles of the antigen-produced protein. Interpret the interval for the scientists.

7.103 Characteristics of a rockfall. Refer to the *Environmental Geology* (Vol. 58, 2009) simulation study of how far a block from a collapsing rock wall will bounce down a soil slope, Exercise 2.59 (p. 57). Rebound lengths (in meters) were estimated for 13 rock bounces. The data are repeated in the table and saved in the **ROCKFALL** file. A MINITAB analysis of the data is shown in the printout below.
  a. Locate a 95% confidence interval for σ^2 on the printout. Interpret the result. [NW]
  b. Locate a 95% confidence interval for σ on the printout. Interpret the result. [NW]
  c. What conditions are required for the intervals, parts **a** and **b,** to be valid?

10.94	13.71	11.38	7.26	17.83	11.92	11.87
5.44	13.35	4.90	5.85	5.10	6.77	

Based on Paronuzzi, P. "Rockfall-induced block propagation on a soil slope, northern Italy." *Environmental Geology*, Vol. 58, 2009 (Table 2).

Test and CI for One Variance: REB-LENGTH

```
Method

The chi-square method is only for the normal distribution.
The Bonett method is for any continuous distribution.


Statistics

Variable      N  StDev  Variance
REB-LENGTH   13   4.09      16.8


95% Confidence Intervals

                          CI for        CI for
Variable    Method         StDev       Variance
REB-LENGTH  Chi-Square  (2.94, 6.76)  (8.6, 45.7)
            Bonett      (2.97, 6.64)  (8.8, 44.1)
```

Applying the Concepts—Intermediate

7.104 Jitter in a water power system. *Jitter* is a term used to describe the variation in conduction time of a water power system. Low throughput jitter is critical to successful waterline technology. An investigation of throughput jitter in the opening switch of a prototype system (*Journal of Applied Physics*) yielded the following descriptive statistics on conduction time for $n = 18$ trials: $\bar{x} = 334.8$ nanoseconds, $s = 6.3$ nanoseconds. (Conduction time is defined as the length of time required for the downstream current to equal 10% of the upstream current.)
  a. Construct a 95% confidence interval for the true standard deviation of conduction times of the prototype system.
  b. Practically interpret the confidence interval, part **a.**
  c. A system is considered to have low throughput jitter if the true conduction time standard deviation is less than 7 nanoseconds. Does the prototype system satisfy this requirement? Explain.

7.105 Cheek teeth of extinct primates. Refer to the *American Journal of Physical Anthropology* (Vol. 142, 2010) study of the characteristics of cheek teeth (e.g., molars) in an extinct primate species, Exercise 2.34 (p. 46). Recall that the researchers recorded the dentary depth of molars (in millimeters) for a sample of 18 cheek teeth extracted from skulls. The data are repeated in the table and saved in the **CHEEKTEETH** file. Estimate the true standard deviation in molar depths for the population of cheek teeth in extinct primates using a 95% confidence interval. Give a practical interpretation of the result. Are the conditions required for a valid confidence interval reasonably satisfied?

18.12	16.55
19.48	15.70
19.36	17.83
15.94	13.25
15.83	16.12
19.70	18.13
15.76	14.02
17.00	14.04
13.96	16.20

Based on Boyer, D. M., Evans, A. R., and Jernvall, J. "Evidence of dietary differentiation among Late Paleocene–Early Eocene Plesiadapids (Mammalia, Primates)." *American Journal of Physical Anthropology*, Vol. 142, 2010 (Table A3).

7.106 Shell lengths of sea turtles. Refer to the *Aquatic Biology* (Vol. 9, 2010) study of green sea turtles inhabiting the Grand Cayman South Sound lagoon, Exercise 7.20 (p. 308). Recall that the data on shell length, measured in centimeters, for 76 captured turtles are saved in the **TURTLES** file. Use the sample data to estimate the true variance in shell lengths of all green sea turtles in the lagoon with 90% confidence. Interpret the result.

7.107 Lobster trap placement. Refer to the *Bulletin of Marine Science* (April 2010) observational study of teams fishing for the red spiny lobster in Baja California Sur, Mexico, Exercise 7.35 (p. 317). Trap spacing measurements (in meters) for a sample of seven teams of red spiny lobster fishermen are repeated in the table (p. 339) and saved in the **TRAPSPACE** file. The researchers want to know how variable the trap spacing measurements are for the population of red spiny lobster fishermen fishing in Baja California Sur, Mexico. Provide the researchers with an estimate of the target parameter using a 99% confidence interval.

| 93 | 99 | 105 | 94 | 82 | 70 | 86 |

Based on Shester, G. G. "Explaining catch variation among Baja California lobster fishers through spatial analysis of trap-placement decisions." *Bulletin of Marine Science*, Vol. 86, No. 2, April 2010 (Table 1), pp. 479-498.

7.108 Is honey a cough remedy? Refer to the *Archives of Pediatrics and Adolescent Medicine* (Dec. 2007) study of honey as a remedy for coughing, Exercise 2.32 (p. 45). Recall that the 105 ill children in the sample were randomly divided into groups. One group received a dosage of an over-the-counter cough medicine (DM); another group received a dosage of honey (H). The coughing improvement scores (as determined by the children's parents) for the patients in the two groups are reproduced in the accompanying table and saved in the **HONEYCOUGH** file. The pediatric researchers desire information on the variation in coughing improvement scores for each of the two groups.

a. Find a 90% confidence interval for the standard deviation in improvement scores for the honey dosage group.
b. Repeat part **a** for the DM dosage group.
c. Based on the results, parts **a** and **b**, what conclusions can the pediatric researchers draw about which group has the smaller variation in improvement scores? (We demonstrate a more statistically valid method for comparing variances in Chapter 9.)

| Honey Dosage: | 12 11 15 11 10 13 10 4 15 16 9 14 10 6 10 8 11 12 12 8 12 9 11 15 10 15 9 13 8 12 10 8 9 5 12 |
| DM Dosage: | 4 6 9 4 7 7 7 9 12 10 11 6 3 4 9 12 7 6 8 12 12 4 12 13 7 10 13 9 4 4 10 15 9 |

Based on Paul, I. M., et al. "Effect of honey, dextromethorphan, and no treatment on nocturnal cough and sleep quality for coughing children and their parents." *Archives of Pediatrics and Adolescent Medicine*, Vol. 161, No. 12, Dec. 2007 (data simulated).

CHAPTER NOTES

Key Terms

Note: Asterisks (*) denote terms from the optional section in this chapter.

*Chi-square (χ^2) distribution 334
Confidence coefficient 303
Confidence interval 300
Confidence level 303
Degrees of freedom (df) 310

Interval estimator 300
Point estimator 300
Sampling error 328
Target parameter 300
t-statistic 310
Wilson adjustment for estimating p 323
z-statistic 301

Key Symbols

θ — General population parameter (theta)
μ — Population mean
σ^2 — Population variance
σ — Population standard deviation
p — Population proportion; P (Success) in binomial trial
q — $1 - p$
$\bar{x}$ — Sample mean (estimator of μ)
s^2 — Sample variance (estimator of σ^2)
$\hat{p}$ — Sample proportion (estimator of p)
$\mu_{\bar{x}}$ — Mean of the population sampling distribution of $\bar{x}$
$\sigma_{\bar{x}}$ — Standard deviation of the sampling distribution of $\bar{x}$
$\sigma_{\hat{p}}$ — Standard deviation of the sampling distribution of $\hat{p}$
SE — Sampling error in estimation
α — $(1 - \alpha)$ represents the confidence coefficient
$z_{\alpha/2}$ — z-value used in a $100(1 - \alpha)\%$ large-sample confidence interval for μ or p
$t_{\alpha/2}$ — Student's *t*-value used in a $100(1 - \alpha)\%$ small-sample confidence interval for μ
$\chi^2_{\alpha/2}$ — *Chi-Square value used in a $100(1 - \alpha)\%$ confidence interval for σ^2

Key Ideas

Confidence Interval: An interval that encloses an unknown population parameter with a certain level of confidence, $(1 - \alpha)$
Confidence Coefficient: The probability $(1 - \alpha)$ that a randomly selected confidence interval encloses the true value of the population parameter.

Population Parameters, Estimators, & Standard Errors

Parameter (θ)	Estimator ($\hat{\theta}$)	Standard Error of Estimator ($\hat{\sigma}_{\hat{\theta}}$)	Estimated Std. Error ($\hat{\sigma}_{\hat{\theta}}$)
Mean, μ	$\bar{x}$	$\sigma/\sqrt{n}$	$s/\sqrt{n}$
Proportion, p	$\hat{p}$	$\sqrt{pq}/n$	$\sqrt{\hat{p}\hat{q}}/n$
*Variance, σ^2	s^2	—	—

Determining the Sample Size n:

Estimating μ: $n = (z_{\alpha/2})^2(\sigma^2)/(\text{SE})^2$
Estimating p: $n = (z_{\alpha/2})^2(pq)/(\text{SE})^2$

Key Words for Identifying the Target Parameter:

μ — Mean, Average
p — Proportion, Fraction, Percentage, Rate, Probability
*σ^2 — Variance, Variation, Spread

Commonly Used z-values for a Large-Sample Confidence Interval for μ or p:

90% CI: $(1 - \alpha) = .10$ $z_{.05} = 1.645$
95% CI: $(1 - \alpha) = .05$ $z_{.025} = 1.96$
99% CI: $(1 - \alpha) = .01$ $z_{.005} = 2.575$

Illustrating the Notion of "95% Confidence"

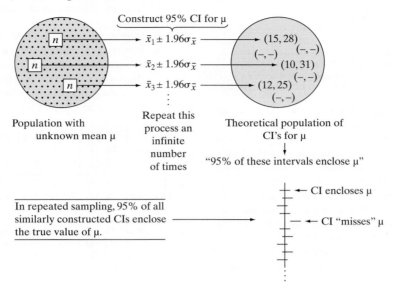

In repeated sampling, 95% of all similarly constructed CIs enclose the true value of μ.

Guide to Forming a Confidence Interval

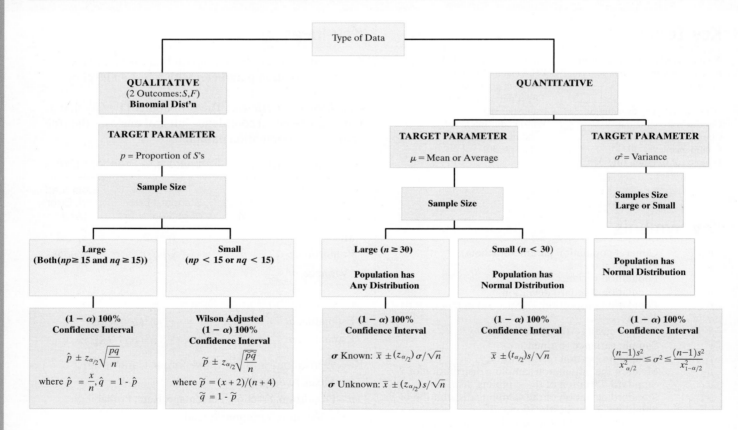

Supplementary Exercises 7.109–7.141

Note: List the assumptions necessary for the valid implementation of the statistical procedures you use in solving all these exercises. Starred (*) exercises are from the optimal section in this chapter.

Understanding the Principles

7.109 For each of the following, identify the target parameter as $\mu, p,$ or σ^2.

a. Average score on the SAT
b. Mean time waiting at a supermarket checkout lane
c. Proportion of voters in favor of legalizing marijuana
d. Percentage of NFL players who have ever made the Pro Bowl
e. Dropout rate of American college students
*f. Variation in IQ scores of sociopaths

7.110 Interpret the phrase "95% confident" in the following statement: "We are 95% confident that the proportion of all PCs with a computer virus falls between .12 and .18."

7.111 In each of the following instances, determine whether you would use a z- or t-statistic (or neither) to form a 95% confidence interval for μ; then look up the appropriate z or t value.
  a. Random sample of size $n = 21$ from a normal distribution with unknown mean μ and standard deviation σ
  b. Random sample of size $n = 175$ from a normal distribution with unknown mean μ and standard deviation σ
  c. Random sample of size $n = 12$ from a normal distribution with unknown mean and standard deviation $\sigma = 5$
  d. Random sample of size $n = 65$ from a distribution about which nothing is known
  e. Random sample of size $n = 8$ from a distribution about which nothing is known

Learning the Mechanics

7.112 Let t_0 represent a particular value of t from Table VI of Appendix A. Find the table values such that the following statements are true:
  a. $P(t \leq t_0) = .05$, where df = 17
  b. $P(t \geq t_0) = .005$, where df = 14
  c. $P(t \leq -t_0 \text{ or } t \geq t_0) = .10$, where df = 6
  d. $P(t \leq -t_0 \text{ or } t \geq t_0) = .01$, where df = 22

7.113 In a random sample of 400 measurements, 227 possess the characteristic of interest, A.
  a. Use a 95% confidence interval to estimate the true proportion p of measurements in the population with characteristic A.
  b. How large a sample would be needed to estimate p to within .02 with 95% confidence?

7.114 A random sample of 225 measurements is selected from a population, and the sample mean and standard deviation are $\bar{x} = 32.5$ and $s = 30.0$, respectively.
  a. Use a 99% confidence interval to estimate the mean of the population, μ.
  b. How large a sample would be needed to estimate μ to within .5 with 99% confidence?
  c. What is meant by the phrase "99% confidence" as it is used in this exercise?
  *d. Find a 99% confidence interval for σ^2.

Applying the Concepts—Basic

7.115 CDC health survey. The Centers for Disease Control and Prevention (CDCP) in Atlanta, Georgia, conduct an annual survey of the general health of the U.S. population as part of their Behavioral Risk Factor Surveillance System. Using random-digit dialing, the CDCP telephones U.S. citizens over 18 years of age and asks them the following four questions:
  1. Is your health generally excellent, very good, good, fair, or poor?
  2. How many days during the previous 30 days was your physical health not good because of injury or illness?
  3. How many days during the previous 30 days was your mental health not good because of stress, depression, or emotional problems?
  4. How many days dur- ing the previous 30 days did your physical or mental health prevent you from performing your usual activities?

Identify the parameter of interest for each question.

7.116 Personal networks of older adults. In sociology, a personal network is defined as the people with whom you make frequent contact. The Living Arrangements and Social Networks of Older Adults (LSN) research program used a stratified random sample of men and women born between 1908 and 1937 to gauge the size of the personal network of older adults. Each adult in the sample was asked to "please name the people (e.g., in your neighborhood) you have frequent contact with and who are also important to you." Based on the number of people named, the personal network size for each adult was determined. The responses of 2,819 adults in the LSN sample yielded the following statistics on network size: $\bar{x} = 14.6$; $s = 9.8$ (*Sociological Methods & Research*, August 2001).
  a. Give a point estimate for the mean personal network size of all older adults.
  b. Form a 95% confidence interval for the mean personal network size of all older adults.
  c. Give a practical interpretation of the interval you found in part **b**.
  d. Give the conditions required for the interval in part **b** to be valid.

7.117 Ancient Greek pottery. Refer to the *Chance* (Fall 2000) study of 837 pieces of pottery found at the ancient Greek settlement at Phylakopi, presented in Exercise 2.14 (p. 36). Of the 837 pieces, 183 were painted with either a curvilinear, geometric, or naturalistic decoration. Find a 90% confidence interval for the population proportion of all pottery artifacts at Phylakopi that are painted. Interpret the resulting interval.

7.118 Tax-exempt charities. Donations to tax-exempt organizations such as the Red Cross, the Salvation Army, the

Organization	Charitable Commitment (%)
American Cancer Society	71
American National Red Cross	89
Big Brothers Big Sisters of America	78
Boy Scouts of America	87
Boys & Girls Clubs of America	80
CARE USA	90
Covenant House	67
Disabled American Veterans	69
Ducks Unlimited	86
Feed The Children	84
Girl Scouts of the USA	82
Goodwill Industries International	88
Habitat for Humanity International	83
Mayo Clinic	97
March of Dimes Foundation	77
Multiple Sclerosis Society	78
Museum of Modern Art	80
Nature Conservancy	80
Paralyzed Veterans of America	62
Planned Parenthood Federation	82
Salvation Army	82
Shriners Hospital for Children	92
Smithsonian Institution	82
Special Olympics	74
Trust for Public Land	89
United States Olympic Committee	71
United Way	86
WGBH Educational Foundation	79
World Trade Center Memorial Foundation	79
YMCA of the USA	81

Based on "The 200 largest U.S. charities." Forbes.com, November 19, 2008.

YMCA, and the American Cancer Society not only go to the stated charitable purpose, but are used to cover fund-raising expenses and overhead. The **CHARITY** file (see table on p. 341) contains the charitable commitments (i.e., the percentage of expenses that goes toward the stated charitable purpose) for a sample of 30 charities.

a. Give a point estimate for the mean charitable commitment of tax-exempt organizations.

b. Construct a 95% confidence interval for the mean charitable commitment.

c. What assumption(s) must hold for the method of estimation used in part **b** to be appropriate?

d. Why is the confidence interval of part **b** a better estimator of the mean charitable commitment than the point estimator of part **a**?

*e. Construct a 95% confidence interval for the variance in charitable commitments for all tax-exempt organizations.

7.119 Cell phone use by drivers. The U.S. Department of Transportation reported the results of the *National Occupant Protection Use Survey.* One focus of the survey was to determine the level of cell phone use by drivers while they are in the act of driving a motor passenger vehicle. Data collected by observers at randomly selected intersections across the country revealed that in a sample of 1,165 drivers, 35 were using their cell phone.

a. Give a point estimate of p, the true driver cell phone use rate (i.e., the true proportion of drivers who are using a cell phone while driving).

b. Compute a 95% confidence interval for p.

c. Give a practical interpretation of the interval you found in part **b.**

d. How many drivers need to be sampled to estimate the true cell phone use rate to within .005 with 95% confidence?

7.120 "Made in the USA" survey. Refer to Exercise 2.179 (p. 99) and the *Journal of Global Business* (Spring 2002) survey to determine what "Made in the USA" means to consumers. Recall that 106 shoppers at a shopping mall in Muncie, Indiana, responded to the question "'Made in the USA' means what percentage of U.S. labor and materials?" Sixty-four shoppers answered "100%."

a. Define the population of interest in the survey.

b. What is the characteristic of interest in the population?

c. Use a 90% confidence interval to estimate the true proportion of consumers who believe that "Made in the USA" means 100% U.S. labor and materials.

d. Give a practical interpretation of the interval you used in part **c.**

e. Explain what the phrase "90% confidence" means for this interval.

7.121 "Made in the USA" survey (cont'd). Refer to Exercise 7.120. Suppose the researchers want to increase the sample size in order to estimate the true proportion p to within .05 of its true value with a 90% confidence interval.

a. What is the confidence level desired by the researchers?

b. What is the sampling error desired by the researchers?

c. Compute the sample size necessary to obtain the desired estimate.

7.122 Water pollution testing. The EPA wants to test a randomly selected sample of n water specimens and estimate μ, the mean daily rate of pollution produced by a mining operation. If the EPA wants a 95% confidence interval estimate with a sampling error of 1 milligram per liter (mg/L), how many water specimens are required in the sample? Assume that prior knowledge indicates that pollution readings in water samples taken during a day are approximately normally distributed with a standard deviation equal to 5 (mg/L).

7.123 Al Qaeda attacks on the United States. Refer to the *Studies in Conflict & Terrorism* (Vol. 29, 2006) analysis of incidents involving suicide terrorist attacks, presented in Exercise 2.173 (p. 98). Data on the number of individual suicide bombings or attacks for each in a sample of 21 incidents involving an attack against the United States by the Al Qaeda terrorist group (see table below) are saved in the **ALQAEDA** file.

a. Find the mean and standard deviation of the sample data.

b. Describe the population from which the sample is selected.

c. Use the information from part **a** to find a 90% confidence interval for the mean μ of the population.

d. Give a practical interpretation of the result you obtained in part **c.**

e. In repeatedly sampling, where intervals similar to the one computed in part **c** are generated, what proportion of the intervals will enclose the true value of μ?

1	1	2	1	2	4	1	1	1	1	2	3	4	5	1	1	1	2	2	2	1

Based on Moghadam, A. "Suicide terrorism, occupation, and the globalization of martyrdom: A critique of *Dying to Win.*" *Studies in Conflict & Terrorism*, Vol. 29, No. 8, 2006 (Table 3).

7.124 Ammonia in car exhaust. Refer to the *Environmental Science & Technology* (September 1, 2000) study on the ammonia levels near the exit ramp of a San Francisco highway tunnel, presented in Exercise 2.184 (p. 100). The ammonia concentration (parts per million) data for eight randomly selected days during the afternoon drive time are reproduced in the accompanying table and saved in the **AMMONIA** file. Find a 99% confidence interval for the population mean daily ammonia level in air in the tunnel. Interpret your result.

1.53	1.50	1.37	1.51	1.55	1.42	1.41	1.48

7.125 Oven cooking study. A group of Harvard University School of Public Health researchers studied the impact of cooking on the size of indoor air particles (*Environmental Science & Technology*, September 1, 2000). The decay rate (measured in μm/hour) for fine particles produced from oven cooking or toasting was recorded on six randomly selected days. The six measurements (shown below) are saved in the **DECAY** file.

.95	.83	1.20	.89	1.45	1.12

Based on Abt, E., et al. "Relative contribution of outdoor and indoor particle sources to indoor concentrations." *Environmental Science & Technology*, Vol. 34, No. 17, Sept. 1, 2000 (Table 3).

a. Find and interpret a 95% confidence interval for the true average decay rate of fine particles produced from oven cooking or toasting.

b. Explain what the phrase "95% confident" implies in the interpretation of part **a.**

***c.** Estimate the true standard deviation of decay rate with a 95% confidence interval. Interpret the result.

d. What must be true about the distribution of the population of decay rates for the inferences you made in parts **a** and **c** to be valid?

e. Suppose that we want to estimate the average decay rate of fine particles produced from oven cooking or toasting to within .04 with 95% confidence. How large a sample should be selected?

7.126 Scary-movie study. According to a University of Michigan study, many adults have experienced lingering "fright" effects from a scary movie or TV show they saw as a teenager (*Tampa Tribune*, March 10, 1999). In a survey of 150 college students, 39 said they still experience "residual anxiety" from a scary TV show or movie.

a. Give a point estimate $\hat{p}$ for the true proportion of college students who experience "residual anxiety" from a scary TV show or movie.

b. Find a 95% confidence interval for p.

c. Interpret the interval you found in part **b.**

Applying the Concepts—Intermediate

7.127 Brown-bag lunches at work. In a study reported in *The Wall Street Journal*, the Tupperware Corporation surveyed 1,007 U.S. workers. Of the people surveyed, 665 indicated that they take their lunch to work with them. Of these 665 taking their lunch, 200 reported that they take it in brown bags.

a. Find a 95% confidence interval estimate of the population proportion of U.S. workers who take their lunch to work with them. Interpret the interval.

b. Consider the population of U.S. workers who take their lunch to work with them. Find a 95% confidence interval estimate of the population proportion who take brown-bag lunches. Interpret the interval.

7.128 Research on brain specimens. In Exercise 2.39 (p. 48), you learned that the postmortem interval (PMI) is the elapsed time between death and the performance of an autopsy on the cadaver. *Brain and Language* (June 1995) reported on the PMIs of 22 randomly selected human brain specimens obtained at autopsy. The data, reproduced in the following table, are saved in the **BRAINPMI** file. A coroner claims that the true mean PMI of human brain specimens obtained at autopsy is 10 days. Do you agree? Use a 95% confidence interval to make an inference.

5.5	14.5	6.0	5.5	5.3	5.8	11.0	6.4
7.0	14.5	10.4	4.6	4.3	7.2	10.5	6.5
3.3	7.0	4.1	6.2	10.4	4.9		

Based on Hayes, T. L., and Lewis, D. A. "Anatomical specialization of the anterior motor speech area: Hemispheric differences in magnopyramidal neurons," *Brain and Language*, Vol. 49, No. 3, p. 292 (Table 1).

7.129 Inbreeding of tropical wasps. Tropical swarm-founding wasps rely on female workers to raise their offspring. One possible explanation for this strange behavior is inbreeding, which increases relatedness among the wasps, presumably making it easier for the workers to pick out their closest relatives as propagators of their own genetic material. To test this theory, 197 swarm-founding wasps were captured in Venezuela, frozen at $-70°C$, and then subjected to a series of genetic tests. The data were used to generate an inbreeding coefficient x for each wasp specimen, with the following results: $\bar{x} = .044$ and $s = .884$.

a. Construct a 99% confidence interval for the mean inbreeding coefficient of this species of wasp.

b. A coefficient of 0 implies that the wasp has no tendency to inbreed. Use the confidence interval you constructed in part **a** to make an inference about the tendency for this species of wasp to inbreed.

7.130 Exercise workout dropouts. Researchers at the University of Florida's Department of Exercise and Sport Sciences conducted a study of variety in exercise workouts (*Journal of Sport Behavior*, 2001). A sample of 120 men and women were randomly divided into three groups, with 40 people per group. Group 1 members varied their exercise routine in workouts, group 2 members performed the same exercise at each workout, and group 3 members had no set schedule or regulations for their workouts.

a. By the end of the study, 15 people had dropped out of the first exercise group. Estimate the dropout rate for exercisers who vary their routine in workouts. Use a 90% confidence interval and interpret the result.

b. By the end of the study, 23 people had dropped out of the third exercise group. Estimate the dropout rate for exercisers who have no set schedule for their workouts. Use a 90% confidence interval and interpret the result.

7.131 Time to solve a math programming problem. *IEEE Transactions* presented a hybrid algorithm for solving a polynomial zero–one mathematical programming problem. The algorithm incorporates a mixture of pseudo-Boolean concepts and time-proven implicit enumeration procedures. Fifty-two random problems were solved by the hybrid algorithm; the times to solution (CPU time in seconds) are listed in the accompanying table and saved in the **MATHCPU** file.

a. Estimate, with 95% confidence, the mean solution time for the hybrid algorithm. Interpret the result.

b. How many problems must be solved to estimate the mean μ to within .25 second with 95% confidence?

***c.** Form a 95% confidence interval for the true standard deviation of the solution times for the hybrid algorithm. Interpret the result.

.045	1.055	.136	1.894	.379	.136	.336	.258	1.070
.506	.088	.242	1.639	.912	.412	.361	8.788	.579
1.267	.567	.182	.036	.394	.209	.445	.179	.118
.333	.554	.258	.182	.070	3.985	.670	3.888	.136
.091	.600	.291	.327	.130	.145	4.170	.227	.064
.194	.209	.258	3.046	.045	.049	.079		

Source: Snyder, W. S., and Chrissis, J. W. "A hybrid algorithm for solving zero–one mathematical programming problems." *IEEE Transactions*, Vol. 22, No. 2, June 1990. Copyright © 1990 IEEE. Reprinted with permission.

7.132 Sentence complexity study. Refer to the *Applied Psycholinguistics* (June 1998) study of language skills of low-income children, presented in Exercise 2.104 (p. 72). Each in a sample of 65 low-income children was administered the Communicative Development Inventory (CDI) exam. The sentence complexity scores had a mean of 7.62 and a standard deviation of 8.91.

a. Construct a 90% confidence interval for the mean sentence complexity score of all low-income children.

b. Interpret the interval you found in part **a** in the words of the problem.

c. Suppose we know that the true mean sentence complexity score of middle-income children is 15.55. Is there evidence that the true mean for low-income children differs from 15.55? Explain.

7.133 Jail suicide study. In Exercise 2.194 (p. 103) you considered data on suicides in correctional facilities. The data were published in the *American Journal of Psychiatry* (July 1995). The researchers wanted to know what factors increase the risk of suicide by those incarcerated in urban jails. The data on 37 suicides that occurred in Wayne County Jail, Detroit, Michigan, are saved in the **SUICIDE** file. Answer each of the questions that follow, using a 95% confidence interval for the target parameter implied in the question. Interpret each confidence interval fully.

a. What proportion of suicides at the jail are committed by inmates charged with murder or manslaughter?

b. What proportion of suicides at the jail are committed at night?

c. What is the average length of time an inmate is in jail before committing suicide?

d. What percentage of suicides are committed by white inmates?

7.134 Training zoo animals. Refer to Exercise 2.65 (p. 59) and the *Teaching of Psychology* (May 1998) study in which students assisted in the training of zoo animals. A sample of 15 psychology students rated "The Training Game" as a "great" method of understanding the animal's perspective during training on a 7–1 point scale (where 1 = strongly disagree and 7 = strongly agree). The mean response was 5.87, with a standard deviation of 1.51.

a. Construct a 95% confidence interval for the true mean response of the students.

b. Suppose you want to reduce the width of the 95% confidence interval to half the size obtained in part **a**. How many students are required in the sample in order to obtain the desired confidence interval width?

7.135 Salmonella in ice cream bars. Recently, a case of salmonella (bacterial) poisoning was traced to a particular brand of ice cream bar, and the manufacturer removed the bars from the market. Despite this response, many consumers refused to purchase *any* brand of ice cream bars for some time after the event (McClave, personal correspondence). One manufacturer conducted a survey of consumers 6 months after the poisoning. A sample of 244 ice cream bar consumers was contacted, and 23 indicated that they would not purchase ice cream bars because of the potential for food poisoning.

a. What is the point estimate of the true fraction of the entire market who refuse to purchase bars 6 months after the poisoning?

b. Is the sample size large enough to use the normal approximation for the sampling distribution of the estimator of the binomial probability? Justify your response.

c. Construct a 95% confidence interval for the true proportion of the market who still refuse to purchase ice cream bars 6 months after the event.

d. Interpret both the point estimate and confidence interval in terms of this application.

7.136 Salmonella in ice cream bars (cont'd). Refer to Exercise 7.135. Suppose it is now 1 year after the poisoning was traced to ice cream bars. The manufacturer wishes to estimate the proportion who still will not purchase bars to within .02, using a 95% confidence interval. How many consumers should be sampled?

7.137 Removing a soil contaminant. A common hazardous compound found in contaminated soil is benzo(a)pyrene [B(a)p]. An experiment was conducted to determine the effectiveness of a method designed to remove B(a)p from soil (*Journal of Hazardous Materials*, June 1995). Three soil specimens contaminated with a known amount of B(a)p were treated with a toxin that inhibits microbial growth. After 95 days of incubation, the percentage of B(a)p removed from each soil specimen was measured. The experiment produced the following summary statistics: $\bar{x} = 49.3$ and $s = 1.5$.

a. Use a 99% confidence interval to estimate the mean percentage of B(a)p removed from a soil specimen in which the toxin was used.

b. Interpret the interval in terms of this application.

c. What assumption is necessary to ensure the validity of this confidence interval?

d. How many soil specimens must be sampled to estimate the mean percentage removed to within .5, using a 99% confidence interval?

*e. Find and interpret a 90% confidence interval for the true variance in the percentages of B(a)p removed.

7.138 Material safety data sheets. For over 20 years, the Occupational Safety & Health Administration has required companies that handle hazardous chemicals to complete material safety data sheets (MSDSs). These sheets have been criticized for being too hard to understand and complete by workers. Although improvements were implemented in 1990, a recent study of 150 MSDSs revealed that only 11% were completed satisfactorily (*Chemical & Engineering News*, Feb. 7, 2005). Find an interval that contains the true proportion of all MSDSs that are completed satisfactorily, with 95% confidence. Would it surprise you if the true proportion was as high as 20%? Explain.

Applying the Concepts—Advanced

7.139 IMA salary survey. Each year, *Management Accounting* reports the results of a salary survey of the members of the Institute of Management Accountants (IMA). One year, the 2,112 members responding had a salary distribution with a 20th percentile of $35,100, a median of $50,000, and an 80th percentile of $73,000.

a. Use this information to determine the minimum sample size that could be used in next year's survey to estimate the mean salary of IMA members to within $2,000 with 98% confidence. [*Hint*: To estimate s, first apply Chebyshev's theorem to find k such that at least 60% of the data fall within k standard deviations of μ. Then find $s \approx$ (80th percentile –20th percentile)/$2k$.]

b. Explain how you estimated the standard deviation required for the calculation of the sample size.

c. List any assumptions you make.

7.140 Air bags pose danger for children. By law, all new cars must be equipped with both driver-side and passenger-side safety air bags. There is concern, however, over whether air bags pose a danger for children sitting on the passenger side. In a National Highway Traffic Safety Administration (NHTSA) study of 55 people killed by the explosive force of air bags, 35 were children seated on the front-passenger side. This study led some car owners with the information about children to disconnect the passenger-side air bag.

a. Use the study to estimate the risk of an air bag fatality on a child seated on the front passenger seat.

b. NHTSA investigators determined that 24 of 35 children killed by the air bags were not wearing seat belts or were improperly restrained. How does this information affect your assessment of the risk of an air bag fatality?

Critical Thinking Challenge

7.141 Scallops, sampling, and the law. *Interfaces* (March–April 1995), discussed the case of a ship that fishes for scallops off the coast of New England. In order to protect baby scallops from being harvested, the U.S. Fisheries and Wildlife Service requires that "the average meat per scallop weigh at least $\frac{1}{36}$ of a pound." The ship was accused of violating this weight standard. Author Arnold Barnett lays out the scenario:

The vessel arrived at a Massachusetts port with 11,000 bags of scallops, from which the harbormaster randomly selected 18 bags for weighing. From each such bag, his agents took a large scoopful of scallops; then, to estimate the bag's average meat per scallop, they divided the total weight of meat in the scoopful by the number of scallops it contained. Based on the 18 [numbers] thus generated, the harbormaster estimated that each of the ship's scallops possessed an average $\frac{1}{39}$ of a pound of meat (that is, they were about seven percent lighter than the minimum requirement). Viewing this outcome as conclusive evidence that the weight standard had been

violated, federal authorities at once confiscated 95 percent of the catch (which they then sold at auction). The fishing voyage was thus transformed into a financial catastrophe for its participants.

The actual scallop weight measurements for each of the 18 sampled bags are listed in the accompanying table and saved in the **SCALLOPS** file. For ease of exposition, Bennett expressed each number as a multiple of $\frac{1}{36}$ of a pound, the minimum permissible average weight per scallop. Consequently, numbers below 1 indicate individual bags that do not meet the standard.

The ship's owner filed a lawsuit against the federal government, declaring that his vessel had fully complied with the weight standard. A Boston law firm was hired to represent the owner in legal proceedings and Bennett was retained by the firm to provide statistical litigation support and, if necessary, expert witness testimony.

a. Recall that the harbormaster sampled only 18 of the ship's 11,000 bags of scallops. One of the questions the lawyers asked Bennett was "Can a reliable estimate of the mean weight of all the scallops be obtained from a sample of size 18?" Give your opinion on this issue.

b. As stated in the article, the government's decision rule is to confiscate a catch if the sample mean weight of the scallops is less than $\frac{1}{36}$ of a pound. Do you see any flaws in this rule?

c. Develop your own procedure for determining whether a ship is in violation of the minimum-weight restriction. Apply your rule to the data. Draw a conclusion about the ship in question.

.93	.88	.85	.91	.91	.84	.90	.98	.88
.89	.98	.87	.91	.92	.99	1.14	1.06	.93

Based on Barnett, A. "Misapplications review: Jail terms." *Interfaces*, Vol. 25, No. 2, Mar.–Apr. 1995, p. 20.

Activity   Conducting a Pilot Study

Choose a population pertinent to your major area of interest—a population that has an unknown mean or, if the population is binomial, that has an unknown probability of success. For example, a marketing major may be interested in the proportion of consumers who prefer a certain product. A sociology major may be interested in estimating the proportion of people in a certain socioeconomic group or the mean income of people living in a particular part of a city. A political science major may wish to estimate the proportion of an electorate in favor of a certain candidate, a certain amendment, or a certain presidential policy. A pre-med student might want to find the average length of time patients stay in the hospital or the average number of people treated daily in the emergency room. We could continue with

examples, but the point should be clear: Choose something of interest to you.

Define the parameter you want to estimate and conduct a *pilot study* to obtain an initial estimate of the parameter of interest and, more importantly, an estimate of the variability associated with the estimator. A pilot study is a small experiment (perhaps 20 to 30 observations) used to gain some information about the population of interest. The purpose of the study is to help plan more elaborate future experiments. Using the results of your pilot study, determine the sample size necessary to estimate the parameter to within a reasonable bound (of your choice) with a 95% confidence interval.

References

Agresti, A., and Coull, B. A. "Approximate is better than 'exact' for interval estimation of binomial proportions." *The American Statistician*, Vol. 52, No. 2, May 1998, pp. 119–126.

Cochran, W. G. *Sampling Techniques*, 3rd ed. New York: Wiley, 1977.

Freedman, D., Pisani, R., and Purves, R. *Statistics*. New York: Norton, 1978.

Kish, L. *Survey Sampling*. New York: Wiley, 1965.

Mendenhall, W., Beaver, R. J., and Beaver, B. *Introduction to Probability and Statistics*, 13th ed. Belmont, CA: Brooks/Cole, 2009.

Wilson, E. G. "Probable inference, the law of succession, and statistical inference." *Journal of the American Statistical Association*, Vol. 22, 1927, pp. 209–212.

USING TECHNOLOGY

MINITAB: Confidence Intervals

MINITAB can be used to obtain one-sample confidence intervals for a population mean, a population proportion, and a population variance.

Confidence Interval for a Mean

Step 1 Access the MINITAB data worksheet that contains the quantitative variable of interest.

Step 2 Click on the "Stat" button on the MINITAB menu bar and then click on "Basic Statistics" and "1-Sample t," as shown in Figure 7.M.1.

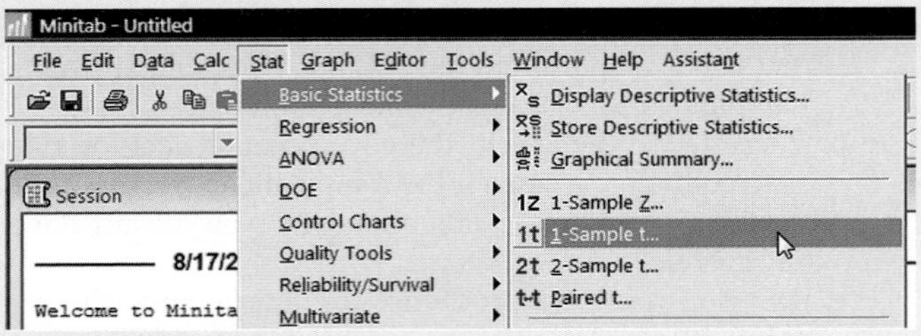

Figure 7.M.1 MINITAB menu options—confidence interval for the mean

Step 3 On the resulting dialog box (shown in Figure 7.M.2), click on "Samples in Columns," and then specify the quantitative variable of interest in the open box.

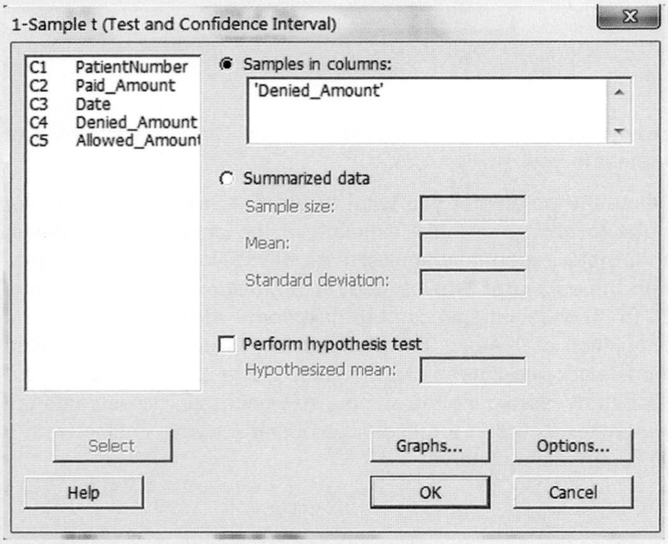

Figure 7.M.2 MINITAB 1-sample t dialog box

Step 4 Click on the "Options" button at the bottom of the dialog box and specify the confidence level in the resulting dialog box, as shown in Figure 7.M.3.

Step 5 Click "OK" to return to the "1-Sample t" dialog box, and then click "OK" again to produce the confidence interval.

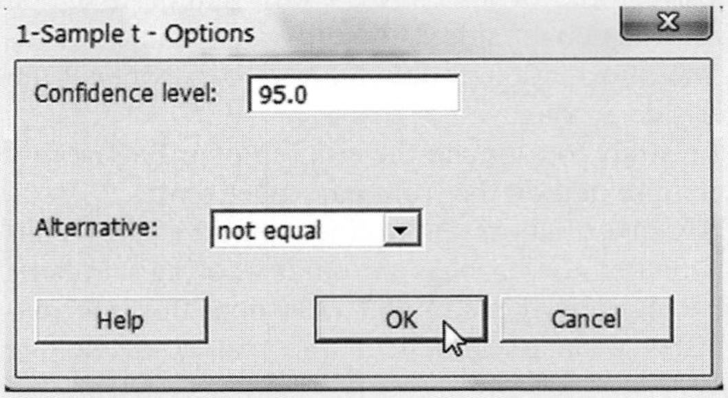

Figure 7.M.3 MINITAB 1-sample t options

Note: If you want to produce a confidence interval for the mean from summary information (e.g., the sample mean, sample standard deviation, and sample size), click on "Summarized data" in the "1-Sample t" dialog box, as shown in Figure 7.M.4. Enter the values of the summary statistics and then click "OK."

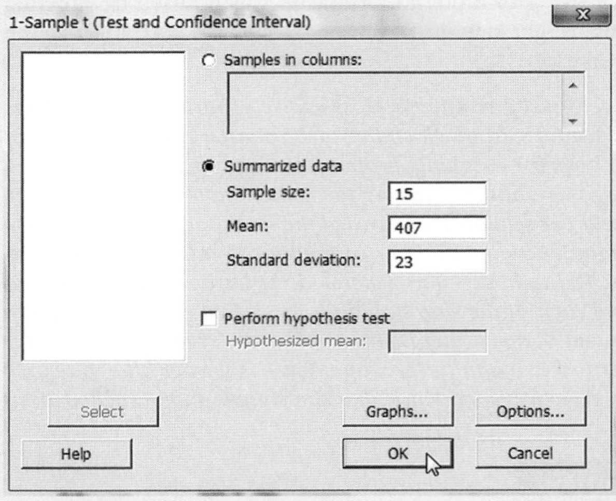

Figure 7.M.4 MINITAB 1-sample t using summary statistics

Important: The MINITAB 1-sample t procedure uses the *t*-statistic to generate the confidence interval. When the sample size n is small, this is the appropriate method. When the sample size n is large, the *t*-value will be approximately equal to the large-sample *z*-value and the resulting interval will still be valid. If you have a large sample and you know the value of the population standard deviation σ (which is rarely the case), select "1-sample Z" from the "Basic Statistics" menu options (see Figure 7.M.1) and make the appropriate selections.

Confidence Interval for a Proportion

Step 1 Access the MINITAB data worksheet that contains the qualitative variable of interest.

Step 2 Click on the "Stat" button on the MINITAB menu bar and then click on "Basic Statistics" and "1-Proportion."

Step 3 On the resulting dialog box (shown in Figure 7.M.5), click on "Samples in Columns" and then specify the qualitative variable of interest in the open box.

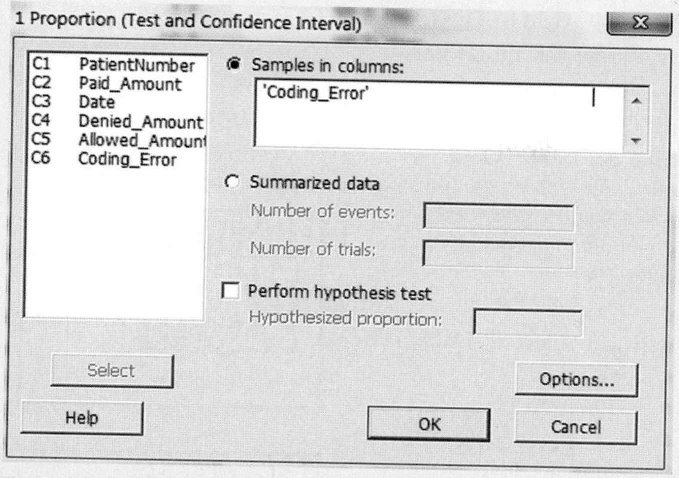

Figure 7.M.5 MINITAB 1 proportion dialog box

Step 4 Click on the "Options" button at the bottom of the dialog box and specify the confidence level in the resulting dialog box, as shown in Figure 7.M.6. Also, check the "Use test and interval based on normal distribution" box at the bottom.

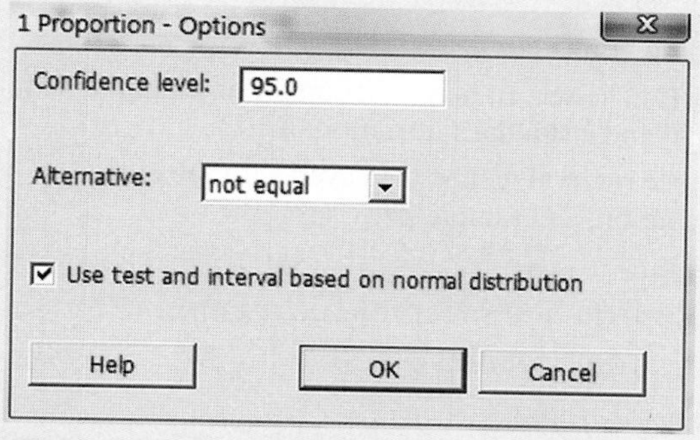

Figure 7.M.6 MINITAB 1 proportion options

Step 5 Click "OK" to return to the "1-Proportion" dialog box and then click "OK" again to produce the confidence interval.

Note: If you want to produce a confidence interval for a proportion from summary information (e.g., the number of successes and the sample size), click on "Summarized data" in the "1-Proportion" dialog box (see Figure 7.M.5). Enter the value for the number of trials (i.e., the sample size) and the number of events (i.e., the number of successes) and then click "OK."

Confidence Interval for a Variance

Step 1 Access the MINITAB data worksheet that contains the quantitative variable of interest.

Step 2 Click on the "Stat" button on the main MINITAB menu bar, then click on "Basic Statistics" and "1 Variance."

Step 3 On the resulting dialog box (shown in Figure 7.M.7), select "Samples in Columns" in the "Data" box, then specify the quantitative variable of interest in the "Columns" box.

Step 4 Click the "Options" button at the bottom of the dialog box and specify the confidence level in the resulting menu, as shown in Figure 7.M.8.

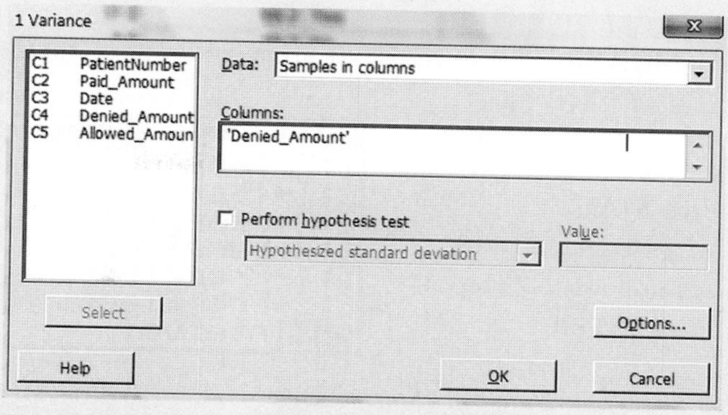

Figure 7.M.7 MINITAB 1 variance dialog box

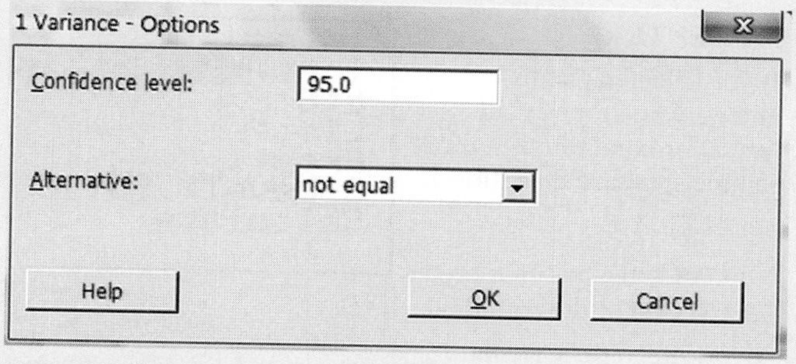

Figure 7.M.8 MINITAB 1 variance options

Step 5 Click "OK" to return to the "1 Variance" dialog box and then click "OK" again to produce the confidence interval.

Note: If you want to produce a confidence interval for a variance from summary information (e.g., the sample variance), select "Sample variance" in the "Data" box of the "1 Variance" dialog box (see Figure 7.M.9). Enter the values of the sample size and sample variance in the appropriate boxes, then click "OK."

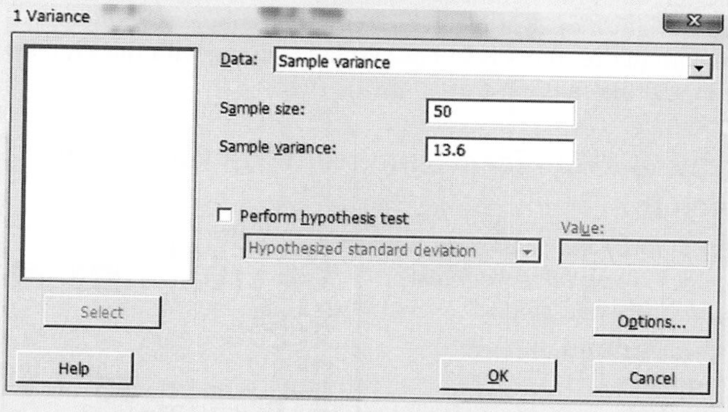

Figure 7.M.9 MINITAB 1 variance dialog box with summary statistics option

TI–83/TI–84 Plus Graphing Calculator: Confidence Intervals

Creating a Confidence Interval for a Population Mean (Known σ or $n \geq 30$)

Step 1 Enter the data (Skip to Step 2 if you have summary statistics, not raw data.)

- Press **STAT** and select **1:Edit**

Note: If the list already contains data, clear the old data. Use the up **ARROW** to highlight "**L1**."

- Press **CLEAR ENTER**
- Use the **ARROW** and **ENTER** kys to enter the data set into **L1**

Step 2 *Access the statistical tests menu*

- Press **STAT**
- Arrow right to **TESTS**
- Arrow down to **Zinterval**
- Press **ENTER**

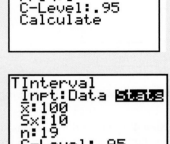

```
EDIT CALC TESTS
1:Z-Test…
2:T-Test…
3:2-SampZTest…
4:2-SampTTest…
5:1-PropZTest…
6:2-PropZTest…
7↓ZInterval…
```

Step 3 *Choose "**Data**," or "**Stats**" ("Data" is selected when you have entered the raw data into a List. "Stats" is selected when you are given only the mean, standard deviation, and sample size.)*

- Press **ENTER**

  If you selected "Data," enter a value for σ. (The best approximation is *s*, the sample standard deviation.)

- Set **List** to **L1**
- Set **Freq** to **1**
- Set **C-Level** to the confidence level
- Arrow down to "**Calculate**"
- Press **ENTER**

```
ZInterval
 Inpt:Data Stats
 σ:8
 List:L₁
 Freq:1
 C-Level:.95
 Calculate
```

If you selected "Stats," enter a value for σ. (The best approximation is *s*, the sample standard deviation.)

- Enter the sample mean and sample size
- Set **C-Level** to the confidence level
- Arrow down to "**Calculate**"
- Press **ENTER**

(The screen at the right is set up for an example with a standard deviation of 20, a mean of 200, and a sample size of 40.) The confidence interval will be displayed along with the sample mean and the sample size.

Creating a Confidence Interval for a Population Mean (n < 30)

Step 1 *Enter the data (Skip to Step 2 if you have summary statistics, not raw data.)*

- Press **STAT** and select **1:Edit**

Note: If the list already contains data, clear the old data. Use the up **ARROW** to highlight "**L1**."

- Press **CLEAR ENTER**
- Use the **ARROW** and **ENTER** keys to enter the data set into **L1**

```
ZInterval
 Inpt:Data Stats
 σ:20
 x̄:200
 n:40
 C-Level:.95
 Calculate
```

Step 2 *Access the statistical tests menu*

- Press **STAT**

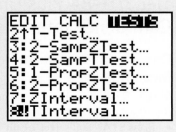

```
EDIT CALC TESTS
2↑T-Test…
3:2-SampZTest…
4:2-SampTTest…
5:1-PropZTest…
6:2-PropZTest…
7:ZInterval…
8↓TInterval…
```

- Arrow right to **TESTS**
- Arrow down to **TInterval**
- Press **ENTER**

Step 3 *Choose "**Data**" or "**Stats**" ("Data" is selected when you have entered the raw data into a List. "Stats" is selected when you are given only the mean, standard deviation, and sample size.)*

- Press **ENTER**
- If you selected "Data," set **List** to **L1**
- Set **Freq** to **1**
- Set **C-Level** to the confidence level
- Arrow down to "**Calculate**"
- Press **ENTER**

  If you selected "Stats" enter the mean, standard deviation, and sample size.

- Set **C-Level** to the confidence level
- Arrow down to "**Calculate**"
- Press **ENTER** (The screen here is set up for an example with a mean of 100 and a standard deviation of 10.)

  The confidence interval will be displayed with the mean, standard deviation, and sample size.

```
TInterval
 Inpt:Data Stats
 List:L₁
 Freq:1
 C-Level:.95
 Calculate
```

```
TInterval
 Inpt:Data Stats
 x̄:100
 Sx:10
 n:19
 C-Level:.95
 Calculate
```

Creating a Confidence Interval for a Population Proportion (Large Samples)

Step 1 *Access the statistical tests menu*

- Press **STAT**
- Arrow right to **TESTS**
- Arrow down to **1-PropZInt**
- Press **ENTER**

```
1-PropZInt
 x:532
 n:1100
 C-Level:.95■
 Calculate
```

Step 2 *Enter the values for* **x, n**, *and* **C-Level**

- where **x** = number of successes
- **n** = sample size
- **C-Level** = level of confidence
- Arrow down to "**Calculate**"
- Press **ENTER**

(The screens at the right are set up for an example with $x = 532$, $n = 1,100$, and confidence level of .95.)

Note: A confidence interval for a population variance is not available in the TI-83/T1–84 plus graphing calculator.

```
1-PropZInt
 (.4541,.51317)
 p̂=.4836363636
 n=1100
■
```

8 Inferences Based on a Single Sample
Tests of Hypothesis

CONTENTS

Where We've Been

- Used sample information to provide a *point estimate* of a population parameter
- Used the sampling distribution of a statistic to assess the reliability of an estimate through a *confidence interval*

Where We're Going

- Introduce the concepts of a *test of hypothesis* (8.1–8.2)
- Provide a measure of reliability for the hypothesis test—called the *significance level* of the test (8.2, 8.4)
- Test a specific value of a population parameter (mean, proportion or variance) (8.3, 8.5–8.6, 8.8)

In 1924, Kimberly-Clark Corporation invented a facial tissue for removing cold cream and began marketing it as KLEENEX® brand tissues. Today, KLEENEX® is recognized as the top-selling brand of tissue in the world. A wide variety of KLEENEX® products is available, ranging from extra-large tissues to tissues with lotion. Over the past 80 years, Kimberly-Clark Corporation has packaged the tissues in boxes of different sizes and shapes and varied the number of tissues packaged in each box. For example, currently a family-size box contains 144 two-ply tissues, a cold-care box contains 70 tissues (coated with lotion), and a convenience pocket pack contains 15 miniature tissues.

How does Kimberly-Clark Corp. decide how many tissues to put in each box? According to the *Wall Street Journal*, marketing experts at the company use the results of a survey of KLEENEX® customers to help determine how many tissues are packed in a box. In the mid-1980s, when Kimberly-Clark Corp. developed the cold-care box, designed especially for people who have a cold, the company conducted its initial survey of customers for this purpose. Hundreds of customers were asked to keep count of their KLEENEX® use in diaries. According to the *Wall Street Journal* report, the survey results left "little doubt that the company should put 60 tissues in each box." The number 60 was "the average number of times people blow their nose during a cold." In 2000, the

company increased the number of tissues packaged in a cold-care box to 70 based on the results of a more recent survey.

From summary information provided in the *Wall Street Journal* article, we constructed a data set that represents the results of a survey similar to the one just described. In the data file named **TISSUES**, we recorded the number of tissues used by each of 250 consumers during a period when they had a cold. We apply the hypothesis-testing methodology presented in this chapter to that data set in several Statistics in Action Revisited examples.

Statistics IN Action Revisited

- Identifying the Key Elements of a Hypothesis Test Relevant to the KLEENEX® Survey (p. 359)
- Testing a Population Mean in the KLEENEX® Survey (p. 371)
- Testing a Population Proportion in the KLEENEX® Survey (p. 384)

Data Set: TISSUES

Suppose you wanted to determine whether the mean level of a driver's blood alcohol exceeds the legal limit after two drinks, or whether the majority of registered voters approve of the president's performance. In both cases, you are interested in making an inference about how the value of a parameter relates to a specific numerical value. Is it less than, equal to, or greater than the specified number? This type of inference, called a **test of hypothesis,** is the subject of this chapter.

We introduce the elements of a test of hypothesis in Sections 8.1 and 8.2. We then show how to conduct tests of hypothesis about a population mean in Sections 8.3–8.5. Large-sample tests about binomial probabilities are the subject of Section 8.6, and some advanced methods for determining the reliability of a test are covered in optional Section 8.7. Finally, we show how to conduct a test about a population variance in optional Section 8.8.

8.1 The Elements of a Test of Hypothesis

Suppose building specifications in a certain city require that the average breaking strength of residential sewer pipe be more than 2,400 pounds per foot of length (i.e., per linear foot). Each manufacturer who wants to sell pipe in that city must demonstrate that its product meets the specification. Note that we are interested in making an inference about the mean μ of a population. However, in this example we are less interested in estimating the value of μ than we are in testing a **hypothesis** about its value—that is, *we want to decide whether the mean breaking strength of the pipe exceeds 2,400 pounds per linear foot.*

A statistical **hypothesis** is a statement about the numerical value of a population parameter.

The method used to reach a decision is based on the rare-event concept explained in earlier chapters. We define two hypotheses: (1) The **null hypothesis** represents the status quo to the party performing the sampling experiment—the hypothesis that will be accepted unless the data provide convincing evidence that it is false. (2) The **alternative, or research, hypothesis** is that which will be accepted only if the data provide convincing evidence of its truth. From the point of view of the city conducting the tests, the null hypothesis is that the manufacturer's pipe does *not* meet specifications unless the tests provide convincing evidence otherwise. The null and alternative hypotheses are therefore

Null hypothesis(H_0): $\mu \leq 2{,}400$

(i.e., the manufacturer's pipe does not meet specifications)

Alternative (research) hypothesis(H_a): $\mu > 2{,}400$

(i.e., the manufacturer's pipe meets specifications)

> The **null hypothesis,** denoted H_0, represents the hypothesis that will be accepted unless the data provide convincing evidence that it is false. This usually represents the "status quo" or some claim about the population parameter that the researcher wants to test.

> The **alternative (research) hypothesis,** denoted H_a, represents the hypothesis that will be accepted only if the data provide convincing evidence of its truth. This usually represents the values of a population parameter for which the researcher wants to gather evidence to support.

How can the city decide when enough evidence exists to conclude that the manufacturer's pipe meets specifications? Because the hypotheses concern the value of the population mean μ, it is reasonable to use the sample mean $\bar{x}$ to make the inference, just as we did when we formed confidence intervals for μ in Sections 7.2 and 7.3. The city will conclude that the pipe meets specifications only when the sample mean $\bar{x}$ convincingly indicates that the population mean exceeds 2,400 pounds per linear foot.

"Convincing" evidence in favor of the alternative hypothesis will exist when the value of $\bar{x}$ exceeds 2,400 by an amount that cannot be readily attributed to sampling variability. To decide, we compute a **test statistic,** i.e., a numerical value computed from the sample. Here, the test statistic is the z-value that measures the distance between the value of $\bar{x}$ and the value of μ specified in the alternative hypothesis. When the null hypothesis contains more than one values of μ, as in this case (H_0: $\mu \leq 2{,}400$), we use the value of μ closest to the values specified in the alternative hypothesis. The idea is that if the hypothesis that μ *equals* 2,400 can be rejected in favor of $\mu > 2{,}400$, then μ *less than or equal to* 2,400 can certainly be rejected. Thus, the test statistic is

$$z = \frac{\bar{x} - 2{,}400}{\sigma_{\bar{x}}} = \frac{\bar{x} - 2{,}400}{\sigma/\sqrt{n}}$$

Note that a value of $z = 1$ means that $\bar{x}$ is 1 standard deviation above $\mu = 2{,}400$, a value of $z = 1.5$ means that $\bar{x}$ is 1.5 standard deviations above $\mu = 2{,}400$, and so on. How large must z be before the city can be convinced that the null hypothesis can be rejected in favor of the alternative and conclude that the pipe meets specifications?

> The **test statistic** is a sample statistic, computed from information provided in the sample, that the researcher uses to decide between the null and alternative hypotheses.

If you examine Figure 8.1, you will note that the chance of observing $\bar{x}$ more than 1.645 standard deviations above 2,400 is only .05—*if in fact the true mean μ is 2,400.* Thus, if the sample mean is more than 1.645 standard deviations above 2,400, either H_0

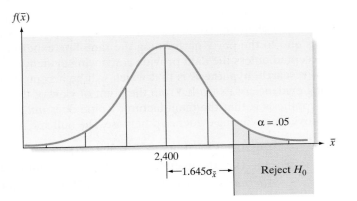

$f(\bar{x})$

$\alpha = .05$

$\bar{x}$

2,400

$\leftarrow 1.645\sigma_{\bar{x}} \rightarrow$ Reject H_0

Figure 8.1

The sampling distribution of $\bar{x}$, assuming $\mu = 2,400$

is true and a relatively rare event has occurred (.05 probability), or H_a is true and the population mean exceeds 2,400. Because we would most likely reject the notion that a rare event has occurred, we would reject the null hypothesis ($\mu \le 2,400$) and conclude that the alternative hypothesis ($\mu > 2,400$) is true. What is the probability that this procedure will lead us to an incorrect decision?

Such an incorrect decision—deciding that the null hypothesis is false when in fact it is true—is called a **Type I error.** As indicated in Figure 8.1, the risk of making a Type I error is denoted by the symbol α—that is,

$$\alpha = P \text{ (Type I error)}$$

$$= P \text{ (Rejecting the null hypothesis when in fact the null hypothesis is true)}$$

> A **Type I error** occurs if the researcher rejects the null hypothesis in favor of the alternative hypothesis when, in fact, H_0 is true. The probability of committing a Type I error is denoted by α.

In our example,

$$\alpha = P \text{ (}z > 1.645 \text{ when in fact } \mu = 2,400\text{)} = .05$$

We now summarize the elements of the test:

$$H_0: \mu \le 2,400 \qquad \text{(Pipe does not meet specifications.)}$$
$$H_a: \mu > 2,400 \qquad \text{(Pipe meets specifications.)}$$
$$\text{Test statistic: } z = \frac{\bar{x} - 2,400}{\sigma_{\bar{x}}}$$

Rejection region: $z > 1.645$, which corresponds to $\alpha = .05$

Note that the **rejection region** refers to the values of the test statistic for which we will *reject the null hypothesis.*

> The **rejection region** of a statistical test is the set of possible values of the test statistic for which the researcher will reject H_0 in favor of H_a.

To illustrate the use of the test, suppose we test 50 sections of sewer pipe and find the mean and standard deviation for these 50 measurements to be

$$\bar{x} = 2,460 \text{ pounds per linear foot}$$
$$s = 200 \text{ pounds per linear foot}$$

As in the case of estimation, we can use s to approximate σ when s is calculated from a large set of sample measurements.

The test statistic is

$$z = \frac{\bar{x} - 2{,}400}{\sigma_{\bar{x}}} = \frac{\bar{x} - 2{,}400}{\sigma/\sqrt{n}} \approx \frac{\bar{x} - 2{,}400}{s/\sqrt{n}}$$

Substituting $\bar{x} = 2{,}460$, $n = 50$, and $s = 200$, we have

$$z \approx \frac{2{,}460 - 2{,}400}{200/\sqrt{50}} = \frac{60}{28.28} = 2.12$$

Therefore, the sample mean lies $2.12\sigma_{\bar{x}}$ above the hypothesized value of μ, 2,400, as shown in Figure 8.2. Because this value of z exceeds 1.645, it falls into the rejection region. That is, we reject the null hypothesis that $\mu = 2{,}400$ and conclude that $\mu > 2{,}400$. Thus, it appears that the company's pipe has a mean strength that exceeds 2,400 pounds per linear foot.

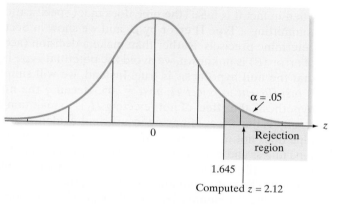

Figure 8.2

Location of the test statistic for a test of the hypothesis $H_0: \mu = 2{,}400$

How much faith can be placed in this conclusion? What is the probability that our statistical test could lead us to reject the null hypothesis (and conclude that the company's pipe meets the city's specifications) when in fact the null hypothesis is true? The answer is $\alpha = .05$—that is, we selected the level of risk, α, of making a Type I error when we constructed the test. Thus, the chance is only 1 in 20 that our test would lead us to conclude the manufacturer's pipe satisfies the city's specifications when in fact the pipe does *not* meet specifications.

Now, suppose the sample mean breaking strength for the 50 sections of sewer pipe turned out to be $\bar{x} = 2{,}430$ pounds per linear foot. Assuming that the sample standard deviation is still $s = 200$, the test statistic is

$$z = \frac{2{,}430 - 2{,}400}{200/\sqrt{50}} = \frac{30}{28.28} = 1.06$$

Therefore, the sample mean $\bar{x} = 2{,}430$ is only 1.06 standard deviations above the null hypothesized value of $\mu = 2{,}400$. As shown in Figure 8.3, this value does not fall into the rejection region ($z > 1.645$). Therefore, we know that we cannot reject H_0 using $\alpha = .05$. Even though the sample mean exceeds the city's specification of 2,400 by 30 pounds per linear foot, it does not exceed the specification by enough to provide *convincing* evidence that the *population mean* exceeds 2,400.

A **Type II error** occurs if the researcher accepts the null hypothesis when, in fact, H_0 is false. The probability of committing a Type II error is denoted by β.

Should we accept the null hypothesis $H_0: \mu \leq 2{,}400$ and conclude that the manufacturer's pipe does not meet specifications? To do so would be to risk a **Type II error**—that of concluding that the null hypothesis is true (the pipe does not meet specifications)

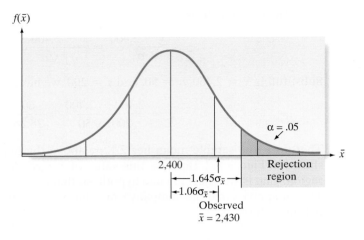

Figure 8.3

Location of test statistic when $\bar{x} = 2{,}430$

when in fact it is false (the pipe does meet specifications). We denote the probability of committing a Type II error by β, and we show in Section 8.7 that β is often difficult to determine precisely. Rather than make a decision (accept H_0) for which the probability of error (β) is unknown, we avoid the potential Type II error by avoiding the conclusion that the null hypothesis is true. Instead, we will simply state that *the sample evidence is insufficient to reject H_0 at $\alpha = .05$.* Because the null hypothesis is the "status-quo" hypothesis, the effect of not rejecting H_0 is to maintain the status quo. In our pipe-testing example, the effect of having insufficient evidence to reject the null hypothesis that the pipe does not meet specifications is probably to prohibit the use of the manufacturer's pipe unless and until there is sufficient evidence that the pipe does meet specifications—that is, until the data indicate convincingly that the null hypothesis is false, we usually maintain the status quo implied by its truth.

Table 8.1 summarizes the four possible outcomes (i.e., **conclusions**) of a test of hypothesis. The "true state of nature" columns in Table 8.1 refer to the fact that either the null hypothesis H_0 is true or the alternative hypothesis H_a is true. Note that the true state of nature is unknown to the researcher conducting the test. The "decision" rows in Table 8.1 refer to the action of the researcher, assuming that he or she will either conclude that H_0 is true or that H_a is true, based on the results of the sampling experiment. Note that a Type I error can be made *only* when the null hypothesis is rejected in favor of the alternative hypothesis, and a Type II error can be made *only* when the null hypothesis is accepted. Our policy will be to make a decision only when we know the probability of making the error that corresponds to that decision. Because α is usually specified by the analyst, we will generally be able to reject H_0 (accept H_a) when the sample evidence supports that decision. However, because β is usually not specified, *we will generally avoid the decision to accept H_0, preferring instead to state that the sample evidence is insufficient to reject H_0 when the test statistic is not in the rejection region.*

Table 8.1	Conclusions and Consequences for a Test of Hypothesis	
	True State of Nature	
Conclusion	H_0 **True**	H_a **True**
Accept H_0 (Assume H_0 True)	Correct decision	Type II error (probability β)
Reject H_0 (Assume H_a True)	Type I error (probability α)	Correct decision

⚠ **CAUTION** Be careful not to "accept H_0" when conducting a test of hypothesis because the measure of reliability, $\beta = P$ (Type II error), is almost always unknown. If the test statistic does not fall into the rejection region, it is better to state the conclusion as "insufficient evidence to reject H_0."* ▲

*In many practical business applications of hypothesis testing, nonrejection leads management to behave as if the null hypothesis were accepted. Accordingly, the distinction between acceptance and nonrejection is frequently blurred in practice. We discuss the issues connected with the acceptance of the null hypothesis and the calculation of β in more detail in Section 8.7

The elements of a test of hypothesis are summarized in the following box. Note that the first four elements are all specified *before* the sampling experiment is performed. In no case will the results of the sample be used to determine the hypotheses; the data are collected to test the predetermined hypotheses, not to formulate them.

Elements of a Test of Hypothesis

1. *Null hypothesis* (H_0): A theory about the specific values of one or more population parameters. The theory generally represents the status quo, which we adopt until it is proven false. The theory is always stated as H_0: parameter $=$ value.

2. *Alternative (research) hypothesis* (H_a): A theory that contradicts the null hypothesis. The theory generally represents that which we will adopt only when sufficient evidence exists to establish its truth.

3. *Test statistic:* A sample statistic used to decide whether to reject the null hypothesis.

4. *Rejection region:* The numerical values of the test statistic for which the null hypothesis will be rejected. The rejection region is chosen so that the probability is α that it will contain the test statistic when the null hypothesis is true, thereby leading to a Type I error. The value of α is usually chosen to be small (e.g., .01, .05, or .10) and is referred to as the **level of significance** of the test.

5. *Assumptions:* Clear statement(s) of any assumptions made about the population(s) being sampled.

6. *Experiment and calculation of test statistic:* Performance of the sampling experiment and determination of the numerical value of the test statistic.

7. *Conclusion:*
   a. If the numerical value of the test statistic falls in the rejection region, we reject the null hypothesis and conclude that the alternative hypothesis is true. We know that the hypothesis-testing process will lead to this conclusion incorrectly (Type I error) only $100\alpha\%$ of the time when H_0 is true.
   b. If the test statistic does not fall in the rejection region, we do not reject H_0. Thus, we reserve judgment about which hypothesis is true. We do not conclude that the null hypothesis is true because we do not (in general) know the probability β that our test procedure will lead to an incorrect acceptance of H_0 (Type II error).

As with confidence intervals, the methodology for testing hypotheses varies depending on the target population parameter. In this chapter, we develop methods for testing a population mean, a population proportion, and (optionally) a population variance. As a reminder, the key words and the type of data associated with these target parameters are again listed in the accompanying box.

Determining the Target Parameter

Parameter	Key Words or Phrases	Type of Data
μ	Mean; average	Quantitative
p	Proportion; percentage; fraction; rate	Qualitative
σ^2	Variance; variability; spread	Quantitative

8.2 Formulating Hypotheses and Setting Up the Rejection Region

In Section 8.1 we learned that the null and alternative hypotheses form the basis for inference using a test of hypothesis. The null and alternative hypotheses may take one of several forms. In the sewer pipe example, we tested the null hypothesis that the population mean strength of the pipe is less than or equal to 2,400 pounds per linear foot against the alternative hypothesis that the mean strength exceeds 2,400—that is, we tested

$$H_0: \mu \leq 2,400 \quad \text{(Pipe does not meet specifications.)}$$
$$H_a: \mu > 2,400 \quad \text{(Pipe meets specifications.)}$$

This is a **one-tailed** (or **one-sided**) **statistical test** because the alternative hypothesis specifies that the population parameter (the population mean μ in this example) is strictly greater than a specified value (2,400 in this example). If the null hypothesis had been $H_0: \mu \geq 2,400$ and the alternative hypothesis had been $H_a: \mu < 2,400$, the test would still be one-sided because the parameter is still specified to be on "one side" of the null hypothesis value. Some statistical investigations seek to show that the population parameter is *either larger or smaller* than some specified value. Such an alternative hypothesis is called a **two-tailed** (or **two-sided**) **hypothesis**.

While alternative hypotheses are always specified as strict inequalities, such as $\mu < 2,400$, $\mu > 2,400$, or $\mu \neq 2,400$, *null hypotheses are usually specified as equalities*, such as $\mu = 2,400$. Even when the null hypothesis is an inequality, such as $\mu \leq 2,400$, we specify $H_0: \mu = 2,400$, reasoning that if sufficient evidence exists to show that $H_a: \mu > 2,400$ is true when tested against $H_0: \mu = 2,400$, then surely sufficient evidence exists to reject $\mu < 2,400$ as well. Therefore, the null hypothesis is specified as the value of μ closest to a one-sided alternative hypothesis and as the only value *not* specified in a two-tailed alternative hypothesis. The steps for selecting the null and alternative hypotheses are summarized in the following box.

Steps for Selecting the Null and Alternative Hypotheses

1. Select the *alternative hypothesis* as that which the sampling experiment is intended to establish. The alternative hypothesis will assume one of three forms:

   a. One-tailed, **upper-tailed** (e.g., $H_a: \mu > 2,400$)
   b. One-tailed, **lower-tailed** (e.g., $H_a: \mu < 2,400$)
   c. Two-tailed (e.g., $H_a: \mu \neq 2,400$)

2. Select the *null hypothesis* as the status quo, that which will be presumed true unless the sampling experiment conclusively establishes the alternative hypothesis. The null hypothesis will be specified as that parameter value closest to the alternative in one-tailed tests and as the complementary (or only unspecified) value in two-tailed tests.

   (e.g., $H_0: \mu = 2,400$)

A **one-tailed test** of hypothesis is one in which the alternative hypothesis is directional and includes the symbol "<" or " > ."

A **two-tailed test** of hypothesis is one in which the alternative hypothesis does not specify departure from H_0 in a particular direction and is written with the symbol " $\neq$."

Example 8.1

Formulating H_0 and H_a for a Test of a Population Mean

Problem A metal lathe is checked periodically by quality control inspectors to determine whether it is producing machine bearings with a mean diameter of .5 inch. If the mean diameter of the bearings is larger or smaller than .5 inch, then the process is out of control and must be adjusted. Formulate the null and alternative hypotheses for a test to determine whether the bearing production process is out of control.

Solution The hypotheses must be stated in terms of a population parameter. Here, we define μ as the true mean diameter (in inches) of all bearings produced by the metal lathe. If either $\mu > .5$ or $\mu < .5$, then the lathe's production process is out of control. Because the inspectors want to be able to detect either possibility (indicating that the process is in need of adjustment), these values of μ represent the alternative (or research) hypothesis. Alternatively, because $\mu = .5$ represents an in-control process (the status quo), this represents the null hypothesis. Therefore, we want to conduct the two-tailed test:

$$H_0: \mu = .5 \text{ (i.e., the process is in control)}$$
$$H_a: \mu \neq .5 \text{ (i.e., the process is out of control)}$$

Look Back Here, the alternative hypothesis is not necessarily the hypothesis that the quality control inspectors desire to support. However, they will make adjustments to the metal lathe settings only if there is strong evidence to indicate that the process is out of control. Consequently, $\mu \neq .5$ must be stated as the alternative hypothesis.

Now Work Exercise 8.11a

Example 8.2

Formulating H_0 and H_a for a Test of a Population Proportion

Problem Cigarette advertisements are required by federal law to carry the following statement: "Warning: The surgeon general has determined that cigarette smoking is dangerous to your health." However, this warning is often located in inconspicuous corners of the advertisements and printed in small type. Suppose the Federal Trade Commission (FTC) claims that 80% of cigarette consumers fail to see the warning. A marketer for a large tobacco firm wants to gather evidence to show that the FTC's claim is too high, i.e., that fewer than 80% of cigarette consumers fail to see the warning. Specify the null and alternative hypotheses for a test of the FTC's claim.

Solution The marketer wants to make an inference about p, the true proportion of all cigarette consumers who fail to see the surgeon general's warning. In particular, the marketer wants to collect data to show that fewer than 80% of cigarette consumers fail to see the warning, i.e., $p < .80$. Consequently, $p < .80$. represents the alternative hypothesis and $p = .80$ (the claim made by the FTC) represents the null hypothesis. That is, the marketer desires the one-tailed (lower-tailed) test:

$$H_0: p = .80 \text{ (i.e., the FTC's claim is true)}$$
$$H_a: p < .80 \text{ (i.e., the FTC's claim is false)}$$

Look Back Whenever a claim is made about the value of a particular population parameter and the researcher wants to test the claim, believing that it is false, the claimed value will represent the null hypothesis.

Now Work Exercise 8.15

The rejection region for a two-tailed test differs from that for a one-tailed test. When we are trying to detect departure from the null hypothesis in *either* direction, we must establish a rejection region in both tails of the sampling distribution of the test statistic. Figures 8.4a and 8.4b show the one-tailed rejection regions for lower- and

Figure 8.4

Rejection regions corresponding to one- and two-tailed tests

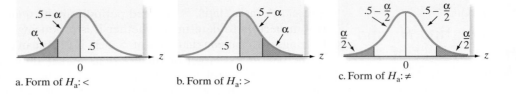

a. Form of $H_a: <$ b. Form of $H_a: >$ c. Form of $H_a: \neq$

upper-tailed tests, respectively. The two-tailed rejection region is illustrated in Figure 8.4c. Note that a rejection region is established in each tail of the sampling distribution for a two-tailed test.

The rejection regions corresponding to typical values selected for α are shown in Table 8.2 for one- and two-tailed tests. Note that the smaller α you select, the more evidence (the larger z) you will need before you can reject H_0.

Table 8.2	Rejection Regions for Common Values of α		
	Alternative Hypotheses		
	Lower-Tailed	Upper-Tailed	Two-Tailed
$\alpha = .10$	$z < -1.28$	$z > 1.28$	$z < -1.645$ or $z > 1.645$
$\alpha = .05$	$z < -1.645$	$z > 1.645$	$z < -1.96$ or $z > 1.96$
$\alpha = .01$	$z < -2.33$	$z > 2.33$	$z < -2.575$ or $z > 2.575$

Example 8.3

Setting Up a Hypothesis Test For μ—Mean Drug Response Time

Problem The effect of drugs and alcohol on the nervous system has been the subject of considerable research. Suppose a research neurologist is testing the effect of a drug on response time by injecting 100 rats with a unit dose of the drug, subjecting each rat to a neurological stimulus, and recording its response time. The neurologist knows that the mean response time for rats not injected with the drug (the "control" mean) is 1.2 seconds. She wishes to test whether the mean response time for drug-injected rats differs from 1.2 seconds. Set up the test of hypothesis for this experiment, using $\alpha = .01$.

Solution The key word *mean* in the statement of the problem implies that the target parameter is μ, the mean response time for all drug-injected rats. Since the neurologist wishes to detect whether μ differs from the control mean of 1.2 seconds in *either* direction—that is, $\mu < 1.2$ or $\mu > 1.2$—we conduct a two-tailed statistical test. Following the procedure for selecting the null and alternative hypotheses, we specify as the alternative hypothesis that the mean differs from 1.2 seconds, since determining whether the drug-injected mean differs from the control mean is the purpose of the experiment. The null hypothesis is the presumption that drug-injected rats have the same mean response time as control rats unless the research indicates otherwise. Thus,

$H_0: \mu = 1.2$ (Mean response time is 1.2 seconds)

$H_a: \mu \neq 1.2$ (Mean response time is less than 1.2 or greater than 1.2 seconds)

The test statistic measures the number of standard deviations between the observed value of $\bar{x}$ and the null-hypothesized value $\mu = 1.2$:

$$\text{Test statistic:} \quad z = \frac{\bar{x} - 1.2}{\sigma_{\bar{x}}}$$

The rejection region must be designated to detect a departure from $\mu = 1.2$ in *either* direction, so we will reject H_0 for values of z that are either too small (negative) or too large (positive). To determine the precise values of z that constitute the rejection region, we first select α, the probability that the test will lead to incorrect rejection of the null hypothesis. Then we divide α equally between the lower and upper tail of the distribution of z, as shown in Figure 8.5. In this example, $\alpha = .01$, so $\alpha/2 = .005$ is placed in each tail. The areas in the tails correspond to $z = -2.575$ and $z = 2.575$, respectively (from Table 8.2), so

$$\text{Rejection region: } z < -2.575 \text{ or } z > 2.575 \qquad \text{(see Figure 8.5)}$$

Assumptions: Since the sample size of the experiment is large enough ($n > 30$), the Central Limit Theorem will apply, and no assumptions need be made about the population of response time measurements. The sampling distribution of the sample mean response of 100 rats will be approximately normal, regardless of the distribution of the individual rats' response times.

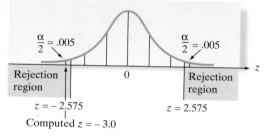

Figure 8.5
Two-tailed rejection region:
$\alpha = .01$

Look Back Note that the test is set up *before* the sampling experiment is conducted. The data are not used to develop the test. Evidently, the neurologist wants to conclude that the mean response time for the drug-injected rats differs from the control mean only when the evidence is very convincing, because the value of α has been set quite low at .01. If the experiment results in the rejection of H_0, she can confidently conclude that the mean response time of the drug-injected rats differs from the control mean because there is only a .01 probability of a Type I error.

Now Work Exercise 8.11b and c

Once the test is set up, the neurologist of Example 8.3 is ready to perform the sampling experiment and conduct the test. The hypothesis test is performed in the next section.

Statistics IN Action   Revisited

Identifying the Key Elements of a Hypothesis Test Relevant to the KLEENEX® Survey

In Kimberly-Clark Corporation's survey of people with colds, each of 250 customers was asked to keep count of his or her use of KLEENEX® tissues in diaries. One goal of the company was to determine how many tissues to package in a cold-care box of KLEENEX®; consequently, the total number of tissues used was recorded for each person surveyed. Since number of tissues is a quantitative variable, the parameter of interest is either μ, the mean number of tissues used by all customers with colds, or σ^2, the variance of the number of tissues used.

Recall that in 2000, the company increased from 60 to 70 the number of tissues it packages in a cold-care box of KLEENEX® tissues. This decision was based on

a claim made by marketing experts that the average number of times a person will blow his or her nose during a cold exceeds the previous mean of 60. The key word *average* implies that the target parameter is μ, and the marketers are claiming that $\mu > 60$. In order to test the claim we set up the following null and alternative hypotheses:

$$H_0: \mu = 60 \qquad H_a: \mu > 60$$

We'll conduct this test in the next Statistics in Action Revisited, on p. 371.

Exercises 8.1–8.19

Understanding the Principles

8.1 Which hypothesis, the null or the alternative, is the status quo hypothesis? Which is the research hypothesis?

8.2 Which element of a test of hypothesis is used to decide whether to reject the null hypothesis in favor of the alternative hypothesis?

8.3 What is the level of significance of a test of hypothesis?

8.4 What is the difference between Type I and Type II errors in hypothesis testing? How do α and β relate to Type I and Type II errors?

8.5 List the four possible results of the combinations of decisions and true states of nature for a test of hypothesis.

8.6 We (generally) reject the null hypothesis when the test statistic falls into the rejection region, but we do not accept the null hypothesis when the test statistic does not fall into the rejection region. Why?

8.7 If you test a hypothesis and reject the null hypothesis in favor of the alternative hypothesis, does your test prove that the alternative hypothesis is correct? Explain.

Learning the Mechanics

8.8 Consider a test of $H_0: \mu = 4$. In each of the following cases, give the rejection region for the test in terms of the z-statistic:
   a. $H_a: \mu > 4, \alpha = .05$
   b. $H_a: \mu > 4, \alpha = .10$
   c. $H_a: \mu < 4, \alpha = .05$
   d. $H_a: \mu \neq 4, \alpha = .05$

8.9 For each of the following rejection regions, sketch the sampling distribution for z and indicate the location of the rejection region.
   a. $z > 1.96$
   b. $z > 1.645$
   c. $z > 2.575$
   d. $z < -1.28$
   e. $z < -1.645$ or $z > 1.645$
   f. $z < -2.575$ or $z > 2.575$
   g. For each of the rejection regions specified in parts **a–f**, what is the probability that a Type I error will be made?

Applet Exercise 8.1

Use the applet entitled *Hypotheses Test for a Mean* to investigate the frequency of Type I and Type II errors. For this exercise, use $n = 100$ and the normal distribution with mean 50 and standard deviation 10.
   a. Set the null mean equal to 50 and the alternative to *not equal*. Run the applet one time. How many times was the null hypothesis rejected at level .05? In this case, the null hypothesis is true. Which type of error occurred each time the true null hypothesis was rejected? What is the probability of rejecting a true null hypothesis at level .05? How does the proportion of times the null hypothesis was rejected compare with this probability?
   b. Clear the applet, then set the null mean equal to 47, and keep the alternative at *not equal*. Run the applet one time. How many times was the null hypothesis *not* rejected at level .05? In this case, the null hypothesis is false. Which type of error occurred each time the

null hypothesis was *not* rejected? Run the applet several more times without clearing. Based on your results, what can you conclude about the probability of failing to reject the null hypothesis for the given conditions?

Applying the Concepts—Basic

8.10 **Infants' listening time.** Researchers writing in *Analysis of Verbal Behavior* (Dec. 2007) reported that the mean listening time of 16-month-old infants exposed to nonmeaningful monosyllabic words (e.g., "giff," "cham," "gack") is 8 seconds. Set up the null and alternative hypotheses for testing the claim.

8.11 **Play Golf America program.** The Professional Golf [NW] Association (PGA) and *Golf Digest* have developed the Play Golf America program, in which teaching professionals at participating golf clubs provide a free 10-minute lesson to new customers. According to *Golf Digest* (July 2008), golf facilities that participate in the program gain, on average, $2,400 in green fees, lessons, or equipment expenditures. A teaching professional at a golf club believes that the average gain in green fees, lessons, or equipment expenditures for participating golf facilities exceeds $2,400.
   a. In order to support the claim made by the teaching professional, what null and alternative hypothesis should you test?
   b. Suppose you select $\alpha = .05$. Interpret this value in the words of the problem.
   c. For $\alpha = .05$, specify the rejection region of a large-sample test.

8.12 **Calories in school lunches.** A University of Florida economist conducted a study of Virginia elementary school lunch menus. During the state-mandated testing period, school lunches averaged 863 calories (*National Bureau of Economic Research*, November 2002). The economist claimed that after the testing period end, the average caloric content of Virginia school lunches dropped significantly. Set up the null and alternative hypothesis to test the economist's claim.

8.13 **A camera that detects liars.** According to *New Scientist* (January 2, 2002), a new thermal imaging camera that detects small temperature changes is now being used as a polygraph device. The United States Department of Defense Polygraph Institute (DDPI) claims that the camera can detect liars correctly 75% of the time by monitoring the temperatures of their faces. Give the null hypothesis for testing the claim made by the DDPI.

8.14 **Effectiveness of online courses.** The Sloan Survey of Online Learning, "Making the Grade: Online Education in the United States, 2006," reported that 60% of college presidents believe that their online education courses are as good as or superior to courses that utilize traditional face-to-face instruction (*Inside Higher Ed*, Nov. 2006). Give the null hypothesis for testing the claim made by the Sloan Survey.

8.15 **DNA-reading tool for quick identification of species.** A [NW] biologist and a zoologist at the University of Florida were the first scientists to test the effectiveness of a high-tech handheld device designed to instantly identify the DNA of an animal species (*PLOS Biology*, Dec. 2005). They used the DNA-reading device on tissue samples collected from mollusks with brightly colored shells. The scientists discovered that the error rate of the device is less than 5 percent. Set up the null and alternative hypotheses if you want to support the findings.

Applying the Concepts—Intermediate

8.16 **Virtual reality hypnosis for pain.** The *International Journal of Clinical and Experimental Hypnosis* (Vol. 58, 2010) investigated using virtual reality hypnosis (VRH) to reduce the pain of trauma patients. Patients reported their pain intensity on the Graphic Rating Scale (GRS) both prior to and one hour after a VRH session. The researchers reported that hypnosis reduced the pain intensity of trauma patients by an average of $\mu = 10$ points on the GRS. Suppose you want to test whether trauma patients who receive normal analgesic care (i.e., medication, but no hypnosis) will have a smaller reduction in average pain intensity than the VRH trauma patients.
a. Set up H_0 and H_a for the test.
b. Describe a Type I error for this test.
c. Describe a Type II error for this test.

8.17 **Mercury levels in wading birds.** According to a University of Florida wildlife ecology and conservation researcher, the average level of mercury uptake in wading birds in the Everglades has declined over the past several years. (*UF News*, December 15, 2000). Ten years ago, the average level was 15 parts per million.
a. Give the null and alternative hypotheses for testing whether the average level today is less than 15 ppm.
b. Describe a Type I error for this test.
c. Describe a Type II error for this test.

Applying the Concepts—Advanced

8.18 **Jury trial outcomes.** Sometimes, the outcome of a jury trial defies the "commonsense" expectations of the general public (e.g., the 1995 O. J. Simpson verdict and the 2011 Casey Anthony verdict). Such a verdict is more acceptable if we understand that the jury trial of an accused murderer is analogous to the statistical hypothesis-testing process. The null hypothesis in a jury trial is that the accused is innocent. (The status quo hypothesis in the U.S. system of justice is innocence, which is assumed to be true until proven *beyond a reasonable doubt*.) The alternative hypothesis is guilt, which is accepted only when sufficient

evidence exists to establish its truth. If the vote of the jury is unanimous in favor of guilt, the null hypothesis of innocence is rejected and the court concludes that the accused murderer is guilty. Any vote other than a unanimous one for guilt results in a "not guilty" verdict. The court never accepts the null hypothesis; that is, the court never declares the accused "innocent." A "not guilty" verdict (as in the O. J. Simpson case) implies that the court could not find the defendant guilty *beyond a reasonable doubt*.
a. Define Type I and Type II errors in a murder trial.
b. Which of the two errors is the more serious? Explain.
c. The court does not, in general, know the values of α and β, but ideally, both should be small. One of these probabilities is assumed to be smaller than the other in a jury trial. Which one, and why?
d. The court system relies on the belief that the value of α is made very small by requiring a unanimous vote before guilt is concluded. Explain why this is so.
e. For a jury prejudiced against a guilty verdict as the trial begins, will the value of α increase or decrease? Explain.
f. For a jury prejudiced against a guilty verdict as the trial begins, will the value of β increase or decrease? Explain.

8.19 **Intrusion detection systems.** Refer to the *Journal of Research of the National Institute of Standards and Technology* (November–December 2003) study of a computer intrusion detection system (IDS), presented in Exercise 3.90 (p. 148). Recall that an IDS is designed to provide an alarm whenever unauthorized access (e.g., an intrusion) to a computer system occurs. The probability of the system giving a false alarm (i.e., providing a warning when, in fact, no intrusion occurs) is defined by the symbol α, while the probability of a missed detection (i.e., no warning given, when, in fact, an intrusion occurs) is defined by the symbol β. These symbols are used to represent Type I and Type II error rates, respectively, in a hypothesis-testing scenario.
a. What is the null hypothesis H_0?
b. What is the alternative hypothesis H_a?
c. According to actual data on the EMERALD system collected by the Massachusetts Institute of Technology Lincoln Laboratory, only 1 in 1,000 computer sessions with no intrusions resulted in a false alarm. For the same system, the laboratory found that only 500 of 1,000 intrusions were actually detected. Use this information to estimate the values of α and β.

8.3 Test of Hypothesis about a Population Mean: Normal (*z*) Statistic

When testing a hypothesis about a population mean μ, the test statistic we use will depend on whether the sample size n is large (say, $n \geq 30$) or small, and whether or not we know the value of the population standard deviation, σ. In this section, we consider the large-sample case.

Because the sample size is large, the Central Limit Theorem guarantees that the sampling distribution of $\bar{x}$ is approximately normal. Consequently, the test statistic for a test based on large samples will be based on the normal *z*-statistic. Although the *z*-statistic requires that we know the true population standard deviation σ, this is rarely, if ever, the case. However, we established in Chapter 7 that when n is large, the sample standard deviation s provides a good approximation to σ, and the z-statistic can be approximated as follows:

$$z = \frac{\overline{x} - \mu_0}{\sigma_{\overline{x}}} = \frac{\overline{x} - \mu_0}{\sigma/\sqrt{n}} \approx \frac{\overline{x} - \mu_0}{s/\sqrt{n}}$$

where μ_0 represents the value of μ specified in the null hypothesis.

The setup of a large-sample test of hypothesis about a population mean is summarized in the following boxes. Both the one- and two-tailed tests are shown.

Large-Sample Test of Hypothesis about μ Based on a Normal (z) Statistic

One-Tailed Test	Two-Tailed Test
$H_0: \mu = \mu_0$	$H_0: \mu = \mu_0$
$H_a: \mu < \mu_0$ (or $H_a: \mu > \mu_0$)	$H_a: \mu \neq \mu_0$
Test Statistic (σ Known): $z = \dfrac{\overline{x} - \mu_0}{\sigma/\sqrt{n}}$	Test statistic (σ Known): $z = \dfrac{\overline{x} - \mu_0}{\sigma/\sqrt{n}}$
Test Statistic (σ unknown): $z \approx \dfrac{\overline{x} - \mu_0}{s/\sqrt{n}}$	Test statistic (σ unknown): $z \approx \dfrac{\overline{x} - \mu_0}{s/\sqrt{n}}$
Rejection region: $z < -z_\alpha$ (or $z > z_\alpha$ when $H_a: \mu > \mu_0$) where z_α is chosen so that $P(z > z_\alpha) = \alpha$	Rejection region: $\lvert z \rvert > z_{\alpha/2}$ where $z_{\alpha/2}$ is chosen so that $P(\lvert z \rvert > z_{\alpha/2}) = \alpha/2$

Note: The symbol for the numerical value assigned to μ under the null hypothesis is μ_0.

Conditions Required for a Valid Large-Sample Hypothesis Test for μ

1. A random sample is selected from the target population.
2. The sample size n is large (i.e., $n \geq 30$). (Due to the Central Limit Theorem, this condition guarantees that the test statistic will be approximately normal regardless of the shape of the underlying probability distribution of the population.)

Once the test has been set up, the sampling experiment is performed and the test statistic calculated. The next box contains possible conclusions for a test of hypothesis, depending on the result of the sampling experiment.

Possible Conclusions for a Test of Hypothesis

1. If the calculated test statistic falls in the rejection region, reject H_0 and conclude that the alternative hypothesis H_a is true. State that you are rejecting H_0 at the α level of significance. Remember that the confidence is in the testing *process,* not the particular result of a single test.

2. If the test statistic does not fall in the rejection region, conclude that the sampling experiment does not provide sufficient evidence to reject H_0 at the α level of significance. [Generally, we will not "accept" the null hypothesis unless the probability β of a Type II error has been calculated (see Section 8.7).]

Example 8.4

Carrying Out a Hypothesis Test for μ—Mean Drug Response Time

Problem Refer to the neurological response-time test set up in Example 8.3 (p. 358). The sample of 100 drug-injected rats yielded the results (in seconds) shown in Table 8.3. At $\alpha = .01$, use these data to conduct the test of hypothesis,

$$H_0: \mu = 1.2$$
$$H_a: \mu \neq 1.2$$

Table 8.3		Drug Response Times for 100 Rats, Example 8.4							
1.90	2.17	0.61	1.17	0.66	1.86	1.41	1.30	0.70	0.56
2.00	1.27	0.98	1.55	0.64	0.60	1.55	0.93	0.48	0.39
0.86	1.19	0.79	1.37	1.31	0.85	0.71	1.21	1.23	0.89
1.84	0.80	0.64	1.08	0.74	0.93	1.71	1.05	1.44	0.42
0.70	0.54	1.40	1.06	0.54	0.17	0.98	0.89	1.28	0.68
0.98	1.14	1.16	1.64	1.16	1.01	1.09	0.77	1.58	0.99
0.57	0.27	0.51	1.27	1.81	0.88	0.31	0.92	0.93	1.66
0.21	0.79	0.94	0.45	1.19	1.60	0.14	0.99	1.08	1.57
0.55	1.65	0.81	1.00	2.55	1.96	1.31	1.88	1.51	1.48
0.61	0.05	1.21	0.48	1.63	1.45	0.22	0.49	1.29	1.40

Data Set: DRUGRAT

Solution To carry out the test, we need to find the values of $\bar{x}$ and s. (In this study, σ is obviously unknown. So we will use s to estimate σ.) These values, $\bar{x} = 1.05$ and $s = .5$, are shown (highlighted) on the MINITAB printout of Figure 8.6. Now we substitute these sample statistics into the test statistic and obtain

$$z = \frac{\bar{x} - 1.2}{\sigma_{\bar{x}}} = \frac{\bar{x} - 1.2}{\sigma/\sqrt{n}} \approx \frac{1.05 - 1.2}{.5/\sqrt{100}} = -3.0$$

The implication is that the sample mean, 1.05, is (approximately) three standard deviations below the null-hypothesized value of 1.2 in the sampling distribution of $\bar{x}$.

Recall from Example 8.3 that the rejection region for the test at $\alpha = .01$ is

Rejection region: $|z| > 2.575$

From Figure 8.5 (p. 359), you can see that $z = -3.0$ falls in the lower-tail rejection region, which consists of all values of $z < -2.575$. Therefore, this sampling experiment provides sufficient evidence to reject H_0 and conclude, at the $\alpha = .01$ level of significance, that the mean response time for drug-injected rats differs from the control mean of 1.2 seconds. It appears that the rats receiving an injection of the drug have a mean response time that is less than 1.2 seconds.

Figure 8.6
MINITAB descriptive statistics for fill amounts, Example 8.4

Descriptive Statistics: TIME

Variable	N	Mean	StDev	Minimum	Q1	Median	Q3	Maximum
TIME	100	1.0517	0.4982	0.0500	0.6650	0.9950	1.4000	2.5500

Look Back Three points about the test of hypothesis in this example apply to all statistical tests:

1. Since z is less than -2.575, it is tempting to state our conclusion at a significance level lower than $\alpha = .01$. We resist this temptation because the level of α is determined *before* the sampling experiment is performed. If we decide that we are willing to tolerate a 1% Type I error rate, the result of the sampling experiment should have no effect on that decision. *In general, the same data should not be used both to set up and to conduct the test.*

2. When we state our conclusion at the .01 level of significance, we are referring to the failure rate of the *procedure*, not the result of this particular test. We know that the test procedure will lead to the rejection of the null hypothesis only 1% of the time when in fact $\mu = 1.2$. *Therefore, when the test statistic falls into the rejection region, we infer that the alternative $\mu \neq 1.2$ is true and express our confidence in the procedure by quoting either the α level of significance or the $100(1 - \alpha)\%$ confidence level.*

3. Although a test may lead to a "statistically significant" result (i.e., rejecting H_0 at significance level α, as in the preceding test), it may not be "practically significant." For example, suppose the neurologist tested $n = 100,000$ drug-injected rats, resulting in

$\bar{x} = 1.1995$ and $s = .05$. Now a two-tailed hypothesis test of H_0: $\mu = 1.2$ results in a test statistic of

$$z = \frac{(1.1995 - 1.2)}{.05/\sqrt{100,000}} = -3.16$$

This result at $\alpha = .01$ leads us to "reject H_0" and conclude that the mean μ is "statistically different" from 1.2. However, for all practical purposes, the sample mean $\bar{x} = 1.1995$ and the hypothesized mean $\mu = 1.2$ are the same. Because the result is not "practically significant," the neurologist is not likely to consider a unit dose of the drug as an inhibitor to response time in rats. Consequently, *not all "statistically significant" results are "practically significant."*

Now Work Exercise 8.28

Exercises 8.20–8.37

Understanding the Principles

8.20 Explain the difference between a one-tailed and a two-tailed test.

8.21 What conditions are required for a valid large-sample test for μ?

8.22 For what values of the test statistic do you reject H_0? fail to reject H_0?

Learning the Mechanics

8.23 A random sample of 100 observations from a population with standard deviation 60 yielded a sample mean of 110.
  a. Test the null hypothesis that $\mu = 100$ against the alternative hypothesis that $\mu > 100$, using $\alpha = .05$. Interpret the results of the test.
  b. Test the null hypothesis that $\mu = 100$ against the alternative hypothesis that $\mu \neq 100$, using $\alpha = .05$. Interpret the results of the test.
  c. Compare the results of the two tests you conducted. Explain why the results differ.

8.24 A random sample of 64 observations produced the following summary statistics: $\bar{x} = .323$ and $s^2 = .034$.
  a. Test the null hypothesis that $\mu = .36$ against the alternative hypothesis that $\mu < .36$, using $\alpha = .10$.
  b. Test the null hypothesis that $\mu = .36$ against the alternative hypothesis that $\mu \neq .36$, using $\alpha = .10$. Interpret the result.

8.25 Suppose you are interested in conducting the statistical test of H_0: $\mu = 200$ against H_a: $\mu > 200$, and you have decided to use the following decision rule: Reject H_0 if the sample mean of a random sample of 100 items is more than 215. Assume that the standard deviation of the population is 80.
  a. Express the decision rule in terms of z.
  b. Find α, the probability of making a Type I error, by using this decision rule.

Applet Exercise 8.2

Use the applet entitled *Hypotheses Test for a Mean* to investigate the effect of the underlying distribution on the proportion of Type I errors. For this exercise, take $n = 100$, mean $= 50$, standard deviation $= 10$, null mean $= 50$, and alternative $<$.
  a. Select the normal distribution and run the applet several times without clearing. What happens to the

proportion of times the null hypothesis is rejected at the .05 level as the applet is run more and more times?
  b. Clear the applet and then repeat part **a**, using the right-skewed distribution. Do you get similar results? Explain.
  c. Describe the effect that the underlying distribution has on the probability of making a Type I error.

Applet Exercise 8.3

Use the applet entitled *Hypotheses Test for a Mean* to investigate the effect of the underlying distribution on the proportion of Type II errors. For this exercise, take $n = 100$, mean $= 50$, standard deviation $= 10$, null mean $= 52$, and alternative $<$.
  a. Select the normal distribution and run the applet several times without clearing. What happens to the proportion of times the null hypothesis is rejected at the .01 level as the applet is run more and more times? Is this what you would expect? Explain.
  b. Clear the applet and then repeat part **a**, using the right-skewed distribution. Do you get similar results? Explain.
  c. Describe the effect that the underlying distribution has on the probability of making a Type II error.

Applet Exercise 8.4

Use the applet entitled *Hypotheses Test for a Mean* to investigate the effect of the null mean on the probability of making a Type II error. For this exercise, take $n = 100$, mean $= 50$, standard deviation $= 10$, and alternative $<$ with the normal distribution. Set the null mean to 55 and run the applet several times without clearing. Record the proportion of Type II errors that occurred at the .01 level. Clear the applet and repeat for null means of 54, 53, 52, and 51. What can you conclude about the probability of a Type II error as the null mean gets closer to the actual mean? Can you offer a reasonable explanation for this behavior?

Applying the Concepts — Basic

8.26 **Speeding and young drivers.** Teenage drivers will inevitably be tempted to drive faster, even to exceed the speed limit, by their friends. Did you resist such temptations when you first began to drive? To gain insight into this phenomenon, psychologists from the United Kingdom conducted a survey of 258 student drivers and reported the results in the *British Journal of Educational Psychology* (Vol. 80,

2010). One of the variables of interest was the response to the question, "Are you confident that you can resist your friends' persuasion to drive faster?" Each response was measured on a 7-point scale, from 1 = "definitely no" to 7 = "definitely yes." The data were collected 5 months after the students had attended a safe-driver presentation. The psychologists reported a sample mean response of 4.98 and a sample standard deviation of 1.62. Suppose it is known that the true mean response of students who *do not* attend a safe-driver presentation is $\mu = 4.7$.

a. Set up the null and alternative hypotheses for testing whether the true mean student-driver response 5 months after a safe-driver presentation is larger than 4.7.

b. Calculate the test statistic for the hypothesis test.

c. Find the rejection region for the hypothesis test, using $\alpha = .05$.

d. State the appropriate conclusion, in the words of the problem.

e. Do the test results indicate that the safe-driver presentation was effective in helping students feel more confident that they can resist their friends' persuasion to drive faster? Explain.

f. The distribution of response scores (on a 7-point scale) for all student drivers is unlikely to be normal. Does this impact the validity of the hypothesis test? Why or why not?

8.27 **Teacher perceptions of child behavior.** *Developmental Psychology* (Mar. 2003) published a study on teacher perceptions of the behavior of elementary school children. Teachers rated the aggressive behavior of a sample of 11,160 New York City public school children by responding to the statement "This child threatens or bullies others in order to get his/her own way." Responses were measured on a scale ranging from 1 (*never*) to 5 (*always*). Summary statistics for the sample of 11,160 children were reported as $\bar{x} = 2.15$ and $s = 1.05$. Let μ represent the mean response for the population of all New York City public school children. Suppose you want to test $H_0: \mu = 3$ against $H_a: \mu \neq 3$.

a. In the words of the problem, define a Type I error and a Type II error.

b. Use the sample information to conduct the test at a significance level of $\alpha = .05$.

c. Conduct the test from part **b** at a significance level of $\alpha = .10$.

8.28 **Latex allergy in health care workers.** Refer to the *Current Allergy & Clinical Immunology* (March 2004) study of $n = 46$ hospital employees who were diagnosed with a latex allergy from exposure to the powder on latex gloves, presented in Exercise 7.13 (p. 307). The number of latex gloves used per week by the sampled workers is summarized as follows: $\bar{x} = 19.3$ and $s = 11.9$. Let μ represent the mean number of latex gloves used per week by all hospital employees. Consider testing $H_0: \mu = 20$ against $H_a: \mu < 20$.

a. Give the rejection region for the test at a significance level of $\alpha = .01$.

b. Calculate the value of the test statistic.

c. Use the results from parts **a** and **b** to draw the appropriate conclusion.

8.29 **Accounting and Machiavellianism.** *Behavioral Research in Accounting* (Jan. 2008) published a study of Machiavellian traits in accountants. (*Machiavellian* describes negative character traits that include manipulation, cunning, duplicity, deception, and bad faith.) A Machiavellian ("Mach")

rating score was determined for each in a sample of 122 purchasing managers with the following results: $\bar{x} = 99.6$, $s = 12.6$. (*Note:* Scores range from a low of 40 to a high of 160, with the theoretical neutral Mach rating score of 100.) A director of purchasing at a major firm claims that the true mean Mach rating score of all purchasing managers is 85.

a. Specify the null and alternative hypothesis for a test of the director's claim.

b. Define a Type I error for this test.

c. Interpret the value, $\alpha = .10$.

d. Give the rejection region for the test using $\alpha = .10$.

e. Find the value of the test statistic.

f. Use the result, part **e**, to make the appropriate conclusion.

g. Do you need to make any assumptions about the distribution of Mach rating scores for the population of all purchasing managers? Explain.

8.30 **Heart rate during laughter.** Laughter is often called "the best medicine," since studies have shown that laughter can reduce muscle tension and increase oxygenation of the blood. In the *International Journal of Obesity* (Jan. 2007), researchers at Vanderbilt University investigated the physiological changes that accompany laughter. Ninety subjects (18–34 years old) watched film clips designed to evoke laughter. During the laughing period, the researchers measured the heart rate (beats per minute) of each subject, with the following summary results: $\bar{x} = 73.5$, $s = 6$. It is well known that the mean resting heart rate of adults is 71 beats per minute.

a. Setup H_0 and H_a for testing whether the true mean heart rate during laughter exceeds 71 beats per minute.

b. If $\alpha = 05$, find the rejection region for the test.

c. Calculate the value of the test statistic.

d. Make the appropriate conclusion.

Applying the Concepts—Intermediate

8.31 **Dating and disclosure.** Refer to the *Journal of Adolescence* (April 2010) study of adolescents' disclosure of their dating and romantic relationships, Exercise 1.29 (p. 21). Recall that a sample of 222 high school students was recruited to participate in the study. One of the variables of interest was the level of disclosure to an adolescent's mother (measured on a 5-point scale, where 1 = "never tell," 2 = "rarely tell," 3 = "sometimes tell," 4 = "almost always tell," and 5 = "always tell"). The sampled high school students had a mean disclosure score of 3.26 and a standard deviation of .93. The researchers hypothesize that the true mean disclosure score of all adolescents will exceed 3. Do you believe the researchers? Conduct a formal test of hypothesis using $\alpha = .01$.

8.32 **Identifying type of urban land cover.** Geographers use remote-sensing data from satellite pictures to identify urban land cover as either grassland, commercial, or residential. In *Geographical Analysis* (Oct. 2006), researchers from Arizona State, Florida State, and Louisiana State Universities collaborated on a new method for analyzing remote-sensing data. A satellite photograph of an urban area was divided into 4 × 4-meter areas (called pixels). Of interest is a numerical measure of the distribution of the sizes of gaps, or holes, in the pixel, a property called *lacunarity*. The mean and standard deviation of the lacunarity measurements for a sample of 100 pixels randomly selected from a specific urban area are 225 and 20, respectively. It is known that the mean lacunarity measurement for all grassland pixels is 220. Do the data suggest that the area sampled is grassland? Test at $\alpha = .01$.

8.33 **Bone fossil study.** Humerus bones from the same species of animal tend to have approximately the same length-to-width ratios. When fossils of humerus bones are discovered, archeologists can often determine the species of animal by examining the length-to-width ratios of the bones. It is known that species A exhibits a mean ratio of 8.5. Suppose 41 fossils of humerus bones were unearthed at an archeological site in East Africa, where species A is believed to have lived. (Assume that the unearthed bones were all from the same unknown species.) The length-to-width ratios of the bones are listed in the following table and saved in the **BONES** file.

10.73	8.89	9.07	9.20	10.33	9.98	9.84	9.59
8.48	8.71	9.57	9.29	9.94	8.07	8.37	6.85
8.52	8.87	6.23	9.41	6.66	9.35	8.86	9.93
8.91	11.77	10.48	10.39	9.39	9.17	9.89	8.17
8.93	8.80	10.02	8.38	11.67	8.30	9.17	12.00
9.38							

a. Test whether the population mean ratio of all bones of this particular species differs from 8.5. Use $\alpha = .01$.
b. What are the practical implications of the test you conducted in part **a**?

8.34 **Cooling method for gas turbines.** During periods of high demand for electricity—especially in the hot summer months—the power output from a gas turbine engine can drop dramatically. One way to counter this drop in power is by cooling the inlet air to the turbine. An increasingly popular cooling method uses high-pressure inlet fogging. The performance of a sample of 67 gas turbines augmented with high-pressure inlet fogging was investigated in the *Journal of Engineering for Gas Turbines and Power* (Jan. 2005). One measure of performance is heat rate (kilojoules per kilowatt per hour). Heat rates for the 67 gas turbines, saved in the **GASTURBINE** file, are listed in the next table. Suppose that a standard gas turbine has, on average, a heat rate of 10,000 kJ/kWh.

a. Conduct a test to determine whether the mean heat rate of gas turbines augmented with high-pressure inlet fogging exceeds 10,000 kJ/kWh. Use $\alpha = .05$.
b. Identify a Type I error for this study. Identify a Type II error.

14622	13196	11948	11289	11964	10526	10387	10592	10460	10086
14628	13396	11726	11252	12449	11030	10787	10603	10144	11674
11510	10946	10508	10604	10270	10529	10360	14796	12913	12270
11842	10656	11360	11136	10814	13523	11289	11183	10951	9722
10481	9812	9669	9643	9115	9115	11588	10888	9738	9295
9421	9105	10233	10186	9918	9209	9532	9933	9152	9295
16243	14628	12766	8714	9469	11948	12414			

8.35 **Salaries of postgraduates.** The *Economics of Education Review* (Vol. 21, 2002) published a paper on the relationship between education level and earnings. The data for the research were obtained from the National Adult Literacy Survey of over 25,000 respondents. The survey revealed that males with a postgraduate degree had a mean salary of \$61,340 (with standard error $s_{\bar{x}} = \$2,185$) while females with a postgraduate degree had a mean of \$32,227 (with standard error $s_{\bar{x}} = \$932$).

a. The article reports that a 95% confidence interval for μ_M, the population mean salary of all males with postgraduate degrees, is (\$57,050, \$65,631). Based on this interval, is there evidence that μ_M differs from \$60,000? Explain.
b. Use the summary information to test the hypothesis that the true mean salary of males with postgraduate degrees differs from \$60,000. Take $\alpha = .05$. (*Note:* $s_{\bar{x}} = s/\sqrt{n}$)
c. Explain why the inferences in parts **a** and **b** agree.
d. The article reports that a 95% confidence interval for μ_F, the population mean salary of all females with postgraduate degrees, is (\$30,396, \$34,058). Based on this interval, is there evidence that μ_F differs from \$33,000? Explain.
e. Use the summary information to test the hypothesis that the true mean salary of females with postgraduate degrees differs from \$33,000. Take $\alpha = .05$ (*Note:* $s_{\bar{x}} = s/\sqrt{n}$.)
f. Explain why the inferences in parts **d** and **e** agree.

8.36 **Dissolved organic compound in lakes.** The level of dissolved oxygen in the surface water of a lake is vital to maintaining the lake's ecosystem. Environmentalists from the University of Wisconsin monitored the dissolved oxygen levels over time for a sample of 25 lakes in the state (*Aquatic Biology*, May 2010). To ensure a representative sample, the environmentalists focused on several lake characteristics, including dissolved organic compound (DOC). The DOC data (measured in grams per cubic meters) for the 25 lakes are listed in the accompanying table and saved in the **WISCLAKES** file. The population of Wisconsin lakes has a mean DOC value of 15 grams/m³. Use a hypothesis test (at $\alpha = .10$) to make an inference about whether the sample is representative of all Wisconsin lakes for the characteristic, dissolved organic compound.

Lake	DOC
Allequash	9.6
Big Muskellunge	4.5
Brown	13.2
Crampton	4.1
Cranberry Bog	22.6
Crystal	2.7
EastLong	14.7
Helmet	3.5
Hiawatha	13.6
Hummingbird	19.8
Kickapoo	14.3
Little Arbor Vitae	56.9
Mary	25.1
Muskellunge	18.4
Northgate Bog	2.7
Paul	4.2
Peter	30.2
Plum	10.3
Reddington Bog	17.6
Sparkling	2.4
Tenderfoot	17.3
Trout Bog	38.8
Trout Lake	3.0
Ward	5.8
West Long	7.6

Based on Langman, O. C., et al. "Control of dissolved oxygen in northern temperate lakes over scales ranging from minutes to days." *Aquatic Biology*, Vol. 9, May 2010 (Table 1).

Applying the Concepts—Advanced

8.37 **Social interaction of mental patients.** The *Community Mental Health Journal* (Aug. 2000) presented the results of a survey of over 6,000 clients of the Department of Mental Health and Addiction Services (DMHAS) in Connecticut. One of the many variables measured for each mental health patient was frequency of social interaction (on a 5-point scale, where 1 = very infrequently, 3 = occasionally, and 5 = very frequently). The 6,681 clients who were evaluated had a mean social interaction score of 2.95 with a standard deviation of 1.10.

a. Conduct a hypothesis test (at $\alpha = .01$) to determine whether the true mean social interaction score of all Connecticut mental health patients differs from 3.

b. Examine the results of the study from a practical view, and then discuss why "statistical significance" does not always imply "practical significance."

c. Because the variable of interest is measured on a 5-point scale, it is unlikely that the population of ratings will be normally distributed. Consequently, some analysts may perceive the test from part **a** to be invalid and search for alternative methods of analysis. Defend or refute this position.

8.4 Observed Significance Levels: *p*-Values

According to the statistical test procedure described in Section 8.2, the rejection region and, correspondingly, the value of α are selected prior to conducting the test and the conclusions are stated in terms of rejecting or not rejecting the null hypothesis. A second method of presenting the results of a statistical test reports the extent to which the test statistic disagrees with the null hypothesis and leaves to the reader the task of deciding whether to reject the null hypothesis. This measure of disagreement is called the *observed significance level* (or *p-value*) for the test.

> The **observed significance level**, or **p-value**, for a specific statistical test is the probability (assuming that H_0 is true) of observing a value of the test statistic that is at least as contradictory to the null hypothesis, and supportive of the alternative hypothesis, as the actual one computed from the sample data.

Recall testing $H_0: \mu = 2{,}400$ versus $H_a: \mu > 2{,}400$, where μ is the mean breaking strength of sewer pipe (Section 8.1). The value of the test statistic computed for the sample of $n = 50$ sections of sewer pipe was $z = 2.12$. Since the test is one tailed—that is, the alternative (research) hypothesis of interest is $H_a: \mu > 2{,}400$—values of the test statistic even more contradictory to H_0 than the one observed would be values larger than $z = 2.12$. Therefore, the observed significance level (*p*-value) for this test is

$$p\text{-value} = P(z \geq 2.12)$$

or, equivalently, the area under the standard normal curve to the right of $z = 2.12$. (See Figure 8.7.)

The area A in Figure 8.7 is given in Table IV in Appendix A as .4830. Therefore, the upper-tail area corresponding to $z = 2.12$ is

$$p\text{-value} = .5 - .4830 = .0170$$

Consequently, we say that these test results are "statistically significant"; that is, they disagree (rather strongly) with the null hypothesis $H_0: \mu = 2{,}400$ and favor $H_a: \mu > 2{,}400$. Hence, the probability of observing a z value as large as 2.12 is only .0170 if in fact the true value of μ is 2,400.

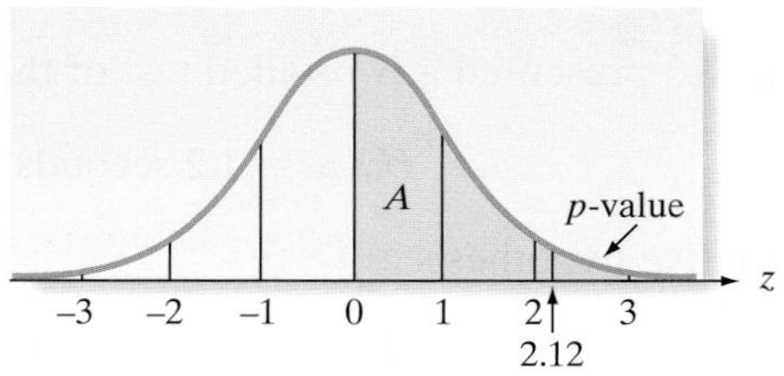

Figure 8.7

Finding the *p*-value for an upper-tailed test when $z = 2.12$

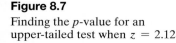

If you are inclined to select $\alpha = .05$ for this test, then you would reject the null hypothesis because the p-value for the test, .0170, is less than .05. In contrast, if you choose $\alpha = .01$, you would not reject the null hypothesis, because the p-value for the test is larger than .01. Thus, the use of the observed significance level is identical to the test procedure described in the preceding sections, except that the choice of α is left to you.

The steps for calculating the p-value corresponding to a test statistic for a population mean are given in the following box:

Steps for Calculating the p-Value for a Test of Hypothesis

1. Determine the value of the test statistic z corresponding to the result of the sampling experiment.

    a. If the test is one-tailed, the p-value is equal to the tail area beyond z in the same direction as the alternative hypothesis. Thus, if the alternative hypothesis is of the form $>$, the p-value is the area to the right of, or above, the observed z-value. Conversely, if the alternative is of the form $<$, the p-value is the area to the left of, or below, the observed z-value. (See Figure 8.8.)

    b. If the test is two tailed, the p-value is equal to twice the tail area beyond the observed z-value in the direction of the sign of z. That is, if z is positive, the p-value is twice the area to the right of, or above, the observed z value. Conversely, if z is negative, the p-value is twice the area to the left of, or below, the observed z-value. (See Figure 8.9.)

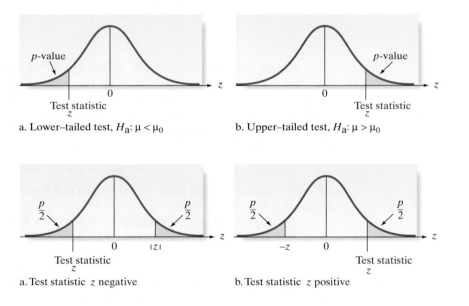

Figure 8.8
Finding the p-value for a one-tailed test

a. Lower–tailed test, H_a: $\mu < \mu_0$

b. Upper–tailed test, H_a: $\mu > \mu_0$

Figure 8.9
Finding the p-value for a two-tailed test: p-value $= 2(p/2)$

a. Test statistic z negative

b. Test statistic z positive

Example 8.5

Computing a p-Value—Test on Mean Drug Response Time

Problem Find the observed significance level for the test of the mean response time for drug-injected rats in Examples 8.3 and 8.4.

Solution Example 8.3 presented a two-tailed test of the hypothesis

$$H_0: \mu = 1.2 \text{ seconds}$$

against the alternative hypothesis

$$H_a: \mu \neq 1.2 \text{ seconds}$$

The observed value of the test statistic in Example 8.4 was $z = -3.0$, and any value of z less than -3.0 or greater than $+3.0$ (because this is a two-tailed test) would be even more contradictory to H_0. Therefore, the observed significance level for the test is

$$p\text{-value} = P(z < -3.0 \text{ or } z > +3.0) = P(|z| > 3.0)$$

Thus, we calculate the area below the observed z-value, $z = -3.0$, and double it. Consulting Table IV in Appendix A, we find that $P(z < -3.0) = .5 - .4987 = .0013$. Therefore, the p-value for this two-tailed test is

$$2P(z < -3.0) = 2(.0013) = .0026$$

This p-value can also be obtained with statistical software. The p-value is shown (highlighted) on the SAS printout of Figure 8.10. [*Note:* The value differs slightly due to rounding in our calculation of the test statistic.]

```
                              One Sample Z Test for a Mean


   Sample Statistics for TIME

        N           Mean          Std. Dev.        Std. Error
   -----------------------------------------------------------------
        100         1.05            0.50              0.05

   Hypothesis Test

      Null hypothesis:    Mean of TIME =  1.2
      Alternative:        Mean of TIME ^= 1.2

      With a specified known standard deviation of 0.5

           Z Statistic        Prob > Z
           -----------        --------
             -2.966            0.0030
```

Figure 8.10
SAS test of mean response time, Example 8.5

Look Back We can interpret this p-value as a strong indication that the mean reaction time of drug-injected rats differs from the control mean ($\mu \neq 1.2$), since we would observe a test statistic this extreme or more extreme only 26 in 10,000 times if the drug-injected mean were equal to the control mean ($\mu = 1.2$). The extent to which the mean differs from 1.2 could be better determined by calculating a confidence interval for μ.

Now Work Exercise 8.44

When publishing the results of a statistical test of hypothesis in journals, case studies, reports, etc., many researchers make use of p-values. Instead of selecting α beforehand and then conducting a test, as outlined in this chapter, the researcher computes (usually with the aid of a statistical software package) and reports the value of the appropriate test statistic and its associated p-value. It is left to the reader of the report to judge the significance of the result (i.e., the reader must determine, on the basis of the reported p-value, whether to reject the null hypothesis in favor of the alternative hypothesis. Usually, the null hypothesis is rejected if the observed significance level is *less than* the fixed significance level α chosen by the reader. The inherent advantage of reporting test results in this manner is twofold: (1) Readers are permitted to select the maximum value of α that they would be willing to tolerate if they actually carried out a standard test of hypothesis in the manner outlined in this chapter, and (2) a measure of the degree of significance of the result (i.e., the p-value) is provided.

Ethics IN Statistics

Selecting the value of α *after* computing the observed significance level (*p*-value) in order to guarantee a preferred conclusion is considered *unethical statistical practice.*

Reporting Test Results as *p*-Values: How to Decide Whether to Reject H_0

a. Choose the maximum value of α that you are willing to tolerate.

b. If the observed significance level (*p*-value) of the test is less than the chosen value of α, reject the null hypothesis. Otherwise, do not reject the null hypothesis.

Example 8.6
Using *p*-Values— Test of Mean Hospital Length of Stay

Problem The lengths of stay (in days) for 100 randomly selected hospital patients, first presented in Table 7.1, are reproduced in Table 8.4. Suppose we want to test the hypothesis that the true mean length of stay (LOS) at the hospital is less than 5 days; that is,

H_0: $\mu = 5$ (Mean LOS is 5 days.)

H_a: $\mu < 5$ (Mean LOS is less than 5 days.)

Assuming that $\sigma = 3.68$, use the data in the table to conduct the test at $\alpha = .05$.

Table 8.4 **Lengths of Stay for 100 Hospital Patients**

2	3	8	6	4	4	6	4	2	5
8	10	4	4	4	2	1	3	2	10
1	3	2	3	4	3	5	2	4	1
2	9	1	7	17	9	9	9	4	4
1	1	1	3	1	6	3	3	2	5
1	3	3	14	2	3	9	6	6	3
5	1	4	6	11	22	1	9	6	5
2	2	5	4	3	6	1	5	1	6
17	1	2	4	5	4	4	3	2	3
3	5	2	3	3	2	10	2	4	2

Data Set: HOSPLOS

Solution The data were entered into a computer and MINITAB was used to conduct the analysis. The MINITAB printout for the lower-tailed test is displayed in Figure 8.11. Both the test statistic, $z = -1.28$, and the *p*-value of the test, $p = .101$, are highlighted on the MINITAB printout. Since the *p*-value exceeds our selected α value, $\alpha = .05$, we cannot reject the null hypothesis. Hence, there is insufficient evidence (at $\alpha = .05$) to conclude that the true mean LOS at the hospital is less than 5 days.

Figure 8.11
MINITAB lower-tailed test of mean LOS, Example 8.6

```
One-Sample Z: LOS

Test of mu = 5 vs < 5
The assumed standard deviation = 3.68


                                        95% Upper
Variable    N    Mean   StDev  SE Mean    Bound      Z      P
LOS        100   4.530  3.678   0.368     5.135   -1.28  0.101
```

Look Back A hospital administrator, desirous of a mean length of stay less than 5 days, may be tempted to select an α level that leads to a rejection of the null hypothesis *after* determining *p*-value = .101. There are two reasons one should resist this temptation. First, the administrator would need to select $\alpha > .101$ (say, $\alpha = .15$) in order to conclude that H_a: $\mu < 5$ is true. A Type I error rate of 15% is considered too large by most researchers. Second, and more importantly, such a strategy is considered unethical statistical practice (See marginal note, p. 369.).

Now Work Exercise 8.51

Note: Some statistical software packages (e.g., SPSS) will conduct only two-tailed tests of hypothesis. For these packages, you obtain the *p*-value for a one-tailed test as shown in the next box.

Converting a Two-Tailed p-Value from a Printout to a One-Tailed p-Value:

$$p = \frac{\text{Reported } p\text{-value}}{2} \text{ if } \begin{cases} H_a \text{ is of form } > \text{ and } z \text{ is positive} \\ H_a \text{ is of form } < \text{ and } z \text{ is negative} \end{cases}$$

$$p = 1 - \left(\frac{\text{Reported } p\text{-value}}{2}\right) \text{ if } \begin{cases} H_a \text{ is of form } > \text{ and } z \text{ is negative} \\ H_a \text{ is of form } < \text{ and } z \text{ is positive} \end{cases}$$

Statistics IN Action | **Revisited** | ## Testing a Population Mean in the KLEENEX® Survey

Refer to Kimberly-Clark Corporation's survey of 250 people who kept a count of their use of KLEENEX® tissues in diaries (p. 359). We want to test the claim made by marketing experts that μ, the average number of tissues used by people with colds, is greater than 60 tissues. That is, we want to test

$$H_0: \mu = 60 \qquad H_a: \mu > 60$$

We will select $\alpha = .05$ as the level of significance for the test.

The survey results for the 250 sampled KLEENEX® users are stored in the **TISSUES** data file. A MINITAB

analysis of the data yielded the printout displayed in Figure SIA8.1.

The observed significance level of the test, highlighted on the printout, is p-value = 000. Since this p-value is less than $\alpha = .05$, we have sufficient evidence to reject H_0; therefore, we conclude that the mean number of tissues used by a person with a cold is greater than 60 tissues. This result supports the company's decision to put 70 tissues in a standard-sized box of KLEENEX.

Data Set: TISSUES

Figure SIA8.1

MINITAB test of $\mu = 60$ for KLEENEX survey

```
Test of mu = 60 vs > 60

                                        95% Lower
Variable    N    Mean   StDev  SE Mean    Bound      T      P
NUMUSED    250  68.68   25.03    1.58     66.06    5.48   0.000
```

Exercises 8.38–8.56

Understanding the Principles

8.38 How does the observed significance level (p-value) of a test differ from the value of α?

8.39 In general, do large p-values or small p-values support the alternative hypothesis H_a?

8.40 If a hypothesis test using $\alpha = .05$ were conducted, for which of the following p-values would the null hypothesis be rejected?
  a. .06   **b.** .10
  c. .01   **d.** .001
  e. .251   **f.** .042

8.41 For each pair consisting of α and an observed significance level (p-value), indicate whether the null hypothesis would be rejected.
  a. $\alpha = .05, p\text{-value} = .10$
  b. $\alpha = .10, p\text{-value} = .05$
  c. $\alpha = .01, p\text{-value} = .001$
  d. $\alpha = .025, p\text{-value} = .05$
  e. $\alpha = .10, p\text{-value} = .45$

Learning the Mechanics

8.42 An analyst tested the null hypothesis $\mu \geq 20$ against the alternative hypothesis $\mu < 20$. The analyst reported a p-value of .06. What is the smallest value of α for which the null hypothesis would be rejected?

8.43 In a test of $H_0: \mu = 100$ against $H_a: \mu > 100$, the sample data yielded the test statistic $z = 2.17$. Find the p-value for the test.

8.44 In a test of $H_0: \mu = 100$ against $H_a: \mu \neq 100$, the sample data yielded the test statistic $z = 2.17$. Find the p-value for the test.
  [NW]

8.45 In a test of the hypothesis $H_0: \mu = 50$ versus $H_a: \mu > 50$, a sample of $n = 100$ observations possessed mean $\bar{x} = 49.4$ and standard deviation $s = 4.1$. Find and interpret the p-value for this test.

8.46 In a test of the hypothesis $H_0: \mu = 10$ versus $H_a: \mu \neq 10$, a sample of $n = 50$ observations possessed mean $\bar{x} = 10.7$ and standard deviation $s = 3.1$. Find and interpret the p-value for this test.

8.47 Consider a test of $H_0: \mu = 75$ performed with the computer. SPSS reports a two-tailed p-value of .1032. Make the appropriate conclusion for each of the following situations:
  a. $H_a: \mu < 75, z = -1.63, \alpha = .05$
  b. $H_a: \mu < 75, z = 1.63, \alpha = .10$
  c. $H_a: \mu > 75, z = 1.63, \alpha = .10$
  d. $H_a: \mu \neq 75, z = -1.63, \alpha = .01$

Applying the Concepts—Basic

8.48 **Speeding and young drivers.** Refer to the *British Journal of Educational Psychology* (Vol. 80, 2010) study of teenage drivers' temptation to exceed the speed limit, Exercise 8.26 (p. 364). Recall that in a survey of 258 student drivers taken 5 months after the students had attended a safe-driver presentation, responses (measured on a 7-point scale) to the question "Are you confident that you can resist your friends' persuasion to drive faster?" had a mean of 4.98 and a standard deviation of 1.62. A MINITAB printout for testing $H_0: \mu = 4.7$ against $H_a: \mu > 4.7$ is shown below.

a. Locate the p-value of the test on the printout. What does the p-value measure?

b. Give the appropriate conclusion for $\alpha = .05$. Does this conclusion agree with your answer to Exercise 8.26d?

```
Test of mu = 4.7 vs > 4.7

                              95% Lower
  N   Mean  StDev  SE Mean    Bound     T      P
 258  4.980  1.620  0.101     4.814    2.78  0.003
```

8.49 **Teacher perceptions of child behavior.** Refer to the *Developmental Psychology* (Mar. 2003) study on the aggressive behavior of elementary school children, presented in Exercise 8.27 (p. 365). Recall that you tested $H_0: \mu = 3$ against $H_a: \mu \neq 3$, where μ is the mean level of aggressiveness for the population of all New York City public school children, as perceived by their teachers and based on the summary statistics $n = 11,160$, $\bar{x} = 2.15$, and $s = 1.05$.

a. Compute the p-value of the test.

b. Compare the p-value with $\alpha = .10$ and make the appropriate conclusion.

8.50 **Latex allergy in health care workers.** Refer to the *Current Allergy & Clinical Immunology* (March 2004) study of latex allergy in health care workers, presented in Exercise 8.28 (p. 365). Recall that you tested $H_0: \mu = 20$ against $H_a: \mu < 20$, where μ is the mean number of latex gloves used per week by all hospital employees, based on the summary statistics $n = 46$, $\bar{x} = 19.3$, and $s = 11.9$.

a. Compute the p-value of the test.

b. Compare the p-value with $\alpha = .01$ and make the appropriate conclusion.

8.51 **Bone fossil study.** In Exercise 8.33 (p. 366), you tested [NW] $H_0: \mu = 8.5$ versus $H_a: \mu \neq 8.5$, where μ is the population mean length-to-width ratio of humerus bones of a particular species of animal. A SAS printout for the hypothesis test is shown below. Locate the p-value on the printout and interpret it.

```
Sample Statistics for LWRATIO

    N         Mean       Std. Dev.      Std. Error
-------------------------------------------------------
    41        9.26         1.20            0.19

Hypothesis Test

    Null hypothesis:     Mean of LWRATIO =  8.5
    Alternative:         Mean of LWRATIO ^= 8.5

With a specified known standard deviation of 1.2

          Z Statistic      Prob > Z
          -----------      --------
            4.042           <.0001
```

8.52 **Emotional empathy in young adults.** According to a theory in psychology, young female adults show more emotional empathy towards others than do males. The *Journal of Moral Education* (June 2010) tested this theory by examining the attitudes of a sample of 30 female college students. Each student completed the Ethic of Care Interview, which consisted of a series of statements on empathy attitudes. For the statement on emotional empathy (e.g., "I often have tender, concerned feelings for people less fortunate than me"), the sample mean response was 3.28. Assume the population standard deviation for females is .5. [*Note*: Empathy scores ranged from 0 to 4, where 0 = "never" and 4 = "always".] Suppose it is known that male college students have an average emotional empathy score of $\mu = 3$.

a. Specify the null and alternative hypothesis for testing whether female college students score higher than 3.0 on the emotional empathy scale.

b. Compute the test statistic.

c. Find the observed significance level (p-value) of the test.

d. At $\alpha = .01$, what is the appropriate conclusion?

e. How small of an α-value can you choose and still have sufficient evidence to reject the null hypothesis?

Applying the Concepts—Intermediate

8.53 **Buy-side vs. sell-side analysts' earnings forecasts.** The *Financial Analysts Journal* (Jul/Aug 2008) presented a study of earnings forecasts of buy-side and sell-side analysts. Buy-side analysts differ from sell-side analysts on a variety of factors, including scope of industry coverage, sources of information used, and target audience. Data were collected on 3,526 forecasts made by buy-side analysts and 58,562 forecasts made by sell-side analysts and the relative absolute forecast error was determined for each. A positive forecast error indicates that the analyst is overestimating earnings, while a negative forecast error implies that the analyst is underestimating earnings. Summary statistics for the forecast errors in the two samples are reproduced in the following table.

	Buy-Side Analysts	Sell-Side Analysts
Mean	0.85	−0.05
Standard Deviation	1.93	0.85

Based on Groysberg, B., Healy, P., and Chapman, C. "Buy-side vs. sell-side analysts' earning forecasts." *Financial Analysts Journal*, Vol. 64, No. 4, Jul/Aug 2008.

a. Conduct a test (at $\alpha = .01$) to determine if the true mean forecast error for buy-side analysts is positive. Use the observed significance level (p-value) of the test to make your decision and state your conclusion in the words of the problem.

b. Conduct a test (at $\alpha = .01$) to determine if the true mean forecast error for sell-side analysts is negative. Use the observed significance level (p-value) of the test to make your decision and state your conclusion in the words of the problem.

8.54 **Colored string preferred by chickens.** Refer to the *Applied Animal Behaviour Science* (October 2000) study of domestic chickens exposed to a pecking stimulus, presented in Exercise 7.21 (p. 308). Recall that the average number of pecks a chicken takes at a white string over a specified time interval is known to be $\mu = 7.5$ pecks. In an experiment in which 72 chickens were exposed to blue string, the average number of pecks was $\bar{x} = 1.13$ pecks, with a standard deviation of $s = 2.21$ pecks.

a. On average, are chickens more apt to peck at white string than at blue string? Conduct the appropriate test of hypothesis, using $\alpha = .05$.

b. Compare your answer to part **a** with your answer to Exercise 7.21**b**.

c. Find the *p*-value for the test and interpret it.

8.55 Feminizing human faces. Research published in *Nature* (August 27, 1998) revealed that people are more attracted to "feminized" faces, regardless of gender. In one experiment, 50 human subjects viewed both a Japanese female and a Caucasian male face on a computer. Using special computer graphics, each subject could morph the faces (by making them more feminine or more masculine) until they attained the "most attractive" face. The level of feminization *x* (measured as a percentage) was measured.

a. For the Japanese female face, $\bar{x} = 10.2\%$ and $s = 31.3\%$. The researchers used this sample information to test the null hypothesis of a mean level of feminization equal to 0%. Verify that the test statistic is equal to 2.3.

b. Refer to part **a.** The researchers reported the *p*-value of the test as $p \approx .02$. Verify and interpret this result.

c. For the Caucasian male face, $\bar{x} = 15.0\%$ and $s = 25.1\%$. The researchers reported the test statistic (for the test of the null hypothesis stated in part **a**) as 4.23, with an associated *p*-value of approximately 0. Verify and interpret these results.

Applying the Concepts—Advanced

8.56 Ages of cable TV shoppers. In a paper presented at the 2000 Conference of the International Association for Time Use Research, professor Margaret Sanik of Ohio State University reported the results of her study on American cable TV viewers who purchase items from one of the home-shopping channels. She found that the average age of these cable TV shoppers was 51 years. Suppose you want to test the null hypothesis $H_0: \mu = 51$, using a sample of $n = 50$ cable TV shoppers.

a. Find the *p*-value of a two-tailed test if $\bar{x} = 52.3$ and $s = 7.1$.

b. Find the *p*-value of an upper-tailed test if $\bar{x} = 52.3$ and $s = 7.1$.

c. Find the *p*-value of a two-tailed test if $\bar{x} = 52.3$ and $s = 10.4$.

d. For each of the tests in parts **a–c**, give a value of α that will lead to a rejection of the null hypothesis.

e. If $\bar{x} = 52.3$, give a value of *s* that will yield a *p*-value of .01 or less for a one-tailed test.

8.5 Test of Hypothesis about a Population Mean: Student's *t*-Statistic

Most water-treatment facilities monitor the quality of their drinking water on an hourly basis. One variable monitored is pH, which measures the degree of alkalinity or acidity in the water. A pH below 7.0 is acidic, one above 7.0 is alkaline, and a pH of 7.0 is neutral. One water-treatment plant has a target pH of 8.5. (Most try to maintain a slightly alkaline level.) The mean and standard deviation of 1 hour's test results, based on 17 water samples at this plant, are

$$\bar{x} = 8.42 \qquad s = .16$$

Does this sample provide sufficient evidence that the mean pH level in the water differs from 8.5?

This inference can be placed in a test-of-hypothesis framework. We establish the target pH as the null hypothesized value and then utilize a two-tailed alternative that the true mean pH differs from the target:

$$H_0: \mu = 8.5 \text{ (Mean pH level is 8.5.)}$$

$$H_a: \mu \neq 8.5 \text{ (Mean pH level differs from 8.5.)}$$

Recall from Section 7.3 that when we are faced with making inferences about a population mean from the information in a small sample, two problems emerge:

1. The normality of the sampling distribution for $\bar{x}$ does not follow from the Central Limit Theorem when the sample size is small. We must assume that the distribution of measurements from which the sample was selected is approximately normally distributed in order to ensure the approximate normality of the sampling distribution of $\bar{x}$.

2. If the population standard deviation σ is unknown, as is usually the case, then we cannot assume that s will provide a good approximation for σ when the sample size is small. Instead, we must use the *t*-distribution rather than the standard normal *z*-distribution to make inferences about the population mean μ.

Therefore, as the test statistic of a small-sample test of a population mean, we use the *t*-statistic:

$$\text{Test statistic: } t = \frac{\bar{x} - \mu_0}{s/\sqrt{n}} = \frac{\bar{x} - 8.5}{s/\sqrt{n}}$$

In this equation, μ_0 is the null-hypothesized value of the population mean μ. In the example here, $\mu_0 = 8.5$.

To find the rejection region, we must specify the value of α, the probability that the test will lead to rejection of the null hypothesis when it is true, and then consult the *t* table (Table VI of Appendix A). With $\alpha = .05$, the two-tailed rejection region is

$$\textit{Rejection region: } t_{\alpha/2} = t_{.025} = 2.120 \text{ with } n - 1 = 16 \text{ degrees of freedom}$$
$$\text{Reject } H_0 \text{ if } t < -2.120 \text{ or } t > 2.120$$

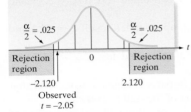

$\frac{\alpha}{2} = .025$ $\frac{\alpha}{2} = .025$

Rejection region   0   Rejection region

−2.120   2.120

Observed
$t = -2.05$

Figure 8.12
Two-tailed rejection region for small-sample *t*-test

The rejection region is shown in Figure 8.12.

We are now prepared to calculate the test statistic and reach a conclusion:

$$t = \frac{\bar{x} - \mu_0}{s/\sqrt{n}} = \frac{8.42 - 8.50}{.16/\sqrt{17}} = \frac{-.08}{.039} = -2.05$$

Since the calculated value of *t* does not fall into the rejection region (Figure 8.12), we cannot reject H_0 at the $\alpha = .05$ level of significance. Thus, the water-treatment plant should not conclude that the mean pH differs from the 8.5 target on the basis of the sample evidence.

It is interesting to note that the calculated *t* value, −2.05, is *less than* the .05-level *z*-value, −1.96. The implication is that if we had *incorrectly* used a *z* statistic for this test, we would have rejected the null hypothesis at the .05 level and concluded that the mean pH level differs from 8.5. The important point is that the statistical procedure to be used must always be closely scrutinized and all the assumptions understood. Many statistical lies are the result of misapplications of otherwise valid procedures.

The technique for conducting a small-sample test of hypothesis about a population mean is summarized in the following boxes:

Small-Sample Test of Hypothesis about μ Based on a Student's *t*-Statistic

One-Tailed Test	**Two-Tailed Test**
$H_0: \mu = \mu_0$	$H_0: \mu = \mu_0$
$H_a: \mu < \mu_0 \text{ (or } H_a: \mu > \mu_0)$	$H_a: \mu \neq \mu_0$
Test statistic: $t = \dfrac{\bar{x} - \mu_0}{s/\sqrt{n}}$	Test statistic: $t = \dfrac{\bar{x} - \mu_0}{s/\sqrt{n}}$
Rejection region: $t < -t_\alpha$ (or $t > t_\alpha$ when $H_a: \mu > \mu_0$)	Rejection region: $t < -t_{\alpha/2}$ or $t > t_{\alpha/2}$

where t_α and $t_{\alpha/2}$ are based on $(n-1)$ degrees of freedom

> **Conditions Required for a Valid Small-Sample Hypothesis Test for μ**
> 1. A random sample is selected from the target population.
> 2. The population from which the sample is selected has a distribution that is approximately normal.

Example 8.7

A Small–Sample Test for μ— Does a New Engine Meet Air–Pollution Standards?

Problem A major car manufacturer wants to test a new engine to determine whether it meets new air-pollution standards. The mean emission μ of all engines of this type must be less than 20 parts per million of carbon. Ten engines are manufactured for testing purposes, and the emission level of each is determined. The data (in parts per million) are listed in Table 8.5. Do the data supply sufficient evidence to allow the manufacturer to conclude that this type of engine meets the pollution standard? Assume that the manufacturer is willing to risk a Type I error with probability $\alpha = .01$.

Table 8.5	**Emission Levels for Ten Engines**								
15.6	16.2	22.5	20.5	16.4	19.4	19.6	17.9	12.7	14.9

Data Set: EMISSIONS

Solution The manufacturer wants to support the research hypothesis that the mean emission level μ for all engines of this type is less than 20 parts per million. The elements of this small-sample one-tailed test are as follows:

$$H_0: \mu = 20 \text{ (Mean emission level is 20 ppm.)}$$
$$H_a: \mu < 20 \text{ (Mean emission level is less than 20 ppm.)}$$

Test statistic: $t = \dfrac{\bar{x} - 20}{s/\sqrt{n}}$

Rejection region: For $\alpha = .01$ and df $= n - 1 = 9$, the one-tailed rejection region (see Figure 8.13) is $t < -t_{.01} = -2.821$.

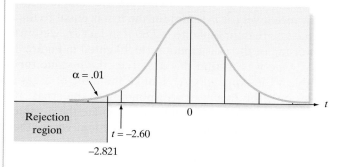

Figure 8.13

A *t*-distribution with 9 df and the rejection region for Example 8.7

Assumption: The relative frequency distribution of the population of emission levels for all engines of this type is approximately normal. Based on the shape of the MINITAB stem-and-leaf display of the data shown in Figure 8.14, this assumption appears to be reasonably satisfied.

To calculate the test statistic, we analyzed the **EMISSIONS** data with MINITAB. The MINITAB printout is shown at the bottom of Figure 8.14. From the printout, we

---

Stem-and-Leaf Display: E-LEVEL

```
Stem-and-leaf of E-LEVEL   N  = 10
Leaf Unit = 1.0


   1    1   2
   3    1   45
  (3)   1   667
   4    1   99
   2    2   0
   1    2   2
```

One-Sample T: E-LEVEL

```
Test of mu = 20 vs < 20
```

					95% Upper		
Variable	N	Mean	StDev	SE Mean	Bound	T	P
E-LEVEL	10	17.5700	2.9522	0.9336	19.2814	-2.60	0.014

Figure 8.14

MINITAB analysis of 10 emission levels, Example 8.7

obtain $\bar{x} = 17.57$ and $s = 2.95$. Substituting these values into the test statistic formula and rounding, we get

$$t = \frac{\bar{x} - 20}{s/\sqrt{n}} = \frac{17.57 - 20}{2.95/\sqrt{10}} = -2.60$$

Since the calculated t falls outside the rejection region (see Figure 8.13), the manufacturer cannot reject H_0. There is insufficient evidence to conclude that $\mu < 20$ parts per million and that the new type of engine meets the pollution standard.

Look Back Are you satisfied with the reliability associated with this inference? The probability is only $\alpha = .01$ that the test would support the research hypothesis if in fact it were false.

Now Work Exercise 8.63a–b

Example 8.8

The *p*-Value for a Small-Sample Test of μ

Problem Find the observed significance level for the test described in Example 8.7. Interpret the result.

Solution The test of Example 8.7 was a lower-tailed test: $H_0: \mu = 20$ versus $H_a: \mu < 20$. Since the value of t computed from the sample data was $t = -2.60$, the observed significance level (or *p*-value) for the test is equal to the probability that t would assume a value less than or equal to -2.60 if in fact H_0 were true. This is equal to the area in the lower tail of the t-distribution (highlighted in Figure 8.15).

One way to find this area (i.e., the *p*-value for the test) is to consult the t-table (Table VI in Appendix A). Unlike the table of areas under the normal curve, Table VI gives only the t values corresponding to the areas .100, .050, .025, .010, .005, .001, and .0005. Therefore, we can only approximate the *p*-value for the test. Since the observed t value was based on nine degrees of freedom, we use the df $= 9$ row in Table VI and move across the row until we reach the t values that are closest to the observed $t = -2.60$. [*Note:* We ignore the minus sign.] The t values corresponding to *p*-values of .010 and .025

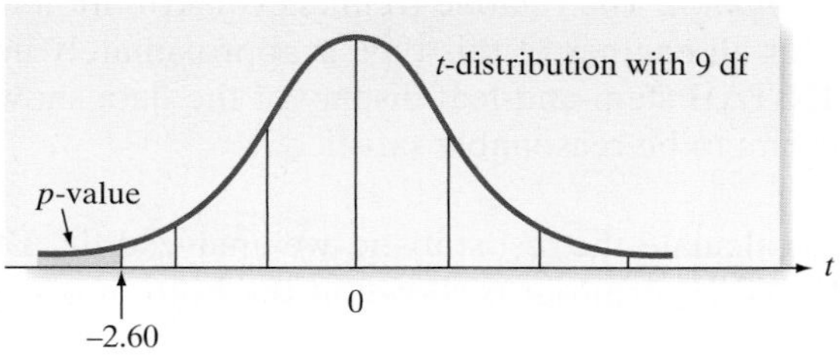

Figure 8.15

The observed significance level for the test of Example 8.7

are 2.821 and 2.262, respectively. Since the observed *t* value falls between $t_{.010}$ and $t_{.025}$, the *p*-value for the test lies between .010 and .025. In other words, $.010 < p\text{-value} < .025$. Thus, we would reject the null hypothesis $H_0: \mu = 20$ parts per million for any value of α larger than .025 (the upper bound of the *p*-value).

A second, more accurate, way to obtain the *p*-value is to use a statistical software package to conduct the test of hypothesis. Both the test statistic (-2.60) and the *p*-value (.014) are highlighted on the MINITAB printout of Figure 8.14.

You can see that the actual *p*-value of the test falls within the bounds obtained from Table VI. Thus, the two methods agree and we will reject $H_0: \mu = 20$ in favor of $H_a: \mu < 20$ for any α level larger than .025.

Now Work Exercise 8.63c

Small-sample inferences typically require more assumptions and provide less information about the population parameter than do large-sample inferences. Nevertheless, the *t*-test is a method of testing a hypothesis about a population mean of a normal distribution when only a small number of observations is available.

> **What Can Be Done If the Population Relative Frequency Distribution Departs Greatly from Normal?**
>
> *Answer:* Use one of the nonparametric statistical methods of Chapter 14.

Exercises 8.57–8.76

Understanding the Principles

8.57 In what ways are the distributions of the *z*-statistic and *t*-statistic alike? How do they differ?

8.58 Under what circumstances should you use the *t*-distribution in testing a hypothesis about a population mean?

Learning the Mechanics

8.59 For each of the following rejection regions, sketch the sampling distribution of *t* and indicate the location of the rejection region on your sketch:
  a. $t > 1.440$, where df $= 6$
  b. $t < -1.782$, where df $= 12$
  c. $t < -2.060$ or $t > 2.060$, where df $= 25$

8.60 For each of the rejection regions defined in Exercise 8.59, what is the probability that a Type I error will be made?

8.61 A random sample of *n* observations is selected from a normal population to test the null hypothesis that $\mu = 10$. Specify the rejection region for each of the following combinations of H_a, α, and *n*:
  a. $H_a: \mu \neq 10; \alpha = .05; n = 14$
  b. $H_a: \mu > 10; \alpha = .01; n = 24$
  c. $H_a: \mu > 10; \alpha = .10; n = 9$
  d. $H_a: \mu < 10; \alpha = .01; n = 12$
  e. $H_a: \mu \neq 10; \alpha = .10; n = 20$
  f. $H_a: \mu < 10; \alpha = .05; n = 4$

8.62 The following sample of six measurements was randomly selected from a normally distributed population: 1, 3, −1, 5, 1, 2.
  a. Test the null hypothesis that the mean of the population is 3 against the alternative hypothesis, $\mu < 3$. Use $\alpha = .05$.

  b. Test the null hypothesis that the mean of the population is 3 against the alternative hypothesis, $\mu \neq 3$. Use $\alpha = .05$.
  c. Find the observed significance level for each test.

8.63 A sample of five measurements, randomly selected from a
  [NW] normally distributed population, resulted in the following summary statistics: $\bar{x} = 4.8, s = 1.3$.
  a. Test the null hypothesis that the mean of the population is 6 against the alternative hypothesis, $\mu < 6$. Use $\alpha = .05$.
  b. Test the null hypothesis that the mean of the population is 6 against the alternative hypothesis, $\mu \neq 6$. Use $\alpha = .05$.
  c. Find the observed significance level for each test.

8.64 Suppose you conduct a *t*-test for the null hypothesis $H_0: \mu = 1,000$ versus the alternative hypothesis $H_a: \mu > 1,000$, based on a sample of 17 observations. The test results are $t = 1.89, p\text{-value} = .038$.
  a. What assumptions are necessary for the validity of this procedure?
  b. Interpret the results of the test.
  c. Suppose the alternative hypothesis had been the two-tailed $H_a: \mu \neq 1,000$. If the *t*-statistic were unchanged, what would the *p*-value be for this test? Interpret the *p*-value for the two-tailed test.

Applying the Concepts—Basic

8.65 **Lobster trap placement.** Refer to the *Bulletin of Marine Science* (April 2010) observational study of lobster trap placement by teams fishing for the red spiny lobster in Baja California Sur, Mexico, Exercise 7.35 (p. 317). Trap spacing measurements (in meters) for a sample of seven teams of red spiny lobster fishermen are reproduced in the accompanying table (and saved in the **TRAPSPACE** file). Let μ represent the average of the trap spacing measurements for the population of red spiny lobster fishermen fishing in Baja California Sur, Mexico. In Exercise 7.35 you computed the mean and standard deviation of the sample measurements to be $\bar{x} = 89.9$ meters and $s = 11.6$ meters, respectively. Suppose you want to determine if the true value of μ differs from 95 meters.

93	99	105	94	82	70	86

Based on Shester, G. G. "Explaining catch variation among Baja California lobster fishers through spatial analysis of trap-placement decisions." *Bulletin of Marine Science*, Vol. 86, No. 2, April 2010 (Table 1), pp. 479–498.

a. Specify the null and alternative hypothesis for this test.
b. Since $\bar{x} = 89.9$ is less than 95, a fisherman wants to reject the null hypothesis. What are the problems with using such a decision rule?
c. Compute the value of the test statistic.
d. Find the approximate p-value of the test.
e. Select a value of α, the probability of a Type I error. Interpret this value in the words of the problem.
f. Give the appropriate conclusion, based on the results of parts **d** and **e**.
g. What conditions must be satisfied for the test results to be valid?
h. In Exercise 7.35 you found a 95% confidence interval for μ. Does the interval support your conclusion in part **f**?

8.66   Reading Japanese books. Refer to the *Reading in a Foreign Language* (April 2004) experiment to improve the Japanese reading comprehension levels of University of Hawaii students presented in Exercise 2.33 (p. 46). Recall that 14 students participated in a 10-week extensive reading program in a second-semester Japanese course. The data on number of books read by each student are saved in the **JAPANESE** file. A MINITAB printout of the data analysis is shown below.

One-Sample T: BOOKS

Test of mu = 25 vs > 25

Variable	N	Mean	StDev	SE Mean	95% Lower Bound	T	P
BOOKS	14	31.64	10.49	2.80	26.68	2.37	0.017

a. State the null and alternative hypotheses for determining whether the average number of books read by all students who participated in the extensive reading program exceeds 25.
b. Find the rejection region for the test, using $\alpha = .05$.
c. Compute the test statistic.
d. State the appropriate conclusion for the test.
e. What conditions are required for the test results to be valid?
f. Locate the p-value on the MINITAB printout and use it to test the hypothesis. Your conclusion should agree with your answer in part **d**.

8.67   Dental anxiety study. Refer to the *BMC Oral Health* (Vol. 9, 2009) study of adults who completed the Dental Anxiety Scale, presented in Exercise 5.37 (p. 242). Recall that scores range from 0 (no anxiety) to 25 (extreme anxiety). Summary statistics for the scores of 15 adults who completed the questionnaire are $\bar{x} = 10.7$ and $s = 3.6$. Conduct a test of hypothesis to determine whether the mean Dental Anxiety Scale score for the population of college students differs from $\mu = 11$. Use $\alpha = .05$.

8.68   Crab spiders hiding on flowers. Refer to the *Behavioral Ecology* (Jan. 2005) experiment on crab spiders' use of camouflage to hide from predators (e.g., birds) on flowers, presented in Exercise 2.38 (p. 47). Researchers at the French Museum of Natural History collected a sample of 10 adult female crab spiders, each sitting on the yellow central part of a daisy, and measured the chromatic contrast between each spider and the flower. The data (for which higher values indicate a greater contrast, and, presumably, an easier detection by predators) are shown in the accompanying table and saved in the **SPIDER** file. The researchers discovered that a contrast of 70 or greater allows birds to see the spider. Of interest is whether or not the true mean chromatic contrast of crab spiders on daisies is less than 70.

57	75	116	37	96	61	56	2	43	32

Based on Thery, M., et al. "Specific color sensitivities of prey and predator explain camouflage in different visual systems." *Behavioral Ecology*, Vol. 16, No. 1, Jan. 2005 (Table 1).

a. Define the parameter of interest, μ
b. Set up the null and alternative hypotheses of interest.
c. Find $\bar{x}$ and s for the sample data, and then use these values to compute the test statistic.
d. Give the rejection region for $\alpha = .10$.
e. State the appropriate conclusion in the words of the problem.

8.69   Cheek teeth of extinct primates. Refer to the *American Journal of Physical Anthropology* (Vol. 142, 2010) study of the characteristics of cheek teeth (e.g., molars) in an extinct primate species, Exercise 2.34 (p. 46). Recall that the researchers recorded the dentary depth of molars (in millimeters) for a sample of 18 cheek teeth extracted from skulls. These depth measurements, saved in the **CHEEKTEETH** file, are listed in the accompanying table. Anthropologists know that the mean dentary depth of molars in an extinct primate species—called Species A—is 15 millimeters. Is there evidence to indicate that the sample of 18 cheek teeth come from some other extinct primate species (i.e., some species other than Species A)? Use the SPSS printout (p. 379) to answer the question.

18.12	16.55
19.48	15.70
19.36	17.83
15.94	13.25
15.83	16.12
19.70	18.13
15.76	14.02
17.00	14.04
13.96	16.20

Based on Boyer, D. M., Evans, A. R., and Jernvall, J. "Evidence of dietary differentiation among Late Paleocene–Early Eocene Plesiadapids (Mammalia, Primates)." *American Journal of Physical Anthropology*, Vol. 142, © 2010.

SPSS Output for Exercise 8.69

One-Sample Statistics

	N	Mean	Std. Deviation	Std. Error Mean
M2Depth	18	16.4994	1.97042	.46443

One-Sample Test

	Test Value = 15					
					95% Confidence Interval of the Difference	
	t	df	Sig. (2-tailed)	Mean Difference	Lower	Upper
M2Depth	3.229	17	.005	1.49944	.5196	2.4793

Applying the Concepts—Intermediate

8.70 **Pitch memory of amusiacs.** Amusia is a congenital disorder that adversely impacts one's perception of music. Refer to the *Advances in Cognitive Psychology* (Vol. 6, 2010) study of the pitch memory of individuals diagnosed with amusia, Exercise 7.41 (p. 318). Recall that each in a sample of 17 amusiacs listened to a series of tone pairs and was asked to determine if the tones were the same or different. In the first trial, the tones were separated by 1 second; in a second trial, the tones were separated by 5 seconds. The difference in accuracy scores for the two trials was determined for each amusiac (where the difference is the score on the first trial minus the score on the second trial). The mean score difference was .11 with a standard deviation of .19.

a. In theory, the longer the delay between tones, the less likely one is to detect a difference between the tones. Consequently, the true mean score difference should exceed 0. Set up the null and alternative hypotheses for testing the theory.

b. Carry out the test, part **a**, using $\alpha = .05$. Is there evidence to support the theory?

8.71 **Radioactive lichen.** Refer to the Lichen Radionuclide Baseline Research project to monitor the level of radioactivity in lichen, presented in Exercise 7.39 (p. 318). Recall that University of Alaska researchers collected nine lichen specimens and measured the amount of the radioactive element cesium-137 (in microcuries per milliliter) in each specimen. (The natural logarithms of the data values are saved in the **LICHEN** file.) Assume that in previous years the mean cesium amount in lichen was $\mu = .003$ microcurie per milliliter. Is there sufficient evidence to indicate that the mean amount of cesium in lichen specimens differs from this value? Use the SAS printout below to conduct a complete test of hypothesis at $\alpha = .10$.

```
Sample Statistics for CESIUM

    N        Mean        Std. Dev.      Std. Error
-------------------------------------------------------
    9       0.0090        0.0049          0.0016

Hypothesis Test

    Null hypothesis:    Mean of CESIUM =  0.003
    Alternative:        Mean of CESIUM ^= 0.003

        t Statistic      Df        Prob > t
    -----------------------------------------
          3.725           8         0.0058
```

8.72 **A new dental bonding agent.** When bonding teeth, orthodontists must maintain a dry field. A new bonding adhesive (called "Smartbond") has been developed to eliminate the necessity of a dry field. However, there is concern that the new bonding adhesive is not as strong as the current standard, a composite adhesive. (*Trends in Biomaterials & Artificial Organs*, Jan. 2003.) Tests on a sample of 10 extracted teeth bonded with the new adhesive resulted in a mean breaking strength (after 24 hours) of $\bar{x} = 5.07$ Mpa and a standard deviation of $s = .46$ Mpa, where Mpa = megapascal, a measure of force per unit area. Orthodontists want to know if the true mean breaking strength of the new bonding adhesive is less than 5.70 Mpa, the mean breaking strength of the composite adhesive. Conduct the appropriate analysis for the orthodontists. Use $\alpha = .01$.

8.73 **Minimizing tractor skidding distance.** Refer to the *Journal of Forest Engineering* (July 1999) study of minimizing tractor skidding distances along a new road in a European forest, presented in Exercise 7.44 (p. 319). The skidding distances (in meters) were measured at 20 randomly selected road sites. The data (saved in the **SKIDDING** file) are repeated in the accompanying table. Recall that a logger working on the road claims that the mean skidding distance is at least 425 meters. Is there sufficient evidence to refute this claim? Use $\alpha = .10$.

488	350	457	199	285	409	435	574	439	546
385	295	184	261	273	400	311	312	141	425

Based on Tujek, J., and Pacola, E. "Algorithms for skidding distance modeling on a raster Digital Terrain Model." *Journal of Forest Engineering*, Vol. 10, No. 1, July 1999 (Table 1).

8.74 **Mongolian desert ants.** Refer to the *Journal of Biogeography* (December 2003) study of ants in Mongolia (central Asia), presented in Exercise 2.68 (p. 59). Recall that botanists placed seed baits at 11 study sites and observed the number of ant species attracted to each site. A portion of the data (saved in the **GOBIANTS** file) is provided in the next table (p. 380). Do these data indicate that the average number of ant species at Mongolian desert sites differs from 5? Conduct the appropriate test at $\alpha = .05$. Are the conditions required for a valid test satisfied?

Data for Exercise 8.74

Site	Region	Number of Ant Species
1	Dry Steppe	3
2	Dry Steppe	3
3	Dry Steppe	52
4	Dry Steppe	7
5	Dry Steppe	5
6	Gobi Desert	49
7	Gobi Desert	5
8	Gobi Desert	4
9	Gobi Desert	4
10	Gobi Desert	5
11	Gobi Desert	4

Based on Pfeiffer, M., et al. "Community organization and species richness of ants in Mongolia along an ecological gradient from steppe to Gobi desert." *Journal of Biogeography,* Vol. 30, No. 12, Dec. 2003.

8.75 **Walking straight into circles.** When people get lost in unfamiliar terrain, do they really walk in circles, as is commonly believed? To answer this question, researchers conducted a field experiment and reported the results in *Current Biology* (Sept. 29, 2009). Fifteen volunteers were blindfolded and asked to walk as straight as possible in a certain direction in a large field. Walking trajectories were monitored every second for 50 minutes using GPS and the average directional bias (degrees per second) recorded for each walker. The data, saved in the **CIRCLES** file, are shown in the table (next column). A strong tendency to veer consistently in the same direction will cause walking in circles. A mean directional bias of 0 indicates that walking trajectories were

random. Consequently, the researchers tested whether the true mean bias differed significantly from 0. A MINITAB printout of the analysis is shown below.

a. Interpret the results of the hypothesis test for the researchers. Use $\alpha = .10$.

b. Although most volunteers showed little overall bias, the researchers produced maps of the walking paths showing that each occasionally made several small circles during the walk. Ultimately, the researchers supported the "walking into circles" theory. Explain why the data in the table is insufficient for testing whether an individual walks into circles.

−4.50	−1.00	−0.50	−0.15	0.00	0.01	0.02	0.05	0.15
0.20	0.50	0.50	1.00	2.00	3.00			

Based on Souman, J. L., Frissen, I., Sreenivasa, M. N., and Ernst, M. O. "Walking straight into circles." *Current Biology,* Vol. 19, No. 18, Sept. 29, 2009 (Figure 2).

Applying the Concepts—Advanced

8.76 **Lengths of great white sharks.** One of the most feared predators in the ocean is the great white shark. It is known that the white shark grows to a mean length of 21 feet; however, one marine biologist believes that great white sharks off the Bermuda coast grow much longer owing to unusual feeding habits. To test this claim, some full-grown great white sharks were captured off the Bermuda coast, measured, and then set free. However, because the capture of sharks is difficult, costly, and very dangerous, only three specimens were sampled. Their lengths were 24, 20, and 22 feet. Do these data support the marine biologist's claim at $\alpha = .10$?

MINITAB Output for Exercise 8.75

One-Sample T: BIAS

Test of mu = 0 vs not = 0

Variable	N	Mean	StDev	SE Mean	95% CI	T	P
BIAS	15	0.085	1.603	0.414	(−0.802, 0.973)	0.21	0.840

8.6 Large-Sample Test of Hypothesis about a Population Proportion

Inferences about population proportions (or percentages) are often made in the context of the probability p of "success" for a binomial distribution. We saw how to use large samples from binomial distributions to form confidence intervals for p in Section 7.4. We now consider tests of hypotheses about p.

Consider, for example, a method currently used by doctors to screen women for breast cancer. The method fails to detect cancer in 20% of the women who actually have the disease. Suppose a new method has been developed that researchers hope will detect cancer more accurately. This new method was used to screen a random sample of 140 women known to have breast cancer. Of these, the new method failed to detect cancer in 12 women. Does this sample provide evidence that the failure rate of the new method differs from the one currently in use?

We first view this experiment as a binomial one with 140 screened women as the trials and failure to detect breast cancer as "Success" (in binomial terminology). Let p represent the probability that the new method fails to detect the breast cancer; on the one hand, if the new method is no better than the current one, then the failure rate is $p = .2$. On the other hand, if the new method is either better or worse than the current method, then the failure rate is either smaller or larger than 20%; that is, $p \neq .2$.

We can now place the problem in the context of a test of hypothesis:

$$H_0: p = .2 \text{ (Cancer detection rate} = .2.)$$
$$H_a: p \neq .2 \text{ (Cancer detection rate differs from .2.)}$$

Recall that the sample proportion $\hat{p}$ is really just the sample mean of the outcomes of the individual binomial trials and, as such, is approximately normally distributed (for large samples) according to the Central Limit Theorem. Thus, for large samples we can use the standard normal z as the test statistic:

$$\textit{Test statistic:} \quad z = \frac{\text{Sample proportion } - \text{ Null hypothesized proportion}}{\text{Standard deviation of sample proportion}}$$
$$= \frac{\hat{p} - p_0}{\sigma_{\hat{p}}}$$

Note that we used the symbol p_0 to represent the null hypothesized value of p.

Rejection region: We use the standard normal distribution to find the appropriate rejection region for the specified value of α. Using $\alpha = .05$, the two-tailed rejection region is

$$z < -z_{\alpha/2} = -z_{.025} = -1.96 \quad \text{or} \quad z > z_{\alpha/2} = z_{.025} = 1.96$$

(See Figure 8.16.)

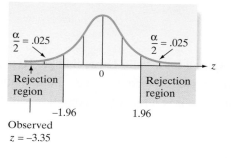

Figure 8.16

Rejection region for breast cancer example

We are now prepared to calculate the value of the test statistic. Before doing so, however, we want to be sure that the sample size is large enough to ensure that the normal approximation for the sampling distribution of $\hat{p}$ is reasonable. Recall from Section 7.4 that we require both np and nq to be at least 15. Because the null hypothesized value, p_0, is assumed to be the true value of p until our test procedure indicates otherwise, then we require $np_0 \geq 15$ and $nq_0 \geq 15$ (where $q_0 = 1 - p_0$). Since $p_0 = .2$ for our test, we have

$$np_0 = (140)(.2) = 28$$

and

$$nq_0 = (140)(.8) = 112$$

Therefore, the normal distribution will provide a reasonable approximation for the sampling distribution of $\hat{p}$.

Returning to the hypothesis test at hand, the proportion of the screenings that failed to detect breast cancer is

$$\hat{p} = \frac{12}{140} = .086$$

Finally, we calculate the number of standard deviations (the z value) between the sampled and hypothesized values of the binomial proportion:

$$z = \frac{\hat{p} - p_0}{\sigma_{\hat{p}}} = \frac{\hat{p} - p_0}{\sqrt{p_0 q_0 / n}} = \frac{.086 - .2}{\sqrt{(.2)(.8)/140}} = \frac{-.114}{.034} = -3.35$$

The implication is that the observed sample proportion is (approximately) 3.35 standard deviations below the null-hypothesized proportion, .2 (Figure 8.16). Therefore, we reject the null hypothesis, concluding at the .05 level of significance that the true failure rate of the new method for detecting breast cancer differs from .20. Since $\hat{p} = .086$, it appears that the new method is better (i.e., has a smaller failure rate) than the method currently in use. (To estimate the magnitude of the failure rate for the new method, a confidence interval can be constructed.)

The test of hypothesis about a population proportion p is summarized in the next box. Note that the procedure is entirely analogous to that used for conducting large-sample tests about a population mean.

Large-Sample Test of Hypothesis about p: Normal (z) Statistic

One-Tailed Test

H_0: $p = p_0$

H_a: $p < p_0$ (or H_a: $p > p_0$)

Test statistic: $z = \dfrac{\hat{p} - p_0}{\sigma_{\hat{p}}}$

Two-Tailed Test

H_0: $p = p_0$

H_a: $p \neq p_0$

Test statistic: $z = \dfrac{\hat{p} - p_0}{\sigma_{\hat{p}}}$

where p_0 = hypothesized value of p, $\sigma_{\hat{p}} = \sqrt{p_0 q_0/n}$, and $q_0 = 1 - p_0$

Rejection region: $z < -z_\alpha$
(or $z > z_\alpha$ when H_a: $p > p_0$)

Rejection region: $z < -z_{\alpha/2}$
or $z > z_{\alpha/2}$

Conditions Required for a Valid Large-Sample Hypothesis Test for p

1. A random sample is selected from a binomial population.
2. The sample size n is large. (This condition will be satisfied if np_0 and nq_0 are both at least 15.)

Example 8.9

A Hypothesis Test for p— Proportion of Defective Batteries

Problem The reputations (and hence sales) of many businesses can be severely damaged by shipments of manufactured items that contain a large percentage of defectives. For example, a manufacturer of alkaline batteries may want to be reasonably certain that less than 5% of its batteries are defective. Suppose 300 batteries are randomly selected from a very large shipment; each is tested and 10 defective batteries are found. Does this outcome provide sufficient evidence for the manufacturer to conclude that the fraction defective in the entire shipment is less than .05? Use $\alpha = .01$.

Solution The objective of the sampling is to determine whether there is sufficient evidence to indicate that the fraction defective, p, is less than .05. Consequently, we will test the null hypothesis that $p = .05$ against the alternative hypothesis that $p < .05$. The elements of the test are

H_0: $p = .05$ (Fraction of defective batteries equals .05.)

H_a: $p < .05$ (Fraction of defective batteries is less than .05.)

Test statistic: $z = \dfrac{\hat{p} - p_0}{\sigma_{\hat{p}}}$

Rejection region: $z < -z_{.01} = -2.33$ (see Figure 8.17)

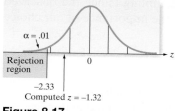

Figure 8.17

Rejection region for Example 8.9

Before conducting the test, we check to determine whether the sample size is large enough to use the normal approximation to the sampling distribution *of* $\hat{p}$. Since $np_0 = (300)(.05) = 15$ and $nq_0 = (300)(.95) = 285$ are both at least 15, the normal approximation will be adequate.

We now calculate the test statistic:

$$z = \frac{\hat{p} - .05}{\sigma_{\hat{p}}} = \frac{(10/300) - .05}{\sqrt{p_0 q_0/n}} = \frac{.03333 - .05}{\sqrt{p_0 q_0/300}}$$

Notice that we use p_0 to calculate $\sigma_{\hat{p}}$ because, in contrast to calculating $\sigma_{\hat{p}}$ for a confidence interval, the test statistic is computed on the assumption that the null hypothesis is true—that is, $p = p_0$. Therefore, substituting the values for $\hat{p}$ and p_0 into the z statistic, we obtain

$$z \approx \frac{-.01667}{\sqrt{(.05)(.95)/300}} = \frac{-.01667}{.0126} = -1.32$$

As shown in Figure 8.17, the calculated z-value does not fall into the rejection region. Therefore, there is insufficient evidence at the .01 level of significance to indicate that the shipment contains less than 5% defective batteries.

Now Work Exercise 8.80a–b

Example 8.10

Finding the *p*-Value for a Test about a Population Proportion *p*

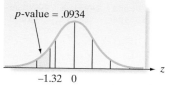

Figure 8.18

The observed significance level for Example 8.10

Problem In Example 8.9, we found that we did not have sufficient evidence at the $\alpha = .01$ level of significance to indicate that the fraction defective, p, of alkaline batteries was less than $p = .05$. How strong was the weight of evidence favoring the alternative hypothesis ($H_a: p < .05$)? Find the observed significance level (*p*-value) for the test.

Solution The computed value of the test statistic z was $z = -1.32$. Therefore, for this lower-tailed test, the observed significance level is

$$p\text{-value} = P(z \leq -1.32)$$

This lower-tail area is shown in Figure 8.18. The area between $z = 0$ and $z = 1.32$ is given in Table IV in Appendix A as .4066. Therefore, the observed significance level is $.5 - .4066 = .0934$.

Note: The *p*-value can also be obtained with statistical software. The MINITAB printout shown in Figure 8.19 gives the *p*-value (highlighted).

Look Back Although we did not reject $H_0: p = .05$ at $\alpha = .01$, the probability of observing a z-value as small as or smaller than -1.32 is only .093 if in fact H_0 is true. Therefore,

Figure 8.19

MINITAB lower-tailed test of p, Example 8.10

Test and CI for One Proportion

Test of p = 0.05 vs p < 0.05

Sample	X	N	Sample p	95% Upper Bound	Z-Value	P-Value
1	10	300	0.033333	0.050380	-1.32	0.093

Using the normal approximation.

we would reject H_0 if we had selected $\alpha = .10$ prior to sampling (since the observed significance level is less than .10). This point reminds us of the importance of our choice of α prior to collecting the sample data.

Now Work Exercise 8.80c

Small-sample test procedures are also available for p, although most surveys use samples that are large enough to employ the large-sample tests presented in this section. A test of proportions that can be applied to small samples is discussed in Chapter 13.

Statistics IN Action Revisited — Testing a Population Proportion in the KLEENEX® Survey

In the previous "Statistics in Action Revisited" (p. 371), we investigated Kimberly-Clark Corporation's assertion that the company should put more than 60 tissues in a cold-care box of KLEENEX® tissues. We did this by testing the claim that the mean number of tissues used by a person with a cold is $\mu = 60$, using data collected from a survey of 250 KLEENEX® users. Another approach to the problem is to consider the proportion of KLEENEX® users who use more than 60 tissues when they have a cold. Now the population parameter of interest is p, the proportion of all KLEENEX® users who use more than 60 tissues when they have a cold.

Kimberly-Clark Corporation's belief that the company should put more than 60 tissues in a cold-care box will be supported if over half of the KLEENEX® users surveyed use more than 60 tissues (i.e., if $p > .5$). Is there evidence to indicate that the population proportion exceeds .5? To answer this question, we set up the following null and alternative hypotheses:

$$H_0: p = .5 \quad H_a: p > .5$$

Recall that the survey results for the 250 sampled KLEENEX® users are stored in the **TISSUES** data file. In addition to the number of tissues used by each person, the file contains a qualitative variable—called USED60—representing whether the person used fewer or more than 60 tissues. (The values of USED60 in the data set are "LESS" or "MORE.") A

MINITAB analysis of this variable yielded the printout displayed in Figure SIA8.2.

On the MINITAB printout, x represents the number of the 250 people with colds that used more than 60 tissues. Note that $x = 154$. This value is used to compute the test statistic $z = 3.67$, highlighted on the printout. The p-value of the test, also highlighted on the printout, is p-value $= .000$. Since this value is less than $\alpha = .05$, there is sufficient evidence (at $\alpha = .05$) to reject H_0; we conclude that the proportion of all KLEENEX® users who use more than 60 tissues when they have a cold exceeds .5. This conclusion supports again, the company's decision to put more than 60 tissues in a pack of KLEENEX.

Data Set: TISSUES

Test and CI for One Proportion: USED60

```
Test of p = 0.5 vs p > 0.5

Event = MORE

                                      95% Lower
Variable    X    N  Sample p            Bound  Z-Value   P-Value
USED60    154  250  0.616000         0.565404     3.67     0.000

Using the normal approximation.
```

Figure SIA8.2
MINITAB test of $p = .5$ for KLEENEX survey

Exercises 8.77–8.95

Understanding the Principles

8.77 What type of data, quantitative or qualitative, is typically associated with making inferences about a population proportion p?

8.78 What conditions are required for a valid large-sample test for p?

Learning the Mechanics

8.79 For the binomial sample sizes and null-hypothesized values of p in each part, determine whether the sample size is large enough to use the normal approximation methodology presented in this section to conduct a test of the null hypothesis $H_0: p = p_0$.
   a. $n = 500, p_0 = .05$
   b. $n = 100, p_0 = .99$
   c. $n = 50, p_0 = .2$
   d. $n = 20, p_0 = .2$
   e. $n = 10, p_0 = .4$

8.80 Suppose a random sample of 100 observations from a bino-
[NW] mial population gives a value of $\hat{p} = .69$ and you wish to

test the null hypothesis that the population parameter p is equal to .75 against the alternative hypothesis that p is less than .75.

a. Noting that $\hat{p} = .69$, what does your intuition tell you? Does the value of $\hat{p}$ appear to contradict the null hypothesis?

b. Use the large-sample z-test to test $H_0: p = .75$ against the alternative hypothesis $H_a: p < .75$. Use $\alpha = .05$. How do the test results compare with your intuitive decision from part **a**?

c. Find and interpret the observed significance level of the test you conducted in part **b**.

8.81 Suppose the sample in Exercise 8.80 has produced $\hat{p} = .84$ and we wish to test $H_0: p = .9$ against the alternative $H_a: p < .9$.

a. Calculate the value of the z statistic for this test.

b. Note that the numerator of the z statistic $(\hat{p} - p_0 = .84 - .90 = -.06)$ is the same as for Exercise 8.80. Considering this, why is the absolute value of z for this exercise larger than that calculated in Exercise 8.80?

c. Complete the test, using $\alpha = .05$, and interpret the result.

d. Find the observed significance level for the test and interpret its value.

8.82 A random sample of 100 observations is selected from a binomial population with unknown probability of success, p. The computed value of $\hat{p}$ is equal to .74.

a. Test $H_0: p = .65$ against $H_a: p > .65$. Use $\alpha = .01$.

b. Test $H_0: p = .65$ against $H_a: p > .65$. Use $\alpha = .10$.

c. Test $H_0: p = .90$ against $H_a: p \neq .90$. Use $\alpha = .05$.

d. Form a 95% confidence interval for p.

e. Form a 99% confidence interval for p.

8.83 Refer to Exercise 7.54 (p. 325), in which 50 consumers taste-tested a new snack food. The data are saved in the **SNACK** file.

a. Test $H_0: p = .5$ against $H_a: p > .5$, where p is the proportion of customers who do not like the snack food. Use $\alpha = .10$.

b. Report the observed significance level of your test.

⬤ Applet Exercise 8.5

Use the applet entitled *Hypotheses Test for a Proportion* to investigate the relationships between the probabilities of Type I and Type II errors occurring at levels .05 and .01. For this exercise, use $n = 100$, true $p = 0.5$, and alternative *not equal*.

a. Set null $p = .5$. What happens to the proportion of times the null hypothesis is rejected at the .05 level and at the .01 level as the applet is run more and more times? What type of error has occurred when the null hypothesis is rejected in this situation? Based on your results, is this type of error more likely to occur at level .05 or at level .01? Explain.

b. Set null $p = .6$. What happens to the proportion of times the null hypothesis is *not* rejected at the .05 level and at the .01 level as the applet is run more and more times? What type of error has occurred when the null hypothesis is *not* rejected in this situation? Based on your results, is this type of error more likely to occur at level .05 or at level .01? Explain.

c. Use your results from parts **a** and **b** to make a general statement about the probabilities of Type I and Type II errors at levels .05 and .01.

⬤ Applet Exercise 8.6

Use the applet entitled *Hypotheses Test for a Proportion* to investigate the effect of the true population proportion p on the probability of a Type I error occurring. For this exercise, use $n = 100$ and alternative *not equal*.

a. Set true $p = .5$ and null $p = .5$. Run the applet several times, and record the proportion of times the null hypothesis is rejected at the .01 level.

b. Clear the applet and repeat part **a** for true $p = .1$ and null $p = .1$. Then repeat one more time for true $p = .01$ and null $p = .01$.

c. Based on your results from parts **a** and **b,** what can you conclude about the probability of a Type I error occurring as the true population proportion gets closer to 0?

Applying the Concepts—Basic

8.84 **Satellite radio in cars.** A spokesperson for the National Association of Broadcasters (NAB) claims that 80% of all satellite radio subscribers have a satellite radio receiver in their car. Recall from Exercise 1.24 (p. 20), that in a June 2007 survey of 501 satellite radio subscribers, 396 had a satellite receiver in their car. Consider a test of the NAB spokesperson's claim.

a. Define the parameter of interest to the NAB spokesperson.

b. Set up the null hypothesis for testing the claim.

c. Specify the alternative hypothesis if you believe that the spokesperson's claim is too high.

d. Compute the value of the test statistic.

e. Determine the rejection region for the test using $\alpha = .10$.

f. Compute the p-value of the test.

g. Make the appropriate conclusion. Show that the decision based on the rejection region agrees with the decision based on the p-value.

8.85 **Accuracy of price scanners at Wal-Mart.** Refer to Exercise 7.81 (p. 332) and the study of the accuracy of checkout scanners at Wal-Mart Stores in California. Recall that the National Institute for Standards and Technology (NIST) mandates that, for every 100 items scanned through the electronic checkout scanner at a retail store, no more than 2 should have an inaccurate price. A study of random items purchased at California Wal-Mart stores found that 8.3% had the wrong price (*Tampa Tribune*, Nov. 22, 2005). Assume that the study included 1,000 randomly selected items.

a. Identify the population parameter of interest in the study.

b. Set up H_0 and H_a for a test to determine whether the true proportion of items scanned at California Wal-Mart stores exceeds the 2% NIST standard.

c. Find the test statistic and rejection region (at $\alpha = .05$) for the test.

d. Give a practical interpretation of the test.

e. What conditions are required for the inference made in part **d** to be valid? Are these conditions met?

8.86 **Gummy bears: red or yellow?** *Chance* (Winter 2010) presented a lesson in hypothesis testing carried out by medical students in a biostatistics class. Students were blind-folded and then given a red-colored or yellow-colored gummy bear to chew. (Half the students were randomly assigned to receive the red gummy bear and half to receive the yellow bear. The students could not see what color gummy bear they were given.) After chewing, the students were asked to guess the color of the candy based on the flavor. Of the 121 students who participated in the study, 97 correctly identified the color of the gummy bear.

a. If there is no relationship between color and gummy bear flavor, what proportion of the population of students will correctly identify the color?

b. Specify the null and alternative hypothesis for testing whether color and flavor are related.

c. Carry out the test and give the appropriate conclusion at $\alpha = .01$. Use the *p*-value of the test to make your decision.

8.87 **Single-parent families.** Examining data collected on 835 males from the National Youth Survey (a longitudinal survey of a random sample of U.S. households), researchers at Carnegie Mellon University found that 401 of the male youths were raised in a single-parent family (*Sociological Methods & Research*, February 2001). Does this information allow you to conclude that more than 45% of male youths are raised in a single-parent family? Test at $\alpha = .05$.

8.88 **Teenagers' use of emoticons in school writing.** Refer to the *Pew Internet and American Life Project* (April 2008) survey of the writing habits of U.S. teenagers, Exercise 7.61 (p. 326). Recall that in a random sample of 700 teenagers, 448 admitted to using at least one informal element in school writing assignments. [*Note*: Emoticons, such as using the symbol ":)" to represent a smile, and abbreviations, such as writing "LOL" for "laughing out loud," are considered informal elements.] Is there evidence to indicate that less than 65% of all U.S. teenagers have used at least one informal element in school writing assignments? Test the relevant hypotheses using $\alpha = .05$.

Applying the Concepts—Intermediate

8.89 **Identifying organisms using a computer.** *National Science Education Standards* recommend that all life science students be exposed to methods of identifying unknown biological specimens. Due to certain limitations of traditional identification methods, biology professors at Slippery Rock University (SRU) developed a computer-aided system for identifying common conifers (deciduous trees) called Confir ID (*The American Biology Teacher*, May 2010). A sample of 171 life science students exposed to both a traditional method of identifying conifers and Confir ID and then asked which method they preferred. As a result, 138 students indicated their preference for Confir ID. In order to change the life sciences curriculum at SRU to include Confir ID, the biology department requires that more than 70% of the students prefer the new, computerized method. Should Confir ID be added to the curriculum at SRU? Explain your reasoning.

8.90 **Graduation rates of student–athletes.** Are student–athletes at Division I universities poorer students than nonathletes? The National Collegiate Athletic Association (NCAA) measures the academic outcomes of student–athletes with the Graduation Success Rate (GSR)—the percentage of eligible athletes who graduate within six years of entering college. According to the NCAA, the GSR for all scholarship athletes at Division I institutions is 63% (*Inside Higher Ed*, Nov. 10, 2006). It is well known that the GSR for all students at Division I colleges is 60%.

a. Suppose the NCAA report was based on a sample of 500 student–athletes, of which 315 graduated within six years. Is this sufficient information to conclude that the GSR for all scholarship athletes at Division I institutions differs from 60%? Test, using $\alpha = .01$.

b. The GSR statistics were also broken down by gender and sport. For example, men's Division I college basketball players had a GSR of 42% (compared with a known GSR of 58% for all male college students). Suppose this statistic was based on a sample of 200 male basketball players, of whom 84 graduated within six years. Is this sufficient information to conclude that the GSR for all male basketball players at Division I institutions differs from 58%? Test, using $\alpha = .01$.

8.91 **Astronomy students and the Big Bang Theory.** Indiana University professors investigated first-year college students' knowledge of astronomy (*Astronomy Education Review*, Vol. 2, 2003). One concept of interest was the Big Bang theory of the creation of the universe. In a sample of 148 freshmen students, 37 believed that the Big Bang theory accurately described the creation of planetary systems. Based on this information, would you be willing to state that more than 20% of all freshmen college students believe in the Big Bang theory? How confident are you of your decision?

8.92 **Study of lunar soil.** *Meteoritics* (March 1995) reported the results of a study of lunar soil evolution. Data were obtained from the *Apollo 16* mission to the moon, during which a 62-cm core was extracted from the soil near the landing site. Monomineralic grains of lunar soil were separated out and examined for coating with dust and glass fragments. Each grain was then classified as coated or uncoated. Of interest is the "coat index"—that is, the proportion of grains that are coated. According to soil evolution theory, the coat index will exceed .5 at the top of the core, equal .5 in the middle of the core, and fall below .5 at the bottom of the core. Use the summary data in the accompanying table to test each part of the three-part theory. Use $\alpha = .05$ for each test.

	Location (depth)		
	Top (4.25 cm)	Middle (28.1 cm)	Bottom (54.5 cm)
Number of grains sampled	84	73	81
Number coated	64	35	29

Based on Basu, A., and McKay, D. S. "Lunar soil evolution processes and Apollo 16 core 60013/60014." *Meteoritics*, Vol. 30, No. 2, Mar. 1995, p. 166 (Table 2).

8.93 **Effectiveness of skin cream.** Pond's has discontinued the production of Age-Defying Complex, a cream with alpha-hydroxy acid, with Age-Defying Towlettes. Pond's advertised that the product could reduce wrinkles and improve the skin. In a study published in *Archives of Dermatology* (June 1996), 33 middle-aged women used a product with alpha-hydroxy acid for 22 weeks. At the end of the study period, a dermatologist judged whether each woman exhibited any improvement in the condition of her skin. The results for the 33 women (where I = improved skin and N = no improvement) are listed in the accompanying table and saved in the **SKINCREAM** file.

a. Do the data provide sufficient evidence to conclude that the cream will improve the skin of more than 60% of middle-aged women? Test, using $\alpha = .05$.

b. Find and interpret the *p*-value of the test.

I	I	N	I	N	N	I	I	I	I	I	I
N	I	I	I	N	I	I	I	N	I	N	I
I	I	I	I	I	N	I	I	N			

Applying the Concepts—Advanced

8.94 Killing insects with low oxygen. A group of Australian entomological toxicologists investigated the impact of exposure to low oxygen on the mortality of insects (*Journal of Agricultural, Biological, and Environmental Statistics*, Sept. 2000). Thousands of adult rice weevils were placed in a chamber filled with wheat grain, and the chamber was exposed to nitrogen gas for 4 days. Insects were assessed as dead or alive 24 hours after exposure. At the conclusion of the experiment, 31,386 weevils were dead and 35 weevils were found alive. Previous studies had shown a 99% mortality rate in adult rice weevils exposed to carbon dioxide for 4 days. What advice would you give to the entomologists regarding the use of nitrogen or carbon dioxide for killing rice weevils?

8.95 Male fetal deaths following 9/11/2001. According to research published in *BMC Public Health* (May 25, 2010), the terrorist attacks of September 11, 2001, may have led to a spike in miscarriages of male babies. Researchers from the University of California, using data from the National Vital Statistics System, reported that the male fetal death rate in September of that year was significantly higher than expected. Suppose it is known that the miscarriage rate for pregnant women expecting a boy in the United States is 3 in 1,000. In a random sample of 2,000 pregnant women expecting a baby boy during September of 2001, how many would need to have experienced a fetal death (miscarriage) in order to support the claim (at $\alpha = .05$) that the male fetal death rate in September 2001 was higher than expected?

8.7 Calculating Type II Error Probabilities: More about β (Optional)

In our introduction to hypothesis testing in Section 8.1, we showed that the probability of committing a Type I error, α, can be controlled by the selection of the rejection region for the test. Thus, when the test statistic falls into the rejection region and we make the decision to reject the null hypothesis, we do so knowing the error rate for incorrect rejections of H_0. The situation corresponding to accepting the null hypothesis, and thereby risking a Type II error, is not generally as controllable. For that reason, we adopted a policy of nonrejection of H_0 when the test statistic does not fall into the rejection region, rather than risking an error of unknown magnitude.

To see how β, the probability of a Type II error, can be calculated for a test of hypothesis, recall the example in Section 8.1 in which a city tests a manufacturer's pipe to see whether it meets the requirement that the mean strength exceed 2,400 pounds per linear foot. The setup for the test is as follows:

$H_0: \mu = 2,400$

$H_a: \mu > 2,400$

Test statistic: $z = \dfrac{\bar{x} - 2,400}{\sigma/\sqrt{n}}$

Rejection region: $z > 1.645$ for $\alpha = .05$

Figure 8.20a shows the rejection region for the **null distribution**—that is, the distribution of the test statistic, assuming that the null hypothesis is true. The area in the rejection region is .05, and this area represents α, the probability that the test statistic leads to rejection of H_0 when in fact H_0 is true.

The Type II error probability β is calculated under the assumption that the null hypothesis is false, because it is defined as the *probability of accepting H_0 when it is false.* Since H_0 is false for any value of μ exceeding 2,400, one value of β exists for each possible value of μ greater than 2,400 (an infinite number of possibilities). Figures 8.20b, 8.20c, and 8.20d show three of the possibilities, corresponding to alternative hypothesis values of μ equal to 2,425, 2,450, and 2,475, respectively. Note that β is the area in the *nonrejection* (or *acceptance*) *region* in each of these distributions and that β decreases as

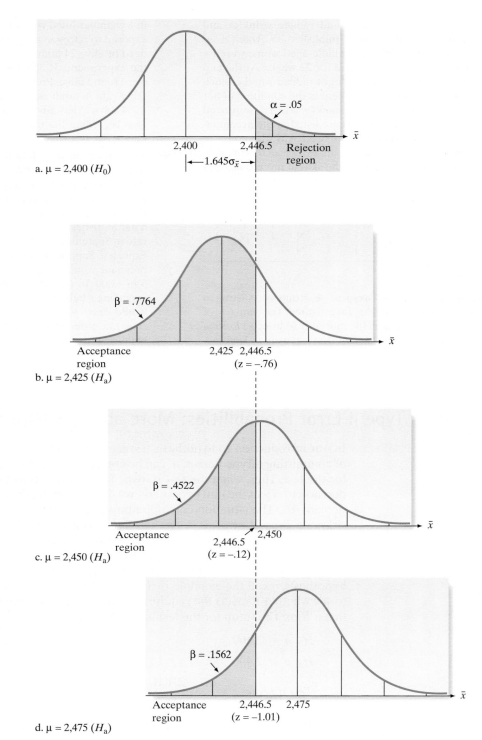

Figure 8.20

Values of α and β for various values of μ

a. $\mu = 2{,}400$ (H_0)

b. $\mu = 2{,}425$ (H_a)

c. $\mu = 2{,}450$ (H_a)

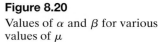

d. $\mu = 2{,}475$ (H_a)

the true value of μ moves farther from the null-hypothesized value of $\mu = 2{,}400$. This behavior is sensible because the probability of incorrectly accepting the null hypothesis should decrease as the distance between the null and alternative values of μ increases.

In order to calculate the value of β for a specific value of μ in H_a, we proceed as follows:

1. We calculate the value of $\bar{x}$ that corresponds to the border between the acceptance and rejection regions. For the sewer pipe example, this is the value of $\bar{x}$ that lies 1.645 standard deviations above $\mu = 2{,}400$ in the sampling distribution of $\bar{x}$. Denoting this value by $\bar{x}_0$, corresponding to the largest value of $\bar{x}$ that supports the null hypothesis, we find (recalling that $s = 200$ and $n = 50$) that

$$\bar{x}_0 = \mu_0 + 1.645\sigma_{\bar{x}} = 2{,}400 + 1.645\left(\frac{\sigma}{\sqrt{n}}\right)$$

$$\approx 2{,}400 + 1.645\left(\frac{s}{\sqrt{n}}\right) = 2{,}400 + 1.645\left(\frac{200}{\sqrt{50}}\right)$$

$$= 2{,}400 + 1.645(28.28) = 2{,}446.5$$

2. For a particular alternative distribution corresponding to a value of μ denoted by μ_a, we calculate the z-value corresponding to $\bar{x}_0$, the border between the rejection and acceptance regions. We then use this z-value and Table IV of Appendix A to determine the area in the *acceptance region* under the alternative distribution. This area is the value of β corresponding to the particular alternative μ_a. For example, for the alternative $\mu_a = 2{,}425$, we calculate

$$z = \frac{\bar{x}_0 - 2{,}425}{\sigma_{\bar{x}}} \approx \frac{\bar{x}_0 - 2{,}425}{s/\sqrt{n}} = \frac{2{,}446.5 - 2{,}425}{28.28} = .76$$

Note in Figure 8.20b that the area in the acceptance region is the area to the left of $z = .76$. This area is

$$\beta = .5 + .2764 = .7764$$

Thus, the probability that the test procedure will lead to an incorrect acceptance of the null hypothesis $\mu = 2{,}400$ when in fact $\mu = 2{,}425$ is about .78. As the average strength of the pipe increases to 2,450, the value of β decreases to .4522 (Figure 8.20c). If the mean strength is further increased to 2,475, the value of β is further decreased to .1562 (Figure 8.20d). Thus, even if the true mean strength of the pipe exceeds the minimum specification by 75 pounds per linear foot, the test procedure will lead to an incorrect acceptance of the null hypothesis (rejection of the pipe) approximately 16% of the time. The upshot is that the pipe must be manufactured so that the mean strength well exceeds the minimum requirement if the manufacturer wants the probability of its acceptance by the city to be large (i.e., β to be small).

The steps for calculating β for a large-sample test about a population mean and a population proportion are summarized in the following boxes:

Steps for Calculating β for a Large-Sample Test about μ

1. Calculate the value(s) of $\bar{x}$ corresponding to the border(s) of the rejection region. There will be one border value for a one-tailed test and two for a two-tailed test. The formula is one of the following, corresponding to a test with level of significance α:

Upper-tailed test: $\quad \bar{x}_0 = \mu_0 + z_\alpha \sigma_{\bar{x}} \approx u_0 + z_\alpha\left(\dfrac{s}{\sqrt{n}}\right)$

$\quad(H_a\colon \mu > \mu_0)$

Lower-tailed test: $\quad \bar{x}_0 = \mu_0 - z_\alpha \sigma_{\bar{x}} \approx u_0 - z_\alpha\left(\dfrac{s}{\sqrt{n}}\right)$

$\quad(H_a\colon \mu < \mu_0)$

Two-tailed test: $\quad \bar{x}_{0,\,L} = \mu_0 - z_{\alpha/2} \sigma_{\bar{x}} \approx u_0 - z_{\alpha/2}\left(\dfrac{s}{\sqrt{n}}\right)$

$\quad(H_a\colon \mu \neq \mu_0)$

$$\bar{x}_{0,\,U} = \mu_0 + z_{\alpha/2} \sigma_{\bar{x}} \approx \mu_0 + z_{\alpha/2}\left(\frac{s}{\sqrt{n}}\right)$$

(continued)

---

2. Specify the value of μ_a in the alternative hypothesis for which the value of β is to be calculated. Then convert the border value(s) of $\bar{x}_0$ to z-value(s), using the alternative distribution with mean μ_a. The general formula for the z value is

$$z = \frac{\bar{x}_0 - \mu_a}{\sigma_{\bar{x}}}$$

3. Sketch the alternative distribution (centered at μ_a), and shade the area in the acceptance (nonrejection) region. Use the z-statistic(s) and Table IV of Appendix A to find the shaded area, which is β.

Lower-tailed test: $\beta = P\left(z > \dfrac{\bar{x}_0 - \mu_a}{\sigma_{\bar{x}}}\right)$

$(H_a: \mu < \mu_0)$

Upper-tailed test: $\beta = P\left(z < \dfrac{\bar{x}_0 - \mu_a}{\sigma_{\bar{x}}}\right)$

$(H_a: \mu > \mu_0)$

Two-tailed test: $\beta = P\left(\dfrac{\bar{x}_{0,L} - \mu_a}{\sigma_{\bar{x}}} < z < \dfrac{\bar{x}_{0,U} - \mu_a}{\sigma_{\bar{x}}}\right)$

$(H_a: \mu \neq \mu_0)$

Steps for Calculating β for a Large-Sample Test about p

1. Calculate the value(s) of $\hat{p}$ corresponding to the border(s) of the rejection region. There will be one border for a one-tailed test and two for a two-tailed test. The formula is one of the following, corresponding to a test with level of significance α.

Upper-tailed test: $\hat{p}_0 = p_0 + z_\alpha \sigma_p = p_0 + z_\alpha \sqrt{\dfrac{p_0 q_0}{n}}$

$(H_a: p > p_0)$

Lower-tailed test: $\hat{p}_0 = p_0 - z_\alpha \sigma_p = p_0 - z_\alpha \sqrt{\dfrac{p_0 q_0}{n}}$

$(H_a: p < p_0)$

Two-tailed test: $\hat{p}_{0,L} = p_0 - z_{\alpha/2} \sigma_p = p_0 - z_{\alpha/2} \sqrt{\dfrac{p_0 q_0}{n}}$

$(H_a: p \neq p_0)$

$\hat{p}_{0,U} = p_0 + z_{\alpha/2} \sigma_p = p_0 + z_{\alpha/2} \sqrt{\dfrac{p_0 q_0}{n}}$

2. Specify the value of p_a in the alternative hypothesis for which the value of β is to be calculated. Then convert the border values of $\hat{p}_0$ to z-value(s), using the alternative distribution with mean p_a. The general formula for the z value is:

$$z = (\hat{p}_0 - p_a)/\sigma_p = \frac{\hat{p}_0 - p_a}{\sqrt{\dfrac{p_0 q_0}{n}}}$$

3. Sketch the alternative distribution (centered at p_a), and shade the area in the acceptance (nonrejection) region. Use the z-statistic(s) and Table IV of Appendix A to find the shaded area, which is β.

$$\text{Upper-tailed test:} \quad \beta = P\left(z < \frac{\hat{p}_0 - p_a}{\sqrt{\frac{p_0 q_0}{n}}}\right)$$

$$(H_a: p > p_0)$$

$$\text{Lower-tailed test:} \quad \beta = P\left(z > \frac{\hat{p}_0 - p_a}{\sqrt{\frac{p_0 q_0}{n}}}\right)$$

$$(H_a: p < p_0)$$

$$\text{Two-tailed test:} \quad \beta = P\left(\frac{\hat{p}_{0,L} - p_a}{\sqrt{\frac{p_0 q_0}{n}}} < z < \frac{p_{0,U} - p_a}{\sqrt{\frac{p_0 q_0}{n}}}\right)$$

$$(H_a: p \neq p_0)$$

Following the calculation of β for a particular value of the parameter in H_a you should interpret the value in the context of the hypothesis-testing application. It is often useful to interpret the value of $1 - \beta$, which is known as the **power of the test** corresponding to a particular alternative, say, μ_a for a population mean. Since β is the probability of accepting the null hypothesis when the alternative hypothesis is true with $\mu = \mu_a$, $1 - \beta$ is the probability of the complementary event, or the probability of rejecting the null hypothesis when the alternative $H_a: \mu = \mu_a$ is true. That is, the power $(1 - \beta)$ measures the likelihood that the test procedure will lead to the correct decision (reject H_0) for a particular value of the mean (or proportion) in the alternative hypothesis.

> The **power of a test** is the probability that the test will correctly lead to the rejection of the null hypothesis for a particular value of μ or p in the alternative hypothesis. The power is equal to $(1 - \beta)$ for the particular alternative considered.

For example, in the sewer pipe example, we found that $\beta = .7764$ when $\mu = 2{,}425$. This is the probability that the test leads to the (incorrect) acceptance of the null hypothesis when $\mu = 2{,}425$. Or, equivalently, the power of the test is $1 - .7764 = .2236$, which means that the test will lead to the (correct) rejection of the null hypothesis only 22% of the time when the pipe exceeds specifications by 25 pounds per linear foot. When the manufacturer's pipe has a mean strength of 2,475 (that is, 75 pounds per linear foot in excess of specifications), the power of the test increases to $1 - .1562 = .8438$. That is, the test will lead to the acceptance of the manufacturer's pipe 84% of the time if $\mu = 2{,}475$.

Example 8.11

Finding the Power of a Test

Problem Recall the drug experiment in Examples 8.3 and 8.4, in which we tested to determine whether the mean response time for rats injected with a drug differs from the control mean response time of $\mu = 1.2$ seconds. The test setup is repeated here:

$$H_0: \mu = 1.2$$

$$H_a: \mu \neq 1.2 \ (\text{i.e., } \mu < 1.2 \text{ or } \mu > 1.2)$$

$$\text{Test statistic: } z = \frac{\bar{x} - 1.2}{\sigma_{\bar{x}}}$$

$$\text{Rejection region: } z < -1.96 \text{ or } z > 1.96 \text{ for } \alpha = .05$$

$$z < -2.575 \text{ or } z > 2.575 \quad \text{for } \alpha = .01$$

Note that two rejection regions have been specified, corresponding to values of $\alpha = .05$ and $\alpha = .01$, respectively. Assume that $n = 100$ and $s = .5$.

a. Suppose drug-injected rats have a mean response time of 1.1 seconds; that is, $\mu = 1.1$. Calculate the values of β corresponding to the two rejection regions. Discuss the relationship between the values of α and β.

b. Calculate the power of the test for each of the rejection regions when $\mu = 1.1$.

Solution

a. We first consider the rejection region corresponding to $\alpha = .05$. The first step is to calculate the border values of $\bar{x}$ corresponding to the two-tailed rejection region, $z < -1.96$ or $z > 1.96$:

$$\bar{x}_{0,L} = \mu_0 - 1.96\sigma_{\bar{x}} \approx \mu_0 - 1.96\left(\frac{s}{\sqrt{n}}\right) = 1.2 - 1.96\left(\frac{.5}{10}\right) = 1.102$$

$$\bar{x}_{0,U} = \mu_0 + 1.96\sigma_{\bar{x}} \approx \mu_0 + 1.96\left(\frac{s}{\sqrt{n}}\right) = 1.2 + 1.96\left(\frac{.5}{10}\right) = 1.298$$

These border values are shown in Figure 8.21.

Next, we convert these values to z-values in the alternative distribution with $\mu_a = 1.1$:

$$z_L = \frac{\bar{x}_{0,L} - \mu_a}{\sigma_{\bar{x}}} \approx \frac{1.102 - 1.1}{.05} = .04$$

$$z_U = \frac{\bar{x}_{0,U} - \mu_a}{\sigma_{\bar{x}}} \approx \frac{1.298 - 1.1}{.05} = 3.96$$

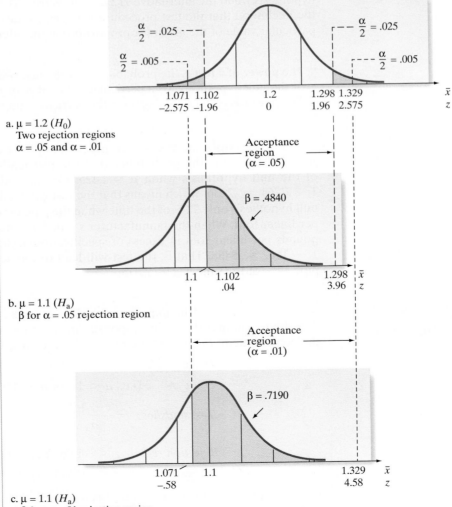

a. $\mu = 1.2$ (H_0)
Two rejection regions
$\alpha = .05$ and $\alpha = .01$

b. $\mu = 1.1$ (H_a)
β for $\alpha = .05$ rejection region

c. $\mu = 1.1$ (H_a)
β for $\alpha = .01$ rejection region

Figure 8.21

Calculation of β for drug-injected rate (Example 8.11)

These z-values are shown in Figure 8.21b. You can see that the acceptance (or non-rejection) region is the area between them. Using Table IV of Appendix A, we find that the area between $z = 0$ and $z = .04$ is .0160 and the area between $z = 0$ and $z = 3.96$ is (approximately) .5 (since $z = 3.96$ is off the scale of Table IV). Then the area between $z = .04$ and $z = 3.96$ is approximately

$$\beta = .5 - .0160 = .4840$$

Thus, the test with $\alpha = .05$ will lead to a Type II error about 48% of the time when the mean reaction time for drug-injected rats is .1 second less than the control mean response time.

For the rejection region corresponding to $\alpha = .01$, $z < -2.575$ or $z > 2.575$, we find that

$$\bar{x}_{0,L} = 1.2 - 2.575\left(\frac{.5}{10}\right) = 1.0712$$

$$\bar{x}_{0,U} = 1.2 + 2.575\left(\frac{.5}{10}\right) = 1.3288$$

These border values of the rejection region are shown in Figure 8.21c.

Converting these two border values to z-values in the alternative distribution with $\mu_a = 1.1$, we find that $z_L = -.58$ and $z_U = 4.58$. The area between these values is approximately

$$\beta = P(-.58 < z < 4.58) = .2190 + .5 = .7190$$

Thus, the chance that the test procedure with $\alpha = .01$ will lead to an incorrect acceptance of H_0 is about 72%.

Note that the value of β increases from .4840 to .7190 when we decrease the value of α from .05 to .01. This is a general property of the relationship between α and β: *As α is decreased (increased), β is increased (decreased).*

b. The power is defined to be the probability of (correctly) rejecting the null hypothesis when the alternative is true. When $\mu = 1.1$ and $\alpha = .05$, we find that

$$\text{Power} = 1 - \beta = 1 - .4840 = .5160$$

When $\mu = 1.1$ and $\alpha = .01$, we get

$$\text{Power} = 1 - \beta = 1 - .7190 = .2810$$

You can see that the power of the test is decreased as the level of α is decreased. This means that as the probability of incorrectly rejecting the null hypothesis is decreased, the probability of correctly accepting the null hypothesis for a given alternative is also decreased.

Look Back A key point in this example is that the value of α must be selected carefully, with the realization that a test is made less capable of detecting departures from the null hypothesis when the value of α is decreased.

Now Work Exercise 8.104a

Most statistical software packages now have options for computing the power of standard tests of hypothesis. Usually, you will need to specify the type of test (z-test or t-test), the form of $H_a (<, >, $ or $\neq)$, the standard deviation, the sample size, and the value of the parameter in H_a (or the difference between the value in H_0 and the value in H_a). The MINITAB power analysis for Example 8.11 when $\alpha = .05$ is displayed in Figure 8.22. The power of the test (.516) is highlighted on the printout.

We have shown that the probability of committing a Type II error, β, is inversely related to α (Example 8.11), and that the value of β decreases as the value of μ_a (or, p_a) moves farther from the null-hypothesis value. Consider again the sewer pipe example. The sample size n also affects β. Remember that the standard deviation of the sampling distribution of $\bar{x}$ is inversely proportional to the square root of the sample size ($\sigma_{\bar{x}} = \sigma/\sqrt{n}$).

Power and Sample Size

```
1-Sample Z Test

Testing mean = null (versus not = null)
Calculating power for mean = null + difference
Alpha = 0.05   Assumed standard deviation = 0.5

                     Sample
      Difference      Size     Power
          0.1          100    0.516005
```

Figure 8.22
MINITAB power analysis for Example 8.11

Thus, as illustrated in Figure 8.23, the variability of both the null and alternative sampling distributions is decreased as n is increased. If the value of α is specified and remains fixed, then the value of β decreases as n increases, as illustrated in Figure 8.23. Conversely, the power of the test for a given alternative hypothesis is increased as the sample size is increased.

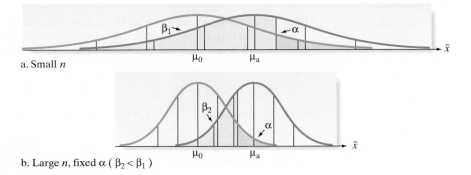

a. Small n

b. Large n, fixed α ($\beta_2 < \beta_1$)

Figure 8.23
Relationship between α, β, and n

The properties of β and power are summarized in the following box:

Properties of β and Power

1. For fixed n and α, the value of β decreases and the power increases as the distance between the specified null value μ_0 (or p_0) and the specified alternative value μ_a (or p_a) increases. (See Figure 8.20.)

2. For fixed n, μ_0, and μ_a (or p_0 and p_a) the value of β increases and the power decreases as the value of α is decreased. (See Figure 8.21.)

3. For fixed α, μ_0, and μ_a, (or p_0 and p_a) the value of β decreases and the power increases as the sample size n is increased. (See Figure 8.23.)

Exercises 8.96–8.109

Understanding the Principles

8.96 Define the power of a test.

8.97 What is the relationship between β (the probability of committing a Type II error) and the power of the test?

8.98 List three factors that will increase the power of the test.

Learning the Mechanics

8.99 Suppose you want to test $H_0: \mu = 1,000$ against $H_a: \mu > 1,000$, using $\alpha = .05$. The population in question is normally distributed with standard deviation 120. A random sample of size $n = 36$ will be used.

a. Sketch the sampling distribution of $\bar{x}$, assuming that H_0 is true.

b. Find the value of $\bar{x}_0$, that value of $\bar{x}$ above which the null hypothesis will be rejected. Indicate the rejection region on your graph of part **a**. Shade the area above the rejection region and label it α.

c. On your graph of part **a**, sketch the sampling distribution of $\bar{x}$ if $\mu = 1,020$. Shade the area under this distribution which corresponds to the probability that $\bar{x}$ falls in the nonrejection region when $\mu = 1,020$. Label this area β.

d. Find β.

e. Compute the power of this test for detecting the alternative H_a: $\mu = 1{,}020$.

8.100 Refer to Exercise 8.99.

a. If $\mu = 1{,}040$ instead of 1,020, what is the probability that the hypothesis test will incorrectly fail to reject H_0? That is, what is β?

b. If $\mu = 1{,}040$, what is the probability that the test will correctly reject the null hypothesis? That is, what is the power of the test?

c. Compare β and the power of the test when $\mu = 1{,}040$ to with values you obtained in Exercise 8.99 for $\mu = 1{,}020$. Explain the differences.

8.101 It is desired to test H_0: $\mu = 50$ against H_a: $\mu < 50$, using $\alpha = .10$. The population in question is uniformly distributed with standard deviation 20. A random sample of size 64 will be drawn from the population.

a. Describe the (approximate) sampling distribution of $\bar{x}$ under the assumption that H_0 is true.

b. Describe the (approximate) sampling distribution of $\bar{x}$ under the assumption that the population mean is 45.

c. If μ were really equal to 45, what is the probability that the hypothesis test would lead the investigator to commit a Type II error?

d. What is the power of this test for detecting the alternative H_a: $\mu = 45$?

8.102 Refer to Exercise 8.101.

a. Find β for each of the following values of the population mean: 49, 47, 45, 43, and 41.

b. Plot each value of β you obtained in part **a** against its associated population mean. Show β on the vertical axis and μ on the horizontal axis. Draw a curve through the five points on your graph.

c. Use your graph from part **b** to find the approximate probability that the hypothesis test will lead to a Type II error when $\mu = 48$.

d. Convert each of the β values you calculated in part **a** to the power of the test at the specified value of μ. Plot the power on the vertical axis against μ on the horizontal axis. Compare the graph of part **b** with the *power curve* you plotted here.

e. Examine the graphs of parts **b** and **d**. Explain what they reveal about the relationships among the distance between the true mean μ and the null-hypothesized mean μ_0, the value of β, and the power.

8.103 Suppose you want to conduct the two-tailed test of H_0: $p = .7$ against H_a: $p \neq .7$ using $\alpha = .05$. A random sample of size 100 will be drawn from the population in question.

a. Describe the sampling distribution of $\hat{p}$ under the assumption that H_0 is true.

b. Describe the sampling distribution of $\hat{p}$ under the assumption that $p = .65$

c. If p were really equal to .65, find the value of β associated with the test.

d. Find the value of β for the alternative H_a: $p = .71$.

Applying the Concepts—Intermediate

8.104 **Latex allergy in health care workers.** Refer to the *Current Allergy & Clinical Immunology* study of health care workers diagnosed with a latex allergy, Exercise 8.28 (p. 365). You tested the null hypothesis of $\mu = 20$ against the alternative hypothesis of $\mu < 20$ using $\alpha = .01$, where μ is the mean number of latex gloves used per week. The sample of $n = 46$ workers had a standard deviation of $s = 11.9$. Assume s is a good estimate of the true standard deviation σ.

NW　**a.** Find the power of the test if the true mean is $\mu = 19$.

b. Repeat part **a** for the true mean values 18, 16, 14, and 12.

c. Plot the power of the test on the vertical axis against the alternative mean on the horizontal axis. Draw a curve through the points. What pattern do you observe?

d. Use the power curve, part **c**, to estimate the power for the mean value $\mu = 15$. Calculate the power for this value of μ, and compare it with your approximation.

e. If the true value of the mean number of latex gloves used per week is really 15, what is the probability that the test will fail to reject H_0: $\mu = 20$?

8.105 **Increasing the sample size.** Refer to Exercise 8.104. Show what happens to the power curve when the sample size is increased from $n = 46$ to $n = 100$. Assume that the standard deviation is $\sigma = 11.9$.

8.106 **Gas mileage of the Honda Civic.** According to the Environmental Protection Agency (EPA) *Fuel Economy Guide*, the 2009 Honda Civic automobile obtains a mean of 36 miles per gallon (mpg) on the highway. Suppose Honda claims that the EPA has underestimated the Civic's mileage. To support its assertion, the company selects $n = 50$ model 2009 Civic cars and records the mileage obtained for each car over a driving course similar to the one used by the EPA. The following data result: $\bar{x} = 38.3$ mpg, $s = 6.4$ mpg.

a. If Honda wishes to show that the mean mpg for 2009 Civic autos is greater than 36 mpg, what should the alternative hypothesis be? the null hypothesis?

b. Do the data provide sufficient evidence to support the auto manufacturer's claim? Test, using $\alpha = .05$. List any assumptions you make in conducting the test.

c. Calculate the power of the test for the mean values of 36.5, 37.0, 37.5, 38.0, and 38.5, assuming that $s = 6.4$ is a good estimate of σ.

d. Plot the power of the test on the vertical axis against the mean on the horizontal axis. Draw a curve through the points.

e. Use the power curve of part **d** to estimate the power for the mean value $\mu = 37.75$. Calculate the power for this value of μ, and compare it with your approximation.

f. Use the power curve to approximate the power of the test when $\mu = 41$. If the true value of the mean mpg for this model is really 41, what (approximately) are the chances that the test will fail to reject the null hypothesis that the mean is 36?

8.107 **Cooling method for gas turbines.** Refer to the *Journal of Engineering for Gas Turbines and Power* (Jan. 2005) study of a new cooling method for gas turbine engines, presented in Exercise 8.34 (p. 366). The heat rates (kilojoules per kilowatt per hour) for 67 gas turbines cooled with the new method are saved in the **GASTURBINE** file. Assume that the standard deviation of the population of heat rates is $\sigma = 1{,}600$. In Exercise 8.34, you tested whether the true mean heat rate μ exceeded 10,000 kJ/kWh at $\alpha = .05$. If the true mean is really $\mu = 10{,}500$, what is the chance that you will make a Type II error?

8.108 Satellite radio in cars. Refer to the National Association of Broadcasters (NAB) survey of 501 satellite radio subscribers, Exercise 8.84 (p. 385). Recall that an NAB spokesperson claims that 80% of all satellite radio subscribers have a satellite radio receiver in their car. You conducted a test to determine if the claimed value is too high using $\alpha = .10$. What is the probability that the test will correctly conclude that the claim is too high, if in fact the true percentage of all satellite radio subscribers who have a satellite radio receiver in their car is 82%?

8.109 Gummy bears: red or yellow? Refer to the *Chance* (Winter 2010) experiment to determine if color of a gummy bear is related to its flavor, Exercise 8.86 (p. 386). You tested the null hypothesis of $p = .5$ against the two-tailed alternative hypothesis of $p \neq .5$ using $\alpha = .01$, where p represents the true proportion of blindfolded students who correctly identify the color of the gummy bear. Recall that of the 121 students who participated in the study, 97 correctly identified the color. Find the power of the test if the true proportion is $p = .65$.

8.8 Test of Hypothesis about a Population Variance (Optional)

Although many practical problems involve inferences about a population mean (or proportion), it is sometimes of interest to make an inference about a population variance σ^2. To illustrate, a quality control supervisor in a cannery knows that the exact amount each can contains will vary, since there are certain uncontrollable factors that affect the amount of fill. The mean fill per can is important, but equally important is the variance of the fill. If σ^2 is large, some cans will contain too little and others too much. Suppose regulatory agencies specify that the standard deviation of the amount of fill in 16-ounce cans should be less than .1 ounce. To determine whether the process is meeting this specification, the supervisor randomly selects 10 cans and weighs the contents of each. The results are given in Table 8.6.

Table 8.6   Fill Weights (ounces) of 10 Cans

16.00	15.95	16.10	16.02	15.99
16.06	16.04	16.05	16.03	16.02

Data Set: FILLAMOUNTS

Do these data provide sufficient evidence to indicate that the variability is as small as desired? To answer this question, we need a procedure for testing a hypothesis about σ^2.

Intuitively, it seems that we should compare the sample variance σ^2 with the hypothesized value of σ^2 (or s with σ) in order to make a decision about the population's variability. The quantity

$$\frac{(n-1)s^2}{\sigma^2}$$

is known to have a **chi-square** (χ^2) **distribution** when the population from which the sample is taken is *normally distributed*. (Chi-square distributions were introduced in optional Section 7.6.)

Since the distribution of $\dfrac{(n-1)s^2}{\sigma^2}$ is known, we can use this quantity as a test statistic in a test of hypothesis for a population variance, as illustrated in the next example.

BIOGRAPHY   **FRIEDRICH R. HELMERT (1843–1917)**

Helmert Transformations

German Friedrich Helmert studied engineering sciences and mathematics at Dresden University, where he earned his Ph.D. Then he accepted a position as a professor of geodesy—the scientific study of the earth's size and shape—at the technical school in Aachen. Helmert's mathematical solutions to geodesy problems led him to several statistics-related discoveries. His greatest statistical contribution occurred in 1876, when he was the first to prove that the sampling distribution of the sample variance s^2 is a chi-square distribution. Helmert used a series of mathematical transformations to obtain the distribution of s^2—transformations that have since been named "Helmert transformations" in his honor. Later in life, Helmert was appointed professor of advanced geodesy at the prestigious University of Berlin and director of the Prussian Geodetic Institute. ■

Example 8.12
A Test for σ^2—Fill Weight Variance

Problem Refer to the fill weights for the sample of ten 16-ounce cans in Table 8.6. Is there sufficient evidence to conclude that the true standard deviation σ of the fill measurements of 16-ounce cans is less than .1 ounce?

Solution Here, we want to test whether $\sigma < .1$. Since the null and alternative hypotheses must be stated in terms of σ^2 (rather than σ), we want to test the null hypothesis that $\sigma^2 = (.1)^2 = .01$ against the alternative that $\sigma^2 < .01$. Therefore, the elements of the test are

$H_0: \sigma^2 = .01$ (Fill variance equals .01—i.e., process specifications are not met.)

$H_a: \sigma^2 < .01$ (Fill variance is less than .01—i.e., process specifications are met.)

Test statistic: $\chi^2 = \dfrac{(n-1)s^2}{\sigma^2}$

Assumption: The distribution of the amounts of fill is approximately normal.

Rejection region: The smaller the value of s^2 we observe, the stronger is the evidence in favor of H_a. Thus, we reject H_0 for "small values" of the test statistic. Recall that the chi-square distribution depends on $(n-1)$ degrees of freedom. With $\alpha = .05$ and $(n-1) = 9$ df, the χ^2 value for rejection is found in Table VII and pictured in Figure 8.24. We will reject H_0 if $\chi^2 < 3.32511$.

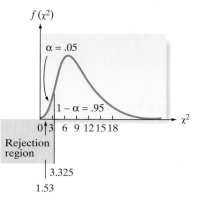

Figure 8.24

Rejection region for Example 8.12

[*Note:* The area given in Table VII is the area to the *right* of the numerical value in the table. Thus, to determine the lower-tail value, which has $\alpha = .05$ to its *left*, we used the $\chi^2_{.95}$ column in Table VII.]

A SAS printout of the analysis is displayed in Figure 8.25. The value of s (highlighted on the printout) is $s = .0412$. Substituting into the formula for the test statistic, we have

$$\chi^2 = \frac{(n-1)s^2}{\sigma^2} = \frac{9(.0412)^2}{.01} = 1.53$$

```
Sample Statistics for FILLAMT

      N      Mean      Std. Dev.      Variance
--------------------------------------------------
     10     16.026       0.0412         0.0017


Hypothesis Test

    Null hypothesis:         Variance of FILLAMT => 0.01
    Alternative:             Variance of FILLAMT <  0.01

        Chi-square         Df         Prob
    ------------------------------------------
          1.524             9         0.0030
```

Figure 8.25

SAS test of fill amount variance, Example 8.12

Conclusion: Since the value of the test statistic is less than 3.32511, the supervisor can conclude that the variance σ^2 of the population of all amounts of fill is less than .01 (i.e., $\sigma < .1$), with probability of a Type I error equal to $\alpha = .05$. If this procedure is repeatedly used, it will incorrectly reject H_0 only 5% of the time. Thus, the quality control supervisor is confident in the decision that the cannery is operating within the desired limits of variability.

Look Back Note that both the test statistic and the one-tailed *p*-value of the test are shown on the SAS printout. Note also that the *p*-value (.003) is less than $\alpha = .05$, thus confirming our conclusion to reject H_0.

Now Work Exercise 8.119

One-tailed and two-tailed tests of hypothesis for σ^2 are given in the following box.

Test of Hypothesis about σ^2

One-Tailed Test

H_0: $\sigma^2 = \sigma_0^2$

H_a: $\sigma^2 < \sigma_0^2$ (or H_a: $\sigma^2 > \sigma_0^2$)

Test statistic: $\chi^2 = \dfrac{(n-1)s^2}{\sigma_0^2}$

Rejection region: $\chi^2 < \chi_{(1-\alpha)}^2$

(or $\chi^2 > \chi_\alpha^2$ when H_a: $\sigma^2 > \sigma_0^2$)

Two-Tailed Test

H_0: $\sigma^2 = \sigma_0^2$

H_a: $\sigma^2 \neq \sigma_0^2$

Test statistic: $\chi^2 = \dfrac{(n-1)s^2}{\sigma_0^2}$

Rejection region: $\chi^2 < \chi_{(1-\alpha/2)}^2$

or $\chi^2 > \chi_{(\alpha/2)}^2$

where σ_0^2 is the hypothesized variance and the distribution of χ^2 is based on $(n-1)$ degrees of freedom.

Conditions Required for a Valid Large-Sample Hypothesis Test for σ^2

1. A random sample is selected from the target population.
2. The population from which the sample is selected has a distribution that is approximately normal.

⚠ **CAUTION** The procedure for conducting a hypothesis test for σ^2 in the preceding examples requires an assumption regardless of whether the sample size n is large or small. We must assume that the population from which the sample is selected has an approximate normal distribution. Unlike small-sample tests for μ based on the *t*-statistic, *slight to moderate departures from normality will render the χ^2 test invalid.* ▲

Exercises 8.110–8.128

Understanding the Principles

8.110 What sampling distribution is used to make inferences about σ^2?

8.111 What conditions are required for a valid test for σ^2?

8.112 *True or False.* The null hypotheses H_0: $\sigma^2 = .25$ and H_0: $\sigma = .5$ are equivalent.

8.113 *True or False.* When the sample size n is large, no assumptions about the population are necessary to test the population variance σ^2.

Learning the Mechanics

8.114 Let χ_0^2 be a particular value of χ^2. Find the value of χ_0^2 such that

a. $P(\chi^2 > \chi_0^2) = .10$ for $n = 12$
b. $P(\chi^2 > \chi_0^2) = .05$ for $n = 9$
c. $P(\chi^2 > \chi_0^2) = .025$ for $n = 5$

8.115 A random sample of n observations is selected from a normal population to test the null hypothesis that $\sigma^2 = 25$. Specify the rejection region for each of the following combinations of H_a, α, and n:

a. H_a: $\sigma^2 \neq 25$; $\alpha = .05$; $n = 16$
b. H_a: $\sigma^2 > 25$; $\alpha = .01$; $n = 23$
c. H_a: $\sigma^2 > 25$; $\alpha = .10$; $n = 15$
d. H_a: $\sigma^2 < 25$; $\alpha = .01$; $n = 13$
e. H_a: $\sigma^2 \neq 25$; $\alpha = .10$; $n = 7$
f. H_a: $\sigma^2 < 25$; $\alpha = .05$; $n = 25$

8.116 A random sample of seven measurements gave $\bar{x} = 9.4$ and $s^2 = 1.84$.

   a. What assumptions must you make concerning the population in order to test a hypothesis about σ^2?

   b. Suppose the assumptions in part **a** are satisfied. Test the null hypothesis $\sigma^2 = 1$ against the alternative hypothesis $\sigma^2 > 1$. Use $\alpha = .05$.

   c. Refer to part **b**. Suppose the test statistic is $\chi^2 = 14.45$. Use Table VII of Appendix A to find the approximate p-value of the test.

   d. Test the null hypothesis $\sigma^2 = 1$ against the alternative hypothesis $\sigma^2 \neq 1$. Use $\alpha = .05$.

8.117 Refer to Exercise 8.116. Suppose we had $n = 100$, $\bar{x} = 9.4$, and $s^2 = 1.84$.

   a. Test the null hypothesis $H_0: \sigma^2 = 1$ against the alternative hypothesis $H_a: \sigma^2 > 1$.

   b. Compare your test result with that of Exercise 8.116.

8.118 A random sample of $n = 7$ observations from a normal population produced the following measurements: 4, 0, 6, 1, 3, 3, 5, 9. Do the data provide sufficient evidence to indicate that $\sigma^2 < 2$? Test, using $\alpha = .05$.

Applying the Concepts—Basic

8.119 **Latex allergy in health care workers.** Refer to the *Current Allergy & Clinical Immunology* (March 2004) study of $n = 46$ hospital employees who were diagnosed with a latex allergy from exposure to the powder on latex gloves, presented in Exercise 8.28 (p. 365). Recall that the number of latex gloves used per week by the sampled workers is summarized as follows: $\bar{x} = 19.3$ and $s = 11.9$. Let σ^2 represent the variance in the number of latex gloves used per week by all hospital employees. Consider testing $H_0: \sigma^2 = 100$ against $H_a: \sigma^2 \neq 100$.

   a. Give the rejection region for the test at a significance level of $\alpha = .01$.

   b. Calculate the value of the test statistic.

   c. Use the results from parts **a** and **b** to draw the appropriate conclusion.

8.120 **A new dental bonding agent.** Refer to the *Trends in Biomaterials & Artificial Organs* (January 2003) study of a new bonding adhesive for teeth called "Smartbond," presented in Exercise 8.72 (p. 379). Recall that tests on a sample of 10 extracted teeth bonded with the new adhesive resulted in a mean breaking strength (after 24 hours) of $\bar{x} = 5.07$ Mpa and a standard deviation of $s = .46$ Mpa. In addition to requiring a good mean breaking strength, orthodontists are concerned about the variability in breaking strength of the new bonding adhesive.

   a. Set up the null and alternative hypothesis for a test to determine whether the breaking strength variance differs from .5 Mpa.

   b. Find the rejection region for the test, using $\alpha = .01$.

   c. Compute the test statistic.

   d. Give the appropriate conclusion for the test.

   e. What conditions are required for the test results to be valid?

8.121 **Characteristics of a rockfall.** Refer to the *Environmental Geology* (Vol. 58, 2009) simulation study of how far a block from a collapsing rockwall will bounce down a soil slope, Exercise 2.59 (p. 57). Rebound lengths (in meters) were estimated for 13 rock bounces. The data are repeated in the table and saved in the **ROCKFALL** file. Descriptive statistics for the rebound lengths are shown on the SAS printout at the bottom of the page. Consider a test of hypothesis for the variation in rebound lengths for the theoretical population of rock bounces from the collapsing rockwall. In particular, a geologist wants to determine if the variance differs from 10 m².

10.94	13.71	11.38	7.26	17.83	11.92	11.87	5.44	13.35
4.90	5.85	5.10	6.77					

Based on Paronuzzi, P. "Rockfall-induced block propagation on a soil slope, northern Italy." *Environmental Geology*, Vol. 58, 2009 (Table 2).

   a. Define the parameter of interest.

   b. Specify the null and alternative hypothesis.

   c. Compute the value of the test statistic.

   d. Determine the rejection region for the test using $\alpha = .10$.

   e. Make the appropriate conclusion.

   f. What condition must be satisfied in order for the inference, part **e**, to be valid?

8.122 **Identifying type of urban land cover.** Refer to the *Geographical Analysis* (Oct. 2006) study of a new method for analyzing remote-sensing data from satellite pixels, presented in Exercise 8.32 (p. 365). Recall that the method uses a numerical measure of the distribution of gaps, or the sizes of holes, in the pixel, called *lacunarity*. Summary statistics for the lacunarity measurements in a sample of 100 grassland pixels are $\bar{x} = 225$ and $s = 20$. As stated in Exercise 8.32, it is known that the mean lacunarity measurement for all grassland pixels is 220. The method will be effective in identifying land cover if the standard deviation of the measurements is 10% (or less) of the true mean (i.e., if the standard deviation is less than 22).

   a. Give the null and alternative hypothesis for a test to determine whether, in fact, the standard deviation of all grassland pixels is less than 22.

   b. A MINITAB analysis of the data is provided on page 400. Locate and interpret the p-value of the test. Use $\alpha = .10$.

Applying the Concepts—Intermediate

8.123 **Point spreads of NFL games.** During the National Football League (NFL) season, Las Vegas oddsmakers establish

SAS Output for Exercise 8.121

The MEANS Procedure

Analysis Variable : LENGTH

N	Mean	Std Dev	Minimum	Maximum
13	9.7169231	4.0947291	4.9000000	17.8300000

MINITAB output for Exercise 8.122

Test for One Standard Deviation

```
Method
Null hypothesis        Sigma = 22
Alternative hypothesis Sigma = < 22

The standard method is only for the normal distribution.

Statistics
N   StDev  Variance
100  20.0      400


Tests
Method     Chi-Square  DF  P-Value
Standard        81.82  99    0.105
```

a point spread on each game for betting purposes. For example, the Champion Green Bay Packers were established as 2.5-point favorites over the Pittsburgh Steelers in the 2011 Super Bowl. The final scores of NFL games were compared against the final point spreads established by the oddsmakers in *Chance* (Fall 1998). The difference between the outcome of the game and the point spread (called a point-spread error) was calculated for 240 NFL games. The mean and standard deviation of the point-spread errors are $\bar{x} = -1.6$ and $s = 13.3$. Suppose the researcher wants to know whether the true standard deviation of the point-spread errors exceeds 15. Conduct the analysis using $\alpha = .05$.

8.124 Mongolian desert ants. Refer to the *Journal of Biogeography* (December 2003) study of ants in Mongolia (central Asia), presented in Exercise 8.74 (p. 379). Data on the number of ant species attracted to 11 randomly selected desert sites are saved in the **GOBIANTS** file. Do these data indicate that the standard deviation of the number of ant species at all Mongolian desert sites exceeds 15 species? Conduct the appropriate test at $\alpha = .05$. Are the conditions required for a valid test satisfied?

8.125 Birth weights of cocaine babies. A group of researchers at the University of Texas-Houston conducted a comprehensive study of pregnant cocaine-dependent women (*Journal of Drug Issues*, Summer 1997). All the women in the study used cocaine on a regular basis (at least three times a week) for more than a year. One of the many variables measured was birth weight (in grams) of the baby delivered. For a sample of 16 cocaine-dependent women, the mean birth weight was 2,971 grams and the standard deviation was 410 grams. Test (at $\alpha = .01$) to determine whether the variance in birth weights of babies delivered by cocaine-dependent women is less than 200,000 grams[2].

8.126 Cooling method for gas turbines. Refer to the *Journal of Engineering for Gas Turbines and Power* (Jan. 2005) study of the performance of augmented gas turbine engines, presented in Exercise 8.34 (p. 366). Recall that the performance of each in a sample of 67 gas turbines was measured by heat rate (kilojoules per kilowatt per hour). The data are saved in the **GASTURBINE** file. Suppose that standard gas turbines have heat rates with a standard deviation of 1,500 kJ/kWh. Is there sufficient evidence to indicate that the heat rates of the augmented gas turbine engine are more variable than the heat rates of the standard gas turbine engine? Test, using $\alpha = 0.5$.

8.127 Jitter in a water power system. Refer to the *Journal of Applied Physics* investigation of throughput jitter in the opening switch of a prototype water power system, Exercise 7.104 (p. 338). Recall that low throughput jitter is critical to successful waterline technology. An analysis of conduction time for a sample of 18 trials of the prototype system yielded $\bar{x} = 334.8$ nanoseconds and $s = 6.3$ nanoseconds. (Conduction time is defined as the length of time required for the downstream current to equal 10% of the upstream current.) A system is considered to have low throughput jitter if the true conduction time standard deviation is less than 7 nanoseconds. Does the prototype system satisfy this requirement? Test using $\alpha = .01$.

Applying the Concepts—Advanced

8.128 Motivation of drug dealers. Refer to the *Applied Psychology in Criminal Justice* (Sept. 2009) study of the personality characteristics of convicted drug dealers, Exercise 7.15 (p. 307). A random sample of 100 drug dealers had a mean Wanting Recognition (WR) score of 39 points, with a standard deviation of 6 points. Recall that the WR score is a quantitative measure of a person's level of need for approval and sensitivity to social situations. (Higher scores indicate a greater need for approval.) A criminal psychologist claims that the range of WR scores for the population of convicted drug dealers is 42 points. Do you believe the psychologist's claim? (*Hint*: Assume the population of WR scores is normally distributed.)

CHAPTER NOTES

Key Terms

Note: Starred (*) terms are from the optional sections in this chapter.

Alternative (research)
  hypothesis 351
*Chi-square (χ^2) distribution 396
Hypothesis 350
Level of significance 355
Lower-tailed test 356
Conclusion 354
*Null distribution 387
Null hypothesis 351
Observed significance level
  (*p*-value) 367

One-tailed (one-sided)
  statistical test 356
*Power of a test 391
Rejection region 352
Test of hypothesis 350
Test statistic 351
Two-tailed (two-sided)
  hypothesis 356
Two-tailed test 356
Type I error 352
Type II error 353
Upper-tailed test 356

Key Symbols

μ	Population mean
p	Population proportion, P(Success), in binomial trial
σ^2	Population variance
$\bar{x}$	Sample mean (estimator of μ)
$\hat{p}$	Sample proportion (estimator of p)
s^2	Sample variance (estimator of σ^2)
H_0	Null hypothesis
H_a	Alternative hypothesis
α	Probability of a Type I error
β	Probability of a Type II error
χ^2	Chi-square (sampling distribution of s^2 for normal data)

Key Ideas

Key Words for Identifying the Target Parameter

μ—Mean, Average
p—Proportion, Fraction, Percentage, Rate, Probability
σ^2—Variance, Variability, Spread

Elements of a Hypothesis Test

1. *Null hypothesis (H_o)*
2. *Alternative hypothesis (H_a)*
3. *Test statistic (z, t, or X^2)*
4. *Significance level (α)*
5. *p-value*
6. *Conclusion*

Forms of Alternative Hypothesis

Lower tailed: H_a: $\mu_0 < 50$

Upper tailed: H_a: $\mu_0 > 50$
Two tailed: H_a: $\mu_0 \neq 50$

Type I Error $=$ Reject H_0 when H_0 is true (occurs with probability α)

Type II Error $=$ Accept H_0 when H_0 is false (occurs with probability β)

Power of a Test $= P$ (Reject H_0 when H_0 is false)
$= 1 - \beta$

Using p-values to Decide

1. Choose significance level (α)
2. Obtain p-value of the test
3. If $\alpha > p$-value, Reject H_0

Guide to Selecting a One-Sample Hypothesis Test

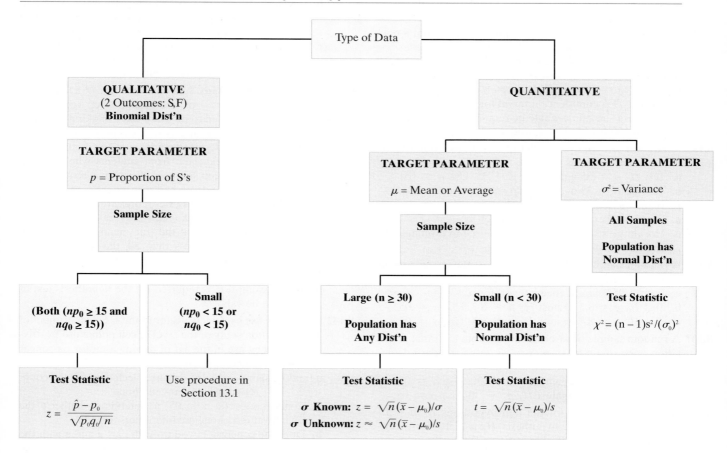

Supplementary Exercises 8.129–8.163

Note: List the assumptions necessary for the valid implementation of the statistical procedures you use in solving all these exercises. Starred (*) exercises refer to the optional sections in this chapter.

Understanding the Principles

8.129 *Complete the following statement:* The smaller the *p*-value associated with a test of hypothesis, the stronger is the support for the _____ hypothesis. Explain your answer.

8.130 Specify the differences between a large-sample and small-sample test of hypothesis about a population mean μ. Focus on the assumptions and test statistics.

8.131 Which of the elements of a test of hypothesis can and should be specified *prior* to analyzing the data that are to be utilized to conduct the test?

8.132 If the rejection of the null hypothesis of a particular test would cause your firm to go out of business, would you want α to be small or large? Explain.

8.133 *Complete the following statement:* The larger the *p*-value associated with a test of hypothesis, the stronger is the support for the _____ hypothesis. Explain your answer.

Learning the Mechanics

8.134 A random sample of 20 observations selected from a normal population produced $\bar{x} = 72.6$ and $s^2 = 19.4$.
 a. Form a 90% confidence interval for μ.
 b. Test $H_0: \mu = 80$ against $H_a: \mu < 80$. Use $\alpha = .05$.
 c. Test $H_0: \mu = 80$ against $H_a: \mu \neq 80$. Use $\alpha = .01$.
 d. Form a 99% confidence interval for μ.
 e. How large a sample would be required to estimate μ to within 1 unit with 95% confidence?

8.135 A random sample of $n = 200$ observations from a binomial population yields $\hat{p} = .29$.
 a. Test $H_0: p = .35$ against $H_a: p < .35$. Use $\alpha = .05$.
 b. Test $H_0: p = .35$ against $H_a: p \neq .35$. Use $\alpha = .05$.
 c. Form a 95% confidence interval for p.
 d. Form a 99% confidence interval for p.
 e. How large a sample would be required to estimate p to within .05 with 99% confidence?

8.136 A random sample of 175 measurements possessed a mean of $\bar{x} = 8.2$ and a standard deviation of $s = .79$.
 a. Form a 95% confidence interval for μ.
 b. Test $H_0: \mu = 8.3$ against $H_a: \mu \neq 8.3$. Use $\alpha = .05$.
 c. Test $H_0: \mu = 8.4$ against $H_a: \mu \neq 8.4$. Use $\alpha = .05$.

*8.137 A random sample of 41 observations from a normal population possessed a mean of $\bar{x} = 88$ and a standard deviation of $s = 6.9$.
 a. Test $H_0: \sigma^2 = 30$ against $H_a: \sigma^2 > 30$. Use $\alpha = .05$.
 b. Test $H_0: \sigma^2 = 30$ against $H_a: \sigma^2 \neq 30$. Use $\alpha = .05$.

8.138 A *t*-test is conducted for the null hypothesis $H_0: \mu = 10$ versus the alternative hypothesis $H_a: \mu > 10$ for a random sample of $n = 17$ observations. The test results are $t = 1.174$ and *p*-value $= .1288$.
 a. Interpret the *p*-value.
 b. What assumptions are necessary for the validity of this test?
 c. Calculate and interpret the *p*-value, assuming that the alternative hypothesis was instead $H_a: \mu \neq 10$.

Applying the Concepts—Basic

8.139 Use of herbal therapy. According to the *Journal of Advanced Nursing* (January 2001), 45% of senior women (i.e., women over the age of 65) use herbal therapies to prevent or treat health problems. Also, senior women who use herbal therapies use an average of 2.5 herbal products in a year.
 a. Give the null hypothesis for testing the first claim by the journal.
 b. Give the null hypothesis for testing the second claim by the journal.

8.140 FDA mandatory new-drug testing. When a new drug is formulated, the pharmaceutical company must subject it to lengthy and involved testing before receiving the necessary permission from the Food and Drug Administration (FDA) to market the drug. The FDA requires the pharmaceutical company to provide substantial evidence that the new drug is safe for potential consumers.
 a. If the new-drug testing were to be placed in a test-of-hypothesis framework, would the null hypothesis be that the drug is safe or unsafe? the alternative hypothesis?
 b. Given the choice of null and alternative hypotheses in part **a**, describe Type I and Type II errors in terms of this application. Define α and β in terms of this application.
 c. If the FDA wants to be very confident that the drug is safe before permitting it to be marketed, is it more important that α or β be small? Explain.

8.141 Sleep deprivation study. In a British study, 12 healthy college students deprived of one night's sleep received an array of tests intended to measure their thinking time, fluency, flexibility, and originality of thought. The overall test scores of the sleep-deprived students were compared with the average score expected from students who received their accustomed sleep. Suppose the overall scores of the 12 sleep-deprived students had a mean of $\bar{x} = 63$ and a standard deviation of 17. (Lower scores are associated with a decreased ability to think creatively.)
 a. Test the hypothesis that the true mean score of sleep-deprived subjects is less than 80, the mean score of subjects who received sleep prior to taking the test. Use $\alpha = .05$.
 b. What assumption is required for the hypothesis test of part **a** to be valid?

8.142 Cell phone use by drivers. Refer to the U.S. Department of Transportation study of the level of cell phone use by drivers while they are in the act of driving a motor passenger vehicle, presented in Exercise 7.119 (p. 342). Recall that in a random sample of 1,165 drivers selected across the country, 35 were found using their cell phone.
 a. Conduct a test (at $\alpha = .05$) to determine whether p, the true driver cell phone use rate, differs from .02.
 b. Does the conclusion, you drew in part **a** agree with the inference you derived from the 95% confidence interval for p in Exercise 7.119? Explain why or why not.

8.143 Al Qaeda attacks on the United States. Refer to the *Studies in Conflict & Terrorism* (Vol. 29, 2006) analysis of recent incidents involving suicide terrorist attacks, presented in Exercise 7.123 (p. 342). Data on the number of individual

MINITAB output for Exercise 8.143

One-Sample T: ATTACKS

Test of mu = 2.5 vs not = 2.5

Variable	N	Mean	StDev	SE Mean	90% CI	T	P
ATTACKS	21	1.857	1.195	0.261	(1.407, 2.307)	-2.46	0.023

suicide bombings that occurred in each of 21 sampled Al Qaeda attacks against the United States are reproduced in the table below and saved in the **ALQAEDA** file.

a. Do the data indicate that the true mean number of suicide bombings for all Al Qaeda attacks against the United States differs from 2.5? Use $\alpha = .10$ and the MINITAB printout above to answer the question.

b. In Exercise 7.123, you found a 90% confidence interval for the mean μ of the population. This interval is also shown on the MINITAB printout. Answer the question in part **a** on the basis of the 90% confidence interval.

c. Do the inferences derived from the test (part **a**) and confidence interval (part **b**) agree? Explain why or why not.

d. What assumption about the data must be true for the inferences to be valid?

e. Use a graph to check whether the assumption you made in part **d** is reasonably satisfied. Comment on the validity of the inference.

1 1 2 1 2 4 1 1 1 1 2 3 4 5 1 1 1 2 2 2 1

Based on Moghadam, A. "Suicide terrorism, occupation, and the globalization of martyrdom: A critique of *Dying to Win*." *Studies in Conflict & Terrorism*, Vol. 29, No. 8, 2006 (Table 3).

8.144 The "Pepsi challenge." "Take the Pepsi Challenge" was a marketing campaign used by the Pepsi-Cola Company. Coca-Cola drinkers participated in a blind taste test in which they tasted unmarked cups of Pepsi and Coke and were asked to select their favorite. Pepsi claimed that "in recent blind taste tests, more than half the Diet Coke drinkers surveyed said they preferred the taste of Diet Pepsi." Suppose 100 Diet Coke drinkers took the Pepsi Challenge and 56 preferred the taste of Diet Pepsi. Test the hypothesis that more than half of all Diet Coke drinkers will select Diet Pepsi in a blind taste test. Use $\alpha = .05$.

8.145 Masculinizing human faces. Refer to the *Nature* (August 27, 1998) study of facial characteristics that are deemed attractive, presented in Exercise 8.55 (p. 373). In another experiment, 67 human subjects viewed side by side an image of a Caucasian male face and the same image 50% masculinized. Each subject was asked to select the facial image they deemed more attractive. Fifty-eight of the 67 subjects felt that masculinization of face shape decreased attractiveness of the male face. The researchers used this sample information to test whether the subjects showed a preference for either the unaltered or the morphed male face.

a. Set up the null and alternative hypotheses for this test.

b. Compute the test statistic.

c. The researchers reported a p-value ≈ 0 for the test. Do you agree?

d. Make the appropriate conclusion in the words of the problem. Use $\alpha = .01$.

*8.146 Time taken to solve a math programming problem. Refer to the *IEEE Transactions* study of a new hybrid algorithm for solving polynomial 0/1 mathematical programs, presented in Exercise 7.131 (p. 343) (Data on solution times are saved in the **MATHCPU** file.) A SAS printout giving descriptive statistics for the sample of 52 solution times is reproduced at the bottom of the page. Use this information to determine whether the variance of the solution times differs from 2. Use $\alpha = .05$.

8.147 Alkalinity of river water. In Exercise 5.125 (p. 264), you learned that the mean alkalinity level of water specimens collected from the Han River in Seoul, Korea, is 50 milligrams per liter (*Environmental Science & Engineering*, September 1, 2000). Consider a random sample of 100 water specimens collected from a tributary of the Han River. Suppose the mean and standard deviation of the alkalinity levels for the sample are, respectively, $\bar{x} = 67.8$ mpl and $s = 14.4$ mpl. Is there sufficient evidence (at $\alpha = .01$) to indicate that the population mean alkalinity level of water in the tributary exceeds 50 mpl?

Applying the Concepts—Intermediate

8.148 Errors in medical tests. Medical tests have been developed to detect many serious diseases. A medical test is designed to minimize the probability that it will produce a "false positive" or a "false negative." A false positive is a positive test result for an individual who does not have the disease, whereas a false negative is a negative test result for an individual who does have the disease.

a. If we treat a medical test for a disease as a statistical test of hypothesis, what are the null and alternative hypotheses for the medical test?

b. What are the Type I and Type II errors for the test? Relate each to false positives and false negatives.

c. Which of these errors has graver consequences? Considering this error, is it more important to minimize α or β? Explain.

SAS output for Exercise 8.146

The MEANS Procedure

Analysis Variable : CPU

Mean	Std Dev	Variance	N	Minimum	Maximum
0.8121923	1.5047603	2.2643035	52	0.0360000	8.7880000

8.149 **Post-traumatic stress of POWs.** *Psychological Assessment* (March 1995) published the results of a study of World War II aviators captured by German forces after having been shot down. Having located a total of 239 World War II aviator POW survivors, the researchers asked each veteran to participate in the study; 33 responded to the letter of invitation. Each of the 33 POW survivors was administered the Minnesota Multiphasic Personality Inventory, one component of which measures level of post-traumatic stress disorder (PTSD). [*Note:* The higher the score, the higher is the level of PTSD.] The aviators produced a mean PTSD score of $\bar{x} = 9.00$ and a standard deviation of $s = 9.32$. Conduct a test to determine if the true mean PTSD score for all World war II aviator POWS is less than 16. [*Note:* The value 16 represents the mean PTSD score established for Vietnam POWS.] Use $\alpha = .10$.

8.150 **Cracks in highway pavement.** Using van-mounted state-of-the-art video technology, the Mississippi Department of Transportation collected data on the number of cracks (called *crack intensity*) in an undivided two-lane highway. (*Journal of Infrastructure Systems*, March 1995). The mean number of cracks found in a sample of eight 50-meter sections of the highway was $\bar{x} = .210$, with a variance of $s^2 = .011$. Suppose the American Association of State Highway and Transportation Officials (AASHTO) recommends a maximum mean crack intensity of .100 for safety purposes.
  a. Test the hypothesis that the true mean crack intensity of the Mississippi highway exceeds the AASHTO recommended maximum. Use $\alpha = .01$.
  b. Define a Type I error and a Type II error for this study.

8.151 **Inbreeding of tropical wasps.** Refer to the *Science* study of inbreeding in tropical swarm-founding wasps, presented in Exercise 7.129 (p. 343). A sample of 197 wasps, captured, frozen, and subjected to a series of genetic tests, yielded a sample mean inbreeding coefficient of $\bar{x} = .044$ with a standard deviation of $s = .884$. Recall that if the wasp has no tendency to inbreed, the true mean inbreeding coefficient μ for the species will equal 0.
  a. Test the hypothesis that the true mean inbreeding coefficient μ for this species of wasp exceeds 0. Use $\alpha = .05$.
  b. Compare the inference you made in part **a** with the inference you obtained in Exercise 7.129, using a confidence interval. Do the inferences agree? Explain.

*8.152 **Weights of parrot fish.** A marine biologist wishes to use parrot fish for experimental purposes due to the belief that their weight is fairly stable (i.e., the variability in weights among parrot fish is small). The biologist randomly samples 10 parrot fish and finds that their mean weight is 4.3 pounds and the standard deviation is 1.4 pounds. The biologist will use the parrot fish only if there is evidence that the variance of their weights is less than 4.
  a. Is there sufficient evidence for the biologist to claim that the variability in weights among parrot fish is small enough to justify their use in the experiment? Test at $\alpha = .05$.
  b. State any assumptions that are needed for the test mentioned in part **a** to be valid.

8.153 **PCB in plant discharge.** The EPA sets a limit of 5 parts per million (ppm) on PCB (polychlorinated biphenyl, a dangerous substance) in water. A major manufacturing firm producing PCB for electrical insulation discharges small amounts from the plant. The company management, attempting to control the PCB in its discharge, has given instructions to halt production if the mean amount of PCB in the effluent exceeds 3 ppm. A random sample of 50 water specimens produced the following statistics: $\bar{x} = 3.1$ ppm and $s = .5$ ppm.
  a. Do these statistics provide sufficient evidence to halt the production process? Use $\alpha = .01$.
  b. If you were the plant manager, would you want to use a large or a small value for α for the test in part **a**?

*8.154 **PCB in plant discharge (cont'd).** Refer to Exercise 8.153.
  a. In the context of the problem, define a Type II error.
  b. Calculate β for the test described in part **a** of Exercise 8.153, assuming that the true mean is $\mu = 3.1$ ppm.
  c. What is the power of the test to detect the effluent's departure from the standard of 3.0 ppm when the mean is 3.1 ppm?
  d. Repeat parts **b** and **c**, assuming that the true mean is 3.2 ppm. What happens to the power of the test as the plant's mean PCB departs farther from the standard?

*8.155 **PCB in plant discharge (cont'd).** Refer to Exercises 8.153 and 8.154.
  a. Suppose an α value of .05 is used to conduct the test. Does this change favor the manufacturer? Explain.
  b. Determine the value of β and the power for the test when $\alpha = .05$ and $\mu = 3.1$.
  c. What happens to the power of the test when α is increased?

8.156 **Federal civil trial appeals.** Refer to the *Journal of the American Law and Economics Association* (Vol. 3, 2001) study of appeals of federal civil trials, presented in Exercise 3.59 (p. 134). A breakdown of 678 civil cases that were originally tried in front of a judge and appealed by either the plaintiff or the defendant is reproduced in the accompanying table. Do the data provide sufficient evidence to indicate that the percentage of civil cases appealed that are actually reversed is less than 25%? Test, using $\alpha = .01$.

Outcome of Appeal	Number of Cases
Plaintiff trial win—reversed	71
Plaintiff trial win—affirmed/dismissed	240
Defendant trial win—reversed	68
Defendant trial win—affirmed/dismissed	299
Total	678

8.157 **Choosing portable grill displays.** Refer to the *Journal of Consumer Research* (March 2003) experiment on influencing the choices of others by offering undesirable alternatives, presented in Exercise 3.27 (p. 122). Recall that each of 124 college students selected three portable grills out of five to display on the showroom floor. The students were instructed to include Grill #2 (a smaller-sized grill) and select the remaining two grills in the display to maximize purchases of Grill #2. If the six possible grill display combinations (1–2–3, 1–2–4, 1–2–5,

2–3–4, 2–3–5, and 2–4–5) are selected at random, then the proportion of students selecting any display will be $1/6 = .167$. One theory tested by the researcher is that the students will tend to choose the three-grill display so that Grill #2 is a compromise between a more desirable and a less desirable grill. Of the 124 students, 85 students selected a three-grill display that was consistent with that theory. Use this information to test the theory proposed by the researcher at $\alpha = .05$.

*8.158 Interocular eye pressure. Ophthalmologists require an instrument that can rapidly measure interocular pressure for glaucoma patients. The device now in general use is known to yield readings of this pressure with a variance of 10.3. The variance of five pressure readings on the same eye by a newly developed instrument is equal to 9.8. Does this sample variance provide sufficient evidence to indicate that the new instrument is more reliable than the instrument currently in use? (Use $\alpha = .05$.)

Applying the Concepts—Advanced

8.159 Polygraph test error rates. In a classic study reported in *Discover* magazine, a group of physicians subjected the *polygraph* (or *lie detector*) to the same careful testing given to medical diagnostic tests. They found that if 1,000 people were subjected to the polygraph and 500 told the truth and 500 lied, the polygraph would indicate that approximately 185 of the truth tellers were liars and that approximately 120 of the liars were truth tellers.

a. In the application of a polygraph test, an individual is presumed to be a truth teller (H_0) until "proven" a liar (H_a). In this context, what is a Type I error? A Type II error?

b. According to the study, what is the probability (approximately) that a polygraph test will result in a Type I error? A Type II error?

8.160 Parents who condone spanking. In Exercise 4.128 (p. 219) you read about a nationwide survey which claimed that 60% of parents with young children condone spanking their child as a regular form of punishment (*Tampa Tribune*, October 5, 2000). In a random sample of 100 parents with young children, how many parents would need to say that they condone spanking as a form of punishment in order to refute the claim?

8.161 Solar joint inspections. X-rays and lasers are used to inspect solder-joint defects on printed circuit boards (PCBs). A particular manufacturer of laser-based inspection equipment claims that its product can inspect at least 10 solder joints per second, on average, when the joints

are spaced .1 inch apart. The equipment was tested by a potential buyer on 48 different PCBs. In each case, the equipment was operated for exactly 1 second. The numbers of solder joints inspected on each run are listed in the table. These data are saved in the **PCB** file.

10	9	10	10	11	9	12	8	8	9	6	10
7	10	11	9	9	13	9	10	11	10	12	8
9	9	9	7	12	6	9	10	10	8	7	9
11	12	10	0	10	11	12	9	7	9	9	10

a. The potential buyer doubts the manufacturer's claim. Do you agree?

b. Assume that the standard deviation of the number of solder joints inspected on each run is 1.2, and the true mean number of solder joints that can be inspected is really equal to 9.5. How likely is the buyer to correctly conclude that the claim is false?

Critical Thinking Challenges

8.162 The Hot Tamale caper. "Hot Tamales" are chewy, cinnamon-flavored candies. A bulk vending machine is known to dispense, on average, 15 Hot Tamales per bag. *Chance* (Fall 2000) published an article on a classroom project in which students were required to purchase bags of Hot Tamales from the machine and count the number of candies per bag. One student group claimed it purchased five bags that had the following candy counts: 25, 23, 21, 21, and 20. There was some question as to whether the students had fabricated the data. Use a hypothesis test to gain insight into whether or not the data collected by the students were fabricated. Use a level of significance that gives the benefit of the doubt to the students.

8.163 Verifying voter petitions. To get their names on the ballot of a local election, political candidates often must obtain petitions bearing the signatures of a minimum number of registered voters. According to the *St. Petersburg Times*, in Pinellas County, Florida, a certain political candidate obtained petitions with 18,200 signatures. To verify that the names on the petitions were signed by actual registered voters, election officials randomly sampled 100 of the signatures and checked each for authenticity. Only 2 were invalid signatures.

a. Is 98 out of 100 verified signatures sufficient to believe that more than 17,000 of the total 18,200 signatures are valid? Use $\alpha = .01$.

b. Repeat part **a** if only 16,000 valid signatures are required.

Activity | Challenging a Claim: Tests of Hypotheses

Use the Internet or a newspaper or magazine to find an example of a claim made by a political or special-interest group about some characteristic (e.g., favor gun control) of the U.S. population. In this activity, you represent a rival group that believes the claim may be false.

1. In your example, what kinds of evidence might exist which would cause one to suspect that the claim might be false and therefore worthy of a statistical study? Be specific. If the claim were false, how would consumers be hurt?

2. Describe what data are relevant and how that data might be collected.

3. Explain the steps necessary to reject the group's claim at level α. State the null and alternative hypotheses. If you reject the claim, does it mean that the claim is false?

4. If you reject the claim when the claim is actually true, what type of error has occurred? What is the probability of this error occurring?

5. If you were to file a lawsuit against the group based on your rejection of its claim, how might the group use your results to defend itself?

References

Snedecor, G. W., and Cochran, W. G. *Statistical Methods*, 7th ed. Ames, IA: Iowa State University Press, 1980.

Wackerly, D., Mendenhall, W., and Scheaffer, R. *Mathematical Statistics with Applications*, 7th ed. Belmont, CA: Thomson, Brooks/Cole, 2008.

USING TECHNOLOGY

MINITAB: Tests of Hypotheses

Testing μ

Step 1 Access the MINITAB data worksheet that contains the sample data.

Step 2 Click on the "Stat" button on the MINITAB menu bar and then click on "Basic Statistics" and "1-Sample t," as shown in Figure 8.M.1.

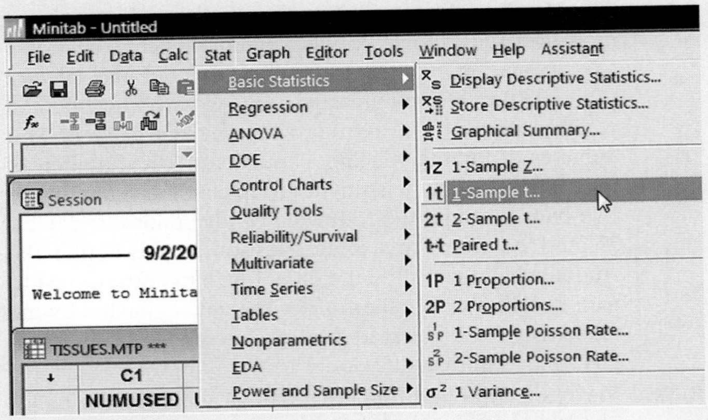

Figure 8.M.1
MINITAB menu options for testing a mean

Step 3 On the resulting dialog box (shown in Figure 8.M.2), click on "Samples in Columns" and then specify the quantitative variable of interest in the open box.

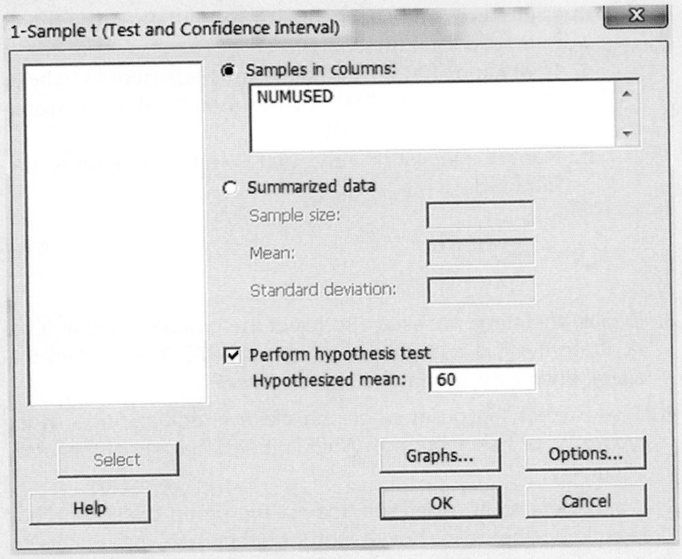

Figure 8.M.2
MINITAB 1-sample *t* test dialog box

Step 4 Check "Perform hypothesis test" and then specify the value of μ_0 for the null hypothesis in the "Hypothesized mean" box.

Step 5 Click on the "Options" button at the bottom of the dialog box and specify the form of the alternative hypothesis, as shown in Figure 8.M.3.

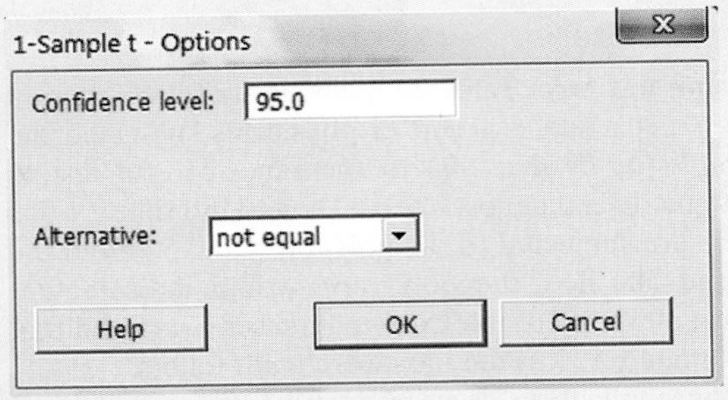

Figure 8.M.3
MINITAB 1-sample t test options

Step 6 Click "OK" to return to the "1-Sample t" dialog box and then click "OK" again to produce the hypothesis test.

Note: If you want to produce a test for the mean from summary information (e.g., the sample mean, sample standard deviation, and sample size), click on "Summarized data" in the "1-Sample t" dialog box, enter the values of the summary statistics and μ_0, and then click "OK."

Important: The MINITAB one-sample *t*-procedure uses the *t*-statistic to generate the hypothesis test. When the sample size *n* is small, this is the appropriate method. When the sample size *n* is large, the *t*-value will be approximately equal to the large-sample *z*-value, and the resulting test will still be valid. If you have a large sample and you know the value of the population standard deviation σ (which is rarely the case), select "1-sample Z" from the "Basic Statistics" menu options (see Figure 8.M.1) and make the appropriate selections.

Testing p

Step 1 Access the MINITAB data worksheet that contains the sample data.

Step 2 Click on the "Stat" button on the MINITAB menu bar and then click on "Basic Statistics" and "1 Proportion" (see Figure 8.M.1).

Step 3 On the resulting dialog box (shown in Figure 8.M.4), click on "Samples in Columns," and then specify the qualitative variable of interest in the open box.

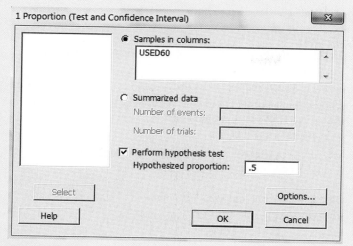

Figure 8.M.4

MINITAB 1-proportion test dialog box

Step 4 Check "Perform hypothesis test" and then specify the null hypothesis value p_0 in the "Hypothesized proportion" box.

Step 5 Click "Options," then specify the form of the alternative hypothesis in the resulting dialog box, as shown in Figure 8.M.5. Also, check the "Use test and interval based on normal distribution" box at the bottom.

Figure 8.M.5

MINITAB 1-proportion test options

Step 6 Click "OK" to return to the "1-Proportion" dialog box and then click "OK" again to produce the test results.

Note: If you want to produce a confidence interval for a proportion from summary information (e.g., the number of successes and the sample size), click on "Summarized data" in the "1 Proportion" dialog box (see Figure 8.M.4). Enter the value for the number of trials (i.e., the sample size) and the number of events (i.e., the number of successes), and then click "OK."

Testing σ^2

Step 1 Access the MINITAB data worksheet that contains the sample data set.

Step 2 Click on the "Stat" button on the MINITAB menu bar and and then click on "Basic Statistics" and "1 Variance" (see Figure 8.M.1).

Step 3 Once the resulting dialog box appears (see Figure 8.M.6), click on "Samples in Columns" and then specify the quantitative variable of interest in the open box.

Figure 8.M.6

MINITAB 1-variance test dialog box

Step 4 Check "Perform hypothesis test" and specify the null hypothesis value of the standard deviation σ_0 in the open box.

Step 5 Click on the "Options" button at the bottom of the dialog box and specify the form of the alternative hypothesis (similar to Figure 8.M.3).

Step 6 Click "OK" twice to produce the hypothesis test.

Note: If you want to produce a test for the variance from summary information (e.g., the sample standard deviation and sample size), click on "Summarized data" in the "1 Variance" dialog box (Figure 8.M.6) and enter the values of the summary statistics.

TI–83/TI–84 Plus Graphing Calculator: Tests of Hypotheses

Note: The TI-83/TI-84 plus graphing calculator cannot currently conduct a test for a population variance.

Hypothesis Test for a Population Mean (Large Sample Case)

Step 1 *Enter the data (Skip to Step 2 if you have summary statistics, not raw data.)*

- Press **STAT** and select **1:Edit**

Note: If the list already contains data, clear the old data. Use the up **ARROW** to highlight "**L1**."

- Press **CLEAR ENTER**
- Use the **ARROW** and **ENTER** keys to enter the data set into **L1**

Step 2 *Access the statistical tests menu*

- Press **STAT**
- Arrow right to **TESTS**
- Press **ENTER** to select either **Z-Test** (if large sample and known) or **T-Test** (if unknown)

Step 3 *Choose "Data" or "Stats" ("Data" is selected when you have entered the raw data into a List. "Stats" is selected when you are given only the mean, standard deviation, and sample size.)*

- Press **ENTER**

If you selected "Data," enter the values for the hypothesis test where $\mu_0 =$ the value for μ in the null hypothesis, $\sigma =$ assumed value of the population standard deviation.

- Set **List** to **L1**
- Set **Freq** to **1**
- Use the **ARROW** to highlight the appropriate alternative hypothesis
- Press **ENTER**
- Arrow down to "**Calculate**"
- Press **ENTER**

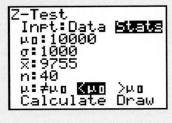

If you selected "Stats," enter the values for the hypothesis test where $\mu_0 =$ the value for μ in the null hypothesis, $\sigma =$ assumed value of the population standard deviation.

- Enter the sample mean, sample standard deviation, and sample size
- Use the **ARROW** to highlight the appropriate alternative hypothesis
- Press **ENTER**
- Arrow down to "**Calculate**"
- Press **ENTER**

```
Z-Test
 μ<10000
 z=-1.549516053
 P=.0606288707
 x̄=9755
 n=40
```

The chosen test will be displayed as well as the z (or t) test statistic, the p-value, the sample mean, and the sample size.

Testing p

Step 1 *Enter the data (Skip to Step 2 if you have summary statistics, not raw data.)*

- Press **STAT** and select **1:Edit**

Note: If the list already contains data, clear the old data. Use the up **ARROW** to highlight "**L1.**"

- Press **CLEAR ENTER**
- Use the **ARROW** and **ENTER** keys to enter the data set into **L1**

Step 2 *Access the statistical tests menu*

- Press **STAT**
- Arrow right to **TESTS**
- Press **ENTER** after selecting **1-Prop Z Test**

Step 3 *Enter the hypothesized proportion p_0, the number of success x, and the sample size n*

- Use the **ARROW** to highlight the appropriate alternative hypothesis
- Press **ENTER**
- Arrow down to "**Calculate**"
- Press **ENTER**

The chosen test will be displayed as well as the z-test statistic and the p-value.

9 Inferences Based on Two Samples *Confidence Intervals and Tests of Hypotheses*

CONTENTS

Where We've Been

- Explored two methods for making statistical inferences: *confidence intervals* and *tests of hypotheses*

- Studied confidence intervals and tests for a single population mean μ, a single population proportion p, and a single population variance σ^2

- Learned how to select the sample size necessary to estimate a population parameter with a specified margin of error

Where We're Going

- Learn how to identify the target parameter for comparing two populations (9.1)

- Learn how to compare two means by using confidence intervals and tests of hypotheses (9.2–9.3)

- Apply these inferential methods to problems in which we want to compare two population proportions, or two population variances (9.4, 9.6)

- Determine the sizes of the samples necessary to estimate the difference between two population parameters with a specified margin of error (9.5)

Statistics IN Action ZixIt Corp. v. Visa USA Inc.—A Libel Case

The National Law Journal (Aug. 26–Sept. 2, 2002) reported on an interesting court case in volving ZixIt Corp., a start-up Internet credit card clearing center. ZixIt claimed that its new online credit card processing system would allow Internet shoppers to make purchases without revealing their credit card numbers. This claim violated the established protocols of most major credit card companies, including Visa. Without the company's knowledge, a Visa vice president for technology research and development began writing e-mails and Web site postings on a Yahoo! message board for ZixIt investors, challenging ZixIt's claim and urging investors to sell their ZixIt stock. The Visa executive posted over 400 e-mail and notes before he was caught. Once it was discovered that a Visa executive was responsible for the postings, ZixIt filed a lawsuit against Visa Corp., alleging that Visa—using the executive as its agent—had engaged in a "malicious two-part scheme to disparage and interfere with ZixIt" and its efforts to market the new online credit card processing system. In the libel case ZixIt asked for $699 million in damages.

Dallas lawyers Jeff Tillotson and Mike Lynn, of the law firm Lynn Tillotson & Pinker, were hired to defend Visa in the lawsuit. The lawyers, in turn, hired Dr. James McClave (co-author of this text) as their expert statistician. McClave testified in court on an "event study" he did matching the Visa executive's e-mail postings with movement of ZixIt's stock price the next business day. McClave's testimony, showing

that there was an equal number of days when the stock went up as went down after a posting, helped the lawyers representing Visa to prevail in the case. *The National Law Journal* reported that, after two and a half days of deliberation, "the jurors found [the Visa executive] was not acting in the scope of his employment and that Visa had not defamed ZixIt or interfered with its business."

In this chapter, we demonstrate several of the statistical analyses McClave used to infer that the Visa executive's postings had no effect on ZixIt's stock price. The daily ZixIt stock prices as well as the timing of the Visa executive's postings are saved in the **ZIXITVISA** file.* We apply the statistical methodology presented in this chapter to this data set in two Statistics in Action Revisited examples.

Statistics IN Action Revisited

- Comparing Mean Price Changes (p. 421)
- Comparing Proportions (p. 443)

Data Set: ZIXITVISA

9.1 Identifying the Target Parameter

Many experiments involve a comparison of two populations. For instance, a sociologist may want to estimate the difference in mean life expectancy between inner-city and suburban residents. Or a consumer group may want to test whether two major brands of food freezers differ in the average amount of electricity they use. Or a political candidate might want to estimate the difference in the proportions of voters in two districts who favor her candidacy. Or a professional golfer might be interested in comparing the variability in the distance that two competing brands of golf balls travel when struck with the same club. In this chapter, we consider techniques for using two samples to compare the populations from which they were selected.

The same procedures that are used to estimate and test hypotheses about a single population can be modified to make inferences about two populations. As in Chapters 7 and 8, the methodology used will depend on the sizes of the samples and the parameter of interest (i.e., the *target parameter*). Some key words and the type of data associated with the parameters covered in this chapter are listed in the following box.

Determining the Target Parameter

Parameter	Key Words or Phrases	Type of Data
$\mu_1 - \mu_2$	Mean difference; difference in averages	Quantitative
$p_1 - p_2$	Difference between proportions, percentages, fractions, or rates; compare proportions	Qualitative
$(\sigma_1)^2/(\sigma_2)^2$	Ratio of variances; difference in variability or spread; compare variation	Quantitative

---

*Data provided (with permission) from Info Tech, Inc., Gainesville, Florida.

You can see that the key words *difference* and *compare* help identify the fact that two populations are to be compared. In the previous examples, the words *mean* in *mean life expectancy* and *average* in *average amount of electricity* imply that the target parameter is the difference in population means, $\mu_1 - \mu_2$. The word *proportions* in *proportions of voters in two districts* indicates that the target parameter is the difference in proportions, $p_1 - p_2$. Finally, the key word *variability* in *variability in the distance* identifies the ratio of population variances, $(\sigma_1)^2/(\sigma_2)^2$, as the target parameter.

As with inferences about a single population, the type of data (quantitative or qualitative) collected on the two samples is also indicative of the target parameter. With quantitative data, you are likely to be interested in comparing the means or variances of the data. With qualitative data with two outcomes (success or failure), a comparison of the proportions of successes is likely to be of interest.

We consider methods for comparing two population means in Sections 9.2 and 9.3. A comparison of population proportions is presented in Section 9.4 and population variances in optional Section 9.6. We show how to determine the sample sizes necessary for reliable estimates of the target parameters in Section 9.5.

9.2 Comparing Two Population Means: Independent Sampling

In this section, we develop both large-sample and small-sample methodologies for comparing two population means. In the large-sample case, we use the z-statistic; in the small-sample case, we use the t-statistic.

Large Samples

Example 9.1

A Large–Sample Confidence Interval for $(\mu_1-\mu_2)$— Comparing Mean Weight Loss for Two Diets

Problem A dietitian has developed a diet that is low in fats, carbohydrates, and cholesterol. Although the diet was initially intended to be used by people with heart disease, the dietitian wishes to examine the effect this diet has on the weights of obese people. Two random samples of 100 obese people each are selected, and one group of 100 is placed on the low-fat diet. The other 100 are placed on a diet that contains approximately the same quantity of food, but is not as low in fats, carbohydrates, and cholesterol. For each person, the amount of weight lost (or gained) in a three-week period is recorded. The data, saved in the **DIETSTUDY** file, are listed in Table 9.1. Form a 95% confidence interval for the difference between the population mean weight losses for the two diets. Interpret the result.

Solution Recall that the general form of a large-sample confidence interval for a single mean μ is $\bar{x} \pm z_{\alpha/2}\sigma_{\bar{x}}$. That is, we add and subtract $z_{\alpha/2}$ standard deviations of the sample estimate $\bar{x}$ to and from the value of the estimate. We employ a similar procedure to form the confidence interval for the difference between two population means.

Let μ_1 represent the mean of the conceptual population of weight losses for all obese people who could be placed on the low-fat diet. Let μ_2 be similarly defined for the other diet. We wish to form a confidence interval for $(\mu_1 - \mu_2)$. An intuitively appealing estimator for $(\mu_1 - \mu_2)$ is the difference between the sample means, $(\bar{x}_1 - \bar{x}_2)$. Thus, we will form the confidence interval of interest with

$$(\bar{x}_1 - \bar{x}_2) \pm z_{\alpha/2}\sigma_{(\bar{x}_1-\bar{x}_2)}$$

Assuming that the two samples are independent, we write the standard deviation of the difference between the sample means (i.e., the *standard error* of $\bar{x}_1-\bar{x}_2$) as

$$\sigma_{(\bar{x}_1-\bar{x}_2)} = \sqrt{\frac{\sigma_1^2}{n_1} + \frac{\sigma_2^2}{n_2}}$$

Table 9.1	Diet Study Data, Example 9.1

Weight Losses for Low-Fat Diet

8	10	10	12	9	3	11	7	9	2
21	8	9	2	2	20	14	11	15	6
13	8	10	12	1	7	10	13	14	4
8	12	8	10	11	19	0	9	10	4
11	7	14	12	11	12	4	12	9	2
4	3	3	5	9	9	4	3	5	12
3	12	7	13	11	11	13	12	18	9
6	14	14	18	10	11	7	9	7	2
16	16	11	11	3	15	9	5	2	6
5	11	14	11	6	9	4	17	20	10

Weight Losses for Regular Diet

6	6	5	5	2	6	10	3	9	11
14	4	10	13	3	8	8	13	9	3
4	12	6	11	12	9	8	5	8	7
6	2	6	8	5	7	16	18	6	8
13	1	9	8	12	10	6	1	0	13
11	2	8	16	14	4	6	5	12	9
11	6	3	9	9	14	2	10	4	13
8	1	1	4	9	4	1	1	5	6
14	0	7	12	9	5	9	12	7	9
8	9	8	10	5	8	0	3	4	8

Data Set: DIETSTUDY

Typically (as in this example), the population variances σ_1^2 and σ_2^2 are unknown. Since the samples are both large ($n_1 = n_2 = 100$), the sample variances s_1^2 and s_2^2 will be good estimators of their respective population variances. Thus, the estimated standard error is

$$\sigma_{(\bar{x}_1 - \bar{x}_2)} \approx \sqrt{\frac{s_1^2}{n_1} + \frac{s_2^2}{n_2}}$$

Summary statistics for the diet data are displayed at the top of the SPSS printout shown in Figure 9.1. Note that $\bar{x}_1 = 9.31$, $\bar{x}_2 = 7.40$, $s_1 = 4.67$, and $s_2 = 4.04$. Using these values and observing that $\alpha = .05$ and $z_{.025} = 1.96$, we find that the 95% confidence interval is, approximately,

$$(9.31 - 7.40) \pm 1.96 \sqrt{\frac{(4.67)^2}{100} + \frac{(4.04)^2}{100}} = 1.91 \pm (1.96)(.62) = 1.91 \pm 1.22$$

or (.69, 3.13). This interval (rounded) is highlighted in Figure 9.1.

Using this estimation procedure over and over again for different samples, we know that approximately 95% of the confidence intervals formed in this manner will enclose the difference in population means ($\mu_1 - \mu_2$). Therefore, we are highly confident that the mean weight loss for the low-fat diet is between .69 and 3.13 pounds more

Group Statistics

DIET		N	Mean	Std. Deviation	Std. Error Mean
WTLOSS	LOWFAT	100	9.31	4.668	.467
	REGULAR	100	7.40	4.035	.404

Independent Samples Test

		Levene's Test for Equality of Variances		t-test for Equality of Means						
									95% Confidence Interval of the Difference	
		F	Sig.	t	df	Sig. (2-tailed)	Mean Difference	Std. Error Difference	Lower	Upper
WTLOSS	Equal variances assumed	1.367	.244	3.095	198	.002	1.910	.617	.693	3.127
	Equal variances not assumed			3.095	193.940	.002	1.910	.617	.693	3.127

Figure 9.1
SPSS analysis of diet study data.

than the mean weight loss for the other diet. With this information, the dietitian better understands the potential of the low-fat diet as a weight-reduction diet.

Look Back If the confidence interval for $(\mu_1 - \mu_2)$ contains 0 [e.g., $(-2.5, 1.3)$], then it is possible for the difference between the population means to be 0 (i.e., $\mu_1 - \mu_2 = 0$). In this case, we could not conclude that a significant difference exists between the mean weight losses for the two diets.

Now Work Exercise 9.6a

The justification for the procedure used in Example 9.1 to estimate $(\mu_1 - \mu_2)$ relies on the properties of the sampling distribution of $(\bar{x}_1 - \bar{x}_2)$. The performance of the estimator in repeated sampling is pictured in Figure 9.2, and its properties are summarized in the following box:

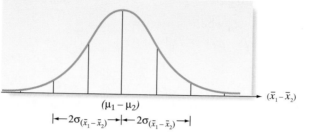

Figure 9.2

Sampling distribution of $(\bar{x}_1 - \bar{x}_2)$

> **Properties of the Sampling Distribution of $(\bar{x}_1 - \bar{x}_2)$**
>
> 1. The mean of the sampling distribution of $(\bar{x}_1 - \bar{x}_2)$ is $(\mu_1 - \mu_2)$.
> 2. If the two samples are independent, the standard deviation of the sampling distribution is
>
> $$\sigma_{(\bar{x}_1-\bar{x}_2)} = \sqrt{\frac{\sigma_1^2}{n_1} + \frac{\sigma_2^2}{n_2}}$$
>
> where σ_1^2 and σ_2^2 are the variances of the two populations being sampled and n_1 and n_2 are the respective sample sizes. We also refer to $\sigma_{(\bar{x}_1-\bar{x}_2)}$ as the **standard error of the statistic** $(\bar{x}_1 - \bar{x}_2)$.
> 3. By the Central Limit Theorem, the sampling distribution of $(\bar{x}_1 - \bar{x}_2)$ is approximately normal *for large samples*.

In Example 9.1, we noted the similarity in the procedures for forming a large-sample confidence interval for one population mean and a large-sample confidence interval for the difference between two population means. When we are testing hypotheses, the procedures are again similar. The general large-sample procedures for forming confidence intervals and testing hypotheses about $(\mu_1 - \mu_2)$ are summarized in the following boxes:

> **Large, Independent Samples Confidence Interval for $(\mu_1 - \mu_2)$: Normal (z) Statistic**
>
> σ_1^2 and σ_2^2 known: $(\bar{x}_1 - \bar{x}_2) \pm z_{\alpha/2}\,\sigma_{(\bar{x}_1-\bar{x}_2)} = (\bar{x}_1 - \bar{x}_2) \pm z_{\alpha/2}\sqrt{\dfrac{\sigma_1^2}{n_1} + \dfrac{\sigma_2^2}{n_2}}$
>
> σ_1^2 and σ_2^2 unknown: $(\bar{x}_1 - \bar{x}_2) \pm z_{\alpha/2}\,\sigma_{(\bar{x}_1-\bar{x}_2)} \approx (\bar{x}_1 - \bar{x}_2) \pm z_{\alpha/2}\sqrt{\dfrac{s_1^2}{n_1} + \dfrac{s_2^2}{n_2}}$

Large, Independent Samples Test of Hypothesis for $(\mu_1 - \mu_2)$: Normal (z) Statistic

One-Tailed Test

H_0: $(\mu_1 - \mu_2) = D_0$
H_a: $(\mu_1 - \mu_2) < D_0$
$\quad$ [or H_a: $(\mu_1 - \mu_2) > D_0$]

Two-Tailed Test

H_0: $(\mu_1 - \mu_2) = D_0$
H_a: $(\mu_1 - \mu_2) \neq D_0$

where D_0 = Hypothesized difference between the means (this difference is often hypothesized to be equal to 0)

Test statistic:

$$z = \frac{(\bar{x}_1 - \bar{x}_2) - D_0}{\sigma_{(\bar{x}_1 - \bar{x}_2)}} \quad \text{where} \quad \sigma_{(\bar{x}_1 - \bar{x}_2)} = \sqrt{\frac{\sigma_1^2}{n_1} + \frac{\sigma_2^2}{n_2}} \quad \text{if both } \sigma_1^2 \text{ and } \sigma_2^2 \text{ are known}$$

$$\approx \sqrt{\frac{s_1^2}{n_1} + \frac{s_2^2}{n_2}} \quad \text{if } \sigma_1^2 \text{ and } \sigma_2^2 \text{ are unknown}$$

Rejection region: $z < -z_\alpha$

$\quad$ [or $z > z_\alpha$ when
H_a: $(\mu_1 - \mu_2) > D_0$]

Rejection region: $|z| > z_{\alpha/2}$

Conditions Required for Valid Large-Sample Inferences about $(\mu_1 - \mu_2)$

1. The two samples are randomly selected in an independent manner from the two target populations.
2. The sample sizes, n_1 and n_2, are both large (i.e., $n_1 \geq 30$ and $n_2 \geq 30$). (By the Central Limit Theorem, this condition guarantees that the sampling distribution of $(\bar{x}_1 - \bar{x}_2)$ will be approximately normal, regardless of the shapes of the underlying probability distributions of the populations. Also, s_1^2 and s_2^2 will provide good approximations to σ_1^2 and σ_2^2 when both samples are large.)

Example 9.2

A Large-Sample Test for $(\mu_1 - \mu_2)$— Comparing Mean Weight Loss for Two Diets

Problem Refer to the study of obese people on a low-fat diet and a regular diet presented in Example 9.1. Another way to compare the mean weight losses for the two different diets is to conduct a test of hypothesis. Use the information on the SPSS printout shown in Figure 9.1 to conduct the test. Take $\alpha = .05$.

Solution Again, we let μ_1 and μ_2 represent the population mean weight losses of obese people on the low-fat diet and regular diet, respectively. If one diet is more effective in reducing the weights of obese people, then either $\mu_1 < \mu_2$ or $\mu_2 < \mu_1$; that is, $\mu_1 \neq \mu_2$. Thus, the elements of the test are as follows:

H_0: $(\mu_1 - \mu_2) = 0$ (i.e., $\mu_1 = \mu_2$; note that $D_0 = 0$ for this hypothesis test)

H_a: $(\mu_1 - \mu_2) \neq 0$ (i.e., $\mu_1 \neq \mu_2$)

Test statistic: $z = \dfrac{(\bar{x}_1 - \bar{x}_2) - D_0}{\sigma_{(\bar{x}_1 - \bar{x}_2)}} = \dfrac{\bar{x}_1 - \bar{x}_2 - 0}{\sigma_{(\bar{x}_1 - \bar{x}_2)}}$

Rejection region: $z < -z_{\alpha/2} = -1.96$ or $z > z_{\alpha/2} = 1.96$ $\qquad$ (see Figure 9.3)

Substituting the summary statistics given in Figure 9.1 into the test statistic, we obtain

$$z = \frac{(\bar{x}_1 - \bar{x}_2) - 0}{\sigma_{(\bar{x}_1 - \bar{x}_2)}} = \frac{9.31 - 7.40}{\sqrt{\dfrac{\sigma_1^2}{n_1} + \dfrac{\sigma_2^2}{n_2}}}$$

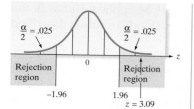

Figure 9.3

Rejection region for Example 9.2

Now, since σ_1^2 and σ_2^2 are unknown, we approximate the test statistic value as follows:

$$z \approx \frac{9.31 - 7.40}{\sqrt{\dfrac{s_1^2}{n_1} + \dfrac{s_2^2}{n_2}}} = \frac{1.91}{\sqrt{\dfrac{(4.67)^2}{100} + \dfrac{(4.04)^2}{100}}} = \frac{1.91}{.617} = 3.09$$

[*Note:* The value of the test statistic is highlighted in the SPSS printout of Figure 9.1.]

As you can see in Figure 9.3, the calculated z-value clearly falls into the rejection region. Therefore, the samples provide sufficient evidence, at $\alpha = .05$, for the dietitian to conclude that the mean weight losses for the two diets differ.

Look Back This conclusion agrees with the inference drawn from the 95% confidence interval in Example 9.1. However, the confidence interval provides more information on the mean weight losses. From the hypothesis test, we know only that the two means differ; that is, $\mu_1 \neq \mu_2$. From the confidence interval in Example 9.1, we found that the mean weight loss μ_1 of the low-fat diet was between .69 and 3.13 pounds more than the mean weight loss μ_2 of the regular diet. In other words, the test tells us that the means differ, but the confidence interval tells us how large the difference is. Both inferences are made with the same degree of reliability—namely, 95% confidence (or at $\alpha = .05$).

Example 9.3

The p-Value for a Test of $(\mu - \mu_2)$

Problem Find the observed significance level for the test in Example 9.2. Interpret the result.

Solution The alternative hypothesis in Example 9.2, $H_a: \mu_1 - \mu_2 \neq 0$, required a two-tailed test using

$$z = \frac{\bar{x}_1 - \bar{x}_2}{\sigma_{(\bar{x}_1 - \bar{x}_2)}}$$

as a test statistic. Since the z-value calculated from the sample data was 3.09, the observed significance level (p-value) for the two-tailed test is the probability of observing a value of z at least as contradictory to the null hypothesis as $z = 3.09$; that is,

$$p\text{-value} = 2 \cdot P(z \geq 3.09)$$

This probability is computed under the assumption that H_0 is true and is equal to the highlighted area shown in Figure 9.4.

The tabulated area corresponding to $z = 3.09$ in Table IV of Appendix A is .4990. Therefore,

$$P(z \geq 3.09) = .5 - .4990 = .0010$$

and the observed significance level for the test is

$$p\text{-value} = 2(.001) = .002$$

Since our selected α value, .05, exceeds this p-value, we have sufficient evidence to reject $H_0: \mu_1 - \mu_2 = 0$.

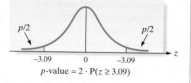

Figure 9.4

The observed significance level for Example 9.2

Look Back The p-value of the test is more easily obtained from a statistical software package. The p-value is highlighted at the bottom of the SPSS printout shown in Figure 9.1. This value agrees with our calculated p-value.

Now Work Exercise 9.6b

Small Samples

In comparing two population means with small samples (say, $n_1 < 30$ and $n_2 < 30$), the methodology of the previous three examples is invalid. The reason? When the sample sizes are small, estimates of σ_1^2 and σ_2^2 are unreliable and the Central Limit Theorem

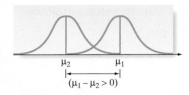

Figure 9.5

Assumptions for the two-sample t: (1) normal populations; (2) equal variances

(which guarantees that the z statistic is normal) can no longer be applied. But as in the case of a single mean (Section 8.4), we use the familiar Student's t-distribution described in Chapter 7.

To use the t-distribution, both sampled populations must be approximately normally distributed with equal population variances, and the random samples must be selected independently of each other. The assumptions of normality and equal variances imply relative frequency distributions for the populations that would appear as shown in Figure 9.5.

Since we assume that the two populations have equal variances ($\sigma_1^2 = \sigma_2^2 = \sigma^2$), it is reasonable to use the information contained in both samples to construct a **pooled sample estimator σ^2** for use in confidence intervals and test statistics. Thus, if s_1^2 and s_2^2 are the two sample variances (each estimating the variance σ^2 common to both populations), the pooled estimator of σ^2, denoted as s_p^2, is

$$s_p^2 = \frac{(n_1 - 1)s_1^2 + (n_2 - 1)s_2^2}{(n_1 - 1) + (n_2 - 1)} = \frac{(n_1 - 1)s_1^2 + (n_2 - 1)s_2^2}{n_1 + n_2 - 2}$$

or

$$s_p^2 = \frac{\overbrace{\sum(x_1 - \bar{x}_1)^2}^{\text{From sample 1}} + \overbrace{\sum(x_2 - \bar{x}_2)^2}^{\text{From sample 2}}}{n_1 + n_2 - 2}$$

where x_1 represents a measurement from sample 1 and x_2 represents a measurement from sample 2. Recall that the term *degrees of freedom* was defined in Section 7.2 as 1 less than the sample size. Thus, in this case, we have $(n_1 - 1)$ degrees of freedom for sample 1 and $(n_2 - 1)$ degrees of freedom for sample 2. Since we are pooling the information on σ^2 obtained from both samples, the number of degrees of freedom associated with the pooled variance s_p^2 is equal to the sum of the numbers of degrees of freedom for the two samples, namely, the denominator of s_p^2; that is, $(n_1 - 1) + (n_2 - 1) = n_1 + n_2 - 2$.

Note that the second formula given for s_p^2 shows that the pooled variance is simply a *weighted average* of the two sample variances s_1^2 and s_2^2. The weight given each variance is proportional to its number of degrees of freedom. If the two variances have the same number of degrees of freedom (i.e., *if the sample sizes are equal*), *then the pooled variance is a simple average of the two sample variances*. The result is an average, or "pooled," variance that is a better estimate of σ^2 than either s_1^2 or s_2^2 alone.

BIOGRAPHY **BRADLEY EFRON (1938–present)**

The Bootstrap Method

Bradley Efron was raised in St. Paul, Minnesota, the son of a truck driver who was the amateur statistician for his bowling and baseball leagues. Efron received a B.S. in mathematics from the California Institute of Technology in 1960, but, by his own admission, had no talent for modern abstract math. His interest in the science of statistics developed after he read a book by Harold Cramer from cover to cover. Efron went to the University of Stanford to study statistics, and he earned his Ph.D there in 1964. He has been a faculty member in Stanford's Department of Statistics since 1966. Over his career, Efron has received numerous awards and prizes for his contributions to modern statistics, including the MacArthur Prize Fellow (1983), the American Statistical Association Wilks Medal (1990), and the Parzen Prize for Statistical Innovation (1998). In 1979, Efron invented a method—called the *bootstrap*—of estimating and testing population parameters in situations in which either the sampling distribution is unknown or the assumptions are violated. The method involves repeatedly taking samples of size n (with replacement) from the original sample and calculating the value of the point estimate. Efron showed that the sampling distribution of the estimator is simply the frequency distribution of the bootstrap estimates. ■

Both the confidence interval and the test-of-hypothesis procedures for comparing two population means with small samples are summarized in the following boxes:

Small, Independent Samples Confidence Interval for $(\mu_1 - \mu_2)$: Student's t-Statistic

$$(\bar{x}_1 - \bar{x}_2) \pm t_{\alpha/2}\sqrt{s_p^2\left(\frac{1}{n_1} + \frac{1}{n_2}\right)}$$

where $s_p^2 = \dfrac{(n_1 - 1)s_1^2 + (n_2 - 1)s_2^2}{n_1 + n_2 - 2}$

and $t_{\alpha/2}$ is based on $(n_1 + n_2 - 2)$ degrees of freedom.

[*Note:* $s_p^2 = \dfrac{s_1^2 + s_2^2}{2}$ when $n_1 = n_2$]

Small, Independent Samples Test of Hypothesis for $(\mu_1 - \mu_2)$: Student's t-Statistic

One-Tailed Test

$H_0: (\mu_1 - \mu_2) = D_0$
$H_a: (\mu_1 - \mu_2) < D_0$
[or $H_a: (\mu_1 - \mu_2) > D_0$]

Two-Tailed Test

$H_0: (\mu_1 - \mu_2) = D_0$
$H_a: (\mu_1 - \mu_2) \neq D_0$

Test statistic: $t = \dfrac{(\bar{x}_1 - \bar{x}_2) - D_0}{\sqrt{s_p^2\left(\dfrac{1}{n_1} + \dfrac{1}{n_2}\right)}}$

Rejection region: $t < -t_\alpha$
or $t > t_\alpha$ when
$H_a: (\mu_1 - \mu_2) > D_0]$

Rejection region: $|t| > t_{\alpha/2}$

where t_α and $t_{\alpha/2}$ are based on $(n_1 + n_2 - 2)$ degrees of freedom.

Conditions Required for Valid Small-Sample Inferences about $(\mu_1 - \mu_2)$

1. The two samples are randomly selected in an independent manner from the two target populations.
2. Both sampled populations have distributions that are approximately normal.
3. The population variances are equal (i.e., $\sigma_1^2 = \sigma_2^2$).

Example 9.4

A Small–Sample Confidence Interval for $(\mu_1 - \mu_2)$– Comparing Two Methods of Teaching

Problem Suppose you wish to compare a new method of teaching reading to "slow learners" with the current standard method. You decide to base your comparison on the results of a reading test given at the end of a learning period of six months. Of a random sample of 22 "slow learners," 10 are taught by the new method and 12 are taught by the standard method. All 22 children are taught by qualified instructors under similar conditions for the designated six-month period. The results of the reading test at the end of this period are given in Table 9.2.

Table 9.2		Reading Test Scores for Slow Learners					
New Method				Standard Method			
80	80	79	81	79	62	70	68
76	66	71	76	73	76	86	73
70	85			72	68	75	66

Data Set: READING

a. Use the data in the table to estimate the true mean difference between the test scores for the new method and the standard method. Use a 95% confidence interval.

b. Interpret the interval you found in part a.

c. What assumptions must be made in order that the estimate be valid? Are they reasonably satisfied?

Solution

a. For this experiment, let μ_1 and μ_2 represent the mean reading test scores of "slow learners" taught with the new and standard methods, respectively. Then the objective is to obtain a 95% confidence interval for $(\mu_1 - \mu_2)$.

The first step in constructing the confidence interval is to obtain summary statistics (e.g., $\bar{x}$ and s) on reading test scores for each method. The data of Table 9.2 were entered into a computer, and SAS was used to obtain these descriptive statistics. The SAS printout appears in Figure 9.6. Note that $\bar{x}_1 = 76.4$, $s_1 = 5.8348$, $\bar{x}_2 = 72.333$, and $s_2 = 6.3437$.

Two Sample t-test for the Means of SCORE within METHOD

Sample Statistics

Group	N	Mean	Std. Dev.	Std. Error
NEW	10	76.4	5.8348	1.8451
STD	12	72.33333	6.3437	1.8313

Hypothesis Test

Null hypothesis: Mean 1 - Mean 2 = 0
Alternative: Mean 1 - Mean 2 ^= 0

If Variances Are	t statistic	Df	Pr > t
Equal	1.552	20	0.1364
Not Equal	1.564	19.77	0.1336

95% Confidence Interval for the Difference between Two Means

Lower Limit	Upper Limit
-1.40	9.53

Figure 9.6
SAS printout for Example 9.4

Next, we calculate the pooled estimate of variance to obtain

$$s_p^2 = \frac{(n_1 - 1)s_1^2 + (n_2 - 1)s_2^2}{n_1 + n_2 - 2}$$

$$= \frac{(10 - 1)(5.8348)^2 + (12 - 1)(6.3437)^2}{10 + 12 - 2} = 37.45$$

where s_p^2 is based on $(n_1 + n_2 - 2) = (10 + 12 - 2) = 20$ degrees of freedom. Also, we find $t_{\alpha/2} = t_{.025} = 2.086$ (based on 20 degrees of freedom) from Table VI of Appendix A.

Finally, the 95% confidence interval for $(\mu_1 - \mu_2)$, the difference between mean test scores for the two methods, is

$$(\bar{x}_1 - \bar{x}_2) \pm t_{\alpha/2}\sqrt{s_p^2\left(\frac{1}{n_1} + \frac{1}{n_2}\right)} = (76.4 - 72.33) \pm t_{.025}\sqrt{37.45\left(\frac{1}{10} + \frac{1}{12}\right)}$$

$$= 4.07 \pm (2.086)(2.62)$$

$$= 4.07 \pm 5.47$$

or $(-1.4, 9.54)$. This interval agrees (except for rounding) with the one shown at the bottom of the SAS printout of Figure 9.6.

b. The interval can be interpreted as follows: With a confidence coefficient equal to .95, we estimate that the difference in mean test scores between using the new method of teaching and using the standard method falls into the interval from -1.4 to 9.54. In other words, we estimate (with 95% confidence) the mean test score for the new method to be anywhere from 1.4 points less than, to 9.54 points more than, the mean test score for the standard method. Although the sample means seem to suggest that the new method is associated with a higher mean test score, there is insufficient evidence to indicate that $(\mu_1 - \mu_2)$ differs from 0 because the interval includes 0 as a possible value for $(\mu_1 - \mu_2)$. To demonstrate a difference in mean test scores (if it exists), you could increase the sample size and thereby narrow the width of the confidence interval for $(\mu_1 - \mu_2)$. Alternatively, you can design the experiment differently. This possibility is discussed in the next section.

c. To use the small-sample confidence interval properly, the following assumptions must be satisfied:

1. The samples are randomly and independently selected from the populations of "slow learners" taught by the new method and the standard method.

2. The test scores are normally distributed for both teaching methods.

3. The variance of the test scores is the same for the two populations; that is, $\sigma_1^2 = \sigma_2^2$.

On the basis of the information provided about the sampling procedure in the description of the problem, the first assumption is satisfied. To check the plausibility of the remaining two assumptions, we resort to graphical methods. Figure 9.7 is a MINITAB printout that gives normal probability plots for the test scores of the two samples of "slow learners." The near straight-line trends on both plots indicate that the distributions of the scores are approximately mound shaped and symmetric.

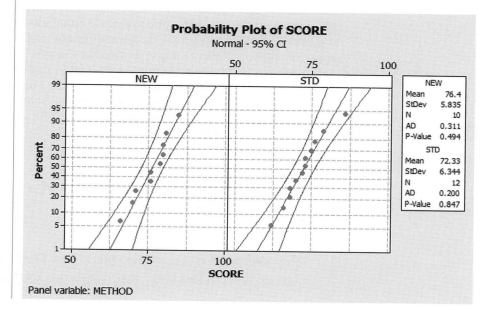

Figure 9.7

MINITAB normal probability plots for Example 9.4

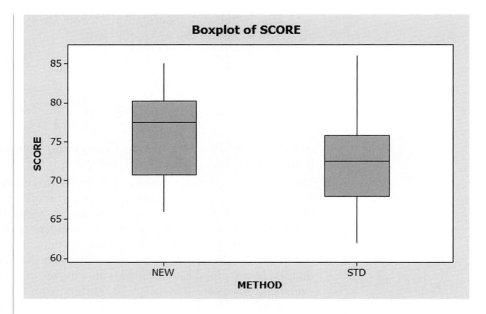

Figure 9.8
MINITAB box plots for
Example 9.4

Consequently, each sample data set appears to come from a population that is approximately normal.

One way to check the third assumption is to test the null hypothesis $H_0: \sigma_1^2 = \sigma_2^2$. This test is covered in Section 9.6. Another approach is to examine box plots of the sample data. Figure 9.8 is a MINITAB printout that shows side-by-side vertical box plots of the test scores in the two samples. Recall from Section 2.9 that the box plot represents the "spread" of a data set. The two box plots appear to have about the same spread; thus, the samples appear to come from populations with approximately the same variance.

Look Back All three assumptions, then, appear to be reasonably satisfied for this application of the small-sample confidence interval.

Now Work Exercise 9.9

The two-sample t-statistic is a powerful tool for comparing population means when the assumptions are satisfied. It has also been shown to retain its usefulness when the sampled populations are only approximately normally distributed. And when the sample sizes are equal, the assumption of equal population variances can be relaxed. That is, if $n_1 = n_2$, then σ_1^2 and σ_2^2 can be quite different, and the test statistic will still possess, approximately, a Student's t-distribution. In the case where $\sigma_1^2 \neq \sigma_2^2$ and $n_1 \neq n_2$, an approximate small-sample confidence interval or test can be obtained by modifying the number of degrees of freedom associated with the t-distribution.

The next box gives the approximate small-sample procedures to use when the assumption of equal variances is violated. The test for the case of "unequal sample sizes" is based on Satterthwaite's (1946) approximation.

Approximate Small-Sample Procedures when $\sigma_1^2 \neq \sigma_2^2$

1. Equal Sample Sizes ($n_1 = n_2 = n$)

Confidence interval: $\qquad (\bar{x}_1 - \bar{x}_2) \pm t_{\alpha/2}\sqrt{(s_1^2 + s_2^2)/n}$

Test statistic for H_0: $(\mu_1 - \mu_2) = 0$: $\qquad t = (\bar{x}_1 - \bar{x}_2)/\sqrt{(s_1^2 + s_2^2)/n}$

where t is based on $\nu = n_1 + n_2 - 2 = 2(n - 1)$ degrees of freedom.

---

2. Unequal Sample Sizes ($n_1 \neq n_2$)

Confidence interval: $$(\bar{x}_1 - \bar{x}_2) \pm t_{\alpha/2}\sqrt{(s_1^2/n_1) + (s_2^2/n_2)}$$

Test statistic for H_0: $(\mu_1 - \mu_2) = 0$: $t = (\bar{x}_1 - \bar{x}_2)/\sqrt{(s_1^2/n_1) + (s_2^2/n_2)}$

where t is based on degrees of freedom equal to

$$\nu = \frac{(s_1^2/n_1 + s_2^2/n_2)^2}{\dfrac{(s_1^2/n_1)^2}{n_1 - 1} + \dfrac{(s_2^2/n_2)^2}{n_2 - 1}}$$

Note: The value of ν will generally not be an integer. Round ν down to the nearest integer to use the t-table.

When the assumptions are not clearly satisfied, you can select larger samples from the populations or you can use other available statistical tests (**nonparametric statistical tests**, described in Chapter 14).

What Should You Do if the Assumptions Are Not Satisfied?

Answer: If you are concerned that the assumptions are not satisfied, use the Wilcoxon rank sum test for independent samples to test for a shift in population distributions. (See Chapter 14).

Statistics IN Action | Revisited | Comparing Mean Price Changes

Refer to the *ZixIt v. Visa* court case described in the Statistics in Action (p. 410). Recall that a Visa executive wrote e-mails and made Web site postings in an effort to undermine a new online credit card processing system developed by ZixIt. ZixIt sued Visa for libel, asking for $699 million in damages. An expert statistician, hired by the defendants (Visa), performed an "event study" in which he matched the Visa executive's e-mail postings with movement of ZixIt's stock price the next business day. The data were collected daily from September 1 to December 30, 1999 (an 83-day period), and are available in the **ZIXITVISA** file. In addition to daily closing price (dollars) of ZixIt stock, the file contains a variable for whether or not the Visa executive posted an e-mail and the change in price of the stock the following business day. During the 83-day period, the executive posted e-mails on 43 days and had no postings on 40 days.

If the daily posting by the Visa executive had a negative impact on ZixIt stock, then the average price change following nonposting days should exceed the average price change following posting days. Consequently, one way to analyze the data is to conduct a comparison of two population means through either a confidence interval or a test of hypothesis. Here, we let μ_1 represent the mean price change of ZixIt stock following all nonposting days and μ_2 represent the mean price change of ZixIt stock following posting days. If, in fact, the charges made by ZixIt are true, then μ_1 will exceed μ_2. However, if the data do not support ZixIt's claim, then we will

not be able to reject the null hypothesis H_0: $(\mu_1 - \mu_2) = 0$ in favor of H_a: $(\mu_1 - \mu_2) > 0$. Similarly, if a confidence interval for $(\mu_1 - \mu_2)$ contains the value 0, then there will be no evidence to support ZixIt's claim.

Because both sample size ($n_1 = 40$ and $n_2 = 43$) are large, we can apply the large-sample z-test or large-sample confidence interval procedure for independent samples. A MINITAB printout for this analysis is shown in Figure SIA9.1. Both the 95% confidence interval and p-value for a two-tailed test of hypothesis are highlighted on the printout. Note that the 95% confidence interval, (−$1.47, $1.09), includes the value $0, and the p-value for the two-tailed hypothesis test (.770) implies that the two population means are not significantly different. Also, interestingly, the sample mean price change after posting days ($\bar{x}_1 = \$.06$) is small and positive, while the sample mean price change after nonposting days ($\bar{x}_2 = -\$.13$) is small and negative, totally contradicting ZixIt's claim.

The statistical expert for the defense presented these results to the jury, arguing that the "average price change following posting days is small and similar to the average price change following nonposting days" and "the difference in the means is not statistically significant."

(continued)

Statistics IN Action
(continued)

Two-Sample T-Test and CI: PriceChange, Posting

```
Two-sample T for PriceChange

Posting  N   Mean   StDev  SE Mean
NO      40  -0.13   3.46    0.55
POST    43   0.06   2.20    0.34


Difference = mu (NO) - mu (POST)
Estimate for difference:  -0.188
95% CI for difference:  (-1.470, 1.093)
T-Test of difference = 0 (vs not =): T-Value = -0.29  P-Value = 0.770  DF = 65
```

Figure SIA9.1
MINITAB comparison of two price change means

Note: The statistician also compared the mean ZixIt trading volume (number of ZixIt stock shares traded) after posting days to the mean trading volume after nonposting days. These results are shown in Figure SIA9.2. You can see that the 95% confidence interval for the difference in mean trading volume (highlighted) includes 0, and the *p*-value for a two-tailed test of hypothesis for a difference in means (also highlighted) is not statistically significant. These results were also presented to the jury in defense of Visa.

● *Data Set:* ZIXITVISA

```
Two-sample T for VolumeAfter

Posting  N    Mean    StDev   SE Mean
NO      40  719645  430837    68121
POST    43  578665  333921    50922


Difference = mu (NO) - mu (POST)
Estimate for difference:  140980
95% CI for difference:  (-28526, 310485)
T-Test of difference = 0 (vs not =): T-Value = 1.66  P-Value = 0.102  DF = 73
```

Figure SIA9.2
MINITAB comparison of two trading volume means

Exercises 9.1–9.29

Understanding the Principles

9.1 Describe the sampling distribution of $(\bar{x}_1 - \bar{x}_2)$ when the samples are large.

9.2 To use the *t*-statistic to test for a difference between the means of two populations, what assumptions must be made about the two populations? About the two samples?

9.3 Two populations are described in each of the cases that follow. In which cases would it be appropriate to apply the small-sample *t*-test to investigate the difference between the population means?
  a. Population 1: Normal distribution with variance σ_1^2 Population 2: Skewed to the right with variance $\sigma_2^2 = \sigma_1^2$
  b. Population 1: Normal distribution with variance σ_1^2 Population 2: Normal distribution with variance $\sigma_2^2 \neq \sigma_1^2$
  c. Population 1: Skewed to the left with variance σ_1^2 Population 2: Skewed to the left with variance $\sigma_2^2 = \sigma_1^2$
  d. Population 1: Normal distribution with variance σ_1^2 Population 2: Normal distribution with variance $\sigma_2^2 = \sigma_1^2$
  e. Population 1: Uniform distribution with variance σ_1^2 Population 2: Uniform distribution with variance $\sigma_2^2 = \sigma_1^2$

9.4 A confidence interval for $(\mu_1 - \mu_2)$ is $(-10, 4)$. Which of the following inferences is correct?
  a. $\mu_1 > \mu_2$
  b. $\mu_1 < \mu_2$
  c. $\mu_1 = \mu_2$
  d. no significant difference between means

9.5 A confidence interval for $(\mu_1 - \mu_2)$ is $(-10, -4)$. Which of the following inferences is correct?
  a. $\mu_1 > \mu_2$
  b. $\mu_1 < \mu_2$
  c. $\mu_1 = \mu_2$
  d. no significant difference between means

Learning the Mechanics

9.6 In order to compare the means of two populations, inde-
[NW] pendent random samples of 400 observations are selected from each population, with the following results:

Sample 1	Sample 2
$\bar{x}_1 = 5{,}275$	$\bar{x}_2 = 5{,}240$
$s_1 = 150$	$s_2 = 200$

a. Use a 95% confidence interval to estimate the difference between the population means $(\mu_1 - \mu_2)$. Interpret the confidence interval.

b. Test the null hypothesis $H_0: (\mu_1 - \mu_2) = 0$ versus the alternative hypothesis $H_a: (\mu_1 - \mu_2) \neq 0$. Give the p-value of the test, and interpret the result.

c. Suppose the test in part **b** were conducted with the alternative hypothesis $H_a: (\mu_1 - \mu_2) > 0$. How would your answer to part **b** change?

d. Test the null hypothesis $H_0: (\mu_1 - \mu_2) = 25$ versus the alternative $H_a: (\mu_1 - \mu_2) \neq 25$. Give the p-value, and interpret the result. Compare your answer with that obtained from the test conducted in part **b**.

e. What assumptions are necessary to ensure the validity of the inferential procedures applied in parts **a–d**?

9.7 Independent random samples of 100 observations each are chosen from two normal populations with the following means and standard deviations:

Population 1	Population 2
$\mu_1 = 14$	$\mu_2 = 10$
$\sigma_1 = 4$	$\sigma_2 = 3$

Let $\bar{x}_1$ and $\bar{x}_2$ denote the two sample means.

a. Give the mean and standard deviation of the sampling distribution of $\bar{x}_1$.

b. Give the mean and standard deviation of the sampling distribution of $\bar{x}_2$.

c. Suppose you were to calculate the difference $(\bar{x}_1 - \bar{x}_2)$ between the sample means. Find the mean and standard deviation of the sampling distribution of $(\bar{x}_1 - \bar{x}_2)$.

d. Will the statistic $(\bar{x}_1 - \bar{x}_2)$ be normally distributed? Explain.

9.8 Assume that $\sigma_1^2 = \sigma_2^2 = \sigma^2$. Calculate the pooled estimator of σ^2 for each of the following cases:

a. $s_1^2 = 200, s_2^2 = 180, n_1 = n_2 = 25$

b. $s_1^2 = 25, s_2^2 = 40, n_1 = 20, n_2 = 10$

c. $s_1^2 = .20, s_2^2 = .30, n_1 = 8, n_2 = 12$

d. $s_1^2 = 2,500, s_2^2 = 1,800, n_1 = 16, n_2 = 17$

e. Note that the pooled estimate is a weighted average of the sample variances. To which of the variances does the pooled estimate fall nearer in each of cases **a–d**?

9.9 Independent random samples from normal populations
NW produced the following results: (saved in the **LM9_9** file).

Sample 1	Sample 2
1.2	4.2
3.1	2.7
1.7	3.6
2.8	3.9
3.0	

a. Calculate the pooled estimate of σ^2.

b. Do the data provide sufficient evidence to indicate that $\mu_2 > \mu_1$? Test, using $\alpha = .10$.

c. Find a 90% confidence interval for $(\mu_1 - \mu_2)$.

d. Which of the two inferential procedures, the test of hypothesis in part **b** or the confidence interval in part **c**, provides more information about $(\mu_1 - \mu_2)$?

9.10 Two independent random samples have been selected, 100 observations from population 1 and 100 from population 2. Sample means $\bar{x}_1 = 70$ and $\bar{x}_2 = 50$ were obtained. From previous experience with these populations, it is known that the variances are $\sigma_1^2 = 100$ and $\sigma_2^2 = 64$.

a. Find $\sigma_{(\bar{x}_1 - \bar{x}_2)}$.

b. Sketch the approximate sampling distribution $(\bar{x}_1 - \bar{x}_2)$, assuming that $(\mu_1 - \mu_2) = 5$.

c. Locate the observed value of $(\bar{x}_1 - \bar{x}_2)$ on the graph you drew in part **b.** Does it appear that this value contradicts the null hypothesis $H_0: (\mu_1 - \mu_2) = 5$?

d. Use the z-table to determine the rejection region for the test of $H_0: (\mu_1 - \mu_2) = 5$ against $H_a: (\mu_1 - \mu_2) \neq 5$. Use $\alpha = .05$.

e. Conduct the hypothesis test of part **d** and interpret your result.

f. Construct a 95% confidence interval for $(\mu_1 - \mu_2)$. Interpret the interval.

g. Which inference provides more information about the value of $(\mu_1 - \mu_2)$, the test of hypothesis in part **e** or the confidence interval in part **f**?

9.11 Independent random samples are selected from two populations and are used to test the hypothesis $H_0: (\mu_1 - \mu_2) = 0$ against the alternative $H_a: (\mu_1 - \mu_2) \neq 0$. An analysis of 233 observations from population 1 and 312 from population 2 yielded a p-value of .115.

a. Interpret the results of the test.

b. If the alternative hypothesis had been $H_a: (\mu_1 - \mu_2) < 0$, how would the p-value change? Interpret the p-value for this one-tailed test.

9.12 Independent random samples selected from two normal populations produced the following sample means and standard deviations:

Sample 1	Sample 2
$n_1 = 17$	$n_2 = 12$
$\bar{x}_1 = 5.4$	$\bar{x}_2 = 7.9$
$s_1 = 3.4$	$s_2 = 4.8$

a. Assuming equal variances, conduct the test $H_0: (\mu_1 - \mu_2) = 0$ against $H_a: (\mu_1 - \mu_2) \neq 0$ using $\alpha = .05$.

b. Find and interpret the 95% confidence interval for $(\mu_1 - \mu_2)$.

Applying the Concepts—Basic

9.13 **Effectiveness of teaching software.** Educational software—ranging from video-game-like programs played on Sony PlayStations to rigorous drilling exercises used on computers—has become very popular in school districts across the country. The U.S. Department of Education (DOE) recently conducted a national study of the effectiveness of educational software. In one phase of the study, a sample of 1,516 first-grade students in classrooms that used educational software was compared to a sample of 1,103 first-grade students in classrooms that did not use the technology. In its *Report to Congress* (March 2007), the DOE concluded that "[mean] test scores [of students on the SAT reading test] were not significantly higher in classrooms using reading ... software products" than in classrooms that did not use educational software.

a. Identify the parameter of interest to the DOE.

b. Specify the null and alternative hypotheses for the test conducted by the DOE.

c. The *p*-value for the test was reported as .62. Based on this value, do you agree with the conclusion of the DOE? Explain.

9.14 Cognitive impairment of schizophrenics. A study of the differences in cognitive function between normal individuals and patients diagnosed with schizophrenia was published in the *American Journal of Psychiatry* (April 2010). The total time (in minutes) a subject spent on the Trail Making Test (a standard psychological test) was used as a measure of cognitive function. The researchers theorize that the mean time on the Trail Making Test for schizophrenics will be larger than the corresponding mean for normal subjects. The data for independent random samples of 41 schizophrenics and 49 normal individuals yielded the following results:

	Schizophrenia	Normal
Sample size	41	49
Mean time	104.23	62.24
Standard deviation	45.45	16.34

Based on Perez-Iglesias, R., et al. "White matter integrity and cognitive impairment in first-episode psychosis." *American Journal of Psychiatry*, Vol. 167, No. 4, April 2010 (Table 1).

a. Define the parameter of interest to the researchers.
b. Set up the null and alternative hypothesis for testing the researchers' theory.
c. The researchers conducted the test, part **b,** and reported a *p*-value of .001. What conclusions can you draw from this result? (Use $\alpha = .01$.)
d. Find a 99% confidence interval for the target parameter. Interpret the result. Does your conclusion agree with that of part **c**?

9.15 Children's recall of TV ads. Marketing professors at Robert Morris and Kent State Universities examined children's recall and recognition of television advertisements (*Journal of Advertising,* Spring 2006). Two groups of children were shown a 60-second commercial for Sunkist FunFruit Rock-n-Roll Shapes. One group (the A/V group) was shown the ad with both audio and video; the second group (the video-only group) was shown only the video portion of the commercial. Following the viewing, the children were asked to recall 10 specific items from the ad. The number of items recalled correctly by each child is summarized in the accompanying table. The researchers theorized that "children who receive an audiovisual presentation will have the same level of mean recall of ad information as those who receive only the visual aspects of the ad."

Video-Only Group	A/V Group
$n_1 = 20$	$n_2 = 20$
$\bar{x}_1 = 3.70$	$\bar{x}_2 = 3.30$
$s_1 = 1.98$	$s_2 = 2.13$

Based on Maher, J. K., Hu, M. Y., and Kolbe, R. H. "Children's recall of television ad elements." *Journal of Advertising*, Vol. 35, No. 1, Spring 2006 (Table 1).

a. Set up the appropriate null and alternative hypotheses to test the researchers' theory.
b. Find the value of the test statistic.
c. Give the rejection region for $\alpha = .10$.

d. Make the appropriate inference. What can you say about the researchers' theory?
e. The researchers reported the *p*-value of the test as *p*-value = .62. Interpret this result.
f. What conditions are required for the inference to be valid?

9.16 Index of Biotic Integrity. The Ohio Environmental Protection Agency used the Index of Biotic Integrity (IBI) to measure the biological condition, or "health," of an aquatic region. The IBI is the sum of metrics that measure the presence, abundance, and health of fish in the region. (Higher values of the IBI correspond to healthier fish populations.) Researchers collected IBI measurements for sites located in different Ohio river basins (*Journal of Agricultural, Biological, and Environmental Sciences*, June 2005). Summary data for two river basins, Muskingum and Hocking, are given in the accompanying table.

River Basin	Sample Size	Mean	Standard Deviation
Muskingum	53	.035	1.046
Hocking	51	.340	.960

Based on Boone, E. L., Keying, Y., and Smith, E. P. "Evaluating the relationship between ecological and habitat conditions using hierarchical models." *Journal of Agricultural, Biological, and Environmental Sciences*, Vol. 10, No. 2, June 2005 (Table 01).

a. Use a 90% confidence interval to compare the mean IBI values of the two river basins. Interpret the interval.
b. Conduct a test of hypothesis (at $\alpha = .10$) to compare the mean IBI values of the two river basins. Explain why the result will agree with the inference you derived from the 90% confidence interval in part **a**.

9.17 Reading Japanese books. Refer to the *Reading in a Foreign Language* (Apr. 2004) experiment to improve the Japanese reading comprehension levels of University of Hawaii students, presented in Exercise 2.33 (p. 46). Recall that 14 students participated in a 10-week extensive reading program in a second-semester Japanese course. The numbers of books read by each student and the student's course grade are repeated in the following table and saved in the **JAPANESE** file.

Number of Books	Course Grade	Number of Books	Course Grade
53	A	30	A
42	A	28	B
40	A	24	A
40	B	22	C
39	A	21	B
34	A	20	B
34	A	16	B

Source: Hitosugi, C. I., and Day, R. R. "Extensive reading in Japanese." *Reading in a Foreign Language*, Vol. 16, No. 1, Apr. 2004 (Table 4). Reprinted with permissions from the National Foreign Language Resource Center, University of Hawaii.

a. Consider two populations of students who participate in the reading program prior to taking a second-semester Japanese course: those who earn an A grade and those who earn a B or C grade. Of interest is the difference in the mean number of books read by the two populations of students. Identify the parameter of interest in words and in symbols.

b. Form a 95% confidence interval for the target parameter identified in part **a.**

c. Give a practical interpretation of the confidence interval you formed in part **b.**

d. Compare the inference in part **c** with the inference you derived from stem-and-leaf plots in Exercise 2.33b.

9.18 **Lobster trap placement.** Refer to the *Bulletin of Marine Science* (April 2010) study of lobster trap placement, Exercise 7.35 (p. 317). Recall that the variable of interest was the average distance separating traps—called *trap spacing*—deployed by teams of fishermen fishing for the red spiny lobster in Baja California Sur, Mexico. The trap spacing measurements (in meters) for a sample of 7 teams from the Bahia Tortugas (BT) fishing cooperative are repeated in the table. In addition, trap spacing measurements for 8 teams from the Punta Abreojos (PA) fishing cooperative are listed. (All these data are saved in the **TRAPSPACE** file). For this problem, we are interested in comparing the mean trap spacing measurements of the two fishing cooperatives.

BT Cooperative:	93	99	105	94	82	70	86	
PA Cooperative:	118	94	106	72	90	66	153	98

Based on Shester, G. G. "Explaining catch variation among Baja California lobster fishers through spatial analysis of trap-placement decisions." *Bulletin of Marine Science*, Vol. 86, No. 2, April 2010 (Table 1), pp. 479–498.

a. Identify the target parameter for this study.

b. Compute a point estimate of the target parameter.

c. What is the problem with using the normal (z) statistic to find a confidence interval for the target parameter?

d. Find a 90% confidence interval for the target parameter.

e. Use the interval, part **d,** to make a statement about the difference in mean trap spacing measurements of the two fishing cooperatives.

f. What conditions must be satisfied for the inference, part **e,** to be valid?

9.19 **Bulimia study.** The "fear of negative evaluation" (FNE) scores for 11 female students known to suffer from the eating disorder bulimia and 14 female students with normal eating habits, first presented in Exercise 2.40 (p. 48), are reproduced in the next table and saved in the **BULIMIA** file. (Recall that the higher the score, the greater is the fear of a negative evaluation.)

Bulimic
students: 21 13 10 20 25 19 16 21 24 13 14
Normal
students: 13 6 16 13 8 19 23 18 11 19 7 10 15 20

Based on Randles, R. H. "On neutral responses (zeros) in the sign test and ties in the Wilcoxon-Mann-Whitney test." *The American Statistician*, Vol. 55, No. 2, May 2001 (Figure 3).

a. Locate a 95% confidence interval for the difference between the population means of the FNE scores for bulimic and normal female students on the MINITAB printout shown at the bottom of the page. Interpret the result.

b. What assumptions are required for the interval of part **a** to be statistically valid? Are these assumptions reasonably satisfied? Explain.

Applying the Concepts—Intermediate

9.20 **Do video game players have superior visual attention skills?** Researchers at Griffin University (Australia) conducted a study to determine whether video game players have superior visual attention skills compared to non–video game players (*Journal of Articles in Support of the Null Hypothesis*, Vol. 6, 2009). Two groups of male psychology students—32 video game players (VGP group) and 28 nonplayers (NVGP group)—were subjected to a series of visual attention tasks that included the *attentional blink* test. A test for the difference between two means yielded $t = -.93$ and $p-value = .358$. Consequently, the researchers reported that "no statistically significant differences in the mean test performances of the two groups were found." Summary statistics for the comparison are provided in the table. Do you agree with the researchers conclusion?

	VGP	NVGP
Sample size	32	28
Mean score	84.81	82.64
Standard deviation	9.56	8.43

Based on Murphy, K., and Spencer, A. "Playing video games does not make for better visual attention skills." *Journal of Articles in Support of the Null Hypothesis*, Vol. 6, No. 1, 2009.

9.21 **Drug content assessment.** Refer to Exercise 5.64 (p. 250) and the *Analytical Chemistry* (Dec. 15, 2009) study in which scientists used high-performance liquid chromatography to determine the amount of drug in a tablet. Twenty-five tablets were produced at each of two different, independent

MINITAB Output for Exercise 9.19

Two-Sample T-Test and CI: FNESCORE, GROUP

```
Two-sample T for FNESCORE

GROUP    N   Mean  StDev  SE Mean
Bulimic  11  17.82  4.92    1.5
Normal   14  14.14  5.29    1.4


Difference = mu (Bulimic) - mu (Normal)
Estimate for difference:  3.68
95% CI for difference:  (-0.60, 7.95)
T-Test of difference = 0 (vs not =): T-Value = 1.78  P-Value = 0.089  DF = 23
Both use Pooled StDev = 5.1303
```

MINITAB Output for Exercise 9.21

Two-Sample T-Test and CI: Content, Site

```
Two-sample T for Content

Site  N   Mean  StDev  SE Mean
1     25  89.55  3.07   0.61
2     25  89.03  3.34   0.67


Difference = mu (1) - mu (2)
Estimate for difference:  0.515
95% CI for difference:  (-1.308, 2.338)
T-Test of difference = 0 (vs not =): T-Value = 0.57  P-Value = 0.573  DF = 48
Both use Pooled StDev = 3.2057
```

sites. Drug concentrations (measured as a percentage) for the tablets produced at the two sites are listed in the accompanying table and saved in the **DRUGCON** file. The scientists want to know whether there is any difference between the mean drug concentration in tablets produced at Site 1 and the corresponding mean at Site 2. Use the MINITAB printout above to help the scientists draw a conclusion.

Site 1								
91.28	92.83	89.35	91.90	82.85	94.83	89.83	89.00	84.62
86.96	88.32	91.17	83.86	89.74	92.24	92.59	84.21	89.36
90.96	92.85	89.39	89.82	89.91	92.16	88.67		

Site 2								
89.35	86.51	89.04	91.82	93.02	88.32	88.76	89.26	90.36
87.16	91.74	86.12	92.10	83.33	87.61	88.20	92.78	86.35
93.84	91.20	93.44	86.77	83.77	93.19	81.79		

Based on Borman, P. J., Marion, J. C., Damjanov, I., and Jackson, P. "Design and analysis of method equivalence studies." *Analytical Chemistry*, Vol. 81, No. 24, December 15, 2009 (Table 3).

9.22 **Patent infringement case.** *Chance* (Fall 2002) described a lawsuit charging Intel Corp. with infringing on a patent for an invention used in the automatic manufacture of computer chips. In response, Intel accused the inventor of adding material to his patent notebook after the patent was witnessed and granted. The case rested on whether a patent witness's signature was written on top of or under key text in the notebook. Intel hired a physicist who used an X-ray beam to measure the relative concentrations of certain elements (e.g., nickel, zinc, potassium) at several spots on the notebook page. The zinc measurements for three notebook locations—on a text line, on a witness line, and on the intersection of the witness and text line—are provided in the following table and saved in the **PATENT** file.

Text line:	.335	.374	.440			
Witness line:	.210	.262	.188	.329	.439	.397
Intersection:	.393	.353	.285	.295	.319	

a. Use a test or a confidence interval (at $\alpha = .05$) to compare the mean zinc measurement for the text line with the mean for the intersection.

b. Use a test or a confidence interval (at $\alpha = .05$) to compare the mean zinc measurement for the witness line with the mean for the intersection.

c. From the results you obtained in parts **a** and **b**, what can you infer about the mean zinc measurements at the three notebook locations?

d. What assumptions are required for the inferences to be valid? Are they reasonably satisfied?

9.23 **How do you choose to argue?** Educators frequently lament weaknesses in students' oral and written arguments. In *Thinking and Reasoning* (April 2007), researchers at Columbia University conducted a series of studies to assess the cognitive skills required for successful arguments. One study focused on whether students would choose to argue by weakening the opposing position or by strengthening the favored position. (For example, suppose you are told you would do better at basketball than soccer, but you like soccer. An argument that weakens the opposing position is "You need to be tall to play basketball." An argument that strengthens the favored position is "With practice, I can become really good at soccer.") A sample of 52 graduate students in psychology was equally divided into two groups. Group 1 was presented with 10 items such that the argument always attempts to strengthens the favored position. Group 2 was presented with the same 10 items, but in this case the argument always attempts to weaken the nonfavored position. Each student then rated the 10 arguments on a five-point scale from very weak (1) to very strong (5). The variable of interest was the sum of the 10 item scores, called the *total rating*. Summary statistics for the data are shown in the accompanying table. Use the methodology of this chapter to compare the mean total ratings for the two groups at $\alpha = .05$. Give a practical interpretation of the results in the words of the problem.

	Group 1 (support favored position)	Group 2 (weaken opposing position)
Sample size	26	26
Mean	28.6	24.9
Standard deviation	12.5	12.2

Based on Kuhn, D., and Udell, W. "Coordinating own and other perspectives in argument." *Thinking and Reasoning*, October 2006.

9.24 **Pig castration study.** Two methods of castrating male piglets were investigated in *Applied Animal Behaviour Science* (Nov. 1, 2000). Method 1 involved an incision in the spermatic cords, while Method 2 involved pulling and severing the cords. Forty-nine male piglets were randomly allocated to one of the two methods. During castration, the researchers measured the number of high-frequency vocal responses (squeals) per second over a 5-second period. The data are summarized in the accompanying table. Conduct a test of hypothesis to determine whether the population mean number of high-frequency vocal responses differs for piglets castrated by the two methods. Use $\alpha = .05$.

	Method 1	Method 2
Sample size	24	25
Mean number of squeals	.74	.70
Standard deviation	.09	.09

Based on Taylor, A. A., and Weary, D. M. "Vocal responses of piglets to castration: Identifying procedural sources of pain." *Applied Animal Behaviour Science,* Vol. 70, No. 1, November 1, 2000.

9.25 Mongolian desert ants. Refer to the *Journal of Biogeography* (Dec. 2003) study of ants in Mongolia (central Asia), presented in Exercise 2.68 (p. 59). Recall that botanists placed seed baits at 5 sites in the Dry Steppe region and 6 sites in the Gobi Desert and observed the number of ant species attracted to each site. These data are listed in the next table and saved in the **GOBIANTS** file. Is there evidence to conclude that a difference exists between the average number of ant species found at sites in the two regions of Mongolia? Draw the appropriate conclusion, using $\alpha = .05$.

Site	Region	Number of Ant Species
1	Dry Steppe	3
2	Dry Steppe	3
3	Dry Steppe	52
4	Dry Steppe	7
5	Dry Steppe	5
6	Gobi Desert	49
7	Gobi Desert	5
8	Gobi Desert	4
9	Gobi Desert	4
10	Gobi Desert	5
11	Gobi Desert	4

Based on Pfeiffer, M., et al. "Community organization and species richness of ants in Mongolia along an ecological gradient from steppe to Gobi desert." *Journal of Biogeography,* Vol. 30, No. 12, Dec. 2003.

9.26 Does rudeness really matter in the workplace? Studies have established that rudeness in the workplace can lead to retaliatory and counterproductive behavior. However, there has been little research on how rude behaviors influence a victim's task performance. Such a study was conducted and the results published in the *Academy of Management Journal* (Oct. 2007). College students enrolled in a management course were randomly assigned to one of two experimental conditions: rudeness condition (45 students) and control group (53 students). Each student was asked to write down as many uses for a brick as possible in five minutes; this value (total number of uses) was used as a performance measure for each student. For those students in the rudeness condition, the facilitator displayed rudeness by berating the students in general for being irresponsible and unprofessional (due to a late-arriving confederate). No comments were made about the late-arriving confederate for students in the control group. The number of different uses of a brick for each of the 98 students was recorded and the data saved in the **RUDE** file, shown in the next table.

Control Group:
```
 1 24  5 16 21  7 20  1  9 20 19 10 23 16  0  4  9 13
17 13  0  2 12 11  7  1 19  9 12 18  5 21 30 15  4  2
12 11 10 13 11  3  6 10 13 16 12 28 19 12 20  3 11
```
Rudeness Condition:
```
 4 11 18 11  9  6  5 11  9 12  7  5  7  3 11  1  9 11
10  7  8  9 10  7 11  4 13  5  4  7  8  3  8 15  9 16
10  0  7 15 13  9  2 13 10
```

Conduct a statistical analysis (at $\alpha = .01$) to determine if the true mean performance level for students in the rudeness condition is lower than the true mean performance level for students in the control group. Use the results shown on the accompanying SAS printout to draw your conclusion   .

```
Sample Statistics

   Group        N      Mean      Std. Dev.    Std. Error
   -----------------------------------------------------------
   Control     53   11.81132     7.3826        1.0141
   Rude        45    8.511111    3.9922        0.5951

Hypothesis Test

   Null hypothesis:     Mean 1 - Mean 2 <= 0
   Alternative:         Mean 1 - Mean 2 >  0

   If Variances Are    t statistic      Df        Pr > t
   -----------------------------------------------------------
   Equal                 2.683          96        0.0043
   Not Equal             2.807          82.43     0.0031

95% Confidence Interval for the Difference between Two Means

         Lower Limit     Upper Limit
         -----------     -----------
            0.86            5.74
```

9.27 Masculinity and crime. The *Journal of Sociology* (July 2003) published a study on the link between the level of masculinity and criminal behavior in men. Using a sample of newly incarcerated men in Nebraska, the researcher identified 1,171 violent events and 532 events in which violence was avoided that the men were involved in. (A violent event involved the use of a weapon, throwing of objects, punching, choking, or kicking. An event in which violence was avoided included pushing, shoving, grabbing, or threats of violence that did not escalate into a violent event.) Each of the sampled men took the Masculinity–Femininity Scale (MFS) test to determine his level of masculinity, based on common male stereotyped traits. MFS scores ranged from 0 to 56 points, with lower scores indicating a more masculine orientation. One goal of the research was to compare the mean MFS scores for two groups of men: those involved in violent events and those who avoided violent events.

a. Identify the target parameter for this study.

b. The sample mean MFS score for the violent-event group was 44.50, while the sample mean MFS score for the avoided-violent-event group was 45.06. Is this sufficient information to make the comparison desired by the researcher? Explain.

c. In a large-sample test of hypothesis to compare the two means, the test statistic was computed to be $z = 1.21$. Compute the two-tailed p-value of the test.

d. Make the appropriate conclusion, using $\alpha = .10$.

9.28 Detection of rigged school milk prices. Each year, the state of Kentucky invites bids from dairies to supply half-pint containers of fluid milk products for its school districts. In several school districts in northern Kentucky (called the "tricounty" market), two suppliers—Meyer Dairy and Trauth Dairy—were accused of price-fixing—that is, conspiring to allocate the districts so that the winning bidder was predetermined and the price per pint was set above the competitive price. These two dairies were the only two bidders on the milk contracts in the tricounty market for eight consecutive years. (In contrast, a large number of different dairies won the milk contracts for school districts in the remainder of the northern Kentucky market, called the "surrounding" market.) Did Meyer and Trauth conspire to rig their bids in the tricounty market? Economic theory states that, if so, the mean winning price in the rigged tricounty market will be higher than the mean winning price in the competitive surrounding market. Data on all bids received from the dairies competing for the milk contracts during the time period in question are saved in the **MILK** file. A MINITAB printout of the comparison of mean prices bid for whole white milk for the two Kentucky milk markets is shown below. Is there support for the claim that the dairies in the tricounty market participated in collusive practices? Explain in detail.

9.29 Ethnicity and pain perception. An investigation of ethnic differences in reports of pain perception was presented at the annual meeting of the American Psychosomatic Society (March 2001). A sample of 55 blacks and 159 whites participated in the study. Subjects rated (on a 13-point scale) the intensity and unpleasantness of pain felt when a bag of ice was placed on their foreheads for two minutes. (Higher ratings correspond to higher pain intensity.) A summary of the results is provided in the following table:

	Blacks	Whites
Sample size	55	159
Mean pain intensity	8.2	6.9

a. Why is it dangerous to draw a statistical inference from the summarized data? Explain.
b. Give values of the missing sample standard deviations that would lead you to conclude (at $\alpha = .05$) that blacks, on average, have a higher pain intensity rating than whites.
c. Give values of the missing sample standard deviations that would lead you to an inconclusive decision (at $\alpha = .05$) regarding whether blacks or whites have a higher mean intensity rating.

MINITAB Output for Exercise 9.28

Two-Sample T-Test and CI: WWBID, Market

```
Two-sample T for WWBID

Market     N     Mean    StDev   SE Mean
SURROUND   254   0.1331  0.0158  0.00099
TRI-COUNTY 100   0.1431  0.0133  0.0013


Difference = mu (SURROUND) - mu (TRI-COUNTY)
Estimate for difference:  -0.009970
95% upper bound for difference:  -0.007232
T-Test of difference = 0 (vs <): T-Value = -6.02  P-Value = 0.000  DF = 213
```

9.3 Comparing Two Population Means: Paired Difference Experiments

In Example 9.4, we compared two methods of teaching reading to "slow learners" by means of a 95% confidence interval. Suppose it is possible to measure the "reading IQs" of the "slow learners" *before* they are subjected to a teaching method. Eight pairs of "slow learners" with similar reading IQs are found, and one member of each pair is randomly assigned to the standard teaching method while the other is assigned to the new method. The data are given in Table 9.3. Do the data support the hypothesis that

Table 9.3	Reading Test Scores for Eight Pairs of "Slow Learners"	
Pair	New Method (1)	Standard Method (2)
1	77	72
2	74	68
3	82	76
4	73	68
5	87	84
6	69	68
7	66	61
8	80	76

Data Set: PAIREDSCORES

the population mean reading test score for "slow learners" taught by the new method is greater than the mean reading test score for those taught by the standard method?

We want to test

$$H_0: (\mu_1 - \mu_2) = 0$$
$$H_a: (\mu_1 - \mu_2) > 0$$

Many researchers mistakenly use the t statistic for two independent samples (Section 9.2) to conduct this test. This invalid analysis is shown on the MINITAB printout of Figure 9.9. The test statistic, $t = 1.26$, and the p-value of the test, $p = .115$., are highlighted on the printout. At $\alpha = .10$, the p-value exceeds α. Thus, from *this* analysis, we might conclude that we do not have sufficient evidence to infer a difference in the mean test scores for the two methods.

Figure 9.9

MINITAB printout of an invalid analysis of reading test scores in Table 9.3

```
Two-sample T for NEW vs STANDARD

              N    Mean   StDev   SE Mean
NEW           8   76.00    6.93      2.4
STANDARD      8   71.63    7.01      2.5


Difference = mu (NEW) - mu (STANDARD)
Estimate for difference:  4.38
95% lower bound for difference:  -1.76
T-Test of difference = 0 (vs >): T-Value = 1.26   P-Value = 0.115   DF = 14
Both use Pooled StDev = 6.9687
```

If you examine the data in Table 9.3 carefully, however, you will find this result difficult to accept. The test score of the new method is larger than the corresponding test score for the standard method *for every one of the eight pairs of "slow learners."* This, in itself, seems to provide strong evidence to indicate that μ_1 exceeds μ_2. Why, then, did the t-test fail to detect the difference? The answer is, *the independent samples t-test is not a valid procedure to use with this set of data.*

The t-test is inappropriate because the assumption of independent samples is invalid. We have randomly chosen *pairs of test scores;* thus, once we have chosen the sample for the new method, we have *not* independently chosen the sample for the standard method. The dependence between observations within pairs can be seen by examining the pairs of test scores, which tend to rise and fall together as we go from pair to pair. This pattern provides strong visual evidence of a violation of the assumption of independence required for the two-sample t-test of Section 9.2. Note also that

$$s_p^2 = \frac{(n_1 - 1)s_1^2 + (n_2 - 1)s_2^2}{n_1 + n_2 - 2} = \frac{(8 - 1)(6.93)^2 + (8 - 1)(7.01)^2}{8 + 8 - 2} = 48.58$$

Hence, there is a *large variation within samples* (reflected by the large value of s_p^2) in comparison to the relatively *small difference between the sample means.* Because s_p^2 is so large, the t-test of Section 9.2 is unable to detect a difference between μ_1 and μ_2.

We now consider a valid method of analyzing the data of Table 9.3. In Table 9.4, we add the column of differences between the test scores of the pairs of "slow learners."

Table 9.4	**Differences in Reading Test Scores**		
Pair	New Method	Standard Method	Difference (New Method − Standard Method)
1	77	72	5
2	74	68	6
3	82	76	6
4	73	68	5
5	87	84	3
6	69	68	1
7	66	61	5
8	80	76	4

We can regard these differences in test scores as a random sample of differences for all pairs (matched on reading IQ) of "slow learners," past and present. Then we can use this sample to make inferences about the mean of the population of differences, μ_d, which is equal to the difference ($\mu_1 - \mu_2$). That is, the mean of the population (and sample) of differences equals the difference between the population (and sample) means. Thus, our test becomes

$$H_0: \mu_d = 0 \quad (\mu_1 - \mu_2 = 0)$$
$$H_a: \mu_d > 0 \quad (\mu_1 - \mu_2 > 0)$$

The test statistic is a one-sample t (Section 8.4), since we are now analyzing a single sample of differences for small n. Thus,

$$Test\ statistic: t = \frac{\bar{x}_d - 0}{s_d/\sqrt{n_d}}$$

where

$$\bar{x}_d = \text{Sample mean difference}$$
$$s_d = \text{Sample standard deviation of differences}$$
$$n_d = \text{Number of differences} = \text{Number of pairs}$$

Assumptions: The population of differences in test scores is approximately normally distributed. The sample differences are randomly selected from the population differences. [*Note:* We do not need to make the assumption that $\sigma_1^2 = \sigma_2^2$.]

Rejection region: At significance level $\alpha = .05$, we will reject H_0 if $t > t_{.05}$, where $t_{.05}$ is based on $(n_d - 1)$ degrees of freedom.

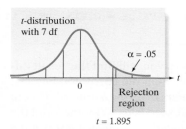

t-distribution with 7 df

$\alpha = .05$

Rejection region

$t = 1.895$

Figure 9.10

Rejection region for Example 9.4

Referring to Table VI in Appendix A, we find the t-value corresponding to $\alpha = .05$ and $n_d - 1 = 8 - 1 = 7$ df to be $t_{.05} = 1.895$. Then we will reject the null hypothesis if $t > 1.895$. (See Figure 9.10.) Note that the number of degrees of freedom decreases from $n_1 + n_2 - 2 = 14$ to 7 when we use the paired difference experiment rather than the two independent random samples design.

Summary statistics for the $n_d = 8$ differences are shown in the MINITAB printout of Figure 9.11. Note that $\bar{x}_d = 4.375$ and $s_d = 1.685$. Substituting these values into the formula for the test statistic, we have

$$t = \frac{\bar{x}_d - 0}{s_d/\sqrt{n_d}} = \frac{4.375}{1.685/\sqrt{8}} = 7.34$$

Because this value of t falls into the rejection region, we conclude (at $\alpha = .05$) that the population mean test score for "slow learners" taught by the new method exceeds the population mean score for those taught by the standard method. We can reach the same conclusion by noting that the p-value of the test, highlighted in Figure 9.11, is much smaller than $\alpha = .05$.

```
Paired T for NEW - STANDARD

              N    Mean   StDev   SE Mean
NEW           8   76.00    6.93      2.45
STANDARD      8   71.63    7.01      2.48
Difference    8   4.375   1.685     0.596

95% lower bound for mean difference: 3.246
T-Test of mean difference = 0 (vs > 0): T-Value = 7.34   P-Value = 0.000
```

Figure 9.11

MINITAB paired difference analysis of reading test scores

Now Work Exercises 9.35a and b

This kind of experiment, in which observations are paired and the differences are analyzed, is called a **paired difference experiment.** In many cases, a paired difference experiment can provide more information about the difference between population means than an independent samples experiment can. The idea is to compare population means by comparing the differences between pairs of experimental units (objects, people, etc.) that were similar prior to the experiment. The differencing removes sources of variation that tend to inflate σ^2. For example, when two children are taught to read by two different methods, the observed difference in achievement may be due to a difference in the effectiveness of the two teaching methods, *or it may be due to differences in the initial reading levels and IQs of the two children (random error)*. To reduce the effect of differences in the children on the observed differences in reading achievement, the two methods of reading are imposed on two children who are more likely to possess similar intellectual capacity, namely, children with nearly equal IQs. The effect of this pairing is to remove the larger source of variation that would be present if children with different abilities were randomly assigned to the two samples. Making comparisons within groups of similar experimental units is called **blocking,** and the paired difference experiment is a simple example of a **randomized block experiment.** In our example, pairs of children with matching IQ scores represent the blocks.

Some other examples for which the paired difference experiment might be appropriate are the following:

1. Suppose you want to estimate the difference $(\mu_1 - \mu_2)$ in mean price per gallon between two major brands of premium gasoline. If you choose two independent random samples of stations for each brand, the variability in price due to geographic location may be large. To eliminate this source of variability, you could choose pairs of stations of similar size, one station for each brand, in close geographic proximity and use the sample of differences between the prices of the brands to make an inference about $(\mu_1 - \mu_2)$.

2. Suppose a college placement center wants to estimate the difference $(\mu_1 - \mu_2)$ in mean starting salaries for men and women graduates who seek jobs through the center. If it independently samples men and women, the starting salaries may vary because of their different college majors and differences in grade point averages. To eliminate these sources of variability, the placement center could match male and female job seekers according to their majors and grade point averages. Then the differences between the starting salaries of each pair in the sample could be used to make an inference about $(\mu_1 - \mu_2)$.

3. Suppose you wish to estimate the difference $(\mu_1 - \mu_2)$ in mean absorption rate into the bloodstream for two drugs that relieve pain. If you independently sample people, the absorption rates might vary because of age, weight, sex, blood pressure, etc. In fact, there are many possible sources of nuisance variability, and pairing individuals who are similar in all the possible sources would be quite difficult. However, it may be possible to obtain two measurements *on the same person.* First, we administer one of the two drugs and record the time until absorption. After a sufficient amount of time, the other drug is administered and a second measurement on absorption time is obtained. The differences between the measurements for each person in the sample could then be used to estimate $(\mu_1 - \mu_2)$. This procedure would be advisable only if the amount of time allotted between drugs is sufficient to guarantee little or no carry-over effect. Otherwise, it would be better to use different people matched as closely as possible on the factors thought to be most important.

Now Work Exercise 9.33

The hypothesis-testing procedures and the method of forming confidence intervals for the difference between two means in a paired difference experiment are summarized in the following boxes for both large and small n:

Paired Difference Confidence Interval for $\mu_d = \mu_1 - \mu_2$

Large Sample, Normal (z) Statistic

$$\bar{x}_d \pm z_{\alpha/2} \frac{\sigma_d}{\sqrt{n_d}} \approx \bar{x}_d \pm z_{\alpha/2} \frac{s_d}{\sqrt{n_d}}$$

Small Sample, Student's t-Statistic

$$\bar{x}_d \pm t_{\alpha/2} \frac{s_d}{\sqrt{n_d}}$$

where $t_{\alpha/2}$ is based on $(n_d - 1)$ degrees of freedom

---

Paired Difference Test of Hypothesis for $\mu_d = \mu_1 - \mu_2$

One-Tailed Test

$H_0: \mu_d = D_0$
$H_a: \mu_d < D_0$
  [or $H_a: \mu_d > D_0$]

Two-Tailed Test

$H_0: \mu_d = D_0$
$H_a: \mu_d \neq D_0$

Large Sample, Normal (z) Statistic

Test statistic: $z = \dfrac{\bar{x}_d - D_0}{\sigma_d/\sqrt{n_d}} \approx \dfrac{\bar{x}_d - D_0}{s_d/\sqrt{n_d}}$

Rejection region: $z < -z_\alpha$ Rejection region: $|z| > z_{\alpha/2}$
[or $z > z_\alpha$ when $H_a: \mu_d > D_0$]

Small Sample, Student's t-Statistic

Test statistic: $t = \dfrac{\bar{x}_d - D_0}{s_d/\sqrt{n_d}}$

Rejection region: $t < -t_\alpha$ Rejection region: $|t| > t_{\alpha/2}$
[or $t > t_\alpha$ when $H_a: \mu_d > D_0$]

where t_α and $t_{\alpha/2}$ are based on $(n_d - 1)$ degrees of freedom

---

Conditions Required for Valid Large-Sample Inferences about μ_d

1. A random sample of differences is selected from the target population of differences.

2. The sample size n_d is large (i.e., $n_d \geq 30$). (By the Central Limit Theorem, this condition guarantees that the test statistic will be approximately normal, regardless of the shape of the underlying probability distribution of the population.)

Conditions Required for Valid Small-Sample Inferences about μ_d

1. A random sample of differences is selected from the target population of differences.

2. The population of differences has a distribution that is approximately normal.

Example 9.5

Confidence Interval For μ_d—Comparing Mean Salaries of Males and Females

Problem An experiment is conducted to compare the starting salaries of male and female college graduates who find jobs. Pairs are formed by choosing a male and a female with the same major and similar grade point averages (GPAs). Suppose a random sample of 10 pairs is formed in this manner and the starting annual salary of each person is recorded. The results are shown in Table 9.5. Compare the mean starting salary μ_1 for males with the mean starting salary μ_2 for females, using a 95% confidence interval. Interpret the results.

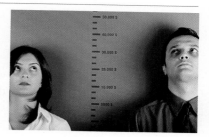

Table 9.5	Data on Annual Salaries for Matched Pairs of College Graduates		
Pair	Male	Female	Difference Male − Female
1	$29,300	$28,800	$ 500
2	41,500	41,600	−100
3	40,400	39,800	600
4	38,500	38,500	0
5	43,500	42,600	900
6	37,800	38,000	−200
7	69,500	69,200	300
8	41,200	40,100	1,100
9	38,400	38,200	200
10	59,200	58,500	700

Data Set: GRADPAIRS

Solution Since the data on annual salary are collected in pairs of males and females matched on GPA and major, a paired difference experiment is performed. To conduct the analysis, we first compute the differences between the salaries, as shown in Table 9.5. Summary statistics for these $n = 10$ differences are displayed at the top of the SAS printout shown in Figure 9.12.

```
                        The MEANS Procedure

                     Analysis Variable : DIFF

        Mean        Std Dev      N      Minimum      Maximum

       400.00       434.61      10      -200.00      1100.00



          Two Sample Paired t-test for the Means of MALE and FEMALE

Sample Statistics

      Group        N      Mean      Std. Dev.    Std. Error
      ------------------------------------------------------
      MALE        10      43930      11665        3688.8
      FEMALE      10      43530      11617        3673.6


Hypothesis Test

      Null hypothesis:      Mean of (MALE - FEMALE) =  0
      Alternative:          Mean of (MALE - FEMALE) ^= 0

          t Statistic      Df       Prob > t
          ------------------------------------
             2.910          9        0.0173


95% Confidence Interval for the Difference between Two Paired Means

         Lower Limit      Upper Limit
         -----------      -----------
            89.10            710.90
```

Figure 9.12

SAS analysis of salary differences

The 95% confidence interval for $\mu_d = (\mu_1 - \mu_2)$ for this small sample is

$$\bar{x}_d \pm t_{\alpha/2}\frac{s_d}{\sqrt{n_d}}$$

where $t_{\alpha/2} = t_{.025} = 2.262$ (obtained from Table VI, Appendix A) is based on $n_d - 1 = 9$ degrees of freedom. Substituting the values of $\bar{x}_d$ and s_d shown on the printout, we obtain

$$\bar{x}_d \pm 2.262\frac{s_d}{\sqrt{n_d}} = 400 \pm 2.262\left(\frac{434.613}{\sqrt{10}}\right)$$

$$= 400 \pm 310.88 \approx 400 \pm 311 = (\$89, \$711)$$

[*Note:* This interval is also shown highlighted at the bottom of the SAS printout of Figure 9.12.] Our interpretation is that the true mean difference between the starting salaries of males and females falls between $89 and $711, with 95% confidence. Since the interval falls above 0, we infer that $\mu_1 - \mu_2 > 0$; that is, the mean salary for males exceeds the mean salary for females.

Look Back Remember that $\mu_d = \mu_1 - \mu_2$. So if $\mu_d > 0$, then $\mu_1 > \mu_2$. Alternatively, if $\mu_d < 0$, then $\mu_1 < \mu_2$.

Now Work Exercise 9.42

To measure the amount of information about $(\mu_1 - \mu_2)$ gained by using a paired difference experiment in Example 9.5 rather than an independent samples experiment, we can compare the relative widths of the confidence intervals obtained by the two methods. A 95% confidence interval for $(\mu_1 - \mu_2)$ obtained from a paired difference experiment is, from Example 9.5, ($89, $711). If we mistakenly analyzed the same data as though this were an independent samples experiment,* we would first obtain the descriptive statistics shown in the SAS printout of Figure 9.13. Then we substitute the sample means and standard deviations shown on the printout into the formula for a 95% confidence interval for $(\mu_1 - \mu_2)$ using independent samples. The result is

$$(\bar{x}_1 - \bar{x}_2) \pm t_{.025}\sqrt{s_p^2\left(\frac{1}{n_1} + \frac{1}{n_2}\right)}$$

where

$$s_p^2 = \frac{(n_1 - 1)s_1^2 + (n_2 - 1)s_2^2}{n_1 + n_2 - 2}$$

Group Statistics

	GENDER	N	Mean	Std. Deviation	Std. Error Mean
SALARY	M	10	43930.00	11665.148	3688.844
	F	10	43530.00	11616.946	3673.601

Independent Samples Test

		Levene's Test for Equality of Variances		t-test for Equality of Means					95% Confidence Interval of the Difference	
		F	Sig.	t	df	Sig. (2-tailed)	Mean Difference	Std. Error Difference	Lower	Upper
SALARY	Equal variances assumed	.000	.991	.077	18	.940	400.000	5206.046	-10537.496	11337.496
	Equal variances not assumed			.077	18.000	.940	400.000	5206.046	-10537.509	11337.509

Figure 9.13

SPSS analysis of salaries, assuming independent samples

*This is done only to provide a measure of the increase in the amount of information obtained by a paired design in comparison to an unpaired design. Actually, if an experiment were designed that used pairing, an unpaired analysis would be invalid because the assumption of independent samples would not be satisfied.

SPSS performed these calculations and obtained the interval ($-10,537.50, $11,337.50), highlighted in Figure 9.13.

Notice that the independent samples interval includes 0. Consequently, if we were to use this interval to make an inference about $(\mu_1 - \mu_2)$, we would incorrectly conclude that the mean starting salaries of males and females do not differ! You can see that the confidence interval for the independent sampling experiment is about 35 times wider than for the corresponding paired difference confidence interval. Blocking out the variability due to differences in majors and grade point averages significantly increases the information about the difference in males' and females' mean starting salaries by providing a much more accurate (a smaller confidence interval for the same confidence coefficient) estimate of $(\mu_1 - \mu_2)$.

Ethics IN Statistics

In a two-group analysis, intentionally pairing observations after the data have been collected in order to produce a desired result is considered *unethical statistcal practice*.

You may wonder whether a paired difference experiment is always superior to an independent samples experiment. The answer is, most of the time, but not always. We sacrifice half the degrees of freedom in the *t*-statistic when a paired difference design is used instead of an independent samples design. This is a loss of information, and unless that loss is more than compensated for by the reduction in variability obtained by blocking (pairing), the paired difference experiment will result in a net loss of information about $(\mu_1 - \mu_2)$. Thus, we should be convinced that the pairing will significantly reduce variability before performing a paired difference experiment. Most of the time, this will happen.

One final note: The pairing of the observations is determined *before* the experiment is performed (i.e., by the *design* of the experiment). A paired difference experiment is *never* obtained by pairing the sample observations *after* the measurements have been acquired.

> **What Do You Do When the Assumption of a Normal Distribution for the Population of Differences Is Not Satisfied?**
>
> *Answer:* Use the Wilcoxon signed rank test for the paired difference design (Chapter 14).

Exercises 9.30–9.50

Understanding the Principles

9.30 What are the advantages of using a paired difference experiment over an independent samples design?

9.31 In a paired difference experiment, when should the observations be paired, before or after the data are collected?

9.32 What conditions are required for valid large-sample inferences about μ_d? small-sample inferences?

Learning the Mechanics

9.33 A paired difference experiment yielded n_d pairs of observations. In each case, what is the rejection region for testing
[NW] $H_0: \mu_d = 2$ against $H_a: \mu_d > 2$?
a. $n_d = 10, \alpha = .05$
b. $n_d = 20, \alpha = .10$
c. $n_d = 5, \alpha = .025$
d. $n_d = 9, \alpha = .01$

9.34 A paired difference experiment produced the following data:

$$n_d = 16 \quad \bar{x}_1 = 143 \quad \bar{x}_2 = 150 \quad \bar{x}_d = -7 \quad s_d^2 = 64$$

a. Determine the values of t for which the null hypothesis $\mu_1 - \mu_2 = 0$ would be rejected in favor of the alternative hypothesis $\mu_1 - \mu_2 < 0$. Use $\alpha = .10$.

b. Conduct the paired difference test described in part **a.** Draw the appropriate conclusions.

c. What assumptions are necessary so that the paired difference test will be valid?

d. Find a 90% confidence interval for the mean difference μ_d.

e. Which of the two inferential procedures, the confidence interval of part **d** or the test of hypothesis of part **b,** provides more information about the difference between the population means?

9.35 The data for a random sample of six paired observations are shown in the following table and saved in the
[NW] LM9_35 file.

Pair	Sample from Population 1	Sample from Population 2
1	7	4
2	3	1
3	9	7
4	6	2
5	4	4
6	8	7

a. Calculate the difference between each pair of observations by subtracting observation 2 from observation 1. Use the differences to calculate $\bar{x}_d$ and s_d^2.

b. If μ_1 and μ_2 are the means of populations 1 and 2, respectively, express μ_d in terms of μ_1 and μ_2.

c. Form a 95% confidence interval for μ_d.

d. Test the null hypothesis $H_0: \mu_d = 0$ against the alternative hypothesis $H_a: \mu_d \neq 0$. Use $\alpha = .05$.

9.36 The data for a random sample of 10 paired observations are shown in the following table and saved in the **LM9_36** file.

Pair	Population 1	Population 2
1	19	24
2	25	27
3	31	36
4	52	53
5	49	55
6	34	34
7	59	66
8	47	51
9	17	20
10	51	55

a. If you wish to test whether these data are sufficient to indicate that the mean for population 2 is larger than that for population 1, what are the appropriate null and alternative hypotheses? Define any symbols you use.

b. Conduct the test from part **a**, using $\alpha = .10$. What is your decision?

c. Find a 90% confidence interval for μ_d. Interpret this interval.

d. What assumptions are necessary to ensure the validity of the preceding analysis?

9.37 A paired difference experiment yielded the following results:

$$n_d = 40, \ \bar{x}_d = 11.7, \ s_d = 6.$$

a. Test $H_0: \mu_d = 10$ against $H_a: \mu_d \neq 10$, where $\mu_d = (\mu_1 - \mu_2)$. Use $\alpha = .05$.

b. Report the p-value for the test you conducted in part **a**. Interpret the p-value.

Applying the Concepts—Basic

9.38 **Summer weight-loss camp.** Camp Jump Start is an 8-week summer camp for overweight and obese adolescents. Counselors develop a weight-management program for each camper that centers on nutrition education and physical activity. In a study published in *Pediatrics* (April 2010), the body mass index (BMI) was measured for each of 76 campers both at the start and end of camp. Summary statistics on BMI measurements are shown in the table.

	Mean	Standard Deviation
Starting BMI	34.9	6.9
Ending BMI	31.6	6.2
Paired Differences	3.3	1.5

Based on Huelsing, J., Kanafani, N., Mao, J., and White, N. H. "Camp Jump Start: Effects of a residential summer weight-loss camp for older children and adolescents." *Pediatrics*, Vol. 125, No. 4, April 2010 (Table 3).

a. Give the null and alternative hypothesis for determining whether the mean BMI at the end of camp is less than the mean BMI at the start of camp.

b. How should the data be analyzed, as an independent-samples t-test or as a paired-difference t-test? Explain.

c. Calculate the test statistic using the formula for an independent-samples t-test. (*Note*: This is *not* how the test should be conducted.)

d. Calculate the test statistic using the formula for a paired-difference t-test.

e. Compare the test statistics, parts **c** and **d**. Which test statistic provides more evidence in support of the alternative hypothesis?

f. The p-value of the test, part **d**, was reported as $p < .0001$. Interpret this result assuming $\alpha = .01$.

g. Do the differences in BMI values need to be normally distributed in order for the inference, part **f**, to be valid? Explain.

h. Find a 99% confidence interval for the true mean change in BMI for Camp Jump Start campers. Interpret the result.

9.39 **Healing potential of handling museum objects.** Does handling a museum object have a positive impact on a sick patient's well-being? To answer this question, researchers at the University College London collected data from 32 sessions with hospital patients (*Museum & Society*, Nov. 2009). Each patient's health status (measured on a 100-point scale) was recorded both before and after handling museum objects such as archaeological artifacts and brass etchings. The data (simulated) are listed in the accompanying table and saved in the **MUSEUM** file.

Session	Before	After	Session	Before	After
1	52	59	17	65	65
2	42	54	18	52	63
3	46	55	19	39	50
4	42	51	20	59	69
5	43	42	21	49	61
6	30	43	22	59	66
7	63	79	23	57	61
8	56	59	24	56	58
9	46	53	25	47	55
10	55	57	26	61	62
11	43	49	27	65	61
12	73	83	28	36	53
13	63	72	29	50	61
14	40	49	30	40	52
15	50	49	31	65	70
16	50	64	32	59	72

a. Explain why the data should be analyzed as paired differences.

b. Compute the difference between the "before" and "after" measurements for each session.

c. Find the mean and standard deviation of the differences, part **b**.

d. Use the summary statistics, part **c**, to find a 90% confidence interval for the true mean difference ("before" minus "after") in health status scale measurements.

e. Interpret the interval, part **d**. Does handling a museum object have a positive impact on a sick patient's well-being?

9.40 Laughter among deaf signers. The *Journal of Deaf Studies and Deaf Education* (Fall 2006) published an article on vocalized laughter among deaf users of American Sign Language (ASL). In videotaped ASL conversations among deaf participants, 28 laughed at least once. The researchers wanted to know if they laughed more as speakers (while signing) or as audience members (while listening). For each of the 28 deaf participants, the number of laugh episodes as a speaker and the number of laugh episodes as an audience member were determined. One goal of the research was to compare the mean numbers of laugh episodes of speakers and audience members.

a. Explain why the data should be analyzed as a paired difference experiment.

b. Identify the study's target parameter.

c. The study yielded a sample mean of 3.4 laughter episodes for speakers and a sample mean of 1.3 laughter episodes for audience members. Is this sufficient evidence to conclude that the population means are different? Explain.

d. A paired difference t-test resulted in $t = 3.14$ and p-value $< .01$. Interpret the results in the words of the problem.

9.41 The placebo effect and pain. According to research published in *Science* (Feb. 20, 2004), the mere belief that you are receiving an effective treatment for pain can reduce the pain you actually feel. Researchers from the University of Michigan and Princeton University tested this placebo effect on 24 volunteers as follows: Each volunteer was put inside a magnetic resonance imaging (MRI) machine for two consecutive sessions. During the first session, electric shocks were applied to their arms and the blood oxygen level–dependent (BOLD) signal (a measure related to neural activity in the brain) was recorded during pain. The second session was identical to the first, except that, prior to applying the electric shocks, the researchers smeared a cream on the volunteer's arms. The volunteers were informed that the cream would block the pain when, in fact, it was just a regular skin lotion (i.e., a placebo). If the placebo is effective in reducing the pain experience, the BOLD measurements should be higher, on average, in the first MRI session than in the second.

a. Identify the target parameter for this study.

b. What type of design was used to collect the data?

c. Give the null and alternative hypotheses for testing the placebo effect theory.

d. The differences between the BOLD measurements in the first and second sessions were computed and summarized in the study as follows: $n_d = 24$, $\bar{x}_d = .21$, $s_d = .47$. Use this information to calculate the test statistic.

e. The p-value of the test was reported as p-value $= .02$. Make the appropriate conclusion at $\alpha = .05$.

9.42 NHTSA new car crash tests. Refer to the National Highway Traffic Safety Administration (NHTSA) crash test data on new cars, saved in the **CRASH** file. Crash test dummies were placed in the driver's seat and front passenger's seat of a new car model, and the car was steered by remote control into a head-on collision with a fixed barrier while traveling at 35 miles per hour. Two of the variables measured for each of the 98 new cars in the data set are (1) the severity of the driver's chest injury and (2) the severity of the passenger's chest injury. (The more points assigned to the chest injury rating, the more severe the injury is.) Suppose the NHTSA wants to determine whether the true mean driver chest injury rating exceeds the true mean passenger chest injury rating and, if so, by how much.

a. State the parameter of interest to the NHTSA.

b. Explain why the data should be analyzed as matched pairs.

c. Find a 99% confidence interval for the true difference between the mean chest injury ratings of drivers and front-seat passengers.

d. Interpret the interval you found in part **c.** Does the true mean driver chest injury rating exceed the true mean passenger chest injury rating? If so, by how much?

e. What conditions are required for the analysis to be valid? Do these conditions hold for these data?

Applying the Concepts—Intermediate

9.43 Acidity of mouthwash. Acid has been found to be a primary cause of dental caries (cavities). It is theorized that oral mouthwashes contribute to the development of caries due to the antiseptic agent oxidizing into acid over time. This theory was tested in the *Journal of Dentistry, Oral Medicine and Dental Education* (Vol. 3, 2009). Three bottles of mouthwash, each of a different brand, were randomly selected from a drugstore. The pH level (where lower pH levels indicate higher acidity) of each bottle was measured on the date of purchase and after 30 days. The data, saved in the **MOUTHWASH** file, are shown in the table. Conduct an analysis to determine if the mean initial pH level of mouthwash differs significantly from the mean pH level after 30 days. Use $\alpha = .05$ as your level of significance.

Mouthwash Brand	Initial pH	Final pH
LMW	4.56	4.27
SMW	6.71	6.51
RMW	5.65	5.58

Based on Chunhye, K. L., and Schmitz, B. C., "Determination of pH, total acid, total ethanol in oral health products: Oxidation of ethanol and recommendations to mitigate its association with dental caries." *Journal of Dentistry, Oral Medicine and Dental Education*, Vol. 3, No. 1, 2009 (Table 1).

9.44 Visual search and memory study. In searching for an item (e.g., a roadside traffic sign, a lost earring, or a tumor in a mammogram), common sense dictates that you will not reexamine items previously rejected. However, researchers at Harvard Medical School found that a visual search has no memory (*Nature*, Aug. 6, 1998). In their experiment, nine subjects searched for the letter "T" mixed among several letters "L." Each subject conducted the search under two conditions: random and static. In the random condition, the locations of the letters were changed every 111 milliseconds; in the static condition, the locations of the letters remained unchanged. In each trial, the reaction time in milliseconds (i.e., the amount of time it took the subject to locate the target letter) was recorded.

a. One goal of the research was to compare the mean reaction times of subjects in the two experimental conditions. Explain why the data should be analyzed as a paired difference experiment.

b. If a visual search has no memory, then the main reaction times in the two conditions will not differ. Specify H_0 and H_a for testing the "no-memory" theory.

c. The test statistic was calculated as $t = 1.52$ with p-value $= .15$. Draw the appropriate conclusion.

9.45 **Linking dementia and leisure activities.** Does participation in leisure activities in your youth reduce the risk of Alzheimer's disease and other forms of dementia? To answer this question, a group of university researchers studied a sample of 107 same-sex Swedish pairs of twins (*Journal of Gerontology: Psychological Sciences and Social Sciences*, Sept. 2003). Each pair of twins was discordant for dementia; that is, one member of each pair was diagnosed with Alzheimer's disease while the other member (the control) was nondemented for at least five years after the sibling's onset of dementia. The level of overall leisure activity (measured on an 80-point scale, where higher values indicate higher levels of leisure activity) of each twin of each pair 20 years prior to the onset of dementia was obtained from the Swedish Twin Registry database. The leisure activity scores (simulated on the basis of summary information presented in the journal article) are saved in the **DEMENTIA** file. The first five and last five observations are shown in the following table:

Pair	Control	Demented
1	27	13
2	57	57
3	23	31
4	39	46
5	37	37
⋮	⋮	⋮
103	22	14
104	32	23
105	33	29
106	36	37
107	24	1

a. Explain why the data should be analyzed as a paired difference experiment.

b. Conduct the appropriate analysis, using $\alpha = .05$. Make an inference about which member of the pair, the demented or control (nondemented) twin, had the largest average level of leisure activity.

9.46 **Ethical sensitivity of teachers towards racial intolerance.** Many high schools have education programs that encourage teachers to embrace racial tolerance, recognize ethnic diversity, and overcome racial stereotypes. To gauge the effectiveness of one such program that utilizes two videos of teachers engaging in racial stereotypes of their students, researchers from New York University and City University of New York recruited 238 high school professionals (including teachers and counselors) to participate in a study (*Journal of Moral Education*, March 2010). Teachers watched the first video, then were given a pretest—the Quick-REST Survey—designed to measure ethical sensitivity towards racial intolerance. The teachers next participated in an all-day workshop on cultural competence. At the end of the workshop, the teachers watched the second video and again were given the Quick-REST Survey (the posttest). To determine whether the program was effective, the researchers compared the mean scores on the Quick-REST Survey using a paired-difference t-test. (*Note:* The higher the score on the Quick-REST Survey, the greater the level of racial tolerance.)

a. The researchers reported the sample means for the pretest and posttest as 75.85 and 80.35, respectively. Why is it dangerous to gauge the effectiveness of the program based only on these summary statistics?

b. The paired-difference t-test (posttest minus pretest) was reported as $t = 4.50$ with an associated observed significance level of p-value $< .001$. Interpret this result.

c. What assumptions, if any, are necessary for the validity of the inference, part **b**?

9.47 **Impact of red light cameras on car crashes.** To combat red-light-running crashes—the phenomenon of a motorist entering an intersection after the traffic signal turns red and causing a crash—many states are adopting photo-red enforcement programs. In these programs, red light cameras installed at dangerous intersections photograph the license plates of vehicles that run the red light. How effective are photo-red enforcement programs in reducing red-light-running crash incidents at intersections? The Virginia Department of Transportation (VDOT) conducted a comprehensive study of its newly adopted photo-red enforcement program and published the results in a June 2007 report. In one portion of the study, the VDOT provided crash data both before and after installation of red light cameras at several intersections. The data (measured as the number of crashes caused by red light running per intersection per year) for 13 intersections in Fairfax County, Virginia, are given in the table and saved in the **REDLIGHT** file. Analyze the data for the VDOT. What do you conclude?

Intersection	Before Camera	After Camera
1	3.60	1.36
2	0.27	0
3	0.29	0
4	4.55	1.79
5	2.60	2.04
6	2.29	3.14
7	2.40	2.72
8	0.73	0.24
9	3.15	1.57
10	3.21	0.43
11	0.88	0.28
12	1.35	1.09
13	7.35	4.92

Based on Virginia Transportation Research Council, "Research report: The impact of red light cameras (photo-red enforcement) on crashes in Virginia." June 2007.

9.48 **Light-to-dark transition of genes.** *Synechocystis,* a type of cyanobacterium that can grow and survive under a wide range of conditions, is used by scientists to model DNA behavior. In the *Journal of Bacteriology* (July 2002), scientists isolated genes of the bacterium responsible for photosynthesis and respiration and investigated the sensitivity of the genes to light. Each gene sample was grown to midexponential phase in a growth incubator in "full light." The lights were then extinguished, and any growth of the sample was measured after 24 hours in the dark ("full dark"). The lights were then turned back on for 90 minutes ("transient light"), followed immediately by an additional 90 minutes in the dark ("transient dark"). Standardized growth measurements in each light–dark condition were obtained for 103 genes. The complete data set is saved in the **GENEDARK** file. Data on the first 10 genes are shown in the following table:

Gene ID	FULL-DARK	TR-LIGHT	TR-DARK
SLR2067	−0.00562	1.40989	−1.28569
SLR1986	−0.68372	1.83097	−0.68723
SSR3383	−0.25468	−0.79794	−0.39719
SLL0928	−0.18712	−1.20901	−1.18618
SLR0335	−0.20620	1.71404	−0.73029
SLR1459	−0.53477	2.14156	−0.33174
SLL1326	−0.06291	1.03623	0.30392
SLR1329	−0.85178	−0.21490	0.44545
SLL1327	0.63588	1.42608	−0.13664
SLL1325	−0.69866	1.93104	−0.24820

Based on Gill, R. T., et al. "Genome-wide dynamic transcriptional profiling of the light to dark transition in *Synechocystis Sp.* PCC6803." *Journal of Bacteriology,* Vol. 184, No. 13, July 2002.

a. Treat the data for the first 10 genes as a random sample collected from the population of 103 genes, and test the hypothesis that there is no difference between the mean standardized growth of genes in the full-dark condition and genes in the transient-light condition. Use $\alpha = .01$.

b. Use a statistical software package to compute the mean difference in standardized growth of the 103 genes in the full-dark condition and the transient-light condition. Did the test you carried out in part **a** detect this difference?

c. Repeat parts **a** and **b** for a comparison of the mean standardized growth of genes in the full-dark condition and genes in the transient-dark condition.

d. Repeat parts **a** and **b** for a comparison of the mean standardized growth of genes in the transient-light condition and genes in the transient-dark condition.

Applying the Concepts—Advanced

9.49 Homophone confusion in Alzheimer's patients. A *homophone* is a word whose pronunciation is the same as that of another word having a different meaning and spelling (e.g., *nun* and *none, doe* and *dough,* etc.). *Brain and Language* (Apr. 1995) reported on a study of homophone spelling in patients with Alzheimer's disease. Twenty Alzheimer's patients were asked to spell 24 homophone pairs given in random order. Then the number of homophone confusions (e.g., spelling *doe* given the context, *bake bread dough*) was recorded for each patient. One year later, the same test was given to the same patients. The data for the study are provided in the next table and saved in the **HOMOPHONE** file. The researchers posed the following question: "Do Alzheimer's patients show a significant increase in mean homophone confusion errors over time?" Perform an analysis of the data to answer the researchers' question. What assumptions are necessary for the procedure used to be valid? Are they satisfied?

Patient	Time 1	Time 2
1	5	5
2	1	3
3	0	0
4	1	1
5	0	1
6	2	1
7	5	6
8	1	2
9	0	9
10	5	8
11	7	10
12	0	3
13	3	9
14	5	8
15	7	12
16	10	16
17	5	5
18	6	3
19	9	6
20	11	8

Based on Neils, J., Roeltgen, D. P., and Constantinidou, F. "Decline in homophone spelling associated with loss of semantic influence on spelling in Alzheimer's disease." *Brain and Language,* Vol. 49, No. 1, pp. 27–49.

9.50 Alcoholic fermentation in wines. Determining alcoholic fermentation in wine is critical to the wine-making process. Must/wine density is a good indicator of the fermentation point, since the density value decreases as sugars are converted into alcohol. For decades, winemakers have measured must/wine density with a hydrometer. Although accurate, the hydrometer employs a manual process that is very time consuming. Consequently, large wineries are searching for more rapid measures of density measurement. An alternative method utilizes the hydrostatic balance instrument (similar to the hydrometer, but digital). A winery in Portugal collected must/wine density measurements on white wine samples randomly selected from the fermentation process for a recent harvest. For each sample, the density of the wine at 20°C was measured with both the hydrometer and the hydrostatic balance. The densities for 40 wine samples are saved in the **WINE40** file. The first five and last five observations are shown in the accompanying table. The winery will use the alternative method of mea-suring wine density only if it can be demonstrated that the mean difference between the density measurements of the two methods does not exceed .002. Perform the analysis for the winery. Provide the winery with a written report of your conclusions.

Sample	Hydrometer	Hydrostatic
1	1.08655	1.09103
2	1.00270	1.00272
3	1.01393	1.01274
4	1.09467	1.09634
5	1.10263	1.10518
⋮	⋮	⋮
36	1.08084	1.08097
37	1.09452	1.09431
38	0.99479	0.99498
39	1.00968	1.01063
40	1.00684	1.00526

Based on Cooperative Cellar of Borba (*Adega Cooperativ a de Borba*), Portugal.

9.4 Comparing Two Population Proportions: Independent Sampling

Suppose a presidential candidate wants to compare the preferences of registered voters in the northeastern United States with those in the southeastern United States. Such a comparison would help determine where to concentrate campaign efforts. The candidate hires a professional pollster to randomly choose 1,000 registered voters in the northeast and 1,000 in the southeast and interview each to learn her or his voting preference. The objective is to use this sample information to make an inference about the difference $(p_1 - p_2)$ between the proportion p_1 of *all* registered voters in the northeast and the proportion p_2 of *all* registered voters in the southeast who plan to vote for the presidential candidate.

The two samples represent independent binomial experiments. (See Section 4.4 for the characteristics of binomial experiments.) The binomial random variables are the numbers x_1 and x_2 of the 1,000 sampled voters in each area who indicate that they will vote for the candidate. The results are summarized in Table 9.6

We can now calculate the sample proportions $\hat{p}_1$ and $\hat{p}_2$ of the voters in favor of the candidate in the northeast and southeast, respectively:

Table 9.6	Results of Poll
Northeast	Southeast
$n_1 = 1{,}000$	$n_2 = 1{,}000$
$x_1 = 546$	$x_2 = 475$

$$\hat{p} = \frac{x_1}{n_1} = \frac{546}{1{,}000} = .546 \quad \hat{p}_2 = \frac{x_2}{n_2} = \frac{475}{1{,}000} = .475$$

The difference between the sample proportions $(\hat{p}_1 - \hat{p}_2)$ makes an intuitively appealing point estimator of the difference between the population $(p_1 - p_2)$. For this example, the estimate is

$$(\hat{p}_1 - \hat{p}_2) = .546 - .475 = .071$$

To judge the reliability of the estimator $(\hat{p}_1 - \hat{p}_2)$, we must observe its performance in repeated sampling from the two populations. That is, we need to know the sampling distribution of $(\hat{p}_1 - \hat{p}_2)$. The properties of the sampling distribution are given in the next box. Remember that $\hat{p}_1$ and $\hat{p}_2$ can be viewed as means of the number of successes per trial in the respective samples, so the Central Limit Theorem applies when the sample sizes are large.

Properties of the Sampling Distribution of $(\hat{p}_1 - \hat{p}_2)$

1. The mean of the sampling distribution of $(\hat{p}_1 - \hat{p}_2)$ is $(p_1 - p_2)$; that is,

$$E(\hat{p}_1 - \hat{p}_2) = p_1 - p_2$$

Thus, $(\hat{p}_1 - \hat{p}_2)$ is an unbiased estimator of $(p_1 - p_2)$.

2. The standard deviation of the sampling distribution of $(\hat{p}_1 - \hat{p}_2)$ is

$$\sigma_{(\hat{p}_1 - \hat{p}_2)} = \sqrt{\frac{p_1 q_1}{n_1} + \frac{p_2 q_2}{n_2}}$$

3. If the sample sizes n_1 and n_2 are large (see Section 7.4 for a guideline), the sampling distribution of $(\hat{p}_1 - \hat{p}_2)$ is approximately normal.

Since the distribution of $(\hat{p}_1 - \hat{p}_2)$ in repeated sampling is approximately normal, we can use the z-statistic to derive confidence intervals for $(p_1 - p_2)$ or to test a hypothesis about $(p_1 - p_2)$.

For the voter example, a 95% confidence interval for the difference $(p_1 - p_2)$ is

$$(\hat{p}_1 - \hat{p}_2) \pm 1.96 \sigma_{(\hat{p}_1 - \hat{p}_2)}, \text{ or } (\hat{p}_1 - \hat{p}_2) \pm 1.96 \sqrt{\frac{p_1 q_1}{n_1} + \frac{p_2 q_2}{n_2}}$$

The quantities $p_1 q_1$ and $p_2 q_2$ must be estimated in order to complete the calculation of the standard deviation $\sigma_{(\hat{p}_1 - \hat{p}_2)}$ and, hence, the calculation of the confidence interval. In Section 7.4, we showed that the value of pq is relatively insensitive to the value chosen to approximate p. Therefore, $\hat{p}_1\hat{q}_1$ and $\hat{p}_2\hat{q}_2$ will provide satisfactory approximations of p_1q_1 and p_2q_2, respectively. Then

$$\sqrt{\frac{p_1q_1}{n_1} + \frac{p_2q_2}{n_2}} \approx \sqrt{\frac{\hat{p}_1\hat{q}_1}{n_1} + \frac{\hat{p}_2\hat{q}_2}{n_2}}$$

and we will approximate the 95% confidence interval by

$$(\hat{p}_1 - \hat{p}_2) \pm 1.96\sqrt{\frac{\hat{p}_1\hat{q}_1}{n_1} + \frac{\hat{p}_2\hat{q}_2}{n_2}}$$

Substituting the sample quantities yields

$$(.546 - .475) \pm 1.96\sqrt{\frac{(.546)(.454)}{1,000} + \frac{(.475)(.525)}{1,000}}$$

or $.071 \pm .044$. Thus, we are 95% confident that the interval from .027 to .115 contains $(p_1 - p_2)$.

We infer that there are between 2.7% and 11.5% more registered voters in the northeast than in the southeast who plan to vote for the presidential candidate. It seems that the candidate should direct a greater campaign effort in the southeast than in the northeast.

Now Work Exercise 9.59

The general form of a confidence interval for the difference $(p_1 - p_2)$ between population proportions is given in the following box:

Large-Sample $100(1 - \alpha)$% Confidence Interval for $(p_1 - p_2)$: Normal (z) Statistic

$$(\hat{p}_1 - \hat{p}_2) \pm z_{\alpha/2}\sigma_{(\hat{p}_1 - \hat{p}_2)} = (\hat{p}_1 - \hat{p}_2) \pm z_{\alpha/2}\sqrt{\frac{p_1q_1}{n_1} + \frac{p_2q_2}{n_2}}$$

$$\approx (\hat{p}_1 - \hat{p}_2) \pm z_{\alpha/2}\sqrt{\frac{\hat{p}_1\hat{q}_1}{n_1} + \frac{\hat{p}_2\hat{q}_2}{n_2}}$$

The z-statistic,

$$z = \frac{(\hat{p}_1 - \hat{p}_2) - (p_1 - p_2)}{\sigma_{(\hat{p}_1 - \hat{p}_2)}}$$

is used to test the null hypothesis that $(p_1 - p_2)$ equals some specified difference, say, D_0. For the special case where $D_0 = 0$—that is, where we want to test the null hypothesis $H_0\colon (p_1 - p_2) = 0$ (or, equivalently, $H_0\colon p_1 = p_2$)—the best estimate of $p_1 = p_2 = p$ is obtained by dividing the total number of successes $(x_1 + x_2)$ for the two samples by the total number of observations $(n_1 + n_2)$, that is,

$$\hat{p} = \frac{x_1 + x_2}{n_1 + n_2}, \quad \text{or} \quad \hat{p} = \frac{n_1\hat{p}_1 + n_2\hat{p}_2}{n_1 + n_2}$$

The second equation shows that $\hat{p}$ is a weighted average of $\hat{p}_1$ and $\hat{p}_2$, with the larger sample receiving more weight. If the sample sizes are equal, then $\hat{p}$ is a simple average of the two sample proportions of successes.

We now substitute the weighted average $\hat{p}$ for both p_1 and p_2 in the formula for the standard deviation of $(\hat{p}_1 - \hat{p}_2)$:

$$\sigma_{(\hat{p} - \hat{p}_2)} = \sqrt{\frac{p_1q_1}{n_1} + \frac{p_2q_2}{n_2}} \approx \sqrt{\frac{\hat{p}\hat{q}}{n_1} + \frac{\hat{p}\hat{q}}{n_2}} = \sqrt{\hat{p}\hat{q}\left(\frac{1}{n_1} + \frac{1}{n_2}\right)}$$

The test is summarized in the following box:

Large-Sample Test of Hypothesis about $(p_1 - p_2)$: Normal (z) Statistic

One-Tailed Test	**Two-Tailed Test**
$H_0: (p_1 - p_2) = 0*$	$H_0: (p_1 - p_2) = 0$
$H_a: (p_1 - p_2) < 0$	$H_a: (p_1 - p_2) \neq 0$
[or $H_a: (p_1 - p_2) > 0$]	

$$\text{Test statistic: } z = \frac{(\hat{p}_1 - \hat{p}_2)}{\sigma_{(\hat{p}_1 - \hat{p}_2)}}$$

Rejection region: $z < -z_\alpha$ Rejection region: $|z| > z_{\alpha/2}$
[or $z > z_\alpha$ when $H_a: (p_1 - p_2) > 0$]

$$\textit{Note: } \sigma_{(\hat{p}_1 - \hat{p}_2)} = \sqrt{\frac{p_1 q_1}{n_1} + \frac{p_2 q_2}{n_2}} \approx \sqrt{\hat{p}\hat{q}\left(\frac{1}{n_1} + \frac{1}{n_2}\right)} \text{ where } \hat{p} = \frac{x_1 + x_2}{n_1 + n_2}$$

Conditions Required for Valid Large-Sample Inferences about $(p_1 - p_2)$

1. The two samples are randomly selected in an independent manner from the two target populations.
2. The sample sizes, n_1 and n_2, are both large, so the sampling distribution of $(\hat{p}_1 - \hat{p}_2)$ will be approximately normal. (This condition will be satisfied if both $n_1\hat{p}_1 \geq 15, n_1\hat{q}_1 \geq 15$, and $n_2\hat{p}_2 \geq 15, n_2\hat{q}_2 \geq 15$.)

Example 9.6

A Large–Sample Test about $(p_1 - p_2)$—Comparing Fractions of Smokers for Two Years

Problem In the past decade, intensive antismoking campaigns have been sponsored by both federal and private agencies. Suppose the American Cancer Society randomly sampled 1,500 adults in 2000 and then sampled 1,750 adults in 2010 to determine whether there was evidence that the percentage of smokers had decreased. The results of the two sample surveys are shown in Table 9.7, where x_1 and x_2 represent the numbers of smokers in the 2000 and 2010 samples, respectively. Do these data indicate that the fraction of smokers decreased over this 10-year period? Use $\alpha = .05$.

Solution If we define p_1 and p_2 as the true proportions of adult smokers in 2000 and 2010, respectively, then the elements of our test are

$$H_0: (p_1 - p_2) = 0$$
$$H_a: (p_1 - p_2) > 0$$

(The test is one tailed, since we are interested only in determining whether the proportion of smokers *decreased*.)

$$\text{Test statistic: } z = \frac{(\hat{p}_1 - \hat{p}_2) - 0}{\sigma_{(\hat{p}_1 - \hat{p}_2)}}$$

Rejection region using $\alpha = .05$: $z > z_\alpha = z_{.05} = 1.645$ (see Figure 9.14)

We now calculate the sample proportions of smokers:

$$\hat{p}_1 = \frac{555}{1,500} = .37 \qquad \hat{p}_2 = \frac{578}{1,750} = .33$$

Table 9.7

Results of Smoking Survey

2000	2010
$n_1 = 1,500$	$n_2 = 1,750$
$x_1 = 555$	$x_2 = 578$

---

*The test can be adapted to test for a difference $D_0 \neq 0$. Because most applications call for a comparison of p_1 and p_2, implying that $D_0 = 0$, we will confine our attention to this case.

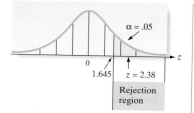

Figure 9.14
Rejection region for Example 9.6

Then

$$z = \frac{(\hat{p}_1 - \hat{p}_2) - 0}{\sigma_{(\hat{p}_1 - \hat{p}_2)}} \approx \frac{(\hat{p}_1 - \hat{p}_2)}{\sqrt{\hat{p}\hat{q}\left(\frac{1}{n_1} + \frac{1}{n_2}\right)}}$$

where

$$\hat{p} = \frac{x_1 + x_2}{n_1 + n_2} = \frac{555 + 578}{1,500 + 1,750} = .349$$

Note that $\hat{p}$ is a weighted average of $\hat{p}_1$ and $\hat{p}_2$, with more weight given to the larger (2010) sample. Thus, the computed value of the test statistic is

$$z = \frac{.37 - .33}{\sqrt{(.349)(.651)\left(\frac{1}{1,500} + \frac{1}{1,750}\right)}} = \frac{.040}{.0168} = 2.38$$

There is sufficient evidence at the $\alpha = .05$ level to conclude that the proportion of adults who smoke has decreased over the 10-year period.

Look Back We could place a confidence interval on $(p_1 - p_2)$ if we were interested in estimating the extent of the decrease.

Now Work Exercise 9.62

Example 9.7

Finding The Observed Significance Level of a Test for $(p_1 - p_2)$

Problem Use a statistical software package to conduct the test presented in Example 9.6. Find and interpret the *p*-value of the test.

Solution We entered the sample sizes (n_1 and n_2) and numbers of successes (x_1 and x_2) into MINITAB and obtained the printout shown in Figure 9.15. The test statistic for this one-tailed test, $z = 2.37$, as well as the *p*-value of the test, are highlighted on the printout. Note that *p*-value = .009 is smaller than $\alpha = .05$. Consequently, we have strong evidence to reject H_0 and conclude that p_1 exceeds p_2.

Test and CI for Two Proportions

```
Sample    X      N   Sample p
1        555   1500   0.370000
2        578   1750   0.330286


Difference = p (1) - p (2)
Estimate for difference:  0.0397143
95% lower bound for difference:  0.0121024
Test for difference = 0 (vs > 0):   Z = 2.37   P-Value = 0.009

Fisher's exact test: P-Value = 0.010
```

Figure 9.15
MINITAB output for test of two proportions

Statistics IN Action | **Revisited** | **Comparing Proportions**

In the first Statistics in Action Revisited in this chapter (p. 421), we demonstrated how the expert statistician used a comparison of two means to defend Visa in a libel case. Recall that ZixIt claims that a Visa executive's e-mail postings had a negative impact on ZixIt's attempt to develop a new online credit card processing system. Here, we demonstrate another way to analyze the data, one successfully presented in court by the statistician.

In addition to daily closing price and trading volume

(continued)

Statistics IN Action
(continued)

of ZixIt stock, the **ZIXITVISA** file also contains a qualitative variable that indicates whether the stock price increased or not (decreased or stayed the same) on the following day. This variable was created by the statistician to compare the proportion of days on which ZixIt stock went up for posting and nonposting days. Let p_1 represent the proportion of days where the ZixIt stock price increased following all nonposting days and p_2 represent the proportion of days where the ZixIt stock price increased following posting days. Then, if the charges made by ZixIt are true (i.e., that postings will have a negative impact on ZixIt stock), p_1 will exceed p_2. Thus, a comparison of two population proportions is appropriate. Recall that during the 83-day period of interest, the executive posted e-mails on 43 days and had no postings on 40 days. Again, both sample sizes ($n_1 = 40$ and $n_2 = 43$) are large, so we can apply the large-sample z-test or large-sample confidence interval procedure for independent samples. (Can you demonstrate this?) A MINITAB printout for this analysis is shown in Figure SIA9.3.

From the printout you can see that following the 40 nonposting days, the price increased on 20 days; following the 43 posting days, the stock price increased on 18 days. Thus, the sample proportions are $p_1 = 20/40 = .5$ and $p_2 = 18/43 = .42$. Are these sample proportions different enough for us to conclude that the population proportions are different and that ZixIt's claim is true? Not according to the statistical analysis. Note that the 95% confidence interval for $(p_1 - p_2)$, $(-.133, .295)$, includes the value 0, and the p-value for the two-tailed test of $H_0: (p_1 - p_2) = 0$, $p - value = .456$, exceeds, say, $\alpha = .05$. Both imply that the two population proportions are not significantly different. Also, neither sample proportion is significantly different from .5. (Can you demonstrate this?) Consequently, in courtroom testimony, the statistical expert used these results to conclude that "the direction of ZixIt's stock price movement following days with postings is random, just like days with no postings."

Data Set: ZIXITVISA

Test and CI for Two Proportions: Up/Down, Posting

```
Event = UP

Posting   X    N   Sample p
NO       20   40   0.500000
POST     18   43   0.418605

Difference = p (NO) - p (POST)
Estimate for difference:  0.0813953
95% CI for difference:  (-0.132500, 0.295291)
Test for difference = 0 (vs not = 0):   Z = 0.75   P-Value = 0.456

Fisher's exact test: P-Value = 0.513
```

Figure SIA9.3
MINITAB comparison of two proportions analysis

Exercises 9.51–9.71

Understanding the Principles

9.51 What conditions are required for valid large-sample inferences about $p_1 - p_2$?

9.52 What is the problem with using the z-statistic to make inferences about $p_1 - p_2$ when the sample sizes are both small?

9.53 Consider making an inference about $p_1 - p_2$, where there are x_1 successes in n_1 binomial trials and x_2 successes in n_2 binomial trials.
   a. Describe the distributions of x_1 and x_2.
   b. For large samples, describe the sampling distribution of $(\hat{p}_1 - \hat{p}_2)$.

Learning the Mechanics

9.54 In each case, determine whether the sample sizes are large enough to conclude that the sampling distribution of $(\hat{p}_1 - \hat{p}_2)$ is approximately normal.
   a. $n_1 = 10, n_2 = 12, \hat{p}_1 = .50, \hat{p}_2 = .50$
   b. $n_1 = 10, n_2 = 12, \hat{p}_1 = .10, \hat{p}_2 = .08$
   c. $n_1 = n_2 = 30, \hat{p}_1 = .20, \hat{p}_2 = .30$
   d. $n_1 = 100, n_2 = 200, \hat{p}_1 = .05, \hat{p}_2 = .09$
   e. $n_1 = 100, n_2 = 200, \hat{p}_1 = .95, \hat{p}_2 = .91$

9.55 Construct a 95% confidence interval for $(p_1 - p_2)$ in each of the following situations:
   a. $n_1 = 400, \hat{p}_1 = .65; n_2 = 400, \hat{p}_2 = .58$
   b. $n_1 = 180, \hat{p}_1 = .31; n_2 = 250, \hat{p}_2 = .25$
   c. $n_1 = 100, \hat{p}_1 = .46; n_2 = 120, \hat{p}_2 = .61$

9.56 Independent random samples, each containing 800 observations, were selected from two binomial populations. The samples from populations 1 and 2 produced 320 and 400 successes, respectively.
   a. Test $H_0: (p_1 - p_2) = 0$ against $H_a: (p_1 - p_2) \neq 0$. Use $\alpha = .05$.
   b. Test $H_0: (p_1 - p_2) = 0$ against $H_a: (p_1 - p_2) \neq 0$. Use $\alpha = .01$.
   c. Test $H_0: (p_1 - p_2) = 0$ against $H_a: (p_1 - p_2) < 0$. Use $\alpha = .01$.
   d. Form a 90% confidence interval for $(p_1 - p_2)$.

9.57 Random samples of size $n_1 = 50$ and $n_2 = 60$ were drawn from populations 1 and 2, respectively. The samples yielded $\hat{p}_1 = .4$ and $\hat{p}_2 = .2$. Test $H_0: (p_1 - p_2) = .1$ against $H_a: (p_1 - p_2) > .1$, using $\alpha = .05$.

9.58 Sketch the sampling distribution of $(\hat{p}_1 - \hat{p}_2)$ based on independent random samples of $n_1 = 100$ and $n_2 = 200$

observations from two binomial populations with probabilities of success $p_1 = .1$ and $p_2 = .5$, respectively.

Applying the Concepts—Basic

9.59 **Bullying behavior study.** School bullying is a form of aggressive behavior that occurs when a student is exposed repeatedly to negative actions (e.g., name-calling, hitting, kicking, spreading slander) from another student. In order to study the effectiveness of an antibullying policy at Dutch elementary schools, a survey of over 2,000 elementary school children was conducted (*Health Education Research*, Feb. 2005). Each student was asked if he or she ever bullied another student. In a sample of 1,358 boys, 746 claimed they had never bullied another student. In a sample of 1,379 girls, 967 claimed they had never bullied another student.

a. Estimate the true proportion of Dutch boys who have never bullied another student.

b. Estimate the true proportion of Dutch girls who have never bullied another student.

c. Estimate the difference in the proportions with a 90% confidence interval.

d. Make a statement about how likely the interval you used in part **c** contains the true difference in proportions.

e. Which group is more likely to bully another student, Dutch boys or Dutch girls?

9.60 **Is steak your favorite barbeque food?** July is National Grilling Month in the United States. On July 1, 2008, *The Harris Poll #70* reported on a survey of Americans' grilling preferences. When asked about their favorite food prepared on a barbeque grill, 662 of 1,250 randomly sampled Democrats preferred steak, as compared to 586 of 930 randomly sampled Republicans.

a. Give a point estimate for the proportion of all Democrats who prefer steak as their favorite barbeque food.

b. Give a point estimate for the proportion of all Republicans who prefer steak as their favorite barbeque food.

c. Give a point estimate for the difference between the proportions of all Democrats and all Republicans who prefer steak as their favorite barbeque food.

d. Construct a 95% confidence interval for the difference between the proportions of all Democrats and all Republicans who prefer steak as their favorite barbeque food.

e. Give a practical interpretation of the interval, part **d**.

f. Explain the meaning of the phrase "95% confident" in your answer to part **e**.

9.61 **Treating depression with St. John's wort.** The *Journal of the American Medical Association* (April 18, 2001) published a study of the effectiveness of using extracts of the herbal medicine St. John's wort in treating major depression. In an eight-week randomized, controlled trial, 200 patients diagnosed with major depression were divided into two groups, one of which ($n_1 = 98$) received St. John's wort extract while the other ($n_2 = 102$) received a placebo (no drug). At the end of the study period, 14 of the St. John's wort patients were in remission, compared with 5 of the placebo patients.

a. Compute the proportion of the St. John's wort patients who were in remission.

b. Compute the proportion of the placebo patients who were in remission.

c. If St. John's wort is effective in treating major depression, then the proportion of St. John's wort patients in remission will exceed the proportion of placebo patients in remission. At $\alpha = .01$, is St. John's wort effective in treating major depression?

d. Repeat part **c**, but use $\alpha = .10$.

e. Explain why the choice of α is critical for this study.

9.62 **Planning-habits survey.** *American Demographics* (Jan. 2002) reported the results of a survey on the planning habits of men and women. In response to the question "What is your preferred method of planning and keeping track of meetings, appointments, and deadlines?" 56% of the men and 46% of the women answered "I keep them in my head." A nationally representative sample of 1,000 adults participated in the survey; therefore, assume that 500 were men and 500 were women.

a. Set up the null and alternative hypotheses for testing whether the percentage of men who prefer keeping track of appointments in their head is larger than the corresponding percentage of women.

b. Compute the test statistic for the test.

c. Give the rejection region for the test, using $\alpha = .01$.

d. Find the p-value for the test.

e. Draw the appropriate conclusion.

9.63 **Racial profiling by the LAPD.** *Racial profiling* is a term used to describe any police action that relies on ethnicity rather than behavior to target suspects engaged in criminal activities. Does the Los Angeles Police Department (LAPD) invoke racial profiling in stops and searches of Los Angeles drivers? This question was addressed in *Chance* (Spring 2006).

a. Data on stops and searches of both African-Americans and white drivers from January through June 2005 are summarized in the accompanying table. Conduct a test (at $\alpha = .05$) to determine whether there is a disparity in the proportions of African-American and white drivers who are searched by the LAPD after being stopped.

Race	Number Stopped	Number Searched	Number of "Hits"
African-American	61,688	12,016	5,134
White	106,892	5,312	3,006

Based on Khadjavi, L. S. "Driving while black in the City of Angels." *Chance*, Vol. 19, No. 2, Spring 2006 (Tables 1 and 2), pp. 43–46.

b. The LAPD defines a "hit rate" as the proportion of searches that result in a discovery of criminal activity. Use the data in the table to estimate the disparity in the hit rates for African-American and white drivers under a 95% confidence interval. Interpret the results.

Applying the Concepts—Intermediate

9.64 **Angioplasty's benefits challenged.** Each year, more than 1 million heart patients undergo an angioplasty. The benefits of an angioplasty were challenged in a recent study of 2,287 patients (2007 Annual Conference of the American College of Cardiology, New Orleans). All the patients had substantial blockage of the arteries, but were medically stable. All were treated with medication such as aspirin and beta-blockers. However, half the patients were randomly assigned to get an angioplasty and half were not. After five years, the researchers found that 211 of the 1,145 patients in the angioplasty group had subsequent heart attacks, compared with 202 of 1,142 patients in the medication-only group. Do you agree with the study's conclusion that "There was no significant difference in the rate of heart attacks for the two groups"? Support your answer with a 95% confidence interval.

9.65 **Killing insects with low oxygen.** Refer to the *Journal of Agricultural, Biological, and Environmental Statistics* (Sept. 2000) study of the mortality of rice weevils exposed to low oxygen, presented in Exercise 8.94 (p. 387). Recall that 31,386 of 31,421 rice weevils were found dead after exposure to nitrogen gas for 4 days. In a second experiment, 23,516 of 23,676 rice weevils were found dead after exposure to nitrogen gas for 3.5 days. Conduct a test of hypothesis to compare the mortality rates of adult rice weevils exposed to nitrogen at the two exposure times. Is there a significant difference (at $\alpha = .10$) in the mortality rates?

9.66 **Effectiveness of drug tests of Olympic athletes.** Erythropoietin (EPO) is a banned drug used by athletes to increase the oxygen-carrying capacity of their blood. New tests for EPO were first introduced prior to the 2000 Olympic Games held in Sydney, Australia. *Chance* (Spring 2004) reported that of a sample of 830 world-class athletes, 159 did not compete in the 1999 World Championships (a year prior to the introduction of the new EPO test). Similarly, 133 of 825 potential athletes did not compete in the 2000 Olympic games. Was the new test effective in deterring an athlete from participating in the 2000 Olympics? If so, then the proportion of nonparticipating athletes in 2000 will be more than the proportion of nonparticipating athletes in 1999. Conduct the analysis (at $\alpha = .10$) and draw the proper conclusion.

9.67 **"Tip-of-the-tongue" study.** Trying to think of a word you know, but can't instantly retrieve, is called the "tip-of-the-tongue" phenomenon. *Psychology and Aging* (Sept. 2001) published a study of this phenomenon in senior citizens. The researchers compared 40 people between 60 and 72 years of age with 40 between 73 and 83 years of age. When primed with the initial syllable of a missing word (e.g., seeing the word *include* to help recall the word *incisor*), the younger seniors had a higher recall rate. Suppose 31 of the 40 seniors in the younger group could recall the word when primed with the initial syllable, while only 22 of the 40 seniors could recall the word. Compare the recall rates of the two groups, using $\alpha = .05$. Does one group of elderly people have a significantly higher recall rate than the other?

9.68 **Detection of rigged school milk prices (cont'd).** Refer to the investigation of collusive bidding in the northern Kentucky school milk market, presented in Exercise 9.28 (p. 428). Market allocation is a common form of collusive behavior in bid-rigging conspiracies. Under collusion, the same dairy usually controls the same school districts year after year. The *incumbency rate* for a market is defined as the proportion of school districts that are won by the vendor that won the previous year. Past experience with milk bids in a competitive environment reveals that a typical incumbency rate is .7. That is, 70% of the school districts are expected to purchase their milk from the dairy that won the previous year. Incumbency rates of .9 or higher are strong indicators of collusive bidding. Over the years, when bid collusion was alleged to have occurred in northern Kentucky, there were 51 potential vendor transitions

(i.e., changes in milk supplier from one year to the next in a district) in the tricounty market and 134 potential vendor transitions in the surrounding market. These values represent the sample sizes ($n_1 = 134$ and $n_2 = 51$) for calculating incumbency rates. Examining the data saved in the **MILK** file, you'll find that in 50 of the 51 potential vendor transitions for the tricounty market, the winning dairy from the previous year won the bid the next year; similarly, you'll find that in 91 of the 134 potential vendor transitions for the surrounding area, the same dairy won the bid the next year.

a. Estimate the incumbency rates for the tricounty and surrounding milk markets.

b. A MINITAB printout comparing the two incumbency rates is shown below. Give a practical interpretation of the results. Do they show further support for the bid collusion theory?

Test and CI for Two Proportions

```
Sample    X     N    Sample p
1        91   134   0.679104
2        50    51   0.980392


Difference = p (1) - p (2)
Estimate for difference:  -0.301288
95% upper bound for difference:  -0.227669
Test for difference = 0 (vs < 0):  Z = -4.30  P-Value = 0.000
```

9.69 **Does sleep improve mental performance?** Are creativity and problem solving linked to adequate sleep? This question was the subject of research conducted by German scientists at the University of Lübeck (*Nature*, Jan. 22, 2004). One hundred volunteers were divided into two equal-sized groups. Each volunteer took a math test that involved transforming strings of eight digits into a new string that fit a set of given rules, as well as a third, hidden rule. Prior to taking the test, one group received eight hours of sleep, while the other group stayed awake all night. The scientists monitored the volunteers to determine whether and when they figured out the third rule. Of the volunteers who slept, 39 discovered the third rule; of the volunteers who stayed awake all night, 15 discovered the third rule. From the study results, what can you infer about the proportions of volunteers in the two groups who discover the third rule? Support your answer with a 90% confidence interval.

Applying the Concepts—Advanced

9.70 **Religious symbolism in TV commercials.** Gonzaga University professors conducted a study of television commercials and published their results in the *Journal of Sociology, Social Work and Social Welfare* (Vol. 2, 2008). The key research question was: "Do television advertisers use religious symbolism to sell goods and services?" In a sample of 797 TV commercials collected in 1998, only 16 commercials used religious symbolism. Of the sample of 1,499 TV commercials examined in the more recent study, 51 commercials used religious symbolism. Conduct an analysis to determine if the percentage of TV commercials that use religious symbolism has changed since the 1998 study. If you detect a change, estimate the magnitude of the difference and attach a measure of reliability to the estimate.

9.71 Teeth defects and stress in prehistoric Japan. Linear enamel hypoplasia (LEH) defects are pits or grooves on the tooth surface that are typically caused by malnutrition, chronic infection, stress, and trauma. A study of LEH defects in prehistoric Japanese cultures was published in the *American Journal of Physical Anthropology* (May 2010). Three groups of Japanese people were studied: Yayoi farmers (early agriculturists), eastern Jomon foragers (broad-based economy), and western Jomon foragers (wet rice economy). LEH defect prevalence was determined from skulls of individuals obtained from each of the three cultures. Of the 182 Yayoi farmers in the study, 63.1% had at least one LEH defect; of the 164 Eastern Jomon foragers, 48.2% had at least one LEH defect; and, of the 122 Western Jomon foragers, 64.8% had at least one LEH defect. Two theories were tested. Theory 1 states that foragers with a broad-based economy will have a lower LEH defect prevalence than early agriculturists. Theory 2 states that foragers with a wet rice economy will not differ in LEH defect prevalence from early agriculturists. Use the results to test both theories, each at $\alpha = .01$.

Based on Temple, D. H. "Patterns of systemic stress during the agricultural transition in prehistoric Japan." *American Journal of Physical Anthropology,* Vol. 142, No. 1, May 2010.

9.5 Determining the Sample Size

You can find the appropriate sample size to estimate the difference between a pair of parameters with a specified sampling error (SE) and degree of reliability by using the method described in Section 7.5. That is, to estimate the difference between a pair of parameters correct to within SE units with confidence level $(1 - \alpha)$, let $z_{\alpha/2}$ standard deviations of the sampling distribution of the estimator equal SE. Then solve for the sample size. To do this, you have to solve the problem for a specific ratio between n_1 and n_2. Most often, you will want to have equal sample sizes—that is, $n_1 = n_2 = n$. We will illustrate the procedure with two examples.

Example 9.8

Finding the Sample Sizes for Estimating $(\mu_1 - \mu_2)$—Comparing Mean Crop Yields

Problem New fertilizer compounds are often advertised with the promise of increased crop yields. Suppose we want to compare the mean yield μ_1 of wheat when a new fertilizer is used with the mean yield μ_2 from a fertilizer in common use. The estimate of the difference in mean yield per acre is to be correct to within .25 bushel with a confidence coefficient of .95. If the sample sizes are to be equal, find $n_1 = n_2 = n$, the number of 1-acre plots of wheat assigned to each fertilizer.

Solution To solve the problem, you need to know something about the variation in the bushels of yield per acre. Suppose that, from past records, you know that the yields of wheat possess a range of approximately 10 bushels per acre. You could then approximate $\sigma_1 = \sigma_2 = \sigma$ by letting the range equal 4σ. Thus,

$$4\sigma \approx 10 \text{ bushels}$$
$$\sigma \approx 2.5 \text{ bushels}$$

The next step is to solve the equation

$$z_{\alpha/2}\sigma_{(\bar{x}_1 - \bar{x}_2)} = \text{SE, or } z_{\alpha/2}\sqrt{\frac{\sigma_1^2}{n_1} + \frac{\sigma_2^2}{n_2}} = \text{SE}$$

for n, where $n = n_1 = n_2$. Since we want our estimate to lie within SE = .25 of $(\mu_1 - \mu_2)$ with confidence coefficient equal to .95, we have $z_{\alpha/2} = z_{.025} = 1.96$. Then, letting $\sigma_1 = \sigma_2 = 2.5$ and solving for n, we get

$$1.96\sqrt{\frac{(2.5)^2}{n} + \frac{(2.5)^2}{n}} = .25$$

$$1.96\sqrt{\frac{2(2.5)^2}{n}} = .25$$

$$n = 768.32 \approx 769 \text{ (rounding up)}$$

Consequently, you will have to sample 769 acres of wheat for each fertilizer to estimate the difference in mean yield per acre to within .25 bushel.

Look Back Since $n = 769$ would necessitate extensive and costly experimentation, you might decide to allow a larger sampling error (say, SE = .50 or SE = 1) in order to reduce the sample size, or you might decrease the confidence coefficient. The point is that we can obtain an idea of the experimental effort necessary to achieve a specified precision in our final estimate by determining the approximate sample size *before* the experiment is begun.

Now Work Exercise 9.76

Example 9.9

Finding the Sample Sizes for Estimating $(p_1 - p_2)$— Comparing Defect Rates of Two Machines

Problem A production supervisor suspects that a difference exists between the proportions p_1 and p_2 of defective items produced by two different machines. Experience has shown that the proportion defective for each of the two machines is in the neighborhood of .03. If the supervisor wants to estimate the difference in the proportions to within .005, using a 95% confidence interval, how many items must be randomly sampled from the output produced by each machine? (Assume that the supervisor wants $n_1 = n_2 = n$.)

Solution In this sampling problem, the sampling error SE = .005, and for the specified level of reliability, $z_{\alpha/2} = z_{.025} = 1.96$. Then, letting $p_1 = p_2 = .03$ and $n_1 = n_2 = n$, we find the required sample size per machine by solving the following equation for n:

$$z_{\alpha/2} \, \sigma_{(\hat{p}_1 - \hat{p}_2)} = \text{SE}$$

or

$$z_{\alpha/2} \sqrt{\frac{p_1 q_1}{n_1} + \frac{p_2 q_2}{n_2}} = \text{SE}$$

$$1.96 \sqrt{\frac{(.03)(.97)}{n} + \frac{(.03)(.97)}{n}} = .005$$

$$1.96 \sqrt{\frac{2(.03)(.97)}{n}} = .005$$

$$n = 8{,}943.2$$

Look Back This large n will likely result in a tedious sampling procedure. If the supervisor insists on estimating $(p_1 - p_2)$ correct to within .005 with 95% confidence, approximately 9,000 items will have to be inspected for each machine.

Now Work Exercise 9.77a

You can see from the calculations in Example 9.9 that $\sigma_{(\hat{p}_1 - \hat{p}_2)}$ (and hence the solution, $n_1 = n_2 = n$) depends on the actual (but unknown) values of p_1 and p_2. In fact, the required sample size $n_1 = n_2 = n$ is largest when $p_1 = p_2 = .5$. Therefore, if you have no prior information on the approximate values of p_1 and p_2, use $p_1 = p_2 = .5$ in the formula for $\sigma_{(\hat{p}_1 - \hat{p}_2)}$. If p_1 and p_2 are in fact close to .5, then the values of n_1 and n_2 that you have calculated will be correct. If p_1 and p_2 differ substantially from .5, then your solutions for n_1 and n_2 will be larger than needed. Consequently, using $p_1 = p_2 = .5$ when solving for n_1 and n_2 is a conservative procedure because the sample sizes n_1 and n_2 will be at least as large as (and probably larger than) needed.

The procedures for determining sample sizes necessary for estimating $(\mu_1 - \mu_2)$ or $(p_1 - p_2)$ for the case $n_1 = n_2$ are given in the following boxes:

Determination of Sample Size for Estimating $(\mu_1 - \mu_2)$

To estimate $(\mu_1 - \mu_2)$ to within a given sampling error SE and with confidence level $(1 - \alpha)$, use the following formula to solve for equal sample sizes that will achieve the desired reliability:

$$n_1 = n_2 = \frac{(z_{\alpha/2})^2(\sigma_1^2 + \sigma_2^2)}{(SE)^2}$$

You will need to substitute estimates for the values of σ_1^2 and σ_2^2 before solving for the sample size. These estimates might be sample variances s_1^2 and s_2^2 from prior sampling (e.g., a pilot study) or from an educated (and conservatively large) guess based on the range—that is, $s \approx R/4$.

Determination of Sample Size for Estimating $p_1 - p_2$

To estimate $(p_1 - p_2)$ to within a given sampling error SE and with confidence level $(1 - \alpha)$, use the following formula to solve for equal sample sizes that will achieve the desired reliability:

$$n_1 = n_2 = \frac{(z_{\alpha/2})^2(p_1 q_1 + p_2 q_2)}{(SE)^2}$$

You will need to substitute estimates for the values of p_1 and p_2 before solving for the sample size. These estimates might be based on prior samples, obtained from educated guesses or, most conservatively, specified as $p_1 = p_2 = .5$.

Exercises 9.72–9.85

Understanding the Principles

9.72 In determining the sample sizes for estimating $\mu_1 - \mu_2$, how do you obtain estimates of the population variances $(\sigma_1)^2$ and $(\sigma_2)^2$ used in the calculations?

9.73 In determining the sample sizes for estimating $p_1 - p_2$, how do you obtain estimates of the binomial proportions p_1 and p_2 used in the calculations?

9.74 If the sample-size calculation yields a value of n that is too large to be practical, how should you proceed?

Learning the Mechanics

9.75 Suppose you want to estimate the difference between two population means correct to within 2.2 with probability .95. If prior information suggests that the population variances are approximately equal to $\sigma_1^2 = \sigma_2^2 = 15$ and you want to select independent random samples of equal size from the populations, how large should the sample sizes, n_1 and n_2, be?

9.76 Find the appropriate values of n_1 and n_2 (assume that $n_1 = n_2$) needed to estimate $(\mu_1 - \mu_2)$ with
  a. A sampling error equal to 3.2 with 95% confidence. From prior experience, it is known that $\sigma_1 \approx 15$ and $\sigma_2 \approx 17$.
  b. A sampling error equal to 8 with 99% confidence. The range of each population is 60.

c. A 90% confidence interval of width 1.0. Assume that $\sigma_1^2 \approx 5.8$ and $\sigma_2^2 \approx 7.5$.

9.77 Assuming that $n_1 = n_2$, find the sample sizes needed to estimate $(p_1 - p_2)$ for each of the following situations:
  a. SE = .01 with 99% confidence. Assume that $p_1 \approx .4$ and $p_2 \approx .7$.
  b. A 90% confidence interval of width .05. Assume there is no prior information available with which to obtain approximate values of p_1 and p_2.
  c. SE = .03 with 90% confidence. Assume that $p_1 \approx .2$ and $p_2 \approx .3$.

9.78 Enough money has been budgeted to collect independent random samples of size $n_1 = n_2 = 100$ from populations 1 and 2 in order to estimate $(p_1 - p_2)$. Prior information indicates that $p_1 = p_2 \approx .6$. Have sufficient funds been allocated to construct a 90% confidence interval for $(p_1 - p_2)$ of width .1 or less? Justify your answer.

Applying the Concepts—Basic

9.79 **Bulimia study.** Refer to the *American Statistician* (May 2001) study comparing the "fear of negative evaluation" (FNE) scores for bulimic and normal female students, presented in Exercise 9.19 (p. 425). Suppose you want to estimate $(\mu_B - \mu_N)$, the difference between the population means of the FNE scores for bulimic and normal female students, using a 95% confidence interval with a sampling error of two points. Find the sample sizes required to obtain such an estimate. Assume equal sample sizes of $\sigma_B^2 = \sigma_N^2 = 25$.

9.80 **Laughter among deaf signers.** Refer to the *Journal of Deaf Studies and Deaf Education* (Fall 2006) paired difference study on vocalized laughter among deaf users of sign language, presented in Exercise 9.40 (p. 437). Suppose you want to estimate $\mu_d = (\mu_S - \mu_A)$, the difference between the population mean number of laugh episodes of deaf speakers and deaf audience members, using a 90% confidence interval with a sampling error of .75. Find the number of pairs of deaf people required to obtain such an estimate, assuming that the variance of the paired differences is $\sigma_d^2 = 3$.

9.81 **Angioplasty's benefits challenged.** Refer to the study of patients with substantial blockage of the arteries presented at the 2007 Annual Conference of the American College of Cardiology, Exercise 9.64 (p. 445). Recall that half the patients were randomly assigned to get an angioplasty and half were not. The researchers compared the proportion of patients with subsequent heart attacks for the two groups and reported no significant difference between the two proportions. Although the study involved over 2,000 patients, the sample size may have been too small to detect a difference in heart attack rates.

a. How many patients must be sampled in each group in order to estimate the difference in heart attack rates to within .015 with 95% confidence? (Use summary data from Exercise 9.64 in your calculation.)

b. Comment on the practicality of carrying out the study with the sample sizes determined in part **a.**

c. Comment on the practical significance of the difference detected in the confidence interval for the study, part **a.**

Applying the Concepts—Intermediate

9.82 **Cable-TV home shoppers.** All cable television companies carry at least one home-shopping channel. Who uses these home-shopping services? Are the shoppers primarily men or women? Suppose you want to estimate the difference in the percentages of men and women who say they have used or expect to use televised home shopping. You want an 80% confidence interval of width .06 or less.

a. Approximately how many people should be included in your samples?

b. Suppose you want to obtain individual estimates for the two percentages of interest. Will the sample size found in part **a** be large enough to provide estimates of each percentage correct to within .02 with probability equal to .90? Justify your response.

9.83 **Do video game players have superior visual attention skills?** Refer to the *Journal of Articles in Support of the Null Hypothesis* (Vol. 6, 2009) study comparing the visual attention skill of video game and non-video game players, Exercise 9.20 (p. 425). Recall that there was no significant difference between the mean score on the *attentional blink* test of video game players and the corresponding mean for non–video game players. It is possible that selecting larger samples would yield a significant difference. How many video game and non–video game players would need to be selected in order to estimate the difference in mean score for the two groups to within 5 points with 95% confidence? (Assume equal sample sizes will be selected from the two groups and that the score standard deviation for both groups is $\sigma \approx 9$.)

9.84 **Rat damage in sugarcane.** Poisons are used to prevent rat damage in sugarcane fields. The U.S. Department of Agriculture is investigating whether the rat poison should be located in the middle of the field or on the outer perimeter. One way to answer this question is to determine where the greater amount of damage occurs. If damage is measured by the proportion of cane stalks that have been damaged by rats, how many stalks from each section of the field should be sampled in order to estimate the true difference between proportions of stalks damaged in the two sections, to within .02 with 95% confidence?

9.85 **Scouting an NFL free agent.** In seeking a free-agent NFL running back, a general manager is looking for a player with high mean yards gained per carry and a small standard deviation. Suppose the GM wishes to compare the mean yards gained per carry for two free agents, on the basis of independent random samples of their yards gained per carry. Data from last year's pro football season indicate that $\sigma_1 = \sigma_2 \approx 5$ yards. If the GM wants to estimate the difference in means correct to within 1 yard with a confidence level of .90, how many runs would have to be observed for each player? (Assume equal sample sizes.)

9.6 Comparing Two Population Variances: Independent Sampling (Optional)

Many times, it is of practical interest to use the techniques developed in this chapter to compare the means or proportions of two populations. However, there are also important instances when we wish to compare two population variances. For example, when two devices are available for producing precision measurements (scales, calipers, thermometers, etc.), we might want to compare the variability of the measurements of the devices before deciding which one to purchase. Or when two standardized tests can be used to rate job applicants, the variability of the scores for both tests should be taken into consideration before deciding which test to use.

For problems like these, we need to develop a statistical procedure to compare population variances. The common statistical procedure for comparing population variances σ_1^2 and σ_2^2 makes an inference about the ratio σ_1^2/σ_2^2. In this section, we will show how to test the null hypothesis that the ratio σ_1^2/σ_2^2 equals 1 (the variances are equal) against the alternative hypothesis that the ratio differs from 1 (the variances differ):

$$H_0: \frac{\sigma_1^2}{\sigma_2^2} = 1 \quad (\sigma_1^2 = \sigma_2^2)$$

$$H_a: \frac{\sigma_1^2}{\sigma_2^2} \neq 1 \quad (\sigma_1^2 \neq \sigma_2^2)$$

To make an inference about the ratio σ_1^2/σ_2^2, it seems reasonable to collect sample data and use the ratio of the sample variances, s_1^2/s_2^2. We will use the test statistic

$$F = \frac{s_1^2}{s_2^2}$$

To establish a rejection region for the test statistic, we need to know the sampling distribution of s_1^2/s_2^2. As you will subsequently see, the sampling distribution of s_1^2/s_2^2 is based on two of the assumptions already required for the t-test:

1. The two sampled populations are normally distributed.

2. The samples are randomly and independently selected from their respective populations.

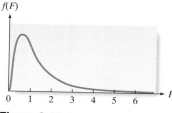

$f(F)$

0 1 2 3 4 5 6 F

Figure 9.16

An F-distribution with 7 numerator and 9 denominator degrees of freedom

When these assumptions are satisfied and when the null hypothesis is true (i.e., when $\sigma_1^2 = \sigma_2^2$), the sampling distribution of $F = s_1^2/s_2^2$ is the **F-distribution** with $(n_1 - 1)$ numerator degrees of freedom and $(n_2 - 1)$ denominator degrees of freedom, respectively. The shape of the F-distribution depends on the number of degrees of freedom associated with s_1^2 and s_2^2—that is, on $(n_1 - 1)$ and $(n_2 - 1)$. An F-distribution with 7 and 9 df is shown in Figure 9.16. As you can see, the distribution is skewed to the right, since s_1^2/s_2^2 cannot be less than 0, but can increase without bound.

BIOGRAPHY **GEORGE W. SNEDECOR (1882–1974)**

Snedecor's F-test

George W. Snedecor's education began at the University of Alabama, where he obtained his bachelor's degree in mathematics and physics. He went on to the University of Michigan for his master's degree in physics and finally earned his Ph.D in mathematics at the University of Kentucky. Snedecor learned of an opening for an assistant professor of mathematics at the University of Iowa, packed his belongings in his car, and began driving to apply for the position. By mistake, he ended up in Ames, Iowa, home of Iowa State University—then an agricultural school that had no need for a mathematics teacher. Nevertheless, Snedecor stayed and founded a statistics laboratory, eventually teaching the first course in statistics at Iowa State in 1915. In 1933, Snedecor turned the statistics laboratory into the first-ever Department of Statistics in the United States. During his tenure as chair of the department, Snedecor published his landmark textbook, *Statistical Methods* (1937). The text contained the first published reference for a test of hypothesis to compare two variances. Although Snedecor named it the *F*-test in honor of statistician R. A. Fisher (who had developed the *F*-distribution a few years earlier) many researchers still refer to it as Snedecor's *F*-test. Now in its ninth edition, *Statistical Methods* (with William Cochran as coauthor) continues to be one of the most frequently cited texts in the statistics field. ■

We need to be able to find F-values corresponding to the tail areas of this distribution in order to establish the rejection region for our test of hypothesis because we expect the ratio F of the sample variances to be either very large or very small when the population variances are unequal. The upper-tail F values for $\alpha = .10, .05, .025$, and $.01$ can be found in Tables VIII, IX, X, and XI of Appendix A. Table IX is partially reproduced in Table 9.8. It gives F values that correspond to $\alpha = .05$ upper-tail areas for different degrees of freedom; the columns correspond to the number of degrees of freedom, v_1, for the numerator sample variance s_1^2, whereas the rows correspond to the number of degrees of freedom, v_2, for the denominator sample variance s_2^2. Thus, if the number of degrees of freedom denoted by the numerator is $v_1 = 7$ and the number of degrees of freedom denoted by the denominator is $v_2 = 9$, we look in the seventh column and ninth row to find $F_{.05} = 3.29$ (highlighted in Table 9.8). As shown in Figure 9.17, $\alpha = .05$

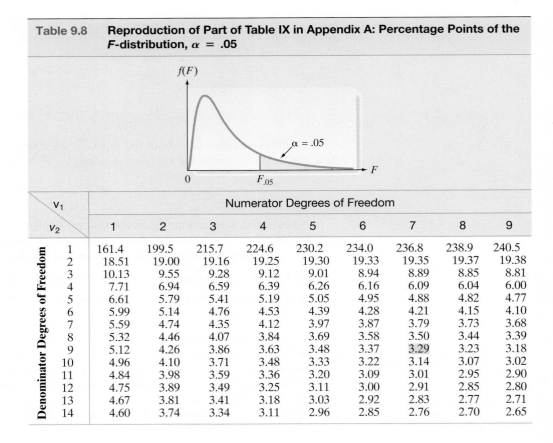

Table 9.8   Reproduction of Part of Table IX in Appendix A: Percentage Points of the F-distribution, $\alpha = .05$

v_1	Numerator Degrees of Freedom								
v_2	1	2	3	4	5	6	7	8	9
1	161.4	199.5	215.7	224.6	230.2	234.0	236.8	238.9	240.5
2	18.51	19.00	19.16	19.25	19.30	19.33	19.35	19.37	19.38
3	10.13	9.55	9.28	9.12	9.01	8.94	8.89	8.85	8.81
4	7.71	6.94	6.59	6.39	6.26	6.16	6.09	6.04	6.00
5	6.61	5.79	5.41	5.19	5.05	4.95	4.88	4.82	4.77
6	5.99	5.14	4.76	4.53	4.39	4.28	4.21	4.15	4.10
7	5.59	4.74	4.35	4.12	3.97	3.87	3.79	3.73	3.68
8	5.32	4.46	4.07	3.84	3.69	3.58	3.50	3.44	3.39
9	5.12	4.26	3.86	3.63	3.48	3.37	3.29	3.23	3.18
10	4.96	4.10	3.71	3.48	3.33	3.22	3.14	3.07	3.02
11	4.84	3.98	3.59	3.36	3.20	3.09	3.01	2.95	2.90
12	4.75	3.89	3.49	3.25	3.11	3.00	2.91	2.85	2.80
13	4.67	3.81	3.41	3.18	3.03	2.92	2.83	2.77	2.71
14	4.60	3.74	3.34	3.11	2.96	2.85	2.76	2.70	2.65

Denominator Degrees of Freedom

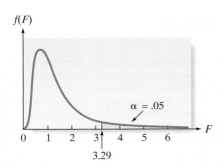

Figure 9.17
An F-distribution for $v_1 = 7$ and $v_2 = 9$ df; $\alpha = .05$

is the tail area to the right of 3.29 in the F-distribution with 7 and 9 df. That is, if $\sigma_1^2 = \sigma_2^2$, then the probability that the F-statistic will exceed 3.29 is $\alpha = .05$.

Now Work Exercise 9.90

Example 9.10

An F-Test Application—Comparing Weight Variations in Mice

Problem An experimenter wants to compare the metabolic rates of white mice subjected to different drugs. The weights of the mice may affect their metabolic rates; thus, the experimenter wishes to obtain mice that are relatively homogeneous with respect to weight. Five hundred mice will be needed to complete the study. Currently, 13 mice from supplier 1 and another 18 mice from supplier 2 are available for comparison. The experimenter weighs these mice  and obtains the data shown in Table 9.9. Do these data provide sufficient evidence to indicate a difference in the variability of weights of mice obtained from the two suppliers? (Use $\alpha = .10$.) From the results of this analysis, what would you suggest to the experimenter?

Table 9.9	Weights (in ounces) of Experimental Mice				
Supplier 1					
4.23	4.35	4.05	3.75	4.41	4.37
4.01	4.06	4.15	4.19	4.52	4.21
4.29					
Supplier 2					
4.14	4.26	4.05	4.11	4.31	4.12
4.17	4.35	4.25	4.21	4.05	4.28
4.15	4.20	4.32	4.25	4.02	4.14

Data Set: MICEWTS

Solution Let

$$\sigma_1^2 = \text{Population variance of weights of white mice from supplier 1}$$
$$\sigma_2^2 = \text{Population variance of weights of white mice from supplier 2}$$

The hypotheses of interest are then

$$H_0: \frac{\sigma_1^2}{\sigma_2^2} = 1 \quad (\sigma_1^2 = \sigma_2^2)$$

$$H_a: \frac{\sigma_1^2}{\sigma_2^2} \neq 1 \quad (\sigma_1^2 \neq \sigma_2^2)$$

The nature of the F-tables given in Appendix A affects the form of the test statistic. To form the rejection region for a two-tailed F-test, we want to make certain that the upper tail is used, because only the upper-tail values of F are shown in Tables VIII, IX, X, and XI. To accomplish that, *we will always place the larger sample variance in the numerator of the F-test statistic.* This has the effect of doubling the tabulated value for α, since we double the probability that the F-ratio will fall in the upper tail by always placing the larger sample variance in the numerator. That is, we establish a one-tailed rejection region by putting the larger variance in the numerator, rather than establishing rejection regions in both tails.

To calculate the value of the test statistic, we require the sample variances. These are shown on the MINITAB printout of Figure 9.18. The sample variances (highlighted) are $s_1^2 = .0409$ and $s_2^2 = .00964$. Therefore,

$$F = \frac{\text{Larger sample variance}}{\text{Smaller sample variance}} = \frac{s_1^2}{s_2^2} = \frac{(.0409)}{(.00964)} = 4.24$$

Descriptive Statistics: WEIGHT

```
Variable  SUPPLIER   N    Mean    StDev   Variance   Minimum   Maximum
WEIGHT       1       13  4.1992   0.2021   0.0409     3.7500    4.5200
             2       18  4.1878   0.0982   0.00964    4.0200    4.3500
```

Test for Equal Variances: WEIGHT versus SUPPLIER

```
95% Bonferroni confidence intervals for standard deviations

SUPPLIER   N     Lower      StDev      Upper
       1  13  0.138579   0.202133   0.361467
       2  18  0.070860   0.098193   0.156836


F-Test (normal distribution)
Test statistic = 4.24, p-value = 0.007


Levene's Test (any continuous distribution)
Test statistic = 4.31, p-value = 0.047
```

Figure 9.18

MINITAB summary statistics and F-test for mice weights

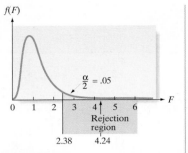

Figure 9.19
Rejection region for
Example 9.10

For our example, we have a numerator s_1^2 with df $= v_1 = n_1 - 1 = 12$ and a denominator s_2^2 with df $= v_2 = n_2 - 1 = 17$. Therefore, we will reject $H_0: \sigma_1^2 = \sigma_2^2$ for $\alpha = .10$ when the calculated value of F exceeds the tabulated value:

$$F_{\alpha/2} = F_{.05} = 2.38 \qquad \text{(see Figure 9.19)}$$

Note that $F = 4.24$ falls into the rejection region. Therefore, the data provide sufficient evidence to indicate that the population variances differ. It appears that the weights of mice obtained from supplier 2 tend to be more homogeneous (less variable) than the weights of mice obtained from supplier 1. On the basis of this evidence, we would advise the experimenter to purchase the mice from supplier 2.

Look Back What would you have concluded if the value of F calculated from the samples had not fallen into the rejection region? Would you conclude that the null hypothesis of equal variances is true? No, because then you risk the possibility of a Type II error (accepting H_0 if H_a is true) without knowing the value of β, the probability of accepting $H_0: \sigma_1^2 = \sigma_2^2$ if in fact it is false. Since we will not consider the calculation of β for specific alternatives in this text, when the F-statistic does not fall into the rejection region we simply conclude that the sample possesses insufficient evidence to refute the null hypothesis that $\sigma_1^2 = \sigma_2^2$.

Now Work Exercise 9.95a

Example 9.11

**The Observed
Significance Level
of an F-Test**

Problem Find the p-value for the test in Example 9.10, using the F-tables in Appendix A. Compare this with the exact p-value obtained from a computer printout.

Solution Since the observed value of the F-statistic in Example 9.10 was found to be 4.24, the observed significance level of the test would equal the probability of observing a value of F at least as contradictory to $H_0: \sigma_1^2 = \sigma_2^2$ as $F = 4.24$, is, if in fact H_0 is true. Since we give the F-tables in Appendix A just for values of α equal to .10, .05, .025, and .01, we can only approximate the observed significance level. Checking Table XI, we find that $F_{.01} = 3.46$. Since the observed value of F is greater than $F_{.01}$, the observed significance level for the test is less than $2(.01) = .02$. (Note that we double the α value shown in Table XI because this is a two-tailed test.) The exact p-value, $p = .007$, is highlighted at the bottom of the MINITAB printout shown in Figure 9.18.

Look Back We double the α value in Table XI because this is a two-tailed test.

Now Work Exercise 9.95b

In addition to applying a hypothesis test for σ_1^2/σ_2^2, we can use the F-statistic to estimate the ratio with a confidence interval. The following boxes summarize the confidence interval and testing procedures:

A $(1 - \alpha) \times 100\%$ Confidence Interval for $(\sigma_1)^2/(\sigma_2)^2$

$$\left(\frac{s_1^2}{s_2^2}\right)\left(\frac{1}{F_{L,\alpha/2}}\right) < \left(\frac{\sigma_1^2}{\sigma_2^2}\right) < \left(\frac{s_1^2}{s_2^2}\right)F_{U,\alpha/2}$$

where $F_{L,\alpha/2}$ is the value of F that places an area $\alpha/2$ in the upper tail of an F-distribution with $v_1 = (n_1 - 1)$ numerator and $v_2 = (n_2 - 1)$ denominator degrees of freedom, and $F_{U,\alpha/2}$ is the value of F that places an area $\alpha/2$ in the upper tail of an F-distribution with $v_1 = (n_2 - 1)$ numerator and $v_2 = (n_1 - 1)$ denominator degrees of freedom.

F-Test for Equal Population Variances*

One-Tailed Test

$H_0: \sigma_1^2 = \sigma_2^2$

$H_a: \sigma_1^2 < \sigma_2^2 \text{ (or } H_a: \sigma_1^2 > \sigma_2^2)$

Test statistic:

$$F = \frac{s_2^2}{s_1^2}$$

$$\left(\text{or } F = \frac{s_1^2}{s_2^2} \text{ when } H_a: \sigma_1^2 > \sigma_2^2 \right)$$

Rejection region:

$F > F_\alpha$

Two-Tailed Test

$H_0: \sigma_1^2 = \sigma_2^2$

$H_a: \sigma_1^2 \neq \sigma_2^2$

Test statistic:

$$F = \frac{\text{Larger sample variance}}{\text{Smaller sample variance}}$$

$$= \frac{s_1^2}{s_2^2} \text{ when } s_1^2 > s_2^2$$

$$\left(\text{or } \frac{s_2^2}{s_1^2} \text{ when } s_2^2 > s_1^2 \right)$$

Rejection region:

$F > F_{\alpha/2}$

where F_α and $F_{\alpha/2}$ are based on ν_1 numerator degrees of freedom and ν_2 denominator degrees of freedom; and ν_1 and ν_2 are the degrees of freedom for the numerator and denominator sample variances, respectively.

Conditions Required for Valid Inferences about $(\sigma_1)^2/(\sigma_2)^2$

1. The samples are random and independent.
2. Both populations are normally distributed.

To conclude this section, we consider the comparison of population variances as a check of the assumption $\sigma_1^2 = \sigma_2^2$ needed for the two-sample t-test. Rejection of the null hypothesis $\sigma_1^2 = \sigma_2^2$ would indicate that the assumption is invalid. [*Note:* Nonrejection of the null hypothesis does *not* imply that the assumption is valid.]

Example 9.12

Checking the Assumption of Equal Variances

Problem In Example 9.4 (Section 9.2), we used the two-sample t-statistic to compare the mean reading scores of two groups of "slow learners" who had been taught to read by two different methods. The data are repeated in Table 9.10 for convenience. The use of the t-statistic was based on the assumption that the population variances of the test scores were equal for the two methods. Conduct a test of hypothesis to check this assumption at $\alpha = .10$.

Table 9.10 **Reading Test Scores for "Slow Learners"**

New Method				Standard Method			
80	80	79	81	79	62	70	68
76	66	71	76	73	76	86	73
70	85			72	68	75	66

Data Set: READING

---

*Although a test of a hypothesis of equality of variances is its most common application, the F-test can also be used to test a hypothesis that the ratio between the population variances is equal to some specified value $H_0: \sigma_1^2/\sigma_2^2 = k$. In that case, the test is conducted in exactly the same way as specified in the box, except that we use the test statistic

$$F = \left(\frac{s_1^2}{s_2^2} \right)\left(\frac{1}{k} \right)$$

Solution We want to test

$$H_0: \sigma_1^2/\sigma_2^2 = 1 \quad (\text{i.e., } \sigma_1^2 = \sigma_2^2)$$
$$H_a: \sigma_1^2/\sigma_2^2 \neq 1 \quad (\text{i.e., } \sigma_1^2 \neq \sigma_2^2)$$

The data were entered into SAS, and the SAS printout shown in Figure 9.20 was obtained. Both the test statistic, $F = .85$, and the two-tailed p-value, .8148, are highlighted on the printout. Since $\alpha = .10$ is less than the p-value, we do not reject the null hypothesis that the population variances of the reading test scores are equal. It is here that the temptation to misuse the F-test is strongest. *We cannot conclude that the data justify the use of the t-statistic.* Doing so would be equivalent to accepting H_0, and we have repeatedly warned against this conclusion because the probability of a Type II error, β, is unknown. The α level of .10 protects us only against rejecting H_0 if it is true. This use of the F-test may prevent us from abusing the t procedure when we obtain a value of F that leads to a rejection of the assumption that $\sigma_1^2 = \sigma_2^2$. But when the F-statistic does not fall into the rejection region, we know little more about the validity of the assumption than before we conducted the test.

```
                    Two Sample Test for Variances of SCORE within METHOD

Sample Statistics

    METHOD
    Group        N        Mean      Std. Dev.    Variance
    ------------------------------------------------------------
    NEW          10        76.4       5.8348      34.04444
    STD          12       72.33333    6.3437      40.24242


Hypothesis Test

    Null hypothesis:      Variance 1 / Variance 2 =  1
    Alternative:          Variance 1 / Variance 2 ^= 1

                   - Degrees of Freedom -
         F          Numer.      Denom.              Pr > F
    ------------------------------------------------------------
       0.85           9           11               0.8148


95% Confidence Interval of the Ratio of Two Variances

         Lower Limit      Upper Limit
         -----------      -----------
           0.2358           3.3096
```

Figure 9.20

SAS F-test for testing assumption of equal variances

Look Back A 95% confidence interval for the ratio $(\sigma_1)^2/(\sigma_2)^2$ is shown at the bottom of the SAS printout of Figure 9.20. Note that the interval, (.2358, 3.3096), includes 1; hence, we cannot conclude that the ratio differs significantly from 1. Thus, the confidence interval leads to the same conclusion that the two-tailed test does: There is insufficient evidence of a difference between the population variances.

Now Work Exercise 9.99

What Do You Do if the Assumption of Normal Population Distributions Is Not Satisfied?

Answer: The F-test is much less robust (i.e., much more sensitive) to departures from normality than the t-test for comparing the population means (Section 9.2). If you have doubts about the normality of the population frequency distributions, use a *nonparametric method* (e.g., Levene's Test) for comparing the two population variances. A method can be found in the nonparametric statistics texts listed in the references for Chapter 14.

Exercises 9.86–9.105

Understanding the Principles

9.86 Describe the sampling distribution of $(s_1)^2/(s_2)^2$ for normal data.

9.87 What conditions are required for valid inferences about $(\sigma_1)^2/(\sigma_2)^2$?

9.88 *True or false.* The F-statistic used for testing $H_0: \sigma_1^2 = \sigma_2^2$ against $H_a: \sigma_1^2 < \sigma_2^2$ is $F = (s_1)^2/(s_2)^2$.

9.89 *True or false.* $H_0: \sigma_1^2 = \sigma_2^2$ is equivalent to $H_0: \sigma_1^2/\sigma_2^2 = 0$.

Learning the Mechanics

9.90 Use Tables VIII, IX, X, and XI of Appendix A to find each
[NW] of the following F-values:
a. $F_{.05}$, where $\nu_1 = 8$ and $\nu_2 = 5$
b. $F_{.01}$, where $\nu_1 = 20$ and $\nu_2 = 14$
c. $F_{.025}$, where $\nu_1 = 10$ and $\nu_2 = 5$
d. $F_{.10}$, where $\nu_1 = 20$ and $\nu_2 = 5$

9.91 Given ν_1 and ν_2, find the following probabilities:
a. $\nu_1 = 2, \nu_2 = 30, P(F \geq 4.18)$
b. $\nu_1 = 24, \nu_2 = 14, P(F < 1.94)$
c. $\nu_1 = 9, \nu_2 = 1, P(F \leq 6,022.0)$
d. $\nu_1 = 30, \nu_2 = 30, P(F > 1.84)$

9.92 For each of the cases that follow, identify the rejection region that should be used to test $H_0: \sigma_1^2 = \sigma_2^2$ against $H_a: \sigma_1^2 > \sigma_2^2$. Assume that $\nu_1 = 20$ and $\nu_2 = 30$.
a. $\alpha = .10$ **b.** $\alpha = .05$
c. $\alpha = .025$ **d.** $\alpha = .01$

9.93 For each of the cases that follow, identify the rejection region that should be used to test $H_0: \sigma_1^2 = \sigma_2^2$ against $H_a: \sigma_1^2 \neq \sigma_2^2$. Assume that $\nu_1 = 8$ and $\nu_2 = 40$.
a. $\alpha = .20$ **b.** $\alpha = .10$
c. $\alpha = .05$ **d.** $\alpha = .02$

9.94 Specify the appropriate rejection region for testing against $H_0: \sigma_1^2 = \sigma_2^2$ in each of the following situations:
a. $H_a: \sigma_1^2 > \sigma_2^2; \alpha = .05, n_1 = 25, n_2 = 20$
b. $H_a: \sigma_1^2 < \sigma_2^2; \alpha = .05, n_1 = 10, n_2 = 15$
c. $H_a: \sigma_1^2 \neq \sigma_2^2; \alpha = .10, n_1 = 21, n_2 = 31$
d. $H_a: \sigma_1^2 < \sigma_2^2; \alpha = .01, n_1 = 31, n_2 = 41$
e. $H_a: \sigma_1^2 \neq \sigma_2^2; \alpha = .05, n_1 = 7, n_2 = 16$

9.95 Independent random samples were selected from each
[NW] of two normally distributed populations, $n_1 = 16$ from population 1 and $n_2 = 25$ from population 2. The means and variances for the two samples are shown in the following table:

Sample 1	Sample 2
$n_1 = 16$	$n_2 = 25$
$\bar{x}_1 = 22.5$	$\bar{x}_2 = 28.2$
$s_1^2 = 2.87$	$s_2^2 = 9.85$

a. Test the null hypothesis $H_0: \sigma_1^2 = \sigma_2^2$ against the alternative hypothesis $H_a: \sigma_1^2 \neq \sigma_2^2$. Use $\alpha = .05$.
b. Find and interpret the p-value of the test.

9.96 Independent random samples were selected from each of two normally distributed populations, $n_1 = 6$ from population 1 and $n_2 = 4$ from population 2. The data are shown in the following table and saved in the **LM9_96** file.

Sample 1	Sample 2
3.1	2.3
4.3	1.4
1.2	3.7
1.7	8.9
.6	
3.4	

a. Test $H_0: \sigma_1^2 = \sigma_2^2$ against $H_a: \sigma_1^2 < \sigma_2^2$. Use $\alpha = .01$.
b. Test $H_0: \sigma_1^2 = \sigma_2^2$ against $H_a: \sigma_1^2 \neq \sigma_2^2$. Use $\alpha = .10$.

Applying the Concepts—Basic

9.97 **Children's recall of TV ads.** Refer to the *Journal of Advertising* (Spring 2006) study of children's recall of television commercials, Exercise 9.15 (p. 424). You used a small-sample t test to test the null hypothesis $H_0: (\mu_1 - \mu_2) = 0$, where $\mu_1 =$ mean number of ads recalled by children in the video-only group and $\mu_2 =$ mean number of ads recalled by children in the A/V group. Summary statistics for the study are reproduced in the table. The validity of the inference derived from the test is based on the assumption of equal group variances, i.e., $\sigma_1^2 = \sigma_2^2$.

Video-Only Group	A/V Group
$n_1 = 20$	$n_2 = 20$
$\bar{x}_1 = 3.70$	$\bar{x}_2 = 3.30$
$s_1 = 1.98$	$s_2 = 2.13$

a. Set up the null and alternative hypothesis for testing this assumption.
b. Compute the test statistic.
c. Find the rejection region for the test using $\alpha = .10$.
d. Make the appropriate conclusion, in the words of the problem.
e. Comment on the validity of the inference derived about the difference in population means in Exercise 9.15.

9.98 **Bulimia study.** Refer to Exercise 9.19 (p. 425). The "fear of negative evaluation" (FNE) scores for the 11 bulimic females and 14 females with normal eating habits are reproduced in the table below and saved in the **BULIMIA** file. The confidence interval you constructed in Exercise 9.19 requires that the variance of the FNE scores of bulimic females be equal to the variance of the FNE scores of normal females. Conduct a test (at $\alpha = .05$) to determine the validity of this assumption.

9.99 **How do you choose to argue?** Refer to the *Thinking and*
[NW] *Reasoning* (Oct. 2006) study of the cognitive skills required for successful arguments, presented in Exercise 9.23 (p. 426). Recall that 52 psychology graduate students were

Table for Exercise 9.98

| Bulimic students: | 21 | 13 | 10 | 20 | 25 | 19 | 16 | 21 | 24 | 13 | 14 | | |
| Normal students: | 13 | 6 | 16 | 13 | 8 | 19 | 23 | 18 | 11 | 19 | 7 | 10 | 15 | 20 |

Based on Randles, R. H. "On neutral responses (zeros) in the sign test and ties in the Wilcoxon-Mann-Whitney test." *The American Statistician,* Vol. 55, No. 2, May 2001 (Figure 3).

equally divided into two groups. Group 1 was presented with arguments that always attempted to strengthen the favored position. Group 2 was presented with arguments that always attempted to weaken the nonfavored position. Summary statistics for the student ratings of the arguments are reproduced in the accompanying table. In Exercise 9.23, you compared the mean ratings for the two groups with a small-sample t-test, assuming equal variances. Determine the validity of this assumption at $\alpha = .05$.

	Group 1 (support favored position)	Group 2 (weaken opposing position)
Sample size	26	26
Mean	28.6	24.9
Standard deviation	12.5	12.2

Based on Kuhn, D., and Udell, W. "Coordinating own and other perspectives in argument." *Thinking and Reasoning*, October 2006.

9.100 Mongolian desert ants. Refer to the *Journal of Biogeography* (Dec. 2003) study of ants in Mongolia (central Asia), presented in Exercise 9.25 (p. 427), in which you compared the mean number of ants at two desert sites. Since the sample sizes were small, the variances of the populations at the two sites must be equal in order for the inference to be valid.

a. Set up H_0 and H_a for determining whether the variances are the same.

b. Use the data in the **GOBIANTS** file to find the test statistic for the test.

c. Give the rejection region for the test if $\alpha = .05$.

d. Find the approximate p-value of the test.

e. Draw the appropriate conclusion in the words of the problem.

f. What conditions are required for the test results to be valid?

9.101 Cognitive impairment of schizophrenics. Refer to the *American Journal of Psychiatry* (April 2010) study of the differences in cognitive function between normal individuals and patients diagnosed with schizophrenia, Exercise 9.14 (p. 424). Recall that the total time (in minutes) a subject spent on the Trail Making Test was used as a measure of cognitive function. Summary data for independent random samples of 41 schizophrenics and 49 normal individuals are reproduced below. Suppose the researchers theorize that schizophrenics will have a wider range in time on the Trail Making Test than will normal subjects. Is there evidence to support this theory? Test using $\alpha = .01$.

	Schizophrenia	Normal
Sample size:	41	49
Mean time:	104.23	62.24
Standard deviation:	45.45	16.34

Based on Perez-Iglesias, R., et al. "White matter integrity and cognitive impairment in first-episode psychosis." *American Journal of Psychiatry*, Vol. 167, No. 4, April 2010 (Table 1).

Applying the Concepts—Intermediate

9.102 Patent infringement case. Refer to the *Chance* (Fall 2002) description of a patent infringement case against Intel Corp., presented in Exercise 9.22 (p. 426). The zinc measurements for three locations listed in the original inventor's notebook—on a text line, on a witness line, and on the intersection of the witness and text line—are reproduced in the following table and saved in the **PATENT** file.

Text line:	.335	.374	.440			
Witness line:	.210	.262	.188	.329	.439	.397
Intersection:	.393	.353	.285	.295	.319	

a. Use a test (at $\alpha = .05$) to compare the variation in zinc measurements for the text line with the corresponding variation for the intersection.

b. Use a test (at $\alpha = .05$) to compare the variation in zinc measurements for the witness line with the corresponding variation for the intersection.

c. From your results in parts **a** and **b**, what can you infer about the variation in zinc measurements at the three notebook locations?

d. What assumptions are required for the inferences to be valid? Are they reasonably satisfied? (You checked these assumptions when you answered Exercise 9.22d.)

9.103 Is honey a cough remedy? Refer to the *Archives of Pediatrics and Adolescent Medicine* (Dec. 2007) study of honey as a children's cough remedy, Exercise 2.32 (p. 45). The data (cough improvement scores) for the 33 children in the DM dosage group and the 35 children in the honey dosage group are reproduced in the accompanying table and saved in the **HONEYCOUGH** file. The researchers want to know if the variability in coughing improvement scores differs for the two groups. Conduct the appropriate analysis, using $\alpha = .10$. If the preferred treatment is the one with the smallest variation in improvement scores, which treatment is preferable?

Honey Dosage:	12 11 15 11 10 13 10 4 15 16 9 14
	10 6 10 8 11 12 12 8 9 5 12
	12 9 11 15 10 15 9 13 8 12 10 8
DM Dosage:	4 6 9 4 7 7 7 9 12 10 11 6
	3 4 9 12 7 6 8 12 12 4 12
	13 7 10 13 9 4 4 10 15 9

Based on Paul, I. M., et al. "Effect of honey, dextromethorphan, and no treatment on nocturnal cough and sleep quality for coughing children and their parents." *Archives of Pediatrics and Adolescent Medicine*, Vol. 161, No. 12, Dec. 2007 (data simulated).

9.104 Detection of rigged school milk prices (cont'd). Refer to the investigation into collusive bidding in the northern Kentucky school milk market, presented in Exercises 9.28 (p. 428) and 9.68 (p. 446). In competitive sealed-bid markets, vendors do not share information about their bids. Consequently, more dispersion or variability among the bids is typically observed than in collusive markets, where vendors communicate about their bids and have a tendency to submit bids in close proximity to one another in an attempt to make the bidding appear competitive. If collusion exists in the tricounty milk market, the variation in winning bid prices in the surrounding ("competitive") market will be significantly larger than the corresponding variation in the tricounty ("rigged") market. A MINITAB analysis of the data on whole white milk in the **MILK** file yielded the printout shown on page 459. Is there evidence that the bid-price variance for the surrounding market exceeds the bid-price variance for the tricounty market?

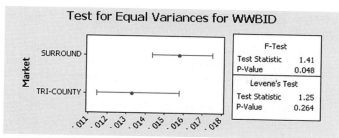

MINITAB output for Exercise 9.104

9.105 Pentagon speeds up order-to-delivery times. Following the initial Persian Gulf War, the Pentagon changed its logistics processes to be more corporate-like. The extravagant "just-in-case" mentality was replaced with "just-in-time" systems. Emulating Federal Express and United Parcel Service, the Pentagon now expedites deliveries from factories to foxholes with the use of bar codes, laser cards, radio tags, and databases to track supplies. The following table contains order-to-delivery times (in days) for a sample of shipments from the United States to the Persian Gulf and a sample of shipments from the United States to Bosnia. These data are saved in the **ORDTIMES** file.

Persian Gulf	Bosnia
28.0	15.1
20.0	6.4
26.5	5.0
10.6	11.4
9.1	6.5
35.2	6.5
29.1	3.0
41.2	7.0
27.5	5.5

Based on Adapted from Crock, S. "The Pentagon goes to B-school." *Business Week*, Dec. 11, 1995, p. 98.

a. Determine whether the variances in order-to-delivery times for Persian Gulf and Bosnia shipments are equal. Use $\alpha = .05$.

b. Given your answer to part a, is it appropriate to construct a confidence interval for the difference between the mean order-to-delivery times? Explain.

CHAPTER NOTES

Key Terms

Note: Starred (*) terms are from the optional section in this chapter.

Blocking 431
F-distribution* 451
Nonparametric statistical tests 421
Paired difference experiment 431
Pooled sample estimator σ^2 416
Randomized block experiment 431
Standard error of the statistic 413

Key Symbols

$\mu_1 - \mu_2$	Difference between population means
μ_d	Paired difference in population means
$p_1 - p_2$	Difference between population proportions
σ_1^2/σ_2^2	Ratio of population variances
D_0	Hypothesized value of difference
$\bar{x}_1 - \bar{x}_2$	Difference between sample means
$\bar{x}_d$	Mean of sample differences
$\hat{p}_1 - \hat{p}_2$	Difference between sample proportions
s_1^2/s_2^2	*Ratio of sample variances
$\sigma_{(\bar{x}_1-\bar{x}_2)}$	Standard error for $\bar{x}_1 - \bar{x}_2$
$\sigma_{\bar{d}}$	Standard error for $\bar{d}$
$\sigma_{(\hat{p}_1 - \hat{p}_2)}$	Standard error for $\hat{p}_1 - \hat{p}_2$
F_a	*Critical value for F-distribution
v_1	*Numerator degrees of freedom for F-distribution
v_2	*Denominator degrees of freedom for F-distribution
SE	Sampling error in estimation

Key Ideas

Key Words for Identifying the Target Parameter

$\mu_1 - \mu_2$	Difference in Means or Averages
μ_d	Paired Difference in Means or Averages
$p_1 - p_2$	Difference in Proportions, Fractions, Percentages, Rates
σ_1^2/σ_2^2	*Ratio (or Difference) in Variances, Spreads

Determining the Sample Size

Estimating $\mu_1 - \mu_2$: $n_1 = n_2 = (z_{a/2})^2(\sigma_1^2 + \sigma_2^2)/(\text{SE})^2$
Estimating $p_1 - p_2$: $n_1 = n_2 = (z_{a/2})^2(p_1q_1 + p_2q_2)/(\text{SE})^2$

Conditions Required for Inferences about $\mu_1 - \mu_2$

Large samples:
1. Independent random samples
2. $n_1 \geq 30, n_2 \geq 30$

Small samples:
1. Independent random samples
2. Both populations normal
3. $\sigma_1^2 = \sigma_2^2$

*Conditions Required for Inferences about σ_1^2/σ_2^2

Large or small samples:
1. Independent random samples
2. Both populations normal

Conditions Required for Inferences about μ_d

Large samples:
1. Random sample of paired differences
2. $n_d \geq 30$

Small samples:
1. Random sample of paired differences
2. Population of differences is normal

Conditions Required for Inferences about $p_1 - p_2$

Large samples:
1. Independent random samples
2. $n_1 p_1 \geq 15, n_1 q_1 \geq 15$
3. $n_2 p_2 \geq 15, n_2 q_2 \geq 15$

Using a Confidence Interval for $(\mu_1 - \mu_2)$ or $(p_1 - p_2)$ to Determine whether a Difference Exists
1. If the confidence interval includes all *positive* numbers $(+, +)$: $\rightarrow$ Infer $\mu_1 > \mu_2$ or $p_1 > p_2$
2. If the confidence interval includes all *negative* numbers $(-, -)$ $\rightarrow$ Infer $\mu_1 < \mu_2$ or $p_1 < p_2$
3. If the confidence interval includes 0 $(-, +)$: $\rightarrow$ Infer "no evidence of a difference."

Guide to Selecting a Two-Sample Hypothesis & Confidence Interval

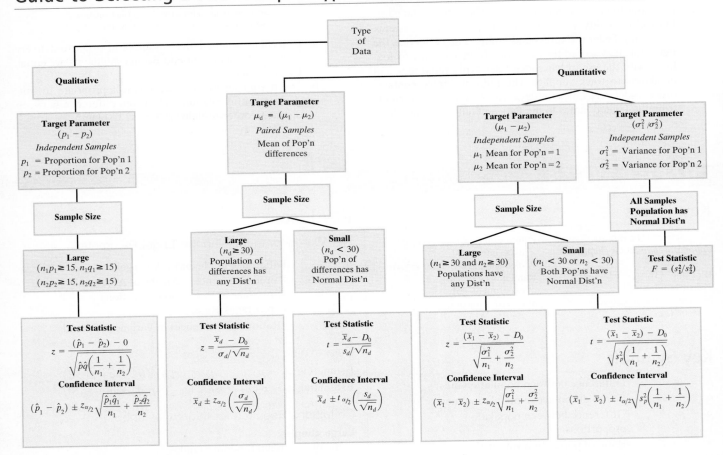

Supplementary Exercises 9.106–9.140

Note: Starred (*) exercises refer to the optional section in this chapter.

Understanding the Principles

9.106 List the assumptions necessary for each of the following inferential techniques:
  a. Large-sample inferences about the difference $(\mu_1 - \mu_2)$ between population means, using a two-sample z-statistic
  b. Small-sample inferences about $(\mu_1 - \mu_2)$, using an independent samples design and a two-sample t-statistic
  c. Small-sample inferences about $(\mu_1 - \mu_2)$, using a paired difference design and a single-sample t-statistic to analyze the differences

  d. Large-sample inferences about the differences $(p_1 - p_2)$ between binomial proportions, using a two-sample z-statistic
  *e. Inferences about the ratio σ_1^2/σ_2^2 of two population variances, using an F-test

9.107 For each of the following, identify the target parameter as $\mu_1 - \mu_2, p_1 - p_2$, or σ_1^2/σ_2^2.
  a. Comparison of average SAT scores of males and females.
  b. Difference between mean waiting times at two supermarket checkout lanes
  c. Comparison of proportions of Democrats and Republicans who favor the legalization of marijuana.

*d. Comparison of variation in salaries of NBA players picked in the first round and the second round.

e. Difference in dropout rates of college student athletes and regular students

Learning the Mechanics

9.108 Two independent random samples were selected from normally distributed populations with means and variances (μ_1, σ_1^2) and (μ_2, σ_2^2), respectively. The sample sizes, means, and variances are shown in the following table:

Sample 1	Sample 2
$n_1 = 20$	$n_2 = 15$
$\bar{x}_1 = 123$	$\bar{x}_2 = 116$
$s_1^2 = 31.3$	$s_2^2 = 120.1$

*a. Test $H_0: \sigma_1^2 = \sigma_2^2$ against $H_a: \sigma_1^2 \neq \sigma_2^2$. Use $\alpha = .05$.

b. Would you be willing to use a t-test to test the null hypothesis $H_0: (\mu_1 - \mu_2) = 0$ against the alternative hypothesis $H_a: (\mu_1 - \mu_2) \neq 0$? Why?

9.109 Independent random samples were selected from two normally distributed populations with means μ_1 and μ_2, respectively. The sample sizes, means, and variances are shown in the following table:

Sample 1	Sample 2
$n_1 = 12$	$n_2 = 14$
$\bar{x}_1 = 17.8$	$\bar{x}_2 = 15.3$
$s_1^2 = 74.2$	$s_2^2 = 60.5$

a. Test $H_0: (\mu_1 - \mu_2) = 0$ against $H_a: (\mu_1 - \mu_2) > 0$. Use $\alpha = .05$.

b. Form a 99% confidence interval for $(\mu_1 - \mu_2)$.

c. How large must n_1 and n_2 be if you wish to estimate $(\mu_1 - \mu_2)$ to within two units with 99% confidence? Assume that $n_1 = n_2$.

9.110 Independent random samples were selected from two binomial populations. The size and number of observed successes for each sample are shown in the following table:

Sample 1	Sample 2
$n_1 = 200$	$n_2 = 200$
$x_1 = 110$	$x_2 = 130$

a. Test $H_0: (p_1 - p_2) = 0$ against $H_a: (p_1 - p_2) < 0$. Use $\alpha = .10$.

b. Form a 95% confidence interval for $(p_1 - p_2)$.

c. What sample sizes would be required if we wish to use a 95% confidence interval of width .01 to estimate $(p_1 - p_2)$?

9.111 Two independent random samples are taken from two populations. The results of these samples are summarized in the following table:

Sample 1	Sample 2
$n_1 = 135$	$n_2 = 148$
$\bar{x}_1 = 12.2$	$\bar{x}_2 = 8.3$
$s_1^2 = 2.1$	$s_2^2 = 3.0$

a. Form a 90% confidence interval for $(\mu_1 - \mu_2)$.

b. Test $H_0: (\mu_1 - \mu_2) = 0$ against $H_a: (\mu_1 - \mu_2) \neq 0$. Use $\alpha = .01$.

c. What sample size would be required if you wish to estimate $(\mu_1 - \mu_2)$ to within .2 with 90% confidence? Assume that $n_1 = n_2$.

9.112 A random sample of five pairs of observations was selected, one observation from a population with mean μ_1, the other from a population with mean μ_2. The data are shown in the following table and saved in the **LM9_112** file.

Pair	Value from Population 1	Value from Population 2
1	28	22
2	31	27
3	24	20
4	30	27
5	22	20

a. Test the null hypothesis $H_0: \mu_d = 0$ against $H_a: \mu_d \neq 0$, where $\mu_d = \mu_1 - \mu_2$. Use $\alpha = .05$.

b. Form a 95% confidence interval for μ_d.

c. When are the procedures you used in parts **a** and **b** valid?

Applying the Concepts—Basic

9.113 **Oil spill impact on seabirds.** Refer to the *Journal of Agricultural, Biological, and Environmental Statistics* (Sept. 2000) study of the impact of a tanker oil spill on the seabird population in Alaska, presented in Exercise 2.189 (p. 102). Recall that for each of 96 shoreline locations (called transects), the number of seabirds found, the length (in kilometers) of the transect, and whether or not the transect was in an oiled area were recorded. (The data are saved in the **EVOS** file.) *Observed seabird density* is defined as the observed count divided by the length of the transect.

MINITAB output for Exercise 9.113

Two-Sample T-Test and CI: Density, Oil

```
Two-sample T for Density

Oil   N   Mean   StDev   SE Mean
no    36  3.27   6.70    1.1
yes   60  3.50   5.97    0.77


Difference = mu (no ) - mu (yes)
Estimate for difference:   -0.221165
95% CI for difference:   (-2.927767, 2.485436)
T-Test of difference = 0 (vs not =): T-Value = -0.16   P-Value = 0.871   DF = 67
```

A comparison of the mean densities of oiled and unoiled transects is displayed in the MINITAB printout on p. 461. Use this information to make an inference about the difference in the population mean seabird densities of oiled and unoiled transects.

9.114 Rating service at five-star hotels. A study published in the *Journal of American Academy of Business, Cambridge* (March 2002) examined whether the perception of the quality of service at five-star hotels in Jamaica differed by gender. Hotel guests were randomly selected from the lobby and restaurant areas and asked to rate 10 service-related items (e.g., "the personal attention you received from our employees"). Each item was rated on a five-point scale (1 = "much worse than I expected," 5 = "much better than I expected"), and the sum of the items for each guest was determined. A summary of the guest scores is provided in the following table:

Gender	Sample Size	Mean Score	Standard Deviation
Males	127	39.08	6.73
Females	114	38.79	6.94

a. Construct a 90% confidence interval for the difference between the population mean service-rating scores given by male and female guests at Jamaican five-star hotels.
b. Use the interval you constructed in part **a** to make an inference about whether the perception of the quality of service at five-star hotels in Jamaica differs by gender.
*c. Is there evidence of a difference in the variation of guest scores for males and females? Test using $\alpha = .10$.

9.115 Effect of altitude on climbers. Dr. Philip Lieberman, a neuroscientist at Brown University, conducted a field experiment to gauge the effect of high altitude on a person's ability to think critically (*New York Times*, Aug. 23, 1995). The subjects of the experiment were five males who took part in an American expedition climbing Mount Everest. At the base camp, Lieberman read sentences to the climbers while they looked at simple pictures in a book. The length of time (in seconds) it took for each climber to match the picture with a sentence was recorded. Using a radio, Lieberman repeated the task when the climbers reached a camp 5 miles above sea level. At this altitude, he noted that the climbers took 50% longer to complete the task.
a. What is the variable measured in this experiment?
b. What are the experimental units?
c. Discuss how the data should be analyzed.

9.116 Executive workout dropouts. Refer to the *Journal of Sport Behavior* (2001) study of variety in exercise workouts, presented in Exercise 7.130 (p. 343). One group of 40 people varied their exercise routine in workouts, while a second group of 40 exercisers had no set schedule or regulations for their workouts. By the end of the study, 15 people had dropped out of the first exercise group and 23 had dropped out of the second group.
a. Find the dropout rates (i.e., the percentage of exercisers who had dropped out of the exercise group) for each of the two groups of exercisers.
b. Find a 90% confidence interval for the difference between the dropout rates of the two groups of exercisers.

c. Give a practical interpretation of the confidence interval you found in part **c**.
d. Suppose you want to reduce the sampling error in the 90% confidence interval to .1. Determine the number of exercisers to be sampled from each group in order to obtain such an estimate. Assume equal sample sizes, and assume that $p_1 \approx .4$ and $p_2 \approx .6$.

9.117 Heights of grade school repeaters. Are children who repeat a grade in elementary school shorter, on average, than their peers? To answer this question, researchers compared the heights of Australian schoolchildren who repeated a grade with the heights of those who did not (*Archives of Disease in Childhood*, Apr. 2000). All height measurements were standardized with the use of *z*-scores. A summary of the results, by gender, is shown in the following table:

Gender/Status	Sample Size	Mean	Standard Deviation
Girls/Repeat Grade	43	.26	.94
Girls/Never Repeated	1,366	.22	1.04
Boys/Repeat Grade	86	−.04	1.17
Boys/Never Repeated	1,349	.30	.97

Reproduced from Wake, M., Coghlan, D., and Hesketh, K. "Does height influence progression through primary school grades?" *The Archives of Disease in Childhood*, Vol. 82, No. 4, April 2000, pp. 297-301. Copyright © 2000 with permission from BMJ Publishing Group, Inc.

a. Set up the null and alternative hypothesis for determining whether the average height of Australian boys who repeated a grade is less than the average height of boys who never repeated.
b. Conduct the test you set up in part **a**, using $\alpha = .05$.
c. Repeat parts **a** and **b** for Australian girls.

*9.118 Bear vs. pig bile study. Bear gallbladder is used in Chinese medicine to treat inflammation. A study in the *Journal of Ethnopharmacology* (June 1995) examined the easier-to-obtain pig gallbladder as an effective substitute for bear gallbladder. Twenty male mice were divided randomly into two groups: Ten were given a dosage of bear bile and 10 were given a dosage of pig bile. All the mice then received an injection of croton oil in the left earlobe to induce inflammation. Four hours later, both the left and right earlobes were weighed, with the difference (in milligrams) representing the degree of swelling. Summary statistics on the degree of swelling are provided in the following table:

Bear Bile	Pig Bile
$n_1 = 10$	$n_2 = 10$
$\bar{x}_1 = 9.19$	$\bar{x}_2 = 9.71$
$s_1 = 4.17$	$s_2 = 3.33$

a. Use a hypothesis test (at $\alpha = .05$) to compare the variation in degree of swelling for mice treated with bear bile and mice treated with pig bile.
b. What assumptions are necessary for the inference you made in part **a** to be valid?

9.119 The "winner's curse" in auction bidding. In auction bidding, the "winner's curse" is the phenomenon of the winning (or highest) bid price being above the expected value of the item being auctioned. *The Review of Economics and Statistics*

(Aug. 2001) published a study on whether experience in bidding affects the likelihood of the winner's curse occurring. Two groups of bidders in a sealed-bid auction were compared: (1) superexperienced bidders and (2) less experienced bidders. In the superexperienced group, 29 of 189 winning bids were above the item's expected value; in the less experienced group, 32 of 149 winning bids were above the item's expected value.

a. Find an estimate of p_1, the true proportion of super-experienced bidders who fall prey to the winner's curse.

b. Find an estimate of p_2, the true proportion of less experienced bidders who fall prey to the winner's curse.

c. Construct a 90% confidence interval for $p_1 - p_2$.

d. Give a practical interpretation of the confidence interval you constructed in part **c**. Make a statement about whether experience in bidding affects the likelihood of the winner's curse occurring.

9.120 Environmental impact study. Some power plants are located near rivers or oceans so that the available water can be used to cool the condensers. Suppose that, as part of an environmental impact study, a power company wants to estimate the difference in mean water temperature between the discharge of its plant and the offshore waters. How many sample measurements must be taken at each site in order to estimate the true difference between means to within .2°C with 95% confidence? Assume that the range in readings will be about 4°C at each site and that the same number of readings will be taken at each site.

9.121 Animal-assisted therapy for heart patients. Refer to the *American Heart Association Conference* (Nov. 2005) study to gauge whether animal-assisted therapy can improve the physiological responses of heart failure patients, presented in Exercise 2.106 (p. 73). Recall that a sample of $n = 26$ heart patients was visited by a human volunteer accompanied by a trained dog; the anxiety level of each patient was measured (in points) both before and after the visits. The drop (before minus after) in anxiety level for patients is summarized as follows: $\bar{x}_d = 10.5, s_d = 7.6$. Does animal-assisted therapy significantly reduce the mean anxiety level of heart failure patients? Support your answer with a 95% confidence interval.

9.122 Life expectancy of Oscar winners. Does winning an Academy of Motion Picture Arts and Sciences award lead to long-term mortality for movie actors? In an article in the *Annals of Internal Medicine* (May 15, 2001), researchers sampled 762 Academy Award winners and matched each one with another actor of the same sex who was in the same winning film and was born in the same era. The life expectancies (ages) of the pairs of actors were compared.

a. Explain why the data should be analyzed as a paired difference experiment.

b. Set up the null hypothesis for a test to compare the mean life expectancies of Academy Award winners and nonwinners.

c. The sample mean life expectancies of Academy Award winners and nonwinners were reported as 79.7 years and 75.8 years, respectively. The *p*-value for comparing the two population means was reported as $p = .003$. Interpret this value in the context of the problem.

Applying the Concepts—Intermediate

9.123 Reading tongue twisters. According to *Webster's New World Dictionary*, a tongue twister is "a phrase that is hard to speak rapidly." Do tongue twisters have an effect on the length of time it takes to read silently? To answer this question, 42 undergraduate psychology students participated in a reading experiment (*Memory & Cognition*, Sept. 1997). Two lists, each composed of 600 words, were constructed. One list contained a series of tongue twisters, and the other list (called the *control*) did not contain any tongue twisters. Each student read both lists, and the length of time (in minutes) required to complete the lists was recorded. The researchers used a test of hypothesis to compare the mean reading response times for the tongue-twister and control lists.

a. Set up the null hypothesis for the test.

b. For each student, the researchers computed the difference between the reading response times for the tongue-twister and control lists. The mean difference was .25 minute with a standard deviation of .78 minute. Use the information to find the test statistic and *p*-value of the test.

c. Give the appropriate conclusion. Use $\alpha = .05$.

Based on Robinson, D. H., and Katayama, A. D. "At-lexical, articulatory interference in silent reading: The 'upstream' tongue-twister effect." Memory & Cognition, Vol. 25, No. 5, Sept. 1997, p. 663.

9.124 Mating habits of snails. Hermaphrodites are animals that possess the reproductive organs of both sexes. *Genetical Research* (June 1995) published a study of the mating systems of hermaphroditic snail species. The mating habits of the snails were classified into two groups: (1) self-fertilizing (selfing) snails that mate with snails of the same sex and (2) cross-fertilizing (outcrossing) snails that mate with snails of the opposite sex. One variable of interest in the study was the effective population size of the snail species. The means and standard deviations of the effective population size for independent random samples of 17 outcrossing snail species and 5 selfing snail species are given in the accompanying table.

Snail Mating System	Effective Population Size		
	Sample Size	Mean	Standard Deviation
Outcrossing	17	4,894	1,932
Selfing	5	4,133	1,890

Based on Jarne, P. "Mating system, bottlenecks, and genetic polymorphism in hermaphroditic animals." *Genetical Research*, Vol. 65, No. 3, June 1995, p. 197 (Table 4).

a. Compare the mean effective population sizes of the two types of snail species with a 90% confidence interval. Interpret the result.

b. Geneticists are often more interested in comparing the variation in population size of the two types of mating systems. Conduct this analysi for the researcher. Interpret the result.

9.125 Children's use of pronouns. Refer to the *Journal of Communication Disorders* (Mar. 1995) study of specifically language-impaired (SLI) children, presented in Exercise 2.188 (p. 101). The data on deviation intelligence quotient (DIQ) for 10 SLI children and 10 younger, normally developing children are reproduced in the accompanying table and saved in the **SLI** file. Use the methodology of this section to compare the mean DIQ of the two groups of children. (Use $\alpha = .10$.) What do you conclude?

	SLI Children			YND Children	
86	87	84	110	90	105
94	86	107	92	92	96
89	98	95	86	100	92
110			90		

9.126 Identical twins reared apart. Because they share an identical genotype, twins make ideal subjects for investigating the degree to which various environmental conditions affect personality. The classical method of studying this phenomenon, nomenon, and the subject of an interesting book by Susan Farber (*Identical Twins Reared Apart*, New York: Basic Books, 1981), is the study of identical twins separated early in life and reared apart. Much of Farber's discussion focuses on a comparison of IQ scores. The data for this analysis appear in the accompanying table and are saved in the **TWINSIQ** file. One member (A) of each of the $n = 32$ pairs of twins was reared by a natural parent; the other member (B) was reared by a relative or some other person. Is there a significant difference between the average IQ scores of identical twins when one member of the pair is reared by the natural parents and the other member of the pair is not? Use $\alpha = .05$ to draw your conclusion.

Pair ID	Twin A	Twin B	Pair ID	Twin A	Twin B
112	113	109	228	100	88
114	94	100	232	100	104
126	99	86	236	93	84
132	77	80	306	99	95
136	81	95	308	109	98
148	91	106	312	95	100
170	111	117	314	75	86
172	104	107	324	104	103
174	85	85	328	73	78
180	66	84	330	88	99
184	111	125	338	92	111
186	51	66	342	108	110
202	109	108	344	88	83
216	122	121	350	90	82
218	97	98	352	79	76
220	82	94	416	97	98

Based on Farber, S. L. *Identical Twins Reared Apart*, © 1981 by Basic Books, Inc.

9.127 Treatments for panic disorder. Inositol is a complex cyclic alcohol found to be effective against clinical depression. Medical researchers believe that inositol may also be used to treat panic disorder. To test this theory, a double-blind, placebo-controlled study of 21 patients diagnosed with panic disorder was conducted (*American Journal of Psychiatry*, July 1995). Patients completed diaries recording the occurrence of their panic attacks. The data (saved in the **INOSITOL** file) for a week in which patients received a glucose placebo and for a week when they were treated with inositol are provided in the next table. [*Note:* Neither the patients nor the treating physicians knew which week the placebo was given.] Analyze the data

and interpret the results. Comment on the validity of the assumptions.

Patient	Placebo	Inositol	Patient	Placebo	Inositol
1	0	0	12	0	2
2	2	1	13	3	1
3	0	3	14	3	1
4	1	2	15	3	4
5	0	0	16	4	2
6	10	5	17	6	4
7	2	0	18	15	21
8	6	4	19	28	8
9	1	1	20	30	0
10	1	0	21	13	0
11	1	3			

Based on Benjamin, J. "Double-blind, placebo-controlled, crossover trial of inositol treatment for panic disorder." *American Journal of Psychiatry*, Vol. 152, No. 7, July 1995, p. 1085 (Table 1).

9.128 Personalities of cocaine abusers. Do cocaine abusers have radically different personalities than nonabusing college students? This was one of the questions researched in *Psychological Assessment* (June 1995). Zuckerman–Kuhlman's Personality Questionnaire (ZKPQ) was administered to a sample of 450 cocaine abusers and a sample of 589 college students. The ZKPQ yields scores (measured on a 20-point scale) on each of five dimensions: impulsive–sensation seeking, sociability, neuroticism–anxiety, aggression–hostility, and activity. The results are summarized in the accompanying table. Compare the mean ZKPQ scores of the two groups on each dimension, using a statistical test of hypothesis. Interpret the results at $\alpha = .01$.

	Cocaine Abusers ($n = 450$)		College Students ($n = 589$)	
ZKPQ Dimension	Mean	Std. Dev.	Mean	Std. Dev.
Impulsive–sensation seeking	9.4	4.4	9.5	4.4
Sociability	10.4	4.3	12.5	4.0
Neuroticism–anxiety	8.6	5.1	9.1	4.6
Aggression–hostility	8.6	3.9	7.3	4.1
Activity	11.1	3.4	8.0	4.1

Based on Ball, S. A. "The validity of an alternative five-factor measure of personality in cocaine abusers." *Psychological Assessment*, Vol. 7, No. 2, June 1995, p. 150 (Table 1). Copyright © 1995 by the American Psychological Association. Reprinted with permission.

9.129 Switching majors in college. When female undergraduates switch from science, mathematics, and engineering (SME) majors into disciplines that are not based on science, are their reasons different from those of their male counterparts? This question was investigated in *Science Education* (July 1995). A sample of 335 junior/senior undergraduates—172 females and 163 males—at two large research universities were identified as "switchers"; that is, they left a declared SME major for a non-SME major. Each student listed one or more factors that contributed to the switching decision.

a. Of the 172 females in the sample, 74 listed lack or loss of interest in SME (i.e., they were "turned off" by science) as a major factor, compared with 72 of the 163 males. Conduct a test (at $\alpha = .10$) to determine whether the proportion of female switchers who give "lack of interest in SME" as a major reason for switching differs from the corresponding proportion of males.

b. Thirty-three of the 172 females in the sample indicated that they because discouraged or lost confidence because of low grades in SME during their early years, compared with 44 of 163 males. Construct a 90% confidence interval for the difference between the proportions of female and male switchers who lost confidence due to low grades in SME. Interpret the result.

9.130 Swim maze study. Merck Research Labs used the single-T swim maze to conduct an experiment to evaluate the effect of a new drug. Nineteen impregnated dam rats were allocated a dosage of 12.5 milligrams of the drug. One male and one female rat pup were randomly selected from each resulting litter to perform in the swim maze. Each pup was placed in the water at one end of the maze and allowed to swim until it escaped at the opposite end. If the pup failed to escape after a certain period of time, it was placed at the beginning of the maze and given another chance. The experiment was repeated until each pup accomplished three successful escapes. The accompanying table (saved in the **RATPUPS** file) reports the number of swims required by each pup to perform three successful escapes. Is there sufficient evidence of a difference between the mean number of swims required by male and female pups? Conduct the test (at $\alpha = .10$). Comment on the assumptions required for the test to be valid.

Litter	Male	Female	Litter	Male	Female
1	8	5	11	6	5
2	8	4	12	6	3
3	6	7	13	12	5
4	6	3	14	3	8
5	6	5	15	3	4
6	6	3	16	8	12
7	3	8	17	3	6
8	5	10	18	6	4
9	4	4	19	9	5
10	4	4			

Based on Thomas E. Bradstreet, "Favorite Data Sets from Early Phases of Drug Research - Part 2." *Proceedings of the Section on Statistical Education of the American Statistical Association.*

9.131 Rating music teachers. Students enrolled in music classes at the University of Texas (Austin) participated in a study to compare the observations and teacher evaluations of music education majors and nonmusic majors (*Journal of Research in Music Education*, Winter 1991). Independent random samples of 100 music majors and 100 nonmajors rated the overall performance of their teacher, using a six-point scale, where 1 was the lowest rating and 6 the highest. Use the information in the accompanying table to compare the mean teacher ratings of the two groups of music students with a 95% confidence interval. Interpret the result.

	Music Majors	Nonmusic Majors
Sample size	100	100
Mean "overall" rating	4.26	4.59
Standard deviation	.81	.78

Based on Duke, R. A., and Blackman, M. D. "The relationship between observers' recorded teacher behavior and evaluation of music instruction." *Journal of Research in Music Education*, Vol. 39, No. 4, Winter 1991 (Table 2).

9.132 Identifying the target parameter. For each of the following studies, give the parameter of interest and state any assumptions that are necessary for the inferences to be invalid.

a. To investigate a possible link between jet lag and memory impairment, a University of Bristol (England) neurologist recruited 20 female flight attendants who worked flights across several time zones. Half of the attendants had only a short recovery time between flights, and half had a long recovery time between flights. The average size of the right temporal lobe of the brain for the short-recovery group was significantly smaller than the average size of the right temporal lobe of the brain for the long-recovery group.

b. In a study presented at a meeting of the Association for the Advancement of Applied Sport Psychology, researchers revealed that the proportion of athletes who have a good self-image of their body is 20% higher than the corresponding proportion of nonathletes.

c. A University of Florida animal sciences professor has discovered that feeding chickens corn oil causes them to produce larger eggs. The weight of eggs produced by each of a sample of chickens on a regular feed diet was recorded. Then the same chickens were fed a diet supplemented by corn oil, and the weight of eggs produced by each was recorded. The mean weight of the eggs produced with corn oil was 3 grams heavier than the mean weight produced with the regular diet.

*9.133 Instrument precision. The quality control department of a paper company measures the brightness (a measure of reflectance) of finished paper on a periodic basis throughout the day. Two instruments that are available to measure the paper specimens are subject to error, but they can be adjusted so that the mean readings for a control paper specimen are the same for both instruments. Suppose you are concerned about the precision of the two instruments and want to compare the variability in the readings of instrument 1 with those of instrument 2. Five brightness measurements were made on a single paper specimen, using each of the two instruments. The data are shown in the following table and saved in the **BRIGHT** file.

Instrument 1	Instrument 2
29	26
28	34
30	30
28	32
30	28

a. Is the variance of the measurements obtained by instrument 1 significantly different from the variance of the measurements obtained by instrument 2?

b. What assumptions must be satisfied for the test in part **a** to be valid?

9.134 Testing electronic circuits. Japanese researchers have developed a compression–depression method of testing electronic circuits based on Huffman coding (*IEICE Transactions on Information & Systems*, Jan. 2005). The new method is designed to reduce the time required for input decompression and output compression—called the compression ratio. Experimental results were obtained by testing a sample of 11 benchmark circuits (all of different sizes) from a SUN Blade 1000 workstation. Each circuit was tested with the standard compression–depression method and the new Huffman-based coding method and the compression ratio recorded. The data are given below and saved in the **CIRCUITS** file.

a. Compare the two methods with a 95% confidence interval. Which method has the smaller mean compression ratio?

b. How many circuits need to be sampled in order to estimate the mean difference in compression ratio to within .03 with 95% confidence.

Circuit	Standard Method	Huffman Coding Method
1	.80	.78
2	.80	.80
3	.83	.86
4	.53	.53
5	.50	.51
6	.96	.68
7	.99	.82
8	.98	.72
9	.81	.45
10	.95	.79
11	.99	.77

Based on Ichihara, H., Shintani, M., and Inoue, T. "Huffman-based test response coding." *IEICE Transactions on Information & Systems*, Vol. E88-D, No. 1, Jan. 2005 (Table 3).

9.135 Kicking the cigarette habit. Can taking an antidepressant drug help cigarette smokers kick their habit? The *New England Journal of Medicine* (Oct. 23, 1997) published a study in which 615 smokers (all of whom wanted to give up smoking) were randomly assigned to receive either Zyban (an antidepressant) or a placebo (a dummy pill) for six weeks. Of the 309 patients who received Zyban, 71 were not smoking one year later. Of the 306 patients who received a placebo, 37 were not smoking one year later. Conduct a test of hypothesis (at $\alpha = .05$) to answer the research question posed in the first sentence of this exercise.

9.136 Accuracy of mental maps. To help students organize global information about people, places, and environments, geographers encourage them to develop "mental maps" of the world. A series of lessons was designed to aid students in the development of mental maps (*Journal of Geography*, May/June 1997). In one experiment, a class of 24 seventh-grade geography students was given mental map lessons, while a second class of 20 students received traditional instruction. All of the students were asked to sketch a map of the world, and each portion of the map was evaluated for accuracy on a five-point scale (1 = low accuracy, 5 = high accuracy).

a. The mean accuracy scores of the two groups of seventh-graders were compared with the use of a test of hypothesis. State H_0 and H_a for a test to determine whether the mental map lessons improve a student's ability to sketch a world map.

b. The observed significance level of the test for comparing the mean accuracy scores for continents drawn is .0507. Interpret this result.

c. The observed significance level of the test for comparing the mean accuracy scores for labeling oceans is .7371. Interpret the result.

d. The observed significance level of the test for comparing the mean accuracy scores for the entire map is .0024. Interpret the result.

e. What assumptions (if any) are required for the tests to be statistically valid? Are they likely to be met? Explain.

Applying the Concepts—Advanced

9.137 Gambling in public high schools. With the rapid growth in legalized gambling in the United States, there is concern that the involvement of youth in gambling activities is also increasing. University of Minnesota professor Randy Stinchfield compared the rates of gambling among Minnesota public school students between 1992 and 1998. (*Journal of Gambling Studies*, Winter 2001). Based on survey data, the following table shows the percentages of ninth-grade boys who gambled weekly or daily on any game (e.g., cards, sports betting, lotteries) for the two years:

	1992	1998
Number of ninth-grade boys in survey	21,484	23,199
Number who gambled weekly/daily	4,684	5,313

a. Are the percentages of ninth-grade boys who gambled weekly or daily on any game in 1992 and 1998 significantly different? (Use $\alpha = .01$.)

b. Professor Stinchfield states that "because of the large sample sizes, even small differences may achieve statistical significance, so interpretations of the differences should include a judgment regarding the magnitude of the difference and its public health significance." Do you agree with this statement? If not, why not? If so, obtain a measure of the magnitude of the difference between 1992 and 1998 and attach a measure of reliability to the difference.

9.138 Feeding habits of sea urchins. The *Florida Scientist* (Summer/Autumn 1991) reported on a study of the feeding habits of sea urchins. A sample of 20 urchins was captured from Biscayne Bay (Miami), placed in marine aquaria, and then starved for 48 hours. Each sea urchin was then fed a 5-cm blade of turtle grass. Ten of the urchins received only green blades, while the other half received only decayed blades. (Assume that the two samples of 10 sea urchins each were randomly and independently selected.) The ingestion time, measured from the time the blade first made contact with the urchin's mouth to the time the urchin had finished ingesting the blade, was recorded. A summary of the results is provided in the following table:

	Green Blades	Decayed Blades
Number of sea urchins	10	10
Mean ingestion time (hours)	3.35	2.36
Standard deviation (hours)	.79	.47

From "Laboratory measurement of ingestion rate for the sea urchin Lytechinus variegatus" by Dr. Jeremy Montague. *Florida Scientist*, Vol. 54, Nos. 3/4, Summer/Autumn 1991. Reprinted with permission from the Florida Academy of Sciences.

According to the researchers, "The difference in rates at which the urchins ingested the blades suggest that green, unblemished turtle grass may not be a particularly palatable food compared with decayed turtle grass. If so, urchins in the field may find it more profitable to selectively graze on decayed portions of the leaves." Do the results support this conclusion?

Critical Thinking Challenges

9.139 Self-managed work teams and family life. To improve quality, productivity, and timeliness, more and more American industries are utilizing self-managed work teams (SMWTs). A team typically consists of 5 to 15 workers who are collectively responsible for making decisions and performing all tasks related to a particular project. Researchers L. Stanley-Stevens (Tarleton State University), D. E. Yeatts, and R. R. Seward (both from the University of North Texas) investigated the connection between SMWTs, work characteristics, and workers' perceptions of positive spillover into family life (*Quality Management Journal*, Summer 1995). Survey data were collected from 114 AT&T employees who worked on 1 of 15 SMWTs at an AT&T technical division. The workers were divided into two groups: (1) those who reported a positive spillover of work skills to family life and (2) those who did not report any such positive work spillover. The two groups were compared on a variety of job and demographic characteristics, several of which are shown in the table (next column). All but the demographic characteristics were measured on a seven-point scale, ranging from 1 = "strongly disagree" to 7 = "strongly agree"; thus, the larger the number, the more the characteristic was indicated. The file named **SPILLOVER** includes the values of the variables listed in the table for each of the 114 survey participants. The researchers' objectives were to compare the two groups of workers on each characteristic. In particular, they wanted to know which job-related

Characteristic	Variable
Information Flow	Use of creative ideas (seven-point scale)
Information Flow	Utilization of information (seven-point scale)
Decision Making	Participation in decisions regarding personnel matters (seven-point scale)
Job	Good use of skills (seven-point scale)
Job	Task identity (seven-point scale)
Demographic	Age (years)
Demographic	Education (years)
Demographic	Gender (male or female)
Comparison	Group (positive spillover or no spillover)

characteristics are most highly associated with positive work spillover. Conduct a complete analysis of the data for the researchers.

9.140 MS and exercise study. A study published in *Clinical Kinesiology* (Spring 1995) was designed to examine the metabolic and cardiopulmonary responses during exercise of persons diagnosed with multiple sclerosis (MS). Leg-cycling and arm-cranking exercises were performed by 10 MS patients and 10 healthy (non-MS) control subjects. Each member of the control group was selected on the basis of gender, age, height, and weight to match (as closely as possible) with one member of the MS group. Consequently, the researchers compared the MS and non-MS groups by matched-pairs t-tests on such outcome variables as oxygen uptake, carbon dioxide output, and peak aerobic power. The data on the matching variables used in the experiment are shown in the table below and saved in the **MSSTUDY** file. Have the researchers successfully matched the MS and non-MS subjects?

	MS Subjects				Non-MS Subjects			
Matched Pair	Gender	Age (years)	Height (cm)	Weight (kg)	Gender	Age (years)	Height (cm)	Weight (kg)
1	M	48	171.0	80.8	M	45	173.0	76.3
2	F	34	158.5	75.0	F	34	158.0	75.6
3	F	34	167.6	55.5	F	34	164.5	57.7
4	M	38	167.0	71.3	M	34	161.3	70.0
5	M	45	182.5	90.9	M	39	179.0	96.0
6	F	42	166.0	72.4	F	42	167.0	77.8
7	M	32	172.0	70.5	M	34	165.8	74.7
8	F	35	166.5	55.3	F	43	165.1	71.4
9	F	33	166.5	57.9	F	31	170.1	60.4
10	F	46	175.0	79.9	F	43	175.0	77.9

From "Maximal aerobic exercise of individuals with multiple sclerosis using three modes of ergometry." *Clinical Kinesiology*, Vol. 49, No. 1, Spring 1995, p. 7. Reprinted with permission from W. Jeffrey Armstrong.

Activity   Paired vs. Unpaired Experiments

We have now discussed two methods of collecting data to compare two population means. In many experimental situations, a decision must be made either to collect two independent samples or to conduct a paired difference experiment. The importance of this decision cannot be overemphasized, since the amount of information obtained and the cost of the experiment are both directly related to the method of experimentation that is chosen.

Choose two populations (pertinent to your school major) that have unknown means and for which you could both collect two independent samples and collect paired observations. Before

conducting the experiment, state which method of sampling you think will provide more information (and why). Compare the two methods, first performing the independent sampling procedure by collecting 10 observations from each population (a total of 20 measurements) and then performing the paired difference experiment by collecting 10 pairs of observations.

Construct two 95% confidence intervals, one for each experiment you conduct. Which method provides the narrower confidence interval and hence more information on this performance of the experiment? Does your result agree with your preliminary expectations?

References

Freedman, D., Pisani, R., and Purves, R. *Statistics*. New York: W. W. Norton and Co., 1978.

Gibbons, J. D. *Nonparametric Statistical Inference*, 2nd ed. New York: McGraw-Hill, 1985.

Hollander, M., and Wolfe, D. A. *Nonparametric Statistical Methods*. New York: Wiley, 1973.

Mendenhall, W., Beaver, R. J., and Beaver, B. *Introduction to Probability and Statistics*, 13th ed. Belmont, CA: Brooks/Cole, 2009.

Satterthwaite, F. W. "An approximate distribution of estimates of variance components." *Biometrics Bulletin*, Vol. 2, 1946, pp. 110–114.

Snedecor, G. W., and Cochran, W. *Statistical Methods*, 7th ed. Ames, IA: Iowa State University Press, 1980.

Steel, R. G. D., and Torrie, J. H. *Principles and Procedures of Statistics*, 2nd ed. New York: McGraw-Hill, 1980.

USING TECHNOLOGY

MINITAB: Two-Sample Inferences

MINITAB can be used to make two-sample inferences about $\mu_1 - \mu_2$ or independent samples, μ_d for paired samples, $p_1 - p_2$ and σ_1^2/σ_2^2.

Comparing Means with Independent Samples

Step 1 Access the MINITAB worksheet that contains the sample data.

Step 2 Click on the "Stat" button on the MINITAB menu bar and then click on "Basic Statistics" and "2-Sample t," as shown in Figure 9.M.1. The resulting dialog box appears as shown in Figure 9.M.2.

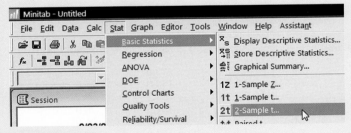

Figure 9.M.1
MINITAB menu options for comparing two means

Step 3a If the worksheet contains data for one quantitative variable (which the means will be computed on) and one qualitative variable (which represents the two groups or populations), select "Samples in one column" and then specify the quantitative variable in the "Samples" area and the qualitative variable in the "Subscripts" area. (See Figure 9.M.2.)

Step 3b If the worksheet contains the data for the first sample in one column and the data for the second sample in another column, select "Samples in different columns" and then specify the "First" and "Second" variables. Alternatively, if you have only summarized data (i.e., sample sizes, sample means, and sample standard deviations), select "Summarized data" and enter these summarized values in the appropriate boxes.

Step 4 Click the "Options" button on the MINITAB "2-Sample T" dialog box. Specify the confidence level for a confidence

Figure 9.M.2
MINITAB 2-sample *t* dialog box

interval, the null-hypothesized value of the difference, $\mu_1 - \mu_2$, and the form of the alternative hypothesis (lower tailed, two tailed, or upper tailed) in the resulting dialog box, as shown in Figure 9.M.3.

Figure 9.M.3
MINITAB options dialog box

Step 5 Click "OK" to return to the "2-Sample T" dialog box and then click "OK" again to generate the MINITAB printout.

Important Note: The MINITAB two-sample *t*-procedure uses the *t*-statistic to conduct the test of hypothesis. When the sample sizes are small, check the "Assume equal variances" box in Figure 9.M.2. When the sample sizes are large, leave the "Assume equal variances" box unchecked; the *t*-value will be approximately equal to the large-sample *z*-value, and the resulting test will still be valid.

Comparing Means with Paired Samples

Step 1 Access the MINITAB worksheet that contains the sample data. The data file should contain two quantitative variables—one with the data values for the first group (or population) and one with the data values for the second group. (*Note:* The sample size should be the same for each group.)

Step 2 Click on the "Stat" button on the MINITAB menu bar and then click on "Basic Statistics" and "Paired *t*" (see Figure 9.M.1).

Step 3 On the resulting dialog box, select the "Samples in columns" option and specify the two quantitative variables of interest in the "First sample" and "Second sample" boxes, as shown in Figure 9.M.4. [Alternatively, if you have only summarized data of the paired differences, select the "Summarized data (differences)" option and enter the sample size, sample mean, and sample standard deviation in the appropriate boxes.]

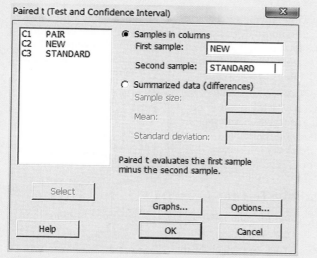

Figure 9.M.4
MINITAB paired-samples *t* dialog box

Step 4 Click the "Options" button and specify the confidence level for a confidence interval, the null-hypothesized value of the difference μ_d, and the form of the alternative hypothesis (lower tailed, two tailed, or upper tailed) in the resulting dialog box. (See Figure 9.M.3.)

Step 5 Click "OK" to return to the "Paired *t*" dialog box and then click "OK" again to generate the MINITAB printout.

Comparing Proportions with Large Independent Samples

Step 1 Access the MINITAB worksheet that contains the sample data.

Step 2 Click on the "Stat" button on the MINITAB menu bar and then click on "Basic Statistics" and "2 Proportions," as shown in Figure 9.M.1.

Step 3 On the resulting dialog box (shown in Figure 9.M.5), select the data option ("Samples in different columns" or "Summarized data") and make the appropriate menu choices. (Figure 9.M.5 shows the menu options when you select "Summarized data.")

Figure 9.M.5
MINITAB 2 proportions dialog box

Step 4 Click the "Options" button and specify the confidence level for a confidence interval, the null-hypothesized value of the difference, and the form of the alternative hypothesis (lower tailed, two tailed, or upper tailed) in the resulting dialog box, as shown in Figure 9.M.6. (If you desire a pooled estimate of *p* for the test, be sure to check the appropriate box.)

Figure 9.M.6
MINITAB 2 proportions options

Step 5 Click "OK" to return to the "2 Proportions" dialog box and then click "OK" again to generate the MINITAB printout.

Comparing Variances with Independent Samples

Step 1 Access the MINITAB worksheet that contains the sample data.

Step 2 Click on the "Stat" buton on the MINITAB menu bar and then click on "Basic Statistics" and "2 Variances" (Figure 9.M.1).

Step 3 On the resulting dialog box (shown in Figure 9.M.7), the menu selections and options are similar to those for the two-sample *t*-test.

2 Variances (Test and Confidence Interval)	
C1 HWCOND	Data: Samples in one column
C2 SCIENCE	
C3 MATH	Samples: SCIENCE
C4 LANGUAGE	
C5 INV-SCI	Subscripts: HWCOND
C6 INV-MATH	
C7 INV-LANG	
Select	Graphs... Options...
Help	OK Cancel

Figure 9.M.7
MINITAB 2 variances dialog box

Step 4 Click "OK" to produce the MINITAB *F*-test printout.

TI-83/TI-84 Plus Graphing Calculator: Two Sample Inferences

The TI-83/ TI-84 plus graphing calculator can be used to conduct tests and form confidence intervals for the difference between two means with independent samples, the difference between two means with matched pairs, the difference between two proportions for large independent samples, and the ratio of two variances.

Confidence Interval for $\mu_1 - \mu_2$

Step 1 *Enter the data (Skip to Step 2 if you have summary statistics, not raw data)*

- Press **STAT** and select **1:Edit**

Note: If the lists already contain data, clear the old data. Use the up **ARROW** to highlight "**L1**."

- Press **CLEAR ENTER**
- Use the up **ARROW** to highlight "**L2**"
- Press **CLEAR ENTER**
- Use the **ARROW** and **ENTER** keys to enter the first data set into **L1**
- Use the **ARROW** and **ENTER** keys to enter the second data set into **L2**

Step 2 *Access the statistical tests menu*

- Press **STAT**
- Arrow right to **TESTS**
- Arrow down to **2-SampTInt**
- Press **ENTER**

Step 3 *Choose "Data" or "Stats" ("Data" is selected when you have entered the raw data into the Lists. "Stats" is selected when you are given only the means, standard deviations, and sample sizes)*

- Press **ENTER**
- If you selected "Data," set **List1** to **L1** and **List2** to **L2**

- Set **Freq1** to **1** and set **Freq2** to **1**
- Set **C-Level** to the confidence level
- If you are assuming that the two populations have equal variances, select **Yes** for **Pooled**

- If you are not assuming equal variances, select **No**
- Press **ENTER**
- Arrow down to "**Calculate**"
- Press **ENTER**
- If you selected "Stats," enter the means, standard deviations, and sample sizes
- Set **C-Level** to the confidence level
- If you are assuming that the two populations have equal variances, select **Yes** for **Pooled**
- If you are not assuming equal variances, select **No**
- Press **ENTER**
- Arrow down to "**Calculate**"
- Press **ENTER**

(The accompanying screen is set up for an example with a mean of 100, a standard deviation of 10, and a sample size of 15 for the first data set and a mean of 105, a standard deviation of 12, and a sample size of 18 for the second data set.)

The confidence interval will be displayed with the number of degrees of freedom, the sample statistics, and the pooled standard deviation (when appropriate).

Hypothesis Test for $\mu_1 - \mu_2$

Step 1 *Enter the data (Skip to Step 2 if you have summary statistics, not raw data)*

- Press **STAT** and select **1:Edit**

Note: If the lists already contain data, clear the old data. Use the up **ARROW** to highlight "**L1**."

- Press **CLEAR ENTER**
- Use the up **ARROW** to highlight "**L2**"
- Press **CLEAR ENTER**
- Use the **ARROW** and **ENTER** keys to enter the first data set into **L1**
- Use the **ARROW** and **ENTER** keys to enter the second data set into **L2**

Step 2 *Access the statistical tests menu*

- Press **STAT**
- Arrow right to **TESTS**
- Arrow down to **2-SampTTest**
- Press **ENTER**

Step 3 *Choose "Data" or "Stats" ("Data" is selected when you have entered the raw data into the Lists. "Stats" is selected when* *you are given only the means, standard deviations, and sample sizes)*

- Press **ENTER**
- If you selected "Data," set **List1** to **L1** and **List2** to **L2**
- Set **Freq1** to **1** and set **Freq2** to **1**
- Use the **ARROW** to highlight the appropriate alternative hypothesis
- Press **ENTER**
- If you are assuming that the two populations have equal variances, select **Yes** for **Pooled**
- If you are not assuming equal variances, select **No**
- Press **ENTER**
- Arrow down to "**Calculate**"
- Press **ENTER**
- If you selected "**Stats**," enter the means, standard deviations, and sample sizes

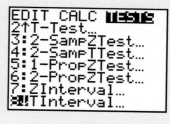

- Use the **ARROW** to highlight the appropriate alternative hypothesis
- Press **ENTER**
- If you are assuming that the two populations have equal variances, select **Yes** for **Pooled**
- If you are not assuming equal variances, select **No**
- Press **ENTER**
- Arrow down to "**Calculate**"
- Press **ENTER**

(The screen that follows is set up for an example with a mean of 100, a standard deviation of 10, and a sample size of 15 for the first data set and a mean of 120, a standard deviation of 12, and a sample size of 18 for the second data set.)

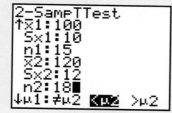

The results of the hypothesis test will be displayed with the p-value, the number of degrees of freedom, the sample statistics, and the pooled standard deviation (when appropriate).

Confidence Interval for a Paired Difference Mean

Note: There is no paired difference option on the calculator. These instructions demonstrate how to calculate the differences and then use the 1-sample *t*-interval.

Step 1 *Enter the data and calculate the differences*

- Press **STAT** and select **1:Edit**

Note: If the lists already contain data, clear the old data. Use the up **ARROW** to highlight "**L1**."

- Press **CLEAR ENTER**
- Use the up **ARROW** to highlight "**L2**"
- Press **CLEAR ENTER**
- Use the **ARROW** and **ENTER** keys to enter the first data set into **L1**

- Use the **ARROW** and **ENTER** keys to enter the second data set into **L2**
- The differences will be calculated in **L3**
- Use the up **ARROW** to highlight "**L3**"
- Press **CLEAR**—This will clear any old data, but **L3** will remain highlighted
- To enter the equation L3 = L1 − L2, use the following keystrokes:
- Press **2**[ND] "**1**" (this will enter L1)
- Press the **MINUS** button
- Press **2**[ND] "**2**" (this will enter L2)

(Notice the equation at the bottom of the screen.)

- Press **ENTER** (the differences should be calculated in **L3**)

Step 2 *Access the statistical tests menu*

- Press **STAT**
- Arrow right to **TESTS**
- Arrow down to **TInterval (even for large sample case)**
- Press **ENTER**

Step 3 *Choose "Data"*

- Press **ENTER**
- Set **List** to **L3**
- Set **Freq** to **1**
- Set **C-Level** to the confidence level
- Arrow down to "**Calculate**"
- Press **ENTER**

The confidence interval will be displayed with the mean, standard deviation, and sample size of the differences.

Hypothesis Test for a Paired Difference Mean

Note: There is no paired difference option on the calculator. These instructions demonstrate how to calculate the differences and then use the 1-sample *t*-test.

Step 1 *Enter the data and calculate the differences*

- Press **STAT** and select **1:Edit**

Note: If the lists already contain data, clear the old data. Use the up **ARROW** to highlight "**L1**."

- Press **CLEAR ENTER**
- Use the up **ARROW** to highlight "**L2**"
- Press **CLEAR ENTER**
- Use the **ARROW** and **ENTER** keys to enter the first data set into **L1**
- Use the **ARROW** and **ENTER** keys to enter the second data set into **L2**

- The differences will be calculated in **L3**
- Use the up **ARROW** to highlight "**L3**"
- Press **CLEAR**—This will clear any old data, but **L3** will remain highlighted
- To enter the equation L3 = L1 − L2, use the following keystrokes:
- Press **2ᴺᴰ "1"** (this will enter L1)
- Press the **MINUS** button
- Press **2ᴺᴰ "2"** (this will enter L2)

(Notice the equation at the bottom of the screen.)

- Press **ENTER** (the differences should be calculated in **L3**)

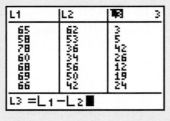

Step 2 *Access the statistical tests menu*

- Press **STAT**
- Arrow right to **TESTS**
- Arrow down to **T-Test (even for a large-sample case)**
- Press **ENTER**

Step 3 *Choose "Data"*

- Press **ENTER**
- Enter the values for the hypothesis test, where μ_0 = the value for μ_d in the null hypothesis
- Set **List** to **L3**
- Set **Freq** to **1**
- Use the **ARROW** to highlight the appropriate alternative hypothesis
- Press **ENTER**
- Arrow down to "**Calculate**"
- Press ENTER

The test statistic and the *p*-value will be displayed, as will the sample mean, standard deviation, and sample size of the differences.

Confidence Interval for $(p_1 - p_2)$

Step 1 *Access the statistical tests menu*

- Press **STAT**
- Arrow right to **TESTS**
- Arrow down to **2-PropZInt**
- Press **ENTER**

Step 2 *Enter the values from the sample information and the confidence level*

where x_1 = number of successes in the first sample (e.g., 53)

n_1 = sample size for the first sample (e.g., 400)

x_2 = number of successes in the second sample (e.g., 78)

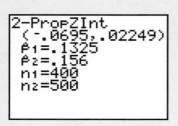

n_2 = sample size for the second sample (e.g., 500)

- Set **C-Level** to the confidence level
- Arrow down to "**Calculate**"
- Press **ENTER**

Hypothesis Test for $(p_1 - p_2)$

Step 1 *Access the statistical tests menu*

- Press **STAT**
- Arrow right to **TESTS**
- Arrow down to **2-PropZTest**
- Press **ENTER**

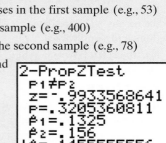

Step 2 *Enter the values from the sample information and select the alternative hypothesis*

where x_1 = number of successes in the first sample (e.g., 53)

n_1 = sample size for the first sample (e.g., 400)

x_2 = number of successes in the second sample (e.g., 78)

n_2 = sample size for the second sample (e.g., 500)

- Use the **ARROW** to highlight the appropriate alternative hypothesis
- Press **ENTER**
- Arrow down to "**Calculate**"
- Press **ENTER**

Hypothesis Test for (σ_1^2/σ_2^2)

Step 1 *Enter the data (Skip to Step 2 if you have summary statistics, not raw data)*

- Press **STAT** and select **1:Edit**

Note: If the lists already contain data, clear the old data. Use the up **ARROW** to highlight "**L1**."

- Press **CLEAR ENTER**
- Use the up **ARROW** to highlight "**L2**"
- Press **CLEAR ENTER**
- Use the **ARROW** and **ENTER** keys to enter the first data set into **L1**
- Use the **ARROW** and **ENTER** keys to enter the second data set into **L2**

Step 2 *Access the statistical tests menu*

- Press **STAT**
- Arrow right to **TESTS**
- Arrow down to **2-SampFTest**
- Press **ENTER**

Step 3 *Choose "Data" or "Stats"* ("Data" is selected when you have entered the raw data into the Lists. "Stats" is selected when you are given only the means, standard deviations, and sample sizes)

- Press **ENTER**
- If you selected "Data"
- Set **List1** to **L1** and **List2** to **L2**
- Set **Freq1** to **1** and set **Freq2** to **1**
- Use the **ARROW** to highlight the appropriate alternative hypothesis
- Press **ENTER**
- Arrow down to "**Calculate**"
- Press **ENTER**

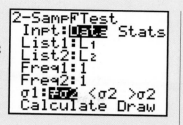

- If you selected "Stats," enter the standard deviations and sample sizes
- Use the **ARROW** to highlight the appropriate alternative hypothesis
- Press **ENTER**
- Arrow down to "**Calculate**"
- Press **ENTER**

The results of the hypothesis test will be displayed with the p-value and the input data used.

10 Analysis of Variance

Comparing More than Two Means

CONTENTS

Where We've Been

- Presented methods for estimating and testing hypotheses about a single population mean
- Presented methods for comparing two population means

Where We're Going

- Discuss the critical elements in the *design* of a sampling experiment (10.1)
- Learn how to set up three of the more popular experimental designs for comparing more than two population means: *completely randomized, randomized block,* and *factorial designs* (10.2, 10.4–10.5)
- Show how to analyze data collected from a designed experiment using a technique called an *analysis of variance* (10.2, 10.4–10.5)
- Present a follow-up analysis to an ANOVA: ranking means (10.3)

On the Trail of the Cockroach:
Do Roaches Travel at Random?

Entomologists have long established that insects such as ants, bees, caterpillars, and termites use chemical or "odor" trails for navigation. These trails are used as highways between sources of food and the insects' nest. Until recently, however, "bug" researchers believed that the navigational behavior of cockroaches scavenging for food was random and not linked to a chemical trail.

One of the first researchers to challenge the "random-walk" theory about cockroaches was professor and entomologist Dini Miller of Virginia Tech University. According to Miller, "The idea that roaches forage randomly means that they would have to come out of their hiding places every night and bump into food and water by accident. But roaches never seem to go hungry." Since cockroaches had never before been evaluated for trail-following behavior, Miller designed an experiment to test a cockroach's ability to follow a trail of their fecal material (*Journal of Economic Entomology*, Aug. 2000).

First, Miller developed a methanol extract from roach feces—called a pheromone. She theorized that "pheromones are communication devices between cockroaches. If you have an infestation and have a lot of fecal material around, it advertises, 'Hey, this is a good cockroach place.'" Then she created a chemical trail with the pheromone on a strip of white chromatography paper and placed the paper at the bottom of a plastic, V-shaped container, 122 square centimeters in area. German cockroaches were released into the container at the beginning of the trail, one at a time, and a video surveillance camera was used to monitor the roaches' movements.

In addition to the trail containing the fecal extract (the treatment), a trail using methanol only was created. This second trail served as a control against which the treated trail could be compared. Because Miller also wanted to determine whether trail-following ability differed among cockroaches of different age, sex, and reproductive status, four roach groups were utilized in the experiment: adult males, adult females, gravid (pregnant) females, and nymphs (immatures). Twenty roaches of each type were randomly assigned to the treatment trail, and 10 of each type were randomly assigned to the control trail. Thus, a total of 120 roaches were used in the experiment.

The movement pattern of each cockroach tested was translated into *xy*-coordinates every one-tenth of a second by the Dynamic Animal Movement Analyzer (DAMA) program. Miller measured the perpendicular distance of each *xy*-coordinate from the trail and then averaged all the distances, or deviations, for each cockroach. The average trail deviations (measured in pixels, where 1 pixel equals approximately 2 centimeters) for each of the 120 cockroaches in the study are stored in the data file named **ROACH**.

We apply the statistical methodology presented in this chapter to the cockroach data in several Statistics in Action Revisited sections.

Data Set: ROACH

Statistics IN Action Revisited

- A One-Way Analysis of the Cockroach Data (p. 491)
- Ranking the Means of the Cockroach Groups (p. 501)
- A Two-Way Analysis of the Cockroach Data (p. 530)

Most of the data analyzed in previous chapters were collected in *observational* sampling experiments rather than *designed* sampling experiments. In *observational experiments*, the analyst has little or no control over the variables under study and merely observes their values. In contrast, in *designed experiments* the analyst attempts to control the levels of one or more variables to determine their effect on a variable; of interest. When properly designed, such experiments allow the analyst to determine whether a change in the controlled variable *causes* a change in the response variable; that is, they allow the analyst to infer *cause and effect*. Although many practical situations do not present the opportunity for such control, it is instructive, even with observational experiments, to have a working knowledge of the analysis and interpretation of data that result from designed experiments and to know the basics of how to design experiments when the opportunity arises.

We first present the basic elements of an experimental design in Section 10.1. We then discuss two of the simpler, and more popular, experimental designs in Sections 10.2 and 10.4. Slightly more complex experiments are discussed in Section 10.5. Methods for ranking means from a designed experiment are presented in Section 10.3.

10.1 Elements of a Designed Study

Certain elements are common to almost all designed experiments, regardless of the specific area of application. For example, the *response* is the variable of interest in the experiment. The response might be the SAT scores of a high school senior, the total sales of a firm last year, or the total income of a particular household this year. We will also refer to the response as the *dependent variable*.

> The **response variable** is the variable of interest to be measured in the experiment. We also refer to the response as the **dependent variable**.

The intent of most statistical experiments is to determine the effect of one or more variables on the response. These variables are usually referred to as the *factors* in a designed experiment. Factors are either *quantitative* or *qualitative*, depending on whether the variable is measured on a numerical or a nonnumerical scale. For example, we might want to explore the effect of the qualitative factor Gender on the response SAT score. In other words, we might want to compare the SAT scores of male and female high school seniors. Or we might wish to determine the effect of the quantitative factor Number of salespeople on the response Total sales for retail firms. Often, two or more factors are of interest. For example, we might want to determine the effect of the quantitative factor Number of wage earners and the qualitative factor Location on the response Household income.

> **Factors** are those variables whose effect on the response is of interest to the experimenter. **Quantitative factors** are measured on a numerical scale, whereas **qualitative factors** are not (naturally) measured on a numerical scale.

Levels are the values of the factors that are utilized in the experiment. The levels of qualitative factors are usually nonnumerical. For example, the levels of Gender are Male and Female, and the levels of Location might be North, East, South, and West.* The levels of quantitative factors are the numerical values of the variable utilized in the experiment. The Number of salespeople in each of a set of companies, the Number of wage earners in each of a set of households, and the GPAs for a set of high school seniors all represent levels of the respective quantitative factors.

> **Factor levels** are the values of the factor utilized in the experiment.

When a *single factor* is employed in an experiment, the *treatments* of the experiment are the levels of the factor. For example, if the effect of the factor Gender on the response SAT score is being investigated, the treatments of the experiment are the two levels of Gender: Female and Male. Or if the effect of the Number of wage earners on Household income is the subject of the experiment, the numerical values assumed by the quantitative factor Number of wage earners are the treatments. If *two or more factors* are utilized in an experiment, the treatments are the factor–level combinations used. For example, if the effects of the factors Gender and Socioeconomic status (SES) on the response SAT score are being investigated, the treatments are the combinations of the levels of Gender and SES used; thus, (Female, high SES) and (Male, low SES) would be treatments.

> The **treatments** of an experiment are the factor–level combinations utilized.

*The levels of a qualitative variable may bear numerical labels. For example, Locations could be numbered 1, 2, 3, and 4. However, in such cases the numerical labels for a qualitative variable will usually be codes representing nonnumerical levels.

The objects on which the response variable and factors are observed are the *experimental units*. For example, SAT score, High school GPA and Gender are all variables that can be observed on the same experimental unit: a high school senior. Similarly, Total sales, Earnings per share, and Number of salespeople can be measured on a particular firm in a particular year, and the firm–year combination is the experimental unit. Likewise, Total income, Number of female wage earners, and Location can be observed for a household at a particular point in time, and the household–time combination is the experimental unit. Every experiment, whether observational or designed, has experimental units on which the variables are observed. However, the identification of the experimental units is more important in designed experiments, when the experimenter must actually sample the experimental units and measure the variables.

> An **experimental unit** is the object on which the response and factors are observed or measured.*

When the specification of the treatments and the method of assigning the experimental units to each of the treatments are controlled by the analyst, the study is said to be *designed*. In contrast, if the analyst is just an observer of the treatments on a sample of experimental units, the study is *observational*. For example, if, on the one hand, you specify the number of female and male high school students within each GPA range to be randomly selected in order to evaluate the effect of gender and GPA on SAT scores, you are designing the experiment. If, on the other hand, you simply observe the SAT scores, gender, and GPA for all students who took the SAT test last month at a particular high school, the study is observational.

> A **designed study** is an experiment in which the analyst controls the specification of the treatments and the method of assigning the experimental units to each treatment. An **observational study** is an experiment in which the analyst simply observes the treatments and the response on a sample of experimental units.

Figure 10.1 provides an overview of the experimental process and a summary of the terminology introduced in this section. Note that the experimental unit is at the core of the process. The method by which the sample of experimental units is selected from the population determines the type of experiment. The level of every factor (the treatment) and the response are all variables that are observed or measured on each experimental unit.

BIOGRAPHY SIR RONALD A. FISHER (1890–1962)

The Founder of Modern Statistics

At a young age, Ronald Fisher demonstrated special abilities in mathematics, astronomy, and biology. (Fisher's biology teacher once divided all his students into two groups on the basis of their "sheer brilliance": Fisher and the rest.) Fisher graduated from prestigious Cambridge University in London in 1912 with a B.A. degree in astronomy. After several years teaching mathematics, he found work at the Rothamsted Agricultural Experiment station, where he began his extraordinary career as a statistician. Many consider Fisher to be the leading founder of modern statistics. His contributions to the field include the notion of unbiased statistics, the development of *p*-values for hypothesis tests, the invention of analysis of variance for designed experiments, the maximum-likelihood estimation theory, and the formulation of the mathematical distributions of several well-known statistics. Fisher's book, *Statistical Methods for Research Workers* (written in 1925), revolutionized applied statistics, demonstrating with very readable and practical examples how to analyze data and interpret the results. In 1935, Fisher wrote *The Design of Experiments*, in which he first described his famous experiment on the "lady tasting tea." (Fisher showed, through a designed experiment, that the lady really could determine whether tea poured into milk tastes better than milk poured into tea.) Before his death, Fisher was elected a Fellow of the Royal Statistical Society, was awarded numerous medals, and was knighted by the Queen of England. ■

*Recall (Chapter 1) that the set of all experimental units is the population.

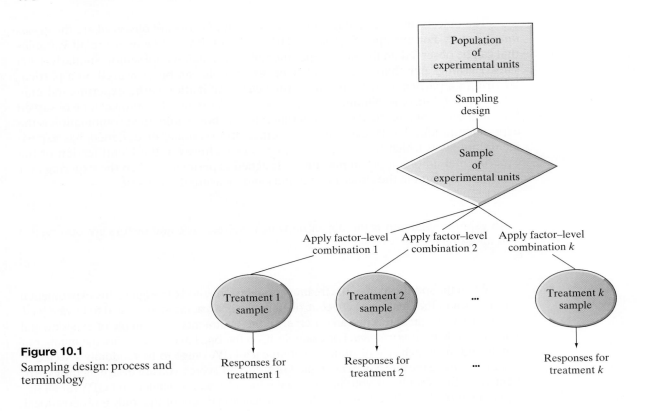

Figure 10.1
Sampling design: process and terminology

Example 10.1
The Key Elements of a Designed Experiment— Testing Golf Ball Brands

Problem The United States Golf Association (USGA) regularly tests golf equipment to ensure that it conforms to the association's standards. Suppose the USGA wishes to compare the mean distances traveled by four different brands of golf balls struck by a driver (the club used to maximize distance). The following experiment is conducted: 10 balls of each brand are randomly selected. Each ball is struck with a driver by "Iron Byron" (the USGA's golf robot named for the famous golfer Byron Nelson), and the distance traveled is recorded. Identify each of the following elements in this study: response, factors, types of factor, levels, treatments, and experimental units.

Solution The response is the variable of interest, Distance traveled. The only factor being investigated is Brand of golf ball, and it is nonnumerical and therefore qualitative. The four brands (say, A, B, C, and D) represent the levels of this factor. Since only one factor is utilized, the treatments are the four levels of that factor—that is, the four brands. The experimental unit is a golf ball; more specifically, it is a golf ball at a particular position in the striking sequence, since the distance traveled can be recorded only when the ball is struck, and we would expect the distance to be different (due to random factors such as wind resistance, landing place, and so forth) if the same ball is struck a second time. Note that 10 experimental units are sampled for each treatment, generating a total of 40 observations.

Look Back This study, like many real applications, is a blend of designed and observational: The analyst cannot control the assignment of the brand to each golf ball (observational), but he or she can control the assignment of each ball to the position in the striking sequence (designed).

Now Work Exercise 10.5

Example 10.2
A Two-Factor Experiment—Testing Golf Ball Brands

Problem Suppose the USGA is also interested in comparing the mean distances the four brands of golf balls travel when struck by a five-iron and by a driver. Ten balls of each brand are randomly selected, five to be struck by the driver and five by the five-iron. Identify the elements of this experiment, and construct a schematic diagram similar to Figure 10.1 to provide an overview of the study.

Solution The response is the same as in Example 10.1, Distance traveled. The experiment now has two factors: Brand of golf ball and Club utilized. There are four levels of Brand (A, B, C, and D) and two of Club (driver and five-iron, or 1 and 5). Treatments are factor–level combinations, so there are $4 \times 2 = 8$ treatments in this experiment: (A, 1), (A, 5), (B, 1), (B, 5), (C, 1), (C, 5), (D, 1), and (D, 5). The experimental units are still the combinations of golf ball and hitting position. Note that five experimental units are sampled per treatment, generating 40 observations. The study is summarized in Figure 10.2.

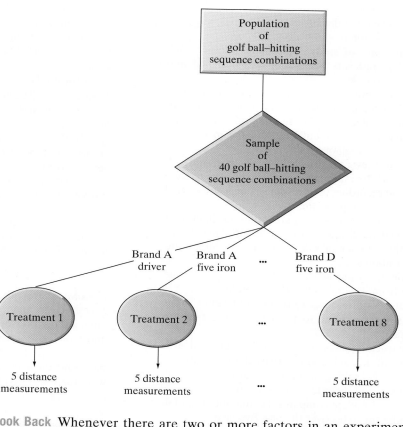

Figure 10.2
Two-factor golf study:
Example 10.2

Look Back Whenever there are two or more factors in an experiment, remember to combine the levels of the factors—one from each factor—to obtain the treatments.

Now Work Exercise 10.12

Our objective in designing a study is usually to maximize the amount of information obtained about the relationship between the treatments and the response. Of course, we are almost always subject to constraints on budget, time, and even the availability of studies units. Nevertheless, designed studies are generally preferred to observational experiments: Not only do we have better control over the amount and quality of the information collected, but we also avoid the biases that are inherent in observational studies in the selection of the experimental units representing each treatment. Inferences based on observational studies always carry the implicit assumption that the sample has no hidden bias which was not considered in the statistical analysis. A better understanding of the potential problems with observational studies is a by-product of our study of experimental design in the remainder of this chapter.

Exercises 10.1–10.14

Understanding the Principles

10.1 What are the treatments for a designed study that utilizes one qualitative factor with four levels: A, B, C, and D?

10.2 What are the treatments for a designed study with two factors, one qualitative with two levels (A and B) and one quantitative with five levels (50, 60, 70, 80, and 90)?

10.3 What is the difference between an observational and a designed study?

Applying the Concepts—Basic

10.4 What are the experimental units on which each of the following responses are observed?
  a. GPA of a college student
  b. Household income
  c. Your time in running the 100-yard dash
  d. A patient's reaction to a new drug

10.5 **Identifying the type of study.** Determine whether each
[NW] of the following studies is observational or designed, and explain your reasoning:
  a. An economist obtains the unemployment rate and gross state product for a sample of states over the past 10 years, with the objective of examining the relationship between the unemployment rate and the gross state product by census region.
  b. A psychologist tests the effects of three different feedback programs by randomly assigning five rats to each program and recording their response times at specified intervals during the program.
  c. A marketer of notebook computers runs ads in each of four national publications for one quarter and keeps track of the number of sales that are attributable to each publication's ad.
  d. An electric utility engages a consultant to monitor the discharge from its smokestack on a monthly basis over a one-year period in order to relate the level of sulfur dioxide in the discharge to the load on the facility's generators.
  e. Intrastate trucking rates are compared before and after governmental deregulation of prices charged, with the comparison also taking into account distance of haul, goods hauled, and the price of diesel fuel.
  f. An agriculture student compares the amount of rainfall in four different states over the past five years.

10.6 **Health risks to beachgoers.** According to a University of Florida veterinary researcher, the longer a beachgoer sits in wet sand or stays in the water, the higher the risk of gastroenteritis (*University of Florida News*, Jan. 29, 2008). The result is based on a study of more than 1,000 adults conducted at three popular Florida beaches. The adults were divided into three groups: (1) beachgoers who were recently exposed to wet sand and water for at least two consecutive hours, (2) beachgoers who were not recently exposed to wet sand and water, and (3) people who had not recently visited a beach. Suppose the researcher wants to compare the mean levels of intestinal bacteria for the three groups. For this study, identify each of the following:
  a. experimental unit
  b. response variable
  c. factor
  d. factor levels

10.7 **Treating depression with a combination of drugs.** Physicians are now experimenting with using a combination of drugs, rather than a single drug, to treat major depression. In a study published in the *American Journal of Psychiatry* (March 2010), a sample of 105 patients diagnosed with a major depressive disorder were randomly assigned to one of four drug combination groups. Group 1 received daily doses of the antidepressant drug fluoxetine and a placebo; group 2 received the antidepressant drug mirtazapine plus fluoxetine; group 3 received mirtazapine plus veniafaxine; and group 4 received mirtazapine plus bupropion. After one month, the score on the Hamilton Depression Rating Scale (HAM-D) was determined for each patient. The researchers compared the mean HAM-D scores of the four groups of depressed patients. For this study, identify each of the following:
  a. experimental unit
  b. response variable
  c. factor
  d. factor levels

10.8 **Extinct New Zealand birds.** Refer to the *Evolutionary Ecology Research* (July 2003) study of extinction in the New Zealand bird population, presented in Exercise 1.18 (p. 19). Recall that biologists measured the body mass (in grams) and type of habitat (aquatic, ground terrestrial, or aerial terrestrial) for each bird species. One objective is to compare the body mass means of birds with the three different types of habitat.
  a. Identify the response variable of the study.
  b. Identify the experimental units of the study.
  c. Identify the factor(s) in the study.
  d. Identify the treatments.

Applying the Concepts—Intermediate

10.9 **Back/knee strength, gender, and lifting strategy.** *Human Factors* (December 2009) investigated whether back and knee strength dictates the load-lifting strategies of males and females. A sample of 32 healthy adults (16 men and 16 women) participated in a series of strength tests on the back and the knees. Following the tests, the participants were randomly divided into two groups, where each group consisted of 8 men and 8 women. One group was provided with knowledge of their strength test results, while the other group was not provided with this knowledge. The final phase of the study required the participants to lift heavy cast iron plates out of a bin. Based on the different angles used to lift the plates, a quantitative measure of posture—called a postural index—was measured for each participant. The goal of the research was to determine the effect of gender and strength knowledge (provided or not provided) on the mean postural index. For this study, identify each of the following:
  a. experimental unit
  b. response variable
  c. factors
  d. levels of each factor
  e. treatments

10.10 **Treatment for tendon pain.** Chronic Achilles tendon pain (i.e., tendinosis) is common among middle-aged recreational athletes. A group of Swedish physicians investigated the use of heavy-load eccentric calf muscle training to treat Achilles tendinosis (*British Journal of Sports Medicine*, Feb. 1, 2004). A sample of 25 patients with chronic Achilles tendinosis undertook the treatment. Data on tendon thickness (measured in millimeters) were collected by ultrasonography both before and following treatment of each patient. The researchers want to compare the mean tendon thickness before treatment with the mean tendon thickness after treatment.

a. Is this a designed experiment or an observational study? Explain.

b. What is the experimental unit of the study?

c. What is the response variable of the study?

d. What are the treatments in this experiment?

e. After reading Section 10.3, you will learn that patients represent a blocking factor in this study. How many levels are in the blocking factor?

10.11 Taste preferences of cockatiels. *Applied Animal Behaviour Science* (Oct. 2000) published a study of the taste preferences of caged cockatiels. A sample of birds bred at the University of California at Davis was randomly divided into three experimental groups. Group 1 was fed purified water in bottles on both sides of the cage. Group 2 was fed water on one side and a liquid sucrose (sweet) mixture on the opposite side of the cage. Group 3 was fed water on one side and a liquid sodium chloride (salty) mixture on the opposite side of the cage. One variable of interest to the researchers was total consumption of liquid by each cockatiel.

a. What is the experimental unit of this study?

b. Is the study a designed experiment? Why?

c. What are the factors in the study?

d. Give the levels of each factor.

e. How many treatments are in the study? Identify them.

f. What is the response variable?

10.12 Exam performance study. In *Teaching of Psychology*
[NW] (Aug. 1998), a study investigated whether final exam performance is affected by whether or not students take a practice test. Students in an introductory psychology class at Pennsylvania State University were initially divided into three groups based on their class standing: Low, Medium, and High. Within each group, the students were randomly assigned to either attend a review session or take a practice test prior to the final exam. Thus, six groups were formed: (Low, Review), (Low, Practice exam), (Medium, Review), (Medium, Practice exam), (High, Review), and (High, Practice exam). One goal of the study was to compare the mean final exam scores of the six groups of students.

a. What is the experimental unit of this study?

b. Is the study a designed experiment? Why?

c. What are the factors in the study?

d. Give the levels of each factor.

e. How many treatments are in the study? Identify them.

f. What is the response variable?

10.13 Baker's versus brewer's yeast. The *Electronic Journal of Biotechnology* (Dec. 15, 2003) published an article comparing two yeast extracts: baker's yeast and brewer's yeast. Brewer's yeast is a surplus by-product obtained from a brewery; hence, it is less expensive than primary-grown baker's yeast. Samples of both yeast extracts were prepared at four different temperatures (45, 48, 51, and 54°C), and the autolysis yield (recorded as a percentage) was measured for each of the yeast–temperature combinations. The goal of the analysis is to investigate the impact of yeast extract and temperature on mean autolysis yield.

a. Identify the factors (and factor levels) in the experiment.

b. Identify the response variable.

c. How many treatments are included in the experiment?

d. What type of experimental design is employed?

Applying the Concepts—Advanced

10.14 Testing a new pain-relief tablet. Paracetamol is the active ingredient in drugs designed to relieve mild to moderate pain and fever. The properties of paracetamol tablets derived from khaya gum were studied in the *Tropical Journal of Pharmaceutical Research* (June 2003). Three factors believed to affect the properties of parcetamol tablets are (1) the nature of the binding agent, (2) the concentration of the binding agent, and (3) the relative density of the tablet. In the experiment, binding agent was set at two levels (khaya gum and PVP), binding concentration at two levels (.5% and 4.0%), and relative density at two levels (low and high). One of the dependent variables investigated in the study was tablet dissolution time (i.e., the amount of time, in minutes, for 50% of the tablet to dissolve). The goal of the study was to determine the effect of binding agent, binding concentration, and relative density on mean dissolution time.

a. Identify the dependent (response) variable in the study.

b. What are the factors investigated in the study? Give the levels of each.

c. How many treatments are possible in the study? List them.

10.2 The Completely Randomized Design: Single Factor

The simplest experimental design, the *completely randomized design*, consists of the *independent random selection* of experimental units representing each treatment. For example, in an experiment with Gender as the only factor, we could independently select random samples of 20 female and 20 male high school seniors in order to compare their mean SAT scores. Or, in an experiment with Experimental Cancer treatment as the single factor at three levels, we could randomly assign cancer patients to receive one of three treatments and then compare the mean pain levels of patients in the treatment groups. In both examples, our objective is to compare treatment means by selecting random, independent samples for each treatment.

> Consider an experiment that involves a single factor with k treatments. The **completely randomized design** is a design in which the k treatments are randomly assigned to the experimental units or in which independent random samples of experimental units are selected for each treatment.*

Example 10.3

Assigning Treatments in a Completely Randomized Design—Comparing Bottled Water Brands

Problem Suppose we want to compare the taste preferences of consumers for three different brands of bottled water (say, Brands A, B, and C), using a random sample of 15 consumers of bottled water. Set up a completely randomized design for this purpose. That is, assign the treatments to the experimental units for this design.

Solution In this study, the experimental units are the 15 consumers, the factor is brand of bottled water, and the treatments are Brands A, B, and C. One way to set up the completely randomized design is to randomly assign one of the three brands to each consumer to taste. Then we could measure (say, on a 1- to 10-point scale) the taste preference of each consumer. A good practice is to assign the same number of consumers to each brand—in this case, five consumers to each of the three brands. (When an equal number of experimental units is assigned to each treatment, we call the design a **balanced design**.)

A random-number table (Table 1, Appendix A) or computer software can be used to make the random assignments. Figure 10.3 is a MINITAB worksheet showing the random assignments made with the MINITAB "Random Data" function. You can see that MINITAB randomly assigned consumers numbered 2, 11, 1, 13, and 3 to taste Brand A; consumers numbered 15, 14, 7, 10, and 8 to taste Brand B; and consumers numbered 6, 5, 12, 9, and 4 to taste Brand C.

Look Back In some experiments, it will not be possible to randomly assign treatments to the experimental units: The units will already be associated with one of the treatments. (For example, if the treatments are "Male" and "Female," you cannot change a person's gender.) In this case, a completely randomized design is a design in which you select independent random samples of experimental units from each treatment.

CRD3brands.MTW *

↓	C1	C2	C3	C4	C5
	Consumer	BrandA	BrandB	BrandC	
1	1	2	15	6	
2	2	11	14	5	
3	3	1	7	12	
4	4	13	10	9	
5	5	3	8	4	
6	6				
7	7				
8	8				
9	9				
10	10				
11	11				
12	12				
13	13				
14	14				
15	15				
16					

Figure 10.3
MINITAB random assignments of consumers to brands

*We use *completely randomized design* to refer to both designed and observational experiments. Thus, the only requirement is that the experimental units to which treatments are applied (designed) or on which treatments are observed (observational) be independently selected for each treatment.

The objective of a completely randomized design is usually to compare the treatment means. If we denote the true, or population, means of the k treatments as $\mu_1, \mu_2, \ldots, \mu_k$, then we will test the null hypothesis that the treatment means are all equal against the alternative that at least two of the treatment means differ:

$$H_0: \mu_1 = \mu_2 = \cdots = \mu_k$$

H_a: At least two of the k treatment means differ

The μ's might represent the means of *all* female and male high school seniors' SAT scores or the means of *all* households' income in each of four census regions.

To conduct a statistical test of these hypotheses, we will use the means of the independent random samples selected from the treatment populations in a completely randomized design. That is, we compare the k sample means $\bar{x}_1, \bar{x}_2, \ldots, \bar{x}_k$.

For example, suppose you select independent random samples of five female and five male high school seniors and record their SAT scores. The data are shown in Table 10.1. A MINITAB analysis of the data, shown in Figure 10.4, reveals that the sample mean SAT scores (shaded) are 590 for females and 550 for males. Can we conclude that the population of female high school students scores 40 points higher, on average, than the population of male students?

Table 10.1

SAT Scores for High School Students

Females	Males
530	490
560	520
590	550
620	580
650	610

Data Set: TAB10_1

Figure 10.4

MINITAB descriptive statistics for data in Table 10.1

Descriptive Statistics: Females, Males

Variable	N	Mean	StDev	Variance	Minimum	Maximum
Females	5	590.0	47.4	2250.0	530.0	650.0
Males	5	550.0	47.4	2250.0	490.0	610.0

To answer this question, we must consider the amount of sampling variability among the experimental units (students). The SAT scores in Table 10.1 are depicted in the dot plot shown in Figure 10.5. Note that the difference between the sample means is small relative to the sampling variability of the scores within the treatments, namely, Female and Male. We would be inclined not to reject the null hypothesis of equal population means in this case.

Figure 10.5

Dot plot of SAT scores: difference between means dominated by sampling variability.

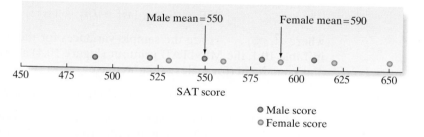

In contrast, if the data are as depicted in the dot plot of Figure 10.6, then the sampling variability is small relative to the difference between the two means. In this case, we would be inclined to favor the alternative hypothesis that the population means differ.

Figure 10.6

Dot plot of SAT scores: difference between means large relative to sampling variability

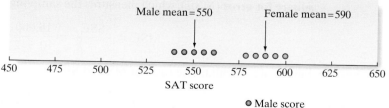

Now Work Exercise 10.21a

You can see that the key is to compare the difference between the treatment means with the amount of sampling variability. To conduct a formal statistical test of the hypothesis requires numerical measures of the difference between the treatment means and the sampling variability within each treatment. The variation between the treatment means is measured by the **sum of squares for treatments** (SST), which is calculated by squaring the distance between each treatment mean and the overall mean of *all* sample measurements, then multiplying each squared distance by the number of sample measurements for the treatment, and, finally, adding the results over all treatments. For the data in Table 10.1, the overall mean is 570. Thus, we have:

$$\text{SST} = \sum_{i=1}^{k} n_i (\bar{x}_i - \bar{x})^2 = 5(550 - 570)^2 + 5(590 - 570)^2 = 4{,}000$$

In this equation, we use $\bar{x}$ to represent the overall mean response of all sample measurements—that is, the mean of the combined samples. The symbol n_i is used to denote the sample size for the ith treatment. You can see that the value of SST is 4,000 for the two samples of five female and five male SAT scores depicted in Figures 10.5 and 10.6.

Next, we must measure the sampling variability within the treatments. We call this the **sum of squares for error** (SSE), because it measures the variability around the treatment means that is attributed to sampling error. The value of SSE is computed by summing the squared distance between each response measurement and the corresponding treatment mean and then adding the squared differences over all measurements in the entire sample:

$$\text{SSE} = \sum_{j=1}^{n_1} (x_{1j} - \bar{x}_1)^2 + \sum_{j=1}^{n_2} (x_{2j} - \bar{x}_2)^2 + \cdots \sum_{j=1}^{n_k} (x_{kj} - \bar{x}_k)^2$$

Here, the symbol x_{1j} is the jth measurement in sample 1, x_{2j} is the jth measurement in sample 2, and so on. This rather complex-looking formula can be simplified by recalling the formula for the sample variance s^2 given in Chapter 2:

$$s^2 = \sum_{i=1}^{n} \frac{(x_i - \bar{x})^2}{n - 1}$$

Note that each sum in SSE is simply the numerator of s^2 for that particular treatment. Consequently, we can rewrite SSE as

$$\text{SSE} = (n_1 - 1)s_1^2 + (n_2 - 1)s_2^2 + \cdots + (n_k - 1)s_k^2$$

where $s_1^2, s_2^2, \ldots, s_k^2$ are the sample variances for the k treatments. For the SAT scores in Table 10.1, the MINITAB printout (Figure 10.4) shows that $s_1^2 = 2{,}250$ (for females) and $s_2^2 = 2{,}250$ (for males); then we have

$$\text{SSE} = (5 - 1)(2{,}250) + (5 - 1)(2{,}250) = 18{,}000$$

To make the two measurements of variability comparable, we divide each by the number of degrees of freedom in order to convert the sums of squares to mean squares. First, the **mean square for treatments** (MST), which measures the variability *among* the treatment means, is equal to

$$\text{MST} = \frac{\text{SST}}{k - 1} = \frac{4{,}000}{2 - 1} = 4{,}000$$

where the number of degrees of freedom for the k treatments is $(k - 1)$. Next, the **mean square for error** (MSE), which measures the sampling variability *within* the treatments, is

$$\text{MSE} = \frac{\text{SSE}}{n - k} = \frac{18{,}000}{10 - 2} = 2{,}250$$

Finally, we calculate the ratio of MST to MSE—an *F*-statistic:

$$F = \frac{\text{MST}}{\text{MSE}} = \frac{4{,}000}{2{,}250} = 1.78$$

These quantities—MST, MSE, and *F*—are shown (highlighted) on the MINITAB printout displayed in Figure 10.7.

One-way ANOVA: Females, Males

Source	DF	SS	MS	F	P
Factor	1	4000	4000	1.78	0.219
Error	8	18000	2250		
Total	9	22000			

Figure 10.7

MINITAB printout with ANOVA results for data in Table 10.1

Values of the *F*-statistic near 1 indicate that the two sources of variation, between treatment means and within treatments, are approximately equal. In this case, the difference between the treatment means may well be attributable to sampling error, which provides little support for the alternative hypothesis that the population treatment means differ. Values of *F* well in excess of 1 indicate that the variation among treatment means well exceeds that within means and therefore support the alternative hypothesis that the population treatment means differ.

When does *F* exceed 1 by enough to reject the null hypothesis that the means are equal? This depends on the degrees of freedom for treatments and for error and on the value of α selected for the test. We compare the calculated *F*-value with an *F*-value taken from a table (see Tables VIII–XI of Appendix A) with $v_1 = (k - 1)$ degrees of freedom in the numerator and $v_2 = (n - k)$ degrees of freedom in the denominator and corresponding to a Type I error probability of α. For the example of the SAT scores, the *F*-statistic has $v_1 = (2 - 1)$ numerator degrees of freedom and $v_2 = (10 - 2) = 8$ denominator degrees of freedom. Thus, for $\alpha = .05$, we find (from Table IX of Appendix A) that

$$F_{.05} = 5.32$$

The implication is that MST would have to be 5.32 times greater than MSE before we could conclude, at the .05 level of significance, that the two population treatment means differ. Since the data yielded $F = 1.78$, our initial impressions of the dot plot in Figure 10.5 are confirmed: There is insufficient information to conclude that the mean SAT scores differ for the populations of female and male high school seniors. The rejection region and the calculated *F* value are shown in Figure 10.8.

Table 10.2

SAT Scores for High School Students Shown in Figure 10.6

Females	Males
580	540
585	545
590	550
595	555
600	560

Data Set: TAB10_2

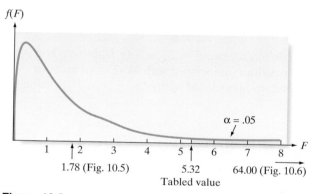

Figure 10.8

Rejection region and calculated *F*-values for SAT score samples

In contrast, consider the dot plot in Figure 10.6. The SAT scores depicted in this dot plot are listed in Table 10.2, followed by MINITAB descriptive statistics in Figure 10.9. Note that the sample means for females and males, 590 and 550, respectively, are the same as in the previous example. Consequently, the variation between the means is the same, namely, MST = 4,000. However, the variation within the two treatments appears to be considerably smaller. In fact, Figure 10.9 shows that $s_1^2 = 62.5$ and $s_2^2 = 62.5$.

Descriptive Statistics: Females, Males

Variable	N	Mean	StDev	Variance	Minimum	Maximum
Females	5	590.00	7.91	62.50	580.00	600.00
Males	5	550.00	7.91	62.50	540.00	560.00

One-way ANOVA: Females, Males

Figure 10.9

MINITAB descriptive statistics and ANOVA results for data in Table 10.2

Source	DF	SS	MS	F	P
Factor	1	4000.0	4000.0	64.00	0.000
Error	8	500.0	62.5		
Total	9	4500.0			

Thus, the variation within the treatments is measured by

$$\text{SSE} = (5 - 1)(62.5) + (5 - 1)(62.5) = 500$$

$$\text{MSE} = \frac{\text{SSE}}{n - k} = \frac{500}{8} = 62.5 \quad (\text{shaded on Figure 10.9})$$

Then the *F*-ratio is

$$F = \frac{\text{MST}}{\text{MSE}} = \frac{4,000}{62.5} = 64.0 \quad (\text{shaded on Figure 10.9})$$

Again, our visual analysis of the dot plot is confirmed statistically: $F = 64.0$ well exceeds the table's *F* value, 5.32, corresponding to the .05 level of significance. We would there-fore reject the null hypothesis at that level and conclude that the SAT mean score of males differs from that of females.

Now Work Exercise 10.21b–h

The **analysis of variance *F*-test** for comparing treatment means is summarized in the following box:

ANOVA *F*-Test to Compare *k* Treatment Means: Completely Randomized Design

$H_0: \mu_1 = \mu_2 = \cdots = \mu_k$

H_a: At least two treatment means differ.

Test statistic: $F = \dfrac{\text{MST}}{\text{MSE}}$

Rejection region: $F > F_\alpha$, where F_α is based on $v_1 = (k - 1)$ numerator degrees of freedom (associated with MST) and $v_2 = (n - k)$ denominator degrees of freedom (associated with MSE), or, $\alpha > p\text{-value}$

Conditions Required for a Valid ANOVA *F*-Test: Completely Randomized Design

1. The samples are randomly selected in an independent manner from the *k* treatment populations. (This can be accomplished by randomly assigning the experimental units to the treatments.)

2. All *k* sampled populations have distributions that are approximately normal.

3. The *k* population variances are equal (i.e., $\sigma_1^2 = \sigma_2^2 = \sigma_3^2 = \cdots = \sigma_k^2$).

Computational formulas for MST and MSE are given in Appendix B. We will rely on statistical software to compute the *F*-statistic, concentrating on the interpretation of the results rather than their calculation.

Example 10.4

Conducting an Anova *F*-Test— Comparing Golf Ball Brands

Problem Suppose the USGA wants to compare the mean distances reached of four different brands of golf balls struck with a driver. A completely randomized design is employed, with Iron Byron, the USGA's robotic golfer, using a driver to hit a random sample of 10 balls of each brand in a random sequence. The distance is recorded for each hit, and the results are shown in Table 10.3, organized by brand.

a. Set up the test to compare the mean distances for the four brands. Use $\alpha = .10$.

b. Use statistical software to obtain the test statistic and *p*-value. Give the appropriate conclusion.

Table 10.3	Results of Completely Randomized Design: Iron Byron Driver			
	Brand A	Brand B	Brand C	Brand D
	251.2	263.2	269.7	251.6
	245.1	262.9	263.2	248.6
	248.0	265.0	277.5	249.4
	251.1	254.5	267.4	242.0
	260.5	264.3	270.5	246.5
	250.0	257.0	265.5	251.3
	253.9	262.8	270.7	261.8
	244.6	264.4	272.9	249.0
	254.6	260.6	275.6	247.1
	248.8	255.9	266.5	245.9
Sample means	250.8	261.1	270.0	249.3

Data Set: GOLFCRD

Solution

a. To compare the mean distances of the $k = 4$ brands, we first specify the hypotheses to be tested. Denoting the population mean of the ith brand by μ_i, we test

$$H_0: \mu_1 = \mu_2 = \mu_3 = \mu_4$$
$$H_a: \text{The mean distances differ for at least two of the brands.}$$

The test statistic compares the variation among the four treatment (Brand) means with the sampling variability within each of the treatments:

$$\text{Test statistic:} \quad F = \frac{\text{MST}}{\text{MSE}}$$

$$\text{Rejection region:} \quad F > F_\alpha = F_{.10} \text{ with } v_1 = (k - 1) = 3 \text{ df}$$
$$\text{and } v_2 = (n - k) = 36 \text{ df}$$

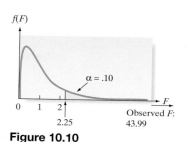

$f(F)$

$\alpha = .10$

0   1   2

2.25

Observed F:
43.99

Figure 10.10

F-test for completely randomized design: golf ball experiment

From Table VIII of Appendix A, we find that $F_{.10} \approx 2.25$ for 3 and 36 df. Thus, we will reject H_0 if $F > 2.25$. (See Figure 10.10.)

The assumptions necessary to ensure the validity of the test are as follows:

1. The samples of 10 golf balls for each brand are selected randomly and independently.

2. The probability distributions of the distances for each brand are normal.

3. The variances of the distance probability distributions for each brand are equal.

b. The MINITAB printout for the data in Table 10.3 resulting from this completely randomized design is given in Figure 10.11. The values, MST = 931.5, MSE = 21.2, and $F = 43.99$, are highlighted on the printout. Since $F > 2.25$, we reject H_0. Also, the p-value of the test (.000) is highlighted on the printout. Since $\alpha = .10$ exceeds the p-value, we draw the same conclusion: Reject H_0. Therefore, at the .10 level of significance, we conclude that at least two of the brands differ with respect to mean distance traveled when struck by the driver.

Figure 10.11

MINITAB ANOVA for completely randomized design

One-way ANOVA: DISTANCE versus BRAND

```
Source   DF      SS      MS      F       P
BRAND     3   2794.4   931.5   43.99   0.000
Error    36    762.3    21.2
Total    39   3556.7
```

Look Ahead Now that we know that mean distances differ, a logical follow-up question is: "Which ball brand travels farther, on average, when hit with a driver?" In Section 10.3 we present a method for ranking treatment means in an ANOVA.

Now Work Exercise 10.24

The results of an **analysis of variance (ANOVA)** can be summarized in a simple tabular format similar to that obtained from the MINITAB program in Example 10.4. The general form of the table is shown in Table 10.4, where the symbols df, SS, and MS stand for degrees of freedom, sum of squares, and mean square, respectively. Note that the two sources of variation, Treatments and Error, add to the total sum of squares, SS(Total). The ANOVA summary table for Example 10.4 is given in Table 10.5, and the partitioning of the total sum of squares into its two components is illustrated in Figure 10.12.

Table 10.4	General ANOVA Summary Table for a Completely Randomized Design			
Source	df	SS	MS	F
Treatments	$k-1$	SST	$MST = \dfrac{SST}{k-1}$	$\dfrac{MST}{MSE}$
Error	$n-k$	SSE	$MSE = \dfrac{SSE}{n-k}$	
Total	$n-1$	SS(Total)		

Table 10.5	ANOVA Summary Table for Example 10.4				
Source	df	SS	MS	F	p-Value
Brands	3	2,794.39	931.46	43.99	.000
Error	36	762.30	21.18		
Total	39	3,556.69			

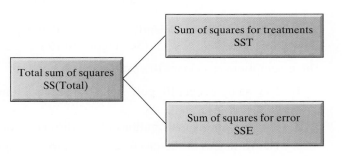

Figure 10.12
Partitioning of the total sum of squares for the completely randomized design

Example 10.5

Checking the ANOVA Assumptions

Problem Refer to the completely randomized ANOVA design conducted in Example 10.4. Are the assumptions required for the test approximately satisfied?

Solution The assumptions for the test are repeated as follows:

1. The samples of golf balls for each brand are selected randomly and independently.

2. The probability distributions of the distances for each brand are normal.

3. The variances of the distance probability distributions for each brand are equal.

Since the sample consisted of 10 randomly selected balls of each brand, and since the robotic golfer Iron Byron was used to drive all the balls, the first assumption of independent random samples is satisfied. To check the next two assumptions, we will employ two graphical methods presented in Chapter 2: histograms and box plots. A MINITAB histogram of driving distances for each brand of golf ball is shown in Figure 10.13, and SAS box plots are shown in Figure 10.14.

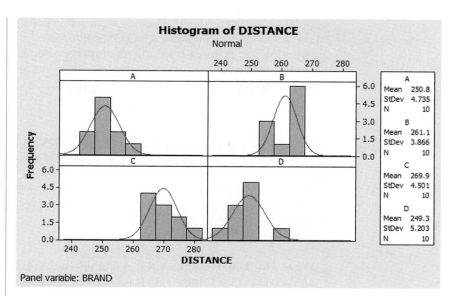

Figure 10.13

MINITAB histograms for golf ball driving distances

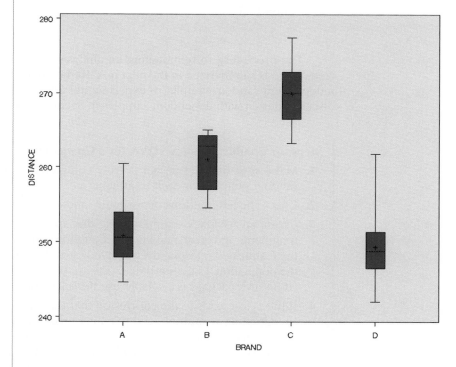

Figure 10.14

SAS box plots for golf ball distances

The normality assumption can be checked by examining the histograms in Figure 10.13. With only 10 sample measurements for each brand, however, the displays are not very informative. More data would need to be collected for each brand before we could assess whether the distances come from normal distributions. Fortunately, analysis of variance has been shown to be a very **robust method** when the assumption of normality is not satisfied exactly. That is, *moderate departures from normality do not have much effect on the significance level of the ANOVA F-test or on confidence coefficients*. Rather than spend the time, energy, or money to collect additional data for this experiment in order to verify the normality assumption, we will rely on the robustness of the ANOVA methodology.

Box plots are a convenient way to obtain a rough check on the assumption of equal variances. With the exception of a possible outlier for Brand D, the box plots in Figure 10.14 show that the spread of the distance measurements is about the same for each brand. Since the sample variances appear to be the same, the assumption of equal population variances for the brands is probably satisfied. Although robust with respect to the normality assumption, ANOVA is *not robust* with respect to the equal-variances assumption. Departures from the assumption of equal population variances can affect the associated measures of reliability (e.g., p-values and confidence levels). Fortunately, the effect is slight when the sample sizes are equal, as in this experiment.

Now Work Exercise 10.32

Although graphs can be used to check the ANOVA assumptions as in Example 10.5, no measures of reliability can be attached to these graphs. When you have a plot that is unclear as to whether or not an assumption is satisfied, you can use formal statistical tests that are beyond the scope of this text. Consult the references at the end of the chapter for information on these tests. When the validity of the ANOVA assumptions is in doubt, nonparametric statistical methods are useful.

What Do You Do When the Assumptions Are Not Satisfied for the Analysis of Variance for a Completely Randomized Design?

Answer: Use a nonparametric statistical method such as the Kruskal–Wallis H-Test of Section 14.5.

The procedure for conducting an analysis of variance for a completely randomized design is summarized in the next box. Remember that the hallmark of this design is independent random samples of experimental units associated with each treatment. We discuss a design with dependent samples in Section 10.4.

Steps for Conducting an ANOVA for a Completely Randomized Design

1. Make sure that the design is truly completely randomized, with independent random samples for each treatment.

2. Check the assumptions of normality and equal variances.

3. Create an ANOVA summary table that specifies the variabilities attributable to treatments and error, making sure that those variabilities lead to the calculation of the F-statistic for testing the null hypothesis that the treatment means are equal in the population. Use a statistical software program to obtain the numerical results. (If no such package is available, use the calculation formulas in Appendix B.)

4. If the F-test leads to the conclusion that the means differ,
    a. Conduct a multiple-comparisons procedure for as many of the pairs of means as you wish to compare. (See Section 10.3.) Use the results to summarize the statistically significant differences among the treatment means.
    b. If desired, form confidence intervals for one or more individual treatment means.

5. If the F-test leads to the nonrejection of the null hypothesis that the treatment means are equal, consider the following possibilities:
    a. The treatment means are equal; that is, the null hypothesis is true.
    b. The treatment means really differ, but other important factors affecting the response are not accounted for by the completely randomized design. These factors inflate the sampling variability, as measured by MSE, resulting in smaller values of the F-statistic. Either increase the sample size for each treatment, or use a different experimental design (as in Section 10.4) that accounts for the other factors affecting the response.

[*Note:* Be careful not to automatically conclude that the treatment means are equal since the possibility of a Type II error must be considered if you accept H_0.]

We conclude this section by making two important points about an analysis of variance. First, recall that we performed a hypothesis test for the difference between two means in Section 9.2 using a two-sample t-statistic for two independent samples. When two independent samples are being compared, the t- and F-tests are equivalent. To see this, apply the formula for t to the two samples of SAT scores in Table 10.2:

$$t = \frac{\bar{x}_1 - \bar{x}_2}{\sqrt{s_p^2\left(\frac{1}{n_1} + \frac{1}{n_2}\right)}} = \frac{590 - 550}{\sqrt{(62.5)\left(\frac{1}{5} + \frac{1}{5}\right)}} = \frac{40}{5} = 8$$

Here, we used the fact that $s_p^2 = $ MSE, which you can verify by comparing the formulas. Recall that the calculated F for the two samples in Table 10.2 is $F = 64$. This value equals the square of the calculated t for the same samples ($t = 8$). Likewise, the critical F-value (5.32) equals the square of the critical t-value at the two-sided .05 level of significance ($t_{.025} = 2.306$ with 8 df). Since both the rejection region and the calculated values are related in the same way, the tests are equivalent. Moreover, the assumptions that must be met to ensure the validity of the t- and F-tests are the same:

1. The probability distributions of the populations of responses associated with each treatment must all be normal.

2. The probability distributions of the populations of responses associated with each treatment must have equal variances.

3. The samples of experimental units selected for the treatments must be random and independent.

In fact, the only real difference between the tests is that the F-test can be used to compare *more than two* treatment means, whereas the t-test is applicable to two samples only.

For our second point, refer to Example 10.4. Our conclusion that at least two of the brands of golf balls have different mean distances traveled when struck with a driver leads naturally to the following questions: Which of the brands differ? and How are the brands ranked with respect to mean distance?

One way to obtain this information is to construct a confidence interval for the difference between the means of any pair of treatments, using the method of Section 9.2. For example, if a 95% confidence interval for $\mu_A - \mu_C$ in Example 10.4 is found to be $(-24, -13)$, we are confident that the mean distance for Brand C exceeds the mean for Brand A (since all differences in the interval are negative). Constructing these confidence intervals for all possible pairs of brands allows you to rank the brand means. A method for conducting these *multiple comparisons*—one that controls for Type I errors—is presented in Section 10.3.

Statistics IN Action | Revisited | A One-Way Analysis of the Cockroach Data

Consider the experiment designed to investigate the trail-following ability of German cockroaches (p. 475). Recall that an entomologist created a chemical trail with either a methanol extract from roach feces or just methanol (the control). Cockroaches were then released into a container at the beginning of the trail, one at a time, and a video surveillance camera was used to monitor the roaches' movements. The movement pattern of each cockroach was measured by its average deviation (in pixels) from the extract trail and the data stored in the **ROACH** file.

For this application, consider only the cockroaches assigned to the fecal extract trail. Four roach groups were utilized in the experiment—adult males, adult females, gravid females, and nymphs—with 20 roaches of each type independently and randomly selected. Is there sufficient evidence to say that the ability to follow the extract trail differs among cockroaches of different age, sex, and reproductive status? In other words, is there evidence to suggest that the mean trail deviation μ differs for the four roach groups?

To answer this question, we conduct a one-way analysis of variance on the **ROACH** data. The dependent (response)

(continued)

Statistics IN Action
(continued)

variable of interest is deviation from the extract trail, while the treatments are the four different roach groups. Thus, we want to test the null hypothesis:

$$H_0: \mu_{\text{Male}} = \mu_{\text{Female}} = \mu_{\text{Gravid}} = \mu_{\text{Nymph}}$$

A MINITAB printout of the ANOVA is displayed in Figure SIA10.1. The *p*-value of the test (highlighted on the printout) is 0. Since this value is less than, say, $\alpha = .05$, we reject the null hypothesis and conclude (at the .05 level of significance) that the mean deviation from the extract trail differs among

the populations of adult male, adult female, gravid, and nymph cockroaches.

The sample means for the four cockroach groups are also highlighted in Figure SIA10.1. Note that adult males have the smallest sample mean deviation (7.38) while gravids have the largest sample mean deviation (44.03). In the next Statistics in Action Revisited application (p. 501), we demonstrate how to rank, statistically, the four population means on the basis of their respective sample means.

> Data Set: ROACH

```
One-way ANOVA: DEVIATE versus GROUP

Source  DF     SS     MS      F      P
GROUP    3  14164   4721  11.61  0.000
Error   76  30918    407
Total   79  45083

S = 20.17    R-Sq = 31.42%    R-Sq(adj) = 28.71%


                              Individual 95% CIs For Mean Based on
                              Pooled StDev
Level    N    Mean  StDev  -+---------+---------+---------+--------
Female  20   21.07  26.13                (-----*-----)
Gravid  20   44.03  24.84                           (-----*-----)
Male    20    7.38   8.61  (-----*-----)
Nymph   20   18.73  15.92             (-----*-----)
                          -+---------+---------+---------+--------
                           0        15        30        45

Pooled StDev = 20.17
```

Figure SIA10.1
MINITAB one-way ANOVA for deviation from extract trail

Exercises 10.15–10.38

Understanding the Principles

10.15 Explain how to collect the data for a completely randomized design.

10.16 Explain the concept of a balanced design.

10.17 What conditions are required for a valid ANOVA *F*-test in a completely randomized design?

10.18 *True or False.* The ANOVA method is robust when the assumption of normality is not exactly satisfied in a completely randomized design.

Learning the Mechanics

10.19 Use Tables VIII, IX, X, and XI of Appendix A to find each of the following *F* values:
  a. $F_{.05}, v_1 = 3, v_2 = 4$
  b. $F_{.01}, v_1 = 3, v_2 = 4$
  c. $F_{.10}, v_1 = 20, v_2 = 40$

  d. $F_{.025}, v_1 = 12, v_2 = 9$

10.20 Find the following probabilities:
  a. $P(F \le 3.48)$ for $v_1 = 5, v_2 = 9$
  b. $P(F > 3.09)$ for $v_1 = 15, v_2 = 20$
  c. $P(F > 2.40)$ for $v_1 = 15, v_2 = 15$
  d. $P(F \le 1.83)$ for $v_1 = 8, v_2 = 40$

10.21 Consider dot plots A and B (shown at the top of p. 493).
  [NW] Assume that the two samples represent independent random samples corresponding to two treatments in a completely randomized design.
  a. In which dot plot is the difference between the sample means small relative to the variability within the sample observations? Justify your answer.
  b. Calculate the treatment means (i.e., the means of samples 1 and 2) for both dot plots.
  c. Use the means to calculate the sum of squares for treatments (SST) for each dot plot.

Dot Plots for Exercise 10.21

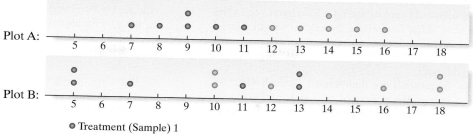

- Treatment (Sample) 1
- Treatment (Sample) 2

d. Calculate the sample variance for each sample and use these values to obtain the sum of squares for error (SSE) for each dot plot.

e. Calculate the total sum of squares [SS(Total)] for the two dot plots by adding the sums of squares for treatment and error. What percentage of SS(Total) is accounted for by the treatments—that is, what percentage of the total sum of squares is the sum of squares for treatment—in each case?

f. Convert the sums of squares for treatment and error to mean squares by dividing each by the appropriate number of degrees of freedom. Calculate the F-ratio of the mean square for treatment (MST) to the mean square for error (MSE) for each dot plot.

g. Use the F-ratios to test the null hypothesis that the two samples are drawn from populations with equal means. Take $\alpha = .05$.

h. What assumptions must be made about the probability distributions corresponding to the responses for each treatment in order to ensure the validity of the F-tests conducted in part **g**?

10.22 Refer to Exercise 10.21. Conduct a two-sample t-test (Section 9.2) of the null hypothesis that the two treatment means are equal for each dot plot. Use $\alpha = .05$ and two-tailed tests. In the course of the test, compare each of the following with the F-tests in Exercise 10.21:

a. The pooled variances and the MSEs

b. The t- and the F-test statistics

c. The tabled values of t and F that determine the rejection regions

d. The conclusions of the t- and F-tests

e. The assumptions that must be made in order to ensure the validity of the t- and F-tests

10.23 Refer to Exercises 10.21 and 10.22. Complete the following ANOVA table for each of the two dot plots:

Source	df	SS	MS	F
Treatments				
Error				
Total				

10.24 A partially completed ANOVA table for a completely randomized design is shown here:

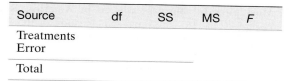

Source	df	SS	MS	F
Treatments	6	18.4		
Error				
Total	41	45.2		

a. Complete the ANOVA table.

b. How many treatments are involved in the experiment?

c. Do the data provide sufficient evidence to indicate a difference among the population means? Test, using $\alpha = .10$.

d. Find the approximate observed significance level for the test in part **c**, and interpret it.

10.25 Suppose the total sum of squares for a completely randomized design with $p = 5$ treatments and $n = 30$ total measurements (6 per treatment) is equal to 500. In each of the following cases, conduct an F-test of the null hypothesis that the mean responses for the 5 treatments are the same. Use $\alpha = .10$.

a. Sum of squares for treatment (SST) is 20% of SS(Total)

b. SST is 50% of SS(Total)

c. SST is 80% of SS(Total)

d. What happens to the F-ratio as the percentage of the total sum of squares attributable to treatments is increased?

10.26 The data in the following table (saved in the **LM10_26** file) resulted from an experiment that utilized a completely randomized design:

Treatment 1	Treatment 2	Treatment 3
3.9	5.4	1.3
1.4	2.0	.7
4.1	4.8	2.2
5.5	3.8	
2.3	3.5	

a. Use statistical software (or the formulas in Appendix B) to complete the following ANOVA table:

Source	df	SS	MS	F
Treatments				
Error				
Total				

b. Test the null hypothesis that $\mu_1 = \mu_2 = \mu_3$, where μ_i represents the true mean for treatment i, against the alternative that at least two of the means differ. Use $\alpha = .01$.

Applying the Concepts—Basic

10.27 **Treating cancer with yoga.** According to a study funded by the National Institutes of Health, yoga classes can help cancer survivors sleep better. The study results were presented at the June 2010 American Society of Clinical Oncology's annual meeting. Researchers randomly assigned 410 cancer patients (who had finished cancer therapy) to receive either their usual follow-up care or attend a 75-minute yoga class twice per week. After four weeks, the researchers measured the level of fatigue and sleepiness experienced by each cancer survivor. Those who took yoga were less fatigued than those who did not.

a. Assume the patients are numbered 1 through 410. Use the random number generator of a statistical software package to randomly assign each patient to either receive the usual follow-up care or to attend yoga classes. Assign 205 patients to each treatment.

b. Consider the following treatment assignment scheme. The patients are ranked according to severity of cancer and the most severe patients are assigned to the yoga class while the others are assigned to receive their usual follow-up care. Comment on the validity of the results obtained from such an assignment.

10.28 Whales entangled in fishing gear. Entanglement of marine mammals (e.g., whales) in fishing gear is considered a significant threat to the species. A study published in *Marine Mammal Science* (April 2010) investigated the type of net most likely to entangle a certain species of whale inhabiting the East Sea of Korea. A sample of 207 entanglements of whales in the area formed the data for the study. These entanglements were caused by one of three types of fishing gear: set nets, pots, and gill nets. One of the variables investigated was body length (in meters) of the entangled whale.

a. Set up the null and alternative hypotheses for determining whether the average body length of entangled whales differs for the three types of fishing gear.

b. An ANOVA *F*-test yielded the following results: $F = 34.81, p - \text{value} < .0001$. Interpret the results for $\alpha = .05$.

10.29 College tennis recruiting with a team Web site. Most university athletic programs have a Web site with information on individual sports and a Prospective Student Athlete Form that allows high school athletes to submit information about their academic and sports achievements directly to the college coach. *The Sport Journal* (Winter 2004) published a study of how important team Web sites are to the recruitment of college tennis players. A survey was conducted of National Collegiate Athletic Association (NCAA) tennis coaches, of which 53 were from Division I schools, 20 were from Division II schools, and 53 were from Division III schools. Coaches were asked to respond to a series of statements, including "The Prospective Student Athlete Form on the Web site contributes very little to the recruiting process." Responses were measured on a seven-point scale (where 1 = strongly disagree and 7 = strongly agree). In order to compare the mean responses of tennis coaches from the three NCAA divisions, the data were analyzed with a completely randomized ANOVA design.

a. Identify the experimental unit, the dependent (response) variable, the factor, and the treatments in this study.

b. Give the null and alternative hypothesis for the ANOVA *F*-test.

c. The observed significance level of the test was found to be *p-value* < .003. What conclusion can you draw if you want to test at $\alpha = .05$?

10.30 A new dental bonding agent. Refer to the *Trends in Biomaterials & Artificial Organs* (Jan. 2003) study of a new bonding adhesive for teeth, presented in Exercise 8.72 (p. 379). Recall that the new adhesive (called "Smartbond") has been developed to eliminate the necessity of a dry field. In one portion of the study, 30 extracted teeth were bonded with Smartbond and each was randomly assigned one of three different bonding times: 1 hour, 24 hours, or 48 hours. At the end of the bonding period, the breaking strength (in Mpa) of each tooth was determined. The data were analyzed with the use of analysis of variance in order to determine whether the true mean breaking strength of the new adhesive varies with the bonding time.

a. Identify the experimental units, treatments, and response variable for this completely randomized design.

b. Set up the null and alternative hypotheses for the ANOVA.

c. Find the rejection region using $\alpha = .01$.

d. The test results were $F = 61.62$ and *p*-value ≈ 0. Give the appropriate conclusion for the test.

e. What conditions are required for the test results to be valid?

10.31 Robots trained to behave like ants. Robotics researchers investigated whether robots could be trained to behave like ants in an ant colony (*Nature*, Aug. 2000). Robots were trained and randomly assigned to "colonies" (i.e., groups) consisting of 3, 6, 9, or 12 robots. The robots were assigned the tasks of foraging for "food" and recruiting another robot when they identified a resource-rich area. One goal of the experiment was to compare the mean energy expended (per robot) of the four different sizes of colonies.

a. What type of experimental design was employed?

b. Identify the treatments and the dependent variable.

c. Set up the null and alternative hypotheses of the test.

d. The following ANOVA results were reported: $F = 7.70$, numerator df = 3, denominator df = 56, *p*-value < .001. Conduct the test at a significance level of $\alpha = .05$ and interpret the result.

10.32 Most powerful business women in America. Refer to *Fortune* (Oct. 16, 2008) magazine's study of the 50 most powerful women in business in America, Exercise 2.60

Data for Exercise 10.32

Rank	Name	Age	Company	Title
1	Indra Nooyi	52	PepsiCo	CEO/Chairman
2	Irene Rosenfeld	55	Kraft Foods	CEO/Chairman
3	Pat Woertz	55	Archer Daniels Midland	CEO/Chairman
4	Anne Mulcahy	55	Xerox	CEO/Chairman
5	Angela Braley	47	Wellpoint	CEO/President
⋮	⋮	⋮	⋮	⋮
49	Cathie Black	64	Hearst Magazines	President
50	Marissa Mayer	33	Google	VP

Source: Fortune, Oct. 16, 2008

SPSS Output for Exercise 10.32

ANOVA

AGE

	Sum of Squares	df	Mean Square	F	Sig.
Between Groups	215.541	2	107.770	2.784	.072
Within Groups	1819.439	47	38.711		
Total	2034.980	49			

AGE

GROUP	Mean	N	Std. Deviation
1	52.50	18	5.044
2	49.58	19	5.357
3	47.23	13	8.516
Total	50.02	50	6.444

(p. 57). Recall that data on age (in years) and title of each of these 50 women are stored in the **WPOWER50** file. (Some of the data set is listed in the table on the bottom of p. 494.) Suppose you want to compare the average ages of the most powerful American women in three groups based on their position (title) within the firm: Group 1 (CEO, CEO/Chairman, or CEO/President), Group 2 (Chairman, President, COO, or COO/President), and Group 3 (Director, EVP, EVP/COO, EVP/President, Executive, SVP, or VP).

a. Give the null and alternative hypotheses to be tested.

b. An SPSS analysis-of-variance printout for the test you stated in part **a** is shown above. The sample means for the 3 groups appear at the bottom of the printout. Why is it insufficient to make a decision about the null hypothesis based solely on these sample means?

c. Locate the test statistic and p-value on the printout. Use this information to make the appropriate conclusion at $\alpha = .10$.

d. Use the data in the **WPOWER50** file to determine whether the ANOVA assumptions are reasonably satisfied.

Applying the Concepts—Intermediate

10.33 Study of recall of TV commercials. Do TV shows with violence and sex impair memory for commercials? To answer this question, Iowa St. researchers conducted a designed experiment in which 324 adults were randomly assigned to one of three viewer groups of 108 participants each (*Journal of Applied Psychology*, June 2002). One group watched a TV program with a violent content code (V) rating, the second group viewed a show with a sex content code (S) rating, and the last group watched a neutral TV program with neither a V nor an S rating. Nine commercials were embedded into each TV show. After viewing the program, each participant was scored on his or her recall of the brand names in the commercial messages, with scores ranging from 0 (no brands recalled) to 9 (all brands recalled). The data (simulated from information provided in the article) are saved in the **TVADRECALL** file. The researchers compared the mean recall scores of the three viewing groups with an analysis of variance for a completely randomized design.

a. Identify the experimental units in the study.

b. Identify the dependent (response) variable in the study.

c. Identify the factor and treatments in the study.

d. The sample mean recall scores for the three groups were $\bar{x}_v = 2.08$, $\bar{x}_s = 1.71$, and $\bar{x}_{Neutral} = 3.17$. Explain why one should not draw an inference about differences in the population mean recall scores on the basis of only these summary statistics.

e. An ANOVA on the data in the **TVADRECALL** file yielded the results shown in the accompanying MINITAB printout. Locate the test statistic and p-value on the printout.

f. Interpret the results from part **e**, using $\alpha = 0.01$. What can the researchers conclude about the three groups of TV ad viewers?

One-way ANOVA: VIOLENT, SEX, NEUTRAL

```
Source    DF        SS       MS       F       P
Factor     2    123.27    61.63   20.45   0.000
Error    321    967.35     3.01
Total    323   1090.62

S = 1.736    R-Sq = 11.30%    R-Sq(adj) = 10.75%
```

10.34 Restoring self-control when intoxicated. Does coffee or some other form of stimulation (e.g., an incentive to stop when seeing a flashing red light on a car) really allow a person suffering from alcohol intoxication to "sober up"? Psychologists from the University of Waterloo investigated the matter in *Experimental and Clinical Psychopharmacology* (February 2005). A sample of 44 healthy male college students participated in the experiment. Each student was asked to memorize a list of 40 words (20 words on a green list and 20 words on a red list). The students were then randomly assigned to one of four different treatment groups (11 students in each group). Students in three of the groups were each given two alcoholic beverages to drink prior to performing a word completion task. Students in Group A received only the alcoholic drinks. Participants in Group AC had caffeine powder dissolved in their drinks. Group AR participants received a monetary award for correct responses on the word completion task. Students in Group P (the placebo group) were told that they would receive alcohol, but instead received two drinks containing a carbonated beverage (with a few drops of alcohol on the surface to provide an alcoholic scent). After consuming their drinks and resting for 25 minutes, the students performed the word completion task. Their scores (simulated on the basis of summary information from the

AR	AC	A	P
.51	.50	.16	.58
.58	.30	.10	.12
.52	.47	.20	.62
.47	.36	.29	.43
.61	.39	− .14	.26
.00	.22	.18	.50
.32	.20	− .35	.44
.53	.21	.31	.20
.50	.15	.16	.42
.46	.10	.04	.43
.34	.02	− .25	.40

Based on Grattan-Miscio, K.E., and Vogel-Sprott, M. "Alcohol, intentional control, and inappropriate behavior: Regulation by caffeine or an incentive." *Experimental and Clinical Psychopharmacology*, Vol. 13, No. 1, February 2005 (Table 1).

article) are reported in the table on the previous page and saved in the **DRINKERS** file. (*Note*: A task score represents the difference between the proportion of correct responses on the green list of words and the proportion of incorrect responses on the red list of words.)

a. What type of experimental design is employed in this study?

b. Analyze the data for the researchers, using $\alpha = .05$. Are there differences among the mean task scores for the four groups?

c. What assumptions must be met in order to ensure the validity of the inference you made in part **b**?

10.35 Is honey a cough remedy? Pediatric researchers at Pennsylvania State University carried out a designed study to test whether a teaspoon of honey before bed calms a child's cough and published their results in *Archives of Pediatrics and Adolescent Medicine* (Dec. 2007). (This experiment was first described in Exercise 2.32, p. 46). A sample of 105 children who were ill with an upper respiratory tract infection and their parents participated in the study. On the first night, the parents rated their children's cough symptoms on a scale from 0 (no problems at all) to 6 (extremely severe) in five different areas. The total symptoms score (ranging from 0 to 30 points) was the variable of interest for the 105 patients. On the second night, the parents were instructed to give their sick child a dosage of liquid "medicine" prior to bedtime. Unknown to the parents, some were given a dosage of dextromethorphan (DM)—an over-the-counter cough medicine—while others were given a similar dose of honey. Also, a third group of parents (the control group) gave their sick children no dosage at all. Again, the parents rated their children's cough symptoms, and the improvement in total cough symptoms score was determined for each child. The data (improvement scores) for the study are shown in the accompanying table and saved in the **HONEYCOUGH** file. The goal of the researchers was to compare the mean improvement scores for the three treatment groups.

a. Identify the type of experimental design employed. What are the treatments?

b. Conduct an analysis of variance on the data and interpret the results.

Honey Dosage:	12 11 15 11 10 13 10 4 15 16 9 14 10 6 10 8 11 12 12 8 12 9 11 15 10 15 9 13 8 12 10 8 9 5 12
DM Dosage:	4 6 9 4 7 7 7 9 12 10 11 6 3 4 9 12 7 6 8 12 12 4 12 13 7 10 13 9 4 4 10 15 9
No Dosage (Control):	5 8 6 1 0 8 12 8 7 7 1 6 7 7 12 7 9 7 9 5 11 9 5 6 8 8 6 7 10 9 4 8 7 3 1 4 3

Based on Paul, I. M., et al. "Effect of honey, dextromethorphan, and no treatment on nocturnal cough and sleep quality for coughing children and their parents," *Archives of Pediatrics and Adolescent Medicine*, Vol. 161, No. 12, Dec. 2007 (data simulated).

10.36 The "name game." Psychologists at Lancaster University (United Kingdom) evaluated three methods of name retrieval in a controlled setting (*Journal of Experimental Psychology—Applied*, June 2000). A sample of 139 students was randomly divided into three groups, and each group of students used a different method to learn the

names of the other students in the group. Group 1 used the "simple name game," in which the first student states his or her full name, the second student announces his or her name and the name of the first student, the third student says his or her name and the names of the first two students, etc. Group 2 used the "elaborate name game," a modification of the simple name game such that the students state not only their names, but also their favorite activity (e.g., sports). Group 3 used "pairwise introductions," according to which students are divided into pairs and each student must introduce the other member of the pair. One year later, all subjects were sent pictures of the students in their group and asked to state the full name of each. The researchers measured the percentage of names recalled by each student respondent. The data (simulated on the basis of summary statistics provided in the research article) are shown in the table and saved in the **NAMEGAME** file. Conduct an analysis of variance to determine whether the mean percentages of names recalled differ for the three name-retrieval methods. Use $\alpha = .05$.

Simple Name Game
24 43 38 65 35 15 44 44 18 27 0 38 50 31 7 46 33 31 0 29 0 0 52 0 29 42 39 26 51 0 42 20 37 51 0 30 43 30 99 39 35 19 24 34 3 60 0 29 40 40

Elaborate Name Game
39 71 9 86 26 45 0 38 5 53 29 0 62 0 1 35 10 6 33 48 9 26 83 33 12 5 0 0 25 36 39 1 37 2 13 26 7 35 3 8 55 50

Pairwise Introductions
5 21 22 3 32 29 32 0 4 41 0 27 5 9 66 54 1 15 0 26 1 30 2 13 0 2 17 14 5 29 0 45 35 7 11 4 9 23 4 0 8 2 18 0 5 21 14

Source: Morris, P. E., and Fritz, C. O. "The name game: Using retrieval practice to improve the learning of names," *Journal of Experimental Psychology—Applied*, Vol. 6, No. 2, June 2000 (data simulated from Figure 1).

10.37 Estimating the age of glacial drifts. Refer to the *American Journal of Science* (Jan. 2005) study of the chemical makeup of buried tills (glacial drifts) in Wisconsin, presented in Exercise 2.136 (p. 88). The ratio of the elements aluminum (Al) and beryllium (Be) in sediment is related to the duration of burial. Recall the Al/Be ratios for a sample of 26 buried till specimens were determined and are saved in the **TILLRATIO** file. The till specimens were obtained from five different boreholes (labeled UMRB-1, UMRB-2, UMRB-3, SWRA, and SD). The data are shown here.

UMRB-1:	3.75	4.05	3.81	3.23	3.13	3.30	3.21
UMRB-2:	3.32	4.09	3.90	5.06	3.85	3.88	
UMRB-3:	4.06	4.56	3.60	3.27	4.09	3.38	3.37
SWRA:	2.73	2.95	2.25				
SD:	2.73	2.55	3.06				

Source: Adapted from *American Journal of Science*, Vol. 305, No. 1, Jan. 2005, p. 16 (Table 2).

Conduct an analysis of variance of the data. Is there sufficient evidence to indicate differences among the mean AI/Be ratios for the five boreholes? Test, using $\alpha = .10$.

Applying the Concepts—Advanced

10.38 Animal-assisted therapy for heart patients. Refer to the *American Heart Association Conference* (Nov. 2005) study to gauge whether animal-assisted therapy can improve the physiological responses of heart failure patients, presented in Exercise 2.106 (p. 73). Recall that 76 heart patients were randomly assigned to one of three groups. Each patient in group T was visited by a human volunteer accompanied by a trained dog, each patient in group V was visited by a volunteer only, and the patients in group C were not visited at all. The anxiety level of each patient was measured (in

	Sample Size	Mean Drop	Std. Dev.
Group T: Volunteer + Trained Dog	26	10.5	7.6
Group V: Volunteer only	25	3.9	7.5
Group C: Control group (no visit)	25	1.4	7.5

Based on Cole, K., et al. "Animal assisted therapy decreases hemodynamics, plasma epinephrine and state anxiety in hospitalized heart failure patients." *American Journal of Critical Care,* 2007, 16: 575–585.

points) both before and after the visits. The accompanying table gives summary statistics for the drop in anxiety level for patients in the three groups. The mean drops in anxiety levels of the three groups of patients were compared with the use of an analysis of variance. Although the ANOVA table was not provided in the article, sufficient information is given to reconstruct it.

a. Compute SST for the ANOVA, using the formula (on p. 484)

$$SST = \sum_{i=1}^{3} n_i(\bar{x}_i - \bar{x})^2$$

where $\bar{x}$ is the overall mean drop in anxiety level of all 76 subjects. [*Hint:* $\bar{x} = (\sum_{i=1}^{3} n_i\,(\bar{x}_i))/76.$]

b. Recall that SSE for the ANOVA can be written as

$$SSE = (n_1 - 1)s_1^2 + (n_2 - 1)s_2^2 + (n_3 - 1)s_3^2$$

where $s_1^2, s_2^2,$ and s_3^2 are the sample variances associated with the three treatments. Compute SSE for the ANOVA.

c. Use the results from parts **a** and **b** to construct the ANOVA table.

d. Is there sufficient evidence (at $\alpha = .01$) of differences among the mean drops in anxiety levels by the patients in the three groups?

e. Comment on the validity of the ANOVA assumptions. How might this affect the results of the study?

10.3 Multiple Comparisons of Means

Consider a completely randomized design with three treatments: A, B, and C. Suppose we determine, via the ANOVA F-test of Section 10.2, that the treatment means are statistically different. To complete the analysis, we want to rank the three treatment means. As mentioned in Section 10.2, we start by placing confidence intervals on the differences between various pairs of treatment means in the experiment. In the three-treatment experiment, for example, we would construct confidence intervals for the following differences: $\mu_A - \mu_B, \mu_A - \mu_C,$ and $\mu_B - \mu_C$.

Determining the Number of Pairwise Comparisons of Treatment Means

In general, if there are k treatment means, there are

$$c = k(k - 1)/2$$

pairs of means that can be compared.

If we want to have $100(1 - \alpha)\%$ confidence that each of the c confidence intervals contains the true difference it is intended to estimate, we must use a smaller value of α for each individual confidence interval than we would use for a single interval. For example, suppose we want to rank the means of the three treatments A, B, and C, with 95% confidence that all three confidence intervals contain the true differences between the treatment means. Then, each individual confidence interval will need to be constructed using a level of significance smaller than $\alpha = .05$ in order to have 95% confidence that the three intervals collectively include the true differences.*

Now Work Exercise 10.43

*The reason each interval must be formed at a higher confidence level than that specified for the collection of intervals can be demonstrated as follows:

$$P\{\text{At least one of } c \text{ intervals fails to contain the true difference}\}$$
$$= 1 - P\{\text{All } c \text{ intervals contain the true differences}\}$$
$$= 1 - (1 - \alpha)^c \geq \alpha$$

Thus, to make this probability of at least one failure equal to α, we must specify the individual levels of significance to be less than α.

To make **multiple comparisons of a set of treatment means**, we can use a number of procedures which, under various assumptions, ensure that the overall confidence level associated with all the comparisons remains at or above the specified $100(1 - \alpha)\%$ level. Three widely used techniques are the Bonferroni, Scheffé, and Tukey methods. For each of these procedures, the risk of making a Type I error applies to the comparisons of the treatment means in the experiment; thus, the value of α selected is called an **experimentwise error rate** (in contrast to a **comparisonwise error rate**).

For a single comparison of two means in a designed experiment, the probability of making a Type I error (i.e., the probability of concluding that a difference in the means exists, given that the means are the same) is called a **comparisonwise error rate (CER)**.

For multiple comparisons of means in a designed experiment, the probability of making at least one Type I error (i.e., the probability of concluding that at least one difference in means exists, given that the means are all the same) is called an **experimentwise error rate (EER)**.

The choice of a multiple-comparison method in ANOVA will depend on the type of experimental design used and the comparisons that are of interest to the analyst. For example, **Tukey** (1949) developed his procedure specifically for pairwise comparisons when the sample sizes of the treatments are equal. The **Bonferroni** method (see Miller, 1981), like the Tukey procedure, can be applied when pairwise comparisons are of interest; however, Bonferroni's method does not require equal sample sizes. **Scheffé** (1953) developed a more general procedure for comparing all possible linear combinations of treatment means (called *contrasts*). Consequently, in making pairwise comparisons, the confidence intervals produced by Scheffé's method will generally be wider than the Tukey or Bonferroni confidence intervals.

BIOGRAPHY CARLO E. BONFERRONI (1892–1960)

Bonferroni Inequalities

During his childhood years in Turin, Italy, Carlo Bonferroni developed an aptitude for mathematics while studying music. He went on to obtain a degree in mathematics at the University of Turin. Bonferroni's first appointment as a professor of mathematics was at the University of Bari in 1923. Ten years later, he became chair of financial mathematics at the University of Florence, where he remained until his death. Bonferroni was a prolific writer, authoring over 65 research papers and books. His interest in statistics included various methods of calculating a mean and a correlation coefficient. Among statisticians, however, Bonferroni is most well known for developing his Bonferroni inequalities in probability theory in 1935. Later, other statisticians proposed using these inequalities to find simultaneous confidence intervals, which led to the development of the Bonferroni multiple-comparison method in ANOVA. Bonferroni balanced these scientific accomplishments with his music, becoming an excellent pianist and composer. ∎

The formulas for constructing confidence intervals for differences between treatment means by the Tukey, Bonferroni, or Scheffé methods are provided in Appendix B. However, since these procedures (and many others) are available in the ANOVA programs of most statistical software packages, we will use the software to conduct the analyses. The programs generate a confidence interval for the difference between two treatment means for all possible pairs of treatments, based on the experimentwise error rate (α) selected by the analyst.

Example 10.6

Ranking Treatment Means—Golf Ball Experiment

Problem Refer to the completely randomized design of Example 10.4, in which we concluded that at least two of the four brands of golf balls are associated with different mean distances traveled when struck with a driver.

a. Use Tukey's multiple comparison procedure to rank the treatment means with an overall confidence level of 95%.

b. Estimate the mean distance traveled for balls manufactured by the brand with the highest rank.

Solution

a. To rank the treatment means with an overall confidence level of .95, we require the experimentwise error rate of $\alpha = .05$. The confidence intervals generated by Tukey's method appear at the bottom of the SAS ANOVA printout, shown in Figure 10.15. Note that for any pair of means μ_i and μ_j, SAS computes two confidence intervals—one for $(\mu_i - \mu_j)$ and one for $(\mu_j - \mu_i)$. Only one of these intervals is necessary to decide whether the means differ significantly.

The ANOVA Procedure

Tukey's Studentized Range (HSD) Test for DISTANCE

NOTE: This test controls the Type I experimentwise error rate.

```
Alpha                                     0.05
Error Degrees of Freedom                    36
Error Mean Square                     21.17503
Critical Value of Studentized Range    3.80880
Minimum Significant Difference          5.5424
```

Comparisons significant at the 0.05 level are indicated by ***.

```
                  Difference
   BRAND           Between        Simultaneous 95%
 Comparison         Means       Confidence Limits

  C - B             8.890       3.348    14.432   ***
  C - A            19.170      13.628    24.712   ***
  C - D            20.630      15.088    26.172   ***
  B - C            -8.890     -14.432    -3.348   ***
  B - A            10.280       4.738    15.822   ***
  B - D            11.740       6.198    17.282   ***
  A - C           -19.170     -24.712   -13.628   ***
  A - B           -10.280     -15.822    -4.738   ***
  A - D             1.460      -4.082     7.002
  D - C           -20.630     -26.172   -15.088   ***
  D - B           -11.740     -17.282    -6.198   ***
  D - A            -1.460      -7.002     4.082
```

Figure 10.15

SAS multiple–comparison printout for Example 10.6

In this example, we have $k = 4$ brand means to compare. Consequently, the number of relevant pairwise comparisons—that is, the number of nonredundant confidence intervals—is $c = 4(3)/2 = 6$. These six intervals, highlighted in Figure 10.15, are given in Table 10.6.

Table 10.6	Pairwise Comparisons for Example 10.6
Brand Comparison	**Confidence Interval**
$(\mu_A - \mu_B)$	$(-15.82, -4.74)$
$(\mu_A - \mu_C)$	$(-24.71, -13.63)$
$(\mu_A - \mu_D)$	$(-4.08, 7.00)$
$(\mu_B - \mu_C)$	$(-14.43, -3.35)$
$(\mu_B - \mu_D)$	$(6.20, 17.28)$
$(\mu_C - \mu_D)$	$(15.09, 26.17)$

We are 95% confident that the intervals *collectively* contain all the differences between the true brand mean distances. Note that intervals containing 0, such as the

Mean: $\overline{249.3}$ $\overline{250.8}$ 261.1   270.0
Brand: D   A   B   C

Figure 10.16

Summary of Tukey multiple comparisons

(Brand A–Brand D) interval from -4.08 to 7.00, do not support a conclusion that the true brand mean distances differ. If both endpoints of the interval are positive, as with the (Brand B–Brand D) interval from 6.20 to 17.28, the implication is that the first brand (B) mean distance exceeds the second (D). Conversely, if both endpoints of the interval are negative, as with the (Brand A–Brand C) interval from -24.71 to -13.63, the implication is that the second brand (C) mean distance exceeds the first brand (A) mean distance.

A convenient summary of the results of the Tukey multiple comparisons is a listing of the brand means from highest to lowest, with a solid line connecting those which are *not* significantly different. This summary is shown in Figure 10.16. The interpretation is that brand C's mean distance exceeds all others, brand B's mean exceeds that of brands A and D, and the means of brands A and D do not differ significantly. All these inferences are made with 95% confidence, the overall confidence level of the Tukey multiple comparisons.

b. Brand C is ranked highest; thus, we want a confidence interval for μ_C. Since the samples were selected independently in a completely randomized design, a confidence interval for an individual treatment mean is obtained with the one-sample t confidence interval of Section 7.3, using the standard deviation, $s = \sqrt{MSE}$, as the measure of sampling variability for the experiment. A 95% confidence interval on the mean distance traveled by brand C (apparently the "longest ball" of those tested) is

$$\bar{x}_C \pm t_{.025}\, s \sqrt{1/n}$$

where $n = 10$, $t_{.025} \approx 2$ (based on 36 degrees of freedom), and $s = \sqrt{MSE} = \sqrt{21.175} = 4.6$ (where MSE is obtained from Figure 10.15). Substituting, we obtain

$$270.0 \pm (2)(4.60)(\sqrt{.1})$$
$$270.0 \pm 2.9 \quad \text{or} \quad (267.1, 272.9)$$

Thus, we are 95% confident that the true mean distance traveled for brand C is between 267.1 and 272.9 yards when the ball is hit with a driver by Iron Byron.

Look Back The easiest way to create a summary table like Figure 10.16 is to first list the treatment means in rank order. Begin with the largest mean and compare it to (in order), the second largest mean, the third largest mean, etc., by examining the appropriate confidence intervals shown on the computer printout. If a confidence interval contains 0, then connect the two means with a line. (These two means are not significantly different.) Continue in this manner by comparing the second largest mean with the third largest, fourth largest, etc., until all possible $c = (k)(k-1)/2$ comparisons are made.

Now Work Exercise 10.45

Many of the available statistical software packages that have multiple-comparison routines will also produce the rankings shown in Figure 10.16. For example, the SAS printout in Figure 10.17 displays the ranking of the mean distances for the four golf ball brands, achieved with Tukey's method.

The experimentwise error rate (.05), MSE value (21.17503), and minimum significant difference (5.5424) used in the analysis are highlighted on the printout. The Tukey rankings of the means are displayed at the bottom of the figure. Instead of a solid line, note that SAS uses a "Tukey Grouping" letter to connect means that are not significantly different. You can see that the mean for brand C is ranked highest, followed by the mean for brand B. These two brand means are significantly different, since they are associated with a different "Tukey Grouping" letter. Brands A and D are ranked lowest, and their means are not significantly different (since they have the same "Tukey Grouping" letter).

```
                    Tukey's Studentized Range (HSD) Test for DISTANCE
    NOTE: This test controls the Type I experimentwise error rate, but it generally has a higher Type
                              II error rate than REGWQ.

                    Alpha                                    0.05
                    Error Degrees of Freedom                   36
                    Error Mean Square                    21.17503
                    Critical Value of Studentized Range   3.80880
                    Minimum Significant Difference         5.5424

             Means with the same letter are not significantly different.

                Tukey Grouping        Mean       N   BRAND
                            A        269.950     10   C

                            B        261.060     10   B

                            C        250.780     10   A
                            C
                            C        249.320     10   D
```

Figure 10.17

Alternative SAS multiple-comparison output for Example 10.6

Remember that the Tukey method—designed for comparing treatments with equal sample sizes—is just one of numerous multiple-comparison procedures available. Another technique may be more appropriate for the experimental design you employ. Consult the references for details on these other methods and when they should be applied. Guidelines for using the Tukey, Bonferroni, and Scheffé methods are given in the following box:

Guidelines for Selecting a Multiple-Comparison Method in ANOVA

Method	Treatment Sample Sizes	Types of Comparisons
Tukey	Equal	Pairwise
Bonferroni	Equal or unequal	Pairwise or general contrasts (number of contrasts known)
Scheffé	Equal or unequal	General contrasts

Note: For equal sample sizes and pairwise comparisons, Tukey's method will yield simultaneous confidence intervals with the smallest width, and the Bonferroni intervals will have smaller widths than the Scheffé intervals.

Ethics IN Statistics Running several multiple comparisons methods and reporting only the one that produces the desired outcome, without regard to the experimental design, is considered *unethical statistical practice.*

Statistics IN Action | Revisited | Ranking the Means of the Cockroach Groups

Refer to the experiment designed to investigate the trail-following ability of German cockroaches. In the previous Statistics in Action Revisited (p. 491), we applied a one-way ANOVA to the **ROACH** data and discovered statistically significant differences among the mean extract trail deviations for the four groups of cockroaches: adult males, adult females, gravid females, and nymphs. In order to determine which group has the highest degree of trail-following ability, we want to rank the population means from largest to smallest. That is, we want to follow up the ANOVA by conducting multiple comparisons of the four treatment means.

Figure SIA10.2 is a MINITAB printout of the ANOVA and the multiple- comparison results. MINITAB uses Tukey's method to compare the means. The experimentwise confidence level (95%) is highlighted on the printout; also highlighted are the Tukey confidence intervals for all possible pairs of means. The information contained in these confidence intervals will enable us to rank the treatment (population) means.

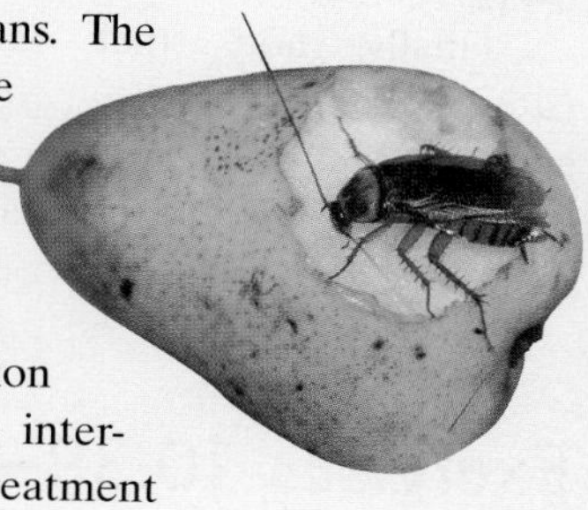

For example, the confidence interval for $(\mu_{Gravid} - \mu_{Female})$ is (6.18, 39.74). Since the endpoints of the interval are both positive, the difference between the means is positive. This implies

(continued)

Statistics IN Action
(continued)

```
Grouping Information Using Tukey Method

GROUP    N    Mean   Grouping
Gravid   20   44.03  A
Female   20   21.07     B
Nymph    20   18.73     B
Male     20    7.38     B

Means that do not share a letter are significantly different.


Tukey 95% Simultaneous Confidence Intervals
All Pairwise Comparisons among Levels of GROUP

Individual confidence level = 98.97%


GROUP = Female subtracted from:

GROUP    Lower   Center   Upper    --------+---------+---------+---------+-
Gravid    6.18    22.96   39.74                      (-----*----)
Male    -30.47   -13.70    3.08          (----*-----)
Nymph   -19.12    -2.34   14.44              (----*-----)
                                  --------+---------+---------+---------+-
                                        -30        0        30       60


GROUP = Gravid subtracted from:

GROUP    Lower   Center   Upper    --------+---------+---------+---------+-
Male    -53.43   -36.65  -19.88   (-----*----)
Nymph   -42.08   -25.30   -8.52      (-----*----)
                                  --------+---------+---------+---------+-
                                        -30        0        30       60


GROUP = Male subtracted from:

GROUP    Lower   Center   Upper    --------+---------+---------+---------+-
Nymph    -5.42    11.36   28.13                  (-----*----)
                                  --------+---------+---------+---------+-
                                        -30        0        30       60
```

Figure SIA10.2
MINITAB Multiple
Comparisons of Extract Trail
Deviation Means

that the population mean deviation from the extract trail for gravid cockroaches is greater than the population mean for adult females (i.e., $\mu_{Gravid} > \mu_{Female}$).

Now consider the confidence interval for $(\mu_{Male} - \mu_{Female})$. The interval shown on the printout is $(-30.47, 3.08)$. Since the value 0 is included in the interval, there is no evidence of a significant difference between the two treatment means. Similar interpretations are made for the confidence intervals for $(\mu_{Nymph} - \mu_{Female})$ and $(\mu_{Nymph} - \mu_{Male})$, since both these intervals contain 0.

Finally, note that the confidence intervals for $(\mu_{Male} - \mu_{Gravid})$ and $(\mu_{Nymph} - \mu_{Gravid})$ both have negative endpoints, implying that the differences between the means is negative. Thus, the population mean deviation from the extract trail for gravids is greater than either the population mean for adult males ($\mu_{Gravid} > \mu_{Male}$) or the population mean for nymphs ($\mu_{Gravid} > \mu_{Nymph}$).

The results of these multiple comparisons are summarized in Table SIA10.1 and at the top of the MINITAB printout. With an overall confidence level of .95, we conclude that gravid cockroaches have a mean extract trail deviation larger than any of the other three groups; there are no significant differences among adult males, adult females, or nymphs.

Table SIA10.1   Ranking of the Cockroach Group Means

Treatment Mean:	7.38	18.73	21.07	44.03
Cockroach Group:	Male	Nymph	Female	Gravid

🔵 *Data Set:* ROACH

Exercises 10.39–10.58

Understanding the Principles

10.39 Define an experimentwise error rate.

10.40 Define a comparisonwise error rate.

10.41 For each of the following confidence intervals for the difference between two means, $(\mu_1 - \mu_2)$, which mean is significantly larger?

a. $(-10, 5)$
b. $(-10, -5)$
c. $(5, 10)$

10.42 Give a situation when it is most appropriate to apply Tukey's multiple-comparison-of-means method.

Learning the Mechanics

10.43 Consider a completely randomized design with k
[NW] treatments. Assume that all pairwise comparisons of treatment means are to be made with the use of a multiple-comparison procedure. Determine the total number of pairwise comparisons for the following values of k:
- **a.** $k = 3$
- **b.** $k = 5$
- **c.** $k = 4$
- **d.** $k = 10$

10.44 Consider a completely randomized design with five treatments: A, B, C, D, and E. The ANOVA F-test revealed significant differences among the means. A multiple-comparison procedure was used to compare all possible pairs of treatment means at $\alpha = .05$. The ranking of the five treatment means is summarized here. Identify which pairs of means are significantly different.

- **a.** $\overline{\text{A}\quad\text{C}\quad\text{E}\quad\text{B}}\ \text{D}$
- **b.** $\text{A}\quad\overline{\text{C}\quad\text{E}\quad\text{B}}\ \text{D}$
- **c.** $\text{A}\quad\overline{\text{C}\quad\text{E}\quad\text{B}\quad\text{D}}$
- **d.** $\text{A}\quad\text{C}\quad\overline{\text{E}\quad\text{B}\quad\text{D}}$

10.45 A multiple-comparison procedure for comparing four
[NW] treatment means produced the confidence intervals shown here. Rank the means from smallest to largest. Which means are significantly different?

$(\mu_1 - \mu_2)$: $(2, 15)$

$(\mu_1 - \mu_3)$: $(4, 7)$

$(\mu_1 - \mu_4)$: $(-10, 3)$

$(\mu_2 - \mu_3)$: $(-5, 11)$

$(\mu_2 - \mu_4)$: $(-12, -6)$

$(\mu_3 - \mu_4)$: $(-8, -5)$

Applying the Concepts—Basic

10.46 Whales entangled in fishing gear. Refer to the *Marine Mammal Science* (April 2010) investigation of whales entangled by fishing gear, Exercise 10.28 (p. 494). The mean body lengths (meters) of whales entangled in each of the three types of fishing gear (set nets, pots, and gill nets) are reported below. Tukey's method was used to conduct multiple comparisons of the means with an experimentwise error rate of .01. Based on the results, which type of fishing gear will entangle the shortest whales, on average? The longest whales, on average?

Mean Length:	4.45	5.28	5.63
Fishing Gear:	Set nets	Gill nets	Pots

10.47 Guilt in decision making. The effect of guilt emotion on how a decision maker focuses on the problem was investigated in the Jan. 2007 issue of the *Journal of Behavioral Decision Making*. A sample of 77 volunteer students participated in one portion of the experiment, where each was randomly assigned to one of three emotional states (guilt, anger, or neutral) through a reading/writing task. Immediately after the task, the students were presented with a decision problem where the stated option has predominantly negative features (e.g., spending money on repairing a very old car). Prior to making the decision, the researchers asked each subject to list possible, more attractive, alternatives. The researchers then compared the mean number of alternatives listed across the three emotional states with an analysis of variance for a completely randomized design. A partial ANOVA summary table is shown next.

Source	df	F-value	p-value
Emotional State	2	22.68	0.001
Error	74		
Total	76		

- **a.** What conclusion can you draw from the ANOVA results?
- **b.** A multiple comparisons of means procedure was applied to the data using an experiment-wise error rate of .05. Explain what the .05 represents.
- **c.** The multiple comparisons yielded the following results. What conclusion can you draw?

Sample mean:	1.90	2.17	4.75
Emotional State:	*Angry*	*Neutral*	*Guilt*

10.48 College tennis recruiting with a team Web site. Refer to *The Sport Journal* (Winter 2004) study comparing the attitudes of Division I, Division II, and Division III college tennis coaches towards team Web sites as recruiting tools, presented in Exercise 10.29 (p. 494). The mean responses (measured on a seven-point scale) to the statement "The Prospective Student-Athlete Form on the Web site contributes very little to the recruiting process" are listed and ranked in the accompanying table. The results were obtained with the use of a multiple-comparison procedure with an experimentwise error rate of .05. Interpret the results practically.

Mean:	4.51	3.60	3.21
Division:	I	II	III

10.49 Chemical properties of whole wheat breads. Whole wheat breads contain a high amount of phytic acid, which tends to lower the absorption of nutrient minerals. The *Journal of Agricultural and Food Chemistry* (Jan. 2005) published the results of a study to determine whether sourdough can increase the solubility of whole wheat bread. Four types of bread were prepared from whole meal flour: (1) yeast added, (2) sourdough added, (3) no yeast or sourdough added (control), and (4) lactic acid added. Data were collected on the soluble magnesium level (percent of total magnesium) during fermentation for samples of each type of bread and were analyzed with the use of a one-way ANOVA. The four mean soluble magnesium means were compared in pairwise fashion with Bonferroni's method. The results are summarized as follows:

Mean:	7%	12.5%	22%	27.5%
Type of Bread:	Control	Yeast	Lactate	Sourdough

- **a.** How many pairwise comparisons are made in the Bonferroni analysis?
- **b.** Which treatment(s) yielded the significantly highest mean soluble magnesium level? The lowest?
- **c.** The experimentwise error rate for the analysis was .05. Interpret this value.

10.50 A new dental bonding agent. Refer to the *Trends in Biomaterials & Artificial Organs* (Jan. 2003) study of a new bonding adhesive for teeth, presented in Exercise 10.30 (p. 494). A completely randomized design was used to compare the mean breaking strengths of teeth bonded for three different bonding times: 1 hour, 24 hours, and 48 hours. The sample mean breaking strengths were $\bar{x}_{1\ \text{hour}} = 3.32$ Mpa,

$\bar{x}_{24 \text{ hours}} = 5.07$ Mpa, and $\bar{x}_{48 \text{ hours}} = 5.03$ Mpa. Using an experimentwise error rate of .05, Tukey's method detected no significant difference between the means at 24 and 48 hours; however, the mean at 1 hour was found to be significantly smaller than the other two means.

a. Illustrate the results of the multiple comparisons of means by ordering the sample means and connecting means that are not significantly different.

b. What practical conclusions can you draw from the analysis?

c. Give a measure of reliability (i.e., overall confidence level) for the inferences drawn in part **b.**

10.51 Robots trained to behave like ants. Refer to the *Nature* (Aug. 2000) study of robots trained to behave like ants, presented in Exercise 10.31 (p. 494). Multiple comparisons of mean energy expended for the four colony sizes were conducted with an experimentwise error rate of .05. The results are summarized as follows:

Sample mean:	.97	.95	.93	.80
Group size:	3	6	9	12

a. How many pairwise comparisons are conducted in this analysis?

b. Interpret the results shown.

Applying the Concepts—Intermediate

10.52 Study of recall of TV commercials. Refer to the *Journal of Applied Psychology* (June 2002) completely randomized design study to compare the mean commercial recall scores of viewers of three TV programs, presented in Exercise 10.33 (p. 495). Recall that one program had a violent content code (V) rating, one had a sex content code (S) rating, and one was a neutral TV program. Using Tukey's method, the researchers conducted multiple comparisons of the three mean recall scores.

a. How many pairwise comparisons were made in this study?

b. The multiple-comparison procedure was applied to the **TVADRECALL** data and the results are shown in the MINITAB printout at the bottom of the page. An experimentwise error rate of .05 was used. Locate the confidence interval for the comparison of the V and S groups. Interpret this result practically.

c. Repeat part **b** for the remaining comparisons. Which of the groups has the largest mean recall score?

d. In the journal article, the researchers concluded that "memory for [television] commercials is impaired after watching violent or sexual programming." Do you agree?

10.53 Dental fear study. Does recalling a traumatic dental experience increase your level of anxiety at the dentist's office? In a study published in *Psychological Reports* (Aug. 1997), researchers at Wittenberg University randomly assigned 74 undergraduate psychology students to one of three experimental conditions. Subjects in the "Slide" condition viewed 10 slides of scenes from a dental office. Subjects in the "Questionnaire" condition completed a full dental history questionnaire; one of the questions asked them to describe their worst dental experience. Subjects in the "Control" condition received no actual treatment. All students then completed the Dental Fear Scale, with scores ranging from 27 (no fear) to 135 (extreme fear). The sample dental fear means for the Slide, Questionnaire, and Control groups were reported as 43.1, 53.8, and 41.8, respectively.

a. A completely randomized ANOVA design was carried out on the data, with the following results: $F = 4.43$, p-value $< .05$. Interpret these results.

b. According to the article, a Bonferroni ranking of the three dental fear means (at $\alpha = .05$) "indicated a

```
Grouping Information Using Tukey Method

           N   Mean   Grouping
NEUTRAL   108   3.167   A
VIOLENT   108   2.083     B
SEX       108   1.713     B

Means that do not share a letter are significantly different.


Tukey 95% Simultaneous Confidence Intervals
All Pairwise Comparisons

Individual confidence level = 98.01%


VIOLENT subtracted from:

          Lower   Center  Upper   -------+---------+---------+---------+--
SEX      -0.923   -0.370  0.183            (----*----)
NEUTRAL   0.530    1.083  1.636                       (----*----)
                                  -------+---------+---------+---------+--
                                     -1.2      0.0       1.2       2.4


SEX subtracted from:

          Lower   Center  Upper   -------+---------+---------+---------+--
NEUTRAL   0.901    1.454  2.007                       (---*----)
                                  -------+---------+---------+---------+--
                                     -1.2      0.0       1.2       2.4
```

MINITAB output for Exercise 10.52

significant difference between the mean scores on the Dental Fear Scale for the Control and Questionnaire groups, but not for the means between the Control and Slide groups." Summarize these results in a chart similar to Figure 10.16 (p. 500).

10.54 Is honey a cough remedy? Refer to the *Archives of Pediatrics and Adolescent Medicine* (Dec. 2007) study of treatments for children's cough symptoms, Exercise 10.35 (p. 496). The data are saved in the **HONEYCOUGH** file. Do you agree with the statement (extracted from the article), "honey may be a preferable treatment for the cough and sleep difficulty associated with childhood upper respiratory tract infection"? Perform a multiple comparisons of means to answer the question.

10.55 Estimating the age of glacial drifts. Refer to the *American Journal of Science* (Jan. 2005) study of the chemical makeup of buried tills (glacial drifts) in Wisconsin, presented in Exercise 10.37 (p. 496). The data is saved in the **TILLRATIO** file. Use a multiple-comparisons procedure to compare the mean Al/Be ratios for the five boreholes (labeled UMRB-1, UMRB-2, UMRB-3, SWRA, and SD), with an experimentwise error rate of .10. Identify the means that appear to differ.

10.56 Effect of scopolamine on memory. The drug scopolamine is often used as a sedative to induce sleep in patients. In *Behavioral Neuroscience* (Feb. 2004), medical researchers examined scopolamine's effects on memory with associated word pairs. A total of 28 human subjects, recruited from a university community, were given a list of related word pairs to memorize. For every word pair in the list (e.g., robber–jail), there was an associated word pair with the same first word, but a different second word (e.g., robber– police). The subjects were then randomly divided into three treatment groups. Group 1 subjects were administered an injection of scopolamine, group 2 subjects were given an injection of glycopyrrolate (an active placebo), and group 3 subjects were not given any drug. Four hours later, subjects were shown 12 word pairs from the list and tested on how many they could recall. The data on number of pairs recalled (simulated on the basis of summary information provided in the research article) are listed below and saved in the **SCOPOLAMINE** file. Prior to the analysis, the researchers theorized that the mean number of word pairs recalled for the scopolamine subjects (group 1) would be less than the corresponding means for the other two groups.

Group 1 (Scopolamine):	5	8	8	6	6	6	6	8	6	4	5	6
Group 2 (Placebo):	8	10	12	10	9	7	9	10				
Group 3 (No drug):	8	9	11	12	11	10	12	12				

a. Explain why this is a completely randomized design.

b. Identify the treatments and response variable.

c. Find the sample means for the three groups. Is this sufficient information to support the researchers' theory? Explain.

d. Conduct an ANOVA *F*-test on the data. Is there sufficient evidence (at $\alpha = .05$) to conclude that the mean number of word pairs recalled differs among the three treatment groups?

e. Conduct multiple comparisons of the three means (using an experimentwise error rate of .05). Do the results support the researchers' theory? Explain.

Applying the Concepts—Advanced

10.57 Restoring self-control while intoxicated. Refer to the *Experimental and Clinical Psychopharmacology* (Feb. 2005) study of restoring self-control while intoxicated, presented in Exercise 10.34 (p. 495). The researchers theorized that if caffeine can really restore self-control, then students in Group AC (the group that drank alcohol plus caffeine) will perform the same as students in Group P (the placebo group) on the word completion task. Similarly, if an incentive can restore self-control, then students in Group AR (the group that drank alcohol and got a reward for correct responses on the word completion task) will perform the same as students in Group P. Finally, the researchers theorized that students in Group A (the alcohol-only group) will perform worse on the word completion task than students in any of the other three groups. Access the data in the **DRINKERS** file and conduct Tukey's multiple comparisons of the means, using an experimentwise error rate of .05. Are the researchers' theories supported?

10.58 Animal-assisted therapy for heart patients. Refer to the *American Heart Association Conference* (Nov. 2005) study to gauge whether animal-assisted therapy can improve the physiological responses of heart failure patients, presented in Exercise 10.38 (p. 497). You found evidence of a difference among the treatment means for the three treatments: Group T (volunteer plus trained dog), Group V (volunteer only), and Group C (control). Conduct a Bonferroni analysis to rank the three treatment means. Use an experimentwise error rate of $\alpha = .03$. Interpret the results for the researchers. [*Hint:* As shown in Appendix B, the Bonferroni formula for a confidence interval for the difference $(\mu_i - \mu_j)$ is

$$(\bar{x}_i - \bar{x}_j) \pm t_{\alpha^*/2}(s)\sqrt{(1/n_i) + (1/n_j)}$$

where $\alpha^* = 2\alpha/[(k)(k-1)]$ is the experimentwise error rate, and k is the total number of treatment means compared.]

10.4 The Randomized Block Design

If the completely randomized design results in nonrejection of the null hypothesis that the treatment means are equal because the sampling variability (as measured by MSE) is large, we may want to consider an experimental design that better controls the variability. In contrast to the selection of independent samples of experimental units specified by the completely randomized design, the *randomized block design* utilizes experimental units that are *matched sets*, assigning one from each set to each treatment. The matched sets of experimental units are called *blocks*. The theory behind the randomized block design is that the sampling variability of the experimental units in each block will be reduced, in turn reducing the measure of error, MSE.

> The **randomized block design** consists of a two-step procedure:
>
> 1. Matched sets of experimental units, called **blocks**, are formed, with each block consisting of k experimental units (where k is the number of treatments). The b blocks should consist of experimental units that are as similar as possible.
>
> 2. One experimental unit from each block is randomly assigned to each treatment, resulting in a total of $n = bk$ responses.

For example, if we wish to compare SAT scores of female and male high school seniors, we could select independent random samples of five females and five males, and analyze the results of the completely randomized design as outlined in Section 10.2. Or we could select matched pairs of females and males according to their scholastic records and analyze the SAT scores of the pairs. For instance, we could select pairs of students with approximately the same GPAs from the same high school. Five such pairs (blocks) are depicted in Table 10.7. Note that this is just a *paired difference experiment*, first discussed in Section 9.3.

Table 10.7	Randomized Block Design: SAT Score Comparison		
Block	**Female SAT Score**	**Male SAT Score**	**Block Mean**
1 (School A, 2.75 GPA)	540	530	535
2 (School B, 3.00 GPA)	570	550	560
3 (School C, 3.25 GPA)	590	580	585
4 (School D, 3.50 GPA)	640	620	630
5 (School E, 3.75 GPA)	690	690	690
Treatment mean	606	594	

Data Set: TAB10_7

As before, the variation between the treatment means is measured by squaring the distance between each treatment mean and the overall mean, multiplying each squared distance by the number of measurements for the treatment, and then summing over treatments:

$$\text{SST} = \sum_{i=1}^{k} b(\bar{x}_{T_i} - \bar{x})^2$$

$$= 5(606 - 600)^2 + 5(594 - 600)^2 = 360$$

Here, $\bar{x}_{T_i}$ represents the sample mean for the ith treatment, b (the number of blocks) is the number of measurements for each treatment, and k is the number of treatments.

The blocks also account for some of the variation among the different responses. That is, just as SST measures the variation between the female and male means, we can calculate a measure of variation among the five block means representing different schools and scholastic abilities. Analogous to the computation of SST, we sum the squares of the differences between each block mean and the overall mean, multiply each squared difference by the number of measurements for each block, and then sum over blocks to calculate the **sum of squares for blocks (SSB):**

$$\text{SSB} = \sum_{i=1}^{b} k(\bar{x}_{B_i} - \bar{x})^2$$

$$= 2(535 - 600)^2 + 2(560 - 600)^2 + 2(585 - 600)^2 +$$
$$2(630 - 600)^2 + 2(690 - 600)^2$$
$$= 30,100$$

Here, $\bar{x}_{B_i}$ represents the sample mean for the ith block and k (the number of treatments) is the number of measurements in each block. As we expect, the variation in SAT scores attributable to Schools and Levels of scholastic achievement is apparently large.

As before, we want to compare the variability attributed to treatments with that which is attributed to sampling. In a randomized block design, the sampling variability is measured by subtracting that portion attributed to treatments and blocks from the total sum of squares, SS(Total). The total variation is the sum of the squared differences of each measurement from the overall mean:

$$SS(Total) = \sum_{i=1}^{n}(x_i - \bar{x})^2$$

$$= (540 - 600)^2 + (530 - 600)^2 + (570 - 600)^2 + (550 - 600)^2 + \cdots + (690 - 600)^2$$

$$= 30,600$$

The variation attributable to sampling error is then found by subtraction:

$$SSE = SS(Total) - SST - SSB = 30,600 - 360 - 30,100 = 140$$

In sum, the total sum of squares, 30,600, is divided into three components: 360 attributed to treatments (Gender), 30,100 attributed to blocks (Scholastic ability), and 140 attributed to sampling error.

The mean squares associated with each source of variability are obtained by dividing the sum of squares by the appropriate number of degrees of freedom. The partitioning of the total sum of squares and the total number of degrees of freedom for a randomized block experiment are summarized in Figure 10.18.

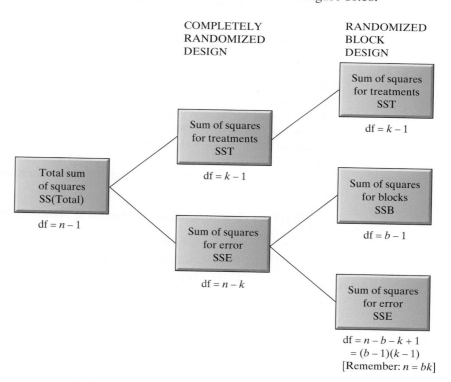

Figure 10.18

Partitioning of the total sum of squares for the randomized block design

To determine whether we can reject the null hypothesis that the treatment means are equal in favor of the alternative that at least two of them differ, we calculate

$$MST = \frac{SST}{k-1} = \frac{360}{2-1} = 360$$

$$MSE = \frac{SSE}{n-b-k+1} = \frac{140}{10-5-2+1} = 35$$

The F-ratio that is used to test the hypothesis is

$$F = \frac{360}{35} = 10.29$$

Comparing this ratio with the tabular F-value corresponding to $\alpha = .05$, $v_1 = (k - 1) = 1$ degree of freedom in the numerator, and $v_2 = (n - b - k + 1) = 4$ degrees of freedom in the denominator, we find that

$$F = 10.29 > F_{.05} = 7.71$$

which indicates that we should reject the null hypothesis and conclude that the mean SAT scores differ for females and males. All of these calculations, of course, can be obtained using statistical software. The output for the MINITAB analysis of the data in Table 10.7 is shown in Figure 10.19. The values of SST, SSE, MST, MSE, and F are highlighted on the printout.

Figure 10.19

MINITAB printout with ANOVA results for the data in Table 10.7

Two-way ANOVA: SAT versus GENDER, BLOCK

Source	DF	SS	MS	F	P
GENDER	1	360	360	10.29	0.033
BLOCK	4	30100	7525	215.00	0.000
Error	4	140	35		
Total	9	30600			

Comment: If you review Section 9.3, you will find that the analysis of a paired difference experiment results in a one-sample t-test on the differences between the treatment responses within each block. Applying the procedure to the differences between female and male scores in Table 10.7, we obtain

$$t = \frac{\bar{x}_d}{s_d/\sqrt{n_d}} = \frac{12}{\sqrt{70}/\sqrt{5}} = 3.207$$

At the .05 level of significance with $(n_d - 1) = 4$ degrees of freedom,

$$t = 3.207 > t_{.025} = 2.776$$

Since $t^2 = (3.207)^2 = 10.29$ and $t_{.025}^2 = (2.776)^2 = 7.71$, we find that the paired difference t-test and the ANOVA F-test are equivalent, with both the calculated test statistics and the rejection region related by the formula $F = t^2$. The difference between the tests is that the paired difference t-test can be used to compare only two treatments in a randomized block design, whereas the F-test can be applied to *two or more* treatments in a randomized block design.

The F-test for a randomized block design is summarized in the following box:

ANOVA F-Test to Compare k Treatment Means: Randomized Block Design

H_0: $\mu_1 = \mu_2 = \cdots = \mu_k$

H_a: At least two treatment means differ

Test statistic: $F = \dfrac{\text{MST}}{\text{MSE}}$

Rejection region: $F > F_\alpha$, where F_α is based on $(k - 1)$ numerator degrees of freedom and $(n - b - k + 1)$ denominator degrees of freedom.

Conditions Required for a Valid ANOVA F-Test: Randomized Block Design

1. The b blocks are randomly selected, and all k treatments are applied (in random order) to each block.
2. The distributions of observations corresponding to all bk block–treatment combinations are approximately normal.
3. The bk block–treatment distributions have equal variances.

Note that the assumptions concern the probability distributions associated with each block–treatment combination. The experimental unit selected for each combination is assumed to have been randomly selected from all possible experimental units for that combination, and the response is assumed to be normally distributed with the same variance for each of the block–treatment combinations. For example, the *F*-test comparing female and male SAT score means requires the scores for each combination of gender and scholastic ability (e.g., females with 3.25 GPA) to be normally distributed with the same variance as the other combinations employed in the experiment.

For those who are interested, the calculation formulas for randomized block designs are given in Appendix B. Throughout this section, we will rely on statistical software packages to analyze randomized block designs and to obtain the necessary ingredients for testing the null hypothesis that the treatment means are equal.

Example 10.7

Experimental Design Principles

Problem Refer to Examples 10.4–10.6. Suppose the USGA wants to compare the mean distances associated with the four brands of golf balls struck by a driver, but wishes to employ human golfers rather than the robot Iron Byron. Assume that 10 balls of each brand are to be utilized in the experiment.

a. Explain how a completely randomized design could be employed.

b. Explain how a randomized block design could be employed.

c. Which design is likely to provide more information about the differences among the brand mean distances?

Solution

a. Since the completely randomized design calls for independent samples, we can employ such a design by randomly selecting 40 golfers and then randomly assigning 10 golfers to each of the four brands. Finally, each golfer will strike the ball of the assigned brand, and the distance will be recorded. This design is illustrated in Figure 10.20.

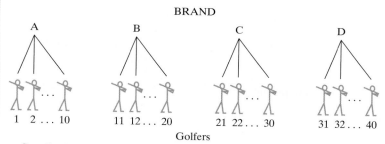

Figure 10.20

Illustration of completely randomized design and randomized block design: comparison of four golf ball brands

a. Completely randomized design

b. Randomized block design

b. The randomized block design employs blocks of relatively homogeneous experimental units. For example, we could randomly select 10 golfers and permit each golfer to hit four balls, one of each brand, in a random sequence. Then each golfer is a block, with each treatment (brand) assigned to each block (golfer). This design is summarized in Figure 10.20.

c. Because we expect much more variability among distances generated by "real" golfers than by Iron Byron, we would expect the randomized block design to control the variability better than the completely randomized design does. That is, with 40 different golfers, we would expect the sampling variability among the measured distances within each brand to be greater than that among the four distances generated by each of 10 golfers hitting one ball of each brand.

Now Work Exercise 10.69 a,b

Example 10.8

A Randomized Block Design ANOVA— Comparing Golf Ball Brands

Problem Refer to Example 10.7. Suppose the randomized block design of part **b** is employed, utilizing a random sample of 10 golfers, with each golfer using a driver to hit four balls, one of each brand, in a random sequence.

a. Set up a test of the research hypothesis that the brand mean distances differ. Use $\alpha = .05$.

b. The data for the experiment are given in Table 10.8. Use statistical software to analyze the data, and conduct the test set up in part **a.**

Table 10.8   Distance Data for Randomized Block Design

Golfer (Block)	Brand A	Brand B	Brand C	Brand D
1	202.4	203.2	223.7	203.6
2	242.0	248.7	259.8	240.7
3	220.4	227.3	240.0	207.4
4	230.0	243.1	247.7	226.9
5	191.6	211.4	218.7	200.1
6	247.7	253.0	268.1	244.0
7	214.8	214.8	233.9	195.8
8	245.4	243.6	257.8	227.9
9	224.0	231.5	238.2	215.7
10	252.2	255.2	265.4	245.2
Sample means	227.0	233.2	245.3	220.7

Data Set: GOLFRBD

Solution

a. We want to test whether the data in Table 10.8 provide sufficient evidence to conclude that the brand mean distances differ. Denoting the population mean of the *i*th brand by μ_i, we test

$$H_0: \mu_1 = \mu_2 = \mu_3 = \mu_4$$
$$H_a: \text{The mean distances differ for at least two of the brands.}$$

The test statistic compares the variation among the four treatment (brand) means with the sampling variability within each of the treatments:

$$\text{Test statistic:} \quad F = \frac{\text{MST}}{\text{MSE}}$$

Rejection region: $F > F_\alpha = F_{.05}$, with $v_1 = (k - 1) = 3$ numerator degrees of freedom and $v_2 = (n - k - b + 1) = 27$ denominator degrees of freedom. From Table IX of Appendix A, we find that $F_{.05} = 2.96$. Thus, we will reject H_0 if $F > 2.96$.

The assumptions necessary to ensure the validity of the test are as follows: (1) The probability distributions of the distances for each brand–golfer combination are normal. (2) The variances of the distance probability distributions for each brand–golfer combination are equal.

b. SPSS was used to analyze the data in Table 10.8, and the result is shown in Figure 10.21. The values of MST and MSE (highlighted on the printout) are 1,099.552 and 20.245, respectively. The F-ratio for Brand (also highlighted on the printout) is $F = 54.312$, which exceeds the tabled value of 2.96. We therefore reject the null hypothesis at the $\alpha = .05$ level of significance, concluding that at least two of the brands differ with respect to mean distance traveled when struck by the driver.

Tests of Between-Subjects Effects

Dependent Variable:DISTANCE

Source	Type III Sum of Squares	df	Mean Square	F	Sig.
Corrected Model	15372.539a	12	1281.045	63.276	.000
Intercept	2145032.910	1	2145032.910	105952.598	.000
BRAND	3298.657	3	1099.552	54.312	.000
GOLFER	12073.882	9	1341.542	66.265	.000
Error	546.621	27	20.245		
Total	2160952.070	40			
Corrected Total	15919.160	39			

a. R Squared = .966 (Adjusted R Squared = .950)

Figure 10.21

SPSS printout for randomized block design ANOVA of data in Table 10.8

Look Back The result of part **b** is confirmed by noting that the observed significance level of the test, highlighted on the printout, is $p \approx 0$.

Now Work Exercise 10.69c

The results of an ANOVA for a randomized block design can be summarized in a simple tabular format similar to that utilized for the completely randomized design in Section 10.2. The general form of the table is shown in Table 10.9, and that for Example 10.8 is given in Table 10.10. Note that the randomized block design is characterized by three sources of variation—treatments, blocks, and error—which sum to the total sum of squares. We hope that employing blocks of experimental units will reduce the error variability, thereby making the test for comparing treatment means more powerful.

When the F-test results in the rejection of the null hypothesis that the treatment means are equal, we will usually want to compare the various pairs of treatment means to determine which specific pairs differ. We can employ a multiple-comparison procedure as in Section 10.3. The number of pairs of means to be compared will again be $c = k(k - 1)/2$, where k is the number of treatment means. In Example 10.8, $c = 4(3)/2 = 6$; that is, there are six pairs of golf ball brand means to be compared.

Table 10.9   General ANOVA Summary Table for a Randomized Block Design

Source	df	SS	MS	F
Treatments	$k - 1$	SST	MST	MST/MSE
Blocks	$b - 1$	SSB	MSB	
Error	$n - k - b + 1$	SSE	MSE	
Total	$n - 1$	SS(Total)		

Table 10.10   ANOVA Table for Example 10.8

Source	df	SS	MS	F	p
Treatment (Brand)	3	3,298.66	1,099.55	54.31	.000
Block (Golfer)	9	12,073.88	1,341.54		
Error	27	546.62	20.25		
Total	39	15,919.16			

Example 10.9

Ranking Treatment Means in a Randomized Block Design—Comparing Golf Ball Brands

Problem Bonferroni's procedure is used to compare the mean distances of the four golf ball brands in Example 10.8. The resulting confidence intervals, with an experimentwise error rate of $\alpha = .05$, are shown in the SPSS printout of Figure 10.22. Interpret the results.

Multiple Comparisons

DISTANCE
Bonferroni

(I) BRAND	(J) BRAND	Mean Difference (I-J)	Std. Error	Sig.	95% Confidence Interval	
					Lower Bound	Upper Bound
A	B	-6.1300*	2.01222	.031	-11.8586	-.4014
	C	-18.2800*	2.01222	.000	-24.0086	-12.5514
	D	6.3200*	2.01222	.024	.5914	12.0486
B	A	6.1300*	2.01222	.031	.4014	11.8586
	C	-12.1500*	2.01222	.000	-17.8786	-6.4214
	D	12.4500*	2.01222	.000	6.7214	18.1786
C	A	18.2800*	2.01222	.000	12.5514	24.0086
	B	12.1500*	2.01222	.000	6.4214	17.8786
	D	24.6000*	2.01222	.000	18.8714	30.3286
D	A	-6.3200*	2.01222	.024	-12.0486	-.5914
	B	-12.4500*	2.01222	.000	-18.1786	-6.7214
	C	-24.6000*	2.01222	.000	-30.3286	-18.8714

Based on observed means.
The error term is Mean Square(Error) = 20.245.
*. The mean difference is significant at the 0.05 level.

Figure 10.22

SPSS printout of Bonferroni confidence intervals for the randomized block design

Solution Note that 12 confidence intervals are shown in Figure 10.22, rather than 6. SPSS (like SAS) computes intervals for both $\mu_i - \mu_j$ and $\mu_j - \mu_i$, $i \neq j$. Only half of these are necessary to conduct the analysis, and these are highlighted on the printout. The intervals (rounded) are summarized as follows:

$$(\mu_A - \mu_B): \quad (-11.9, -.4)$$
$$(\mu_A - \mu_C): \quad (-24.0, -12.6)$$
$$(\mu_A - \mu_D): \quad (.6, 12.0)$$
$$(\mu_B - \mu_C): \quad (-17.9, -6.4)$$
$$(\mu_B - \mu_D): \quad (6.7, 18.2)$$
$$(\mu_C - \mu_D): \quad (18.9, 30.3)$$

Note that we are 95% confident that all the brand means differ, because none of the intervals contains 0. The listing of the brand means in Figure 10.23 has no lines connecting them, because there are no nonsignificant differences at the .05 level.

Mean:	220.7	227.0	233.2	245.3
Brand:	D	A	B	C

Figure 10.23

Listing of brand means for randomized block design [*Note:* All differences are statistically significant.]

Now Work Exercise 10.69d

Unlike the completely randomized design, the randomized block design cannot, in general, be used to estimate individual treatment means. Whereas the completely randomized design employs a random sample for each treatment, the randomized block design does not necessarily do so: The experimental units within the blocks are assumed to be randomly selected, but the blocks themselves may not be randomly selected.

We can, however, test the hypothesis that the block means are significantly different. We simply compare the variability attributable to differences among the block means with that associated with sampling variability. The ratio of MSB to MSE is an F-ratio similar to that formed in testing treatment means. The F-statistic is compared with a tabular value for a specific value of α, with $(b - 1)$ numerator degrees of freedom and $(n - k - b + 1)$ denominator degrees of freedom. The test is usually given on the same printout as the test for treatment means. Note in the SPSS printout in Figure 10.21 that the test statistic for comparing the block means is

$$F = \frac{\text{MSB}}{\text{MSE}} = \frac{\text{MS(Golfers)}}{\text{MS(Error)}} = \frac{1{,}341.54}{20.25} = 66.27$$

with a p-value of .000. Since $\alpha = .05$ exceeds this p-value, we conclude that the block means are different. The results of the test are summarized in Table 10.11.

Table 10.11 ANOVA Table for Randomized Block Design: Test for Blocks Included

Source	df	SS	MS	F	p
Treatments (Brands)	3	3,298.66	1,099.55	54.31	.000
Blocks (Golfers)	9	12,073.88	1,341.54	66.27	.000
Error	27	546.62	20.25		
Total	39	15,919.16			

In the golf example, the test for block means confirms our suspicion that the golfers vary significantly; therefore, the use of the block design was a good decision. However, be careful not to conclude that the block design was a mistake if the F-test for blocks does not result in rejection of the null hypothesis that the block means are the same. Remember that the possibility of a Type II error exists, and we are not controlling its probability as we are the probability α of a Type I error. If the experimenter believes that the experimental units are more homogeneous within blocks than between blocks, then he or she should use the randomized block design regardless of the results of a single test comparing the block means.

The procedure for conducting an analysis of variance for a randomized block design is summarized in the next box. Remember that the hallmark of this design is the utilization of blocks of homogeneous experimental units in which each treatment is represented.

Steps for Conducting an ANOVA for a Randomized Block Design

1. Be sure that the design consists of blocks (preferably, blocks of homogeneous experimental units) and that each treatment is randomly assigned to one experimental unit in each block.

2. If possible, check the assumptions of normality and equal variances for all block–treatment combinations. [*Note:* This may be difficult to do, since the design will likely have only one observation for each block–treatment combination.]

3. Create an ANOVA summary table that specifies the variability attributable to treatments, blocks, and error, and that leads to the calculation of the F-statistic to test the null hypothesis that the treatment means are equal in the population. Use a statistical software package or the calculation formulas in Appendix B to obtain the necessary numerical ingredients.

4. If the F-test leads to the conclusion that the means differ, employ the Bonferroni or Tukey procedure, or a similar procedure, to conduct multiple comparisons of as many of the pairs of means as you wish. Use the results to summarize the statistically significant differences among the treatment means. Remember that, in general, the randomized block design cannot be employed to form confidence intervals for individual treatment means.

5. If the F-test leads to the nonrejection of the null hypothesis that the treatment means are equal, several possibilities exist:
   a. The treatment means *are* equal: that is, the null hypothesis is true.
   b. The treatment means really differ, but other important factors affecting the response are not accounted for by the randomized block design. These

(continued)

factors inflate the sampling variability, as measured by MSE, resulting in smaller values of the F-statistic. Either increase the sample size for each treatment, or conduct an experiment that accounts for the other factors affecting the response (as is to be done in Section 10.5). Do not automatically reach the former conclusion, since the possibility of a Type II error must be considered if you accept H_0.

6. If desired, conduct the F-test of the null hypothesis that the block means are equal. Rejection of this hypothesis lends statistical support to the utilization of the randomized block design.

Note: It is often difficult to check whether the assumptions for a randomized block design are satisfied. When you feel that these assumptions are likely to be violated, a nonparametric procedure is advisable.

What Do You Do When the Assumptions Are Not Satisfied for the Analysis of Variance for a Completely Randomized Design?

Answer: Use a nonparametric statistical method such as the Friedman F_r test of Section 14.6.

Exercises 10.59–10.76

Understanding the Principles

10.59 Explain the difference between a randomized block design and a paired difference experiment.

10.60 When is it advantageous to use a randomized block design over a completely randomized design?

10.61 What conditions are required for valid inferences from a randomized block ANOVA design?

Learning the Mechanics

10.62 A randomized block design yielded the ANOVA table shown here:

Source	df	SS	MS	F
Treatments	4	501	125.25	9.109
Blocks	2	225	112.50	8.182
Error	8	110	13.75	
Total	14	836		

a. How many blocks and treatments were used in this experiment?

b. How many observations were collected in the experiment?

c. Specify the null and alternative hypotheses you would use to compare the treatment means.

d. What test statistic should be used to conduct the hypothesis test of part **c**?

e. Specify the rejection region for the test of parts **c** and **d**. Use $\alpha = .01$.

f. Conduct the test of parts **c–e,** and state the proper conclusion.

g. What assumptions are necessary to ensure the validity of the test you conducted in part **f**?

10.63 An experiment was conducted that used a randomized block design. The data from the experiment are displayed in the following table and saved in the **LM10_63** file.

Treatment	Block 1	Block 2	Block 3
1	2	3	5
2	8	6	7
3	7	6	5

a. Fill in the missing entries in the following ANOVA table:

Source	df	SS	MS	F
Treatments		21.5555		
Blocks				
Error				
Total		30.2222		

b. Specify the null and alternative hypotheses you would use to investigate whether a difference exists among the treatment means.

c. What test statistic should be used in conducting the test of part **b**?

d. Describe the Type I and Type II errors associated with the hypothesis test of part **b**.

e. Conduct the hypothesis test of part **b**, using $\alpha = .05$.

10.64 Suppose an experiment utilizing a randomized block design has four treatments and nine blocks, for a total of $4 \times 9 = 36$ observations. Assume that the total sum of squares for the response is SS(Total) = 500. For each of the following partitions of SS(Total), test the null hypothesis that the treatment means are equal and the null hypothesis that the block means are equal (use $\alpha = .05$ for each test):

a. The sum of squares for treatments (SST) is 20% of SS(Total), and the sum of squares for blocks (SSB) is 30% of SS(Total).

b. SST is 50% of SS(Total), and SSB is 20% of SS(Total).

c. SST is 20% of SS(Total), and SSB is 50% of SS(Total).

d. SST is 40% of SS(Total), and SSB is 40% of SS(Total).

e. SST is 20% of SS(Total), and SSB is 20% of SS(Total).

10.65 A randomized block design was used to compare the mean responses for three treatments. Four blocks of three homogeneous experimental units were selected, and each treatment was randomly assigned to one experimental unit within each block. The data (saved in the **LM10_65** file) are shown in the following table, and a MINITAB ANOVA printout for this experiment is displayed below.

	Block			
Treatment	1	2	3	4
A	3.4	5.5	7.9	1.3
B	4.4	5.8	9.6	2.8
C	2.2	3.4	6.9	.3

a. Use the printout to fill in the entries in the following ANOVA table:

Source	df	SS	MS	F
Treatments				
Blocks				
Error				
Total				

b. Do the data provide sufficient evidence to indicate that the treatment means differ?

c. Do the data provide sufficient evidence to indicate that blocking was effective in reducing the experimental error? Use $\alpha = .05$.

d. Use the printout to rank the treatment means at $\alpha = .05$.

e. What assumptions are necessary to ensure the validity of the inferences made in parts **b, c,** and **d**?

Applying the Concepts—Basic

10.66 Making high-stakes insurance decisions. The *Journal of Economic Psychology* (Sept. 2008) published a study on high-stakes insurance decisions. In part A of the experiment, 84 subjects were informed of the hazards (both fire and theft) of owning a valuable painting, but were not told the exact probabilities of the hazards occurring. The subjects then provided an amount they were willing to pay (WTP) for insuring the painting. In part B of the experiment, these same subjects were informed of the exact probabilities of the hazards (fire and theft) of owning a valuable sculpture.

MINITAB output for Exercise 10.65

```
Analysis of Variance for X

Source       DF      SS       MS       F       P
Treatment     2   12.032    6.016   50.96   0.000
Block         3   71.749   23.916  202.59   0.000
Error         6    0.708    0.118
Total        11   84.489


S = 0.343592    R-Sq = 99.16%    R-Sq(adj) = 98.46%


Grouping Information Using Tukey Method and 95.0% Confidence

Treatment   N   Mean   Grouping
B           4   5.7    A
A           4   4.5      B
C           4   3.2        C

Means that do not share a letter are significantly different.


Tukey 95.0% Simultaneous Confidence Intervals
Response Variable X
All Pairwise Comparisons among Levels of Treatment
Treatment = A   subtracted from:

Treatment   Lower   Center   Upper    -+---------+---------+---------+-----
B           0.379   1.125   1.8706                            (----*---)
C          -2.071  -1.325  -0.5794                (----*----)
                                      -+---------+---------+---------+-----
                                    -3.0      -1.5       0.0       1.5


Treatment = B   subtracted from:

Treatment   Lower   Center   Upper    -+---------+---------+---------+-----
C          -3.196  -2.450  -1.704     (----*----)
                                      -+---------+---------+---------+-----
                                    -3.0      -1.5       0.0       1.5
```

The subjects then provided a WTP amount for insuring the sculpture. The researchers were interested in comparing the mean WTP amounts for the painting and the sculpture.

a. Explain why the experimental design employed is a randomized block design.

b. Identify the dependent (response) variable, treatments, and blocks for the design.

c. Give the null and alternative hypotheses of interest to the researchers.

10.67 **"Topsy-turvy" seasons in college football.** Each week during the college football season the Associated Press (AP) ranks the top 25 Division I college football teams based on voting by sportswriters. Many recent upsets of top-rated teams have led to major changes in the weekly AP Poll over the past several seasons. Statisticians A. K. Kaw and A. Yalcin have created a formula for determining a weekly "topsy-turvy" (TT) index, designed to measure the degree to which the top 25 ranked teams changed from the previous week (*Chance*, Summer 2009). The greater the TT index, the greater the changes in the ranked teams. The statisticians calculated the TT index each week of the 15-week college football season for 6 recent seasons. In order to determine whether any of the 15 weeks in a season tends to be more or less topsy-turvy than others, they conducted an ANOVA on the data using a randomized block design. Here, weeks were considered the treatments and seasons were the blocks. The ANOVA summary table is shown below.

Source	df	SS	MS	F	p-Value
Weeks	14	3,562.3	254.4	2.57	.0044
Seasons	5	2,936.5	587.3	5.94	.0001
Error	70	6,914.4	98.8		
Total	89	13,413.1			

Source: Kaw, A. K., and Yalcin, A. "A metric to quantify college football's topsy-turvy season." *Chance*, Vol. 22, No. 3, Summer 2009 (Table 2). Reprinted with permission from *Chance*. Copyright 2009 by the American Statistical Association. All rights reserved.

a. Is there a significant difference (at $\alpha = .01$) in the mean TT index across the 15 weeks of the college football season?

b. Is there evidence (at $\alpha = .01$) that blocking on seasons was effective in removing an extraneous source of variation in the data?

c. Tukey's multiple comparisons method was applied to compare the mean TT index across the 15 weeks. How many pairwise comparisons are involved in this analysis?

d. Using an experimentwise error rate of .05, Tukey's analysis revealed that the last week of the regular season (week 14), when only conference championships are played, had a significantly smaller mean TT index than week 6. All other pairs of means were not significantly different. Use this result to make a comment about whether any of the 15 weeks in a college football season tend to be more or less topsy-turvy than others.

10.68 **Treatment for tendon pain.** Refer to the *British Journal of Sports Medicine* (Feb. 1, 2004) study of chronic Achilles tendon pain, presented in Exercise 10.10 (p. 480). Recall that each in a sample of 25 patients with chronic Achilles tendinosis was treated with heavy-load eccentric calf muscle training. Tendon thickness (in millimeters) was measured both before and following the treatment of each

patient. These experimental data are shown in the accompanying table and saved in the **TENDON** file.

a. What experimental design was employed in this study?

b. How many treatments are used in this study? How many blocks?

c. Give the null and alternative hypothesis for determining whether the mean tendon thickness before treatment differs from the mean after treatment.

d. Compute the test statistic for the test in part **c**, using the paired difference *t* formula of Section 9.3.

e. Use statistical software (or the formulas in Appendix B) to compute the ANOVA *F*-statistic for the test in part **c**.

f. Compare the test statistics you computed in parts **d** and **e**. You should find that $F = t^2$.

g. Show that the two tests from parts **d** and **e** yield identical *p*-values.

h. Give the appropriate conclusion, using $\alpha = .05$.

Patient	Before Thickness (millimeters)	After Thickness (millimeters)
1	11.0	11.5
2	4.0	6.4
3	6.3	6.1
4	12.0	10.0
5	18.2	14.7
6	9.2	7.3
7	7.5	6.1
8	7.1	6.4
9	7.2	5.7
10	6.7	6.5
11	14.2	13.2
12	7.3	7.5
13	9.7	7.4
14	9.5	7.2
15	5.6	6.3
16	8.7	6.0
17	6.7	7.3
18	10.2	7.0
19	6.6	5.3
20	11.2	9.0
21	8.6	6.6
22	6.1	6.3
23	10.3	7.2
24	7.0	7.2
25	12.0	8.0

Based on Ohberg, L., et al. "Eccentric training in patients with chronic Achilles tendinosis: Normalized tendon structure and decreased thickness at follow up." *British Journal of Sports Medicine*, Vol. 38, No. 1, Feb. 1, 2004 (Table 2).

10.69 **Endangered dwarf shrubs.** Rugel's pawpaw (yellow squirrel banana) is an endangered species of a dwarf shrub. Biologists from Stetson University conducted an experiment to determine the effects of fire on the shrub's growth (*Florida Scientist*, Spring 1997). Twelve experimental plots of land were selected in a pasture where the shrub is abundant. Within each plot, three pawpaws were randomly selected and treated as follows: One shrub was subjected to fire, another to clipping, and the third left unmanipulated (a control). After five months, the number of flowers produced by each of the 36 shrubs was determined. The objective of the study was to compare the mean number of flowers produced by pawpaws for the three treatments (fire, clipping, and control).

a. Identify the type of experimental design employed, including the treatments, response variable, and experimental units.

b. Illustrate the layout of the design, using a graphic similar to Figure 10.20.

c. The ANOVA of the data resulted in a test statistic of $F = 5.42$ for treatments with p-value $= .009$. Interpret this result.

d. The three treatment means were compared by Tukey's method at $\alpha = .05$. Interpret the results, shown as follows:

Mean number of flowers:	1.17	10.58	17.08
Treatment:	Control	Clipping	Burning

10.70 A new method of evaluating health care research reports. When evaluating research reports in health care, a popular tool is the Assessment of Multiple Systematic Reviews (AMSTAR). AMSTAR, which incorporates 11 items (questions), has been widely accepted by professional health associations. A group of dental researchers have revised the assessment tool and named it R-AMSTAR (*The Open Dentistry Journal*, Vol. 4, 2010). The revised assessment tool was validated on five systematic reviews (named R1,

R2, R3, R4, and R5) on rheumatoid arthritis. For each review, scores on the 11 R-AMSTAR items (all measured on a 4-point scale) were obtained. The data, saved in the **RAMSTAR** file, are shown in the accompanying table.

a. One goal of the study was to compare the mean item scores of the five reviews. Set up the null and alternative hypotheses for this test.

b. Examine the data in the table and explain why a randomized block ANOVA is appropriate to apply.

c. The SAS output for a randomized block ANOVA of the data (with Review as treatments and Item as blocks) appears below. Interpret the p-values of the tests shown.

d. The SAS printout also reports the results of a Tukey multiple-comparison analysis of the five Review means. Which pairs of means are significantly different? Do these results agree with your conclusion in part c?

e. The experimentwise error rate used in the analysis in part d is .05. Interpret this value.

Applying the Concepts—Intermediate

10.71 Stress in cows prior to slaughter. What is the level of stress (if any) that cows undergo prior to being slaughtered? To answer this question, researchers designed an experiment involving cows bred in Normandy, France (*Applied*

Data for Exercise 10.70

Review	Item 1	Item 2	Item 3	Item 4	Item 5	Item 6	Item 7	Item 8	Item 9	Item 10	Item 11
R1	4.0	1.0	4.0	2.0	3.5	3.5	3.5	3.5	1.0	1.0	1.0
R2	3.5	2.5	4.0	4.0	3.5	4.0	3.5	2.5	3.5	1.5	1.0
R3	4.0	4.0	3.5	4.0	1.5	2.5	3.5	3.5	2.5	1.5	1.0
R4	3.5	2.0	4.0	4.0	2.0	4.0	3.5	3.0	3.5	1.0	1.0
R5	3.5	4.0	4.0	3.0	2.5	4.0	4.0	4.0	2.5	1.0	2.5

Source: Kung, J., Chiappelli, F., Cajulis, O., Avezona, R., Kossan, G., Chew, L., and Maida, C. A. "From systematic reviews to clinical recommendations to clinical-based health care: Validation of revised assessment of multiple-systematic reviews (R-AMSTAR) for grading of clinical relevance," *The Open Dentistry Journal*, Vol. 4, 2010, pp. 84–91.

SAS output for Exercise 10.70

```
                              The ANOVA Procedure
Dependent Variable: SCORE

        Source              DF       Sum of         Mean Square    F Value    Pr > F
                                     Squares
        Model               14       46.01818182    3.28701299     6.36       <.0001
        Error               40       20.68181818    0.51704545
        Corrected Total     54       66.70000000


              R-Square        Coeff Var       Root MSE      SCORE Mean
              0.689928        24.79513        0.719059      2.900000


        Source              DF       Anova SS       Mean Square    F Value    Pr > F
        REVIEW              4        2.51818182     0.62954545     1.22       0.3186
        ITEM                10       43.50000000    4.35000000     8.41       <.0001


    Means with the same letter are not significantly different.

    Tukey Grouping          Mean      N     REVIEW
                 A          3.1818     11    R5
                 A
                 A          3.0455     11    R2
                 A
                 A          2.8636     11    R3
                 A
                 A          2.8636     11    R4
                 A
                 A          2.5455     11    R1
```

Animal Behaviour Science, June 2010). The heart rate (beats per minute) of a cow was measured at four different pre-slaughter phases—(1) first phase of visual contact with pen mates, (2) initial isolation from pen mates for prepping, (3) restoration of visual contact with pen mates, and (4) first contact with human prior to slaughter. Data for eight cows (simulated from information provided in the article) are shown in the accompanying table and saved in the **COWSTRESS** file. The researchers analyzed the data using an analysis of variance for a randomized block design. Their objective was to determine whether the mean heart rate of cows differed in the four pre-slaughter phases.

a. Identify the treatments and blocks for this experimental design.
b. Conduct the appropriate analysis using a statistical software package. Summarize the results in an ANOVA table.
c. Is there evidence of differences among the mean heart rates of cows in the four pre-slaughter phases? Test using $\alpha = .05$.
d. If warranted, conduct a multiple comparisons procedure to rank the four treatment means. Use an experiment wise error rate of $\alpha = .05$.

Cow	Phase			
	1	2	3	4
1	124	124	109	107
2	100	98	98	99
3	103	98	100	106
4	94	91	98	95
5	122	109	114	115
6	103	92	100	106
7	98	80	99	103
8	120	84	107	110

10.72 Effect of massage on boxers. Eight amateur boxers participated in an experiment to investigate the effect of massage on boxing performance (*British Journal of Sports Medicine,* Apr. 2000). The punching power of each boxer (measured in newtons) was recorded in the round following each of four different interventions: (M1) in round 1, following a pre-bout sports massage;

Boxer	Intervention			
	M1	R1	M5	R5
1	1,243	1,244	1,291	1,262
2	1,147	1,053	1,169	1,177
3	1,247	1,375	1,309	1,321
4	1,274	1,235	1,290	1,285
5	1,177	1,139	1,233	1,238
6	1,336	1,313	1,366	1,362
7	1,238	1,279	1,275	1,261
8	1,261	1,152	1,289	1,266

Based on Hemmings, B., Smith, M., Graydon, J., and Dyson, R. "Effects of massage on physiological restoration, perceived recovery, and repeated sports performance." *British Journal of Sports Medicine,* Vol. 34, No. 2, Apr. 2000 (adapted from Table 3).

(R1) in round 1, following a pre-bout period of rest; (M5) in round 5, following a sports massage between rounds; and (R5) in round 5, following a period of rest between rounds. Based on information provided in the article, the data in the accompanying table were obtained and saved in the **BOXING** file. The main goal of the experiment was to compare the punching power means of the four interventions.

a. Set up H_0 and H_a for this analysis.
b. Identify the treatments and blocks in the experiment.
c. Conduct the test set up in part **a**. What conclusions can you draw regarding the effect of massage on punching power?

10.73 Plants and stress reduction. Plant therapists believe that plants can reduce the stress levels of humans. A Kansas State University study was conducted to investigate this phenomenon. Two weeks prior to final exams, 10 undergraduate students took part in an experiment to determine what effect the presence of a live plant, a photo of a plant, or the absence of a plant has on the student's ability to relax while isolated in a dimly lit room. Each student participated in three sessions: one with a live plant, one with a photo of a plant, and one with no plant (the control).* During each session, finger temperature was measured at one-minute intervals for 20 minutes. Since increasing finger temperature indicates an increased level of relaxation, the maximum temperature (in degrees) was used as the response variable. For example, one student's finger measured 95.6° in the "Live Plant" condition, 92.6° in the "Plant Photo" condition, and 96.6° in the "No Plant" condition. The temperatures under the three conditions for the other nine students follow: Student 2 (95.6°, 94.8°, 96.0°); Student 3 (96.0°, 97.2°, 96.2°); Student 4 (95.2°, 94.6°, 95.7°); Student 5 (96.7°, 95.5°, 94.8°); Student 6 (96.0°, 96.6°, 93.5°); Student 7 (93.7, 96.2°, 96.7°); Student 8 (97.0°, 95.8°, 95.4°); Student 9 (94.9°, 96.6°, 90.5°); Student 10 (91.4°, 93.5°, 96.6°). These data are saved in the **PLANTS** file. Conduct an ANOVA and make the proper inferences at $\alpha = .10$.

Based on data from Elizabeth Schreiber, Department of Statistics, Kansas State University, Manhattan, Kansas.

10.74 Absentee rates at a jeans plant. A plant that manufactures denim jeans in the United Kingdom recently introduced a computerized automated handling system. The new system delivers garments to the assembly-line operators by means of an overhead conveyor. Although the automated system minimizes operator handling time, it inhibits operators from working ahead and taking breaks to be away from their machine. A study in *New Technology, Work, and Employment* (July 2001) investigated the impact of the new handling system on worker absentee rates at the jeans plant. One theory is that the mean absentee rate will vary by day of the week, as operators decide to indulge in one-day absences to relieve work pressure. Nine weeks were randomly selected, and the absentee rate (percentage of workers

*The experiment is simplified for this exercise. The actual experiment involved 30 students who participated in 12 sessions.

absent) was determined for each day (Monday through Friday) of the workweek. For example, the absentee rates for the five days of the first week selected are: 5.3, .6, 1.9, 1.3, and 1.6, respectively. The data for all nine weeks are saved in the **JEANS** file. Conduct a complete analysis of the data to determine whether the mean absentee rate differs across the five days of the workweek.

Based on Boggis, J. J. "The eradication of leisure." *New Technology, Work, and Employment*, Vol. 16, No. 2, July 2001 (Table 3).

10.75 Light-to-dark transition of genes. Refer to the *Journal of Bacteriology* (July 2002) study of the sensitivity of bacteria genes to light, presented in Exercise 9.48 (p. 438). Recall that scientists isolated 103 genes of the bacterium responsible for photosynthesis and respiration. Each gene was grown to midexponential phase in a growth incubator in "full light" and then was exposed to three alternative light/dark conditions: "full dark" (lights extinguished for 24 hours), "transient light" (lights kept on for another minutes), and "transient dark" (lights turned off for 90 minutes). At the end of each light/dark condition, the standardized growth measurement was determined for each of the 103 genes. The complete data set is saved in the **GENEDARK** file. (Data for the first 10 genes are shown in the accompanying table.) Assume that the goal of the experiment is to compare the mean standardized growth measurements for the three light/dark conditions.

Gene ID	FULL-DARK	TR-LIGHT	TR-DARK
SLR2067	−0.00562	1.40989	−1.28569
SLR1986	−0.68372	1.83097	−0.68723
SSR3383	−0.25468	−0.79794	−0.39719
SLL0928	−0.18712	−1.20901	−0.18618
SLR0335	−0.20620	1.71404	−0.73029
SLR1459	−0.53477	2.14156	−0.33174
SLL1326	−0.06291	1.03623	−0.30392
SLR1329	−0.85178	−0.21490	0.44545
SLL1327	0.63588	1.42608	−0.13664
SLL1325	−0.69866	1.93104	−0.24820

Based on Gill, R. T., et al. "Genome-wide dynamic transcriptional profiling of the light to dark transition in *Synechocystis Sp.* PCC6803." *Journal of Bacteriology*, Vol. 184, No. 13, July 2002.

a. Explain why the data should be analyzed as a randomized block design.

b. Specify the null and alternative hypotheses for comparing the light/dark condition means.

c. Using a statistical software package, conduct the test you set up in part **b.** Interpret the results at $\alpha = .05$.

d. If necessary, employ a multiple-comparison procedure to rank the light/dark condition means. Use an experimentwise error rate of .05.

Applying the Concepts—Advanced

10.76 Listening ability of infants. *Science* (Jan. 1, 1999) reported on the ability of seven-month-old infants to learn an unfamiliar language. In one experiment, 16 infants were trained in an artificial language. Then each infant was presented with two 3-word sentences that consisted entirely of new words (e.g., "wo fe wo"). One sentence was consistent (i.e., constructed from the same grammar the infants got in the training session), and one sentence was inconsistent (i.e., constructed from grammar in which the infant was not trained). The variable measured in each trial was the time (in seconds) the infant spent listening to the speaker, with the goal being to compare the mean listening times of consistent and inconsistent sentences.

a. The data were analyzed as a randomized block design with the 16 infants representing the blocks and the two types of sentences (consistent and inconsistent) representing the treatments. Do you agree with this data analysis method? Explain.

b. Refer to part **a.** The test statistic for testing treatments was $F = 25.7$ with an associated observed significance level of $p < .001$. Interpret this result.

c. Explain why the data could also be analyzed as a paired difference experiment with a test statistic of $t = 5.07$.

d. The mean listening times and standard deviations for the two treatments are given here. Use this information to calculate the F-statistic for comparing the treatment means in a completely randomized ANOVA design. Explain why this test statistic provides weaker evidence of a difference between treatment means than the test in part **b.** provides.

	Consistent Sentences	Inconsistent Sentences
Mean	6.3	9.0
Standard deviation	2.6	2.16

e. Explain why there is no need to control the experimentwise error rate in ranking the treatment means for this experiment.

10.5 Factorial Experiments: Two Factors

All the experiments discussed in Sections 10.2 and 10.4 were **single-factor experiments**. The treatments were levels of a single factor, with the sampling of experimental units performed with either a completely randomized or a randomized block design. However, most responses are affected by more than one factor, and we will therefore often wish to design experiments involving more than one factor.

Consider an experiment in which the effects of two factors on the response are being investigated. Assume that factor *A* is to be investigated at *a* levels and factor *B* at *b* levels. Recalling that treatments are factor–level combinations, you can see that the experiment has, potentially, ab treatments that could be included in it. A *complete factorial experiment* is a factorial experiment in which all possible ab treatments are utilized.

A **complete factorial experiment** is a factorial experiment in which every factor–level combination is utilized. That is, the number of treatments in the experiment equals the total number of factor–level combinations.

For example, suppose the USGA wants to determine the relationship not only between distance and brand of golf ball, but also between distance and the club used to hit the ball. If the USGA decides to use four brands and two clubs (say, driver and five-iron) in the experiment, then a complete factorial experiment would call for utilizing all $4 \times 2 = 8$ Brand–Club combinations. This experiment is referred to more specifically as a *complete* 4×2 *factorial*. A layout for a two-factor factorial experiment (henceforth, we use the term *factorial* to refer to a *complete factorial*) is given in Table 10.12. The factorial experiment is also referred to as a **two-way classification**, because it can be arranged in the row–column format exhibited in Table 10.12 .

Table 10.12 Schematic Layout of Two-Factor Factorial Experiment

		Factor B at b Levels				
	Level	1	2	3	$\cdots$	b
Factor A at a levels	1	Trt. 1	Trt. 2	Trt. 3	$\cdots$	Trt. b
	2	Trt. $b + 1$	Trt. $b + 2$	Trt. $b + 3$	$\cdots$	Trt. $2b$
	3	Trt. $2b + 1$	Trt. $2b + 2$	Trt. $2b + 3$	$\cdots$	Trt. $3b$
	$\vdots$	$\vdots$	$\vdots$	$\vdots$	$\cdots$	$\vdots$
	a	Trt. $(a - 1)b + 1$	Trt. $(a - 1)b + 2$	Trt. $(a - 1)b + 3$	$\cdots$	Trt. ab

In order to complete the specification of the experimental design, the treatments must be assigned to the experimental units. If the assignment of the ab treatments in the factorial experiment is random and independent, the design is completely randomized. For example, if the machine Iron Byron is used to hit 80 golf balls, 10 for each of the eight Brand–Club combinations, in a random sequence, the design would be completely randomized. By contrast, if the assignment is made within homogeneous blocks of experimental units, then the design is a randomized block design. For example, if 10 golfers are employed to hit each of the eight golf balls, and each golfer hits all eight Brand–Club combinations in a random sequence, then the design is a randomized block design, with the golfers serving as blocks. In the remainder of this section, we confine our attention to factorial experiments utilizing completely randomized designs.

If we utilize a completely randomized design to conduct a factorial experiment with ab treatments, we can proceed with the analysis in exactly the same way as we did in Section 10.2. That is, we calculate (or let the computer calculate) the measure of treatment mean variability (MST) and the measure of sampling variability (MSE) and use the F-ratio of these two quantities to test the null hypothesis that the treatment means are equal. However, if this hypothesis is rejected, because we conclude that some differences exist among the treatment means, important questions remain: Are both factors affecting the response, or only one? If both, do they affect the response independently, or do they interact to affect the response?

For example, suppose the distance data indicate that at least two of the eight treatment (Brand–Club combinations) means differ in the golf experiment. Does the brand

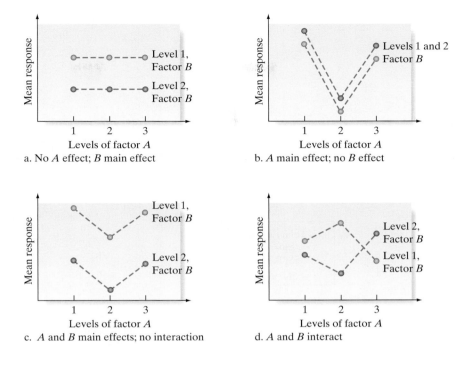

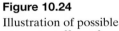

Figure 10.24

Illustration of possible treatment effects: factorial experiment

of ball (factor *A*) or the club utilized (factor *B*) affect mean distance, or do both affect it? Several possibilities are shown in Figure 10.24. In Figure 10.24a, the brand means are equal (only three are shown for the purpose of illustration), but the distances differ for the two levels of factor *B* (Club). Thus, there is no effect of Brand on distance, but a Club main effect is present. In Figure 10.24b, the Brand means differ, but the Club means are equal for each Brand. Here, a Brand main effect is present, but no effect of Club is present.

Figure 10.24c and Figure 10.24d illustrate cases in which both factors affect the response. In Figure 10.24c, the mean distance between clubs does not change for the three Brands, so the effect of Brand on distance is independent of Club. That is, the two factors Brand and Club *do not interact*. In contrast, Figure 10.24d shows that the difference between mean distances between clubs varies with Brand. Thus, the effect of Brand on distance depends on Club, and the two factors *do interact*.

In order to determine the nature of the treatment effect, if any, on the response in a factorial experiment, we need to break the treatment variability into three components: Interaction between factors *A* and *B*, Main effect of factor *A*, and Main effect of factor *B*. The **Factor interaction** component is used to test whether the factors combine to affect the response, while the **Factor main effect** components are used to determine whether the factors affect the response separately.

Now Work Exercise 10.91

The partitioning of the total sum of squares into its various components is illustrated in Figure 10.25. Notice that at stage 1 the components are identical to those in the one-factor, completely randomized designs of Section 10.2: The sums of squares for treatment and error sum to the total sum of squares. The degrees of freedom for treatments is equal to ($ab - 1$), one less than the number of treatments. The degrees of freedom for error is equal to ($n - ab$), the total sample size minus the number of treatments. Only at stage 2 of the partitioning does the factorial experiment differ from those previously discussed. Here, we divide the treatment sum of squares into its three components: interaction and the two main effects. These components can then be used to test the nature of the differences, if any, among the treatment means.

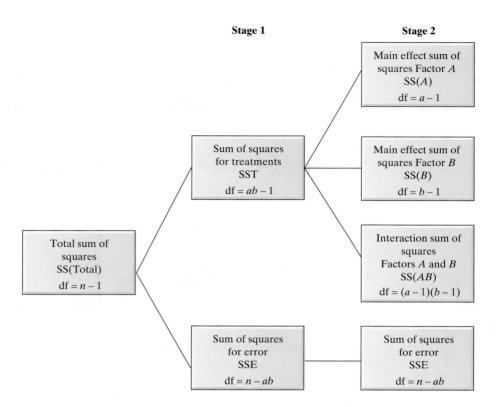

Figure 10.25

Partitioning the total sum of squares for a two-factor factorial

There are a number of ways to proceed in the testing and estimation of factors in a factorial experiment. We present one approach in the following box:

Procedure for Analysis of Two-Factor Factorial Experiment

1. Partition the total sum of squares into the treatment and error components (stage 1 of Figure 10.25). Use either a statistical software package or the calculation formulas in Appendix B to accomplish the partitioning.

2. Use the F-ratio of the mean square for treatments to the mean square for error to test the null hypothesis that the treatment means are equal.*
   a. If the test results in nonrejection of the null hypothesis, consider refining the experiment by increasing the number of replications or introducing other factors. Also, consider the possibility that the response is unrelated to the two factors.
   b. If the test results in rejection of the null hypothesis, then proceed to step 3.

3. Partition the treatment sum of squares into the main effect and the interaction sum of squares (stage 2 of Figure 10.25). Use either a statistical software package or the calculation formulas in Appendix B to accomplish the partitioning.

4. Test the null hypothesis that factors A and B do not interact to affect the response by computing the F-ratio of the mean square for interaction to the mean square for error.
   a. If the test results in nonrejection of the null hypothesis, proceed to step 5.
   b. If the test results in rejection of the null hypothesis, conclude that the two factors interact to affect the mean response. Then proceed to step 6a.

5. Conduct tests of two null hypotheses that the mean response is the same at each level of factor A and factor B. Compute two F-ratios by comparing the mean square for each factor main effect with the mean square for error.

---

*Some analysts prefer to proceed directly to test the interaction and main-effect components, skipping the test of treatment means. We begin with this test to be consistent with our approach in the one-factor completely randomized design.

> **a.** If one or both tests result in rejection of the null hypothesis, conclude that the factor affects the mean response. Proceed to step 6b.
>
> **b.** If both tests result in nonrejection, an apparent contradiction has occurred. Although the treatment means seemingly differ (step 2 test), the interaction (step 4) and main-effect (step 5) tests have not supported that result. Further experimentation is advised.
>
> **6.** Compare the means:
>
>   **a.** If the test for interaction (step 4) is significant, use a multiple-comparison procedure to compare any or all pairs of the treatment means.
>
>   **b.** If the test for one or both main effects (step 5) is significant, use a multiple-comparison procedure to compare the pairs of means corresponding to the levels of the significant factor(s).

We assume that the completely randomized design is a **balanced design,** meaning that the same number of observations are made for each treatment. That is, we assume that r experimental units are randomly and independently selected for each treatment. The numerical value of r must exceed 1 in order to have any degrees of freedom with which to measure the sampling variability. [Note that if $r = 1$, then $n = ab$, and the number of degrees of freedom associated with error (Figure 10.25) is df $= n - ab = 0$.] The value of r is often referred to as the number of **replicates of the factorial experiment,** since we assume that all ab treatments are repeated, or replicated, r times. Whatever approach is adopted in the analysis of a factorial experiment, several tests of hypotheses are usually conducted. The tests are summarized in the following box.

> **Tests Conducted in Analyses of Factorial Experiments: Completely Randomized Design, r Replicates per Treatment**
>
> **Test for Treatment Means**
>
> H_0: No difference among the ab treatment means
>
> H_a: At least two treatment means differ
>
> *Test statistic:* $F = \dfrac{\text{MST}}{\text{MSE}}$
>
> *Rejection region:* $F \geq F_\alpha$, based on $(ab - 1)$ numerator and $(n - ab)$ denominator degrees of freedom [*Note:* $n = abr$.]
>
> **Test for Factor Interaction**
>
> H_0: Factors A and B do not interact to affect the response mean
>
> H_a: Factors A and B do interact to affect the response mean
>
> *Test statistic:* $F = \dfrac{\text{MS}(AB)}{\text{MSE}}$
>
> *Rejection region:* $F \geq F_\alpha$, based on $(a - 1)(b - 1)$ numerator and $(n - ab)$ denominator degrees of freedom
>
> **Test for Main Effect of Factor A**
>
> H_0: No difference among the a mean levels of factor A
>
> H_a: At least two factor A mean levels differ
>
> *Test statistic:* $F = \dfrac{\text{MS}(A)}{\text{MSE}}$
>
> *Rejection region:* $F \geq F_\alpha$, based on $(a - 1)$ numerator and $(n - ab)$ denominator degrees of freedom

(continued)

> **Test for Main Effect of Factor B**
>
> H_0: No difference among the b mean levels of factor B
>
> H_a: At least two factor B mean levels differ
>
> Test statistic: $F = \dfrac{MS(B)}{MSE}$
>
> Rejection region: $F \geq F_\alpha$, based on $(b - 1)$ numerator and $(n - ab)$ denominator degrees of freedom

> **Conditions Required for Valid F-Tests in Factorial Experiments**
>
> 1. The response distribution for each factor–level combination (treatment) is normal.
> 2. The response variance is constant for all treatments.
> 3. Random and independent samples of experimental units are associated with each treatment.

Example 10.10

Conducting a Factorial ANOVA— Golf Ball Driving Distances

Problem Suppose the USGA tests four different brands (A, B, C, D) of golf balls and two different clubs (driver, five-iron) in a 4 × 2 factorial design. Each of the eight Brand–Club combinations (treatments) is randomly and independently assigned to four experimental units, each consisting of a specific position in the sequence of hits by Iron Byron. The distance response is recorded for each of the 32 hits, and the results are shown in Table 10.13.

Table 10.13 **Distance Data for 4 × 2 Factorial Golf Experiment**

Club	Brand			
	A	B	C	D
Driver	226.4	238.3	240.5	219.8
	232.6	231.7	246.9	228.7
	234.0	227.7	240.3	232.9
	220.7	237.2	244.7	237.6
Five-iron	163.8	184.4	179.0	157.8
	179.4	180.6	168.0	161.8
	168.6	179.5	165.2	162.1
	173.4	186.2	156.5	160.3

Data Set: GOLFFAC1

a. Use statistical software to partition the total sum of squares into the components necessary to analyze this 4 × 2 factorial experiment.

b. Conduct the appropriate ANOVA tests and interpret the results of your analysis. Use $\alpha = .10$ for the tests you conduct.

c. If appropriate, conduct multiple comparisons of the treatment means. Use an experimentwise error rate of .10. Illustrate the comparisons with a graph.

Solution

a. The SAS printout that partitions the total sum of squares for this factorial experiment is given in Figure 10.26. The partitioning takes place in two stages. First, the total sum of squares is partitioned into the model (treatment) and error sums of squares at the top of the printout. Note that SST is 33,659.8 with 7 degrees of freedom and SSE is 822.2 with 24 degrees of freedom, adding to 34,482.0 and 31 degrees of freedom. In the second stage of partitioning, the treatment sum of

```
Dependent Variable: DISTANCE

                                         Sum of
          Source              DF         Squares    Mean Square   F Value   Pr > F

          Model                7     33659.80875    4808.54411    140.35   <.0001

          Error               24       822.24000      34.26000

          Corrected Total     31     34482.04875


                    R-Square    Coeff Var    Root MSE    DISTANCE Mean

                    0.976155    2.896461     5.853204       202.0813


          Source              DF    Type III SS   Mean Square   F Value   Pr > F

          CLUB                 1    32093.11125   32093.11125    936.75   <.0001
          BRAND                3      800.73625     266.91208      7.79   0.0008
          CLUB*BRAND           3      765.96125     255.32042      7.45   0.0011
```

Figure 10.26

SAS printout for factorial ANOVA of golf data

squares is further divided into the main-effect and interaction sums of squares. From the highlighted values at the bottom of the printout, we see that SS(Club) is 32,093.1 with 1 degree of freedom, SS(Brand) is 800.7 with 3 degrees of freedom, and SS(Club × Brand) is 766.0 with 3 degrees of freedom, adding to 33,659.8 and 7 degrees of freedom.

b. Once partitioning is accomplished, our first test is

H_0: The eight treatment means are equal.

H_a: At least two of the eight means differ.

Test statistic: $F = \dfrac{MST}{MSE} = 140.35$ (highlighted, top of printout)

Observed significance level: $p < .0001$ (highlighted, top of printout)

Since $\alpha = .10$ exceeds p, we reject this null hypothesis and conclude that at least two of the Brand–Club combinations differ in mean distance.

After accepting the hypothesis that the treatment means differ, and therefore that the factors Brand and/or Club somehow affect the mean distance, we want to determine how the factors affect the mean response. We begin with a test of interaction between Brand and Club:

H_0: The factors Brand and Club do not interact to affect the mean response.

H_a: Brand and Club interact to affect mean response.

Test statistic: $F = \dfrac{MS(AB)}{MSE} = \dfrac{MS(\text{Brand} \times \text{Club})}{MSE}$

$= \dfrac{255.32}{34.26} = 7.45$ (highlighted, bottom of printout)

Observed significance level: $p = .0011$ (highlighted, bottom of printout)

Since $\alpha = .10$ exceeds the p-value, we conclude that the factors Brand and Club interact to affect mean distance.

Because the factors interact, we do not test the main effects for Brand and Club. Instead, we compare the treatment means in an attempt to learn the nature of the interaction.

c. Rather than compare all $8(7)/2 = 28$ pairs of treatment means, we test for differences only between pairs of brands within each club. That differences exist *between* clubs can be assumed. Therefore, only $4(3)/2 = 6$ pairs of means need to be compared for each club, or a total of 12 comparisons for the two clubs. The results of these comparisons, obtained from Tukey's method with an experimentwise error rate

--- CLUB=5 IRON --

The ANOVA Procedure

Tukey's Studentized Range (HSD) Test for DISTANCE

NOTE: This test controls the Type I experimentwise error rate, but it generally has a higher Type II error rate than REGWQ.

Alpha	0.1
Error Degrees of Freedom	12
Error Mean Square	36.10792
Critical Value of Studentized Range	3.62071
Minimum Significant Difference	10.878

Means with the same letter are not significantly different.

Tukey Grouping		Mean	N	BRAND
	A	182.675	4	B
B				
B		171.300	4	A
B		167.175	4	C
B				
B		160.500	4	D

--- CLUB=DRIVER --

The ANOVA Procedure

Tukey's Studentized Range (HSD) Test for DISTANCE

NOTE: This test controls the Type I experimentwise error rate, but it generally has a higher Type II error rate than REGWQ.

Alpha	0.1
Error Degrees of Freedom	12
Error Mean Square	32.41208
Critical Value of Studentized Range	3.62071
Minimum Significant Difference	10.307

Means with the same letter are not significantly different.

Tukey Grouping		Mean	N	BRAND
	A	243.100	4	C
	A			
B	A	233.725	4	B
B				
B		229.750	4	D
B				
B		228.425	4	A

Figure 10.27

SAS printout of Tukey rankings of golf ball brand means for each club

of $\alpha = .10$ for each club, are displayed in the SAS printout shown in Figure 10.27. For each club, the brand means are listed in descending order in the Figure 10.23, and those not significantly different are connected by the same letter in the "Tukey Grouping" column.

As shown in Figure 10.27, the picture is unclear with respect to Brand means. For the five-iron (top of figure), the brand B mean significantly exceeds all other brands. However, when the brand B golf balls are hit with a driver (bottom of figure), brand B's mean is not significantly different from any of the other brands. The Club × Brand interaction can be seen in the SPSS plot of means shown in Figure 10.28. Note that the difference between the mean distances of the two clubs (driver and five-iron) varies with the brand. The biggest difference appears for Brand C, while the smallest difference is for Brand B.

Look Back Note the nontransitive nature of the multiple comparisons. For example, for the driver, the brand C mean can be "the same" as the brand B mean, and the brand B mean can be "the same" as the brand D mean, yet the brand C mean can significantly exceed the brand D mean. The reason lies in the definition of "the same": We must be careful not to conclude that two means are equal simply because they are connected by a line or a letter. The line indicates only that *the connected means are not significantly different.* You should conclude (at the overall α level of significance) only that means which are *not* connected are different, while withholding judgment on those which are connected. The picture of which means differ and by how much will become clearer as we increase the number of replicates of the factorial experiment.

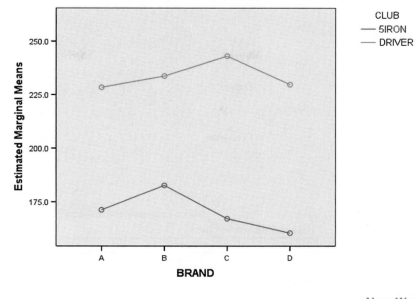

Figure 10.28
SPSS means plot for factorial golf ball experiment

Now Work Exercise 10.94

As with completely randomized and randomized block designs, the results of a factorial ANOVA are typically presented in an ANOVA summary table. Table 10.14 gives the general form of the ANOVA table, while Table 10.15 gives the ANOVA table for the golf ball data analyzed in Example 10.10. A two-factor factorial is characterized by four sources of variation—factor A, factor B, $A \times B$ interaction, and error—that sum to the total sum of squares.

Table 10.14   General ANOVA Summary Table for a Two-Factor Factorial Experiment with r Replicates, Where Factor A Has a Levels and Factor B Has b Levels

Source	df	SS	MS	F
A	$a - 1$	SSA	MSA	MSA/MSE
B	$b - 1$	SSB	MSB	MSB/MSE
AB	$(a - 1)(b - 1)$	SSAB	MSAB	MSAB/MSE
Error	$ab(r - 1)$	SSE	MSE	
Total	$n - 1$	SS(Total)		

Table 10.15   ANOVA Summary Table for Example 10.10

Source	df	SS	MS	F
Brand	1	32,093.11	32,093.11	936.75
Club	3	800.74	266.91	7.79
Interaction	3	765.96	255.32	7.45
Error	24	822.24	34.26	
Total	31	34,482.05		

Example 10.11

More Practice on Conducting a Factorial ANOVA— Golf Ball Driving Distances

Problem Refer to Example 10.10. Suppose the same factorial experiment is performed on four other brands (E, F, G, and H), and the results are as shown in Table 10.16. Repeat the factorial analysis and interpret the results.

Solution An SPSS printout for the second factorial experiment is shown in Figure 10.29. We conduct several tests, as outlined in the box on page 523.

Table 10.16 Distance Data for Second Factorial Golf Experiment

Club	Brand			
	E	F	G	H
Driver	238.6	261.4	264.7	235.4
	241.9	261.3	262.9	239.8
	236.6	254.0	253.5	236.2
	244.9	259.9	255.6	237.5
Five-iron	165.2	179.2	189.0	171.4
	156.9	171.0	191.2	159.3
	172.2	178.0	191.3	156.6
	163.2	182.7	180.5	157.4

Data Set: GOLFFAC2

Tests of Between-Subjects Effects

Dependent Variable:DISTANCE

Source	Type III Sum of Squares	df	Mean Square	F	Sig.
Corrected Model	49959.375ᵃ	7	7137.054	290.120	.000
Intercept	1423532.828	1	1423532.828	57866.453	.000
BRAND	3410.316	3	1136.772	46.210	.000
CLUB	46443.900	1	46443.900	1887.939	.000
BRAND * CLUB	105.158	3	35.053	1.425	.260
Error	590.407	24	24.600		
Total	1474082.610	32			
Corrected Total	50549.782	31			

a. R Squared = .988 (Adjusted R Squared = .985)

Figure 10.29

SPSS printout of second factorial ANOVA

Test for Equality of Treatment Means

The F-ratio for Treatments is $F = 290.1$ (highlighted at the top of the printout), which exceeds the tabular value of $F_{.10} = 1.98$ for seven numerator and 24 denominator degrees of freedom. (Note that the same rejection regions will apply in this example as in Example 10.10, since the factors, treatments, and replicates are the same.) We conclude that at least two of the Brand–Club combinations have different mean distances.

Test for Interaction

We next test for interaction between Brand and Club. The F-value (highlighted on the SPSS printout) is

$$F = \frac{\text{MS(Brand} \times \text{Club)}}{\text{MSE}} = 1.425$$

Since this F-ratio does *not* exceed the tabled value of $F_{.10} = 2.33$ with 3 and 24 df, we cannot conclude at the .10 level of significance that the factors interact. In fact, note that the observed significance level (highlighted on the SPSS printout) for the test of interaction is .260. Thus, at any level of significance lower than $\alpha = .26$, we could not conclude that the factors interact. We therefore test the main effects for Brand and Club.

Test for Brand Main Effect

We first test for the Brand main effect:

H_0: No difference exists among the true Brand mean distances.

H_a: At least two Brand mean distances differ.

Test statistic:

$$F = \frac{\text{MS(Brand)}}{\text{MSE}} = \frac{1,136.772}{24.600} = 46.210 \quad \text{(highlighted on printout)}$$

Observed significance level: $p = .000$ (highlighted on printout)

Since $\alpha = .10$ exceeds the p-value, we conclude that at least two of the brand means differ. We will subsequently determine which brand means differ by using Tukey's multiple-comparison procedure. But first, we want to test for the Club main effect:

Test for Club Main Effect

H_0: No differences exist between the Club mean distances.

H_a: The Club mean distances differ.

Test statistic:

$$F = \frac{MS(Club)}{MSE} = \frac{46,443.900}{24.600} = 1,887.94 \text{ (highlighted on printout)}$$

Observed significance level: $p = .000$ (highlighted on printout)

Since $\alpha = .10$ exceeds the p-value, we conclude that the two clubs are associated with different mean distances. Since only two levels of Club were utilized in the experiment, this F-test leads to the inference that the mean distance differs for the two clubs. It is no surprise (to golfers) that the mean distance for balls hit with the driver is significantly greater than the mean distance for those hit with the five-iron.

Ranking of Means

To determine which of the Brands' mean distances differ, we wish to compare the four Brand means by using Tukey's method at $\alpha = .10$. The results of these multiple comparisons are displayed in the SPSS printout shown in Figure 10.30. The Brand means are shown, grouped by subset, in the figure, with means that are not significantly different in the same subset. Brands G and F (subset 2) are associated with significantly greater mean distances than brands E and H (subset 1). However, since G and F are in the same Tukey subset, and since E and H are in the same subset, we cannot distinguish between brands G and F or between brands E and H by means of these data. The means that do not differ are illustated with lines in Figure 10.31.

Look Back Since the interaction between Brand and Club was not significant, we conclude that this difference among brands applies to both clubs. The sample means for all Club–Brand combinations are shown in the MINITAB graph of Figure 10.32 and appear to support the conclusions of the tests and comparisons. Note that the Brand means maintain their relative positions for each Club: Brands F and G dominate brands E and H for both the driver and the five-iron.

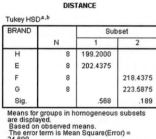

Figure 10.30

SPSS printout of Tukey rankings of brand means

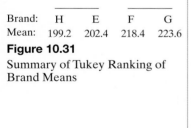

Figure 10.31

Summary of Tukey Ranking of Brand Means

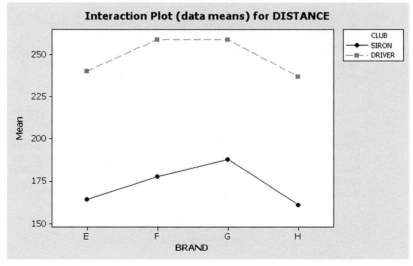

Figure 10.32

MINITAB means plot for second factorial golf ball experiment

Now Work Exercise 10.89

Analysis-of-factorial experiments can become complex if the number of factors is increased. Even the two-factor experiment becomes more difficult to analyze if some factor combinations have different numbers of observations than others. We have provided an introduction to these important experiments by using two-factor factorials with equal numbers of observations for each treatment. Although similar principles apply to most factorial experiments, you should consult the references at the end of this chapter if you need to design and analyze more complex factorials.

Statistics IN Action | **Revisited**   A Two-Way Analysis of the Cockroach Data

We now return to the study of the trail-following ability of German cockroaches (p. 475). Recall that an entomologist created a chemical trail with either a methanol extract from roach feces or just methanol (control). Cockroaches from one of four age–sex groups were released into a container at the beginning of the trail, one at a time, and the movement pattern of each cockroach—the deviation from the trail (in pixels)—was measured. The layout for the experimental design is shown in Figure SIA10.3. You can see that 20 roaches of each type were randomly assigned to the treatment trail and 10 of each type were randomly assigned to the control trail. Thus, a total of 120 roaches were used in the experiment. The design is a factorial with two factors: Trail (extract or control) and Group (adult males, adult females, gravid females, or nymphs).

The entomologist wants to determine whether cockroaches in different age–sex groups differ in their ability to follow either the extract trail or the control trail. In other words, how do the two factors age–sex group and type of trail affect the cockroaches' mean deviation from the trail? To answer this question, we conduct a two-way factorial analysis of variance on the data saved in the **ROACH** file. A MINITAB printout of the ANOVA is displayed in Figure SIA10.4.

		Roach Type			
		Adult Male	Adult Female	Gravid Female	Nymph
Trail	**Extract**	$n = 20$	$n = 20$	$n = 20$	$n = 20$
	Control	$n = 10$	$n = 10$	$n = 10$	$n = 10$

Figure SIA10.3
Layout of experimental design for cockroach study

```
General Linear Model: DEVIATE versus TRAIL, GROUP

Factor   Type    Levels  Values
TRAIL    fixed        2  Control, Extract
GROUP    fixed        4  Female, Gravid, Male, Nymph


Analysis of Variance for DEVIATE, using Adjusted SS for Tests

Source         DF     Seq SS    Adj SS    Adj MS      F      P
TRAIL           1    46445.5   46445.5   46445.5  63.34  0.000
GROUP           3    16271.2   13000.6    4333.5   5.91  0.001
TRAIL*GROUP     3     2245.2    2245.2     748.4   1.02  0.386
Error         112    82131.7   82131.7     733.3
Total         119   147093.6


S = 27.0799   R-Sq = 44.16%   R-Sq(adj) = 40.67%
```

Figure SIA10.4
MINITAB two-way ANOVA for extract trail deviation

Statistics IN Action
(continued)

First, note that the *p*-value for the test for factor interaction (highlighted on the printout) is .386. Thus, there is insufficient evidence (at $\alpha = .05$) of interaction between the two factors. This implies that the impact of one factor (say, age–sex group) on mean trail deviation does not depend on the level of the other factor (type of trail). With no evidence of interaction, it is appropriate to conduct main-effect tests on the two factors.

The *p*-values for the main effects of type of trail and age–sex group (both highlighted on the printout) are .000 and .001, respectively. Since both *p*-values are less than $\alpha = .05$, there is sufficient evidence of (1) a difference between the mean deviations of cockroaches following the fecal extract trail and the control, and (2) differences in mean deviations among the four age–sex groups. To determine which main effect means are largest, we perform multiple comparisons of the means for both main effects.

Figure SIA10.5 is an SAS printout showing (at the top) the Bonferroni rankings of the age–sex group means and (at the bottom) the Bonferroni rankings of the trail-type means. Both comparisons are made with an experimentwise error rate of .05. On the basis of the "Bon Grouping" letters shown at the top of the printout, the only significant difference between age–sex group means is for the adult male and gravid cockroaches: Adult males have a significantly smaller mean deviation from the trail than gravids have. No other pair of age–sex group means is significantly different. The bottom of Figure SIA10.5 shows that the mean deviation for cockroaches following the fecal extract trail is significantly smaller than the mean for cockroaches following the control.

The final conclusions of the entomologist were as follows: There is evidence that cockroaches exhibit more of an ability to follow a fecal extract trail than a control (methanol) trail. Also, adult males appear to have more of an ability to follow a trail than gravid (pregnant female) cockroaches have.

Data Set: ROACH

```
            Bonferroni (Dunn) t Tests for DEVIATE

NOTE: This test controls the Type I experimentwise error rate, but it generally has a higher Type
                          II error rate than REGWQ.


                Alpha                                  0.05
                Error Degrees of Freedom                112
                Error Mean Square                   733.3187
                Critical Value of t                 2.68593
                Minimum Significant Difference        18.78


        Means with the same letter are not significantly different.


          Bon Grouping           Mean      N    GROUP

                        A       53.747     30    Gravid
                        A
                 B      A       37.077     30    Nymph
                 B      A
                 B      A       35.110     30    Female
                 B
                 B              20.917     30    Male


            Bonferroni (Dunn) t Tests for DEVIATE

NOTE: This test controls the Type I experimentwise error rate, but it generally has a higher Type
                          II error rate than REGWQ.


                Alpha                                  0.05
                Error Degrees of Freedom                112
                Error Mean Square                   733.3187
                Critical Value of t                 1.98137
                Minimum Significant Difference        10.39
                Harmonic Mean of Cell Sizes        53.33333

                    NOTE: Cell sizes are not equal.


        Means with the same letter are not significantly different.


          Bon Grouping           Mean      N    TRAIL

                 A              64.535     40    Control

                 B              22.801     80    Extract
```

Figure SIA10.5
SAS Bonferroni rankings of group means and trail means

Exercises 10.77–10.99

Understanding the Principles

10.77 Describe how the treatments are formed in a complete factorial experiment.

10.78 What is a balanced factorial design?

10.79 What conditions are required for valid inferences from a factorial ANOVA?

10.80 Describe what is meant by factor interaction.

10.81 Suppose you conduct a 3×5 factorial experiment.
  a. How many factors are used in this experiment?
  b. Can you determine the type(s) of factors—qualitative or quantitative—from the information given? Explain.
  c. Can you determine the number of levels used for each factor? Explain.
  d. Describe a treatment for the experiment, and determine the number of treatments used.
  e. What problem is caused by using a single replicate of this experiment? How is the problem solved?

Learning the Mechanics

10.82 The partially complete ANOVA table given here is for a two-factor factorial experiment:

Source	df	SS	MS	F
A	3		.75	
B	1	.95		
AB			.30	
Error				
Total	23	6.5		

  a. Give the number of levels for each factor.
  b. How many observations were collected for each factor–level combination?
  c. Complete the ANOVA table.
  d. Test to determine whether the treatment means differ. Use $\alpha = .10$.
  e. Conduct the tests of factor interaction and mean effects, each at the $\alpha = .10$ level of significance. Which of the tests are warranted as part of the factorial analysis? Explain.

10.83 Following is a partially completed ANOVA table for a 3×4 factorial experiment with two replications:

Source	df	SS	MS	F
A		.8		
B		5.3		
AB		9.6		
Error				
Total		18.1		

  a. Complete the ANOVA table.

  b. Which sums of squares are combined to find the sum of squares for treatment? Do the data provide sufficient evidence to indicate that the treatment means differ? Use $\alpha = .05$.
  c. Does the result of the test in part **b** warrant further testing? Explain.
  d. What is meant by factor interaction, and what is the practical implication if it exists?
  e. Test to determine whether these factors interact to affect the response mean. Use $\alpha = .05$, and interpret the result.
  f. Does the result of the interaction test warrant further testing? Explain.

10.84 The following two-way table gives data for a 2×3 factorial experiment with two observations for each factor–level combination. The data are saved in the **LM10_84** file.

	Factor B		
Level	1	2	3
Factor A 1	3.1, 4.0	4.6, 4.2	6.4, 7.1
2	5.9, 5.3	2.9, 2.2	3.3, 2.5

  a. Identify the treatments for this experiment. Calculate and plot the treatment means, using the response variable as the y-axis and the levels of factor B as the x-axis. Use the levels of factor A as plotting symbols. Do the treatment means appear to differ? Do the factors appear to interact?
  b. The MINITAB ANOVA printout for this experiment is shown below. Test to determine whether the treatment means differ at the $\alpha = .05$ level of significance. Does the test support your visual interpretation from part **a**?

Two-way ANOVA: Response versus A, B

Source	DF	SS	MS	F	P
A	1	4.4408	4.44083	18.06	0.005
B	2	4.1267	2.06333	8.39	0.018
Interaction	2	18.0067	9.00333	36.62	0.000
Error	6	1.4750	0.24583		
Total	11	28.0492			

S = 0.4958   R-Sq = 94.74%   R-Sq(adj) = 90.36%

  c. Does the result of the test in part **b** warrant a test for interaction between the two factors? If so, perform it, using $\alpha = .05$.
  d. Do the results of the previous tests warrant tests of the two factor main effects? If so, perform them, using $\alpha = .05$.
  e. Interpret the results of the tests. Do they support your visual interpretation from part **a**?

10.85 Suppose a 3×3 factorial experiment is conducted with three replications. Assume that SS(Total) = 1,000. For each of the following scenarios, form an ANOVA table, conduct the appropriate tests, and interpret the results:

a. The sum of squares of factor A main effect [SS(A)] is 20% of SS(Total), the sum of squares of factor B main effect [SS(B)] is 10% of SS(Total), and the sum of squares of interaction [SS(AB)] is 10% of SS(Total).

b. SS(A) is 10%, SS(B) is 10%, and SS(AB) is 50% of SS(Total).

c. SS(A) is 40%, SS(B) is 10%, and SS(AB) is 20% of SS(Total).

d. SS(A) is 40%, SS(B) is 40%, and SS(AB) is 10% of SS(Total).

10.86 The following two-way table gives data for a 2×2 factorial experiment with two observations per factor–level combination: The data are saved in the **LM10_86** file.

	Factor B	
Level	**1**	**2**
Factor A 1	29.6, 35.2	47.3, 42.1
2	12.9, 17.6	28.4, 22.7

a. Identify the treatments for this experiment. Calculate and plot the treatment means, using the response variable as the y-axis and the levels of factor B as the x-axis. Use the levels of factor A as plotting symbols. Do the treatment means appear to differ? Do the factors appear to interact?

b. Construct an ANOVA table for this experiment.

c. Test to determine whether the treatment means differ at the $\alpha = .05$ level of significance. Does the test support your visual interpretation from part **a**?

d. Does the result of the test in part **c** warrant a test for interaction between the two factors? If so, perform it, using

e. Do the results of the previous tests warrant tests of the two factor main effects? If so, perform them, using $\alpha = .05$.

f. Interpret the results of the tests. Do they support your visual interpretation from part **a**?

g. Given the results of your tests, which pairs of means, if any, should be compared?

Applying the Concepts—Basic

10.87 Egg shell quality in laying hens. Introducing calcium into a hen's diet can improve the shell quality of the eggs laid. One way to do this is with a limestone diet. In *Animal Feed Science and Technology* (June 2010) researchers investigated the effect of hen's age and limestone diet on eggshell quality. Two different diets were studied—fine limestone (FL) and coarse limestone (CL). Hens were classified as either younger hens (24–36 weeks old) or older hens (56–68 weeks old). The study used 120 younger hens and 120 older hens. Within each age group, half the hens were fed a fine limestone diet and the other half a coarse limestone diet. Thus, there were 60 hens in each of the four combinations of age and diet. The characteristics of the eggs produced from the laying hens were recorded, including shell thickness.

a. Identify the type of experimental design employed by the researchers.

b. Identify the factors and the factor levels (treatments) for this design.

c. Identify the experimental unit.

d. Identify the dependent variable.

e. The researchers found no evidence of factor interaction. Interpret this result, practically.

f. The researchers found no evidence of a main effect for hen's age. Interpret this result, practically.

g. The researchers found statistical evidence of a main effect for limestone diet. Interpret this result, practically. (*Note:* The mean shell thickness for eggs produced by hens on a CL diet was larger than the corresponding mean for hens on an FL diet.)

10.88 Insomnia and education. Refer to the *Journal of Abnormal Psychology* (Feb. 2005) study of whether insomnia is related to education status, presented in Exercise 1.26 (p. 20). A random-digit telephone dialing procedure was employed to collect data on 575 study participants. In addition to insomnia status (normal sleeper or chronic insomnia), the researchers classified each participant into one of four education categories (college graduate, some college, high school graduate, and high school dropout). One dependent variable of interest to the researchers was a quantitative measure of daytime functioning called the Fatigue Severity Scale (FSS). The data were analyzed as a 2×4 factorial experiment, with Insomnia status and Education level as the two factors.

a. Determine the number of treatments for this study. List them.

b. The researchers reported that "the Insomnia $\times$ Education interaction was not statistically significant." Interpret this result practically. (Illustrate with a graph.)

c. The researchers discovered that the sample mean FSS for people with insomnia was greater than the sample mean FSS for normal sleepers and that this difference was statistically significant. Interpret this result practically.

d. The researchers reported that the main effect of Education was statistically significant. Interpret this result practically.

e. Refer to part **d**. In a follow-up analysis, the sample mean FSS values for the four Education levels were compared with the use of Tukey's method ($\alpha = .05$), with the results shown in the accompanying table. What do you conclude?

Mean:	3.3	3.6	3.7	4.2
Education:	College graduate	Some college	High school graduate	High school dropout

10.89 Mussel settlement patterns on algae. Mussel larvae are in great abundance in the drift material that washes up on Ninety Mile Beach in New Zealand. These larvae tend to settle on algae. Environmentalists at the University of Auckland investigated the impact of the type of algae on the abundance of mussel larvae in drift material (*Malacologia*, Feb. 8, 2002). Drift material from three different wash-up events on Ninety Mile Beach was collected;

for each washup, the algae was separated into four strata: coarse-branching, medium-branching, fine-branching, and hydroid algae. Two samples were randomly selected for each of the $3 \times 4 = 12$ event–strata combinations, and the mussel density (percent per square centimeter) was measured for each. The data were analyzed as a complete 3×4 factorial design. The ANOVA summary table is as follows:

Source	df	F	p-Value
Event	2	.35	>.05
Strata	3	217.33	<.05
Interaction	6	1.91	>.05
Error	12		
Total	23		

a. Identify the factors (and levels) in this experiment.
b. How many treatments are included in the experiment?
c. How many replications are included in the experiment?
d. What is the total sample size for the experiment?
e. What is the response variable measured?
f. Which ANOVA F-test should be conducted first? Conduct this test (at $\alpha = .05$) and interpret the results.
g. If appropriate, conduct the F-tests (at $\alpha = .05$) for the main effects. Interpret the results.
h. Tukey multiple comparisons of the four algae strata means (at $\alpha = .05$) are summarized in the accompanying table. Which means are significantly different?

Mean (%/cm^2):	9	10	27	55
Algae stratum:	Coarse	Medium	Fine	Hydroid

10.90 Do video game players have superior visual attention skills? Refer to the *Journal of Articles in Support of the Null Hypothesis* (Vol. 6, 2009) study of video game players, Exercise 9.20 (p. 425). In a second experiment, the two groups of male psychology students—32 video game players (VGP group) and 28 nonplayers (NVGP group)—were subjected to the *inattentional blindness* test. This test gauges students' ability to detect the appearance of unexpected or task-irrelevant objects in their visual field. The students were required to silently count the number of times moving block letters touch the sides of a display window while focusing on a small square in the center of the window. Within each group, half the students were assigned a task with 4 moving letters (low load) and half were assigned a task with 8 moving letters (high load). At the end of the task, the difference between the student's count of letter touches and the actual count (recorded as a percentage error) was determined. The data on percentage error were subjected to an ANOVA for a factorial design, with Group (VGP or NVGP) as one factor and Load (low or high) as the other factor. The results are summarized in the accompanying table.

Source	df	F	p-Value
Group	1	0.58	.449
Load	1	35.04	< .0005
Group × Load	1	1.37	.247
Error	56		
Total	59		

Based on Murphy, K., and Spencer, A. "Playing video games does not make for better visual attention skills." *Journal of Articles in Support of the Null Hypothesis*, Vol. 6, No. 1, 2009.

a. Interpret the F-test for Group × Load interaction. Use $\alpha = .10$.
b. Interpret the F-test for Group main effect. Use $\alpha = .10$.
c. Interpret the F-test for Load main effect. Use $\alpha = .10$.
d. Based on your answers to parts **a–c**, do you believe video game players perform better on the inattentional blindness test than non–video game players?

10.91 Impact of paper color on exam scores. A study published in *Teaching Psychology* (May 1998) examined how external clues influence student performance. Introductory psychology students were randomly assigned to one of four different midterm examinations. Form 1 was printed on blue paper and contained difficult questions, while form 2 was also printed on blue paper, but contained simple questions. Form 3 was printed on red paper with difficult questions; form 4 was printed on red paper with simple questions. The researchers were interested in the impact that Color (red or blue) and Question (simple or difficult) had on mean exam score.
a. What experimental design was employed in this study? Identify the factors and treatments.
b. The researchers conducted an ANOVA and found a significant interaction between Color and Question (p-value $< .03$). Interpret this result.
c. The sample mean scores (percentage correct) for the four exam forms are listed in the accompanying table. Plot the four means on a graph to illustrate the Color × Question interaction.

Form	Color	Question	Mean Score
1	Blue	Difficult	53.3
2	Blue	Simple	80.0
3	Red	Difficult	39.3
4	Red	Simple	73.6

10.92 Virtual-reality–based rehabilitation systems. In *Robotica* (Vol. 22, 2004), researchers described a study of the effectiveness of display devices for three virtual-reality (VR)-based hand rehabilitation systems. Display device A is a projector, device B is a desktop computer monitor, and device C is a head-mounted display. Twelve nondisabled right-handed male subjects were randomly assigned to the three VR display devices, four subjects in each group. In addition, within each group two subjects were randomly assigned to use an auxiliary lateral image and two subjects were not. Consequently, a 3×2 factorial design was employed, with the VR display device at three levels (A, B, or C) and an auxiliary lateral image at two levels (yes or no). Using the assigned VR system, each subject carried out a "pick-and-place" procedure and the collision frequency (number of collisions between moved objects) was measured.

a. Give the sources of variation and associated degrees of freedom in an ANOVA summary table for this design.

b. How many treatments are investigated in this experiment?

c. The factorial ANOVA resulted in the following p-values: Display main effect (.045), Auxiliary lateral image main effect (.003), and Interaction (.411). Interpret these results practically. Use $\alpha = .05$ for each test you conduct.

Applying the Concepts—Intermediate

10.93 The thrill of a close game. Do women enjoy the thrill of a close basketball game as much as men do? To answer this question, male and female undergraduate students were recruited to participate in an experiment (*Journal of Sport & Social Issues*, Feb. 1997). The students watched one of eight live televised games of a recent NCAA basketball tournament. (None of the games involved a home team to which the students could be considered emotionally committed.) The "suspense" of each game was classified into one of four categories according to the closeness of scores at the game's conclusions: minimal (15-point or greater differential), moderate (10–14-point differential), substantial (5–9-point differential), and extreme (1–4-point differential). After the game, each student rated his or her enjoyment on an 11-point scale ranging from 0 (not at all) to 10 (extremely). The enjoyment rating data were analyzed as a 4×2 factorial design, with suspense (four levels) and gender (two levels) as the two factors. The $4 \times 2 = 8$ treatment means are shown in the following table:

Suspense	Gender Male	Gender Female
Minimal	1.77	2.73
Moderate	5.38	4.34
Substantial	7.16	7.52
Extreme	7.59	4.92

Based on Gan, Su-lin, et al. "The thrill of a close game: Who enjoys it and who doesn't?" *Journal of Sport & Social Issues*, Vol. 21, No. 1, Feb. 1997, pp. 59–60.

a. Plot the treatment means in a graph similar to Figure 10.28. Does the pattern of means suggest interaction between suspense and gender? Explain.

b. The ANOVA F-test for interaction yielded the following results: numerator df = 3, denominator df = 68, $F = 4.42$, p-value = .007. What can you infer from these results?

c. On the basis of the test carried out in part **b**, is the difference between the mean enjoyment levels of males and females the same, regardless of the suspense level of the game?

10.94 Baker's versus brewer's yeast. The *Electronic Journal of Biotechnology* (Dec. 15, 2003) published an article on a comparison of two yeast extracts: baker's yeast and brewer's yeast. Brewer's yeast is a surplus by-product obtained from a brewery; hence, it is less expensive than primary-grown baker's yeast. Samples of both yeast extracts were prepared at four different temperatures (45, 48, 51, and 54°C); thus, a 2×4 factorial design with yeast extract at two levels and temperature at four levels was employed. The response variable was the autolysis yield (recorded as a percentage).

a. How many treatments are included in the experiment?

b. An ANOVA found sufficient evidence of factor interaction at $\alpha = .05$. Interpret this result practically.

c. Give the null and alternative hypotheses for testing the main effects of yeast extract and temperature.

d. Explain why the tests referred to in part **c** should not be conducted.

e. Multiple comparisons of the four temperature means were conducted for each of the two yeast extracts. Interpret the results, shown as follows:

Baker's yeast:	Mean yield (%):	41.1	47.5	48.6	50.3
	Temperature (°C):	54	45	48	51
Brewer's yeast:	Mean yield (%):	39.4	47.3	49.2	49.6
	Temperature (°C):	54	51	48	45

10.95 Learning from picture book reading. *Developmental Psychology* (Nov. 2006) published an article that examined toddlers' ability to learn from reading picture books. The experiment involved 36 children at each of three different ages: 18, 24, and 30 months. The children were randomly assigned into one of three different reading book conditions: book with color photographs (Photos), book with colored pencil drawings (Drawings), and book with no photographs or drawing (Control). Thus, a 3×3 factorial experiment was employed (with age at three levels and reading book condition at three levels). After a book-reading session, the children were scored on their ability to reenact the target actions in the book. Scores ranged from 0 (low) to 3 (high). An ANOVA of the reenactment scores is summarized in the following table:

Source	df	F	p-value
Age	–	11.93	< .001
Book	–	23.64	< .001
Age × Book	–	2.99	< .05
Error	–		
Total	107		

a. Fill in the missing degrees-of-freedom (df) values in the table.

b. How many treatments are investigated in this experiment? List them.

c. Conduct a test for Age × Book interaction at $\alpha = .05$. Interpret the result practically.

d. On the basis of the test you conducted in part **c**, do you need to conduct tests for Age and Book main effects? Explain.

e. At each age level, the researchers performed multiple comparisons of the reading book condition means at $\alpha = .05$. The results are summarized in the accompanying table. What can you conclude from this analysis? Support your answer with a plot of the means.

Age = 18 months:	.40 Control	.75 Drawings	1.20 Photos
Age = 24 months:	.60 Control	1.61 Drawings	1.63 Photos
Age = 30 months:	.50 Control	2.20 Drawings	2.21 Photos

10.96 Violent lyrics and aggressiveness. In the *Journal of Personality and Social Psychology* (May 2003), psychologists investigated the potentially harmful effects of violent music lyrics. The researchers theorized that listening to a song with violent lyrics will lead to more violent thoughts and actions. A total of 60 undergraduate college students participated in an experiment designed by the researchers. Half of the students were volunteers, and half were required to participate as part of their introductory psychology class. Each student listened to a song by the group "Tool," with half the students randomly assigned a song with violent lyrics and half assigned a song with nonviolent lyrics. Consequently, the experiment used a 2×2 factorial design with the factors Song (violent, nonviolent) and Pool (volunteer, psychology class). After listening to the song, each student was given a list of word pairs and asked to rate the similarity of each word in the pair on a seven-point scale. One word in each pair was aggressive in meaning (e.g., *choke*) and the other was ambiguous (e.g., *night*). An aggressive cognition score was assigned on the basis of the average word-pair scores. (The higher the score, the more the subject associated an ambiguous word with a violent word.) The data (simulated) are shown in the accompanying table and saved in the **LYRICS** file. Conduct a complete analysis of variance on the data.

	Volunteer	Psychology Class
Violent Song	4.1 3.5 3.4 4.1 3.7 2.8 3.4 4.0 2.5 3.0 3.4 3.5 3.2 3.1 3.6	3.4 3.9 4.2 3.2 4.3 3.3 3.1 3.2 3.8 3.1 3.8 4.1 3.3 3.8 4.5
Nonviolent Song	2.4 2.4 2.5 2.6 3.6 4.0 3.3 3.7 2.8 2.9 3.2 2.5 2.9 3.0 2.4	2.5 2.9 2.9 3.0 2.6 2.4 3.5 3.3 3.7 3.3 2.8 2.5 2.8 2.0 3.1

10.97 Eyewitnesses and mug shots. When an eyewitness to a crime examines a set of mug shots at a police station, the photos are usually presented in groups (e.g., 6 mug shots at a time). Criminologists at Niagara University investigated whether mug shot group size has an effect on the selections made by eyewitnesses (*Applied Psychology in Criminal Justice*, April 2010). A sample of 90 college students was shown a video of a simulated theft. Shortly thereafter, each student was shown 180 mug shots and asked to select a photo which most closely resembled the thief. (Multiple photos could be selected.) The students were randomly assigned to view either 3, 6, or 12 mug shots at a time. Within each mug shot group size, the students were further randomly divided into three sets. In the first set, the researchers focused on the selections made in the first 60 photos shown; in the second set, the focus was on selections

made in the middle 60 photos shown; and, in the third set, selections made in the last 60 photos were recorded. The dependent variable of interest was the number of mug shot selections. Simulated data for this 3×3 factorial ANOVA, with Mug Shot Group Size at 3 levels (3, 6, or 12 photos) and Photo Set at 3 levels (first 60, middle 60, and last 60), are saved in the **MUGSHOT** file. Fully analyze the data for the researchers. In particular, the researchers want to know if mug shot group size has an effect on the mean number of selections, and, if so, which group size leads to the most selections. Also, is a higher number of selections made in the first 60, middle 60, or last 60 photos viewed?

Applying the Concepts—Advanced

10.98 Testing a new pain-relief tablet. Refer to the *Tropical Journal of Pharmaceutical Research* (June 2003) study of the impact of binding agent, binding concentration, and relative density on the mean dissolution time of pain-relief tablets, presented in Exercise 10.14 (p. 481). Recall that binding agent was set at two levels (khaya gum and PVP), binding concentration at two levels (.5% and 4.0%), and relative density at two levels (low and high); thus, a $2 \times 2 \times 2$ factorial design was employed. The sample mean dissolution times for the treatments associated with the factors binding agent and relative density when the other factor (binding concentration) is held fixed at .5% are $\bar{x}_{Gum/Low} = 4.70$, $\bar{x}_{Gum/High} = 7.95$, $\bar{x}_{PVP/Low} = 3.00$, and $\bar{x}_{PVP/High} = 4.10$. Do the results suggest that there is an interaction between binding agent and relative density? Explain.

10.99 Impact of flavor name on consumer choice. Do consumers react favorably to products with ambiguous colors or names? Marketing professors E. G. Miller and B. E. Kahn investigated this phenomenon in the *Journal of Consumer Research* (June 2005). As a "reward" for participating in an unrelated experiment, 100 consumers were told that they could have some jelly beans available in several cups on a table. Half the consumers were assigned to take jelly beans with common descriptive flavor names (e.g., watermelon green), while the other half were assigned to take jelly beans with ambiguous flavor names (e.g., monster green). Within each group, half of the consumers took the jelly beans and left (low cognitive load condition), while the other half were asked questions designed to distract them while they were taking their jelly beans (high cognitive load condition). Consequently, a 2×2 factorial experiment was employed—with Flavor name (common or ambiguous) and Cognitive load (low or high) as the two factors—with 25 consumers assigned to each of four treatments. The dependent variable of interest was the number of jelly beans taken by each consumer. The means and standard deviations of the four treatments are shown in the following table:

	Ambiguous		Common	
	Mean	Std. Dev.	Mean	Std. Dev.
Low Load	18.0	15.0	7.8	9.5
High Load	6.1	9.5	6.3	10.0

Based on Miller, E. G., and Kahn, B. E. "Shades of meaning: The effect of color and flavor names on consumer choice." *Journal of Consumer Research*, Vol. 32, June 2005 (Table 1).

a. Calculate the total of the $n = 25$ measurements for each of the four categories in the 2×2 factorial experiment.

b. Calculate the correction for mean, CM. (See Appendix B for computational formulas.)

c. Use the results of parts **a** and **b** to calculate the sums of squares for Load, Name, and Load $\times$ Name interaction.

d. Calculate the sample variance for each treatment. Then calculate the sum of squares of deviations within each sample for the four treatments.

e. Calculate SSE. (*Hint*: SSE is the pooled sum of squares for the deviations calculated in part **d**.)

f. Now that you know SS(Load), SS(Name), SS(Load $\times$ Name), and SSE, find SS(Total).

g. Summarize the calculations in an ANOVA table.

h. The researchers reported the F-value for Load $\times$ Name interaction as $F = 5.34$. Do you agree?

i. Conduct a complete analysis of these data. Use $\alpha = .05$ for any inferential techniques you employ. Illustrate your conclusions graphically.

j. What assumptions are necessary to ensure the validity of the inferential techniques you utilized? State them in terms of this experiment.

CHAPTER NOTES

Key Terms

Analysis of variance F-test 486
Analysis of variance (ANOVA) 488
Balanced design 482, 523
Blocks 506
Bonferroni multiple-comparison procedure 498
Comparisonwise error rate (CER) 498
Complete factorial experiment 520

Completely randomized design 482
Dependent variable 476
Designed study 477
Experimental unit 477
Experimentwise error rate (EER) 498
Factor interaction 521
Factor levels 476
Factor main effect 521
Factors 476

F-statistic 484
F-test 486
Mean square for error 484
Mean square for treatments 484
Multiple comparisons of a set of treatment means 498
Observational study 477
Qualitative factors 476
Quantitative factors 476
Randomized block design 506
Replicates of the factorial experiment 523
Response variable 476

Robust method 489
Scheffé multiple-comparison procedure 498
Single-factor experiments 519
Sum of squares for blocks (SSB) 506
Sum of squares for error 484
Sum of squares for treatments 484
Treatments 476
Tukey multiple-comparison procedure 498
Two-way classification 520

Guide to Conducting ANOVA F-Tests

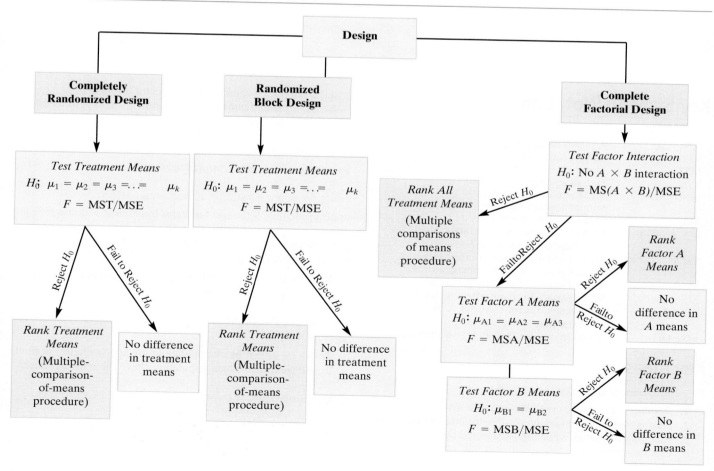

Guide to Selecting an Experimental Design

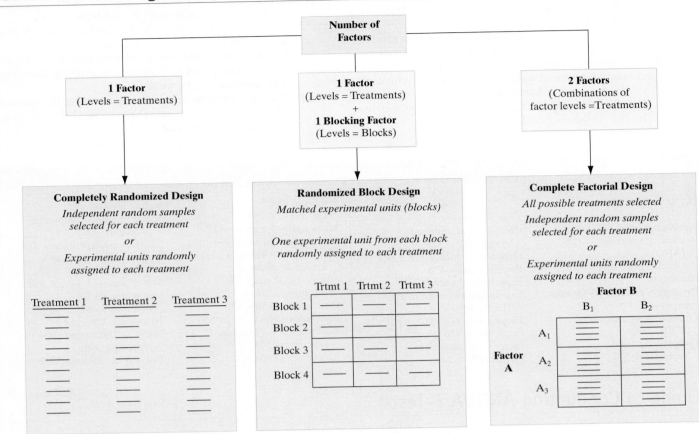

Key Symbols/Notation

ANOVA	Analysis of variance
SST	Sum of squares for treatments
MST	Mean square for treatments
SSB	Sum of squares for blocks
MSB	Mean square for blocks
SSE	Sum of squares for error
MSE	Mean square for error
$a \times b$ factorial	Factorial design with one factor at a levels and the other factor at b levels
SS(A)	Sum of squares for main-effect factor A
MS(A)	Mean square for main-effect factor A
SS(B)	Sum of squares for main-effect factor B
MS(B)	Mean square for main-effect factor B
SS(AB)	Sum of squares for factor $A \times B$ interaction
MS(AB)	Mean square for factor $A \times B$ interaction

Key Ideas

Key Elements of a Designed Experiment

1. *Response (dependent) variable*—quantitative
2. *Factors (independent variables)*—quantitative or qualitative
3. *Factor levels (values of factors)*—selected by the experimenter

4. *Treatments*—combinations of factor levels
5. *Experimental units*—assign treatments to experimental units and measure response for each

Balanced Design

Sample sizes for each treatment are equal.

Tests for Main Effects in a Factorial Design

appropriate only if the test for factor interaction is nonsignificant

Conditions Required for Valid *F*-Test in a Completely Randomized Design

1. All k treatment populations are approximately normal.
2. $\sigma_1^2 = \sigma_2^2 = \ldots = \sigma_k^2$

Conditions Required for Valid *F*-Tests in a Randomized Block Design

1. All treatment-block populations are approximately normal.
2. All treatment-block populations have the same variance.

Conditions Required for Valid *F*-Tests in a Complete Factorial Design

1. All treatment populations are approximately normal.
2. All treatment populations have the same variance.

Robust Method

Slight to moderate departures from normality have no impact on the validity on the ANOVA results.

Experimentwise Error Rate

Risk of making at least one Type I error when making multiple comparisons of means in ANOVA

Number of Pairwise Comparisons with *k* Treatment Means

$$c = k(k-1)/2$$

Multiple-Comparison-of-Means Methods

Tukey
1. Balanced design
2. Pairwise comparisons of means

Bonferroni
1. Either balanced or unbalanced design
2. Pairwise comparisons of means

Scheffé
1. Either balanced or unbalanced design
2. General contrasts of means

Supplementary Exercises 10.100–10.128

Understanding the Principles

10.100 What is the difference between a one-way ANOVA and a two-way ANOVA?

10.101 Explain the difference between an experiment that utilizes a completely randomized design and one that utilizes a randomized block design.

10.102 What are the treatments in a two-factor experiment with factor *A* at three levels and factor *B* at two levels?

10.103 Why does the experimentwise error rate of a multiple-comparison procedure differ from the significance level for each comparison (assuming that the experiment has more than two treatments)?

Learning the Mechanics

10.104 A completely randomized design is utilized to compare four treatment means. The data are shown in the accompanying table and saved in the **LM10_104** file.
 a. Given that SST = 36.95 and SS(Total) = 62.55, complete an ANOVA table for this experiment.
 b. Is there evidence that the treatment means differ? Use $\alpha = .10$.

Treatment 1	Treatment 2	Treatment 3	Treatment 4
8	6	9	12
10	9	10	13
9	8	8	10
10	8	11	11
11	7	12	11

10.105 An experiment utilizing a randomized block design was conducted to compare the mean responses for four treatments, A, B, C, and D. The treatments were randomly assigned to the four experimental units in each of five blocks. The data are shown in the table and saved in the **LM10_105** file:

Treatment	Block 1	2	3	4	5
A	8.6	7.5	8.7	9.8	7.4
B	7.3	6.3	7.3	8.4	6.3
C	9.1	8.3	9.0	9.9	8.2
D	9.3	8.2	9.2	10.0	8.4

 a. Given that SS(Total) = 22.31, SS(Block) = 10.688, and SSE = .288, complete an ANOVA table for the experiment.
 b. Do the data provide sufficient evidence to indicate a difference among treatment means? Test, using $\alpha = .05$
 c. Does the result of the test in part **b** warrant further comparison of the treatment means? If so, how many pairwise comparisons need to be made?
 d. Is there evidence that the block means differ? Use $\alpha = .05$.

10.106 The following table shows a partially completed ANOVA table for a two-factor factorial experiment:

Source	df	SS	MS	F
A	3	2.6		
B	5	9.2		
A × B			3.1	
Error		18.7		
Total	47			

 a. Complete the ANOVA table.
 b. How many levels were used for each factor? How many treatments were used? How many replications were performed?
 c. Find the value of the sum of squares for treatments. Test to determine whether the data provide evidence that the treatment means differ. Use $\alpha = .05$.
 d. Is further testing of the nature of factor effects warranted? If so, test to determine whether the factors interact. Use $\alpha = .05$. Interpret the result.

Applying the Concepts—Basic

10.107 **Accounting and Machiavellianism.** A study of Machiavellian traits in accountants was published in *Behavioral Research in Accounting* (Jan. 2008). *Machiavellian* describes negative character traits such as manipulation, cunning, duplicity, deception, and bad faith. A Machiavellian ("Mach") rating score was determined for each in a sample of accounting alumni of a large southwestern university. The accountants were then classified as having high, moderate, or low Mach rating scores. For one portion of the study, the researcher investigated the impact of both Mach score classification and gender on the average income of an accountant. For this experiment, identify each of the following:

a. experimental unit
b. response variable
c. factors
d. levels of each factor
e. treatments

10.108 Rotary oil rigs. An economist wants to compare the average monthly number of rotary oil rigs running in three states: California, Utah, and Alaska. In order to account for month-to-month variation, three months were randomly selected over a two-year period and the number of oil rigs running in each state in each month was obtained from data provided from *World Oil* (Jan. 2002) magazine. The data, reproduced in the accompanying table and saved in the **OILRIGS** file, were analyzed by means of a randomized block design. The MINITAB printout is shown below (following the data).

Month	California	Utah	Alaska
1	27	17	11
2	34	20	14
3	36	15	14

Two-way ANOVA: NumRigs versus State, Month/Year

```
Source       DF      SS       MS       F      P
State         2  617.556  308.778  38.07  0.002
Month/Year    2   30.889   15.444   1.90  0.262
Error         4   32.444    8.111
Total         8  680.889

S = 2.848   R-Sq = 95.23%   R-Sq(adj) = 90.47%


                   Individual 95% CIs For Mean Based on
                   Pooled StDev
State    Mean   ---------+---------+---------+---------+
AL     13.0000  (----*-----)
CAL    32.3333                        (----*-----)
UT     17.3333      (-----*----)
                   ---------+---------+---------+---------+
                        16.0      24.0      32.0      40.0
```

a. Why is a randomized block design preferred over a completely randomized design for comparing the mean number of oil rigs running monthly in California, Utah, and Alaska?
b. Identify the treatments for the experiment.

NUMRIGS

Tukey HSD[a,b]

STATE	N	Subset 1	Subset 2
AL	3	13.00	
UT	3	17.33	
CAL	3		32.33
Sig.		.262	1.000

Means for groups in homogeneous subsets are displayed.
Based on Type III Sum of Squares
The error term is Mean Square(Error) = 8.111.
  a. Uses Harmonic Mean Sample Size = 3.000.
  b. Alpha = .05.

c. Identify the blocks for the experiment.
d. State the null hypothesis for the ANOVA *F*-test.
e. Locate the test statistic and *p*-value on the MINITAB printout at the bottom of the previous column. Interpret the results.
f. A Tukey multiple comparison of means (at $\alpha = .05$) is summarized in the SPSS printout at the bottom of the left column. Which state(s) have the significantly largest mean number of oil rigs running monthly?

10.109 Strength of fiberboard boxes. The *Journal of Testing and Evaluation* (July 1992) published an investigation of the mean compression strength of corrugated fiberboard shipping containers. Comparisons were made for boxes of five different sizes: A, B, C, D, and E. Twenty identical boxes of each size were tested, and the peak compression strength (in pounds) was recorded for each box. The accompanying figure shows the sample means for the five types of box, as well as the variation around each sample mean.
  a. Explain why the data are collected as a completely randomized design.
  b. Refer to box types B and D. On the basis of the graph, does it appear that the mean compressive strengths of these two types of box are significantly different? Explain.
  c. On the basis of the graph, does it appear that the mean compressive strengths of all five types of box are significantly different? Explain.

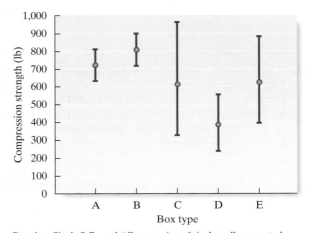

Based on Singh, S. P., et al. "Compression of single-wall corrugated shipping containers using fixed and floating test platens." *Journal of Testing and Evaluation*, Vol. 20, No. 4, July 1992, p. 319 (Figure 10.3).

10.110 Alcohol-and-marriage study. An experiment was conducted to examine the effects of alcohol on the marital interactions of husbands and wives (*Journal of Abnormal Psychology*, Nov. 1998). A total of 135 couples participated in the experiment. The husband in each couple was classified as aggressive (60 husbands) or nonaggressive (75 husbands), on the basis of an interview and his response to a questionnaire. Before the marital interactions of the couples were observed, each husband was randomly assigned to one of three groups: receive no alcohol, receive several alcoholic mixed drinks, or receive placebos (nonalcoholic drinks disguised as mixed drinks). Consequently, a 2 × 3 factorial design was employed, with husband's aggressiveness at two levels (aggressive or nonaggressive) and husband's alcohol condition at three levels (no alcohol, alcohol, and placebo). The response variable observed during the marital interaction was severity of conflict (measured on a 100-point scale).

a. A partial ANOVA table is shown. Fill in the missing degrees of freedom.

Source	df	F
Aggressiveness (A)	–	16.43 ($p < .001$)
Alcohol Condition (C)	–	6.00 ($p < .01$)
A × C	–	–
Error	129	
Total	–	

b. Interpret the p-value of the F-test for Aggressiveness.

c. Interpret the p-value of the F-test for Alcohol Condition.

d. The F-test for interaction was omitted from the article. Discuss the dangers of making inferences based on the tests in parts **b** and **c** without knowing the result of the interaction test.

10.111 Heights of grade school repeaters. Refer to the Archives of Disease in Childhood (Apr. 2000) study of whether height influences a child's progression through elementary school, presented in Exercise 9.117 (p. 462). Within each grade, Australian schoolchildren were divided into equal thirds (tertiles) based on age (youngest third, middle third, and oldest third). The researchers compared the average heights of the three groups, using an analysis of variance. (All height measurements were standardized with z-scores.) A summary of the results for all grades combined, by gender, is shown in the table at the bottom of the page.

a. What is the null hypothesis for the ANOVA of the boys' data?

b. Interpret the results of the test, part **a**. Use $\alpha = .05$.

c. Repeat parts **a** and **b** for the girls' data.

d. Summarize the results of the hypothesis tests in the words of the problem.

e. The three height means for boys were ranked with the Bonferroni method at $\alpha = .05$. The line in the summary table connects those means which are not significantly different. Which tertile has the smallest mean height?

f. What is the experimentwise error rate for the inference made in parts **e**? Interpret this value.

g. The researchers did not perform a Bonferroni analysis of the height means for the three groups of girls. Explain why not.

10.112 Contaminated fish in the Tennessee River. Refer to the U.S. Army Corps of Engineers data on contaminated fish, saved in the **FISHDDT** file. The results of an ANOVA and Tukey

multiple-comparison analysis to compare the three fish species (channel catfish, largemouth bass, and smallmouth buffalo) on the dependent variable length (in centimeters) are shown in the following MINITAB printout.

```
One-way ANOVA: LENGTH versus SPECIES

Source     DF      SS      MS      F       P
SPECIES     2   3533.1  1766.5  76.88   0.000
Error     141   3239.9    23.0
Total     143   6772.9

S = 4.794   R-Sq = 52.16%   R-Sq(adj) = 51.49%


Level             N     Mean    StDev
CHANNELCATFISH   96   44.729    4.581
LARGEMOUTHBASS   12   26.542    4.480
SMALLMOUTHBUFF   36   43.125    5.413

Pooled StDev = 4.794


Tukey 95% Simultaneous Confidence Intervals
All Pairwise Comparisons among Levels of SPECIES

Individual confidence level = 98.08%


SPECIES = CHANNELCATFISH subtracted from:

SPECIES            Lower    Center    Upper
LARGEMOUTHBASS   -21.664   -18.188  -14.711
SMALLMOUTHBUFF    -3.823    -1.604    0.615


SPECIES = LARGEMOUTHBASS subtracted from:

SPECIES            Lower    Center    Upper
SMALLMOUTHBUFF    12.798    16.583   20.368
```

a. Is there sufficient evidence (at $\alpha = .01$) to conclude that the mean lengths differ among the three fish species?

b. What is the experimentwise error rate for the Tukey multiple comparisons?

c. Locate the confidence interval for the difference $(\mu_{bass} - \mu_{catfish})$. Are the two means significantly different? If so, which mean is significantly larger?

d. Locate the confidence interval for the difference $(\mu_{buffalo} - \mu_{catfish})$. Are the two means significantly different? If so, which mean is significantly larger?

e. Locate the confidence interval for the difference $(\mu_{buffalo} - \mu_{bass})$. Are the two means significantly different? If so, which mean is significantly larger?

10.113 Are you lucky? Parapsychologists define "lucky" people as individuals who report that seemingly chance events consistently tend to work out in their favor. A team of British psychologists designed a study to examine the effects of luckiness and competition on performance in a guessing task (*The Journal of Parapsychology*, Mar. 1997). Each in a sample of 56 college students was

Summary information for Exercise 10.111

	Sample Size	Youngest Tertile Mean Height	Middle Tertile Mean Height	Oldest Tertile Mean Height	F-Value	p-Value
Boys	1439	0.33	0.33	0.16	4.57	0.01
Girls	1409	0.27	0.18	0.21	0.85	0.43

Based on Wake, M., Coghlan, D., and Hesketh, K. "Does height influence progression through primary school grades?" *The Archives of Disease in Childhood*, Vol. 82, No. 4, Apr. 2000 (Table 2), pp. 297–301.

classified as lucky, unlucky, or uncertain on the basis of their responses to a Luckiness Questionnaire. In addition, the participants were randomly assigned to either a competitive or a noncompetitive condition. All students were then asked to guess the outcomes of 50 flips of a coin. The response variable measured was percentage of coin flips correctly guessed.

a. A 2×3 factorial ANOVA design was conducted on the data. Identify the factors and their levels for this design.

b. The results of the ANOVA are summarized in the accompanying table. Interpret the results fully.

Source	df	F	p-Value
Luckiness (L)	2	1.39	.26
Competition (C)	1	2.84	.10
L × C	2	0.72	.72
Error	50		
Total	55		

Applying the Concepts—Intermediate

10.114 Income and road rage. The phenomenon of road rage has received much media attention in recent years. Is a driver's propensity to engage in road rage related to his or her income? Researchers at Mississippi State University attempted to answer this question by conducting a survey of a representative sample of over 1,000 U.S. adult drivers (*Accident Analysis and Prevention*, Vol. 34, 2002). Based on how often each driver engaged in certain road rage behaviors (e.g., making obscene gestures at, tailgating, and thinking about physically hurting another driver), a road rage score was assigned. (Higher scores indicate a greater pattern of road rage behavior.) The drivers were also grouped by annual income: under \$30,000, between \$30,000 and \$60,000, and over \$60,000. The data were subjected to an analysis of variance, with the results summarized in the following table.

Income Group	Sample Size	Mean Road Rage Score
Under \$30,000	379	4.60
\$30,000 to \$60,000	392	5.08
Over \$60,000	267	5.15
ANOVA results:	*F*-value = 3.90	*p*-value < .01

a. Is a driver's propensity to engage in road rage related to his or her income?

b. An experimentwise error rate of .01 was used to rank the three means. Give a practical interpretation of this error rate.

c. How many pairwise comparisons are necessary to compare the three means? List them.

d. A multiple-comparisons procedure revealed that the means for the two income groups Between \$30 and \$60 thousand and Over \$60 thousand were not significantly different. All other pairs of means were found to be significantly different. Summarize these results in tabular form.

e. Which of the comparisons of part **c** will yield a confidence interval that does not contain 0?

10.115 Prompting walkers to walk. A study was conducted to investigate the effect of prompting in a walking program (*Health Psychology*, Mar. 1995). Five groups of walkers—27 in each group—agreed to participate by walking for 20 minutes at least one day per week over a 24-week period. The participants were prompted to walk each week via telephone calls, but different prompting schemes were used for each group. Walkers in the control group received no prompting phone calls, walkers in the "frequent/low" group received a call once a week with low structure (e.g., "just touching base"), walkers in the "frequent/high" group received a call once a week with high structure (i.e., goals are set), walkers in the "infrequent/low" group received a call once every 3 weeks with low structure, and walkers in the "infrequent/high" group received a call once every 3 weeks with high structure. The table below lists the number of participants in each group who actually walked the minimum requirement each week for weeks 1, 4, 8, 12, 16, and 24. The data, saved in the **WALKERS** file, were subjected to an analysis of variance for a randomized block design, with the five walker groups representing the treatments and the six periods (weeks) representing the blocks.

a. What is the purpose of blocking on weeks in this study?

b. Use statistical software (or the formulas in Appendix B) to construct an ANOVA summary table.

c. Is there sufficient evidence of a difference in the mean number of walkers per week among the five walker groups? Use $\alpha = .05$.

d. Use Tukey's technique to compare all pairs of treatment means with an experimentwise error rate of $\alpha = .05$. Interpret the results.

e. What assumptions must hold to ensure the validity of the inferences you made in parts **c** and **d**?

10.116 Exposure to low-frequency sound. *Infrasound* refers to sound frequencies below the audibility range of the human ear. A study of the physiological effects of infrasound was published in the *Journal of Low Frequency Noise, Vibration and Active Control* (Mar. 2004). In the experiment, one group of five university students (Group A) was exposed to infrasound at 4 hertz and 120 decibels

Data for Exercise 10.115

Week	Control	Frequent/Low	Frequent/High	Infrequent/Low	Infrequent/High
1	7	23	25	21	19
4	2	19	25	10	12
8	2	18	19	9	9
12	2	7	20	8	2
16	2	18	18	8	7
24	1	17	17	7	6

Source: Lombard, D. N., et al. "Walking to meet health guidelines: The effect of prompting frequency and prompt structure." Health Psychology, Vol. 14, No. 2, Mar. 1995, p. 167 (Table 2). Copyright © 1995 by the American Psychological Association. Reprinted with permission.

for 1 hour, and a second group of five students (Group B) was exposed to infrasound at 2 hertz and 110 decibels, also for 1 hour. The heart rate (beats/minute) of each student was measured both before and after exposure. The experimental data are provided in the accompanying table and saved in the **INFRASOUND** file. To determine the impact of infrasound, the researchers compared the mean heart rate before exposure to the mean heart rate after exposure.

Group A Students	Before Exposure	After Exposure	Group B Students	Before Exposure	After Exposure
A1	70	70	B1	73	79
A2	69	80	B2	68	60
A3	76	84	B3	61	69
A4	77	86	B4	72	77
A5	64	76	B5	61	66

Based on Qibai, C. Y. H., and Shi, H. "An investigation on the physiological and psychological effects of infrasound on persons." *Journal of Low Frequency Noise, Vibration and Active Control*, Vol. 23, No. 1, March 2004 (Tables I–IV).

a. Analyze the data on Group A students with an ANOVA for a randomized block design. Conduct the ANOVA test of interest with $\alpha = .05$.

b. Repeat part **a** for Group B students.

c. Now analyze the data via a paired difference *t*-test. Show that, for both groups of students, the results are equivalent to the randomized block ANOVA.

10.117 Commercial eggs produced from different housing systems. In the production of commercial eggs, four different types of housing systems for the chickens are used: cage, barn, free range, and organic. The characteristics of eggs produced from the four housing systems were investigated in *Food Chemistry* (Vol. 106, 2008). Twenty-eight commercial grade A eggs were randomly selected from supermarkets—10 of which were produced in cages, 6 in barns, 6 with free range, and 6 organically. A number of quantitative characteristics were measured for each egg, including shell thickness (millimeters), whipping capacity (percent overrun), and penetration strength (Newtons). The data (simulated from summary statistics provided in the journal article) are saved in the **EGGS** file. For each characteristic, the researchers compared the means of the four housing systems. MINITAB descriptive statistics and ANOVA printouts for each characteristic are shown in the next column. Fully interpret the results. Identify the characteristics for which housing systems differ.

10.118 Testing the ability to perform left-handed tasks. Most people are right handed due to the propensity of the left hemisphere of the brain to control sequential movement. Similarly, the fact that some tasks are performed better with the left hand is likely due to the superiority of the right hemisphere of the brain in processing the necessary information. Does such cerebral specialization in spatial processing occur in adults with Down syndrome? A 2 × 2 factorial experiment was conducted to answer this question (*American Journal on Mental Retardation*, May 1995). A sample of adults with Down syndrome was compared with a control group of individuals of a similar age, but not affected by the condition. Thus, one factor was Group at two levels (Down syndrome and control), and the second factor was Handedness (left or right) of

MINITAB Output for Exercise 10.117

Descriptive Statistics: THICKNESS, OVERRUN, STRENGTH

Variable	HOUSING	N	Mean	StDev	Minimum	Maximum
THICKNESS	BARN	6	0.50000	0.01414	0.48000	0.52000
	CAGE	10	0.4230	0.0350	0.3700	0.4700
	FREE	6	0.5017	0.0279	0.4700	0.5500
	ORGANIC	6	0.4817	0.0387	0.4300	0.5200
OVERRUN	BARN	6	513.33	8.38	501.00	526.00
	CAGE	10	480.60	12.91	462.00	502.00
	FREE	6	517.50	8.17	510.00	531.00
	ORGANIC	6	529.17	10.65	511.00	544.00
STRENGTH	BARN	6	39.333	1.120	37.600	40.300
	CAGE	10	37.320	2.127	33.000	40.200
	FREE	6	37.17	3.79	31.50	40.60
	ORGANIC	6	35.97	3.04	32.60	40.20

One-way ANOVA: THICKNESS versus HOUSING

Source	DF	SS	MS	F	P
HOUSING	3	0.034291	0.011430	11.74	0.000
Error	24	0.023377	0.000974		
Total	27	0.057668			

One-way ANOVA: OVERRUN versus HOUSING

Source	DF	SS	MS	F	P
HOUSING	3	10788	3596	31.36	0.000
Error	24	2752	115		
Total	27	13540			

One-way ANOVA: STRENGTH versus HOUSING

Source	DF	SS	MS	F	P
HOUSING	3	35.12	11.71	1.70	0.193
Error	24	164.82	6.87		
Total	27	199.94			

the subject. All the subjects performed a task that typically yields a left-hand advantage. The response variable was "laterality index," measured on a −100- to 100-point scale. (A large positive index indicates a right-hand advantage, a large negative index a left-hand advantage.)

a. Identify the treatments in this experiment.
b. Construct a graph that would support a finding of no interaction between the two factors.
c. Construct a graph that would support a finding of interaction between the two factors.
d. The *F*-test for factor interaction yielded an observed significance level of $p < .05$. Interpret this result.
e. Multiple comparisons of all pairs of treatment means yielded the rankings shown in the table below. Interpret the results.
f. The experimentwise error rate for part **e** was .05. Interpret this value.

Mean laterality index:	−30	−4	−.5	+.5
Group/Handed:	Down/ Left	Control/ Right	Control/ Left	Down/ Right

10.119 Facial expression study. What do people infer from facial expressions of emotion? This was the research question of interest in an article published in the *Journal of Nonverbal Behavior* (Fall 1996). A sample of 36 introductory psychology students was randomly divided into six groups. Each group was assigned to view one of six slides showing a person making a facial expression.* The six expressions were (1) angry, (2) disgusted, (3) fearful, (4) happy, (5) sad, and (6) neutral. After viewing the slides, the students rated the degree of dominance they inferred from the facial expression (on a scale ranging from −15 to +15). The data (simulated from summary information provided in the article) are listed in the accompanying table and saved in the **FACES** file.

a. Conduct an analysis of variance to determine whether the mean dominance ratings differ among the six facial expressions. Use $\alpha = .10$.

b. Use Tukey's method to rank the six dominance rating means. (Use $\alpha = .05$.)

Angry	Disgusted	Fearful	Happy	Sad	Neutral
2.10	.40	.82	1.71	.74	1.69
.64	.73	−2.93	−.04	−1.26	−.60
.47	−.07	−.74	1.04	−2.27	−.55
.37	−.25	.79	1.44	−.39	.27
1.62	.89	−.77	1.37	−2.65	−.57
−.08	1.93	−1.60	.59	−.44	−2.16

10.120 **Effectiveness of geese decoys.** Using decoys is a common method of hunting waterfowl. A study in the *Journal of Wildlife Management* (July 1995) compared the effectiveness of three different types of decoy— taxidermy-mounted decoys, plastic shell decoys, and full-bodied plastic decoys—in attracting Canada geese to sunken pit blinds. In order to account for an extraneous source of variation, three pit blinds were used as blocks in the experiment. Thus, a randomized block design with three treatments (types of decoy) and three blocks (pit blinds) was employed. The response variable was the percentage of a goose flock to approach within 46 meters of the pit blind on a given day. The data are given in the following table and saved in the **DECOY** file:[†]

Blind	Shell	Full Bodied	Taxidermy Mounted
1	7.3	13.6	17.8
2	12.6	10.4	17.0
3	16.4	23.4	13.6

Based on Harrey, W. F., Hindman, L. J., and Rhodes, W. E. "Vulnerability of Canada geese to taxidermy-mounted decoys." *Journal of Wildlife Management*, Vol. 59, No. 3, July 1995, p. 475 (Table 1).

a. Use statistical software (or the formulas in Appendix B) to construct an ANOVA table.

b. Interpret the *F*-statistic for comparing the response means of the three types of decoy.

c. What assumptions are necessary for the validity of the inference made in part **a**?

d. Why is it not necessary to conduct multiple comparisons of the response means for the three types of decoy?

10.121 **Impact of vitamin-B supplement.** In the *Journal of Nutrition* (July 1995), University of Georgia researchers

examined the impact of a vitamin-B supplement (nicotinamide) on the kidney. The experimental "subjects" were 28 Zucker rats—a species that tends to develop kidney problems. Half of the rats were classified as obese and half as lean. Within each group, half were randomly assigned to receive a vitamin-B-supplemented diet and half were not. Thus, a 2×2 factorial experiment was conducted, with seven rats assigned to each of the four combinations of size (lean or obese) and diet (supplemental or not). One of the response variables measured was weight (in grams) of the kidney at the end of a 20-week feeding period. The data (simulated from summary information provided in the journal article) are shown in the accompanying table and saved in the **VITAMINB** file.

Rat Size		Diet			
		Regular		Vitamin-B Supplement	
Lean		1.62	1.47	1.51	1.63
		1.80	1.37	1.65	1.35
		1.71	1.71	1.45	1.66
		1.81		1.44	
Obese		2.35	2.84	2.93	2.63
		2.97	2.05	2.72	2.61
		2.54	2.82	2.99	2.64
		2.93		2.19	

a. Conduct an analysis of variance on the data. Summarize the results in an ANOVA table.

b. Conduct the appropriate ANOVA *F*-tests at $\alpha = .01$. Interpret the results.

10.122 **Mosquito insecticide study.** A species of Caribbean mosquito is known to be resistant against certain insecticides. The effectiveness of five different types of insecticides— temephos, malathion, fenitrothion, fenthion, and chlorpyrifos—in controlling this mosquito species was investigated in the *Journal of the American Mosquito Control Association* (March 1995). Mosquito larvae were collected from each of seven Caribbean locations. In a laboratory, the larvae from each location were divided into five batches and each batch was exposed to one of the five insecticides. The dosage of insecticide required to kill 50% of the larvae was recorded and divided by the known dosage for a susceptible mosquito strain. The resulting value is called the *resistance ratio*. (The higher the ratio, the more resistant the mosquito species is to the insecticide relative to the susceptible mosquito strain.) The resistance ratios for the study are listed in the next table (top of p. 545) and saved in the **MOSQUITO** file. The researchers want to compare the mean resistance ratios of the five insecticides.

a. Explain why the experimental design is a randomized block design. Identify the treatments and the blocks.

b. Conduct a complete analysis of the data. Are any of the insecticides more effective than any of the others?

10.123 **Short-day traits of lemmings.** Many temperate-zone animal species exhibit physiological and morphological changes when the hours of daylight begin to decrease during the autumn months. A study was conducted to investigate the "short-day" traits of collared lemmings (*Journal of Experimental Zoology*, Sept. 1993). A total of 124 lemmings were bred in a colony maintained with a photoperiod of 22

---

*In the actual experiment, each group viewed all six facial expression slides and the design employed was a Latin Square (beyond the scope of this text).

[†]The actual design employed in the study was more complex than the randomized block design shown here. In the actual study, each number in the table represented the mean daily percentage of goose flocks attracted to the blind, averaged over 13–17 days.

Data for Exercise 10.122

Location	Insecticide				
	Temephos	Malathion	Fenitrothion	Fenthion	Chlorpyrifos
Anguilla	4.6	1.2	1.5	1.8	1.5
Antigua	9.2	2.9	2.0	7.0	2.0
Dominica	7.8	1.4	2.4	4.2	4.1
Guyana	1.7	1.9	2.2	1.5	1.8
Jamaica	3.4	3.7	2.0	1.5	7.1
St. Lucia	6.7	2.7	2.7	4.8	8.7
Suriname	1.4	1.9	2.0	2.1	1.7

Source: Rawlins, S. C., and Oh Hing Wan, J. "Resistance in some Caribbean population of Aedes aegypti to several insecticides." *Journal of the American Mosquito Control Associations*, Vol. 11, No. 1, Mar. 1995 (Table 1). Reprinted with permission.

hours of light per day. At weaning (19 days of age), the lemmings were weighed and randomly assigned to live under one of two photoperiods: 16 hours or less of light per day and more than 16 hours of light per day. (Each group was assigned the same number of males and females.) After 10 weeks, the lemmings were weighed again. The response variable of interest was the gain in body weight (measured in grams) over the 10-week experimental period. The researchers analyzed the data by means of a 2×2 factorial ANOVA design with the two factors being Photoperiod (at two levels) and Gender (at two levels).

a. Construct an ANOVA table for the experiment, listing the sources of variation and associated degrees of freedom.

b. The F-test for interaction was not significant. Interpret this result practically.

c. The p-values for testing for Photoperiod and Gender main effects were both smaller than .001. Interpret these results practically.

10.124 Ranging behavior of Spanish cattle. The cattle inhabiting the Biological Reserve of Doñana (Spain), live under free-range conditions, with virtually no human interference. The cattle population is organized into four herds (LGN, MTZ, PLC, and QMD). The *Journal of Zoology* (July 1995) investigated the ranging behavior of the four herds across the four seasons. Thus, a 4×4 factorial experiment was employed, with Herd and Season representing the two factors. Three animals from each herd during each season were sampled and the home range of each individual was measured (in square kilometers). The data were subjected to an ANOVA, with the results shown in the following table.

Source	df	F	p-Value
Herd (H)	3	17.2	$p < .001$
Season (S)	3	3.0	$p < .05$
H × S	9	1.2	$p > .05$
Error	32		
Total	47		

a. Conduct the appropriate ANOVA F-tests and interpret the results.

b. The researcher conducting the experiment ranked the four herd means independently of season. Do you agree with this strategy? Explain.

c. Refer to part **b** The Bonferroni rankings of the four herd means (at $\alpha = .05$) are shown in the next table. Interpret the results.

Mean (km)^2:	.75	1.0	2.7	3.8
Herd:	PLC	LGN	QMD	MTZ

Applying the Concepts—Advanced

10.125 Testing a new insecticide. Traditionally, people protect themselves from mosquito bites by applying insect repellent to their skin and clothing. Recent research suggests that peremethrin, an insecticide with low toxicity to humans, can provide protection from mosquitoes. A study in the *Journal of the American Mosquito Control Association* (Mar. 1995) investigated whether a tent sprayed with a commercially available 1% peremethrin formulation would protect people, both inside and outside the tent, against biting mosquitoes. Two canvas tents—one treated with peremethrin, the other untreated—were positioned 25 meters apart on flat dry ground in an area infested with mosquitoes. Eight people participated in the experiment, with four randomly assigned to each tent. Of the four stationed at each tent, two were randomly assigned to stay inside the tent (at opposite corners) and two to stay outside the tent (at opposite corners). During a specified 20-minute period during the night, each person kept count of the number of mosquito bites received. The goal of the study was to determine the effect of both Tent type (treated or untreated) and Location (inside or outside the tent) on the mean mosquito bite count.

a. What type of design was employed in the study?

b. Identify the factors and treatments.

c. Identify the response variable.

d. The study found statistical evidence of interaction between Tent type and Location. Give a practical interpretation of this result.

10.126 Therapy for binge eaters. Do you experience episodes of excessive eating accompanied by being overweight? If so, you may suffer from binge eating disorder. Cognitive-behavioral therapy (CBT), in which patients are taught how to make changes in specific behavior patterns (e.g., exercise, eat only low-fat foods), can be effective in treating the disorder. A group of Stanford University researchers investigated the effectiveness of interpersonal therapy (IPT) as a second level of treatment for binge eaters (*Journal of Consulting and Clinical Psychology*, June 1995). The researchers employed a design that randomly assigned a sample of 41 overweight individuals diagnosed with binge eating disorder to either a treatment group (30 subjects) or a control group (11 subjects). Subjects in the treatment group received 12 weeks of CBT and then were subdivided into two groups. Those who responded successfully to CBT (17 subjects) were assigned to a weight-loss therapy (WLT) program for the next 12 weeks. Those CBT subjects who did not respond to treatment (13 subjects) received 12 weeks of IPT. The subjects in the control group received no therapy of any type. Thus, the study ultimately consisted of three groups of overweight binge eaters: the CBT-WLT group, the CBT-IPT group, and the control group. One outcome (response) variable measured for each subject was the number x of binge eating episodes per week. Summary statistics for each of the three groups at the end of the 24-week period are shown in the accompanying table. The data were analyzed as a completely randomized design with three treatments (CBT-WLT, CBT-IPT, and Control). Although the ANOVA tables were not provided in the article, sufficient information is given in the table to reconstruct them. [See Exercise 10.38 (p. 497).] Is CBT effective in reducing the mean number of binges experienced per week?

	CBT-WLT	CBT-IPT	Control
Sample size	17	13	11
Mean number of binges per week	0.2	1.9	2.9
Standard deviation	0.4	1.7	2.0

Based on Agras, W. S., et al. "Does interpersonal therapy help patients with binge eating disorder who fail to respond to cognitive–behavioral therapy?" *Journal of Consulting and Clinical Psychology*, Vol. 63, No. 3, June 1995, p. 358 (Table 1).

Critical Thinking Challenges

10.127 Anticorrosive behavior of steel coated with epoxy. Organic coatings that use epoxy resins are widely used to protect steel and metal against weathering and corrosion. Researchers at National Technical University in Athens, Greece, examined the steel anticorrosive behavior of different epoxy coatings formulated with zinc pigments in an attempt to find the epoxy coating with the best resistance to corrosion (*Pigment & Resin Technology*, Vol. 32, 2003). The experimental units were flat, rectangular panels cut from steel sheets. Each panel was coated with one of four different coating systems: S1, S2, S3, and S4. Three panels, labeled, S1-A, S1-B, S1-C, S2-A, S2-B, ..., S4-C, were prepared for each coating system. The characteristics of the four coating systems are listed in the following table:

Coating System	First Layer	Second Layer
S1	Zinc dust	Epoxy paint, 100 micrometers thick
S2	Zinc phosphate	Epoxy paint, 100 micrometers thick
S2	Zinc phosphate with mica	Finish layer, 100 micrometers thick
S4	Zinc phosphate with mica	Finish layer, 200 micrometers thick

Each coated panel was immersed in deionized and deaerated water and then tested for corrosion. Since exposure time is likely to have a strong influence on anticorrosive behavior, the researchers attempted to remove this extraneous source of variation through the experimental design. Exposure times were fixed at 24 hours, 60 days, and 120 days. For each of the coating systems, one panel was exposed to water for 24 hours, one for 60 days, and

one for 120 days, in random order. The design is illustrated in the following diagram:

Exposure Time	Coating System/Panel Exposed
24 Hours	S1-A, S2-C, S3-C, S4-B
60 Days	S1-C, S2-A, S3-B, S4-A
120 Days	S1-B, S2-B, S3-A, S4-C

Following exposure, the corrosion rate (in nanoamperes per square centimeter) was determined for each panel. The lower the corrosion rate, the greater the anticorrosion performance of the coating system. The data are shown in the next table and saved in the **EPOXY** file. Are there differences among the epoxy treatment means? If so, which of the epoxy coating systems yields the lowest corrosion rate?

Exposure Time	System S1	System S2	System S3	System S4
24 Hours	6.7	7.5	8.2	6.1
60 Days	8.7	9.1	10.5	8.3
120 Days	11.8	12.6	14.5	11.8

Source: Kouloumbi, N., et al. "Anticorrosion performance of epoxy coatings on steel surface exposed to de-ionized water," *Pigment & Resin Technology*, Vol. 32, No. 2, 2003 (Table II).

10.128 Exam performance study. Refer to the *Teaching of Psychology* (Aug. 1998) study of whether a practice test helps students prepare for a final exam, presented in Exercise 10.12 (p. 481). Recall that students in an introductory psychology class were grouped according to their class standing and whether they attended a review session or took a practice test prior to the final exam. The experimental design was a 3×2 factorial design, with Class Standing at 3 levels (low, medium, high) and Exam Preparation at 2 levels (practice exam, review session). There were 22 students in each of the $3 \times 2 = 6$ treatment groups. After completing the final exam, each student rated her or his exam preparation on an 11-point scale ranging from 0 (not helpful at all) to 10 (extremely helpful). The data for this experiment (simulated from summary statistics provided in the article) are saved in the **PRACEXAM** file. The first five and last five observations in the data set are listed in the accompanying table. Conduct a complete analysis of variance of the helpfulness ratings data, including (if warranted) multiple comparisons of means. Do your findings support the article's conclusion that "Students at all levels of academic ability benefit from a ... practice exam"?

Exam Preparation	Class Standing	Helpfulness Rating
PRACTICE	LOW	6
PRACTICE	LOW	7
PRACTICE	LOW	7
PRACTICE	LOW	5
PRACTICE	LOW	3
⋮	⋮	⋮
REVIEW	HI	5
REVIEW	HI	2
REVIEW	HI	5
REVIEW	HI	4
REVIEW	HI	3

Based on Balch, W. R. "Practice versus review exams and final exam performance." *Teaching of Psychology*, Vol. 25, No. 3, Aug. 1998 (Table 1).

Activity · Comparing Supermarket Food Prices

Due to ever-increasing food costs, consumers are becoming more discerning in their choice of supermarkets. It is usually more convenient to shop at just one market, as opposed to buying different items at different markets. Thus, it would be useful to compare the mean food expenditure for a market basket of food items from store to store. Since there is a great deal of variability in the prices of products sold at any supermarket, we will consider an experiment that blocks on products.

Choose three (or more) supermarkets in your area that you want to compare. Then choose approximately 10 (or more) food products you typically purchase. For each food item, record the price each store charges in the following manner:

Food Item 1	Food Item 2	⋯	Food Item 10
Price store 1	Price store 1	⋯	Price store 1
Price store 2	Price store 2	⋯	Price store 2
Price store 3	Price store 3	⋯	Price store 3

Use the data you obtain to test

H_0: Mean expenditures at the stores are the same.

H_a: Mean expenditures for at least two of the stores are different.

Also, test to determine whether blocking on food items is advisable in this kind of experiment. Interpret the results of your analysis fully.

References

Cochran, W. G., and Cox, G. M. *Experimental Designs*, 2nd ed. New York: Wiley, 1957.

Hsu, J. C. *Multiple Comparisons: Theory and Methods*. London: Chapman & Hall, 1996.

Kramer, C. Y. "Extension of multiple range tests to group means with unequal number of replications." *Biometrics*, Vol. 12, 1956, pp. 307–310.

Kutner, M., Nachtsheim, C., Neter, J., and Li, W. *Applied Linear Statistical Models*, 5th ed. New York: McGraw-Hill/Irwin, 2005.

Mason, R. L., Gunst, R. F., and Hess, J. L. *Statistical Design and Analysis of Experiments*. New York: Wiley, 1989.

Mendenhall, W. *Introduction to Linear Models and the Design and Analysis of Experiments*. Belmont, CA: Wadsworth, 1968.

Miller, R. G., Jr. *Simultaneous Statistical Inference*. New York: Springer-Verlag, 1981.

Scheffé, H. "A method for judging all contrasts in the Analysis of Variance," *Biometrica*, Vol. 40, 1953, pp. 87–104.

Scheffé, H. *The Analysis of Variance*. New York: Wiley, 1959.

Snedecor, G. W., and Cochran, W. G. *Statistical Methods*, 7th ed. Ames, IA: Iowa State University Press, 1980.

Steele, R. G. D., and Torrie, J. H. *Principles and Procedures of Statistics: A Biometrical Approach*, 2nd ed. New York: McGraw-Hill, 1980.

Tukey, J. "Comparing individual means in the Analysis of Variance," *Biometrics*, Vol. 5, 1949, pp. 99–114.

Winer, B. J. *Statistical Principles in Experimental Design*, 2nd ed. New York: McGraw-Hill, 1971.

USING TECHNOLOGY

MINITAB: Analysis of Variance

MINITAB can conduct ANOVAs for all three types of experimental designs discussed in this chapter: completely randomized, randomized block, and factorial designs.

Completely Randomized Design

Step 1 Access the MINITAB worksheet file that contains the sample data. The data file should contain one quantitative variable (the response, or dependent, variable) and one factor variable with at least two levels.

Figure 10.M.1
MINITAB menu options for one-way ANOVA

Step 2 Click on the "Stat" button on the MINITAB menu bar and then click on "ANOVA" and "One-Way," as shown in Figure 10.M.1.

Step 3 On the resulting dialog screen Figure 10.M.2, specify the response variable in the "Response" box and the factor variable in the "Factor" box.

Figure 10.M.2
MINITAB one-way ANOVA dialog box

Step 4 Click the "Comparisons" button and select a multiple comparisons method and experimentwise error rate in the resulting dialog box (see Figure 10.M.3).

One-Way Multiple Comparisons

☑ Tukey's, family error rate: 5

☐ Fisher's, individual error rate: 5

☐ Dunnett's, family error rate: 5

  Control group level:

☐ Hsu's MCB, family error rate: 5
   ◉ Largest is best
   ○ Smallest is best

Help   OK   Cancel

Figure 10.M.3
MINITAB multiple comparisons dialog box

Step 5 Click "OK" to return to the "One-Way ANOVA" dialog screen and then click "OK" to generate the MINITAB printout.

Randomized Block and Factorial Designs

Step 1 Access the MINITAB worksheet file that contains the sample data. The data file should contain one quantitative variable (the response, or dependent, variable) and two other variables that represent the factors and/or blocks.

Step 2 Click on the "Stat" button on the MINITAB menu bar, and then click on "ANOVA" and "Two-Way" (see Figure 10.M.1). The resulting dialog screen appears as shown in Figure 10.M.4.

Two-Way Analysis of Variance

Response: DISTANCE

Row factor: BRAND   ☐ Display means

Column factor: GOLFER   ☐ Display means

☐ Store residuals
☐ Store fits

Confidence level: 95.0

Select   ☑ Fit additive model   Graphs...

Help   OK   Cancel

Figure 10.M.4
Minitab two-way ANOVA dialog box

Step 3 Specify the response variable in the "Response" box, the first factor variable in the "Row factor" box, and the second factor or block variable in the "Column factor" box. If the design is a randomized block, select the "Fit additive model" option, as shown in Figure 10.M.4. If the design is factorial, leave the "Fit additive model" option unselected.

Step 4 Click "OK" to generate the MINITAB printout.

Note: Multiple comparisons of treatment means are obtained by selecting "Stat," then "ANOVA," and then "General Linear Models." Specify the factors in the "Model" box and then select "Comparisons" and put the factor of interest in the "Terms" box. Press "OK" twice.

TI-83/TI-84 Plus Graphing Calculator: Analysis of Variance

The TI-83/TI-84 plus graphing calculator can be used to compute a one-way ANOVA for a completely randomized design but not a two-way ANOVA for either a randomized block or factorial design.

Completely Randomized Design

Step 1 *Enter each data set into its own list (i.e., sample 1 into L1, sample 2 into L2, sample 3 into L3, etc.)*

Step 2 *Access the statistical test menu*

- Press **STAT**
- Arrow right to **TESTS**
- Arrow down to **ANOVA**
- Press **ENTER**
- Type in each List name, separated by commas (*e.g.,* L1, L2, L3, L4)
- Press **ENTER**

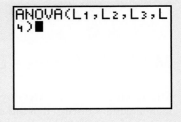

Step 3 *View display*

The calculator will display the *F*-test statistic, as well as the *p*-value, the factor degrees of freedom, sum of squares, mean square, and by arrowing down, the Error degrees of freedom, sum of squares, mean square, and the pooled standard deviation.

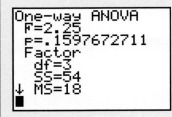

11 Simple Linear Regression

CONTENTS

Where We've Been

- Presented methods for estimating and testing population parameters (e.g., the mean, proportion, and variance) for a single sample
- Extended these methods to allow for a comparison of population parameters for multiple samples

Where We're Going

- Introduce the straight-line (*simple linear regression*) model as a means of relating one quantitative variable to another quantitative variable (11.1)
- Assess how well the simple linear regression model fits the sample data (11.2–11.4)
- Introduce the *correlation coefficient* as a means of relating one quantitative variable to another quantitative variable (11.5)
- Utilize the simple linear regression model to predict the value of one variable from a specified value of another variable (11.6, 11.7)

Statistics in Action Can "Dowsers" Really Detect Water?

The act of searching for and finding underground supplies of water with the use of nothing more than a divining rod is commonly known as "dowsing." Although widely regarded among scientists as no more than a superstitious relic from medieval times, dowsing remains popular in folklore, and to this day, there are individuals who claim to have this mysterious skill.

Many dowsers in Germany claim that they respond to "earthrays" which emanate from the water source. Earthrays, say the dowsers, are a subtle form of radiation that is potentially hazardous to human health. As a result of these claims, in the mid-1980s the German government conducted a two-year experiment to investigate the possibility that dowsing is a genuine skill. If such a skill could be demonstrated, reasoned government officials, then dangerous levels of radiation in Germany could be detected, avoided, and disposed of.

A group of university physicists in Munich, Germany, was provided a grant of 400,000 marks (about $250,000) to conduct the study. Approximately 500 candidate dowsers were recruited to participate in preliminary tests of their skill. To avoid fraudulent claims, the 43 individuals who seemed to be the most successful in the preliminary tests were selected for the final, carefully controlled, experiment.

The researchers set up a 10-meter-long line on the ground floor of a vacant barn, along which a small wagon could be moved. Attached to the wagon was a short length of pipe, perpendicular to the test line, that was connected by hoses to a pump with running water. The location of the pipe along the line for each trial of the experiment was assigned by a computer-generated random number. On the upper floor of the barn, directly above the experimental line, a 10-meter test line was painted. In each trial, a dowser was admitted to this upper level and required, with his or her rod, stick, or other tool of choice, to ascertain where the pipe with running water on the ground floor was located.

Each dowser participated in at least one test series constituting a sequence of from 5 to 15 trials (typically, 10), with the pipe randomly repositioned after each trial. (Some dowsers undertook only 1 test series, whereas selected others underwent more than 10 test series.) Over the two-year experimental period, the 43 dowsers participated in a total of 843 tests. The experiment was "double blind" in that neither the observer (researcher) on the top floor nor the dowser knew the pipe's location, even after a guess was made. [*Note:* Before the experiment began, a professional magician inspected the entire arrangement for potential deception or cheating by the dowsers.]

For each trial, two variables were recorded: the actual location of the pipe (in decimeters from the beginning of the line) and the dowser's guess (also measured in decimeters). On the basis of an examination of these data, the German physicists concluded in their final report that although most dowsers did not do particularly well in the experiments, "some few dowsers, in particular tests, showed an extraordinarily high rate of success, which can scarcely if at all be explained as due to chance . . . a real core of dowser-phenomena can be regarded as empirically proven . . ." (Wagner, Betz, and König, 1990. Final Report 01 KB8602, Federal Ministry for Research and Technology).

This conclusion was critically assessed by Professor J. T. Enright of the University of California at San Diego (*Skeptical Inquirer*, Jan./Feb. 1999). In the Statistics in Action Revisited sections of this chapter, we demonstrate how Enright concluded the exact opposite of the German physicists.

Statistics in Action Revisited

- Estimating a Straight-Line Regression Model for the Dowsing Data (p. 559)
- Assessing How Well the Straight-Line Model Fits the Dowsing Data (p. 574)
- Using the Coefficients of Correlation and Determination to Assess the Dowsing Data (p. 584)
- Using the Straight-Line Model to Predict Pipe Location for the Dowsing Data (p. 592)

In Chapters 7–10, we described methods for making inferences about population means. The mean of a population has been treated as a *constant*, and we have shown how to use sample data to estimate or to test hypotheses about this constant mean. In many applications, the mean of a population is not viewed as a constant, but rather as a variable. For example, the mean sale price of residences in a large city might be treated as a variable that depends on the number of square feet of living space in the residence. The relationship might be

$$\text{Mean sale price} = \$30{,}000 + \$60\,(\text{Square feet})$$

This formula implies that the mean sale price of 1,000-square-foot homes is $90,000, the mean sale price of 2,000-square-foot homes is $150,000, and the mean sale price of 3,000-square-foot homes is $210,000.

In this chapter, we discuss situations in which the mean of the population is treated as a variable, dependent on the value of another variable. The dependence of the residential sale price on the number of square feet of living space is one illustration. Other examples include the dependence of the mean reaction time on the amount of a drug in the bloodstream, the dependence of the mean starting salary of a college graduate on the student's GPA, and the dependence of the mean number of years to which a criminal is sentenced on the number of previous convictions.

Here, we present the simplest of all models relating a populating mean to another variable: the *straight-line model*. We show how to use the sample data to estimate the straight-line relationship between the mean value of one variable, y, as it relates to a second variable, x. The methodology of estimating and using a straight-line relationship is referred to as *simple linear regression analysis*.

11.1 Probabilistic Models

An important consideration in taking a drug is how it may affect one's perception or general awareness. Suppose you want to model the length of time it takes to respond to a stimulus (a measure of awareness) as a function of the percentage of a certain drug in the bloodstream. The first question to be answered is this: "Do you think that an exact relationship exists between these two variables?" That is, do you think that it is possible to state the exact length of time it takes an individual (subject) to respond if the amount of the drug in the bloodstream is known? We think that you will agree with us that this is *not* possible, for several reasons: The reaction time depends on many variables other than the percentage of the drug in the bloodstream—for example, the time of day, the amount of sleep the subject had the night before, the subject's visual acuity, the subject's general reaction time without the drug, and the subject's age. Even if many variables are included in a model (the topic of Chapter 12), it is still unlikely that we would be able to predict the subject's reaction time *exactly*. There will almost certainly be some variation in response times due strictly to *random phenomena* that cannot be modeled or explained.

If we were to construct a model that hypothesized an exact relationship between variables, it would be called a **deterministic model**. For example, if we believe that y, the reaction time (in seconds), will be exactly one-and-one-half times x, the amount of drug in the blood, we write

$$y = 1.5x$$

This represents a **deterministic relationship** between the variables y and x. It implies that y can always be determined exactly when the value of x is known. *There is no allowance for error in this prediction.*

If, however, we believe that there will be unexplained variation in reaction times—perhaps caused by important, but unincluded, variables or by random phenomena—we discard the deterministic model and use a model that accounts for this **random error.** Our **probabilistic model** will include both a deterministic component and a random-error component. For example, if we hypothesize that the response time y is related to the percentage x of drug by

$$y = 1.5x + \text{Random error}$$

we are hypothesizing a **probabilistic relationship** between y and x. Note that the deterministic component of this probabilistic model is $1.5x$.

Figure 11.1a shows the possible responses for five different values of x, the percentage of drug in the blood, when the model is deterministic. All the responses must fall exactly on the line, because a deterministic model leaves no room for error.

Figure 11.1b shows a possible set of responses for the same values of x when we are using a probabilistic model. Note that the deterministic part of the model (the straight line itself) is the same. Now, however, the inclusion of a random-error component allows the response times to vary from this line. Since we know that the response time does vary randomly for a given value of x, the probabilistic model for y is more realistic than the deterministic model.

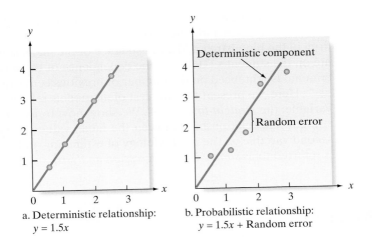

Figure 11.1
Possible reaction times y for five different drug percentages x

a. Deterministic relationship: $y = 1.5x$

b. Probabilistic relationship: $y = 1.5x +$ Random error

General Form of Probabilistic Models

$$y = \text{Deterministic component} + \text{Random error}$$

where y is the variable of interest. We always assume that the mean value of the random error equals 0. This is equivalent to assuming that the mean value of y, $E(y)$, equals the deterministic component of the model; that is,

$$E(y) = \text{Deterministic component}$$

BIOGRAPHY **FRANCIS GALTON (1822–1911)**

The Law of Universal Regression

Francis Galton was the youngest of seven children born to a middle-class English family of Quaker faith. A cousin of Charles Darwin, Galton attended Trinity College (Cambridge, England) to study medicine. Due to the death of his father, Galton was unable to obtain his degree. His competence in both medicine and mathematics, however, led Galton to pursue a career as a scientist. He made major contributions to the fields of genetics, psychology, meteorology, and anthropology. Some consider Galton to be the first social scientist for his applications of the novel statistical concepts of the time—in particular, regression and correlation. While studying natural inheritance in 1886, Galton collected data on heights of parents and adult children. He noticed the tendency for tall (or short) parents to have tall (or short) children, but that the children were not as tall (or short), on average as their parents. Galton called this phenomenon the "law of universal regression," for the average heights of adult children tended to "regress" to the mean of the population. With the help of his friend and disciple, Karl Pearson, Galton applied the straight-line model to the height data, and the term *regression model* was coined. ■

In this chapter, we present the simplest of probabilistic models—the **straight-line model**—which gets its name from the fact that the deterministic portion of the model graphs as a straight line. Fitting this model to a set of data is an example of **regression analysis**, or **regression modeling**. The elements of the straight-line model are summarized in the following box:

A First-Order (Straight-Line) Probabilistic Model

$$y = \beta_0 + \beta_1 x + \varepsilon$$

where

$y =$ **Dependent** *or* **response variable** (variable to be modeled)

$x =$ **Independent** *or* **predictor variable** (variable used as a predictor of y)*

$\beta_0 + \beta_1 x = E(y) =$ Deterministic component

ε (epsilon) $=$ Random error component

---

*The word *independent* should not be interpreted in a probabilistic sense as defined in Chapter 3. The phrase *independent variable* is used in regression analysis to refer to a predictor variable for the response y.

β_0 (beta zero) = **y-intercept of the line**—that is, the point at which the line intersects, or cuts through, the y-axis (see Figure 11.2)

β_1 (beta one) = **Slope of the line**—that is, the change (amount of increase or decrease) in the deterministic component of y for every one-unit increase in x.

[*Note:* A *positive* slope implies that $E(y)$ *increases* by the amount β_1. (See Figure 11.2.) A *negative* slope implies that $E(y)$ *decreases* by the amount β_1.]

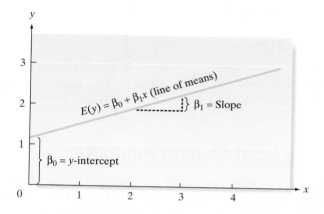

Figure 11.2
The straight-line model

In the probabilistic model, the deterministic component is referred to as the **line of means**, because the mean of y, $E(y)$, is equal to the straight-line component of the model. That is,

$$E(y) = \beta_0 + \beta_1 x$$

Note that the Greek symbols β_0 and β_1 respectively represent the y-intercept and slope of the model. They are population parameters that will be known only if we have access to the entire population of (x, y) measurements. Together with a specific value of the independent variable x, they determine the mean value of y, which is just a specific point on the line of means (Figure 11.2).

The values of β_0 and β_1 will be unknown in almost all practical applications of regression analysis. The process of developing a model, estimating the unknown parameters, and using the model can be viewed as the five-step procedure shown in the following box:

Conducting a Simple Linear Regression:

Step 1 Hypothesize the deterministic component of the model that relates the mean $E(y)$ to the independent variable x (Section 11.2).

Step 2 Use the sample data to estimate unknown parameters in the model (Section 11.2).

Step 3 Specify the probability distribution of the random-error term and estimate the standard deviation of this distribution (Section 11.3).

Step 4 Statistically evaluate the usefulness of the model (Sections 11.4 and 11.5).

Step 5 When satisfied that the model is useful, use it for prediction, estimation, and other purposes (Section 11.6).

Exercises 11.1–11.10

Understanding the Principles

11.1 Why do we generally prefer a probabilistic model to a deterministic model? Give examples for which the two types of models might be appropriate.

11.2 What is the difference between a dependent variable and an independent variable in a probabilistic model?

11.3 What is the line of means?

11.4 If a straight-line probabilistic relationship relates the mean $E(y)$ to an independent variable x, does it imply that every value of the variable y will always fall exactly on the line of means? Why or why not?

Learning the Mechanics

11.5 In each case, graph the line that passes through the given points.
  a. $(1, 1)$ and $(5, 5)$
  b. $(0, 3)$ and $(3, 0)$
  c. $(-1, 1)$ and $(4, 2)$
  d. $(-6, -3)$ and $(2, 6)$

11.6 Give the slope and y-intercept for each of the lines graphed in Exercise 11.5.

11.7 The equation (deterministic) for a straight line is

$$y = \beta_0 + \beta_1 x$$

If the line passes through the point $(-2, 4)$, then $x = -2, y = 4$ must satisfy the equation; that is,

$$4 = \beta_0 + \beta_1(-2)$$

Similarly, if the line passes through the point $(4, 6)$, then $x = 4, y = 6$ must satisfy the equation; that is,

$$6 = \beta_0 + \beta_1(4)$$

Use these two equations to solve for β_0 and β_1; then find the equation of the line that passes through the points $(-2, 4)$ and $(4, 6)$.

11.8 Refer to Exercise 11.7. Find the equations of the lines that pass through the points listed in Exercise 11.5.

11.9 Plot the following lines:
  a. $y = 4 + x$
  b. $y = 5 - 2x$
  c. $y = -4 + 3x$
  d. $y = -2x$
  e. $y = x$
  f. $y = .50 + 1.5x$

11.10 Give the slope and y-intercept for each of the lines defined in Exercise 11.9.

11.2 Fitting the Model: The Least Squares Approach

After the straight-line model has been hypothesized to relate the mean $E(y)$ to the independent variable x, the next step is to collect data and to estimate the (unknown) population parameters, the y-intercept β_0 and the slope β_1.

To begin with a simple example, suppose an experiment involving five subjects is conducted to determine the relationship between the percentage of a certain drug in the bloodstream and the length of time it takes to react to a stimulus. The results are shown in Table 11.1. (The number of measurements and the measurements themselves are unrealistically simple in order to avoid arithmetic confusion in this introductory example.) This set of data will be used to demonstrate the five-step procedure of regression modeling given in the previous section. In the current section, we hypothesize the deterministic component of the model and estimate its unknown parameters (steps 1 and 2). The model's assumptions and the random-error component (step 3) are the subjects of Section 11.3, whereas Sections 11.4 and 11.5 assess the utility of the model (step 4). Finally, we use the model for prediction and estimation (step 5) in Section 11.6.

Table 11.1	Reaction Time versus Drug Percentage	
Subject	Percent x of Drug	Reaction Time y (seconds)
1	1	1
2	2	1
3	3	2
4	4	2
5	5	4

Data Set: STIMULUS

Step 1 *Hypothesize the deterministic component of the probabilistic model.* As stated before, we will consider only straight-line models in this chapter. Thus, the complete model relating mean response time $E(y)$ to drug percentage x is given by

$$E(y) = \beta_0 + \beta_1 x$$

Step 2 *Use sample data to estimate unknown parameters in the model.* This step is the subject of this section—namely, how can we best use the information in the sample of five observations in Table 11.1 to estimate the unknown y-intercept β_0 and slope β_1?

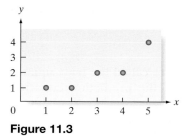

Figure 11.3
Scatterplot for data in Table 11.1

To determine whether a linear relationship between y and x is plausible, it is help-ful to plot the sample data in a **scatterplot** (or **scattergram**). Recall (Section 2.9) that a scatterplot locates each data point on a graph, as shown in Figure 11.3 for the five data points of Table 11.4. Note that the scatterplot suggests a general tendency for y to increase as x increases. If you place a ruler on the scatterplot, you will see that a line may be drawn through three of the five points, as shown in Figure 11.4. To obtain the equation of this visually fitted line, note that the line intersects the y-axis at $y = -1$, so the y-intercept is -1. Also, y increases exactly one unit for every one-unit increase in x, indicating that the slope is $+1$. Therefore, the equation is

$$\tilde{y} = -1 + 1(x) = -1 + x$$

where $\tilde{y}$ is used to denote the y that is predicted from the visual model.

One way to decide quantitatively how well a straight line fits a set of data is to note the extent to which the data points deviate from the line. For example, to evaluate the model in Figure 11.4, we calculate the magnitude of the *deviations* (i.e., the differences between the observed and the predicted values of y). These deviations, or **errors of prediction**, are the vertical distances between observed and predicted values (see Figure 11.4).* The observed and predicted values of y, their differences, and their squared differences are shown in Table 11.2. Note that the *sum of errors* equals 0 and the *sum of squares of the errors* (SSE), which places a greater emphasis on large deviations of the points from the line, is equal to 2.

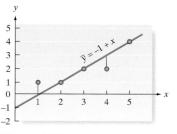

Figure 11.4
Visual straight line fitted to the data in Figure 11.3

Table 11.2   Comparing Observed and Predicted Values for the Visual Model

x	y	$\tilde{y} = -1 + x$	$(y - \tilde{y})$	$(y - \tilde{y})^2$
1	1	0	$(1 - 0) = 1$	1
2	1	1	$(1 - 1) = 0$	0
3	2	2	$(2 - 2) = 0$	0
4	2	3	$(2 - 3) = -1$	1
5	4	4	$(4 - 4) = 0$	0
			Sum of errors $= 0$	Sum of squared errors (SSE) $= 2$

You can see by shifting the ruler around the graph that it is possible to find many lines for which the sum of errors is equal to 0, but it can be shown that there is one (and only one) line for which the SSE is a *minimum*. This line is called the **least squares line**, the **regression line,** or the **least squares prediction equation.** The methodology used to obtain that line is called the **method of least squares.**

Now Work Exercise 11.16a–d

To find the least squares prediction equation for a set of data, assume that we have a sample of n data points consisting of pairs of values of x and y, say, $(x_1, y_1), (x_2, y_2), \dots, (x_n, y_n)$. For example, the $n = 5$ data points shown in Table 11.2 are $(1, 1), (2, 1), (3, 2), (4, 2),$ and $(5, 4)$. The fitted line, which we will calculate on the basis of the five data points, is written as

$$\hat{y} = \hat{\beta}_0 + \hat{\beta}_1 x$$

The "hats" indicate that the symbols below them are estimates: $\hat{y}$ (y-hat) is an estimator of the mean value of y, $E(y)$, and is a predictor of some future value of y; and $\hat{\beta}_0$ and $\hat{\beta}_1$ are estimators of β_0 and β_1, respectively.

For a given data point—say, the point (x_i, y_i),—the observed value of y is y_i and the predicted value of y would be obtained by substituting x_i into the prediction equation:

$$\hat{y}_i = \hat{\beta}_0 + \hat{\beta}_1 x_i$$

The deviation of the ith value of y from its predicted value is

$$(y_i - \hat{y}_i) = [y_i - (\hat{\beta}_0 + \hat{\beta}_1 x_i)]$$

*In Chapter 12, we refer to these errors of prediction as **regression residuals**. There, we learn that an analysis of residuals is essential in establishing a useful regression model.

Then the sum of the squares of the deviations of the y-values about their predicted values for all the n data points is

$$SSE = \sum [\, y_i - (\hat{\beta}_0 + \hat{\beta}_1 x_i)\,]^2$$

The quantities $\hat{\beta}_0$ and $\hat{\beta}_1$ that make the SSE a minimum are called the **least squares estimates** of the population parameters β_0 and β_1, and the prediction equation $\hat{y} = \hat{\beta}_0 + \hat{\beta}_1 x$ is called the *least squares line*.

The **least squares line** $\hat{y} = \hat{\beta}_0 + \hat{\beta}_1 x$ is the line that has the following two properties:

1. The sum of the errors equals 0, i.e., mean error of prediction $= 0$.
2. The sum of squared errors (SSE) is smaller than that for any other straight-line model

---

The values of $\hat{\beta}_0$ and $\hat{\beta}_1$ that minimize the SSE are given by the formulas in the following box (proof omitted):*

Formulas for the Least Squares Estimates

Slope: $\hat{\beta}_1 = \dfrac{SS_{xy}}{SS_{xx}}$

y-intercept: $\hat{\beta}_0 = \bar{y} - \hat{\beta}_1 \bar{x}$

where

$$SS_{xy} = \sum (x_i - \bar{x})(y_i - \bar{y}) = \sum x_i y_i - \frac{\left(\sum x_i\right)\left(\sum y_i\right)}{n}$$

$$SS_{xx} = \sum (x_i - \bar{x})^2 = \sum x_i^2 - \frac{\left(\sum x_i\right)^2}{n}$$

n = Sample size

Example 11.1

Applying the Method of Least Squares— Drug Reaction Data

Problem Refer to the reaction data presented in Table 11.1. Consider the straight-line model $E(y) = \beta_0 + \beta_1 x$, where y = reaction time (in seconds) and x = percent of drug received.

a. Use the method of least squares to estimate the values of β_0 and β_1.
b. Predict the reaction time when $x = 2\%$.
c. Find the SSE for the analysis.
d. Give practical interpretations of $\hat{\beta}_0$ and $\hat{\beta}_1$.

Solution

a. Preliminary computations for finding the least squares line for the drug reaction example are presented in Table 11.3. We can now calculate

$$SS_{xy} = \sum x_i y_i - \frac{\left(\sum x_i\right)\left(\sum y_i\right)}{5} = 37 - \frac{(15)(10)}{5} = 37 - 30 = 7$$

$$SS_{xx} = \sum x_i^2 - \frac{\left(\sum x_i\right)^2}{5} = 55 - \frac{(15)^2}{5} = 55 - 45 = 10$$

*Students who are familiar with calculus should note that the values of β_0 and β_1 that minimize SSE $= \Sigma (y_i - \hat{y}_i)^2$ are obtained by setting the two partial derivatives $\partial SSE/\partial \beta_0$ and $\partial SSE/\partial \beta_1$ equal to 0. The solutions of these two equations yield the formulas shown in the box. Furthermore, we denote the *sample* solutions of the equations by $\hat{\beta}_0$ and $\hat{\beta}_1$, where the "hat" denotes that these are sample estimates of the true population intercept β_0 and slope β_1.

Table 11.3 **Preliminary Computations for the Drug Reaction Example**

x_i	y_i	x_i^2	$x_i y_i$
1	1	1	1
2	1	4	2
3	2	9	6
4	2	16	8
5	4	25	20
Totals $\sum x_i = 15$	$\sum y_i = 10$	$\sum x_i^2 = 55$	$\sum x_i y_i = 37$

Then the slope of the least squares line is

$$\hat{\beta}_1 = \frac{SS_{xy}}{SS_{xx}} = \frac{7}{10} = .7$$

and the y-intercept is

$$\hat{\beta}_0 = \bar{y} - \hat{\beta}_1\bar{x} = \frac{\sum y_i}{5} - \hat{\beta}_1\frac{\sum x_i}{5}$$

$$= \frac{10}{5} - (.7)\left(\frac{15}{5}\right) = 2 - (.7)(3) = 2 - 2.1 = -.1$$

The least squares line is thus

$$\hat{y} = \hat{\beta}_0 + \hat{\beta}_1 x = -.1 + .7x$$

The graph of this line is shown in Figure 11.5.

b. The predicted value of y for a given value of x can be obtained by substituting into the formula for the least squares line. Thus, when $x = 2$, we predict y to be

$$\hat{y} = -.1 + .7x = -.1 + .7(2) = 1.3$$

We show how to find a prediction interval for y in Section 11.6.

c. The observed and predicted values of y, the deviations of the y values about their predicted values, and the squares of these deviations are shown in Table 11.4. Note that the sum of the squares of the deviations, SSE, is 1.10 and (as we would expect) this is less than the SSE = 2.0 obtained in Table 11.2 for the visually fitted line.

d. The estimated y-intercept, $\hat{\beta}_0 = -.1$, appears to imply that the estimated mean reaction time is equal to $-.1$ second when the percent x of drug is equal to 0%. Since negative reaction times are not possible, this seems to make the model nonsensical. However, *the model parameters should be interpreted only within the sampled range of the independent variable*—in this case, for amounts of drug in the bloodstream between 1% and 5%. Thus, the y-intercept—which is, by definition, at $x = 0$ (0% drug)—is not within the range of the sampled values of x and is not subject to meaningful interpretation.

The slope of the least squares line, $\hat{\beta}_1 = .7$, implies that for every unit increase in x, the mean value of y is estimated to increase by .7 unit. In terms of this example, for every 1% increase in the amount of drug in the bloodstream, the mean reaction time

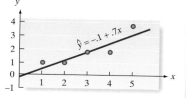
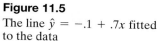

Figure 11.5

The line $\hat{y} = -.1 + .7x$ fitted to the data

Table 11.4 **Comparing Observed and Predicted Values for the Least Squares Prediction Equation**

x	y	$\hat{y} = -.1 + .7x$	$(y - \hat{y})$	$(y - \hat{y})^2$
1	1	.6	$(1 - .6) = .4$	.16
2	1	1.3	$(1 - 1.3) = -.3$	.09
3	2	2.0	$(2 - 2.0) = 0$	.00
4	2	2.7	$(2 - 2.7) = -.7$	.49
5	4	3.4	$(4 - 3.4) = .6$	.36
			Sum of errors = 0	SSE = 1.10

is estimated to increase by .7 second *over the sampled range of drug amounts from 1% to 5%*. Thus, the model does not imply that increasing the drug amount from 5% to 10% will result in an increase in mean reaction time of 3.5 seconds, because the range of x in the sample does not extend to 10% $(x = 10)$. In fact, 10% might be such a high concentration that the drug would kill the subject! Be careful to interpret the estimated parameters only within the sampled range of x.

Look Back The calculations required to obtain $\hat{\beta}_0$, $\hat{\beta}_1$, and SSE in simple linear regression, although straightforward, can become rather tedious. Even with the use of a pocket calculator, the process is laborious and susceptible to error, especially when the sample size is large. Fortunately, the use of statistical computer software can significantly reduce the labor involved in regression calculations. The SAS, SPSS, and MINITAB outputs for the simple linear regression of the data in Table 11.1 are displayed in Figure 11.6a–c. The values of $\hat{\beta}_0$ and $\hat{\beta}_1$ are highlighted on the printouts. These values, $\hat{\beta}_0 = -.1$ and $\hat{\beta}_1 = .7$, agree exactly with our hand-calculated values. The value of SSE $= 1.10$ is also highlighted on the printouts.

Dependent Variable: TIME_Y

Number of Observations Read	5
Number of Observations Used	5

Analysis of Variance

Source	DF	Sum of Squares	Mean Square	F Value	Pr > F
Model	1	4.90000	4.90000	13.36	0.0354
Error	3	1.10000	0.36667		
Corrected Total	4	6.00000			

Root MSE	0.60553	R-Square	0.8167
Dependent Mean	2.00000	Adj R-Sq	0.7556
Coeff Var	30.27650		

Parameter Estimates

Variable	DF	Parameter Estimate	Standard Error	t Value	Pr > |t|
Intercept	1	-0.10000	0.63509	-0.16	0.8849
DRUG_X	1	0.70000	0.19149	3.66	0.0354

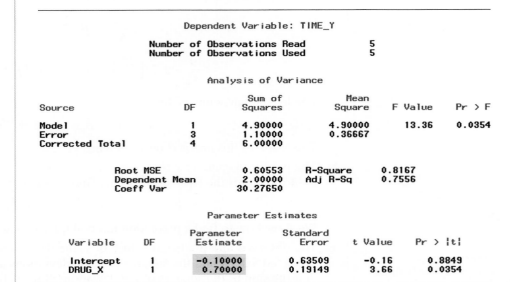

Figure 11.6a
SAS printout for the time–drug regression

Model Summary

Model	R	R Square	Adjusted R Square	Std. Error of the Estimate
1	.904[a]	.817	.756	.606

a. Predictors: (Constant), DRUG_X

ANOVA[b]

Model		Sum of Squares	df	Mean Square	F	Sig.
1	Regression	4.900	1	4.900	13.364	.035[a]
	Residual	1.100	3	.367		
	Total	6.000	4			

a. Predictors: (Constant), DRUG_X
b. Dependent Variable: TIME_Y

Coefficients[a]

Model		Unstandardized Coefficients		Standardized Coefficients		
		B	Std. Error	Beta	t	Sig.
1	(Constant)	-.100	.635		-.157	.885
	DRUG_X	.700	.191	.904	3.656	.035

a. Dependent Variable: TIME_Y

Figure 11.6b
SPSS printout for the time–drug regression

Regression Analysis: TIME_Y versus DRUG_X

```
The regression equation is
TIME_Y = - 0.100 + 0.700 DRUG_X


Predictor        Coef   SE Coef       T       P
Constant      -0.1000    0.6351   -0.16   0.885
DRUG_X         0.7000    0.1915    3.66   0.035


S = 0.605530    R-Sq = 81.7%    R-Sq(adj) = 75.6%


Analysis of Variance

Source          DF        SS        MS       F       P
Regression       1    4.9000    4.9000   13.36   0.035
Residual Error   3    1.1000    0.3667
Total            4    6.0000
```

Figure 11.6c
MINITAB printout for the
time–drug regression

Now Work Exercise 11.23

Interpreting the Estimates of β_0 and β_1 in Simple Linear Regression

y-intercept: $\hat{\beta}_0$ represents the predicted value of y when $x = 0$. (*Caution*: This value will not be meaningful if the value $x = 0$ is nonsensical or outside the range of the sample data.)

slope: $\hat{\beta}_1$ represents the increase (or decrease) in y for every 1-unit increase in x. (*Caution*: This interpretation is valid only for x-values within the range of the sample data.)

Even when the interpretations of the estimated parameters in a simple linear regression are meaningful, we need to remember that they are only estimates based on the sample. As such, their values will typically change in repeated sampling. How much confidence do we have that the estimated slope $\hat{\beta}_1$ accurately approximates the true slope β_1? Determining this requires statistical inference, in the form of confidence intervals and tests of hypotheses, which we address in Section 11.4.

To summarize, we defined the best-fitting straight line to be the line that minimizes the sum of squared errors around it, and we called it the least squares line. We should interpret the least squares line only within the sampled range of the independent variable. In subsequent sections, we show how to make statistical inferences about the model.

Statistics IN Action | **Revisited**

Estimating a Straight–Line Regression Model for the Dowsing Data

After conducting a series of experiments in a Munich barn, a group of German physicists concluded that dowsing (i.e., the ability to find underground water with a divining rod) "can be regarded as empirically proven." This observation was based on the data collected on 3 (of the participating 500) dowsers who had particularly impressive results. All of these "best" dowsers (numbered 99, 18, and 108) performed the experiment multiple times, and the best test series (sequence of trials) for each of them was identified. These data, saved in the **DOWSING** file, are listed in Table SIA11.1.

Recall (p. 550) that for various hidden pipe locations, each dowser guessed where the pipe with running water

was located. Let x = dowser's guess (in meters) and y = pipe location (in meters) for each trial. One way to determine whether the "best" dowsers are effective is to fit the straight-line model $E(y) = \beta_0 + \beta_1 x$ to the data in Table SIA11.1.

A MINITAB scatterplot of the data is shown in Figure SIA11.1. The least squares line, obtained from the MINITAB regression printout shown in Figure SIA11.2, is also displayed on the scatterplot. Although the least squares line

(continued)

Statistics IN Action
(continued)

Table SIA11.1 Dowsing Trial Results: Best Series for the Three Best Dowsers

Trial	Dowser Number	Pipe Location	Dowser's Guess
1	99	4	4
2	99	5	87
3	99	30	95
4	99	35	74
5	99	36	78
6	99	58	65
7	99	40	39
8	99	70	75
9	99	74	32
10	99	98	100
11	18	7	10
12	18	38	40
13	18	40	30
14	18	49	47
15	18	75	9
16	18	82	95
17	108	5	52
18	108	18	16
19	108	33	37
20	108	45	40
21	108	38	66
22	108	50	58
23	108	52	74
24	108	63	65
25	108	72	60
26	108	95	49

Based on Enright, J. T. "Testing dowsing: The failure of the Munich experiments." *Skeptical Inquirer*, Jan./Feb. 1999, p. 45 (Figure 6a).

Data Set: DOWSING

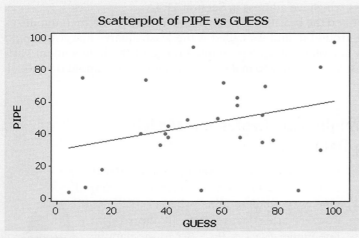

Figure SIA11.1
MINITAB scatterplot of dowsing data

Regression Analysis: PIPE versus GUESS

```
The regression equation is
PIPE = 30.1 + 0.308 GUESS


Predictor      Coef   SE Coef      T       P
Constant      30.07     11.41   2.63   0.015
GUESS        0.3079    0.1900   1.62   0.118


S = 26.0298    R-Sq = 9.9%    R-Sq(adj) = 6.1%


Analysis of Variance

Source          DF       SS       MS      F       P
Regression       1   1778.9   1778.9   2.63   0.118
Residual Error  24  16261.2    677.6
Total           25  18040.2
```

Figure SIA11.2
MINITAB simple linear regression for dowsing data

has a slight upward trend, the variation of the data points around the line is large. It does not appear that a dowser's guess (x) will be a very good predictor of the actual pipe location (y). In fact, the estimated slope (obtained from Figure SIA11.2) is $\hat{\beta}_1 = .31$. Thus, for every 1-meter increase in a dowser's guess, we estimate that the actual pipe location will increase only .31 meter. In the Statistics in Action Revisited sections that follow, we will provide a measure of reliability for this inference and investigate the phenomenon of dowsing further.

Exercises 11.11–11.31

Understanding the Principles

11.11 In regression, what is an error of prediction?

11.12 Give two properties of the line estimated with the method of least squares.

11.13 *True or False*. The estimates of β_0 and β_1 should be interpreted only within the sampled range of the independent variable, x.

Learning the Mechanics

11.14 The accompanying table is similar to Table 11.3. It is used to make the preliminary computations for finding the least squares line for the given pairs of x and y values.
a. Complete the table.   b. Find SS_{xy}.
c. Find SS_{xx}.   d. Find $\hat{\beta}_1$.
e. Find $\bar{x}$ and $\bar{y}$.   f. Find $\hat{\beta}_0$.
g. Find the least squares line.

x_i	y_i	x_i^2	$x_i y_i$
7	2	—	—
4	4	—	—
6	2	—	—
2	5	—	—
1	7	—	—
1	6	—	—
3	5	—	—
Totals $\sum x_i =$	$\sum y_i =$	$\sum x_i^2 =$	$\sum x_i y_i =$

11.15 Refer to Exercise 11.14. After the least squares line has been obtained, the following table (which is similar to Table 11.4) can be used (1) to compare the observed and the predicted values of y and (2) to compute SSE.

x	y	$\hat{y}$	$(y - \hat{y})$	$(y - \hat{y})^2$
7	2	—	—	—
4	4	—	—	—
6	2	—	—	—
2	5	—	—	—
1	7	—	—	—
1	6	—	—	—
3	5	—	—	—
		$\sum(y - \hat{y}) =$	SSE $= \sum(y - \hat{y})^2 =$	

a. Complete the table.
b. Plot the least squares line on a scatterplot of the data. Plot the following line on the same graph:
$$\hat{y} = 14 - 2.5x$$
c. Show that SSE is larger for the line in part **b** than it is for the least squares line.

11.16 Construct a scatterplot of the following data.

x	.5	1	1.5
y	2	1	3

a. Plot the following two lines on your scatterplot:
$$y = 3 - x \quad \text{and} \quad y = 1 + x$$

b. Which of these lines would you choose to characterize the relationship between x and y? Explain.
c. Show that the sum of errors for both of these lines equals 0.
d. Which of these lines has the smaller SSE?
e. Find the least squares line for the data, and compare it with the two lines described in part **a**.

11.17 Consider the following pairs of measurements, saved in the LM11_17 file:

x	5	3	−1	2	7	6	4
y	4	3	0	1	8	5	3

a. Construct a scatterplot of these data.
b. What does the scatterplot suggest about the relationship between x and y?
c. Given that $SS_{xx} = 43.4286$, $SS_{xy} = 39.8571$, $\bar{y} = 3.4286$, and $\bar{x} = 3.7143$, calculate the least squares estimates of β_0 and β_1.
d. Plot the least squares line on your scatterplot. Does the line appear to fit the data well? Explain.
e. Interpret the y-intercept and slope of the least squares line. Over what range of x are these interpretations meaningful?

Applet Exercise 11.1

Use the applet entitled *Regression by Eye* to explore the relationship between the pattern of data in a scatterplot and the corresponding least squares model.
a. Run the applet several times. For each time, attempt to move the green line into a position that appears to minimize the vertical distances of the points from the line. Then click *Show regression line* to see the actual regression line. How close is your line to the actual line? Click *New data* to reset the applet.
b. Click the trash can to clear the graph. Use the mouse to place five points on the scatterplot that are approximately in a straight line. Then move the green line to approximate the regression line. Click *Show regression line* to see the actual regression line. How close were you this time?
c. Continue to clear the graph, and plot sets of five points with different patterns among the points. Use the green line to approximate the regression line. How close do you come to the actual regression line each time?
d. On the basis of your experiences with the applet, explain why we need to use more reliable methods of finding the regression line than just "eyeing" it.

Applying the Concepts—Basic

11.18 Do nice guys really finish last? In baseball, there is an old saying that "nice guys finish last." Is this true in the competitive corporate world? Researchers at Harvard University attempted to answer this question and reported their results in *Nature* (March 20, 2008). In the study, Boston-area college students repeatedly played a version of the game "prisoner's dilemma," where competitors choose cooperation, defection, or costly punishment. (Cooperation meant paying 1 unit for the opponent to receive 2 units; defection meant gaining 1 unit at a cost of

1 unit for the opponent; and punishment meant paying 1 unit for the opponent to lose 4 units.) At the conclusion of the games, the researchers recorded the average payoff and the number of times punishment was used for each player. A graph of the data is shown in the accompanying scatterplot.

a. Consider punishment use (x) as a predictor of average payoff (y). Based on the scatterplot, is there evidence of a linear trend?

b. Refer to part **a**. Is the slope of the line relating punishment use (x) to average payoff (y) positive or negative?

c. The researchers concluded that "winners don't punish"? Do you agree? Explain.

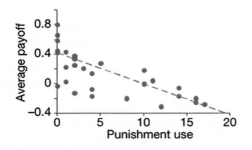

11.19 New method for blood typing. Refer to the *Analytical Chemistry* (May 2010) study in which medical researchers tested a new method of typing blood using lost cost paper, Exercise 2.151 (p. 90). The researchers applied blood drops to the paper and recorded the rate of absorption (called *blood wicking*). The table gives the wicking lengths (millimeters) for six blood drops, each at a different antibody concentration. The data are saved in the **BLOODTYPE** file. Let y = wicking length and x = antibody concentration.

Droplet	Length (mm)	Concentration
1	22.50	0.0
2	16.00	0.2
3	13.50	0.4
4	14.00	0.6
5	13.75	0.8
6	12.50	1.0

Based on Khan, M. S., et al. "Paper diagnostic for instant blood typing." *Analytical Chemistry*, Vol. 82, No. 10, May 2010 (Figure 4b).

a. Give the equation of the straight-line model relating y to x.

b. An SPSS printout of the simple linear regression analysis is shown below. Give the equation of the least squares line.

c. Give practical interpretations (if possible) of the estimated y-intercept and slope of the line.

Coefficients[a]

Model		Unstandardized Coefficients		Standardized Coefficients		
		B	Std. Error	Beta	t	Sig.
1	(Constant)	19.393	1.708		11.357	.000
	ABConc	-8.036	2.820	-.819	-2.849	.046

a. Dependent Variable: WickLength

11.20 Quantitative models of music. Writing in *Chance* (Fall 2004), University of Konstanz (Germany) statistics professor Jan Beran demonstrated that certain aspects of music can be described by quantitative models. For example, the information content of a musical composition (called *entropy*) can be quantified by determining how many times a certain pitch occurs. In a sample of 147 famous compositions ranging from the 13th to the 20th century, Beran computed the Z12-note entropy (y) and plotted it against the year of birth (x) of the composer. The graph is reproduced here.

a. Do you observe a trend, especially since the year 1400?

b. The least squares line for the data since 1400 is shown on the graph. Is the slope of the line positive or negative? What does this imply?

c. Explain why the line shown is not the true line of means.

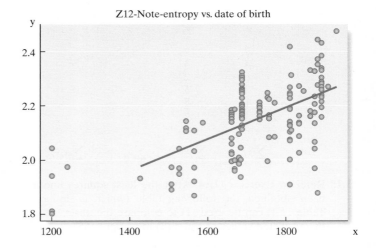

11.21 Wind turbine blade stress. Mechanical engineers at the University of Newcastle (Australia) investigated the use of timber in high-efficiency small wind turbine blades (*Wind Engineering*, Jan. 2004). The strengths of two types of timber—radiata pine and hoop pine—were compared. Twenty specimens (called "coupons") of each timber blade were fatigue tested by measuring the stress (in MPa) on the blade after various numbers of blade cycles. A simple linear regression analysis of the data, one conducted for each type of timber, yielded the following results (where y = stress and x = natural logarithm of number of cycles):

$$\text{Radiata Pine: } \hat{y} = 97.37 - 2.50x$$

$$\text{Hoop Pine: } \hat{y} = 122.03 - 2.36x$$

a. Interpret the estimated slope of each line.

b. Interpret the estimated y-intercept of each line.

c. On the basis of these results, which type of timber blade appears to be stronger and more fatigue resis-tant? Explain.

11.22 Mongolian desert ants. Refer to the *Journal of Biogeography* (Dec. 2003) study of ants in Mongolia, presented in Exercise 2.155 (p. 91). Data on annual rainfall, maximum daily temperature, and number of ant species recorded at each of 11 study sites are listed in the table and saved in the **GOBIANTS** file.

Site	Region	Annual Rainfall (mm)	Max. Daily Temp. (°C)	Number of Ant Species
1	Dry Steppe	196	5.7	3
2	Dry Steppe	196	5.7	3
3	Dry Steppe	179	7.0	52
4	Dry Steppe	197	8.0	7
5	Dry Steppe	149	8.5	5
6	Gobi Desert	112	10.7	49
7	Gobi Desert	125	11.4	5
8	Gobi Desert	99	10.9	4
9	Gobi Desert	125	11.4	4
10	Gobi Desert	84	11.4	5
11	Gobi Desert	115	11.4	4

Based on Pfeiffer, M., et al. "Community organization and species richness of ants in Mongolia along an ecological gradient from steppe to Gobi desert." *Journal of Biogeography*, Vol. 30, No. 12, Dec. 2003 (Tables 1 and 2).

a. Consider a straight-line model relating annual rainfall (y) and maximum daily temperature (x). A MINITAB printout of the simple linear regression is shown below. Give the least squares prediction equation.

Regression Analysis: Rain versus Temp

```
The regression equation is
Rain = 295 - 16.4 Temp

Predictor      Coef   SE Coef      T      P
Constant     295.25     22.41  13.18  0.000
Temp         -16.364     2.346  -6.97  0.000

S = 17.5111   R-Sq = 84.4%   R-Sq(adj) = 82.7%

Analysis of Variance
Source          DF     SS     MS      F      P
Regression       1  14915  14915  48.64  0.000
Residual Error   9   2760    307
Total           10  17675
```

---

b. Construct a scatterplot for the analysis you performed in part **a**. Include the least square line on the plot. Does the line appear to be a good predictor of annual rainfall?

c. Now consider a straight-line model relating number of ant species (y) to annual rainfall (x). On the basis of the MINITAB printout below. repeat parts **a** and **b**.

Regression Analysis: AntSpecies versus Rain

```
The regression equation is
AntSpecies = 10.5 + 0.016 Rain

Predictor   Coef   SE Coef     T      P
Constant   10.52     22.03  0.48  0.644
Rain      0.0160    0.1480  0.11  0.916

S = 19.6726   R-Sq = 0.1%   R-Sq(adj) = 0.0%

Analysis of Variance
Source          DF      SS     MS      F      P
Regression       1     4.5    4.5   0.01  0.916
Residual Error   9  3483.1  387.0
Total           10  3487.6
```

11.23 Redshifts of quasi-stellar objects. Astronomers call a shift in the spectrum of galaxies a "redshift." A correlation between redshift level and apparent magnitude (i.e., brightness on a logarithmic scale) of a quasi-stellar object was discovered and reported in the *Journal of Astrophysics & Astronomy* (Mar./Jun. 2003). Physicist D. Basu (Carleton University, Ottawa) applied simple linear regression to data collected for a sample of over 6,000 quasi-stellar objects with confirmed redshifts. The analysis yielded the following results for a specific range of magnitudes: $\hat{y} = 18.13 + 6.21x$, where y = magnitude and x = redshift level.

a. Graph the least squares line. Is the slope of the line positive or negative?

b. Interpret the estimate of the y-intercept in the words of the problem.

c. Interpret the estimate of the slope in the words of the problem.

Applying the Concepts—Intermediate

11.24 Extending the life of an aluminum smelter pot. An investigation of the properties of bricks used to line aluminum smelter pots was published in *The American Ceramic Society Bulletin* (Feb. 2005). Six different commercial bricks were evaluated. The life span of a smelter pot depends on the porosity of the brick lining (the less porosity, the longer is the life); consequently, the researchers measured the apparent porosity of each brick specimen, as well as the mean pore diameter of each brick. The data are given in the next table and saved in the **SMELTPOT** file.

Brick	Apparent Porosity (%)	Mean Pore Diameter (micrometers)
A	18.8	12.0
B	18.3	9.7
C	16.3	7.3
D	6.9	5.3
E	17.1	10.9
F	20.4	16.8

Based on Bonadia, P., et al. "Aluminosilicate refractories for aluminum cell linings." *The American Ceramic Society Bulletin*, Vol. 84, No. 2, Feb. 2005 (Table II).

a. Find the least squares line relating porosity (y) to mean pore diameter (x).

b. Interpret the y-intercept of the line.

c. Interpret the slope of the line.

d. Predict the apparent percentage of porosity for a brick with a mean pore diameter of 10 micrometers.

11.25 Ranking driving performance of professional golfers. Refer to *The Sport Journal* (Winter 2007) study of a new method for ranking the total driving performance of golfers on the Professional Golf Association (PGA) tour, presented in Exercise 2.64 (p. 59). Recall that the method computes a driving performance index based on a golfer's average driving distance (yards) and driving accuracy (percent of drives that land in the fairway). Data for the top 40 PGA golfers (as ranked by the new method) are saved in the **PGADRIVER** file. (The first five and last five observations are listed in the next table.)

a. Write the equation of a straight-line model relating driving accuracy (y) to diving distance (x).

Rank	Player	Driving Distance (yards)	Driving Accuracy (%)	Driving Performance Index
1	Woods	316.1	54.6	3.58
2	Perry	304.7	63.4	3.48
3	Gutschewski	310.5	57.9	3.27
4	Wetterich	311.7	56.6	3.18
5	Hearn	295.2	68.5	2.82
⋮	⋮	⋮	⋮	⋮
36	Senden	291	66	1.31
37	Mickelson	300	58.7	1.30
38	Watney	298.9	59.4	1.26
39	Trahan	295.8	61.8	1.23
40	Pappas	309.4	50.6	1.17

Based on Frederick Wiseman, Ph.D., Mohamed Habibullah, Ph.D., and Mustafa Yilmaz, Ph.D, *Sports Journal*, Vol. 10, No. 1.

b. Use simple linear regression to fit the model you found in part **a** to the data. Give the least squares prediction equation.

c. Interpret the estimated y-intercept of the line.

d. Interpret the estimated slope of the line.

e. In Exercise 2.157 (p. 91), you were informed that a professional golfer practicing a new swing to increase his average driving distance is concerned that his driving accuracy will be lower. Which of the two estimates, y-intercept or slope, will help you determine whether the golfer's concern is a valid one? Explain.

11.26 FCAT scores and poverty. In the state of Florida, elementary school performance is based on the average score obtained by students on a standardized exam, called the Florida Comprehensive Assessment Test (FCAT). An analysis of the link between FCAT scores and sociodemographic factors was published in the *Journal of Educational*

Elementary School	FCAT—Math	FCAT—Reading	% Below Poverty
1	166.4	165.0	91.7
2	159.6	157.2	90.2
3	159.1	164.4	86.0
4	155.5	162.4	83.9
5	164.3	162.5	80.4
6	169.8	164.9	76.5
7	155.7	162.0	76.0
8	165.2	165.0	75.8
9	175.4	173.7	75.6
10	178.1	171.0	75.0
11	167.1	169.4	74.7
12	177.0	172.9	63.2
13	174.2	172.7	52.9
14	175.6	174.9	48.5
15	170.8	174.8	39.1
16	175.1	170.1	38.4
17	182.8	181.4	34.3
18	180.3	180.6	30.3
19	178.8	178.0	30.3
20	181.4	175.9	29.6
21	182.8	181.6	26.5
22	186.1	183.8	13.8

Based on Tekwe, C. D., et al. "An empirical comparison of statistical models for value-added assessment of school performance." *Journal of Educational and Behavioral Statistics*, Vol. 29, No. 1, Spring 2004 (Table 2).

and Behavioral Statistics (Spring 2004). Data on average math and reading FCAT scores of third graders, as well as the percentage of students below the poverty level, for a sample of 22 Florida elementary schools are listed in the accompanying table and saved in the **FCAT** file.

a. Propose a straight-line model relating math score (y) to percentage (x) of students below the poverty level.

b. Use the method of least squares to fit the model to the data in the FCAT file.

c. Graph the least squares line on a scatterplot of the data. Is there visual evidence of a relationship between the two variables? Is the relationship positive or negative?

d. Interpret the estimates of the y-intercept and slope in the words of the problem.

e. Now consider a model relating reading score (y) to percentage (x) of students below the poverty level. Repeat parts **a–d** for this model.

11.27 Sound waves from a basketball. Refer to the *American Journal of Physics* (June 2010) study of sound waves in a spherical cavity, Exercise 2.37 (p. 47). The frequencies of sound waves (estimated using a mathematical formula) resulting from the first 24 resonances (echoes) after striking a basketball with a metal rod are reproduced in the following table and saved in the **BBALL** file. Recall that the researcher expects the sound wave frequency to increase as the number of resonances increases.

Resonance	Frequency
1	979
2	1572
3	2113
4	2122
5	2659
6	2795
7	3181
8	3431
9	3638
10	3694
11	4038
12	4203
13	4334
14	4631
15	4711
16	4993
17	5130
18	5210
19	5214
20	5633
21	5779
22	5836
23	6259
24	6339

Based on Russell, D. A. "Basketballs as spherical acoustic cavities." *American Journal of Physics*, Vol. 48, No. 6, June 2010 (Table I).

a. Hypothesize a model for frequency (y) as a function of number of resonances (x) that proposes a linearly increasing relationship.

b. According to the researcher's theory, will the slope of the line be positive or negative?

c. Estimate the beta parameters of the model and (if possible) give a practical interpretation of each.

11.28 Sweetness of orange juice The quality of the orange juice produced by a manufacturer is constantly monitored. There are numerous sensory and chemical components that combine to make the best-tasting orange juice. For example, one manufacturer has developed a quantitative index of the "sweetness" of orange juice. (The higher the index, the sweeter is the juice.) Is there a relationship between the sweetness index and a chemical measure such as the amount of water-soluble pectin (parts per million) in the orange juice? Data collected on these two variables during 24 production runs at a juice-manufacturing plant are shown in the table and saved in the **OJUICE** file. Suppose a manufacturer wants to use simple linear regression to predict the sweetness (y) from the amount of pectin (x).

a. Find the least squares line for the data.
b. Interpret $\hat{\beta}_0$ and $\hat{\beta}_1$ in the words of the problem.
c. Predict the sweetness index if the amount of pectin in the orange juice is 300 ppm. [*Note:* A measure of reliability of such a prediction is discussed in Section 11.6.]

Run	Sweetness Index	Pectin (ppm)
1	5.2	220
2	5.5	227
3	6.0	259
4	5.9	210
5	5.8	224
6	6.0	215
7	5.8	231
8	5.6	268
9	5.6	239
10	5.9	212
11	5.4	410
12	5.6	256
13	5.8	306
14	5.5	259
15	5.3	284
16	5.3	383
17	5.7	271
18	5.5	264
19	5.7	227
20	5.3	263
21	5.9	232
22	5.8	220
23	5.8	246
24	5.9	241

Note: The data in the table are authentic. For reasons of confidentiality, the name of the manufacturer cannot be disclosed.

11.29 Ideal height of your mate. Anthropologists theorize that humans tend to choose mates who are similar to themselves. This includes choosing mates who are similar in height. To test this theory, a study was conducted on 147 Cornell University students (*Chance*, Summer 2008). Each student was asked to select the height of his or her ideal spouse or life partner. The researchers fit the simple linear regression model, $E(y) = \beta_0 + \beta_1 x$, where y = ideal partner's height (in inches) and x = student's height (in inches). The data for the study (simulated from information provided in a scatterplot) are saved in the **IDHEIGHT** file. The next table lists selected observations from the full data set.

a. The researchers found the estimated slope of the line to be negative. Fit the model to the data in the **IDHEIGHT** file using statistical software and verify this result.

Gender	Actual Height	Ideal Height
F	59	66
F	60	70
F	60	72
F	61	65
F	61	67
⋮	⋮	⋮
M	73.5	66
M	74	67
M	74	68
M	74	69
M	74	70

Based on Lee, G., Velleman, P., and Wainer, H. "Giving the finger to dating services." *Chance*, Vol. 21, No. 3, Summer 2008 (adapted from Figure 3).

b. The negative slope was interpreted as follows: "The taller the respondent was, the shorter they felt their ideal partner ought to be." Do you agree?
c. The result, part **b**, contradicts the theory developed by anthropologists. To gain insight into this phenomenon, use a scatterplot to graph the full data set. Use a different plotting symbol for male and female students. Now focus on just the data for the female students. What trend do you observe? Repeat for male students.
d. Fit the straight-line model to the data for the female students. Interpret the estimated slope of the line.
e. Repeat part **d** for the male students.
f. Based on the results, parts **d** and **e**, comment on whether the study data support the theory developed by anthropologists.

11.30 The "name game." Refer to the *Journal of Experimental Psychology—Applied* (June 2000) study in which the "name game" was used to help groups of students learn the names of other students in the group, presented in Exercise 10.36 (p. 496). Recall that the "name game" requires the first student in the group to state his or her full name, the second student to say his or her name and the name of the first student, the third student to say his or her name and the names of the first two students, etc. After making their introductions, the students listened to a seminar speaker for 30 minutes. At the end of the seminar, all students were asked to remember the full name of each of the other students in their group, and the researchers measured the proportion of names recalled for each. One goal of the study was to investigate the linear trend between y = proportion of names recalled and x = position (order) of the student during the game. The data (simulated on the basis of summary statistics provided in the research article) for 144 students in the first eight positions are saved in the **NAMEGAME2** file. The first five and last five observations in the data set are listed in the table on the next page. [*Note:* Since the student in position 1 actually must recall the names of all the other students, he or she is assigned position number 9 in the data set.] Use the method of least squares to estimate the line $E(y) = \beta_0 + \beta_1 x$. Interpret the β estimates in the words of the problem.

Data for Exercise 11.30

Position	Recall
2	0.04
2	0.37
2	1.00
2	0.99
2	0.79
⋮	⋮
9	0.72
9	0.88
9	0.46
9	0.54
9	0.99

Based on Morris, P. E., and Fritz, C. O. "The name game: Using retrieval practice to improve the learning of names." *Journal of Experimental Psychology—Applied*, Vol. 6, No. 2, June 2000 (data simulated from Figure 2).

Applying the Concepts—Advanced

11.31 Spreading rate of spilled liquid. Refer to the *Chemical Engineering Progress* (Jan. 2005) study of the rate at which a spilled volatile liquid will spread across a surface, presented in Exercise 2.158 (p. 91). Recall that a DuPont Corp. engineer calculated the mass (in pounds) of a 50-gallon methanol spill after a period ranging from 0 to 60 minutes. Do the data shown in the accompanying table (saved in the **LIQUIDSPILL** file) indicate that the mass of the spill tends to diminish as time increases? If so, how much will the mass diminish each minute?

Time (minutes)	Mass (pounds)
0	6.64
1	6.34
2	6.04
4	5.47
6	4.94
8	4.44
10	3.98
12	3.55
14	3.15
16	2.79
18	2.45
20	2.14
22	1.86
24	1.60
26	1.37
28	1.17
30	0.98
35	0.60
40	0.34
45	0.17
50	0.06
55	0.02
60	0.00

Based on Barry, J. "Estimating rates of spreading and evaporation of volatile liquids." *Chemical Engineering Progress*, Vol. 101, No. 1, Jan. 2005.

11.3 Model Assumptions

In Section 11.2, we assumed that the probabilistic model relating the drug reaction time y to the percentage x of drug in the bloodstream is

$$y = \beta_0 + \beta_1 x + \varepsilon$$

We also recall that the least squares estimate of the deterministic component of the model, $\beta_0 + \beta_1 x$, is

$$\hat{y} = \hat{\beta}_0 + \hat{\beta}_1 x = -.1 + .7x$$

Now we turn our attention to the random component ε of the probabilistic model and its relation to the errors in estimating β_0 and β_1. We will use a probability distribution to characterize the behavior of ε. We will see how the probability distribution of ε determines how well the model describes the relationship between the **dependent variable** y and the independent variable x.

Step 3 in a regression analysis requires us to specify the probability distribution of the random error ε. We will make four basic assumptions about the general form of this probability distribution:

Assumption 1 The mean of the probability distribution of ε is 0. That is, the average of the values of ε over an infinitely long series of experiments is 0 for each setting of the independent variable x. This assumption implies that the mean value of y, for a given value of x, is $E(y) = \beta_0 + \beta_1 x$.

Assumption 2 The variance of the probability distribution of ε is constant for all settings of the independent variable x. For our straight-line model, this assumption means that the variance of ε is equal to a constant—say, σ^2—for all values of x.

Assumption 3 The probability distribution of ε is normal.

Assumption 4 The values of ε associated with any two observed values of y are independent. That is, the value of ε associated with one value of y has no effect on any of the values of ε associated with any other y values.

The implications of the first three assumptions can be seen in Figure 11.7, which shows distributions of errors for three values of x, namely, 5, 10, and 15. Note that the relative frequency distributions of the errors are normal with a mean of 0 and a constant variance σ^2. (All of the distributions shown have the same amount of spread or variability.) The straight line shown in Figure 11.7 is the line of means; it indicates the mean value of y for a given value of x. We denote this mean value as $E(y)$. Then the line of means is given by the equation

$$E(y) = \beta_0 + \beta_1 x$$

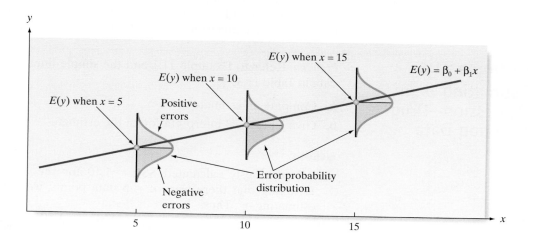

Figure 11.7

The probability distribution of ε

These assumptions make it possible for us to develop measures of reliability for the least squares estimators and to devise hypothesis tests for examining the usefulness of the least squares line. We have various techniques for checking the validity of these assumptions, and we have remedies to apply when they appear to be invalid. Several remedies are discussed in Chapter 12. Fortunately, the assumptions need not hold exactly in order for least squares estimators to be useful. The assumptions will be satisfied adequately for many applications encountered in practice.

It seems reasonable to assume that the greater the variability of the random error ε (which is measured by its variance σ^2), the greater will be the errors in the estimation of the model parameters β_0 and β_1 and in the error of prediction when $\hat{y}$ is used to predict y for some value of x. Consequently, you should not be surprised, as we proceed through this chapter, to find that σ^2 appears in the formulas for all confidence intervals and test statistics that we will be using.

Estimation of σ^2 for a (First-Order) Straight-Line Model

$$s^2 = \frac{\text{SSE}}{\text{Degrees of freedom for error}} = \frac{\text{SSE}}{n-2}$$

where $\text{SSE} = \sum(y_i - \hat{y}_i)^2 = \text{SS}_{yy} - \hat{\beta}_1 \text{SS}_{xy}$

in which

$$\text{SS}_{yy} = \sum(y_i - \bar{y})^2 = \sum y_i^2 - \frac{(\sum y_i)^2}{n}$$

To estimate the standard deviation σ of ε, we calculate

$$s = \sqrt{s^2} = \sqrt{\frac{\text{SSE}}{n-2}}$$

We will refer to s as the **estimated standard error of the regression model**.

In most practical situations, σ^2 is unknown and we must use our data to estimate its value. The best estimate of σ^2, denoted by s^2, is obtained by dividing the sum of the squares of the deviations of the y values from the prediction line, or

$$\text{SSE} = \sum (y_i - \hat{y}_i)^2$$

by the number of degrees of freedom associated with this quantity. We use 2 df to estimate the two parameters β_0 and β_1 in the straight-line model, leaving $(n - 2)$ df for the estimation of the error variance.

⚠ **CAUTION** When performing these calculations, you may be tempted to round the calculated values of SS_{yy}, $\hat{\beta}_1$, and SS_{xy}. Be certain to carry at least six significant figures for each of these quantities, to avoid substantial errors in calculating the SSE. ▲

Example 11.2

Estimating σ in Regression—Drug Reaction Data

Problem Refer to Example 11.1 and the simple linear regression of the drug reaction data in Table 11.3.

a. Compute an estimate of σ.

b. Give a practical interpretation of the estimate.

Solution

a. We previously calculated SSE = 1.10 for the least squares line $\hat{y} = -.1 + .7x$. Recalling that there were $n = 5$ data points, we have $n - 2 = 5 - 2 = 3$ df for estimating σ^2. Thus,

$$s^2 = \frac{\text{SSE}}{n - 2} = \frac{1.10}{3} = .367$$

is the estimated variance, and

$$s = \sqrt{.367} = .61$$

is the standard error of the regression model.

b. You may be able to grasp s intuitively by recalling the interpretation of a standard deviation given in Chapter 2 and remembering that the least squares line estimates the mean value of y for a given value of x. Since s measures the spread of the distribution of y values about the least squares line and these errors of prediction are assumed to be normally distributed, we should not be surprised to find that most (about 95%) of the observations lie within $2s$, or $2(.61) = 1.22$, of the least squares line. For this simple example (only five data points), all five data points fall within $2s$ of the least squares line. In Section 11.6, we use s to evaluate the error of prediction when the least squares line is used to predict a value of y to be observed for a given value of x.

Dependent Variable: TIME_Y

Number of Observations Read		5
Number of Observations Used		5

Analysis of Variance

Source	DF	Sum of Squares	Mean Square	F Value	Pr > F
Model	1	4.90000	4.90000	13.36	0.0354
Error	3	1.10000	0.36667		
Corrected Total	4	6.00000			

Root MSE	0.60553	R-Square	0.8167	
Dependent Mean	2.00000	Adj R-Sq	0.7556	
Coeff Var	30.27650			

Parameter Estimates

Variable	DF	Parameter Estimate	Standard Error	t Value	Pr > |t|
Intercept	1	-0.10000	0.63509	-0.16	0.8849
DRUG_X	1	0.70000	0.19149	3.66	0.0354

Figure 11.8

SAS printout for the time–drug regression

Look Back The values of s^2 and s can also be obtained from a simple linear regression print-out. The SAS printout for the drug reaction example is reproduced in Figure 11.8. The value of s^2 is highlighted on the printout (in the **Mean Square** column in the row labeled **Error**). The value $s^2 = .36667$, rounded to three decimal places, agrees with the one calculated by hand. The value of s is also highlighted in Figure 11.8 (next to the heading **Root MSE**). This value, $s = .60553$, agrees (except for rounding) with our hand-calculated value.

Now Work Exercise 11.36a–b

> **Interpretation of s, the Estimated Standard Deviation of ε**
>
> We expect most ($\approx 95\%$) of the observed y values to lie within $2s$ of their respective least squares predicted values, $\hat{y}$.

Exercises 11.30–11.42

Understanding the Principles

11.32 What are the four assumptions made about the probability distribution of ε in regression?

11.33 Illustrate the assumptions of Exercise 11.32 with a graph.

11.34 Visually compare the scatterplots shown below. If a least squares line were determined for each data set, which do you think would have the smallest variance s^2? Explain.

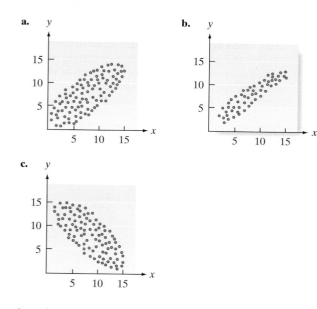

Learning the Mechanics

11.35 Calculate SSE and s^2 for each of the following cases:
   a. $n = 20$, $SS_{yy} = 95$, $SS_{xy} = 50$, $\hat{\beta}_1 = .75$
   b. $n = 40$, $\sum y^2 = 860$, $\sum y = 50$,

   $SS_{xy} = 2{,}700$, $\hat{\beta}_1 = .2$

   c. $n = 10$, $\sum (y_i - \bar{y})^2 = 58$,

   $SS_{xy} = 91$, $SS_{xx} = 170$

11.36 Suppose you fit a least squares line to 12 data points and the calculated value of SSE is .429.
[NW]
   a. Find s^2, the estimator of σ^2 (the variance of the random error term ε).
   b. Find s, the estimate of σ.

   c. What is the largest deviation that you might expect between any one of the 12 points and the least squares line?

11.37 Refer to Exercises 11.14 and 11.17 (p. 561). Calculate SSE, s^2, and s for the least squares lines obtained in those exercises. Interpret the standard errors of the regression model for each.

Applying the Concepts—Basic

11.38 Do nice guys really finish last? Refer to the *Nature* (March 20, 2008) study of whether "nice guys finish last," Exercise 11.18 (p. 561). Recall that Boston-area college students repeatedly played a version of the game "prisoner's dilemma," where competitors choose cooperation, defection, or costly punishment. At the conclusion of the games, the researchers recorded the average payoff and the number of times punishment was used for each player. Based on a scatterplot of the data, the simple linear regression relating average payoff (y) to punishment use (x) resulted in SSE = 1.04.
   a. Assuming a sample size of $n = 28$, compute the estimated standard deviation of the error distribution, s.
   b. Give a practical interpretation of s.

11.39 New method for blood typing. Refer to the *Analytical Chemistry* (May 2010) study in which medical researchers tested a new method of typing blood using lost cost paper, Exercise 11.19 (p. 562). The data in the **BLOODTYPE** file were used to fit the straight-line model relating $y =$ wicking length to $x =$ antibody concentration. The SPSS printout follows.

Model Summary

Model	R	R Square	Adjusted R Square	Std. Error of the Estimate
1	.819[a]	.670	.587	2.35944

a. Predictors: (Constant), ABConc

ANOVA[b]

Model		Sum of Squares	df	Mean Square	F	Sig.
1	Regression	45.201	1	45.201	8.119	.046[a]
	Residual	22.268	4	5.567		
	Total	67.469	5			

a. Predictors: (Constant), ABConc

b. Dependent Variable: WickLength

a. Give the values of SSE, s^2, and s shown on the print-out.

b. Give a practical interpretation of s. Recall that wicking length is measured in millimeters.

11.40 Quantitative models of music. Refer to the *Chance* (Fall 2004) study on modeling a certain pitch of a musical composition, presented in Exercise 11.20 (p. 562). Recall that the number of times (y) a certain pitch occurs—called entropy—was modeled as a straight-line function of year of birth (x) of the composer. On the basis of the scatter-plot of the data, the standard deviation σ of the model is estimated to be $s = .1$. For a given year (x), about 95% of the actual entropy values (y) will fall within d units of their predicted values. Find the value of d.

11.41 Mongolian desert ants. Refer to the *Journal of Biogeography* (Dec. 2003) study of ant sites in Mongolia, presented in Exercise 11.22 (p. 562). The data in the **GOBIANTS** file was used to estimate the straight-line model relating annual rainfall (y) to maximum daily temperature (x).

a. Give the values of SSE, s^2, and s, shown on the MINITAB printout (p. 563).

b. Give a practical interpretation of the value of s.

Applying the Concepts—Intermediate

11.42 Extending the life of an aluminum smelter pot. Refer to *The American Ceramic Society Bulletin* (Feb. 2005) study of bricks that line aluminum smelter pots, presented in Exercise 11.24 (p. 563). You fit the simple linear regression model relating brick porosity (y) to mean pore diameter (x) to the data in the **SMELTPOT** file.

a. Find an estimate of the standard deviation σ of the model.

b. In Exercise 11.24d, you predicted brick porosity percentage when $x = 10$ micrometers. Use the result of part **a** to estimate the error of prediction.

11.43 FCAT scores and poverty. Refer to the *Journal of Educational and Behavioral Statistics* (Spring 2004) study of scores on the Florida Comprehensive Assessment Test (FCAT), presented in Exercise 11.26 (p. 564). The data are saved in the **FCAT** file.

a. Consider the simple linear regression relating math score (y) to percentage (x) of students below the poverty level. Find and interpret the value of s for this regression.

b. Consider the simple linear regression relating reading score (y) to percentage (x) of students below the poverty level. Find and interpret the value of s for this regression.

c. Which dependent variable, math score or reading score, can be more accurately predicted by percentage (x) of students below the poverty level? Explain.

11.44 Sweetness of orange juice. Refer to the study of the quality of orange juice produced at a juice manufacturing plant, Exercise 11.28 (p. 565). The data are saved in the **OJUICE** file. Recall that simple linear regression was used to predict the sweetness index (y) from the amount of pectin (x) in orange juice manufactured during a production run.

a. Give the values of SSE, s^2, and s for this regression.

b. Explain why it is difficult to give a practical interpretation to s^2.

c. Use the value of s to derive a range within which most (about 95%) of the errors of prediction of sweetness index fall.

11.45 Ideal height of your mate. Refer to the *Chance* (Summer 2008) study of the height of the ideal mate, Exercise 11.29 (p. 565). The data in the **IDHEIGHT** file were used to fit the simple linear regression model, $E(y) = \beta_0 + \beta_1 x$, where y = ideal partner's height (in inches) and x = student's height (in inches).

a. Fit the straight-line model to the data for the male students. Find an estimate for σ, the standard deviation of the error term, and interpret its value practically.

b. Repeat part **a** for the female students.

c. For which group, males or females, is student's height the more accurate predictor of ideal partner's height?

Applying the Concepts—Advanced

11.46 Life tests of cutting tools. To improve the quality of the output of any production process, it is necessary first to understand the capabilities of the process (For example, see Gitlow, H., *Quality Management Systems: A Practical Guide*, 2000.) In a particular manufacturing process, the useful life of a cutting tool is linearly related to the speed at which the tool is operated. The data in the accompanying table, saved in the **CUTTOOL** file, were derived from life tests for the two different brands of cutting tools currently used in the production process. For which brand would you feel more confident using the least squares line to predict useful life for a given cutting speed? Explain.

Cutting Speed (meters per minute)	Useful Life (hours)	
	Brand A	Brand B
30	4.5	6.0
30	3.5	6.5
30	5.2	5.0
40	5.2	6.0
40	4.0	4.5
40	2.5	5.0
50	4.4	4.5
50	2.8	4.0
50	1.0	3.7
60	4.0	3.8
60	2.0	3.0
60	1.1	2.4
70	1.1	1.5
70	.5	2.0
70	3.0	1.0

11.4 Assessing the Utility of the Model: Making Inferences about the Slope β_1

Now that we have specified the probability distribution of ε and found an estimate of the variance σ^2, we are ready to make statistical inferences about the linear model's usefulness in predicting the response y. This is step 4 in our regression modeling procedure.

Refer again to the data of Table 11.1, and suppose the reaction times are *completely unrelated* to the percentage of drug in the bloodstream. What could then be said about the values of β_0 and β_1 in the hypothesized probabilistic model

$$y = \beta_0 + \beta_1 x + \varepsilon$$

if x contributes no information for the prediction of y? The implication is that the mean of y—that is, the deterministic part of the model $E(y) = \beta_0 + \beta_1 x$—does not change as x changes. In the straight-line model, this means that the true slope, β_1, is equal to 0. (See Figure 11.9.) Therefore, to test the null hypothesis that the linear model contributes no information for the prediction of y against the alternative hypothesis that the linear model is useful in predicting y, we test

$$H_0: \beta_1 = 0$$
$$H_a: \beta_1 \neq 0$$

If the data support the alternative hypothesis, we will conclude that x does contribute information for the prediction of y with the straight-line model [although the true relationship between $E(y)$ and x could be more complicated than a straight line]. In effect, then, this is a test of the usefulness of the hypothesized model.

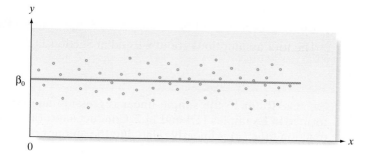

Figure 11.9
Graph of the straight-line model when the slope is zero, i.e., $y = \beta_0 + \varepsilon$

The appropriate test statistic is found by considering the sampling distribution of $\hat{\beta}_1$, the least squares estimator of the slope β_1, as shown in the following box:

Sampling Distribution of $\hat{\beta}_1$

If we make the four assumptions about ε (see Section 11.3), the sampling distribution of the least squares estimator $\hat{\beta}_1$ of the slope will be normal with mean β_1 (the true slope) and standard deviation

$$\sigma_{\hat{\beta}_1} = \frac{\sigma}{\sqrt{SS_{xx}}} \qquad \text{(see Figure 11.10)}$$

We estimate $\sigma_{\hat{\beta}_1}$ by $s_{\hat{\beta}_1} = \dfrac{s}{\sqrt{SS_{xx}}}$ and refer to $s_{\hat{\beta}_1}$ as the **estimated standard error of the least squares slope $\hat{\beta}_1$.**

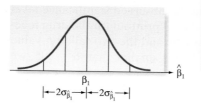

Figure 11.10
Sampling distribution of $\hat{\beta}_1$

Since σ is usually unknown, the appropriate test statistic is a *t*-statistic, formed as:

$$t = \frac{\hat{\beta}_1 - \text{Hypothesized value of } \beta_1}{s_{\hat{\beta}_1}} \qquad \text{where} \qquad s_{\hat{\beta}_1} = \frac{s}{\sqrt{SS_{xx}}}$$

Thus,

$$t = \frac{\hat{\beta}_1 - 0}{s/\sqrt{SS_{xx}}}$$

Note that we have substituted the estimator s for σ and then formed the estimated standard error $s_{\hat{\beta}_1}$ by dividing s by $\sqrt{SS_{xx}}$. The number of degrees of freedom associated with this t statistic is the same as the number of degrees of freedom associated with s. Recall that this number is $(n - 2)$ df when the hypothesized model is a straight line. (See Section 11.3.) The setup of our test of the usefulness of the straight-line model is summarized in the following two boxes.

A Test of Model Utility: Simple Linear Regression

One-Tailed Test	**Two-Tailed Test**
$H_0: \beta_1 = 0$	$H_0: \beta_1 = 0$
$H_a: \beta_1 < 0$ (or $H_a: \beta_1 > 0$)	$H_a: \beta_1 \neq 0$

Test statistic: $t = \dfrac{\hat{\beta}_1}{s_{\hat{\beta}_1}} = \dfrac{\hat{\beta}_1}{s/\sqrt{SS_{xx}}}$

Rejection region: $t < -t_\alpha$	Rejection region: $|t| > t_{\alpha/2}$

(or $t > t_\alpha$ when $H_a: \beta_1 > 0$)

where t_α and $t_{\alpha/2}$ are based on $(n - 2)$ degrees of freedom

---

Conditions Required for a Valid Test: Simple Linear Regression

The four assumptions about ε listed in Section 11.3.

Example 11.3

Testing the Regression Slope, β_1—Drug Reaction Model

Problem Refer to the simple linear regression analysis of the drug reaction data performed in Examples 11.1 and 11.2. Conduct a test (at $\alpha = .05$) to determine whether the reaction time (y) is linearly related to the amount of drug (x).

Solution For the drug reaction example, $n = 5$. Thus, t will be based on $n - 2 = 3$ df, and the rejection region t (at $\alpha = .05$) will be

$$|t| > t_{.025} = 3.182$$

We previously calculated $\hat{\beta}_1 = .7$, $s = .61$, and $SS_{xx} = 10$. Thus,

$$t = \frac{\hat{\beta}_1}{s/\sqrt{SS_{xx}}} = \frac{.7}{.61/\sqrt{10}} = \frac{.7}{.19} = 3.7$$

Since this calculated t-value falls into the upper-tail rejection region (see Figure 11.11), we reject the null hypothesis and conclude that the slope β_1 is not 0. The sample evidence indicates that the percentage x of drug in the bloodstream contributes information for the prediction of the reaction time y when a linear model is used.

[*Note:* We can reach the same conclusion by using the observed significance level (p-value) of the test from a computer printout. The MINITAB printout for the drug reaction example is reproduced in Figure 11.12. The test statistic and the two-tailed p-value are highlighted on the printout. Since the p-value $= .035$ is smaller than $\alpha = .05$, we will reject H_0.]

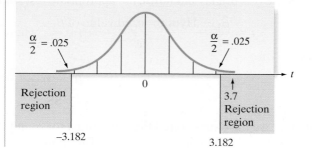

Figure 11.11

Rejection region and calculated t value for testing $H_0: \beta_1 = 0$ versus $H_a: \beta_1 \neq 0$

Regression Analysis: TIME_Y versus DRUG_X

```
The regression equation is
TIME_Y = - 0.100 + 0.700 DRUG_X


Predictor     Coef  SE Coef      T      P
Constant   -0.1000   0.6351  -0.16  0.885
DRUG_X      0.7000   0.1915   3.66  0.035


S = 0.605530   R-Sq = 81.7%   R-Sq(adj) = 75.6%


Analysis of Variance

Source           DF      SS      MS      F      P
Regression        1  4.9000  4.9000  13.36  0.035
Residual Error    3  1.1000  0.3667
Total             4  6.0000
```

Figure 11.12

MINITAB printout for the time–drug regression

Look Back What conclusion can be drawn if the calculated t-value does not fall into the rejection region or if the observed significance level of the test exceeds α? We know from previous discussions of the philosophy of hypothesis testing that such a t-value does *not* lead us to accept the null hypothesis. That is, we do not conclude that $\beta_1 = 0$. Additional data might indicate that β_1 differs from 0, or a more complicated relationship may exist between x and y, requiring the fitting of a model other than the straight-line model. We discuss several such models in Chapter 12.

Now Work Exercise 11.54

Interpreting p-Values for β Coefficients in Regression

Almost all statistical computer software packages report a *two-tailed* p-value for each of the β parameters in the regression model. For example, in simple linear regression, the p-value for the two-tailed test $H_0: \beta_1 = 0$ versus $H_a: \beta_1 \neq 0$ is given on the printout. If you want to conduct a *one-tailed* test of hypothesis, you will need to adjust the p-value reported on the printout as follows:

$$Upper-tailed\ test(H_a: \beta_1 > 0):\quad p\text{-value} = \begin{cases} p/2 & \text{if } t > 0 \\ 1 - p/2 & \text{if } t < 0 \end{cases}$$

$$Lower-tailed\ test(H_a: \beta_1 < 0):\quad p\text{-value} = \begin{cases} p/2 & \text{if } t < 0 \\ 1 - p/2 & \text{if } t > 0 \end{cases}$$

where p is the p-value reported on the printout and t is the value of the test statistic.

Another way to make inferences about the slope β_1 is to estimate it with a confidence interval, formed as shown in the following box:

A $100(1 - \alpha)\%$ Confidence Interval for the Simple Linear Regression Slope β_1

$$\hat{\beta}_1 \pm (t_{\alpha/2})s_{\hat{\beta}_1}$$

where the estimated standard error of $\hat{\beta}_1$ is calculated by

$$s_{\hat{\beta}_1} = \frac{s}{\sqrt{\text{SS}_{xx}}}$$

and $t_{\alpha/2}$ is based on $(n - 2)$ degrees of freedom.

Conditions Required for a Valid Confidence Interval: Simple Linear Regression

The four assumptions about ε listed in Section 11.3.

For the simple linear regression for the drug reaction (Examples 11.1–11.3), $t_{\alpha/2}$ is based on $(n - 2) = 3$ degrees of freedom. Therefore, a 95% confidence interval for the slope β_1, the expected change in reaction time for a 1% increase in the amount of drug in the bloodstream, is

$$\hat{\beta}_1 \pm t_{.025}s_{\hat{\beta}_1} = .7 \pm 3.182\left(\frac{s}{\sqrt{SS_{xx}}}\right) = .7 \pm 3.182\left(\frac{.61}{\sqrt{10}}\right) = .7 \pm .61$$

Thus, the estimate of the interval for the slope parameter β_1 is from .09 to 1.31. [*Note:* This interval can also be obtained with statistical software and is highlighted on the SPSS printout shown in Figure 11.13.] In terms of this example, the implication is that we can be 95% confident that the *true* mean increase in reaction time per additional 1% of the drug is between .09 and 1.31 seconds. This inference is meaningful only over the sampled range of x—that is, from 1% to 5% of the drug in the bloodstream.

Figure 11.13

SPSS printout with 95% confidence intervals for the time–drug regression betas

Coefficients^a

Model		Unstandardized Coefficients		Standardized Coefficients	t	Sig.	95% Confidence Interval for B	
		B	Std. Error	Beta			Lower Bound	Upper Bound
1	(Constant)	-.100	.635		-.157	.885	-2.121	1.921
	TIME_X	.700	.191	.904	3.656	.035	.091	1.309

a. Dependent Variable: DRUG_Y

Now Work Exercise 11.59

Since all the values in this interval are positive, it appears that β_1 is positive and that the mean of y, $E(y)$, increases as x increases. However, the rather large width of the confidence interval reflects the small number of data points (and, consequently, a lack of information) used in the experiment. We would expect a narrower interval if the sample size were increased.

We conclude this section with a comment on the other β-parameter in the straight-line model—the y-intercept, β_0. Why not conduct a test of hypothesis on β_0? For example, we could conduct the test $H_0: \beta_0 = 0$ against $H_a: \beta_0 \neq 0$. The p-value for this test appears on the printouts for SAS, SPSS, MINITAB, and most other statistical software packages. The answer lies in the interpretation of β_0. In the previous section, we learned that the y-intercept represents the mean value of y when $x = 0$. Thus, the test $H_0: \beta_0 = 0$ is equivalent to testing whether $E(y) = 0$ when $x = 0$. In the drug reaction simple linear regression, we would be testing whether the mean reaction time (y) is 0 seconds when the amount of drug in the blood (x) is 0%. The value $x = 0$ is typically not a meaningful value (as in the drug reaction example), or, $x = 0$ is typically outside the range of the sample data. In either of these cases, the test $H_0: \beta_0 = 0$ is not meaningful and should be avoided.

For those regression analyses where $x = 0$ is a meaningful value, one may desire to predict the value of y when $x = 0$. We discuss a confidence interval for such a prediction in Section 11.6.

⚠ **CAUTION** In simple linear regression, the test $H_0: \beta_0 = 0$ is only meaningful if the value $x = 0$ makes sense and is within the range of the sample data. ▲

Statistics ɪɴ Action | **Revisited** | **Assessing How Well the Straight–Line Model Fits the Dowsing Data**

In the previous Statistics in Action Revisited, we fit the straight-line model $E(y) = \beta_0 + \beta_1 x$, where $x =$ dowser's guess (in meters) and $y =$ pipe location (in meters) for each trial. The MINITAB regression printout is reproduced in Figure SIA11.3. The two-tailed p-value for testing the null hypothesis $H_0: \beta_1 = 0$ (highlighted on the printout) is p-value = .118. Even for an α-level as high as $\alpha = .10$, there is insufficient evidence to reject H_0. Consequently, the dowsing data in Table SIA11.1 provide no statistical support for the German researchers' claim that the three best dowsers have an ability to find underground water with a divining rod.

This lack of support for the dowsing theory is made clearer with a confidence interval for the slope of the line. When

Statistics IN Action
(continued)

$n = 26$, df $= (n - 2) = 24$ and $t_{.025} = 2.064$. Substituting the latter value and the relevant values shown on the MINITAB printout, we find that a 95% confidence interval for β_1 is

$$\hat{\beta}_1 \pm t_{.025}(s_{\hat{\beta}_1}) = .31 \pm (2.064)(.19)$$
$$= .31 \pm .39, \text{ or } (-.08, .70)$$

Thus, for every 1-meter increase in a dowser's guess, we estimate (with 95% confidence) that the change in the actual pipe location will range anywhere from a decrease of .08 meter to an increase of .70 meter. In other words, we're not sure whether the pipe location will increase or decrease along the 10-meter pipeline! Keep in mind also that the data in Table SIA11.1 represent the "best" performances of the three dowsers (i.e., the outcome of the dowsing experiment in its most favorable light). When the data for all trials are considered and plotted, there is not even a hint of a trend.

Regression Analysis: PIPE versus GUESS

```
The regression equation is
PIPE = 30.1 + 0.308 GUESS


Predictor     Coef   SE Coef      T      P
Constant     30.07     11.41   2.63  0.015
GUESS       0.3079    0.1900   1.62  0.118


S = 26.0298    R-Sq = 9.9%    R-Sq(adj) = 6.1%


Analysis of Variance

Source         DF       SS       MS      F      P
Regression      1   1778.9   1778.9   2.63  0.118
Residual Error 24  16261.2    677.6
Total          25  18040.2
```

Figure SIA11.3
MINITAB simple linear regression for dowsing data

Exercises 11.47–11.66

Understanding the Principles

11.47 In the equation $E(y) = \beta_0 + \beta_1 x$, what is the value of β_1 if x has no linear relationship to y?

11.48 What conditions are required for valid inferences about the β's in simple linear regression?

11.49 How do you adjust the p-value obtained from a computer printout when you perform a one-tailed test of β_1 in simple linear regression?

11.50 For each of the following 95% confidence intervals for β_1 in simple linear regression, decide whether there is evidence of a positive or negative linear relationship between y and x:
a. $(22, 58)$ b. $(-30, 111)$ c. $(-45, -7)$

Learning the Mechanics

11.51 Construct both a 95% and a 90% confidence interval for β_1 for each of the following cases:
a. $\hat{\beta}_1 = 31, s = 3, SS_{xx} = 35, n = 12$
b. $\hat{\beta}_1 = 64, SSE = 1,960, SS_{xx} = 30, n = 18$
c. $\hat{\beta}_1 = -8.4, SSE = 146, SS_{xx} = 64, n = 24$

11.52 Consider the following pairs of observations:

x	1	5	3	2	6	6	0
y	1	3	3	1	4	5	1

a. Construct a scatterplot of the data.
b. Use the method of least squares to fit a straight line to the seven data points in the table.
c. Plot the least squares line on your scatterplot of part a.
d. Specify the null and alternative hypotheses you would use to test whether the data provide sufficient evidence to indicate that x contributes information for the (linear) prediction of y.
e. What is the test statistic that should be used in conducting the hypothesis test of part d? Specify the number of degrees of freedom associated with the test statistic.
f. Conduct the hypothesis test of part d, using $\alpha = .05$.
g. Construct a 95% confidence interval for β_1.

11.53 Consider the following pairs of observations:

y	4	2	5	3	2	4
x	1	4	5	3	2	4

a. Construct a scatterplot of the data.
b. Use the method of least squares to fit a straight line to the six data points.
c. Graph the least squares line on the scatterplot of part a.
d. Compute the test statistic for determining whether x and y are linearly related.
e. Carry out the test you set up in part d, using $\alpha = .01$.
f. Find a 99% confidence interval for β_1.

Applying the Concepts—Basic

11.54 English as a second language reading ability. What are
NW   the factors that allow a native Spanish-speaking person to understand and read English? A study published in the *Bilingual Research Journal* (Summer 2006) investigated the relationship of Spanish (first-language) grammatical knowledge to English (second-language) reading. The study involved a sample of $n = 55$ native Spanish-speaking adults who were students in an English as a second language (ESL) college class. Each student took four standardized exams: Spanish grammar (SG), Spanish reading (SR), English grammar (EG), and English reading (ESLR). Simple linear regression was used to model the ESLR score (y) as a function of each of the other exam scores (x). The results are summarized in the next table (p. 576).
a. At $\alpha = .05$, is there sufficient evidence to indicate that ESLR score is linearly related to SG score?

Independent variable (x)	p-value for testing H_0: $\beta_1 = 0$
SG score	.739
SR score	.012
ER score	.022

b. At $\alpha = .05$, is there sufficient evidence to indicate that ESLR score is linearly related to SR score?

c. At $\alpha = .05$, is there sufficient evidence to indicate that ESLR score is linearly related to ER score?

11.55 Lobster fishing study. Refer to the *Bulletin of Marine Science* (April 2010) study of teams of fishermen fishing for the red spiny lobster in Baja California Sur, Mexico, Exercise 9.18 (p. 425). Two variables measured for each of 8 teams from the Punta Abreojos (PA) fishing cooperative were $y = $ total catch of lobsters (in kilograms) during the season and $x = $ average percentage of traps allocated per day to exploring areas of unknown catch (called *search frequency*). These data, saved in the **TRAPSPACE** file, are listed in the table.

Total Catch	Search Frequency
2,785	35
6,535	21
6,695	26
4,891	29
4,937	23
5,727	17
7,019	21
5,735	20

Source: From Shester, G. G. "Explaining catch variation among Baja California lobster fishers through spatial analysis of trap-placement decisions." *Bulletin of Marine Science*, Vol. 86, No. 2, April 2010 (Table 1). Reprinted with permission from the University of Miami-*Bulletin of Marine Science*.

a. Graph the data in a scatterplot. What type of trend, if any, do you observe?

b. A simple linear regression analysis was conducted using SAS. Find the least squares prediction equation on the accompanying SAS printout. Interpret the slope of the least squares line.

c. Give the null and alternative hypothesis for testing whether total catch (y) is negatively linearly related to search frequency (x).

d. Find the p-value of the test, part **c**, on the SAS printout.

e. Give the appropriate conclusion of the test, part **c**, using $\alpha = .05$.

11.56 Ranking driving performance of professional golfers. Refer to *The Sport Journal* (Winter 2007) study of a new method for ranking the total driving performance of golfers on the Professional Golf Association (PGA) tour, presented in Exercise 11.25 (p. 563). You fit a straight-line model relating driving accuracy (y) to driving distance (x) to the data saved in the **PGADRIVER** file.

a. Give the null and alternative hypotheses for testing whether driving accuracy (y) decreases linearly as driving distance (x) increases.

b. Find the test statistic and p-value of the test you set up in part **a**.

c. Make the appropriate conclusion at $\alpha = .01$.

11.57 FCAT scores and poverty. Refer to the *Journal of Educational and Behavioral Statistics* (Spring 2004) study of scores on the Florida Comprehensive Assessment Test (FCAT), first presented in Exercise 11.26 (p. 564). Consider the simple linear regression relating math score (y) to percentage (x) of students below the poverty level. The data are saved in the **FCAT** file.

a. Test whether y is negatively related to x. Use $\alpha = .01$.

b. Construct a 99% confidence interval for β_1. Interpret the result practically.

11.58 Ideal height of your mate. Refer to the *Chance* (Summer 2008) study of the height of the ideal mate, Exercise 11.45 (p. 570). You used the data in the **IDHEIGHT** file to fit the simple linear regression model $E(y) = \beta_0 + \beta_1 x$, where $y = $ ideal partner's height (in inches) and $x = $ student's height (in inches), for both males and females.

a. Find a 90% confidence interval for β_1 in the model for the male students. Give a practical interpretation of the result.

b. Repeat part **a** for the female students.

c. Which group, males or females, has the greater increase in ideal partner's height for every 1 inch increase in student's height?

11.59 Sweetness of orange juice. Refer to Exercise 11.28 (p. 565) and the simple linear regression relating the sweetness index (y) of an orange juice sample to the amount of

SAS Output for Exercise 11.55

Dependent Variable: CATCH

Number of Observations Read	8
Number of Observations Used	8

Analysis of Variance

Source	DF	Sum of Squares	Mean Square	F Value	Pr > F
Model	1	6888299	6888299	6.81	0.0402
Error	6	6071019	1011837		
Corrected Total	7	12959318			

Root MSE	1005.90086	R-Square	0.5315
Dependent Mean	5540.50000	Adj R-Sq	0.4535
Coeff Var	18.15542		

Parameter Estimates

Variable	DF	Parameter Estimate	Standard Error	t Value	Pr > |t|
Intercept	1	9658.24359	1617.76208	5.97	0.0010
SEARCHFREQ	1	-171.57265	65.75780	-2.61	0.0402

water-soluble pectin (x) in the juice. The data are saved in the **OJUICE** file. Find a 95% confidence interval for the true slope of the line. Interpret the result.

Applying the Concepts—Intermediate

11.60 Effect of massage on boxers. Refer to the *British Journal of Sports Medicine* (Apr. 2000) study of the effect of massage on boxing performance, presented in Exercise 10.72 (p. 518). Two other variables measured on the boxers were blood lactate concentration (in mM) and the boxer's perceived recovery (on a 28-point scale). On the basis of information provided in the article, the data shown in the accompanying table (and saved in the **BOXING2** file) were obtained for 16 five-round boxing performances in which a massage was given to the boxer between rounds. Conduct a test to determine whether blood lactate level (y) is linearly related to perceived recovery (x). Use $\alpha = .10$.

Blood Lactate Level	Perceived Recovery
3.8	7
4.2	7
4.8	11
4.1	12
5.0	12
5.3	12
4.2	13
2.4	17
3.7	17
5.3	17
5.8	18
6.0	18
5.9	21
6.3	21
5.5	20
6.5	24

Based on Hemmings, B., Smith, M., Graydon, J., and Dyson, R. "Effects of massage on physiological restoration, perceived recovery, and repeated sports performance." *British Journal of Sports Medicine*, Vol. 34, No. 2, Apr. 2000 (data adapted from Figure 3).

11.61 Forest fragmentation study. Refer to the *Conservation Ecology* (Dec. 2003) study on the causes of fragmentation of 54 South American forests, presented in Exercise 2.156 (p. 91). Recall that researchers developed two fragmentation indexes for each forest—one index for anthropogenic (human development activities) fragmentation and one for fragmentation from natural causes. Data on 5 of the 54 forests saved in the **FORFRAG** file are listed in the following table:

Ecoregion (forest)	Anthropogenic Index, y	Natural Origin Index, x
Araucaria moist forests	34.09	30.08
Atlantic Coast *restingas*	40.87	27.60
Bahia coastal forests	44.75	28.16
Bahia interior forests	37.58	27.44
Bolivian *Yungas*	12.40	16.75

Based on Wade, T. G., et al. "Distribution and causes of global forest fragmentation." *Conservation Ecology*, Vol. 72, No. 2, Dec. 2003 (Table 6).

a. Ecologists theorize that a linear relationship exists between the two fragmentation indexes. Write the model relating y to x.

b. Fit the model to the data in the **FORFRAG** file, using the method of least squares. Give the equation of the least squares prediction equation.

c. Interpret the estimates of β_0 and β_1 in the context of the problem.

d. Is there sufficient evidence to indicate that the natural origin index (x) and the anthropogenic index (y) are positively linearly related? Test, using $\alpha = .05$.

e. Find and interpret a 95% confidence interval for the change in the anthropogenic index (y) for every 1-point increase in the natural origin index (x).

11.62 Pain empathy and brain activity. *Empathy* refers to being able to understand and vicariously feel what others actually feel. Neuroscientists at University College of London investigated the relationship between brain activity and pain-related empathy in persons who watch others in pain (*Science*, Feb. 20, 2004). Sixteen couples participated in the experiment. The female partner watched while painful stimulation was applied to the finger of her male partner. Two variables were measured for each female: y = pain-related brain activity (measured on a scale ranging from -2 to 2) and x = score on the Empathic Concern Scale (0 to 25 points). The data are listed in the accompanying table and saved in the **BRAINPAIN** file. The research question of interest was "Do people scoring higher in empathy show higher pain-related brain activity?" Use simple linear regression analysis to answer this question.

Couple	Brain Activity (y)	Empathic Concern (x)
1	.05	12
2	−.03	13
3	.12	14
4	.20	16
5	.35	16
6	0	17
7	.26	17
8	.50	18
9	.20	18
10	.21	18
11	.45	19
12	.30	20
13	.20	21
14	.22	22
15	.76	23
16	.35	24

Based on Singer, T. et al. "Empathy for pain involves the affective but not sensory components of pain." *Science*, Vol. 303, Feb. 20, 2004. (Adapted from Figure 4.)

11.63 Relation of eye and head movements. How do eye and head movements relate to body movements when a person reacts to a visual stimulus? Scientists at the California Institute of Technology designed an experiment to answer this question and reported their results in *Nature* (Aug. 1998). Adult male rhesus monkeys were exposed to a visual stimulus (i.e., a panel of light-emitting diodes), and their eye, head, and body movements were electronically recorded. In one variation of the experiment, two variables were measured: active head movement (x, percent per degree) and body-plus-head rotation (y, percent per degree). The data for $n = 39$ trials were subjected to a simple linear regression analysis, with the following results: $\hat{\beta}_1 = .88$, $s_{\hat{\beta}_1} = .14$

a. Conduct a test to determine whether the two variables, active head movement x and body-plus-head rotation y are positively linearly related. Use $\alpha = .05$.

b. Construct and interpret a 90% confidence interval for β_1.

c. The scientists want to know whether the true slope of the line differs significantly from 1. On the basis of your answer to part **b,** make the appropriate inference.

11.64 The "name game." Refer to the *Journal of Experimental Psychology—Applied* (June 2000) name-retrieval study, presented in Exercise 11.30 (p. 565). Recall that the goal of the study was to investigate the linear trend between proportion of names recalled (y) and position (order) of the student (x) during the "name game." Is there sufficient evidence (at $\alpha = .01$) of a linear trend? Answer the question by analyzing the data for 144 students saved in the **NAMEGAME2** file.

Applying the Concepts—Advanced

11.65 Does elevation affect hitting performance in baseball? Refer to the *Chance* (Winter 2006) investigation of the effects of elevation on slugging percentage in Major League Baseball, Exercise 2.148 (p. 89). Data were compiled on players' composite slugging percentages at each of 29 cities for the 2003 season, as well as on each city's elevation (feet above sea level.) The data are saved in the **MLBPARKS** file. (Selected observations are shown in the table in the next column.) Consider a straight-line model relating slugging percentage (y) to elevation (x).

a. The model was fit to the data with the use of MINITAB, with the results shown in the printout below. Locate the estimates of the model parameters on the printout.

Regression Analysis: SLUGPCT versus ELEVATION

```
The regression equation is
SLUGPCT = 0.515 + 0.000021 ELEVATION

Predictor        Coef      SE Coef       T       P
Constant      0.515140    0.007954    64.76   0.000
ELEVATION    0.00002074  0.00000719    2.89   0.008

S = 0.0369803    R-Sq = 23.6%    R-Sq(adj) = 20.7%

Analysis of Variance

Source          DF      SS         MS       F      P
Regression       1   0.011390   0.011390   8.33   0.008
Residual Error  27   0.036924   0.001368
Total           28   0.048314
```

City	Slug Pct.	Elevation
Anaheim	.480	160
Arlington	.605	616
Atlanta	.530	1,050
Baltimore	.505	130
Boston	.505	20
⋮	⋮	⋮
Denver	.625	5,277
⋮	⋮	⋮
Seattle	.550	350
San Francisco	.510	63
St. Louis	.570	465
Tampa	.500	10
Toronto	.535	566

Based on Schaffer, J., and Heiny, E. L. "The effects of elevation on slugging percentage in Major League Baseball." *Chance*, Vol. 19, No. 1, Winter 2006 (adapted from Figure 2).

b. Is there sufficient evidence (at $\alpha = .01$) of a positive linear relationship between elevation (x) and slugging percentage (y)? Use the p-value shown on the printout to make the inference.

c. Construct a scatterplot of the data and draw the least squares line on the graph. Locate the data point for Denver on the graph. What do you observe?

d. Recall that the Colorado Rockies, who play their home games in Denver, are annually among the league leaders in slugging percentage. Baseball experts attribute this to the "thin air" of Denver—called the Mile High city due to its elevation. Remove the data point for Denver from the data set and refit the straight-line model to the remaining data. Repeat parts **a** and **b.** What conclusions can you draw about the "thin air" theory from this analysis?

11.66 Spreading rate of spilled liquid. Refer to the *Chemical Engineering Progress* (Jan. 2005) study of the rate at which a spilled volatile liquid will spread across a surface, Exercise 11.31 (p. 566). Recall that the data on mass of the spill and elapsed time is saved in the **LIQUIDSPILL** file. Is there sufficient evidence (at $\alpha = .05$) to indicate that the mass of the spill tends to diminish linearly as elapsed time increases? If so, give an interval estimate (with 95% confidence) of the decrease in spill mass for each minute of elapsed time.

11.5 The Coefficients of Correlation and Determination

In this section, we present two statistics that describe the adequacy of a model: the *coefficient of correlation* and the *coefficient of determination*.

Coefficient of Correlation

Recall (from optional Section 2.9) that a **bivariate relationship** describes a relationship—or correlation—between two variables x and y. Scatterplots are used to describe a bivariate relationship graphically. In this section, we will discuss the concept of **correlation** and how it can be used to measure the linear relationship between two variables x and y. A numerical descriptive measure of correlation is provided by the *coefficient of correlation, r.*

> The **coefficient of correlation**,* r, is a measure of the strength of the *linear* relationship between two variables x and y. It is computed (for a sample of n measurements on x and y) as follows:
>
> $$r = \frac{SS_{xy}}{\sqrt{SS_{xx} \, SS_{yy}}}$$
>
> where
>
> $$SS_{xy} = \sum (x - \bar{x})(y - \bar{y}) = \sum xy - \frac{(\sum x)(\sum y)}{n}$$
>
> $$SS_{xx} = \sum (x - \bar{x})^2 = \sum x^2 - \frac{(\sum x)^2}{n}$$
>
> $$SS_{yy} = \sum (y - \bar{y})^2 = \sum y^2 - \frac{(\sum y)^2}{n}$$

Note that the computational formula for the correlation coefficient r given above involves the same quantities that were used in computing the least squares prediction equation. In fact, since the numerators of the expressions for $\hat{\beta}_1$ and r are identical, it is clear that $r = 0$ when $\hat{\beta}_1 = 0$ (the case where x contributes no information for the prediction of y) and that r is when the slope is positive and negative when the slope is negative. Unlike $\hat{\beta}_1$, the correlation coefficient r is *scaleless* and assumes a value between -1 and $+1$, regardless of the units of x and y.

A value of r near or equal to 0 implies little or no linear relationship between y and x. In contrast, the closer r comes to 1 or -1, the stronger is the linear relationship between y and x. And if $r = 1$ or $r = -1$, all the sample points fall exactly on the least squares line. Positive values of r imply a positive linear relationship between y and x; that is, y increases as x increases. Negative values of r imply a negative linear relationship between y and x; that is, y decreases as x increases. Each of these situations is portrayed in Figure 11.14.

Now Work Exercise 11.69

We use the data in Table 11.1 for the drug reaction example to demonstrate how to calculate the coefficient of correlation, r. The quantities needed to calculate r are SS_{xy}, SS_{xx}, and SS_{yy}. The first two quantities have been calculated previously and are repeated here for convenience:

$$SS_{xy} = 7, \quad SS_{xx} = 10, \quad SS_{yy} = \sum y^2 - \frac{(\sum y)^2}{n}$$

$$= 26 - \frac{(10)^2}{5} = 26 - 20 = 6$$

We now find the coefficient of correlation:

$$r = \frac{SS_{xy}}{\sqrt{SS_{xx} \, SS_{yy}}} = \frac{7}{\sqrt{(10)(6)}} = \frac{7}{\sqrt{60}} = .904$$

The fact that r is positive and near 1 indicates that the reaction time tends to increase as the amount of drug in the bloodstream increases—*for the given sample of five subjects.* This is the same conclusion we reached when we found the calculated value of the least squares slope to be positive.

---

*The value of r is often called the *Pearson correlation coefficient* to honor its developer, Karl Pearson. (See Biography, p. 729).

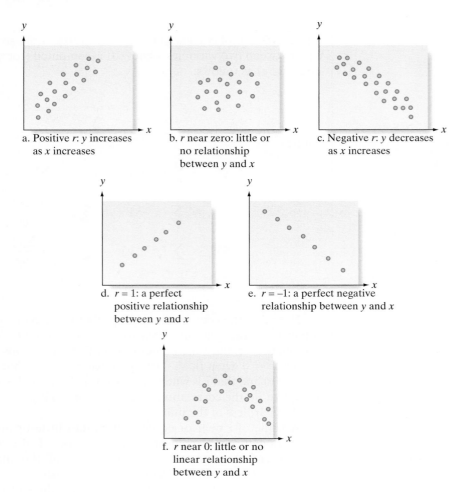

Figure 11.14

Values of r and their implications

Example 11.4

Using the Correlation Coefficient— Relating Crime Rate and Casino Employment

Problem Legalized gambling is available on several riverboat casinos operated by a city in Mississippi. The mayor of the city wants to know the correlation between the number of casino employees and the yearly crime rate. The records for the past 10 years are examined, and the results listed in Table 11.5 are obtained. Calculate the coefficient of correlation, r, for the data. Interpret the result.

Table 11.5	Data on Casino Employees and Crime Rate, Example 11.4	
Year	Number x of Casino Employees (thousands)	Crime Rate y (number of crimes per 1,000 population)
2001	15	1.35
2002	18	1.63
2003	24	2.33
2004	22	2.41
2005	25	2.63
2006	29	2.93
2007	30	3.41
2008	32	3.26
2009	35	3.63
2010	38	4.15

Data Set: CASINO

Solution Rather than use the computing formula given earlier, we resort to a statistical software package. The data of Table 11.5 were entered into a computer and MINITAB was used to compute r. The MINITAB printout is shown in Figure 11.15.

Correlations: EMPLOYEES, CRIMERAT

Pearson correlation of EMPLOYEES and CRIMERAT = 0.987
P-Value = 0.000

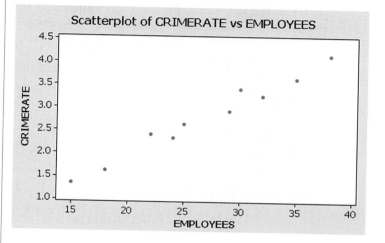

Figure 11.15
MINITAB correlation printout and scatterplot for Example 11.4

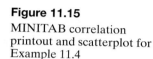

Ethics IN Statistics

Intentionally using the correlation coefficient only to make an inference about the relationship between two variables in situations where a nonlinear relationship may exist is considered *unethical statistical practice.*

The coefficient of correlation, highlighted at the top of the printout, is $r = .987$. Thus, the size of the casino workforce and crime rate in this city are very highly correlated—at least over the past 10 years. The implication is that a strong positive linear relationship exists between these variables. (See Figure 11.15.) We must be careful, however, not to jump to any unwarranted conclusions. For instance, the mayor may be tempted to conclude that hiring more casino workers next year will increase the crime rate—that is, that there is a *causal relationship* between the two variables. However, high correlation does not imply causality. The fact is, many things have probably contributed both to the increase in the casino workforce and to the increase in crime rate. The city's tourist trade has undoubtedly grown since riverboat casinos were legalized, and it is likely that the casinos have expanded both in services offered and in number. *We cannot infer a causal relationship on the basis of high sample correlation. When a high correlation is observed in the sample data, the only safe conclusion is that a linear trend may exist between x and y.*

Look Back Another variable, such as the increase in tourism, may be the underlying cause of the high correlation between x and y.

Now Work Exercise 11.79a

⚠ **CAUTION** Two caveats apply in using the sample correlation coefficient r to infer the nature of the relationship between x and y: (1) A *high correlation* does not necessarily imply that a causal relationship exists between x and y—only that a linear trend may exist; (2) a *low correlation* does not necessarily imply that x and y are unrelated—only that x and y are not strongly linearly related. ▲

Keep in mind that the correlation coefficient r measures the linear correlation between x values and y values in the sample, and a similar linear coefficient of correlation exists for the population from which the data points were selected. The **population correlation coefficient** is denoted by the symbol ρ (rho). As you might expect, ρ is estimated by the corresponding sample statistic r. Or, instead of estimating ρ, we might want to test the null hypothesis $H_0: \rho = 0$ against $H_a: \rho \neq 0$; that is, we can test the hypothesis that x contributes no information for the prediction of y by using the straight-line model against the alternative that the two variables are at least linearly related.

However, we already performed this *identical* test in Section 11.4 when we tested $H_0: \beta_1 = 0$ against $H_a: \beta_1 \neq 0$. That is, the null hypothesis $H_0: \rho = 0$ is equivalent to the hypothesis $H_0: \beta_1 = 0$.* When we tested the null hypothesis $H_0: \beta_1 = 0$ in connection

*The two tests are equivalent in simple linear regression only.

with the drug reaction example, the data led to a rejection of the null hypothesis at the $\alpha = .05$ level. This rejection implies that the null hypothesis of a 0 linear correlation between the two variables (drug and reaction time) can also be rejected at the $\alpha = .05$ level. The only real difference between the least squares slope $\hat{\beta}_1$ and the coefficient of correlation, r, is the measurement scale. Therefore, the information they provide about the usefulness of the least squares model is to some extent redundant. For this reason, we will use the slope to make inferences about the existence of a positive or negative linear relationship between two variables.

For the sake of completeness, a summary of the test for linear correlation is provided in the following boxes.

A Test for Linear Correlation

One-Tailed Test

H_0: $\rho = 0$

H_a: $\rho > 0$ (or H_a: $\rho < 0$)

Two-Tailed Test

H_0: $\rho = 0$

H_a: $\rho \neq 0$

Test statistic: $t = \dfrac{r\sqrt{n-2}}{\sqrt{1 - r^2}} = \dfrac{\hat{\beta}}{s_{\hat{\beta}_1}}$

Rejection region: $t > t_\alpha$ (or $t < -t_\alpha$)

Rejection region: $|t| > t_{\alpha/2}$

where the distribution of t depends on $(n-2)$ df.

Condition Required for a Valid Test of Correlation

The sample of (x, y) values is randomly selected from a normal population.

Coefficient of Determination

Another way to measure the usefulness of a linear model is to measure the contribution of x in predicting y. To accomplish this, we calculate how much the errors of prediction of y were reduced by using the information provided by x. To illustrate, consider the sample shown in the scatterplot of Figure 11.16a. If we assume that x contributes no information for the prediction of y, the best prediction for a value of y is the sample mean $\bar{y}$, which is shown as the horizontal line in Figure 11.16b. The vertical line segments in Figure 11.16b are the deviations of the points about the mean $\bar{y}$. Note that the sum of the squares of the deviations for the prediction equation $\hat{y} = \bar{y}$ is

$$SS_{yy} = \sum (y_i - \bar{y})^2$$

Now suppose you fit a least squares line to the same set of data and locate the deviations of the points about the line, as shown in Figure 11.16c. Compare the deviations about the prediction lines in Figures 11.16b and 11.16c You can see that

1. If x contributes little or no information for the prediction of y, the sums of the squares of the deviations for the two lines

$$SS_{yy} = \sum (y_i - \bar{y})^2 \quad \text{and} \quad SSE = \sum (y_i - \hat{y}_i)^2$$

will be nearly equal.

2. If x does contribute information for the prediction of y, the SSE will be smaller than SS_{yy}. In fact, if all the points fall on the least squares line, then SSE = 0.

Consequently, the reduction in the sum of the squares of the deviations that can be attributed to x, expressed as a proportion of SS_{yy}, is

$$\frac{SS_{yy} - SSE}{SS_{yy}}$$

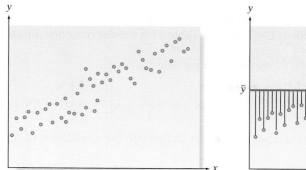

a. Scatterplot of data

b. Assumption: x contributes no information for predicting y, $\hat{y} = \bar{y}$

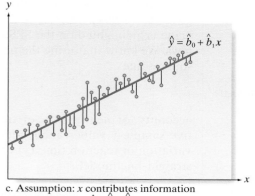

Figure 11.16

A comparison of the sum of squares of deviations for two models

c. Assumption: x contributes information for predicting y, $\hat{y} = \hat{\beta}_0 + \hat{\beta}_1 x$

Note that SS_{yy} is the "total sample variability" of the observations around the mean $\bar{y}$ and that SSE is the remaining "unexplained sample variability" after fitting the line $\hat{y}$. Thus, the difference $(SS_{yy} - SSE)$ is the "explained sample variability" attributable to the linear relationship with x. Thus, a verbal description of the proportion is

$$\frac{SS_{yy} - SSE}{SS_{yy}} = \frac{\text{Explained sample variability}}{\text{Total sample variability}}$$

$$= \text{Proportion of total sample variability explained by the linear relationship}$$

In simple linear regression, it can be shown that this proportion—called the *coefficient of determination*—is equal to the square of the simple linear coefficient of correlation, r.

The **coefficient of determination** is

$$r^2 = \frac{SS_{yy} - SSE}{SS_{yy}} = 1 - \frac{SSE}{SS_{yy}}$$

and represents the proportion of the total sample variability around $\bar{y}$ that is explained by the linear relationship between y and x. (In simple linear regression, it may also be computed as the square of the coefficient of correlation, r.)

Note that r^2 is always between 0 and 1, because r is between -1 and $+1$. Thus, an r^2 of .60 means that the sum of the squares of the deviations of the y values about their predicted values has been reduced 60% by the use of the least squares equation $\hat{y}$, instead of $\bar{y}$, to predict y.

Example 11.5
Obtaining the Value of r^2—Drug Reaction Regression

Problem Calculate the coefficient of determination for the drug reaction example. The data are repeated in Table 11.6 for convenience. Interpret the result.

Solution From previous calculations,

$$SS_{yy} = 6 \quad \text{and} \quad SSE = \sum(y - \hat{y})^2 = 1.10$$

Then, from our earlier definition, the coefficient of determination is

$$r^2 = \frac{SS_{yy} - SSE}{SS_{yy}} = \frac{6.0 - 1.1}{6.0} = \frac{4.9}{6.0} = .817$$

Another way to compute r^2 is to recall from Section 11.5 that $r = .904$. Then we have $r^2 = (.904)^2 = .817$. A third way to obtain r^2 is from a computer printout. Its value is highlighted on the SPSS printout in Figure 11.17. Our interpretation is as follows: We know that using the percent x of drug in the blood to predict y with the least squares line

$$\hat{y} = -.1 + .7x$$

accounts for nearly 82% of the total sum of the squares of the deviations of the five sample y values about their mean. Or, stated another way, 82% of the sample variation in reaction time (y) can be "explained" by using the percent x of drug in a straight-line model.

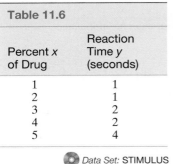

Table 11.6

Percent x of Drug	Reaction Time y (seconds)
1	1
2	1
3	2
4	2
5	4

Data Set: STIMULUS

Figure 11.17
Portion of SPSS printout for time-drug regression

Model Summary

Model	R	R Square	Adjusted R Square	Std. Error of the Estimate
1	.904ᵃ	.817	.756	.606

a. Predictors: (Constant), DRUG_X

Now Work Exercise 11.79b

> **Practical Interpretation of the Coefficient of Determination, r^2**
>
> $100(r^2)\%$ of the sample variation in y (measured by the total sum of the squares of the deviations of the sample y values about their mean $\bar{y}$) can be explained by (or attributed to) using x to predict y in the straight-line model.

Statistics in Action **Revised**

Using the Coefficients of Correlation and Determination to Assess the Dowsing Data

In the previous Statistics in Action Revisited, we discovered that using a dowser's guess (x) in a straight-line model was not statistically useful in predicting actual pipe location (y). Both the coefficient of correlation and the coefficient of determination (highlighted on the MINITAB printouts in Figure SIA11.4) also support this conclusion. The value of the correlation coefficient, $r = .314$, indicates a fairly weak positive linear relationship between the variables. This value, however, is not statistically significant (p-value $= .118$). In other words, there is no evidence to indicate that the population correlation coefficient is different from 0. The coefficient of determination, $r^2 = .099$, implies that only about 10% of the sample variation in pipe location values can be explained by the simple linear model.

Statistics IN Action
(continued)

Regression Analysis: PIPE versus GUESS

```
The regression equation is
PIPE = 30.1 + 0.308 GUESS

Predictor     Coef    SE Coef      T       P
Constant     30.07     11.41     2.63   0.015
GUESS        0.3079    0.1900    1.62   0.118

S = 26.0298    R-Sq = 9.9%    R-Sq(adj) = 6.1%

Analysis of Variance

Source          DF      SS       MS      F      P
Regression       1    1778.9   1778.9   2.63  0.118
Residual Error  24   16261.2    677.6
Total           25   18040.2
```

Correlations: PIPE, GUESS

```
Pearson correlation of PIPE and GUESS = 0.314
P-Value = 0.118
```

Figure SIA11.4
MINITAB printouts with coefficients of correlation and determination for the dowsing data

Exercises 11.67–11.88

Understanding the Principles

11.67 *True or False.* The correlation coefficient is a measure of the strength of the linear relationship between x and y.

11.68 Describe the slope of the least squares line if
  a. $r = .7$ **b.** $r = -.7$ **c.** $r = 0$ **d.** $r^2 = .64$

11.69 Explain what each of the following sample correlation
[NW] coefficients tells you about the relationship between the x and y values in the sample:
  a. $r = 1$ **b.** $r = -1$
  c. $r = 0$ **d.** $r = .90$
  e. $r = .10$ **f.** $r = -.88$

11.70 *True or False.* A value of the correlation coefficient near 1 or near −1 implies a causal relationship between x and y.

Learning the Mechanics

11.71 Construct a scatterplot for each data set. Then calculate r and r^2 for each data set.

a.

x	−2	−1	0	1	2
y	−2	1	2	5	6

b.

x	−2	−1	0	1	2
y	6	5	3	2	0

c.

x	1	2	2	3	3	3	4
y	2	1	3	1	2	3	2

d.

x	0	1	3	5	6
y	0	1	2	1	0

11.72 Calculate r^2 for the least squares line in Exercise 11.14 (p. 561).

11.73 Calculate r^2 for the least squares line in Exercise 11.17 (p. 561).

Applet Exercise 11.2

Use the applet entitled *Correlation by the Eye* to explore the relationship between the pattern of data in a scatterplot and the corresponding correlation coefficient.

  a. Run the applet several times. Each time, guess the value of the correlation coefficient. Then click *Show r* to see the actual correlation coefficient. How close is your value to the actual value of r? Click *New data* to reset the applet.

  b. Click the trash can to clear the graph. Use the mouse to place five points on the scatterplot that are approximately in a straight line. Then guess the value of the correlation coefficient. Click *Show r* to see the actual correlation coefficient. How close were you this time?

  c. Continue to clear the graph and plot sets of five points with different patterns among the points. Guess the value of r. How close do you come to the actual value of r each time?

  d. On the basis of your experiences with the applet, explain why we need to use more reliable methods of finding the correlation coefficient than just "eyeing" it.

Applying the Concepts—Basic

11.74 RateMyProfessors.com. A popular Web site among college students is RateMyProfessors.com (RMP). Established over 10 years ago, RMP allows students to post quantitative ratings of their instructors. In *Practical Assessment, Research & Evaluation* (May 2007), University of Maine researchers investigated whether instructor ratings posted on RMP are correlated with the formal in-class student evaluations of teaching (SET) that all universities are required to administer at the end of the semester. Data collected for $n = 426$ University of Maine instructors yielded a correlation between RMP and SET ratings of .68.

a. Give the equation of a linear model relating SET rating (y) to RMP rating (x).

b. Give a practical interpretation of the value $r = .68$.

c. Is the estimated slope of the line, part **a**, positive or negative? Explain.

d. A test of the null hypothesis $H_0: \rho = 0$ yielded a p-value of .001. Interpret this result.

e. Compute the coefficient of determination, r^2, for the regression analysis. Interpret the result.

11.75 Going for it on fourth down in the NFL. Each week coaches in the National Football League (NFL) face a decision during the game. On fourth down, should the team punt the ball or go for a first down? To aid in the decision-making process, statisticians at California State University, Northridge, developed a regression model for predicting the number of points scored (y) by a team that has a first down with a given number of yards (x) from the opposing goal line (*Chance*, Winter 2009). One of the models fit to data collected on five NFL teams from a recent season was the simple linear regression model, $E(y) = \beta_0 + \beta_1 x$. The regression yielded the following results: $\hat{y} = 4.42 - .048\,x, r^2 = .18$.

a. Give a practical interpretation of the coefficient of determination, r^2.

b. Compute the value of the coefficient of correlation, r, from the value of r^2. Is the value of r positive or negative? Why?

11.76 Lobster fishing study. Refer to the *Bulletin of Marine Science* (April 2010) study of teams of fishermen fishing for the red spiny lobster in Baja California Sur, Mexico, Exercise 11.55 (p. 576). Recall that simple linear regression was used to model $y = $ total catch of lobsters (in kilograms) during the season as a function of $x = $ average percentage of traps allocated per day to exploring areas of unknown catch (called *search frequency*).

a. Locate and interpret the coefficient of determination, r^2, on the SAS printout shown on p. 576.

b. Note that the coefficient of correlation, r, is not shown on the SAS printout. Is there information on the printout to determine whether total catch (y) is negatively linearly related to search frequency (x)? Explain.

11.77 Physical activity of obese young adults. Refer to the *International Journal of Obesity* (Jan. 2007) study of the physical activity of obese young adults, presented in Exercise 6.35 (p. 290). For two groups of young adults—13 obese and 15 of normal weight—researchers recorded the total number of registered movements (counts) of each young adult over a period of time. *Baseline* physical activity was then computed as the number of counts per minute (cpm). Four years later, physical activity measurements were taken again—called physical activity *at follow-up*.

a. For the 13 obese young adults, the researchers reported a correlation of $r = .50$ between baseline and follow-up physical activity, with an associated p-value of .07. Give a practical interpretation of this correlation coefficient and p-value.

b. Refer to part **a**. Construct a scatterplot of the 13 data points that would yield a value of $r = .50$.

c. For the 15 young adults of normal weight, the researchers reported a correlation of $r = -.12$ between baseline and follow-up physical activity, with an associated p-value of .66. Give a practical interpretation of this correlation coefficient and p-value.

d. Refer to part **c**. Construct a scatterplot of the 15 data points that would yield a value of $r = -12$.

11.78 Wind turbine blade stress. Refer to the *Wind Engineering* (Jan. 2004) study of two types of timber—radiata pine and hoop pine—used in high-efficiency small wind turbine blades, presented in Exercise 11.21 (p. 562). Data on stress (y) and the natural logarithm of the number of blade cycles (x) for each type of timber were analyzed by means of simple linear regression. The results are as follows, with additional information on the coefficient of determination:

Radiata Pine: $\hat{y} = 97.37 - 2.50x, r^2 = .84$
Hoop Pine: $\hat{y} = 122.03 - 2.36x, r^2 = .90$

Interpret the value of r^2 for each type of timber.

11.79 Sports news on local TV broadcasts. *The Sports Journal* (Winter 2004) published the results of a study conducted to assess the factors that affect the time allotted to sports news on local television news broadcasts. Information on total time (in minutes) allotted to sports and on audience ratings of the TV news broadcast (measured on a 100-point scale) was obtained from a national sample of 163 news directors. A correlation analysis of the data yielded $r = .43$.

a. Interpret the value of the correlation coefficient r.

b. Find and interpret the value of the coefficient of determination r^2.

11.80 Redshifts of quasi-stellar objects. Refer to the *Journal of Astrophysics & Astronomy* (Mar./Jun. 2003) study of redshifts in quasi-stellar objects presented in Exercise 11.23 (p. 563). Recall that simple linear regression was used to model the magnitude (y) of a quasi-stellar object as a function of the redshift level (x). In addition to the least squares line, $\hat{y} = 18.13 + 6.21x$, the coefficient of correlation was determined to be $r = .84$.

a. Interpret the value of r in the words of the problem.

b. What is the relationship between r and the estimated slope of the line?

c. Find and interpret the value of r^2.

Applying the Concepts—Intermediate

11.81 Performance in online courses. Florida State University information scientists assessed the impact of online courses on student performance (*Educational Technology & Society*, Jan. 2005). Each in a sample of 24 graduate students enrolled in an online advanced Web application course was asked, "How many courses per semester (on average) do you take online?" Each student's performance on weekly quizzes was also recorded. The information scientists found that the number of online courses and the weekly quiz grade were negatively correlated at $r = -.726$.

a. Give a practical interpretation of r.

b. The researchers concluded that there was a "significant negative correlation" between the number of online courses and the weekly quiz grade. Do you agree?

11.82 Salary linked to height. Are short people shortchanged when it comes to salary? According to business professors T. A. Judge (University of Florida) and D. M. Cable (University of North Carolina), tall people tend to earn more money over their career than short people earn (*Journal of Applied Psychology*, June 2004). Using data collected from participants in the National Longitudinal Surveys, the researchers computed the correlation between average earnings (in dollars) from 1985 to 2000 and height (in inches) for several occupations. The results are given in the following table:

Occupation	Correlation, r	Sample Size, n
Sales	.41	117
Managers	.35	455
Blue Collar	.32	349
Service Workers	.31	265
Professional/Technical	.30	453
Clerical	.25	358
Crafts/Forepersons	.24	250

Source: Judge, T. A., and Cable, D. M. "The effect of physical height on work-place success and income: Preliminary test of a theoretical model." *Journal of Applied Psychology*, Vol. 89, No. 3, June 2004 (Table 5). Copyright © 2004 by the American Psychological Association. Reprinted with permission.

a. Interpret the value of r for people in sales occupations.
b. Compute r^2 for people in sales occupations. Interpret the result.
c. Give H_0 and H_a for testing whether average earnings and height are positively correlated.
d. Compute the test statistic for testing H_0 and H_a in part **c** for people in sales occupations.
e. Use the result you obtained in part **d** to conduct the test at $\alpha = .01$. State the appropriate conclusion.
f. Select another occupation and repeat parts **a–e**.

11.83 View of rotated objects. *Perception & Psychophysics* (July 1998) reported on a study of how people view three-dimensional objects projected onto a rotating two-dimensional image. Each in a sample of 25 university students viewed various depth-rotated objects (e.g., a hairbrush, a duck, and a shoe) until they recognized the object. The recognition exposure time—that is, the minimum time (in milliseconds) required for the subject to recognize the object—was recorded for each object. In addition, each subject rated the "goodness of view" of the object on a numerical scale, with lower scale values corresponding to better views. The following table gives the correlation coefficient r between recognition exposure time and goodness of view for several different rotated objects:

Object	r	t
Piano	.447	2.40
Bench	−.057	.27
Motorbike	.619	3.78
Armchair	.294	1.47
Teapot	.949	14.50

a. Interpret the value of r for each object.
b. Calculate and interpret the value of r^2 for each object.
c. The table also includes the t-value for testing the null hypothesis of no correlation (i.e., for testing $H_0: \beta_1 = 0$). Interpret these results.

11.84 Snow geese feeding trial. Botanists at the University of Toronto conducted a series of experiments to investigate the feeding habits of baby snow geese (*Journal of Applied Ecology*, Vol. 32, 1995). Goslings were deprived of food until their guts were empty and then were allowed to feed for 6 hours on a diet of plants or Purina® Duck Chow®. For each feeding trial, the change in the weight of the gosling after 2.5 hours was recorded as a percentage of the bird's initial weight. Two other variables recorded were digestion efficiency (measured as a percentage) and amount of acid-detergent fiber in the digestive tract (also measured as a percentage). Data on 42 feeding trials are saved in the **SNOWGEESE** file. The first and last 5 observations are listed in the table below.

Feeding Trial	Diet	Weight Change (%)	Digestion Efficiency (%)	Acid-Detergent Fiber (%)
1	Plants	−6	0	28.5
2	Plants	−5	2.5	27.5
3	Plants	−4.5	5	27.5
4	Plants	0	0	32.5
5	Plants	2	0	32
⋮	⋮	⋮	⋮	⋮
38	Duck Chow	9	59	8.5
39	Duck Chow	12	52.5	8
40	Duck Chow	8.5	75	6
41	Duck Chow	10.5	72.5	6.5
42	Duck Chow	14	69	7

Based on Gadallah, F. L., and Jefferies, R. L. "Forage quality in brood rearing areas of the lesser snow goose and the growth of captive goslings." *Journal of Applied Biology*, Vol. 32, No. 2, 1995, pp. 281–282 (adapted from Figures 2 and 3).

a. The botanists were interested in the correlation between weight change (y) and digestion efficiency (x). Plot the data for these two variables in a scatterplot. Do you observe a trend?
b. Find the coefficient of correlation relating weight change y to digestion efficiency x. Interpret this value.
c. Conduct a test to determine whether weight change y is correlated with digestion efficiency x. Use $\alpha = .01$.
d. Repeat parts **b** and **c**, but exclude the data for trials that used duck chow. What do you conclude?
e. The botanists were also interested in the correlation between digestion efficiency y and acid-detergent fibe x. Repeat parts **a–d** for these two variables.

11.85 Do nice guys finish first or last? Refer to the *Nature* (March 20, 2008) study of the use of punishment in cooperation games, Exercise 11.18 (p. 561). Recall that college students repeatedly played a version of the game "prisoner's dilemma" and the researchers recorded the average payoff (y) and the number of times punishment was used (x) for each player. A negative correlation was discovered between x and y.

a. Give the null and alternative hypothesis for testing whether average payoff and punishment use are negatively correlated.
b. The test, part **a**, yielded a p-value of .001. Interpret this result.
c. Does the result, part **b**, imply that increasing punishment causes your payoff to decrease? Explain.

11.86 The "name game." Refer to the *Journal of Experimental Psychology—Applied* (June 2000) name-retrieval study, first presented in Exercise 11.30 (p. 565). The data for the study are saved in the **NAMEGAME2** file. Find and interpret the values of r and r^2 for the simple linear regression relating the proportion of names recalled (y) and the position (order) of the student (x) during the "name game."

11.87 Effect of massage on boxing. Refer to the *British Journal of Sports Medicine* (April 2000) study of the effect of massage on boxing performance, presented in Exercise 11.60 (p. 577). The data for the study are saved in the **BOXING2** file. Find and interpret the values of r and r^2 for the simple linear regression relating the blood lac-tate concentration and the boxer's perceived recovery.

Applying the Concepts—Advanced

11.88 Pain tolerance study. A study published in *Psychosomatic Medicine* (Mar./Apr. 2001) explored the relationship between reported severity of pain and actual pain tolerance in 337 patients who suffer from chronic pain. Each patient reported his or her severity of chronic pain on a seven-point scale (1 = no pain, 7 = extreme pain). To obtain a pain tolerance level, a tourniquet was applied to the arm of each patient and twisted. The maximum pain level tolerated was measured on a quantitative scale.

a. According to the researchers, "Correlational analysis revealed a small but significant inverse relationship between [actual] pain tolerance and the reported severity of chronic pain." On the basis of this statement, is the value of r for the 337 patients positive or negative?

b. Suppose that the result reported in part **a** is significant at $\alpha = .05$. Find the approximate value of r for the sample of 337 patients.

11.6 Using the Model for Estimation and Prediction

If we are satisfied that a useful model has been found to describe the relationship between reaction time and percent of drug in the bloodstream, we are ready for step 5 in our regression modeling procedure: using the model for estimation and prediction.

> *The most common uses of a probabilistic model for making inferences can be divided into two categories. The first is the use of the model for estimating the mean value of y, E(y), for a specific value of x.*

For our drug reaction example, we may want to estimate the mean response time for all people whose blood contains 4% of the drug.

> *The second use of the model entails predicting a new individual y value for a given x.*

That is, we may want to predict the reaction time for a specific person who possesses 4% of the drug in the bloodstream.

In the first case, we are attempting to estimate the mean value of y for a very large number of experiments at the given x value. In the second case, we are trying to predict the outcome of a single experiment at the given x value. Which of these uses of the model—estimating the mean value of y or predicting an individual new value of y (for the same value of x)—can be accomplished with the greater accuracy?

Before answering this question, we first consider the problem of choosing an estimator (or predictor) of the mean (or a new individual) y value. We will use the least squares prediction equation

$$\hat{y} = \hat{\beta}_0 + \hat{\beta}_1 x$$

both to estimate the mean value of y and to predict a specific new value of y for a given value of x. For our example, we found that

$$\hat{y} = -.1 + .7x$$

so the estimated mean reaction time for all people when $x = 4$ (the drug is 4% of the blood content) is

$$\hat{y} = -.1 + .7(4) = 2.7 \text{ seconds}$$

The same value is used to predict a new y value when $x = 4$. That is, both the estimated mean and the predicted value of y are $\hat{y} = 2.7$ when $x = 4$, as shown in Figure 11.18. The difference between these two uses of the model lies in the accuracies of the estimate and the prediction, best measured by the sampling errors of the least squares line when it is used as an estimator and as a predictor, respectively. These errors are reflected in the standard deviations given in the following box:

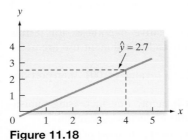

Figure 11.18

Estimated mean value and predicted individual value of reaction time y for $x = 4$

Sampling Errors for the Estimator of the Mean of y and the Predictor of an Individual New Value of y

1. The standard deviation of the sampling distribution of the estimator $\hat{y}$ of the mean value of y at a specific value of x, say x_p, is

$$\sigma_y = \sigma\sqrt{\frac{1}{n} + \frac{(x_p - \bar{x})^2}{SS_{xx}}}$$

where σ is the standard deviation of the random error ε. We refer to $\sigma_{\hat{y}}$ as the standard error of $\hat{y}$.

2. The standard deviation of the prediction error for the predictor $\hat{y}$ of an individual new y value at a specific value of x is

$$\sigma_{(y-\hat{y})} = \sigma\sqrt{1 + \frac{1}{n} + \frac{(x_p - \bar{x})^2}{SS_{xx}}}$$

where σ is the standard deviation of the random error ε. We refer to $\sigma_{(y-\hat{y})}$ as the standard error of prediction.

The true value of σ is rarely known, so we estimate σ by s and calculate the estimation and prediction intervals as shown in the next two boxes:

A $100(1 - \alpha)\%$ Confidence Interval for the Mean Value of y at $x = x_p$

$$\hat{y} + t_{\alpha/2}(\text{Estimated standard error of } \hat{y})$$

or

$$\hat{y} \pm t_{\alpha/2}s\sqrt{\frac{1}{n} + \frac{(x_p - \bar{x})^2}{SS_{xx}}}$$

where $t_{\alpha/2}$ is based on $(n - 2)$ degrees of freedom.

A $100(1 - \alpha)\%$ Prediction Interval* for an Individual New Value of y at $x = x_p$

$$\hat{y} \pm t_{\alpha/2}(\text{Estimated standard error of prediction})$$

or

$$\hat{y} \pm t_{\alpha/2}s\sqrt{1 + \frac{1}{n} + \frac{(x_p - \bar{x})^2}{SS_{xx}}}$$

where $t_{\alpha/2}$ is based on $(n - 2)$ degrees of freedom.

Example 11.6

Estimating the Mean of y—Drug Reaction Regression

Problem Refer to the simple linear regression on drug reaction. Find a 95% confidence interval for the mean reaction time when the concentration of the drug in the bloodstream is 4%.

Solution For a 4% concentration, $x = 4$ and the confidence interval for the mean value of y is

$$\hat{y} \pm t_{\alpha/2}s\sqrt{\frac{1}{n} + \frac{(x_p - \bar{x})^2}{SS_{xx}}} = \hat{y} \pm t_{.025}s\sqrt{\frac{1}{5} + \frac{(4 - \bar{x})^2}{SS_{xx}}}$$

*The term *prediction interval* is used when the interval formed is intended to enclose the value of a random variable. The term *confidence interval* is reserved for the estimation of population parameters (such as the mean).

where $t_{.025}$ is based on $n - 2 = 5 - 2 = 3$ degrees of freedom. Recall that $\hat{y} = 2.7$, $s = .61, \bar{x} = 3$, and $SS_{xx} = 10$. From Table VI in Appendix A, $t_{.025} = 3.182$. Thus, we have

$$2.7 \pm (3.182)(.61)\sqrt{\frac{1}{5} + \frac{(4-3)^2}{10}} = 2.7 \pm (3.182)(.61)(.55)$$

$$= 2.7 \pm (3.182)(.34)$$

$$= 2.7 \pm 1.1$$

Therefore, when the percentage of drug in the bloodstream is 4%, we can be 95% confident that the mean reaction time for all possible subjects will range from 1.6 to 3.8 seconds.

Look Back Note that we used a small amount of data (a small sample size) for purposes of illustration in fitting the least squares line. The interval would probably be narrower if more information had been obtained from a larger sample.

Now Work Exercise 11.94a–d

Example 11.7

Predicting an Individual Value of y— Drug Reaction Regression

Problem Refer again to the drug reaction regression. Predict the reaction time for the next performance of the experiment for a subject with a drug concentration of 4%. Use a 95% prediction interval.

Solution To predict the response time for an individual new subject for whom $x = 4$, we calculate the 95% prediction interval as

$$\hat{y} \pm t_{\alpha/2}s\sqrt{1 + \frac{1}{n} + \frac{(x_p - \bar{x})^2}{SS_{xx}}} = 2.7 \pm (3.182)(.61)\sqrt{1 + \frac{1}{5} + \frac{(4-3)^2}{10}}$$

$$= 2.7 \pm (3.182)(.61)(1.14)$$

$$= 2.7 \pm (3.182)(.70)$$

$$= 2.7 \pm 2.2$$

Therefore, when the drug concentration for an individual is 4%, we predict with 95% confidence that the reaction time for this new individual will fall into the interval from .5 to 4.9 seconds.

Look Back Like the confidence interval for the mean value of y, the prediction interval for y is quite large. This is because we have chosen a simple example (one with only five data points) to fit the least squares line. The width of the prediction interval could be reduced by using a larger number of data points.

Now Work Exercise 11.94e

Both the confidence interval for $E(y)$ and the prediction interval for y can be obtained from a statistical software package. Figure 11.19 is a MINITAB printout showing the confidence interval and prediction interval, respectively, for the data in the drug example.

The 95% confidence interval for $E(y)$ when $x = 4$, highlighted under "95% CI" in Figure 11.19, is (1.645, 3.755). The 95% prediction interval for y when $x = 4$, highlighted in Figure 11.19 under "95% PI," is (.503, 4.897). These agree with the ones computed in Examples 11.6 and 11.7.

Note that the prediction interval for an individual new value of y is *always* wider than the corresponding confidence interval for the mean value of y. Will this always be true? The answer is "Yes." The error in estimating the mean value of y, $E(y)$, for a given value of x, say, x_p, is the distance between the least squares line and the true line of means, $E(y) = \beta_0 + \beta_1 x$. This error, $[\hat{y} - E(y)]$, is shown in Figure 11.20. In contrast, *the error* $(y_p - \hat{y})$ *in predicting some future value of y is the sum of two errors*: the error

```
Predicted Values for New Observations

New
Obs    Fit   SE Fit      95% CI              95% PI
  1  2.700   0.332    (1.645, 3.755)     (0.503, 4.897)
```

Figure 11.19

MINITAB printout giving 95% confidence interval for $E(y)$ and 95% prediction interval for y

```
Values of Predictors for New Observations

New
Obs    DRUG_X
  1     4.00
```

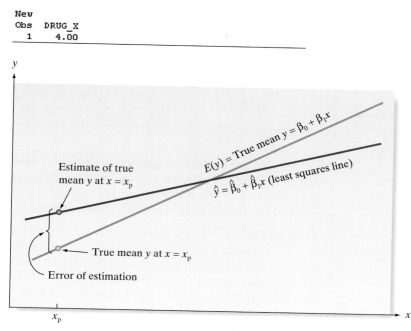

Figure 11.20

Error in estimating the mean value of y for a given value of x

in estimating the mean of y, $E(y)$, shown in Figure 11.20, plus the random error that is a component of the value of y that is to be predicted. (See Figure 11.21.) Consequently, the error in predicting a particular value of y will be larger than the error in estimating the mean value of y for a particular value of x. Note from their formulas that both the error of estimation and the error of prediction take their smallest values when $x_p = \bar{x}$. The farther x_p lies from $\bar{x}$, the larger will be the errors of estimation and prediction. You can see why this is true by noting the deviations for different values of x_p between the actual line of means $E(y) = \beta_0 + \beta_1 x$ and the predicted line of means $\hat{y} = \hat{\beta}_0 + \hat{\beta}_1 x$ shown in Figure 11.21. The deviation is larger at the extremes of the interval, where the largest and smallest values of x in the data set occur.

Both the confidence intervals for mean values and the prediction intervals for new values are depicted over the entire range of the regression line in Figure 11.22. You

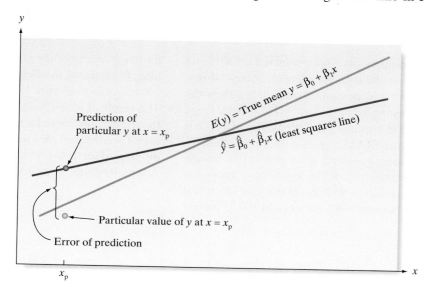

Figure 11.21

Error in predicting a future value of y for a given value of x

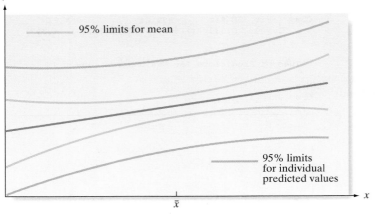

Figure 11.22

Confidence intervals for mean values and prediction intervals for new values

can see that the confidence interval is always narrower than the prediction interval and that they are both narrowest at the mean $\bar{x}$, increasing steadily as the distance $|x - \bar{x}|$ increases. In fact, when x is selected far enough away from $\bar{x}$ so that it falls outside the range of the sample data, it is dangerous to make any inferences about $E(y)$ or y.

⚠ **CAUTION** Using the least squares prediction equation to estimate the mean value of y or to predict a particular value of y for values of x that fall outside the range of the values of x contained in your sample data may lead to errors of estimation or prediction that are much larger than expected. Although the least squares model may provide a very good fit to the data over the range of x values contained in the sample, it could give a poor representation of the true model for values of x outside that region. ▲

The width of the confidence interval grows smaller as n is increased; thus, in theory, you can obtain as precise an estimate of the mean value of y as desired (at any given x) by selecting a large enough sample. The prediction interval for a new value of y also grows smaller as n increases, but there is a lower limit on its width. If you examine the formula for the prediction interval, you will see that the interval can get no smaller than $\hat{y} \pm z_{\alpha/2}\sigma$.* Thus, the only way to obtain more accurate predictions for new values of y is to reduce the standard deviation σ of the regression model. This can be accomplished only by improving the model, either by using a curvilinear (rather than linear) relationship with x or by adding new independent variables to the model (or both). Methods of improving the model are discussed in Chapter 12.

Now Work Exercise 11.94f

Statistics IN Action | **Revisited** | **Using the Straight–Line Model to Predict Pipe Location for the Dowsing Data**

The group of German physicists who conducted the dowsing experiments stated that the data for the three "best" dowsers empirically support the dowsing theory. If so, then the straight-line model relating a dowser's guess (x) to actual pipe location (y) should yield accurate predictions. The MINITAB printout shown in Figure SIA11.5 gives a 95% prediction interval for y when a dowser guesses $x = 50$ meters (the middle of the 100-meter-long waterpipe). The highlighted interval is $(-9.3, 100.23)$. Thus, we can be 95% confident that the actual pipe location will fall between -9.3 meters and 100.23 meters

for this guess. Since the pipe is only 100 meters long, the interval in effect ranges from 0 to 100 meters—the entire length of the pipe! This result, of course, is due to the fact that the straight-line model is not a statistically useful predictor of pipe location, a fact we discovered in the previous Statistics in Action Revisited sections.

---

*The result follows from the facts that, for large n, $t_{\alpha/2} \approx z_{\alpha/2}$, $s \approx \sigma$, and the last two terms under the radical in the standard error of the predictor are approximately 0.

(continued)

```
Predicted Values for New Observations

New
Obs    Fit   SE Fit     95% CI          95% PI
 1   45.47    5.15   (34.83, 56.10)   (-9.30, 100.23)


Values of Predictors for New Observations

New
Obs   GUESS
 1    50.0
```

Figure SIA11.5
MINITAB prediction interval
for dowsing data

Exercises 11.89–11.107

Understanding the Principles

11.89 Explain the difference between y and $E(y)$ for a given x.

11.90 *True or False.* For a given x, a confidence interval for $E(y)$ will always be wider than a prediction interval for y.

11.91 *True or False.* The greater the deviation between x and $\bar{x}$, the wider the prediction interval for y will be.

11.92 For each of the following, decide whether the proper inference is a prediction interval for y or a confidence interval for $E(y)$:
   a. A jeweler wants to predict the selling price of a diamond stone on the basis of its size (number of carats).
   b. A psychologist wants to estimate the average IQ of all patients who have a certain income level.

Learning the Mechanics

11.93 In fitting a least squares line to $n = 10$ data points, the following quantities were computed:

$$SS_{xx} = 32, \bar{x} = 3, SS_{yy} = 26, \bar{y} = 4, SS_{xy} = 28$$

   a. Find the least squares line.
   b. Graph the least squares line.
   c. Calculate SSE.
   d. Calculate s^2.
   e. Find a 95% confidence interval for the mean value of y when $x_p = 2.5$.
   f. Find a 95% prediction interval for y when $x_p = 4$.

11.94 Consider the following pairs of measurements saved in the **LM11_94** file.

x	1	2	3	4	5	6	7
y	3	5	4	6	7	7	10

   a. Construct a scatterplot of these data.
   b. Find the least squares line, and plot it on your scatterplot.
   c. Find s^2.
   d. Find a 90% confidence interval for the mean value of y when $x = 4$. Plot the upper and lower bounds of the confidence interval on your scatterplot.
   e. Find a 90% prediction interval for a new value of y when $x = 4$. Plot the upper and lower bounds of the prediction interval on your scatterplot.

f. Compare the widths of the intervals you constructed in parts **d** and **e**. Which is wider and why?

11.95 Consider the following pairs of measurements, saved in the **LM11_95** file.

x	4	6	0	5	2	3	2	6	2	1
y	3	5	−1	4	3	2	0	4	1	1

For these data, $SS_{xx} = 38.900$, $SS_{yy} = 33.600$, $SS_{xy} = 32.8$, and $\hat{y} = -.414 + .843x$.
   a. Construct a scatterplot of the data.
   b. Plot the least squares line on your scatterplot.
   c. Use a 95% confidence interval to estimate the mean value of y when $x_p = 6$. Plot the upper and lower bounds of the interval on your scatterplot.
   d. Repeat part **c** for $x_p = 3.2$ and $x_p = 0$.
   e. Compare the widths of the three confidence intervals you constructed in parts **c** and **d**, and explain why they differ.

11.96 Refer to Exercise 11.95.
   a. Using no information about x, estimate and calculate a 95% confidence interval for the mean value of y. [*Hint:* Use the one-sample t methodology of Section 7.3.]
   b. Plot the estimated mean value and the confidence interval as horizontal lines on your scatterplot.
   c. Compare the confidence intervals you calculated in parts **c** and **d** of Exercise 11.95 with the one you calculated in part **a** of this exercise. Does x appear to contribute information about the mean value of y?
   d. Check the answer you gave in part **c** with a statistical test of the null hypothesis $H_0: \beta_1 = 0$ against $H_a: \beta_1 \neq 0$. Use $\alpha = .05$.

Applying the Concepts—Basic

11.97 **Do nice guys finish first or last?** Refer to the *Nature* (March 20, 2008) study of the use of punishment in cooperation games, Exercise 11.18 (p. 561). Recall that simple linear regression was used to model a player's average payoff (y) as a straight-line function of the number of times punishment was used (x) by the player.
   a. If the researchers want to predict average payoff for a single player who used punishment 10 times, how should they proceed?

SAS output for Exercise 11.99

Predictions

Obs	Region	Rain	Temp	Predicted Rain	Lower prediction limit of Rain	Upper prediction limit of Rain
1	DryStepp	196	5.7	201.977	179.525	224.430
2	DryStepp	196	5.7	201.977	179.525	224.430
3	DryStepp	179	7.0	180.704	163.694	197.714
4	DryStepp	197	8.0	164.340	150.594	178.085
5	DryStepp	149	8.5	156.157	143.513	168.802
6	GobiDese	112	10.7	120.156	106.038	134.274
7	GobiDese	125	11.4	108.701	92.298	125.104
8	GobiDese	99	10.9	116.883	102.172	131.595
9	GobiDese	125	11.4	108.701	92.298	125.104
10	GobiDese	84	11.4	108.701	92.298	125.104
11	GobiDese	115	11.4	108.701	92.298	125.104

b. If the researchers want to estimate the mean of the average payoffs for all players who used punishment 10 times, how should they proceed?

11.98 English as a second language reading ability. Refer to the *Bilingual Research Journal* (Summer 2006) study of the relationship of Spanish (first-language) grammatical knowledge to English (second-language) reading, presented in Exercise 11.54 (p. 575). Recall that three simple linear regressions were used to model the English reading (ESLR) score (y) as a function of Spanish grammar (SG), Spanish reading (SR), and English grammar (EG), respectively.
  a. If the researchers want to predict the ELSR score (y) of a native Spanish-speaking adult who scored 50% in Spanish grammar (x), how should they proceed?
  b. If the researchers want to estimate the mean ELSR score $E(y)$ of all native Spanish-speaking adults who scored 70% in Spanish grammar (x), how should they proceed?

11.99 Mongolian desert ants. Refer to the *Journal of Biogeography* (Dec. 2003) study of ant sites in Mongolia, presented in Exercise 11.22 (p. 562). You applied the method of least squares to the data in the **GOBIANTS** file to estimate the straight-line model relating annual rainfall (y) and maximum daily temperature (x). A SAS printout giving 95% prediction intervals for the amount of rainfall at each of the 11 sites is shown above. Select the interval associated with site (observation) 7 and interpret it practically.

11.100 Ranking driving performance of professional golfers. Refer to *The Sport Journal* (Winter 2007) study of a new method for ranking the total driving performance of golfers on the Professional Golf Association (PGA) tour, presented in Exercise 11.25 (p. 563). You fit a straight-line model relating diving accuracy (y) to driving distance (x) to the data saved in the **PGADRIVER** file. A MINITAB printout with prediction and confidence intervals for a driving distance of $x = 300$ yards is shown below.

```
Predicted Values for New Observations

New
Obs    Fit    SE Fit      95% CI           95% PI
  1  61.309  0.357  (60.586, 62.032)  (56.724, 65.894)


Values of Predictors for New Observations

New
Obs  DISTANCE
  1    300
```

a. Locate the 95% prediction interval for driving accuracy (y) on the printout, and give a practical interpretation of the result.
b. Locate the 95% prediction interval for mean driving accuracy (y) on the printout, and give a practical interpretation of the result.
c. If you are interested in knowing the average driving accuracy of all PGA golfers who have a driving distance of 300 yards, which of the intervals is relevant? Explain.

11.101 Sweetness of orange juice. Refer to the simple linear regression of sweetness index y and amount of pectin, x, for $n = 24$ orange juice samples, presented in Exercise 11.28 (p. 565). The SPSS printout of the analysis is shown at the top of page 595. A 90% confidence interval for the mean sweetness index $E(y)$ for each value of x is shown on the SPSS spreadsheet on the next page. Select an observation and interpret this interval.

Applying the Concepts—Intermediate

11.102 Sound waves from a basketball. Refer to the *American Journal of Physics* (June 2010) study of sound waves in a spherical cavity, Exercise 11.27 (p. 564). The frequencies of sound waves resulting from the first 24 resonances (echoes) after striking a basketball with a metal rod are saved in the **BBALL** file. You fit a straight-line model relating frequency (y) to number of resonances (x) in Exercise 11.27.
  a. Use the model to predict the sound wave frequency for the 10th resonance.
  b. Form a 90% confidence interval for the prediction, part **a**. Interpret the result.
  c. Suppose you want to predict the sound wave frequency for the 30th resonance. What are the dangers in making this prediction with the fitted model?

11.103 Ideal height of your mate. Refer to the *Chance* (Summer 2008) study of the height of the ideal mate, Exercise 11.29 (p. 565). The data in the **IDHEIGHT** file was used to fit the simple linear regression model, $E(y) = \beta_0 + \beta_1 x$, where $y =$ ideal partner's height (in inches) and $x =$ student's height (in inches). One model was fitted for male students and one model was fitted for female students. Consider a student who is 66 inches tall.
  a. If the student is a female, use the model to predict the height of her ideal mate. Form a 95% confidence interval for the prediction and interpret the result.

	run	sweet	pectin	lower90m	upper90m
1	1	5.2	220	5.64898	5.83848
2	2	5.5	227	5.63898	5.81613
3	3	6.0	259	5.57819	5.72904
4	4	5.9	210	5.66194	5.87173
5	5	5.8	224	5.64337	5.82560
6	6	6.0	215	5.65564	5.85493
7	7	5.8	231	5.63284	5.80379
8	8	5.6	268	5.55553	5.71011
9	9	5.6	239	5.61947	5.78019
10	10	5.9	212	5.65946	5.86497
11	11	5.4	410	5.05526	5.55416
12	12	5.6	256	5.58517	5.73592
13	13	5.8	306	5.43785	5.65219
14	14	5.5	259	5.57819	5.72904
15	15	5.3	284	5.50957	5.68213
16	16	5.3	383	5.15725	5.57694
17	17	5.7	271	5.54743	5.70434
18	18	5.5	264	5.56591	5.71821
19	19	5.7	227	5.63898	5.81613
20	20	5.3	263	5.56843	5.72031
21	21	5.9	232	5.63125	5.80075
22	22	5.8	220	5.64898	5.83848
23	23	5.8	246	5.60640	5.76091
24	24	5.9	241	5.61587	5.77454

SPSS output for Exercise 11.101

b. If the student is a male, use the model to predict the height of his ideal mate. Form a 95% confidence interval for the prediction and interpret the result.

c. Which of the two inferences, parts **a** and **b,** may be invalid? Why?

11.104 The "name game." Refer to the *Journal of Experimental Psychology—Applied* (June 2000) name-retrieval study, presented in Exercise 11.30, (p. 565). The data for the study are saved in the **NAMEGAME2** file.

a. Find a 99% confidence interval for the mean recall proportion for students in the fifth position during the "name game." Interpret the result.

b. Find a 99% prediction interval for the recall proportion of a particular student in the fifth position during the "name game." Interpret the result.

c. Compare the intervals you found in parts **a** and **b.** Which interval is wider? Will this always be the case? Explain.

11.105 Spreading rate of spilled liquid. Refer to the *Chemicial Engineering Progress* (Jan. 2005) study of the rate at which a spilled volatile liquid will spread across a surface, presented in Exercise 11.31 (p. 566). Recall that simple linear regression was used to model y = mass of the spill as a function of y = elapsed time of the spill. The data for the study are saved in the **LIQUIDSPILL** file.

a. Find a 99% confidence interval for the mean mass of all spills with an elapsed time of 15 minutes. Interpret the result.

b. Find a 99% prediction interval for the mass of a single spill with an elapsed time of 15 minutes. Interpret the result.

c. Compare the intervals you found in parts **a** and **b.** Which interval is wider? Will this always be the case? Explain.

11.106 Feeding habits of snow geese. Refer to the *Journal of Applied Ecology* feeding study of the relationship between the weight change y of baby snow geese and their digestion efficiency x, presented in Exercise 11.84 (p. 587). The data for the study are saved in the **SNOWGEESE** file.

a. Fit the simple linear regression model to the data.

b. Do you recommend using the model to predict weight change y? Explain.

c. Use the model to form a 95% confidence interval for the mean weight change of all baby snow geese with a digestion efficiency of $x = 15\%$. Interpret the interval.

Applying the Concepts—Advanced

11.107 Life tests of cutting tools. Refer to the data saved in the **CUTTOOL** file of Exercise 11.46 (p. 570).

a. Use a 90% confidence interval to estimate the mean useful life of a brand-A cutting tool when the cutting speed is 45 meters per minute. Repeat for brand B. Compare the widths of the two intervals and comment on the reasons for any difference.

b. Use a 90% prediction interval to predict the useful life of a brand-A cutting tool when the cutting speed is 45 meters per minute. Repeat for brand B. Compare the widths of the two intervals with each other and with the two intervals you calculated in part **a.** Comment on the reasons for any differences.

c. Note that the estimation and prediction you performed in parts **a** and **b** were for a value of x that was not included in the original sample. That is, the value $x = 45$ was not part of the sample. However, the value is within the range of x values in the sample, so that the regression model spans the x value for which the estimation and prediction were made. In such situations, estimation and prediction represent **interpolations.**

Suppose you were asked to predict the useful life of a brand-A cutting tool for a cutting speed of $x = 100$ meters per minute. Since the given value of x is outside the range of the sample x values, the prediction is an example of **extrapolation.** Predict the useful life of a brand-A cutting tool that is operated at 100 meters per minute, and construct a 95% confidence interval for the actual useful life of the tool. What additional assumption do you have to make in order to ensure the validity of an extrapolation?

11.7 A Complete Example

In the previous sections, we presented the basic elements necessary to fit and use a straight-line regression model. In this section, we will assemble these elements by applying them in an example with the aid of computer software.

Suppose a fire insurance company wants to relate the amount of fire damage in major residential fires to the distance between the burning house and the nearest fire station. The study is to be conducted in a large suburb of a major city; a sample of 15 recent fires in this suburb is selected. The amount of damage, y, and the distance between the fire and the nearest fire station, x, are recorded for each fire. The results are given in Table 11.7 and saved in the **FIREDAM** file.

Step 1 First, we hypothesize a model to relate fire damage, y, to the distance from the nearest fire station, x. We hypothesize a straight-line probabilistic model:

$$y = \beta_0 + \beta_1 x + \varepsilon$$

Step 2 Next, we open the **FIREDAM** file and use statistical software to estimate the unknown parameters in the deterministic component of the hypothesized model. The SAS printout for the simple linear regression analysis is shown in Figure 11.23. The least squares estimate of the slope β_1 and intercept β_0, highlighted on the printout, are

$$\hat{\beta}_1 = 4.91933$$
$$\hat{\beta}_0 = 10.27793$$

Table 11.7
Fire Damage Data

Distance from Fire Station, x (miles)	Fire Damage y (thousands of dollars)
3.4	26.2
1.8	17.8
4.6	31.3
2.3	23.1
3.1	27.5
5.5	36.0
.7	14.1
3.0	22.3
2.6	19.6
4.3	31.3
2.1	24.0
1.1	17.3
6.1	43.2
4.8	36.4
3.8	26.1

Data Set: FIREDAM

Dependent Variable: DAMAGE

Analysis of Variance

Source	DF	Sum of Squares	Mean Square	F Value	Pr > F
Model	1	841.76636	841.76636	156.89	<.0001
Error	13	69.75098	5.36546		
Corrected Total	14	911.51733			

Root MSE	2.31635	R-Square	0.9235
Dependent Mean	26.41333	Adj R-Sq	0.9176
Coeff Var	8.76961		

Parameter Estimates

Variable	DF	Parameter Estimate	Standard Error	t Value	Pr > |t|	95% Confidence Limits	
Intercept	1	10.27793	1.42028	7.24	<.0001	7.20960	13.34625
DISTANCE	1	4.91933	0.39275	12.53	<.0001	4.07085	5.76781

Output Statistics

Obs	DISTANCE	Dep Var DAMAGE	Predicted Value	Std Error Mean Predict	95% CL Predict		Residual
1	3.4	26.2000	27.0037	0.5999	21.8344	32.1729	-0.8037
2	1.8	17.8000	19.1327	0.8340	13.8141	24.4514	-1.3327
3	4.6	31.3000	32.9068	0.7915	27.6186	38.1951	-1.6068
4	2.3	23.1000	21.5924	0.7112	16.3577	26.8271	1.5076
5	3.1	27.5000	25.5279	0.6022	20.3573	30.6984	1.9721
6	5.5	36.0000	37.3342	1.0573	31.8334	42.8351	-1.3342
7	0.7	14.1000	13.7215	1.1766	8.1087	19.3342	0.3785
8	3	22.3000	25.0359	0.6081	19.8622	30.2097	-2.7359
9	2.6	19.6000	23.0682	0.6550	17.8678	28.2686	-3.4682
10	4.3	31.3000	31.4311	0.7198	26.1908	36.6713	-0.1311
11	2.1	24.0000	20.6085	0.7566	15.3442	25.8729	3.3915
12	1.1	17.3000	15.6892	1.0444	10.1999	21.1785	1.6108
13	6.1	43.2000	40.2858	1.2587	34.5906	45.9811	2.9142
14	4.8	36.4000	33.8907	0.8450	28.5640	39.2175	2.5093
15	3.8	26.1000	28.9714	0.6320	23.7843	34.1585	-2.8714
16	3.5	.	27.4956	0.6043	22.3239	32.6672	.

Figure 11.23

SAS printout for fire damage regression

and the least squares equation is (rounded)

$$\hat{y} = 10.28 + 4.92x$$

This prediction equation is graphed by MINITAB in Figure 11.24, along with a plot of the data points.

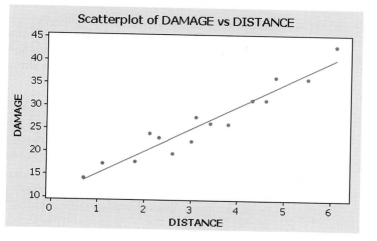

Figure 11.24

MINITAB scatterplot with least squares line for fire damage regression analysis

The least squares estimate of the slope, $\hat{\beta}_1 = 4.92$, implies that the estimated mean damage increases by \$4,920 for each additional mile from the fire station. This interpretation is valid over the range of x, or from .7 to 6.1 miles from the station. The estimated y-intercept, $\hat{\beta}_0 = 10.28$, has the interpretation that a fire 0 miles from the fire station has an estimated mean damage of \$10,280. Although this would seem to apply to the fire station itself, remember that the y-intercept is meaningfully interpretable only if $x = 0$ is within the sampled range of the independent variable. Since $x = 0$ is outside the range, β_0 has no practical interpretation.

Step 3   Now we specify the probability distribution of the random-error component ε. The assumptions about the distribution are identical to those listed in Section 11.3. Although we know that these assumptions are not completely satisfied (they rarely are for practical problems), we are willing to assume that they are approximately satisfied for this example. The estimate of the standard deviation σ of ε, highlighted on the SAS printout, is

$$s = 2.31635$$

This implies that most of the observed fire damage (y) values will fall within approximately $2s = 4.64$ thousand dollars of their respective predicted values when the least squares line is used.

Step 4   We can now check the usefulness of the hypothesized model—in other words, whether x really contributes information for the prediction of y by the straight-line model. First, test the null hypothesis that the slope β_1 is 0—that is, that there is no linear relationship between fire damage and the distance from the nearest fire station—against the alternative hypothesis that fire damage increases as the distance increases. We test

$$H_0: \beta_1 = 0$$
$$H_a: \beta_1 > 0$$

The two-tailed observed significance level for testing $H_a: \beta_1 \neq 0$, highlighted on the SAS printout, is less than .0001. Thus, the p-value for our one-tailed test is less than half of this value (.00005). This small p-value leaves little doubt that mean fire damage and distance between the fire and the fire station are at least linearly related, with mean fire damage increasing as the distance increases.

We gain additional information about the relationship by forming a confidence interval for the slope β_1. A 95% confidence interval, highlighted on the

SAS printout, is (4.071, 5.768). Thus, with 95% confidence, we estimate that the interval from $4,071 to $5,768 encloses the mean increase (β_1) in fire damage per additional mile in distance from the fire station.

Another measure of the utility of the model is the coefficient of determination, r^2. The value (also highlighted on the printout) is $r^2 = .9235$, which implies that about 92% of the sample variation in fire damage (y) is explained by the distance (x) between the fire and the fire station.

The coefficient of correlation, r, that measures the strength of the linear relationship between y and x is not shown on the SAS printout and must be calculated. Using the facts that $r = \sqrt{r^2}$ in simple linear regression and that r and $\hat{\beta}_1$ have the same sign, we calculate

$$r = +\sqrt{r^2} = \sqrt{.9235} = .96$$

The high correlation confirms our conclusion that β_1 is greater than 0; it appears that fire damage and distance from the fire station are positively correlated. All signs point to a strong linear relationship between y and x.

Step 5 We are now prepared to use the least squares model. Suppose the insurance company wants to predict the fire damage if a major residential fire were to occur 3.5 miles from the nearest fire station. The predicted value (highlighted at the bottom of the SAS printout) is $\hat{y} = 27.496$, while the 95% prediction interval (also highlighted) is (22.324, 32.667). Therefore, with 95% confidence, we predict fire damage in a major residential fire 3.5 miles from the nearest station to be between $22,324 and $32,667.

⚠ **CAUTION** We would not use this model to make predictions for homes less than .7 mile or more than 6.1 miles from the nearest fire station. A look at the data in Table 11.7 reveals that all the x-values fall between .7 and 6.1. It is dangerous to use the model to make predictions outside the region in which the sample data fall. A straight line might not provide a good model for the relationship between the mean value of y and the value of x when stretched over a wider range of x-values. ▲

Exercises 11.108–11.109

Applying the Concepts—Intermediate

11.108 An MBA's work-life balance. The importance of having employees with a healthy work-life balance has been recognized by U.S. companies for decades. Many business schools offer courses that assist MBA students with developing good work-life balance habits and most large companies have developed work-life balance programs for their employees. In April 2005, the Graduate Management Admission Council (GMAC) conducted a survey of over 2,000 MBA alumni to explore the work-life balance issue. (For example, one question asked alumni to state their level of agreement with the statement, "My personal and work demands are overwhelming.") Based on these responses, the GMAC determined a work-life balance scale score for each MBA alumni. Scores ranged from 0 to 100, with lower scores indicating a higher imbalance between work and life. Many other variables, including average number of hours worked per week, were also measured. The data for the work-life balance study are saved in the **GMAC** file. (The first 15 observations are listed in the accompanying table.) Let x = average number of hours worked per week and y = work-life balance scale score for each MBA alumnus. Investigate the link between these two variables by conducting a complete simple linear regression analysis of the data. Summarize your findings in a professional report.

WLB Score	Hours
75.22	50
64.98	45
49.62	50
44.51	55
70.10	50
54.74	60
55.98	55
21.24	60
59.86	50
70.10	50
29.00	70
64.98	45
36.75	40
35.45	40
45.75	50

Based on "Work-life balance: An MBA alumni report."
Graduate Management Admission Council (GMAC)
Research Report (Oct. 13, 2005).

11.109 Legal advertising—does it pay? According to the American Bar Association, there are over one million lawyers competing for your business. To gain a competitive edge, these lawyers are aggressively advertising their services. In fact, Erickson Marketing, Inc., reports that "attorneys are the #1 category of advertising in the Yellow Pages." Does legal advertising really pay? To partially answer this question, consider the case of an actual law firm that specializes in personal injury (PI) cases. The firm spends thousands of dollars each month on advertising. The accompanying table shows the firm's new personal injury cases each month over a 42-month period. Also shown is the total expenditure on advertising each month, and over the previous 6 months. These data are saved in the **LEGALADV** file. Do these data provide support for the hypothesis that increased advertising expenditures are associated with more personal injury cases? Conduct a complete simple linear regression analysis of the data, letting y = new PI cases and x = 6-month cumulative advertising expenditure. Summarize your findings in a professional report.

Month	New PI Cases	6 Months Cumulative Adv. Exp.
7	11	$41,632.74
8	7	$38,227.39
9	13	$39,779.77
10	7	$37,490.22
11	9	$52,225.71
12	8	$56,249.15
13	18	$59,938.03
14	9	$65,250.59
15	25	$66,071.85
16	26	$81,765.94
17	27	$66,895.46
18	12	$71,426.16
19	14	$75,346.40
20	5	$81,589.97
21	22	$78,828.68

Month	New PI Cases	6 Months Cumulative Adv. Exp.
22	15	$78,415.73
23	12	$90,802.77
24	18	$95,689.44
25	20	$83,099.55
26	38	$82,703.75
27	13	$90,484.38
28	18	$102,084.54
29	21	$84,976.99
30	7	$95,314.29
31	16	$115,858.28
32	12	$108,557.00
33	15	$127,693.57
34	18	$122,761.67
35	30	$123,545.67
36	12	$119,388.26
37	30	$134,675.68
38	20	$133,812.93
39	19	$142,417.13
40	29	$149,956.61
41	58	$165,204.46
42	42	$156,725.72
43	24	$146,397.56
44	47	$197,792.64
45	24	$198,460.28
46	14	$206,662.87
47	31	$253,011.27
48	26	$249,496.28

Source: Info Tech, Inc., Gainesville, Florida.

CHAPTER NOTES

Key Terms

Bivariate relationship 578
Coefficient of correlation 579
Coefficient of determination 583
Confidence interval for
  mean of y 589
Correlation 578
Dependent (or response)
  variable 552
Deterministic model 551
Deterministic relationship 551
Errors of prediction 555
Estimated standard error of the
  least squares slope β_1 571
Estimated standard error of the
  regression model 567
Extrapolation 596
Independent (or predictor)
  variable 552
Interpolation 595
Least squares estimates 556

Least squares line (or regression
  line or least squares prediction
  equation) 555
Line of means 553
Method of least squares 555
Population correlation
  coefficient 581
Prediction interval for y 589
Probabilistic model 551
Probabilistic relationship 551
Random error 551
Regression analysis
  (modeling) 552
Regression residuals 555
Scatterplot or scattergram 555
Slope 553
Straight-line (first-order)
  model 552
y-intercept 555

Key Symbols/Notation

y	Dependent variable (variable to be predicted)
x	Independent variable (variable used to predict)
$E(y)$	Expected (mean) of y
β_0	y-intercept of true line
β_1	slope of true line
$\hat{\beta}_0$	Least squared estimate of y-intercept
$\hat{\beta}_1$	Least squares estimate of slope
ε	Random error
$\hat{y}$	Predicted value of y for a given x-value
$(y - \hat{y})$	Estimated error of prediction
SSE	Sum of squared errors of prediction
r	Coefficient of correlation
r^2	Coefficient of determination
x_p	Value of x used to predict y
$r^2 = \dfrac{SS_{yy} - SSE}{SS_{yy}}$	Coefficient of determination
$\hat{y} \pm t_{\alpha/2}s\sqrt{\dfrac{1}{n} + \dfrac{(x_p - \bar{x})^2}{SS_{xx}}}$	$(1 - \alpha)100\%$ confidence interval *for* $E(y)$ when $x = x_p$
$\hat{y} \pm t_{\alpha/2}s\sqrt{1 + \dfrac{1}{n} + \dfrac{(x_p - \bar{x})^2}{SS_{xx}}}$	$(1 - \alpha)100\%$ prediction *interval for* y when $x = x_p$

Key Ideas

Simple Linear Regression Variables

y = **Dependent** variable (quantitative)
x = **Independent** variable (quantitative)

Method of Least Squares Properties

1. average error of prediction $= 0$
2. sum of squared errors is minimum

First-order (straight-line) Model

$$E(y) = \beta_0 + \beta_1 x$$

where $E(y) = $ mean of y

$\beta_0 = $ **y-intercept** of line (point where line intercepts y-axis)

$\beta_1 = $ **slope** of line (change in y for every one-unit change in x)

Practical Interpretation of y-intercept

Predicted y-value when $x = 0$
(no practical interpretation if $x = 0$ is either nonsensical or outside range of sample data)

Practical Interpretation of Slope

Increase (or decrease) in y for every one-unit increase in x

Coefficient of Correlation, r

1. ranges between -1 and $+1$
2. measures strength of *linear relationship* between y and x

Coefficient of Determination, r^2

1. ranges between 0 and 1
2. measures proportion of sample variation in y "explained" by the model.

Practical Interpretation of Model Standard Deviation s

Ninety-five percent of y-values fall within $2s$ of their respective predicted values

Comparing Intervals in Step 5

Width of *confidence interval for $E(y)$* will always be **narrower** than width of *prediction interval for y.*

Guide to Simple Linear Regression

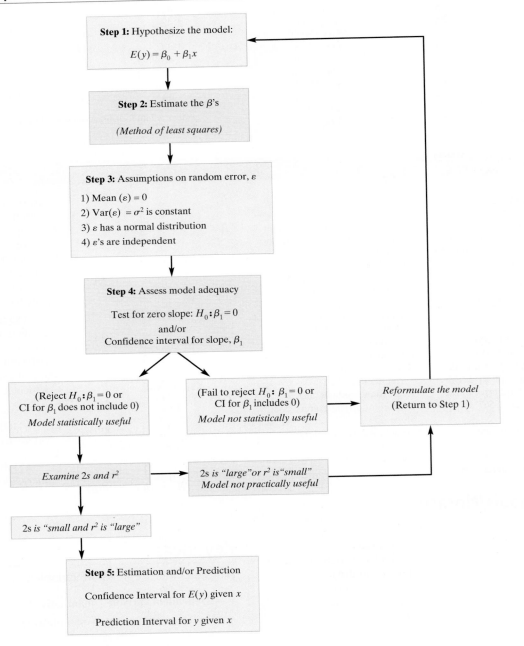

Step 1: Hypothesize the model:

$$E(y) = \beta_0 + \beta_1 x$$

Step 2: Estimate the β's

(Method of least squares)

Step 3: Assumptions on random error, ε

1) Mean $(\varepsilon) = 0$
2) Var$(\varepsilon) = \sigma^2$ is constant
3) ε has a normal distribution
4) ε's are independent

Step 4: Assess model adequacy

Test for zero slope: $H_0\text{:}\,\beta_1 = 0$
and/or
Confidence interval for slope, β_1

(Reject $H_0\text{:}\,\beta_1 = 0$ or CI for β_1 does not include 0)
Model statistically useful

(Fail to reject $H_0\text{:}\,\beta_1 = 0$ or CI for β_1 includes 0)
Model not statistically useful

Reformulate the model
(Return to Step 1)

Examine $2s$ and r^2

$2s$ is "large" or r^2 is "small"
Model not practically useful

$2s$ is "small and r^2 is "large"

Step 5: Estimation and/or Prediction

Confidence Interval for $E(y)$ given x

Prediction Interval for y given x

Supplementary Exercises 11.110–11.136

Understanding the Principles

11.110 Explain the difference between a probabilistic model and a deterministic model.

11.111 Give the general form of a straight-line model for $E(y)$.

11.112 Outline the five steps in a simple linear regression analysis.

11.113 *True or False.* In simple linear regression, about 95% of the y-values in the sample will fall within $2s$ of their respective predicted values.

Learning the Mechanics

11.114 In fitting a least squares line to $n = 15$ data points, the following quantities were computed: $SS_{xx} = 55$, $SS_{yy} = 198$, $SS_{xy} = -88$, $\bar{x} = 1.3$, and $\bar{y} = 35$.
  a. Find the least squares line.
  b. Graph the least squares line.
  c. Calculate SSE.
  d. Calculate s^2.
  e. Find a 90% confidence interval for β_1. Interpret this estimate.
  f. Find a 90% confidence interval for the mean value of y when $x = 15$.
  g. Find a 90% prediction interval for y when $x = 15$.

11.115 Consider the following sample data:

y	5	1	3
x	5	1	3

  a. Construct a scatterplot for the data.
  b. It is possible to find many lines for which $\Sigma (y - \hat{y}) = 0$. For this reason, the criterion $\Sigma (y - \hat{y}) = 0$ is not used to identify the "best-fitting" straight line. Find two lines that have $\Sigma (y - \hat{y}) = 0$.
  c. Find the least squares line.
  d. Compare the value of SSE for the least squares line with that of the two lines you found in part **b**. What principle of least squares is demonstrated by this comparison?

11.116 Consider the following 10 data points saved in the **LM11_116** file.

x	3	5	6	4	3	7	6	5	4	7
y	4	3	2	1	2	3	3	5	4	2

  a. Plot the data on a scatterplot.
  b. Calculate the values of r and r^2.
  c. Is there sufficient evidence to indicate that x and y are linearly correlated? Test at the $\alpha = .10$ level of significance.

Applying the Concepts—Basic

11.117 Arsenic in soil. In Denver, Colorado, environmentalists have discovered a link between high arsenic levels in soil and a crabgrass killer used in the 1950s and 1960s (*Environmental Science & Technology*, Sept. 1, 2000). The recent discovery was based, in part, on the accompanying scatterplots. The graphs plot the level of the metals cadmium and arsenic, respectively, against the distance from a former smelter plant for samples of soil taken from Denver residential properties.
  a. Normally, the metal level in soil decreases as distance from the source (e.g., a smelter plant) increases.

Propose a straight-line model relating metal level y to distance x from the plant. On the basis of the theory, would you expect the slope of the line to be positive or negative?
  b. Examine the scatterplot for cadmium. Does the plot support the theory you set forth in part **a**?
  c. Examine the scatterplot for arsenic. Does the plot support the theory of part **a**? (*Note:* This finding led investigators to discover the link between high arsenic levels and the use of the crabgrass killer.)

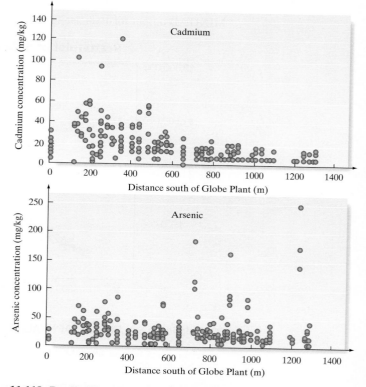

11.118 Predicting sale prices of homes. Real-estate investors, home buyers, and homeowners often use the appraised value of property as a basis for predicting the sale of that property. Data on sale prices and total appraised value of 78 residential properties sold recently in an upscale Tampa, Florida, neighborhood named Hunter's Green are saved in the **HUNGREEN** file. Selected observations are listed in the accompanying table.

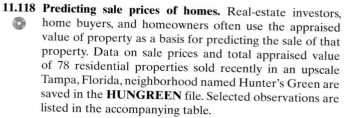

Property	Sale Price	Appraised Value
1	$489,900	$418,601
2	1,825,000	1,577,919
3	890,000	687,836
4	250,00	191,620
5	1,275,000	1,063,901
⋮	⋮	⋮
74	325,000	292,702
75	516,000	407,449
76	309,300	272,275
77	370,000	347,320
78	580,000	511,359

Based on data from Hillsborough Country (Florida) Property Appraiser's Officer.

a. Propose a straight-line model to relate the appraised property value (x) to the sale price (y) for residential properties in this neighborhood.

b. A MINITAB scatterplot of the data with the least squared line is shown at the top of the printout below. Does it appear that a straight-line model will be an appropriate fit to the data?

c. A MINITAB simple linear regression printout is also shown at the bottom of the printout below. Find the equation of the least squared line. Interpret the estimated slope and y-intercept in the words of the problem.

d. Locate the test statistic and p-value for testing $H_0: \beta_1 = 0$ against $H_a: \beta_1 > 0$. Is there sufficient evidence (at $\alpha = .01$) of a positive linear relationship between apprised property value (x) and sale price (y)?

e. Locate and interpret practically the values of r and r^2 on the printout.

f. Locate and interpret practically the 95% prediction interval for sale price (y) on the printout.

11.119 Baseball batting averages versus wins. Is the number of games won by a major league baseball team in a season related to the team's batting average? Consider data from

MINITAB output for Exercise 11.118

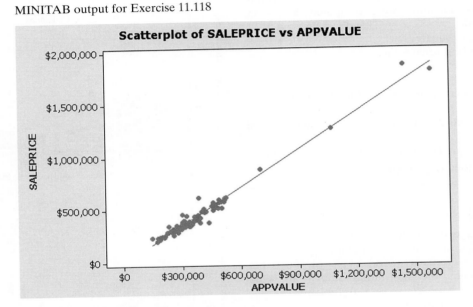

Correlations: SALEPRICE, APPVALUE

```
Pearson correlation of SALEPRICE and APPVALUE = 0.987
P-Value = 0.000
```

Regression Analysis: SALEPRICE versus APPVALUE

```
The regression equation is
SALEPRICE = 184 + 1.20 APPVALUE


Predictor     Coef  SE Coef      T      P
Constant       184     9834   0.02  0.985
APPVALUE   1.19956  0.02234  53.70  0.000


S = 44859.7   R-Sq = 97.4%   R-Sq(adj) = 97.4%


Analysis of Variance

Source          DF          SS           MS        F      P
Regression       1  5.80234E+12  5.80234E+12  2883.31  0.000
Residual Error  76  1.52942E+11   2012390103
Total           77  5.95528E+12


Predicted Values for New Observations

New
Obs     Fit  SE Fit        95% CI              95% PI
  1  480007    5105  (469839, 490175)  (390085, 569930)


Values of Predictors for New Observations

New
Obs  APPVALUE
  1    400000
```

the *Baseball Almanac* on the number of games won and the batting averages for the 14 teams in the American League for the 2010 Major League Baseball season. The data are listed in the next table and saved in the **ALWINS** file.

Team	Games Won	Batting Avg. (average number of hits per 1,000 at bats)
New York	95	.267
Toronto	85	.248
Baltimore	66	.259
Boston	89	.268
Tampa Bay	96	.247
Cleveland	69	.248
Detroit	81	.268
Chicago	88	.268
Kansas City	67	.274
Minnesota	94	.273
Los Angeles	80	.248
Texas	90	.276
Seattle	61	.236
Oakland	81	.256

Based on data from *Baseball Almanac*, 2010; www.mlb.com.

a. If you were to model the relationship between the mean (or expected) number of games won by a major league team and the team's batting average x, using a straight line, would you expect the slope of the line to be positive or negative? Explain.

b. Construct a scatterplot of the data. Does the pattern revealed by the scatterplot agree with your answer to part **a**?

c. An SAS printout of the simple linear regression is shown below. Find the estimates of the β's on the printout and write the equation of the least squares line.

d. Graph the least squares line on your scatterplot. Does your least squares line seem to fit the points on your scatterplot?

e. Interpret the estimates of β_0 and β_1 in the words of the problem.

f. Conduct a test (at $\alpha = .05$) to determine whether the mean (or expected) number of games won by a major

league baseball team is positively linearly related to the team's batting average.

g. Find the coefficient of determination, r^2, and interpret its value.

h. Do you recommend using the model to predict the number of games won by a team during the 2010 season?

11.120 College protests of labor exploitation. Refer to the *Journal of World-Systems Research* (Winter 2004) study of student "sit-ins" for a "sweat-free campus" at universities, presented in Exercise 2.153 (p. 90). Recall that the **SITIN** file contains data on the duration (in days) of each sit-in, as well as the number of student arrests. The data for 5 sit-ins in which there was at least one arrest are shown in the table. Let y = number of arrests and x = duration.

Sit-In	University	Duration (days)	Number of Arrests
12	Wisconsin	4	54
14	SUNY Albany	1	11
15	Oregon	3	14
17	Iowa	4	16
18	Kentucky	1	12

Based on Ross, R. J. S. "From antisweatshop to global justice to antiwar: How the new new left is the same and different from the old new left." *Journal of Word-Systems Research*, Vol. X, No. 1, Winter 2004 (Tables 1 and 3).

a. Give the equation of a straight-line model relating y to x.

b. SPSS was used to fit the model to the data for the 5 sitins. The printout is shown on page 604. Give the least squares prediction equation.

c. Interpret the estimates of β_0 and β_1 in the context of the problem.

d. Find and interpret the value of s on the printout.

e. Find and interpret the value of r^2 on the printout.

f. Conduct a test to determine whether number of arrests is positively linearly related to duration. (Use $\alpha = .10$.)

11.121 Feeding habits of fish. Refer to the *Brain and Behavior Evolution* (Apr. 2000) study of the feeding behavior of

SAS output for Exercise 11.119

```
                    Dependent Variable: WINS

                Number of Observations Read        14
                Number of Observations Used        14


                         Analysis of Variance

                               Sum of         Mean
     Source          DF       Squares        Square     F Value    Pr > F

     Model            1     270.31115     270.31115       2.18     0.1660
     Error           12    1491.11742     124.25979
     Corrected Total 13    1761.42857


          Root MSE           11.14719      R-Square     0.1535
          Dependent Mean     81.57143      Adj R-Sq     0.0829
          Coeff Var          13.66555


                         Parameter Estimates

                        Parameter      Standard
     Variable    DF      Estimate         Error     t Value    Pr > |t|

     Intercept    1     -12.62556      63.93554       -0.20      0.8468
     BATAVG       1       0.36269       0.24591        1.47      0.1660
```

SPSS output for Exercise 11.120

Model Summary

Model	R	R Square	Adjusted R Square	Std. Error of the Estimate
1	.601[a]	.361	.148	16.913

a. Predictors: (Constant), DURATION

ANOVA[b]

Model		Sum of Squares	df	Mean Square	F	Sig.
1	Regression	485.026	1	485.026	1.696	.284[a]
	Residual	858.174	3	286.058		
	Total	1343.200	4			

a. Predictors: (Constant), DURATION

b. Dependent Variable: ARRESTS

Coefficients[a]

Model		Unstandardized Coefficients		Standardized Coefficients	t	Sig.
		B	Std. Error	Beta		
1	(Constant)	2.522	16.352		.154	.887
	DURATION	7.261	5.576	.601	1.302	.284

a. Dependent Variable: ARRESTS

black-bream fish, presented in Exercise 2.150 (p. 89). Recall that the zoologists recorded the number of aggressive strikes of two black-bream fish feeding at the bottom of an aquarium in the 10-minute period following the addition of food. The table listing the weekly number of strikes and the age of the fish (in days) is reproduced below. These data are saved in the **BLACKBREAM** file.

Week	Number of Strikes	Age of Fish (days)
1	85	120
2	63	136
3	34	150
4	39	155
5	58	162
6	35	169
7	57	178
8	12	184
9	15	190

Based on Shand, J., et al. "Variability in the location of the retinal ganglion cell area centralis is correlated with ontogenetic changes in feeding behavior in the Blackbream, Acanthopagrus 'butcher'." *Brain and Behavior*, Vol. 55, No. 4, Apr. 2000 (Figure H).

a. Write the equation of a straight-line model relating number of strikes (y) to age of fish (x).

b. Fit the model to the data by the method of least squares and give the least squares prediction equation.

c. Give a practical interpretation of the value of $\hat{\beta}_0$ if possible.

d. Give a practical interpretation of the value of $\hat{\beta}_1$ if possible.

e. Test $H_0: \beta_1 = 0$ versus $H_a: \beta_1 < 0$, using $\alpha = .10$. Interpret the result.

11.122 English as a second language reading ability. Refer to the *Bilingual Research Journal* (Summer 2006) study of the relationship of Spanish (first-language) grammatical

knowledge to English (second-language) reading, presented in Exercise 11.54 (p. 575). Recall that each in a sample of $n = 55$ native Spanish-speaking adults took four standardized exams: Spanish grammar (SG), Spanish reading, (SR), English grammar (EG), and English reading (ESLR). Simple linear regressions were used to model the ESLR score (y) as a function of each of the other exam scores (x). The coefficient of determination, r^2, for each model is listed in the accompanying table. Give a practical interpretation of each of these values.

Independent Variable (x)	r^2
SG score	.002
SR score	.099
EG score	.078

11.123 Removing metal from water. In the *Electronic Journal of Biotechnology* (Apr. 15, 2004), Egyptian scientists studied a new method for removing heavy metals from water. Metal solutions were prepared in glass vessels, and then biosorption was used to remove the metal ions. Two variables were measured for each test vessel: y = metal uptake (milligrams of metal per gram of biosorbent) and x = final concentration of metal in the solution (milligrams per liter).

a. Write a simple linear regression model relating y to x.]

b. For one metal, a simple linear regression analysis yielded $r^2 = .92$. Interpret this result.

Applying the Concepts—Intermediate

11.124 New method of estimating rainfall. Accurate measurements of rainfall are critical for many hydrological and meteorological projects. Two standard methods of monitoring rainfall use rain gauges and weather radar. Both, however, can be contaminated by human and environmental interference. In the *Journal of Data Science* (Apr. 2004), researchers employed artificial neural networks (i.e., computer-based mathematical models) to estimate rainfall at a meteorological station in Montreal. Rainfall estimates were made every 5 minutes over a 70-minute period by each of the three methods. The data (in millimeters) are listed in the table and saved in the **RAINFALL** file.

Time	Radar	Rain Gauge	Neural Network
8:00 a.m.	3.6	0	1.8
8:05	2.0	1.2	1.8
8:10	1.1	1.2	1.4
8:15	1.3	1.3	1.9
8:20	1.8	1.4	1.7
8:25	2.1	1.4	1.5
8:30	3.2	2.0	2.1
8:35	2.7	2.1	1.0
8:40	2.5	2.5	2.6
8:45	3.5	2.9	2.6
8:50	3.9	4.0	4.0
8:55	3.5	4.9	3.4
9:00 a.m.	6.5	6.2	6.2
9:05	7.3	6.6	7.5
9:10	6.4	7.8	7.2

Based on Hessami, M. et al. "Selection of an artificial neural network model for the post-calibration of weather radar rainfall estimation." *Journal of Data Science*, Vol. 2, No. 2, Apr. 2004. (Adapted from Figures 2 and 4.)

a. Propose a straight-line model relating rain gauge amount (y) to weather radar rain estimate (x).

b. Use the method of least squares to fit the model to the data in the **RAINFALL** file.

c. Graph the least squares line on a scatterplot of the data. Is there visual evidence of a relationship between the two variables? Is the relationship positive or negative?

d. Interpret the estimates of the y-intercept and slope in the words of the problem.

e. Find and interpret the value of s for this regression.

f. Test whether y is linearly related to x. Use $\alpha = .01$.

g. Construct a 99% confidence interval for β_1. Interpret the result practically.

h. Now consider a model relating rain gauge amount (y) to the artificial neural network rain estimate (x). Repeat parts **a–g** for this model.

11.125 Are geography journals worth their cost? Refer to the *Geoforum* (Vol. 37, 2006) study of whether the price of a geography journal is correlated with quality, presented in Exercise 2.154 (p. 90). Several quantitative variables were recorded for each in a sample of 28 geography journals: cost of a one-year subscription (dollars); journal impact factor (JIF), the average number of times articles from the journal have been cited; number of citations for the journals over the past five years; and relative price index (RPI). The data for the 28 journals are saved in the **GEOJRNL** file. Selected observations are listed in the next table.

a. Fit a straight-line model relating cost (y) to JIF (x). Give a practical interpretation of the estimated slope of the line.

b. Within how many dollars can you expect to predict cost?

c. Find and interpret a 95% confidence interval for the slope.

Journal	Cost ($)	JIF	Citations	RPI
J. Econ. Geogr	468	3.139	207	1.16
Prog. Hum. Geog	624	2.943	544	0.77
T. I. Brit. Geogr	499	2.388	249	1.11
Econ. Geogr.	90	2.325	173	0.30
A. A. A. Geogr.	698	2.115	377	0.93
⋮	⋮	⋮	⋮	⋮
Geogr.Anal.	213	0.902	106	0.88
Geogr.J.	223	0.857	81	0.94
Appl.Geogr	646	0.853	74	3.38

Based on Blomley, N. "Is this journal worth US$1118?" *Geoforum*, Vol. 37, No. 6, p. 118.

d. Repeat parts a–c for a line relating cost (y) to number of citations (x).

e. Repeat parts a–c for a line relating cost (y) to RPI (x).

11.126 Organic chemistry experiment. Chemists at Kyushu University (Japan) examined the linear relationship between the maximum absorption rate y (in nanomoles) and the Hammett substituent constant x for metacyclophane compounds (*Journal of Organic Chemistry*, July 1995). The data for variants of two compounds are given in the accompanying table and saved in the **ORGCHEM** file. The variants of compound 1 are labeled 1a, 1b, 1d, 1e, 1f, 1g, and 1h; the variants of compound 2 are 2a, 2b, 2c, and 2d.

Compound	Maximum Absorption y	Hammett Constant x
1a	298	0.00
1b	346	.75
1d	303	.06
1e	314	−.26
1f	302	.18
1g	332	.42
1h	302	−.19
2a	343	.52
2b	367	1.01
2c	325	.37
2d	331	.53

Source: Tsuge, A., et al. "Preparation and spectral properties of disubstituted [2-2] metacyclophanes." *Journal of Organic Chemistry*, Vol. 60, No. 15, July 1995, pp. 4390–4391 (Table 1 and Figure 1). Reprinted with permission from the American Chemical Society.

a. Plot the data in a scatterplot. Use two different plotting symbols for the two compounds. What do you observe?

b. Using only the data for compound 1, fit the model $E(y) = \beta_0 + \beta_1 x$.

c. Assess the adequacy of the model you fit in part **b**. Use $\alpha = .01$.

d. Repeat parts **b** and **c**, using only the data for compound 2.

11.127 Mortality of predatory birds. Two species of predatory birds—collard flycatchers and tits—compete for nest holes during breeding season on the island of Gotland, Sweden. Frequently, dead flycatchers are found in nest boxes occupied by tits. A field study examined whether the risk of mortality to flycatchers is related to the degree of competition between the two bird species for nest sites (*The Condor*, May 1995). The table (p. 606) gives data on the number y of flycatchers killed at each of 14 discrete locations (plots) on the island, as well as on the nest box tit occupancy x (i.e., the percentage of nest boxes occupied by tits) at each plot. These data are saved in the **CONDOR2** file. Consider the simple linear regression model $E(y) = \beta_0 + \beta_1 x$.

Plot	Number of Flycatchers Killed y	Nest Box Tit Occupancy x (%)
1	0	24
2	0	33
3	0	34
4	0	43
5	0	50
6	1	35
7	1	35
8	1	38
9	1	40
10	2	31
11	2	43
12	3	55
13	4	57
14	5	64

Based on Merila, J., and Wiggins, D. A. "Interspecific competition for nest holes causes adult mortality in the collard flycatcher." *The Condor*, Vol. 97, No. 2, May 1995, p. 449 (Figure 2), Cooper Ornithological Society.

a. Plot the data in a scatterplot. Does the frequency of flycatcher casualties per plot appear to increase linearly with increasing proportion of nest boxes occupied by tits?

b. Use the method of least squares to find the estimates of β_0 and β_1. Interpret their values.

c. Test the utility of the model, using $\alpha = .05$.

d. Find r and r^2 and interpret their values.

e. Find s and interpret the result.

f. Do you recommend using the model to predict the number of flycatchers killed? Explain.

11.128 Winning marathon times. In *Chance* (Winter 2000), statistician Howard Wainer and two students compared men's and women's winning times in the Boston Marathon. One of the graphs used to illustrate gender differences is reproduced below. The scatterplot graphs the winning times (in minutes) against the year in which the race was run. Men's times are represented by purple dots and women's times by red dots.

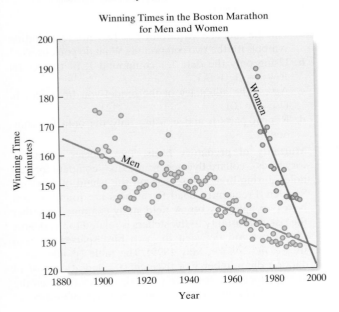

Winning Times in the Boston Marathon for Men and Women

a. Consider only the winning times for men. Is there evidence of a linear trend? If so, propose a straight-line model for predicting winning time (y) based on year (x). Would you expect the slope of this line to be positive or negative?

b. Repeat part **b** for women's times.

c. Which slope, the men's or the women's, will be greater in absolute value?

d. Would you recommend using the straight-line models to predict the winning time in the 2020 Boston Marathon? Why or why not?

e. Which model, the men's or the women's, is likely to have the smallest estimate of σ?

11.129 Quantum tunneling. At temperatures approaching absolute zero ($-273°C$), helium exhibits traits that seem to defy many laws of Newtonian physics. An experiment has been conducted with helium in solid form at various temperatures near absolute zero. The solid helium is placed in a dilution refrigerator along with a solid impure substance, and the fraction (in weight) of the impurity passing through the solid helium is recorded. (This phenomenon of solids passing directly through solids is known as *quantum tunneling*.) The data are given in the next table and saved in the **HELIUM** file.

Temperature x (°C)	Proportion of Impurity
−262.0	.315
−265.0	.202
−256.0	.204
−267.0	.620
−270.0	.715
−272.0	.935
−272.4	.957
−272.7	.906
−272.8	.985
−272.9	.987

a. Find the least squares estimates of the intercept and slope. Interpret them.

b. Use a 95% confidence interval to estimate the slope β_1. Interpret the interval in terms of this application. Does the interval support the hypothesis that temperature contributes information about the proportion of impurity passing through helium?

c. Interpret the coefficient of determination for this model.

d. Find a 95% prediction interval for the percentage of impurity passing through solid helium at $-273°C$. Interpret the result.

e. Note that the value of x in part **d** is outside the experimental region. Why might this lead to an unreliable prediction?

11.130 Dance/movement therapy. In cotherapy, two or more therapists lead a group. An article in the *American Journal of Dance Therapy* (Spring/Summer 1995) examined the use of cotherapy in dance/movement therapy. Two of several variables measured on each of a sample of 136 professional dance/movement therapists were years x of formal training and reported success rate y (measured as a percentage) of coleading dance/movement therapy groups.

a. Propose a linear model relating y to x.

b. The researcher hypothesized that dance/movement therapists with more years in formal dance training will report higher perceived success rates in cotherapy relationships. State the hypothesis in terms of the parameter of the model you proposed in part **a**.

c. The correlation coefficient for the sample data was reported as $r = -.26$. Interpret this result.

d. Does the value of r in part **c** support the hypothesis in part **b**? Test, using $\alpha = .05$.

11.131 Conversing with the hearing impaired. A study was conducted to investigate how people with a hearing impairment communicate with their conversational partners. (*Journal of the Academy of Rehabilitative Audiology*, Vol. 27, 1994). Each of 13 hearing-impaired subjects, all fitted with a cochlear implant, participated in a structured communication interaction with a familiar conversational partner (a family member) and with an unfamiliar conversational partner (who was instructed not to take the initiative to repair breakdowns in communication). The total number of words used by the subject in each of the two conversations is given in the accompanying table and saved in the **HEARAID** file.

Subject	Words with Familiar Partner x	Words with Unfamiliar Partner y
1	65	47
2	160	78
3	55	90
4	83	75
5	0	6
6	140	101
7	49	40
8	164	215
9	62	29
10	56	75
11	207	121
12	207	139
13	93	83

Based on Tye-Murray, N., et al. "Communication breakdowns: Partner contingencies and partner reactions." *Journal of the Academy of Rehabilitative Audiology*, Vol. 27, 1994, pp. 116–117 (Tables 6, 7).

a. Plot the data in a scatterplot. Is there visual evidence of a linear relationship between x and y? If so, is it positive or negative?

b. Propose a straight-line model relating y to x.

c. Use the method of least squares to find the estimates of β_0 and β_1.

d. Interpret the values of $\hat{\beta}_0$ and $\hat{\beta}_1$.

11.132 Sports participation survey. The Sasakawa Sports Foundation conducted a national survey to assess the physical activity patterns of Japanese adults. The next table lists the frequency (average number of days in the past year) and the amount of time (average number of minutes per single activity) Japanese adults spent participating in a sample of 11 sports activities. The data are saved in the **JAPANSPORTS** file.

a. Write the equation of a straight-line model relating duration (y) to frequency (x).

b. Find the least squares prediction equation.

c. Is there evidence of a linear relationship between y and x? Test, using $\alpha = .05$.

d. Use the least squares line to predict the amount of time Japanese adults participate in a sport that they play 25 times a year. Form a 95% confidence interval around the prediction and interpret the result.

Activity	Frequency x (days/years)	Amount of Time y (minutes)
Jogging	135	43
Cycling	68	99
Aerobics	44	61
Swimming	39	60
Volleyball	30	80
Tennis	21	100
Softball	16	91
Baseball	19	127
Skating	7	115
Skiing	10	249
Golf	5	262

Based on J. Bennett, ed. *Statistics in Sport*. London: Arnold, 1998 (adapted from Figure 11.6).

Applying the Concepts—Advanced

11.133 Regression through the origin. Sometimes it is known from theoretical considerations that the straight-line relationship between two variables x and y passes through the origin of the xy-plane. Consider the relationship between the total weight y of a shipment of 50-pound bags of flour and the number x of bags in the shipment. Since a shipment containing $x = 0$ bags (i.e., no shipment at all) has a total weight of $y = 0$, a straight-line model of the relationship between x and y should pass through the point $x = 0, y = 0$. In such a case, you could assume that $\beta_0 = 0$ and characterize the relationship between x and y with the following model:

$$y = \beta_1 x + \varepsilon$$

The least squares estimate of β_1 for this model is

$$\hat{\beta}_1 = \frac{\Sigma x_i y_i}{\Sigma x_i^2}$$

From the records of past flour shipments, 15 shipments were randomly chosen and the data shown in the following table were recorded. These data are saved in the **FLOUR** file.

Weight of Shipment	Number of 50-Pound Bags in Shipment
5,050	100
10,249	205
20,000	450
7,420	150
24,685	500
10,206	200
7,325	150
4,958	100
7,162	150
24,000	500
4,900	100
14,501	300
28,000	600
17,002	400
16,100	400

a. Find the least squares line for the given data under the assumption that $\beta_0 = 0$. Plot the least squares line on a scatterplot of the data.

b. Find the least squares line for the given data, using the model

$$y = \beta_0 + \beta_1 x + \varepsilon$$

(i.e., do not restrict β_0 to equal 0). Plot this line on the same scatterplot you constructed in part a.

c. Refer to part b. Why might $\hat{\beta}_0$ be different from 0 even though the true value of β_0 is known to be 0?

d. The estimated standard error of $\hat{\beta}_0$ is equal to

$$s\sqrt{\frac{1}{n} + \frac{\bar{x}^2}{SS_{xx}}}$$

Use the t-statistic

$$t = \frac{\hat{\beta}_0 - 0}{s\sqrt{(1/n) + (\bar{x}^2/SS_{xx})}}$$

to test the null hypothesis $H_0: \beta_0 = 0$ against the alternative $H_a: \beta_0 \neq 0$. Take $\alpha = .10$. Should you include β_0 in your model?

11.134 Long-jump "takeoff error." The long jump is a track-and-field event in which a competitor attempts to jump a maximum distance into a sandpit after a running start. At the edge of the pit is a takeoff board. Jumpers usually try to plant their toes at the front edge of this board to maximize their jumping distance. The absolute distance between the front edge of the takeoff board and the spot where the toe actually lands on the board prior to jumping is called "takeoff error." Is takeoff error in the long jump linearly related to best jumping distance? To answer this question, kinesiology researchers videotaped the performances of 18 novice long jumpers at a high school track meet. (*Journal of Applied Biomechanics*, May 1995). The average takeoff

Jumper	Best Jumping Distance y (meters)	Average Takeoff Error x (meters)
1	5.30	.09
2	5.55	.17
3	5.47	.19
4	5.45	.24
5	5.07	.16
6	5.32	.22
7	6.15	.09
8	4.70	.12
9	5.22	.09
10	5.77	.09
11	5.12	.13
12	5.77	.16
13	6.22	.03
14	5.82	.50
15	5.15	.13
16	4.92	.04
17	5.20	.07
18	5.42	.04

Based on Berg, W. P., and Greer, N. L. "A kinematic profile of the approach run of novice long jumpers." *Journal of Applied Biomechanics*, Vol. 11, No. 2, May 1995, p. 147 (Table 1).

error x and the best jumping distance y (out of three jumps) for each jumper are recorded in the accompanying table and saved in the **LONGJUMP** file. If a jumper can reduce his or her average takeoff error by .1 meter, how much would you estimate the jumper's best jumping distance to change? On the basis of your answer, comment on the usefulness of the model for predicting best jumping distance.

Critical Thinking Challenges

11.135 Study of fertility rates. The fertility rate of a country is defined as the number of children a woman citizen bears, on average, in her lifetime. *Scientific American* (Dec. 1993) reported on the declining fertility rate in developing countries. The researchers found that family planning can have a great effect on fertility rate. The accompanying table gives the fertility rate y and contraceptive prevalence x (measured as the percentage of married women who use contraception) for each of 27 developing countries. These data are saved in the **FERTRATE** file.

a. According to the researchers, "The data reveal that differences in contraceptive prevalence explain about 90% of the variation in fertility rates." Do you concur?

b. The researchers also concluded that "if contraceptive use increases by 18 percent, women bear, on average, one fewer child." Is this statement supported by the data? Explain.

Country	Contraceptive Prevalence x	Fertility Rate y
Mauritius	76	2.2
Thailand	69	2.3
Colombia	66	2.9
Costa Rica	71	3.5
Sri Lanka	63	2.7
Turkey	62	3.4
Peru	60	3.5
Mexico	55	4.0
Jamaica	55	2.9
Indonesia	50	3.1
Tunisia	51	4.3
El Salvador	48	4.5
Morocco	42	4.0
Zimbabwe	46	5.4
Egypt	40	4.5
Bangladesh	40	5.5
Botswana	35	4.8
Jordan	35	5.5
Kenya	28	6.5
Guatemala	24	5.5
Cameroon	16	5.8
Ghana	14	6.0
Pakistan	13	5.0
Senegal	13	6.5
Sudan	10	4.8
Yemen	9	7.0
Nigeria	7	5.7

Based on Robey, B., et al. "The fertility decline in developing countries." *Scientific American*, Dec. 1993, p. 62. [*Note:* The data values are estimated from a scatterplot.]

11.136 Spall damage in bricks. A recent civil suit revolved around a five-building brick apartment complex located in the Bronx, New York, which began to suffer *spalling* damage (i.e., a separation of some portion of the face of a brick from its body). The owner of the complex alleged that the bricks were manufactured defectively. The brick manufacturer countered that poor design and shoddy management led to the damage. To settle the suit, an estimate of the rate of damage per 1,000 bricks, called the spall rate, was required (*Chance*, Summer 1994). The owner estimated the spall rate by using several *scaffold-drop* surveys. (With this method, an engineer lowers a scaffold down at selected places on building walls and counts the number of visible spalls for every 1,000 bricks in the observation area.) The brick manufacturer conducted its own survey by dividing the walls of the complex into 83 wall segments and taking a photograph of each one. (The number of spalled bricks that could be made out from each photo was recorded, and the sum over all 83 wall segments was used as an estimate of total spall damage.) In this court case, the jury was faced with the following dilemma: On the one hand, the scaffold-drop survey provided the most accurate estimate of spall rates in a given wall segment. Unfortunately, however, the drop areas were not selected at random from the entire complex; rather, drops were made at areas with high spall concentrations, leading to an overestimate of the total damage. On the other hand, the photo survey was complete in that all 83 wall segments in the complex were checked for spall damage. But the spall rate estimated by the photos, at least in areas of high spall concentration, was biased low (spalling damage cannot always be seen from a photo), leading to an underestimate of the total damage.

The data in the table (saved in the **BRICKS** file) are the spall rates obtained from the two methods at 11 drop locations. Use the data, as did expert statisticians who testified in the case, to help the jury estimate the true spall rate at a given wall segment. Then explain how this information, coupled with the data (not given here) on all 83 wall segments, can provide a reasonable estimate of the total spall damage (i.e., total number of damaged bricks).

Drop Location	Drop Spall Rate (per 1,000 bricks)	Photo Spall Rate (per 1,000 bricks)
1	0	0
2	5.1	0
3	6.6	0
4	1.1	.8
5	1.8	1.0
6	3.9	1.0
7	11.5	1.9
8	22.1	7.7
9	39.3	14.9
10	39.9	13.9
11	43.0	11.8

Based on Fairley, W. B., et al. "Bricks, buildings, and the Bronx: Estimating masonry deterioration." *Chance*, Vol. 7. No. 3, Summer 1994, p. 36 (Figure 3). [*Note:* The data points are estimated from the points shown on a scatterplot.]

Activity Applying Simple Linear Regression to Your Favorite Data

Many dependent variables in all areas of research serve as the subjects of regression-modeling efforts. We list five such variables here:

1. Crime rate in various communities

2. Daily maximum temperature in your town

3. Grade point average of students who have completed one academic year at your college

4. Gross domestic product of the United States

5. Points scored by your favorite football team in a single game

Choose one of these dependent variables, or choose some other dependent variable, for which you want to construct a prediction model. There may be a large number of independent variables that should be included in a prediction equation for the dependent variable you choose. List three potentially important independent variables, x_1, x_2, and x_3, that you think might be (individually)

strongly related to your dependent variable. Next, obtain 10 data values, each of which consists of a measure of your dependent variable y and the corresponding values of x_1, x_2, and x_3.

a. Use the least squares formulas given in this chapter to fit three straight-line models—one for each independent variable—for predicting y.

b. Interpret the sign of the estimated slope coefficient $\hat{\beta}_1$ in each case, and test the utility of each model by testing $H_0: \beta_1 = 0$ against $H_a: \beta_1 \neq 0$. What assumptions must be satisfied to ensure the validity of these tests?

c. Calculate the coefficient of determination, r^2, for each model. Which of the independent variables predicts y best for the 10 sampled sets of data? Is this variable necessarily best in general (i.e., for the entire population)? Explain.

Be sure to keep the data and the results of your calculations, since you will need them for the Activity section in Chapter 12.

References

Chatterjee, S., and Price, B. *Regression Analysis by Example*, 2nd ed. New York: Wiley, 1991.

Draper, N., and Smith, H. *Applied Regression Analysis*, 3rd ed. New York: Wiley, 1987.

Graybill, F. *Theory and Application of the Linear Model*. North Scituate, MA: Duxbury, 1976.

Kleinbaum, D., and Kupper, L. *Applied Regression Analysis and Other Multivariable Methods*, 2nd ed. North Scituate, MA: Duxbury, 1997.

Kutner, M., Nachtsheim, C., Neter, J., and Li, W. *Applied Linear Statistical Models*, 5th ed. New York: McGraw-Hill/Irwin, 2006.

Mendenhall, W. *Introduction to Linear Models and the Design and Analysis of Experiments*. Belmont, CA.: Wadsworth, 1968.

Mendenhall, W., and Sincich, T. *A Second Course in Statistics: Regression Analysis*, 7th ed. Upper Saddle River, NJ: Prentice Hall, 2011.

Montgomery, D., Peck, E., and Vining, G. *Introduction to Linear Regression Analysis*, 3rd ed. New York: Wiley, 2001.

Mosteller, F., and Tukey, J. W. *Data Analysis and Regression: A Second Course in Statistics*. Reading, MA: Addison-Wesley, 1977.

Rousseeuw, P. J., and Leroy, A. M. *Robust Regression and Outlier Detection*. New York: Wiley, 1987.

Weisburg, S. *Applied Linear Regression*, 2nd ed. New York: Wiley, 1985.

USING TECHNOLOGY

MINITAB: Simple Linear Regression

Regression Analysis

Step 1 Access the MINITAB worksheet file that contains the two quantitative variables (dependent and independent variables).

Step 2 Click on the "Stat" button on the MINITAB menu bar, and then click on "Regression" and "Regression" again, as shown in Figure 11.M.1.

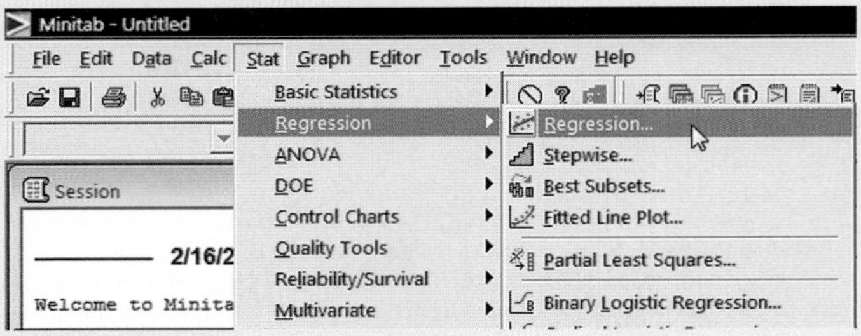

Figure 11.M.1
MINITAB menu options for regression

Step 3 On the resulting dialog box (see Figure 11.M.2), specify the dependent variable in the "Response" box and the independent variable in the "Predictors" box.

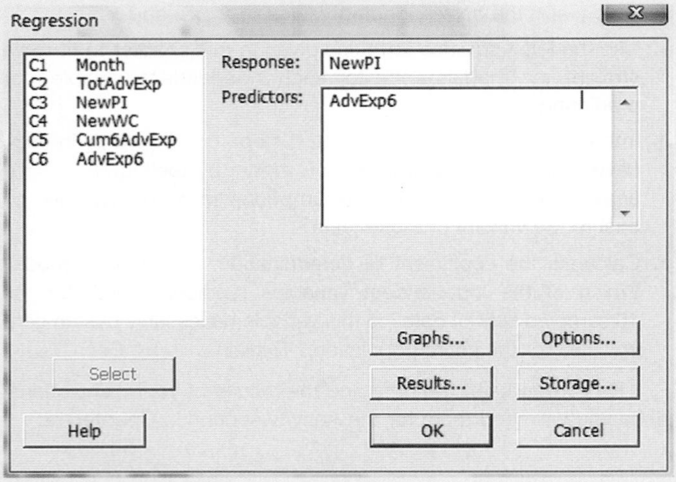

Figure 11.M.2
MINITAB regression dialog box

Step 4 To produce prediction intervals for y and confidence intervals for $E(y)$, click the "Options" button. The resulting dialog box is shown in Figure 11.M.3.

Step 5 Check "Confidence limits" and/or "Prediction limits", specify the "Confidence level," and enter the value of x in the "Prediction intervals for new observations" box.

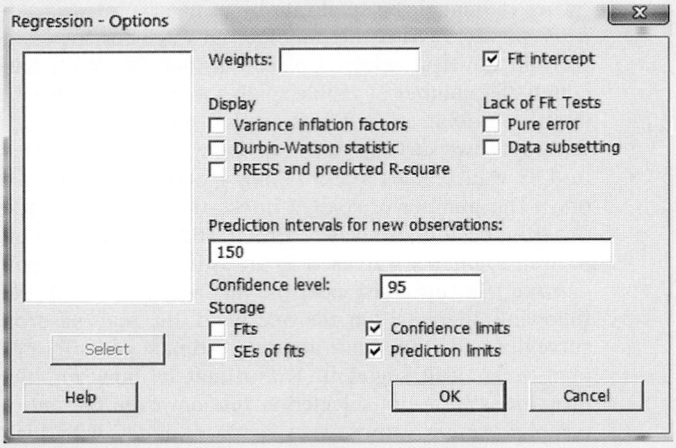

Figure 11.M.3
MINITAB regression options

Step 6 Click "OK" to return to the main Regression dialog box and then click "OK" again to produce the MINITAB simple linear regression printout.

Correlation Analysis

Step 1 Click on the "Stat" button on the MINITAB main menu bar, then click on "Basic Statistics," and then click on "Correlation," as shown in Figure 11.M.4.

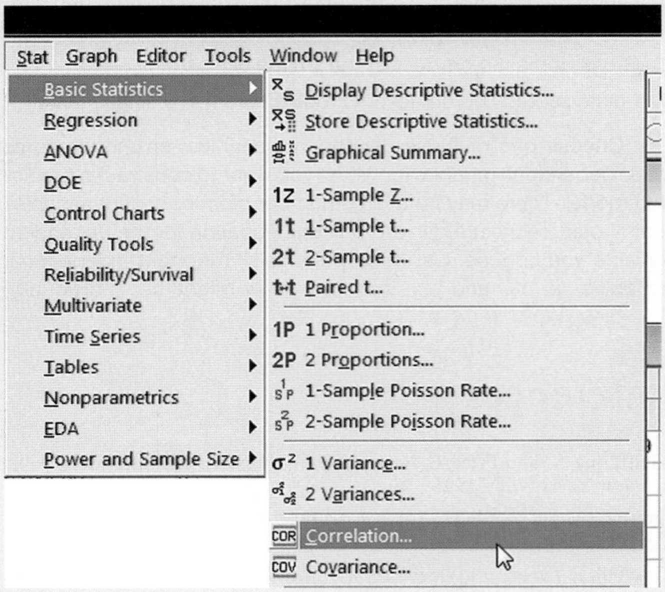

Figure 11.M.4
MINITAB menu options for correlation

Step 2 On the resulting dialog box (see Figure 11.M.5), enter the two variables of interest in the "Variables" box.

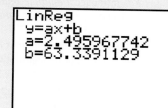

Figure 11.M.5
MINITAB correlation dialog box

Step 3 Click "OK" to obtain a printout of the correlation.

TI–83/TI–84 Plus Graphing Calculator: Simple Linear Regression

Finding the Least Squares Regression Equation

Step 1 *Enter the data*

- Press **STAT** and select **1:Edit**

Note: If a list already contains data, clear the old data.

- Use the up arrow to highlight the list name, "**L1**" or "**L2**"
- Press **CLEAR ENTER**
- Enter your *x*-data in **L1** and your *y*-data in **L2**

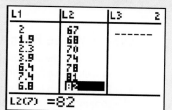

Step 2 *Find the equation*

- Press **STAT** and highlight **CALC**
- Press **4** for **LinReg(ax + b)**
- Press **ENTER**
- The screen will show the values for *a* and *b* in the equation $y = ax + b$

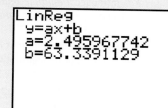

Finding *r* and r^2

Use this procedure if *r* and r^2 do not already appear on the LinReg screen from part I:

Step 1 *Turn the diagnostics feature on*

- Press **2nd 0** for **CATALOG**
- Press the **ALPHA** key and x^{-1} for **D**
- Press the down **ARROW** until **DiagnosticsOn** is highlighted
- Press **ENTER** twice

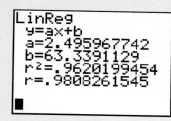

Step 2 *Find the regression equation as shown in part I above*
The values for *r* and r^2 will appear on the screen as well.

Graphing the Least Squares Line with the Scatterplot

Step 1 *Enter the data as shown in part I above*

Step 2 *Set up the data plot*

- Press **Y =** and **CLEAR** all functions from the **Y** register
- Press **2ndY =** for **STAT PLOT**
- Press **1** for **Plot1**
- Set the cursor so that **ON** is flashing and press **ENTER**
- For **Type**, use the **ARROW** and **ENTER** keys to highlight and select the scatterplot (first icon in the first row)
- For **Xlist**, choose the column containing the *x*-data
- For **Ylist**, choose the column containing the *y*-data

Step 3 *Find the regression equation and store the equation in Y1*

- Press **STAT** and highlight **CALC**
- Press **4** for **LinReg(ax + b)** (*Note*: Don't press **ENTER** here because you want to store the regression equation in Y1.)
- Press **VARS**
- Use the right arrow to highlight **Y-VARS**
- Press **ENTER** to select **1:Function**
- Press **ENTER** to select **1:Y1**
- Press **ENTER**

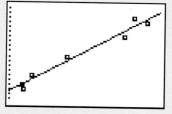

Step 4 *View the scatterplot and regression line*

- Press **ZOOM** and then press **9** to select **9:ZoomStat**

You should see the data graphed along with the regression line.

12 Multiple Regression and Model Building

CONTENTS

Where We've Been

- Introduced the straight-line model relating a dependent variable y to a single independent variable x
- Demonstrated how to estimate the parameters of the straight-line model by the method of least squares
- Showed how to statistically assess the adequacy of the model
- Showed how to use the model to estimate $E(y)$ and predict y for a given value of x

Where We're Going

- Introduce a *multiple-regression* model as a means of relating a dependent variable y to two or more independent variables. (12.1)
- Present several different multiple-regression models involving both quantitative and qualitative independent variables. (12.2, 12.5–12.8)
- Assess how well the multiple-regression model fits the sample data. (12.3)
- Demonstrate how to use the model for prediction (12.4)
- Present some model building techniques (12.9–12.10)
- Show how an analysis of a model's *residuals* can aid in detecting violations of the model's assumptions and in identifying modifications required by the model. (12.11)
- Alert the analyst to some regression pitfalls (12.12)

Statistics IN Action

Modeling Condominium Sales: What Factors Affect Auction Price?

This application involves an investigation of the factors that affect the sale price of oceanside condominium units. It represents an extension of an analysis of data collected by Herman Kelting. Although condo sale prices have increased dramatically since the time of the study, the relationship between these factors and sale prices remain about the same. Consequently, the data provide valuable insight into today's condominium sales market.

Sales data were obtained for a newly built oceanside condominium complex consisting of two adjacent, connected eight-floor buildings. The complex contains 209 units of equal size (approximately 500 square feet each). The locations of the buildings relative to the ocean, the swimming pool, the parking lot, etc., are shown in Figure SIA12.1.

Among the features of the complex that you should note are the following:

1. The units facing south, called *oceanview*, face the beach and ocean. In addition, units in building 1 have a good view of the pool. Units to the rear of the building, called *bayview*, face the parking lot and an area of land that, ultimately, borders a bay. The view from the upper floors of these units is primarily of wooded, sandy terrain. The bay is very distant and barely visible.

2. The only elevator in the complex is located at the east end of building 1, next to the office and the game room. People moving to or from the higher floor units in building 2 would probably use the elevator, then move through the passages to their units. Thus, units on the higher floors and at a greater distance from the elevator would be less convenient for their occupants, who would expend greater effort in moving baggage, groceries, etc., and would be

farther away from the game room, the office, and the swimming pool. These units also possess an advantage: Because traffic through the hallways in the area would be minimal, these units would be the most private.

3. Because lower-floor oceanside units open onto the beach, ocean, and pool, they are most suited to active people. They are within easy reach of the game room, and they are also easily reached from the parking area.

4. Checking Figure SIA12.1, you will see that the views in some of the units at the center of the complex—units ending in numbers 11 and 14—are partially blocked.

5. The condominium complex was completed at the time of the 1975 recession; sales were slow, and the developer was forced to sell about half of the units at auction approximately 18 months after the complex opened. Many unsold units were furnished by the developer and rented prior to the auction.

This condominium complex is particularly suited to our study. Because the single elevator is located at one end of the complex, it is the source of a remarkably high level of both inconvenience and privacy for the people occupying units on the top floors in building 2. Consequently, the data provide a good opportunity to investigate the relationship that might exist between sale price, height of the unit (floor number), distance of the unit from the elevator, and presence or absence of an ocean view. In addition, the presence or absence of furniture in each of the units permits an investigation of the

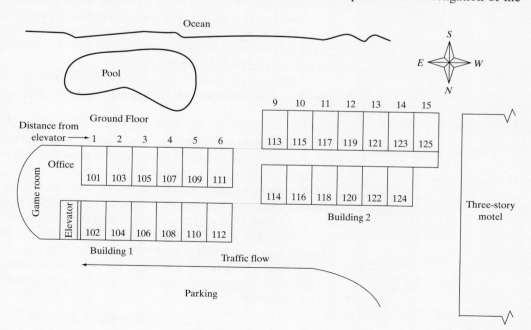

Figure SIA12.1

Layout of condominium complex

(*continued*)

Statistics IN Action
(continued)

effect of the availability of furniture on the sale price. Finally, the units sold at auction are completely specified by the buyer and hence are consumer oriented, in contrast to most other real estate units, which are, to a high degree, seller oriented and specified by the broker.

The **CONDO** file contains data on each of the 209 units sold—106 at public auction and 103 at the developer's fixed price. The variables measured for each condominium unit are listed in Table SIA12.1. We want to build a model for the auction price and, ultimately, to use the model to predict the sale prices of future units.

In several Statistics in Action Revisited sections, we show how to analyze the data by means of a multiple-regression analysis.

Statistics IN Action Revisited

- A First-Order Model for Condominium Sale Price (p. 635)
- Building a Model for Condominium Sale Price (p. 676)
- A Residual Analysis for the Condominium Sale Price Model (p. 700)

Table SIA12.1   Variables in the CONDO Data File

Variable Name	Type	Description
PRICE100	Quantitative	Sales price (hundreds of dollars)
FLOOR	Quantitative	Floor height $(1, 2, 3, \ldots, 8)$
DIST	Quantitative	Distance, in units, from the elevator $(1, 2, 3, \ldots, 15)$
VIEW	Qualitative	View $(1 =$ ocean view, $0 =$ non–ocean view$)$
END	Qualitative	Location of unit $(1 =$ end of complex, $0 =$ not an end unit$)$
FURNISH	Qualitative	Furniture status $(1 =$ furnished, $0 =$ nonfurnished$)$
AUCTION	Qualitative	Method of sale $(1 =$ public auction, $0 =$ fixed price$)$

Data Set: CONDO

12.1 Multiple–Regression Models

Most practical applications of regression analysis utilize models that are more complex than the simple straight-line model. For example, a realistic probabilistic model for reaction time would include more than just the amount of a particular drug in the bloodstream. Factors such as age, a measure of visual perception, and sex of the subject are a few of the many variables that might be related to reaction time. Thus, we would want to incorporate these and other potentially important independent variables into the model in order to make accurate predictions.

Probabilistic models that include more than one independent variable are called **multiple-regression models.** The general form of these models is

$$y = \beta_0 + \beta_1 x_1 + \beta_2 x_2 + \cdots + \beta_k x_k + \varepsilon$$

The dependent variable y is now written as a function of k independent variables $x_1, x_2, \ldots, x_k$. The random-error term is added to make the model probabilistic rather than deterministic. The value of the coefficient β_i determines the contribution of the independent variable x_i, and β_0 is the y-intercept. The coefficients $\beta_0, \beta_1, \ldots, \beta_k$ are usually unknown because they represent population parameters.

At first glance, it might appear that the regression model just described would not allow for anything other than straight-line relationships between y and the independent variables, but this is not true. Actually, $x_1, x_2, \ldots, x_k$ can be functions of variables, as long as the functions do not contain unknown parameters. For example, the reaction time y of a subject to a visual stimulus could be a function of the independent variables

$$x_1 = \text{Age of the subject}$$
$$x_2 = (\text{Age})^2 = x_1^2$$
$$x_3 = 1 \text{ if male subject, } 0 \text{ if female subject}$$

The x_2-term is called a **higher order term,** since it is the value of a quantitative variable (x_1) squared (i.e., raised to the second power). The x_3-term is a (qualitative) **coded variable** representing a quality (gender). The **multiple-regression model** is quite versatile and can be made to model many different types of response variables.

The Multiple-Regression Model*

$$y = \beta_0 + \beta_1 x_1 + \beta_2 x_2 + \cdots + \beta_k x_k + \varepsilon$$

where

y is the dependent variable

$x_1, x_2, \ldots, x_k$ are the independent variables

$E(y) = \beta_0 + \beta_1 x_1 + \beta_2 x_2 + \cdots + \beta_k x_k$ is the deterministic portion of the model

β_i determines the contribution of the independent variable x_i

Note: The symbols $x_1, x_2, \ldots, x_k$ may represent higher order terms for quantitative predictors or terms that represent qualitative predictors.

As shown in the following box, the steps used to develop the multiple-regression model are similar to those used for the simple regression model:

Analyzing a Multiple-Regression Model

Step 1 Hypothesize the deterministic component of the model. This component relates the mean $E(y)$ to the independent variables $x_1, x_2, \ldots, x_k$. Involved here is the choice of the independent variables to be included in the model (Sections 12.2, 12.5–12.10).

Step 2 Use the sample data to estimate the unknown parameters $\beta_0, \beta_1, \beta_2, \ldots, \beta_k$ in the model (Section 12.2).

Step 3 Specify the probability distribution of the random-error term ε, and estimate the standard deviation σ of this distribution (Section 12.3).

Step 4 Check that the assumptions about ε are satisfied, and make modifications to the model if necessary (Section 12.11).

Step 5 Statistically evaluate the usefulness of the model (Section 12.3).

Step 6 When you are satisfied that the model is useful, use it for prediction, estimation, and other purposes (Section 12.4).

The assumptions we make about the random error ε of the multiple-regression model are similar to those we make about the random error in a simple linear regression and are summarized as follows:

Assumptions about Random Error ε

For any given set of values of $x_1, x_2, \ldots, x_k$, the random error ε has a probability distribution with the following properties:

1. The mean is equal to 0.

2. The variance is equal to σ^2.

3. The probability distribution is a normal distribution.

4. Random errors are independent (in a probabilistic sense).

---

*Technically, this model is referred to as a multiple *linear* regression model, since the equation is a linear function of the β's.

Throughout this chapter, we introduce several different types of models that form the foundation of **model building** (or useful model construction). In the next several sections, we consider the most basic multiple-regression model, called the *first-order model*.

PART I: FIRST–ORDER MODELS WITH QUANTITATIVE INDEPENDENT VARIABLES

12.2 Estimating and Making Inferences about the β Parameters

A model that includes only terms denoting *quantitative* independent variables, called a **first-order model,** is described in the next box. Note that the first-order model does not include any higher order terms (such as x_1^2).

A First-Order Model in Five Quantitative Independent Variables*

$$E(y) = \beta_0 + \beta_1 x_1 + \beta_2 x_2 + \beta_3 x_3 + \beta_4 x_4 + \beta_5 x_5$$

where $x_1, x_2, \ldots, x_5$ are all quantitative variables that *are not* functions of other independent variables.

Note: β_i represents the slope of the line relating y to x_i when all the other x's are held fixed.

The method of fitting first-order models—and multiple-regression models in general—is identical to that of fitting the simple straight-line model: the *method of least squares.* That is, we choose the estimated model

$$\hat{y} = \hat{\beta}_0 + \hat{\beta}_1 x_1 + \cdots + \hat{\beta}_k x_k$$

that (1) has an average error of prediction of 0, i.e., $\Sigma(y - \hat{y}) = 0$, and (2) minimizes $\text{SSE} = \sum(y - \hat{y})^2$. As in the case of the simple linear model, the sample estimates $\hat{\beta}_0, \hat{\beta}_1, \ldots, \hat{\beta}_k$ are obtained as a solution of a set of simultaneous linear equations.[†]

The primary difference between fitting the simple- and multiple-regression models is computational difficulty. The $(k + 1)$ simultaneous linear equations that must be solved to find the $(k + 1)$ estimated coefficients $\hat{\beta}_0, \hat{\beta}_1, \ldots, \hat{\beta}_k$ are difficult (sometimes nearly impossible) to solve with a calculator. Consequently, we resort to the use of computers. Instead of presenting the tedious hand calculations required to fit the models, we present output from SAS, SPSS, and MINITAB.

BIOGRAPHY GEORGE U. YULE (1871–1951)

Yule Processes

Born on a small farm in Scotland, George Yule received an extensive childhood education. After graduating from University College (London), where he studied civil engineering, Yule spent a year employed in engineering workshops. However, he made a career change in 1893, accepting a teaching position back at University College under the guidance of statistician Karl Pearson (see p. 729). Inspired by Pearson's work, Yule produced a series of important articles on the statistics of regression and correlation. Yule is considered the first to have applied the method of least squares in regression analysis and he developed the theory of multiple regression. He eventually was appointed a lecturer in statistics at Cambridge University and later became the president of the prestigious Royal Statistical Society. Yule made many other contributions to the field, including the invention of time-series analysis and the development of "Yule" processes and the "Yule" distribution. ∎

---

*The terminology "first-order" is derived from the fact that each x in the model is raised to the first power.

[†]Students who are familiar with calculus should note that $\hat{\beta}_0, \hat{\beta}_1, \ldots, \hat{\beta}_k$ are the solutions of the set of equations $\partial \text{SSE}/\partial\hat{\beta}_0 = 0, \partial \text{SSE}/\partial\hat{\beta}_1 = 0, \ldots, \partial \text{SSE}/\partial\hat{\beta}_k = 0$. The solution is usually given in matrix form, but we do not present the details here. (See the references for details.)

Example 12.1

Fitting a First-Order Model—Price of an Antique Clock

Problem A collector of antique grandfather clocks sold at auction believes that the price received for the clocks depends on both the age of the clocks and the number of bidders at the auction. Thus, he hypothesizes the first-order model

$$y = \beta_0 + \beta_1 x_1 + \beta_2 x_2 + \varepsilon$$

where

$$y = \text{Auction price (dollars)}$$
$$x_1 = \text{Age of clock (years)}$$
$$x_2 = \text{Number of bidders}$$

A sample of 32 auction prices of grandfather clocks, along with their age and the number of bidders, is given in Table 12.1.

a. Use scatterplots to plot the sample data. Interpret the plots.

b. Use the method of least squares to estimate the unknown parameters β_0, β_1, and β_2 of the model.

c. Find the value of SSE that is minimized by the least squares method.

d. Estimate σ, the standard deviation of the model, and interpret the result.

Table 12.1 Auction Price Data

Age x_1	Number of Bidders x_2	Auction Price y	Age x_1	Number of Bidders x_2	Auction Price y
127	13	$1,235	170	14	$2,131
115	12	1,080	182	8	1,550
127	7	845	162	11	1,884
150	9	1,522	184	10	2,041
156	6	1,047	143	6	845
182	11	1,979	159	9	1,483
156	12	1,822	108	14	1,055
132	10	1,253	175	8	1,545
137	9	1,297	108	6	729
113	9	946	179	9	1,792
137	15	1,713	111	15	1,175
117	11	1,024	187	8	1,593
137	8	1,147	111	7	785
153	6	1,092	115	7	744
117	13	1,152	194	5	1,356
126	10	1,336	168	7	1,262

Data Set: GFCLOCKS

Solution

a. MINITAB side-by-side scatterplots for examining the bivariate relationships between y and x_1 and between y and x_2 are shown in Figure 12.1. Of the two variables, age (x_1) appears to have the strongest linear relationship with auction price (y).

b. The model hypothesized is fit to the data of Table 12.1 with MINITAB. A portion of the printout is reproduced in Figure 12.2. The least squares estimates of the β parameters (highlighted) are $\hat{\beta}_0 = -1,339$, $\hat{\beta}_1 = 12.74$, and $\hat{\beta}_2 = 85.95$. Therefore, the equation that minimizes SSE for this data set (i.e., the **least squares prediction equation**) is

$$\hat{y} = -1,339 + 12.74x_1 + 85.95x_2$$

c. The minimum value of the sum of the squared errors, also highlighted in Figure 12.2, is SSE = 516,727.

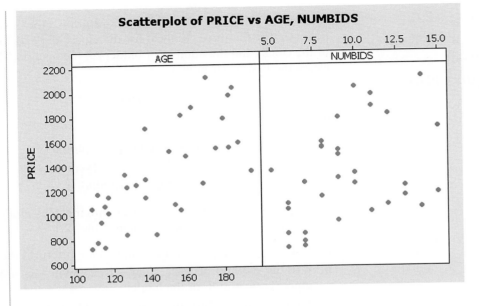

Figure 12.1
MINITAB side-by-side scatterplots for the data of Table 12.1

Figure 12.2
MINITAB analysis of the auction price model

Regression Analysis: PRICE versus AGE, NUMBIDS

```
The regression equation is
PRICE = - 1339 + 12.7 AGE + 86.0 NUMBIDS


Predictor     Coef   SE Coef       T       P
Constant    -1339.0    173.8   -7.70   0.000
AGE         12.7406   0.9047   14.08   0.000
NUMBIDS      85.953    8.729    9.85   0.000


S = 133.485    R-Sq = 89.2%    R-Sq(adj) = 88.5%


Analysis of Variance

Source           DF       SS       MS        F       P
Regression        2  4283063  2141531   120.19   0.000
Residual Error   29   516727    17818
Total            31  4799790
```

d. Recall that the estimator of σ^2 for the straight-line model is $s^2 = \text{SSE}/(n - 2)$, and note that the denominator is ($n -$ Number of estimated β parameters), which is ($n - 2$) in the straight-line model. Since we must estimate the three parameters β_0, β_1, and β_2, for the first-order model, the estimator of σ^2 is

$$s^2 = \frac{\text{SSE}}{n - 3} = \frac{\text{SSE}}{32 - 3} = \frac{516{,}727}{29} = 17{,}818$$

This value, often called the **mean square for error (MSE)** is also highlighted at the bottom of the MINITAB printout in Figure 12.2. The estimate of σ, then, is

$$s = \sqrt{17{,}818} = 133.5$$

which is highlighted in the middle of the printout in Figure 12.2. One useful interpretation of the estimated standard deviation s is that the interval $\pm 2s$ will provide a rough approximation to the accuracy with which the model will predict future values of y for given values of x. Thus, we expect the model to provide predictions of auction price to within about $\pm 2s = \pm 2(133.5) = \pm 267$ dollars.*

*The $\pm 2s$ approximation improves as the sample size is increased. We will provide a more precise methodology for the construction of prediction intervals in Section 12.4.

Look Back As with simple linear regression, we will use the estimator of σ^2 both to check the utility of the model (Section 12.3) and to provide a measure of the reliability of predictions and estimates when the model is used for those purposes (Section 12.4). Thus, you can see that the estimation of σ^2 plays an important part in the development of a regression model.

Now Work Exercise 12.6a–c

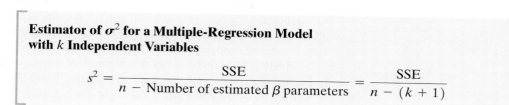

Estimator of σ^2 for a Multiple-Regression Model with k Independent Variables

$$s^2 = \frac{\text{SSE}}{n - \text{Number of estimated } \beta \text{ parameters}} = \frac{\text{SSE}}{n - (k + 1)}$$

After obtaining the least squares prediction equation, the analyst will usually want to make meaningful interpretations of the β estimates. Recall that in the straight-line model (Chapter 11)

$$y = \beta_0 + \beta_1 x + \varepsilon$$

β_0 represents the y-intercept of the line and β_1 represents the slope of the line. From our discussion in Chapter 11, β_1 has a practical interpretation: the mean change in y for every 1-unit increase in x. When the independent variables are quantitative, the β parameters in the first-order model specified in Example 12.1 have similar interpretations. The difference is that when we interpret the β that multiplies one of the variables (e.g., x_1), we must be certain to hold the values of the remaining independent variables (e.g., x_2, x_3) fixed.

To see this, suppose that the mean $E(y)$ of a response y is related to two quantitative independent variables x_1 and x_2 by the first-order model

$$E(y) = 1 + 2x_1 + x_2$$

In other words, $\beta_0 = 1$, $\beta_1 = 2$, and $\beta_2 = 1$.

Now, when $x_2 = 0$, the relationship between $E(y)$ and x_1 is given by

$$E(y) = 1 + 2x_1 + (0) = 1 + 2x_1$$

A MINITAB graph of this relationship (a straight line) is shown in Figure 12.3. Similar graphs of the relationship between $E(y)$ and x_1 for $x_2 = 1$, namely,

$$E(y) = 1 + 2x_1 + (1) = 2 + 2x_1$$

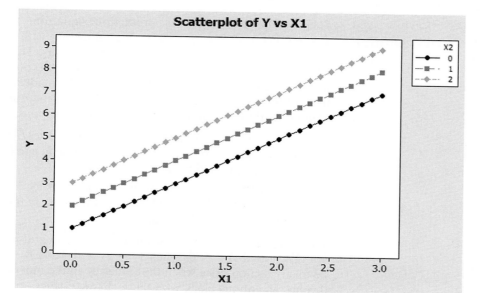

Figure 12.3

MINITAB graph of $E(y) = 1 + 2x_1 + x_2$ for $x_2 = 0, 1, 2$

and for $x_2 = 2$, that is,

$$E(y) = 1 + 2x_1 + (2) = 3 + 2x_1$$

also are shown in Figure 12.3. Note that the slopes of the three lines are all equal to $\beta_1 = 2$, the coefficient that multiplies x_1.

Figure 12.3 exhibits a characteristic of all first-order models: If you graph $E(y)$ versus any one variable—say, x_1—for fixed values of the other variables, the result will always be a *straight line* with slope equal to β_1. If you repeat the process for other values of the fixed independent variables, you will obtain a set of *parallel* straight lines. This indicates that the effect of the independent variable x_i on $E(y)$ is independent of all the other independent variables in the model, and this effect is measured by the slope β_i (see note in box on p. 615).

A MINITAB three-dimensional graph of the model $E(y) = 1 + 2x_1 + x_2$ is shown in Figure 12.4. Note that the graph is a plane. If you slice the plane at a particular value of x_2 (say, $x_2 = 0$), you obtain a straight line relating $E(y)$ to x_1 (e.g., $E(y) = 1 + 2x_1$). Similarly, if you slice the plane at a particular value of x_2 (say, $x_2 = 0$), you obtain a straight line relating $E(y)$ to x_1 (e.g., $E(y) = 1 + 2x_1$). Similarly, if you slice the plane at a particular value of x_1, you obtain a straight line relating $E(y)$ to x_2. Since it is more difficult to visualize three-dimensional and, in general, k-dimensional surfaces, we will graph all the models presented in this chapter in two dimensions. *The key to obtaining these graphs is to hold fixed all but one of the independent variables in the model.*

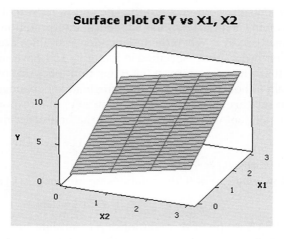

Figure 12.4
MINITAB 3-dimensional graph
of $E(y) = 1 + 2x_1 + x_2$

Example 12.2

Interpreting the β Estimates—Clock Auction Price Model

Problem Refer to the first-order model for auction price y considered in Example 12.1. Interpret the estimates of the β parameters in the model.

Solution The least squares prediction equation, as given in Example 12.1, is $\hat{y} = -1,339 + 12.74x_1 + 85.95x_2$. We know that with first-order models β_1 represents the slope of the line relating y to x_1 for fixed x_2. That is, β_1 measures the change in $E(y)$ for every one-unit increase in x_1 when the other independent variable in the model is held fixed. A similar statement can be made about β_2: β_2 measures the change in $E(y)$ for every one-unit increase in x_2 when the other x in the model is held fixed. Consequently, we obtain the following interpretations:

$\hat{\beta}_1 = 12.74$: We estimate the mean auction price $E(y)$ of an antique clock to increase \$12.74 for every $1 -$ year increase in age (x_1) when the number of bidders (x_2) is held fixed.

$\hat{\beta}_2 = 85.95$: We estimate the mean auction price $E(y)$ of an antique clock to increase \$85.95 for every $1 -$ bidder increase in the number of bidders (x_2) when age (x_1) is held fixed.

The value $\hat{\beta}_0 = -1,339$ does not have a meaningful interpretation in this example. To see this, note that $\hat{y} = \hat{\beta}_0$ when $x_1 = x_2 = 0$. Thus, $\hat{\beta}_0 = -1,339$ represents the estimated mean auction price when the values of all the independent variables are set equal to 0. Since an antique clock with these characteristics—an age of 0 years and 0 bidders on the clock—is not practical, the value of $\hat{\beta}_0$ has no meaningful interpretation. In general, $\hat{\beta}_0$ will not have a practical interpretation unless it makes sense to set the values of the x's simultaneously equal to 0.

Look Back In general, $\hat{\beta}_0$ will not have a practical interpretation unless it makes sense to set the values of the x's simultaneously equal to 0.

Now Work Exercise 12.17a–b

⚠️ **CAUTION** The interpretation of the β parameters in a multiple-regression model will depend on the terms specified in the model. The interpretations in Example 12.2 are for a first-order linear model only. In practice, you should be sure that a first-order model is the correct model for $E(y)$ before making β interpretations. [We discuss alternative models for $E(y)$ in Sections 12.5–12.8.] ▲

Inferences about the individual β parameters in a model are obtained with the use of either a confidence interval or a test of hypothesis, as outlined in the following boxes:*

A 100$(1 - \alpha)$% Confidence Interval for a β Parameter

$$\hat{\beta}_i \pm (t_{\alpha/2})s_{\hat{\beta}_1}$$

where $t_{\alpha/2}$ is based on $n - (k + 1)$ degrees of freedom and

$$n = \text{Number of observations}$$
$$k + 1 = \text{Number of } \beta \text{ parameters in the model}$$

Test of an Individual Parameter Coefficient in the Multiple-Regression Model

One-Tailed Test

$H_0: \beta_i = 0$

$H_a: \beta_i < 0$ [or $H_a: \beta_i > 0$]

Two-Tailed Test

$H_0: \beta_i = 0$

$H_a: \beta_i \neq 0$

$$\textit{Test statistic: } t = \frac{\hat{\beta}_i}{s_{\hat{\beta}_i}}$$

Rejection region: $t < -t_\alpha$

[or $t > t_\alpha$ when $H_a: \beta_i > 0$]

Rejection region: $|t| > t_{\alpha/2}$

where t_α and $t_{\alpha/2}$ are based on $n - (k + 1)$ degrees of freedom and

$$n = \text{Number of observations}$$
$$k + 1 = \text{Number of } \beta \text{ parameters in the model}$$

Conditions Required for Valid Inferences about Individual β Parameters

The four assumptions about the probability distribution for the random error ε (p. 615).

We illustrate these methods with another example.

*The formulas for computing $\hat{\beta}_i$ and its standard error are so complicated that the only reasonable way to present them is by using matrix algebra. We do not assume a prerequisite of matrix algebra for this text, and in any case, we think that the formulas can be omitted in an introductory course without serious loss. They are programmed into almost all statistical software packages with multiple-regression routines and are presented in some of the texts listed in the references.

Example 12.3
Inferences about the β Parameters— Auction Price Model

Problem Refer to Examples 12.1 and 12.2. The collector of antique grandfather clocks knows that the price (y) received for the clocks increases linearly with the age (x_1) of the clocks. Moreover, the collector hypothesizes that the auction price (y) of the clocks will increase linearly as the number of bidders (x_2) increases. Use the information on the MINITAB printout shown in Figure 12.2 (p. 618) to

a. Test the hypothesis that the mean auction price of a clock increases as the number of bidders increases when age is held constant (i.e., when $\beta_2 > 0$). (Use $\alpha = .05$.)

b. Find a 90% confidence interval for β_1 and interpret the result.

Solution

a. The hypotheses of interest concern the parameter β_2. Specifically,

$$H_0: \beta_2 = 0$$
$$H_a: \beta_2 > 0$$

The test statistic is a t-statistic formed by dividing the sample estimate $\hat{\beta}_2$ of the parameter β_2 by the estimated standard error of $\hat{\beta}_2$ (denoted $s_{\hat{\beta}_2}$). These estimates, $\hat{\beta}_2 = 85.953$ and $s_{\hat{\beta}_2} = 8.729$, as well as the calculated t-value,

$$\text{Test statistic: } t = \frac{\hat{\beta}_2}{s_{\hat{\beta}_2}} = \frac{85.953}{8.729} = 9.85$$

are highlighted on the MINITAB printout in Figure 12.2.

The rejection region for the test is found in exactly the same way as the rejection regions for the t-tests in previous chapters. That is, we consult Table VI in Appendix A to obtain an upper-tail value of t. This is a value t_α such that $P(t > t_\alpha) = \alpha$. We can then use this value to construct rejection regions for either one-tailed or two-tailed tests.

For $\alpha = .05$ and $n - (k + 1) = 32 - (2 + 1) = 29$ df, the critical t-value obtained from Table VI is $t_{.05} = 1.699$. Therefore,

Rejection region: $t > 1.699$ (see Figure 12.5)

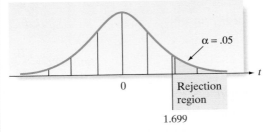

$\alpha = .05$

0

Rejection region

1.699

Figure 12.5
Rejection region for
$H_0: \beta_2 = 0$ vs. $H_a: \beta_2 > 0$

Since the test statistic value, $t = 9.85$, falls into the rejection region, we have sufficient evidence to reject H_0. Thus, the collector can conclude that the mean auction price of a clock increases as the number of bidders increases when age is held constant. Note that the two-tailed observed significance level of the test (highlighted on the printout) is approximately .000. Since the one-tailed p-value (half this value) is also .000, any nonzero α will lead us to reject H_0.

b. From the box, a 90% confidence interval for β_1 is

$$\hat{\beta}_1 \pm t_{\alpha/2} s_{\hat{\beta}_1} = \hat{\beta}_1 \pm t_{.05} s_{\hat{\beta}_1}$$

Substituting $\hat{\beta}_1 = 12.74$, $s_{\hat{\beta}_i} = .905$ (both obtained from the MINITAB printout in Figure 12.2), and $t_{.05} = 1.699$ (from part **a**) into the equation, we obtain

$$12.74 \pm 1.699(.905) = 12.74 \pm 1.54$$

or (11.20, 14.28). Thus, we are 90% confident that β_1 falls between 11.20 and 14.28. Since β_1 is the slope of the line relating the auction price (y) to the age of the clock

(x_1), we conclude that the price increases between \$11.20 and \$14.28 for every 1-year increase in age, holding number of bidders (x_2) constant.

Look Back When interpreting the β multiplied by one x, be sure to hold fixed the values of the other x's in the model.

Now Work Exercise 12.17c–d

12.3 Evaluating Overall Model Utility

In Section 12.2, we demonstrated the use of t-tests in making inferences about β parameters in a multiple-regression model. There are caveats, however, to conducting these t-tests for the purposes of determining which x's are useful for predicting y. Several such caveats are listed in the following box:

Use Caution When Conducting t-Tests on the β Parameters

It is dangerous to conduct t-tests on the individual β parameters in a *first-order linear model* for the purpose of determining which independent variables are useful for predicting y and which are not. If you fail to reject H_0: $\beta_i = 0$, several conclusions are possible:

1. There is no relationship between y and x_i.
2. A straight-line relationship between y and x_i exists (holding the other x's in the model fixed), but a Type II error occurred.
3. A relationship between y and x_i (holding the other x's in the model fixed) exists, but is more complex than a straight-line relationship (e.g., a curvilinear relationship may be appropriate). The most you can say about a β parameter test is that there is either sufficient (if you reject H_0: $\beta_i = 0$) or insufficient (if you do not reject H_0: $\beta_i = 0$) evidence of a *linear* (*straight-line*) relationship between y and x_i.

In addition, conducting t-tests on each β parameter in a model is *not* the best way to determine whether the *overall* model is contributing information relevant to the prediction of y. If we were to conduct a series of t-tests to determine whether the independent variables are contributing to the predictive relationship, we would be very likely to make one or more errors in deciding which terms to retain in the model and which to exclude.

For example, suppose you fit a first-order model in 10 quantitative x variables and decide to conduct t-tests on all 10 of the individual β's in the model, each at $\alpha = .05$. Even if all the β parameters (except β_0) are equal to 0, approximately 40% of the time you will incorrectly reject the null hypothesis at least once and conclude that some β parameter differs from 0.* Thus, in multiple-regression models for which a large number of independent variables are being considered, conducting a series of t-tests may include a large number of insignificant variables and exclude some useful ones. If we want to test the utility of a multiple-regression model, we will need a *global test* (one that encompasses all the β parameters). We would also like to find some statistical quantity that measures how well the model fits the data.

*The proof of this result (assuming independence of tests) proceeds as follows:

$P(\text{Reject } H_0 \text{ at least once} \mid \beta_1 = \beta_2 = \cdots = \beta_{10} = 0)$

$= 1 - P(\text{Reject } H_0 \text{ no times} \mid \beta_1 = \beta_2 = \cdots = \beta_{10} = 0)$

$\leq 1 - [P(\text{Accept } H_0\text{: } \beta_1 = 0 \mid \beta_1 = 0) \cdot P(\text{Accept } H_0\text{: } \beta_2 = 0 \mid \beta_2 = 0)] \cdots \cdot P(\text{Accept } H_0\text{: } \beta_{10} = 0 \mid \beta_{10} = 0)]$

$= 1 - [(1 - \alpha)^{10}] = 1 - (.95)^{10} = .401.$

For dependent tests, the Bonferroni inequality states that

$$P(\text{Reject } H_0 \text{ at least once} \mid \beta_1 = \beta_2 = \cdots = \beta_{10} = 0) \leq 10(\alpha) = 10(.05) = .50$$

We commence with the easier problem: finding a measure of how well a linear model fits a set of data. For this, we use the multiple-regression equivalent of r^2, the coefficient of determination for the straight-line model (Chapter 11), as given in the next definition.

The **multiple coefficient of determination**, R^2, is defined as

$$R^2 = 1 - \frac{SSE}{SS_{yy}} = \frac{SS_{yy} - SSE}{SS_{yy}} = \frac{\text{Explained variability}}{\text{Total variability}}$$

and represents the proportion of the total sample variation in y that can be "explained" by the multiple-regression model.

Just like r^2 in the simple linear model, R^2 represents the fraction of the sample variation of the y-values (measured by SS_{yy}) that is explained by the least squares prediction equation. Thus, $R^2 = 0$ implies a complete lack of fit of the model to the data, and $R^2 = 1$ implies a perfect fit, with the model passing through every data point. In general, the larger the value of R^2, the better the model fits the data.

To illustrate, consider the first-order model for the grandfather clock auction price, presented in Examples 12.1–12.3. A SAS printout of the analysis is shown in Figure 12.6. The value $R^2 = .8923$ is highlighted in Figure 12.6. This high value of R^2 implies that using the independent variables age and number of bidders in a first-order model explains 89.2% of the total *sample variation* (measured by SS_{yy}) in auction price y. Thus, R^2 is a sample statistic that tells how well the model fits the data and thereby represents a measure of the usefulness of the entire model.

```
                           Dependent Variable: PRICE

                      Number of Observations Read          32
                      Number of Observations Used          32

                               Analysis of Variance

                                       Sum of         Mean
          Source             DF       Squares       Square    F Value    Pr > F

          Model               2       4283063      2141531     120.19    <.0001
          Error              29        516727        17818
          Corrected Total    31       4799790

                  Root MSE           133.48467    R-Square     0.8923
                  Dependent Mean    1326.87500    Adj R-Sq     0.8849
                  Coeff Var           10.06008

                               Parameter Estimates

                      Parameter      Standard
          Variable   DF   Estimate      Error    t Value  Pr > |t|    90% Confidence Limits

          Intercept   1  -1338.95134  173.80947    -7.70    <.0001   -1634.27571  -1043.62697
          AGE         1     12.74057    0.90474    14.08    <.0001      11.20331     14.27784
          NUMBIDS     1     85.95298    8.72852     9.85    <.0001      71.12211    100.78385
```

Figure 12.6

SAS analysis of the auction price model

A large value of R^2 computed from the *sample* data does not necessarily mean that the model provides a good fit to all of the data points in the *population*. For example, a first-order linear model that contains three parameters will provide a perfect fit to a sample of three data points, and R^2 will equal 1. Likewise, you will always obtain a perfect fit ($R^2 = 1$) to a set of n data points if the model contains exactly n parameters. Consequently, if you want to use the value of R^2 as a measure of how useful the model will be in predicting y, it should be based on a sample that contains substantially more data points than the number of parameters in the model.

⚠ **CAUTION** In a multiple-regression analysis, use the value of R^2 as a measure of how useful a linear model will be in predicting y only if the sample contains substantially more data points than the number of β parameters in the model. ▲

As an alternative to using R^2 as a measure of model adequacy, the *adjusted multiple coefficient of determination*, denoted R_a^2, is often reported. The formula for R_a^2 is given in the next definition.

The **adjusted multiple coefficient of determination** is given by

$$R_a^2 = 1 - \left[\frac{(n-1)}{n-(k+1)}\right]\left(\frac{\text{SSE}}{\text{SS}_{yy}}\right)$$

$$= 1 - \left[\frac{(n-1)}{n-(k+1)}\right](1 - R^2)$$

Note: $R_a^2 \leq R^2$.

R^2 and R_a^2 have similar interpretations. However, unlike R^2, R_a^2 takes into account ("adjusts" for) both the sample size n and the number of β parameters in the model. R_a^2 will always be smaller than R^2 and, more importantly, cannot be "forced" to 1 simply by adding more and more independent variables to the model. Consequently, analysts prefer the more conservative R_a^2 in choosing a measure of model adequacy. The value of R_a^2 is also highlighted in Figure 12.6. Note that $R_a^2 = .8849$, a value only slightly smaller than R^2.

Despite their utility, R^2 and R_a^2 are only sample statistics. Therefore, it is dangerous to judge the global usefulness of a model solely on the basis of these values. A better method is to conduct a test of hypothesis involving *all* the β parameters (except β_0) in a model. In particular, for the general multiple-regression model $E(y) = \beta_0 + \beta_1 x_1 + \beta_2 x_2 + \cdots + \beta_k x_k$, we would test

$$H_0: \beta_1 = \beta_2 = \cdots = \beta_k = 0$$
$$H_a: \text{At least one of the coefficients is nonzero}$$

The test statistic used to test this hypothesis is an *F*-statistic, and several equivalent versions of the formula can be used (although we will usually rely on the computer to calculate the *F*-statistic):

$$\textit{Test statistic: } F = \frac{(\text{SS}_{yy} - \text{SSE})/k}{\text{SSE}/[n-(k+1)]} = \frac{\text{Mean square (Model)}}{\text{Mean square (Error)}}$$

$$= \frac{R^2/k}{(1-R^2)/[n-(k+1)]}$$

Both formulas indicate that the *F*-statistic is the ratio of the *explained* variability divided by the number of degrees of freedom in the model to the *unexplained* variability divided by the number of degrees of freedom associated with the error. (For this reason, the test is often called the "analysis-of-variance" *F*-test.) Thus, the larger the proportion of the total variability accounted for by the model, the larger is the *F*-statistic.

To determine when the ratio becomes large enough that we can confidently reject the null hypothesis and conclude that the model is more useful than no model at all in predicting y, we compare the calculated *F*-statistic with a tabulated *F*-value with k df in the numerator and $[n-(k+1)]$ df in the denominator. Recall that tabulations of the *F*-distribution for various values of α are given in Tables VIII, IX, X, and XI of Appendix A. We thus have

Rejection region: $F > F_\alpha$, where F is based on k numerator and
$n - (k + 1)$ denominator degrees of freedom

The analysis-of-variance F-test for testing the overall utility of the multiple-regression model is summarized in the following box:

Testing the Global Usefulness of the Model: The Analysis-of-Variance F-Test

$H_0: \beta_1 = \beta_2 = \cdots = \beta_k = 0$ (All model terms are unimportant in predicting y.)

H_a: At least one $\beta_i \neq 0$ (At least one model term is useful in predicting y.)

$$\text{Test statistic: } F = \frac{(SS_{yy} - SSE)/k}{SSE/[n - (k + 1)]} = \frac{R^2/k}{(1 - R^2)/[n - (k + 1)]}$$

$$= \frac{\text{Mean square (Model)}}{\text{Mean square (Error)}}$$

where n is the sample size and k is the number of terms in the model.

Rejection region: $F > F_\alpha$, with k numerator degrees of freedom and $[n - (k + 1)]$ denominator degrees of freedom.

Conditions Required for the Global F-Test to Be Valid

The standard regression assumptions about the random error component (Section 12.1).

⚠ **CAUTION** A rejection of the null hypothesis $H_0: \beta_1 = \beta_2 = \cdots = \beta_k = 0$ in the **global F-test** leads to the conclusion [with $100(1 - \alpha)\%$ confidence] that the model is statistically useful. However, "useful" does not necessarily mean "best." Another model may prove even more useful in terms of providing more reliable estimates and predictions. This global F-test is usually regarded as a test that the model *must* pass to merit further consideration. ⚠

Example 12.4

Assessing Overall Model Adequacy—Antique Clock Auction Price Model

Problem Refer to Example 12.3, in which an antique collector modeled the auction price y of grandfather clocks as a function of the age x_1 of the clock and the number x_2 of bidders. Recall that the hypothesized first-order model is

$$y = \beta_0 + \beta_1 x_1 + \beta_2 x_2 + \varepsilon$$

a. Find and interpret the adjusted coefficient of determination, R_a^2, for this example.

b. Conduct the global F-test of model usefulness at the $\alpha = .05$ level of significance.

Solution

a. The R_a^2 value (highlighted in the SAS printout shown in Figure 12.6) is .8849. This implies that the least squares model has explained about 88.5% of the total sample variation in y values (auction prices), after adjusting for sample size and number of independent variables in the model.

b. The elements of the global test of the model are as follows:

$H_0: \beta_1 = \beta_2 = 0$ [*Note:* $k = 2$]

H_a: At least one of the two model coefficients is nonzero

$$\text{Test statistic: } F = \frac{\text{MS(Model)}}{\text{MSE}} = \frac{2,141,531}{17,818} = 120.19 \qquad \text{(see Figure 12.6)}$$

p-value: less than .0001

Conclusion: Since $\alpha = .05$ exceeds the observed significance level, $p < .0001$, the data provide strong evidence that at least one of the model coefficients is nonzero. The overall model appears to be statistically useful in predicting auction prices.

Look Back Can we be sure that the best model for prediction has been found if the global F-test indicates that a model is useful? Unfortunately, we cannot. The addition of other independent variables may improve the usefulness of the model (see Caution on page 626). We consider more complex multiple regression models in Sections 12.5–12.8.

Now Work Exercise 12.16

In this section, we discussed several different statistics for assessing the utility of a multiple-regression model: t-tests on the individual β parameters, R^2, R_a^2, and the global F-test. Both R^2 and R_a^2 are indicators of how well the prediction equation fits the data. Intuitive evaluations of the contribution of the model based on R^2 must be examined with care. Unlike R_a^2, the value of R^2 increases as more and more variables are added to the model. Consequently, you could force R^2 to take a value very close to 1 even though the model contributes no information relevant to the prediction of y. In fact, R^2 equals 1 when the number of terms in the model (including β_0) equals the number of data points. Therefore, you should not rely solely on the value of R^2 (or even R_a^2) to tell you whether a model is useful in predicting y.

Conducting t-tests on all the β parameters is also not the best method of testing the global utility of a model, since these multiple tests result in a high probability of making at least one Type I error. Use the F-test for testing the global utility of the model.

After we have used the F-test and determined that the overall model is useful in predicting y, we may elect to conduct one or more t-tests on the individual β parameters. However, the test (or tests) to be conducted should be decided *a priori*—that is, prior to fitting the model. Also, we should limit the number of t-tests conducted, to avoid the potential problem of making too many Type I errors. Generally, the regression analyst will conduct t-tests only on the "most important" β's. We provide insight into identifying the most important β's in a linear model in Sections 12.5–12.8.

Recommendation for Checking the Utility of a Multiple-Regression Model

1. First, use the F-test to conduct a test of the adequacy of the overall model; that is, test $H_0: \beta_1 = \beta_2 = \cdots = \beta_k = 0$. If the model is deemed adequate (i.e., if you reject H_0), then proceed to step 2. Otherwise, you should hypothesize and fit another model. The new model may include more independent variables or higher order terms.

2. Conduct t-tests on those β parameters in which you are particularly interested (i.e., the "most important" β's). These usually involve only the β's associated with higher order terms (x_2, x_1x_2, etc.). However, it is a safe practice to limit the number of β's that are tested. Conducting a series of t-tests leads to a high overall Type I error rate α.

3. Examine the values of R_a^2 and $2s$ to evaluate how well, numerically, the model fits the data.

Exercises 12.1–12.28

Understanding the Principles

12.1 Write a first-order model relating $E(y)$ to
  a. two quantitative independent variables
  b. four quantitative independent variables
  c. five quantitative independent variables

12.2 List the four assumptions about the random error ε required for a multiple-regression analysis.

12.3 Outline the six steps in a multiple-regression analysis.

12.4 What are the caveats to conducting t-tests on all of the individual β parameters in a multiple-regression model?

12.5 How should you test the overall adequacy of a multiple-regression model?

Learning the Mechanics

12.6 MINITAB was used to fit the model $y = \beta_0 + \beta_1x_1 +$
 [NW] $\beta_2x_2 + \varepsilon$ to $n = 20$ data points, and the printout (top of page 628) was obtained.
  a. What are the sample estimates of β_0, β_1, and β_2?
  b. What is the least squares prediction equation?
  c. Find SSE, MSE, and s. Interpret the standard deviation in the context of the problem.
  d. Test $H_0: \beta_1 = 0$ against $H_a: \beta_1 \neq 0$. Use $\alpha = .05$.
  e. Use a 95% confidence interval to estimate β_2.
  f. Find R^2 and R_a^2 and interpret these values.
  g. Use the two formulas given in this section to calculate the test statistic for the null hypothesis $H_0: \beta_1 = \beta_2 = 0$.

MINITAB output for Exercise 12.6

```
The regression equation is
Y = 506.35 - 941.9 X1 - 429.1 X2


Predictor    Coef   SE Coef      T      P
Constant  506.346    45.17   11.21  0.000
X1       -941.900   275.08   -3.42  0.003
X2       -429.060   379.83   -1.13  0.274


S = 94.251    R-Sq = 45.9%    R-Sq(adj) = 39.6%


Analysis of Variance

Source          DF      SS      MS      F      P
Regression       2  128329   64165   7.22  0.005
Residual Error  17  151016    8883
Total           19  279345
```

Compare your results with the test statistic shown on the printout.

h. Find the observed significance level of the test you conducted in part **g.** Interpret the value.

12.7 Suppose you fit the model

$$y = \beta_0 + \beta_1 x_1 + \beta_2 x_2 + \beta_3 x_3 + \varepsilon$$

to $n = 30$ data points and obtain the following result:

$$\hat{y} = 3.4 - 4.6x_1 + 2.7x_2 + .93x_3$$

The estimated standard errors of $\hat{\beta}_2$ and $\hat{\beta}_3$ are 1.86 and .29, respectively.

a. Test the null hypothesis $H_0: \beta_2 = 0$ against the alternative hypothesis $H_a: \beta_2 \neq 0$. Use $\alpha = .05$.
b. Test the null hypothesis $H_0: \beta_3 = 0$ against the alternative hypothesis $H_a: \beta_3 \neq 0$. Use $\alpha = .05$.
c. The null hypothesis $H_0: \beta_2 = 0$ is not rejected. In contrast, the null hypothesis $H_0: \beta_3 = 0$ is rejected. Explain how this can happen even though $\hat{\beta}_2 > \hat{\beta}_3$.

12.8 Suppose you fit the first-order multiple-regression model

$$y = \beta_0 + \beta_1 x_1 + \beta_2 x_2 + \varepsilon$$

to $n = 25$ data points and obtain the prediction equation

$$\hat{y} = 6.4 + 3.1x_1 + .92x_2$$

The estimated standard deviations of the sampling distributions of $\hat{\beta}_1$ and $\hat{\beta}_2$ are 2.3 and .27, respectively.
a. Test $H_0: \beta_1 = 0$ against $H_a: \beta_1 > 0$. Use $\alpha = .05$.
b. Test $H_0: \beta_2 = 0$ against $H_a: \beta_2 \neq 0$. Use $\alpha = .05$.
c. Find a 90% confidence interval for β_1. Interpret the interval.
d. Find a 99% confidence interval for β_2. Interpret the interval.

12.9 How is the number of degrees of freedom available for estimating σ^2 (the variance of ε) related to the number of independent variables in a regression model?

12.10 Consider the following first-order model equation in three quantitative independent variables:

$$E(y) = 1 + 2x_1 + x_2 - 3x_3$$

a. Graph the relationship between y and x_1 for $x_2 = 1$ and $x_3 = 3$.

b. Repeat part **a** for $x_2 = -1$ and $x_3 = 1$.
c. How do the graphed lines in parts **a** and **b** relate to each other? What is the slope of each line?
d. If a linear model is first order in three independent variables, what type of geometric relationship will you obtain when you graph $E(y)$ as a function of one of the independent variables for various combinations of values of the other independent variables?

12.11 Suppose you fit the first-order model

$$y = \beta_0 + \beta_1 x_1 + \beta_2 x_2 + \beta_3 x_3 + \beta_4 x_4 + \beta_5 x_5 + \varepsilon$$

to $n = 30$ data points and obtain

$$SSE = .33 \quad R^2 = .92$$

a. Do the values of SSE and R^2 suggest that the model provides a good fit to the data? Explain.
b. Is the model of any use in predicting y? Test the null hypothesis $H_0: \beta_1 = \beta_2 = \cdots = \beta_5 = 0$ against the alternative hypothesis H_a: At least one of the parameters $\beta_1, \beta_2, \cdots, \beta_5$ is nonzero. Use $\alpha = .05$.

12.12 If the analysis-of-variance F-test leads to the conclusion that at least one of the model parameters is nonzero, can you conclude that the model is the best predictor for the dependent variable y? Can you conclude that all of the terms in the model are important in predicting y? What is the appropriate conclusion?

Applying the Concepts—Basic

12.13 **Characteristics of lead users.** During new product development, companies often involve "lead users," i.e., creative individuals who are on the leading edge of an important market trend. *Creativity and Innovation Management* (Feb. 2008) published an article on identifying the social network characteristics of lead users of children's computer games. Data were collected for $n = 326$ children and the following variables measured: lead-user rating (y, measured on a 5-point scale), gender ($x_1 = 1$ if female, 0 if male), age (x_2, years), degree of centrality (x_3, measured as the number of direct ties to other peers in the network), and betweenness centrality (x_4, measured as the number of shortest paths between peers). A first-order model for y was fit to the data, yielding the following least squares prediction equation:

$$\hat{y} = 3.58 + .01x_1 - .06x_2 - .01x_3 + .42x_4$$

a. Give two properties of the errors of prediction that result from using the method of least squares to obtain the parameter estimates.
b. Give a practical interpretation of the estimate of β_4 in the model.
c. A test of $H_0: \beta_4 = 0$ resulted in a p-value of .002. Make the appropriate conclusion at $\alpha = .05$.

12.14 **Dating and disclosure.** Refer to the *Journal of Adolescence* (April 2010) study of adolescents' disclosure of their dating and romantic relationships, Exercise 8.31 (p. 365). Data collected for a sample of 222 high school students were used to determine the level of disclosure of the date's identity to an adolescent's mother (measured on a 5-point scale, where 1 = "never tell," 2 = "rarely tell," 3 = "sometimes tel," 4 = "almost always tell," and 5 = "always tell). Multiple regression was used to model level of disclosure (y) to several independent variables, including gender ($x_1 = 1$ if female, 0 if male), age (x_2, years), dating experience (x_3, years), and level of trust in parents (x_4, 5-point scale).

a. Give the equation of a first-order model for y as a function of the four independent variables.

b. The coefficient of determination for the model, part **a**, was reported as $R^2 = .24$. Give a practical interpretation of this value.

c. Give the null hypothesis for testing the overall adequacy of the model, part **a**.

d. The test, part **c**, resulted in $F = 56.60$ with p-value $< .001$. Interpret this result using $\alpha = .05$.

e. The estimate of the beta coefficient for age (x_2) was reported as $-.09$. Give a practical interpretation of this value.

12.15 Whales entangled in fishing gear. Refer to the *Marine Mammal Science* (April 2010) study of whales entangled in fishing gear, Exercise 10.28 (p. 494). Data collected for a sample of 207 entanglements in the East Sea of Korea were used to model the length (y) of an entangled whale (in meters). Two independent variables used to predict whale length were water depth of the entanglement (x_1, in meters) and distance of the entanglement from land (x_2, in miles).

a. Give the equation of a first-order model for length (y) as a function of the two independent variables.

b. The researchers theorize that the length of an entangled whale will increase linearly as the water depth increases, for entanglements that are a fixed distance from land. Explain how to use the model, part **a**, to test this theory.

c. The p-value for testing $H_0: \beta_2 = 0$ in the model, part **a**, was reported as .013. Interpret this result using $\alpha = .05$.

12.16 Estimation of urban population by means of satellite images. Can the population of an urban area be estimated without taking a census? In *Geographical Analysis* (Jan. 2007) geography professors at the University of Wisconsin at Milwaukee and Ohio State University demonstrated the use of satellite image maps in estimating urban population. A portion of Columbus, Ohio, was partitioned into $n = 125$ census block groups, and satellite imagery was obtained. For each census block, the following variables were measured: population density (y), proportion of block with low-density residential areas (x_1), and proportion of block with high-density residential areas (x_2). A first-order model for y was fitted to the data and produced the following results:

$$\hat{y} = -.0304 + 2.006x_1 + 5.006x_2, R^2 = .686$$

a. Give a practical interpretation of each β-estimate in the model.

b. Give a practical interpretation of the coefficient of determination, R^2.

c. State H_0 and H_a for a test of the overall adequacy of the model.

d. Refer to part **c**. Compute the value of the test statistic.

e. Refer to parts **c** and **d**. Make the appropriate conclusion at $\alpha = .01$.

12.17 Predicting runs scored in baseball. In *Chance* (Fall 2000), statistician Scott Berry built a multiple-regression model for predicting the total number of runs scored by a Major League Baseball team during a season. Using data on all

teams from 1990 to 1998 (a sample with $n = 234$), the results in the following table were obtained.

Independent Variable	β Estimate	Standard Error
Intercept	3.70	15.00
Walks (x_1)	.34	.02
Singles (x_2)	.49	.03
Doubles (x_3)	.72	.05
Triples (x_4)	1.14	.19
Home Runs (x_5)	1.51	.05
Stolen Bases (x_6)	.26	.05
Caught Stealing (x_7)	$-.14$	.14
Strikeouts (x_8)	$-.10$	.01
Outs (x_9)	$-.10$	.01

Source: Berry, S. M. "A statistician reads the sports pages: Modeling offensive ability in baseball." *Chance,* Vol. 13, No. 4, Fall 2000 (Table 2). Reprinted with permission from *Chance.* Copyright 2000 by the American Statistical Association. All rights reserved.

a. Write the least squares prediction equation for y = total number of runs scored by a team in a season.

b. Give practical interpretations of the β estimates.

c. Conduct a test of $H_0: \beta_7 = 0$ against $H_a: \beta_7 < 0$ at $\alpha = .05$. Interpret the results.

d. Form a 95% confidence interval for β_5. Interpret the results.

e. Predict the number of runs scored by your favorite Major League Baseball team last year. How close is the predicted value to the actual number of runs scored by your team? (*Note*: You can find data on your favorite team on the Internet at *www.majorleaguebaseball.com*.)

12.18 Growth of Japanese beetles. In the *Journal of Insect Behavior* (Nov. 2001), biologists at Eastern Illinois University published the results of their study on Japanese beetles. The biologists collected beetles over a period of $n = 13$ summer days in a soybean field. For one portion of the study, the biologists modeled y, the average size (in millimeters) of female beetles as a function of the average daily temperature x_1 (degrees) and Julian date x_2.

a. Write a first-order model for $E(y)$ as a function of x_1 and x_2.

b. The model was fit to the data, with the results shown in the accompanying table. Interpret the estimate of β_1.

Variable	Parameter Estimate	t-Value	p-Value
Intercept	6.51	26.0	<.0001
Temperature (x_1)	$-.002$	-0.72	.49
Date (x_2)	$-.010$	-3.30	.008

c. Conduct a test to determine whether the average size of female Japanese beetles decreases linearly as the temperature increases. Use $\alpha = .05$.

12.19 Study of adolescents with ADHD. Children with attention-deficit/hyperactivity disorder (ADHD) were monitored to evaluate their risk for substance (e.g., alcohol, tobacco, illegal drug) use (*Journal of Abnormal Psychology*, Aug. 2003). The following data were collected on 142 adolescents diagnosed with ADHD:

y = frequency of marijuana use the past six months

x_1 = severity of inattention (5-point scale)

x_2 = severity of impulsivity–hyperactivity (5-point scale)

x_3 = level of oppositional-defiant and conduct disorder (5-point scale)

a. Write the equation of a first-order model for $E(y)$.

b. The coefficient of determination for the model is $R^2 = .08$. Interpret this value.

c. The global F-test for the model yielded a p-value less than .01. Interpret this result.

d. The t-test for $H_0: \beta_1 = 0$ resulted in a p-value less than .01. Interpret this result.

e. The t-test for $H_0: \beta_2 = 0$ resulted in a p-value greater than .05. Interpret this result.

f. The t-test for $H_0: \beta_3 = 0$ resulted in a p-value greater than .05. Interpret this result.

Applying the Concepts—Intermediate

12.20 Novelty of a vacation destination. Many tourists choose a vacation destination on the basis of the newness or uniqueness (i.e., the novelty) of the itinerary. Texas A&M University professor J. Petrick investigated the relationship between novelty and vacationing golfers' demographics (*Annals of Tourism Research*, April 2002). Data were obtained from a mail survey of 393 golf vacationers to a large coastal resort in the southeastern United States. Several measures of novelty level (on a numerical scale) were obtained for each vacationer, including "change from routine," "thrill," "boredom-alleviation," and "surprise." The researcher employed four independent variables in a regression model to predict each measure of novelty. The independent variables were x_1 = number of rounds of golf per year, x_2 = total number of golf vacations taken, x_3 = number of years the respondent played golf, and x_4 = average golf score.

a. Give the hypothesized equation of a first-order model for y = change from routine.

b. A test of $H_0: \beta_3 = 0$ versus $H_a: \beta_3 < 0$ yielded a p-value of .005. Interpret this result if $\alpha = .01$.

c. The estimate of β_3 was found to be negative. On the basis of this result (and the result of part **b**), the researcher concluded that "those who have played golf for more years are less apt to seek change from their normal routine in their golf vacations." Do you agree with this statement? Explain.

d. The regression results for the three other dependent measures of novelty are summarized in the accompanying table. Give the null hypothesis for testing the overall adequacy of each first-order regression model.

Dependent Variable	F-Value	p-Value	R^2
Thrill	5.56	< .001	.055
Change from routine	3.02	.018	.030
Surprise	3.33	.011	.023

Based on Petrick, J. F. "An examination of golf vacationers' novelty." *Annals of Tourism Research*, Vol. 29, No. 2, April 2002.

e. Give the rejection region for the test mentioned in part **d**. Use $\alpha = .01$.

f. Use the test statistics reported in the table and the rejection region from part **e** to conduct the test for each of the dependent measures of novelty.

g. Verify that the p-values in the table support the conclusions you drew in part **f**.

h. Interpret the values of R^2 reported in the table.

12.21 Arsenic in groundwater. *Environmental Science & Technology* (Jan. 2005) reported on a study of the reliability of a commercial kit designed to test for arsenic in groundwater. The field kit was used to test a sample of 328 groundwater wells in Bangladesh. In addition to the arsenic level (in micrograms per liter), the latitude (degrees), longitude (degrees), and depth (feet) of each well were measured. The data are saved in the **ASWELLS** file. The first and last five observations are listed in the following table:

Wellid	Latitude	Longitude	Depth	Arsenic
10	23.7887	90.6522	60	331
14	23.7886	90.6523	45	302
30	23.7880	90.6517	45	193
59	23.7893	90.6525	125	232
85	23.7920	90.6140	150	19
⋮	⋮	⋮	⋮	⋮
7353	23.7949	90.6515	40	48
7357	23.7955	90.6515	30	172
7890	23.7658	90.6312	60	175
7893	23.7656	90.6315	45	624
7970	23.7644	90.6303	30	254

a. Write a first-order model for arsenic level (y) as a function of latitude, longitude, and depth.

b. Use the method of least squares to fit the model to the data.

c. Give practical interpretations of the β estimates.

d. Find the standard deviation s of the model, and interpret its value.

e. Find and interpret the values of R^2 and R_a^2.

f. Conduct a test of overall model utility at $\alpha = .05$.

g. On the basis of the results you obtained in parts **d–f**, would you recommend using the model to predict arsenic level (y)? Explain.

12.22 Study of contaminated fish. Refer to the U.S. Army Corps of Engineers data on fish contaminated from the toxic discharges of a chemical plant located on the banks of the Tennessee River in Alabama, shown below. Recall that the engineers measured the length (in centimeters), weight (in grams), and DDT level (in parts per million) for 144 captured fish. In addition, the number of miles upstream from the river was recorded. The data are saved in the **FISHDDT** file. (The first and last five observations are shown in the table, p. 631.)

a. Fit the first-order model $E(y) = \beta_0 + \beta_1 x_1 + \beta_2 x_2 + \beta_3 x_3$ to the data, where y = DDT level, x_1 = mile, x_2 = length, and x_3 = weight. Report the least squares prediction equation.

b. Find the estimate of the standard deviation of ε for the model, and give a practical interpretation of its value.

c. Do the data provide sufficient evidence to conclude that DDT level increases as length increases? Report the observed significance level of the test, and reach a conclusion. Use $\alpha = .05$.

d. Find and interpret a 95% confidence interval for β_3.

e. Test the overall adequacy of the model, using $\alpha = .05$.

Data for Exercise 12.22

River	Mile	Species	Length	Weight	DDT
FC	5	CHANNELCATFISH	42.5	732	10.00
FC	5	CHANNELCATFISH	44.0	795	16.00
FC	5	CHANNELCATFISH	41.5	547	23.00
FC	5	CHANNELCATFISH	39.0	465	21.00
FC	5	CHANNELCATFISH	50.5	1252	50.00
⋮	⋮	⋮	⋮	⋮	⋮
TR	345	LARGEMOUTHBASS	23.5	358	2.00
TR	345	LARGEMOUTHBASS	30.0	856	2.20
TR	345	LARGEMOUTHBASS	29.0	793	7.40
TR	345	LARGEMOUTHBASS	17.5	173	0.35
TR	345	LARGEMOUTHBASS	36.0	1433	1.90

12.23 Reality TV and cosmetic surgery. How much influence do the media, especially reality television programs, have on one's decision to undergo cosmetic surgery? This was the question of interest to psychologists who published an article in *Body Image: An International Journal of Research* (March 2010). In the study, 170 college students answered questions about their impression of reality TV shows featuring cosmetic surgery, level of self-esteem, satisfaction with their own body, and desire to have cosmetic surgery to alter their body. The variables analyzed in the study were measured as follows: DESIRE—scale ranging from 5 to 25, where the higher the value, the greater the interest in having cosmetic surgery; GENDER—1 if male, 0 if female; SELFESTM—scale ranging from 4 to 40, where the higher the value, the greater the level of self-esteem; BODYSAT—scale ranging from 1 to 9, where the higher the value, the greater the satisfaction with one's own body; and IMPREAL—scale ranging from 1 to 7, where the higher the value, the more one believes reality television shows featuring cosmetic surgery are realistic. The data for the study (simulated based on statistics reported in the journal article) are saved in the **BODYIMAGE** file. Selected observations are listed in the table (next column). The psychologists used multiple regression to model desire to have cosmetic surgery (y) as a function of gender (x_1), self-esteem (x_2), body satisfaction (x_3), and impression of reality TV (x_4).

a. Fit the first-order model, $E(y) = \beta_0 + \beta_1 x_1 + \beta_2 x_2 + \beta_3 x_3 + \beta_4 x_4$, to the data in the **BODYIMAGE** file. Give the least squares prediction equation.

b. Interpret the β-estimates in the words of the problem.

c. Is the overall model statistically useful for predicting desire to have cosmetic surgery? Test using $\alpha = .01$.

d. Which statistic, R^2 or R_a^2, is the preferred measure of model fit? Practically interpret the value of this statistic.

Data for Exercise 12.23

Student	DESIRE	GENDER	SELFESTM	BODYSAT	IMPREAL
1	11	0	24	3	4
2	13	0	20	3	4
3	11	0	25	4	5
4	11	1	22	9	4
5	18	0	8	1	6
⋮	⋮	⋮	⋮	⋮	⋮
166	18	0	25	3	5
167	13	0	26	4	5
168	9	1	13	5	6
169	14	0	20	3	2
170	6	1	27	8	3

e. Conduct a test to determine whether desire to have cosmetic surgery decreases linearly as level of body satisfaction increases. Use $\alpha = .05$.

f. Find a 95% confidence interval for β_4. Practically interpret the result.

12.24 Deep-space survey of quasars. A quasar is a distant celestial object (at least 4 billion light-years away) that provides a powerful source of radio energy. The *Astronomical Journal* (July 1995) reported on a study of 90 quasars detected by a deep-space survey. The survey enabled astronomers to measure several different quantitative characteristics of each quasar, including redshift range, line flux (erg/cm² · · · s), line luminosity (erg/s), AB_{1450} magnitude, absolute magnitude, and rest-frame equivalent width. The data for a sample of 25 large (redshift) quasars are saved in the **QUASAR** file. (Several quasars are listed in the table.)

a. Hypothesize a first-order model for equivalent width y as a function of the first four variables shown in the table.

b. Fit the first-order model to the data. Give the least squares prediction equation.

c. Interpret the β estimates in the model.

Data for Exercise 12.24

Quasar	Redshift (x_1)	Line Flux (x_2)	Line Luminosity (x_3)	AB_{1450} x_4	Absolute Magnitude (x_5)	Rest-Frame Equivalent Width (y)
1	2.81	−13.48	45.29	19.50	−26.27	117
2	3.07	−13.73	45.13	19.65	−26.26	82
3	3.45	−13.87	45.11	18.93	−27.17	33
4	3.19	−13.27	45.63	18.59	−27.39	92
5	3.07	−13.56	45.30	19.59	−26.32	114

Based on Schmidt, M., Schneider, D. P., and Gunn, J. E. "Spectroscopic CCD surveys for quasars at large redshift." *The Astronomical Journal*, Vol. 110, No. 1, July 1995, p. 70 (Table 1).

d. Test the overall adequacy of the model, using $\alpha = .05$.

e. Test to determine whether redshift (x_1) is a useful linear predictor of equivalent width (y). Use $\alpha = .05$.

12.25 Cooling method for gas turbines. Refer to the *Journal of Engineering for Gas Turbines and Power* (Jan. 2005) study of a high-pressure inlet fogging method for a gas turbine engine, presented in Exercise 8.34 (p. 366). Recall that the heat rate (kilojoules per kilowatt per hour) was measured for each in a sample of 67 gas turbines augmented with high-pressure inlet fogging. In addition, several other variables were measured, including cycle speed (revolutions per minute), inlet temperature (°C), exhaust gas temperature (°C), cycle pressure ratio, and air mass flow rate (kilograms per second). The data are saved in the **GASTURBINE** file. The first and last five observations are listed in the table below.

Rpm	Cpratio	Inlet-Temp	Exh-Temp	Airflow	Heatrate
27245	9.2	1134	602	7	14622
14000	12.2	950	446	15	13196
17384	14.8	1149	537	20	11948
11085	11.8	1024	478	27	11289
14045	13.2	1149	553	29	11964
⋮	⋮	⋮	⋮	⋮	⋮
18910	14.0	1066	532	8	12766
3600	35.0	1288	448	152	8714
3600	20.0	1160	456	84	9469
16000	10.6	1232	560	14	11948
14600	13.4	1077	536	20	12414

Based on Bhargava, R., and Meher-Homji, C. B. "Parametric analysis of existing gas turbines with inlet evaporative and overspray fogging." *Journal of Engineering for Gas Turbines and Power*, Vol. 127, No. 1, Jan. 2005. Table from pp. 156–157.

a. Write a first-order model for heat rate (y) as a function of speed, inlet temperature, exhaust temperature, cycle pressure ratio, and air mass flow rate.

b. Use the method of least squares to fit the model to the data.

c. Give practical interpretations of the β estimates.

d. Find the standard deviation s of the model, and interpret its value.

e. Find R_a^2 and interpret its value.

f. Is the overall model statistically useful in predicting heat rate (y)? Test, using $\alpha = .01$.

12.26 R^2 and model fit. Because the coefficient of determination, R^2, always increases when a new independent variable is added to a model, it is tempting to include many variables in the model in order to force R^2 to be near 1. However, doing so reduces the number of degrees of freedom available for estimating σ^2, which adversely affects our ability to make reliable inferences. Suppose you want to use 18 psychological and sociological factors to predict a student's Scholastic Assessment Test (SAT) score. You fit the model

$$y = \beta_0 + \beta_1 x_1 + \beta_2 x_2 + \cdots + \beta_{17} x_{17} + \beta_{18} x_{18} + \varepsilon$$

where y = SAT score and $x_1, x_2, \ldots, x_{18}$ are the psychological and sociological factors. Only 20 years of data ($n = 20$) are used to fit the model, and you obtain $R^2 = .95$. Test to see whether this impressive-looking R^2 is large enough for you to infer that the model is useful—that is, that at least one term in the model is important in predicting SAT scores. Use $\alpha = .05$.

12.27 Highway crash data analysis. Researchers at Montana State University have written a tutorial on an empirical method for analyzing before and after highway crash data (Montana Department of Transportation, Research Report, May 2004). The initial step in the methodology is to develop a safety performance function (SPF)—a mathematical model that estimates the probability of occurrence of a crash for a given segment of roadway. Using data on over 100 segments of roadway, the researchers fit the model $E(y) = \beta_0 + \beta_1 x_1 + \beta_2 x_2$, where y = number of crashes per three years, x_1 = roadway length (miles), and x_2 = average annual daily traffic (number of vehicles) = AADT. The results are shown in the following tables:

Interstate Highways

Variable	Parameter Estimate	Standard Error	t-Value
Intercept	1.81231	.50568	3.58
Length (x_1)	.10875	.03166	3.44
AADT (x_2)	.00017	.00003	5.19

Non-Interstate Highways

Variable	Parameter Estimate	Standard Error	t-Value
Intercept	1.20785	.28075	4.30
Length (x_1)	.06343	.01809	3.51
AADT (x_2)	.00056	.00012	4.86

a. Give the least squares prediction equation for the interstate highway model.

b. Give practical interpretations of the β estimates you made in part **a**.

c. Refer to part **a**. Find a 99% confidence interval for β_1 and interpret the result.

d. Refer to part **a**. Find a 99% confidence interval for β_2 and interpret the result.

e. Repeat parts **a–d** for the non-interstate-highway model.

Applying the Concepts—Advanced

12.28 Bordeaux wine sold at auction. The vineyards in the Bordeaux region of France are known for producing excellent red wines. However, the uncertainty of the weather during the growing season, the phenomenon that wine tastes better with age, and the fact that some Bordeaux vineyards produce better wines than others encourage speculation concerning the value of a case of wine produced by a certain vineyard during a certain year (or of a certain vintage). As a result, many wine experts attempt to predict the auction price of a case of Bordeaux wine. The publishers of a newsletter titled *Liquid Assets: The International Guide to Fine Wine* discussed a multiple-regression approach to predicting the London auction price of red Bordeaux wine in *Chance* (Fall 1995). The natural logarithm of the price y (in dollars) of a case containing a dozen bottles of red wine was modeled as a function of weather during the growing season and age of vintage. Data collected for the vintages of 1952 to 1980 were used. Three models were fit to the data. The results of the regressions are summarized in the table (next page).

a. For each model, conduct a t-test for each of the β parameters in the model. Interpret the results.

Results for Exercise 12.28

Independent Variables	Beta Estimates (Standard Errors)		
	Model 1	Model 2	Model 3
x_1 = Vintage year	.0354 (.0137)	.0238 (.00717)	.0240 (.00747)
x_2 = Average growing-season temperature (°C)	(not included)	.616 (.0952)	.608 (.116)
x_3 = Sept.–Aug. rainfall (cm)	(not included)	−.00386 (.00081)	−.00380 (.00095)
x_4 = Rainfall in months preceding vintage (cm)	(not included)	.0001173 (.000482)	.00115 (.000505)
x_5 = Average Sept. temperature (°C)	(not included)	(not included)	.00765 (.565)
	$R^2 = .212, s = .575$	$R^2 = .828, s = .287$	$R^2 = .828, s = .293$

b. When the natural logarithm of y is used as a dependent variable, the antilogarithm of a β coefficient minus 1 (i.e., $e^{\beta_i} - 1$) represents the percentage change in y for every one-unit increase in the associated x value.* Use this information to interpret the β estimates of each model.

c. On the basis of the values of R^2 and s shown, which of the three models would you use to predict red Bordeaux wine prices? Explain.

12.4 Using the Model for Estimation and Prediction

In Section 11.6, we discussed the use of the least squares line in estimating the mean value of y, $E(y)$, for some particular value of x, say, $x = x_p$. We also showed how to use the same fitted model to predict, when $x = x_p$, some new value of y to be observed in the future. Recall that the least squares line yielded the same value for both the estimate of $E(y)$ and the prediction of some future value of y. That is, both are the result of substituting x_p into the prediction equation $\hat{y} = \hat{\beta}_0 + \hat{\beta}_1 x$ and calculating $\hat{y}_p$. There the equivalence ends, however: The confidence interval for the mean $E(y)$ is narrower than the prediction interval for y because of the additional uncertainty attributable to the random error ϵ that arises in predicting some future value of y.

These same concepts carry over to the multiple-regression model. Consider a first-order model relating sale price (y) of a residential property to land value (x_1), improvements (x_2), and home size (x_3). Suppose we want to estimate the mean sale price for a given property with $x_1 = \$15,000$, $x_2 = \$50,000$, and $x_3 = 1,800$ square feet. Assuming that the first-order model represents the true relationship between the sale price and the three independent variables, we want to estimate

$$E(y) = \beta_0 + \beta_1 x_1 + \beta_2 x_2 + \beta_3 x_3$$
$$= \beta_0 + \beta_1(15,000) + \beta_2(50,000) + \beta_3(1,800)$$

After obtaining the least squares estimates, $\beta_0, \beta_1, \beta_2,$ and β_3, we find the estimate of $E(y)$ to be

$$\hat{y} = \hat{\beta}_0 + \hat{\beta}_1(15,000) + \hat{\beta}_2(50,000) + \hat{\beta}_3(1,800)$$

To form a confidence interval for the mean (or for an individual value of y), we need to know the standard deviation of the sampling distribution for the estimator $\hat{y}$. For multiple-regression models, the form of this standard deviation is rather complicated. However, the regression routines of statistical software packages allow us to obtain the confidence intervals for mean values or individual values of y for any given combination of values of the independent variables. We illustrate with an example.

---

*This result is derived by expressing the percentage change in price y as $(y_1 - y_0)/y_0$, where y_1 = the value of y when, say, $x = 1$ and y_0 = the value of y when $x = 0$. Now let $y^* = \ln(y)$, and assume that the model is $y^* = \beta_0 + \beta_1 x$. Then

$$y = e^{y^*} = e^{\beta_0}e^{\beta_1 x} = \begin{cases} e^{\beta_0} & \text{when } x = 0 \\ e^{\beta_0}e^{\beta_1} & \text{when } x = 1 \end{cases}$$

Substituting, we have

$$\frac{y_1 - y_0}{y_0} = \frac{e^{\beta_0}e^{\beta_1} - e^{\beta_0}}{e^{\beta_0}} = e^{\beta_1} - 1$$

Example 12.5

Estimating $E(y)$ and Predicting y—Auction Price Model

Problem Refer to Examples 12.1–12.4 and the first-order model $E(y) = \beta_0 + \beta_1 x_1 + \beta_2 x_2$, where y = auction price of a grandfather clock, x_1 = age of the clock, and x_2 = number of bidders.

a. Estimate the average auction price for all 150-year-old clocks sold at an auction with 10 bidders. Use a 95% confidence interval. Interpret the result.

b. Predict the auction price for a single 150-year-old clock sold at an auction with 10 bidders. Use a 95% prediction interval. Interpret the result.

c. Suppose you want to predict the auction price for one clock that is 50 years old and has 2 bidders. How should you proceed?

Solution

a. Here, the key words *average* and *for all* imply that we want to estimate the mean of y, $E(y)$. We want a 95% confidence interval for $E(y)$ when $x_1 = 150$ years and $x_2 = 10$ bidders. A MINITAB printout for this analysis is shown in Figure 12.7. The confidence interval (highlighted under "**95% CI**") is (1381.4, 1481.9). Thus, we are 95% confident that the mean auction price for all 150-year-old clocks sold at an auction with 10 bidders lies between $1,381.4 and $1,481.9.

b. The key words *predict* and *for a single* imply that we want a 95% prediction interval for y when $x_1 = 150$ years and $x_2 = 10$ bidders. This interval (highlighted under "**95% PI**" on the MINITAB printout shown in Figure 12.7) is (1154.1, 1709.3). We say, with 95% confidence, that the auction price for a single 150-year-old clock sold at an auction with 10 bidders falls between $1,154.1 and $1,709.3.

c. Now we want to predict the auction price y for a single (*one*) grandfather clock when $x_1 = 50$ years and $x_2 = 2$ bidders. Consequently, we desire a 95% prediction interval for y. However, before we form this prediction interval, we should check to make sure that the selected values of the independent variables, $x_1 = 50$ and $x_2 = 2$, are both reasonable and within their respective sample ranges. If you examine the sample data shown in Table 12.1 (p. 617), you will see that the range for age is $108 \le x_1 \le 194$ and the range for number of bidders is $5 \le x_2 \le 15$. Thus, both selected values fall well *outside* their respective ranges. Recall the *Caution* box in

---

Regression Analysis: PRICE versus AGE, NUMBIDS

```
The regression equation is
PRICE = - 1339 + 12.7 AGE + 86.0 NUMBIDS

Predictor      Coef   SE Coef       T       P
Constant   -1339.0     173.8   -7.70   0.000
AGE        12.7406    0.9047   14.08   0.000
NUMBIDS     85.953     8.729    9.85   0.000

S = 133.485   R-Sq = 89.2%   R-Sq(adj) = 88.5%

Analysis of Variance

Source           DF        SS       MS       F       P
Regression        2   4283063  2141531  120.19   0.000
Residual Error   29    516727    17818
Total            31   4799790

Predicted Values for New Observations

New
Obs     Fit   SE Fit        95% CI              95% PI
  1  1431.7     24.6  (1381.4, 1481.9)   (1154.1, 1709.3)

Values of Predictors for New Observations

New
Obs   AGE   NUMBIDS
  1   150      10.0
```

Figure 12.7

MINITAB printout with 95% confidence intervals for grandfather clock model

Section 11.6 (p. 592) warning about the dangers of using the model to predict y for a value of an independent variable that is not within the range of the sample data. Doing so may lead to an unreliable prediction.

Look Back If we want to make the prediction requested in part **c**, we would need to collect additional data on clocks with the requested characteristics (i.e., $x_1 = 50$ years and $x_2 = 2$ bidders) and then refit the model.

Now Work Exercise 12.35

Statistics IN Action | Revisited   A First–Order Model for Condominium Sale Price

The developer of a Florida condominium complex wants to build a model for the sale price (y) of a condo unit (recorded in hundreds of dollars) and then use the model to predict the prices of future units sold. In addition to sale price, the **CONDO** file contains data on six potential predictor variables for a sample of 209 units sold. (See Table SIA12.1 on p. 614.) These independent variables are defined as follows:

$x_1 =$ floor location (i.e., floor height) of the unit (1, 2, 3, . . . , or 8)

$x_2 =$ distance, in units, from the elevator (1, 2, 3 . . . , or 15)

$x_3 =$ 1 if ocean view, 0 if non–ocean view

$x_4 =$ 1 if end unit, 0 if not an end unit

$x_5 =$ 1 if furnished unit, 0 if unfurnished

$x_6 =$ 1 if sold at public auction, 0 if sold at developer's price

MINITAB scatterplots (with the dependent variable, PRICE100, plotted against each of the potential predictors) for the data are shown in Figure SIA12.2. From the scatterplots, it appears that DISTANCE (x_2) and VIEW (x_3) may be the best predictors, since they both show stronger trends with sales price than the other independent variables. However, it would not be sound statistical practice to discard the other predictors based on graphs alone. Consequently, in this section, we will use all six independent variables in a multiple-regression model for sales price.

Consider the first-order multiple regression model

$$E(y) = \beta_0 + \beta_1 x_1 + \beta_2 x_2 + \beta_3 x_3 + \beta_4 x_4 + \beta_5 x_5 + \beta_6 x_6$$

The MINITAB printout for the regression analysis is shown in Figure SIA12.3. The global F-statistic ($F = 49.99$) and associated p-value (.000) shown on the printout indicate that the overall model is statistically useful in predicting auction price. The value of the adjusted R^2, however, indicates that the model explains only about 59% of the sample variation in price, and the standard deviation of the model ($s = 21.8$) implies that the model can predict price to within about

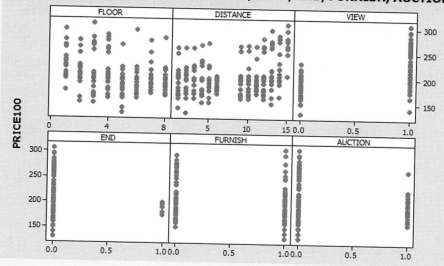

Figure SIA12.2

MINITAB scatterplots for CONDO data

(continued)

Statistics IN Action
(continued)

```
The regression equation is
PRICE100 = 187 - 1.16 FLOOR + 0.917 DISTANCE + 48.3 VIEW - 23.6 END
           + 6.95 FURNISH - 28.0 AUCTION


Predictor      Coef   SE Coef      T      P
Constant    186.990     4.693  39.84  0.000
FLOOR        -1.1599    0.7636  -1.52  0.130
DISTANCE      0.9168    0.3544   2.59  0.010
VIEW         48.250     3.281   14.70  0.000
END         -23.633     8.661   -2.73  0.007
FURNISH       6.945     3.243    2.14  0.033
AUCTION     -27.970     3.886   -7.20  0.000


S = 21.8163    R-Sq = 59.8%    R-Sq(adj) = 58.6%


Analysis of Variance

Source            DF      SS     MS      F      P
Regression         6  142750  23792  49.99  0.000
Residual Error   202   96142    476
Total            208  238893
```

Figure SIA12.3
MINITAB regression output for first-order model of condominium unit sale price

$2s = 43.6$ hundred dollars (i.e., \$4,360). While the model appears to be "statistically" useful in predicting auction price, the moderate value of R_a^2 and relatively large $2s$ value indicate that the model may not yield accurate predictions.

[*Note*: Not all of the independent variables have statistically significant *t*-values. However, we caution against dropping the insignificant variables from the model at this stage. One reason (discussed in Section 12.3) is that performing a large number of *t*-tests will yield an inflated probability of at least one Type I error. In later sections of this chapter, we develop other reasons for why the multiple *t*-test approach is not a good strategy for determining which independent variables to keep in the model.]

The MINITAB printout shown in Figure SIA12.4 gives a 95% prediction interval for auction price and a 95% confidence interval for the mean price for the following *x*-values:

$x_1 = 5$ (i.e., unit on the fifth floor)

$x_2 = 9$ (i.e., distance of nine units from the elevator)

$x_3 = 1$ (i.e., ocean view)

$x_4 = 0$ (i.e., not an end unit)

$x_5 = 0$ (i.e., an unfurnished unit)

$x_6 = 1$ (i.e., a unit sold at public auction)

```
Predicted Values for New Observations

New Obs     Fit  SE Fit       95% CI              95% PI
   1     209.72    2.87  (204.05, 215.39)  (166.33, 253.11)


Values of Predictors for New Observations

New Obs  FLOOR  DISTANCE  VIEW      END   FURNISH  AUCTION
   1      5.00      9.00  1.00  0.000000  0.000000     1.00
```

Figure SIA12.4
MINITAB printout with 95% confidence and prediction intervals

The 95% confidence interval of (204.05, 215.39) implies that, for all condo units with these *x*-values, the mean auction price falls between 204.05 and 215.39 hundred dollars, with 95% confidence. The 95% prediction interval of (166.33, 253.11) implies that, for an individual condo unit with these *x*-values, the auction price falls between 166.33 and 253.11 hundred dollars, with 95% confidence. Note the wide range of the prediction interval. This is due to the large magnitude of the model's standard deviation $s = 21.8$ hundred dollars. Again, although the model is deemed statistically useful in predicting auction price, it may not be "practically" useful. To reduce the magnitude of s, we will need to improve the model's predictive ability. (We consider such a model in the next Statistics in Action Revisited section.)

Data Set: CONDO

Exercises 12.29–12.39

Understanding the Principles

12.29 Explain why we use $\hat{y}$ as an estimate of $E(y)$ and to predict y.

12.30 Which interval will be narrower, a 95% confidence interval for $E(y)$ or a 95% prediction interval for y? (Assume that the values of the x's are the same for both intervals.)

Applying the Concepts—Basic

12.31 Characteristics of lead users. Refer to the *Creativity and Innovation Management* (Feb. 2008) study of lead users of children's computer games, Exercise 12.13 (p. 628). Recall that the researchers modeled lead-user rating (y, measured on a 5-point scale) as a function of gender ($x_1 = 1$ if female, 0 if male), age (x_2, years), degree of centrality (x_3, measured as the number of direct ties to other peers in the network), and betweenness centrality (x_4, measured as the number of shortest paths between peers). The least squares prediction equation was $\hat{y} = 3.58 + .01x_1 - .06x_2 - .01x_3 + .42x_4$

a. Compute the predicted lead-user rating of a 10-year-old female child with 5 direct ties to other peers in her social network and with 2 shortest paths between peers.

b. Compute an estimate for the mean lead-user rating of all 8-year-old male children with 10 direct ties to other peers and with 4 shortest paths between peers.

12.32 Predicting runs scored in baseball. Refer to the *Chance* (Fall 2000) study of runs scored in Major League Baseball games, Exercise 12.17 (p. 629). Multiple regression was used to model total number of runs scored (y) of a team during the season as a function of number of walks (x_1), number of singles (x_2), number of doubles (x_3), number of triples (x_4), number of home runs (x_5), number of stolen bases (x_6), number of times caught stealing (x_7), number of strikeouts (x_8), and total number of outs (x_9). Using the β-estimates given in Exercise 12.17, predict the number of runs scored by your favorite Major League Baseball team last year. How close is the predicted value to the actual number of runs scored by your team? [*Note:* You can find data on your favorite team on the Internet at *www.major-leaguebaseball.com.*]

12.33 Reality TV and cosmetic surgery. Refer to the *Body Image: An International Journal of Research* (March 2010) study of the impact of reality TV shows on one's desire to undergo cosmetic surgery, Exercise 12.23 (p. 631). Recall that psychologists used multiple regression to model desire to have cosmetic surgery (y) as a function of gender (x_1), self-esteem (x_2), body satisfaction (x_3), and impression of reality TV (x_4). The SAS printout below shows a confidence interval for $E(y)$ for each of the first five students in the study.

a. Interpret the confidence interval for $E(y)$ for student 1.

b. Interpret the confidence interval for $E(y)$ for student 4.

12.34 Deep-space survey of quasars. Refer to *The Astronomical Journal* study of quasars, presented in Exercise 12.24 (p. 631). Recall that a first-order model was used to relate a quasar's equivalent width (y) to its redshift (x_1), line flux (x_2), line luminosity (x_3), and AB_{1450}(x_4). A portion of the SPSS spreadsheet showing 95% prediction intervals for y for the first five observations in the data set is reproduced below. Interpret the interval corresponding to the fifth observation.

12.35 Cooling method for gas turbines. Refer to the *Journal of Engineering for Gas Turbines and Power* (Jan. 2005) study of a high-pressure inlet fogging method for a gas turbine engine, presented in Exercise 12.25 (p. 632). Recall that you fitted a first-order model for heat rate (y) as a function of speed (x_1), inlet temperature (x_2), exhaust temperature (x_3), cycle pressure ratio (x_4), and air mass flow rate (x_5). A MINITAB printout with both a 95% confidence interval

SAS Output for Exercise 12.33

Dependent Variable: DESIRE

Output Statistics

Obs	GENDER	SELFESTM	BODYSAT	IMPREAL	Dependent Variable	Predicted Value	Std Error Mean Predict	95% CL Mean	
1	0	24	3	4	11.0000	13.8655	0.2235	13.4243	14.3067
2	0	20	3	4	13.0000	14.0573	0.2519	13.5598	14.5547
3	0	25	4	5	11.0000	13.9883	0.2899	13.4159	14.5607
4	1	22	9	4	11.0000	9.8409	0.5318	8.7908	10.8910
5	0	8	1	6	18.0000	16.2634	0.5472	15.1830	17.3438

SPSS Output for Exercise 12.34

	REDSHIFT	LINEFLUX	LINELUM	AB1450	RFEWIDTH	LCL 95	UCL 95
1	2.81	-13.48	45.29	19.50	117	101.29	172.22
2	3.07	-13.73	45.13	19.65	82	59.62	125.89
3	3.45	-13.87	45.11	18.93	33	-35.09	37.04
4	3.19	-13.27	45.63	18.59	92	63.76	139.25
5	3.07	-13.56	45.30	19.59	114	90.69	158.57

MINITAB output for Exercise 12.35

```
Predicted Values for New Observations

New
Obs     Fit   SE Fit       95% CI                95% PI
  1  12632.5   237.3  (12157.9, 13107.1)   (11599.6, 13665.5)


Values of Predictors for New Observations

New
Obs   RPM   INLET-TEMP   EXH-TEMP   CPRATIO   AIRFLOW
  1  7500        1000        525      13.5      10.0
```

Data for Exercise 12.36

Station	Avg. Annual Precipitation y (inches)	Altitude x_1 (feet)	Latitude x_2 (degrees)	Distance from Coast x_3 (miles)
1. Eureka	39.57	43	40.8	1
2. Red Bluff	23.27	341	40.2	97
3. Thermal	18.20	4152	33.8	70
4. Fort Bragg	37.48	74	39.4	1
5. Soda Springs	49.26	6752	39.3	150
⋮	⋮	⋮	⋮	⋮
26. San Diego	9.94	19	32.7	5
27. Daggett	4.25	2105	34.1	85
28. Death Valley	1.66	−178	36.5	194
29. Crescent City	74.87	35	41.7	1
30. Colusa	15.95	60	39.2	91

Source: Taylor, P. I. "A pedagogic application of multiple regression analysis." *Geography*, July 1980, Vol. 65, pp. 203–212. Reprinted with permission from the author.

for $E(y)$ and a prediction interval for y, for selected values of the x's, is shown at the bottom of page 637.

a. Interpret the 95% prediction interval for y in the words of the problem.

b. Interpret the 95% confidence interval for $E(y)$ in the words of the problem.

c. Will the confidence interval for $E(y)$ always be narrower than the prediction interval for y? Explain.

Applying the Concepts—Intermediate

12.36 California rain levels. An article published in *Geography* (July 1980) used multiple regression to predict annual rainfall levels in California. Data on the average annual precipitation (y), altitude (x_1), latitude (x_2), and distance from the Pacific coast (x_3) for 30 meteorological stations scattered throughout California are saved in the **CALIRAIN** file. (Selected observations are listed in the table above.) Consider the first-order model $y = \beta_0 + \beta_1 x_1 + \beta_2 x_2 + \beta_3 x_3 + \varepsilon$.

a. Fit the model to the data and give the least squares prediction equation.

b. Is there evidence that the model is useful in predicting annual precipitation y? Test, using $\alpha = .05$.

c. Find a 95% prediction interval for y for the Giant Forest meteorological station (station 9). Interpret the interval.

12.37 Study of contaminated fish. Refer to Exercise 12.22 (p. 630) and the U.S. Army Corps of Engineers data on contaminated fish. You fit the first-order model $E(y) = \beta_0 + \beta_1 x_1 + \beta_2 x_2 + \beta_3 x_3$ to the data saved in the **FISHDDT** file, where y = DDT level (parts per million), x_1 = number of miles upstream, x_2 = length (centimeters), and x_3 = weight (in grams). Predict, with 95% confidence, the DDT level of a fish caught 100 miles upstream with a length of 40 centimeters and a weight of 800 grams. Interpret the result.

12.38 Boiler drum production. In a production facility, an accurate estimate of hours needed to complete a task is crucial to management in making such decisions as hiring the proper number of workers, quoting an accurate deadline for a client, or performing cost analyses regarding budgets. A manufacturer of boiler drums wants to use regression to predict the number of hours needed to erect the drums in future projects. To accomplish this task, data on 36 boilers were collected. In addition to hours (y), the variables measured were boiler capacity (x_1 = lb/hr), boiler design pressure (x_2 = pounds per square inch, or psi), boiler type ($x_3 = 1$ if industry field erected, 0 if utility field erected), and drum type ($x_4 = 1$ if steam, 0 if mud). The data are saved in the **BOILERS** file. (Selected observations are shown in the table below.)

a. Fit the model $E(y) = \beta_0 + \beta_1 x_1 + \beta_2 x_2 + \beta_3 x_3 + \beta_4 x_4$ to the data and give the prediction equation.

b. Conduct a test for the global utility of the model. Use $\alpha = .01$.

c. Find a 95% confidence interval for $E(y)$ when $x_1 = 150,000$, $x_2 = 500$, $x_3 = 1$, and $x_4 = 0$. Interpret the result.

d. What type of interval would you use if you want to estimate the average number of hours required to erect all industrial mud boilers with a capacity of 150,000 lb/hr and a design pressure of 500 psi?

12.39 Arsenic in groundwater. Refer to the *Environmental Science & Technology* (Jan. 2005) study of the reliability of a commercial kit designed to test for arsenic in groundwater, presented in Exercise 12.21 (p. 630). Using the data in the **ASWELLS** file, you fit a first-order model for arsenic level (y) as a function of latitude, longitude, and depth. On the basis of the model statistics, the researchers concluded that the arsenic level is highest at a low latitude, high longitude, and low depth. Do you agree? If so, find a 95% prediction interval for arsenic level for the lowest latitude, highest longitude, and lowest depth that are within the range of the sample data. Interpret the result.

Data for Exercise 12.38

Hours y	Boiler Capacity x_1	Design Pressure x_2	Boiler Type x_3	Drum Type x_4
3,137	120,000	375	1	1
3,590	65,000	750	1	1
4,526	150,000	500	1	1
10,825	1,073,877	2,170	0	1
4,023	150,000	325	1	1
⋮	⋮	⋮	⋮	⋮
4,206	441,000	410	1	0
4,006	441,000	410	1	0
3,728	627,000	1,525	0	0
3,211	610,000	1,500	0	0
1,200	30,000	325	1	0

Based on data provided by Dr. Kelly Uscategui, University of Connecticut.

PART II: MODEL BUILDING IN MULTIPLE REGRESSION

12.5 Interaction Models

In Section 12.2, we demonstrated the relationship between $E(y)$ and the independent variables in a first-order model. When $E(y)$ is graphed against any one variable (say, x_1) for fixed values of the other variables, the result is a set of *parallel* straight lines. (See Figure 12.3, p. 619). When this situation occurs (as it always does for a first-order model), we say that the relationship between $E(y)$ and any one independent variable *does not depend* on the values of the other independent variables in the model.

However, if the relationship between $E(y)$ and x_1 does, in fact, depend on the values of the remaining x's held fixed, then the first-order model is not appropriate for predicting y. In this case, we need another model that will take into account this dependence. Such a model includes the *cross products* of two or more x's.

For example, suppose that the mean value $E(y)$ of a response y is related to two quantitative independent variables x_1 and x_2 by the model

$$E(y) = 1 + 2x_1 - x_2 + x_1x_2$$

A graph of the relationship between $E(y)$ and x_1 for $x_2 = 0, 1$, and 2 is displayed in the MINITAB graph shown in Figure 12.8.

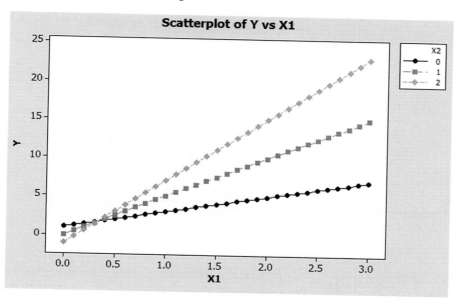

Figure 12.8

MINITAB graphs of $1 + 2x_1 - x_2 + 3x_1x_2$ for $x_2 = 0, 1, 2$

Note that the graph shows three nonparallel straight lines. You can verify that the slopes of the lines differ by substituting each of the values $x_2 = 0, 1$, and 2 into the equation. For $x_2 = 0$,

$$E(y) = 1 + 2x_1 - (0) + x_1(0) = 1 + 2x_1 \qquad \text{(slope} = 2)$$

For $x_2 = 1$,

$$E(y) = 1 + 2x_1 - (1) + x_1(1) = 3x_1 \qquad \text{(slope} = 3)$$

For $x_2 = 2$,

$$E(y) = 1 + 2x_1 - (2) + x_1(2) = -1 + 4x_1 \qquad \text{(slope} = 4)$$

Note that the slope of each line is represented by $\beta_1 + \beta_3x_2 = 2 + x_2$. Thus, the effect on $E(y)$ of a change in x_1 (i.e., the slope) now *depends* on the value of x_2. When this situation occurs, we say that x_1 and x_2 **interact**. The cross-product term, x_1x_2, is called an **interaction term**, and the model $E(y) = \beta_0 + \beta_1x_1 + \beta_2x_2 + \beta_3x_1x_2$ is called an **interaction model** with two quantitative variables.

> **An Interaction Model Relating $E(y)$ to Two Quantitative Independent Variables**
>
> $$E(y) = \beta_0 + \beta_1 x_1 + \beta_2 x_2 + \beta_3 x_1 x_2$$
>
> where
>
> $(\beta_1 + \beta_3 x_2)$ represents the change in $E(y)$ for every one-unit increase in x_1, holding x_2 fixed
>
> $(\beta_2 + \beta_3 x_1)$ represents the change in $E(y)$ for every one-unit increase in x_2, holding x_1 fixed

A three-dimensional graph (generated by MINITAB) of an interaction model in two quantitative x's is shown in Figure 12.9. The interaction model traces a ruled surface (twisted plane) in three-dimensional space. Unlike the flat planar surface displayed in Figure 12.4 (p. 620). If we slice the twisted plane at a fixed value of x_2, we obtain a straight line relating $E(y)$ to x_1; however, the slope of the line will change as we change the value of x_2. Consequently, *an interaction model is appropriate when the linear relationship between y and one independent variable depends on the value of the other independent variable.* The next example illustrates this idea.

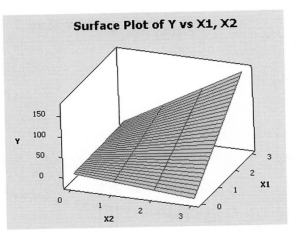

Figure 12.9

MINITAB 3-dimensional graph of $1 + 2x_1 - x_2 + 3x_1 x_2$

Example 12.6

Evaluating an Interaction Model— Clock Auction Prices

Problem Refer to Examples 12.1–12.4. Suppose the collector of grandfather clocks, having observed many auctions, believes that the *rate of increase* in the auction price with age will be driven upward by a large number of bidders. Thus, instead of a relationship like that shown in Figure 12.10a, in which the rate of increase in price with age is the same for any number of bidders, the collector believes that the relationship is like that shown in Figure 12.10b. Note that as the number of bidders increases from 5 to 15, the slope of the price-versus-age line increases.

Consequently, the following interaction model is proposed:

$$y = \beta_0 + \beta_1 x_1 + \beta_2 x_2 + \beta_3 x_1 x_2 + \varepsilon$$

The 32 data points listed in Table 12.1 were used to fit the model with interaction. A portion of the MINITAB printout is shown in Figure 12.11.

a. Use the global F-test at $\alpha = .05$ to test the overall utility of the model.

b. Test the hypothesis (at $\alpha = .05$) that the price–age slope increases as the number of bidders increases—that is, that age and number of bidders, x_2, interact positively.

c. Estimate the change in auction price y of a 150-year-old grandfather clock for each additional bidder.

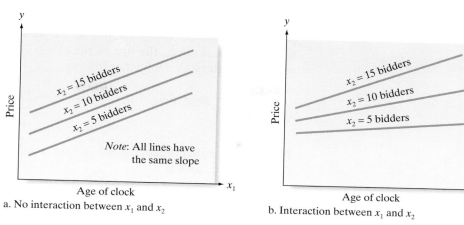

Figure 12.10

Examples of no-interaction and interaction models

a. No interaction between x_1 and x_2

b. Interaction between x_1 and x_2

Regression Analysis: PRICE versus AGE, NUMBIDS, AGEBID

```
The regression equation is
PRICE = 320 + 0.88 AGE - 93.3 NUMBIDS + 1.30 AGEBID


Predictor    Coef   SE Coef      T      P
Constant    320.5    295.1    1.09  0.287
AGE         0.878    2.032    0.43  0.669
NUMBIDS    -93.26    29.89   -3.12  0.004
AGEBID     1.2978   0.2123    6.11  0.000


S = 88.9145    R-Sq = 95.4%    R-Sq(adj) = 94.9%


Analysis of Variance

Source          DF      SS       MS       F      P
Regression       3  4578427  1526142  193.04  0.000
Residual Error  28   221362     7906
Total           31  4799790
```

Figure 12.11

MINITAB printout of interaction model of auction price

Solution

a. The global F-test is used to test the null hypothesis

$$H_0: \beta_1 = \beta_2 = \beta_3 = 0$$

The test statistic and p-value of the test (highlighted on the MINITAB printout) are $F = 193.04$ and $p = 0$, respectively. Since $\alpha = .05$ exceeds the p-value, there is sufficient evidence to conclude that the model fit is a statistically useful predictor of the auction price y.

b. The hypotheses of interest to the collector concern the interaction parameter β_3. Specifically,

$$H_0: \beta_3 = 0$$
$$H_a: \beta_3 > 0$$

Since we are testing an individual β parameter, a t-test is required. The test statistic and the two-tailed p-value (highlighted on the printout) are $t = 6.11$ and $p = 0$, respectively. The upper-tailed p-value, obtained by dividing the two-tailed p-value in half, is $0/2 = 0$. Since $\alpha = .05$ exceeds the p-value, the collector can reject H_0 and conclude that the rate of change of the mean price of the clocks with age increases as the number of bidders increases; that is, x_1 and x_2 interact positively. Thus, it appears that the interaction term should be included in the model.

c. To estimate the change in auction price y for every one-unit increase in number of bidders, x_2, we need to estimate the slope of the line relating y to x_2 when the age of the clock, x_1, is 150 years old. An analyst who is not careful may estimate this slope as $\hat{\beta}_2 = -93.26$. Although the coefficient of x_2 is negative, this does *not* imply that the auction price decreases as the number of bidders increases. Since interaction is present, the rate of change (slope) of the mean auction price with the number of bidders *depends* on x_1, the age of the clock. For a fixed value of age (x_1), we can rewrite the interaction model as follows:

$$E(y) = \beta_0 + \beta_1 x_1 + \beta_2 x_2 + \beta_3 x_1 x_2 = \underbrace{\{\beta_0 + \beta_1 x_1\}}_{y\text{-intercept}} + \underbrace{\{\beta_2 + \beta_3 x_1\}}_{\text{slope}} x_2$$

Thus, the estimated rate of change of y for a unit increase in x_2 (one new bidder) for a 150-year-old clock is

$$\text{Estimated } x_2 \text{ slope} = \hat{\beta}_2 + \hat{\beta}_3 x_1 = -93.26 + 1.30(150) = 101.74$$

In other words, we estimate that the auction price of a 150-year-old clock will *increase* by about $101.74 for every additional bidder.

Look Back Although the rate of increase will vary as x_1 is changed, it will remain positive for the range of values of x_1 included in the sample. Extreme care is needed in interpreting the signs and sizes of coefficients in a multiple-regression model.

Now Work Exercise 12.46

Example 12.6 illustrates an important point about conducting t-tests on the β parameters in the interaction model. The "most important" β parameter in this model is the interaction β, β_3. [Note that this β is also the one associated with the highest-order term in the model, $x_1 x_2$.*] Consequently, we will want to test $H_0: \beta_3 = 0$ after we have determined that the overall model is useful in predicting y. Once interaction is detected (as in Example 12.6), however, tests on the first-order terms x_1 and x_2 should *not* be conducted, since they are meaningless tests; the presence of interaction implies that both x's are important.

⚠ **CAUTION** Once interaction has been deemed important in the model $E(y) = \beta_0 + \beta_1 x_1 + \beta_2 x_2 + \beta_3 x_1 x_2$, do not conduct t-tests on the β coefficients of the first-order terms x_1 and x_2. These terms should be kept in the model regardless of the magnitude of their associated p-values shown on the printout. ▲

We close this section with a comment: You will probably never know *a priori* whether interaction exists between two independent variables; consequently, you will need to fit and test the interaction term to determine its importance.

Exercises 12.40–12.55

Understanding the Principles

12.40 If two variables x_1 and x_2 do not interact, how would you describe their effect on the mean response $E(y)$?

12.41 Write an interaction model relating the mean value of y, $E(y)$, to
  a. two quantitative independent variables
  b. three quantitative independent variables [*Hint:* Include all possible two-way cross-product terms.]

Learning the Mechanics

12.42 Suppose the true relationship between $E(y)$ and the quantitative independent variables x_1 and x_2 is

$$E(y) = 3 + x_1 + 2x_2 - x_1 x_2$$

  a. Describe the corresponding three-dimensional response surface.
  b. Plot the linear relationship between y and x_2 for $x_2 = 0, 1, 2$, where $0 \le x_2 \le 5$.
  c. Explain why the lines you plotted in part **b** are not parallel.
  d. Use the lines you plotted in part **b** to explain how changes in the settings of x_1 and x_2 affect $E(y)$.
  e. Use your graph from part **b** to determine how much $E(y)$ changes when x_1 is changed from 2 to 0 and x_2 is simultaneously changed from 4 to 5.

12.43 Suppose you fit the interaction model

$$y = \beta_0 + \beta_1 x_1 + \beta_2 x_2 + \beta_3 x_1 x_2 + \varepsilon$$

*The order of a term is equal to the sum of the exponents of the quantitative variables included in the term. Thus, when x_1 and x_2 are both quantitative variables, the cross product, $x_1 x_2$, is a second-order term.

to $n = 32$ data points and obtain the following results:

$$SS_{yy} = 479 \qquad SSE = 21 \qquad \hat{\beta}_3 = 10 \qquad s_{\hat{\beta}_3} = 4$$

a. Find R^2 and interpret its value.

b. Is the model adequate for predicting y? Test at $\alpha = .05$.

c. Use a graph to explain the contribution of the x_1x_2 term to the model.

d. Is there evidence that x_1 and x_2 interact? Test at $\alpha = .05$.

12.44 MINITAB was used to fit the model

$$y = \beta_0 + \beta_1x_1 + \beta_2x_2 + \beta_3x_1x_2 + \varepsilon$$

to $n = 15$ data points. The resulting printout is shown below.

a. What is the prediction equation for the response surface?

b. Describe the geometric form of the response surface of part **a**.

c. Plot the prediction equation for the case when $x_2 = 1$. Do this twice more on the same graph for the cases when $x_2 = 3$ and $x_2 = 5$.

d. Explain what it means to say that x_1 and x_2 interact. Explain why the graph you plotted in part **c** suggests that x_1 and x_2 interact.

e. Specify the null and alternative hypotheses you would use to test whether x_1 and x_2 interact.

f. Conduct the hypothesis test of part **e**, using $\alpha = .01$.

```
The regression equation is
Y = -2.55 + 3.82 X1 + 2.63 X2 -1.29 X1X2


Predictor    Coef   SE Coef     T      P
Constant   -2.550    1.142   -2.23  0.043
X1          3.815    0.529    7.22  0.000
X2          2.630    0.344    7.64  0.000
X1X2       -1.285    0.159   -8.06  0.000


S = 0.713    R-Sq = 85.6%    R-Sq(adj) = 81.6%


Analysis of Variance

Source          DF      SS      MS      F      P
Regression       3  33.149  11.050  21.75  0.000
Residual Error  11   5.587   0.508
Total           14  38.736
```

Applying the Concepts — Basic

12.45 Whales entangled in fishing gear. Refer to the *Marine Mammal Science* (April 2010) study of whales entangled in fishing gear, Exercise 12.15 (p. 629). Recall that the length (y) of an entangled whale (in meters) was modeled as a function of water depth of the entanglement (x_1, in meters) and distance of the entanglement from land (x_2, in miles).

a. Give the equation of an interaction model for length (y) as a function of the two independent variables.

b. The researchers theorize that the length of an entangled whale will increase linearly as the water depth increases. In terms of the parameters in the model, part **a**, write the slope of the line relating length (y) to water depth (x_1) for a distance of $x_2 = 10$ miles from land.

c. Repeat part **b** for a distance of $x_2 = 25$ miles from land.

12.46 Role of retailer interest in shopping behavior. Retail
[NW] interest is defined by marketers as the level of interest a consumer has in a given retail store. Marketing professors at the University of Tennessee at Chattanooga and the University of Alabama investigated the role of retailer interest in consumers' shopping behavior (*Journal of Retailing*, Summer 2006). Using survey data collected on $n = 375$ consumers, the professors developed an interaction model for $y =$ willingness of the consumer to shop at a retailer's store in the future (called "repatronage intentions") as a function of $x_1 =$ consumer satisfaction and $x_2 =$ retailer interest. The regression results are shown below.

Variable	$\hat{\beta}$	t-Value	p-Value
Satisfaction (x_1)	.426	7.33	$< .01$
Interest (x_2)	.044	0.85	$> .10$
Satisfaction $\times$ Interest (x_1x_2)	-.157	-3.09	$< .01$

$R^2 = .65, F = 226.35, p$-value $< .001$

a. Is the overall model statistically useful in predicting y? Test, using $\alpha = .05$.

b. Conduct a test for interaction at $\alpha = .05$.

c. Use the β-estimates to sketch the estimated relationship between repatronage intentions (y) and satisfaction (x_1) when retailer interest is $x_2 = 1$ (a low value).

d. Repeat part **c** for the case when retailer interest is $x_2 = 7$ (a high value).

e. Put the two lines you sketched in parts **c** and **d** on the same graph to illustrate the nature of the interaction.

12.47 Defects in nuclear missile housing parts. The technique of multivariable testing was discussed in *The Journal of the Reliability Analysis Center* (First Quarter, 2004). Multivariable testing was shown to improve the quality of carbon-foam rings used in nuclear missile housings. The rings are produced via a casting process that involves mixing ingredients, oven curing, and carving the finished part. One type of defect analyzed was the number y of black streaks in the manufactured ring. Two variables found to affect the number of defects were turntable speed (revolutions per minute) x_1 and cutting blade position (inches from center) x_2.

a. The researchers discovered "an interaction between blade position and turntable speed." Hypothesize a regression model for $E(y)$ that incorporates this interaction.

b. Interpret what it means practically to say that "blade position and turntable speed interact."

c. The researchers reported a positive linear relationship between number of defects (y) and turntable speed (x_1), but found that the slope of the relationship was much steeper for lower values of cutting blade position (x_2). What does this imply about the interaction term in the model you hypothesized in part **a**? Explain.

12.48 Psychology of waiting in line. While waiting in a long line for service (e.g., to use an ATM or at the post office), at some point you may decide to leave the line. The *Journal of Consumer Research* (Nov. 2003) published a study of consumer behavior while waiting in a line. College students (sample size $n = 148$) were asked to imagine that they were waiting in line at a post office to mail a package and that the estimated waiting time was 10 minutes or less. After a 10-minute wait, students were asked about their level of negative feelings (annoyed, anxious) on a scale of 1 (strongly disagree) to 9 (strongly agree). Before answering, however, the students were informed about how many people were ahead of them and behind them in the line. The researchers used regression to relate negative feelings score (y) to number ahead in line (x_1) and number behind in line (x_2).

a. The researchers fit an interaction model to the data. Write the hypothesized equation of this model.
b. In the words of the problem, explain what it means to say that "x_1 and x_2 interact to affect y."
c. A t-test for the interaction β resulted in a p-value greater than .25. Interpret this result.
d. From their analysis, the researchers concluded that "the greater the number of people ahead, the higher [is] the negative feeling score" and "the greater the number of people behind, the lower [is] the negative feeling score." Use this information to determine the signs of $\hat{\beta}_1$ and $\hat{\beta}_2$ in the model.

12.49 Eye movement and spatial distortion. A study of how eye movement behavior can distort one's judgment of the location of an object was published in *Advances in Cognitive Psychology* (Vol. 6, 2010). The medical term for fast voluntary movement of the eyes when focusing on an object is *saccadic* eye movement. The researchers designed several experiments in which volunteers fixated their eyes on a cross in the middle of a computer screen. A probe was then spatially extended near the cross and each volunteer was asked to judge the location of the probe. Saccadic eye movement was monitored during each session. The researchers used spatial position of the probe (x_1, measured in degrees) and position of the cross (x_2, degrees) to predict the amplitude (y) of saccadic eye movement. The following model was fit to the data:

$$E(y) = \beta_0 + \beta_1 x_1 + \beta_2 x_2 + \beta_3 x_1 x_2$$

a. The model yielded $R^2 = .994$. Interpret this result.
b. In the words of the problem, what does it mean to say that "x_1 and x_2 interact"?
c. The least squares prediction equation was determined as: $\hat{y} = .91 + .70x_1 - .06x_2 - .03x_1 x_2$. Illustrate interaction by graphing the relationship between predicted amplitude ($\hat{y}$) and cross position (x_2) for probe positions $x_1 = 3.5$ and $x_1 = 6.5$.

Applying the Concepts—Intermediate

12.50 Reality TV and cosmetic surgery. Refer to the *Body Image: An International Journal of Research* (March 2010) study of the influence of reality TV shows on one's desire to undergo cosmetic surgery, Exercise 12.23 (p. 631). Recall that psychologists modeled desire to have cosmetic surgery (y) as a function of gender (x_1), self-esteem (x_2), body satisfaction (x_3), and impression of reality TV (x_4). For this exercise, consider only the independent variables gender (x_1) and impression of reality TV (x_4).

a. The research psychologists theorize that the impact of one's impression of reality TV on level of desire for cosmetic surgery will be greater for females than for males. Does this theory imply that the independent variables x_1 and x_4 interact, or that there is no interaction? Explain.
b. Fit the interaction model, $E(y) = \beta_0 + \beta_1 x_1 + \beta_2 x_4 + \beta_3 x_1 x_4$, to the simulated data saved in the **BODYIMAGE** file.
c. Use the results, part **b**, to carry out a test for interaction. Make your conclusion using $\alpha = .05$.

12.51 Child abuse report. Licensed therapists are mandated by law to report child abuse by their clients. This law requires the therapist to breach confidentiality and possibly lose the client's trust. A national survey of licensed psychotherapists was conducted to investigate clients' reactions to legally mandated child-abuse reports (*American Journal of Orthopsychiatry*, Jan. 1997). The sample consisted of 303 therapists who had filed a child-abuse report against at least one of their clients. The researchers were interested in finding the best predictors of a client's reaction (y) to the report, where y is measured on a 30-point scale. (The higher the value, the more favorable was the client's response to the report.) The independent variables found to have the most predictive power are as follows:

x_1: Therapist's age (years)
x_2: Therapist's gender (1 if male, 0 if female)
x_3: Degree of therapist's role strain (25 − point scale)
x_4: Strength of client − therapist relationship (40 − point scale)
x_5: Type of case (1 if family, 0 if not)
$x_1 x_2$: Age × Gender interaction

a. Hypothesize a first-order model relating y to each of the five independent variables and the interaction term.
b. Give the null hypothesis for testing the contribution of x_4, strength of client–therapist relationship, to the model.
c. The test statistic for the test suggested in part **b** was $t = 4.408$, with an associated p-value of .001. Interpret this result.
d. The estimated β coefficient for the $x_1 x_2$ interaction term was positive and highly significant ($p < .001$). According to the researchers, "This interaction suggests that ... as the age of the therapist increased, ... male therapists were less likely to get negative client reactions than were female therapists." Do you agree?
e. For the model presented here, $R^2 = .2946$. Interpret this value.

12.52 Unconscious self-esteem study. Psychologists define *implicit* self-esteem as unconscious evaluations of one's worth or value. In contrast, *explicit* self-esteem refers to the extent to which a person consciously considers oneself as valuable and worthy. An article published in *Journal of Articles in Support of the Null Hypothesis* (March 2006) investigated whether implicit self-esteem is really unconscious. A sample of 257 college undergraduate students completed a questionnaire designed to measure implicit self-esteem and explicit self-esteem. Thus, an implicit self-esteem score (x_1) and explicit self-esteem score (x_2) were obtained for each. (*Note*: Higher scores indicate higher levels of self-esteem.) Also, a second questionnaire was administered in order to obtain each subject's estimate of his or her level of implicit self-esteem. The score obtained from this questionnaire was called an estimated implicit self-esteem score (x_3). Finally, the researchers computed two measures of accuracy in estimating implicit self-esteem: $y_1 = (x_3 - x_1)$ and $y_2 = |x_3 - x_1|$.

a. The researchers fit the interaction model $E(y_1) = \beta_0 + \beta_1 x_1 + \beta_2 x_2 + \beta_3 x_1 x_2$. The t-test of the interaction term, β_3, was "nonsignificant," with a p-value $> .10$. However, both t-tests of β_1 and β_2 were statistically significant (p-value $< .001$). Interpret these results practically.

b. The researchers also fit the interaction model $E(y_2) = \beta_0 + \beta_1 x_1 + \beta_2 x_2 + \beta_3 x_1 x_2$. The t-test on the interaction term, β_3, was "significant," with a p-value $< .001$. Interpret this result practically.

12.53 Factors that affect an auditor's judgment. A study was conducted to determine the effects of linguistic delivery style and client credibility on auditors' judgments (*Advances in Accounting and Behavioral Research*, 2003). Each of 200 auditors from "Big 5" accounting firms were asked to assume that he or she was an audit team supervisor of a new manufacturing client and was performing an analytical review of the client's financial statement. The researchers gave the auditors different information on the client's credibility and the linguistic delivery style of the client's explanation. Each auditor then provided an assessment of the likelihood that the client's explanation accounts for the fluctuation in the financial statement. The three variables of interest—credibility (x_1), linguistic delivery style (x_2), and likelihood (y)—were all measured on a numerical scale. Regression analysis was used to fit the interaction model $y = \beta_0 + \beta_1 x_1 + \beta_2 x_2 + \beta_3 x_1 x_2 + \varepsilon$. The results are summarized in the table below.

a. Interpret the phrase "client credibility and linguistic delivery style interact" in the words of the problem.

b. Give the null and alternative hypotheses for testing the overall adequacy of the model.

c. Conduct the test suggested in part **b,** using the information in the table.

d. Give the null and alternative hypotheses for testing whether client credibility and linguistic delivery style interact.

e. Conduct the test suggested in part **d,** using the information in the table.

f. The researchers estimated the slope of the likelihood–linguistic delivery style line at a low level of client credibility $(x_1 = 22)$. Obtain this estimate and interpret it in the words of the problem.

g. The researchers also estimated the slope of the likelihood–linguistic delivery style line at a high level of client credibility $(x_1 = 46)$. Obtain this estimate and interpret it in the words of the problem.

12.54 Arsenic in groundwater. Refer to the *Environmental Science & Technology* (Jan. 2005) study of the reliability of a commercial kit to test for arsenic in groundwater, presented in Exercise 12.21 (p. 630). Recall that you fit a first-order model for arsenic level (y) as a function of latitude (x_1), longitude (x_2), and depth (x_3) to data saved in the **ASWELLS** file.

a. Write a model for arsenic level (y) that includes first-order terms for latitude, longitude, and depth, as well as terms for interaction between latitude and depth and interaction between longitude and depth.

b. Use statistical software to fit the interaction model you wrote in part **a** to the data in the **ASWELLS** file. Give the least squares prediction equation.

c. Conduct a test (at $\alpha = .05$) to determine whether latitude and depth interact to affect arsenic level.

d. Conduct a test (at $\alpha = .05$) to determine whether longitude and depth interact to affect arsenic level.

e. Interpret practically the results of the tests you conducted in parts **c** and **d.**

12.55 Cooling method for gas turbines. Refer to the *Journal of Engineering for Gas Turbines and Power* (Jan. 2005) study of a high-pressure inlet fogging method for a gas turbine engine, presented in Exercise 12.25 (p. 632). Recall that you fit a first-order model for heat rate (y) as a function of speed (x_1), inlet temperature (x_2), exhaust temperature (x_3), cycle pressure ratio (x_4), and air mass flow rate (x_5) to data saved in the **GASTURBINE** file.

a. Researchers hypothesize that the linear relationship between heat rate (y) and temperature (both inlet and exhaust) depends on air mass flow rate. Write a model for heat rate that incorporates the researchers' theories.

b. Use statistical software to fit the interaction model you wrote in part **a** to the data in the **GASTURBINE** file. Give the least squares prediction equation.

c. Conduct a test (at $\alpha = .05$) to determine whether inlet temperature and air mass flow rate interact to affect heat rate.

d. Conduct a test (at $\alpha = .05$) to determine whether exhaust temperature and air mass flow rate interact to affect heat rate.

e. Interpret practically the results of the tests you conducted in parts **c** and **d.**

Results for Exercise 12.53

	Beta Estimate	Std Error	t-statistic	p-value
Constant	15.865	10.980	1.445	0.150
Client credibility (x_1)	0.037	0.339	0.110	0.913
Linguistic delivery style (x_2)	−0.678	0.328	−2.064	0.040
Interaction $(x_1 x_2)$	0.036	0.009	4.008	<0.005

$F = 55.35$ $(p < 0.0005)$; Adjusted $R^2 = .450$

12.6 Quadratic and Other Higher Order Models

All of the models discussed in the previous sections proposed straight-line relationships between $E(y)$ and each of the independent variables in the model. In this section, we consider models that allow for curvature in the relationships. Each of these models is a **second-order model,** because it includes an x^2-term.

First, we consider a model that includes only one independent variable x. The form of this model, called the **quadratic model,** is

$$y = \beta_0 + \beta_1 x + \beta_2 x^2 + \varepsilon$$

The term involving x^2, called a **quadratic term** (or **second-order term**), enables us to hypothesize curvature in the graph of the response model relating y to x. Graphs of the quadratic model for two different values of β_2 are shown in Figure 12.12. When the curve opens upward, the sign of β_2 is positive (see Figure 12.12a); when the curve opens downward, the sign of β_2 is negative (see Figure 12.12b).

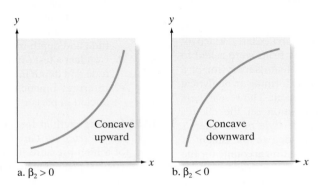

Figure 12.12
Graphs of two quadratic models

a. $\beta_2 > 0$ b. $\beta_2 < 0$

A Quadratic (Second-Order) Model in a Single Quantitative Independent Variable

$$E(y) = \beta_0 + \beta_1 x + \beta_2 x^2$$

where

β_0 is the y-intercept of the curve

β_1 is a shift parameter

β_2 is the rate of curvature

Example 12.7

Analyzing a Quadratic Model— Predicting Electrical Usage

Problem In all-electric homes, the amount of electricity expended is of interest to consumers, builders, and groups involved with energy conservation. Suppose we wish to investigate the monthly electrical usage y in all-electric homes and its relationship to the size x of the home. Moreover, suppose we think that monthly electrical usage in all-electric homes is related to the size of the home by the quadratic model

$$y = \beta_0 + \beta_1 x + \beta_2 x^2 + \varepsilon$$

To fit the model, the values of y and x are collected for 15 homes during a particular month. The data are shown in Table 12.2.

a. Construct a scatterplot of the data. Is there evidence to support the use of a quadratic model?

b. Use the method of least squares to estimate the unknown parameters β_0, β_1, and β_2 in the quadratic model.

Table 12.2

Home Size–Electrical Usage Data

Size of Home x (sq. ft)	Monthly Usage, y (kilowatt-hours)
1,290	1,182
1,350	1,172
1,470	1,264
1,600	1,493
1,710	1,571
1,840	1,711
1,980	1,804
2,230	1,840
2,400	1,956
2,710	2,007
2,930	1,984
3,000	1,960
3,210	2,001
3,240	1,928
3,520	1,945

Data Set: ELECTRIC

c. Graph the prediction equation and assess how well the model fits the data, both visually and numerically.

d. Interpret the β estimates.

e. Is the overall model useful (at $\alpha = .01$) for predicting electrical usage y?

f. Is there sufficient evidence of concave-downward curvature in the electrical usage–home size relationship? Test, using $\alpha = .01$.

Solution

a. A scatterplot of the data of Table 12.2, produced with MINITAB, is shown in Figure 12.13. The figure illustrates that the electrical usage appears to increase in a curvilinear manner with the size of the home. This relationship provides some support for the inclusion of the quadratic term x^2 in the model.

b. We used SAS to fit the model to the data in Table 12.2. Part of the SAS regression output is displayed in Figure 12.14. The least squares estimates of the β parameters

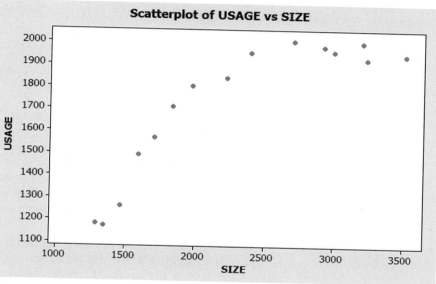

Figure 12.13

MINITAB scatterplot for electrical usage data

```
                    Dependent Variable: USAGE

            Number of Observations Read        15
            Number of Observations Used        15


                        Analysis of Variance

                             Sum of        Mean
Source              DF      Squares       Square    F Value    Pr > F

Model                2      1300900       650450     258.11    <.0001
Error               12        30240    2520.01514
Corrected Total     14      1331140


            Root MSE           50.19975    R-Square    0.9773
            Dependent Mean   1721.20000    Adj R-Sq    0.9735
            Coeff Var           2.91656


                        Parameter Estimates

                        Parameter    Standard
Variable     DF          Estimate       Error    t Value    Pr > |t|

Intercept     1        -806.71666   166.87209      -4.83      0.0004
SIZE          1           1.96162     0.15252      12.86     <.0001
SIZESQ        1        -0.00034044   0.00003212    -10.60     <.0001
```

Figure 12.14

SAS regression output for electrical usage model

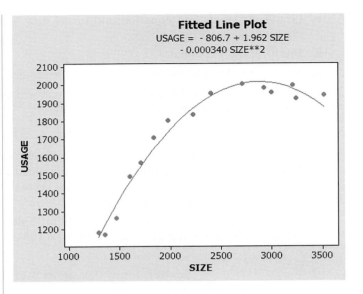

Figure 12.15

MINITAB plot of least squares model for electrical usage

(highlighted) are $\hat{\beta}_0 = -806.7, \hat{\beta}_1 = 1.9616,$ and $\hat{\beta}_2 = -.00034.$ Therefore, the equation that minimizes the SSE for the data is

$$\hat{y} = -806.7 + 1.9616x - .00034x^2$$

c. Figure 12.15 is a MINITAB graph of the least squares prediction equation. Note that the graph provides a good fit to the data of Table 12.2. A numerical measure of fit is obtained with the adjusted coefficient of determination, R_a^2. From the SAS printout, $R_a^2 = .9735$. This implies that over 97% of the sample variation in electrical usage (y) can be explained by the quadratic model (after adjusting for sample size and number of degrees of freedom).

d. The interpretation of the estimated coefficients in a quadratic model must be undertaken cautiously. First, the estimated y-intercept, $\hat{\beta}_0$, can be meaningfully interpreted only if the range of the independent variable includes zero—that is, if $x = 0$ is included in the sampled range of x. Although $\hat{\beta}_0 = -806.7$ seems to imply that the estimated electrical usage is negative when $x = 0$, this zero point is not in the range of the sample (the lowest value of x is 1,290 square feet), and the value is nonsensical (a home with 0 square feet); thus, the interpretation of $\hat{\beta}_0$ is not meaningful.

The estimated coefficient of x is $\hat{\beta}_1 = 1.9616$, but it no longer represents a slope in the presence of the quadratic term x^2.* The estimated coefficient of the first-order term x will not, in general, have a meaningful interpretation in the quadratic model.

The sign of the coefficient, $\hat{\beta}_2 = -.00034$, of the quadratic term, x^2, is the indicator of whether the curve is concave downward (mound shaped) or concave upward (bowl shaped). A negative $\hat{\beta}_2$ implies downward concavity, as in this example (Figure 12.15), and a positive $\hat{\beta}_2$ implies upward concavity. Rather than interpreting the numerical value of $\hat{\beta}_2$ itself, we utilize a graphical representation, as in Figure 12.15, to describe the model.

Note that Figure 12.15 implies that the estimated electrical usage levels off as the home sizes increase beyond 2,500 square feet. In fact, the concavity of the model would lead to decreasing usage estimates if we were to display the model out to 4,000 square feet and beyond. (See Figure 12.16.) However, model interpretations are not meaningful outside the range of the independent variable, which has a maximum

---

*For students with a knowledge of calculus, note that the slope of the quadratic model is the first derivative $\partial y/\partial x = \beta_1 + 2\beta_2 x$. Thus, the slope varies as a function of x, unlike the constant slope associated with the straight-line model.

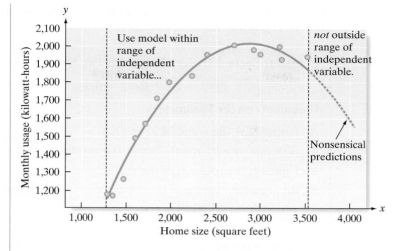

Figure 12.16

Potential misuse of quadratic model

value of 3,520 square feet in this example. Thus, although the model appears to support the hypothesis that the *rate of increase* per square foot *decreases* for home sizes near the high end of the sampled values, the conclusion that usage will actually begin to decrease for very large homes would be a *misuse* of the model, since no homes of 3,600 square feet or more were included in the sample.

e. To test whether the quadratic model is statistically useful, we conduct the global *F*-test:

$$H_0: \beta_1 = \beta_2 = 0$$

$$H_a: \text{At least one of the preceding coefficients is nonzero}$$

From the SAS printout shown in Figure 12.14, the test statistic (highlighted) is $F = 258.11$, with an associated *p*-value $< .0001$. Thus, for any reasonable α, we reject H_0 and conclude that the overall model is a useful predictor of electrical usage y.

f. Figure 12.15 shows concave-downward curvature in the relationship between the size of a home and electrical usage in the sample of 15 data points. To determine whether this type of curvature exists in the population, we want to test

$$H_0: \beta_2 = 0 \text{ (no curvature exists in the response curve)}$$

$$H_a: \beta_2 < 0 \text{ (downward concavity exists in the response curve)}$$

The test statistic for testing β_2, highlighted on Figure 12.14, is $t = -10.60$, and the associated two-tailed *p*-value is less than .0001. Since this is a one-tailed test, the appropriate *p*-value is less than $.0001/2 = .00005$. Now, $\alpha = .01$ exceeds this *p*-value. Thus, there is very strong evidence of downward curvature in the population; that is, electrical usage increases more slowly per square foot for large homes than for small homes.

Look Back Note that the SAS printout in Figure 12.14 also provides the *t*-test statistic and corresponding two-tailed *p*-values for the tests of $H_0: \beta_0 = 0$ and $H_0: \beta_1 = 0$. Since the interpretation of these parameters is not meaningful for this model, the tests are not of interest.

Now Work Exercise 12.68

When two or more quantitative independent variables are included in a second-order model, we can incorporate squared terms for each x in the model, as well as the interaction between the two independent variables. A model that includes all possible second-order terms in two independent variables—called a **complete second-order model**—is given in the following box:

Complete Second-Order Model with Two Quantitative Independent Variables

$$E(y) = \beta_0 + \beta_1 x_1 + \beta_2 x_2 + \beta_3 x_1 x_2 + \beta_4 x_1^2 + \beta_5 x_2^2$$

Comments on the Parameters

β_0: y-intercept, the value of $E(y)$ when $x_1 = x_2 = 0$

β_1, β_2: Changing β_1 and β_2 causes the surface to shift along the x_1- and x_2-axes.

β_3: Controls the rotation of the surface

β_4, β_5: Signs and values of these parameters control the type of surface and the rates of curvature.

Three types of surfaces are produced by a second-order model:* a **paraboloid** that opens upward (Figure 12.17a), a paraboloid that opens downward (Figure 12.17b), and a **saddle-shaped surface** (Figure 12.17c).

A complete second-order model is the three-dimensional equivalent of a quadratic model in a single quantitative variable. Instead of tracing parabolas, however, it traces paraboloids and saddle-shaped surfaces. Since only a portion of the complete surface is used to fit the data, this model provides a very large variety of gently curving surfaces that can be used to fit data. It is a good choice for a model if you expect curvature in the response surface relating $E(y)$ to x_1 and x_2.

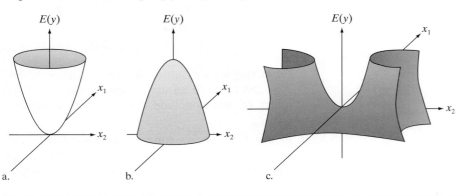

Figure 12.17
Graphs for three second-order surfaces

a.   b.   c.

Example 12.8

A More Complex Second–Order Model—Predicting Hours Worked per Week

Problem A social scientist would like to relate the number of hours worked per week (outside the home) by a married woman to the number of years of formal education she has completed and the number of children in her family.

a. Identify the dependent variable and the independent variables.

b. Write the first-order model for this example.

c. Modify the model in part **b** so that it includes an interaction term.

d. Write a complete second-order model for $E(y)$.

Solution

a. The dependent variable is

    y = Number of hours worked per week by a married woman

---

*The paraboloid opens upward (Figure 12.17a) when $\beta_4 + \beta_5 > 0$ and opens downward (Figure 12.17b) when $\beta_4 + \beta_5 < 0$; the saddle-shaped surface (Figure 12.17c) is produced when $\beta_3^2 > 4\beta_4\beta_5$.

The two independent variables, both quantitative in nature, are

x_1 = Number of years of formal education completed by the woman

x_2 = Number of children in the family

b. The first-order model is

$$E(y) = \beta_0 + \beta_1 x_1 + \beta_2 x_2$$

This model would probably not be appropriate in the current situation because x_1 and x_2 may interact or curvature terms corresponding to x_1^2 and x_2^2 may be needed to obtain a good model for $E(y)$.

c. Adding the interaction term, we obtain

$$E(y) = \beta_0 + \beta_1 x_1 + \beta_2 x_2 + \beta_3 x_1 x_2$$

This model should be better than the model in part **b,** since we have now allowed for interaction between x_1 and x_2.

d. The complete second-order model is

$$E(y) = \beta_0 + \beta_1 x_1 + \beta_2 x_2 + \beta_3 x_1 x_2 + \beta_4 x_1^2 + \beta_5 x_2^2$$

Since it would not be surprising to find curvature in the response surface, the complete second-order model would be preferred to the models in parts **b** and **c.** How can we tell whether the complete second-order model really does provide better predictions of hours worked than the models in parts **b** and **c**? The answers to these and similar questions are examined in Section 12.9.

Most relationships between $E(y)$ and two or more quantitative independent variables are second order and require the use of either the interactive or the complete second-order model to obtain a good fit to a data set. As in the case of a single quantitative independent variable, however, the curvature in the response surface may be very slight over the range of values of the variables in the data set. When this happens, a first-order model may provide a good fit to the data.

Exercises 12.56–12.74

Understanding the Principles

12.56 In the model $E(y) = \beta_0 + \beta_1 x + \beta_2 x^2$,
  a. Which β represents the y-intercept?
  b. Which β represents the shift?
  c. Which β represents the rate of curvature?

12.57 Write a second-order model relating the mean of y, $E(y)$, to
  a. one quantitative independent variable
  b. two quantitative independent variables
  c. three quantitative independent variables [*Hint:* Include all possible two-way cross-product terms and squared terms.]

Learning the Mechanics

12.58 Suppose you fit the quadratic model

$$E(y) = \beta_0 + \beta_1 x + \beta_2 x^2$$

to a set of $n = 20$ data points and find that $R^2 = .91$, $SS_{yy} = 29.94$, and $SSE = 2.63$.
  a. Is there sufficient evidence to indicate that the model contributes information relevant to predicting y? Test using $\alpha = .05$.

  b. What null and alternative hypotheses would you test to determine whether upward curvature exists?
  c. What null and alternative hypotheses would you test to determine whether downward curvature exists?

12.59 Suppose you fit the second-order model

$$y = \beta_0 + \beta_1 x + \beta_2 x^2 + \varepsilon$$

to $n = 25$ data points. Your estimate of β_2 is $\hat{\beta}_2 = .47$, and the estimated standard error of the estimate is .15.
  a. Test $H_0: \beta_2 = 0$ against $H_a: \beta_2 \neq 0$. Use $\alpha = .05$.
  b. Suppose you want to determine only whether the quadratic curve opens upward; that is, as x increases, the slope of the curve would increase. Give the test statistic and the rejection region for the test for $\alpha = .05$. Do the data support the theory that the slope of the curve increases as x increases? Explain.

12.60 MINITAB was used to fit the complete second-order model

$$E(y) = \beta_0 + \beta_1 x_1 + \beta_2 x_2 + \beta_3 x_1 x_2 + \beta_4 x_1^2 + \beta_5 x_2^2$$

to $n = 39$ data points. (See the MINITAB printout on page 652.)
  a. Is there sufficient evidence to indicate that at least one of the parameters β_1, β_2, β_3, β_4, and β_5 is nonzero? Test, using $\alpha = .05$.

```
The regression equation is
Y = -24.56 + 1.12 X1 + 27.99 X2 - 0.54 X1X2 - 0.004 X1SQ + 0.002 X2SQ


Predictor          Coef      SE Coef       T      P
Constant        -24.563        6.531   -3.76  0.001
X1               1.19848       0.1103   10.86  0.000
X2              27.988        79.489     0.35  0.727
X1X2            -0.5397        1.0338   -0.52  0.605
X1SQ            -0.0043        0.0004  -10.74  0.000
X2SQ             0.0020        0.0033    0.60  0.550


S = 2.762     R-Sq = 79.7%   R-Sq(adj) = 76.6%

Analysis of Variance

Source          DF       SS       MS       F      P
Regression       5   989.30   197.86   25.93  0.000
Residual Error  33   251.81     7.63
Total           38  1241.11
```

MINITAB output for
Exercise 12.60

b. Test $H_0: \beta_4 = 0$ against $H_a: \beta_4 \neq 0$. Use $\alpha = .01$.

c. Test $H_0: \beta_5 = 0$ against $H_a: \beta_5 \neq 0$. Use $\alpha = .01$.

d. Use graphs to explain the consequences of the tests in parts **b** and **c**.

12.61 Consider the following quadratic models:

$$(1)\ y = 1 - 2x + x^2$$
$$(2)\ y = 1 + 2x + x^2$$
$$(3)\ y = 1 + x^2$$
$$(4)\ y = 1 - x^2$$
$$(5)\ y = 1 + 3x^2$$

a. Graph each of these quadratic models, side by side, on the same sheet of graph paper.

b. What effect does the first-order term $(2x)$ have on the graph of the curve?

c. What effect does the second-order term (x^2) have on the graph of the curve?

Applying the Concepts—Basic

12.62 Childhood obesity study. The eating patterns of families of overweight preschool children was the subject of an article published in the *Journal of Education and Human Development* (Vol. 3, 2009). A sample of 10 overweight children living in a rural area of the United States was selected. A portion of the research focused on the body mass index of each child and his/her parent. (Body mass index—or BMI—is determined by dividing weight by height squared.) These data are provided in the accompanying table and saved in the **BMI** file. The researchers were interested in determining whether parent BMI could be used as a predictor of child BMI for overweight children.

Child	Parent	Child	Parent
17.10	24.62	17.30	26.30
17.15	24.70	17.32	26.60
17.20	25.70	17.40	26.80
17.24	25.80	17.60	27.20
17.25	26.20	17.80	27.35

Based on Seal, N., and Seal, J. "Eating patterns of the rural families of overweight preschool children: A pilot study." *Journal of Education and Human Development*, Vol. 3, No. 1, 2009 (Figure 1).

a. For this study, identify the dependent and independent variables.

b. Construct a scatterplot for the data. What trend do you observe?

c. A quadratic model was fit to the data in the **BMI** file, with the results shown in the SPSS printout below. Give the least squares prediction equation.

d. Is the overall model statistically useful? Test using $\alpha = .05$.

e. Is there evidence of upward curvature in the relationship between child BMI and parent BMI? Test using $\alpha = .05$.

f. Do you agree with the statement, "for obese children, child BMI increases at an increasing rate as parent BMI increases"?

Model Summary

Model	R	R Square	Adjusted R Square	Std. Error of the Estimate
1	.970[a]	.942	.925	.05887

a. Predictors: (Constant), ParBMIsq, ParentBMI

ANOVA[b]

Model		Sum of Squares	df	Mean Square	F	Sig.
1	Regression	.392	2	.196	56.620	.000[a]
	Residual	.024	7	.003		
	Total	.416	9			

a. Predictors: (Constant), ParBMIsq, ParentBMI
b. Dependent Variable: ChildBMI

Coefficients[a]

Model		Unstandardized Coefficients		Standardized Coefficients	t	Sig.
		B	Std. Error	Beta		
1	(Constant)	99.834	18.520		5.423	.001
	ParentBMI	-5.763	1.375	-25.280	-4.535	.003
	ParBMIsq	.115	.025	26.185	4.692	.002

a. Dependent Variable: ChildBMI

12.63 Going for it on fourth down in the NFL. Refer to the *Chance* (Winter 2009) study of fourth-down decisions by coaches in the National Football League (NFL), Exercise 11.75

(p. 586). Recall that statisticians at California State University, Northridge, fit a straight-line model for predicting the number of points scored (y) by a team that has a first down with a given number of yards (x) from the opposing goal line. A second model fit to data collected on five NFL teams from a recent season was the quadratic regression model, $E(y) = \beta_0 + \beta_1 x + \beta_2 x^2$. The regression yielded the following results: $\hat{y} = 6.13 + .141x - .0009x^2$, $R^2 = .226$.

a. If possible, give a practical interpretation of each of the β-estimates in the model.

b. Give a practical interpretation of the coefficient of determination, R^2.

c. In Exercise 11.75, the coefficient of correlation for the straight-line model was reported as $R^2 = .18$. Does this statistic alone indicate that the quadratic model is a better fit than the straight-line model? Explain.

d. What test of hypothesis would you conduct to determine if the quadratic model is a better fit than the straight-line model?

12.64 Assertiveness and leadership. Management professors at Columbia University examined the relationship between assertiveness and leadership (*Journal of Personality and Social Psychology*, Feb. 2007). The sample comprised 388 people enrolled in a full-time master's in business administration (MBA) program. On the basis of answers to a questionnaire, the researchers measured two variables for each subject: assertiveness score (x) and leadership ability score (y). A quadratic regression model was fit to the data, with the following results:

Independent Variable	β Estimate	t-Value	p-Value
x	.57	2.55	.01
x^2	$-.88$	-3.97	$<.01$

Model $R^2 = .12$

a. Conduct a test of overall model utility. Use $\alpha = .05$.

b. The researchers hypothesized that leadership ability will increase at a decreasing rate with assertiveness. Set up the null and alternative hypotheses to test this theory.

c. Use the reported results to conduct the test you set up in part **b.** Give your conclusion (at $\alpha = .05$) in the words of the problem.

12.65 Testing tires for wear. Underinflated or overinflated tires can increase tire wear. A new tire was tested for wear at different pressures, with the results shown in the following table and saved in the **TIRES** file.

Pressure x (pounds per square inch)	Mileage y (thousands)
30	29
31	32
32	36
33	38
34	37
35	33
36	26

a. Plot the data on a scatterplot.

b. If you were given only the information for $x = 30, 31, 32,$ and 33, what kind of model would you suggest? For $x = 33, 34, 35,$ and 36? For all the data?

12.66 Goal congruence in top management teams. Do chief executive officers (CEOs) and their top managers always agree on the goals of the company? Goal importance congruence between CEOs and vice presidents (VPs) was studied in the *Academy of Management Journal* (Feb 2008). The researchers used regression to model a VP's attitude toward the goal of improving efficiency (y) as a function of the two quantitative independent variables, level of CEO leadership (x_1) and level of congruence between the CEO and the VP (x_2). A complete second-order model in x_1 and x_2 was fit to data collected for $n = 517$ top management team members at U.S. credit unions.

a. Write the complete second-order model for $E(y)$.

b. The coefficient of determination for the model, part **a,** was reported as $R^2 = .14$. Interpret this value.

c. The estimate of the β-value for the $(x_2)^2$ term in the model was found to be negative. Interpret this result, practically.

d. A t-test on the β-value for the interaction term in the model, $x_1 x_2$, resulted in a p-value of .02. Practically interpret this result, using $\alpha = .05$.

12.67 Estimation of urban population by means of satellite images. Refer to the *Geographical Analysis* (Jan. 2007) study that demonstrated the use of satellite image maps for estimating urban population, presented in Exercise 12.16 (p. 629). A first-order model for census block population density (y) was fit as a function of the proportion of a block with low-density residential areas (x_1) and the proportion of a block with high-density residential areas (x_2). Now consider a complete second-order model for y.

a. Write the equation of the model.

b. Identify the terms in the model that allow for curvilinear relationships.

12.68 Violent behavior in children. Refer to the *Development Psychology* (Mar. 2003) study of the behavior of elementary school children, presented in Exercise 8.27 (p. 365). The researchers used a quadratic equation to model the level (y) of aggressive fantasies experienced by a child as a function of the child's age (x). [*Note:* Level y was measured as an average of responses to six questions (e.g., "Do you sometimes have daydreams about hitting or hurting someone you don't like?"). Responses were measured on a scale ranging from 1 (*never*) to 5 (*always*).]

a. Write the equation of the hypothesized model for $E(y)$.

b. Research psychologists theorized that a child's aggressive fantasies increase with age, but at a slower rate of acceleration in older children. Sketch the curve hypothesized by the researchers.

c. Set up H_0 and H_a for testing the researchers' theory.

d. The model was fitted to data collected for over 11,000 elementary school children, with the following results: $\hat{y} = 1.926 + .097x - .003x^2$, standard error of $\hat{\beta}_2 = .001$. Compute the test statistic for the test you set up in part **c.**

e. Use the result you found in part **d** to make the appropriate conclusion. Take $\alpha = .05$.

Applying the Concepts—Intermediate

12.69 Estimating change-point dosage. A standard method for studying toxic substances and their effects on humans is to observe the responses of rodents exposed to various doses of the substance over time. In the *Journal of Agricultural, Biological, and Environmental Statistics* (June 2005), researchers used least squares regression to estimate the "change-point" dosage, defined as the largest dose level that has no adverse effects. Data were obtained from a dose-response study of rats exposed to the toxic substance aconiazide. A sample of 50 rats was evenly divided into five dosage groups: 0, 100, 200, 500, and 750 milligrams per kilograms of body weight. The dependent variable y measured was the weight change (in grams) after a 2-week exposure. The researchers fit the quadratic model $E(y) = \beta_0 + \beta_1 x + \beta_2 x^2$, where x = dosage level, with the following results: $\hat{y} = 10.25 + .0053x - .0000266x^2$.

a. Construct a rough sketch of the least squares prediction equation. Describe the nature of the curvature in the estimated model.

b. Estimate the weight change (y) for a rat given a dosage of 500 mg/kg of aconiazide.

c. Estimate the weight change (y) for a rat given a dosage of 0 mg/kg of aconiazide. (This dosage is called the "control" dosage level.)

d. Of the five groups in the study, find the largest dosage level x which yields an estimated weight change that is closest to, but below, the estimated weight change for the control group. This value is the change-point dosage.

12.70 Revenues of popular movies. The *Internet Movie Database* (www.imdb.com) monitors the gross revenues of all major motion pictures. The table below gives both the domestic (United States and Canada) and international gross revenues for a sample of 20 popular movies. The data are saved in the **IMDB** file.

a. Write a first-order model for international gross revenues y as a function of domestic gross revenues x.

b. Write a second-order model for international gross revenues y as a function of domestic gross revenues x.

c. Construct a scatterplot of these data. Which of the models appears to be a better choice for explaining variation in international gross revenues?

d. Fit the model of part **b** to the data and investigate its usefulness. Is there evidence of a curvilinear relationship between international and domestic gross revenues? Test, using $\alpha = .05$.

e. On the basis of your analysis in part **d**, which of the two models better explains the variation in international gross revenues?

12.71 Satisfaction with membership in a new religious movement. How satisfied are people who have recently joined a new religious movement? To answer this question, German researchers collected data for a sample of 58 believers who had recently joined a new religious group (*Applied Psychology: An International Review*, April 2010). The dependent variable of interest was *satisfaction level* (y), measured quantitatively on an 11-point scale (where 0 = totally dissatisfied and 10 = totally satisfied). Two independent variables were used to predict satisfaction level: *Needs* (x_1)—a measure of the level of needs one requires in a religion, and, *Supplies* (x_2)—a measure of the level of supplies provided by the religion. In theory, if the level of needs matches the level of supplies, one will be highly satisfied with the religion.

a. The researchers fitted a complete second-order model for $E(y)$ as a function of x_1 and x_2. Write the equation of this model.

b. The regression results are reported in the table on page 655. Interpret the value of R^2.

c. Use the R^2 statistic to conduct a test of overall model adequacy. Test using $\alpha = .10$.

d. Conduct a test to determine whether needs (x_1) is curvilinearly related to satisfaction (y). Test using $\alpha = .10$.

e. Conduct a test to determine whether supplies (x_2) is curvilinearly related to satisfaction (y). Test using $\alpha = .10$.

Movie (Year)	Domestic Gross ($ millions)	International Gross ($ millions)
Avatar (2009)	760.5	2,021.0
Titanic (1997)	600.8	1,234.6
The Dark Knight (2008)	533.3	464.0
Pirates of the Caribbean (2006)	423.0	642.9
E.T. (1982)	434.9	321.8
Spider Man (2002)	403.7	417.9
Jurassic Park (1993)	356.8	563.0
Lion King (1994)	328.4	455.0
Harry Potter Sorcerer's Stone (2001)	317.6	651.1
Inception (2010)	291.4	468.2
Sixth Sense (1999)	293.5	368.0
The Hangover (2009)	277.3	201.6
Jaws (1975)	260.0	210.6
Ghost (1990)	217.6	300.0
Saving Private Ryan (1998)	216.1	263.2
Gladiator (2000)	187.7	258.3
Dances with Wolves (1990)	184.2	240.0
The Exorcist (1973)	204.6	153.0
My Big Fat Greek Wedding (2002)	241.4	115.1
Rocky IV (1985)	127.9	172.6

Based on The Internet Movie Database (www.imdb.com)

Results for Exercise 12.71

Independent Variable	Estimated Beta	Standard Error
Needs (x_1)	.952	.780
Supplies (x_2)	1.198	.766
Needs × Supplies (x_1x_2)	.356	.429
Needs × Needs (x_1^2)	−.181	.302
Supplies × Supplies (x_2^2)	−.755	.413

$R^2 = .402$

12.72 Failure times of silicon wafer microchips. Researchers at National Semiconductor experimented with tin-lead solder bumps used to manufacture silicon wafer integrated circuit chips (*International Wafer Level Packaging Conference*, Nov. 3–4, 2005). The failure times of the microchips (in hours) were determined at different solder temperatures (degrees Centigrade). The data for one experiment are given in the table and saved in the **WAFER** file. The researchers want to predict failure time (y) based on solder temperature (x).

a. Construct a scatterplot for the data. What type of relationship, linear or curvilinear, appears to exist between failure time and solder temperature?

b. Fit the model, $E(y) = \beta_0 + \beta_1 x + \beta_2 x^2$, to the data. Give the least squares prediction equation.

c. Conduct a test to determine if there is upward curvature in the relationship between failure time and solder temperature. (Use $\alpha = .05$.)

Temperature (°C)	Time to Failure (hours)
165	200
162	200
164	1,200
158	500
158	600
159	750
156	1,200
157	1,500
152	500
147	500
149	1,100
149	1,150
142	3,500
142	3,600
143	3,650
133	4,200
132	4,800
132	5,000
134	5,200
134	5,400
125	8,300
123	9,700

Based on Gee, S., and Nguyen, L. "Mean time to failure in wafer level–CSP packages with SnPb and SnAgCu solder bmps." International Wafer Level Packaging Conference, San Jose, CA, Nov. 3–4, 2005 (adapted from Figure 7).

12.73 Public perceptions of health risks. In the *Journal of Experimental Psychology: Learning, Memory, and Cognition* (July 2005), University of Basel (Switzerland) psychologists tested the ability of people to judge the risk of an infectious disease. The researchers asked German college students to estimate the number of people who are infected with a certain disease in a typical year. The median estimates, as well as the actual incidence of the disease for each in a sample of 24 infections, are listed in the table and saved in the **INFECTION** file. Consider the quadratic model $E(y) = \beta_0 + \beta_1 x + \beta_2 x^2$, where y = actual incidence rate and x = estimated rate.

Infection	Actual Incidence	Estimate
Polio	0.25	300
Diphtheria	1	1,000
Trachoma	1.75	691
Rabbit Fever	2	200
Cholera	3	17.5
Leprosy	5	0.8
Tetanus	9	1,000
Hemorrhagic Fever	10	150
Trichinosis	22	326.5
Undulant Fever	23	146.5
Well's Disease	39	370
Gas Gangrene	98	400
Parrot Fever	119	225
Typhoid	152	200
Q Fever	179	200
Malaria	936	400
Syphilis	1,514	1,500
Dysentery	1,627	1,000
Gonorrhea	2,926	6,000
Meningitis	4,019	5,000
Tuberculosis	12,619	1,500
Hepatitis	14,889	10,000
Gastroenteritis	203,864	37,000
Botulism	15	37,500

Based on Hertwig, R., Pachur, T., and Kurzenhauser, S. "Judgments of risk frequencies: Tests of possible cognitive mechanisms." *Journal of Experimental Psychology: Learning, Memory, and Cognition*, Vol. 31, No. 4, July 2005 (Table 1).

a. Fit the quadratic model to the data, and then conduct a test to determine whether the actual incidence is curvilinearly related to the estimated incidence. (Use $\alpha = .05$.)

b. Construct a scatterplot of the data. Locate the data point for botulism on the graph. What do you observe?

c. Repeat part **a**, but omit the data point for botulism from the analysis. Has the fit of the model improved? Explain.

Applying the Concepts—Advanced

12.74 Tree frog study. The optomotor responses of tree frogs were studied in the *Journal of Experimental Zoology* (Sept. 1993). Microspectrophotometry was used to measure the threshold quantal flux (the light intensity at which the optomotor response was first observed) of tree frogs tested at different spectral wavelengths. The data revealed the relationship between the logarithm of quantal flux (y) and wavelength (x), shown in the following graph:

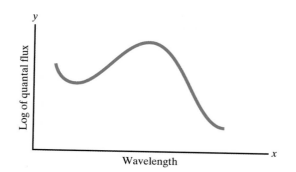

a. Explain why a first-order model would not be appropriate for modeling $E(y)$.

b. Explain why a second-order model would not be appropriate for modeling $E(y)$.

c. Demonstrate that the third-order model $E(y) = \beta_0 + \beta_1 x + \beta_2 x^2 + \beta_3 x^3$ may be the most appropriate model for $E(y)$.

12.7 Qualitative (Dummy) Variable Models

Multiple-regression models can also be written to include **qualitative** (or **categorical**) independent variables. **Qualitative variables,** unlike quantitative variables, cannot be measured on a numerical scale. Therefore, we must code the values of the qualitative variable (called **levels**) as numbers before we can fit the model. These coded qualitative variables are called **dummy** (or **indicator**) **variables,** since the numbers assigned to the various levels are arbitrarily selected.

To illustrate, suppose a female executive at a certain company claims that male executives earn higher salaries, on average, than female executives with the same education, experience, and responsibilities. To support her claim, she wants to model the salary y of an executive, using a qualitative independent variable representing the gender of the executive (male or female).

A convenient method of coding the values of a qualitative variable at two levels involves assigning a value of 1 to one of the levels and a value of 0 to the other. For example, the **dummy variable** used to describe gender could be coded as follows:

$$x = \begin{cases} 1 & \text{if male} \\ 0 & \text{if female} \end{cases}$$

The choice of which level is assigned to 1 and which is assigned to 0 is arbitrary. The model then takes the following form:

$$E(y) = \beta_0 + \beta_1 x$$

The advantage of using a 0–1 coding scheme is that the β coefficients are easily interpreted. The foregoing model allows us to compare the mean executive salary $E(y)$ for males with the corresponding mean for females:

$$\textit{Males } (x = 1): \qquad E(y) = \beta_0 + \beta_1(1) = \beta_0 + \beta_1$$
$$\textit{Females } (x = 0): \qquad E(y) = \beta_0 + \beta_1(0) = \beta_0$$

These two means are illustrated in the bar graph in Figure 12.18.

First note that β_0 represents the mean salary for females (say, μ_F). When a 0–1 coding convention is used, β_0 will always represent the mean response associated with the level of the qualitative variable assigned the value 0 (called the **base level**). The difference between the mean salary for males and the mean salary for females, $\mu_M - \mu_F$, is represented by β_1; that is,

$$\mu_M - \mu_F = (\beta_0 + \beta_1) - (\beta_0) = \beta_1$$

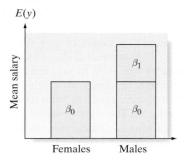

Figure 12.18

Bar chart comparing $E(y)$ for males and females

This difference is shown in Figure 12.18.* With a 0–1 coding convention, β_1 will always represent the difference between the mean response for the level assigned the value 1 and the mean for the base level. Thus, for the executive salary model, we have

$$\beta_0 = \mu_F$$
$$\beta_1 = \mu_M - \mu_F$$

The model relating a mean response $E(y)$ to a qualitative independent variable at two levels is summarized in the following box.

---

*Note that β_1 could be negative. If β_1 were negative, the height of the bar corresponding to males would be reduced (rather than increased) from the height of the bar for females by the amount β_1. Figure 12.18 is constructed under the assumption that β_1 is a positive quantity.

A Model Relating $E(y)$ to a Qualitative Independent Variable with Two Levels

$$E(y) = \beta_0 + \beta_1 x$$

where

$$x = \begin{cases} 1 & \text{if level A} \\ 0 & \text{if level B} \end{cases}$$

Interpretation of β's:

$$\beta_0 = \mu_B \text{ (mean for base level)}$$
$$\beta_1 = \mu_A - \mu_B$$

Note: μ_i represents the value of $E(y)$ for level i.

Now carefully examine the model with a single qualitative independent variable at two levels, because we will use exactly the same pattern for any number of levels. Moreover, the interpretation of the parameters will always be the same.

One level (say, level A) is selected as the **base level.** Then, for the 0–1 coding[†] for the dummy variables,

$$\mu_A = \beta_0$$

The coding for all dummy variables is as follows: To represent the mean value of y for a particular level, let that dummy variable equal 1; otherwise, the dummy variable is set equal to 0. Using this system of coding, we obtain

$$\mu_B = \beta_0 + \beta_1$$
$$\mu_C = \beta_0 + \beta_2$$

and so on. Because $\mu_A = \beta_0$, any other model parameter will represent the difference between means for that level and the base level; that is,

$$\beta_1 = \mu_B - \mu_A$$
$$\beta_2 = \mu_C - \mu_A$$

and so on. Consequently, each β multiplied by a dummy variable represents the difference between $E(y)$ at one level of the qualitative variable and $E(y)$ at the base level.

A Model Relating $E(y)$ to One Qualitative Independent Variable with k Levels

Always use a number of dummy variables that is one less than the number of levels of the qualitative variable. Thus, for a qualitative variable with k levels, use $k - 1$ dummy variables:

$$E(y) = \beta_0 + \beta_1 x_1 + \beta_2 x_2 + \cdots + \beta_{k-1} x_{k-1}$$

where x_i is the dummy variable for level $i + 1$ and

$$x_i = \begin{cases} 1 & \text{if } y \text{ is observed at level } i + 1 \\ 0 & \text{otherwise} \end{cases}$$

Then, for this system of coding,

$$\mu_1 = \beta_0 \qquad\qquad \text{and} \qquad\qquad \beta_1 = \mu_2 - \mu_1$$
$$\mu_2 = \beta_0 + \beta_1 \qquad\qquad\qquad\qquad\qquad \beta_2 = \mu_3 - \mu_1$$
$$\mu_3 = \beta_0 + \beta_2 \qquad\qquad\qquad\qquad\qquad \beta_3 = \mu_4 - \mu_1$$
$$\mu_4 = \beta_0 + \beta_3$$
$$\vdots \qquad\qquad \vdots \qquad\qquad\qquad\qquad\qquad \vdots \qquad \vdots$$
$$\mu_k = \beta_0 + \beta_{k-1} \qquad\qquad\qquad\qquad\qquad \beta_{k-1} = \mu_k - \mu_1$$

Note: μ_i represents the value of $E(y)$ for level i.

[†]You do not have to use a 0–1 system of coding for the dummy variables. *Any* two-value system will work, but the interpretation given to the model parameters will depend on the code. Using the 0–1 system makes the model parameters easy to interpret.

Example 12.9

A Model with One Qualitative Independent Variable: Golf Ball Driving Distances

Problem Refer to Example 10.4 (p. 486). Recall that the USGA wants to compare the mean driving distances of four different golf ball brands (A, B, C, and D). Iron Byron, the USGA's robotic golfer, is used to hit a sample of 10 balls of each brand. The distance data are reproduced in Table 12.3.

a. Hypothesize a regression model for driving distance y, using Brand as an independent variable.

b. Interpret the β's in the model.

c. Fit the model to the data and give the least squares prediction equation. Show that the β-estimates can also be obtained from the sample means.

d. Use the model to determine whether the mean driving distances for the four brands are significantly different at $\alpha = .05$.

Table 12.3 Driving Distances (in feet) for Four Golf Ball Brands

Brand A	Brand B	Brand C	Brand D
251.2	263.2	269.7	251.6
245.1	262.9	263.2	248.6
248.0	265.0	277.5	249.4
251.1	254.5	267.4	242.0
260.5	264.3	270.5	246.5
250.0	257.0	265.5	251.3
253.9	262.8	270.7	261.8
244.6	264.4	272.9	249.0
254.6	260.6	275.6	247.1
248.8	255.9	266.5	245.9

Data Set: GOLFCRD

Solution

a. Note that golf ball brand (A, B, C, and D) is a qualitative variable (measured on a nominal scale). According to the previous box, for a four-level qualitative variable we require three dummy variables in the regression model. The model relating $E(y)$, where y is the distance the ball is driven by Iron Byron, to this single qualitative variable, golf ball Brand, is

$$E(y) = \beta_0 + \beta_1 x_1 + \beta_2 x_2 + \beta_3 x_3$$

where

$$x_1 = \begin{cases} 1 & \text{if Brand B} \\ 0 & \text{if not} \end{cases} \quad x_2 = \begin{cases} 1 & \text{if Brand C} \\ 0 & \text{if not} \end{cases} \quad x_3 = \begin{cases} 1 & \text{if Brand D} \\ 0 & \text{if not} \end{cases}$$

We arbitrarily choose Brand A to be the base level.

b. Since Brand A is the base level, β_0 represents the mean driving distance for Brand A (i.e., $\beta_0 = \mu_A$). The other β's are differences in means, namely,

$$\beta_1 = \mu_B - \mu_A$$
$$\beta_2 = \mu_C - \mu_A$$
$$\beta_3 = \mu_D - \mu_A$$

where μ_A, μ_B, μ_C, and μ_D are the mean distances for Brands A, B, C, and D, respectively.

c. The MINITAB printout of the regression analysis is shown in Figure 12.19. The least squares estimates of the β's are highlighted on the printout, yielding the following least squares prediction equation:

$$\hat{y} = 250.78 + 10.28x_1 + 19.17x_2 - 1.46x_3$$

Regression Analysis: DISTANCE versus X1, X2, X3

```
The regression equation is
DISTANCE = 251 + 10.3 X1 + 19.2 X2 - 1.46 X3

Predictor      Coef   SE Coef       T       P
Constant    250.780     1.455  172.34   0.000
X1           10.280     2.058    5.00   0.000
X2           19.170     2.058    9.32   0.000
X3           -1.460     2.058   -0.71   0.483

S = 4.60163   R-Sq = 78.6%   R-Sq(adj) = 76.8%

Analysis of Variance

Source           DF       SS      MS       F       P
Regression        3  2794.39  931.46   43.99   0.000
Residual Error   36   762.30   21.18
Total            39  3556.69
```

Descriptive Statistics: DISTANCE

```
Variable  BRAND   N    Mean  StDev  Minimum  Maximum
DISTANCE  A      10  250.78   4.74   244.60   260.50
          B      10  261.06   3.87   254.50   265.00
          C      10  269.95   4.50   263.20   277.50
          D      10  249.32   5.20   242.00   261.80
```

Figure 12.19

MINITAB output for model with dummy variables

The interpretations of the β's in part **b** allow us to obtain the β estimates from the sample means associated with the different levels of the qualitative variable.*

Since $\beta_0 = \mu_A$, then the estimate of β_0 is the estimated mean driving distance for Brand A (base level). This sample mean, highlighted at the bottom of the MINITAB printout, is 250.78; thus $\hat{\beta}_0 = 250.78$.

Now $\beta_1 = \mu_B - \mu_A$; therefore, the estimate of β_1 is the difference between the sample mean driving distances for Brands B and A. Based on the sample means highlighted at the bottom of the MINITAB printout, we have $\hat{\beta}_1 = 261.06 - 250.78 = 10.28$.

Similarly, the estimate of $\beta_2 = \mu_C - \mu_A$ is the difference between the sample mean distances for Brands C and A. From the sample means highlighted at the bottom of the MINITAB printout, we have $\hat{\beta}_2 = 269.95 - 250.78 = 19.17$.

Finally, the estimate of $\beta_3 = \mu_D - \mu_A$ is the difference between the sample mean distances for Brands D and A. Using the sample means highlighted at the bottom of the MINITAB printout, we have $\hat{\beta}_3 = 249.32 - 250.78 = -1.46$.

d. Testing the null hypothesis that the means for the four brands are equal (i.e., $\mu_A = \mu_B = \mu_C = \mu_D$), is equivalent to testing

$$H_0: \beta_1 = \beta_2 = \beta_3 = 0$$

You can see this by observing that if $\beta_1 = \mu_B - \mu_A = 0$, then $\mu_A = \mu_B$. Similarly, $\beta_2 = \mu_C - \mu_A = 0$ implies that $\mu_A = \mu_C$, and $\beta_3 = \mu_D - \mu_A = 0$ implies that $\mu_A = \mu_D$. The alternative hypothesis is

$$H_a: \text{At least one of the parameters } \beta_1, \beta_2, \text{ and } \beta_3 \text{ differs from 0}$$

which implies that at least two of the four means ($\mu_A, \mu_B, \mu_C,$ and μ_D) differ.

To test this hypothesis, we conduct the global F-test on the model. The value of the F-statistic for testing the adequacy of the model, $F = 43.99$, and the observed significance level of the test, $p = .000$, are both highlighted in Figure 12.19. Since $\alpha = .05$ exceeds the p-value, we reject H_0 and conclude that at least one of the parameters differs from 0. Or, equivalently, we conclude that the data provide sufficient evidence to indicate that the mean driving distance does vary from one golf ball brand to another.

*The least squares method and the sample means method will yield equivalent β estimates when the sample sizes associated with the different levels of the qualitative variable are equal.

> **Look Back** This global F-test is equivalent to the analysis-of-variance F-test in Chapter 10 for a completely randomized design.
>
> Now Work Exercise 12.78

⚠ **CAUTION** A common mistake by regression analysts is the use of a single dummy variable x for a qualitative variable at k levels, where $x = 1, 2, 3, \ldots, k$. Such a regression model will have unestimable β's and β's that are difficult to interpret. Remember, in modeling $E(y)$ with a single qualitative independent variable, the number of 0–1 dummy variables to include in the model will always be one less than the number of levels of the qualitative variable. ▲

Exercises 12.75–12.91

Understanding the Principles

12.75 Write a regression model relating the mean value of y to a qualitative independent variable that can assume two levels. Interpret all the terms in the model.

12.76 Write a regression model relating $E(y)$ to a qualitative independent variable that can assume three levels. Interpret all the terms in the model.

Learning the Mechanics

12.77 The model $E(y) = \beta_0 + \beta_1 x_1 + \beta_2 x_2 + \beta_3 x_3$, where

$$x_1 = \begin{cases} 1 & \text{if level 2} \\ 0 & \text{if not} \end{cases}$$

$$x_2 = \begin{cases} 1 & \text{if level 3} \\ 0 & \text{if not} \end{cases}$$

$$x_3 = \begin{cases} 1 & \text{if level 4} \\ 0 & \text{if not} \end{cases}$$

was used to relate $E(y)$ to a single qualitative variable with four levels. This model was fitted to $n = 30$ data points and the following result was obtained:

$$\hat{y} = 10.2 - 4x_1 + 12x_2 + 2x_3$$

a. Use the least squares prediction equation to find the estimate of $E(y)$ for each level of the qualitative independent variable.

b. Specify the null and alternative hypotheses you would use to test whether $E(y)$ is the same for all four levels of the independent variable.

12.78 MINITAB was used to fit the model
[NW]

$$y = \beta_0 + \beta_1 x_1 + \beta_2 x_2 + \varepsilon$$

where

$$x_1 = \begin{cases} 1 & \text{if level 2} \\ 0 & \text{if not} \end{cases}$$

$$x_2 = \begin{cases} 1 & \text{if level 3} \\ 0 & \text{if not} \end{cases}$$

to $n = 15$ data points. The results are shown in the accompanying MINITAB printout (top of next column).

a. Report the least squares prediction equation.

b. Interpret the values of β_1 and β_2.

MINITAB output for Exercise 12.78

```
The regression equation is
Y = 80.0 + 16.8 X1 + 40.4 X2


Predictor    Coef   SE Coef      T      P
Constant   80.000    4.082   19.60  0.000
X1         16.800    5.774    2.91  0.013
X2         40.400    5.774    7.00  0.000


S = 9.129    R-Sq = 80.5%    R-Sq(adj) = 77.2%

Analysis of Variance

Source         DF      SS      MS      F      P
Regression      2   4118.9  2059.5  24.72  0.000
Residual Error 12   1000.0    83.3
Total          14   5118.9
```

c. Interpret the following hypotheses in terms of μ_1, μ_2, and μ_3:

$$H_0: \beta_1 = \beta_2 = 0$$

$$H_a: \text{At least one of the parameters } \beta_1 \text{ and } \beta_2 \text{ differs from 0}$$

d. Conduct the hypothesis test of part **c.**

Applying the Concepts—Basic

12.79 **Whales entangled in fishing gear.** Refer to the *Marine Mammal Science* (April 2010) study of whales entangled in fishing gear, Exercise 12.15 (p. 629). These entanglements involved one of three types of fishing gear: set nets, pots, and gill nets. Consequently, the researchers used gear type as a predictor of the body length (y, in meters) of the entangled whale. Consider the regression model, $E(y) = \beta_0 + \beta_1 x_1 + \beta_2 x_2$, where $x_1 = \{1 \text{ if set net, } 0 \text{ if not}\}$ and $x_2 = \{1 \text{ if pots, } 0 \text{ if not}\}$. [*Note:* Gill nets is the "base" level of gear type.]

a. The researchers want to know the mean body length of whales entangled in gill nets. Give an expression for this value in terms of the β's in the model.

b. Practically interpret the value of β_1 in the model.

c. In terms of the β's in the model, how would you test to determine if the mean body lengths of entangled whales differ for the three types of fishing gear?

12.80 **Production technologies, terroir, and quality of Bordeaux wine.** In addition to state-of-the-art technologies, the production of quality wine is strongly influenced by the natural endowments of the grape-growing region—called the "terroir." *The Economic Journal* (May 2008) published an empirical study of the factors that yield a quality Bordeaux wine. A quantitative measure of wine quality (y) was modeled as a function of several qualitative independent variables, including grape-picking method (manual or automated), soil type (clay, gravel, or sand), and slope orientation (east, south, west, southeast, or southwest).

a. Create the appropriate dummy variables for each of the qualitative independent variables.

b. Write a model for wine quality (y) as a function of grape-picking method. Interpret the β's in the model.

c. Write a model for wine quality (y) as a function of soil type. Interpret the β's in the model.

d. Write a model for wine quality (y) as a function of slope orientation. Interpret the β's in the model.

12.81 Impact of race on football card values. University of Colorado sociologists investigated the impact of race on the value of professional football players' "rookie" cards (*Electronic Journal of Sociology*, 2007). The sample consisted of 148 rookie cards of National Football League (NFL) players who were inducted into the Football Hall of Fame. The price of a card (in dollars) was modeled as a function of several qualitative independent variables: race of player (black or white), availability of the card (high or low), and position of the player (quarterback, running back, wide receiver, tight end, defensive lineman, linebacker, defensive back, or offensive lineman).

a. Create the appropriate dummy variables for each of the qualitative independent variables.

b. Write a model for price (y) as a function of race. Interpret the β's in the model.

c. Write a model for price (y) as a function of the availability of the card. Interpret the β's in the model.

d. Write a model for price (y) as a function of the player's position. Interpret the β's in the model.

12.82 Chemical composition of rainwater. Researchers at the University of Aberdeen (Scotland) developed a statistical model for estimating the chemical composition of water (*Journal of Agricultural, Biological, and Environmental Statistics*, March 2005). For one application, the nitrate concentration y (milligrams per liter) in a water sample collected after a heavy rainfall was modeled as a function of water source (groundwater, subsurface flow, or overground flow).

a. Write a model for E(y) as a function of the qualitative independent variable.

b. Give an interpretation of each of the β parameters in the model you wrote in part **a**.

12.83 Detecting quantitative traits in genes. In gene therapy, it is important to know the location of a gene for a disease on the genome (genetic map). Although many genes yield a specific trait (e.g., disease or not), others cannot be categorized, since they are quantitative in nature (e.g., extent of disease). Researchers at the University of North Carolina at Wilmington developed statistical models that link quantitative genetic traits to locations on the genome (*Chance*, Summer 2006). The extent of a certain disease is determined by the absence (A) or presence (B) of a gene marker at each of two locations, L1 and L2, on the genome.

For example, AA represents absence of the marker at both locations, while AB represents absence at location L1, but presence at location L2.

a. How many different gene marker combinations are possible at the two locations?

b. Using dummy variables, write a model for extent of the disease, y, as a function of gene marker combination.

c. Interpret the β-values in the model you wrote in part **b**.

d. Give the null hypothesis for testing whether the overall model from part **b** is statistically useful for predicting extent of the disease, y.

12.84 Improving SAT scores. Refer to the *Chance* (Winter 2001) study of students who paid a private tutor (or coach) to help them improve their SAT scores, presented in Exercise 2.105 (p. 73). Multiple regression was used to estimate the effect of coaching on SAT-Mathematics scores. Data on 3,492 students (573 of whom were coached) were used to fit the model $E(y) = \beta_0 + \beta_1 x_1 + \beta_2 x_2$, where y = SAT-Math score, x_1 = score on PSAT, and $x_2 = \{1$ if student was coached, 0 if not $\}$.

a. The fitted model had an adjusted R^2 value of .76. Interpret this result.

b. The estimate of β_2 in the model was 19, with a standard error of 3. Use this information to form a 95% confidence interval for β_2. Interpret the interval.

c. On the basis of the interval you found in part **b**, what can you say about the effect of coaching on SAT-Math scores?

Applying the Concepts—Intermediate

12.85 Major depression and personality disorders. When psychiatric patients have an episode of depression, they are often diagnosed with several personality disorders. A team of physicians, psychiatrists, and psychologists investigated whether these patients exhibit more or fewer personality disorder symptoms than nondepressed patients in the *American Journal of Psychiatry* (May 2010). A study group of over 400 psychiatric patients was monitored over a six-year period. At the start of the study, each was diagnosed as having (1) major depression only, (2) personality disorder only, or (3) both major depression and personality disorder. Of interest to the researchers was the number of personality disorder criteria met at the end of the study. Consider a regression model for the number of personality disorders (y).

a. Write a model for E(y) as a function of the qualitative variable, patient diagnosis group.

b. If there are no differences among the mean number of personality disorders for the three patient groups, what are the values of the β's in the model, part **a**?

c. How could you test to determine if the mean number of personality disorders for the major depression–only patients is less than the corresponding mean for the patients with both major depression and personality disorder?

12.86 Study of recall of TV commercials. Refer to the *Journal of Applied Psychology* (June 2002) study of recall of television commercials, presented in Exercise 10.33 (p. 495). Participants were assigned to watch one of three types of TV programs, with nine commercials embedded in each show. Group V watched a TV program with a violent-content code rating, Group S viewed a show with a sex-content code rating, and Group N watched a neutral TV program with neither a V nor an S rating. The dependent variable measured for each participant was the score (y) on his or her recall of the brand names mentioned in the commercial messages, with scores ranging from 0 (no brands recalled) to 9 (all brands recalled). The data are saved in the **TVADRECALL** file.

a. Write a model for $E(y)$ as a function of viewer group.

b. Fit the model you wrote in part **a** to the data saved in the **TVADRECALL** file. Give the least squares prediction equation.

c. Conduct a test of overall model utility at $\alpha = .01$. Interpret the results. Show that the results agree with the analysis performed in Exercise 10.33.

d. The sample mean recall scores for the three groups were $\bar{y}_V = 2.08, \bar{y}_S = 1.71$, and $\bar{y}_N = 3.17$. Show how to find these sample means by using only the β-estimates obtained in part **b**.

12.87 Expert testimony in homicide trials of battered women. For over 20 years, courts have accepted evidence of "battered woman syndrome" as a defense in homicide cases. An article published in the *Duke Journal of Gender Law & Policy* (Summer 2003) examined the impact of expert testimony on the outcome of homicide trials that involve battered woman syndrome. On the basis of data collected on individual juror votes from past trials, the article reported that "when expert testimony was present, women jurors were more likely than men to change a verdict from not guilty to guilty after deliberations." Assume that when no expert testimony was present, male jurors were more likely than women to change a verdict from not guilty to guilty after deliberations. These results were obtained from a multiple-regression model for likelihood of changing a verdict from not guilty to guilty after deliberations, y, as a function of juror gender (male or female) and expert testimony (yes or no). Give the model for $E(y)$ that hypothesizes the relationships reported in the article. Illustrate the model with a sketch.

12.88 Homework assistance for college students. Do college professors who provide their students with assistance on homework help improve student grades? This was the research question of interest in the *Journal of Accounting Education* (Vol. 25, 2007). A sample of 175 accounting students took a pretest on a topic not covered in class, then each was given a homework problem to solve on the same topic. The students were assigned to one of three homework assistance groups. Some students received the completed solution, some were given check figures at various steps of the solution, and some received no help at all. After finishing the homework, the students were all given a posttest on the subject. The dependent variable of interest was the knowledge gain (or, test score improvement). These data are saved in the **ACCHW** file.

a. Propose a model for the knowledge gain (y) as a function of the qualitative variable, homework assistance group.

b. In terms of the β's in the model, give an expression for the difference between the mean knowledge gains of students in the "completed solution" and "no help" groups.

c. Fit the model to the data and give the least squares prediction equation.

d. Conduct the global F-test for model utility using $\alpha = .05$. Interpret the results, practically.

12.89 Extinct New Zealand birds. Refer to the *Evolutionary Ecology Research* (July 2003) study of the patterns of extinction in the New Zealand bird population, presented in Exercise 2.20 (p. 37). Recall that the **NZBIRDS** file contains qualitative data on flight capability (volant or flightless), habitat (aquatic, ground terrestrial, or aerial terrestrial), nesting site (ground, cavity within ground, tree, or cavity above ground), nest density (high or low), diet (fish, vertebrates, vegetables, or invertebrates), and extinct status (extinct, absent from island, present), and quantitative data on body mass (grams) and egg length (millimeters) for 132 bird species at the time of the Maori colonization of New Zealand.

a. Write a model for mean body mass as a function of flight capability.

b. Write a model for mean body mass as a function of diet.

c. Write a model for mean egg length as a function of nesting site.

d. Fit the model you wrote in part **a** to the data and interpret the estimates of the β's.

e. Conduct a test to determine whether the model from part **a** is statistically useful (at $\alpha = .01$) for estimating mean body mass.

f. Fit the model you wrote in part **b** to the data and interpret the estimates of the β's.

g. Conduct a test to determine whether the model from part **b** is statistically useful (at $\alpha = .01$) for estimating mean body mass.

h. Fit the model you wrote in part **c** to the data and interpret the estimates of the β's.

i. Conduct a test to determine whether the model from part **c** is statistically useful (at $\alpha = .01$) for estimating mean egg length.

12.90 Heights of grade school repeaters. Refer to *The Archives of Disease in Childhood* (Apr. 2000) study of whether height influences a child's progression through elementary school, presented in Exercise 10.111 (p. 541). Recall that Australian schoolchildren were divided into equal thirds (tertiles) based on age (youngest third, middle third, and oldest third). The average heights of the three groups (for which all height measurements were standardized by using z-scores), by gender, are shown in the accompanying table.

	Youngest Tertile Mean Height	Middle Tertile Mean Height	Oldest Tertile Mean Height
Boys	0.33	0.33	0.16
Girls	0.27	0.18	0.21

Based on Wake, M., Coghlan, D., and Hesketh, K. "Does height influence progression through primary school grades?" *The Archives of Disease in Childhood*, Vol. 82, Apr. 2000, pp. 297–301 (Table 2).

a. Propose a regression model that will enable you to compare the average heights of the three age groups for boys.

b. Find the estimates of the β's in the model you proposed in part **a**.

c. Repeat parts **a** and **b** for girls.

Applying the Concepts—Advanced

12.91 Community responses to a violent crime. How communities respond to a disaster or a violent crime was the subject of research published in the *American Journal of*

Community Psychology (Vol. 44, 2009). Psychologists at the University of California tracked monthly violent crime incidents in two Texas cities, Jasper and Center, before and after the murder of a Jasper citizen that had racial overtones and heavy media coverage. (Center, Texas, was selected as comparison city since it had roughly the same population and racial makeup as Jasper.) Using monthly data on violent crimes, the researchers fit the regression model:

$$E(y) = \beta_0 + \beta_1 x_1 + \beta_2 x_2 + \beta_3 x_1 x_2,$$

where y = violent crime rate (number of crimes per 1,000 population), x_1 = {1 if Jasper, 0 if Center}, and x_2 = {1 if after the murder, 0 if before the murder}.

a. In terms of the β's in the model, what is the mean violent crime rate for months following the murder in Center, Texas?

b. In terms of the β's in the model, what is the mean violent crime rate for months following the murder in Jasper, Texas?

c. For months following the murder, find the difference between the mean violent crime rate for Jasper and Center. (Use your answers to parts **a** and **b**.)

d. Repeat part **c** for months before the murder.

e. Note that the differences, parts **c** and **d**, are not the same. Explain why this illustrates the notion of interaction between x_1 and x_2.

f. A test for $H_0: \beta_3 = 0$ yielded a p-value<.001. Using α = .01, interpret this result.

g. The regression resulted in the following β-estimates: $\hat{\beta}_1 = -429, \hat{\beta}_2 = -169, \hat{\beta}_3 = 255$. Use these estimates to illustrate that average monthly violent crime decreased in Center after the murder, but increased in Jasper.

12.8 Models with Both Quantitative and Qualitative Variables (Optional)

Suppose you want to relate the mean monthly sales $E(y)$ of a company to the monthly advertising expenditure x for three different advertising media (say, newspaper, radio, and television) and you wish to use first-order (straight-line) models to model the responses for all three media. Graphs of these three relationships might appear as shown in Figure 12.20.

Since the lines in Figure 12.20 are hypothetical, a number of practical questions arise. Is one advertising medium as effective as any other? That is, do the three mean sales lines differ for the three advertising media? Do the increases in mean sales per dollar input in advertising differ for the three advertising media? That is, do the *slopes* of the three lines differ? Note that the two practical questions have been rephrased into questions about the parameters that define the three lines of Figure 12.20. To answer these questions, we must write a single regression model that will characterize the three lines of the figure and that, by testing hypotheses about the lines, will answer the questions.

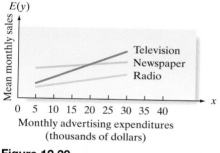

Figure 12.20

Graphs of the relationship between mean sales $E(y)$ and advertising expenditure x

The response described previously, monthly sales, is a function of *two* independent variables, one quantitative (advertising expenditure x_1) and one qualitative (type of medium). We will proceed in stages to build a model relating $E(y)$ to these variables and will show graphically the interpretation we would give to the model at each stage. This approach will help you see the contributions of the various terms in the model.

1. The straight-line relationship between mean sales $E(y)$ and advertising expenditure is the same for all three media; that is, a single line will describe the relationship between $E(y)$ and advertising expenditure x_1 for all the media. (See Figure 12.21.) Thus,

$$E(y) = \beta_0 + \beta_1 x_1,$$

where

$$x_1 = \text{Advertising expenditure}$$

2. The straight lines relating mean sales $E(y)$ to advertising expenditure x_1

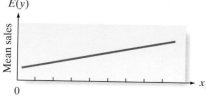

Figure 12.21

The relationship between $E(y)$ and x_1 is the same for all media

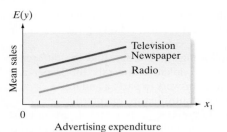

$E(y)$

Mean sales

Television
Newspaper
Radio

0

Advertising expenditure

x_1

Figure 12.22
Parallel response lines for the three media

differ from one medium to another, but the rate of increase in mean sales per increase in dollar advertising expenditure x_1 is the same for all media; that is, the lines are parallel, but possess different y-intercepts. (See Figure 12.22.) Hence,

$$E(y) = \beta_0 + \beta_1 x_1 + \beta_2 x_2 + \beta_3 x_3$$

where

$$x_1 = \text{Advertising expenditure}$$

$$x_2 = \begin{cases} 1 & \text{if radio medium} \\ 0 & \text{if not} \end{cases}$$

$$x_3 = \begin{cases} 1 & \text{if television medium} \\ 0 & \text{if not} \end{cases}$$

Notice that this model is essentially a combination of a first-order model with a single quantitative variable and a model with a single qualitative variable. That is,

First-order model with a single
quantitative variable: $\qquad E(y) = \beta_0 + \beta_1 x_1$

Model with single qualitative
variable at three levels: $\qquad E(y) = \beta_0 + \beta_2 x_2 + \beta_3 x_3$

where x_1, x_2, and x_3 are as just defined. The model described here implies no interaction between the two independent variables, which are advertising expenditure x_1 and the qualitative variable (type of advertising medium). The change in $E(y)$ for a one-unit increase in x_1 is identical (the slopes of the lines are equal) for all three advertising media. The terms corresponding to each of the independent variables are called **main-effect terms,** because they imply no interaction.

3. The straight lines relating mean sales $E(y)$ to advertising expenditure x_1 differ for the three advertising media; that is, both the line intercepts and the slopes differ. (See Figure 12.23.) This interaction model is obtained by adding terms involving the cross-product terms, one each from each of the two independent variables:

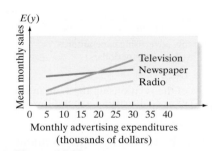

$E(y)$

Mean monthly sales

Television
Newspaper
Radio

0 5 10 15 20 25 30 35 40

Monthly advertising expenditures
(thousands of dollars)

x

Figure 12.23
Different response lines for the three media

$$E(y) = \beta_0 \;+\; \underbrace{\beta_1 x_1}_{\substack{\text{Main effect,}\\\text{advertising}\\\text{expenditure}}} \;+\; \underbrace{\beta_2 x_2 + \beta_3 x_3}_{\substack{\text{Main effect,}\\\text{type of}\\\text{medium}}} \;+\; \underbrace{\beta_4 x_1 x_2 + \beta_5 x_1 x_3}_{\text{Interaction}}$$

Note that each of the preceding models is obtained by adding terms to Model 1, the single first-order model used to model the responses for all three media. Model 2 is obtained by adding the main-effect terms for type of medium, the qualitative variable. Model 3 is obtained by adding the interaction terms to Model 2.

Example 12.10

Interpreting the β's in a Model with Mixed Variables

Problem Substitute the appropriate values of the dummy variables in Model 3 to obtain the equations of the three response lines in Figure 12.23.

Solution The complete model that characterizes the three lines in Figure 12.23 is

$$E(y) = \beta_0 + \beta_1 x_1 + \beta_2 x_2 + \beta_3 x_3 + \beta_4 x_1 x_2 + \beta_5 x_1 x_3$$

where

$$x_1 = \text{Advertising expenditure}$$

$$x_2 = \begin{cases} 1 & \text{if radio medium} \\ 0 & \text{if not} \end{cases}$$

$$x_3 = \begin{cases} 1 & \text{if television medium} \\ 0 & \text{if not} \end{cases}$$

Examining the coding, you can see that $x_2 = x_3 = 0$ when the advertising medium is newspaper. Substituting these values into the expression for $E(y)$, we obtain the newspaper medium line:

$$E(y) = \beta_0 + \beta_1 x_1 + \beta_2(0) + \beta_3(0) + \beta_4 x_1(0) + \beta_5 x_1(0) = \beta_0 + \beta_1 x_1$$

Similarly, we substitute the appropriate values of x_2 and x_3 into the expression for $E(y)$ to obtain the radio medium line ($x_2 = 1, x_3 = 0$),

$$E(y) = \beta_0 + \beta_1 x_1 + \beta_2(1) + \beta_3(0) + \beta_4 x_1(1) + \beta_5 x_1(0)$$
$$= \underbrace{(\beta_0 + \beta_2)}_{y\text{-intercept}} + \underbrace{(\beta_1 + \beta_4)x_1}_{\text{Slope}}$$

and the television medium line ($x_2 = 0, x_3 = 1$),

$$E(y) = \beta_0 + \beta_1 x_1 + \beta_2(0) + \beta_3(1) + \beta_4 x_1(0) + \beta_5 x_1(1)$$
$$= \underbrace{(\beta_0 + \beta_3)}_{y\text{-intercept}} + \underbrace{(\beta_1 + \beta_5)x_1}_{\text{Slope}}$$

Look Back If you were to fit Model 3, obtain estimates of $\beta_0, \beta_1, \beta_2, \ldots, \beta_5$, and substitute them into the equations for the three media lines, you would obtain exactly the same prediction equations as you would if you were to fit three separate straight lines, one to each of the three sets of media data. You may ask why we would not fit the three lines separately. Why bother fitting a model that combines all three lines (Model 3) into the same equation? The answer is that you need to use this procedure if you wish to use statistical tests to compare the three media lines. We need to be able to express a practical question about the lines in terms of a hypothesis that a set of parameters in the model equals 0. (We demonstrate this procedure in the next section.) You could not do that if you were to perform three separate regression analyses and fit a line to each set of media data.

Now Work Exercise 12.95

Example 12.11

Testing for Two Different Slopes—Worker Productivity Data

Problem An industrial psychologist conducted an experiment to investigate the relationship between worker productivity and a measure of salary incentive for two manufacturing plants; one plant operates under "disciplined management practices" and the other plant uses a traditional management style. The productivity y per worker was measured by recording the number of machined castings that a worker could produce in a four-week period of 40 hours per week. The incentive was the amount x_1 of bonus (in cents per casting) paid for all castings produced in excess of 1,000 per worker for the four-week period. Nine workers were selected from each plant, and three from each group of nine were assigned to receive a 20¢ bonus per casting, three a 30¢ bonus, and three a 40¢ bonus. The productivity data for the 18 workers, three for each type of plant and incentive combination, are shown in Table 12.4.

Table 12.4 Productivity Data (Number of Castings) for Example 12.11

Management Style	Incentive								
	20¢/casting			30¢/casting			40¢/casting		
Traditional	1,435	1,512	1,491	1,583	1,529	1,610	1,601	1,574	1,636
Disciplined	1,575	1,512	1,488	1,635	1,589	1,661	1,645	1,616	1,689

Data Set: CASTING

a. Write a model for mean productivity $E(y)$, assuming that the relationship between $E(y)$ and incentive x_1 is first order.

b. Fit the model, and graph the prediction equations for the traditional and disciplined plants.

c. Do the data provide sufficient evidence to indicate that the rate of increase in worker productivity is different for disciplined and traditional plants? Test at $\alpha = .10$.

Solution

a. If we assume that a first-order model* is adequate to detect a change in mean productivity as a function of incentive x_1, then the model that produces two straight lines, one for each management style, is

$$E(y) = \beta_0 + \beta_1 x_1 + \beta_2 x_2 + \beta_3 x_1 x_2$$

where

$$x_1 = \text{Incentive} \qquad x_2 = \begin{cases} 1 & \text{if disciplined management style} \\ 0 & \text{if traditional management style} \end{cases}$$

b. A MINITAB printout for the regression analysis is shown in Figure 12.24. Reading the parameter estimates highlighted on the printout, you can see that

$$\hat{y} = 1{,}365.83 + 6.217 x_1 + 47.78 x_2 + .033 x_1 x_2$$

The prediction equation for the plant using traditional management style can be obtained (see the coding) by substituting $x_2 = 0$ into the general prediction equation. Then

$$\hat{y} = \hat{\beta}_0 + \hat{\beta}_1 x_1 + \hat{\beta}_2(0) + \hat{\beta}_3 x_1(0) = \hat{\beta}_0 + \hat{\beta}_1 x_1$$
$$= 1{,}365.83 + 6.217 x_1$$

Similarly, the prediction equation for the plant with a disciplined management style can be obtained by substituting $x_2 = 1$ into the general prediction equation. Then

$$\hat{y} = \hat{\beta}_0 + \hat{\beta}_1 x_1 + \hat{\beta}_2 x_2 + \hat{\beta}_3 x_1 x_2$$
$$= \hat{\beta}_0 + \hat{\beta}_1 x_1 + \hat{\beta}_2(1) + \hat{\beta}_3 x_1(1)$$
$$= \underbrace{(\hat{\beta}_0 + \hat{\beta}_2)}_{y\text{-intercept}} + \underbrace{(\hat{\beta}_1 + \hat{\beta}_3)}_{\text{Slope}} x_1$$
$$= (1{,}365.83 + 47.78) + (6.217 + .033) x_1$$
$$= 1{,}413.61 + 6.250 x_1$$

```
The regression equation is
CASTINGS = 1366 + 6.22 INCENTIVE + 47.8 SDUMMY + 0.03 INC_SDUM


Predictor      Coef   SE Coef       T       P
Constant    1365.83     51.84   26.35   0.000
INCENTIVE     6.217     1.667    3.73   0.002
SDUMMY        47.78     73.31    0.65   0.525
INC_SDUM      0.033     2.358    0.01   0.989


S = 40.8387    R-Sq = 71.1%    R-Sq(adj) = 64.9%


Analysis of Variance

Source            DF     SS      MS       F      P
Regression         3  57332   19111   11.46  0.000
Residual Error    14  23349    1668
Total             17  80682
```

Figure 12.24

MINITAB printout of the complete model for the casting data

*Although the model contains a term involving $x_1 x_2$, it is first-order (graphs as a straight line) in the quantitative variable x_1. The variable x_2 is a dummy variable that introduces or deletes terms in the model. The order of a model is determined only by the quantitative variables that appear in the model.

A MINITAB graph of these prediction equations is shown in Figure 12.25. Note that the slopes of the two lines are nearly identical (6.217 for traditional and 6.250 for disciplined).

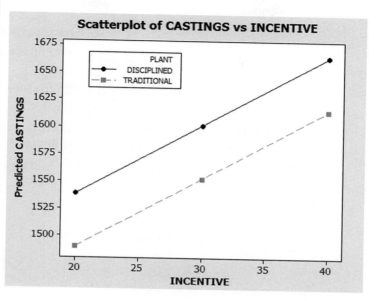

Figure 12.25

MINITAB plot of prediction equations for two plants

c. If the rate of increase in productivity with incentive (i.e, the slope) for the disciplined management style plant is different from the corresponding slope for the traditional plant, then the interaction β (i.e., β_3) will differ from 0. Consequently, we want to test

$$H_0: \beta_3 = 0$$
$$H_a: \beta_3 \neq 0$$

This test is conducted with the use of the t-test of Section 12.2. The test statistic and the corresponding p-value are highlighted on the MINITAB printout:

$$t = .014 \qquad p\text{-value} = .989$$

Since $\alpha = .10$ is less than the p-value, we fail to reject H_0, meaning that there is insufficient evidence to conclude that the traditional and disciplined shapes differ. Thus, the test supports our observation of two nearly identical slopes in part **b.**

Look Back Since interaction is not significant, we will drop the $x_1 x_2$ term from the model and use the simpler model $E(y) = \beta_0 + \beta_1 x_1 + \beta_2 x_2$ to predict productivity.

Now Work Exercise 12.99

Models with both quantitative and qualitative x's may also include higher order (e.g., second-order) terms. In the problem of relating mean monthly sales $E(y)$ of a company to monthly advertising expenditure x_1 and type of medium, suppose we think that the relationship between $E(y)$ and x_1 is curvilinear. We will construct the model stage by stage to enable you to compare the procedure with the stage-by-stage construction of the first-order model at the beginning of this section. The graphical interpretations will help you understand the contributions of the model terms.

1. The mean sales curves are identical for all three advertising media; that is, a single second-order curve will suffice to describe the relationship between $E(y)$ and x_1 for all the media. (See Figure 12.26.) Thus,

$$E(y) = \beta_0 + \beta_1 x_1 + \beta_2 x_1^2$$

where $x_1 = $ Advertising expenditure

Figure 12.26

The relationship between $E(y)$ and x_1 is the same for all media

2. The response curves possess the same shapes, but different y-intercepts. (See Figure 12.27.) Hence,

$$E(y) = \beta_0 + \beta_1 x_1 + \beta_2 x_1^2 + \beta_3 x_2 + \beta_4 x_3$$

where

$$x_1 = \text{Advertising expenditure}$$

$$x_2 = \begin{cases} 1 & \text{if radio medium} \\ 0 & \text{if not} \end{cases}$$

$$x_3 = \begin{cases} 1 & \text{if television medium} \\ 0 & \text{if not} \end{cases}$$

3. The response curves for the three advertising media are different (i.e., Advertising expenditure and Type of medium interact), as shown in Figure 12.28. Then

$$E(y) = \beta_0 + \beta_1 x_1 + \beta_2 x_1^2 + \beta_3 x_2 + \beta_4 x_3 + \beta_5 x_1 x_2 + \beta_6 x_1 x_3 + \beta_7 x_1^2 x_2 + \beta_8 x_1^2 x_3$$

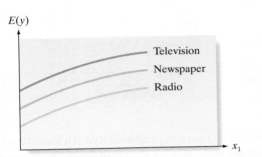

Figure 12.27
The response curves have the same shapes, but different y-intercepts

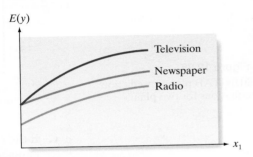

Figure 12.28
The response curves for the three media differ

Now that you know how to write a model with two independent variables—one qualitative and one quantitative—we ask a question: Why do it? Why not write a separate second-order model for each type of medium where $E(y)$ is a function of only advertising expenditure? As stated earlier, one reason we wrote the single model representing all three response curves is so that we can test to determine whether the curves are different. We illustrate this procedure in optional Section 12.9. A second reason for writing a single model is that we obtain a pooled estimate of σ^2, the variance of the random-error component ε. If the variance of ε is truly the same for each type of medium, the pooled estimate is superior to three separate estimates calculated by fitting a separate model for each type of medium.

Exercises 12.92–12.106

Understanding the Principles

12.92 Consider a multiple-regression model for a response y with one quantitative independent variable x_1 and one qualitative variable at three levels.
  a. Write a first-order model that relates the mean response $E(y)$ to the quantitative independent variable.
  b. Add the main-effect terms for the qualitative independent variable to the model of part **a**. Specify the coding scheme you use.
  c. Add terms to the model of part **b** to allow for interaction between the quantitative and qualitative independent variables.
  d. Under what circumstances will the response lines of the model in part **c** be parallel?
  e. Under what circumstances will the model in part **c** have only one response line?

12.93 Refer to Exercise 12.92.
  a. Write a complete second-order model that relates $E(y)$ to the quantitative variable.
  b. Add the main-effect terms for the qualitative variable (at three levels) to the model of part **a**.
  c. Add terms to the model of part **b** to allow for interaction between the quantitative and qualitative independent variables.
  d. Under what circumstances will the response curves of the model have the same shape, but different y-intercepts?
  e. Under what circumstances will the response curves of the model be parallel lines?
  f. Under what circumstances will the response curves of the model be identical?

12.94 Write a model that relates $E(y)$ to two independent variables, one quantitative and one qualitative (at four levels).

Construct a model that allows the associated response curves to be second order but does not allow for interaction between the two independent variables.

Learning the Mechanics

12.95 Consider the model

[NW]

$$y = \beta_0 + \beta_1 x_1 + \beta_2 x_2 + \beta_3 x_3 + \varepsilon$$

where x_1 is a quantitative variable and x_2 and x_3 are dummy variables describing a qualitative variable at three levels, using the coding scheme

$$x_2 = \begin{cases} 1 & \text{if level 2} \\ 0 & \text{otherwise} \end{cases} \quad x_3 = \begin{cases} 1 & \text{if level 3} \\ 0 & \text{otherwise} \end{cases}$$

The resulting least squares prediction equation is

$$\hat{y} = 44.8 + 2.2x_1 + 9.4x_2 + 15.6x_3$$

a. What is the response line (equation) for $E(y)$ when $x_2 = x_3 = 0$? When $x_2 = 1$ and $x_3 = 0$? When $x_2 = 0$ and $x_3 = 1$?

b. What is the least squares prediction equation associated with level 1? Level 2? Level 3? Plot these on the same graph.

12.96 Consider the model

$$y = \beta_0 + \beta_1 x_1 + \beta_2 x_1^2 + \beta_3 x_2 + \beta_4 x_3 + \beta_5 x_1 x_2 + \beta_6 x_1 x_3 + \beta_7 x_1^2 x_2 + \beta_8 x_1^2 x_3 + \varepsilon$$

where x_1 is a quantitative variable and

$$x_2 = \begin{cases} 1 & \text{if level 2} \\ 0 & \text{otherwise} \end{cases} \quad x_3 = \begin{cases} 1 & \text{if level 3} \\ 0 & \text{otherwise} \end{cases}$$

The resulting least squares prediction equation is

$$\hat{y} = 48.8 - 3.4x_1 + .07x_1^2 - 2.4x_2 - 7.5x_3 + 3.7x_1x_2 + 2.7x_1x_3 - .02x_1^2x_2 - .04x_1^2x_3$$

a. What is the equation of the response curve for $E(y)$ when $x_2 = 0$ and $x_3 = 0$? When $x_2 = 1$ and $x_3 = 0$? When $x_2 = 0$ and $x_3 = 1$?

b. On the same graph, plot the least squares prediction equation associated with level 1, level 2, and level 3.

Applying the Concepts—Basic

12.97 Reality TV and cosmetic surgery. Refer to the *Body Image: An International Journal of Research* (March 2010) study of the impact of reality TV shows on a college student's decision to undergo cosmetic surgery, Exercise 12.23 (p. 631). Recall that the data for the study (simulated based on statistics reported in the journal article) are saved in the **BODYIMAGE** file. Consider the interaction model, $E(y) = \beta_0 + \beta_1 x_1 + \beta_2 x_4 + \beta_3 x_1 x_4$, where y = desire to have cosmetic surgery (25-point scale), x_1 = {1 if male, 0 if female}, and x_4 = impression of reality TV (7-point scale). The model was fit to the data and the resulting SAS printout appears below.

a. Give the least squares prediction equation.

b. Find the predicted level of desire (y) for a male college student with an impression-of-reality-TV-scale score of 5.

c. Conduct a test of overall model adequacy. Use $\alpha = .10$.

d. Give a practical interpretation of R_a^2.

e. Give a practical interpretation of s.

f. Conduct a test (at $\alpha = .10$) to determine if gender (x_1) and impression of reality TV show (x_4) interact in the prediction of level of desire for cosmetic surgery (y).

g. Give an estimate of the change in desire (y) for every 1-point increase in impression of reality TV show (x_4) for female students.

h. Repeat part **g** for male students.

12.98 Impact of race on football card values. Refer to the *Electronic Journal of Sociology* (2007) study of the impact of race on the value of professional football players' "rookie" cards, presented in Exercise 12.81 (p. 661). Recall that the sample consisted of 148 rookie cards of National Football League (NFL) players who were inducted into the Football Hall of Fame. The researchers modeled the natural logarithm of card price (y) as a function of the following independent variables:

Dependent Variable: DESIRE

Number of Observations Read	170	
Number of Observations Used	170	

Analysis of Variance

Source	DF	Sum of Squares	Mean Square	F Value	Pr > F
Model	3	747.00146	249.00049	45.09	<.0001
Error	166	916.78678	5.52281		
Corrected Total	169	1663.78824			

Root MSE	2.35007	R-Square	0.4490
Dependent Mean	12.56471	Adj R-Sq	0.4390
Coeff Var	18.70371		

Parameter Estimates

Variable	DF	Parameter Estimate	Standard Error	t Value	Pr > |t|
Intercept	1	11.77891	0.67363	17.49	<.0001
GENDER	1	-1.97223	1.17922	-1.67	0.0963
IMPREAL	1	0.58462	0.16163	3.62	0.0004
IMPREAL_GENDER	1	-0.55330	0.27611	-2.00	0.0467

SAS output for Exercise 12.97

Race: $x_1 = 1$ if black, 0 if white
Card availability: $x_2 = 1$ if high, 0 if low
Card vintage: $x_3 = $ year card printed
Finalist: $x_4 = $ natural logarithm of number of times player was on final Hall of Fame ballot
Position-QB: $x_5 = 1$ if quarterback, 0 if not
Position-RB: $x_7 = 1$ if running back, 0 if not
Position-WR: $x_8 = 1$ if wide receiver, 0 if not
Position-TE: $x_9 = 1$ if tight end, 0 if not
Position-DL: $x_{10} = 1$ if defensive lineman, 0 if not
Position-LB: $x_{11} = 1$ if linebacker, 0 if not
Position-DB: $x_{12} = 1$ if defensive back, 0 if not
[*Note:* For Position, offensive lineman is the base level]

a. The model $E(y) = \beta_0 + \beta_1 x_1 + \beta_2 x_2 + \beta_3 x_3 + \beta_4 x_4 + \beta_5 x_5 + \beta_6 x_6 + \beta_7 x_7 + \beta_8 x_8 + \beta_9 x_9 + \beta_{10} x_{10} + \beta_{11} x_{11} + \beta_{11} x_{11} + \beta_{12} x_{12}$ was fit to the data, with the following results: $R^2 = .705$, adj-$R^2 = .681$, $F = 26.9$. Interpret the results practically. Make an inference about the overall adequacy of the model.

b. Refer to part **a**. Statistics for the race variable were reported as follows: $\hat{\beta}_1 = -147$, $s_{\hat{\beta}_1} = .145$, $t = -1.014$, p-value = .312. Use this information to make an inference about the impact of race on the value of professional football players' rookie cards.

c. Refer to part **a**. Statistics for the card vintage variable were reported as follows: $\hat{\beta}_3 = -.074$, $s_{\hat{\beta}_2} = .007$, $t = -10.92$, p-value = .000. Use this information to make an inference about the impact of card vintage on the value of professional football players' rookie cards.

d. Write a first-order model for $E(y)$ as a function of card vintage (x_4) and position (x_5–x_{12}) that allows for the relationship between price and vintage to vary with position.

12.99 **Smoking and resting energy.** The influence of cigarette smoking on resting energy expenditure (REE) in normal-weight and obese smokers was investigated (*Health Psychology*, Mar. 1995). The researchers hypothesized that the relationship between a smoker's REE and length of time since smoking differs for these two types of smokers. Consequently, they examined the interaction model

$$E(y) = \beta_0 + \beta_1 x_1 + \beta_2 x_2 + \beta_3 x_1 x_2$$

where

$y = $ REE, measured in kilocalories per day

$x_1 = $ Time, in minutes, after smoking, of metabolic energy reading (levels = 10, 20, and 30 minutes)

$x_2 = \begin{cases} 1 & \text{if normal weight} \\ 0 & \text{if obese} \end{cases}$

a. Give the equation of the hypothesized line relating mean REE to time after smoking for obese smokers. What is the slope of the line?

b. Repeat part **a** for normal-weight smokers.

c. A test for interaction resulted in an observed significance level of .044. Interpret this value.

12.100 **Winning marathon times.** Refer to the *Chance* (Winter 2000) study of men's and women's winning times in the Boston Marathon, presented in Exercise 11.128 (p. 606). Suppose the researchers want to build a model for predicting winning time (y) of the marathon as a function of year (x_1) in which race is run and gender of winning runner (x_2).

a. Set up the appropriate dummy variables (if necessary) for x_1 and x_2.

b. Write the equation of a model that proposes parallel straight-line relationships between winning time (y) and year (x_1), one line for each gender.

c. Write the equation of a model that proposes nonparallel straight-line relationships between winning time (y) and year (x_1), one line for each gender.

d. Which of the models do you think will provide the best predictions of winning time (y)? Base your answer on the graph displayed in Exercise 11.128.

12.101 **Whales entangled in fishing gear.** Refer to the *Marine Mammal Science* (April 2010) study of whales entangled in fishing gear, Exercises 12.15 and 12.79 (pp. 629 and 660). Now consider a model for the length (y) of an entangled whale (in meters) that is a function of water depth of the entanglement (in meters) and gear type (set nets, pots, or gill nets).

a. Write a main-effects-only model for $E(y)$.

b. Sketch the relationships hypothesized by the model, part **a**. (*Hint:* Plot length on the vertical axis and water depth on the horizontal axis.)

c. Add terms to the model, part **a**, that include interaction between water depth and gear type. (*Hint:* Be sure to interact each dummy variable for gear type with water depth.)

d. Sketch the relationships hypothesized by the model, part **c**.

e. In terms of the β's in the model of part **c**, give the rate of change of whale length with water depth for set nets.

f. Repeat part **e** for pots.

g. Repeat part **e** for gill nets.

h. In terms of the β's in the model of part **c**, how would you test to determine if the rate of change of whale length with water depth is the same for all three types of fishing gear?

Applying the Concepts—Intermediate

12.102 **RNA analysis of wheat genes.** Engineers from the Department of Crop and Soil Sciences at Washington State University used regression to estimate the number of copies of a gene transcript in an aliquot of ribonucleic acid (RNA) extracted from a wheat plant (*Electronic Journal of Biotechnology*, April 15, 2004). The proportion (x_1) of RNA extracted from a wheat plant exposed to the cold was varied, and the transcript copy number (y, in thousands) was measured for each of two cloned genes: Mn superoxide dismutase (MnSOD) and phospholipase D (PLD). The data are listed in the next table (p. 671) and saved in the **WHEATRNA** file.

a. Write a first-order model for number of copies (y) as a function of proportion (x_1) of RNA extracted and gene type (MnSOD or PLD). Assume that proportion of RNA and gene type interact to affect y.

b. Fit the model you wrote in part **a** to the data. Give the least squares prediction equation for y.

c. Conduct a test to determine whether, in fact, proportion of RNA and gene type interact. Test, using $\alpha = .01$.

d. Use the results from part **b** to estimate the rate of increase of number of copies (y) with proportion (x_1) of RNA extracted for the MnSOD gene type.

e. Repeat part **d** for the PLD gene type.

Data for Exercise 12.102

RNA Proportion (x_1)	Number of Copies (y, thousands)	
	MnSOD	PLD
0.00	401	80
0.00	336	83
0.00	337	75
0.33	711	132
0.33	637	148
0.33	602	115
0.50	985	147
0.50	650	142
0.50	747	146
0.67	904	146
0.67	1,007	150
0.67	1,047	184
0.80	1,151	173
0.80	1,098	201
0.80	1,061	181
1.00	1,261	193
1.00	1,272	187
1.00	1,256	199

Based on Baek K. H., and Skinner, D. Z. "Quantitative real-time PCR method to detect changes in specific transcript and total RNA amounts." *Electronic Journal of Biotechnology*, Vol. 7, No. 1, April 15, 2004 (adapted from Figure 2).

12.103 Workplace bullying and intention to leave. Workplace bullying (e.g., work-related harassment, persistent criticism, withholding of key information, spreading of rumors, intimidation) has been shown to have a negative psychological effect on victims, often leading the victim to quit or resign. In *Human Resource Management Journal* (Oct. 2008), researchers employed multiple regression to examine whether perceived organizational support would moderate the relationship between workplace bullying and victims' intention to leave the firm. The dependent variable in the analysis, intention to leave (y), was measured on a quantitative scale. The two key independent variables in the study were bullying (x_1, measured on a quantitative scale) and perceived organizational support (measured qualitatively as "low," "neutral," or "high").

a. Set up the dummy variables required to represent perceived organizational support (POS) in the regression model.

b. Write a model for $E(y)$ as a function of bullying and POS that hypothesizes three parallel straight lines, one for each level of POS.

c. Write a model for $E(y)$ as a function of bullying and POS that hypothesizes three nonparallel straight lines, one for each level of POS.

d. The researchers discovered that the effect of bullying on intention to leave was greater at the low level of POS than at the high level of POS. Which of the two models, parts **b** and **c**, support these findings?

12.104 Lead levels in mountain moss. A study of the atmospheric pollution on the slopes of the Blue Ridge Mountains (in Tennessee) was conducted. The file **LEADMOSS** contains the levels of lead found in 70 fern moss specimens (in micrograms of lead per gram of moss tissue) collected from the mountain slopes, as well as the elevation of the moss specimen (in feet) and the direction (1 if east, 0 if west) of the slope face. The first five and last five observations of the data set are listed in the following table:

Specimen	Lead Level	Elevation	Slope Face
1	3.475	2,000	0
2	3.359	2,000	0
3	3.877	2,000	0
4	4.000	2,500	0
5	3.618	2,500	0
⋮	⋮	⋮	⋮
66	5.413	2,500	1
67	7.181	2,500	1
68	6.589	2,500	1
69	6.182	2,000	1
70	3.706	2,000	1

Based on Schilling, J. "Bioindication of atmospheric heavy metal deposition in the Blue Ridge using the moss, *Thuidium delicatulum*." master-of-science thesis, spring 2000.

a. Write the equation of a first-order model relating mean lead level $E(y)$ to elevation (x_1) and slope face (x_2). Include interaction between elevation and slope face in the model.

b. Graph the relationship between mean lead level and elevation for the different slope faces that is hypothesized by the model you wrote in part **a.**

c. In terms of the β's of the model from part **a,** give the change in lead level for every 1-foot increase in elevation for moss specimens on the east slope.

d. Fit the model from part **a** to the data, using an available statistical software package. Is the overall model statistically useful in predicting lead level? Test, using $\alpha = .10$.

12.105 Chemical composition of rainwater. Refer to the *Journal of Agricultural, Biological, and Environmental Statistics* (March 2005) study of the chemical composition of rainwater, presented in Exercise 12.82 (p. 661). Recall that the nitrate concentration y (milligrams per liter) in a sample of rainwater was modeled as a function of water source (groundwater, subsurface flow, or overground flow). Now consider adding a second independent variable, silica concentration (milligrams per liter), to the model.

a. Write a first-order model for $E(y)$ as a function of the independent variables. Assume that the rate of increase of nitrate concentration with silica concentration is the same for all three water sources. Sketch the relationships hypothesized by the model on a graph.

b. Write a first-order model for $E(y)$ as a function of the independent variables, but now assume that the rate of increase of nitrate concentration with silica concentration differs for the three water sources. Sketch the relationships hypothesized by the model on a graph.

Applying the Concepts—Advanced

12.106 Iron supplement for anemia. Many women suffer from anemia. A female physician who is also an avid jogger wanted to know if women who exercise regularly have a different mean red blood cell count than women who do not. She also wanted to know if the amount of a particular iron supplement a woman takes has any effect and whether the effect is the same for both groups. Write a model that will reflect the relationship between red blood cell count and the two independent variables just described, assuming that

a. the effect of the iron supplement on mean blood cell count is the same regardless of whether a woman exercises regularly.

b. the effect of the iron supplement on mean blood cell count depends on whether a woman exercises regularly.

12.9 Comparing Nested Models (Optional)

To be successful model builders, we require a statistical method that will allow us to determine (with a high degree of confidence) which one among a set of candidate models best fits the data. In this section, we present such a technique for *nested models*.

> Two models are **nested** if one model contains all the terms of the second model and at least one additional term. The more complex of the two models is called the **complete** model, and the simpler of the two is called the **reduced** model.

To illustrate the concept of nested models, consider the straight-line interaction model for the mean auction price $E(y)$ of a grandfather clock as a function of two quantitative variables: age of the clock (x_1) and number of bidders (x_2). The interaction model fit in Example 12.6 is

$$E(y) = \beta_0 + \beta_1 x_1 + \beta_2 x_2 + \beta_3 x_1 x_2$$

If we assume that the relationship between auction price (y), age (x_1), and bidders (x_2) is curvilinear, then the complete second-order model is more appropriate:

$$E(y) = \overbrace{\beta_0 + \beta_1 x_1 + \beta_2 x_2 + \beta_3 x_1 x_2}^{\text{Terms in interaction model}} + \overbrace{\beta_4 x_1^2 + \beta_5 x_2^2}^{\text{Quadratic terms}}$$

Note that the curvilinear model contains quadratic terms for x_1 and x_2, as well as the terms in the interaction model. Therefore, the models are nested models. In this case, the interaction model is nested within the more inclusive curvilinear model. Thus, the curvilinear model is the *complete* model and the interaction model is the *reduced* model.

Suppose we want to know whether the curvilinear model contributes more information relevant to the prediction of y than the straight-line interaction model does. This is equivalent to determining whether the quadratic betas β_4 and β_5 should be retained in the model. To test whether these terms should be retained, we set up the null and alternative hypotheses as follows:

H_0: $\beta_4 = \beta_5 = 0$ (i.e., quadratic terms are not important in predicting y).

H_a: At least one of the parameters β_4 and β_5 is nonzero (i.e., at least one of the quadratic terms is useful in predicting y).

Note that the terms being tested are those additional terms in the complete (curvilinear) model that are not in the reduced (straight-line interaction) model.

We presented the t-test for a single β coefficient (Section 12.2) and the global F-test for *all* the β parameters (except β_0) in the model (Section 12.3). We now need a test for a *subset* of the β parameters in the complete model. The test procedure is intuitive. First, we use the method of least squares to fit the reduced model and calculate the corresponding sum of squares for error, SSE_R (the sum of squares of the deviations between the observed and the predicted y-values). Next, we fit the complete model and calculate its sum of squares for error, SSE_C. Then we compare SSE_R with SSE_C by calculating the difference, $SSE_R - SSE_C$. If the additional terms in the complete model are significant, then SSE_C should be much smaller than SSE_R, and the difference $SSE_R - SSE_C$ will be large.

Since SSE will always decrease when new terms are added to the model, the question is whether the difference $SSE_R - SSE_C$ is large enough to conclude that it is due to more than just an increase in the number of model terms and to chance. The formal statistical test utilizes an F-statistic, as shown in the following box:

F-Test for Comparing Nested Models

Reduced model: $E(y) = \beta_0 + \beta_1 x_1 + \cdots + \beta_g x_g$

Complete model: $E(y) = \beta_0 + \beta_1 x_1 + \cdots + \beta_g x_g + \beta_{g+1} x_{g+1} + \cdots + \beta_k x_k$

$H_0: \beta_{g+1} = \beta_{g+2} = \cdots = \beta_k = 0$

H_a: At least one of the β parameters specified in H_0 is nonzero.

Test statistic:
$$F = \frac{(SSE_R - SSE_C)/(k - g)}{SSE_C/[n - (k + 1)]}$$

$$= \frac{(SSE_R - SSE_C)/\# \, \beta\text{'s tested in } H_0}{MSE_C}$$

where

SSE_R = Sum of squared errors for the reduced model

SSE_C = Sum of squared errors for the complete model

MSE_C = Mean square error (s^2) for the complete model

$k - g$ = Number of β parameters specified in H_0 (i.e., number of β's tested)

$k + 1$ = Number of β parameters in the complete model (including β_0)

n = Total sample size

Rejection region: $F > F_\alpha$

where F is based on $\nu_1 = k - g$ numerator degrees of freedom and $\nu_2 = n - (k + 1)$ denominator degrees of freedom

When the assumptions listed in Section 12.1 about the random-error term are satisfied, this F-statistic has an F-distribution with ν_1 and ν_2 df. Note that ν_1 is the number of β parameters being tested and ν_2 is the number of degrees of freedom associated with s^2 in the complete model.

Example 12.12

Analyzing a Complete Second–Order Model— Carnation Growth Data

Problem A botanist conducted an experiment to study the growth of carnations as a function of the temperature x_1 (°F) in a greenhouse and the amount of fertilizer x_2 [kilograms (kg) per plot] applied to the soil. Twenty-seven plots of equal size were treated with fertilizer in amounts varying between 50 and 60 kg per plot and were mechanically kept at constant temperatures between 80 and 100°F. Small carnation plants [approximately 15 centimeters (cm) in height] were planted in each plot, and their height y (cm) was measured after a six-week growing period. The resulting data are shown in Table 12.5.

a. Fit a complete second-order model to the data.

b. Sketch the fitted model in three dimensions.

Table 12.5 **Temperature (x_1), Amount of Fertilizer (x_2), and Height (y) of Carnations**

x_1	x_2	y	x_1	x_2	y	x_1	x_2	y
80	50	50.8	90	50	63.4	100	50	46.6
80	50	50.7	90	50	61.6	100	50	49.1
80	50	49.4	90	50	63.4	100	50	46.4
80	55	93.7	90	55	93.8	100	55	69.8
80	55	90.9	90	55	92.1	100	55	72.5
80	55	90.9	90	55	97.4	100	55	73.2
80	60	74.5	90	60	70.9	100	60	38.7
80	60	73.0	90	60	68.8	100	60	42.5
80	60	71.2	90	60	71.3	100	60	41.4

Data Set: CARNATIONS

c. Do the data provide sufficient evidence to indicate that the second-order terms β_3, β_4, and β_5 contribute information relevant to the prediction of y?

Solution

a. The complete second-order model is

$$E(y) = \beta_0 + \beta_1 x_1 + \beta_2 x_2 + \beta_3 x_1 x_2 + \beta_4 x_1^2 + \beta_5 x_2^2$$

The data in Table 12.5 were used to fit this model, and a portion of the SAS output is shown in Figure 12.29.

```
                    Dependent Variable: HEIGHT

                  Number of Observations Read        27
                  Number of Observations Used        27

                          Analysis of Variance

                                Sum of        Mean
Source              DF         Squares       Square    F Value    Pr > F
Model                5      8402.26454   1680.45291     596.32    <.0001
Error               21        59.17843      2.81802
Corrected Total     26      8461.44296


        Root MSE                 1.67870    R-Square     0.9930
        Dependent Mean          66.96296    Adj R-Sq     0.9913
        Coeff Var                2.50690

                          Parameter Estimates

                        Parameter      Standard
Variable      DF         Estimate         Error    t Value    Pr > |t|

Intercept      1      -5127.89907     110.29601     -46.49     <.0001
TEMP           1         31.09639       1.34441      23.13     <.0001
FERT           1        139.74722       3.14005      44.50     <.0001
TEM_FERT       1         -0.14550       0.00969     -15.01     <.0001
TEMPSQ         1         -0.13339       0.00685     -19.46     <.0001
FERTSQ         1         -1.14422       0.02741     -41.74     <.0001


        Test HIGHORD Results for Dependent Variable HEIGHT

                                    Mean
Source             DF            Square    F Value    Pr > F

Numerator           3        2204.11003     782.15    <.0001
Denominator        21           2.81802
```

Figure 12.29

SAS printout of complete second-order model for height.

The least squares prediction equation (rounded) is

$$\hat{y} = -5{,}127.90 + 31.10x_1 + 139.75x_2 - .146x_1 x_2 - .133x_1^2 - 1.14x_2^2$$

b. A three-dimensional graph of this prediction model, called a **response surface,** is shown in Figure 12.30. Note that the height seems to be greatest for temperatures of about 85–90°F and for applications of about 55–57 kg of fertilizer per plot.* Further experimentation in these ranges might lead to a more precise determination of the optimal temperature–fertilizer combination.

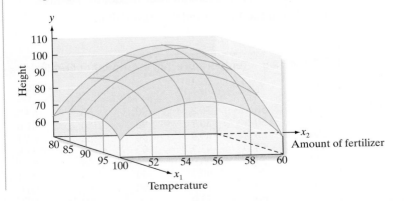

Figure 12.30

Plot of second-order least squares model for Example 12.12

*Students with a knowledge of calculus should note that we can solve for the exact temperature and amount of fertilizer that maximize height in the least squares model by solving $\partial \hat{y} / \partial x_1 = 0$ and $\partial \hat{y} / \partial x_2 = 0$ for x_1 and x_2. Sample estimates of these estimated optimal values are $x_1 = 86.25°F$ and $x_2 = 55.58$ kg per plot.

c. To determine whether the data provide sufficient information to indicate that the second-order terms contribute information for the prediction of y, we wish to test

$$H_0: \beta_3 = \beta_4 = \beta_5 = 0$$

against the alternative hypothesis

H_a: At least one of the parameters β_3, β_4, and β_5 differs from 0.

The first step in conducting the test is to drop the second-order terms out of the complete (second-order) model and fit the reduced model

$$E(y) = \beta_0 + \beta_1 x_1 + \beta_2 x_2$$

to the data. The SAS printout for this model is shown in Figure 12.31.

The sums of squares for error, highlighted in Figures 12.29 and 12.31 for the complete and reduced models, respectively, are

$$\text{SSE}_C = 59.17843$$
$$\text{SSE}_R = 6{,}671.50852$$

and s^2 for the complete model (highlighted on Figure 12.29) is

$$s^2 = \text{MSE}_C = 2.81802$$

Recall that $n = 27, k = 5$, and $g = 2$. Therefore, the calculated value of the F-statistic, based on $\nu_1 = (k - g) = 3$ numerator df and $\nu_2 = [n - (k + 1)] = 21$ denominator df, is

$$F = \frac{(\text{SSE}_R - \text{SSE}_C)/(k - g)}{\text{SSE}_C/[n - (k + 1)]} = \frac{(\text{SSE}_R - \text{SSE}_C)/(k - g)}{\text{MSE}_C}$$

where $\nu_1 = (k - g)$ is equal to the number of parameters involved in H_0. Therefore,

$$\textit{Test statistic: } F = \frac{(6{,}671.50852 - 59.17843)/3}{2.81802} = 782.15$$

The final step in the test is to compare this computed value of F with the tabulated value based on $\nu_1 = 3$ and $\nu_2 = 21$ df. If we choose $\alpha = .05$, then $F_{.05} = 3.07$ and the rejection region is

$$\textit{Rejection region: } F > 3.07.$$

Since the computed value of F falls in the rejection region (i.e., it exceeds $F_{.05} = 3.07$), we reject H_0 and conclude that at least one of the second-order terms contributes information relevant to the prediction of y. Thus, the second-order model appears to provide better predictions of y than does a first-order model.

Figure 12.31

SAS printout of first-order model for height

Dependent Variable: HEIGHT

Number of Observations Read	27
Number of Observations Used	27

Analysis of Variance

Source	DF	Sum of Squares	Mean Square	F Value	Pr > F
Model	2	1789.93444	894.96722	3.22	0.0577
Error	24	6671.50852	277.97952		
Corrected Total	26	8461.44296			

Root MSE	16.67272	R-Square	0.2115	
Dependent Mean	66.96296	Adj R-Sq	0.1458	
Coeff Var	24.89842			

Parameter Estimates

Variable	DF	Parameter Estimate	Standard Error	t Value	Pr > |t|
Intercept	1	106.08519	55.94500	1.90	0.0700
TEMP	1	-0.91611	0.39298	-2.33	0.0285
FERT	1	0.78778	0.78596	1.00	0.3262

Look Back Using special commands, you can get SAS to perform the desired nested-model F-test. The test statistic and p-value for the preceding test are highlighted at the bottom of the SAS printout in Figure 12.29.

Now Work Exercise 12.110

The nested-model F-test can be used to determine whether *any* subset of terms should be included in a complete model by testing the null hypothesis that a particular set of β parameters simultaneously equals 0. For example, we may want to test whether a set of interaction terms for quantitative variables or a set of main-effect terms for a qualitative variable should be included in a model. If we reject H_0, the complete model is the better of the two nested models.

Suppose the F-test in Example 12.12 yielded a test statistic that did not fall into the rejection region. Although we must be cautious about accepting H_0, most practitioners of regression analysis adopt the principle of *parsimony*. That is, in situations where two competing models are found to have essentially the same predictive power (as in this case), the model with the lesser number of β's (i.e., the more parsimonious model) is selected. On the basis of this principle, we would drop the three second-order terms and select the straight-line (reduced) model over the second-order (complete) model.

> A **parsimonious model** is a general linear model with a small number of β parameters. In situations where two competing models have essentially the same predictive power (as determined by an F-test), choose the more parsimonious of the two.

Guidelines for Selecting Preferred Model in a Nested Model F-Test

Conclusion	Preferred Model
Reject H_0	→ Complete Model
Fail to reject H_0	→ Reduced Model

When the candidate models in model building are nested models, the F-test developed in this section is the appropriate procedure to apply to compare the models. However, if the models are not nested, this F-test is not applicable. In such a situation, the analyst must base the choice of the best model on statistics such as R_a^2 and s. It is important to remember that decisions based on these and other numerical descriptive measures of the adequacy of a model cannot be supported with a measure of reliability and are often highly subjective in nature.

Statistics IN Action | **Revisited** | ### Building a Model for Condominium Sale Price

In the previous Statistics in Action Revisited section (p. 635), we used the six independent variables listed in Table SIA12.1 to fit a first-order model for the auction price (y) of a condominium unit. Although the model was deemed statistically useful in predicting y, the standard deviation of the model ($s = 21.8$ hundred dollars) was probably too large for the model to be "practically" useful. A more complicated model—one involving higher order terms (interactions and squared terms)—needs to be considered. We start with a second-order model involving only the two quantitative independent variables FLOOR (x_1) and DISTANCE (x_2). The model is given by the equation

$$E(y) = \beta_0 + \beta_1 x_1 + \beta_2 x_2 + \beta_3 x_1 x_2$$
$$+ \beta_4 (x_1)^2 + \beta_5 (x_2)^2$$

The SAS printout for this model is shown in Figure SIA12.5. Note that the global F-test for the model is statistically significant (p-value $<.0001$).

Are the higher (second)-order terms in the model, namely, $\beta_3 x_1 x_2, \beta_4(x_1)^2$, and $\beta_5(x_2)^2$, necessary? If not, we can simplify the model by dropping these curvature terms. The hypothesis of interest is $H_0: \beta_3 = \beta_4 = \beta_5 = 0$. To test this subset of β's, we compare the second-order model with a model that lacks the interaction and curvilinear terms. The reduced model takes the form

```
                          Dependent Variable: PRICE100

                       Number of Observations Read        209
                       Number of Observations Used        209

                              Analysis of Variance

                                     Sum of          Mean
Source                  DF          Squares         Square      F Value    Pr > F

Model                    5            60858          12172       13.88     <.0001
Error                  203           178035      877.01813
Corrected Total        208           238893


              Root MSE             29.61449     R-Square      0.2548
              Dependent Mean      201.28708     Adj R-Sq      0.2364
              Coeff Var            14.71256


                             Parameter Estimates

                       Parameter       Standard
Variable        DF      Estimate          Error     t Value    Pr > |t|

Intercept        1     229.80506       13.17675       17.44     <.0001
FLOOR            1      -8.76315        4.49100       -1.95      0.0524
DISTANCE         1      -7.33316        2.31456       -3.17      0.0018
FLR_DIST         1      -0.17739        0.20153       -0.88      0.3798
FLOORSQ          1       0.76065        0.44691        1.70      0.0903
DISTSQ           1       0.66948        0.13221        5.06     <.0001


           Test HIORDER Results for Dependent Variable PRICE100

                                    Mean
Source                  DF         Square       F Value    Pr > F

Numerator                3     8500.32594         9.69     <.0001
Denominator            203      877.01813
```

Figure SIA12.5

SAS printout of the second-order model for condo sale price—quantitative variables only

$$E(y) = \beta_0 + \beta_1 x_1 + \beta_2 x_2$$

The results of this nested model (or partial) F-test are shown at the bottom of the SAS printout in Figure SIA12.5. The p-value of the test (highlighted) is less than .0001. Since this p-value is smaller than $\alpha = .01$, there is sufficient evidence to reject H_0. That is, there is evidence to indicate that at least one of the three higher order terms is a useful predictor of auction price.

To improve the model, we now add terms for the qualitative variables VIEW (x_3), END (x_4), FURNISH (x_5), and AUCTION (x_6). The developer theorizes that the impact of floor height and distance from the elevator on price will vary with the unit's view. Consequently, we also add interaction between floor and view and between distance and view. The complete model takes the form

$$E(y) = \beta_0 + \beta_1 x_1 + \beta_2 x_2 + \beta_3 x_1 x_2 + \beta_4 (x_1)^2$$
$$+ \beta_5 (x_2)^2 + \beta_6 x_3 + \beta_7 x_3 x_1 + \beta_8 x_3 x_2$$
$$+ \beta_9 x_3 x_1 x_2 + \beta_{10} x_3 (x_1)^2 + \beta_{11} x_3 (x_2)^2$$
$$+ \beta_{12} x_4 + \beta_{13} x_5 + \beta_{14} x_6$$

The SAS printout for this complete model is shown in Figure SIA12.6. The overall model is statistically useful (p-value <.0001 for global F-test), explaining about 68%

(adjusted $R^2 = .6815$) of the sample variation in auction prices. The model standard deviation, $s = 19$, implies that we can predict price to within about 38 hundred dollars. Both the adjusted R^2 and $2s$ values are improvements over the corresponding values for the first-order model of the previous Statistics in Action Revisited (p. 635).

To test the developer's theory of how the view affects the sales price relationship, we conduct a nested-model F-test of all the VIEW (x_3) interaction terms. The null hypothesis of interest is $H_0: \beta_7 = \beta_8 = \beta_9 = \beta_{10} = \beta_{11} = 0$, and the reduced model takes the form

$$E(y) = \beta_0 + \beta_1 x_1 + \beta_2 x_2 + \beta_3 x_1 x_2$$
$$+ \beta_4 (x_1)^2 + \beta_5 (x_2)^2 + \beta_6 x_3 + \beta_{12} x_4$$
$$+ \beta_{13} x_5 + \beta_{14} x_6$$

The p-value of the test (highlighted at the bottom of the SAS printout in Figure SIA12.6) is less than .0001. Since this p-value is smaller than $\alpha = .01$, there is sufficient evidence to conclude that at least one of the view interaction terms is useful in predicting the auction price. This implies, as theorized by the developer, that the price–floor and price–distance relationships depend on the unit's view (ocean view or not).

(continued)

```
                    Dependent Variable: PRICE100

                Number of Observations Read        209
                Number of Observations Used        209

                       Analysis of Variance

                             Sum of          Mean
    Source            DF     Squares        Square    F Value    Pr > F

    Model             14      167924         11995      32.79    <.0001
    Error            194       70968     365.81593
    Corrected Total  208      238893


            Root MSE              19.12632     R-Square    0.7029
            Dependent Mean       201.28708     Adj R-Sq    0.6815
            Coeff Var              9.50201

                       Parameter Estimates

                       Parameter     Standard
    Variable       DF   Estimate        Error    t Value    Pr > |t|

    Intercept       1   188.72646     13.15224      14.35    <.0001
    FLOOR           1    -4.61416      4.49901      -1.03     0.3064
    DISTANCE        1    -2.35297      2.43410      -0.97     0.3349
    FLR_DIST        1    -0.33458      0.20177      -1.66     0.0989
    FLOORSQ         1     1.01858      0.42699       2.39     0.0180
    DISTSQ          1     0.29095      0.14111       2.06     0.0406
    VIEW            1    74.33636     17.73275       4.19    <.0001
    VU_FLOOR        1    -4.77034      5.90393      -0.81     0.4201
    VU_DIST         1    -1.62826      3.17204      -0.51     0.6083
    VU_FLR_DIST     1     0.07496      0.27422       0.27     0.7849
    VU_FLRSQ        1    -0.26670      0.57992      -0.46     0.6461
    VU_DISTSQ       1     0.13340      0.18392       0.73     0.4691
    END             1   -16.27750      7.90038      -2.06     0.0407
    FURNISH         1     7.96051      2.99043       2.66     0.0084
    AUCTION         1   -25.59314      3.66294      -6.99    <.0001


      Test VUINT Results for Dependent Variable PRICE100

                              Mean
    Source          DF       Square    F Value    Pr > F

    Numerator        5   2565.21064       7.01    <.0001
    Denominator    194    365.81593
```

Figure SIA12.6
SAS regression printout for the complete second-order model of condo sale price—qualitative variables added

Exercises 12.107–12.120

Understanding the Principles

12.107 Determine which pairs of models that follow are nested models. For each pair of nested models, identify the complete and reduced model.
   a. $E(y) = \beta_0 + \beta_1 x_1 + \beta_2 x_2$
   b. $E(y) = \beta_0 + \beta_1 x_1$
   c. $E(y) = \beta_0 + \beta_1 x_1 + \beta_2 x_1^2$
   d. $E(y) = \beta_0 + \beta_1 x_1 + \beta_2 x_2 + \beta_3 x_1 x_2$
   e. $E(y) = \beta_0 + \beta_1 x_1 + \beta_2 x_2 + \beta_3 x_1 x_2 + \beta_4 x_1^2 + \beta_5 x_2^2$

12.108 Explain why the F-test used to compare complete and reduced models is a one-tailed, upper-tailed test.

12.109 What is a parsimonious model?

Learning the Mechanics

12.110 Suppose you fit the regression model

$$y = \beta_0 + \beta_1 x_1 + \beta_2 x_2 + \beta_3 x_1 x_2 + \beta_4 x_1^2 + \beta_5 x_2^2 + \varepsilon$$

to $n = 30$ data points and you wish to test

$$H_0: \beta_3 = \beta_4 = \beta_5 = 0$$

   a. State the alternative hypothesis H_a.
   b. Give the reduced model appropriate for conducting the test.

   c. What are the numerator and denominator degrees of freedom associated with the F-statistic?
   d. Suppose the SSE's for the complete and reduced models are $\text{SSE}_R = 1,250.2$ and $\text{SSE}_C = 1,125.2$, respectively. Conduct the hypothesis test and interpret the results. Use $\alpha = .05$.

12.111 The complete model

$$y = \beta_0 + \beta_1 x_1 + \beta_2 x_2 + \beta_3 x_3 + \beta_4 x_4 + \varepsilon$$

was fitted to $n = 20$ data points, with SSE = 152.66. The independent variables x_3 and x_4 were dropped from the model, yielding SSE = 160.44.
   a. How many β parameters are in the complete model? The reduced model?
   b. Specify the null and alternative hypotheses you would use to investigate whether the complete model contributes more information relevant to the prediction of y than the reduced model does.
   c. Conduct the hypothesis test of part **b.** Use $\alpha = .05$.

Applying the Concepts—Basic

12.112 Mental health of a community. An article in the *Community Mental Health Journal* (Aug. 2000) used multiple-regression analysis to model the level of community adjustment of clients of the Department of Mental Health and Addiction Services in Connecticut. The dependent variable, community adjustment (y), was measured quantitatively on the basis of staff ratings of the clients. (Lower scores indicate better adjustment.) The complete model was a first-order model with 21 independent variables. The independent variables were categorized as demographic (four variables), diagnostic (seven variables), treatment (four variables), and community (six variables).

a. Write the equation of $E(y)$ for the complete model.

b. Give the null hypothesis for testing whether the seven diagnostic variables contribute information relevant to the prediction of y.

c. Give the equation of the reduced model appropriate for the test suggested in part **b.**

d. The test in part **b** was carried out and resulted in a test statistic of $F = 59.3$ and p-value $<.0001$. Interpret this result in the words of the problem.

12.113 Workplace bullying and intention to leave. Refer to the *Human Resource Management Journal* (Oct. 2008) study of workplace bullying, Exercise 12.103 (p. 671). Recall that multiple regression was used to model an employee's intention to leave (y) as a function of bullying (x_1, measured on a quantitative scale) and perceived organizational support (measured qualitatively as "low POS," "neutral POS," or "high POS"). In Exercise 12.103b, you wrote a model for $E(y)$ as a function of bullying and POS that hypothesizes three parallel straight lines, one for each level of POS. In Exercise 12.103c, you wrote a model for $E(y)$ as a function of bullying and POS that hypothesizes three nonparallel straight lines, one for each level of POS.

a. Explain why the two models are nested. Which is the complete model? Which is the reduced model?

b. Give the null hypothesis for comparing the two models.

c. If you reject H_0 in part **b,** which model do you prefer? Why?

d. If you fail to reject H_0 in part **b,** which model do you prefer? Why?

12.114 Cooling method for gas turbines. Refer to the *Journal of Engineering for Gas Turbines and Power* (Jan. 2005) study of a high-pressure inlet fogging method for a gas turbine engine, presented in Exercise 12.25 (p. 632). Consider a model for the heat rate (kilojoules per kilowatt per hour) produced by a gas turbine as a function of cycle speed (revolutions per minute) and cycle pressure ratio. The data are saved in the **GASTURBINE** file.

a. Write a complete second-order model for heat rate (y).

b. Give the null and alternative hypotheses for determining whether the curvature terms in the complete second-order model are statistically useful in predicting the heat rate (y).

c. For the test in part **b,** identify the complete and reduced models.

d. Portions of the MINITAB printouts for the two models are shown below. Find the values of SSE_R, SSE_C, and MSE_C on the printouts.

e. Compute the value of the test statistics for the test of part **b.**

f. Find the rejection region for the test of part **b.** Use $\alpha = .10$.

g. State the conclusion of the test in the words of the problem.

12.115 Study of supervisor-targeted aggression. "Moonlighters" are workers who hold two jobs at the same time. What are the factors that affect the likelihood of a moonlighting worker becoming aggressive toward his or her supervisor? This was the research question of interest in the *Journal of Applied Psychology* (July 2005). Completed questionnaires were obtained from $n = 105$ moonlighters, and the data were used to fit several multiple-regression models for supervisor-targeted aggression score (y). Two of the models (with R^2 values in parentheses) are shown in the next table (p. 680).

MINITAB output for Exercise 12.114

Complete Model

```
The regression equation is
HEATRATE = 15583 + 0.078 RPM - 523 CPRATIO + 0.00445 RPM_CPR - 0.000000 RPMSQ
           + 8.84 CPRSQ

S = 563.513    R-Sq = 88.5%    R-Sq(adj) = 87.5%

Analysis of Variance

Source          DF        SS         MS        F      P
Regression       5  148526859   29705372   93.55  0.000
Residual Error  61   19370350     317547
Total           66  167897208
```

Reduced Model

```
The regression equation is
HEATRATE = 12065 + 0.170 RPM - 146 CPRATIO - 0.00242 RPM_CPR

S = 633.842    R-Sq = 84.9%    R-Sq(adj) = 84.2%

Analysis of Variance

Source          DF        SS         MS        F      P
Regression       3  142586570   47528857  118.30  0.000
Residual Error  63   25310639     401756
Total           66  167897208
```

Models for Exercise 12.115

Model 1: $E(y) = \beta_0 + \beta_1(\text{Age}) + \beta_2(\text{Gender})$

$\qquad + \beta_3(\text{Interaction injustice at secondary job}) + \beta_4(\text{Abusive supervisor at secondary job})$

$(R^2 = .101)$

Model 2: $E(y) = \beta_0 + \beta_1(\text{Age}) + \beta_2(\text{Gender})$

$\qquad + \beta_3(\text{Interactional injustice at secondary job}) + \beta_4(\text{Abusive supervisor at secondary job})$

$\qquad + \beta_5(\text{Self-esteem}) + \beta_6(\text{History of aggression}) + \beta_7(\text{Interactional injustice at primary job})$

$\qquad + \beta_8(\text{Abusive supervisor at primary job})$

$(R^2 = .555)$

---

a. Interpret the R^2 values for the models.

b. Give the null and alternative hypotheses for comparing the fits of Models 1 and 2.

c. Are the two models nested? Explain.

d. The nested F-test for comparing the two models resulted in $F = 42.13$ and p-value $< .001$. What can you conclude from these results?

e. A third model was fit, one that hypothesizes all possible pairs of interactions between self-esteem, history of aggression, interactional injustice at primary job, and abusive supervisor at primary job. Give the equation of this model (Model 3).

f. A nested F-test to compare Models 2 and 3 resulted in p-value $> .10$. What can you conclude from this result?

Applying the Concepts—Intermediate

12.116 Reality TV and cosmetic surgery. Refer to the *Body Image: An International Journal of Research* (March 2010) study of the influence of reality TV shows on one's desire to undergo cosmetic surgery, Exercise 12.23 (p. 631). Recall that psychologists modeled desire to have cosmetic surgery (y) as a function of gender (x_1), self-esteem (x_2), body satisfaction (x_3), and impression of reality TV (x_4). The psychologists theorize that one's impression of reality TV will "moderate" the impact that each of the first three independent variables has on one's desire to have cosmetic surgery. If so, then x_4 will interact with each of the other independent variables.

a. Give the equation of the model for $E(y)$ that matches the theory.

b. Fit the model, part **a,** to the simulated data saved in the **BODYIMAGE** file. Evaluate the overall utility of the model.

c. Give the null hypothesis for testing the psychologists theory.

d. Conduct a nested model F-test to test the theory. What do you conclude?

12.117 Improving SAT scores. Refer to the *Chance* (Winter 2001) study of students who paid a private tutor (or coach) to help them improve their SAT scores, presented in Exercise 12.84 (p. 661). Recall that the baseline model, $E(y) = \beta_0 + \beta_1 x_1 + \beta_2 x_2$, where $y =$ SAT-Math score, $x_1 =$ score on PSAT, and $x_2 = \{1$ if student was coached, 0 if not$\}$, had the following results: $R_a^2 = .76, \hat{\beta}_2 = 19,$ and $s_{\hat{\beta}_2} = 3$. As an alternative model, the researcher added several "control" variables, including dummy variables for student ethnicity ($x_3, x_4,$ and x_5), a socioeconomic status index variable (x_6), two variables that measured high school performance

(x_7 and x_8), the number of math courses taken in high school (x_9), and the overall GPA for the math courses (x_{10}).

a. Write the hypothesized equation for $E(y)$ for the alternative model.

b. Give the null hypothesis for a nested-model F-test comparing the initial and alternative models.

c. The nested model F-test from part **b,** was statistically significant at $\alpha = .05$. Interpret this result practically.

d. The alternative model from part **a** resulted in $R_a^2 = .79, \hat{\beta}_2 = 14,$ and $s_{\hat{\beta}_2} = 3$. Interpret the value of R_a^2.

e. Refer to part **d.** Find and interpret a 95% confidence interval for β_2.

f. The researcher concluded that "the estimated effect of SAT coaching decreases from the baseline model when control variables are added to the model." Do you agree? Justify your answer.

g. As a modification to the model of part **a,** the researcher added all possible interactions between the coaching variable (x_2) and the other independent variables in the model. Write the equation for $E(y)$ for this modified model.

h. Give the null hypothesis for comparing the models from parts **a** and **g.** How would you perform this test?

12.118 Glass as a waste encapsulant. Since glass is not subject to radiation damage, the encapsulation of waste in glass is considered to be one of the most promising solutions to the problem of low-level nuclear waste in the environment. However, chemical reactions may weaken the glass. This concern led to a study undertaken jointly by the Department of Materials Science and Engineering at the University of Florida and the U.S. Department of Energy to assess the utility of glass as a waste encapsulant.[*] Corrosive chemical solutions (called corrosion baths) were prepared and applied directly to glass samples containing one of three types of waste (TDS-3A, FE, and AL); the chemical reactions were observed over time. A few of the key variables measured were

$y =$ Amount of silicon (in parts per million) found in solution at end of experiment (This is both a measure of the degree of breakdown in the glass and a proxy for the amount of radioactive species released into the environment.)

$x_1 =$ Temperature (°C) of the corrosion bath

$x_2 = 1$ if waste type TDS-3A, 0 if not

$x_3 = 1$ if waste type FE, 0 if not

[*]The background information for this exercise was provided by Dr. David Clark, Department of Materials Science and Engineering, University of Florida.

(Waste type AL is the base level.) Suppose we want to model amount y of silicon as a function of temperature (x_1) and type of waste (x_2, x_3).

a. Write a model that proposes parallel straight-line relationships between amount of silicon and temperature, one line for each of the three types of waste.

b. Add terms for the interaction between temperature and waste type to the model from part **a.**

c. Refer to the model from part **b.** For each type of waste, give the slope of the line relating amount of silicon to temperature.

d. Explain how you could test for the presence of temperature–type-of-waste interaction.

12.119 Whales entangled in fishing gear. Refer to the *Marine Mammal Science* (April 2010) study of whales entangled in fishing gear, Exercise 12.101 (p. 670). A first-order model for the length (y) of an entangled whale that is a function of water depth of the entanglement (x_1) and gear type (set nets, pots, or gill nets) is written as follows: $E(y) = \beta_0 + \beta_1 x_1 + \beta_2 x_2 + \beta_3 x_3 + \beta_4 x_1 x_2 + \beta_5 x_1 x_3$, where $x_2 = \{1$ if set net, 0 if not$\}$ and $x_2 = \{1$ if pot, 0 if not$\}$. Consider this model the complete model in a nested model F-test.

a. Suppose you want to determine if there are any differences in the mean lengths of entangled whales for the three gear types. Give the appropriate null hypothesis to test.

b. Refer to part **a.** Give the reduced model for the test.

c. Refer to parts **a** and **b.** If you reject the null hypothesis, what would you conclude?

d. Suppose you want to determine if the rate of change of whale length (y) with water depth (x_1) is the same for all three types of fishing gear. Give the appropriate null hypothesis to test.

e. Refer to part **d.** Give the reduced model for the test.

f. Refer to parts **d** and **e.** If you fail to reject the null hypothesis, what would you conclude?

Applying the Concepts—Advanced

12.120 Emotional distress in firefighters. The *Journal of Human Stress* (Summer 1987) reported on a study of the "psychological response of firefighters to chemical fire." It is thought that the complete second-order model

$$E(y) = \beta_0 + \beta_1 x_1 + \beta_2 x_1^2 + \beta_3 x_2 + \beta_4 x_1 x_2 + \beta_5 x_1^2 x_2$$

where

$$y = \text{Emotional distress}$$
$$x_1 = \text{Experience (years)}$$
$$x_2 = 1 \text{ if exposed to chemical fire, 0 if not}$$

will be adequate to describe the relationship between emotional distress and years of experience for two groups of firefighters: those exposed to a chemical fire and those not exposed.

a. How would you determine whether the *rate* of increase of emotional distress with experience is different for the two groups of firefighters?

b. How would you determine whether there are differences in mean emotional distress *levels* that are attributable to exposure group?

12.10 Stepwise Regression (Optional)

Consider the problem of predicting the salary y of an executive. Perhaps the biggest problem in building a model to describe executive salaries is choosing the important independent variables to be included. The list of potentially important independent variables is extremely long (e.g., age, experience, tenure, education level, etc.), and we need some objective method of screening out those which are not important.

The problem of deciding which of a large set of independent variables to include in a model is a common one. Trying to determine which variables influence the profit of a firm, affect the blood pressure of humans, or are related to a student's performance in college are only a few examples.

A systematic approach to building a model with a large number of independent variables is difficult because the interpretation of multivariable interactions and higher order terms is tedious. We therefore turn to a screening procedure, available in most statistical software packages, known as *stepwise regression*.

The most commonly used **stepwise regression** procedure works as follows: The user first identifies the response y and the set of potentially important independent variables $x_1, x_2, \ldots, x_k$, where k is generally large. [*Note:* This set of variables could include both first-order and higher order terms. However, we often include only the main effects of both quantitative variables (first-order terms) and qualitative variables (dummy variables), since the inclusion of second-order terms greatly increases the number of independent variables.] The response and independent variables are then entered into the computer software, and the stepwise procedure begins.

Step 1 The software program fits all possible one-variable models of the form

$$E(y) = \beta_0 + \beta_1 x_i$$

to the data, where x_i is the ith independent variable, $i = 1, 2, \ldots, k$. For each model, the t-test (or the equivalent F-test) for a single β parameter is conducted to test the null hypothesis

$$H_0: \beta_1 = 0$$

against the alternative hypothesis

$$H_a: \beta_1 \neq 0$$

The independent variable that produces the largest (absolute) t-value is then declared the best one-variable predictor of y.[*] Call this independent variable x_1.

Step 2 The stepwise program now begins to search through the remaining $(k - 1)$ independent variables for the best two-variable model of the form

$$E(y) = \beta_0 + \beta_1 x_1 + \beta_2 x_i$$

This is done by fitting all two-variable models containing x_1 and each of the other $(k - 1)$ options for the second variable x_i. The t-values for the test $H_0: \beta_2 = 0$ are computed for each of the $(k - 1)$ models (corresponding to the remaining independent variables $x_i, i = 2, 3, \ldots, k$), and the variable having the largest t is retained. Call this variable x_2. At this point, some software packages diverge in methodology. The better packages now go back and check the t-value of $\hat{\beta}_1$ after $\hat{\beta}_2 x_2$ has been added to the model. If the t-value has become nonsignificant at some specified α level (say, $\alpha = .05$), the variable x_1 is removed and a search is made for the independent variable with a β parameter that will yield the most significant t-value in the presence of $\hat{\beta}_2 x_2$. Other packages do not recheck the significance of $\hat{\beta}_1$, but proceed directly to step 3.[†]

The reason the t-value for x_1 may change from step 1 to step 2 is that the meaning of the coefficient $\hat{\beta}_1$ changes. In step 2, we are approximating a complex response surface in two variables by a plane. The best-fitting plane may yield a different value for $\hat{\beta}_1$ than that obtained in step 1. Thus, both the value of $\hat{\beta}_1$ and its significance usually change from step 1 to step 2. For this reason, the software packages that recheck the t-values at each step are preferred.

Step 3 The stepwise procedure now checks for a third independent variable to include in the model with x_1 and x_2. That is, we seek the best model of the form

$$E(y) = \beta_0 + \beta_1 x_1 + \beta_2 x_2 + \beta_3 x_i$$

To do this, we fit all the $(k - 2)$ models using x_1, x_2, and each of the $(k - 2)$ remaining variables x_i as a possible x_3. The criterion is again to include the independent variable with the largest t-value. Call this best third variable x_3. The better programs now recheck the t-values corresponding to the x_1- and x_2-coefficients, replacing the variables that yield nonsignificant t-values. This procedure is continued until no further independent variables can be found that yield significant t-values (at the specified α level) in the presence of the variables already in the model.

The result of the stepwise procedure is a model containing only those terms with t-values that are significant at the specified α level. Thus, in most practical situations, only several of the large number of independent variables remain. However, it is very important *not* to jump to the conclusion that all the independent variables which are important in predicting y have been identified or that the unimportant independent variables have been eliminated. Remember, the stepwise procedure is using only *sample estimates* of the true model coefficients (β's) to select the important variables. An extremely large number of single β parameter t-tests have been conducted, and the probability is very high

---

[*]Note that the variable with the largest t-value is also the one with the largest (absolute) Pearson product moment correlation r (Section 11.6) with y.

[†]*Forward selection* is the name given to stepwise routines that *do not* recheck the significance of each previously entered independent variable. This is in contrast to *stepwise selection* routines that recheck the significance of each entered term. A third approach is to use *backward selection*, where initially all terms are entered then eliminated one by one.

that one or more errors have been made in including or excluding variables. That is, we have very likely included some unimportant independent variables in the model (Type I errors) and eliminated some important ones (Type II errors).

There is a second reason we might not have arrived at a good model. When we choose the variables to be included in the stepwise regression, we often omit high-order terms (to keep the number of variables manageable). Consequently, we may have initially omitted several important terms from the model. Thus, we should recognize stepwise regression for what it is: an **objective variable-screening procedure.**

Successful model builders will now consider second-order terms (for quantitative variables) and other interactions among variables screened by the stepwise procedure. Indeed, it would be best to develop this response surface model with a second set of data independent of those used for the screening, so that the results of the stepwise procedure can be partially verified with new data. This is not always possible, however, because in many modeling situations only a small amount of data is available.

Do not be deceived by the impressive-looking t-values that result from the stepwise procedure: it has retained only the independent variables with the largest t-values. Also, be certain to consider second-order terms in systematically developing the prediction model. Finally, if you have used a first-order model for your stepwise procedure, remember that it may be greatly improved by the addition of higher order terms.

⚠ **CAUTION** Be wary of using the results of stepwise regression to make inferences about the relationship between $E(y)$ and the independent variables in the resulting first-order model. First, an extremely large number of t-tests have been conducted, leading to a high probability of making one or more Type I or Type II errors. Second, the stepwise model does not include any higher order or interaction terms. Stepwise regression should be used only when necessary—that is, when you want to determine which of a large number of potentially important independent variables should be used in the model-building process. ▲

Example 12.13
Stepwise Regression— Modeling Executive Salary
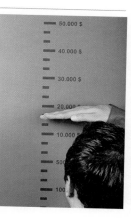

Problem An international management consulting company develops multiple-regression models for executive salaries of its client firms. The consulting company has found that models which use the natural logarithm of salary as the dependent variable have better predictive power than those using salary as the dependent variable.* A preliminary step in the construction of these models is the determination of the most important independent variables. For one firm, 10 potential independent variables (7 quantitative and 3 qualitative) were measured in a sample of 100 executives. The data, described in Table 12.6, are saved in the **EXECSAL** file. Since it would be very difficult to construct a complete second-order model with all of the 10 independent variables, use stepwise regression to decide which of the 10 variables should be included in the building of the final model for the logarithm of executive salaries.

Solution We will use stepwise regression with the main effects of the 10 independent variables to identify the most important variables. The dependent variable y is the natural logarithm of the executive salaries. The MINITAB stepwise regression printout is shown in Figure 12.32.

Note that the first variable included in the model is x_1, years of experience. At the second step, x_3, a dummy variable for the qualitative variable bonus eligibility is entered into the model. In steps 3, 4, and 5, the variables x_4 (number of employees supervised), x_2 (years of education), and x_5 (corporate assets), respectively, are selected for inclusion in

---

*This is probably because salaries tend to be incremented in *percentages* rather than dollar values. When a response variable undergoes percentage changes as the independent variables are varied, the logarithm of the response variable will be more suitable as a dependent variable.

Table 12.6 **Independent Variables in the Executive Salary Example**

Independent Variable	Description
x_1	Experience (years)—quantitative
x_2	Education (years)—quantitative
x_3	Bonus eligibility (1 if yes, 0 if no)—qualitative
x_4	Number of employees supervised—quantitative
x_5	Corporate assets (millions of dollars)—quantitative
x_6	Board member (1 if yes, 0 if no)—qualitative
x_7	Age (years)—quantitative
x_8	Company profits (past 12 months, millions of dollars)—quantitative
x_9	Has international responsibility (1 if yes, 0 if no)—qualitative
x_{10}	Company's total sales (past 12 months, millions of dollars)—quantitative

Data Set: EXECSAL

Figure 12.32
MINITAB stepwise
regression printout for
executive salary data

```
Alpha-to-Enter: 0.15   Alpha-to-Remove: 0.15


Response is Y on 10 predictors, with N = 100


Step                1        2        3        4        5
Constant        11.091   10.968   10.783   10.278    9.962

X1              0.0278   0.0273   0.0273   0.0273   0.0273
T-Value          12.62    15.13    18.80    24.68    26.50
P-Value          0.000    0.000    0.000    0.000    0.000

X3                        0.197    0.233    0.232    0.225
T-Value                    7.10    10.17    13.30    13.74
P-Value                    0.000    0.000    0.000    0.000

X4                                0.00048  0.00055  0.00052
T-Value                             7.32    10.92    11.06
P-Value                            0.000    0.000    0.000

X2                                         0.0300   0.0291
T-Value                                      8.38     8.72
P-Value                                     0.000    0.000

X5                                                  0.00196
T-Value                                                3.95
P-Value                                              0.000

S                0.161    0.131    0.106   0.0807   0.0751
R-Sq             61.90    74.92    83.91    90.75    92.06
R-Sq(adj)        61.51    74.40    83.41    90.36    91.64
Mallows C-p      343.9    195.5     93.8     16.8      3.6
```

the model. MINITAB stops after five steps, because no other independent variables met the criterion for admission into the model. As a default, MINITAB uses $\alpha = .15$ in the t-tests conducted. In other words, if the p-value associated with a β coefficient exceeds $\alpha = .15$, the variable is *not* included in the model.

The results of the stepwise regression suggest that we should concentrate on the preceding five independent variables. Models with second-order terms and interactions should be proposed and evaluated to determine the best model for predicting executive salaries.

Now Work Exercise 12.123

We conclude this section with some advice on the use of stepwise regression.

RECOMMENDATION Do *not* use the stepwise regression model as the final model for predicting y. Recall that the stepwise procedure tends to perform a large number of t-tests, inflating the overall probability of a Type I error, and does not automatically include higher order terms (e.g., interactions and squared terms) in the final model. Use stepwise regression as a variable-screening tool when there exists a large number of potentially important independent variables. Then begin building models for y, using the variables identified by stepwise regression.

Exercises 12.121–12.129

Understanding the Principles

12.121 Explain the difference between a stepwise model and a standard regression model.

12.122 Give two caveats associated with using the stepwise regression results as the final model for predicting y.

Learning the Mechanics

12.123 Suppose there are six independent variables
NW $x_1, x_2, x_3, x_4, x_5,$ and x_6 that might be useful in predicting a response y. A total of $n = 50$ observations is available, and it is decided to employ stepwise regression to help in selecting the independent variables that appear to be useful. The computer fits all possible one-variable models of the form

$$E(y) = \beta_0 + \beta_1 x_i$$

where x_i is the ith independent variable, $i = 1, 2, \ldots, 6$. The information in the following table is provided from the computer printout:

Independent Variable	$\hat{\beta}_i$	$s_{\hat{\beta}_i}$
x_1	1.6	.42
x_2	−.9	.01
x_3	3.4	1.14
x_4	2.5	2.06
x_5	−4.4	.73
x_6	.3	.35

a. Which independent variable is declared the best one-variable predictor of y? Explain.

b. Would this variable be included in the model at this stage? Explain.

c. Describe the next phase that a stepwise procedure would execute.

Applying the Concepts — Basic

12.124 Accuracy of software effort estimates. Periodically, software engineers must provide estimates of their effort in developing new software. In the *Journal of Empirical Software Engineering* (Vol. 9, 2004), multiple regression was used to predict the accuracy of these effort estimates. The dependent variable, defined as the relative error in estimating effort,

$$y = (\text{Actual effort} - \text{Estimated effort})/(\text{Actual effort})$$

was determined for each in a sample of $n = 49$ software development tasks. Eight independent variables were

evaluated as potential predictors of relative error using stepwise regression. Each of these was formulated as a dummy variable, as shown in the table.

Company role of estimator: $x_1 = 1$ if developer, 0 if project leader
Task complexity: $x_2 = 1$ if low, 0 if medium/high
Contract type: $x_3 = 1$ if fixed price, 0 if hourly rate
Customer importance: $x_4 = 1$ if high, 0 if low/medium
Customer priority: $x_5 = 1$ if time of delivery, 0 if cost or quality
Level of knowledge: $x_6 = 1$ if high, 0 if low/medium
Participation: $x_7 = 1$ if estimator participates in work, 0 if not
Previous accuracy: $x_8 = 1$ if more than 20% accurate, 0 if less than 20% accurate

---

a. In step 1 of the stepwise regression, how many different one-variable models are fitted to the data?

b. In step 1, the variable x_1 is selected as the "best" one-variable predictor. How is this determined?

c. In step 2 of the stepwise regression, how many different two-variable models (where x_1 is one of the variables) are fitted to the data?

d. The only two variables selected for entry into the stepwise regression model were x_1 and x_8. The stepwise regression yielded the following prediction equation:

$$\hat{y} = .12 - .28x_1 + .27x_8$$

Give a practical interpretation of the β estimates multiplied by x_1 and x_8.

e. Why should a researcher be wary of using the model, part **d**, as the final model for predicting effort (y)?

12.125 An analysis of footprints in sand. Fossilized human footprints provide a direct source of information on the gait dynamics of extinct species. How paleontologists and anthropologists interpret these prints, however, may vary. To gain insight into this phenomenon, a group of scientists used human subjects (16 young adults) to generate footprints in sand (*American Journal of Physical Anthropology*, April 2010). One dependent variable of interest was *heel depth* (y) of the footprint (in millimeters). The scientists wanted to find the best predictors of depth from among six possible independent variables. Three variables were related to the human subject (*foot mass, leg length,* and *foot type*) and three variables were related to walking in sand (*velocity, pressure,* and *impulse*). A stepwise regression run on these six variables yielded the following results:

Selected independent variables: *pressure* and *leg length*

$R^2 = .771$, Global F-test p-value $< .001$

---

a. Write the hypothesized equation of the final stepwise regression model.

b. Interpret the value of R^2 for the model.

c. Conduct a test of the overall utility of the final stepwise model.

d. At minimum, how many t-tests on individual β's were conducted to arrive at the final stepwise model?

e. Based on your answer to part **d,** comment on the probability of making at least one Type I error during the stepwise analysis.

12.126 Yield strength of steel alloy. Industrial engineers at the University of Florida used regression modeling as a tool to reduce the time and cost associated with developing new metallic alloys (*Modelling and Simulation in Materials Science and Engineering*, Vol. 13, 2005). To illustrate, the engineers build a regression model for the tensile yield strength (y) of a new steel alloy. The potential important predictors of yield strength are listed in the following table:

---

x_1 = Carbon amount (% weight)
x_2 = Manganese amount (% weight)
x_3 = Chromium amount (% weight)
x_4 = Nickel amount (% weight)
x_5 = Molybdenum amount (% weight)
x_6 = Copper amount (% weight)
x_7 = Nitrogen amount (% weight)
x_8 = Vanadium amount (% weight)
x_9 = Plate thickness (millimeters)
x_{10} = Solution treating (millimeters)
x_{11} = Ageing temperature (degrees Celsius)

---

a. The engineers used stepwise regression to search for a parsimonious set of predictor variables. Do you agree with this decision? Explain.

b. The stepwise regression selected the following independent variables: x_1 = Carbon, x_2 = Manganese, x_3 = Chromium, x_5 = Molybdenum, x_6 = Copper, x_8 = Vanadium, x_9 = Plate thickness, x_{10} = Solution treating, and x_{11} = Ageing temperature. On the basis of this information, determine the total number of first-order models that were fit in the stepwise routine.

c. Refer to part **b.** All the variables listed there were statistically significant in the stepwise model, with $R^2 = .94$. Consequently, the engineers used the estimated stepwise model to predict yield strength. Do you agree with this decision? Explain.

Applying the Concepts—Intermediate

12.127 Bus rapid-transit study. Bus rapid transit (BRT) is a rapidly growing trend in the provision of public transportation in America. The Center for Urban Transportation Research (CUTR) at the University of South Florida conducted a survey of BRT customers in Miami (*Transportation Research Board* Annual Meeting, Jan. 2003). Data on the following variables (all measured on a five-point scale, where 1 = "very unsatisfied" and 5 = "very satisfied") were collected for a sample of over

500 bus riders: overall satisfaction with BRT (y), safety on bus (x_1), seat availability (x_2), dependability (x_3), travel time (x_4), cost (x_5), information/maps (x_6), convenience of routes (x_7), traffic signals (x_8), safety at bus stops (x_9), hours of service (x_{10}), and frequency of service (x_{11}). CUTR analysts used stepwise regression to model overall satisfaction (y).

a. How many models are fitted at step 1 of the stepwise regression?

b. How many models are fitted at step 2 of the stepwise regression?

c. How many models are fitted at step 11 of the stepwise regression?

d. The stepwise regression selected the following eight variables to include in the model (in order of selection): $x_{11}, x_4, x_2, x_7, x_{10}, x_1, x_9$, and x_3. Write the equation for $E(y)$ that results.

e. The model in part **d** was tested and resulted in $R^2 = .677$. Interpret this value.

f. Explain why the CUTR analysts should be cautious in concluding that the "best" model for $E(y)$ has been found.

12.128 Modeling species abundance. A marine biologist was hired by the EPA to determine whether the hot-water runoff from a particular power plant located near a large gulf is having an adverse effect on the marine life in the area. The biologist's goal is to acquire a prediction equation for the number of marine animals located at certain designated areas, or stations, in the gulf. On the basis of past experience, the EPA considered the following environmental factors as predictors for the number of animals at a particular station:

x_1 = Temperature of water (TEMP)
x_2 = Salinity of water (SAL)
x_3 = Dissolved oxygen content of water (DO)
x_4 = Turbidity index, a measure of the turbidity of the water (TI)
x_5 = Depth of the water at the station (ST_DEPTH)
x_6 = Total weight of sea grasses in sampled area (TGRSWT)

As a preliminary step in the construction of this model, the biologist used a stepwise regression procedure to identify the most important of these six variables. A total of 716 samples was taken at different stations in the gulf, producing the SPSS printout shown on page 687. (The response measured was y, the logarithm of the number of marine animals found in the sampled area.)

a. According to the printout, which of the independent variables should be used in the model?

b. Are we able to assume that the marine biologist has identified all the important independent variables for the prediction of y? Why?

c. Using the variables identified in part **a,** write the first-order model with interaction that may be used to predict y.

d. How would the marine biologist determine whether the model specified in part **c** is better than the first-order model?

e. Note the small value of R^2. What action might the biologist take to improve the model?

12.129 Using corn in a duck diet. Corn is high in starch content; consequently, it is considered excellent feed for domestic

SPSS output for Exercise 12.128

Variables Entered/Removed[a]

Model	Variables Entered	Variables Removed	Method
1	ST_DEPTH	.	Stepwise (Criteria: Probability-of-F-to-enter <= .050, Probability-of-F-to-remove >= .100).
2	TGRSWT	.	Stepwise (Criteria: Probability-of-F-to-enter <= .050, Probability-of-F-to-remove >= .100).
3	TI	.	Stepwise (Criteria: Probability-of-F-to-enter <= .050, Probability-of-F-to-remove >= .100).

a. Dependent Variable: LOGNUM

Model Summary

Model	R	R Square	Adjusted R Square	Std. Error of the Estimate
1	.329[a]	.122	.121	.7615773
2	.427[b]	.182	.180	.7348470
3	.432[c]	.187	.184	.7348469

a. Predictors: (Constant), ST_DEPTH

b. Predictors: (Constant), ST_DEPTH, TGRSWT

c. Predictors: (Constant), ST_DEPTH, TGRSWT, TI

chickens. Does corn possess the same potential in feeding ducks bred for broiling? This was the subject of research published in *Animal Feed Science and Technology* (April 2010). The objective of the study was to establish a prediction model for the true metabolizable energy (TME) of corn regurgitated from ducks. The researchers considered 11 potential predictors of TME: dry matter (DM), crude protein (CP), ether extract (EE), ash (ASH), crude fiber (CF), neutral detergent fiber (NDF), acid detergent fiber (ADF), gross energy (GE), amylose (AM), amylopectin (AP), and amylopectin/amylose (AMAP). Stepwise regression was used to find the best subset of predictors. The final stepwise model yielded the following results:

$$\widehat{\text{TME}} = 7.70 + 2.14(\text{AMAP}) + .16(\text{NDF}),$$
$$R^2 = .988, s = .07, \text{global } F \text{ } p\text{-value} = .001$$

a. Determine the number of t-tests performed in step 1 of the stepwise regression.

b. Determine the number of t-tests performed in step 2 of the stepwise regression.

c. Give a full interpretation of the final stepwise model regression results.

d. Explain why it is dangerous to use the final stepwise model as the "best" model for predicting TME.

e. Using the independent variables selected by the stepwise routine, write a complete second-order model for TME.

f. Refer to part **e**. How would you determine if the terms in the model that allow for curvature are statistically useful for predicting TME?

PART III: MULTIPLE REGRESSION DIAGNOSTICS

12.11 Residual Analysis: Checking the Regression Assumptions

When we apply regression analysis to a set of data, we never know for certain whether the assumptions of Section 12.1 are satisfied. How far can we deviate from the assumptions and still expect regression analysis to yield results that will have the reliability stated in this chapter? How can we detect departures (if they exist) from the assumptions, and what can we do about them? We provide some answers to these questions in this section.

Recall from Section 12.1 that, for any given set of values of $x_1, x_2, \ldots, x_k$, we assume that the random-error term ε has the following properties:

1. mean equal to 0

2. constant variance (σ^2)

3. normal propability distribution

4. probabilistically independent

It is unlikely that these assumptions are ever satisfied exactly in a practical application of regression analysis. Fortunately, experience has shown that least squares regression analysis produces reliable statistical tests, confidence intervals, and prediction intervals, as long as the departures from the assumptions are not too great. In this section, we present some methods for determining whether the data indicate significant departures from the assumptions.

Because the assumptions all concern the random-error component ε of the model, the first step is to estimate the random error. Since the actual random error associated with a particular value of y is the difference between the actual y value and its unknown mean, we estimate the error by the difference between the actual y value and the *estimated* mean. This estimated error is called the *regression residual*, or simply the **residual**, and is denoted by $\hat{\varepsilon}$. The actual error ε and residual $\hat{\varepsilon}$ are shown in Figure 12.33.

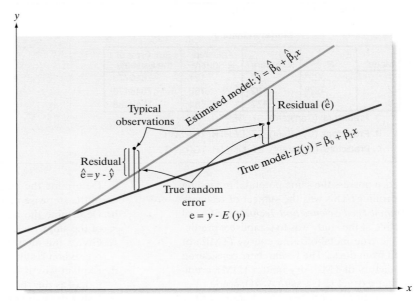

Figure 12.33
Actual random error ε and regression residual $\hat{\varepsilon}$

A **regression residual** $\hat{\varepsilon}$ is defined as the difference between an observed y-value and its corresponding predicted value:

$$\hat{\varepsilon} = (y - \hat{y}) = y - (\hat{\beta}_0 + \hat{\beta}_1 x_1 + \hat{\beta}_2 x_2 + \cdots + \hat{\beta}_k x_k)$$

Since the true mean of y (i.e., the true regression model) is not known, the actual random error cannot be calculated. However, because the residual is based on the estimated mean (the least squares regression model), it can be calculated and used to estimate the random error and to check the regression assumptions. Such checks are generally referred to as **residual analyses.** Two useful properties of residuals are given in the next box.

Properties of Regression Residuals

1. *The mean of the residuals is equal to 0.* This property follows from the fact that the sum of the differences between the observed y values and their least squares predicted $\hat{y}$ values is equal to 0:

$$\sum (\text{Residuals}) = \sum (y - \hat{y}) = 0$$

2. *The standard deviation of the residuals is equal to the standard deviations of the fitted regression model.* This property follows from the fact that the sum of the squared residuals is equal to SSE, which, when divided by the error degrees of freedom, is equal to the variance of the fitted regression model. The square root of the variance is both the standard deviation of the residuals and the standard deviation of the regression model:

$$\sum(\text{Residuals})^2 = \sum(y - \hat{y})^2 = \text{SSE}$$

$$s = \sqrt{\frac{\sum(\text{Residuals})^2}{n - (k+1)}} = \sqrt{\frac{\text{SSE}}{n - (k+1)}} = \sqrt{\text{MSE}}$$

BIOGRAPHY FRANCIS J. ANSCOMBE (1918–2001)

Anscombe's Data

British citizen Frank Anscombe grew up in a small town near the English Channel. He attended Trinity College in Cambridge, England, on a merit scholarship, graduating with first-class honors in mathematics in 1939. He earned his master's degree in 1943. During World War II, Anscombe worked for the British Ministry of Supply, developing a mathematical solution for aiming antiair-craft rockets at German bombers and buzz bombs. Following the war, Anscombe worked at the Rothamsted Experimental Station, applying statistics to agriculture. There, he formed his appre-ciation for solving problems with social relevance. During his career as a professor of statistics, Anscombe served on the faculty of Cambridge, Princeton, and Yale Universities. He was a pioneer in the application of computers to statistical analysis and was one of the original developers of residual analysis in regression. Anscombe is famous for a paper he wrote in 1973 in which he showed that one regression model could be fitted by four very different data sets ("Anscombe's data"). Although Anscombe published 50 research articles on statistics, he also had serious interests in classical music, poetry, and art. ∎

The examples that follow show how a graphical analysis of regression residuals can be used to verify the assumptions associated with a model and to support improvements to the model when the assumptions do not appear to be satisfied. Although the residuals can be calculated and plotted by hand, we rely on statistical software for these tasks in the examples and exercises.

Checking Assumption #1: Mean $\varepsilon = 0$

First, we demonstrate how a residual plot can detect a model in which the hypothesized relationship between $E(y)$ and an independent variable x is misspecified. The assump-tion of mean error of 0 is violated in these types of models.*

Example 12.14

Analyzing Residuals: Electrical Usage Model

Problem Refer to the problem of modeling the relationship between home size (x) and electrical usage (y) in Example 12.7 (p. 646). The data for $n = 15$ homes are repeated in Table 12.7. MINITAB printouts for a straight-line model and a quadratic model fit-ted to the data are shown in Figures 12.34a and 12.34b, respectively. The residuals from these models are highlighted in the printouts. The residuals are then plotted on the vertical axis against the variable x, size of home, on the horizontal axis in Figures 12.35a and 12.35b.

a. Verify that each residual is equal to the difference between the observed y-value and the estimated mean value $\hat{y}$.

b. Analyze the plots of the residuals.

Solution

a. Consider the first usage value in the data set, $y = 1,182$. For the straight-line model, the predicted value for this observation is $\hat{y} = 1,362.2$. Consequently, the residual is

$$\hat{\varepsilon} = (y - \hat{y}) = 1,182 - 1,362.2 = -180.2 \qquad \text{(highlighted in Figure 12.34a)}$$

*For a misspecified model, the hypothesized mean of y, denoted by $E_h(y)$, will not equal the true mean of y, $E(y)$. Since $y = E_h(y) + \varepsilon$, then $\varepsilon = y - E_h(y)$ and $E(\varepsilon) = E\{y - E_h(y)\} = E(y) - E_h(y) \neq 0$.

Table 12.7

Home Size–Electrical Usage Data

Size of Home, x (sq. ft)	Monthly Usage, y (kilowatt-hours)
1,290	1,182
1,350	1,172
1,470	1,264
1,600	1,493
1,710	1,571
1,840	1,711
1,980	1,804
2,230	1,840
2,400	1,956
2,710	2,007
2,930	1,984
3,000	1,960
3,210	2,001
3,240	1,928
3,520	1,945

⊙ *Data Set:* ELECTRIC

Similarly, the residual for the first y-value from the quadratic model (Figure 12.34b) is

$$\hat{\varepsilon} = 1{,}182 - 1{,}157.2 = 24.8 \qquad \text{(highlighted in Figure 12.34b)}$$

The two residuals agree (after rounding) with the first values given in the column labeled "**Residual**" in Figures 12.34a and 12.34b, respectively. Although the residuals both correspond to the same observed y value, 1,182, they differ because the predicted mean value changes, depending on whether the straight-line model or quadratic model is used. Similar calculations produce the remaining residuals.

b. The MINITAB plot of the residuals for the straight-line model (Figures 12.35a) reveals a nonrandom pattern. The residuals exhibit a curved shape, with the residuals for the small values of x below the horizontal 0 (mean of the residuals) line, the residuals corresponding to the middle values of x above the 0 line, and the residuals for the largest values of x again below the 0 line. The indication is that the mean value of the random error *within* each of these ranges of x (small, medium, large) may not be equal to 0. Such a pattern usually indicates that curvature needs to be added to the model.

When the second-order term is added to the model, the nonrandom pattern disappears. In Figure 12.35b, the residuals appear to be randomly distributed around the 0 line, as expected. Also, the variation of the residuals around the 0 line appears to be much smaller for the quadratic model. In fact, $s \approx 50$ for the quadratic model is much smaller than $s \approx 155$ for the straight-line model. The implication is that the quadratic model provides a considerably better model for predicting electrical usage.

Regression Analysis: USAGE versus SIZE

```
The regression equation is
USAGE = 903 + 0.356 SIZE

Predictor      Coef   SE Coef      T       P
Constant      903.0     132.1    6.83   0.000
SIZE        0.35594   0.05477    6.50   0.000

S = 155.251    R-Sq = 76.5%    R-Sq(adj) = 74.7%

Analysis of Variance

Source          DF       SS        MS       F       P
Regression       1  1017803   1017803   42.23   0.000
Residual Error  13   313338     24103
Total           14  1331140
```

Obs	SIZE	USAGE	Fit	SE Fit	Residual	St Resid
1	1290	1182.0	1362.2	68.3	-180.2	-1.29
2	1350	1172.0	1383.5	65.6	-211.5	-1.50
3	1470	1264.0	1426.2	60.6	-162.2	-1.13
4	1600	1493.0	1472.5	55.4	20.5	0.14
5	1710	1571.0	1511.7	51.4	59.3	0.41
6	1840	1711.0	1557.9	47.3	153.1	1.04
7	1980	1804.0	1607.8	43.7	196.2	1.32
8	2230	1840.0	1696.8	40.3	143.2	0.96
9	2400	1956.0	1757.3	40.5	198.7	1.33
10	2710	2007.0	1867.6	46.0	139.4	0.94
11	2930	1984.0	1945.9	52.9	38.1	0.26
12	3000	1960.0	1970.8	55.5	-10.8	-0.07
13	3210	2001.0	2045.6	64.0	-44.6	-0.32
14	3240	1928.0	2056.3	65.3	-128.3	-0.91
15	3520	1945.0	2155.9	78.0	-210.9	-1.57

Figure 12.34a
MINITAB printout for straight-line model of electrical usage

Regression Analysis: USAGE versus SIZE, SIZESQ

```
The regression equation is
USAGE = - 807 + 1.96 SIZE - 0.000340 SIZESQ

Predictor         Coef       SE Coef        T       P
Constant        -806.7         166.9    -4.83   0.000
SIZE           1.9616        0.1525    12.86   0.000
SIZESQ    -0.00034044    0.00003212   -10.60   0.000

S = 50.1998    R-Sq = 97.7%    R-Sq(adj) = 97.3%

Analysis of Variance

Source          DF       SS       MS       F       P
Regression       2  1300900   650450   258.11   0.000
Residual Error  12    30240     2520
Total           14  1331140
```

Obs	SIZE	USAGE	Fit	SE Fit	Residual	St Resid
1	1290	1182.0	1157.2	29.3	24.8	0.61
2	1350	1172.0	1221.0	26.2	-49.0	-1.14
3	1470	1264.0	1341.2	21.2	-77.2	-1.70
4	1600	1493.0	1460.3	18.0	32.7	0.70
5	1710	1571.0	1552.2	17.1	18.8	0.40
6	1840	1711.0	1650.1	17.6	60.9	1.30
7	1980	1804.0	1742.6	19.0	61.4	1.32
8	2230	1840.0	1874.7	21.2	-34.7	-0.76
9	2400	1956.0	1940.2	21.7	15.8	0.35
10	2710	2007.0	2009.0	20.0	-2.0	-0.04
11	2930	1984.0	2018.2	18.4	-34.2	-0.73
12	3000	1960.0	2014.1	18.4	-54.1	-1.16
13	3210	2001.0	1982.1	21.5	18.9	0.42
14	3240	1928.0	1975.1	22.5	-47.1	-1.05
15	3520	1945.0	1880.0	36.2	65.0	1.87

Figure 12.34b
MINITAB printout for quadratic model of electrical usage

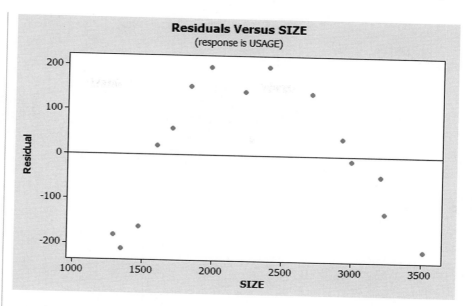

Figure 12.35a

MINITAB residual plot for straight-line model of electrical usage

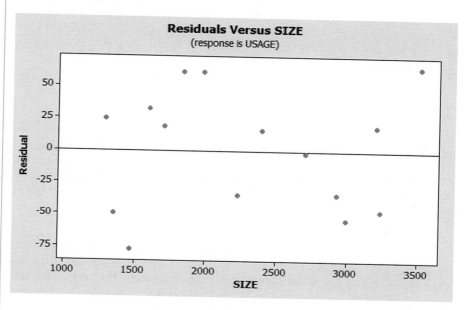

Figure 12.35b

MINITAB residual plot for quadratic model of electrical usage

Look Back The residual analysis verifies our conclusions from Example 12.7, where we found the quadratic term, $\beta_2 x^2$, to be statistically significant.

Now Work Exercise 12.138a

Checking Assumption #2: Constant Error Variance

Residual plots can also be used to detect violations of the assumption of constant error variance. For example, a plot of the residuals versus the predicted value $\hat{y}$ may display one of the patterns shown in Figure 12.36. In these figures, the range in values of the residuals increases (or decreases) as $\hat{y}$ increases, thus indicating that the variance of the random error, ε, becomes larger (or smaller) as the estimate of $E(y)$ increases in value. Because $E(y)$ depends on the x values in the model, this implies that the variance of ε is not constant for all settings of the x's.

In the next example, we demonstrate how to use this plot to detect a nonconstant variance and suggest a useful remedy.

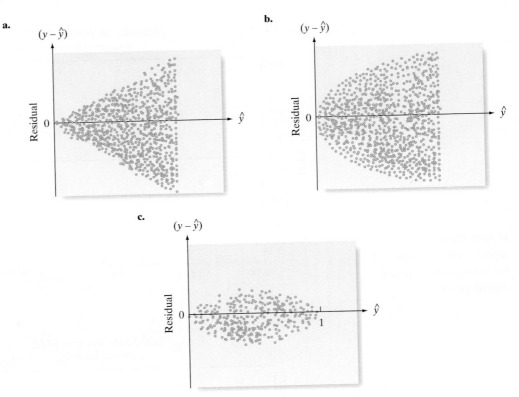

Figure 12.36
Residual plots showing changes
in the variance of ε

Example 12.15

**Using Residuals
to Check Equal
Variances**

Problem The data in Table 12.8 are the salaries, y, and years of experience, x, for a sample of 50 social workers. The first-order model $E(y) = \beta_0 + \beta_1 x$ was fitted to the data using by SPSS. The SPSS printout is shown in Figure 12.37, followed by a plot of the residuals versus $\hat{y}$ in Figure 12.38. Interpret the results. Make model modifications, if necessary.

Solution The SPSS printout, Figure 12.37, suggests that the first-order model provides an adequate fit to the data. The R^2-value indicates that the model explains about 78.7% of the sample variation in salaries. The t-value for testing β_1, 13.31, is highly significant (p-value ≈ 0) and indicates that the model contributes information for the prediction of y. However, an examination of the residuals plotted against $\hat{y}$ (Figure 12.38) reveals a potential problem. Note the "cone" shape of the residual variability; the size of the residuals increases as the estimated mean salary increases, implying that the constant variance assumption is violated.

One way to stabilize the variance of ε is to refit the model using a transformation on the dependent variable y. With economic data (e.g., salaries), a useful **variance-stabilizing transformation** is the natural logarithm of y, denoted $\ln(y)$.* We fit the model

$$\ln(y) = \beta_0 + \beta_1 x + \varepsilon$$

to the data of Table 12.8. Figure 12.39 shows the SPSS regression analysis printout for the $n = 50$ measurements, while Figure 12.40 shows a plot of the residuals from the log model.

You can see that the logarithmic transformation has stabilized the error variances. Note that the cone shape is gone; there is no apparent tendency of the residual variance

---

*Other variance-stabilizing transformations that are used successfully in practice are $\sqrt{y}$ and $\sin^{-1}\sqrt{y}$. Consult the references for more details on these transformations.

Table 12.8

Salary Data for Example 12.15

Years of Experience, x	Salary, y
7	$26,075
28	79,370
23	65,726
18	41,983
19	62,308
15	41,154
24	53,610
13	33,697
2	22,444
8	32,562
20	43,076
21	56,000
18	58,667
7	22,210
2	20,521
18	49,727
11	33,233
21	43,628
4	16,105
24	65,644
20	63,022
20	47,780
15	38,853
25	66,537
25	67,447
28	64,785
26	61,581
27	70,678
20	51,301
18	39,346
1	24,833
26	65,929
20	41,721
26	82,641
28	99,139
23	52,624
17	50,594
25	53,272
26	65,343
19	46,216
16	54,288
3	20,844
12	32,586
23	71,235
20	36,530
19	52,745
27	67,282
25	80,931
12	32,303
11	38,371

Data Set: SOCWORK

Model Summary[b]

Model	R	R Square	Adjusted R Square	Std. Error of the Estimate
1	.887[a]	.787	.782	8642.441

a. Predictors: (Constant), ESP

b. Dependent Variable: SALARY

ANOVA[b]

Model		Sum of Squares	df	Mean Square	F	Sig.
1	Regression	1.3E+10	1	1.324E+10	177.257	.000[a]
	Residual	3.6E+09	48	74691793.28		
	Total	1.7E+10	49			

a. Predictors: (Constant), ESP

b. Dependent Variable: SALARY

Coefficients[a]

Model		Unstandardized Coefficients		Standardized Coefficients	t	Sig.
		B	Std. Error	Beta		
1	(Constant)	11368.72	3160.317		3.597	.001
	ESP	2141.381	160.839	.887	13.314	.000

a. Dependent Variable: SALARY

Figure 12.37

SPSS regression printout for first-order model of salary

Scatterplot

Dependent Variable: SALARY

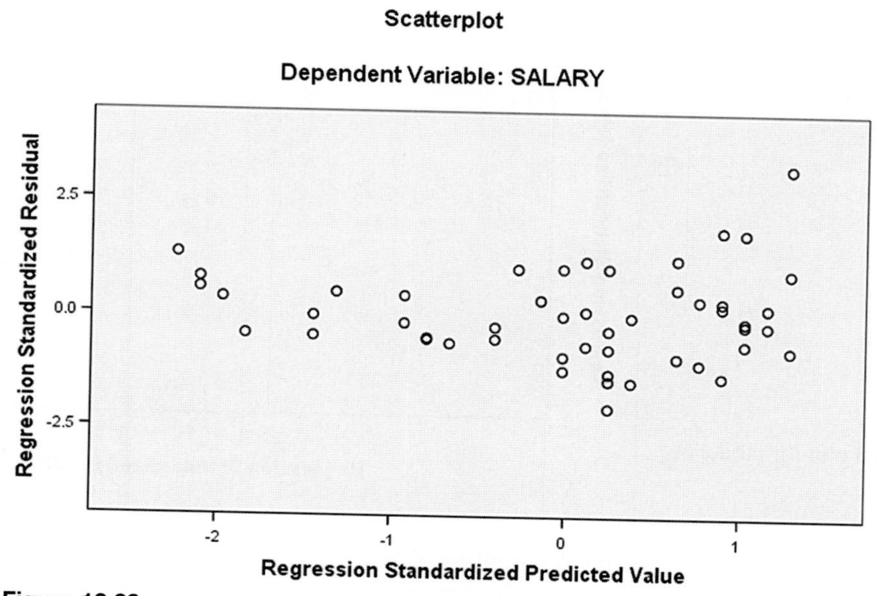

Figure 12.38

SPSS residual plot for first-order model of salary

to increase as mean salary increases. We therefore are confident that inferences using the ln(y) model are more reliable than those using the untransformed model.

Look Back With transformed models, the analyst should be wary when interpreting model statistics such as $\hat{\beta}_1$ and s. These interpretations must take into account that the dependent variable is not some function of y. For example, the antilogarithm of $\hat{\beta}_1$ in the

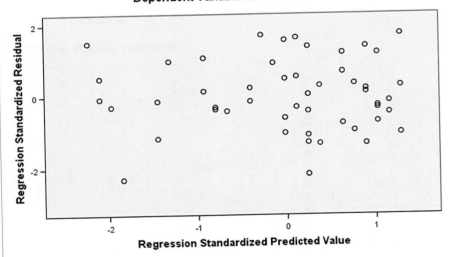

Model Summary[b]

Model	R	R Square	Adjusted R Square	Std. Error of the Estimate
1	.929[a]	.864	.861	.1541127

a. Predictors: (Constant), ESP
b. Dependent Variable: LNSALARY

ANOVA[b]

Model		Sum of Squares	df	Mean Square	F	Sig.
1	Regression	7.212	1	7.212	303.660	.000[a]
	Residual	1.140	48	.024		
	Total	8.352	49			

a. Predictors: (Constant), ESP
b. Dependent Variable: LNSALARY

Coefficients[a]

Model		Unstandardized Coefficients		Standardized Coefficients	t	Sig.
		B	Std. Error	Beta		
1	(Constant)	9.841	.056		174.631	.000
	ESP	.050	.003	.929	17.426	.000

a. Dependent Variable: LNSALARY

Figure 12.39

SPSS regression printout for model of log salary

Scatterplot

Dependent Variable: LNSALARY

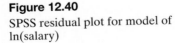

Figure 12.40

SPSS residual plot for model of ln(salary)

$\ln(y)$ model for salary (y) represents the percentage change in salary for every 1 year increase in experience (x).

Now Work Exercise 12.138b

Checking Assumption #3: Errors Normally Distributed

Several graphical methods are available for assessing whether the random error ε has an approximate normal distribution. Recall (Section 5.4) that stem-and-leaf displays, histograms, and normal probability plots are useful for checking whether the data are normally distributed. We illustrate these techniques in an example. But first, we discuss a related problem—using residuals to check for *outliers*.

If the assumption of normally distributed errors is satisfied, then we expect approximately 95% of the residuals to fall within 2 standard deviations of the mean of 0, and almost all of the residuals to lie within 3 standard deviations of the mean of 0. Residuals that are extremely far from 0 and disconnected from the bulk of the other residuals are called **regression outliers** and should receive special attention from the analyst.

A **regression outlier** is a residual that is larger than $3s$ (in absolute value).

Example 12.16

Identifying Outliers— Grandfather Clock Price Model

Problem Refer to Example 12.6 (p. 640), in which we modeled the auction price y of a grandfather clock as a function of age x_1 and number of bidders, x_2. The data for this example are repeated in Table 12.9, with one important difference: The auction price of the clock at the top of the second column has been changed from \$2,131 to \$1,131 (highlighted in the table). The interaction model

$$E(y) = \beta_0 + \beta_1 x_1 + \beta_2 x_2 + \beta_3 x_1 x_2$$

is again fit to these (modified) data, with the MINITAB printout shown in Figure 12.41. The residuals are shown highlighted in the printout and then are plotted against the number of bidders, x_2, in Figure 12.42. Analyze the residual plot.

Table 12.9   Altered Auction Price Data

Age x_1	Number of Bidders, x_2	Auction Price y	Age x_1	Number of Bidders, x_2	Auction Price y
127	13	\$1,235	170	14	\$1,131
115	12	1,080	182	8	1,550
127	7	845	162	11	1,884
150	9	1,522	184	10	2,041
156	6	1,047	143	6	845
182	11	1,979	159	9	1,483
156	12	1,822	108	14	1,055
132	10	1,253	175	8	1,545
137	9	1,297	108	6	729
113	9	946	179	9	1,792
137	15	1,713	111	15	1,175
117	11	1,024	187	8	1,593
137	8	1,147	111	7	785
153	6	1,092	115	7	744
117	13	1,152	194	5	1,356
126	10	1,336	168	7	1,262

Data Set: GFCLOCKALT

Solution The residual plot dramatically reveals the one altered measurement. Note that one of the two residuals at $x_2 = 14$ bidders falls more than three standard deviations below 0. (This observation is highlighted on Figure 12.41.) Note also that no other residual falls more than two standard deviations from 0.

What do we do with outliers once we identify them? First, we try to determine the cause. Were the data entered into the computer incorrectly? Was the observation recorded incorrectly when the data were collected? If so, we correct the observation and rerun the analysis. Another possibility is that the observation is not representative of the conditions we are trying to model. For example, in this case the low price may be attributable to extreme damage to the clock or to a clock of inferior quality compared with the others. In these cases, we probably would exclude the observation from the analysis. In many cases, you may not be able to determine the cause of the outlier. Even so, you may want to rerun the regression analysis excluding the outlier, in order to assess the effect of that observation on the results of the analysis.

```
The regression equation is
PRICE = - 513 + 8.17 AGE + 19.9 NUMBIDS + 0.320 AGE_BIDS


Predictor      Coef   SE Coef       T       P
Constant     -512.8     665.9   -0.77   0.448
AGE           8.165     4.585    1.78   0.086
NUMBIDS       19.89     67.44    0.29   0.770
AGE_BIDS     0.3196    0.4790    0.67   0.510


S = 200.598    R-Sq = 72.9%    R-Sq(adj) = 70.0%


Analysis of Variance

Source          DF       SS        MS       F       P
Regression       3  3033587   1011196   25.13   0.000
Residual Error  28  1126703     40239
Total           31  4160290


Obs   AGE    PRICE      Fit   SE Fit   Residual   St Resid
  1   127   1235.0   1310.4     59.3      -75.4      -0.39
  2   115   1080.0   1105.9     62.1      -25.9      -0.14
  3   127    845.0    947.5     61.1     -102.5      -0.54
  4   150   1522.0   1322.5     37.1      199.5       1.01
  5   156   1047.0   1179.5     60.3     -132.5      -0.69
  6   182   1979.0   1831.9     82.9      147.1       0.81
  7   156   1822.0   1598.0     61.9      224.0       1.17
  8   132   1253.0   1185.8     39.7       67.2       0.34
  9   137   1297.0   1178.9     39.0      118.1       0.60
 10   113    946.0    913.9     58.6       32.1       0.17
 11   137   1713.0   1561.0     78.4      152.0       0.82
 12   117   1024.0   1072.6     53.1      -48.6      -0.25
 13   137   1147.0   1115.2     44.3       31.8       0.16
 14   153   1092.0   1149.2     59.0      -57.2      -0.30
 15   117   1152.0   1187.2     69.7      -35.2      -0.19
 16   126   1336.0   1117.6     43.4      218.4       1.12
 17   170   1131.0   1914.4    116.7     -783.4      -4.80R
 18   182   1550.0   1597.7     62.8      -47.7      -0.25
 19   162   1884.0   1598.3     57.0      285.7       1.49
 20   184   2041.0   1776.6     70.7      264.4       1.41
 21   143    845.0   1048.4     58.9     -203.4      -1.06
 22   159   1483.0   1421.8     40.6       61.2       0.31
 23   108   1055.0   1130.7     97.9      -75.7      -0.43
 24   175   1545.0   1522.7     55.4       22.3       0.12
 25   108    729.0    695.5     99.6       33.5       0.19
 26   179   1792.0   1642.7     57.6      149.3       0.78
 27   111   1175.0   1224.0    107.2      -49.0      -0.29
 28   187   1593.0   1651.3     68.6      -58.3      -0.31
 29   111    785.0    781.1     80.9        3.9       0.02
 30   115    744.0    822.7     75.5      -78.7      -0.42
 31   194   1356.0   1480.7    133.6     -124.7      -0.83 X
 32   168   1262.0   1374.0     57.7     -112.0      -0.58

R denotes an observation with a large standardized residual.
X denotes an observation whose X value gives it large influence.
```

Figure 12.41

MINITAB regression printout for altered grandfather clock data

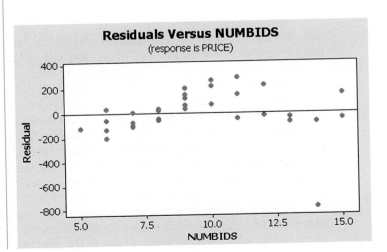

Figure 12.42

MINITAB residual plot of altered grandfather clock data

```
The regression equation is
PRICE = 474 - 0.46 AGE - 114 NUMBIDS + 1.48 AGE_BIDS


Predictor      Coef   SE Coef       T      P
Constant      474.0     298.2    1.59  0.124
AGE          -0.465     2.107   -0.22  0.827
NUMBIDS     -114.12     31.23   -3.65  0.001
AGE_BIDS     1.4781    0.2295    6.44  0.000


S = 85.8286   R-Sq = 95.2%   R-Sq(adj) = 94.7%


Analysis of Variance

Source           DF        SS        MS       F      P
Regression        3   3933417   1311139  177.99  0.000
Residual Error   27    198897      7367
Total            30   4132314
```

Figure 12.43

MINITAB regression printout when outlier is deleted

Figure 12.43 shows the printout when the outlier observation is excluded from the grandfather clock analysis, and Figure 12.44 shows the new plot of the residuals against the number of bidders. Now only one of the residuals lies beyond two standard deviations from 0, and none lies beyond three standard deviations. Also, the statistics indicate a much better model without the outlier. Most notably, the standard deviation (s) has decreased from 200.6 to 85.83, indicating a model that will provide more precise estimates and predictions (narrower confidence and prediction intervals) for clocks that are similar to those in the reduced sample.

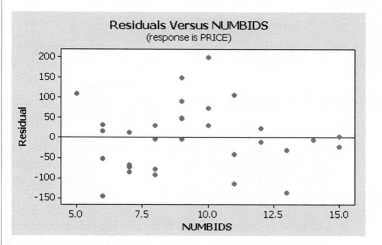

Figure 12.44

MINITAB residual plot when outlier is deleted

Look Back Remember that if the outlier is removed from the analysis when in fact it belongs to the same population as the rest of the sample, the resulting model may provide misleading estimates and predictions.

Now Work Exercise 12.143

Ethics IN Statistics

Removing observations from a sample data set for the sole purpose of improving the fit of a regression model without investigating whether the observations are outliers or legitimate data points is considered *unethical statistical practice.*

The next example checks the assumption of the normality of the random error component.

Example 12.17
Using Residuals to Check for Normal Errors

Problem Refer to Example 12.16. Analyze the distribution of the residuals in the grandfather clock example, both before and after the outlier residual is removed. Determine whether the assumption of a normally distributed error term is reasonable.

Solution A histogram and normal probability plot for the two sets of residuals are constructed using MINITAB and are shown in Figure 12.45 and 12.46. Note that the outlier appears to skew the histogram in Figure 12.45, whereas the histogram in Figure 12.46

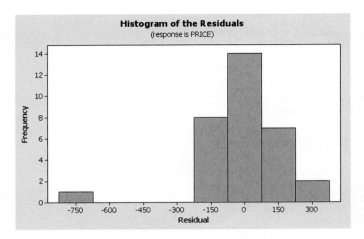

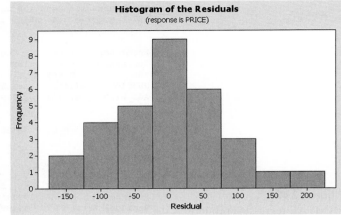

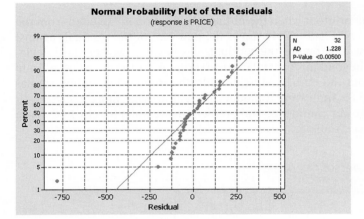

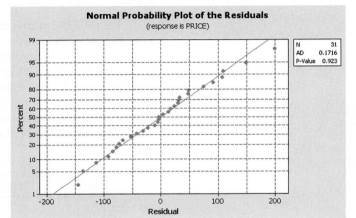

Figure 12.45

MINITAB graphs of regression residuals for grandfather clock model (outlier included)

Figure 12.46

MINITAB graphs of regression residuals for grandfather clock model (outlier deleted)

appears to be more mound shaped. Similarly, the pattern of residuals in the normal probability plot in Figure 12.46 (outlier deleted) is more nearly a straight line than the pattern in Figure 12.45 (outlier included). Thus, the normality assumption appears to be more plausible after the outlier is removed.

Look Back Although graphs do not provide formal statistical tests of normality, they do provide a descriptive display. Consult the references for methods to conduct statistical tests of normality using the residuals.

Now Work Exercise 12.138d

Of the four assumptions, the assumption that the random error is normally distributed is the least restrictive when we apply regression analysis in practice—that is, moderate departures from a normal distribution have very little effect on the validity of the statistical tests, confidence intervals, and prediction intervals presented in this chapter. In this case, we say that regression analysis is **robust** with respect to nonnormal errors. However, great departures from normality cast doubt on any inferences derived from the regression analysis.

Checking Assumption #4: Errors Independent

The assumption of independent errors is violated when successive errors are correlated. This typically occurs when the data for both the dependent and independent variables are observed sequentially over a period of time—called **time-series data.** Time-series data have a unique characteristic; the experimental unit represents a unit of time (e.g., a year, a month, a quarter). There are both graphical and formal statistical tests available for checking the assumption of independent regression errors. For example, a simple graph is to plot the residuals against time. If the residuals tend to group alternately into positive and negative clusters (as shown in Figure 12.47), then it is likely that the errors are correlated and the assumption is violated. If **correlated errors** are detected, one solution is to construct a **time-series model** for $E(y)$. These methods are beyond the scope of this text. (Consult the references for examples of time-series models.)

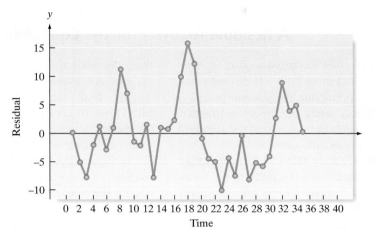

Figure 12.47

Hypothetical residual plot for time-series data

Summary

Residual analysis is a useful tool for the regression analyst, not only to check the assumptions but also to provide information about how the model can be improved. A summary of the residual analyses presented in this section to check the assumption that the random error ε is normally distributed with mean 0 and constant variance is presented in the next box.

Steps in a Residual Analysis

1. Check for a misspecified model by plotting the residuals against each of the quantitative independent variables. Analyze each plot, looking for a curvilinear trend. This shape signals the need for a quadratic term in the model. Try a second-order term in the variable against which the residuals are plotted.

2. Examine the residual plots for outliers. Draw lines on the plots at two and three standard deviations below and above the 0 line. Examine residuals outside the three-standard-deviation lines as potential outliers, and check to see that approximately 5% of the residuals exceed the two-standard-deviation lines. Determine whether each outlier can be explained as an error in data collection or transcription, whether it corresponds to a member of a population different from that of the remainder of the sample, or whether it simply represents an unusual observation. If the observation is determined to be an error, fix it or remove it. Even if you can't determine the cause, you may want to rerun the regression analysis without the observation to determine its effect on the analysis.

3. Check for nonnormal errors by plotting a frequency distribution of the residuals, using a stem-and-leaf display, histogram, or normal probability plot. Check to see if obvious departures from normality exist. Extreme skewness of the frequency distribution may be due to outliers or could indicate the need for a transformation of the dependent variable. (Normalizing transformations are beyond the scope of this book, but you can find discussions of the procedure in the references.)

(continued)

> **4.** Check for unequal error variances by plotting the residuals against the predicted values $\hat{y}$. If you detect a cone-shaped pattern or some other pattern which indicates that the variance of ε is not constant, refit the model, using an appropriate variance-stabilizing transformation on y, such as $\ln(y)$. (Consult the references for other useful variance-stabilizing transformations.)

Statistics IN Action Revisited A Residual Analysis for the Condominium Sale Price Model

In the previous Statistics in Action Revisited section (p. 676), we found a second-order model to be both a statistically and practically useful model for predicting the auction price (y) of a condominium unit. Before using the model in practice, we need to examine the residuals to be sure that the standard regression assumptions are reasonably satisfied.

Figures SIA12.7 and SIA12.8 are MINITAB graphs of the residuals from the model. Except for a few outliers, the histogram shown in Figure SIA12.7 appears to be approximately normally distributed; consequently, the assumption of normal errors is reasonably satisfied. The scatterplot of the residuals against $\hat{y}$ shown in Figure SIA12.8 reveals no distinct pattern;

thus, the assumption of a constant error variance is reasonably satisfied.

To find the outliers, we used MINITAB to compute the residuals of the model. A list of the unusual residual values is shown in the MINITAB printout in Figure SIA12.9. Four condo units (numbers 111, 184, 185, and 188; the values of the standardized residuals are highlighted on the printout) have residual values that are greater than 3 in absolute value. (Research reveals that

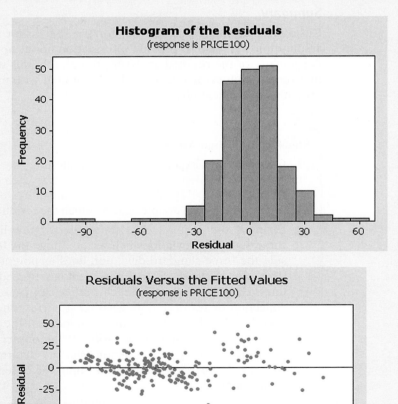

Figure SIA12.7

MINITAB histogram of residuals from second-order model of condo sale price

Figure SIA12.8

MINITAB plot of residuals versus predicted values from second-order model of condo sale price

Statistics IN Action
(continued)

```
Unusual Observations

Obs   FLOOR   PRICE100     Fit   SE Fit   Residual   St Resid
 11    8.00    279.00    240.71    7.62      38.29      2.18R
 26    3.00    269.00    228.48    4.75      40.52      2.19R
 37    6.00    294.00    246.22    5.34      47.78      2.60R
111    5.00    265.00    202.98    3.69      62.02      3.30R
183    3.00    150.00    197.54    4.36     -47.54     -2.55R
184    5.00    140.00    238.04    6.02     -98.04     -5.40R
185    5.00    130.00    192.19    4.37     -62.19     -3.34R
186    6.00    182.00    224.70    5.17     -42.70     -2.32R
188    5.00    164.00    252.16    4.66     -88.16     -4.75R

R denotes an observation with a large standardized residual.
```

Figure SIA12.9

MINITAB list of outliers for the second-order model of condo sale price

all of these units are located on the fifth floor.) If we delete these observations from the data set and refit the model, we obtain the results shown in the MINITAB printout in Figure SIA12.10. You can see that the adjusted-R^2 value is now .793, an increase of about 9% from the previous value. Also, the standard deviation of the model is now $s = 15.1$, a reduction

of about four hundred dollars from the previous value. Obviously, removing the outliers from the analysis yields a "better" prediction equation. If we can justify removing these four outliers (except for number 111, they were all undersold units), then we can use the model shown in Figure SIA12.10 to obtain more accurate predictions of future condo sale prices.

```
The regression equation is
PRICE100 = 185 - 2.59 FLOOR - 1.98 DISTANCE - 0.387 FLR_DST + 0.912 FLOORSQ
            + 0.264 DISTSQ + 83.9 VIEW - 4.01 VU_FLR - 4.61 VU_DIST
            + 0.154 VU_FLR_DST - 0.426 VU_FLRSQ + 0.304 VU_DISTSQ - 13.1 END
            + 11.5 FURNISH - 31.8 AUCTION

Predictor      Coef   SE Coef       T       P
Constant     184.93     10.39   17.81   0.000
FLOOR        -2.591      3.570   -0.73   0.469
DISTANCE     -1.977      1.923   -1.03   0.305
FLR_DST     -0.3871     0.1593   -2.43   0.016
FLOORSQ      0.9120     0.3384    2.70   0.008
DISTSQ       0.2644     0.1115    2.37   0.019
VIEW          83.88      14.03    5.98   0.000
VU_FLR       -4.007      4.705   -0.85   0.396
VU_DIST      -4.608      2.529   -1.82   0.070
VU_FLR_DST   0.1535     0.2166    0.71   0.479
VU_FLRSQ    -0.4261     0.4624   -0.92   0.358
VU_DISTSQ    0.3041     0.1465    2.08   0.039
END         -13.142      6.244   -2.10   0.037
FURNISH      11.549      2.405    4.80   0.000
AUCTION     -31.816      2.954  -10.77   0.000

S = 15.0925   R-Sq = 80.7%   R-Sq(adj) = 79.3%

Analysis of Variance

Source           DF      SS      MS      F      P
Regression       14  181271   12948  56.84  0.000
Residual Error  190   43279     228
Total           204  224550
```

Figure SIA12.10

MINITAB printout of second-order model of condo sale price with outliers deleted

12.12 Some Pitfalls: Estimability, Multicollinearity, and Extrapolation

You should be aware of several potential problems when you construct a prediction model for some response y. A few of the most important ones are discussed in this final section.

Problem 1: Parameter Estimability

Suppose you want to fit a model relating annual crop yield y to the total expenditure for fertilizer, x. We propose the first-order model

$$E(y) = \beta_0 + \beta_1 x$$

Now suppose we have three years of data and $1,000 is spent on fertilizer each year. The data are shown in Figure 12.48. You can see the problem: The parameters of the model cannot be estimated when all the data are concentrated at a single x value. Recall that it takes two points (x-values) to fit a straight line. Thus, the parameters are not estimable when only one x is observed.

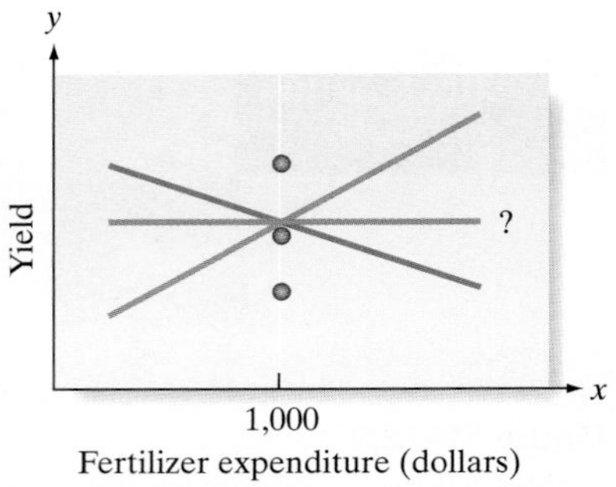

A similar problem would occur if we attempted to fit the quadratic model

$$E(y) = \beta_0 + \beta_1 x + \beta_2 x^2$$

to a set of data for which only one or two different x-values were observed. (See Figure 12.49.) At least three different x-values must be observed before a quadratic model can be fitted to a set of data (i.e., before all three parameters are estimable).

In general, the number of levels of observed x-values must be one more than the order of the polynomial in x that you want to fit.

For controlled experiments, the researcher can select one of the experimental designs in Chapter 10 that will permit estimation of the model parameters. Even when the values of the independent variables cannot be controlled by the researcher, the independent variables are almost always observed at a sufficient number of levels to permit estimation of the model parameters. When the statistical software you use suddenly refuses to fit a model, however, the problem is probably inestimable parameters.

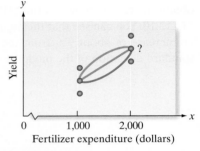

Figure 12.48
Yield and fertilizer expenditure data: three years

Figure 12.49
Only two x values observed: quadratic model is not estimable

Problem 2: Multicollinearity

Often, two or more of the independent variables used in a regression model contribute redundant information. That is, the independent variables are correlated with each other. For example, suppose we want to construct a model to predict the gas mileage rating of a truck as a function of its load x_1 (in tons) and the horsepower x_2 (in foot-pounds per second) of its engine. We would expect heavy loads to require greater horsepower and to result in lower mileage ratings. Thus, although both x_1 and x_2 contribute information for the prediction of mileage rating y, some of the information is overlapping because x_1 and x_2 are correlated.

When the independent variables are correlated, we say *multicollinearity* exists. In practice, it is not uncommon to observe correlations among the independent variables. However, a few problems arise when serious multicollinearity is present in the regression variables.

> **Multicollinearity** exists when two or more of the independent variables used in a regression are correlated.

First, high correlations among the independent variables increase the likelihood of rounding errors in the calculations of the β estimates, standard errors, and so forth. Second, and more important, the regression results may be confusing and misleading. Consider the model for gasoline mileage rating (y) of a truck,

$$E(y) = \beta_0 + \beta_1 x_1 + \beta_2 x_2$$

where $x_1 =$ load and $x_2 =$ horsepower. Fitting the model to a sample data set, we might find that the t-tests for testing β_1 and β_2 are both nonsignificant at the $\alpha = .05$ level, while the F-test for $H_0: \beta_1 = \beta_2 = 0$ is highly significant ($p = .001$). The tests may seem to be contradictory, but really they are not. One the one hand, the t-tests indicate that the contribution of one variable, say, x_1 = load, is not significant after the effect of x_2 = horsepower has been accounted for (because x_2 is also in the model). On the other hand, the significant F-test tells us that at least one of the two variables is making a contribution to the prediction of y (i.e., either β_1 or β_2 or both differ from 0). In fact, both are probably contributing, but the contribution of one overlaps with that of the other.

Multicollinearity can also have an effect on the signs of the parameter estimates. More specifically, a value of $\hat{\beta}_i$ may have the opposite sign from what is expected. In the truck gasoline mileage example, we expect heavy loads to result in lower mileage ratings and we expect higher horsepowers to result in lower mileage ratings; consequently, we expect the signs of both the parameter estimates to be negative. Yet, we may actually see a positive value of $\hat{\beta}_1$ and be tempted to claim that heavy loads result in *higher* mileage ratings. This is the danger of interpreting a β coefficient when the independent variables are correlated. Because the variables contribute redundant information, the effect of x_1 = load on y = mileage rating is measured only partially by β_1.

How can you avoid the problems of multicollinearity in regression analysis? One way is to conduct a designed experiment (Chapter 10), so that the levels of the x variables are uncorrelated. Unfortunately, time and cost constraints may prevent you from collecting data in this manner. Consequently, most data are collected observationally. Since observational data frequently consist of correlated independent variables, you will need to recognize when multicollinearity is present and, if necessary, make modifications in the regression analysis.

Several methods are available for detecting multicollinearity in regression. A simple technique is to calculate the coefficient of correlation, r, between each pair of independent variables in the model and use the procedure outlined in Section 11.5 to test for significantly correlated variables. If one or more of the r values is statistically different from 0, the variables in question are correlated and a multicollinearity problem may exist.* The degree of multicollinearity will depend on the magnitude of the value of r, as shown in the box.

Using the Correlation Coefficient r to Detect Multicollinearity

Extreme Multicollinearity: $|r| \geq .8$

Moderate Multicollinearity: $.2 \leq |r| < .8$

Low Multicollinearity: $|r| < .2$

Other indications of the presence of multicollinearity include those just mentioned, namely, nonsignificant t-tests for the individual parameter estimates when the F-test for overall model adequacy is significant, and estimates with opposite signs from what is expected.[†]

Detecting Multicollinearity in the Regression Model

Multicollinearity may exist if there are

   1. Significant correlations between pairs of independent variables

   2. Nonsignificant t-tests for all (or nearly all) of the individual β parameters when the F-test for overall model adequacy is significant

   3. Signs opposite from what is expected in the estimated β parameters

---

*Remember that r measures only the pairwise correlation between x values. Three variables, $x_1, x_2,$ and x_3, may be highly correlated as a group, but may not exhibit large pairwise correlations. Thus, multicollinearity may be present even when all pairwise correlations are not significantly different from 0.

[†]More formal methods for detecting multicollinearity, such as variance-inflation factors (VIFs), are available. Independent variables with a VIF of 10 or above are usually considered to be highly correlated with one or more of the other independent variables in the model. Calculation of VIFs is beyond the scope of this introductory text. (Consult the chapter references for a discussion of VIFs and other formal methods of detecting multicollinearity.)

Example 12.18
Detecting Multicollinearity— Modeling Carbon Monoxide Content

Problem The Federal Trade Commission (FTC) annually ranks varieties of domestic cigarettes according to their tar, nicotine, and carbon monoxide content. The U.S. surgeon general considers each of these three substances hazardous to a smoker's health. Past studies have shown that increases in the tar and nicotine content of a cigarette are accompanied by an increase in the carbon monoxide emitted from the cigarette smoke. Table 12.10 presents data on tar, nicotine, and carbon monoxide content (in milligrams) and weight (in grams) for a sample of 25 (filter) brands tested in a recent year. Suppose we want to model carbon monoxide content y as a function of tar content x_1, nicotine content x_2, and weight x_3, using the model

$$E(y) = \beta_0 + \beta_1 x_1 + \beta_2 x_2 + \beta_3 x_3$$

The model is fitted to the 25 data points in Table 12.10, and a portion of the SAS printout is shown in Figure 12.50. Examine the printout. Do you detect any signs of multicollinearity?

Table 12.10 FTC Cigarette Data for Example 12.18

Tar (x_1)	Nicotine (x_2)	Weight (x_3)	Carbon Monoxide (y)
14.1	.86	.9853	13.6
16.0	1.06	1.0938	16.6
29.8	2.03	1.1650	23.5
8.0	.67	.9280	10.2
4.1	.40	.9462	5.4
15.0	1.04	.8885	15.0
8.8	.76	1.0267	9.0
12.4	.95	.9225	12.3
16.6	1.12	.9372	16.3
14.9	1.02	.8858	15.4
13.7	1.01	.9643	13.0
15.1	.90	.9316	14.4
7.8	.57	.9705	10.0
11.4	.78	1.1240	10.2
9.0	.74	.8517	9.5
1.0	.13	.7851	1.5
17.0	1.26	.9186	18.5
12.8	1.08	1.0395	12.6
15.8	.96	.9573	17.5
4.5	.42	.9106	4.9
14.5	1.01	1.0070	15.9
7.3	.61	.9806	8.5
8.6	.69	.9693	10.6
15.2	1.02	.9496	13.9
12.0	.82	1.1184	14.9

Source: Federal Trade Commission.   *Data Set:* FTC

Solution First, note that the F-test for overall model utility is highly significant. The test statistic ($F = 78.98$) and the observed significance level (p-value $< .0001$) are highlighted on the SAS printout shown in Figure 12.50. Therefore, we can conclude at, say, $\alpha = .01$, that at least one of the parameters β_1, β_2, and β_3 in the model is nonzero. The t-tests for two of three individual β's, however, are nonsignificant. (The p-values for these tests are highlighted on the printout.) Unless tar (x_1) is the only one of the three variables that is useful in predicting carbon monoxide content, these results are the first indication of a potential multicollinearity problem.

The negative values for $\hat{\beta}_2$ and $\hat{\beta}_3$ (highlighted on the printout) are a second clue to the presence of multicollinearity. From past studies, the FTC expects carbon monoxide content (y) to increase when either nicotine content (x_2) or weight (x_3) increases; that is, the FTC expects *positive* relationships between y and x_2 and between y and x_3, not negative ones.

All signs indicate that a serious multicollinearity problem exists.*

*Note also that the variance-inflation factors (VIFs) for both tar and nicotine, given on the SAS printout, Figure 12.49, exceed 10.

```
                        Dependent Variable: CO

                    Number of Observations Read        25
                    Number of Observations Used        25

                            Analysis of Variance

                                  Sum of        Mean
        Source           DF      Squares       Square    F Value   Pr > F

        Model             3    495.25781    165.08594      78.98   <.0001
        Error            21     43.89259      2.09012
        Corrected Total  24    539.15040


                    Root MSE            1.44573    R-Square    0.9186
                    Dependent Mean     12.52800    Adj R-Sq    0.9070
                    Coeff Var          11.53996


                           Parameter Estimates

                        Parameter     Standard                      Variance
        Variable    DF    Estimate       Error    t Value  Pr > |t|  Inflation

        Intercept    1     3.20219     3.46175       0.93    0.3655          0
        TAR          1     0.96257     0.24224       3.97    0.0007   21.63071
        NICOTINE     1    -2.63166     3.90056      -0.67    0.5072   21.89992
        WEIGHT       1    -0.13048     3.88534      -0.03    0.9735    1.33386


                Pearson Correlation Coefficients, N = 25
                     Prob > |r| under H0: Rho=0

                           TAR      NICOTINE     WEIGHT

        TAR            1.00000       0.97661    0.49077
                                      <.0001     0.0127

        NICOTINE       0.97661       1.00000    0.50018
                        <.0001                   0.0109

        WEIGHT         0.49077       0.50018    1.00000
                        0.0127        0.0109
```

Figure 12.50

SAS printout for model of CO
content, Example 12.18

Look Back To confirm our suspicions, we had SAS produce the coefficient of correlation, r, for each of the three pairs of independent variables in the model. The resulting output is shown (highlighted) at the bottom of Figure 12.50. You can see that tar (x_1) and nicotine (x_2) are highly correlated $(r = .9766)$, while weight (x_3) is moderately correlated with the other two x's $(r \approx .5)$. All three correlations have p-values less than .05; consequently, all three are significantly different from 0 at $\alpha = .05$.

Now Work Exercise 12.140

Once you have detected multicollinearity, you can choose from among several alternative measures available for solving the problem. Several of these are outlined in the next box. The appropriate measure to take depends on the severity of the multicollinearity and the ultimate goal of the regression analysis. Some researchers, when confronted with highly correlated independent variables, choose to include only one of the correlated variables in the final model. If you are interested only in using the model for estimation and prediction (step 6), you may decide not to drop any of the independent variables from the model. We have seen that it is dangerous to interpret the individual β parameters in the presence of multicollinearity. However, confidence intervals for $E(y)$ and prediction intervals for y generally remain unaffected *as long as the values of the x's used to predict y follow the same pattern of multicollinearity exhibited in the sample data.* That is, you must take strict care to ensure that the values of the x-variables fall within the range of the sample data.

Solutions to Some Problems Created by Multicollinearity in Regression*

1. Drop one or more of the correlated independent variables from the model. One way to decide which variables to keep in the model is to employ stepwise regression (Section 12.10).

(continued)

---

*Several other solutions are available. For example, in the case where higher order regression models are fitted, the analyst may want to code the independent variables so that higher order terms (e.g., x^2) for a particular x-variable are not highly correlated with x. One transformation that works is $z = (x - \bar{x})/s$. Other, more sophisticated procedures for addressing multicollinearity (such as *ridge regression*) are beyond the scope of this text. (Consult the references at the end of the chapter.)

2. If you decide to keep all the independent variables in the model,

    a. Avoid making inferences about the individual β parameters on the basis of the t-tests.

    b. Restrict inferences about $E(y)$ and future y values to values of the x's that fall within the range of the sample data.

Problem 3: Prediction Outside the Experimental Region

Many research economists have developed highly technical models to relate the state of the economy to various economic indexes and other independent variables. Many of these models are multiple-regression models, in which, for example, the dependent variable y might be next year's gross domestic product (GDP) and the independent variables might include this year's rate of inflation, this year's Consumer Price Index (CPI), etc. In other words, the model might be constructed to predict next year's economy using this year's knowledge.

    Unfortunately, models such as these were almost all unsuccessful in predicting the recessions of the early 1970s, the late 1990s, and the mid 2000s. What went wrong? One of the problems was that many of the regression models were used to **extrapolate** (i.e., predict y for values of the independent variables that were outside the region in which the model was developed). For example, the inflation rate in the late 1960s, when many of the models were developed, ranged from 6% to 8%. When the double-digit inflation of the early 1970s became a reality, some researchers attempted to use the same models to predict future growth in GDP. As you can see in Figure 12.51, the model may be highly accurate in predicting y when x is in the range of experimentation, but the use of the model outside that range is a dangerous practice.

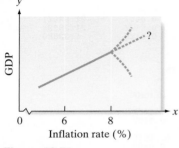

Figure 12.51

Using a regression model outside the experimental region

Exercises 12.130–12.150

Understanding the Principles

12.130 Define a regression residual.

12.131 Define an outlier.

12.132 Give two properties of the regression residuals from a model.

12.133 *True or False.* Regression models fit to time-series data typically result in uncorrelated errors.

12.134 Define multicollinearity in regression.

12.135 Give three indicators of a multicollinearity problem.

12.136 Define extrapolation.

Learning the Mechanics

12.137 Consider fitting the multiple regression model

$$E(y) = \beta_0 + \beta_1 x_1 + \beta_2 x_2 + \beta_3 x_3 + \beta_4 x_4 + \beta_5 x_5$$

A matrix of correlations for all pairs of independent variables is shown. Do you detect a multicollinearity problem? Explain.

	x_1	x_2	x_3	x_4	x_5
x_1	—	.17	.02	−.23	.19
x_2		—	.45	.93	.02
x_3			—	.22	−.01
x_4				—	.86
x_5					—

12.138 Identify the problem(s) in each of the residual plots [NW] shown on page 707.

Applying the Concepts—Basic

12.139 **Dating and disclosure.** Refer to the *Journal of Adolescence* (April 2010) study of adolescents' disclosure of their dating and romantic relationships, Exercise 12.14 (p. 628). Recall that multiple regression was used to model y = level of an adolescent's disclosure of a date's identity to his/her mother (measured on a 5-point scale). The independent variables in the study were gender (x_1 = 1 if female, 0 if male), age (x_2, years), dating experience (x_3, years), and level of trust in parents (x_4, 5-point scale). The highest correlation (in absolute value) for any pair of independent variables was $r = -.16$ for gender and level of trust. Do you believe that the regression analysis will exhibit multicollinearity problems? Explain.

12.140 **Women in top management.** The *Journal of Organizational* [NW] *Culture, Communications and Conflict* (July 2007) published a study on women in upper management positions at U.S. firms. Monthly data ($n = 252$ months) were collected for several variables in an attempt to model the number of females in managerial positions (y). The independent variables included the number of females with a college degree (x_1), the number of female high school graduates with no college degree (x_2), the number of males in managerial positions (x_3), the number of males with a college degree (x_4), and the number of male high school

Residual plots for Exercise 12.138

a.

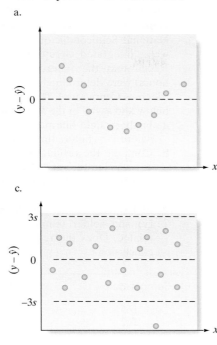

b.

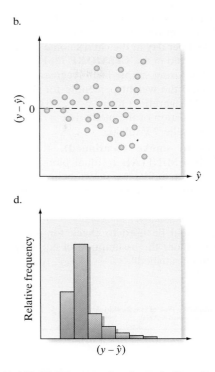

c.

d.

graduates with no college degree (x_5). Determine which of the correlations reported in parts **a-d** results in a potential multicollinearity problem for the regression analysis.

a. The correlation relating number of females in managerial positions and number of females with a college degree: $r = .983$.

b. The correlation relating number of females in managerial positions and number of female high school graduates with no college degree: $r = .074$.

c. The correlation relating number of males in managerial positions and number of males with a college degree: $r = .722$.

d. The correlation relating number of males in managerial positions and number of male high school graduates with no college degree: $r = .528$.

12.141 Personality and aggressive behavior. *Psychological Bulletin* (Vol. 132, 2006) reported on a study linking personality and aggressive behavior. Four of the variables measured in the study were aggressive behavior, irritability, trait anger, and narcissism. Pairwise correlations for these four variables are given in the following table:

Aggressive behavior–irritability: .80
Aggressive behavior–trait anger: .48
Aggressive behavior–narcissism: .50
Irritability–trait anger: .57
Irritability–narcissism: .16
Trait anger–narcissism: .13

a. Suppose aggressive behavior is the dependent variable in a regression model and the other variables are independent variables. Is there evidence of extreme multicollinearity? Explain.

b. Suppose narcissism is the dependent variable in a regression model and the other variables are independent variables. Is there evidence of extreme multicollinearity? Explain.

12.142 Yield strength of steel alloy. Refer to Exercise 12.126 (p. 686) and the *Modelling and Simulation in Materials Science and Engineering* (Vol. 13, 2005) study in which engineers built a regression model for the tensile yield strength (y) of a new steel alloy. The engineers discovered that the independent variable nickel (x_4) was highly correlated with the other 10 potential independent variables. Consequently, nickel was dropped from the model. Do you agree with this decision? Explain.

12.143 Passive exposure to smoke. Passive exposure to environmental tobacco smoke has been associated with suppression of growth and an increased frequency of respiratory tract infections in normal children. Is this association more pronounced in children with cystic fibrosis? To answer this question, 43 children (18 girls and 25 boys) attending a two-week summer camp for cystic fibrosis

Weight Percentile y	No. of Cigarettes Smoked per Day x	Weight Percentile y	No. of Cigarettes Smoked per Day x
6	0	43	0
6	15	49	0
2	40	50	0
8	23	49	22
11	20	46	30
17	7	54	0
24	3	58	0
25	0	62	0
17	25	66	0
25	20	66	23
25	15	83	0
31	23	87	44
35	10		

Based on Rubin, B. K. "Exposure of children with cystic fibrosis to environmental tobacco smoke." *The New England Journal of Medicine*, Sept. 20, 1990. Vol. 323, No. 12, p. 85 (data extracted from Figure 3).

patients were studied (*New England Journal of Medicine*, Sept. 20, 1990). Researchers investigated the correlation between a child's weight percentile (y) and the number of cigarettes smoked per day in the child's home (x). The table on page 707 (saved in the **CFSMOKE** file) lists the data on the 25 boys. Using simple linear regression, the researchers predicted the weight percentile for the last observation ($x = 44$ cigarettes) to be $\hat{y} = 29.63$. Given that the standard deviation of the model is $s = 24.68$, is this observation an outlier? Explain.

12.144 Passive exposure to smoke (continued). Refer to Exercise 12.143. Two MINITAB residual plots for the simple linear regression model are shown below.
 a. Which graph should be used to check for normal errors? Does the assumption of normality appear to be satisfied?
 b. Which graph should be used to check for unequal error variances? Does the assumption of equal variances appear to be satisfied?

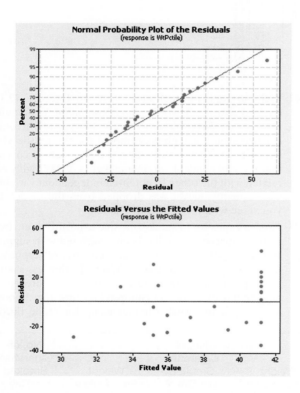

Applying the Concepts—Intermediate

12.145 Accuracy of software effort estimates. Refer to the *Journal of Empirical Software Engineering* (Vol. 9, 2004) study of the accuracy of new software effort estimates, Exercise 12.124 (p. 685). Recall that stepwise regression was used to develop a model for the relative error in estimating effort (y) as a function of company role of estimator ($x_1 = 1$ if developer, 0 if project leader) and previous accuracy ($x_8 = 1$ if more than 20% accurate, 0 if less than 20% accurate). The stepwise regression yielded the prediction equation $\hat{y} = .12 - .28x_1 + .27x_8$. The researcher is concerned that the sign of the estimated β multiplied by x_1 is the opposite from what is expected. (The researcher expects a project leader to have a smaller relative error

of estimation than a developer.) Give at least one reason why this phenomenon occurred.

12.146 Failure times of silicon wafer microchips. Refer to the National Semiconductor study of manufactured silicon wafer integrated circuit chips, Exercise 12.72 (p. 655). Recall that the failure times of the microchips (in hours) were determined at different solder temperatures (degrees Centigrade). The data are repeated in the table below and saved in the **WAFER** file.
 a. Fit the straight-line model $E(y) = \beta_0 + \beta_1 x$ to the data, where y = failure time and x = solder temperature.
 b. Compute the residual for a microchip manufactured at a temperature of 152°C.
 c. Plot the residuals against solder temperature (x). Do you detect a trend?
 d. In Exercise 12.72c, you determined that failure time (y) and solder temperature (x) were curvilinearly related. Does the residual plot, part **c**, support this conclusion?

Temperature (°C)	Time to Failure (hours)
165	200
162	200
164	1,200
158	500
158	600
159	750
156	1,200
157	1,500
152	500
147	500
149	1,100
149	1,150
142	3,500
142	3,600
143	3,650
133	4,200
132	4,800
132	5,000
134	5,200
134	5,400
125	8,300
123	9,700

Based on Gee, S., and Nguyen, L. "Mean time to failure in wafer level–CSP packages with SnPb and SnAgCu solder bmps." International Wafer Level Packaging Conference, San Jose, CA, Nov. 3–4, 2005 (adapted from Figure 7).

12.147 Arsenic in groundwater. Refer to the *Environmental Science & Technology* (Jan. 2005) study of the reliability of a commercial kit to test for arsenic in groundwater, Exercise 12.21 (p. 630). Recall that you fit a first-order model for arsenic level (y) as a function of latitude (x_1), longitude (x_2), and depth (x_3) to data saved in the **ASWELLS** file. Conduct a residual analysis of the data. Based on the results, comment on each of the following:
 a. assumption of mean error = 0
 b. assumption of constant error variance
 c. outliers
 d. assumption of normally distributed errors
 e. multicollinearity

12.148 Contamination of fish in the Tennessee River. Refer to the U.S. Army Corps of Engineers data on fish contaminated from the toxic discharges of a chemical plant located on the banks of the Tennessee River in Alabama, presented in Exercise 12.22 (p. 630). In that exercise, you fitted the first-order model $E(y) = \beta_0 + \beta_1 x_1 + \beta_2 x_2 + \beta_3 x_3$, where y = DDT level in captured fish, x_1 = miles captured upstream, x_2 = fish length, and x_3 = fish weight. Conduct a complete residual analysis of the model using the data in the **FISHDDT** file. Do you recommend any modifications to be made to the model? Explain.

12.149 **Reality TV and cosmetic surgery.** Refer to the *Body Image: An International Journal of Research* (March 2010) study of the influence of reality TV shows on one's desire to undergo cosmetic surgery, Exercise 12.23 (p. 631). Simulated data for the study are saved in the **BODYIMAGE** file. In Exercise 12.23, you fit the first-order model, $E(y) = \beta_0 + \beta_1 x_1 + \beta_2 x_2 + \beta_3 x_3 + \beta_4 x_4$, where

y = desire to have cosmetic surgery, x_1 is a dummy variable for gender, x_2 = level of self-esteem, x_3 = level of body satisfaction, and x_4 = impression of reality TV. Conduct a complete residual analysis for the model. Do you detect any violations of the assumptions?

12.150 **Cooling method for gas turbines.** Refer to the *Journal of Engineering for Gas Turbines and Power* (Jan. 2005) study of a high-pressure inlet fogging method for a gas turbine engine, presented in Exercise 12.25 (p. 632). Now consider the interaction model $E(y) = \beta_0 + \beta_1 x_1 + \beta_2 x_2 + \beta_3 x_1 x_2$ for heat rate (y) of a gas turbine as a function of cycle speed (x_1) and cycle pressure ratio (x_2). Use the data saved in the **GASTURBINE** file to conduct a complete residual analysis of the model. Do you recommend making modifications to the model?

Guide to Multiple Regression

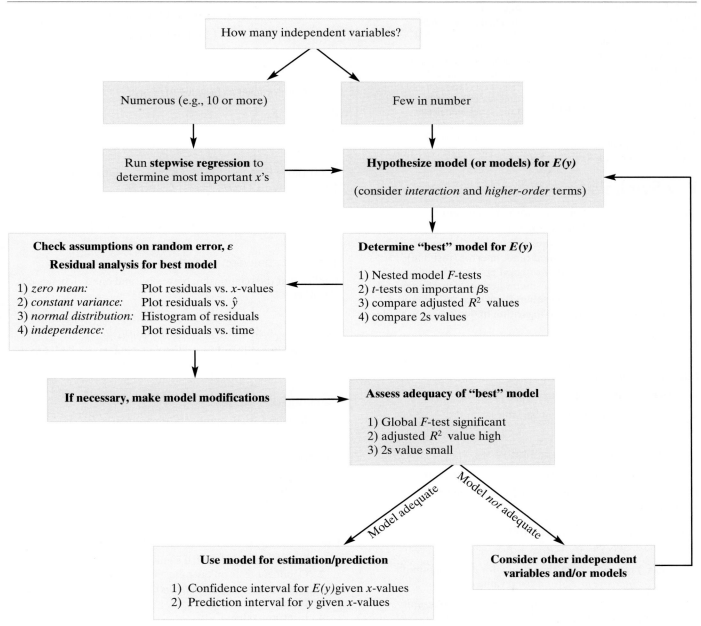

CHAPTER NOTES

Key Terms

[Note:Starred (*) items are from the optional sections in this chapter.]

Adjusted multiple coefficient of determination 625
Base level 656
Coded variable 615
*Complete (nested) model 672
Complete second-order model 650
Correlated errors 699
Dummy (or indicator) variables 656
Extrapolate 706
First-order model 616
Global *F*-test 626
Higher order term 615
Interact 639
Interaction model 639
Interaction term 639
Least squares prediction equation 617
Level of a variable 656
*Main-effect terms 664
Mean square for error (MSE) 618
Model building 616
Multicollinearity 702
Multiple coefficient of determination 624

Multiple-regression model 615
*Nested model 673
*Nested-model *F*-test 676
Objective variable-screening procedure 683
Paraboloid 650
Parameter estimability 701
*Parsimonious model 676
Quadratic model 646
Quadratic term (or second-order term) 646
Qualitative (or categorical) independent variable 656
Regression outlier 695
*Reduced (nested) model 672
Regression residual 688
Residual 688
Residual analysis 688
*Response surface 674
Robust 698
Saddle-shaped surface 650
Second-order model 646
*Stepwise regression 681
Time-series data 698
Time-series model 699
Variance-stabilizing transformation 692

Key Symbols

x_1^2	Quadratic form for a quantitative x
$x_1 x_2$	Interaction term
MSE	Mean square for error (estimates σ^2)
$\hat{\varepsilon}$	Estimated random error (residual)
SSE_R	*Sum of squared errors, reduced model
SSE_C	*Sum of squared errors, complete model
MSE_C	*Mean squared error, complete model
$\ln(y)$	Natural logarithm of dependent variable

Key Ideas

Multiple-Regression Variables

y = **Dependent** variable (quantitative)

$x_1, x_2, \ldots, x_k$ are **independent** variables (*quantitative or qualitative*)

First-Order Model in *k* Quantitative *x*'s

$$E(y) = \beta_0 + \beta_1 x_1 + \beta_2 x_2 + \ldots + \beta_k x_k$$

Each β_1 represents the change in y for every one-unit increase in x_1, holding all other x's fixed.

Interaction Model in 2 Quantitative *x*'s

$$E(y) = \beta_0 + \beta_1 x_1 + \beta_2 x_2 + \beta_3 x_1 x_2.$$

$(\beta_1 + \beta_3 x_2)$ represents the change in y for every one-unit *increase in* x_1, for a fixed value of x_2.

$(\beta_2 + \beta_3 x_1)$ represents the change in y for every one-unit *increase in* x_2, for a fixed value of x_1.

Quadratic Model in 1 Quantitative *x*

$$E(y) = \beta_0 + \beta_1 x + \beta_2 x^2$$

β_2 represents the rate of curvature of x.

($\beta_2 > 0$ implies *upward* curvature.)

($\beta_2 < 0$ implies *downward* curvature.)

Complete Second-Order Model in 2 Quantitative *x*'s

$$E(y) = \beta_0 + \beta_1 x_1 + \beta_2 x_2 + \beta_3 x_1 x_2 + \beta_4 x_1^2 + \beta_5 x_2^2$$

β_4 represents the rate of curvature of x_1, holding x_2 fixed.

β_5 represents the rate of curvature of x_2, holding x_1 fixed.

Dummy Variable Model for 1 Qualitative *x*

$$E(y) = \beta_0 + \beta_1 x_1 + \beta_2 x_2 + \cdots + \beta_{k-1} x_{k-1}$$
$$x_1 = \{1 \text{ if level 1, 0 if not}\}$$
$$x_2 = \{1 \text{ if level 2, 0 if not}\}$$
$$\vdots$$
$$x_{k-1} = \{1 \text{ if level } k - 1, 0 \text{ if not}\}$$

$\beta_0 = E(y)$ for level k (base level) $= \mu_k$

$\beta_1 = \mu_1 - \mu_k$

$\beta_2 = \mu_2 - \mu_k$

*Complete Second-Order Model in 1 Quantitative *x* and 1 Qualitative *x* (Two Levels, A and B)

$$E(y) = \beta_0 + \beta_1 x_1 + \beta_2 x_1^2 + \beta_3 x_2 + \beta_4 x_1 x_2 + \beta_5 x_1^2 x_2$$
$$x_2 = \{1 \text{ if level A, 0 if level B}\}$$

Interaction between x_1 and x_2

Implies that the relationship between y and one x depends on the other x.

Adjusted Coefficient of Determination, R_a^2

Cannot be "forced" to 1 by adding independent variables to the model.

Parsimonious Model

A model with a small number of β parameters.

*Nested Models

Are models such that one model (the *complete model*) contains all the terms of another model (the *reduced model*) plus at least one additional term.

Recommendation for Assessing Model Adequacy

1. Conduct global *F*-test; if significant, then
2. Conduct *t*-tests on only the "most important" β's (*interaction* or *squared terms*).
3. Interpret value of 2s.
4. Interpret value of R_a^2.

Recommendation for Testing Individual β's

1. If *curvature* (x^2) is deemed important, do not conduct test for first-order (x) term in the model.
2. If *interaction* (x_1x_2) is deemed important, do not conduct tests for first-order terms $(x_1$ and $x_2)$ in the model.

*Problems with Using Stepwise Regression Model as the "Final" Model

1. *Extremely large number of t-tests* inflats overall probability of at least one Type I error.
2. *No higher order terms* (interactions or squared terms) are included in the model.

Analysis of Residuals

1. Detect **misspecified model:** *plot residuals vs. quantitative x* [look for trends (e.g., curvilinear trend)]
2. Detect **nonconstant error variance:** *plot residuals vs.* $\hat{y}$ [look for patterns (e.g., cone shape)]
3. Detect **nonnormal errors:** *histogram, stem–leaf, or normal probability plot of residuals* (look for strong departures from normality)
4. Identify **outliers:** *residuals greater than 3s in absolute value* (investigate outliers before deleting)

Multicollinearity

Occurs when two or more x's are correlated.
Indicators of multicollinearity:
1. Highly correlated x's
2. Significant global F-test, but all t-tests nonsignificant
3. Signs on β's opposite from what is expected

Extrapolation

Occurs when you predict y for values of x's that are outside of range of sample data.

Key Formulas

$s^2 = \text{MSE} = \dfrac{\text{SSE}}{n - (k + 1)}$ — Estimator of σ^2 for a model with k independent variables

$t = \dfrac{\hat{\beta}_i}{s_{\hat{\beta}_i}}$ — Test statistic for testing H_0: β_i

$\hat{\beta}_i \pm (t_{\alpha/2}) s_{\hat{\beta}_i}$, where $t_{\alpha/2}$ depends on $n - (k + 1)$ df — $100(1 - \alpha)\%$ confidence interval for β_i

$R^2 = \dfrac{\text{SS}_{yy} - \text{SSE}}{\text{SS}_{yy}}$ — Multiple coefficient of determination

$R_a^2 = 1 - \left[\dfrac{(n - 1)}{n - (k + 1)}\right](1 - R^2)$ — Adjusted multiple coefficient of determination

$F = \dfrac{\text{MS(Model)}}{\text{MSE}} = \dfrac{R^2/k}{(1 - R^2)/[n - (k + 1)]}$ — Test statistic for testing H_0: $\beta_1 = \beta_2 = \ldots = \beta_k = 0$

$F = \dfrac{(\text{SSE}_R - \text{SSE}_C)/\text{number of }\beta\text{'s tested}}{\text{MSE}_C}$ — *Test statistic for comparing reduced and complete models

$y - \hat{y}$ — Regression residual

Supplementary Exercises 12.151–12.185

Understanding the Principles

12.151 Write a model relating $E(y)$ to one qualitative independent variable that is at four levels. Define all the terms in your model.

12.152 Explain why stepwise regression is used. What is its value in the model-building process?

12.153 It is desired to relate $E(y)$ to a quantitative variable x_1 and a qualitative variable at three levels.
   a. Write a first-order model.
   b. Write a model that will graph as three different second-order curves, one for each level of the qualitative variable.

12.154 a. Write a first-order model relating $E(y)$ to two quantitative independent variables x_1 and x_2.
   b. Write a complete second-order model.

Learning the Mechanics

12.155 Suppose you fit the model

$$y = \beta_0 + \beta_1 x_1 + \beta_2 x_1^2 + \beta_3 x_2 + \beta_4 x_1 x_2 + \varepsilon$$

to $n = 25$ data points and find that

$\hat{\beta}_0 = 1.26 \quad \hat{\beta}_1 = -2.43 \quad \hat{\beta}_2 = .05 \quad \hat{\beta}_3 = .62 \quad \hat{\beta}_4 = 1.81$
$s_{\hat{\beta}_1} = 1.21 \quad s_{\hat{\beta}_2} = .16 \quad s_{\hat{\beta}_3} = .26 \quad s_{\hat{\beta}_4} = 1.49$

$\text{SSE} = .41 \quad R^2 = .83$

   a. Is there sufficient evidence to conclude that at least one of the parameters $\beta_1, \beta_2, \beta_3$, and β_4 is nonzero? Test using $\alpha = .05$.
   b. Test H_0: $\beta_1 = 0$ against H_a: $\beta_1 < 0$. Use $\alpha = .05$.
   c. Test H_0: $\beta_2 = 0$ against H_a: $\beta_2 > 0$. Use $\alpha = .05$.
   d. Test H_0: $\beta_3 = 0$ against H_a: $\beta_3 \neq 0$. Use $\alpha = .05$.

12.156 Suppose you used MINITAB to fit the model

$$y = \beta_0 + \beta_1 x_1 + \beta_2 x_2 + \varepsilon$$

to $n = 15$ data points and you obtained the printout below.

```
The regression equation is
Y  = 90.1 - 1.84 X1 + .285 X2


Predictor    Coef   SE Coef       T      P
Constant    90.10     23.10    3.90  0.002
X1          -1.836    0.367   -5.01  0.001
X2           0.285    0.231    1.24  0.465


S = 10.68     R-Sq = 91.6%    R-Sq(adj) = 90.2%

Analysis of Variance

Source          DF      SS     MS      F      P
Regression       2   14801   7400  64.91  0.001
Residual Error  12    1364    114
Total           14   16165
```

a. What is the least squares prediction equation?
b. Find R^2 and interpret its value.
c. Is there sufficient evidence to indicate that the model is useful in predicting y? Conduct an F-test, using $\alpha = .05$.
d. Test the null hypothesis $H_0: \beta_1 = 0$ against the alternative hypothesis $H_a: \beta_1 \neq 0$. Use $\alpha = .05$. Draw the appropriate conclusions.
e. Find the standard deviation of the regression model and interpret it.

12.157 Suppose you have developed a regression model to explain the relationship between y and $x_1, x_2,$ and x_3. The ranges of the variables you observed were as follows: $10 \leq y \leq 100, 5 \leq x_1 \leq 55, .5 \leq x_2 \leq 1,$ and $1,000 \leq x_3 \leq 2,000$. Will the error of prediction be smaller when you use the least squares equation to predict y when $x_1 = 30, x_2 = .6,$ and $x_3 = 1,300$, or when $x_1 = 60, x_2 = .4,$ and $x_3 = 900$? Why?

12.158 The first-order model $E(y) = \beta_0 + \beta_1 x_1$ was fit to $n = 19$ data points. A plot of the residuals of the model is shown. Is the need for a quadratic term in the model evident from the plot? Explain.

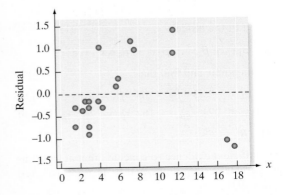

12.159 To model the relationship between y (a dependent variable) and x (an independent variable), a researcher has taken one measurement of y at each of three different

x values. Drawing on his mathematical expertise, the researcher realizes that he can fit the second-order model

$$E(y) = \beta_0 + \beta_1 x + \beta_2 x^2$$

and it will pass exactly through all three points, yielding SSE = 0. The researcher, delighted with the "excellent" fit of the model, eagerly sets out to use it to make inferences. What problems will he encounter in attempting to make inferences?

12.160 Suppose you fit the regression model

$$E(y) = \beta_0 + \beta_1 x_1 + \beta_2 x_2 + \beta_3 x_2^2 + \beta_4 x_1 x_2 + \beta_5 x_1 x_2^2$$

to $n = 35$ data points and wish to test the null hypothesis $H_0: \beta_4 = \beta_5 = 0$.
a. State the alternative hypothesis.
b. Explain in detail how to compute the F-statistic needed to test the null hypothesis.
c. What are the numerator and denominator degrees of freedom associated with the F-statistic in part b?
d. Give the rejection region for the test if $\alpha = .05$.

Applying the Concepts—Basic

12.161 Global warming and foreign investments. Scientists believe that a major cause of global warming is higher levels of carbon dioxide (CO_2) in the atmosphere. In the *Journal of World-Systems Research* (Summer 2003), sociologists examined the impact of a dependence on foreign investment on CO_2 emissions in $n = 66$ developing countries. In particular, the researchers modeled the level of CO_2 emissions in a particular year on the basis of foreign investments made 16 years earlier and several other independent variables. The variables and the model results are listed in the following table:

$y = \ln$ (level of CO_2 emissions incurrent year)	β Estimate	t-Value	p-Value
$x_1 = \ln$ (foreign investments)	.79	2.52	<.05
$x_2 =$ gross domestic investment	.01	.13	>.10
$x_3 =$ trade exports	-.02	-1.66	>.10
$x_4 = \ln$(GNP)	-.44	-.97	>.10
$x_5 =$ agricultural production	-.03	-.66	>.10
$x_6 = 1$ if African country, 0 if not	-1.19	-1.52	>.10
$x_7 = \ln$(level of CO_2 emissions)	.56	3.35	<.001

$R^2 = .31$

Based on Grimes, P., and Kentor, J. "Exporting the greenhouse: Foreign capital penetration and CO_2 emissions 1980–1996." *Journal of World-Systems Research*, Vol. IX, No. 2, Summer 2003 (Table 1).

a. Interpret the value of R^2.
b. Use the value of R^2 to test the null hypothesis, $H_0: \beta_1 = \beta_2 = \cdots = \beta_7 = 0$ at $\alpha = .01$. Give the appropriate conclusion.
c. Do you advise conducting t-tests on each of the independent variables to test the overall adequacy of the model? Explain.

Correlations for Exercise 12.161

Independent Variable	x_2	x_3	x_4	x_5	x_6	$x_7 = $ ln (level of CO_2 emissions)
$x_1 = $ ln(foreign investments)	.13	.57	.30	−.38	.14	−.14
$x_2 = $ gross domestic investment)		.49	.36	−.47	−.14	.25
$x_3 = $ trade exports			.43	−.47	−.06	−.07
$x_4 = $ ln(GNP)				−.84	−.53	.42
$x_5 = $ agricultural production						−.50
$x_6 = $ 1 if African country, 0 if not						−.47

Based on Grimes, P., and Kentor, J. "Exporting the greenhouse: Foreign capital pentration and CO2 emissions 1980–1996." *Journal of World-Systems Research*, Vol. IX, No. 2, Summer 2003 (Appendix B).

d. What null hypothesis would you test to determine whether the number of foreign investments made 16 years earlier is a statistically useful predictor of CO_2 emissions the current year?

e. Conduct the test mentioned in part **d** at $\alpha = .05$. Give the appropriate conclusion.

f. A matrix giving the correlation (r) for each pair of independent variables is shown in the table above. Identify the independent variables that are highly correlated. What problems may result from including these highly correlated variables in the regression model?

12.162 Students' ability in science. An article published in the *American Educational Research Journal* (Fall 1998) used multiple regression to model the students' perceptions of their ability in science classes. The sample consisted of 165 Grade 5–Grade 8 students in six performance-based science classrooms, all of which use hands-on activities as the main teaching tool. The dependent variable of interest, the student's perception of his or her ability (y), was measured on a four-point scale (where 1 = little or no ability and 4 = high ability). Two types of independent variables were included in the model, *control variables* and *performance behavior variables*. The control variables are: prior science attitude (measured on a 4-point scale), score on standardized science test, gender, and classroom (1, 3, 4, 5, or 6). The performance behavior variables (all measured on a numerical scale between 0 and 1) are: active-leading behavior, passive-assisting behavior, and active-manipulating behavior.

a. Identify the independent variables as quantitative or qualitative.

b. Individual β-tests on the independent variables all had p-values greater than .10 except for prior science attitute, gender, and active-leading behavior. Which variables appear to contribute to the prediction of a student's perception of his or her ability in science?

c. The estimated β-value for the active-leading behavior variable is .88 with a standard error of .34. Use this information to construct a 95% confidence interval for this β. Interpret the interval.

d. The following statistics for evaluating the overall predictive power of the model were reported: $R^2 = .48$, $F = 12.84$, $p < .001$. Interpret the results.

e. Hypothesize the equation of the first-order main effects model for $E(y)$

f. The researchers also considered a model that included all possible interactions between the control variables

and the performance behavior variables. Write the equation for this model for $E(y)$.

g. The researchers determined that the interaction terms in the model formulated in part **b** were not significant; therefore, they used the model from part **a** to make inferences. Explain the best way to conduct this test for interaction. Give the null hypothesis of the test.

Based on Jovanovic, J., and King, S. S. "Boys and girls in the performance-based science classroom: Who's doing the performing." *American Educational Research Journal*, Vol. 35, No. 3, Fall 1998, p. 489 (Table 8).

12.163 Distress in EMS workers. The *Journal of Consulting and Clinical Psychology* (June 1995) reported on a study of emergency service (EMS) rescue workers who responded to the I-880 freeway collapse during a San Francisco earthquake. The goal of the study was to identify the predictors of symptomatic distress in the workers. One of the distress variables studied was the Global Symptom Index (GSI). Several models for GSI (y) based on the following independent variables were considered:

$x_1 = $ Critical Incident Exposure scale (CIE)
$x_2 = $ Hogan Personality Inventory-Adjustment scale (HPI-A)
$x_3 = $ Years of experience (EXP)
$x_4 = $ Locus of Control scale (LOC)
$x_5 = $ Social Support scale (SS)
$x_6 = $ Dissociative Experiences scale (DES)
$x_7 = $ Peritraumatic Dissociation Experiences Questionnaire, self-report (PDEQ-SR)

a. Write a first-order model for $E(y)$ as a function of the first five independent variables, x_1–x_5.

b. The model from part **a,** fitted to data collected on $n = 147$ EMS workers, yielded the following results: $R^2 = .469$, $F = 34.47$, p-value $< .001$. Interpret these results.

c. Write a first-order model for $E(y)$ as a function of all seven independent variables, x_1–x_7.

d. The model from part **c** yielded $R^2 = .603$. Interpret this result.

e. The t-tests for testing the DES and PDEQ-SR variables both yielded a p-value of .001. Interpret this result.

12.164 Listen-and-look study. Where do you look when you are listening to someone speak? Researchers have discovered that listeners tend to gaze at the eyes or mouth of the speaker. In a study published in *Perception & Psychophysics* (Aug. 1998), subjects watched a videotape of a speaker giving a series of short monologues at a social gathering (e.g., a party). The level of background noise (multilingual voices and music) was varied during the listening sessions. Each subject wore a pair of clear plastic goggles on which an infrared corneal detection system was mounted, enabling the researchers to monitor the subject's eye movements. One response variable of interest was the proportion y of times the subject's eyes fixated on the speaker's mouth.

a. The researchers wanted to estimate $E(y)$ for four different noise levels: none, low, medium, and high. Hypothesize a model that will allow the researchers to obtain these estimates.

b. Interpret the β's in the model you hypothesized in part **a.**

c. Explain how to test the hypothesis of no differences in the mean proportions of mouth fixations for the four background noise levels.

12.165 Genetics of a brain disease. Spinocerebellar ataxia type 1 (SCA1) is an inherited neurodegenerative disorder characterized by dysfunction of the brain. From a deoxyribonucleic acid (DNA) analysis of SCA1 chromosomes, researchers discovered the presence of repeat gene sequences (*Cell Biology*, Feb. 1995). In general, the more repeat sequences observed, the earlier was the onset of the disease (in years of the person's age). The scatterplot (next column) shows this relationship for data collected on 113 individuals diagnosed with SCA1.

a. Suppose you want to model the age y of onset of the disease as a function of number x of repeat gene sequences in SCA1 chromosomes. Propose a quadratic model for y.

b. Will the sign of β_2 in the model you proposed in part **a** be positive or negative? Base your decision on the results shown in the scatterplot.

c. The researchers reported a correlation of $r = -.815$ between age and number of repeats. Since $r^2 = (-.815)^2 = .664$, they concluded that about "66% of the variability in the age of onset can be accounted for by the number of repeats." Does this statement apply to the quadratic model $E(y) = \beta_0 + \beta_1 x + \beta_2 x^2$? If not, give the equation of the model for which it does apply.

12.166 Frequency of drinking alcohol. To what degree do the attitudes of your peers influence your behavior? A study presented in *Social Psychology Quarterly* (Vol. 50, 1987) included a sample of $n = 143$ adult drinkers in an urban setting characterized by a high physical availability of alcoholic beverages. The goal of the study was to build a model relating frequency of drinking alcoholic beverages (y) to attitude toward drinking (x_1) and social support (x_2). Consider the interaction model

$$E(y) = \beta_0 + \beta_1 x_1 + \beta_2 x_2 + \beta_3 x_1 x_2$$

a. Interpret the phrase "x_1 and x_2 interact" in terms of the problem.

Scatterplot for Exercise 12.165

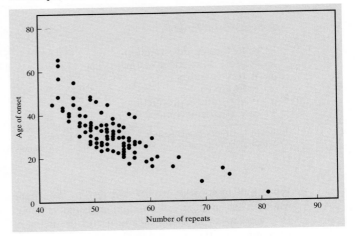

b. Write the null and alternative hypotheses for determining whether attitude (x_1) and social support (x_2) interact.

c. The reported p-value for the test suggested in part **b** was $p < .001$. Interpret this result.

12.167 Density of mosquito larvae. A field experiment was conducted to assess the effect of organic enrichment on the mean density of mosquito larvae (*Journal of the American Mosquito Control Association*, June 1995). Larval specimens were collected from a pond three days after the pond was flooded with canal water. A second sample of specimens was collected three weeks after flooding and enriching the pond with rabbit pellets. All specimens were returned to the laboratory and the number y of mosquito larvae in each specimen was counted.

a. Write a model that will allow you to compare the mean number of mosquito larvae found in the enriched pond with the corresponding mean for the natural pond.

b. Interpret the β coefficients in the model you wrote in part **a.**

c. Set up the null and alternative hypotheses for testing whether the mean larval density for the enriched pond exceeds the mean for the natural pond.

d. The p-value associated with the global F-test for the model from part **a** was determined to be .004. Interpret this result.

12.168 Factors identifying urban counties. The *Professional Geographer* (Feb. 2000) published a study of urban and rural counties in the western United States. The researchers used six independent variables—total county population (x_1), population density (x_2), population concentration (x_3), population growth (x_4), proportion of county land in farms (x_5), and five-year change in agricultural land base (x_6)—to model the urban/rural rating (y) of a county on a scale of 1 (most rural) to 10 (most urban). Prior to running the multiple-regression analysis, the researchers were concerned about possible multicollinearity in the data. Following is a MINITAB printout of correlations between all pairs of the independent variables:

Correlations: X1, X2, X3, X4, X5, X6

	X1	X2	X3	X4	X5
X2	0.20				
X3	0.45	0.43			
X4	-0.05	-0.14	-0.01		
X5	-0.16	-0.15	-0.07	-0.20	
X6	-0.12	-0.12	-0.22	-0.06	-0.06

Cell Contents: Pearson correlation

a. On the basis of the correlation printout, is there any evidence of extreme multicollinearity?

b. The first-order model with all six independent variables was fit to the data. The multiple-regression results are shown in the accompanying MINITAB printout. On the basis of the reported tests, is there any evidence of extreme multicollinearity?

Predictor	Coef	T	P
Constant	--	-	-
X1	0.110	2.01	0.045
X2	0.065	1.20	0.230
X3	0.540	3.59	0.000
X4	-0.009	-0.18	0.860
X5	-0.150	-3.00	0.003
X6	-0.027	-0.55	0.580

R-Sq = 44.0%   R-Sq(adj) = 43.0%

Analysis of Variance

Source	DF	SS	MS	F	P
Regression	6	--	--	32.47	0.000
Residual Error	249	--	--		
Total	255	--			

Based on Berry, K. A. et al., "Interpreting what is rural and urban for western U. S. counties." *Professional Geographer*, Vol. 52, No. 1, Feb. 2000 (Table 2).

12.169 Occupational safety study. An important goal in occupational safety is "active caring." Employees demonstrate active caring about the safety of their coworkers when they identify environmental hazards and unsafe work practices and then implement appropriate corrective actions for these unsafe conditions or behaviors. Three factors hypothesized to increase the propensity of an employee to actively care for safety are (1) high self-esteem, (2) optimism, and (3) group cohesiveness. *Applied & Preventive Psychology* (Winter 1995) attempted to establish empirical support for the active-caring hypothesis by fitting the model $E(y) = \beta_0 + \beta_1 x_1 + \beta_2 x_2 + \beta_3 x_3$, where

> y = active-caring score (measuring active caring on a 15-point scale)
>
> x_1 = Self-esteem score
> x_2 = Optimism score
> x_3 = Group cohesion score

The regression analysis, based on data collected for $n = 31$ hourly workers at a large fiber-manufacturing plant, yielded a multiple coefficient of determination of $R^2 = .362$.

a. Interpret the value of R^2.

b. Use the R^2 value to test the global utility of the model. Take $\alpha = .05$.

12.170 Deferred tax allowance study. A study was conducted to identify accounting choice variables that influence a manager's decision to change the level of the deferred tax asset allowance at a firm (*The Engineering Economist*, Jan./Feb. 2004). Data were collected on a sample of 329 firms that reported deferred tax assets. The dependent variable of interest (DTVA) is measured as the change in the deferred tax asset valuation allowance divided by the deferred tax asset. The independent variables used as predictors of DTVA are as follows:

LEVERAGE: x_1 = ratio of debt book value to shareholder's equity

BONUS: x_2 = 1 if firm maintains a management bonus plan, 0 if not

MVALUE: x_3 = market value of common stock

BBATH: x_4 = 1 if operating earnings are negative and lower than last year, 0 if not

EARN: x_5 = change in operating earnings divided by total assets

A first-order model was fitted to the data with the following results (*p*-values are in parantheses):

$$\hat{y} = .044 + .006x_1 - .035x_2 - .001x_3 + .296x_4, + .010x_5, R_a^2 = .280$$
$$(.070) \quad (.228) \quad (.157) \quad (.678) \quad (.001) \quad (.869)$$

a. Interpret the estimate of the β-coefficient for x_4.

b. The "Big Bath" theory proposed by the researchers states that the mean DTVA for firms with negative earnings and earnings lower than last year will exceed the mean DTVA of other firms. Is there evidence to support this theory? Test, using $\alpha = .05$.

c. Interpret the value of R_a^2.

Applying the Concepts—Intermediate

12.171 Snow geese feeding trial. Refer to the *Journal of Applied Ecology* (Vol. 32, 1995) study of the feeding habits of baby snow geese, presented Exercise 11.84 (p. 587). Data on gosling weight change, digestion efficiency, acid-detergent fiber (all measured as percentages), and diet (plants or duck chow) for 42 feeding trials are saved in the **SNOWGEESE** file. Selected observations are shown in the table (p. 716). The botanists were interested in predicting weight change (y) as a function of the other variables. Consider the first-order model $E(y) = \beta_0 + \beta_1 x_1 + B_2 x_2$ where x_1 is digestion efficiency and x_2 is acid-detergent fiber.

a. Find the least squares prediction equation for weight change y.

b. Interpret the β-estimates in the equation you found in part **a.**

c. Conduct a test to determine whether digestion efficiency, x_1, is a useful linear predictor of weight change. Use $\alpha = .01$.

d. Form a 99% confidence interval for β_2. Interpret the result.

Data for Exercise 12.171

Feeding Trial	Diet	Weight Change (%)	Digestion Efficiency (%)	Acid-Detergent Fibre (%)
1	Plants	−6	0	28.5
2	Plants	−5	2.5	27.5
3	Plants	−4.5	5	27.5
4	Plants	0	0	32.5
5	Plants	2	0	32
⋮	⋮	⋮	⋮	⋮
38	Duck Chow	9	59	8.5
39	Duck Chow	12	52.5	8
40	Duck Chow	8.5	75	6
41	Duck Chow	10.5	72.5	6.5
42	Duck Chow	14	69	7

Based on Gadallah, F. L., and Jefferies, R. L. "Forage quality in brood rearing areas of the lesser snow goose and the growth of captive goslings." *Journal of Applied Biology*, Vol. 32, No. 2, 1995, pp. 281–282 (adapted from Figures 2 and 3).

Rate	Time
1.00	0.1
0.80	0.3
0.40	0.5
0.20	0.7
0.05	0.9
0.00	1.1
−0.05	1.3
−0.02	1.5
0.00	1.7
−0.10	1.9
−0.15	2.1
−0.05	2.3
−0.13	2.5
−0.08	2.7
0.00	2.9

Based on Takizawa, K., et al. "Characteristics of C_3 radicals in high-density C_4F_8 plasmas studied by laser-induced fluorescence spectroscopy." *Journal of Applied Physics*, Vol. 88, No. 11, Dec. 1, 2000 (Figure 7).

e. Find and interpret R^2 and R_a^2. Which statistic is the preferred measure of model fit? Explain.

f. Is the overall model statistically useful in predicting weight change? Test, using $\alpha = .05$.

g. Write a first-order model relating gosling weight change (y) to digestion efficiency (x_1) and diet (plants or duck chow) that allows for different slopes for each diet.

h. Fit the model you wrote in part g to the data saved in the **SNOWGEESE** trial. Give the least squares prediction equation.

i. Refer to part g. Find the estimated slope of the line for goslings fed a diet of plants. Interpret its value.

j. Refer to part g. Find the estimated slope of the line for goslings fed a diet of duck chow. Interpret its value.

k. Refer to part g. Conduct a test to determine whether the slopes associated with the two diets are significantly different. Use $\alpha = .05$.

12.172 Optimizing semiconductor material processing. Fluorocarbon plasmas are used in the production of semiconductor materials. In the *Journal of Applied Physics* (Dec. 1, 2000), electrical engineers at Nagoya University (Japan) studied the kinetics of fluorocarbon plasmas in order to optimize material processing. In one portion of the study, the surface production rate of fluorocarbon radicals emitted from the production process was

measured at various points in time (in milliseconds) after the radio frequency power was turned off. The data are given in the table above and saved in the **RADICALS** file. Consider a model relating surface production rate (y) to time (x).

a. Graph the data in a scatterplot. What trend do you observe?

b. Fit a quadratic model to the data. Give the least squares prediction equation.

c. Is there sufficient evidence of upward curvature in the relationship between surface production rate and time after turnoff? Use $\alpha = .05$.

12.173 Socialization of graduate students. *Teaching Sociology* (July 1995) developed a model for the professional socialization of graduate students working toward a Ph.D. in sociology. One of the dependent variables modeled was professional confidence y, measured on a five-point scale. The model included over 20 independent variables and was fitted to data collected on a sample of 309 sociology graduate students. One concern was whether multicollinearity existed in the data. A matrix of Pearson product moment correlations for 10 of the independent variables is shown below. [*Note:* Each entry in the table is the correlation coefficient r between the variable in the corresponding row and column.]

a. Examine the correlation matrix, and find the independent variables that are moderately or highly correlated.

Table for Exercise 12.173

Independent Variable	(1)	(2)	(3)	(4)	(5)	(6)	(7)	(8)	(9)	(10)
(1) Father's occupation	1.000	.363	.099	−.110	−.047	−.053	−.111	.178	.078	.049
(2) Mother's education	.363	1.000	.228	−.139	−.216	.084	−.118	.192	.125	.068
(3) Race	.099	.228	1.000	.036	−.515	.014	−.120	.112	.117	.337
(4) Sex	−.110	−.139	.036	1.000	.165	−.256	.173	−.106	−.117	.073
(5) Foreign status	−.047	−.216	−.515	.165	1.000	−.041	.159	−.130	−.165	−.171
(6) Undergraduate GPA	−.053	.084	.014	−.256	−.041	1.000	.032	.028	−.034	.092
(7) Year GRE taken	−.111	−.118	−.120	.173	.159	.032	1.000	−.086	−.602	.016
(8) Verbal GRE score	.178	.192	.112	−.106	−.130	.028	−.086	1.000	.132	.087
(9) Years in graduate program	.078	.125	.117	−.117	−.165	−.034	−.602	.132	1.000	−.071
(10) First-year graduate GPA	.049	.068	.337	.073	−.171	.092	.016	.087	−.071	1.000

Based on Keith, B., and Moore, H. A. "Training sociologists: An assessment of professional socialization and the emergence of career aspirations." *Teaching Sociology*, Vol. 23, No. 3, July 1995, p. 205 (Table 1).

b. What modeling problems may occur if the variables you found in part **a** are left in the model? Explain.

12.174 Comparing mosquito repellents. Which insect repellents protect best against mosquitoes? Periodically, consumer groups conduct tests to compare insect repellents (e.g., *Consumer Reports*, June 2000). Consider a similar study of 14 popular mosquito repellents. Each product was classified as either an aerosol spray or as a lotion. The cost of the product (in dollars) was divided by the amount of the repellent needed to cover exposed areas of the skin (about 1/3 ounce) to obtain a cost-per-use value. Effectiveness was measured as the maximum number of hours of protection (in half-hour increments) provided when human testers exposed their arms to 200 mosquitoes. The simulated data are listed in the table at the bottom of the page and saved in the **REPELLENT** file.

a. Suppose you want to use repellent type to model the cost per use (y). Create the appropriate number of dummy variables for repellent type, and write the model.

b. Fit the model you wrote in part **a** to the data.

c. Give the null hypothesis for testing whether repellent type is a useful predictor of cost per use (y).

d. Conduct the test suggested in part **c**, and give the appropriate conclusion. Use $\alpha = .10$.

e. Repeat parts **a–d** if the dependent variable is maximum number of hours of protection (y).

12.175 Rating funny cartoons. Newspaper cartoons, although designed to be funny, often invoke hostility, pain, or aggression in readers, especially when those cartoons depict violence. A study was undertaken to determine how violence in cartoons is related to aggression or pain (*Motivation and Emotion*, Vol. 10, 1986). A group of volunteers (psychology students) rated each of 32 violent newspaper cartoons (16 "Herman" and 16 "Far Side" cartoons) on three dimensions:

$y = $ Funniness ($0 = $ not funny, ... , $9 = $ very funny)

$x_1 = $ Pain ($0 = $ none, ... , $9 = $ a very great deal)

$x_2 = $ Aggression/hostility ($0 = $ none, ... , $9 = $ a very great deal)

The ratings of the students on each dimension were averaged, and the resulting $n = 32$ observations were subjected to a multiple-regression analysis. On the basis of the underlying theory (called the inverted-U theory) that the funniness of a joke will increase at low levels of aggression or pain, level off, and then decrease at high levels of aggression or pain, the following quadratic models were proposed:

Model 1: $E(y) = \beta_0 + \beta_1 x_1 + \beta_2 x_1^2, R^2 = .099, F = 1.60$

Model 2: $E(y) = \beta_0 + \beta_1 x_2 + \beta_2 x_2^2, R^2 = .100, F = 1.61$

a. According to the theory, what is the expected sign of β_2 in each model?

b. Is there sufficient evidence to indicate that the quadratic model relating pain to funniness rating is useful? Test at $\alpha = .05$.

c. Is there sufficient evidence to indicate that the quadratic model relating aggression/hostility to funniness rating is useful? Test at $\alpha = .05$.

12.176 "Sun safety" study. Excessive exposure to solar radiation is known to increase the risk of developing skin cancer, yet many people do not practice "sun safety." A group of University of Arizona researchers examined the feasibility of educating preschool (four- to five-year-old) children about sun safety (*American Journal of Public Health*, July 1995). A sample of 122 preschool children was divided into two groups: the control group and the intervention group. Children in the intervention group received a *Be Sun Safe* curriculum in preschool, while the control group did not. All children were tested for their knowledge, comprehension, and application of sun safety at two points in time: prior to the sun safety curriculum (pretest, x_1) and seven weeks following the curriculum (posttest, y).

a. Write a first-order model for mean posttest score $E(y)$ as a function of pretest score x_1 and group. Assume that no interaction exists between pretest score and group.

b. For the model you wrote in part **a**, show that the slope of the line relating posttest score to pretest score is the same for both groups of children.

c. Repeat part **a**, but assume that pretest score and group interact.

d. For the model of part **c**, show that the slope of the line relating posttest score to pretest score differs for the two groups of children.

e. Assuming that interaction exists, give the reduced model for testing whether the mean posttest scores differ for the intervention and control groups.

Data for Exercise 12.174

Repellent	Type	Cost/Use ($)	Maximum Protection (hours)
1	Aerosol	0.61	6.5
2	Aerosol	0.72	3.5
3	Aerosol	0.69	6.0
4	Aerosol	0.74	7.0
5	Aerosol	0.77	1.5
6	Aerosol	2.27	14.5
7	Lotion	1.17	3.5
8	Lotion	0.86	7.5
9	Aerosol	3.25	24.5
10	Lotion	2.58	14.0
11	Aerosol	1.17	1.0
12	Lotion	1.50	2.5
13	Lotion	1.25	7.5
14	Lotion	0.96	3.5

Based on "Comparing mosquito repellents," from "Buzz off: Insect repellents: Which keep bugs at bay?" *Consumer Reports*, June 2000.

f. With sun safety knowledge as the dependent variable, the test presented in part **e** was carried out and resulted in a *p*-value of .03. Interpret this result.

g. With sun safety comprehension as the dependent variable, the test presented in part **e** was carried out and resulted in a *p*-value of .033. Interpret this result.

h. With sun safety application as the dependent variable, the test presented in part **e** was carried out and resulted in a *p*-value of .322. Interpret this result.

12.177 Soil loss during rainfall. Phosphorus used in soil fertilizers can contaminate freshwater sources during rainfall runoff. Consequently, it is important for water quality engineers to estimate the amount of dissolved phosphorus in the water. *Geoderma* (June 1995) presented an investigation of the relationship between soil loss and percentage of dissolved phosphorus in water samples collected at 20 fertilized watersheds in Oklahoma. The data are given in the accompanying table and saved in the **PHOSPHOR** file.

a. Plot the data in a scatterplot. Do you detect a linear or curvilinear trend?

b. Fit the quadratic model $E(y) = \beta_0 + \beta_1 x + \beta_2 x^2$ to the data.

c. Conduct a test to determine whether a curvilinear relationship exists between dissolved phosphorus percentage (*y*) and soil loss (*x*). Use $\alpha = .05$.

Watershed	Soil Loss *x* (kilometers per half-acre)	Dissolved Phosphorus Percentage *y*
1	18	42.3
2	17	50.2
3	35	52.7
4	16	77.1
5	14	36.8
6	54	17.5
7	153	66.4
8	81	67.5
9	183	28.9
10	284	15.1
11	767	20.1
12	148	38.3
13	649	5.6
14	479	8.6
15	1,371	5.5
16	9,150	4.6
17	15,022	2.2
18	69	77.9
19	4,392	7.8
20	312	42.9

Based on Sharpley, A. N., Robinson, J. S., and Smith, S. J. "Bioavailable phosphorus dynamics in agricultural soils and effects on water quality." *Geoderma*, Vol. 67, No. 1–2, June 1995, p. 11 (Table 4).

12.178 Habitats of grizzly bears. Do grizzly bears segregate on the basis of sex? One hypothesis is that female grizzlies avoid male-occupied habitats because of competition for food and cannibalism. A competing theory is that females do not avoid males, but simply have different habitats available to them. These hypotheses were investigated in the *Journal of Wildlife Management* (July 1995). Grizzly bears were trapped, fitted with a radio collar, and released in the Highwood trapping zone (HTZ) in Alberta, Canada. The percentage of time, *y*,

each bear used the HTZ as a habitat over a four-year period was recorded. The researchers modeled $E(y)$ as a function of reproductive class at five levels: estrous adult females, adult females with offspring, independent subadult females, adult males, and independent subadult males. One goal was to compare the mean percentage use of HTZ for the five classes of grizzly bears.

Grizzly Class	*n*	Mean Percentage of Use
Estrous adult females	5	38
Adult females with offspring	7	16
Independent subadult females	2	89
Adult males	7	43
Independent subadult males	8	58

Based on Wielgus, R. B., and Bunnell, F. L. "Tests of hypotheses for sexual segregation in grizzly bears." *Journal of Wildlife Management*, Vol. 59, No. 3, July 1995, p. 555 (Table 1).

a. Write a model for $E(y)$ that will enable the researchers to carry out the comparison.

b. The sample sizes and sample means for the five classes are shown in the table above. Use this information to find estimates of the β's in the model you wrote in part **a**.

c. Give the null hypothesis for a test to determine whether the mean percentage of use of HTZ differs among the grizzly bear classes.

d. The *p*-value for the test mentioned in part **c** was reported as .15. Interpret this result.

12.179 Sale prices of apartments. A Minneapolis, Minnesota, real-estate appraiser used regression analysis to explore the relationship between the sale prices of apartment buildings sold in Minneapolis and various characteristics of the properties. Twenty-five apartment buildings were randomly sampled from all apartment buildings that were sold during a recent year. The table on page 719 (saved in the **MNSALES** file) lists the data collected by the appraiser. [*Note:* The physical condition of each apartment building is coded E (excellent), G (good), or F (fair).]

a. Write a model that describes the relationship between sale price and number of apartment units as three parallel lines, one for each level of physical condition. Be sure to specify the dummy-variable coding scheme you use.

b. Plot *y* against x_1 (number of apartment units) for all buildings in excellent condition. On the same graph, plot *y* against x_1 for all buildings in good condition. Do this again for all buildings in fair condition. Does it appear that the model you specified in part **a** is appropriate? Explain.

c. Fit the model from part **a** to the data. Report the least squares prediction equation for each of the three building condition levels.

d. Plot the three prediction equations of part **c** on a scatterplot of the data.

e. Do the data provide sufficient evidence to conclude that the relationship between sale price and number of units varies with the physical condition of the apartments? Test, using $\alpha = .05$.

f. Check the data set for multicollinearity. How does your result affect your choice of independent variables to use in a model for sale price?

Data for Exercise 12.179

Code No.	Sale Price y ($)	No. of Apartments, x_1	Age of Structure, x_2 (years)	Lot Size x_3 (sq. ft.)	No. of On-Site Parking Spaces, x_4	Gross Building Area x_5 (sq. ft.)	Condition of Apartment Building
0229	90,300	4	82	4,635	0	4,266	F
0094	384,000	20	13	17,798	0	14,391	G
0043	157,500	5	66	5,913	0	6,615	G
0079	676,200	26	64	7,750	6	34,144	E
0134	165,000	5	55	5,150	0	6,120	G
0179	300,000	10	65	12,506	0	14,552	G
0087	108,750	4	82	7,160	0	3,040	G
0120	276,538	11	23	5,120	0	7,881	G
0246	420,000	20	18	11,745	20	12,600	G
0025	950,000	62	71	21,000	3	39,448	G
0015	560,000	26	74	11,221	0	30,000	G
0131	268,000	13	56	7,818	13	8,088	F
0172	290,000	9	76	4,900	0	11,315	E
0095	173,200	6	21	5,424	6	4,461	G
0121	323,650	11	24	11,834	8	9,000	G
0077	162,500	5	19	5,246	5	3,828	G
0060	353,500	20	62	11,223	2	13,680	F
0174	134,400	4	70	5,834	0	4,680	E
0084	187,000	8	19	9,075	0	7,392	G
0031	155,700	4	57	5,280	0	6,030	E
0019	93,600	4	82	6,864	0	3,840	F
0074	110,000	4	50	4,510	0	3,092	G
0057	573,200	14	10	11,192	0	23,704	E
0104	79,300	4	82	7,425	0	3,876	F
0024	272,000	5	82	7,500	0	9,542	E

g. Consider the first-order model $E(y) = \beta_0 + \beta_1 x_1 + \cdots + \beta_5 x_5$. Conduct a complete residual analysis for the model to check the assumptions on ε.

12.180 Physical characteristics of boys. A physiologist wishes to investigate the relationship between the physical characteristics of preadolescent boys and their maximum oxygen uptake (measured in milliliters of oxygen per kilogram of body weight). The data below (saved in the **BOYS10** file) were collected on a random sample of 10 preadolescent boys.

a. Fit the regression model

$$y = \beta_0 + \beta_1 x_1 + \beta_2 x_2 + \beta_3 x_3 + \beta_4 x_4 + \varepsilon$$

to the data, and give the least squares prediction equation.

b. It seems reasonable to assume that the greater a child's weight, the greater should be the maximum oxygen uptake. Is $\hat{\beta}_3$, the estimated coefficient of weight x_3, positive as expected? Give an explanation for this result.

c. It would seem that the chest depth of a child should be positively correlated with lung volume and hence to maximum oxygen uptake. Is $\hat{\beta}_4$ significantly different from 0, as expected? If not, explain why.

d. Calculate the correlation coefficients between all pairs of the independent variables x_1–x_4. Do these correlations provide an explanation for the confusing signs and small t-values associated with the estimated regression coefficients of the model?

*12.181 Entry-level job preferences. Benefits Quarterly (First Quarter, 1995) published a study of entry-level job preferences. A number of independent variables were used to model the job preferences (measured on a 10-point scale) of 164 business school graduates. Suppose stepwise regression is used to build a model for job preference score (y) as a function of the following independent variables:

$$x_1 = \begin{cases} 1 & \text{if flextime position} \\ 0 & \text{if not} \end{cases}$$

Maximum Oxygen Uptake y	Age x_1 (years)	Height x_2 (centimeters)	Weight x_3 (kilograms)	Chest Depth x_4 (centimeters)
1.54	8.4	132.0	29.1	14.4
1.74	8.7	135.5	29.7	14.5
1.32	8.9	127.7	28.4	14.0
1.50	9.9	131.1	28.8	14.2
1.46	9.0	130.0	25.9	13.6
1.35	7.7	127.6	27.6	13.9
1.53	7.3	129.9	29.0	14.0
1.71	9.9	138.1	33.6	14.6
1.27	9.3	126.6	27.7	13.9
1.50	8.1	131.8	30.8	14.5

$$x_2 = \begin{cases} 1 & \text{if day care support required} \\ 0 & \text{if not} \end{cases}$$

$$x_3 = \begin{cases} 1 & \text{if spousal transfer support required} \\ 0 & \text{if not} \end{cases}$$

x_4 = Number of sick days allowed

$$x_5 = \begin{cases} 1 & \text{if applicant married} \\ 0 & \text{if not} \end{cases}$$

x_6 = Number of children of applicant

$$x_7 = \begin{cases} 1 & \text{if male applicant} \\ 0 & \text{if female applicant} \end{cases}$$

a. How many models are fitted to the data in step 1? Give the general form of these models.

b. How many models are fitted to the data in step 2? Give the general form of these models.

c. How many models are fitted to the data in step 3? Give the general form of these models.

d. Explain how the procedure determines when to stop adding independent variables to the model.

e. Describe two major drawbacks to using the final stepwise model as the "best" model for job preference score (y).

12.182 **Characteristics of sea-ice melt ponds.** Surface albedo is defined as the ratio of solar energy directed upward from a surface over energy incident upon the surface. Surface albedo is a critical climatological parameter of sea ice. The National Snow and Ice Data Center (NSIDC) collects data on the albedo, depth, and physical characteristics of ice-melt ponds in the Canadian Arctic. Data on 504 ice-melt ponds located in the Barrow Strait in the Canadian Arctic are saved in the **PONDICE** file. Environmental engineers want to examine the relationship between the broadband surface albedo level y of the ice and the pond depth x (in meters).

a. Construct a scatterplot of the **PONDICE** data. On the basis of the scatterplot, hypothesize a model for $E(y)$ as a function of x.

b. Fit the model you hypothesized in part **a** to the data in the **PONDICE** file. Give the least squares prediction equation.

c. Conduct a test of the overall adequacy of the model. Use $\alpha = .01$.

d. Conduct tests (at $\alpha = .01$) on any important β parameters in the model.

e. Find and interpret the values of adjusted R^2 and s.

f. Do you detect any outliers in the data? Explain.

Applying the Concepts—Advanced

12.183 **Abundance of bird species.** Multiple-regression analysis was used to model the abundance y of an individual bird species in transects in the United Kingdom (*Journal of Applied Ecology*, Vol. 32, 1995). Three of the independent variables used in the model, all field boundary attributes, are

1. Transect location (small pasture field, small arable field, or large arable field)

2. Land use (pasture or arable) adjacent to the transect

3. Total number of trees in the transect

a. Identify each of the independent variables as a quantitative or qualitative variable.

b. Write a first-order model for $E(y)$ as a function of the total number of trees.

c. Add main-effect terms for transect location to the model you wrote in part **b.** Graph the hypothesized relationships of the new model.

d. Add main-effect terms for land use to the model you came up with in part **c.** In terms of the β's of the new model, what is the slope of the relationship between $E(y)$ and number of trees for any combination of transect location and land use?

e. Add terms for interaction between transect location and land use to the model you arrived at in part **d.** Do these interaction terms affect the slope of the relationship between $E(y)$ and number of trees? Explain.

f. Add terms for interaction between number of trees and all coded dummy variables to the model you formulated in part **e.** In terms of the β's of the new model, give the slope of the relationship between $E(y)$ and number of trees for each combination of transect location and land use.

Critical Thinking Challenges

12.184 **IQ and *The Bell Curve*.** In Exercise 5.153 (p. 267), we introduced *The Bell Curve* (New York: Free Press, 1994) by Richard Herrnstein and Charles Murray (H&M), a controversial book about race, genes, IQ, and economic mobility. The book heavily employs statistics and statistical methodology in an attempt to support the authors' positions on the relationships among these variables and their social consequences. The main theme of *The Bell Curve* can be summarized as follows:

1. Measured intelligence (IQ) is largely genetically inherited.

2. IQ is correlated positively with a variety of socioeconomic status success measures, such as a prestigious job, a high annual income, and high educational attainment.

3. From 1 and 2, it follows that socioeconomic successes are largely genetically caused and therefore resistant to educational and environmental interventions (such as affirmative action).

The statistical methodology (regression) employed by the authors and the inferences derived from the statistics were critiqued in *Chance* (Summer 1995) and *The Journal of the American Statistical Association* (Dec. 1995). The following are just a few of the problems with H&M's use of regression that have been identified:

Problem 1 H&M consistently use a trio of independent variables—IQ, socioeconomic status, and age—in a series of first-order models designed to predict dependent social outcome variables such as income and unemployment. (Only on a single occasion are interaction terms incorporated.) Consider, for example, the model

$$E(y) = \beta_0 + \beta_1 x_1 + \beta_2 x_2 + \beta_3 x_3$$

where y = income, x_1 = IQ, x_2 = socioeconomic status, and x_3 = age. H&M utilize t-tests on the individual β parameters to assess the importance of the independent variables. As with most of the models considered in *The Bell Curve*, the estimate of β_1 in the income model is positive and statistically significant at $\alpha = .05$, and the associated t-value is larger (in absolute value) than the t-values associated with the other independent variables.

Consequently, *H&M claim that IQ is a better predictor of income than the other two independent variables.* No attempt was made to determine whether the model was properly specified or whether the model provides an adequate fit to the data.

Problem 2   In an appendix, the authors describe multiple regression as a "mathematical procedure that yields coefficients for each of [the independent variables], indicating how much of a change in [the dependent variable] can be anticipated for a given change in any particular [independent] variable, with all the others held constant." Armed with this information and the fact that the estimate of β_1 in the model just described is positive, *H&M infer that a high IQ necessarily implies (or causes) a high income, and a low IQ inevitably leads to a low income.* (Cause-and-effect inferences like this are made repeatedly throughout the book.)

Problem 3   The title of the book refers to the normal distribution and its well-known "bell-shaped" curve. There is a misconception among the general public that scores on intelligence tests (IQS) are normally distributed. In fact, most IQ scores have distributions that are decidedly skewed. Traditionally, psychologists and psychometricians have transformed these scores so that the resulting numbers have a precise normal distribution. H&M make a special point to do this. Consequently, *the measure of IQ used in all the regression models is normalized (i.e., transformed so that the resulting distribution is normal), despite the fact that regression methodology does not require predictor (independent) variables to be normally distributed.*

Problem 4   A variable that is not used as a predictor of social outcome in any of the models in *The Bell Curve* is level of education. H&M purposely omit education from the models, arguing that IQ causes education, not the other way around. Other researchers who have examined H&M's data report that *when education is included as an independent variable in the model, the effect of IQ on the dependent variable (say, income) is diminished.*

a. Comment on each of the problems identified. Why do these problems cast a shadow on the inferences made by the authors?

b. Using the variables specified in the model presented, describe how you would conduct the multiple-regression analysis. (Propose a more complicated model and describe the appropriate model tests, including a residual analysis.)

12.185 FLAG study of bid collusion. Road construction contracts in the state of Florida are awarded on the basis of competitive, sealed bids; the contractor who bids the lowest price wins the contract. During the 1980s, the Office of the Florida Attorney General (FLAG) suspected numerous contractors of practicing bid collusion (i.e., setting the winning bid price above the fair, or competitive, price in order to increase their own profit margin). By comparing the prices bid (and other important bid variables) of the fixed (i.e., rigged) contracts with the competitively bid contracts, FLAG was able to establish invaluable benchmarks for detecting future bid rigging. FLAG collected data on 279 road construction contracts. For each contract, the following variables were measured (the data are saved in the **FLAG** file):

1. Price of contract ($) bid by lowest bidder
2. Department of Transportation (DOT) engineer's estimate of fair contract price ($)
3. Ratio of low (winning) bid price to DOT engineer's estimate of fair price
4. Status of contract (1 if fixed, 0 if competitive)
5. District (1, 2, 3, 4, or 5) in which construction project is located
6. Number of bidders on contract
7. Estimated number of days required to complete work
8. Length of road project (miles)
9. Percentage of costs allocated to liquid asphalt
10. Percentage of costs allocated to base material
11. Percentage of costs allocated to excavation
12. Percentage of costs allocated to mobilization
13. Percentage of costs allocated to structures
14. Percentage of costs allocated to traffic control
15. Subcontractor utilization (1 if yes, 0 if no)

Use the methodology of this chapter to build a model for low-bid contract price (y). Comment on how the status of the bid affects the price.

Activity   Collecting Data and Fitting a Multiple-Regression Model

Note: The use of statistical software is required for this project. This is a continuation of the Activity section in Chapter 11, in which you selected three independent variables as predictors of a dependent variable of your choice and obtained at least 10 data values. Now, by means of an available software package, fit the multiple-regression model

$$y = \beta_0 + \beta_1 x_1 + \beta_2 x_2 + \beta_3 x_3 + \varepsilon$$

where

y = Dependent variable you chose

x_1 = First independent variable you chose

x_2 = Second independent variable you chose

x_3 = Third independent variable you chose

a. Compare the coefficients $\hat{\beta}_1$, $\hat{\beta}_2$, and $\hat{\beta}_3$ with their corresponding slope coefficients in the Activity of Chapter 11, where you fit three separate straight-line models. How do you account for the differences?

b. Calculate the coefficient of determination, R^2, and conduct the F-test of the null hypothesis $H_0: \beta_1 = \beta_2 = \beta_3 = 0$. What is your conclusion?

c. Check the data for multicollinearity. If multicollinearity exists, how should you proceed?

d. Now increase your list of 3 variables to include approximately 10 that you think would be useful in predicting the dependent variable. With the aid of statistical software, employ a stepwise regression program to choose the important variables among those you have listed. To test your intuition, list the variables in the order you think they will be selected before you conduct the analysis. How does your list compare with the stepwise regression results?

e. After the group of 10 variables has been narrowed to a smaller group of variables by the stepwise analysis, try to improve the model by including interactions and quadratic terms. Be sure to consider the meaning of each interaction or quadratic term

before adding it to the model. (A quick sketch can be very helpful.) See if you can systematically construct a useful model for prediction. If you have a large data set, you might want to hold out the last observations to test the predictive ability of your model after it is constructed. (As noted in Section 12.10, using the same data to construct *and* to evaluate predictive ability can lead to invalid statistical tests and a false sense of security.)

References

Barnett, V., and Lewis, T. *Outliers in Statistical Data*. New York: Wiley, 1978.

Belsley, D. A., Kuh, E., and Welsch, R. E. *Regression Diagnostics: Identifying Influential Data and Sources of Collinearity*. New York: Wiley, 1980.

Chatterjee, S., and Price, B. *Regression Analysis by Example*, 2nd ed. New York: Wiley, 1991.

Draper, N., and Smith, H. *Applied Regression Analysis*, 2nd ed. New York: Wiley, 1981.

Graybill, F. *Theory and Application of the Linear Model*. North Scituate, MA: Duxbury, 1976.

Kelting, H. "Investigation of condominium sale prices in three market scenarios: Utility of stepwise, interactive, multiple regression

analysis and implications for design and appraisal methodology." Unpublished paper, University of Florida, Gainesville, FL, 1979.

Kutner, M., Nachtsheim, C., Neter, J., and Li, W. *Applied Linear Statistical Models*, 5th ed. New York: McGraw-Hill/Irwin, 2005.

Mendenhall, W. *Introduction to Linear Models and the Design and Analysis of Experiments*. Belmont, CA: Wadsworth, 1968.

Mendenhall, W., and Sincich, T. *A Second Course in Statistics: Regression Analysis*, 7th ed. Upper Saddle River, NJ: Prentice Hall, 2011.

Mosteller, F., and Tukey, J. W. *Data Analysis and Regression: A Second Course in Statistics*. Reading, MA: Addison-Wesley, 1977.

Rousseeuw, P. J., and Leroy, A. M. *Robust Regression and Outlier Detection*. New York: Wiley, 1987.

Weisberg, S. *Applied Linear Regression*, 2nd ed. New York: Wiley, 1985.

USING TECHNOLOGY

MINITAB: Multiple Regression

Multiple Regression

Step 1 Access the MINITAB worksheet file that contains the dependent and independent variables.

Step 2 Click on the "Stat" button on the MINITAB menu bar, and then click on "Regression" and "Regression" again, as shown in Figure 12.M.1.

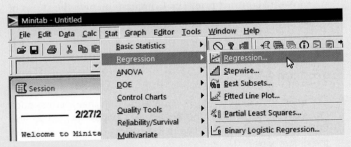

Figure 12.M.1
MINITAB menu options for regression

Step 3 The resulting dialog box appears as shown in Figure 12.M.2 Specify the dependent variable in the "Response" box and the independent variables in the "Predictors" box. [*Note:* If your model includes interaction and/or squared terms, you must create and add these higher-order variables to the MINITAB worksheet *prior* to running a regression analysis. You can do this by clicking the "Calc" button on the MINITAB main menu and selecting the "Calculator" option.]

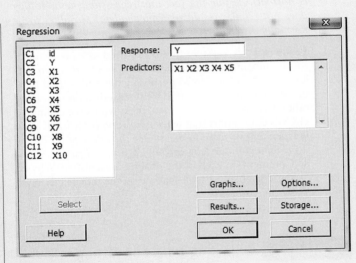

Figure 12.M.2
MINITAB regression dialog box

Step 4 To produce prediction intervals for *y* and confidence intervals for $E(y)$, click the "Options" button and select the appropriate menu items in the resulting menu list. (See Figure 12.M.3.)

Step 5 Residual plots are obtained by clicking the "Graphs" button and making the appropriate selections on the resulting menu. (See Figure 12.M.4.)

Step 6 To return to the main Regression dialog box from any of these optional screens, click "OK."

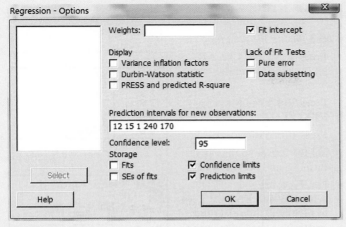

Figure 12.M.3
MINITAB regression options

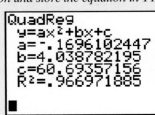

Figure 12.M.4
MINITAB regression graphs options

Step 7 When you have made all your selections, click "OK" on the main Regression dialog box to produce the MINITAB multiple-regression printout.

Stepwise Regression

Step 1 Click on the "Stat" button on the main menu bar; then click on "Regression," and click on "Stepwise." (See Figure 12.M.1.) The resulting dialog box appears like the one in Figure 12.M.2.

Step 2 Specify the dependent variable in the "Response" box and the independent variables in the stepwise model in the "Predictors" box.

Step 3 As an option, you can select the value of α to use in the analysis by clicking on the "Methods" button and specifying the value. (The default is $\alpha = .15$.)

Step 4 Click "OK" to view the stepwise regression results.

TI-83/TI-84 Plus Graphing Calculator: Multiple Regression

Note: Only simple linear and quadratic regression models can be fit using the TI-83/ TI-84 plus graphing calculator.

Quadratic Regression
I. Finding the Quadratic Regression Equation

Step 1 *Enter the data*

- Press **STAT** and select **1:Edit**

  [*Note:* If the list already contains data, clear the old data. Use the up arrow to highlight "**L1**" or "**L2**."]

- Press **CLEAR ENTER**

- Use the **ARROW** and **ENTER** keys to enter the data set into **L1** and **L2**

Step 2 *Find the quadratic regression equation*

- Press **STAT** and highlight **CALC**

- Press **5** for **QuadReg**

- Press **ENTER**

- The screen will show the values for a, b, and c in the equation

- If the diagnostics are on, the screen will also give the value for r^2

- To turn the diagnostics feature on:

- Press **2nd 0** for **CATALOG**

- Press the **ALPHA** key and x^{-1} for **D**

- Press the down **ARROW** until **DiagnosticsOn** is highlighted

- Press **ENTER** twice

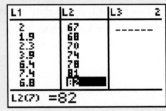

II. Graphing the Quadratic Curve with the Scatterplot

Step 1 *Enter the data as shown in part I above*

Step 2 *Set up the data plot*

- Press **Y =** and **CLEAR** all functions from the Y registers

- Press **2nd Y =** for **STAT PLOT**

- Press **1** for **Plot**

- Set the cursor so that **ON** is flashing, and press **ENTER**

- For **Type**, use the **ARROW** and **ENTER** keys to highlight and select the scatterplot (first icon in the first row)

- For **Xlist,** choose the column containing the *x*-data

- For **Ylist,** choose the column containing the *y*-data

Step 3 *Find the regression equation and store the equation in Y1*

- Press **STAT** and highlight **CALC**

- Press **5** for **Quad Reg**

  (*Note:* Don't press ENTER here, because you want to store the regression equation in Y1.)

- Press **VARS**

- Use the right arrow to highlight **Y-VARS**
- Press **ENTER** to select **1:Function**
- Press **ENTER** to select **1:Y1**
- Press **ENTER**

Step 4 *View the scatterplot and regression line*

- Press **ZOOM** and then press **9** to select **9:ZoomStat**

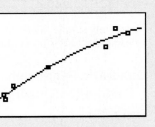

III. Plotting Residuals

When computing a regression equation on the graphing calculator, the residuals are automatically computed and saved to a list called **RESID. RESID** can be found under the **LIST menu (2nd STAT).**

Step 1 *Enter the data*

- Press **STAT** and select **1:Edit**

[*Note:* If the list already contains data, clear the old data. Use the up arrow to highlight "**L1**" or "**L2**."]

- Press **CLEAR ENTER**
- Use the **ARROW** and **ENTER** keys to enter the data set into **L1** and **L2**

Step 2 *Compute the regression equation*

- Press **STAT** and highlight **CALC**
- Press **4** for **LinReg(ax + b)**

- Press **ENTER**

Step 3 *Set up the data plot*

- Press **Y =** and **CLEAR** all functions from the Y registers
- Press **2nd Y =** for **STATPLOT**
- Press **1** for **Plot1**
- Set the cursor so that **ON** is flashing, and press **ENTER**

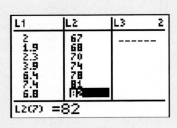

- For **Type**, use the **ARROW** and **ENTER** keys to highlight and select the scatterplot (first icon in the first row)
- Move the cursor to **Xlist** and choose the column containing the *x*-data
- Move the cursor to **Ylist**
- Press **2nd STAT** for **LIST**
- Use the down arrow to highlight the listname **RESID** and press **ENTER**

Step 4 *View the scatterplot of the residuals*

- Press **ZOOM 9** for **ZoomStat**

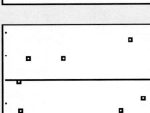

13 Categorical Data Analysis

CONTENTS

Where We've Been

- Presented methods for making inferences about the population proportion associated with a two-level qualitative variable (i.e., a binomial variable)
- Presented methods for making inferences about the difference between two binomial proportions

Where We're Going

- Discuss qualitative (i.e., categorical) data with more than two outcomes (13.1)
- Present a *chi-square* hypothesis test for comparing the category proportions associated with a single qualitative variable—called a *one-way analysis* (13.2)
- Present a chi-square hypothesis test for relating two qualitative variables—called a *two-way analysis* (13.3)
- Caution about the misuse of chi-square tests (13.4)

Statistics IN Action College Students and Alcohol: Is Amount Consumed Related to Drinking Frequency?

Traditionally, a common social activity on American college campuses is drinking alcohol. Despite laws on underage drinking, fraternities, sororities, and other campus groups often have alcohol available at their weekend parties. For some students, this activity leads to binge drinking and excessive alcohol use, often resulting in academic failure, physical violence, accidental injury, and even death. In fact, the Journal of Studies on Alcohol (Vol. 63, 2002) recently reported that about 1,400 alcohol-related deaths occur each year on American college campuses.

To gain insight into the alcohol consumption behavior of college students, professors Soyeon Shim (University of Arizona) and Jennifer Maggs (Pennsylvania State University) designed a study and reported their results in *Family and Consumer Sciences Research Journal* (Mar. 2005). Among the researchers' main objectives were (1) to segment college students on the basis of their rates of alcohol consumption and (2) to establish a statistical link between the frequency of drinking and the amount of alcohol consumed. They collected survey data from undergraduate students enrolled in a variety of courses at the University of Arizona, a large state university in the Southwest. To increase the likelihood of obtaining a representative sample, the researchers balanced the sample with both lower and upper division students, as well as students with majors in the social sciences, humanities, business, engineering, and the natural sciences. A total of 657 students completed usable surveys.

The survey consisted of a six-page booklet that took approximately 10 minutes to complete. Two of the many questions on the survey (and the subject of this *Statistics in Action*) pertained to the frequency with which the student drank alcohol (beer, wine, or liquor) during the previous one-month period and the average number of drinks the student consumed per occasion. From this information, the researchers categorized students according to Type of drinker. Responses for the three variables of interest were classified qualitatively as shown in Table SIA13.1. The data for the 657 students are saved in the **COLLDRINKS** file.

In an attempt to help the researchers achieve their objectives, we apply the statistical methodology presented in this chapter to this data set in two *Statistics in Action Revisited* examples.

Statistics IN Action Revisited

- Testing Category Proportions for Type of College Drinker (p. 732)

- Testing whether Frequency of Drinking Is Related to Amount of Alcohol Consumed (p. 744)

Table SIA13.1	Qualitative Variables Measured in the Drinking Study
Variable Name	Levels (possible values)
AMOUNT	None, 1 drink, 2–3 drinks, 4–6 drinks, 7–9 drinks, 10 or more drinks
FREQUENCY	None, Once a month, Once or twice per week, More
TYPE	Non/Seldom, Social, Typical binge, Heavy binge

Data Set: COLLDRINKS

13.1 Categorical Data and the Multinomial Experiment

Recall from Section 1.4 (p. 9) that observations on a qualitative variable can only be categorized. For example, consider the highest level of education attained by a professional hockey player. Level of education is a qualitative variable with several categories, including some high school, high school diploma, some college, college undergraduate degree, and graduate degree. If we were to record education level for all professional hockey players, the result of the categorization would be a count of the numbers of players falling into the respective categories.

When the qualitative variable of interest results in one of two responses (e.g., yes or no, success or failure, favor or do not favor), the data—called *counts*—can be analyzed with the binomial probability distribution discussed in Section 4.4. However, qualitative variables, such as level of education, that allow for more than two categories for a response are much more common, and these must be analyzed by a different method.

Qualitative data with more than two levels often result from a **multinomial experiment**. The characteristics for a multinomial experiment with k outcomes are described in the next box. You can see that the binomial experiment of Chapter 4 is a multinomial experiment with $k = 2$.

Properties of the Multinomial Experiment

1. The experiment consists of n identical trials.
2. There are k possible outcomes to each trial. These outcomes are sometimes called **classes, categories,** or **cells.**
3. The probabilities of the k outcomes, denoted by $p_1, p_2, \ldots, p_k$, where $p_1 + p_2 + \cdots + p_k = 1$, remain the same from trial to trial.
4. The trials are independent.
5. The random variables of interest are the **cell counts** $n_1, n_2, \ldots, n_k$ of the number of observations that fall into each of the k categories.

Example 13.1

Identifying a Multinomial Experiment

Problem Consider the problem of determining the highest level of education attained by each of a sample of $n = 40$ National Hockey League (NHL) players. Suppose we categorize level of education into one of five categories—some high school, high school diploma, some college, college undergraduate degree, and graduate degree—and count the number of the 40 players that fall into each category. Is this a multinomial experiment, to a reasonable degree of approximation?

Solution Checking the five properties of a multinomial experiment shown in the box, we have the following:

1. The experiment consists of $n = 40$ identical trials, each of which is undertaken to determine the education level of an NHL player.
2. There are $k = 5$ possible outcomes to each trial, corresponding to the five education-level responses.
3. The probabilities of the $k = 5$ outcomes p_1, p_2, p_3, p_4, and p_5, where p_i represents the true probability that an NHL player attains level-of-education category i, remain the same from trial to trial (to a reasonable degree of approximation).
4. The trials are independent; that is, the education level attained by one NHL player does not affect the level attained by any other player.
5. We are interested in the count of the number of hockey players who fall into each of the five education-level categories. These five cell counts are denoted n_1, n_2, n_3, n_4, and n_5.

Thus, the properties of a multinomial experiment are satisfied.

In this chapter, we are concerned with the analysis of categorical data—specifically, the data that represent the counts for each category of a multinomial experiment. In Section 13.2, we learn how to make inferences about the probabilities of categories for data classified according to a single qualitative (or categorical) variable. Then, in Sections 13.3 and 13.4 we consider inferences about categorical probabilities for data classified according to two qualitative variables. The statistic used for these inferences is one that possesses, approximately, the familiar chi-square distribution.

13.2 Testing Categorical Probabilities: One–Way Table

In this section, we consider a multinomial experiment with k outcomes that correspond to the categories of a *single* qualitative variable. The results of such an experiment are summarized in a **one-way table.** The term *one-way* is used because only one variable is classified. Typically, we want to make inferences about the true percentages that occur in the k categories on the basis of the sample information in the one-way table.

To illustrate, suppose three political candidates are running for the same elective position. Prior to the election, we conduct a survey to determine the voting preferences of a random sample of 150 eligible voters. The qualitative variable of interest is *preferred candidate*, which has three possible outcomes: candidate 1, candidate 2, and candidate 3. Suppose the number of voters preferring each candidate is tabulated and the resulting count data appear as in Table 13.1.

Table 13.1	**Results of Voter Preference Survey**	
	Candidate	
1	**2**	**3**
61 votes	53 votes	36 votes

Note that our voter preference survey satisfies the properties of a multinomial experiment for the qualitative variable, preferred candidate. The experiment consists of randomly sampling $n = 150$ voters from a large population of voters containing an unknown proportion p_1 that favors candidate 1, a proportion p_2 that favors candidate 2, and a proportion p_3 that favors candidate 3. Each voter sampled represents a single trial that can result in one of three outcomes: The voter will favor candidate 1, 2, or 3 with probabilities $p_1, p_2,$ and p_3, respectively. (Assume that all voters will have a preference.) The voting preference of any single voter in the sample does not affect the preference of any other; consequently, the trials are independent. Finally, you can see that the recorded data are the numbers of voters in each of the three preference categories. Thus, the voter preference survey satisfies the five properties of a multinomial experiment.

In this survey, and in most practical applications of the multinomial experiment, the k outcome probabilities $p_1, p_2, \ldots, p_k$ are unknown and we want to use the survey data to make inferences about their values. The unknown probabilities in the voter preference survey are

$$p_1 = \text{Proportion of all voters who favor candidate 1}$$
$$p_2 = \text{Proportion of all voters who favor candidate 2}$$
$$p_3 = \text{Proportion of all voters who favor candidate 3}$$

To decide whether the voters, in total, have a preference for any one of the candidates, we will test the null hypothesis that the candidates are equally preferred (i.e., $p_1 = p_2 = p_3 = \frac{1}{3}$) against the alternative hypothesis that one candidate is preferred (i.e., at least one of the probabilities $p_1, p_2,$ and p_3 exceeds $\frac{1}{3}$). Thus, we want to test

$H_0: p_1 = p_2 = p_3 = \frac{1}{3}$ (no preference)

$H_a:$ At least one of the proportions exceeds $\frac{1}{3}$ (a preference exists)

If the null hypothesis is true and $p_1 = p_2 = p_3 = \frac{1}{3}$, then the expected value (mean value) of the number of voters who prefer candidate 1 is given by

$$E_1 = np_1 = (n)\tfrac{1}{3} = (150)\tfrac{1}{3} = 50$$

Similarly, $E_2 = E_3 = 50$ if the null hypothesis is true and no preference exists.

The **chi-square test** measures the degree of disagreement between the data and the null hypothesis:

$$\chi^2 = \frac{[n_1 - E_1]^2}{E_1} + \frac{[n_2 - E_2]^2}{E_2} + \frac{[n_3 - E_3]^2}{E_3}$$
$$= \frac{(n_1 - 50)^2}{50} + \frac{(n_2 - 50)^2}{50} + \frac{(n_3 - 50)^2}{50}$$

Note that the farther the observed numbers $n_1, n_2,$ and n_3 are from their expected value (50), the larger χ^2 will become. That is, large values of χ^2 imply that the null hypothesis is false.

We have to know the distribution of χ^2 in repeated sampling before we can decide whether the data indicate that a preference exists. When H_0 is true, χ^2 can be shown to have (approximately) the familiar chi-square distribution of Section 8.7. For this one-way classification, the χ^2 distribution has $(k - 1)$ degrees of freedom.* The rejection region for the voter preference survey for $\alpha = .05$ and $k - 1 = 3 - 1 = 2$ df is

Rejection region: $\chi^2 > \chi^2_{.05}$

This value of $\chi^2_{.05}$ (found in Table VII) is 5.99147. (See Figure 13.1.) The computed value of the test statistic is

$$\chi^2 = \frac{(n_1 - 50)^2}{50} + \frac{(n_2 - 50)^2}{50} + \frac{(n_3 - 50)^2}{50}$$
$$= \frac{(61 - 50)^2}{50} + \frac{(53 - 50)^2}{50} + \frac{(36 - 50)^2}{50} = 6.52$$

Since the computed $\chi^2 = 6.52$ exceeds the critical value of 5.99147, we conclude at the $\alpha = .05$ level of significance that there does exist a voter preference for one or more of the candidates.

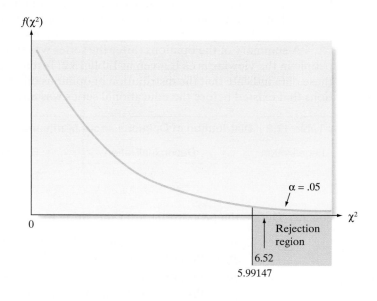

Figure 13.1

Rejection region for voter preference survey

Now that we have evidence to indicate that the proportions $p_1, p_2,$ and p_3 are unequal, we can use the methods of Section 7.4 to make inferences concerning their individual values. [*Note:* We cannot use the methods of Section 9.4 to compare two proportions, because the cell counts are dependent random variables.] The general

---

*The derivation of the number of degrees of freedom for χ^2 involves the number of linear restrictions imposed on the count data. In the present case, the only constraint is that $\Sigma n_i = n$, where n (the sample size) is fixed in advance. Therefore, df $= k - 1$. For other cases, we will give the number of degrees of freedom for each usage of χ^2 and refer the interested reader to the references for more detail.

form for a test of hypothesis concerning multinomial probabilities is shown in the following box:

A Test of a Hypothesis about Multinomial Probabilities: One-Way Table

H_0: $p_1 = p_{1,0}$, $p_2 = p_{2,0}, \ldots, p_k = p_{k,0}$

where $p_{1,0}, p_{2,0}, \ldots, p_{k,0}$ represent the hypothesized values of the multinomial probabilities

H_a: At least one of the multinomial probabilities does not equal its hypothesized value

Test statistic: $\chi^2 = \sum \dfrac{[n_i - E_i]^2}{E_i}$

where $E_i = np_{i,0}$ is the **expected cell count**—that is, the expected number of outcomes of type i, assuming that H_0 is true. The total sample size is n.

Rejection region: $\chi^2 > \chi_\alpha^2$, where χ_α^2 has $(k - 1)$ df

---

Conditions Required for a Valid χ^2 Test: One-Way Table

1. A multinomial experiment has been conducted. This is generally satisfied by taking a random sample from the population of interest.

2. The sample size n will be large enough so that, for every cell, the expected cell count $E(n_i)$ will be equal to 5 or more.*

Example 13.2
A One-Way χ^2 Test—Effectiveness of a TV Program on Marijuana

Problem Suppose an educational television station has broadcast a series of programs on the physiological and psychological effects of smoking marijuana. Now that the series is finished, the station wants to see whether the citizens within the viewing area have changed their minds about how the possession of marijuana should be considered legally. Before the series was shown, it was determined that 7% of the citizens favored legalization, 18% favored decriminalization, 65% favored the existing law (an offender could be fined or imprisoned), and 10% had no opinion.

A summary of the opinions (after the series was shown) of a random sample of 500 people in the viewing area is given in Table 13.2. Test at the $\alpha = .01$ level to see whether these data indicate that the distribution of opinions differs significantly from the proportions that existed before the educational series was aired.

Table 13.2   Distribution of Opinions about Marijuana Possession

Legalization	Decriminalization	Existing Laws	No Opinion
39	99	336	26

Data Set: MARIJUANA

Solution Define the proportions after the airing to be

p_1 = Proportion of citizens favoring legalization
p_2 = Proportion of citizens favoring decriminalization
p_3 = Proportion of citizens favoring existing laws
p_4 = Proportion of citizens with no opinion

Then the null hypothesis representing no change in the distribution of percentages is

$$H_0: p_1 = .07, p_2 = .18, p_3 = .65, p_4 = .10$$

and the alternative is

H_a: At least one of the proportions differs from its null hypothesized value

---

*The assumption that all expected cell counts are at least 5 is necessary in order to ensure that the χ^2 approximation is appropriate. Exact methods for conducting the test of hypothesis exist and may be used for small expected cell counts, but these methods are beyond the scope of this text.

Thus, we have

$$\text{Test statistic: } \chi^2 = \sum \frac{[n_i - E_i]^2}{E_i}$$

where

$$E_1 = np_{1,0} = 500(.07) = 35$$
$$E_2 = np_{2,0} = 500(.18) = 90$$
$$E_3 = np_{3,0} = 500(.65) = 325$$
$$E_4 = np_{4,0} = 500(.10) = 50$$

Since all these values are larger than 5, the χ^2 approximation is appropriate. Also, if the citizens in the sample were randomly selected, then the properties of the multinomial probability distribution are satisfied.

Rejection region: For $\alpha = .01$ and df $= k - 1 = 3$, reject H_0 if $\chi^2 > \chi^2_{.01}$, where (from Table VII in Appendix A) $\chi^2_{.01} = 11.3449$.

We now calculate the test statistic:

$$\chi^2 = \frac{(39 - 35)^2}{35} + \frac{(99 - 90)^2}{90} + \frac{(336 - 325)^2}{325} + \frac{(26 - 50)^2}{50} = 13.249$$

Since this value exceeds the table value of χ^2 (11.3449), the data provide sufficient evidence ($\alpha = .01$) that the opinions on the legalization of marijuana have changed since the series was aired.

The χ^2 test can also be conducted with the use of an available statistical software package. Figure 13.2 is an SPSS printout of the analysis of the data in Table 13.2. The test statistic and p-value of the test are highlighted on the printout. Since $\alpha = .01$ exceeds $p = .004$, there is sufficient evidence to reject H_0.

OPINION

	Observed N	Expected N	Residual
LEGAL	39	35.0	4.0
DECRIM	99	90.0	9.0
EXISTLAW	336	325.0	11.0
NONE	26	50.0	-24.0
Total	500		

Test Statistics

	OPINION
Chi-Square[a]	13.249
df	3
Asymp. Sig.	.004

a. 0 cells (.0%) have expected frequencies less than 5. The minimum expected cell frequency is 35.0.

Figure 13.2

SPSS analysis of data in Table 13.2

Look Back If the conclusion for the χ^2 test is "fail to reject H_0," then there is insufficient evidence to conclude that the distribution of opinions differs from the proportions stated in H_0. Be careful not to "accept H_0" and conclude that $p_1 = .07$, $p_2 = .18$, $p_3 = .65$, and $p_4 = .10$. The probability (β) of a Type II error is unknown.

Now Work Exercise 13.9

If we focus on one particular outcome of a multinomial experiment, we can use the methods developed in Section 7.4 for a binomial proportion to establish a confidence interval for any one of the multinomial probabilities.* For example, if we want a 95%

*Note that focusing on one outcome has the effect of lumping the other $(k - 1)$ outcomes into a single group. Thus, we obtain, in effect, two outcomes—or a binomial experiment.

confidence interval for the proportion of citizens in the viewing area who have no opinion about the issue, we calculate

$$\hat{p}_4 \pm 1.96\sigma_{\hat{p}_4}$$

where

$$\hat{p}_4 = \frac{n_4}{n} = \frac{26}{500} = .052 \quad \text{and} \quad \sigma_{\hat{p}_4} \approx \sqrt{\frac{\hat{p}_4(1 - \hat{p}_4)}{n}}$$

Thus, we get

$$.052 \pm 1.96\sqrt{\frac{(.052)(.948)}{500}} = .052 \pm .019$$

or (.033, .071). Consequently, we estimate that between 3.3% and 7.1% of the citizens now have no opinion on the issue of the legalization of marijuana. The series of programs may have helped citizens who formerly had no opinion on the issue to form an opinion, since it appears that the proportion of "no opinions" is now less than 10%.

Statistics IN Action | Revisited | Testing Category Proportions for Type of College Drinker

In the *Family and Consumer Sciences Research Journal* (Mar. 2005) study of college students and drinking (p. 726), one of the researchers' main objectives was to segment college students according to their rates of alcohol consumption. A segmentation was developed on the basis of the students' responses to the questions on frequency of drinking and average number of drinks per occasion. Four types, or groups, of college drinkers emerged: non/seldom drinkers, social drinkers, typical binge drinkers, and heavy binge drinkers. What are the proportions of students in each of these groups, and are these proportions statistically different?

To answer these questions, we used SPSS to analyze the type-of-drinker variable in the **COLLDRINKS** file. Figure SIA13.1 shows summary statistics and a graph describing

TYPE

		Frequency	Percent	Valid Percent	Cumulative Percent
Valid	Non/Seldom	118	18.0	18.0	18.0
	Social	282	42.9	42.9	60.9
	Typical Binge	163	24.8	24.8	85.7
	Heavy Binge	94	14.3	14.3	100.0
	Total	657	100.0	100.0	

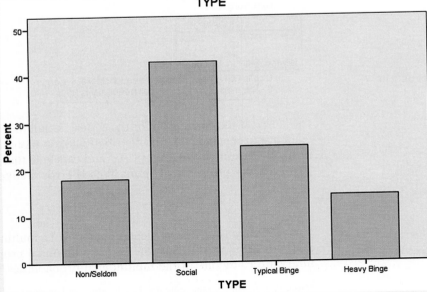

Figure SIA13.1

SPSS descriptive statistics and graph for type of drinker

Statistics IN Action
(continued)

the four categories. From the summary table at the top of the printout, you can see that 118 (or 18%) of the students are non/seldom drinkers, 282 (or 43%) are social drinkers, 163 (or 25%) are typical binge drinkers, and 94 (or 14%) are heavy binge drinkers. These sample percentages are illustrated in the bar graph in Figure SIA13.1. In this sample of students, the largest percentage (43%) consists of social drinkers.

Is this sufficient evidence to indicate that the true proportions in the population of college students are different? Letting $p_1, p_2, p_3,$ and p_4 represent the true proportions for non/seldom, social, typical binge, and heavy binge drinkers, respectively, we tested $H_0: p_1 = p_2 = p_3 = p_4 = .25$, using the chi-square test in SPSS. The printout is displayed in Figure SIA13.2. The cell frequencies and expected numbers are shown in the top table of the figure, while the chi-square test statistic (127.5) and p-value (.000) are shown in the bottom table. At any reasonably selected α-level (say, $\alpha = .01$), the small p-value indicates that there is sufficient evidence to reject the null hypothesis; thus, we conclude that the true proportions associated with the four type-of-drinker categories are indeed statistically different.

Chi-Square Test

Frequencies

TYPE

	Observed N	Expected N	Residual
Non/Seldom	118	164.3	-46.3
Social	282	164.3	117.8
Typical Binge	163	164.3	-1.3
Heavy Binge	94	164.3	-70.3
Total	657		

Test Statistics

	TYPE
Chi-Square[a]	127.493
df	3
Asymp. Sig.	.000

a. 0 cells (.0%) have expected frequencies less than 5. The minimum expected cell frequency is 164.3.

Figure SIA13.2
SPSS chi-square test for type-of-drinker categories

Exercises 13.1–13.19

Understanding the Principles

13.1 What are the characteristics of a multinomial experiment? Compare the characteristics with those of a binomial experiment.

13.2 What conditions must n satisfy to make the χ^2 test for a one-way table valid?

Learning the Mechanics

13.3 Use Table VII of Appendix A to find each of the following χ^2 values:
a. $\chi^2_{.05}$ for df = 10
b. $\chi^2_{.990}$ for df = 50
c. $\chi^2_{.10}$ for df = 16
d. $\chi^2_{.005}$ for df = 50

13.4 Use Table VII of Appendix A to find the following probabilities:
a. $P(\chi^2 \leq 1.063623)$ for df = 4
b. $P(\chi^2 > 30.5779)$ for df = 15
c. $P(\chi^2 \geq 82.3581)$ for df = 100
d. $P(\chi^2 < 18.4926)$ for df = 30

13.5 Find the rejection region for a one-dimensional χ^2-test of a null hypothesis concerning $p_1, p_2, \ldots, p_k$ if
a. $k = 3; \alpha = .05$
b. $k = 5; \alpha = .10$
c. $k = 4; \alpha = .01$

13.6 A multinomial experiment with $k = 3$ cells and $n = 320$ produced the data shown in the accompanying table. Do these data provide sufficient evidence to contradict the null hypothesis that $p_1 = .25, p_2 = .25,$ and $p_3 = .50$? Test, using $\alpha = .05$.

	Cell		
	1	2	3
n_i	78	60	182

13.7 A multinomial experiment with $k = 4$ cells and $n = 205$ produced the data shown in the following table:

	Cell			
	1	2	3	4
n_i	43	56	59	47

a. Is there sufficient evidence to conclude that the multinomial probabilities differ? Test, using $\alpha = .05$.
b. What are the Type I and Type II errors associated with the test of part **a**?
c. Construct a 95% confidence interval for the multinomial probability associated with cell 3.

13.8 A multinomial experiment with $k = 4$ cells and $n = 400$ produced the data shown in the accompanying table. Do these data provide sufficient evidence to contradict the null hypothesis that $p_1 = .2, p_2 = .4, p_3 = .1,$ and $p_4 = .3$? Test, using $\alpha = .05$.

	Cell			
	1	2	3	4
n_i	70	196	46	88

Applying the Concepts—Basic

13.9 Jaw dysfunction study. A report on dental patients with
[NW] temporomandibular (jaw) joint dysfunction (TMD) was
published in *General Dentistry* (Jan/Feb. 2004). A random
sample of 60 patients was selected for an experimental
treatment of TMD. Prior to treatment, the patients filled
out a survey on two nonfunctional jaw habits—bruxism
(teeth grinding) and teeth clenching—that have been
linked to TMD. Of the 60 patients, 3 admitted to bruxism,
11 admitted to teeth clenching, 30 admitted to both habits,
and 16 claimed they had neither habit.

  a. Describe the qualitative variable of interest in the
  study. Give the levels (categories) associated with the
  variable.
  b. Construct a one-way table for the sample data.
  c. Give the null and alternative hypotheses for testing
  whether the percentages associated with the admitted
  habits are the same.
  d. Calculate the expected numbers for each cell of the
  one-way table.
  e. Calculate the appropriate test statistic.
  f. Give the rejection region for the test at
  $\alpha = .05$
  g. Give the appropriate conclusion in the words of the
  problem.
  h. Find and interpret a 95% confidence interval for the
  true proportion of dental patients who admit to both
  habits.

13.10 Beetles and slime molds. Myxomycetes are mushroom-
like slime molds that are a food source for insects. The
Journal of Natural History (May 2010) published the
results of a study that investigated which of six species
of slime molds are most attractive to beetles inhabiting
an Atlantic rain forest. A sample of 19 beetles feeding on
slime mold was obtained and the species of slime mold
was determined for each beetle. The numbers of beetles
captured on each of the six species are given in the accompanying table. These data are saved in the **SLIMEMOLD**
file. The researchers want to know if the relative frequency
of occurrence of beetles differs for the six slime mold
species.

Slime mold species:	LE	TM	AC	AD	HC	HS
Number of beetles:	3	2	7	3	1	3

  a. Identify the categorical variable (and its levels) of interest in this study.
  b. Set up the null and alternative hypotheses of interest to
  the researchers.
  c. Find the test statistic and corresponding *p*-value.
  d. The researchers found "no significant differences in the
  relative frequencies of occurrence" using $\alpha = .05$. Do
  you agree?
  e. Comment on the validity of the inference, part **d.**
  (Determine the expected cell counts.)

13.11 Excavating ancient pottery. Refer to the *Chance* (Fall 2000)
study of ancient Greek pottery, presented in Exercise 2.14
(p. 36). Recall that 837 pottery pieces were uncovered at
the excavation site. The table describing the types of pottery found is reproduced in the next table and the information saved in the **POTTERY** file.

Pot Category	Number Found
Burnished	133
Monochrome	460
Painted	183
Other	61
Total	837

Based on Berg, I., and Bliedon, S. "The pots of Phyiakopi:
Applying statistical techniques to archaeology." *Chance*,
Vol. 13, No. 4, Fall 2000.

  a. Describe the qualitative variable of interest in the study.
  Give the levels (categories) associated with the variable.
  b. Assume that the four types of pottery occur with equal
  probability at the excavation site. What are the values
  of $p_1, p_2, p_3,$ and p_4, the probabilities associated with the
  four pottery types?
  c. Give the null and alternative hypotheses for testing whether one type of pottery is more likely to occur at the site
  than any of the other types.
  d. Find the test statistic for testing the hypotheses stated in
  part **c.**
  e. Find and interpret the *p*-value of the test. State the
  conclusion in the words of the problem if you use
  $\alpha = .10.$

13.12 Museum management. Refer to the *Museum Management
and Curatorship* (June 2010) worldwide survey of 30 leading museums of contemporary art, Exercise 2.19 (p. 37).
Recall that each museum manager was asked to provide the
performance measure used most often for internal evaluation. A summary of the results is provided in the table and
saved in the **MUSEUM2** file. The data were analyzed using
a chi-square test for a multinomial experiment. The results
are shown in the MINITAB printout (top of page 735).

  a. Is there evidence to indicate that one performance
  measure is used more often than any of the others? Test
  using $\alpha = .10.$
  b. Find a 90% confidence interval for the proportion of
  museums worldwide that use total visitors as their performance measure. Interpret the result.

Performance Measure	Number of Museums
Total visitors	8
Paying visitors	5
Big shows	6
Funds raised	7
Members	4

Applying the Concepts—Intermediate

13.13 Gender in two-child families. Refer to the *Human Biology*
(Feb. 2009) study on the gender of children in two-child
families, Exercise 4.25 (p. 188). The article reported on the
results of the National Health Interview Survey (NHIS)
of 42,888 two-child families. The table below (saved in the
BOYGIRL file) gives the number of families with each
gender configuration.

Gender Configuration	Number of Families
Girl-girl (GG)	9,523
Boy-girl (BG)	11,118
Girl-boy (GB)	10,913
Boy-boy (BB)	11,334

MINITAB Output for Exercise 13.12

Chi-Square Goodness-of-Fit Test for Observed Counts in Variable: NUMBER

Using category names in PERFORM

Category	Observed	Test Proportion	Expected	Contribution to Chi-Sq
Total visitors	8	0.2	6	0.666667
Paying visitors	5	0.2	6	0.166667
Big shows	6	0.2	6	0.000000
Funds raised	7	0.2	6	0.166667
Members	4	0.2	6	0.666667

N	DF	Chi-Sq	P-Value
30	4	1.66667	0.797

a. If it is just as likely to have a boy as a girl, find the probability of each of the gender configurations for a two-child family.

b. Use the probabilities, part **a**, to determine the expected number of families for each gender configuration.

c. Compute the chi-square test statistic for testing the hypothesis that it is just as likely to have a boy as a girl.

d. Interpret the result, part **c**, if you conduct the test using $\alpha = .10$.

e. Recent research indicates that the ratio of boys to girls in the world population is not 1 to 1, but instead higher (e.g., 1.06 to 1). Using a ratio of 1.06 to 1, the researchers showed that the probabilities of the different gender configurations are: GG—.23795, BG—.24985, GB—.24985, and BB—.26235. Repeat parts **b–d** using these probabilities.

13.14 Sociology fieldwork methods. Refer to the *Teaching Sociology* (July 2006) study of the fieldwork methods used by qualitative sociologists, presented in Exercise 2.16 (p. 36). Recall that fieldwork methods can be categorized as follows: Interview, Observation plus Participation, Observation Only, and Grounded Theory. The accompanying table (saved in the **FIELDWORK** file) shows the number of papers published over the past seven years in each category. Suppose a sociologist claims that 70%, 15%, 10%, and 5% of the fieldwork methods involve interview, observation plus participation, observation only, and grounded theory, respectively. Do the data support or refute the claim? Explain.

Fieldwork Method	Number of Papers
Interview	5,079
Observation + Participation	1,042
Observation Only	848
Grounded Theory	537

Based on Hood, J. C. "Teaching against the text: The case of qualitative methods." *Teaching Sociology*, Vol. 34, July 2006 (Exhibit 2).

13.15 Do you believe in the Bible? Refer to the General Social Survey (GSS) and the question pertaining to a person's belief in the Bible, presented in Exercise 2.18 (p. 36). Recall that approximately 2,800 Americans selected from one of the following answers: (1) The Bible is the actual word of God and is to be taken literally; (2) the Bible is the inspired word of God, but not everything is to be taken literally; (3)

the Bible is an ancient book of fables; or (4) the Bible has some other origin, but is recorded by men. The variable "Bible1" in the **BIBLE** file contains the responses.

a. Summarize the responses in a one-way table.

b. State the null and alternative hypotheses for testing whether the true proportions in each category are equal.

c. Find the expected number of responses in each answer category for the test mentioned in part **b**.

d. Compute the chi-square statistic for the test.

e. Give the appropriate conclusion for the test if $\alpha = .10$.

f. A more realistic null hypothesis is that 30% of Americans believe that the Bible is the actual word of God; 50% believe that it is inspired by God, but not to be taken literally; 15% believe that it is an ancient book of fables; and 5% believe that the Bible has some other origin. Repeat parts **b–e** for this hypothesis.

13.16 Characteristics of ice-melt ponds. Refer to the National Snow and Ice Data Center (NSIDC) collection of data on 504 ice-melt ponds in the Canadian Arctic, presented in Exercise 12.182 (p. 720). The data are saved in the **PONDICE** file. One variable of interest to environmental engineers studying the ponds is the type of ice observed in each. Ice type is classified as first-year ice, multiyear ice, or landfast ice. The SAS summary table for the types of ice of the 504 ice-melt ponds is reproduced below.

a. Use a 90% confidence interval to estimate the proportion of ice-melt ponds in the Canadian Arctic that have first-year ice.

b. Suppose environmental engineers hypothesize that 15% of Canadian Arctic ice-melt ponds have first-year ice, 40% have landfast ice, and 45% have multiyear ice. Test the engineers' theory, using $\alpha = .01$.

The FREQ Procedure				
ICETYPE	Frequency	Percent	Cumulative Frequency	Cumulative Percent
First-year	88	17.46	88	17.46
Landfast	196	38.89	284	56.35
Multi-year	220	43.65	504	100.00

13.17 Detecting Alzheimer's disease at an early age. Geneticists at Australian National University are studying whether the cognitive effects of Alzheimer's disease can be detected at an early age (*Neuropsychology*, Jan. 2007). One portion of the study focused on a particular strand of DNA extracted from each in a sample of 2,097 young adults between the ages of 20 and 24. The DNA strand was classified into one of three genotypes: E4⁺/E4⁺, E4⁺/E4⁻, and E4⁻/E4⁻. The number of young adults with each genotype is shown in the table and the data are saved in the **E4E4** file. Suppose that in adults who are not afflicted with Alzheimer's disease, the distribution of genotypes for this strand of DNA is 2% with E4⁺/E4⁺, 25% with E4⁺/E4⁻, and 73% with E4⁻/E4⁻. If differences in this distribution are detected, then this strand of DNA could lead researchers to an early test for the onset of Alzheimer's. Conduct a test (at $\alpha = .05$) to determine if the distribution of E4/E4 genotypes for the population of young adults differs from the norm.

Genotype:	E4⁺/E4⁺	E4⁺/E4⁻	E4⁻/E4⁻
Number of young adults:	56	517	1,524

Applying the Concepts—Advanced

13.18 Political representation of religious groups. Do those elected to the U.S. House of Representatives really "represent" their constituents demographically? This was a question of interest in *Chance* (Summer 2002). One of several demographics studied was religious affiliation. The accompanying table (saved in the **USHOUSE** file) gives the proportion of the U.S. population for several religions, as well as the number of the 435 seats in the House of Representatives affiliated with that religion. Give your opinion on whether or not the members of the House of Representatives are statistically representative of the religious affiliation of their constituents in the United States.

Religion	Proportion of U.S. Population	Number of Seats in House
Catholic	.28	117
Methodist	.04	61
Jewish	.02	30
Other	.66	227
Totals	1.00	435

13.19 Analysis of a Scrabble game. In the board game Scrabble™, a player initially draws a "hand" of seven tiles at random from 100 tiles. Each tile has a letter of the alphabet, and the player attempts to form a word from the letters in his or her hand. In *Chance* (Winter 2002), scientist C. J. Robinove investigated whether a handheld electronic version of the game, called ScrabbleExpress™, produces too few vowels

in the 7-letter draws. For each of the 26 letters (and "blank" for any letter), the accompanying table gives the true relative frequency of the letter in the board game, as well as the frequency of occurrence of the letter in a sample of 700 tiles (i.e., 100 "hands") randomly drawn in the electronic game. These data are saved in the **SCRABBLE** file.

a. Do the data support the scientist's contention that ScrabbleExpress™ "presents the player with unfair word selection opportunities" that are not the same as in the Scrabble™ board game? Test, using $\alpha = .05$.

b. Use a 95% confidence interval to estimate the true proportion of letters drawn in the electronic game that are vowels. Compare the results with the true relative frequency of a vowel in the board game.

Letter	Relative Frequency in Board Game	Frequency in Electronic Game
A	.09	39
B	.02	18
C	.02	30
D	.04	30
E	.12	31
F	.02	21
G	.03	35
H	.02	21
I	.09	25
J	.01	17
K	.01	27
L	.04	18
M	.02	31
N	.06	36
O	.08	20
P	.02	27
Q	.01	13
R	.06	27
S	.04	29
T	.06	27
U	.04	21
V	.02	33
W	.02	29
X	.01	15
Y	.02	32
Z	.01	14
# (blank)	.02	34
Total		700

Source: Robinove, C. J. "Letter-frequency bias in an electronic Scrabble game." *Chance*, Vol. 15, No. 1, Winter 2002, p. 31 (Table 3). Reprinted with permission from *Chance*. Copyright 2002 by the American Statistical Association. All rights reserved.

13.3 Testing Categorical Probabilities: Two-Way (Contingency) Table

In Section 13.2, we introduced the multinomial probability distribution and considered data classified according to a single criterion. We now consider multinomial experiments in which the data are classified according to two criteria—that is, *classification with respect to two qualitative factors*.

Consider a study similar to one in the *Journal of Marketing* on the impact of using celebrities in television advertisements. The researchers investigated the relationship between the gender of a viewer and the viewer's brand awareness. Three hundred TV viewers were randomly selected and each asked to identify products advertised by male celebrity spokespersons. The data are summarized in the **two-way table** shown in Table 13.3. This table, called a **contingency table,** presents multinomial count data classified on two scales, or **dimensions, of classification:** gender of viewer and brand awareness.

Table 13.3   Contingency Table for Marketing Example

		Gender		
		Male	Female	Totals
Brand Awareness	**Could Identify Product**	95	41	136
	Could Not Identify Product	50	114	164
	Totals	145	155	300

Data Set: CELEBRITY

The symbols representing the cell counts for the multinomial experiment in Table 13.3 are shown in Table 13.4a, and the corresponding cell, row, and column probabilities are shown in Table 13.4b. Thus, n_{11} represents the number of viewers who are male and could identify the brand, and p_{11} represents the corresponding cell probability. Note the symbols for the row and column totals and also the symbols for the probability totals. The latter are called **marginal probabilities** for each row and column. The marginal probability p_{r1} is the probability that a TV viewer identifies the product; the marginal probability p_{c1} is the probability that a TV viewer is male. Thus,

$$p_{r1} = p_{11} + p_{12} \text{ and } p_{c1} = p_{11} + p_{21}$$

Table 13.4a   Observed Counts for Contingency Table 13.3

		Gender		
		Male	Female	Totals
Brand Awareness	**Could Identify Product**	n_{11}	n_{12}	R_1
	Could Not Identify Product	n_{21}	n_{22}	R_2
	Totals	C_1	C_2	n

Table 13.4b   Probabilities for Contingency Table 13.3

		Gender		
		Male	Female	Totals
Brand Awareness	**Could Identify Product**	p_{11}	p_{12}	p_{r1}
	Could Not Identify Product	p_{21}	p_{22}	p_{r2}
	Totals	p_{c1}	p_{c2}	1

We can see, then, that this really is a multinomial experiment with a total of 300 trials, $(2)(2) = 4$ cells or possible outcomes, and probabilities for each cell as shown in Table 13.4b. Since the 300 TV viewers are randomly chosen, the trials are considered independent and the probabilities are viewed as remaining constant from trial to trial.

Suppose we want to know whether the two classifications of gender and brand awareness are dependent. That is, if we know the gender of the TV viewer, does that information give us a clue about the viewer's brand awareness? In a probabilistic sense, we know (Chapter 3) that the independence of events A and B implies that $P(AB) = P(A)P(B)$. Similarly, in the contingency table analysis, if the **two classifications are independent,** the probability that an item is classified into any particular cell of the table is the product of the corresponding marginal probabilities. Thus, under the hypothesis of independence, in Table 13.4b we must have

$$p_{11} = p_{r1}p_{c1} \quad p_{12} = p_{r1}p_{c2}$$
$$p_{21} = p_{r2}p_{c1} \quad p_{22} = p_{r2}p_{c2}$$

To test the hypothesis of independence, we use the same reasoning employed in the one-dimensional tests of Section 13.2. First, we calculate the *expected*, or *mean, count in each cell*, assuming that the null hypothesis of independence is true. We do this by noting that the expected count in a cell of the table is just the total number of multinomial trials, n, times the cell probability. Recall that n_{ij} represents the **observed count** in the cell located in the ith row and jth column. Then the expected cell count for the upper left-hand cell (first row, first column) is

$$E_{11} = np_{11}$$

or, when the null hypothesis (the classifications are independent) is true,

$$E_{11} = np_{r1}p_{c1}$$

Since these true probabilities are not known, we estimate p_{r1} and p_{c1} by the same proportions $\hat{p}_{r1} = R_1/n$ and $\hat{p}_{c1} = C_1/n$. Thus, the estimate of the expected value E_{11} is

$$\hat{E}_{11} = n\left(\frac{R_1}{n}\right)\left(\frac{C_1}{n}\right) = \frac{R_1C_1}{n}$$

Similarly, for each i, j,

$$\hat{E}_{ij} = \frac{(\text{Row total})(\text{Column total})}{\text{Total sample size}}$$

Hence,

$$\hat{E}_{12} = \frac{R_1C_2}{n}$$

$$\hat{E}_{21} = \frac{R_2C_1}{n}$$

$$\hat{E}_{22} = \frac{R_2C_2}{n}$$

Finding Expected Cell Counts for a Two-Way Contingency Table

The estimate of the expected number of observations falling into the cell in row i and column j is given by

$$\hat{E}_{ij} = \frac{R_iC_j}{n}$$

where R_i = total for row i, C_j = total for column j, and n = sample size.

Using the data in Table 13.3, we find that

$$\hat{E}_{11} = \frac{R_1C_1}{n} = \frac{(136)(145)}{300} = 65.73$$

$$\hat{E}_{12} = \frac{R_1C_2}{n} = \frac{(136)(155)}{300} = 70.27$$

$$\hat{E}_{21} = \frac{R_2C_1}{n} = \frac{(164)(145)}{300} = 79.27$$

$$\hat{E}_{22} = \frac{R_2C_2}{n} = \frac{(164)(155)}{300} = 84.73$$

Tabulated statistics: AWARE, GENDER

```
Using frequencies in NUMBER


Rows: AWARE    Columns: GENDER

               Male   Female    All

ID-Product       95       41    136
               65.7     70.3  136.0

No-ID            50      114    164
               79.3     84.7  164.0

All             145      155    300
              145.0    155.0  300.0


Cell Contents:        Count
                      Expected count
```

Figure 13.3

MINITAB contingency table analysis of data in Table 13.3

```
Pearson Chi-Square = 46.135, DF = 1, P-Value = 0.000
Likelihood Ratio Chi-Square = 47.362, DF = 1, P-Value = 0.000
```

These estimated expected values are more easily obtained using computer software. Figure 13.3 is a MINITAB printout of the analysis, with the expected values highlighted.

We now use the χ^2 statistic to compare the observed and expected (estimated) counts in each cell of the contingency table:

$$\chi^2 = \frac{[n_{11} - \hat{E}_{11}]^2}{\hat{E}_{11}} + \frac{[n_{12} - \hat{E}_{12}]^2}{\hat{E}_{12}} + \frac{[n_{21} - \hat{E}_{21}]^2}{\hat{E}_{21}} + \frac{[n_{22} - \hat{E}_{22}]^2}{\hat{E}_{22}}$$

$$= \sum \frac{[n_{ij} - \hat{E}_{ij}]^2}{\hat{E}_{ij}}$$

(*Note:* The use of $\sum$ in the context of a contingency table analysis refers to a sum over all cells in the table.)

Substituting the data of Table 13.3 and the expected values into this expression, we get

$$\chi^2 = \frac{(95 - 65.73)^2}{65.73} + \frac{(41 - 70.27)^2}{70.27} + \frac{(50 - 79.27)^2}{79.27} + \frac{(114 - 84.73)^2}{84.73} = 46.14$$

Note that this value is also shown (highlighted) in Figure 13.3.

Large values of χ^2 imply that the observed counts do not closely agree and hence that the hypothesis of independence is false. To determine how large χ^2 must be before it is too large to be attributed to chance, we make use of the fact that the sampling distribution of χ^2 is approximately a χ^2 probability distribution when the classifications are independent.

When testing the null hypothesis of independence in a two-way contingency table, the appropriate degrees of freedom will be $(r - 1)(c - 1)$, where r is the number of rows and c is the number of columns in the table. For the brand awareness example, the number of degrees of freedom for χ^2 is $(r - 1)(c - 1) = (2 - 1)(2 - 1) = 1$. Then, for $\alpha = .05$, we reject the hypothesis of independence when

$$\chi^2 > \chi^2_{.05} = 3.84146$$

Since the computed $\chi^2 = 46.14$ exceeds the value 3.84146, we conclude that viewer gender and brand awareness are dependent events. This result may also be obtained by noting that the *p*-value of the test (highlighted on Figure 13.3) is approximately 0.

The pattern of **dependence** can be seen more clearly by expressing the data as percentages. We first select one of the two classifications to be used as the base variable. In the preceding example, suppose we select gender of the TV viewer as the classificatory variable to be the base. Next, we represent the responses for each level of the second categorical variable (brand awareness here) as a percentage of the subtotal for the base variable. For example, from Table 13.3, we convert the response for males who identify the brand (95) to a percentage of the total number of male viewers (145). That is,

$$(^{95}/_{145})100\% = 65.5\%$$

All of the entries in Table 13.3 are similarly converted, and the values are shown in Table 13.5. The value shown at the right of each row is the row's total, expressed as a percentage of the total number of responses in the entire table. Thus, the percentage of TV viewers who identify the product is $(\frac{136}{300})100\% = 45.3\%$ (rounded to the nearest percent).

Table 13.5   Percentage of TV Viewers Who Identify Brand, by Gender

		Gender		Totals
		Male	Female	
Brand Awareness	**Could Identify Product**	65.5	26.5	45.3
	Could Not Identify Product	34.5	73.5	54.7
	Totals	100	100	100

If the gender and brand awareness variables are independent, then the percentages in the cells of the table are expected to be approximately equal to the corresponding row percentages. Thus, we would expect the percentage of viewers who identify the brand for each gender to be approximately 45% if the two variables are independent. The extent to which each gender's percentage departs from this value determines the dependence of the two classifications, with greater variability of the row percentages meaning a greater degree of dependence. A plot of the percentages helps summarize the observed pattern. In the SPSS bar graph in Figure 13.4, we show the gender of the viewer (the base variable) on the horizontal axis and the percentage of TV viewers who identify the brand (green bars) on the vertical axis. The "expected" percentage under the assumption of independence is shown as a horizontal line.

Figure 13.4 clearly indicates the reason that the test resulted in the conclusion that the two classifications in the contingency table are dependent. The percentage of male

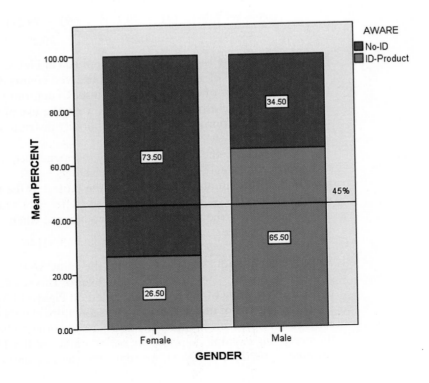

Figure 13.4

SPSS bar graph showing percent of viewers who identify TV product

TV viewers who identify the brand promoted by a male celebrity is more than twice as high as the percentage of female TV viewers who identify the brand. Statistical measures of the degree of dependence and procedures for making comparisons of pairs of levels for classifications are beyond the scope of this text, but can be found in the references. We will utilize descriptive summaries such as Figure 13.4 to examine the degree of dependence exhibited by the sample data.

The general form of a two-way contingency table containing r rows and c columns (called an $r \times c$ contingency table) is shown in Table 13.6. Note that the observed count in the ijth cell is denoted by n_{ij}, the ith row total is r_i, the jth column total is c_j, and the total sample size is n. Using this notation, we give the general form of the contingency table test for independent classifications in the box.

Table 13.6 General $r \times c$ Contingency Table

		Column				Row Totals
		1	2	...	c	
Row	1	n_{11}	n_{12}	...	n_{1c}	R_1
	2	n_{21}	n_{22}	...	n_{2c}	R_2
	:	:	:		:	:
	r	n_{r1}	n_{r2}	...	n_{rc}	R_r
Column Totals		C_1	C_2	...	C_c	n

General Form of a Two-Way (Contingency) Table Analysis: A Test for Independence

H_0: The two classifications are independent

H_a: The two classifications are dependent

Test statistic: $\chi^2 = \Sigma \dfrac{[n_{ij} - \hat{E}_{ij}]^2}{\hat{E}_{ij}}$

where $\hat{E}_{ij} = \dfrac{R_i C_j}{n}$

Rejection region: $\chi^2 > \chi^2_\alpha$, where χ^2_α has $(r - 1)(c - 1)$ df

---

Conditions Required for a Valid χ^2 Test: Contingency Tables

1. The n observed counts are a random sample from the population of interest. We may then consider this to be a multinomial experiment with $r \times c$ possible outcomes.

2. The sample size n will be large enough so that, for every cell, the expected count $\hat{E}(n_{ij})$ will be equal to 5 or more.

Example 13.3

Conducting a Two-Way Analysis— Marital Status and Religion

Problem A social scientist wants to determine whether the marital status (divorced or not divorced) of U.S. men is independent of their religious affiliation (or lack thereof). A sample of 500 U.S. men is surveyed, and the results are tabulated as shown in Table 13.7 and saved in the **MARREL** file.

a. Test to see whether there is sufficient evidence to indicate that the marital status of men who have been or are currently married is dependent on religious affiliation. Take $\alpha = .01$.

b. Graph the data and describe the patterns revealed. Is the result of the test supported by the graph?

Table 13.7 Survey Results (Observed Counts), Example 13.3

		Religious Affiliation					
		A	B	C	D	None	Totals
Marital Status	**Divorced**	39	19	12	28	18	116
	Married, never divorced	172	61	44	70	37	384
	Totals	211	80	56	98	55	500

Data Set: MARREL

Solution

a. The first step is to calculate estimated expected cell frequencies under the assumption that the classifications are independent. Rather than compute these values by hand, we resort to a computer. The SAS printout of the analysis of Table 13.7 is displayed in Figure 13.5, each cell of which contains the observed (top) and expected (bottom) frequency in that cell. Note that $\hat{E}_{11}$, the estimated expected count for the Divorced, A cell, is 48.952. Similarly, the estimated expected count for the Divorced, B cell, is $\hat{E}_{12} = 18.56$. Since all the estimated expected cell frequencies are greater than 5, the χ^2 approximation for the test statistic is appropriate. Assuming that the men chosen were randomly selected from all married or previously married American men, the characteristics of the multinomial probability distribution are satisfied.

```
                    The FREQ Procedure

               Table of MARITAL by RELIGION

   MARITAL      RELIGION

   Frequency
   Expected   A       B       C       D      NONE    | Total

   DIVORCED       39      19      12      28      18  |  116
              48.952  18.56  12.992  22.736  12.76   |

   NEVER         172      61      44      70      37  |  384
              162.05  61.44  43.008  75.264  42.24   |

   Total         211      80      56      98      55     500


            Statistics for Table of MARITAL by RELIGION

   Statistic                        DF      Value      Prob

   Chi-Square                        4      7.1355     0.1289
   Likelihood Ratio Chi-Square       4      6.9854     0.1367
   Mantel-Haenszel Chi-Square        1      6.4943     0.0108
   Phi Coefficient                          0.1195
   Contingency Coefficient                  0.1186
   Cramer's V                               0.1195


                    Fisher's Exact Test

            Table Probability (P)     6.936E-06
            Pr <= P                   0.1251

               Sample Size = 500
```

Figure 13.5

SAS contingency table printout for Example 13.3

The null and alternative hypotheses we want to test are

H_0: The marital status of U.S. men and their religious affiliation are independent

H_a: The marital status of U.S. men and their religious affiliation are dependent

The test statistic, $\chi^2 = 7.135$, is highlighted at the bottom of the printout, as is the observed significance level (*p*-value) of the test. Since $\alpha = .01$ is less than $p = .129$, we fail to reject H_0; that is, we cannot conclude that the marital status of U.S. men depends on their religious affiliation. (Note that we could not reject H_0 even with $\alpha = .10$.)

b. The marital status frequencies can be expressed as percentages of the number of men in each religious affiliation category. The expected percentage of divorced men under the assumption of independence is $(^{116}/_{500})100\% = 23\%$. A SAS graph of the percentages is shown in Figure 13.6. Note that the percentages of divorced men (see the bars in the "DIVORCED" block of the SAS graph) deviate only slightly from

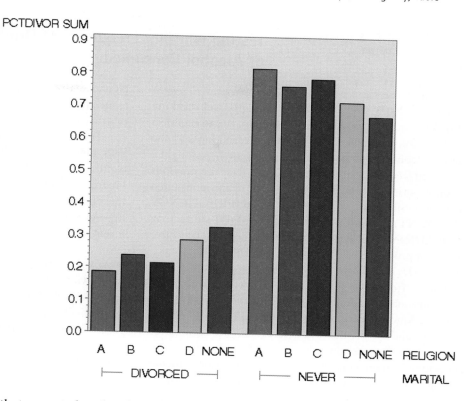

Figure 13.6

SAS side-by-side bar graphs showing percentage of divorced and never divorced males by religion

that expected under the assumption of independence, supporting the result of the test in part **a.** That is, neither the descriptive bar graph nor the statistical test provides evidence that the male divorce rate depends on (varies with) religious affiliation.

Now Work Exercise 13.29

Contingency Tables with Fixed Marginals

In the *Journal of Marketing* study on celebrities in TV ads, a single random sample was selected from the target population of all TV viewers and the outcomes—values of gender and brand awareness—were recorded for each viewer. For this type of study, the researchers had no a *priori* knowledge of how many observations would fall into the categories of the qualititative variables. In other words, prior to obtaining the sample, the researchers did not know how many males or how many brand identifiers would make up the sample. Often times, it is advantageous to select a random sample from each of the levels of one of the qualitative variables.

For example, in the *Journal of Marketing* study, the researchers may want to be sure of an equivalent number of males and females in their sample. Consequently, they will select independent random samples of 150 males and 150 females. (In fact, this was the sampling plan for the actual study.) Summary data for this type of study yield a **contingency table with fixed marginals** since the column totals for one qualitative variable (e.g., gender) are known in advance.* The goal of the analysis does not change—determine whether the two qualitative variables (e.g., gender and brand awareness) are dependent.

The procedure for conducting a chi-square analysis for a contingency table with fixed marginals is identical to the one outlined above, since it can be shown (proof omitted) that the χ^2 test statistic for this type of sampling also has an approximate chi-square distribution with $(r - 1)(c - 1)$ degrees of freedom. One reason why you might choose this alternative sampling plan is to obtain sufficient observations in each cell of the contingency table to ensure that the chi-square approximation is valid. Remember, this will usually occur when the expected cell counts are all greater than or equal to 5. By selecting a large sample (150 observations) for each gender in the *Journal of Marketing* study, the researchers improved the odds of obtaining large expected cell counts in the contingency table.

*Data from this type of study are also known as *product binomial data.*

Statistics IN Action | Revisited

Testing whether Frequency of Drinking Is Related to Amount of Alcohol Consumed

Refer again to the *Family and Consumer Sciences Research Journal* (Mar. 2005) study of college students and drinking (p. 726). A second objective of the researchers was to establish a statistical link between the frequency of drinking and the amount of alcohol consumed. That is, the researchers sought a link between frequency of drinking alcohol over the previous one-month period and average number of drinks consumed per occasion. Since both of these variables (FREQUENCY and AMOUNT) measured on the sample of 657 students in the **COLLDRINKS** file are qualitative, a contingency table analysis is appropriate.

Figure SIA13.3 shows the SPSS contingency table analyses relating frequency of drinking to average amount of alcohol consumed. The null hypothesis for the test is H_0: Frequency and Amount are independent. The chi-square test statistic (756.6) and the *p*-value of the test (.000) are highlighted on the printout. If we conduct the test at $\alpha = .01$, there is sufficient evidence to reject H_0. That is, the data provide evidence indicating that, for college students, the average amount of alcohol consumed per occasion is associated with the frequency of drinking.

The row percentages highlighted in the contingency table of Figure SIA13.3 reveal the differences in drinking amounts for the different levels of drinking frequency. For frequency of drinking "None" and "Once a month", 0% drink heavily (7–9 or 10 or more drinks per occasion). However, for frequency of drinking "Twice a week" and "More," 12.7% and 17.1%, respectively, have 7–9 drinks per occasion, while 4.0% and 11.2%, respectively, have 10 or more drinks per occasion. These results led the researchers to report that "The frequent drinkers were more likely to consume more [alcohol] on each occasion, a tendency that clearly makes them heavy drinkers."

Data Set: COLLDRINKS

FREQNCY * AMOUNT Crosstabulation

			AMOUNT						Total
			None	1 drink	2-3 drinks	4-6 drinks	7-9 drinks	More	
FREQNCY	None	Count	118	0	0	0	0	0	118
		Expected Count	21.2	11.0	51.2	17.8	11.5	5.4	118.0
		% within FREQNCY	100.0%	.0%	.0%	.0%	.0%	.0%	100.0%
	OnceMonth	Count	0	30	43	20	0	0	93
		Expected Count	16.7	8.6	40.3	14.0	9.1	4.2	93.0
		% within FREQNCY	.0%	32.3%	46.2%	21.5%	.0%	.0%	100.0%
	TwiceWeek	Count	0	26	158	46	35	11	276
		Expected Count	49.6	25.6	119.7	41.6	26.9	12.6	276.0
		% within FREQNCY	.0%	9.4%	57.2%	16.7%	12.7%	4.0%	100.0%
	More	Count	0	5	84	33	29	19	170
		Expected Count	30.5	15.8	73.7	25.6	16.6	7.8	170.0
		% within FREQNCY	.0%	2.9%	49.4%	19.4%	17.1%	11.2%	100.0%
Total		Count	118	61	285	99	64	30	657
		Expected Count	118.0	61.0	285.0	99.0	64.0	30.0	657.0
		% within FREQNCY	18.0%	9.3%	43.4%	15.1%	9.7%	4.6%	100.0%

Chi-Square Tests

	Value	df	Asymp. Sig. (2-sided)
Pearson Chi-Square	756.606[a]	15	.000
Likelihood Ratio	706.412	15	.000
Linear-by-Linear Association	322.813	1	.000
N of Valid Cases	657		

a. 1 cells (4.2%) have expected count less than 5. The minimum expected count is 4.25.

Figure SIA13.3
SPSS contingency table analysis: frequency of drinking vs. average amount

Exercises 13.20–13.44

Understanding the Principles

13.20 What is a two-way (contingency) table?

13.21 What is a contingency table with fixed marginals?

13.22 *True or False.* One goal of a contingency table analysis is to determine whether the two classifications are independent or dependent.

13.23 What conditions are required for a valid chi-square test of data from a contingency table?

Learning the Mechanics

13.24 Find the rejection region for a test of independence of two classifications for which the contingency table contains r rows and c columns and
  a. $r = 5$, $c = 5$, $\alpha = .05$
  b. $r = 3$, $c = 6$, $\alpha = .10$
  c. $r = 2$, $c = 3$, $\alpha = .01$

13.25 Consider the following 2×3 (i.e., $r = 2$ and $c = 3$) contingency table:

		Column		
		1	2	3
Row	1	9	34	53
	2	16	30	25

  a. Specify the null and alternative hypotheses that should be used in testing the independence of the row and column classifications.
  b. Specify the test statistic and the rejection region that should be used in conducting the hypothesis test of part **a.** Use $\alpha = .01$.
  c. Assuming that the row classification and the column classification are independent, find estimates for the expected cell counts.
  d. Conduct the hypothesis test of part **a.** Interpret your result.

13.26 Refer to Exercise 13.25.
  a. Convert the frequency responses to percentages by calculating the percentage of each column total falling in each row. Also, convert the row totals to percentages of the total number of responses. Display the percentages in a table.
  b. Create a bar graph with row 1 percentage on the vertical axis and column number on the horizontal axis. Show the row 1 total percentage as a horizontal line on the graph.
  c. What pattern do you expect to see if the rows and columns are independent? Does the plot support the result of the test of independence in Exercise 13.25?

13.27 Test the null hypothesis of independence of the two classifications A and B of the 3×3 contingency table shown here. Use $\alpha = .05$.

		B		
		B_1	B_2	B_3
	A_1	40	72	42
A	A_2	63	53	70
	A_3	31	38	30

13.28 Refer to Exercise 13.27. Convert the responses to percentages by calculating the percentage of each B class total falling into each A classification. Also, calculate the percentage of the total number of responses that constitute each of the A classification totals.
  a. Create a bar graph with row A_1 percentage on the vertical axis and B classification on the horizontal axis. Does the graph support the result of the test of hypothesis in Exercise 13.27? Explain.
  b. Repeat part **a** for the row A_2 percentages.
  c. Repeat part **a** for the row A_3 percentages.

Applying the Concepts—Basic

13.29 **Children's perceptions of their neighborhood.** In *Health* NW *Education Research* (Feb. 2005), nutrition scientists at Deakin University (Australia) investigated children's perceptions of their environments. Each in a sample of 147 ten-year-old children drew maps of their home and neighborhood environment. The researchers examined the maps for certain themes (e.g., presence of a pet, television in the bedroom, opportunities for physical activity). The results, broken down by gender, for two themes (presence of a dog and TV in the bedroom) are shown in the tables below and saved in the **MAPDOG** and **MAPTV** files, respectively.
  a. Find the sample proportion of boys who drew a dog on their maps.
  b. Find the sample proportion of girls who drew a dog on their maps.
  c. Compare the proportions you found in parts **a** and **b.** Does it appear that the likelihood of drawing a dog on the neighborhood map depends on gender?
  d. Give the null hypothesis for testing whether the likelihood of a drawing a dog on the neighborhood map depends on gender.
  e. Use the MINITAB printout (p. 746) to conduct the test mentioned in part **d** at $\alpha = .05$.
  f. Conduct a test to determine whether the likelihood of drawing a TV in the bedroom is different for boys and girls. Use $\alpha = .05$.

Presence of a Dog	Number of Boys	Number of Girls
Yes	6	11
No	71	59
Total	77	70

Presence of TV in Bedroom	Number of Boys	Number of Girls
Yes	11	9
No	66	61
Total	77	70

Based on Hume, C., Salmon, J., and Ball, K. "Children's perceptions of their home and neighborhood environments, and their association with objectively measured physical activity: A qualitative and quantitative study." *Health Education Research*, Vol. 20, No. 1, February 2005 (Table III).

13.30 **Eyewitnesses and mug shots.** Refer to the *Applied Psychology in Criminal Justice* (April 2010) study of mug shot choices by eyewitnesses to a crime, Exercise 10.97

MINITAB output for Exercise 13.29

Tabulated statistics: DOG, GENDER

```
Using frequencies in NUMBER


Rows: DOG    Columns: GENDER

         Boy    Girl    All

No        71     59     130
        68.10  61.90  130.00

Yes        6     11      17
         8.90   8.10   17.00

All       77     70     147
        77.00  70.00  147.00


Cell Contents:        Count
                      Expected count


Pearson Chi-Square = 2.250, DF = 1, P-Value = 0.134
Likelihood Ratio Chi-Square = 2.268, DF = 1, P-Value = 0.132
```

(p. 536). Recall that a sample of 96 college students was shown a video of a simulated theft, then asked to select the mug shot that most closely resembled the thief. The students were randomly assigned to view either 3, 6, or 12 mug shots at a time, with 32 students in each group. The number of students in the 3-, 6-, or 12-photos-per-page groups who selected the target mugshot were 19, 19, and 15, respectively.

a. For each photo group, compute the proportion of students who selected the target mug shot. Which group yielded the lowest proportion?

b. Create a contingency table for these data, with photo group in the rows and whether or not the target mug shot was selected in the columns.

c. Refer to, part **b.** Are there differences in the proportions who selected the target mug shot among the three photo groups? Test using $\alpha = .10$.

13.31 Stereotyping deceptive and authentic news stories. Major newspapers lose their credibility (and subscribers) when they are found to have published deceptive or misleading news stories. In *Journalism and Mass Communication Quarterly* (Summer 2007), University of Texas researchers investigated whether certain stereotypes (e.g., negative references to certain nationalities) occur more often in deceptive news stories than in authentic news stories. The researchers analyzed 183 news stories that were proven to be deceptive in nature and 128 news stories that were considered authentic. Specifically, the researchers determined whether each story was negative, neutral, or positive in tone. The accompanying table (saved in the **NEWSSTORY** file) gives the number of news stories found in each tone category.

	Authentic News Stories	Deceptive News Stories
Negative Tone	59	111
Neutral Tone	49	61
Positive Tone	20	11
Total	128	183

Based on Lasorsa, D., and Dai, J. "When news reporters deceive: The production of stereotypes." *Journalism and Mass Communication Quarterly*, Vol. 84, No. 2, Summer 2007 (Table 2).

SPSS output for Exercise 13.31

TONE * STORY Crosstabulation

			STORY		
			Authentic	Deceptive	Total
TONE	Negative	Count	59	111	170
		Expected Count	70.0	100.0	170.0
	Neutral	Count	49	61	110
		Expected Count	45.3	64.7	110.0
	Positive	Count	20	11	31
		Expected Count	12.8	18.2	31.0
Total		Count	128	183	311
		Expected Count	128.0	183.0	311.0

Chi-Square Tests

	Value	df	Asymp. Sig. (2-sided)
Pearson Chi-Square	10.427[a]	2	.005
Likelihood Ratio	10.348	2	.006
N of Valid Cases	311		

a. 0 cells (.0%) have expected count less than 5. The minimum expected count is 12.76.

a. Find the sample proportion of negative tone news stories that is deceptive.

b. Find the sample proportion of neutral news stories that is deceptive.

c. Find the sample proportion of positive news stories that is deceptive.

d. Compare the sample proportions, parts **a–c.** Does it appear that the proportion of news stories that is deceptive depends on story tone?

e. Give the null hypothesis for testing whether the authenticity of a news story depends on tone.

f. Use the SPSS printout above to conduct the test, part **e.** Test at $\alpha = .05$.

13.32 Healing heart patients with music, imagery, touch, and prayer. "Frontier medicine" is a term used to describe medical therapies (e.g., energy healing, therapeutic prayer, spiritual healing) for which there is no plausible explanation. *The Lancet* (July 16, 2005) published the results of a study designed to test the effectiveness of two types of frontier medicine—music, imagery, and touch (MIT) therapy and therapeutic prayer—in healing cardiac care patients. Patients were randomly assigned to receive one of four types of treatment: (1) prayer, (2) MIT, (3) prayer and MIT, and (4) standard care (no prayer and no MIT). Six months after therapy, the patients were evaluated for a major adverse cardiovascular event (e.g., a heart attack). The results of the study are summarized in the accompanying table and saved in the **HEALING** file.

Therapy	Number of Patients with Major Cardiovascular Events	Number of Patients with No Events	Total
Prayer	43	139	182
MIT	47	138	185
Prayer and MIT	39	150	189
Standard	50	142	192

Based on Krucoff, M. W., et al. "Music, imagery, touch, and prayer as adjuncts to interventional cardiac care: The Monitoring and Actualization of Noetic Trainings (MANTRA) II randomized study." *The Lancet*, Vol. 366, July 16, 2005 (Table 4).

MINITAB output for Exercise 13.32

Tabulated statistics: THERAPY, EVENT

```
Using frequencies in NUMBER


Rows: THERAPY     Columns: EVENT

               No       Yes      All

MIT           138        47      185
             140.7      44.3    185.0

Prayer        139        43      182
             138.4      43.6    182.0

Prayer&MIT    150        39      189
             143.8      45.2    189.0

Standard      142        50      192
             146.1      45.9    192.0

All           569       179      748
             569.0     179.0    748.0

Cell Contents:       Count
                     Expected count


Pearson Chi-Square = 1.828, DF = 3, P-Value = 0.609
Likelihood Ratio Chi-Square = 1.855, DF = 3, P-Value = 0.603
```

a. Identify the two qualitative variables (and associated levels) measured in the study.
b. State H_o and H_a for testing whether a major adverse cardiovascular event depends on type of therapy.
c. Use the MINITAB printout above to conduct the test mentioned in part **b** at $\alpha = .10$. On the basis of this test, what can the researchers infer about the effectiveness of music, imagery, and touch therapy and the effectiveness of healing prayer in heart patients?

13.33 Masculinity and crime. Refer to the *Journal of Sociology* (July 2003) study on the link between the level of masculinity and criminal behavior in men, presented in Exercise 9.27 (p. 427). The researcher identified events that a sample of newly incarcerated men were involved in and classified each event as "violent" (involving the use of a weapon, the throwing of objects, punching, choking, or kicking) or "avoided-violent" (involving pushing, shoving, grabbing, or threats of violence that did not escalate into a violent event). Each man (and corresponding event) was also classified as possessing "high-risk masculinity" (scored high on the Masculinity–Femininity Scale test and low on the Traditional Outlets of Masculinity Scale test) or "low-risk masculinity." The data on 1,507 events are summarized in the following table and saved in the **HRM** file.

	Violent Events	Avoided-Violent Events	Totals
High-Risk Masculinity	236	143	379
Low-Risk Masculinity	801	327	1,128
Totals	1,037	470	1,507

Based on Krienert, J. L. "Masculinity and crime: A quantitative exploration of Messerschmidt's hypothesis." *Journal of Sociology*, Vol. 7, No. 2, July 2003 (Table 4).

a. Identify the two categorical variables measured (and their levels) in the study.
b. Identify the experimental units.
c. If the type of event (violent or avoided-violent) is independent of high- low-risk masculinity, how many of the 1,507 events would you expect to be violent and involve a high-risk-masculine man?
d. Repeat part **c** for the other combinations of event type and high- low-risk masculinity.
e. Calculate the χ^2 statistic for testing whether event type depends on high- low-risk masculinity.
f. Give the appropriate conclusion of the test mentioned in part **e**, using $\alpha = .05$.

Applying the Concepts—Intermediate

13.34 "Cry wolf" effect in air traffic controlling. Researchers at Alion Science Corporation and New Mexico State University collaborated on a study of how air traffic controllers respond to false alarms (*Human Factors*, Aug. 2009). The researchers theorize that the high rate of false alarms regarding midair collisions leads to the "cry wolf" effect, i.e., the tendency for air traffic controllers to ignore true alerts in the future. The investigation examined data on a random sample of 437 conflict alerts. Each alert was first classified as a "true" or "false" alert. Then, each was classified according to whether or not there was a human controller response to the alert. The number of the 437 alerts that fall into each of the combined categories is given as follows: True alert/No response–3; True alert/Response–231; False alert/No response–37; False alert/Response–166. This summary information is saved in the **ATC** file. Do the data indicate that the response rate of air traffic controllers to midair collision alarms differs for true and false alerts? Test using $\alpha = .05$. What inference can you make concerning the "cry wolf" effect?

Based on Wickens, C. D., Rice, S., Keller, D., Hutchins, S., Hughes, J., and Clayton, K., "False alerts in air traffic control conflict alerting system: Is there a 'cry wolf' effect?" *Human Factors*, Vol. 51, Issue 4, August 2009 (Table 2).

13.35 IQ and mental deficiency. A person is diagnosed with a mental deficiency if, before the age of 18, his or her score on a standard IQ test is no higher than 70 (two standard deviations below the mean of 100). Researchers at Cornell and West Virginia Universities examined the impact of rising IQ scores on diagnoses of mental deficiency (MD) (*American Psychologist*, October, 2003). IQ data were collected from different school districts across the United States, and the students were tested with either the Wechsler Intelligence Scale for Children—Revised (WISC-R) or the Wechsler Intelligence Scale for Children—Third Revision (WISC-III) IQ tests. The researchers focused on those students with IQs just above the mental deficiency cutoff (between 70 and 85), based on the original IQ test. These "borderline" MD students

Test/Retest	Diagnosed with MD	Above MD Cutoff IQ	Total
WISC-R / WISC-R	25	167	192
WISC-R / WISC-III	54	103	157
WISC-III / WISC-III	36	141	177

Source: Kanaya, T., Scullin, M. H., and Ceci, S. J. "The Flynn effect and U.S. Policies." *American Psychologist*, Vol. 58, No. 10, Oct. 2003 (Figure 1). Copyright © 2003 by the American Psychological Association. Reprinted with permission.

were then retested one year later with one of the IQ tests. The accompanying table gives the number of students diagnosed with mental deficiency on the basis of the retest. These data are saved in the **MDIQ** file. Conduct a chi-square test for independence to determine whether the proportion of students diagnosed with MD depends on the IQ test/retest method. Use $\alpha = .01$.

13.36 Creating menus to influence others. Refer to the *Journal of Consumer Research* (Mar. 2003) study on influencing the choices of others by offering undesirable alternatives, presented in Exercise 8.157 (p. 404). In another experiment conducted by the researcher, 96 subjects were asked to imagine that they had just moved to an apartment with two others and that they were shopping for a new appliance (e.g., a television, a microwave oven). Each subject was asked to create a menu of three brand choices for his or her roommates; then subjects were randomly assigned (in equal numbers) to one of three different "goal" conditions: (1) Create the menu in order to influence roommates to buy a preselected brand, (2) create the menu in order to influence roommates to buy a brand of your choice, and (3) create the menu with no intent to influence roommates. The researcher theorized that the menus created to influence others would likely include undesirable alternative brands. Consequently, the number of menus in each goal condition that was consistent with the theory was determined. The data are summarized in the accompanying table and saved in the **MENU3** file. Analyze the data for the purpose of determining whether the proportion of subjects who select menus consistent with the theory depends on the goal condition. Use $\alpha = .01$.

Goal Condition	Number Consistent with Theory	Number Not Consistent with Theory	Totals
Influence/preselected brand	15	17	32
Influence/own brand	14	18	32
No influence	3	29	32

Based on Hamilton, R. W. "Why do people suggest what they do not want? Using context effects to influence others' choices." *Journal of Consumer Research*, Vol. 29, March 2003 (Table 2).

13.37 Detecting Alzheimer's disease at an early age. Refer to the *Neuropsychology* (Jan. 2007) study of whether the cognitive effects of Alzheimer's disease can be detected at an early age, Exercise 13.17 (p. 735). Recall that a particular strand of DNA was classified into one of three genotypes: $E4^+/E4^+$, $E4^+/E4^-$, and $E4^-/E4^-$. In addition to a sample of 2,097 young adults (20–24 years), two other age groups were studied: a sample of 2,182 middle-aged adults (40–44 years) and a sample of 2,281 elderly adults (60–64 years). The accompanying table gives a breakdown of the number of adults with the three genotypes in each age category for

Age Group	$E4^+/E4^+$ Genotype	$E4^+/E4^-$ Genotype	$E4^-/E4^-$ Genotype	Sample Size
20–24	56	517	1,524	2,097
40–44	45	566	1,571	2,182
60–64	48	564	1,669	2,281

Source: Jorm, A. F., et al. "APOE genotype and cognitive functioning in a large age-stratified population sample." *Neuropsychology*, Vol. 21, No. 1, January 2007 (Table 1). Copyright © 2007 by the American Psychological Association. Reprinted with permission.

the total sample of 6,560 adults. (These data are saved in the **E4E4ALL** file.) The researchers concluded that "there were no significant genotype differences across the three age groups" using $\alpha = .05$. Do you agree?

13.38 Trapping grain moths. In an experiment described in the *Journal of Agricultural, Biological, and Environmental Statistics* (Dec. 2000), bins of corn were stocked with various parasites (e.g., grain moths) in late winter. In early summer (June), three bowl-shaped traps were placed on the surface of the grain in order to capture the moths. All three traps were baited with a sex pheromone lure; however, one trap used an unmarked sticky adhesive, one was marked with a fluorescent red powder, and one was marked with a fluorescent blue powder. The traps were set on a Wednesday, and the catch was collected the following Thursday and Friday. The accompanying table (saved in the **MOTHTRAP** file) shows the number of moths captured in each trap on each day. Conduct a test (at $\alpha = .10$) to determine whether the percentages of moths caught by the three traps depends on the day of the week.

	Adhesive, No Mark	Red Mark	Blue Mark
Thursday	136	41	17
Friday	101	50	18

Based on Wileyto, E. P. et al. "Self-marking recapture models for estimating closed insect populations." *Journal of Agricultural, Biological, and Environmental Statistics*, Vol. 5, No. 4, December 2000 (Table 5A).

13.39 Classifying air threats with heuristics. The *Journal of Behavioral Decision Making* (Jan. 2007) published a study on the use of heuristics to classify the threat level of approaching aircraft. Of special interest was the use of a fast and frugal heuristic—a computationally simple procedure for making judgments with limited information—named "Take-the-Best-for-Classification" (TTB-C). The subjects were 48 men and women, some from a Canadian Forces reserve unit, others university students. Each subject was presented with a radar screen on which simulated approaching aircraft were identified with asterisks. By using the computer mouse to click on the asterisk, one could receive further information about the aircraft. The goal was to identify the aircraft as "friend" or "foe" as fast as possible. Half the subjects were given cue-based instructions for determining the type of aircraft, while the other half were given pattern-based instructions. The researcher also classified the heuristic strategy used by the subject

Instruction	Strategy
Pattern	Other
Pattern	Other
Pattern	Other
Cue	TTBC
Cue	TTBC
⋮	⋮
Pattern	TTBC
Cue	Guess
Cue	TTBC
Cue	Guess
Pattern	Guess

Based on Bryant, D. J. "Classifying simulated air threats with fast and frugal heuristics." *Journal of Behavioral Decision Making*. Vol. 20, January 2007 (Appendix C).

as TTB-C, Guess, or Other. Data on the two variables Instruction type and Strategy, measured for each of the 48 subjects, are saved in the **AIRTHREAT** file. (Data on the first and last five subjects are shown in the table on p. 748.) Do the data provide sufficient evidence at $\alpha = .05$ to indicate that choice of heuristic strategy depends on type of instruction provided? How about at $\alpha = .01$?

13.40 Subarctic plant study. The traits of seed-bearing plants indigenous to subarctic Finland were studied in *Arctic, Antarctic, and Alpine Research* (May 2004). Plants were categorized according to *type* (dwarf shrub, herb, or grass), *abundance of seedlings* (no seedlings, rare seedlings, or abundant seedlings), *regenerative group* (no vegetative reproduction, vegetative reproduction possible, vegetative reproduction ineffective, or vegetative reproduction effective), *seed weight class* (0–.1, .1–.5, .5–1.0, 1.0–5.0, and > 5.0 milligrams), and *diaspore morphology* (no structure, pappus, wings, fleshy fruits, or awns/hooks). The data on a sample of 73 plants are saved in the **SEEDLING** file.

a. A contingency table for plant type and seedling abundance, produced by MINITAB, is shown below. (*Note:* NS = no seedlings, SA = seedlings abundant, and SR = seedlings rare.) Suppose you want to perform a chi-square test of independence to determine whether seedling abundance depends on plant type. Find the expected cell counts for the contingency table. Are the assumptions required for the test satisfied?

```
Tabulated statistics: Abundance, Type

Rows: Abundance   Columns: Type

        DwarfShrub  Grasses  Herbs  All

NS          3          1       1      5
SA          5         14      32     51
SR          5          2      10     17
All        13         17      43     73

Cell Contents:      Count
```

b. Reformulate the contingency table by combining the NS and SR categories of seedling abundance. Find the expected cell counts for this new contingency table. Are the assumptions required for the test satisfied?

c. Reformulate the contingency table of part **b** by combining the dwarf shrub and grasses categories of plant type. Find the expected cell counts for this contingency table. Are the assumptions required for the test satisfied?

d. Carry out the chi-square test for independence on the contingency table you came up with in part **c,** using $\alpha = .10$. What do you conclude?

13.41 Susceptibility to hypnosis. A standardized procedure for determining a person's susceptibility to hypnosis is the Stanford Hypnotic Susceptibility Scale, Form C (SHSS:C). Recently, a new method called the Computer-Assisted Hypnosis Scale (CAHS), which uses a computer as a facilitator of hypnosis, has been developed. Each scale classifies a person's hypnotic susceptibility as low, medium, high, or very high. Researchers at the University of Tennessee compared the two scales by administering both tests to each of 130 undergraduate volunteers. (*Psychological Assessment*, Mar. 1995). The hypnotic classifications are summarized in the table at the bottom of the page and saved in the **HYPNOSIS** file. A contingency table analysis will be performed to determine whether CAHS level and SHSS level are independent.

a. Check to see if the assumption of expected cell counts of 5 or more is satisfied. Should you proceed with the analysis? Explain.

b. One way to satisfy the assumption of part **a** is to combine the data for two or more categories (e.g., high and very high) in the contingency table. Form a new contingency table by combining the data for the high and very high categories in both the rows and the columns.

c. Calculate the expected cell counts in the new contingency table you formed in part **c.** Is the assumption now satisfied?

d. Perform the chi-square test on the new contingency table. Use $\alpha = .05$. Interpret the results.

13.42 Guilt in decision making. The effect of guilt emotion on how a decision maker focuses on the problem was investigated in the Jan. 2007 issue of the *Journal of Behavioral Decision Making*. A total of 171 volunteer students participated in the experiment, where each was randomly assigned to one of three emotional states (guilt, anger, or neutral) through a reading/writing task. Immediately after the task, the students were presented with a decision problem where the stated option has predominantly negative features (e.g., spending money on repairing a very old car). The results (number responding in each category) are summarized in the accompanying table and saved in the **GUILT** file. Is there sufficient evidence (at $\alpha = .10$) to claim that the option choice depends on emotional state?

Emotional State	Choose Stated Option	Do Not Choose Stated Option	Totals
Guilt	45	12	57
Anger	8	50	58
Neutral	7	49	56
Totals	60	111	171

Based on Gangemi, A., and Mancini, F. "Guilt and focusing in decision-making." *Journal of Behavioral Decision Making*, Vol. 20, Jan. 2007 (Table 2).

Table for Exercise 13.41

		CAHS Level				
		Low	Medium	High	Very High	Totals
SHSS: C Level	**Low**	32	14	2	0	48
	Medium	11	14	6	0	31
	High	6	14	19	3	42
	Very High	0	2	4	3	9
	Totals	49	44	31	6	130

Applying the Concepts—Advanced

13.43 Efficacy of an HIV vaccine. New, effective AIDS vaccines are now being developed through the process of "sieving"—that is, sifting out infections with some strains of HIV. Harvard School of Public Health statistician Peter Gilbert demonstrated how to test the efficacy of an HIV vaccine in *Chance* (Fall 2000). As an example, using the 2 × 2 table shown below, Gilbert reported the results of VaxGen's preliminary HIV vaccine trial. The vaccine was designed to eliminate a particular strain of the virus called the "MN strain." The trial consisted of 7 AIDS patients vaccinated with the new drug and 31 AIDS patients who were treated with a placebo (no vaccination). The first table (saved in the **HIVVAC1** file) shows the number of patients who tested positive and negative for the MN strain in the trial follow-up period.

		MN Strain		
		Positive	Negative	Totals
Patient Group	**Unvaccinated**	22	9	31
	Vaccinated	2	5	7
	Totals	24	14	38

Source: Gilbert, P. "Developing an AIDS vaccine by sieving." *Chance,* Vol. 13, No. 4, Fall 2000. Reprinted with permission from *Chance.* Copyright 2000 by the American Statistical Association. All rights reserved.

a. Conduct a test to determine whether the vaccine is effective in treating the MN strain of HIV. Use $\alpha = .05$.

b. Are the assumptions for the test you carried out in part **a** satisfied? What are the consequences if the assumptions are violated?

SAS output for Exercise 13.43

The FREQ Procedure

Table of GROUP by MNSTRAIN

GROUP   MNSTRAIN

Frequency Expected	NEG	POS	Total
UN	9 11.421	22 19.579	31
V	5 2.5789	2 4.4211	7
Total	14	24	38

Statistics for Table of GROUP by MNSTRAIN

Statistic	DF	Value	Prob
Chi-Square	1	4.4112	0.0357
Likelihood Ratio Chi-Square	1	4.2893	0.0384
Continuity Adj. Chi-Square	1	2.7773	0.0956
Mantel-Haenszel Chi-Square	1	4.2952	0.0382
Phi Coefficient		-0.3407	
Contingency Coefficient		0.3225	
Cramer's V		-0.3407	

WARNING: 50% of the cells have expected counts less than 5. Chi-Square may not be a valid test.

Fisher's Exact Test

Cell (1,1) Frequency (F)	9
Left-sided Pr <= F	0.0498
Right-sided Pr >= F	0.9940
Table Probability (P)	0.0438
Two-sided Pr <= P	0.0772

Sample Size = 38

c. In the case of a 2 × 2 contingency table, R. A. Fisher (1935) developed a procedure for computing the exact *p*-value for the test (called *Fisher's exact test*). The method utilizes the hypergeometric probability distribution of Chapter 4 (p. 214). Consider the hypergeometric probability

$$\frac{\binom{7}{2}\binom{31}{22}}{\binom{38}{24}}$$

which represents the probability that 2 out of 7 vaccinated AIDS patients test positive and 22 out of 31 unvaccinated patients test positive—that is, the probability of the result shown in the table, given that the null hypothesis of independence is true. Compute this probability (called the *probability of the contingency table*).

d. Refer to part **c**. Two contingency tables (with the same marginal totals as the original table) that are more unsupportive of the null hypothesis of independence than the observed table are shown below (These data are saved in the **HIVVAC2** and **HIVVAC3** files respectively.). First, explain why these tables provide more evidence to reject H_0 than the original table does. Then compute the probability of each table, using the hypergeometric formula.

e. The *p*-value of Fisher's exact test is the probability of observing a result at least as unsupportive of the null hypothesis as is the observed contingency table, given the same marginal totals. Sum the probabilities of parts **c** and **d** to obtain the *p*-value of Fisher's exact test. (To verify your calculations, check the *p*-value labeled **Left-sided Pr <= F** at the bottom of the SAS printout shown on the left. Interpret this value in the context of the vaccine trial.

		MN Strain		
		Positive	Negative	Totals
Patient Group	**Unvaccinated**	23	8	31
	Vaccinated	1	6	7
	Totals	24	14	38

		MN Strain		
		Positive	Negative	Totals
Patient Group	**Unvaccinated**	24	7	31
	Vaccinated	0	7	7
	Totals	24	14	38

13.44 Examining the "Monty Hall Dilemma." In Exercise 3.197 (p. 176) you solved the game show problem of whether or not to switch your choice of three doors, one of which hides a prize, after the host reveals what is behind a door that is not chosen. (Despite the natural inclination of many to keep one's first choice, the correct answer is that you should switch your choice of doors.) This problem is sometimes called the "Monty Hall Dilemma," named for Monty Hall, the host of the popular TV game show *Let's Make a*

Deal. In *Thinking & Reasoning* (July, 2007), Wichita State University professors set up an experiment designed to influence subjects to switch their original choice of doors. Each subject participated in 23 trials. In trial 1, three (boxes) representing doors were presented on a computer screen; only one box hid a prize. In each subsequent trial, an additional box was presented, so that in trial 23, twenty-five boxes were presented. In each trial, after a box was selected, all of the remaining boxes except for one either (1) were shown to be empty (*Empty* condition), (2) disappeared (*Vanish* condition), (3) disappeared, and the chosen box was enlarged (*Steroids* condition), or (4) disappeared, and the remaining box not chosen was enlarged (*Steroids2* condition). Twenty-seven subjects were assigned to each condition. The number of subjects who ultimately switched boxes is tallied, by condition, in the following table for both the first trial and the last trial, These data are saved in the **MONTYHALL** file.

	First Trial (1)		Last Trial (23)	
Condition	Switch Boxes	No Switch	Switch Boxes	No Switch
Empty	10	17	23	4
Vanish	3	24	12	15
Steroids	5	22	21	6
Steroids2	8	19	19	8

Based on Howard, J. N., Lambdin, C. G., and Datteri, D. L. "Let's make a deal: Quality and availability of second-stage information as a catalyst for change." *Thinking & Reasoning*, Vol. 13, No. 3, July 2007 (Table 2).

a. For a selected trial, does the likelihood of switching boxes depend on condition?

b. For a given condition, does the likelihood of switching boxes depend on trial number?

c. On the basis of the results you obtained in parts **a** and **b**, what factors influence a subject to switch choices?

13.4 A Word of Caution about Chi-Square Tests

Because the χ^2 statistic for testing hypotheses about multinomial probabilities is one of the most widely applied statistical tools, it is also one of the most abused statistical procedures. Consequently, the user should always be certain that the experiment satisfies the assumptions underlying each procedure. Furthermore, the user should be certain that the sample is drawn from the correct population—that is, from the population about which the inference is to be made.

The use of the χ^2 probability distribution as an approximation to the sampling distribution for χ^2 should be avoided when the expected counts are very small. The approximation can become very poor when these expected counts are small; thus, the true α level may be quite different from the tabular value. As a rule of thumb, an expected cell count of at least 5 means that the χ^2 probability distribution can be used to determine an approximate critical value.

If the χ^2 value does not exceed the established critical value of χ^2, *do not accept the hypothesis of independence.* You would be risking a Type II error (accepting H_0 when it is false), and the probability β of committing such an error is unknown. The usual alternative hypothesis is that the classifications are dependent. Because the number of ways in which two classifications can be dependent is virtually infinite, it is difficult to calculate one or even several values of β to represent such a broad alternative hypothesis. Therefore, we avoid concluding that two classifications are independent, even when χ^2 is small.

Finally, if a contingency table χ^2 value does exceed the critical value, we must be careful to avoid inferring that a *causal* relationship exists between the classifications. Our alternative hypothesis states that the two classifications are statistically dependent—and a statistical dependence does not imply causality. Therefore, *the existence of a causal relationship cannot be established by a contingency table analysis.*

CHAPTER NOTES

Key Terms

Categories 727
Cells 727
Cell counts 727
Chi-square test 729
Classes 727
Contingency table 737
Contingency table with fixed marginals 743
Dependence 739

Dimensions of classification 737
Expected cell count 730
Independence of two classifications 737
Marginal probabilities 737
Multinomial experiment 727
Observed cell count 738
One-way table 728
Two-way table 736

Key Symbols/Notation

$p_{i,0}$	Value of multinomial probability p_1 hypothesized in H_0
χ^2	Chi-square test statistic used in analysis of categorical data
n_i	Number of observed outcomes in cell i of a one-way table
E_i	Expected number of outcomes in cell i of a one-way table

p_{ij} Probability of an outcome in row i and column j of a two-way table

n_{ij} Number of observed outcomes in row i and column j of a two-way table

E_{ij} Expected number of outcomes in row i and column j of a two-way table

R_i Total number of outcomes in row i of a two-way table

C_j Total number of outcomes in column j of a two-way table

Key Ideas

Multinomial Data

Qualitative data that fall into more than two categories (or classes)

Properties of a Multinomial Experiment

1. n identical trials
2. k possible outcomes to each trial
3. probabilities of the k outcomes $(p_1, p_2, \ldots, p_k)$ where $p_1 + p_2 + \ldots + p_k = 1$, remain the same from trial to trial

4. trials are independent
5. variables of interest: *cell counts* (i.e., number of observations falling into each outcome category), denoted $n_1, n_2, \ldots, n_k$

One-Way Table

Summary table for a *single* qualitative variable

Two-Way (Contingency) Table

Summary table for *two* qualitative variables

Chi-Square (χ^2) Statistic

used to test category probabilities in one-way and two-way tables

Chi-square tests for independence

should **not** be used to *infer a causal relationship between two QLs*

Conditions Required for Valid χ^2 Tests

1. multinomial experiment
2. sample size n is large (expected cell counts are all greater than or equal to 5)

Categorical Data Analysis Guide

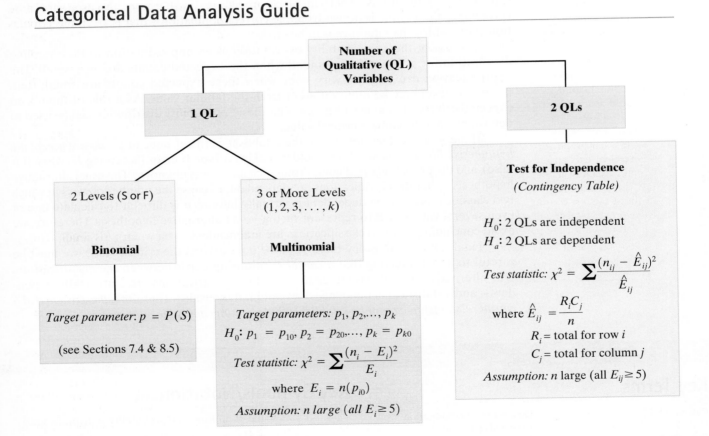

Supplementary Exercises 13.45–13.69

Understanding the Principles

13.45 *True or False.* Rejecting the null hypothesis in a chi-square test for independence implies that a causal relationship exists between the two categorical variables.

13.46 What is the difference between a one-way chi-square analysis and a two-way chi-square analysis?

Learning the Mechanics

13.47 A random sample of 250 observations was classified according to the row and column categories shown in the following table:

		Column		
		1	2	3
	1	20	20	10
Row	2	10	20	70
	3	20	50	30

a. Do the data provide sufficient evidence to conclude that the rows and columns are dependent? Test, using $\alpha = .05$.

b. Would the analysis change if the row totals were fixed before the data were collected?

c. Do the assumptions required for the analysis to be valid differ according to whether the row (or column) totals are fixed? Explain.

d. Convert the table entries to percentages by using each column total as a base and calculating each row response as a percentage of the corresponding column total. In addition, calculate the row totals and convert them to percentages of all 250 observations.

e. Create a bar graph with the percentages from row 1 on the vertical axis and the column number on the horizontal axis. Draw a horizontal line corresponding to the total percentage for row 1. Does the graph support the result of the test conducted in part **a**?

f. Repeat part **e** for the percentages from row 2.

g. Repeat part **e** for the percentages from row 3.

13.48 A random sample of 150 observations was classified into the categories shown in the following table:

	Category				
	1	2	3	4	5
n_i	28	35	33	25	29

a. Do the data provide sufficient evidence that the categories are not equally likely? Use $\alpha = .10$.

b. Form a 90% confidence interval for p_2, the probability that an observation will fall into category 2.

Applying the Concepts—Basic

13.49 Location of major sports venues. There has been a recent trend for professional sports franchises in Major League Baseball (MLB), the National Football League (NFL), the National Basketball Association (NBA), and the National Hockey League (NHL) to build new stadiums and ballparks in urban, downtown venues. An article in *Professional Geographer* (Feb. 2000) investigated whether there has been a significant suburban-to-urban shift in the location of major sport facilities. In 1985, 40% of all major sport facilities were located downtown, 30% in central cities, and 30% in suburban areas. In contrast, of the 113 major sports franchises that existed in 1997, 58 were built downtown, 26 in central cities, and 29 in suburban areas.

a. Describe the qualitative variable of interest in the study. Give the levels (categories) associated with the variable.

b. Give the null hypothesis for a test to determine whether the proportions of major sports facilities in downtown, central city, and suburban areas in 1997 are the same as in 1985.

c. If the null hypothesis of part **b** is true, how many of the 113 sports facilities in 1997 would you expect to be located in downtown, central city, and suburban areas, respectively?

d. Find the value of the chi-square statistic for testing the null hypothesis of part **b**.

e. Find the (approximate) *p*-value of the test, and give the appropriate conclusion in the words of the problem. Assume that $\alpha = .05$.

13.50 Scanning Internet messages. *Inc. Technology* (Mar. 18, 1997) reported the results of an Equifax/Harris Consumer Privacy Survey in which 328 Internet users indicated their level of agreement with the following statement: "The government needs to be able to scan Internet messages and user communications to prevent fraud and other crimes." The number of users in each response category is summarized as follows:

Agree Strongly	Agree Somewhat	Disagree Somewhat	Disagree Strongly
59	108	82	79

a. Specify the null and alternative hypotheses you would use to determine whether the opinions of Internet users are evenly divided among the four categories.

b. Conduct the test of part **a**, using $\alpha = .05$.

c. In the context of this exercise, what is a Type I error? A Type II error?

d. What assumptions must hold in order to ensure the validity of the test you conducted in part **b**?

13.51 Risk factor for lumbar disease. One of the most common musculoskeletal disorders is lumbar disk disease (LDD). Medical researchers reported finding a common genetic risk factor for LDD (*Journal of the American Medical Association*, Apr. 11, 2001). The study included 171 Finnish patients diagnosed with LDD (the patient group) and 321 without LDD (the control group). Of the 171 LDD patients, 21 were discovered to have the genetic trait. Of the 321 people in the control group, 15 had the genetic trait.

a. Consider the two categorical variables group and presence/absence of genetic trait. Form a 2 × 2 contingency table for these variables.

b. Conduct a test to determine whether the genetic trait occurs at a higher rate in LDD patients than in the controls. Use $\alpha = .01$.

c. Construct a bar graph that will visually support your conclusion in part **b**.

13.52 Late-emerging reading disabilities. Studies of children with reading disabilities typically focus on "early-emerging" difficulties identified prior to the fourth grade. Psychologists at Haskins Laboratories recently studied children with "late-emerging" reading difficulties (i.e., children who appeared to undergo a fourth-grade "slump" in reading achievement) and published their findings in the *Journal of Educational Psychology* (June 2003). A sample of 161 children was selected from fourth and fifth graders at elementary schools in Philadelphia. In addition to recording the grade level, the researchers determined whether each child had a previously undetected reading disability. Sixty-six children were diagnosed with a reading disability. Of these children, 32 were fourth graders and 34 were fifth graders. Similarly, of the 95 children with normal reading achievement, 55 were fourth graders and 40 were fifth graders.

   a. Identify the two qualitative variables (and corresponding levels) measured in the study.

   b. From the information provided, form a contingency table.

   c. Assuming that the two variables are independent, calculate the expected cell counts.

   d. Find the test statistic for determining whether the proportions of fourth and fifth graders with reading disabilities differs from the proportions of fourth and fifth graders with normal reading skills.

   e. Find the rejection region for the test if $\alpha = .10$.

   f. Is there a link between reading disability and grade level? Give the appropriate conclusion of the test.

13.53 Politics and religion. University of Maryland professor Ted R. Gurr examined the political strategies used by ethnic groups worldwide in their fight for minority rights. (*Political Science & Politics*, June 2000). Each in a sample of 275 ethnic groups was classified according to world region and highest level of political action reported. The data are summarized in the contingency table below and saved in the **ETHNIC** file. Conduct a test at $\alpha = .10$ to determine whether political strategy of ethnic groups depends on world region. Support your answer with a graph.

13.54 "Made in the USA" survey. Refer to the *Journal of Global Business* (Spring 2002) study of what "Made in the USA" on product labels means to the typical consumer, presented in Exercise 2.179 (p. 99). Recall that 106 shoppers participated in the survey. Their responses, given as a percentage of U.S. labor and materials in four categories, are summarized as follows: 64 shoppers responded 100%; 20 shoppers stated 75 to 99%; 18 shoppers stated 50 to 74%; and 4 shoppers said less than 50%. Suppose a consumer advocate group claims that half of all consumers believe that "Made in the USA" means "100%" of labor and materials are produced in the United States, one-fourth believe that "75 to 99%" are produced in the United States, one-fifth believe that "50 to 74%" are produced in the United States, and 5 percent believe that "less than 50%" are produced in the United States.

   a. Describe the qualitative variable of interest in the study. Give the levels (categories) associated with the variable.

   b. What are the values of $p_1, p_2, p_3,$ and p_4, the probabilities associated with the four response categories hypothesized by the consumer advocate group?

   c. Give the null and alternative hypotheses for testing the consumer advocate group's claim.

   d. Compute the test statistic for testing the hypotheses stated in part **c.**

   e. Find the rejection region of the test at $\alpha = .10$.

   f. State the conclusion in the words of the problem.

   g. Find and interpret a 90% confidence interval for the true proportion of consumers who believe that "Made in the USA" means that "100%" of labor and materials are produced in the United States.

   Based on "'Made in the USA': Consumer perceptions, deception and policy alternatives." *Journal of Global Business*, Vol. 13, No. 24, Spring 2002 (Table 3).

13.55 Hearing impairment study. The *Journal of Intellectual Disability Research* (Feb. 1995) published a longitudinal study of hearing impairment in a group of elderly patients with intellectual disability. The hearing function of each patient was screened each year over a 10-year period. At the study's conclusion, the hearing loss of each patient was categorized as severe, moderate, mild, or none. The classifications of the 28 surviving patients are summarized in the table below and saved in the **HEARIMP** file.

Hearing Loss	Number of Patients
None	7
Mild	7
Moderate	9
Severe	5
Total	28

Table for Exercise 13.53

		Political Strategy		
		No Political Action	Mobilization, Mass Action	Terrorism, Rebellion, Civil War
World Region	**Latin American**	24	31	7
	Post-Communist	32	23	4
	South, Southeast, East Asia	11	22	26
	Africa/Middle East	39	36	20

Table 1 from "Nonviolence ethnopolitics: Strategies for the attainment of group rights and autonomy" by Ted Robert Gurr. *Political Science & Politics*, Vol. 33, No. 2, June 2000. Copyright © 2000 The American Political Science Association. Reprinted with the permission of Cambridge University Press.

a. Conduct a test to determine whether the true proportions of intellectually disabled elderly patients in each of the hearing-loss categories differ. Use $\alpha = .05$.

b. Use a 90% confidence interval to estimate the proportion of disabled elderly patients with severe hearing loss.

13.56 Butterfly hot spots. *Nature* (Sept. 1993) reported on a study of animal and plant species "hot spots" in Great Britain. A hot spot is defined as a 10-km² area that is species rich—that is, heavily populated by a species of interest. Analogously, a cold spot is a 10-km² area that is species poor. The accompanying table gives the number of butterfly hot spots and the number of butterfly cold spots in a sample of 2,588 10-km² areas. In theory, 5% of the areas should be butterfly hot spots and 5% should be butterfly cold spots, while the remaining areas (90%) are neutral. Test the theory, using $\alpha = .01$.

Butterfly hot spots	123
Butterfly cold spots	147
Neutral areas	2,318
Total	2,588

Source: Prendergast, J. R., et al. "Rare species, the coincidence of diversity hotspots and conservation strategies." *Nature*, Vol. 365, No. 6444, Sept. 23, 1993, p. 335 (Table 1), copyright 1993. Adapted by permission from Macmillan Publishers Ltd.

Applying the Concepts—Intermediate

13.57 Iraq War survey. The Pew Internet & American Life Project commissioned Princeton Survey Research Associates to develop and carry out a survey of what Americans think about the War in Iraq. Some of the results of the March 2003 survey of over 1,400 American adults are saved in the **IRAQWAR** file. Responses to the following questions were recorded:

1. Do you support or oppose the Iraq War? (1 = Support, 2 = Oppose)

2. Do you ever go online to access the Internet or World Wide Web? (1 = Yes, 2 = No)

3. Do you consider yourself a Republican, Democrat, or Independent? (1 = Rep., 2 = Dem., 3 = Ind.)

4. Have you or anyone in your household served in the U.S. military? (1 = Yes, I have; 2 = Yes, other; 3 = Yes, both; 4 = No)

5. In general, would you describe your political views as very conservative, conservative, moderate, liberal, or very liberal? (1 = Very conservative, 2 = Conservative, 3 = Moderate, 4 = Liberal, 5 = Very liberal)

6. What is your race? (1 = White, 2 = African-American, 3 = Asian, 4 = Mixed, 5 = Native American, 6 = Other)

7. What is your income range? (1 = < 10K, 2 = 10–20K, 3 = 20–30K, 4 = 30–40K, 5 = 40–50K, 6 = 50–75K, 7 = 75–100K, 8 = > 100K)

8. Do you live in a suburban, rural, or urban community? (1 = urban, 2 = suburban, 3 = rural)

Conduct a series of contingency table analyses to determine whether support for the Iraq War depends on one or more of the other categorical variables measured in the March 2003 survey.

13.58 Pig farmer study. An article in *Sociological Methods & Research* (May 2001) analyzed the data presented in the accompanying table. A sample of 262 Kansas pig farmers was classified according to their education level (college or not) and size of their pig farm (number of pigs). The data are saved in the **PIGFARM** file. Conduct a test to determine whether a pig farmer's education level has an impact on the size of the pig farm. Use $\alpha = .05$ and support your answer with a graph.

		Education Level		
		No College	College	Totals
Farm Size	**<1,000 pigs**	42	53	95
	1,000–2,000 pigs	27	42	69
	2,001–5,000 pigs	22	20	42
	>5,000 pigs	27	29	56
	Totals	118	144	262

Based on Agresti, A., and Liu, I. "Strategies for modeling a categorical variable allowing multiple category choices." *Sociological Methods & Research*, Vol. 29, No. 4, May 2001 (Table I).

13.59 Multiple-sclerosis drug. Interferons are proteins produced naturally by the human body that help fight infections and regulate the immune system. A drug developed from interferons, called Avonex, is now available for treating patients with multiple sclerosis (MS). In a clinical study, 85 MS patients received weekly injections of Avonex over a two-year period. The number of exacerbations (i.e., flare-ups of symptoms) was recorded for each patient and is summarized in the accompanying table. For MS patients who take a placebo (no drug) over a similar two-year period, it is known from previous studies that 26% will experience no exacerbations, 30% one exacerbation, 11% two exacerbations, 14% three exacerbations, and 19% four or more exacerbations.

Number of Exacerbations	Number of Patients
0	32
1	26
2	15
3	6
4 or more	6

Based on data from Biogen, Inc.

a. Conduct a test to determine whether the exacerbation distribution of MS patients who take Avonex differs from the percentages reported for placebo patients. Use $\alpha = .05$.

b. Find a 95% confidence interval for the true percentage of Avonex MS patients who remain free of exacerbations during a two-year period.

c. Refer to part **b.** Is there evidence that Avonex patients are more likely to have no exacerbations than placebo patients? Explain.

13.60 Flight response of geese to helicopter traffic. Offshore oil drilling near an Alaskan estuary has led to increased air traffic—mostly large helicopters—in the area. The U.S. Fish and Wildlife Service commissioned a study to investigate the impact these helicopters have on the flocks of Pacific brant geese that inhabit the estuary in the fall before migrating (*Statistical Case Studies: A Collaboration between Academe and Industry*, 1998). Two large helicopters were flown repeatedly over the estuary at different altitudes and lateral distances from the flock. The flight responses of the geese (recorded as "low" or "high"), the altitude (in hundreds of meters), and the lateral distance (also in hundreds of meters) for each of 464 helicopter overflights were recorded and are saved in the **PACGEESE** file. The data for the first 10 overflights are shown in the following table:

Overflight	Altitude	Lateral Distance	Flight Response
1	0.91	4.99	HIGH
2	0.91	8.21	HIGH
3	0.91	3.38	HIGH
4	9.14	21.08	LOW
5	1.52	6.60	HIGH
6	0.91	3.38	HIGH
7	3.05	0.16	HIGH
8	6.10	3.38	HIGH
9	3.05	6.60	HIGH
10	12.19	6.60	HIGH

Source: From Erickson, W., Nick, T., and Ward, D. "Investigating Flight Response of Pacific Brant to Helicopters at Izembek Lagoon, Alaska by Using Logistic Regression." *Statistical Case Studies: A Collaboration between Academe and Industry,* ASA-SIAM Series on Statistics and Applied Probability, 1998. Copyright © 1998 Society for Industrial and Applied Mathematics. Reprinted with permission. All rights reserved.

a. The researchers categorized altitude as follows: less than 300 meters, 300–600 meters, and 600 or more meters. Summarize the data in the **PACGEESE** file by creating a contingency table for altitude category and flight response.

b. Conduct a test to determine whether flight response of the geese depends on altitude of the helicopter. Test, using $\alpha = .01$.

c. The researchers categorized lateral distance as follows: less than 1,000 meters, 1,000–2,000 meters, 2,000–3,000 meters, and 3,000 or more meters. Summarize the data in the **PACGEESE** file by creating a contingency table for lateral distance category and flight response.

d. Conduct a test to determine whether flight response of the geese depends on lateral distance of helicopter from the flock. Test, using $\alpha = .01$.

e. The current Federal Aviation Authority (FAA) minimum altitude standard for flying over the estuary is 2,000 feet (approximately 610 meters). On the basis of the results obtained in parts **a–d,** what changes to the FAA regulations do you recommend in order to minimize the effects to Pacific brant geese?

13.61 Birds feeding on gypsy moths. A field study was conducted to identify the natural predators of the gypsy moth. (*Environmental Entomology,* June 1995). For one part of the study, 24 black-capped chickadees (common wintering birds) were captured in mist nets and individually caged. Each bird was offered a mass of gypsy moth eggs attached to a piece of bark. Half the birds were offered no other food (no choice), and half were offered a variety of other naturally occurring foods such as spruce and pine seeds (choice). The numbers of birds that did and did not feed on the gypsy moth egg mass are given in the next table and saved in the **MOTH** file. Analyze the data in the table to determine whether a relationship exists between food

choice and feeding or not feeding on gypsy moth eggs. Use $\alpha = .10$.

	Fed on Egg Mass	
	Yes	No
Choice of foods	2	10
No choice	8	4

13.62 Gangs and homemade weapons. The National Gang Crime Research Center (NGCRC) has developed a six-level gang classification system for both adults and juveniles. The six categories are shown in the accompanying table. The classification system was developed as a potential predictor of a gang member's propensity for violence in prison, jail, or a correctional facility. To test the system, the NGCRC collected data on approximately 10,000 confined offenders and assigned each a score from the gang classification system. (*Journal of Gang Research,* Winter 1997). One of several other variables measured by the NGCRC was whether or not the offender had ever carried a homemade weapon (e.g., knife) while in custody. The data on gang score and homemade weapon are summarized in the table below and saved in the **GANGS** file. Conduct a test to determine whether carrying a homemade weapon in custody depends on gang classification score. (Use $\alpha = .01$.) Support your conclusion with a graph.

Gang Classification Score	Homemade Weapon Carried	
	Yes	No
0 (Never joined a gang, no close friends in a gang)	255	2,551
1 (Never joined a gang, 1–4 close friends in a gang)	110	560
2 (Never joined a gang, 5 or more friends in a gang)	151	636
3 (Inactive gang member)	271	959
4 (Active gang member, no position of rank)	175	513
5 (Active gang member, holds position of rank)	476	831

Source: From Knox, G. W., et al. "A gang classification system for corrections." *Journal of Gang Research,* Vol. 4, No. 2, Winter 1997, p. 54 (Table 4). Reprinted with permission from National Gang Crime Research Center.

13.63 Top Internet search engines. Nielsen/NetRatings is a global leader in Internet media and market research. In May 2006, the firm reported on the "search" shares (i.e., the percentage of all Internet searches) for the most popular search engines available on the Web. Google Search accounted for 50% of all searches, Yahoo! Search for 22%, MSN Search for 11%, and all other search engines for 17%. Suppose that, in a random sample of 1,000 recent Internet searches, 487 used Google Search, 245 used Yahoo! Search, 121 used MSN Search, and 147 used another search engine.

a. Do the sample data disagree with the percentages reported by Nielsen/NetRatings? Test, using $\alpha = .05$.

b. Find and interpret a 95% confidence interval for the percentage of all Internet searches that use the Google Search engine.

13.64 Orientation clue experiment. *Human Factors* (Dec. 1988) published a study of color brightness as a body orientation clue. Ninety college students reclining on their backs in the dark were disoriented when positioned on a rotating platform under a slowly rotating disk that blocked their field of vision. The subjects were asked to say "Stop" when they felt as if they were right-side up. The position of the brightness pattern on the disk in relation to each student's body orientation was then recorded. Subjects selected only three disk brightness patterns as subjective vertical clues: (1) brighter side up, (2) darker side up, and (3) brighter and darker sides aligned on either side of the subjects' heads. The frequency counts for the experiment are given in the accompanying table and saved in the **BODYCLUE** file. Conduct a test to compare the proportions of subjects who fall into the three disk-orientation categories. Assume that you want to determine whether the three proportions differ. Use $\alpha = .05$.

Disk Orientation		
Brighter Side Up	Darker Side Up	Bright and Dark Sides Aligned
58	15	17

13.65 Coupon usage study. A hot topic in marketing research is the exploration of a technology-based self-service (TBSS) encounter, in which various technologies (e.g., ATMs, online banking, self-scanning at retail stores) allow the customer to perform all or part of the service. Marketing professor Dan Ladik of the University of Suffolk investigated whether there were differences in customer characteristics and customer satisfaction between users of discount coupons distributed through the mail (nontechnology users) and users of coupons distributed via the Internet (TBSS users). A questionnaire measured several qualitative variables (defined in the accompanying table) for each of 440 coupon users. The data are saved in the **COUPONS** file.

Variable Name	Levels (Possible Values)
Coupon User Type	Mail, Internet, or Both
Gender	Male or Female
Education	High School, Vo-Tech/College, 4-year College Degree, or Graduate School
Work Status	Full Time, Part Time, Not Working, Retired
Coupon Satisfaction	Satisfied, Unsatisfied, Indifferent

a. Consider the variable Coupon User Type. Conduct a test (at $\alpha = .05$) to determine whether the proportions of mail-only users, Internet-only users, and users of both media are statistically different. Illustrate the results with a graph.

b. The researcher wants to know whether there are differences in customer characteristics (i.e., Gender, Education, Work Status, and Coupon Satisfaction) among the three types of coupon users. For each characteristic, conduct a contingency table analysis (at $\alpha = .05$) to determine whether Coupon User Type is related to that characteristic. Illustrate your results with graphs.

13.66 Battle simulation trials. In order to evaluate their situational awareness, fighter aircraft pilots participate in battle simulations. At a random point in the trial, the simulator is frozen and data on situational awareness are immediately collected. The simulation is then continued until, ultimately, performance (e.g., number of kills) is measured. A study reported in *Human Factors* (Mar. 1995) investigated whether temporarily stopping the simulation results in any change in pilot performance. Trials were designed so that some simulations were stopped to collect situational awareness data while others were not. Each trial was then classified according to the number of kills made by the pilot. The data for 180 trials are summarized in the accompanying contingency table and saved in the **SIMKILLS** file. Conduct a contingency table analysis and interpret the results fully.

	Number of Kills					
	0	1	2	3	4	Totals
Stops	32	33	19	5	2	91
No Stops	24	36	18	8	3	89
Totals	56	69	37	13	5	180

Applying the Concepts—Advanced

13.67 Goodness-of-fit test. A statistical analysis is to be done on a set of data consisting of 1,000 monthly salaries. The analysis requires the assumption that the sample was drawn from a normal distribution. A preliminary test, called the χ^2 *goodness-of-fit test*, can be used to help determine whether it is reasonable to assume that the sample is from a normal distribution. Suppose the mean and standard deviation of the 1,000 salaries are hypothesized to be $1,200 and $200, respectively. Using the standard normal table, we can approximate the probability of a salary being in the intervals listed in the accompanying table. The third column represents the expected number of the 1,000 salaries to be found in each interval if the sample was drawn from a normal distribution with $\mu = \$1,200$ and $\sigma = \$200$. Suppose the last column contains the actual observed frequencies in the sample. Large differences between the observed and expected frequencies cast doubt on the normality assumption.

Interval	Probability	Expected Frequency	Observed Frequency
Less than $800	.023	23	26
$800 < \$1,000$	.136	136	146
$1,000 < \$1,200$	.341	341	361
$1,200 < \$1,400$	.341	341	311
$1,400 < \$1,600$	.136	136	143
$1,600 or above	.023	23	13

a. Compute the χ^2 statistic on the basis of the observed and expected frequencies.

b. Find the tabulated χ^2 value when $\alpha = .05$ and there are five degrees of freedom. (There are $k - 1 = 5$ df associated with this χ^2 statistic.)

c. On the basis of the χ^2 statistic and the tabulated χ^2 value, is there evidence that the salary distribution is nonnormal?

d. Find the approximate observed significance level for the test in part c.

13.68 Testing normality. Suppose a random variable is hypothesized to be normally distributed with a mean of 0 and a standard deviation of 1. A random sample of 200 observations of the variable yields frequencies in the intervals listed in the table shown below. Do the data provide sufficient evidence to contradict the hypothesis that x is normally distributed with $\mu = 0$ and $\sigma = 1$? Use the technique developed in Exercise 13.67.

Critical Thinking Challenge

13.69 A "rigged" election? *Chance* (Spring 2004) presented data from a recent election held to determine the board of directors of a local community. There were 27 candidates for the board, and each of 5,553 voters was allowed to choose 6 candidates. The claim was that "a fixed vote with fixed percentages [was] assigned to each and every candidate, making it impossible to participate in an honest election." Votes were tallied in six time slots: after 600 total votes were in, after 1,200, after 2,444, after 3,444, after 4,444, and, finally, after 5,553 votes. The data on three of the candidates (Smith, Coppin, and Montes) are shown in the accompanying table and saved in the **RIGVOTE** file. A residential organization believes that "there was nothing random about the count and tallies for each time slot, and specific unnatural or rigged percentages were being assigned to each and every candidate." Give your opinion. Is the probability of a candidate receiving votes independent of the time slot, and if so, does this imply a rigged election?

Time Slot	1	2	3	4	5	6
Votes for Smith	208	208	451	392	351	410
Votes for Coppin	55	51	109	98	88	104
Votes for Montes	133	117	255	211	186	227
Total Votes	600	600	1,244	1,000	1,000	1,109

Based on Gelman, A. "55,000 residents desperately need your help!" *Chance*, Vol. 17, No. 2, Spring 2004 (Figures 1 and 5).

Table for Exercise 13.68

Interval	$x < -2$	$-2 \le x < -1$	$-1 \le x < 0$	$0 \le x < 1$	$1 \le x < 2$	$x \ge 2$
Frequency	7	20	61	77	26	9

Activity   Binomial versus Multinomial Experiments

In this activity, you will study the difference between binomial and multinomial experiments.

1. A television station has hired an independent research group to determine whether television viewers in the area prefer its local news program to the news programs of two other stations in the same city. Explain why a multinomial experiment would be appropriate, and design a poll that satisfies the five properties of a multinomial experiment. State the null and alternative hypotheses for the corresponding χ^2 test.

2. Suppose the television station believes that a majority of local viewers prefers its news program to those of its two competitors. Explain why a binomial experiment would be appropriate to support this claim, and design a poll that satisfies the five properties of a binomial experiment. State the null and alternative hypotheses for the corresponding test.

3. Generalize the situations in Exercises 1 and 2 in order to describe conditions under which a multinomial experiment can be rephrased as a binomial experiment. Is there any advantage in doing so? Explain.

References

Agresti, A. *Categorical Data Analysis*. New York: Wiley, 1990.

Cochran, W. G. "The χ^2 test of goodness of fit." *Annals of Mathematical Statistics*, 1952, 23.

Conover, W. J. *Practical Nonparametric Statistics*, 2nd ed. New York: Wiley, 1980.

Fisher, R. A. "The logic of inductive inference (with discussion)." *Journal of the Royal Statistical Society*, Vol. 98, 1935, pp. 39–82.

Hollander, M., and Wolfe, D. A. *Nonparametric Statistical Methods*. New York: Wiley, 1973.

Savage, I. R. "Bibliography of nonparametric statistics and related topics." *Journal of the American Statistical Association*, 1953, 48.

USING TECHNOLOGY

MINITAB: Chi-Square Analyses

MINITAB can conduct chi-square tests for both one-way and two-way (contingency) tables.

One-Way Table

Step 1 Access the MINITAB worksheet file that contains the sample data for the qualitative variable of interest. [*Note:* The data file can have actual values (levels) of the variable for each observation, or, alternatively, two columns—one listing the levels of the qualitative variable and the other column with the observed counts for each level.]

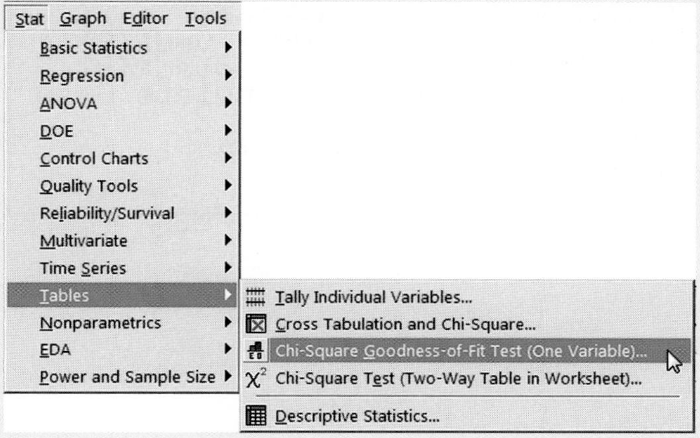

Figure 13.M.1
MINITAB menu options for a one-way chi-square analysis

Step 2 Click on the "Stat" button on the MINITAB menu bar, and then click on "Tables" and "Chi-Square Goodness-of-Fit Test (One Variable)," as shown in Figure 13.M.1. The resulting dialog box appears as shown in Figure 13.M.2.

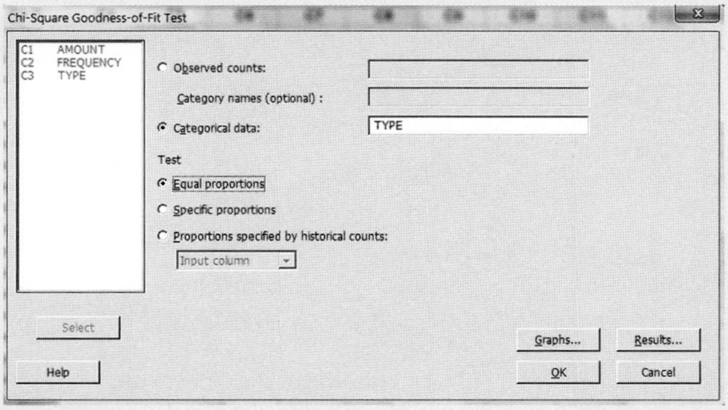

Figure 13.M.2
MINITAB one-way chi-square dialog box

Step 3 If your data have one column of values for your qualitative variable, select "Categorical data" and specify the variable name (or column) in the box. If your data have summary information in two columns (see above), select "Observed counts" and specify the column with the counts and the column with the variable names in the respective boxes.

Step 4 Select "Equal proportions" for a test of equal proportions, or select "Specific proportions" and enter the hypothesized proportion next to each level in the resulting box.

Step 5 Click "OK" to generate the MINITAB printout.

Two-Way Table

Step 1 Access the MINITAB worksheet file that contains the sample data. The data file should contain two qualitative variables, with category values for each of the *n* observations in the data set. Alternatively, the worksheet can contain the cell counts for each of the categories of the two qualitative variables.

Step 2 Click on the "Stat" button on the MINITAB menu bar and then click on "Tables" and "Cross Tabulation and Chi-Square," (see Figure 13.M.1). The resulting dialog box appears as shown in Figure 13.M.3.

Step 3 Specify one qualitative variable in the "For rows" box and the other qualitative variable in the "For columns" box. [*Note:* If your worksheet contains cell counts for the categories, enter the variable with the cell counts in the "Frequencies are in" box.]

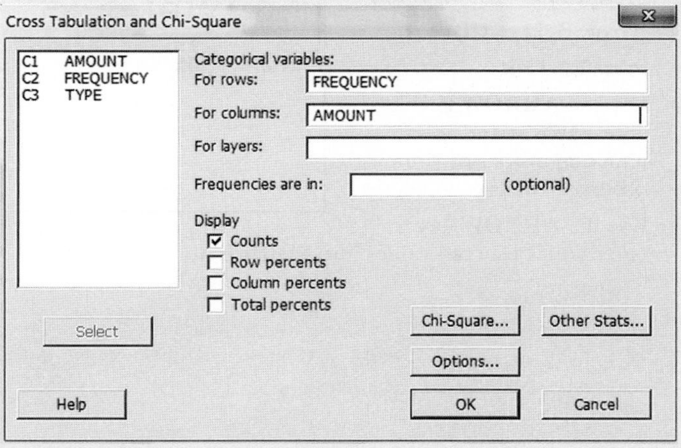

Figure 13.M.3
MINITAB cross tabulation dialog box

Step 4 Select the summary statistics (e.g., counts, percentages) you want to display in the contingency table.

Step 5 Click the "Chi-square" button. The resulting dialog box is shown in Figure 13.M.4.

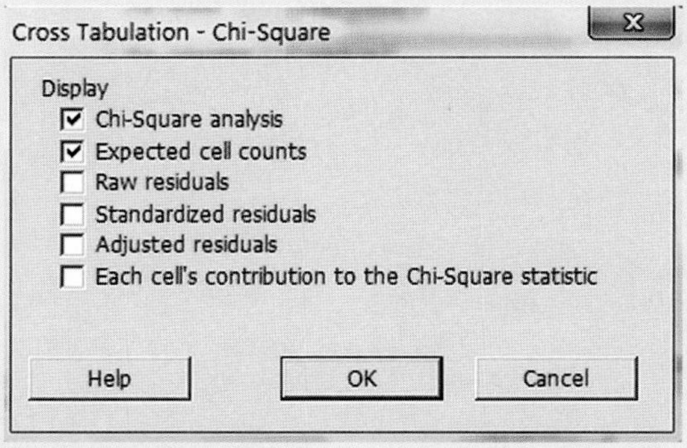

Figure 13.M.4
MINITAB chi-square dialog box

Step 6 Select "Chi-Square analysis" and "Expected cell counts" and click "OK."

Step 7 When you return to the "Cross Tabulation" menu screen, click "OK" to generate the MINITAB printout. *Note:* If your MINITAB worksheet contains only the cell counts for the contingency table in columns, click the "Chi-Square Test (Table in Worksheet)" menu option (see Figure 13.M.1) and specify the columns in the "Columns containing the table" box. Click "OK" to produce the MINITAB printout.

TI-83/TI-84 Plus Graphing Calculator: Chi-Square Analyses

The TI-83/TI-84 plus graphing calculator can be used to conduct a chi-square test for a two-way (contingency) table but cannot conduct a chi-square test for a one-way table.

Two-Way (Contingency) Table

Step 1 *Access the matrix menu to enter the observed values*

- Press **2nd x⁻¹** for **MATRX**
- Arrow right to **EDIT**
- Press **ENTER**
- Use the **ARROW** key to enter the row and column dimensions of your observed Matrix
- Use the **ARROW** key to enter your observed values into Matrix [A]

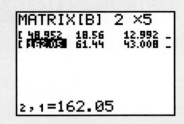

Step 2 *Access the matrix menu to enter the expected values*

- Press **2nd x⁻¹** for **MATRX**
- Arrow right to **EDIT**
- Arrow down to **2:[B]**
- Press **ENTER**
- Use the **ARROW** key to enter the row and column dimensions of your expected matrix (The dimensions will be the same as in Matrix A)
- Use the **ARROW** key to enter your expected values into Matrix [B]

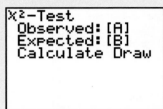

Step 3 *Access the statistical tests menu and perform the chi-square test*

- Press **STAT**
- Arrow right to **TESTS**
- Arrow down to χ^2 **Test**
- Press **ENTER**
- Arrow down to **Calculate**
- Press **ENTER**

Appendix A: Tables

Table I Random Numbers

Row \ Column	1	2	3	4	5	6	7	8	9	10	11	12	13	14
1	10480	15011	01536	02011	81647	91646	69179	14194	62590	36207	20969	99570	91291	90700
2	22368	46573	25595	85393	30995	89198	27982	53402	93965	34095	52666	19174	39615	99505
3	24130	48360	22527	97265	76393	64809	15179	24830	49340	32081	30680	19655	63348	58629
4	42167	93093	06243	61680	07856	16376	39440	53537	71341	57004	00849	74917	97758	16379
5	37570	39975	81837	16656	06121	91782	60468	81305	49684	60672	14110	06927	01263	54613
6	77921	06907	11008	42751	27756	53498	18602	70659	90655	15053	21916	81825	44394	42880
7	99562	72905	56420	69994	98872	31016	71194	18738	44013	48840	63213	21069	10634	12952
8	96301	91977	05463	07972	18876	20922	94595	56869	69014	60045	18425	84903	42508	32307
9	89579	14342	63661	10281	17453	18103	57740	84378	25331	12566	58678	44947	05585	56941
10	85475	36857	53342	53988	53060	59533	38867	62300	08158	17983	16439	11458	18593	64952
11	28918	69578	88231	33276	70997	79936	56865	05859	90106	31595	01547	85590	91610	78188
12	63553	40961	48235	03427	49626	69445	18663	72695	52180	20847	12234	90511	33703	90322
13	09429	93969	52636	92737	88974	33488	36320	17617	30015	08272	84115	27156	30613	74952
14	10365	61129	87529	85689	48237	52267	67689	93394	01511	26358	85104	20285	29975	89868
15	07119	97336	71048	08178	77233	13916	47564	81056	97735	85977	29372	74461	28551	90707
16	51085	12765	51821	51259	77452	16308	60756	92144	49442	53900	70960	63990	75601	40719
17	02368	21382	52404	60268	89368	19885	55322	44819	01188	65255	64835	44919	05944	55157
18	01011	54092	33362	94904	31273	04146	18594	29852	71585	85030	51132	01915	92747	64951
19	52162	53916	46369	58586	23216	14513	83149	98736	23495	64350	94738	17752	35156	35749
20	07056	97628	33787	09998	42698	06691	76988	13602	51851	46104	88916	19509	25625	58104
21	48663	91245	85828	14346	09172	30168	90229	04734	59193	22178	30421	61666	99904	32812
22	54164	58492	22421	74103	47070	25306	76468	26384	58151	06646	21524	15227	96909	44592
23	32639	32363	05597	24200	13363	38005	94342	28728	35806	06912	17012	64161	18296	22851
24	29334	27001	87637	87308	58731	00256	45834	15398	46557	41135	10367	07684	36188	18510
25	02488	33062	28834	07351	19731	92420	60952	61280	50001	67658	32586	86679	50720	94953
26	81525	72295	04839	96423	24878	82651	66566	14778	76797	14780	13300	87074	79666	95725
27	29676	20591	68086	26432	46901	20849	89768	81536	86645	12659	92259	57102	80428	25280
28	00742	57392	39064	66432	84673	40027	32832	61362	98947	96067	64760	64584	96096	98253
29	05366	04213	25669	26422	44407	44048	37937	63904	45766	66134	75470	66520	34693	90449
30	91921	26418	64117	94305	26766	25940	39972	22209	71500	64568	91402	42416	07844	69618
31	00582	04711	87917	77341	42206	35126	74087	99547	81817	42607	43808	76655	62028	76630
32	00725	69884	62797	56170	86324	88072	76222	36086	84637	93161	76038	65855	77919	88006
33	69011	65795	95876	55293	18988	27354	26575	08625	40801	59920	29841	80150	12777	48501
34	25976	57948	29888	88604	67917	48708	18912	82271	65424	69774	33611	54262	85963	03547
35	09763	83473	73577	12908	30883	18317	28290	35797	05998	41688	34952	37888	38917	88050
36	91576	42595	27958	30134	04024	86385	29880	99730	55536	84855	29080	09250	79656	73211
37	17955	56349	90999	49127	20044	59931	06115	20542	18059	02008	73708	83517	36103	42791
38	46503	18584	18845	49618	02304	51038	20655	58727	28168	15475	56942	53389	20562	87338
39	92157	89634	94824	78171	84610	82834	09922	25417	44137	48413	25555	21246	35509	20468
40	14577	62765	35605	81263	39667	47358	56873	56307	61607	49518	89656	20103	77490	18062
41	98427	07523	33362	64270	01638	92477	66969	98420	04880	45585	46565	04102	46880	45709

(continued)

Table I (continued)

Column Row	1	2	3	4	5	6	7	8	9	10	11	12	13	14
42	34914	63976	88720	82765	34476	17032	87589	40836	32427	70002	70663	88863	77775	69348
43	70060	28277	39475	46473	23219	53416	94970	25832	69975	94884	19661	72828	00102	66794
44	53976	54914	06990	67245	68350	82948	11398	42878	80287	88267	47363	46634	06541	97809
45	76072	29515	40980	07391	58745	25774	22987	80059	39911	96189	41151	14222	60697	59583
46	90725	52210	83974	29992	65831	38857	50490	83765	55657	14361	31720	57375	56228	41546
47	64364	67412	33339	31926	14883	24413	59744	92351	97473	89286	35931	04110	23726	51900
48	08962	00358	31662	25388	61642	34072	81249	35648	56891	69352	48373	45578	78547	81788
49	95012	68379	93526	70765	10592	04542	76463	54328	02349	17247	28865	14777	62730	92277
50	15664	10493	20492	38391	91132	21999	59516	81652	27195	48223	46751	22923	32261	85653
51	16408	81899	04153	53381	79401	21438	83035	92350	36693	31238	59649	91754	72772	02338
52	18629	81953	05520	91962	04739	13092	97662	24822	94730	06496	35090	04822	86774	98289
53	73115	35101	47498	87637	99016	71060	88824	71013	18735	20286	23153	72924	35165	43040
54	57491	16703	23167	49323	45021	33132	12544	41035	80780	45393	44812	12512	98931	91202
55	30405	83946	23792	14422	15059	45799	22716	19792	09983	74353	68668	30429	70735	25499
56	16631	35006	85900	98275	32388	52390	16815	69290	82732	38480	73817	32523	41961	44437
57	96773	20206	42559	78985	05300	22164	24369	54224	35083	19687	11052	91491	60383	19746
58	38935	64202	14349	82674	66523	44133	00697	35552	35970	19124	63318	29686	03387	59846
59	31624	76384	17403	53363	44167	64486	64758	75366	76554	31601	12614	33072	60332	92325
60	78919	19474	23632	27889	47914	02584	37680	20801	72152	39339	34806	08930	85001	87820
61	03931	33309	57047	74211	63445	17361	62825	39908	05607	91284	68833	25570	38818	46920
62	74426	33278	43972	10110	89917	15665	52872	73823	73144	88662	88970	74492	51805	99378
63	09066	00903	20795	95452	92648	45454	09552	88815	16553	51125	79375	97596	16296	66092
64	42238	12426	87025	14267	20979	04508	64535	31355	86064	29472	47689	05974	52468	16834
65	16153	08002	26504	41744	81959	65642	74240	56302	00033	67107	77510	70625	28725	34191
66	21457	40742	29820	96783	29400	21840	15035	34537	33310	06116	95240	15957	16572	06004
67	21581	57802	02050	89728	17937	37621	47075	42080	97403	48626	68995	43805	33386	21597
68	55612	78095	83197	33732	05810	24813	86902	60397	16489	03264	88525	42786	05269	92532
69	44657	66999	99324	51281	84463	60563	79312	93454	68876	25471	93911	25650	12682	73572
70	91340	84979	46949	81973	37949	61023	43997	15263	80644	43942	89203	71795	99533	50501
71	91227	21199	31935	27022	84067	05462	35216	14486	29891	68607	41867	14951	91696	85065
72	50001	38140	66321	19924	72163	09538	12151	06878	91903	18749	34405	56087	82790	70925
73	65390	05224	72958	28609	81406	39147	25549	48542	42627	45233	57202	94617	23772	07896
74	27504	96131	83944	41575	10573	08619	64482	73923	36152	05184	94142	25299	84387	34925
75	37169	94851	39117	89632	00959	16487	65536	49071	39782	17095	02330	74301	00275	48280
76	11508	70225	51111	38351	19444	66499	71945	05422	13442	78675	84081	66938	93654	59894
77	37449	30362	06694	54690	04052	53115	62757	95348	78662	11163	81651	50245	34971	52924
78	46515	70331	85922	38329	57015	15765	97161	17869	45349	61796	66345	81073	49106	79860
79	30986	81223	42416	58353	21532	30502	32305	86482	05174	07901	54339	58861	74818	46942
80	63798	64995	46583	09785	44160	78128	83991	42865	92520	83531	80377	35909	81250	54238
81	82486	84846	99254	67632	43218	50076	21361	64816	51202	88124	41870	52689	51275	83556
82	21885	32906	92431	09060	64297	51674	64126	62570	26123	05155	59194	52799	28225	85762

(continued)

Table I (continued)

Row \ Column	1	2	3	4	5	6	7	8	9	10	11	12	13	14
83	60336	98782	07408	53458	13564	59089	26445	29789	85205	41001	12535	12133	14645	23541
84	43937	46891	24010	25560	86355	33941	25786	54990	71899	15475	95434	98227	21824	19585
85	97656	63175	89303	16275	07100	92063	21942	18611	47348	20203	18534	03862	78095	50136
86	03299	01221	05418	38982	55758	92237	26759	86367	21216	98442	08303	56613	91511	75928
87	79626	06486	03574	17668	07785	76020	79924	25651	83325	88428	85076	72811	22717	50585
88	85636	68335	47539	03129	65651	11977	02510	26113	99447	68645	34327	15152	55230	93448
89	18039	14367	64337	06177	12143	46609	32989	74014	64708	00533	35398	58408	13261	47908
90	08362	15656	60627	36478	65648	16764	53412	09013	07832	41574	17639	82163	60859	75567
91	79556	29068	04142	16268	15387	12856	66227	38358	22478	73373	88732	09443	82558	05250
92	92608	82674	27072	32534	17075	27698	98204	63863	11951	34648	88022	56148	34925	57031
93	23982	25835	40055	67006	12293	02753	14827	23235	35071	99704	37543	11601	35503	85171
94	09915	96306	05908	97901	28395	14186	00821	80703	70426	75647	76310	88717	37890	40129
95	59037	33300	26695	62247	69927	76123	50842	43834	86654	70959	79725	93872	28117	19233
96	42488	78077	69882	61657	34136	79180	97526	43092	04098	73571	80799	76536	71255	64239
97	46764	86273	63003	93017	31204	36692	40202	35275	57306	55543	53203	18098	47625	88684
98	03237	45430	55417	63282	90816	17349	88298	90183	36600	78406	06216	95787	42579	90730
99	86591	81482	52667	61582	14972	90053	89534	76036	49199	43716	97548	04379	46370	28672
100	38534	01715	94964	87288	65680	43772	39560	12918	86537	62738	19636	51132	25739	56947

Table II   **Binomial Probabilities**

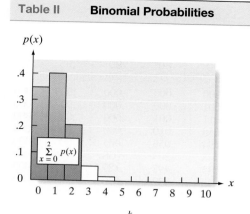

Tabulated values are $\sum_{x=0}^{k} p(x)$. *(Computations are rounded at the third decimal place.)*

a. $n = 5$

k \ p	.01	.05	.10	.20	.30	.40	.50	.60	.70	.80	.90	.95	.99
0	.951	.774	.590	.328	.168	.078	.031	.010	.002	.000	.000	.000	.000
1	.999	.977	.919	.737	.528	.337	.188	.087	.031	.007	.000	.000	.000
2	1.000	.999	.991	.942	.837	.683	.500	.317	.163	.058	.009	.001	.000
3	1.000	1.000	1.000	.993	.969	.913	.812	.663	.472	.263	.081	.023	.001
4	1.000	1.000	1.000	1.000	.998	.990	.969	.922	.832	.672	.410	.226	.049

b. $n = 6$

k \ p	.01	.05	.10	.20	.30	.40	.50	.60	.70	.80	.90	.95	.99
0	.941	.735	.531	.262	.118	.047	.016	.004	.001	.000	.000	.000	.000
1	.999	.967	.886	.655	.420	.233	.109	.041	.011	.002	.000	.000	.000
2	1.000	.998	.984	.901	.744	.544	.344	.179	.070	.017	.001	.000	.000
3	1.000	1.000	.999	.983	.930	.821	.656	.456	.256	.099	.016	.002	.000
4	1.000	1.000	1.000	.998	.989	.959	.891	.767	.580	.345	.114	.033	.001
5	1.000	1.000	1.000	1.000	.999	.996	.984	.953	.882	.738	.469	.265	.059

c. $n = 7$

k \ p	.01	.05	.10	.20	.30	.40	.50	.60	.70	.80	.90	.95	.99
0	.932	.698	.478	.210	.082	.028	.008	.002	.000	.000	.000	.000	.000
1	.998	.956	.850	.577	.329	.159	.063	.019	.004	.000	.000	.000	.000
2	1.000	.996	.974	.852	.647	.420	.227	.096	.029	.005	.000	.000	.000
3	1.000	1.000	.997	.967	.874	.710	.500	.290	.126	.033	.003	.000	.000
4	1.000	1.000	1.000	.995	.971	.904	.773	.580	.353	.148	.026	.004	.000
5	1.000	1.000	1.000	1.000	.996	.981	.937	.841	.671	.423	.150	.044	.002
6	1.000	1.000	1.000	1.000	1.000	.998	.992	.972	.918	.790	.522	.302	.068

(continued)

Table II   (continued)

d. n = 8

k \ p	.01	.05	.10	.20	.30	.40	.50	.60	.70	.80	.90	.95	.99
0	.923	.663	.430	.168	.058	.017	.004	.001	.000	.000	.000	.000	.000
1	.997	.943	.813	.503	.255	.106	.035	.009	.001	.000	.000	.000	.000
2	1.000	.994	.962	.797	.552	.315	.145	.050	.011	.001	.000	.000	.000
3	1.000	1.000	.995	.944	.806	.594	.363	.174	.058	.010	.000	.000	.000
4	1.000	1.000	1.000	.990	.942	.826	.637	.406	.194	.056	.005	.000	.000
5	1.000	1.000	1.000	.999	.989	.950	.855	.685	.448	.203	.038	.006	.000
6	1.000	1.000	1.000	1.000	.999	.991	.965	.894	.745	.497	.187	.057	.003
7	1.000	1.000	1.000	1.000	1.000	.999	.996	.983	.942	.832	.570	.337	.077

e. n = 9

k \ p	.01	.05	.10	.20	.30	.40	.50	.60	.70	.80	.90	.95	.99
0	.914	.630	.387	.134	.040	.010	.002	.000	.000	.000	.000	.000	.000
1	.997	.929	.775	.436	.196	.071	.020	.004	.000	.000	.000	.000	.000
2	1.000	.992	.947	.738	.463	.232	.090	.025	.004	.000	.000	.000	.000
3	1.000	.999	.992	.914	.730	.483	.254	.099	.025	.003	.000	.000	.000
4	1.000	1.000	.999	.980	.901	.733	.500	.267	.099	.020	.001	.000	.000
5	1.000	1.000	1.000	.997	.975	.901	.746	.517	.270	.086	.008	.001	.000
6	1.000	1.000	1.000	1.000	.996	.975	.910	.768	.537	.262	.053	.008	.000
7	1.000	1.000	1.000	1.000	1.000	.996	.980	.929	.804	.564	.225	.071	.003
8	1.000	1.000	1.000	1.000	1.000	1.000	.998	.990	.960	.866	.613	.370	.086

f. n = 10

k \ p	.01	.05	.10	.20	.30	.40	.50	.60	.70	.80	.90	.95	.99
0	.904	.599	.349	.107	.028	.006	.001	.000	.000	.000	.000	.000	.000
1	.996	.914	.736	.376	.149	.046	.011	.002	.000	.000	.000	.000	.000
2	1.000	.988	.930	.678	.383	.167	.055	.012	.002	.000	.000	.000	.000
3	1.000	.999	.987	.879	.650	.382	.172	.055	.011	.001	.000	.000	.000
4	1.000	1.000	.998	.967	.850	.633	.377	.166	.047	.006	.000	.000	.000
5	1.000	1.000	1.000	.994	.953	.834	.623	.367	.150	.033	.002	.000	.000
6	1.000	1.000	1.000	.999	.989	.945	.828	.618	.350	.121	.013	.001	.000
7	1.000	1.000	1.000	1.000	.998	.988	.945	.833	.617	.322	.070	.012	.000
8	1.000	1.000	1.000	1.000	1.000	.998	.989	.954	.851	.624	.264	.086	.004
9	1.000	1.000	1.000	1.000	1.000	1.000	.999	.994	.972	.893	.651	.401	.096

(continued)

Table II   (continued)

g. $n = 15$

k \ p	.01	.05	.10	.20	.30	.40	.50	.60	.70	.80	.90	.95	.99
0	.860	.463	.206	.035	.005	.000	.000	.000	.000	.000	.000	.000	.000
1	.990	.829	.549	.167	.035	.005	.000	.000	.000	.000	.000	.000	.000
2	1.000	.964	.816	.398	.127	.027	.004	.000	.000	.000	.000	.000	.000
3	1.000	.995	.944	.648	.297	.091	.018	.002	.000	.000	.000	.000	.000
4	1.000	.999	.987	.838	.515	.217	.059	.009	.001	.000	.000	.000	.000
5	1.000	1.000	.998	.939	.722	.403	.151	.034	.004	.000	.000	.000	.000
6	1.000	1.000	1.000	.982	.869	.610	.304	.095	.015	.001	.000	.000	.000
7	1.000	1.000	1.000	.996	.950	.787	.500	.213	.050	.004	.000	.000	.000
8	1.000	1.000	1.000	.999	.985	.905	.696	.390	.131	.018	.000	.000	.000
9	1.000	1.000	1.000	1.000	.996	.966	.849	.597	.278	.061	.002	.000	.000
10	1.000	1.000	1.000	1.000	.999	.991	.941	.783	.485	.164	.013	.001	.000
11	1.000	1.000	1.000	1.000	1.000	.998	.982	.909	.703	.352	.056	.005	.000
12	1.000	1.000	1.000	1.000	1.000	1.000	.996	.973	.873	.602	.184	.036	.000
13	1.000	1.000	1.000	1.000	1.000	1.000	1.000	.995	.965	.833	.451	.171	.010
14	1.000	1.000	1.000	1.000	1.000	1.000	1.000	1.000	.995	.965	.794	.537	.140

h. $n = 20$

k \ p	.01	.05	.10	.20	.30	.40	.50	.60	.70	.80	.90	.95	.99
0	.818	.358	.122	.012	.001	.000	.000	.000	.000	.000	.000	.000	.000
1	.983	.736	.392	.069	.008	.001	.000	.000	.000	.000	.000	.000	.000
2	.999	.925	.677	.206	.035	.004	.000	.000	.000	.000	.000	.000	.000
3	1.000	.984	.867	.411	.107	.016	.001	.000	.000	.000	.000	.000	.000
4	1.000	.997	.957	.630	.238	.051	.006	.000	.000	.000	.000	.000	.000
5	1.000	1.000	.989	.804	.416	.126	.021	.002	.000	.000	.000	.000	.000
6	1.000	1.000	.998	.913	.608	.250	.058	.006	.000	.000	.000	.000	.000
7	1.000	1.000	1.000	.968	.772	.416	.132	.021	.001	.000	.000	.000	.000
8	1.000	1.000	1.000	.990	.887	.596	.252	.057	.005	.000	.000	.000	.000
9	1.000	1.000	1.000	.997	.952	.755	.412	.128	.017	.001	.000	.000	.000
10	1.000	1.000	1.000	.999	.983	.872	.588	.245	.048	.003	.000	.000	.000
11	1.000	1.000	1.000	1.000	.995	.943	.748	.404	.113	.010	.000	.000	.000
12	1.000	1.000	1.000	1.000	.999	.979	.868	.584	.228	.032	.000	.000	.000
13	1.000	1.000	1.000	1.000	1.000	.994	.942	.750	.392	.087	.002	.000	.000
14	1.000	1.000	1.000	1.000	1.000	.998	.979	.874	.584	.196	.011	.000	.000
15	1.000	1.000	1.000	1.000	1.000	1.000	.994	.949	.762	.370	.043	.003	.000
16	1.000	1.000	1.000	1.000	1.000	1.000	.999	.984	.893	.589	.133	.016	.000
17	1.000	1.000	1.000	1.000	1.000	1.000	1.000	.996	.965	.794	.323	.075	.001
18	1.000	1.000	1.000	1.000	1.000	1.000	1.000	.999	.992	.931	.608	.264	.017
19	1.000	1.000	1.000	1.000	1.000	1.000	1.000	1.000	.999	.988	.878	.642	.182

(continued)

Table II	(continued)

i. $n = 25$

k \ p	.01	.05	.10	.20	.30	.40	.50	.60	.70	.80	.90	.95	.99
0	.778	.277	.072	.004	.000	.000	.000	.000	.000	.000	.000	.000	.000
1	.974	.642	.271	.027	.002	.000	.000	.000	.000	.000	.000	.000	.000
2	.998	.873	.537	.098	.009	.000	.000	.000	.000	.000	.000	.000	.000
3	1.000	.966	.764	.234	.033	.002	.000	.000	.000	.000	.000	.000	.000
4	1.000	.993	.902	.421	.090	.009	.000	.000	.000	.000	.000	.000	.000
5	1.000	.999	.967	.617	.193	.029	.002	.000	.000	.000	.000	.000	.000
6	1.000	1.000	.991	.780	.341	.074	.007	.000	.000	.000	.000	.000	.000
7	1.000	1.000	.998	.891	.512	.154	.022	.001	.000	.000	.000	.000	.000
8	1.000	1.000	1.000	.953	.677	.274	.054	.004	.000	.000	.000	.000	.000
9	1.000	1.000	1.000	.983	.811	.425	.115	.013	.000	.000	.000	.000	.000
10	1.000	1.000	1.000	.994	.902	.586	.212	.034	.002	.000	.000	.000	.000
11	1.000	1.000	1.000	.998	.956	.732	.345	.078	.006	.000	.000	.000	.000
12	1.000	1.000	1.000	1.000	.983	.846	.500	.154	.017	.000	.000	.000	.000
13	1.000	1.000	1.000	1.000	.994	.922	.655	.268	.044	.002	.000	.000	.000
14	1.000	1.000	1.000	1.000	.998	.966	.788	.414	.098	.006	.000	.000	.000
15	1.000	1.000	1.000	1.000	1.000	.987	.885	.575	.189	.017	.000	.000	.000
16	1.000	1.000	1.000	1.000	1.000	.996	.946	.726	.323	.047	.000	.000	.000
17	1.000	1.000	1.000	1.000	1.000	.999	.978	.846	.488	.109	.002	.000	.000
18	1.000	1.000	1.000	1.000	1.000	1.000	.993	.926	.659	.220	.009	.000	.000
19	1.000	1.000	1.000	1.000	1.000	1.000	.998	.971	.807	.383	.033	.001	.000
20	1.000	1.000	1.000	1.000	1.000	1.000	1.000	.991	.910	.579	.098	.007	.000
21	1.000	1.000	1.000	1.000	1.000	1.000	1.000	.998	.967	.766	.236	.034	.000
22	1.000	1.000	1.000	1.000	1.000	1.000	1.000	1.000	.991	.902	.463	.127	.002
23	1.000	1.000	1.000	1.000	1.000	1.000	1.000	1.000	.998	.973	.729	.358	.026
24	1.000	1.000	1.000	1.000	1.000	1.000	1.000	1.000	1.000	.996	.928	.723	.222

Table III   Poisson Probabilities

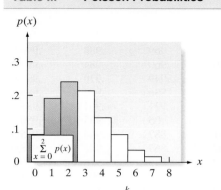

Tabulated values are $\sum_{x=0}^{k} p(x)$. (Computations are rounded at the third decimal place.)

λ \ k	0	1	2	3	4	5	6	7	8	9
.02	.980	1.000								
.04	.961	.999	1.000							
.06	.942	.998	1.000							
.08	.923	.997	1.000							
.10	.905	.995	1.000							
.15	.861	.990	.999	1.000						
.20	.819	.982	.999	1.000						
.25	.779	.974	.998	1.000						
.30	.741	.963	.996	1.000						
.35	.705	.951	.994	1.000						
.40	.670	.938	.992	.999	1.000					
.45	.638	.925	.989	.999	1.000					
.50	.607	.910	.986	.998	1.000					
.55	.577	.894	.982	.998	1.000					
.60	.549	.878	.977	.997	1.000					
.65	.522	.861	.972	.996	.999	1.000				
.70	.497	.844	.966	.994	.999	1.000				
.75	.472	.827	.959	.993	.999	1.000				
.80	.449	.809	.953	.991	.999	1.000				
.85	.427	.791	.945	.989	.998	1.000				
.90	.407	.772	.937	.987	.998	1.000				
.95	.387	.754	.929	.981	.997	1.000				
1.00	.368	.736	.920	.981	.996	.999	1.000			
1.1	.333	.699	.900	.974	.995	.999	1.000			
1.2	.301	.663	.879	.966	.992	.998	1.000			
1.3	.273	.627	.857	.957	.989	.998	1.000			
1.4	.247	.592	.833	.946	.986	.997	.999	1.000		
1.5	.223	.558	.809	.934	.981	.996	.999	1.000		
1.6	.202	.525	.783	.921	.976	.994	.999	1.000		
1.7	.183	.493	.757	.907	.970	.992	.998	1.000		
1.8	.165	.463	.731	.891	.964	.990	.997	.999	1.000	
1.9	.150	.434	.704	.875	.956	.987	.997	.999	1.000	
2.0	.135	.406	.677	.857	.947	.983	.995	.999	1.000	
2.2	.111	.355	.623	.819	.928	.975	.993	.998	1.000	
2.4	.091	.308	.570	.779	.904	.964	.988	.997	.999	1.000
2.6	.074	.267	.518	.736	.877	.951	.983	.995	.999	1.000
2.8	.061	.231	.469	.692	.848	.935	.976	.992	.998	.999
3.0	.050	.199	.423	.647	.815	.916	.966	.988	.996	.999
3.2	.041	.171	.380	.603	.781	.895	.955	.983	.994	.99
3.4	.033	.147	.340	.558	.744	.871	.942	.977	.992	.9

(co

Table III	(continued)								
λ \ k	0	1	2	3	4	5	6	7	8	9
3.6	.027	.126	.303	.515	.706	.844	.927	.969	.988	.996
3.8	.022	.107	.269	.473	.668	.816	.909	.960	.984	.994
4.0	.018	.092	.238	.433	.629	.785	.889	.949	.979	.992
4.2	.015	.078	.210	.395	.590	.753	.867	.936	.972	.989
4.4	.012	.066	.185	.359	.551	.720	.844	.921	.964	.985
4.6	.010	.056	.163	.326	.513	.686	.818	.905	.955	.980
4.8	.008	.048	.143	.294	.476	.651	.791	.887	.944	.975
5.0	.007	.040	.125	.265	.440	.616	.762	.867	.932	.968
5.2	.006	.034	.109	.238	.406	.581	.732	.845	.918	.960
5.4	.005	.029	.095	.213	.373	.546	.702	.822	.903	.951
5.6	.004	.024	.082	.191	.342	.512	.670	.797	.886	.941
5.8	.003	.021	.072	.170	.313	.478	.638	.771	.867	.929
6.0	.002	.017	.062	.151	.285	.446	.606	.744	.847	.916

	10	11	12	13	14	15	16			
2.8	1.000									
3.0	1.000									
3.2	1.000									
3.4	.999	1.000								
3.6	.999	1.000								
3.8	.998	.999	1.000							
4.0	.997	.999	1.000							
4.2	.996	.999	1.000							
4.4	.994	.998	.999	1.000						
4.6	.992	.997	.999	1.000						
4.8	.990	.996	.999	1.000						
5.0	.986	.995	.998	.999	1.000					
5.2	.982	.993	.997	.999	1.000					
5.4	.977	.990	.996	.999	1.000					
5.6	.972	.988	.995	.998	.999	1.000				
5.8	.965	.984	.993	.997	.999	1.000				
6.0	.957	.980	.991	.996	.999	.999	1.000			

	0	1	2	3	4	5	6	7	8	9
6.2	.002	.015	.054	.134	.259	.414	.574	.716	.826	.902
6.4	.002	.012	.046	.119	.235	.384	.542	.687	.803	.886
6.6	.001	.010	.040	.105	.213	.355	.511	.658	.780	.869
6.8	.001	.009	.034	.093	.192	.327	.480	.628	.755	.850
7.0	.001	.007	.030	.082	.173	.301	.450	.599	.729	.830
7.2	.001	.006	.025	.072	.156	.276	.420	.569	.703	.810
7.4	.001	.005	.022	.063	.140	.253	.392	.539	.676	.788
7.6	.001	.004	.019	.055	.125	.231	.365	.510	.648	.765
7.8	.000	.004	.016	.048	.112	.210	.338	.481	.620	.741
8.0	.000	.003	.014	.042	.100	.191	.313	.453	.593	.717
8.5	.000	.002	.009	.030	.074	.150	.256	.386	.523	.653
9.0	.000	.001	.006	.021	.055	.116	.207	.324	.456	.587
9.5	.000	.001	.004	.015	.040	.089	.165	.269	.392	.522
10.0	.000	.000	.003	.010	.029	.067	.130	.220	.333	.458

	10	11	12	13	14	15	16	17	18	19
6.2	.949	.975	.989	.995	.998	.999	1.000			
6.4	.939	.969	.986	.994	.997	.999	1.000			
6.6	.927	.963	.982	.992	.997	.999	.999	1.000		
6.8	.915	.955	.978	.990	.996	.998	.999	1.000		
7.0	.901	.947	.973	.987	.994	.998	.999	1.000		

(*continued*)

Table III (continued)

λ \ k	10	11	12	13	14	15	16	17	18	19
7.2	.887	.937	.967	.984	.993	.997	.999	.999	1.000	
7.4	.871	.926	.961	.980	.991	.996	.998	.999	1.000	
7.6	.854	.915	.954	.976	.989	.995	.998	.999	1.000	
7.8	.835	.902	.945	.971	.986	.993	.997	.999	1.000	
8.0	.816	.888	.936	.966	.983	.992	.996	.998	.999	1.000
8.5	.763	.849	.909	.949	.973	.986	.993	.997	.999	.999
9.0	.706	.803	.876	.926	.959	.978	.989	.995	.998	.999
9.5	.645	.752	.836	.898	.940	.967	.982	.991	.996	.998
10.0	.583	.697	.792	.864	.917	.951	.973	.986	.993	.997

λ \ k	20	21	22
8.5	1.000		
9.0	1.000		
9.5	.999	1.000	
10.0	.998	.999	1.000

λ \ k	0	1	2	3	4	5	6	7	8	9
10.5	.000	.000	.002	.007	.021	.050	.102	.179	.279	.397
11.0	.000	.000	.001	.005	.015	.038	.079	.143	.232	.341
11.5	.000	.000	.001	.003	.011	.028	.060	.114	.191	.289
12.0	.000	.000	.001	.002	.008	.020	.046	.090	.155	.242
12.5	.000	.000	.000	.002	.005	.015	.035	.070	.125	.201
13.0	.000	.000	.000	.001	.004	.011	.026	.054	.100	.166
13.5	.000	.000	.000	.001	.003	.008	.019	.041	.079	.135
14.0	.000	.000	.000	.000	.002	.006	.014	.032	.062	.109
14.5	.000	.000	.000	.000	.001	.004	.010	.024	.048	.088
15.0	.000	.000	.000	.000	.001	.003	.008	.018	.037	.070

λ \ k	10	11	12	13	14	15	16	17	18	19
10.5	.521	.639	.742	.825	.888	.932	.960	.978	.988	.994
11.0	.460	.579	.689	.781	.854	.907	.944	.968	.982	.991
11.5	.402	.520	.633	.733	.815	.878	.924	.954	.974	.986
12.0	.347	.462	.576	.682	.772	.844	.899	.937	.963	.979
12.5	.297	.406	.519	.628	.725	.806	.869	.916	.948	.969
13.0	.252	.353	.463	.573	.675	.764	.835	.890	.930	.957
13.5	.211	.304	.409	.518	.623	.718	.798	.861	.908	.942
14.0	.176	.260	.358	.464	.570	.669	.756	.827	.883	.923
14.5	.145	.220	.311	.413	.518	.619	.711	.790	.853	.901
15.0	.118	.185	.268	.363	.466	.568	.664	.749	.819	.875

λ \ k	20	21	22	23	24	25	26	27	28	29
10.5	.997	.999	.999	1.000						
11.0	.995	.998	.999	1.000						
11.5	.992	.996	.998	.999	1.000					
12.0	.988	.994	.997	.999	.999	1.000				
12.5	.983	.991	.995	.998	.999	.999	1.000			
13.0	.975	.986	.992	.996	.998	.999	1.000			
13.5	.965	.980	.989	.994	.997	.999	1.000			
14.0	.952	.971	.983	.991	.995	.998	.999	1.000		
14.5	.936	.960	.976	.986	.992	.997	.999	.999	1.000	
15.0	.917	.947	.967	.981	.989	.994	.997	.998	.999	1.000

(continued)

Table III (continued)

λ \ k	4	5	6	7	8	9	10	11	12	13
16	.000	.001	.004	.010	.022	.043	.077	.127	.193	.275
17	.000	.001	.002	.005	.013	.026	.049	.085	.135	.201
18	.000	.000	.001	.003	.007	.015	.030	.055	.092	.143
19	.000	.000	.001	.002	.004	.009	.018	.035	.061	.098
20	.000	.000	.000	.001	.002	.005	.011	.021	.039	.066
21	.000	.000	.000	.000	.001	.003	.006	.013	.025	.043
22	.000	.000	.000	.000	.001	.002	.004	.008	.015	.028
23	.000	.000	.000	.000	.000	.001	.002	.004	.009	.017
24	.000	.000	.000	.000	.000	.000	.001	.003	.005	.011
25	.000	.000	.000	.000	.000	.000	.001	.001	.003	.006

λ \ k	14	15	16	17	18	19	20	21	22	23
16	.368	.467	.566	.659	.742	.812	.868	.911	.942	.963
17	.281	.371	.468	.564	.655	.736	.805	.861	.905	.937
18	.208	.287	.375	.469	.562	.651	.731	.799	.855	.899
19	.150	.215	.292	.378	.469	.561	.647	.725	.793	.849
20	.105	.157	.221	.297	.381	.470	.559	.644	.721	.787
21	.072	.111	.163	.227	.302	.384	.471	.558	.640	.716
22	.048	.077	.117	.169	.232	.306	.387	.472	.556	.637
23	.031	.052	.082	.123	.175	.238	.310	.389	.472	.555
24	.020	.034	.056	.087	.128	.180	.243	.314	.392	.473
25	.012	.022	.038	.060	.092	.134	.185	.247	.318	.394

λ \ k	24	25	26	27	28	29	30	31	32	33
16	.978	.987	.993	.996	.998	.999	.999	1.000		
17	.959	.975	.985	.991	.995	.997	.999	.999	1.000	
18	.932	.955	.972	.983	.990	.994	.997	.998	.999	1.000
19	.893	.927	.951	.969	.980	.988	.993	.996	.998	.999
20	.843	.888	.922	.948	.966	.978	.987	.992	.995	.997
21	.782	.838	.883	.917	.944	.963	.976	.985	.991	.994
22	.712	.777	.832	.877	.913	.940	.959	.973	.983	.989
23	.635	.708	.772	.827	.873	.908	.936	.956	.971	.981
24	.554	.632	.704	.768	.823	.868	.904	.932	.953	.969
25	.473	.553	.629	.700	.763	.818	.863	.900	.929	.950

λ \ k	34	35	36	37	38	39	40	41	42	43
19	.999	1.000								
20	.999	.999	1.000							
21	.997	.998	.999	.999	1.000					
22	.994	.996	.998	.999	.999	1.000				
23	.988	.993	.996	.997	.999	.999	1.000			
24	.979	.987	.992	.995	.997	.998	.999	.999	1.000	
25	.966	.978	.985	.991	.991	.997	.998	.999	.999	1.000

| Table IV | **Normal Curve Areas** |

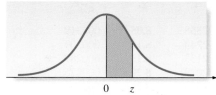

z	.00	.01	.02	.03	.04	.05	.06	.07	.08	.09
.0	.0000	.0040	.0080	.0120	.0160	.0199	.0239	.0279	.0319	.0359
.1	.0398	.0438	.0478	.0517	.0557	.0596	.0636	.0675	.0714	.0753
.2	.0793	.0832	.0871	.0910	.0948	.0987	.1026	.1064	.1103	.1141
.3	.1179	.1217	.1255	.1293	.1331	.1368	.1406	.1443	.1480	.1517
.4	.1554	.1591	.1628	.1664	.1700	.1736	.1772	.1808	.1844	.1879
.5	.1915	.1950	.1985	.2019	.2054	.2088	.2123	.2157	.2190	.2224
.6	.2257	.2291	.2324	.2357	.2389	.2422	.2454	.2486	.2517	.2549
.7	.2580	.2611	.2642	.2673	.2704	.2734	.2764	.2794	.2823	.2852
.8	.2881	.2910	.2939	.2967	.2995	.3023	.3051	.3078	.3106	.3133
.9	.3159	.3186	.3212	.3238	.3264	.3289	.3315	.3340	.3365	.3389
1.0	.3413	.3438	.3461	.3485	.3508	.3531	.3554	.3577	.3599	.3621
1.1	.3643	.3665	.3686	.3708	.3729	.3749	.3770	.3790	.3810	.3830
1.2	.3849	.3869	.3888	.3907	.3925	.3944	.3962	.3980	.3997	.4015
1.3	.4032	.4049	.4066	.4082	.4099	.4115	.4131	.4147	.4162	.4177
1.4	.4192	.4207	.4222	.4236	.4251	.4265	.4279	.4292	.4306	.4319
1.5	.4332	.4345	.4357	.4370	.4382	.4394	.4406	.4418	.4429	.4441
1.6	.4452	.4463	.4474	.4484	.4495	.4505	.4515	.4525	.4535	.4545
1.7	.4554	.4564	.4573	.4582	.4591	.4599	.4608	.4616	.4625	.4633
1.8	.4641	.4649	.4656	.4664	.4671	.4678	.4686	.4693	.4699	.4706
1.9	.4713	.4719	.4726	.4732	.4738	.4744	.4750	.4756	.4761	.4767
2.0	.4772	.4778	.4783	.4788	.4793	.4798	.4803	.4808	.4812	.4817
2.1	.4821	.4826	.4830	.4834	.4838	.4842	.4846	.4850	.4854	.4857
2.2	.4861	.4864	.4868	.4871	.4875	.4878	.4881	.4884	.4887	.4890
2.3	.4893	.4896	.4898	.4901	.4904	.4906	.4909	.4911	.4913	.4916
2.4	.4918	.4920	.4922	.4925	.4927	.4929	.4931	.4932	.4934	.4936
2.5	.4938	.4940	.4941	.4943	.4945	.4946	.4948	.4949	.4951	.4952
2.6	.4953	.4955	.4956	.4957	.4959	.4960	.4961	.4962	.4963	.4964
2.7	.4965	.4966	.4967	.4968	.4969	.4970	.4971	.4972	.4973	.4974
2.8	.4974	.4975	.4976	.4977	.4977	.4978	.4979	.4979	.4980	.4981
2.9	.4981	.4982	.4982	.4983	.4984	.4984	.4985	.4985	.4986	.4986
3.0	.4987	.4987	.4987	.4988	.4988	.4989	.4989	.4989	.4990	.4990

Source: A bridged from Table I of A. Hald, *Statistical Tables and Formulas* (New York: Wiley), 1952. Reproduced by permission of A. Hald

Table V	Exponentials								
λ	$e^{-\lambda}$	λ	$e^{-\lambda}$	λ	$e^{-\lambda}$	λ	$e^{-\lambda}$	λ	$e^{-\lambda}$
.00	1.000000	2.05	.128735	4.05	.017422	6.05	.002358	8.05	.000319
.05	.951229	2.10	.122456	4.10	.016573	6.10	.002243	8.10	.000304
.10	.904837	2.15	.116484	4.15	.015764	6.15	.002133	8.15	.000289
.15	.860708	2.20	.110803	4.20	.014996	6.20	.002029	8.20	.000275
.20	.818731	2.25	.105399	4.25	.014264	6.25	.001930	8.25	.000261
.25	.778801	2.30	.100259	4.30	.013569	6.30	.001836	8.30	.000249
.30	.740818	2.35	.095369	4.35	.012907	6.35	.001747	8.35	.000236
.35	.704688	2.40	.090718	4.40	.012277	6.40	.001661	8.40	.000225
.40	.670320	2.45	.086294	4.45	.011679	6.45	.001581	8.45	.000214
.45	.637628	2.50	.082085	4.50	.011109	6.50	.001503	8.50	.000204
.50	.606531	2.55	.078082	4.55	.010567	6.55	.001430	8.55	.000194
.55	.576950	2.60	.074274	4.60	.010052	6.60	.001360	8.60	.000184
.60	.548812	2.65	.070651	4.65	.009562	6.65	.001294	8.65	.000175
.65	.522046	2.70	.067206	4.70	.009095	6.70	.001231	8.70	.000167
.70	.496585	2.75	.063928	4.75	.008652	6.75	.001171	8.75	.000158
.75	.472367	2.80	.060810	4.80	.008230	6.80	.001114	8.80	.000151
.80	.449329	2.85	.057844	4.85	.007828	6.85	.001059	8.85	.000143
.85	.427415	2.90	.055023	4.90	.007447	6.90	.001008	8.90	.000136
.90	.406570	2.95	.052340	4.95	.007083	6.95	.000959	8.95	.000130
.95	.386741	3.00	.049787	5.00	.006738	7.00	.000912	9.00	.000123
1.00	.367879	3.05	.047359	5.05	.006409	7.05	.000867	9.05	.000117
1.05	.349938	3.10	.045049	5.10	.006097	7.10	.000825	9.10	.000112
1.10	.332871	3.15	.042852	5.15	.005799	7.15	.000785	9.15	.000106
1.15	.316637	3.20	.040762	5.20	.005517	7.20	.000747	9.20	.000101
1.20	.301194	3.25	.038774	5.25	.005248	7.25	.000710	9.25	.000096
1.25	.286505	3.30	.036883	5.30	.004992	7.30	.000676	9.30	.000091
1.30	.272532	3.35	.035084	5.35	.004748	7.35	.000643	9.35	.000087
1.35	.259240	3.40	.033373	5.40	.004517	7.40	.000611	9.40	.000083
1.40	.246597	3.45	.031746	5.45	.004296	7.45	.000581	9.45	.000079
1.45	.234570	3.50	.030197	5.50	.004087	7.50	.000553	9.50	.000075
1.50	.223130	3.55	.028725	5.55	.003887	7.55	.000526	9.55	.000071
1.55	.212248	3.60	.027324	5.60	.003698	7.60	.000501	9.60	.000068
1.60	.201897	3.65	.025991	5.65	.003518	7.65	.000476	9.65	.000064
1.65	.192050	3.70	.024724	5.70	.003346	7.70	.000453	9.70	.000061
1.70	.182684	3.75	.023518	5.75	.003183	7.75	.000431	9.75	.000058
1.75	.173774	3.80	.022371	5.80	.003028	7.80	.000410	9.80	.000056
1.80	.165299	3.85	.021280	5.85	.002880	7.85	.000390	9.85	.000053
1.85	.157237	3.90	.020242	5.90	.002739	7.90	.000371	9.90	.000050
1.90	.149569	3.95	.019255	5.95	.002606	7.95	.000353	9.95	.000048
1.95	.142274	4.00	.018316	6.00	.002479	8.00	.000336	10.00	.000045
2.00	.135335								

Table VI   Critical Values of *t*

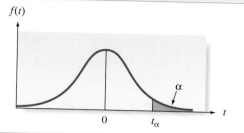

$f(t)$

α

0 t_α t

Degrees of Freedom	$t_{.100}$	$t_{.050}$	$t_{.025}$	$t_{.010}$	$t_{.005}$	$t_{.001}$	$t_{.0005}$
1	3.078	6.314	12.706	31.821	63.657	318.31	636.62
2	1.886	2.920	4.303	6.965	9.925	22.326	31.598
3	1.638	2.353	3.182	4.541	5.841	10.213	12.924
4	1.533	2.132	2.776	3.747	4.604	7.173	8.610
5	1.476	2.015	2.571	3.365	4.032	5.893	6.869
6	1.440	1.943	2.447	3.143	3.707	5.208	5.959
7	1.415	1.895	2.365	2.998	3.499	4.785	5.408
8	1.397	1.860	2.306	2.896	3.355	4.501	5.041
9	1.383	1.833	2.262	2.821	3.250	4.297	4.781
10	1.372	1.812	2.228	2.764	3.169	4.144	4.587
11	1.363	1.796	2.201	2.718	3.106	4.025	4.437
12	1.356	1.782	2.179	2.681	3.055	3.930	4.318
13	1.350	1.771	2.160	2.650	3.012	3.852	4.221
14	1.345	1.761	2.145	2.624	2.977	3.787	4.140
15	1.341	1.753	2.131	2.602	2.947	3.733	4.073
16	1.337	1.746	2.120	2.583	2.921	3.686	4.015
17	1.333	1.740	2.110	2.567	2.898	3.646	3.965
18	1.330	1.734	2.101	2.552	2.878	3.610	3.922
19	1.328	1.729	2.093	2.539	2.861	3.579	3.883
20	1.325	1.725	2.086	2.528	2.845	3.552	3.850
21	1.323	1.721	2.080	2.518	2.831	3.527	3.819
22	1.321	1.717	2.074	2.508	2.819	3.505	3.792
23	1.319	1.714	2.069	2.500	2.807	3.485	3.767
24	1.318	1.711	2.064	2.492	2.797	3.467	3.745
25	1.316	1.708	2.060	2.485	2.787	3.450	3.725
26	1.315	1.706	2.056	2.479	2.779	3.435	3.707
27	1.314	1.703	2.052	2.473	2.771	3.421	3.690
28	1.313	1.701	2.048	2.467	2.763	3.408	3.674
29	1.311	1.699	2.045	2.462	2.756	3.396	3.659
30	1.310	1.697	2.042	2.457	2.750	3.385	3.646
40	1.303	1.684	2.021	2.423	2.704	3.307	3.551
60	1.296	1.671	2.000	2.390	2.660	3.232	3.460
120	1.289	1.658	1.980	2.358	2.617	3.160	3.373
∞	1.282	1.645	1.960	2.326	2.576	3.090	3.291

80%   90%   95%

Table VII   Critical Values of χ^2

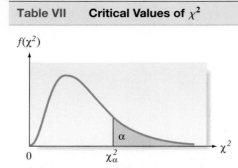

$f(\chi^2)$

α

0 χ^2_α χ^2

Degrees of Freedom	$\chi^2_{.995}$	$\chi^2_{.990}$	$\chi^2_{.975}$	$\chi^2_{.950}$	$\chi^2_{.900}$
1	.0000393	.0001571	.0009821	.0039321	.0157908
2	.0100251	.0201007	.0506356	.102587	.210720
3	.0717212	.114832	.215795	.351846	.584375
4	.206990	.297110	.484419	.710721	1.063623
5	.411740	.554300	.831211	1.145476	1.61031
6	.675727	.872085	1.237347	1.63539	2.20413
7	.989265	1.239043	1.68987	2.16735	2.83311
8	1.344419	1.646482	2.17973	2.73264	3.48954
9	1.734926	2.087912	2.70039	3.32511	4.16816
10	2.15585	2.55821	3.24697	3.94030	4.86518
11	2.60321	3.05347	3.81575	4.57481	5.57779
12	3.07382	3.57056	4.40379	5.22603	6.30380
13	3.56503	4.10691	5.00874	5.89186	7.04150
14	4.07468	4.66043	5.62872	6.57063	7.78953
15	4.60094	5.22935	6.26214	7.26094	8.54675
16	5.14224	5.81221	6.90766	7.96164	9.31223
17	5.69724	6.40776	7.56418	8.67176	10.0852
18	6.26481	7.01491	8.23075	9.39046	10.8649
19	6.84398	7.63273	8.90655	10.1170	11.6509
20	7.43386	8.26040	9.59083	10.8508	12.4426
21	8.03366	8.89720	10.28293	11.5913	13.2396
22	8.64272	9.54249	10.9823	12.3380	14.0415
23	9.26042	10.19567	11.6885	13.0905	14.8479
24	9.88623	10.8564	12.4011	13.8484	15.6587
25	10.5197	11.5240	13.1197	14.6114	16.4734
26	11.1603	12.1981	13.8439	15.3791	17.2919
27	11.8076	12.8786	14.5733	16.1513	18.1138
28	12.4613	13.5648	15.3079	16.9279	18.9392
29	13.1211	14.2565	16.0471	17.7083	19.7677
30	13.7867	14.9535	16.7908	18.4926	20.5992
40	20.7065	22.1643	24.4331	26.5093	29.0505
50	27.9907	29.7067	32.3574	34.7642	37.6886
60	35.5346	37.4848	40.4817	43.1879	46.4589
70	43.2752	45.4418	48.7576	51.7393	55.3290
80	51.1720	53.5400	57.1532	60.3915	64.2778
90	59.1963	61.7541	65.6466	69.1260	73.2912
100	67.3276	70.0648	74.2219	77.9295	82.3581

(continued)

Table VII	(continued)				
Degrees of Freedom	$\chi^2_{.100}$	$\chi^2_{.050}$	$\chi^2_{.025}$	$\chi^2_{.010}$	$\chi^2_{.005}$
1	2.70554	3.84146	5.02389	6.63490	7.87944
2	4.60517	5.99147	7.37776	9.21034	10.5966
3	6.25139	7.81473	9.34840	11.3449	12.8381
4	7.77944	9.48773	11.1433	13.2767	14.8602
5	9.23635	11.0705	12.8325	15.0863	16.7496
6	10.6446	12.5916	14.4494	16.8119	18.5476
7	12.0170	14.0671	16.0128	18.4753	20.2777
8	13.3616	15.5073	17.5346	20.0902	21.9550
9	14.6837	16.9190	19.0228	21.6660	23.5893
10	15.9871	18.3070	20.4831	23.2093	25.1882
11	17.2750	19.6751	21.9200	24.7250	26.7569
12	18.5494	21.0261	23.3367	26.2170	28.2995
13	19.8119	22.3621	24.7356	27.6883	29.8194
14	21.0642	23.6848	26.1190	29.1413	31.3193
15	22.3072	24.9958	27.4884	30.5779	32.8013
16	23.5418	26.2962	28.8454	31.9999	34.2672
17	24.7690	27.5871	30.1910	33.4087	35.7185
18	25.9894	28.8693	31.5264	34.8053	37.1564
19	27.2036	30.1435	32.8523	36.1908	38.5822
20	28.4120	31.4104	34.1696	37.5662	39.9968
21	29.6151	32.6705	35.4789	38.9321	41.4010
22	30.8133	33.9244	36.7807	40.2894	42.7956
23	32.0069	35.1725	38.0757	41.6384	44.1813
24	33.1963	36.4151	39.3641	42.9798	45.5585
25	34.3816	37.6525	40.6465	44.3141	46.9278
26	35.5631	38.8852	41.9232	45.6417	48.2899
27	36.7412	40.1133	43.1944	46.9630	49.6449
28	37.9159	41.3372	44.4607	48.2782	50.9933
29	39.0875	42.5569	45.7222	49.5879	52.3356
30	40.2560	43.7729	46.9792	50.8922	53.6720
40	51.8050	55.7585	59.3417	63.6907	66.7659
50	63.1671	67.5048	71.4202	76.1539	79.4900
60	74.3970	79.0819	83.2976	88.3794	91.9517
70	85.5271	90.5312	95.0231	100.425	104.215
80	96.5782	101.879	106.629	112.329	116.321
90	107.565	113.145	118.136	124.116	128.299
100	118.498	124.342	129.561	135.807	140.169

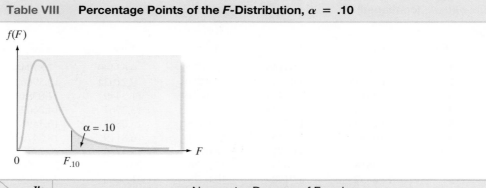

Table VIII **Percentage Points of the *F*-Distribution, $\alpha = .10$**

	ν_1	Numerator Degrees of Freedom								
ν_2		1	2	3	4	5	6	7	8	9
	1	39.86	49.50	53.59	55.83	57.24	58.20	58.91	59.44	59.86
	2	8.53	9.00	9.16	9.24	9.29	9.33	9.35	9.37	9.38
	3	5.54	5.46	5.39	5.34	5.31	5.28	5.27	5.25	5.24
	4	4.54	4.32	4.19	4.11	4.05	4.01	3.98	3.95	3.94
	5	4.06	3.78	3.62	3.52	3.45	3.40	3.37	3.34	3.32
	6	3.78	3.46	3.29	3.18	3.11	3.05	3.01	2.98	2.96
	7	3.59	3.26	3.07	2.96	2.88	2.83	2.78	2.75	2.72
	8	3.46	3.11	2.92	2.81	2.73	2.67	2.62	2.59	2.56
	9	3.36	3.01	2.81	2.69	2.61	2.55	2.51	2.47	2.44
	10	3.29	2.92	2.73	2.61	2.52	2.46	2.41	2.38	2.35
	11	3.23	2.86	2.66	2.54	2.45	2.39	2.34	2.30	2.27
	12	3.18	2.81	2.61	2.48	2.39	2.33	2.28	2.24	2.21
	13	3.14	2.76	2.56	2.43	2.35	2.28	2.23	2.20	2.16
	14	3.10	2.73	2.52	2.39	2.31	2.24	2.19	2.15	2.12
Denominator Degrees of Freedom	15	3.07	2.70	2.49	2.36	2.27	2.21	2.16	2.12	2.09
	16	3.05	2.67	2.46	2.33	2.24	2.18	2.13	2.09	2.06
	17	3.03	2.64	2.44	2.31	2.22	2.15	2.10	2.06	2.03
	18	3.01	2.62	2.42	2.29	2.20	2.13	2.08	2.04	2.00
	19	2.99	2.61	2.40	2.27	2.18	2.11	2.06	2.02	1.98
	20	2.97	2.59	2.38	2.25	2.16	2.09	2.04	2.00	1.96
	21	2.96	2.57	2.36	2.23	2.14	2.08	2.02	1.98	1.95
	22	2.95	2.56	2.35	2.22	2.13	2.06	2.01	1.97	1.93
	23	2.94	2.55	2.34	2.21	2.11	2.05	1.99	1.95	1.92
	24	2.93	2.54	2.33	2.19	2.10	2.04	1.98	1.94	1.91
	25	2.92	2.53	2.32	2.18	2.09	2.02	1.97	1.93	1.89
	26	2.91	2.52	2.31	2.17	2.08	2.01	1.96	1.92	1.88
	27	2.90	2.51	2.30	2.17	2.07	2.00	1.95	1.91	1.87
	28	2.89	2.50	2.29	2.16	2.06	2.00	1.94	1.90	1.87
	29	2.89	2.50	2.28	2.15	2.06	1.99	1.93	1.89	1.86
	30	2.88	2.49	2.28	2.14	2.05	1.98	1.93	1.88	1.85
	40	2.84	2.44	2.23	2.09	2.00	1.93	1.87	1.83	1.79
	60	2.79	2.39	2.18	2.04	1.95	1.87	1.82	1.77	1.74
	120	2.75	2.35	2.13	1.99	1.90	1.82	1.77	1.72	1.68
	∞	2.71	2.30	2.08	1.94	1.85	1.77	1.72	1.67	1.63

(continued)

Table VIII   (continued)

ν_2 \ ν_1	Numerator Degrees of Freedom									
	10	12	15	20	24	30	40	60	120	∞
1	60.19	60.71	61.22	61.74	62.00	62.26	62.53	62.79	63.06	63.33
2	9.39	9.41	9.42	9.44	9.45	9.46	9.47	9.47	9.48	9.49
3	5.23	5.22	5.20	5.18	5.18	5.17	5.16	5.15	5.14	5.13
4	3.92	3.90	3.87	3.84	3.83	3.82	3.80	3.79	3.78	3.76
5	3.30	3.27	3.24	3.21	3.19	3.17	3.16	3.14	3.12	3.10
6	2.94	2.90	2.87	2.84	2.82	2.80	2.78	2.76	2.74	2.72
7	2.70	2.67	2.63	2.59	2.58	2.56	2.54	2.51	2.49	2.47
8	2.54	2.50	2.46	2.42	2.40	2.38	2.36	2.34	2.32	2.29
9	2.42	2.38	2.34	2.30	2.28	2.25	2.23	2.21	2.18	2.16
10	2.32	2.28	2.24	2.20	2.18	2.16	2.13	2.11	2.08	2.06
11	2.25	2.21	2.17	2.12	2.10	2.08	2.05	2.03	2.00	1.97
12	2.19	2.15	2.10	2.06	2.04	2.01	1.99	1.96	1.93	1.90
13	2.14	2.10	2.05	2.01	1.98	1.96	1.93	1.90	1.88	1.85
14	2.10	2.05	2.01	1.96	1.94	1.91	1.89	1.86	1.83	1.80
15	2.06	2.02	1.97	1.92	1.90	1.87	1.85	1.82	1.79	1.76
16	2.03	1.99	1.94	1.89	1.87	1.84	1.81	1.78	1.75	1.72
17	2.00	1.96	1.91	1.86	1.84	1.81	1.78	1.75	1.72	1.69
18	1.98	1.93	1.89	1.84	1.81	1.78	1.75	1.72	1.69	1.66
19	1.96	1.91	1.86	1.81	1.79	1.76	1.73	1.70	1.67	1.63
20	1.94	1.89	1.84	1.79	1.77	1.74	1.71	1.68	1.64	1.61
21	1.92	1.87	1.83	1.78	1.75	1.72	1.69	1.66	1.62	1.59
22	1.90	1.86	1.81	1.76	1.73	1.70	1.67	1.64	1.60	1.57
23	1.89	1.84	1.80	1.74	1.72	1.69	1.66	1.62	1.59	1.55
24	1.88	1.83	1.78	1.73	1.70	1.67	1.64	1.61	1.57	1.53
25	1.87	1.82	1.77	1.72	1.69	1.66	1.63	1.59	1.56	1.52
26	1.86	1.81	1.76	1.71	1.68	1.65	1.61	1.58	1.54	1.50
27	1.85	1.80	1.75	1.70	1.67	1.64	1.60	1.57	1.53	1.49
28	1.84	1.79	1.74	1.69	1.66	1.63	1.59	1.56	1.52	1.48
29	1.83	1.78	1.73	1.68	1.65	1.62	1.58	1.55	1.51	1.47
30	1.82	1.77	1.72	1.67	1.64	1.61	1.57	1.54	1.50	1.46
40	1.76	1.71	1.66	1.61	1.57	1.54	1.51	1.47	1.42	1.38
60	1.71	1.66	1.60	1.54	1.51	1.48	1.44	1.40	1.35	1.29
120	1.65	1.60	1.55	1.48	1.45	1.41	1.37	1.32	1.26	1.19
∞	1.60	1.55	1.49	1.42	1.38	1.34	1.30	1.24	1.17	1.00

Denominator Degrees of Freedom

Table IX	Percentage Points of the *F*-Distribution, $\alpha = .05$

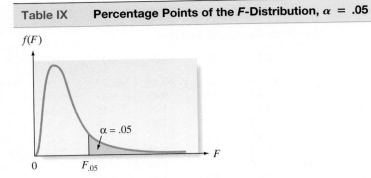

ν_1	Numerator Degrees of Freedom								
ν_2	1	2	3	4	5	6	7	8	9
1	161.4	199.5	215.7	224.6	230.2	234.0	236.8	238.9	240.5
2	18.51	19.00	19.16	19.25	19.30	19.33	19.35	19.37	19.38
3	10.13	9.55	9.28	9.12	9.01	8.94	8.89	8.85	8.81
4	7.71	6.94	6.59	6.39	6.26	6.16	6.09	6.04	6.00
5	6.61	5.79	5.41	5.19	5.05	4.95	4.88	4.82	4.77
6	5.99	5.14	4.76	4.53	4.39	4.28	4.21	4.15	4.10
7	5.59	4.74	4.35	4.12	3.97	3.87	3.79	3.73	3.68
8	5.32	4.46	4.07	3.84	3.69	3.58	3.50	3.44	3.39
9	5.12	4.26	3.86	3.63	3.48	3.37	3.29	3.23	3.18
10	4.96	4.10	3.71	3.48	3.33	3.22	3.14	3.07	3.02
11	4.84	3.98	3.59	3.36	3.20	3.09	3.01	2.95	2.90
12	4.75	3.89	3.49	3.26	3.11	3.00	2.91	2.85	2.80
13	4.67	3.81	3.41	3.18	3.03	2.92	2.83	2.77	2.71
14	4.60	3.74	3.34	3.11	2.96	2.85	2.76	2.70	2.65
15	4.54	3.68	3.29	3.06	2.90	2.79	2.71	2.64	2.59
16	4.49	3.63	3.24	3.01	2.85	2.74	2.66	2.59	2.54
17	4.45	3.59	3.20	2.96	2.81	2.70	2.61	2.55	2.49
18	4.41	3.55	3.16	2.93	2.77	2.66	2.58	2.51	2.46
19	4.38	3.52	3.13	2.90	2.74	2.63	2.54	2.48	2.42
20	4.35	3.49	3.10	2.87	2.71	2.60	2.51	2.45	2.39
21	4.32	3.47	3.07	2.84	2.68	2.57	2.49	2.42	2.37
22	4.30	3.44	3.05	2.82	2.66	2.55	2.46	2.40	2.34
23	4.28	3.42	3.03	2.80	2.64	2.53	2.44	2.37	2.32
24	4.26	3.40	3.01	2.78	2.62	2.51	2.42	2.36	2.30
25	4.24	3.39	2.99	2.76	2.60	2.49	2.40	2.34	2.28
26	4.23	3.37	2.98	2.74	2.59	2.47	2.39	2.32	2.77
27	4.21	3.35	2.96	2.73	2.57	2.46	2.37	2.31	2.25
28	4.20	3.34	2.95	2.71	2.56	2.45	2.36	2.29	2.24
29	4.18	3.33	2.93	2.70	2.55	2.43	2.35	2.28	2.22
30	4.17	3.32	2.92	2.69	2.53	2.42	2.33	2.27	2.21
40	4.08	3.23	2.84	2.61	2.45	2.34	2.25	2.18	2.12
60	4.00	3.15	2.76	2.53	2.37	2.25	2.17	2.10	2.04
120	3.92	3.07	2.68	2.45	2.29	2.17	2.09	2.02	1.96
∞	3.84	3.00	2.60	2.37	2.21	2.10	2.01	1.94	1.88

(continued)

Table IX	(continued)									
ν_1		Numerator Degrees of Freedom								
ν_2	10	12	15	20	24	30	40	60	120	∞
1	241.9	243.9	245.9	248.0	249.1	250.1	251.1	252.2	253.3	254.3
2	19.40	19.41	19.43	19.45	19.45	19.46	19.47	19.48	19.49	19.50
3	8.79	8.74	8.70	8.66	8.64	8.62	8.59	8.57	8.55	8.53
4	5.96	5.91	5.86	5.80	5.77	5.75	5.72	5.69	5.66	5.63
5	4.74	4.68	4.62	4.56	4.53	4.50	4.46	4.43	4.40	4.36
6	4.06	4.00	3.94	3.87	3.84	3.81	3.77	3.74	3.70	3.67
7	3.64	3.57	3.51	3.44	3.41	3.38	3.34	3.30	3.27	3.23
8	3.35	3.28	3.22	3.15	3.12	3.08	3.04	3.01	2.97	2.93
9	3.14	3.07	3.01	2.94	2.90	2.86	2.83	2.79	2.75	2.71
10	2.98	2.91	2.85	2.77	2.74	2.70	2.66	2.62	2.58	2.54
11	2.85	2.79	2.72	2.65	2.61	2.57	2.53	2.49	2.45	2.40
12	2.75	2.69	2.62	2.54	2.51	2.47	2.43	2.38	2.34	2.30
13	2.67	2.60	2.53	2.46	2.42	2.38	2.34	2.30	2.25	2.21
14	2.60	2.53	2.46	2.39	2.35	2.31	2.27	2.22	2.18	2.13
15	2.54	2.48	2.40	2.33	2.29	2.25	2.20	2.16	2.11	2.07
16	2.49	2.42	2.35	2.28	2.24	2.19	2.15	2.11	2.06	2.01
17	2.45	2.38	2.31	2.23	2.19	2.15	2.10	2.06	2.01	1.96
18	2.41	2.34	2.27	2.19	2.15	2.11	2.06	2.02	1.97	1.92
19	2.38	2.31	2.23	2.16	2.11	2.07	2.03	1.98	1.93	1.88
20	2.35	2.28	2.20	2.12	2.08	2.04	1.99	1.95	1.90	1.84
21	2.32	2.25	2.18	2.10	2.05	2.01	1.96	1.92	1.87	1.81
22	2.30	2.23	2.15	2.07	2.03	1.98	1.94	1.89	1.84	1.78
23	2.27	2.20	2.13	2.05	2.01	1.96	1.91	1.86	1.81	1.76
24	2.25	2.18	2.11	2.03	1.98	1.94	1.89	1.84	1.79	1.73
25	2.24	2.16	2.09	2.01	1.96	1.92	1.87	1.82	1.77	1.71
26	2.22	2.15	2.07	1.99	1.95	1.90	1.85	1.80	1.75	1.69
27	2.20	2.13	2.06	1.97	1.93	1.88	1.84	1.79	1.73	1.67
28	2.19	2.12	2.04	1.96	1.91	1.87	1.82	1.77	1.71	1.65
29	2.18	2.10	2.03	1.94	1.90	1.85	1.81	1.75	1.70	1.64
30	2.16	2.09	2.01	1.93	1.89	1.84	1.79	1.74	1.68	1.62
40	2.08	2.00	1.92	1.84	1.79	1.74	1.69	1.64	1.58	1.51
60	1.99	1.92	1.84	1.75	1.70	1.65	1.59	1.53	1.47	1.39
120	1.91	1.83	1.75	1.66	1.61	1.55	1.50	1.43	1.35	1.25
∞	1.83	1.75	1.67	1.57	1.52	1.46	1.39	1.32	1.22	1.00

Denominator degrees of freedom

Table X	Percentage Points of the *F*-Distribution, $\alpha = .025$

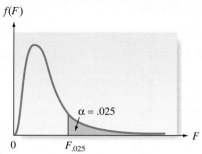

ν_1	Numerator Degrees of Freedom								
ν_2	1	2	3	4	5	6	7	8	9
1	647.8	799.5	864.2	899.6	921.8	937.1	948.2	956.7	963.3
2	38.51	39.00	39.17	39.25	39.30	39.33	39.36	39.37	39.39
3	17.44	16.04	15.44	15.10	14.88	14.73	14.62	14.54	14.47
4	12.22	10.65	9.98	9.60	9.36	9.20	9.07	8.98	8.90
5	10.01	8.43	7.76	7.39	7.15	6.98	6.85	6.76	6.68
6	8.81	7.26	6.60	6.23	5.99	5.82	5.70	5.60	5.52
7	8.07	6.54	5.89	5.52	5.29	5.12	4.99	4.90	4.82
8	7.57	6.06	5.42	5.05	4.82	4.65	4.53	4.43	4.36
9	7.21	5.71	5.08	4.72	4.48	4.32	4.20	4.10	4.03
10	6.94	5.46	4.83	4.47	4.24	4.07	3.95	3.85	3.78
11	6.72	5.26	4.63	4.28	4.04	3.88	3.76	3.66	3.59
12	6.55	5.10	4.47	4.12	3.89	3.73	3.61	3.51	3.44
13	6.41	4.97	4.35	4.00	3.77	3.60	3.48	3.39	3.31
14	6.30	4.86	4.24	3.89	3.66	3.50	3.38	3.29	3.21
15	6.20	4.77	4.15	3.80	3.58	3.41	3.29	3.20	3.12
16	6.12	4.69	4.08	3.73	3.50	3.34	3.22	3.12	3.05
17	6.04	4.62	4.01	3.66	3.44	3.28	3.16	3.06	2.98
18	5.98	4.56	3.95	3.61	3.38	3.22	3.10	3.01	2.93
19	5.92	4.51	3.90	3.56	3.33	3.17	3.05	2.96	2.88
20	5.87	4.46	3.86	3.51	3.29	3.13	3.01	2.91	2.84
21	5.83	4.42	3.82	3.48	3.25	3.09	2.97	2.87	2.80
22	5.79	4.38	3.78	3.44	3.22	3.05	2.93	2.84	2.76
23	5.75	4.35	3.75	3.41	3.18	3.02	2.90	2.81	2.73
24	5.72	4.32	3.72	3.38	3.15	2.99	2.87	2.78	2.70
25	5.69	4.29	3.69	3.35	3.13	2.97	2.85	2.75	2.68
26	5.66	4.27	3.67	3.33	3.10	2.94	2.82	2.73	2.65
27	5.63	4.24	3.65	3.31	3.08	2.92	2.80	2.71	2.63
28	5.61	4.22	3.63	3.29	3.06	2.90	2.78	2.69	2.61
29	5.59	4.20	3.61	3.27	3.04	2.88	2.76	2.67	2.59
30	5.57	4.18	3.59	3.25	3.03	2.87	2.75	2.65	2.57
40	5.42	4.05	3.46	3.13	2.90	2.74	2.62	2.53	2.45
60	5.29	3.93	3.34	3.01	2.79	2.63	2.51	2.41	2.33
120	5.15	3.80	3.23	2.89	2.67	2.52	2.39	2.30	2.22
∞	5.02	3.69	3.12	2.79	2.57	2.41	2.29	2.19	2.11

(continued)

Table X   (continued)

ν_2 \ ν_1	Numerator Degrees of Freedom									
	10	12	15	20	24	30	40	60	120	∞
1	968.6	976.7	984.9	993.1	997.2	1,001	1,006	1,010	1,014	1,018
2	39.40	39.41	39.43	39.45	39.46	39.46	39.47	39.48	39.49	39.50
3	14.42	14.34	14.25	14.17	14.12	14.08	14.04	13.99	13.95	13.90
4	8.84	8.75	8.66	8.56	8.51	8.46	8.41	8.36	8.31	8.26
5	6.62	6.52	6.43	6.33	6.28	6.23	6.18	6.12	6.07	6.02
6	5.46	5.37	5.27	5.17	5.12	5.07	5.01	4.96	4.90	4.85
7	4.76	4.67	4.57	4.47	4.42	4.36	4.31	4.25	4.20	4.14
8	4.30	4.20	4.10	4.00	3.95	3.89	3.84	3.78	3.73	3.67
9	3.96	3.87	3.77	3.67	3.61	3.56	3.51	3.45	3.39	3.33
10	3.72	3.62	3.52	3.42	3.37	3.31	3.26	3.20	3.14	3.08
11	3.53	3.43	3.33	3.23	3.17	3.12	3.06	3.00	2.94	2.88
12	3.37	3.28	3.18	3.07	3.02	2.96	2.91	2.85	2.79	2.72
13	3.25	3.15	3.05	2.95	2.89	2.84	2.78	2.72	2.66	2.60
14	3.15	3.05	2.95	2.84	2.79	2.73	2.67	2.61	2.55	2.49
15	3.06	2.96	2.86	2.76	2.70	2.64	2.59	2.52	2.46	2.40
16	2.99	2.89	2.79	2.68	2.63	2.57	2.51	2.45	2.38	2.32
17	2.92	2.82	2.72	2.62	2.56	2.50	2.44	2.38	2.32	2.25
18	2.87	2.77	2.67	2.56	2.50	2.44	2.38	2.32	2.26	2.19
19	2.82	2.72	2.62	2.51	2.45	2.39	2.33	2.27	2.20	2.13
20	2.77	2.68	2.57	2.46	2.41	2.35	2.29	2.22	2.16	2.09
21	2.73	2.64	2.53	2.42	2.37	2.31	2.25	2.18	2.11	2.04
22	2.70	2.60	2.50	2.39	2.33	2.27	2.21	2.14	2.08	2.00
23	2.67	2.57	2.47	2.36	2.30	2.24	2.18	2.11	2.04	1.97
24	2.64	2.54	2.44	2.33	2.27	2.21	2.15	2.08	2.01	1.94
25	2.61	2.51	2.41	2.30	2.24	2.18	2.12	2.05	1.98	1.91
26	2.59	2.49	2.39	2.28	2.22	2.16	2.09	2.03	1.95	1.88
27	2.57	2.47	2.36	2.25	2.19	2.13	2.07	2.00	1.93	1.85
28	2.55	2.45	2.34	2.23	2.17	2.11	2.05	1.98	1.91	1.83
29	2.53	2.43	2.32	2.21	2.15	2.09	2.03	1.96	1.89	1.81
30	2.51	2.41	2.31	2.20	2.14	2.07	2.01	1.94	1.87	1.79
40	2.39	2.29	2.18	2.07	2.01	1.94	1.88	1.80	1.72	1.64
60	2.27	2.17	2.06	1.94	1.88	1.82	1.74	1.67	1.58	1.48
120	2.16	2.05	1.94	1.82	1.76	1.69	1.61	1.53	1.43	1.31
∞	2.05	1.94	1.83	1.71	1.64	1.57	1.48	1.39	1.27	1.00

Denominator Degrees of Freedom

Table XI   Percentage Points of the *F*-distribution, α = .01

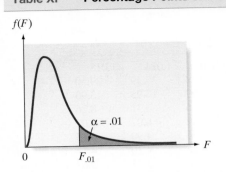

$f(F)$

$\alpha = .01$

$F_{.01}$

0 F

		Numerator Degrees of Freedom							
ν_1	1	2	3	4	5	6	7	8	9
ν_2									
1	4,052	4,999.5	5,403	5,625	5,764	5,859	5,928	5,982	6,022
2	98.50	99.00	99.17	99.25	99.30	99.33	99.36	99.37	99.39
3	34.12	30.82	29.46	28.71	28.24	27.91	27.67	27.49	27.35
4	21.20	18.00	16.69	15.98	15.52	15.21	14.98	14.80	14.66
5	16.26	13.27	12.06	11.39	10.97	10.67	10.46	10.29	10.16
6	13.75	10.92	9.78	9.15	8.75	8.47	8.26	8.10	7.98
7	12.25	9.55	8.45	7.85	7.46	7.19	6.99	6.84	6.72
8	11.26	8.65	7.59	7.01	6.63	6.37	6.18	6.03	5.91
9	10.56	8.02	6.99	6.42	6.06	5.80	5.61	5.47	5.35
10	10.04	7.56	6.55	5.99	5.64	5.39	5.20	5.06	4.94
11	9.65	7.21	6.22	5.67	5.32	5.07	4.89	4.74	4.63
12	9.33	6.93	5.95	5.41	5.06	4.82	4.64	4.50	4.39
13	9.07	6.70	5.74	5.21	4.86	4.62	4.44	4.30	4.19
14	8.86	6.51	5.56	5.04	4.69	4.46	4.28	4.14	4.03
15	8.68	6.36	5.42	4.89	4.56	4.32	4.14	4.00	3.89
16	8.53	6.23	5.29	4.77	4.44	4.20	4.03	3.89	3.78
17	8.40	6.11	5.18	4.67	4.34	4.10	3.93	3.79	3.68
18	8.29	6.01	5.09	4.58	4.25	4.01	3.84	3.71	3.60
19	8.18	5.93	5.01	4.50	4.17	3.94	3.77	3.63	3.52
20	8.10	5.85	4.94	4.43	4.10	3.87	3.70	3.56	3.46
21	8.02	5.78	4.87	4.37	4.04	3.81	3.64	3.51	3.40
22	7.95	5.72	4.82	4.31	3.99	3.76	3.59	3.45	3.35
23	7.88	5.66	4.76	4.26	3.94	3.71	3.54	3.41	3.30
24	7.82	5.61	4.72	4.22	3.90	3.67	3.50	3.36	3.26
25	7.77	5.57	4.68	4.18	3.85	3.63	3.46	3.32	3.22
26	7.72	5.53	4.64	4.14	3.82	3.59	3.42	3.29	3.18
27	7.68	5.49	4.60	4.11	3.78	3.56	3.39	3.26	3.15
28	7.64	5.45	4.57	4.07	3.75	3.53	3.36	3.23	3.12
29	7.60	5.42	4.54	4.04	3.73	3.50	3.33	3.20	3.09
30	7.56	5.39	4.51	4.02	3.70	3.47	3.30	3.17	3.07
40	7.31	5.18	4.31	3.83	3.51	3.29	3.12	2.99	2.89
60	7.08	4.98	4.13	3.65	3.34	3.12	2.95	2.82	2.72
120	6.85	4.79	3.95	3.48	3.17	2.96	2.79	2.66	2.56
∞	6.63	4.61	3.78	3.32	3.02	2.80	2.64	2.51	2.41

Denominator Degrees of Freedom

(*continued*)

Table XI **(continued)**

ν_2	ν_1 Numerator Degrees of Freedom									
	10	12	15	20	24	30	40	60	120	∞
1	6,056	6,106	6,157	6,209	6,235	6,261	6,287	6,313	6,339	6,366
2	99.40	99.42	99.43	99.45	99.46	99.47	99.47	99.48	99.49	99.50
3	27.23	27.05	26.87	26.69	26.60	26.50	26.41	26.32	26.22	26.13
4	14.55	14.37	14.20	14.02	13.93	13.84	13.75	13.65	13.56	13.46
5	10.05	9.89	9.72	9.55	9.47	9.38	9.29	9.20	9.11	9.02
6	7.87	7.72	7.56	7.40	7.31	7.23	7.14	7.06	6.97	6.88
7	6.62	6.47	6.31	6.16	6.07	5.99	5.91	5.82	5.74	5.65
8	5.81	5.67	5.52	5.36	5.28	5.20	5.12	5.03	4.95	4.86
9	5.26	5.11	4.96	4.81	4.73	4.65	4.57	4.48	4.40	4.31
10	4.85	4.71	4.56	4.41	4.33	4.25	4.17	4.08	4.00	3.91
11	4.54	4.40	4.25	4.10	4.02	3.94	3.86	3.78	3.69	3.60
12	4.30	4.16	4.01	3.86	3.78	3.70	3.62	3.54	3.45	3.36
13	4.10	3.96	3.82	3.66	3.59	3.51	3.43	3.34	3.25	3.17
14	3.94	3.80	3.66	3.51	3.43	3.35	3.27	3.18	3.09	3.00
15	3.80	3.67	3.52	3.37	3.29	3.21	3.13	3.05	2.96	2.87
16	3.69	3.55	3.41	3.26	3.18	3.10	3.02	2.93	2.84	2.75
17	3.59	3.46	3.31	3.16	3.08	3.00	2.92	2.83	2.75	2.65
18	3.51	3.37	3.23	3.08	3.00	2.92	2.84	2.75	2.66	2.57
19	3.43	3.30	3.15	3.00	2.92	2.84	2.76	2.67	2.58	2.49
20	3.37	3.23	3.09	2.94	2.86	2.78	2.69	2.61	2.52	2.42
21	3.31	3.17	3.03	2.88	2.80	2.72	2.64	2.55	2.46	2.36
22	3.26	3.12	2.98	2.83	2.75	2.67	2.58	2.50	2.40	2.31
23	3.21	3.07	2.93	2.78	2.70	2.62	2.54	2.45	2.35	2.26
24	3.17	3.03	2.89	2.74	2.66	2.58	2.49	2.40	2.31	2.21
25	3.13	2.99	2.85	2.70	2.62	2.54	2.45	2.36	2.27	2.17
26	3.09	2.96	2.81	2.66	2.58	2.50	2.42	2.33	2.23	2.13
27	3.06	2.93	2.78	2.63	2.55	2.47	2.38	2.29	2.20	2.10
28	3.03	2.90	2.75	2.60	2.52	2.44	2.35	2.26	2.17	2.06
29	3.00	2.87	2.73	2.57	2.49	2.41	2.33	2.23	2.14	2.03
30	2.98	2.84	2.70	2.55	2.47	2.39	2.30	2.21	2.11	2.01
40	2.80	2.66	2.52	2.37	2.29	2.20	2.11	2.02	1.92	1.80
60	2.63	2.50	2.35	2.20	2.12	2.03	1.94	1.84	1.73	1.60
120	2.47	2.34	2.19	2.03	1.95	1.86	1.76	1.66	1.53	1.38
∞	2.32	2.18	2.04	1.88	1.79	1.70	1.59	1.47	1.32	1.00

Table XII **Critical Values of T_L and T_U for the Wilcoxon Rank Sum Test: Independent Samples**

Test statistic is the rank sum associated with the smaller sample (if equal sample sizes, either rank sum can be used).

a. $\alpha = .025$ one-tailed; $\alpha = .05$ two-tailed

n_2 \ n_1	3 T_L	3 T_U	4 T_L	4 T_U	5 T_L	5 T_U	6 T_L	6 T_U	7 T_L	7 T_U	8 T_L	8 T_U	9 T_L	9 T_U	10 T_L	10 T_U
3	5	16	6	18	6	21	7	23	7	26	8	28	8	31	9	33
4	6	18	11	25	12	28	12	32	13	35	14	38	15	41	16	44
5	6	21	12	28	18	37	19	41	20	45	21	49	22	53	24	56
6	7	23	12	32	19	41	26	52	28	56	29	61	31	65	32	70
7	7	26	13	35	20	45	28	56	37	68	39	73	41	78	43	83
8	8	28	14	38	21	49	29	61	39	73	49	87	51	93	54	98
9	8	31	15	41	22	53	31	65	41	78	51	93	63	108	66	114
10	9	33	16	44	24	56	32	70	43	83	54	98	66	114	79	131

b. $\alpha = .05$ one-tailed; $\alpha = .10$ two-tailed

n_2 \ n_1	3 T_L	3 T_U	4 T_L	4 T_U	5 T_L	5 T_U	6 T_L	6 T_U	7 T_L	7 T_U	8 T_L	8 T_U	9 T_L	9 T_U	10 T_L	10 T_U
3	6	15	7	17	7	20	8	22	9	24	9	27	10	29	11	31
4	7	17	12	24	13	27	14	30	15	33	16	36	17	39	18	42
5	7	20	13	27	19	36	20	40	22	43	24	46	25	50	26	54
6	8	22	14	30	20	40	28	50	30	54	32	58	33	63	35	67
7	9	24	15	33	22	43	30	54	39	66	41	71	43	76	46	80
8	9	27	16	36	24	46	32	58	41	71	52	84	54	90	57	95
9	10	29	17	39	25	50	33	63	43	76	54	90	66	105	69	111
10	11	31	18	42	26	54	35	67	46	80	57	95	69	111	83	127

Source: From F. Wilcoxon and R. A. Wilcox, "Some Rapid Approximate Statistical Procedures," 1964, 20–23.

Table XIII	Critical Values of T_0 in the Wilcoxon Paired Difference Signed Rank Test					
One-Tailed	Two-Tailed	$n = 5$	$n = 6$	$n = 7$	$n = 8$	$n = 9$	$n = 10$
$\alpha = .05$	$\alpha = .10$	1	2	4	6	8	11
$\alpha = .025$	$\alpha = .05$		1	2	4	6	8
$\alpha = .01$	$\alpha = .02$			0	2	3	5
$\alpha = .005$	$\alpha = .01$				0	2	3
		$n = 11$	$n = 12$	$n = 13$	$n = 14$	$n = 15$	$n = 16$
$\alpha = .05$	$\alpha = .10$	14	17	21	26	30	36
$\alpha = .025$	$\alpha = .05$	11	14	17	21	25	30
$\alpha = .01$	$\alpha = .02$	7	10	13	16	20	24
$\alpha = .005$	$\alpha = .01$	5	7	10	13	16	19
		$n = 17$	$n = 18$	$n = 19$	$n = 20$	$n = 21$	$n = 22$
$\alpha = .05$	$\alpha = .10$	41	47	54	60	68	75
$\alpha = .025$	$\alpha = .05$	35	40	46	52	59	66
$\alpha = .01$	$\alpha = .02$	28	33	38	43	49	56
$\alpha = .005$	$\alpha = .01$	23	28	32	37	43	49
		$n = 23$	$n = 24$	$n = 25$	$n = 26$	$n = 27$	$n = 28$
$\alpha = .05$	$\alpha = .10$	83	92	101	110	120	130
$\alpha = .025$	$\alpha = .05$	73	81	90	98	107	117
$\alpha = .01$	$\alpha = .02$	62	69	77	85	93	102
$\alpha = .005$	$\alpha = .01$	55	61	68	76	84	92
		$n = 29$	$n = 30$	$n = 31$	$n = 32$	$n = 33$	$n = 34$
$\alpha = .05$	$\alpha = .10$	141	152	163	175	188	201
$\alpha = .025$	$\alpha = .05$	127	137	148	159	171	183
$\alpha = .01$	$\alpha = .02$	111	120	130	141	151	162
$\alpha = .005$	$\alpha = .01$	100	109	118	128	138	149
		$n = 35$	$n = 36$	$n = 37$	$n = 38$	$n = 39$	
$\alpha = .05$	$\alpha = .10$	214	228	242	256	271	
$\alpha = .025$	$\alpha = .05$	195	208	222	235	250	
$\alpha = .01$	$\alpha = .02$	174	186	198	211	224	
$\alpha = .005$	$\alpha = .01$	160	171	183	195	208	
		$n = 40$	$n = 41$	$n = 42$	$n = 43$	$n = 44$	$n = 45$
$\alpha = .05$	$\alpha = .10$	287	303	319	336	353	371
$\alpha = .025$	$\alpha = .05$	264	279	295	311	327	344
$\alpha = .01$	$\alpha = .02$	238	252	267	281	297	313
$\alpha = .005$	$\alpha = .01$	221	234	248	262	277	292
		$n = 46$	$n = 47$	$n = 48$	$n = 49$	$n = 50$	
$\alpha = .05$	$\alpha = .10$	389	408	427	446	466	
$\alpha = .025$	$\alpha = .05$	361	379	397	415	434	
$\alpha = .01$	$\alpha = .02$	329	345	362	380	398	
$\alpha = .005$	$\alpha = .01$	307	323	339	356	373	

Source: From F. Wilcoxon and R. A. Wilcox, "Some Rapid Approximate Statistical Procedures," 1964, p. 28.

Table XIV   Critical Values of Spearman's Rank Correlation Coefficient

The α values correspond to a one-tailed test of H_0: $\rho = 0$. The value should be doubled for two-tailed tests.

n	$\alpha = .05$	$\alpha = .025$	$\alpha = .01$	$\alpha = .005$	n	$\alpha = .05$	$\alpha = .025$	$\alpha = .01$	$\alpha = .005$
5	.900	–	–	–	18	.399	.476	.564	.625
6	.829	.886	.943	–	19	.388	.462	.549	.608
7	.714	.786	.893	–	20	.377	.450	.534	.591
8	.643	.738	.833	.881	21	.368	.438	.521	.576
9	.600	.683	.783	.833	22	.359	.428	.508	.562
10	.564	.648	.745	.794	23	.351	.418	.496	.549
11	.523	.623	.736	.818	24	.343	.409	.485	.537
12	.497	.591	.703	.780	25	.336	.400	.475	.526
13	.475	.566	.673	.745	26	.329	.392	.465	.515
14	.457	.545	.646	.716	27	.323	.385	.456	.505
15	.441	.525	.623	.689	28	.317	.377	.448	.496
16	.425	.507	.601	.666	29	.311	.370	.440	.487
17	.412	.490	.582	.645	30	.305	.364	.432	.478

Table XV   Critical Values of the Studentized Range, $\alpha = .05$

v \ k	2	3	4	5	6	7	8	9	10	11	12	13	14	15	16	17	18	19	20
1	17.97	26.98	32.82	37.08	40.41	43.12	45.40	47.36	49.07	50.59	51.96	53.20	54.33	55.36	56.32	57.22	58.04	58.83	59.56
2	6.08	8.33	9.80	10.88	11.74	12.44	13.03	13.54	13.99	14.39	14.75	15.08	15.38	15.65	15.91	16.14	16.37	16.57	16.77
3	4.50	5.91	6.82	7.50	8.04	8.48	8.85	9.18	9.46	9.72	9.95	10.15	10.35	10.52	10.69	10.84	10.98	11.11	11.24
4	3.93	5.04	5.76	6.29	6.71	7.05	7.35	7.60	7.83	8.03	8.21	8.37	8.52	8.66	8.79	8.91	9.03	9.13	9.23
5	3.64	4.60	5.22	5.67	6.03	6.33	6.58	6.80	6.99	7.17	7.32	7.47	7.60	7.72	7.83	7.93	8.03	8.12	8.21
6	3.46	4.34	4.90	5.30	5.63	5.90	6.12	6.32	6.49	6.65	6.79	6.92	7.03	7.14	7.24	7.34	7.43	7.51	7.59
7	3.34	4.16	4.68	5.06	5.36	5.61	5.82	6.00	6.16	6.30	6.43	6.55	6.66	6.76	6.85	6.94	7.02	7.10	7.17
8	3.26	4.04	4.53	4.89	5.17	5.40	5.60	5.77	5.92	6.05	6.18	6.29	6.39	6.48	6.57	6.65	6.73	6.80	6.87
9	3.20	3.95	4.41	4.76	5.02	5.24	5.43	5.59	5.74	5.87	5.98	6.09	6.19	6.28	6.36	6.44	6.51	6.58	6.64
10	3.15	3.88	4.33	4.65	4.91	5.12	5.30	5.46	5.60	5.72	5.83	5.93	6.03	6.11	6.19	6.27	6.34	6.40	6.47
11	3.11	3.82	4.26	4.57	4.82	5.03	5.20	5.35	5.49	5.61	5.71	5.81	5.90	5.98	6.06	6.13	6.20	6.27	6.33
12	3.08	3.77	4.20	4.51	4.75	4.95	5.12	5.27	5.39	5.51	5.61	5.71	5.80	5.88	5.95	6.02	6.09	6.15	6.21
13	3.06	3.73	4.15	4.45	4.69	4.88	5.05	5.19	5.32	5.43	5.53	5.63	5.71	5.79	5.86	5.93	5.99	6.05	6.11
14	3.03	3.70	4.11	4.41	4.64	4.83	4.99	5.13	5.25	5.36	5.46	5.55	5.64	5.71	5.79	5.85	5.91	5.97	6.03
15	3.01	3.67	4.08	4.37	4.60	4.78	4.94	5.08	5.20	5.31	5.40	5.49	5.57	5.65	5.72	5.78	5.85	5.90	5.96
16	3.00	3.65	4.05	4.33	4.56	4.74	4.90	5.03	5.15	5.26	5.35	5.44	5.52	5.59	5.66	5.73	5.79	5.84	5.90
17	2.98	3.63	4.02	4.30	4.52	4.70	4.86	4.99	5.11	5.21	5.31	5.39	5.47	5.54	5.61	5.67	5.73	5.79	5.84
18	2.97	3.61	4.00	4.28	4.49	4.67	4.82	4.96	5.07	5.17	5.27	5.35	5.43	5.50	5.57	5.63	5.69	5.74	5.79
19	2.96	3.59	3.98	4.25	4.47	4.65	4.79	4.92	5.04	5.14	5.23	5.31	5.39	5.46	5.53	5.59	5.65	5.70	5.75
20	2.95	3.58	3.96	4.23	4.45	4.62	4.77	4.90	5.01	5.11	5.20	5.28	5.36	5.43	5.49	5.55	5.61	5.66	5.71
24	2.92	3.53	3.90	4.17	4.37	4.54	4.68	4.81	4.92	5.01	5.10	5.18	5.25	5.32	5.38	5.44	5.49	5.55	5.59
30	2.89	3.49	3.85	4.10	4.30	4.46	4.60	4.72	4.82	4.92	5.00	5.08	5.15	5.21	5.27	5.33	5.38	5.43	5.47
40	2.86	3.44	3.79	4.04	4.23	4.39	4.52	4.63	4.73	4.82	4.90	4.98	5.04	5.11	5.16	5.22	5.27	5.31	5.36
60	2.83	3.40	3.74	3.98	4.16	4.31	4.44	4.55	4.65	4.73	4.81	4.88	4.94	5.00	5.06	5.11	5.15	5.20	5.24
120	2.80	3.36	3.68	3.92	4.10	4.24	4.36	4.47	4.56	4.64	4.71	4.78	4.84	4.90	4.95	5.00	5.04	5.09	5.13
∞	2.77	3.31	3.63	3.86	4.03	4.17	4.29	4.39	4.47	4.55	4.62	4.68	4.74	4.80	4.85	4.89	4.93	4.97	5.01

Table XVI   Critical Values of the Studentized Range, $\alpha = .01$

v \ k	2	3	4	5	6	7	8	9	10	11	12	13	14	15	16	17	18	19	20
1	90.03	135.0	164.3	185.6	202.2	215.8	227.2	237.0	245.6	253.2	260.0	266.2	271.8	277.0	281.8	286.3	290.0	294.3	298.0
2	14.04	19.02	22.29	24.72	26.63	28.20	29.53	30.68	31.69	32.59	33.40	34.13	34.81	35.43	36.00	36.53	37.03	37.50	37.95
3	8.26	10.62	12.17	13.33	14.24	15.00	15.64	16.20	16.69	17.13	17.53	17.89	18.22	18.52	18.81	19.07	19.32	19.55	19.77
4	6.51	8.12	9.17	9.96	10.58	11.10	11.55	11.93	12.27	12.57	12.84	13.09	13.32	13.53	13.73	13.91	14.08	14.24	14.40
5	5.70	6.98	7.80	8.42	8.91	9.32	9.67	9.97	10.24	10.48	10.70	10.89	11.08	11.24	11.40	11.55	11.68	11.81	11.93
6	5.24	6.33	7.03	7.56	7.97	8.32	8.61	8.87	9.10	9.30	9.48	9.65	9.81	9.95	10.08	10.21	10.32	10.43	10.54
7	4.95	5.92	6.54	7.01	7.37	7.68	7.94	8.17	8.37	8.55	8.71	8.86	9.00	9.12	9.24	9.35	9.46	9.55	9.65
8	4.75	5.64	6.20	6.62	6.96	7.24	7.47	7.68	7.86	8.03	8.18	8.31	8.44	8.55	8.66	8.76	8.85	8.94	9.03
9	4.60	5.43	5.96	6.35	6.66	6.91	7.13	7.33	7.49	7.65	7.78	7.91	8.03	8.13	8.23	8.33	8.41	8.49	8.57
10	4.48	5.27	5.77	6.14	6.43	6.67	6.87	7.05	7.21	7.36	7.49	7.60	7.71	7.81	7.91	7.99	8.08	8.15	8.23
11	4.39	5.15	5.62	5.97	6.25	6.48	6.67	6.84	6.99	7.13	7.25	7.36	7.46	7.56	7.65	7.73	7.81	7.88	7.95
12	4.32	5.05	5.50	5.84	6.10	6.32	6.51	6.67	6.81	6.94	7.06	7.17	7.26	7.36	7.44	7.52	7.59	7.66	7.73
13	4.26	4.96	5.40	5.73	5.98	6.19	6.37	6.53	6.67	6.79	6.90	7.01	7.10	7.19	7.27	7.35	7.42	7.48	7.55
14	4.21	4.89	5.32	5.63	5.88	6.08	6.26	6.41	6.54	6.66	6.77	6.87	6.96	7.05	7.13	7.20	7.27	7.33	7.39
15	4.17	4.84	5.25	5.56	5.80	5.99	6.16	6.31	6.44	6.55	6.66	6.76	6.84	6.93	7.00	7.07	7.14	7.20	7.26
16	4.13	4.79	5.19	5.49	5.72	5.92	6.08	6.22	6.35	6.46	6.56	6.66	6.74	6.82	6.90	6.97	7.03	7.09	7.15
17	4.10	4.74	5.14	5.43	5.66	5.85	6.01	6.15	6.27	6.38	6.48	6.57	6.66	6.73	6.81	6.87	6.94	7.00	7.05
18	4.07	4.70	5.09	5.38	5.60	5.79	5.94	6.08	6.20	6.31	6.41	6.50	6.58	6.65	6.72	6.79	6.85	6.91	6.97
19	4.05	4.67	5.05	5.33	5.55	5.73	5.89	6.02	6.14	6.25	6.34	6.43	6.51	6.58	6.65	6.72	6.78	6.84	6.89
20	4.02	4.64	5.02	5.29	5.51	5.69	5.84	5.97	6.09	6.19	6.28	6.37	6.45	6.52	6.59	6.65	6.71	6.77	6.82
24	3.96	4.55	4.91	5.17	5.37	5.54	5.69	5.81	5.92	6.02	6.11	6.19	6.26	6.33	6.39	6.45	6.51	6.56	6.61
30	3.89	4.45	4.80	5.05	5.24	5.40	5.54	5.65	5.76	5.85	5.93	6.01	6.08	6.14	6.20	6.26	6.31	6.36	6.41
40	3.82	4.37	4.70	4.93	5.11	5.26	5.39	5.50	5.60	5.69	5.76	5.83	5.90	5.96	6.02	6.07	6.12	6.16	6.21
60	3.76	4.28	4.59	4.82	4.99	5.13	5.25	5.36	5.45	5.53	5.60	5.67	5.73	5.78	5.84	5.89	5.93	5.97	6.01
120	3.70	4.20	4.50	4.71	4.87	5.01	5.12	5.21	5.30	5.37	5.44	5.50	5.56	5.61	5.66	5.71	5.75	5.79	5.83
∞	3.64	4.12	4.40	4.60	4.76	4.88	4.99	5.08	5.16	5.23	5.29	5.35	5.40	5.45	5.49	5.54	5.57	5.61	5.65

Appendix B: Calculation Formulas for Analysis of Variance

B.1 Completely Randomized Design

$$
\begin{aligned}
\text{CM} &= \text{Correction for mean} \\
&= \frac{(\text{Total of all observations})^2}{\text{Total number of abservations}} = \frac{(\Sigma y_i)^2}{n} \\
\text{SS(Total)} &= \text{Total sum of squares} \\
&= (\text{Sum of squares of all observations}) - \text{CM} = \Sigma y_i^2 - \text{CM} \\
\text{SST} &= \text{Sum of squares for treatments} \\
&= \left(\begin{array}{c}\text{Sum of squares of treatments totals with} \\ \text{each square divided by the number of} \\ \text{observations for that treatment}\end{array}\right) - \text{CM} \\
&= \frac{T_1^2}{n_1} + \frac{T_2^2}{n_2} + \cdots + \frac{T_k^2}{n_k} - \text{CM} \\
\text{SSE} &= \text{Sum of squares for error} = \text{SS(Total)} - \text{SST} \\
\text{MST} &= \text{Mean square for treatments} = \frac{\text{SST}}{k-1} \\
\text{MSE} &= \text{Mean square for error} = \frac{\text{SSE}}{n-k} \\
F &= \text{Test statistic} = \frac{\text{MST}}{\text{MSE}}
\end{aligned}
$$

where

$$
\begin{aligned}
n &= \text{Total number of observations} \\
k &= \text{Number of treatments} \\
T_i &= \text{Total for treatment } i \ (i = 1, 2, \ldots, k)
\end{aligned}
$$

B.2 Randomized Block Design

$$CM = \text{Correction for mean}$$

$$= \frac{(\text{Total of all observations})^2}{\text{Total number of observations}} = \frac{(\Sigma y_i)^2}{n}$$

$$SS(\text{Total}) = \text{Total sum of squares}$$

$$= (\text{Sum of squares of all observations}) - CM = \Sigma y_i^2 - CM$$

$$SST = \text{Sum of squares for treatments}$$

$$= \left(\begin{array}{c} \text{Sum of squares of treatment totals with} \\ \text{each square divided by } b, \text{ the number of} \\ \text{observations for that treatment} \end{array} \right) - CM$$

$$= \frac{T_1^2}{b} + \frac{T_2^2}{b} + \cdots + \frac{T_k^2}{b} - CM$$

$$SST = \text{Sum of squares for blocks}$$

$$= \left(\begin{array}{c} \text{Sum of squares of block totals with} \\ \text{each square divided by } k, \text{ the number} \\ \text{of observations in that block} \end{array} \right) - CM$$

$$= \frac{B_1^2}{k} + \frac{B_2^2}{k} + \cdots + \frac{B_b^2}{k} - CM$$

$$SSE = \text{Sum of squares for error} = SS(\text{Total}) - SST - SSB$$

$$MST = \text{Mean square for treatments} = \frac{SST}{k-1}$$

$$MSB = \text{Mean square for blocks} = \frac{SSB}{b-1}$$

$$MSE = \text{Mean square for error} = \frac{SSE}{n-k-b+1}$$

$$F = \text{Test statistic} = \frac{MST}{MSE}$$

where

$$n = \text{Total number of observations}$$
$$b = \text{Number of blocks}$$
$$k = \text{Number of treatments}$$
$$T_i = \text{Total for treatment } i \ (i = 1, 2, \ldots, k)$$
$$B_i = \text{Total for block } i \ (i = 1, 2, \ldots, b)$$

B.3 Two-Factor Factorial Experiment

$$CM = \text{Correction for mean}$$

$$= \frac{(\text{Total of all } n \text{ measurements})^2}{n} = \frac{\left(\sum_{i=1}^{n} y_i \right)^2}{n}$$

$$SS(\text{Total}) = \text{Total sum of squares}$$

$$= (\text{Sum of squares of all } n \text{ measurements}) - CM = \sum_{i=1}^{n} y_i^2 - CM$$

$$SS(A) = \text{Sum of squares for main effects, factor } A$$

$$= \left(\begin{array}{c} \text{Sum of squares of the totals } A_1, A_2, \ldots, A_a \\ \text{divided by the number of measurements} \\ \text{in a single total, namely } br \end{array} \right) - \text{CM}$$

$$= \frac{\sum\limits_{i=1}^{a} A_i^2}{br} - \text{CM}$$

$\text{SS}(B) = $ Sum of squares for main effects, factor B

$$= \left(\begin{array}{c} \text{Sum of squares of the totals } B_1, B_2, \ldots, B_b \\ \text{divided by the number of measurements} \\ \text{in a single total, namely } ar \end{array} \right) - \text{CM}$$

$$= \frac{\sum\limits_{i=1}^{b} B_i^2}{ar} - \text{CM}$$

$\text{SS}(AB) = $ Sum of squares for AB interaction

$$= \left(\begin{array}{c} \text{Sum of squares of the cell totals} \\ AB_{11}, AB_{12}, \ldots, AB_{ab} \text{ divided by} \\ \text{the number of measurements in} \\ \text{a single total, namely } r \end{array} \right) - \text{SS}(A) - \text{SS}(B) - \text{CM}$$

$$= \frac{\sum\limits_{j=1}^{b} \sum\limits_{i=1}^{a} AB_{ij}^2}{r} - \text{SS}(A) - \text{SS}(B) - \text{CM}$$

where

$a = $ Number of levels of factor A

$b = $ Number of levels of factor B

$r = $ Number of replicates (observations per treatment)

$A_i = $ Total for level i of factor A ($i = 1, 2, \ldots, a$)

$B_i = $ Total for level i of factor B ($i = 1, 2, \ldots, b$)

$AB_{ij} = $ Total for treatment (ij), i.e., for ith level of factor A and ith level of factor B

B.4 Tukey's Multiple Comparisons Procedure (Equal Sample Sizes)

Step 1 Select the desired experimentwise error rate, α

Step 2 Calculate

$$\omega = q_\alpha(k, v)\frac{s}{\sqrt{n_t}}$$

where

$k = $ Number of sample means (i.e., number of treatments)

$s = \sqrt{\text{MSE}}$

$v = $ Number of degrees of freedom associated with MSE

$n_t = $ Number of observations in each of the k samples (i.e., number of observations per treatment)

$q_\alpha(k, v) = $ Critical value of the Studentized range (Tables XV and XVI of Appendix A)

Step 3 Calculate and rank the k sample means.

Step 4 Place a bar over those pairs of treatment means that differ by less than ω. A pair of treatments not connected by an overbar (i.e., differing by more than ω) implies a difference in the corresponding population means.

Note: The confidence level associated with all inferences drawn from the analysis is $(1 - \alpha)$.

B.5 Bonferroni Multiple Comparisons Procedure (Pairwise Comparisons)

Step 1 Calculate for each treatment pair (i, j)

$$B_{ij} = t_{\alpha/(2c)} s \sqrt{\frac{1}{n_i} + \frac{1}{n_j}}$$

where

k = Number of sample (treatment) means in the experiment

c = Number of pairwise comparisons

   [*Note:* If all pairwise comparisons are to be made, then $c = k(k - 1)/2$]

$s = \sqrt{MSE}$

v = Number of degrees of freedom associated with MSE

n_i = Number of observations in sample for treatment i

n_j = Number of observations in sample for treatment j

$t_{\alpha/(2c)}$ = Critical value of t distribution with v df and tail area $\alpha/(2c)$ (Table VI in Appendix A)

Step 2 Rank the sample means and place a bar over any treatment pair (i, j) whose sample means differ by less than B_{ij}. Any pair of means not connected by an overbar implies a difference in the corresponding population means.

Note: The level of confidence associated with all inferences drawn from the analysis is at least $(1 - \alpha)$.

B.6 Scheffé's Multiple Comparisons Procedure (Pairwise Comparisons)

Step 1 Calculate Scheffé's critical difference for each pair of treatments (i, j):

$$S_{ij} = \sqrt{(k - 1)(F_\alpha)(MSE)\left(\frac{1}{n_1} + \frac{1}{n_j}\right)}$$

where

k = Number of sample (treatment) means

MSE = Mean squared error

n_i = Number of observations in sample for treatment i

n_j = Number of observations in sample for treatment j

F_α = Critical value of F distribution with $k - 1$ numerator df and v denominator df (Tables XIII, IX, X, and XI of Appendix A)

v = Number of degrees of freedom associated with MSE

Step 2 Rank the k sample means and place a bar over any treatment pair (i, j) that differs by less than S_{ij}. Any pair of sample means not connected by an overbar implies a difference in the corresponding population means.

Short Answers to Selected Odd Exercises

Chapter 1

1.3 population; variable(s); sample; inference; measure of reliability **1.11** qualitative; qualitative **1.13 a.** earthquake sites
b. sample **c.** ground motion (qualitative); magnitude (quantitative); ground acceleration (quantitative) **1.15 a.** (1) qualitative;
(2) quantitative; (3) qualitative **b.** sample **1.17** descriptive **1.19** Town, Type of water supply, and Presence of hydrogen sulphide
are qualitative; all others are quantitative **1.21 a.** quantitative **b.** quantitative **c.** qualitative **d.** quantitative **e.** qualitative **f.** quantitative
g. qualitative **1.23 a.** designed experiment **b.** smokers **c.** quantitative **d.** population: all smokers in the U.S.; sample: 50,000 smokers
in trial **e.** the difference in mean age at which each of the scanning methods first detects a tumor. **1.25. a.** designed experiment
b. amateur boxers **c.** heart rate (quantitative); blood lactate level (quantitative) **d.** no difference between the two groups of boxers
e. no **1.27 a.** population; all students; sample: 155 volunteer students **b.** designed experiment **c.** higher proportion of students in
guilty-state group chose not to repair the car than those in neutral-state and anger-state groups **d.** representativeness of sample
1.29 a. quantitative: age and dating experience; qualitative: gender and willingness to tell **b.** possible nonrepresentative sample
1.31 a. survey **b.** qualitative **c.** nonrepresentative sample **1.33 a.** results valid if only eat oat bran **b.** observational (survey) **c.** only
most positive responses reported **d.** children not surveyed about their level of hunger

Chapter 2

2.5 a. X—8, Y—9, Z—3 **b.** X—.40, Y—.45, Z—.15 **2.7 a.** qualitative **b.** Unknown—5, Unworn—2, Slight—4, Light/Mod—2, Mod—3,
Mod/Heavy—1, Heavy—1 **c.** Unknown—.278, Unworn—.111, Slight—.222, Light/Mod—.111, Mod—.167, Mod/Heavy—.056,
Heavy—.056 **d.** Unknown **2.9 a.** 39/266 = .147 **b.** level 2—.286, level 3—.188, level 4—.327, level 5—.041, level 6—.011
e. level 4 **2.11 a.** relative frequencies: Black—.203; White—.637; Sumatran—.017; Javan—.003; Indian—.140 **c.** .839; .161
2.13 a. .389 **b.** yes **c.** multiyear ice is most common **2.15** LEO—most government owned (43.7%); GEO—most commercially
owned (69.1%) **2.17** 50% of sampled CEOs had advanced degrees **2.19** No **2.21 b.** public—40% contaminated; private—21.4%
contaminated **c.** bedrock—31.3% contaminated; unconsolidated—31.8% contaminated **2.27 a.** 23 **2.29** frequencies: 50, 75, 125, 100,
25, 50, 50, 25 **2.31 a.** histogram **b.** 38 **c.** .475 **d.** .1375

2.33 a.

Stem	Leaf
1	6
2	0 1 2 4 8
3	0 4 4 9
4	0 0 2
5	3

b. A students tend to read the most books **2.35. b.** .962 **2.37.** 67% of frequencies exceed 3,500 hertz **2.39.** most of the PMIs
range from 3 to 7.5

2.41 No;

Stem	Leaf
1	0000000
2	0
3	00
4	000
5	
6	
7	
8	
9	0
10	0
11	0
12	00

2.43 histogram looks similar to graph; "inside job" not likely **2.45 a.** 33 **b.** 175 **c.** 20 **d.** 71 **e.** 1,089 **2.47 a.** 6 **b.** 50 **c.** 42.8 **2.49** mean,
median, mode **2.51** sample size and variability of the data **2.53 a.** mean < median **b.** mean > median **c.** mean = median
2.55 mode = 15; mean = 14.545; median = 15 **2.57 a.** 8.5 **b.** 25 **c.** .78 **d.** 13.44 **2.59** mean = 9.72; median = 10.94
2.61 a. mean = 31.6; median = 32; mode = 34 and 40 **b.** approx. symmetric **2.63 a.** mean = −4.86; median = −4.85;
mode = −5.00 **2.65 b.** probably none **2.67 a.** 16.5; increase **b.** 16.16; no change **c.** no mode **2.69 a.** mean = −.15; median = −.11;
4 modes **c.** mean = −.12; median = −.105; 4 modes **2.71** largest value minus smallest value **2.75** more variable **2.77 a.** 5, 3.7, 1.92

b. 99, 1949.25, 44.15 **c.** 98, 1307.84, 36.16 **2.79** data set 1: 1, 1, 2, 2, 3, 3, 4, 4, 5, 5; data set 2: 1, 1, 1, 1, 1, 5, 5, 5, 5, 5 **2.81 a.** 3, 1.3, 1.14 **b.** 3, 1.3, 1.14 **c.** 3, 1.3, 1.14 **2.83 a.** 51.26 **b.** 128.57 **c.** 11.34 **d.** σ^2 and σ **2.85 a.** 2.86 **b.** 3.26 **c.** 2.94 **d.** DM; honey **2.87 a.** 6.45; increase **b.** 3.883; increase **c.** 1.97; increase **2.89 a.** 10, 7.67, 2.77 **b.** 8, 5.15, 2.27; less variable **c.** 8, 5.06, 2.25; less variable **2.91 a.** dollars; quantitative **b.** at least 3/4; at least 8/9; nothing; nothing **2.93 a.** $\approx 68\%$ **b.** $\approx 95\%$ **c.** $\approx$ all **2.95** range/6 = 104.17, range/4 = 156.25; no **2.97 a.** $\bar{x}$ = 95.7, s = 4.96 **b.** (90.74, 100.66); (85.77, 105.63); (80.81, 110.59) **c.** 89.2%, 96.2%, 97.8%; agree (Chebychev) **2.99 a.** unknown **b.** $\approx 84\%$ **2.101 a.** (0, 153) **b.** (0, 281) **c.** hand rubbing appears to be more effective **2.103 a.** at least 8/9 of the velocities will fall within 936 ± 30 **b.** No **2.105 a.** 19 ± 195 **b.** 7 ± 147 **c.** SAT-Math **2.107** do not purchase **2.109 a.** 25%; 75% **b.** 50%; 50% **c.** 80%; 20% **d.** 16%; 84% **2.111 a.** 2 **b.** .5 **c.** 0 **d.** -2.5 **e.** sample: a, d; population: b, c **f.** above: a, b; below: d **2.113** $\mu = 60$, $\sigma = 10$ **2.115** 26th percentile **2.117 a.** $z = -1.5$ **b.** .50 **2.119 a.** 23 **b.** $z = 3.28$ **2.121 a.** $z = 5.08$ **b.** $z = -.81$ **c.** yes **2.123 a.** $z = 2.0$: 3.7; $z = -1.0$: 2.2; $z = .5$: 2.95; $z = -2.5$: 1.45 **b.** 1.9 **c.** $z = 1.0$ and 2.0; GPA = 3.2 and 3.7; mound shaped; symmetric **2.129 a.** no, $z = .73$ **b.** yes, $z = -3.27$ **c.** no, $z = 1.36$ **d.** yes, $z = 3.73$ **2.131 a.** 4 **b.** $Q_U \approx 6$, $Q_L \approx 3$ **c.** 3 **d.** skewed right **e.** 50%; 75% **f.** 12, 13, and 16 **2.133 a.** 10, 15, 27.5 **b.** 3.5, 5, 7.5 **c.** effective **2.135 a.** -1.26 **b.** No **2.137 a.** 69, 73, 74, 78, 83, 84, and 86 **b.** 69, 73, 74, and 78 **c.** no; highly skewed data **2.139** outliers: 28 and 33 **2.141 b.** $\sigma_{2005} = 67.9$; $\sigma_{2009} = 81.4$ **c.** no **2.147** slight positive linear trend **2.149** positive linear trend **2.151** negative trend, nonlinear **2.153 a.** no **b.** yes, slight positive **c.** reliability is suspect (only 5 data points) **2.155 a.** negative trend **b.** positive trend with both plant coverage and diversity **2.157** Yes; accuracy decreases as driving distance increases **2.165 a.** $-1, 1, 2$ **b.** $-2, 2, 4$ **c.** $1, 3, 4$ **d.** .1, .3, .4 **2.167 a.** 3.12 **b.** 9.02 **c.** 9.79 **2.169 a.** $\bar{x} = 5.67$, $s^2 = 1.07$, $s = 1.03$ **b.** $\bar{x} = -\$1.5$, $s^2 = 11.5$, $s = \$3.39$ **c.** $\bar{x} = .413\%$, $s^2 = .088$, $s = .297\%$ **d.** 3; \$10; .7375% **2.171** yes, positive **2.173** skewed to the right **2.175 b.** $z = -1.06$ **2.177 b.** 5.4% **c.** average player rating is 1068 **2.179 a.** survey **b.** quantitative **c.** about 60% **2.181 a.** "favorable/recommended"; .635 **b.** yes **2.183 a.** no **b.** unknown; at least 84%; at least 93.75% **c.** 55%; 97.5; 100% **d.** 87.5%; 95%; 100% **2.185** Over half of the whistle types were "Type a" **2.187 a.** no outliers **2.189 a.** Seabirds—quantitative; length—quantitative; oil—qualitative **b.** transect **c.** oiled: 37.5%; unoiled: 62.5% **e.** distributions are similar **f.** 3.27 ± 13.4 **g.** 3.50 ± 11.94 **h.** unoiled **2.191 b.** yes **c.** A1775A: = 19,462.2, s = 532.29; A1775B: = 22,838.5, s = 560.98 **d.** cluster A1775A **2.193** yes; $z = -2.5$ **2.195 a.** median **b.** mean **2.197** results not reliable; sample not representative

Chapter 3

3.9 a. .5 **b.** .3 **c.** .6 **3.11** $P(A) = .55$; $P(B) = .50$; $P(C) = .70$ **3.13 a.** 10 **b.** 20 **c.** 15,504 **3.15 a.** (R_1R_2), (R_1R_3), (R_2R_3), (R_1B_1), (R_1B_2), (R_2B_1), (R_2B_2), (R_3B_1), (R_3B_2), (B_1B_2) **b.** 1/10 for each sample point **c.** $P(A) = 1/10$, $P(B) = 6/10$, $P(C) = 3/10$ **3.17 a.** Blue, orange, green, yellow, brown, and red **b.** $P(\text{blue}) = .24$, $P(\text{orange}) = .20$, $P(\text{green}) = .16$, $P(\text{yellow}) = .14$, $P(\text{brown}) = .13$, $P(\text{red}) = .13$ **c.** .13 **d.** .43 **e.** .76 **3.19 a.** .01 **b.** yes **3.21 a.** (None), (1 or 2), (3–5), (6–9), (10 or more) **b.** $P(\text{None}) = .25$, $P(1 \text{ or } 2) = .31$, $P(3-5) = .25$, $P(6-9) = .05$, $P(10 \text{ or more}) = .14$ **c.** .44 **3.23** .748 **3.25** .389 **3.27 a.** 6 **b.** .282, .065, .339, .032, .008, .274 **c.** .686 **3.29 a.** 28 **b.** 1/28 **3.31** 693/1,686,366 = .000411 **3.33 a.** 15 **b.** 20 **c.** 15 **d.** 6 **3.39** $P(A \cup B) = P(A) + P(B) - P(A \cap B)$ **3.41** $P(A \cup B) = P(A) + P(B)$ **3.43 b.** $P(A) = 7/8$, $P(B) = 1/2$, $P(A \cup B) = 7/8$, $P(A^c) = 1/8$, $P(A \cap B) = 1/2$ **c.** 7/8 **d.** no **3.45 a.** $^3/_4$ **b.** 13/20 **c.** 1 **d.** 2/5 **e.** $^1/_4$ **f.** 7/20 **g.** 1 **h.** 1/4 **3.47 a.** .65 **b.** .72 **c.** .25 **d.** .08 **e.** .35 **f.** .72 **g.** .65 **h.** A and C, B and C, C and D **3.49. b.** .06 **c.** .94 **3.51 a.** School laboratory, In transit, Chemical plant, Non chemical plant, Other **b.** .06, .26, .21, .35, .12 **c.** .06 **d.** .56 **e.** .74 **3.53 b.** .43 **c.** .57 **3.55 a.** .333 **b.** .351 **c.** .263 **d.** .421 **3.57 a.** .156 **b.** .617 **c.** .210 **d.** .449 **e.** yes **3.59 a.** .684 **b.** .124 **c.** no **d.** .316 **e.** .717 **f.** .091 **3.61 a.** .19 **b.** .405 **c.** .595 **3.65** $P(A|B) = P(A \cap B)/P(B)$ **3.67 a.** .5 **b.** .25 **c.** no **3.69 a.** .08 **b.** .40 **c.** .52 **3.71 a.** $P(A) = .4$; $P(B) = .4$; $P(A \cap B) = .3$ **b.** $P(E_1|A) = .25$, $P(E_2|A) = .25$, $P(E_3|A) = .5$ **c.** .75 **3.73** no **3.75** 1/3, 0, 1/14, 1/7, 1 **3.77 a.** .26 **b.** .013 **3.79** .40 **3.81 a.** .055 **b.** .214 **3.83** .013; independent events **3.85 a.** .23 **b.** .729 **3.87 a.** 38/132 = .2879 **b.** 29/123 = .236 **3.89 a.** .406 **b.** .294 **3.91** $P(A|B) = 0$ **3.93 a.** (A defeats B, C defeats D, A defeats C), (A defeats B, C defeats D, C defeats A), (A defeats B, D defeats C, A defeats D), (A defeats B, D defeats C, D defeats A), (B defeats A, C defeats D, B defeats C), (B defeats A, C defeats D, C defeats B), (B defeats A, D defeats C, B defeats D), (B defeats A, D defeats C, D defeats B) **b.** $P(A \text{ wins}) = 2/8 = .25$ **c.** .576 **3.95 a.** .1 **b.** .522 **c.** The person is guessing **3.97 b.** Worst—.250, 2nd worst—.200, 3rd worst—.157, 4th worst—.120, 5th worst—.089, 6th worst—.064, 7th worst—.044, 8th worst—.029, 9th worst—.018, 10th worst—.011, 11th worst—.007, 12th worst—.006, 13th worst—.005 **c.** 250/800 = .313 **d.** .297 **e.** .288 **3.103 a.** 35, 820, 200 **b.** 1/35, 820, 200 **c.** highly unlikely **3.117 a.** 4 **b.** 8 **c.** 32 **d.** 2^n **3.119 a.** 35 **b.** 15 **c.** 435 **d.** 45 **e.** $\binom{q}{r} = \dfrac{q!}{r!(q!-r!)}$ **3.121 a.** 10 **b.** 6 **c.** 36 **3.123 a.** 56 **b.** 1,680 **c.** 6,720 **3.125 a.** 24 **b.** 12 **3.127** 1,500,625 **3.129 a.** 30 **b.** 1,050 **3.131 a.** 6 **b.** 1/3 **3.133 a.** 18 **b.** 4/18 **3.135 a.** 21/252 **b.** 21/252 **c.** 105/252 **3.137 a.** 2,598,960 **b.** .002 **c.** .00394 **d.** .0000154 **3.141 a.** .225 **b.** .125 **c.** .35 **d.** .643 **e.** .357 **3.143** .2 **3.145 a.** .5 **b.** .99 **c.** .847 **3.147** 1/159 **3.149** $P(\text{dolomite}|\text{reading} > 60) = .1082$ **3.151** .966 **3.153 a.** $P(T^c|E) > P(T|E)$ **3.155 a.** $A \cup B$ **b.** B^c **c.** $A \cap B$ **d.** $A^c|B$ **3.157 a.** 0 **b.** no **3.159** .5 **3.161 c.** $P(A) = 1/4$; $P(B) = 1/2$ **e.** $P(A^c) = 3/4$; $P(B^c) = 1/2$; $P(A \cap B) = 1/4$; $P(A \cup B) = 1/2$; $P(A|B) = 1/2$; $P(B|A) = 1$ **f.** no, no **3.163 a.** no **b.** .3, .1 **c.** .37 **3.165** .05 **3.167 a.** false **b.** true **c.** true **d.** false **3.169** $A = \{8th \ grader \ scores \ above \ 655\}$; $P(A^c) = .95$ **3.171 a.** .261 **b.** Trunk—.85, Leaves—.10, Branch—.05 **3.173 a.** {Single, shore parallel; Other; Planar} **b.** 1/3, 1/3, 1/3 **c.** 2/3 **d.** {No dunes/flat; Bluff/scarp; Single dune; Not observed} **e.** 1/3, 1/6, 1/3, 1/6 **f.** 2/3 **3.175 a.** $\{11, 13, 15, 17, 29, 31, 33, 35\}$ **b.** $\{2, 4, 6, 8, 10, 11, 13, 15, 17, 20, 22, 24, 26, 28, 29, 31, 33, 35, 1, 3, 5, 7, 9, 19, 21, 23, 25, 27\}$ **c.** $P(A) = 9/19$, $P(B) = 9/19$, $P(A \cap B) = 4/19$, $P(A \cup B) = 14/19$, $P(C) = 9/19$ **d.** $\{11, 13, 15, 17\}$ **e.** 14/19, no **f.** 2/19 **g.** $\{1, 2, 3, \cdots, 29, 31, 33, 35\}$ **h.** 16/19 **3.177 a.** .158 **b.** .316 **c.** .526 **d.** #3 **3.179 a.** .64, .32, .04 **b.** .72, .22, .06 **c.** dependent **3.181 a.** .116 **b.** .728 **3.183 a.** .7127 **b.** .2873 **c.** .9639 **d.** .3078 **e.** .0361 **f.** at least 3 **3.185 a.** #4 **b.** #4 or #6 **3.187 a.** .006 **b.** .0007 **c.** .001 **d.** .018 **e.** .006 **3.189 a.** 1 to 2 **b.** .5 **c.** .4 **3.191 a.** .25 **b.** .0156 **c.** .4219 **d.** .001, .000000001, .9970 **3.193** .993 **3.195** 4.4739×10^{-28} **3.197** Marilyn **3.199** yes

Chapter 4

4.3 a. discrete **b.** continuous **c.** continuous **d.** discrete **e.** continuous **f.** continuous   **4.5 a.** continuous **b.** discrete **c.** discrete **d.** discrete **e.** discrete **f.** continuous   **4.7** 0, 1, 2, 3, ...; discrete   **4.9** continuous; discrete   **4.11** gender; IQ score   **4.13** allergy to penicillin; blood pressure   **4.15** table, graph, formula   **4.17 a.** .25 **b.** .35 **c.** .8   **4.19 a.** .7 **b.** .3 **c.** 1 **d.** .2 **e.** .8   **4.21 b.** $P(0) = 1/8$, $P(1) = 3/8, P(2) = 3/8, P(3) = 1/8$ **d.** 1/2   **4.23 a.** 1 **b.** .24 **c.** .39   **4.25 a.** BB, BG, GB, GG **b.** 1/4, 1/4, 1/4, 1/4, **c.** $P(0) = .25$, $P(1) = .5, P(2) = .25$ **d.** $P(0) = .222, P(1) = .513, P(2) = .265$ **4.27 a.** .23 **b.** .081 **c.** .77 **4.29 a.** $P(6) = .282$, $P(7) = .065, P(8) = .339, P(9) = .032, P(10) = .008, P(11) = .274$ **b.** .274   **4.31** 7/8   **4.33 a.** 0, 1, 2 **b.** $P(0) = .625$, $P(1) = .250, P(2) = .125$ **4.35** no   **4.37 a.** 3.8 **b.** 10.56 **c.** 3.25 **e.** no **f.** yes   **4.39 a.** $\mu_x = 1, \mu_y = 1$ **b.** distribution of x **c.** $\mu_x = 1, \sigma^2 = .6; \mu_y = 1, \sigma^2 = .2$ **4.41** 1.04   **4.43 a.** 1.8 **b.** .9899 **c.** .96   **4.45 a.** MC, MS, MB, MO, ML, CS, CB, CO, CL, SB, SO, SL, BO, BL, OL **b.** equally likely, with $p = 1/15$ **d.** $P(0) = 6/15, P(1) = 8/15, P(2) = 1/15$ **e.** .667   **4.47** $-\$0.263$ **4.49** $\$0.25$

4.51 $p(x) = \binom{7}{x}.2^x.8^{7-x}$ $(x = 0, 1, 2, \ldots, 7)$ **4.53 a.** 15 **b.** 10 **c.** 1 **d.** 1 **e.** 4   **4.55 a.** .4096 **b.** .3456 **c.** .027 **d.** .0081 **e.** .3456

f. .027   **4.57 a.** $\mu = 12.5, \sigma^2 = 6.25, \sigma = 2.5$ **b.** $\mu = 16, \sigma^2 = 12.8, \sigma = 3.578$ **c.** $\mu = 60, \sigma^2 = 24, \sigma = 4.899$ **d.** $\mu = 63, \sigma^2 = 6.3$, $\sigma = 2.510$ **e.** $\mu = 48, \sigma^2 = 9.6, \sigma = 3.098$ **f.** $\mu = 40, \sigma^2 = 38.4, \sigma = 6.197$ **4.59 a.** 0 **b.** .998 **c.** .137 **4.61 a.** $p = .5$ **b.** $p < .5$ **c.** $p > .5$ **4.63 b.** 40 **c.** 24 **d.** (30.2, 49.8)   **4.65 b.** $n = 20, p = .8$ **c.** .174 **d.** .804 **e.** 16   **4.67 b.** .4752 **c.** .0769 **4.69 a.** .375 **b.** .5   **4.71 a.** .791 **b.** .056 **c.** rare event if $p = .1$ **4.73 a.** .1 **b.** .7 **c.** .4783 **d.** .0078 **e.** yes   **4.75 b.** $\mu = 2.4, \sigma = 1.47$ **c.** $p = .9, q = .1, n = 24, \mu = 21.6$, $\sigma = 1.47$ **4.79** 3   **4.81 b.** $\mu = 3, \sigma = 1.7321$ **4.83 a.** .934 **b.** .191 **c.** .125 **d.** .223 **e.** .777 **f.** .001   **4.85 b.** $\mu = 3, \sigma = 1.7321$ **c.** .966 **4.87 a.** .368 **b.** .264 **c.** .920   **4.89 a.** .301 **b.** .337   **4.91 a.** .202 **b.** .323 **c.** $\mu = 1.6, \sigma = 1.26$ **4.93 a.** .125 **b.** 5 **c.** no; $p(0) = .007$ **4.95 a.** $p(2) = .039, p(6) = .160, p(10) = .047$ **c.** $\mu = 6.2, \sigma = 2.49$ **d.** very unlikely   **4.97** yes, .96   **4.101 a.** .3 **b.** .119 **c.** .167 **d.** .167   **4.103 b.** $\mu = 4, \sigma = .853$ **d.** .939   **4.105 a.** hypergeometric **b.** binomial   **4.107 a.** hypergeometric **b.** $r = 20, N = 100$, and $n = 3$ **c.** .508 **d.** .391 **e.** .094 **f.** .007   **4.109 a.** .383 **b.** .0002   **4.111** .2693 **4.113 a.** .0883 **b.** .1585   **4.115** $P(x = 1) = .25$ **4.117 a.** $\mu = 113.24, \sigma = 4.19$ **b.** $z = 6.38$ **4.119 a.** Poisson **b.** binomial **c.** binomial   **4.121 a.** .192 **b.** .228 **c.** .772 **d.** .987 **e.** .960 **f.** 14; 4.2; 2.05 **g.** .975   **4.123 a.** .243 **b.** .131 **c.** .36 **d.** .157 **e.** .128 **f.** .121   **4.125 a.** .180 **b.** .015 **c.** .076   **4.127 b.** 4.65   **4.129 b.** .05 **c.** 10 **4.131** binomial   **4.133 a.** yes **b.** .051 **c.** .757 **d.** .192 **e.** 3.678   **4.135 a.** .001 **b.** .322 **c.** .994   **4.137 a.** $\mu = 520, \sigma = 13.49$ **b.** no, $z = -8.90$ **4.139** .642   **4.141 a.** 2 **b.** no, $p = .003$ **4.143 a.** .006 **b.** insecticide is less effective than claimed   **4.145** ≈ 292 **4.147** .109

Chapter 5

5.3 b. $\mu = 20, \sigma = 5.774$ **c.** (8.452, 31.548)   **5.5 b.** $\mu = 3, \sigma = .577$ **c.** .577 **d.** .61 **e.** .65 **f.** 0   **5.7 a.** 0 **b.** 1 **c.** 1   **5.9 a.** 2 **b.** .25 **c.** .375 **5.11 a.** .8 **b.** .3   **5.13 a.** .1333, .5714 **b.** .2667, 0   **5.15** $\mu = .5, \sigma = .2887$, 10th percentile $= .1, Q_L = .25, Q_U = .75$ **5.17 a.** continuous **c.** $\mu = 7, \sigma = .2887$ **d.** .5 **e.** 0 **f.** .75 **g.** .0002   **5.19** .4444   **5.21** standard normal   **5.23 a.** .4772 **b.** .3413 **c.** .4987 **d.** .2190   **5.25 a.** 0 **b.** .8413 **c.** .8413 **d.** .1587 **e.** .6826 **f.** .9544 **g.** .6934 **h.** .6378   **5.27 a.** 1 **b.** -1 **c.** 0 **d.** -2.5 **e.** 3   **5.29 a.** 1.645 **b.** 1.96 **c.** -1.96 **d.** 1.28 **e.** 1.28   **5.31 a.** .3830 **b.** .3023 **c.** .1525 **d.** .7333 **e.** .1314 **f.** .9545   **5.33 a.** 19.76 **b.** 36.72 **c.** 48.64   **5.35** 182   **5.37 a.** $-.25$ **b.** .44 **c.** .0122   **5.39 a.** .1583 **b.** .2434 **c.** .0096 **d.** .9406 **e.** no; $P(x \le 25) \approx 0$ **5.41 a.** .8413 **b.** .7528   **5.43 a.** .4107 **b.** .1508 **c.** .9066 **d.** .0162 **e.** .841.8   **5.45 a.** no; $P(x \le 9) = .7527$ **b.** no; $P(x \le 2) = .0139$ **5.47 a.** (62.26, 65.74); (61.01,66.99); (59.72,68.28); (58.90,69.10); (57.30,70.70)   **5.49** .7019   **5.51 a.** $z_L = -.675, z_U = .675$ **b.** $-2.68, 2.68$ **c.** $-4.69, 4.69$ **d.** .0074, 0   **5.55 a.** .68 **b.** .95 **c.** .997 **5.57** plot c   **5.59 a.** not normal (skewed right) **b.** $Q_L = 37.5, Q_U = 152.8, s = 95.8$ **c.** IQR/s $= 1.204$ **5.61 a.** 7 **b.** 6.444 **c.** IQR/s $= 1.09$ **5.63** no; IQR/s $= .82$ **5.65 a.** histogram too peaked **b.** more than 95%   **5.67** both distributions approx. normal **5.69** IQR/$s = 1.3$, histogram approx. normal   **5.71** lowest possible time (0 minutes) is less than 1 standard deviation below the mean **5.75 a.** yes **b.** $\mu = 10, \sigma^2 = 6$ **c.** .726 **d.** .7291   **5.77 a.** .345; .3446 **b.** .115; .1151 **c.** .924; .9224   **5.79 a.** .1788 **b.** .5236 **c.** .6950 **5.81 a.** 16.25 **b.** 3.49 **c.** 1.07 **d.** .1762   **5.83 a.** $\mu = 1,500, \sigma^2 = 1,275$ **b.** .0026 **c.** no   **5.85** ≈ 0 **5.87 a.** no **b.** yes **c.** yes   **5.89** no; $P(x > 110) = .0018$ **5.91 a.** 300 **b.** 800 **c.** 1   **5.97 a.** .367879 **b.** .950213 **c.** .223130 **d.** .993262 **5.99** .950213   **5.101 a.** .449329 **b.** 864665 **5.103 a.** .283 **b.** yes   **5.105 a.** .279543 **b.** .1123887   **5.107 a.** 17 **b.** .5862   **5.109 a.** .7534 **b.** .6667 **c.** .81   **5.113 a.** exponential **b.** uniform **c.** normal   **5.115 a.** .9821 **b.** .0179 **c.** .9505 **d.** .3243 **e.** .9107 **f.** .0764   **5.117 a.** .6915 **b.** .0228 **c.** .5328 **d.** .3085 **e.** .0 **f.** .9938 **5.119 a.** .3821 **b.** .5398 **c.** .0 **d.** .1395 **e.** .0045 **f.** .4602   **5.121** no   **5.123** 83.15%   **5.125 a.** .9406 **b.** .9406 **c.** .1140   **5.127 a.** 0 **b.** 1   **5.129 a.** .667 **b.** .333 **c.** 82.5°F   **5.131** IQR/$s = 1.56$; approx. normal   **5.133 a.** .0735 **b.** .3651 **c.** 7.29   **5.135** .3125 **5.137** .0154; very unlikely   **5.139** ≈ 0 **5.141** no   **5.143 a.** $d = 56.24$cm **b.** .434598   **5.145 a.** (i) .384, (ii) .49, (iii) .212, (iv) .84 **b.** (i).3849, (ii) .4938, (iii) .2119, (iv) .8413 **c.** (i).0009, (ii) .0038, (iii) .0001, (iv) .0013   **5.147** 52 minutes   **5.149** 5.068   **5.151** 0; unlikely   **5.153 a.** .05, .20, .50, .20, .05 **b.** z-scores **c.** identical

Chapter 6

6.3 c. 1/16   **6.5 c.** .05 **d.** no
6.7 a.

M	1	1.5	2	2.5	3	3.5	4	4.5	5
P(M)	.04	.12	.17	.20	.20	.14	.08	.04	.01

6.13 unbiased, minimum variance   **6.15** a. 5 **b.** $E(\bar{x}) = 5$ **c.** $E(M) = 4.778$ **d.** $\bar{x}$ **6.17** a. yes **b.** median   **6.19 b.** 1.61 **c.** $E(s^2) = 1.61$ **d.** $E(s) = 1.004$ **6.21** mean and standard deviation of sampling distribution of $\bar{x}$ **6.23** smaller   **6.27 a.** $\mu = 100, \sigma = 5$ **b.** $\mu = 100, \sigma = 2$ **c.** $\mu = 100, \sigma = 1$ **d.** $\mu = 100, \sigma = 1.414$ **e.** $\mu = 100, \sigma = .447$ **f.** $\mu = 100, \sigma = .316$ **6.29 a.** $\mu = 2.9, \sigma^2 = 3.29$,

$\sigma = 1.814$ **c.** $\mu_{\bar{x}} = 2.9, \sigma_{\bar{x}} = 1.283$ **6.31 a.** $\mu_{\bar{x}} = 30, \sigma_{\bar{x}} = 1.6$ **b.** approx. normal **c.** .8944 **d.** .0228 **e.** .1303 **f.** .9699
6.33 As n increases, variance decreases. **6.35 a.** $\mu_{\bar{x}} = 320, \sigma_{\bar{x}} = 10$, approx. normal **b.** .1359 **c.** ≈ 0 **6.37 a.** 79 **b.** 2.3 **c.** approx. normal
d. .43 **e.** .3336 **6.39** No, $P(\bar{x} < 103) = .0107$ **6.41 a.** 10; .0002 **b.** Central Limit Theorem **c.** .0170 **6.43 a.** $\mu_{\bar{x}} = .53, \sigma_{\bar{x}} = .0273$,
approx. normal **b.** .0336 **c.** after **6.45 a.** .0034 **b.** $\mu > 6$ **6.47** hand rubbing: $P(\bar{x} < 30) = .2743$; hand washing: $P(\bar{x} < 30) = .0047$;
sample used hand rubbing **6.49** false **6.51** true **6.53 b.** $E(A) = \alpha$ **c.** choose estimator with smallest variance **6.55 a.** .5 **b.** .0606
c. .0985 **d.** .8436 **6.61 a.** 106 **b.** 2.73 **c.** approx. normal **d.** -2.20 **e.** .0136 **6.63** .0838 **6.65** .9772 **6.67 a.** No, prob. ≈ 0 **b.** likely that
$\mu > 4.59$ **6.69 a.** .3264 **b.** 1.881 **c.** valid **6.71** .9332 **6.73 a.** .0031 **b.** more likely if $\mu = 156$; less likely if $\mu = 158$ **c.** less likely if
$\sigma = 2$; more likely if $\sigma = 6$

Chapter 7

7.5 yes **7.7 a.** 1.645 **b.** 2.58 **c.** 1.96 **d.** 1.28 **7.9 a.** 28 ± 4.53 **b.** 102 ± 2.77 **c.** 15 ± 3.92 **d.** 4.05 ± 3.92 **e.** no
7.11 a. 83.2 ± 1.25 **c.** 83.2 ± 1.65 **d.** increases **e.** yes **7.13 a.** 19.3 **b.** 19.3 ± 3.44 **d.** random sample, large n
7.15 39 ± 1.55 **7.17 a.** $.36 \pm .0099$ **c.** first-year: $.303 \pm .0305$; landfast: $.362 \pm .0177$; multiyear: $.381 \pm .0101$ **7.19 a.** μ
b. no; apply Central Limit Theorem **c.** $(0.4786, 0.8167)$ **d.** yes **7.21 a.** $1.13 \pm .67$ **b.** yes **7.23 a.** 19 ± 7.83 **b.** 7 ± 5.90
c. SAT-Math **7.25 a.** males: 16.79 ± 2.35; females: 10.79 ± 1.67 **b.** .0975 **c.** males **7.27** Central Limit Theorem no longer
applies; σ unknown **7.29 a.** large: normal; small: t-distribution **b.** large: normal; small: unknown **7.31 a.** 2.228 **b.** 2.567
c. -3.707 **d.** -1.771 **7.33. a.** 5 ± 1.88 **b.** 5 ± 2.39 **c.** 5 ± 3.75 **d.** (a) $5 \pm .78$, (b) $5 \pm .94$, (c) 5 ± 1.28; width
decreases **7.35. a.** $\mu =$ population mean trap spacing **b.** 89.86 **c.** population of trap spacings has an unknown distribution
d. 89.86 ± 10.76 **e.** random sample; population of trap spacings is normally distributed **7.37 a.** $(652.76, 817.40)$ **d.** yes
7.39 a. .009 **b.** 2.306 **c.** $.009 \pm .0037$ **7.41** $.11 \pm .08$ **7.43 a.** both untreated: 20.9 ± 1.06; male treated: 20.3 ± 1.25; female
treated: 22.9 ± 1.79; both treated: $18.6 \pm .79$ **b.** female treated **7.45 a.** $1.43 \pm .15$ **c.** 90% of all similarly constructed intervals
contain μ **7.47 a.** 37.3 ± 7.70 **b.** One hour before: $\mu > 25.5$ **7.49** mean of sampling distribution of $\hat{p}$ is equal to p
7.51 a. yes **b.** $.64 \pm .07$ **7.53 a.** yes **b.** no **c.** no **d.** no **7.55 a.** all American adults **b.** 1,000 adults surveyed
c. $p =$ proportion of adults who say Starbucks coffee is overpriced **d.** $.73 \pm .03$ **7.57 a.** set of all gun ownership status (yes/no)
values for all U.S. adults **b.** true percentage of all adults who own a gun **c.** .26 **d.** $.26 \pm .02$ **7.59 a.** $.63 \pm .03$ **b.** yes
7.61 $.64 \pm .05$ **7.63 a.** $.338 \pm .025$ **7.65 a.** no **b.** $.009 \pm .010$ **7.67** 95% confident that proportion of health care workers
with latex allergy who suspect they have allergy is between .327 and .541 **7.69** true **7.71** 519 **7.73 a.** 482 **b.** 214 **7.75** 34
7.77 a. $n = 16$: $W = .98$; $n = 25$: $W = .784$; $n = 49$: $W = .56$; $n = 100$: $W = .392$; $n = 400$: $W = .196$ **7.79** 21 **7.81 a.** small sample
b. 125 **7.83** 21 **7.85** 40 **7.87** 14,735 **7.89** $n = 1,041$ **7.91** 271 **7.93** chi-square **7.95** $n-1$ **7.97 a.** $(4.54, 8.81)$ **b.** $(.00024,$
$.00085)$ **c.** $(641.86, 1809.09)$ **d.** $(.95, 12.66)$ **7.99** $(4.27, 65.90)$ **7.101 a.** $(308.17, 537.93)$ **b.** yes **7.103 a.** $(8.6, 45.7)$ **b.** $(2.94,$
$6.76)$ **c.** random sample from a normal population **7.105** $(1.4785, 2.9533)$ **7.107** $(43.76, 1200.99)$ **7.109 a.** μ **b.** μ **c.** p
d. p **e.** p **f.** σ^2 **7.111 a.** $t = 2.086$ **b.** $z = 1.96$ **c.** $z = 1.96$ **d.** $z = 1.96$ **e.** neither t nor z **7.113 a.** $.57 \pm .05$
b. 2,358 **7.115** (1) p; (2) μ; (3) μ; (4) μ **7.117** $.219 \pm .024$ **7.119 a.** .03 **b.** $.03 \pm .01$ **7.121 a.** .90 **b.** .05 **c.** 259
7.123 a. $\bar{x} = 1.86, s = 1.195$ **b.** skewed right **c.** $1.86 \pm .45$ **e.** 90% **7.125 a.** $1.07 \pm .24$ **c.** $(.145, .568)$ **d.** normal **e.** 129
7.127 a. $.660 \pm .029$ **b.** $.301 \pm .035$ **7.129 a.** $.044 \pm .162$ **b.** no evidence of species inbreeding **7.131 a.** $.81 \pm .41$
b. 140 **c.** $(1.27, 1.89)$ **7.133 a.** $.390 \pm .149$ **b.** $.683 \pm .142$ **c.** 41.405 ± 21.493 **d.** $.390 \pm .149$ **7.135 a.** .094 **b.** yes
c. $.094 \pm .037$ **7.137 a.** 49.3 ± 8.6 **c.** population is normal **d.** 60 **e.** $(.75, 43.87)$ **7.139. a.** 196 **7.141 a.** yes **b.** missing
measure of reliability **c.** 95% CI for μ: $.932 \pm .037$

Chapter 8

8.1 null; alternative **8.3** α **8.5** reject H_0 when H_0 is true; accept H_0 when H_0 is true; reject H_0 when H_0 is false; accept H_0
when H_0 is false **8.7** no **8.9 g.** (a) .025, (b) .05, (c) $\approx .005$, (d) $\approx .10$, (e) .10, (f) $\approx .01$ **8.11 a.** $H_0 : \mu = 2400, H_a : \mu > 2400$
b. the probability of concluding that the mean gain in fees is greater than \$2,400 when, in fact, it is equal to \$2,400 is .05
c. $z > 1.645$ **8.13** $H_0 : p = .75$ **8.15** $H_0 : p = .05, H_a : p < .05$ **8.17 a.** $H_0 : \mu = 15, H_a : \mu < 15$ **b.** conclude mean mercury level is
less than 15 ppm when mean equals 15 ppm **c.** conclude mean mercury level equals 15 ppm when mean is less than 15 ppm
8.19 a. H_0: No intrusion occurs **b.** H_a: Intrusion occurs **c.** $\alpha = .001, \beta = .5$ **8.21** random sample, large n **8.23 a.** $z = 1.67$, reject
H_0 **b.** $z = 1.67$, fail to reject H_0 **8.25 a.** $z > 1.875$ **b.** $\approx .03$ **8.27 a.** Type I: conclude true mean response for all New York
City public school children is not 3 when mean equals 3; Type II: conclude true mean response equals 3 when mean is not equal
to 3 **b.** $z = -85.52$, reject H_0 **c.** $z = -85.52$, reject H_0 **8.29 a.** $H_0 : \mu = 85, H_a : \mu \neq 85$ **b.** conclude that the true mean Mach rating
score is different from 85 when, in fact, it is equal to 85 **c.** probability of concluding that the true mean Mach rating score is dif-
ferent from 85 when, in fact, it is equal to 85 is .10 **d.** $|z| > 1.645$ **e.** $z = 12.80$ **f.** reject H_0 **g.** no **8.31** $z = 4.17$, reject H_0
8.33 a. $z = 4.03$, reject H_0 **8.35 a** no **b.** $z = .61$, do not reject H_0 **d.** no **e.** $z = -.83$, do not reject H_0 **8.37 a.** $z = -3.72$, reject
H_0 **8.39** small p-values **8.41 a.** fail to reject H_0 **b.** reject H_0 **c.** reject H_0 **d.** fail to reject H_0 **e.** fail to reject H_0 **8.43** .0150
8.45 p-value $= .9279$, fail to reject H_0 **8.47 a.** fail to reject H_0 **b.** fail to reject H_0 **c.** reject H_0 **d.** fail to reject H_0
8.49 a. p-value ≈ 0 **b.** reject H_0 **8.51** p-value $< .0001$; reject H_0 at $\alpha = .01$ **8.53 a.** $z = 26.15$, p-value ≈ 0; reject H_0
b. $z = -14.24$, p-value ≈ 0; reject H_0 **8.55 b.** reject H_0 at $\alpha = .05$ **c.** reject H_0 at $\alpha = .05$ **8.57** mound-shaped, symmetric;
t-distribution is flatter than z-distribution **8.61 a.** $|t| > 2.160$ **b.** $t > 2.5$ **c.** $t > 1.397$ **d.** $t < -2.718$ **e.** $|t| > 1.729$ **f.** $t < -2.353$

8.63 a. $t = -2.064$; fail to reject H_0 **b.** $t = -2.064$; fail to reject H_0 **c.** (a)$.05 < p$-value $< .10$; (b)$.10 < p$-value $< .20$
8.65 a. $H_0{:}\mu = 95$, $H_a{:}\mu \neq 95$ **b.** variation in sample data not taken into account **c.** $t = -1.163$ **d.** p-value $= .289$ **e.** $\alpha = .10$; probability of concluding mean strap spacing differs from 95 when, in fact, it is equal to 95 is .10 **f.** fail to reject H_0 **g.** random sample from a normal population of strap spacing measurements **h.** yes   **8.67** $t = -.32$; fail to reject H_0 **8.69** yes; $z = 3.23$, p-value $= .005$, reject $H_0{:}\,\mu = 15$ **8.71** $t = 3.725$, p-value $= .0058$, reject H_0 **8.73** yes, $t = -2.53$
8.75 a. $t = .21$, p-value $= .84$, fail to reject $H_0{:}\,\mu = 0$ **b.** average of positive and negative scores will tend to cancel   **8.77** qualitative   **8.79 a.** yes **b.** no **c.** no **d.** no **e.** no   **8.81 a.** $z = -2.00$ **c.** reject H_0 **d.** .0228   **8.83 a.** $z = 1.13$, fail to reject H_0 **b.** .1292
8.85 a. p **b.** $H_0{:}\,p = .02$, $H_a{:}\,p > .02$ **c.** $z = 14.23$; $z > 1.645$ **d.** reject H_0 **e.** large n; yes   **8.87** yes, $z = 1.74$ **8.89** yes; $z = 3.14$, reject $H_0{:}\,p = .70$ **8.91** no, $z = 1.52$ **8.93 a.** no, $z = 1.49$ (but inadequate sample size) **b.** p-value $\approx .07$ **8.95** 10
8.97 power $= 1-\beta$ **8.99 b.** 1,032.9 **d.** .7422 **e.** .2578   **8.101 a.** approx. normal, $\mu_{\bar{x}} = 50$, $\sigma_{\bar{x}} = 2.5$ **b.** approx. normal, $\mu_{\bar{x}} = 45$, $\sigma_{\bar{x}} = 2.5$ **c.** .2358 **d.** .7642   **8.103 a.** approx. normal, $\mu_{\hat{p}} = .7$, $\sigma_{\hat{p}} = .0458$ **b.** approx. normal, $\mu_{\hat{p}} = .65$, $\sigma_{\hat{p}} = .0477$ **c.** .7950
d. .9342   **8.105** power increases   **8.107** .1814   **8.109** .7764   **8.111** random sample from a normal population   **8.113** false
8.115 a. $\chi^2 < 6.26214$ or $\chi^2 > 27.4884$ **b.** $\chi^2 > 40.2894$ **c.** $\chi^2 > 21.0642$ **d.** $\chi^2 < 3.57056$ **e.** $\chi^2 < 1.63539$ or $\chi^2 > 12.5916$
f. $\chi^2 < 13.8484$ **8.117 a.** $\chi^2 = 182.16$, reject H_0 **8.119 a.** $\chi^2 < 20.7065$ or $\chi^2 > 66.7659$ **b.** $\chi^2 = 63.72$ **c.** do not reject H_0
8.121 a. σ^2, the variance of the population of rock bounces **b.** $H_0{:}\,\sigma^2 = 10$, $H_a{:}\,\sigma^2 \neq 10$ **c.** $\chi^2 = 20.12$ **d.** $\chi^2 > 21.0261$ or $\chi^2 < 5.22603$
e. do not reject H_0 **f.** population of rock bounces is normally distributed   **8.123** $\chi^2 = 187.90$, do not reject $H_0{:}\,\sigma^2 = 225$
8.125 $\chi^2 = 12.61$, fail to reject H_0 **8.127** no; $\chi^2 = 13.77$, fail to reject H_0 **8.129** alternative   **8.131** H_0, H_a, α **8.133** null
8.135 a. $z = -1.78$, reject H_0 **b.** $z = -1.78$, fail to reject H_0 **c.** $.29 \pm .063$ **d.** $.29 \pm .083$ **e.** 549   **8.137 a.** $\chi^2 = 63.48$, reject H_0
b. $\chi^2 = 63.48$, reject H_0 **8.139 a.** $H_0{:}\,p = .45$ **b.** $H_0{:}\,\mu = 2.5$ **8.141 a.** $t = -3.46$, reject H_0 **b.** normal population   **8.143 a.** yes; $t = -2.46$, p-value $= .023$ **c.** yes **d.** numbers of suicide bombings are normal **e.** no; data highly skewed to the right
8.145 a. $H_0{:}\,p = .5$, $H_a{:}\,p \neq .5$ **b.** $z = 5.99$ **c.** yes **d.** reject H_0 **8.147** yes, $z = 12.36$ **8.149 a.** $H_0{:}\mu = 16$, $H_a{:}\mu < 16$
b. $z = -4.31$, reject H_0 **8.151 a.** $z = .70$, fail to reject H_0 **b.** yes   **8.153 a.** no, $z = 1.41$ **b.** small α **8.155 a.** no
b. $\beta = .5910$, power $= .4090$ **c.** power increases   **8.157** $z = 15.46$, reject $H_0{:}\,p = .167$ **8.159 b.** .37, .24   **8.161 a.** yes, $z = -2.33$
b. .8925   **8.163 a.** no at $\alpha = .01$, $z = 1.85$ **b.** yes at $\alpha = .01$, $z = 3.10$

Chapter 9

9.1 normally distributed with mean $\mu_1 - \mu_2$ and standard deviation $\sqrt{\left(\dfrac{\sigma_1^2}{n_1} + \dfrac{\sigma_2^2}{n_2}\right)}$

9.3 a. no **b.** no **c.** no **d.** yes **e.** no   **9.5** b   **9.7 a.** 14; .4 **b.** 10; .3 **c.** 4; .5 **d.** yes   **9.9 a.** .5989 **b.** $t = -2.39$, reject H_0,
c. $-1.24 \pm .98$ **9.11 a.** fail to reject H_0 at $\alpha = .10$ **b.** p-value $= .0575$, fail to reject H_0 at $\alpha = .10$ **9.13 a.** $(\mu_1 - \mu_2)$
b. $H_0{:}\,(\mu_1 - \mu_2) = 0$, $H_a{:}\,(\mu_1 - \mu_2) > 0$ **c.** yes; fail to reject H_0 **9.15 a.** $H_0{:}\,(\mu_1 - \mu_2) = 0$, $H_a{:}\,(\mu_1 - \mu_2) \neq 0$ **b.** $t = .62$ **c.** $|t| > 1.684$
d. fail to reject H_0; supports theory   **9.17 a.** $\mu_1 - \mu_2$ **b.** 12.5 ± 10.2 **9.19 a.** $(-.60, 7.95)$ **b.** independent random samples from
normal populations with equal variances   **9.21** $t = .57$, p-value $= .573$, fail to reject $H_0{:}\,(\mu_1 - \mu_2) = 0$ **9.23** $t = 1.08$, fail to reject
H_0 **9.25** no, $t = .18$ **9.27 a.** $\mu_1 - \mu_2$ **b.** no **c.** .2262 **d.** fail to reject H_0 **9.29 a.** no standard deviations reported **b.** $s_1 = s_2 = 5$
c. $s_1 = s_2 = 6$ **9.31** before   **9.33 a.** $t > 1.833$ **b.** $t > 1.328$ **c.** $t > 2.776$ **d.** $t > 2.896$ **9.35 a.** $\bar{x}_d = 2$, $s_d^2 = 2$ **b.** $\mu_d = \mu_1 - \mu_2$
c. 2 ± 1.484 **d.** $t = 3.46$, reject H_0 **9.37 a.** $z = 1.79$, fail to reject H_0 **b.** .0734   **9.39 a.** two measurements for each patient—before
and after **c.** $\bar{x}_d = -7.63$, $s_d = 5.27$ **d.** -7.63 ± 1.53 **e.** yes; evidence that $\mu_{\text{before}} < \mu_{\text{after}}$ **9.41 a.** μ_d **b.** paired difference **c.** $H_0{:}\,\mu_d = 0$,
$H_a{:}\,\mu_d > 0$ **d.** $t = 2.19$ **e.** reject H_0 **9.43** $t = 2.92$, fail to reject H_0 no evidence of a difference   **9.45 a.** two scores for each twin pair
b. 95% CI for μ_d: 1.95 ± 1.91; control group has larger mean   **9.47** $t = 3.00$, reject H_0; after camera mean is smaller than before
camera mean   **9.49** yes; $t = -2.31$, reject H_0 **9.51** large, independent samples   **9.53 a.** binomial **b.** normal   **9.55 a.** $.07 \pm .067$
b. $.06 \pm .086$ **c.** $-.15 \pm .131$ **9.57** $z = 1.16$, fail to reject H_0 **9.59 a.** .55 **b.** .70 **c.** $-.15 \pm .03$ **d.** 90% confidence **e.** Dutch boys
9.61 a. .143 **b.** .049 **c.** $z = 2.27$, fail to reject H_0 **d.** reject H_0 **9.63 a.** $z = 94.35$, reject H_0 **b.** $-.139 \pm .016$ **9.65** yes, $z = 11.05$
9.67 yes; $z = 2.13$, reject H_0 at $\alpha = .05$ **9.69** $.48 \pm .144$; proportion greater for those who slept   **9.71** Theory 1: $z = 2.79$,
reject H_0, support theory; Theory 2: $z = -.30$, fail to reject H_0, support theory   **9.73** if no prior information, use $p_1 = p_2 = .5$
9.75 $n_1 = n_2 = 24$ **9.77 a.** $n_1 = n_2 = 29{,}954$ **b.** $n_1 = n_2 = 2{,}165$ **c.** $n_1 = n_2 = 1{,}113$ **9.79** $n_1 = n_2 = 49$ **9.81 a.** $n_1 = n_2 = 5051$
b. may be impractical to obtain such a large sample **c.** difference almost meaningless   **9.83** $n_1 = n_2 = 25$ **9.85** $n_1 = n_2 = 136$
9.87 normal, independent populations   **9.89** false   **9.91 a.** .025 **b.** .90 **c.** .99 **d.** .05   **9.93 a.** $F > 2.36$ **b.** $F > 3.04$ **c.** $F > 3.84$
d. $F > 5.12$ **9.95 a.** $F = 3.43$, reject H_0 **b.** p-value $< .02$, reject H_0 **9.97 a.** $H_0{:}\,\sigma_1^2 = \sigma_2^2$, $H_a{:}\,\sigma_1^2 \neq \sigma_2^2$ **b.** $F = 1.16$ **c.** $F > 2.16$ **d.** fail
to reject H_0 **e.** valid   **9.99** $F = 1.05$, fail to reject H_0; assumption is valid   **9.101** yes; $F = 7.74$, reject H_0 **9.103** $F = 1.30$, fail to
reject H_0; no evidence of a difference in variation   **9.105 a.** $F = 8.29$, reject H_0 **b.** no   **9.107 a.** $\mu_1 - \mu_2$ **b.** $\mu_1 - \mu_2$ **c.** $p_1 - p_2$ **d.** σ_1^2/σ_2^2
e. $p_1 - p_2$ **9.109 a.** $t = .78$, fail to reject H_0 **b.** 2.5 ± 8.99 **c.** $n_1 = n_2 = 225$ **9.111 a.** $3.9 \pm .31$ **b.** $z = 20.60$, reject H_0
c. $n_1 = n_2 = 346$ **9.113** p-value $= .871$, fail to reject $H_0{:}\,\mu_{\text{no}} - \mu_{\text{yes}} = 0$ **9.115 a.** time needed **b.** climbers **c.** paired experiment
9.117 a. $z = -2.64$, reject H_0 **b.** $z = .27$, fail to reject H_0 **9.119 a.** .153 **b.** .215 **c.** $-.062 \pm .070$ **9.121** yes; $(7.43, 13.57)$
9.123 a. $H_0{:}\,\mu_d = 0$, $H_a{:}\,\mu_d \neq 0$ **b.** $z = 2.08$, p-value $= .0376$ **c.** reject H_0 **9.125** $t = -.46$, fail to reject H_0 **9.127** $t = 1.77$, fail to
reject H_0 at $\alpha = .01$ **9.129 a.** $z = -.22$, fail to reject H_0 **b.** $-.078 \pm .076$ **9.131** $-.33 \pm .22$ **9.133 a.** yes, $F = 10$ **b.** both populations normal   **9.135** $z = -3.55$, reject $H_0{:}\,p_1 - p_2 = 0$ **9.137 a.** yes; $z = -2.79$, reject H_0 **b.** yes   **9.139** use of creative ideas
$(z = 8.85)$; good use of job skills $(z = 4.76)$

Chapter 10

10.1 A, B, C, D   **10.3** designed study—values of independent variables controlled   **10.5 a.** observational **b.** designed **c.** designed **d.** observational **e.** observational **f.** observational   **10.7 a.** patient **b.** HAM-D score **c.** drug combination group **d.** 1, 2, 3, and 4   **10.9 a.** healthy adult **b.** postural index **c.** gender and strength knowledge **d.** gender: male, female; strength knowledge: yes, no **e.** male/yes, male/no, female/yes, female/no   **10.11 a.** cockatiel **b.** yes **c.** experimental group **d.** 1, 2, 3 **e.** 3 **f.** total consumption   **10.13 a.** Temperature (45, 48, 51, and 54°C); Type of yeast (baker's, brewer's) **b.** autolysis yield **c.** 8   **10.15** independent random samples from treatment populations, or, randomly assign treatments to experimental units   **10.17** normal treatment populations, with equal variances   **10.19 a.** 6.59 **b.** 16.69 **c.** 1.61 **d.** 3.87   **10.21 a.** plot b **b.** 9; 14 **c.** 75; 75 **d.** 20; 144 **e.** 95 (78.95%); 219 (34.25%) **f.** MST $=$ 75, MSE $=$ 2, F $=$ 37.5; MST $=$ 75, MSE $=$ 14.4, F $=$ 5.21 **g.** reject H_0; reject H_0 **h.** both populations normal with equal variances   **10.23** plot a: df(T) $=$ 1, df(E) $=$ 10, df(Total) $=$ 11, SST $=$ 75, SSE $=$ 20, SS(Total) $=$ 95, MST $=$ 75, MSE $=$ 2, F $=$ 37.5; plot b: df(T) $=$ 1, df(E) $=$ 10, df(Total) $=$ 11, SST $=$ 75, SSE $=$ 144, SS(Total) $=$ 219, MST $=$ 75, MSE $=$ 14.4, F $=$ 5.21   **10.25 a.** F $=$ 1.56; do not reject H_0 **b.** F $=$ 6.25; reject H_0; **c.** F $=$ 25; reject H_0; **d.** increases   **10.27 b.** not valid   **10.29 a.** exp. units: coaches; dep. variable: 7-point rating; factor: division; treatments: I, II, III **b.** H_0: μ_{I} $=$ μ_{II} $=$ μ_{III} **c.** reject H_0;   **10.31 a.** completely randomized **b.** treatments: 3, 6, 9, 12 robots; dep. variable: energy expended **c.** H_0: μ_3 $=$ μ_6 $=$ μ_9 $=$ μ_{12}, H_a: At least 2 μ's differ **d.** reject H_0 **10.33 a.** TV viewers **b.** recall score **c.** program rating; V, S, neutral **d.** variances not taken into account **e.** F $=$ 20.45, p-value $=$ .000 **f.** reject H_0; mean recall scores differ among program groups   **10.35 a.** completely randomized; honey dosage, DM dosage, no dosage **b.** F $=$ 17.51, p-value $=$ .000, reject H_0: μ_{Honey} $=$ μ_{DM} $=$ μ_{Control} **10.37** yes, F $=$ 7.25   **10.39** probability of at least one Type I error   **10.41 a.** no significant difference **b.** μ_2 **c.** μ_1 **10.43 a.** 3 **b.** 10 **c.** 6 **d.** 45   **10.45** μ_1 $>$ μ_2, μ_1 $>$ μ_3, μ_4 $>$ μ_2, μ_4 $>$ μ_3 **10.47 a.** reject H_0: μ_{Angry} $=$ μ_{Guilt} $=$ μ_{Neutral} **b.** P(at least two means differ | none are different) $=$ .05 **c.** μ_{Guilt} $>$ μ_{Angry}, μ_{Guilt} $>$ μ_{Neutral} **10.49 a.** 6 **b.** sourdough; control and yeast   **10.51 a.** 6 **b.** μ_{12} is the smallest mean; μ_3, μ_6, and μ_9 are not significantly different   **10.53 a.** reject H_0 **b.** Control and Slide not significantly different   **10.55** $\mu_{\mathrm{UMRB-2}}$ $>$ μ_{SD}, $\mu_{\mathrm{UMRB-2}}$ $>$ μ_{SWRA}, $\mu_{\mathrm{UMRB-3}}$ $>$ μ_{SD}, $\mu_{\mathrm{UMRB-3}}$ $>$ μ_{SWRA}, $\mu_{\mathrm{UMRB-1}}$ $>$ μ_{SWRA} **10.57** yes; F $=$ 10.29, p-value $=$ .000, reject H_0: μ_{A} $=$ μ_{AR} $=$ μ_{AC} $=$ μ_{P}; μ_{A} $<$ (μ_{P}, μ_{AR}, μ_{AC})   **10.59** paired difference design has only 2 treatments   **10.61** all block-treatment combinations have normal populations with equal variances   **10.63 a.** df(T) $=$ 2, df(B) $=$ 2, df(E) $=$ 4, df(Total) $=$ 8, SSB $=$ .8889, SSE $=$ 7.7778, MST $=$ 10.7778, MSB $=$ .4444, MSE $=$ 1.9444, F(T) $=$ 5.54, F(B) $=$ .23 **b.** H_0: μ_1 $=$ μ_2 $=$ μ_3 **c.** F $=$ 5.54 **d.** Type I error $=$ conclude means differ when the means are equal; Type II error $=$ conclude means are equal when the means differ **e.** do not reject H_0 **10.65 a.** df(T) $=$ 2, df(B) $=$ 3, df(E) $=$ 6, df(Total) $=$ 11, SST $=$ 12.03, SSB $=$ 71.75, SSE $=$ .71, SS(Total) $=$ 84.49, MST $=$ 6.02, MSB $=$ 23.92, MSE $=$ .12, F(T) $=$ 50.96, F(B) $=$ 202.59 **b.** yes, p-value $=$ .000 **c.** yes, p-value $=$ .000 **d.** μ_{C} $<$ μ_{A} $<$ μ_{B} **10.67 a.** yes; F $=$ 2.57, p-value $=$ .0044 **b.** yes; F $=$ 5.94, p-value $=$ .0001 **c.** 105 **d.** only weeks 6 and 14 are more topsy-turvy than other weeks   **10.69 a.** randomized block design **c.** reject H_0 at α $>$ .009 **d.** μ_{control} $<$ (μ_{burning}, μ_{clipping})   **10.71 a.** 4 pre-slaughter phases **b.** df(T) $=$ 3, df(B) $=$ 7, df(E) $=$ 21, df(Total) $=$ 31, SST $=$ 521, SSB $=$ 1923, SSE $=$ 1005, SS(Total) $=$ 3449, MST $=$ 173.7, MSB $=$ 274.7, MSE $=$ 47.85, F(T) $=$ 3.63, F(B) $=$ 5.74 **c.** yes; F $=$ 3.63, p-value $=$ .030 **d.** $\mu_{\mathrm{Phase-1}}$ $>$ $\mu_{\mathrm{Phase-2}}$ **10.73** do not reject H_0, F $=$ .02   **10.75 b.** H_0: $\mu_{\mathrm{Full-Dark}}$ $=$ $\mu_{\mathrm{TR-Light}}$ $=$ $\mu_{\mathrm{TR-Dark}}$ **c.** F $=$ 5.33, reject H_0 **d.** $\mu_{\mathrm{Full-Dark}}$ $<$ $\mu_{\mathrm{TR-Light}}$ **10.77** all factor-level combinations   **10.79** all treatments have normal populations with equal variances   **10.81 a.** 2 **b.** no **c.** yes; 3 and 5 **d.** 15 **e.** df(Error) $=$ 0; replication   **10.83 a.** df(A) $=$ 2, df(B) $=$ 3, df(AB) $=$ 6, df(E) $=$ 12, df(Total) $=$ 23, SSE $=$ 2.4, MS(A) $=$ .40, MS(B) $=$ 1.77, MS(AB) $=$ 1.60, MSE $=$ .20, F(A) $=$ 2.00, F(B) $=$ 8.83, F(AB) $=$ 8.00 **b.** SSA, SSB, SSAB; yes F $=$ 7.14 **c.** yes **d.** effects of one factor on the dependent variable are not the same at different levels of the second factor **e.** F $=$ 8.00, reject H_0 **f.** no   **10.85 a.** F(AB) $=$ .75; F(A) $=$ 3; F(B) $=$ 1.5 **b.** F(AB) $=$ 7.5; F(A) $=$ 3; F(B) $=$ 3 **c.** F(AB) $=$ 3; F(A) $=$ 12; F(B) $=$ 3 **d.** F(AB) $=$ 4.5; F(A) $=$ 36; F(B) $=$ 36   **10.87 a.** complete factorial design **b.** Age (young, old); Diet (fine limestone, coarse limestone) **c.** hen **d.** shell thickness **e.** effect of diet on thickness is the same for each age **f.** mean thickness is not different for young and old hens **g.** shell thickness is affected by diet   **10.89 a.** event (3 wash-ups), strata (coarse, medium, fine, hydroid) **b.** 12 **c.** 2 **d.** 24 **e.** Mussel density **f.** interaction; do not reject H_0 **g.** F(Event) $=$ .35, do not reject H_0; F(Strata) $=$ 217.33, reject H_0 **h.** μ_{Hydroid} $>$ μ_{Fine} $>$ (μ_{Medium}, μ_{Coarse})   **10.91 a.** 2 $\times$ 2 factorial; Color (blue, red), question (difficult, simple) **b.** evidence of interaction (α $=$ .05)   **10.93 b.** evidence of interaction (α $=$ .01) **c.** no   **10.95 a.** 2; 2; 4; 99 **b.** 9 **c.** evidence of interaction **d.** no **e.** 18 months: μ_{control} $<$ μ_{photos}; 24 months: μ_{control} $<$ μ_{drawings}, μ_{control} $<$ μ_{photos}; 30 months: μ_{control} $<$ μ_{drawings}, μ_{control} $<$ μ_{photos} **10.97** interaction nonsignificant (F $=$ 1.77, p-value $=$ .142); Group main effect significant (F $=$ 7.59, p-value $=$ .001); Set main effect significant (F $=$ 31.11, p-value $=$ 000); mean for group size 3 is largest; mean for first photo set is largest   **10.99 a.** Low/Ambig: 450; Low/Common: 195; High/Ambig: 152.5; High/Common: 157.5 **b.** 9,120.25 **c.** SS(Load) $=$ 1,122.25, SS(Name) $=$ 625, SS(Load $\times$ Name) $=$ 676 **d.** Low/Ambig: 225, 5400; Low/Common: 90.25, 2166; High/Ambig: 90.25, 2166; High/Common: 100, 2400 **e.** 12,132 **f.** 14,555.25

g.

Source	df	SS	MS	F
LOAD	1	1,122.25	1,122.25	8.88
NAME	1	625	625	4.95
LOAD $\times$ NAME	1	676	676	5.35
Error	96	12,132	126.375	
Total	99	14,555.25		

h. yes **i.** significant interaction

10.105 a.

Source	df	SS	MS	F
Treatment	3	11.334	3.778	157.42
Block	4	10.688	2.672	111.33
Error	12	.288	.024	
Total	19	22.31		

b. yes, reject H_0 **c.** yes; 6 **d.** yes, reject H_0 **10.107 a.** accountant **b.** income **c.** Mach rating and Gender **d.** Mach rating (high, moderate, low); Gender (male, female) **e.** high/male, moderate/male, low/male, high/female, moderate/female, low/female   **10.109 b.** yes **c.** no   **10.111 a.** H_0: $\mu_{young} = \mu_{middle} = \mu_{old}$ **b.** reject H_0 **c.** H_0: $\mu_{young} = \mu_{middle} = \mu_{old}$; fail to reject H_0 **e.** oldest **f.** .05 **g.** no differences in means for girls   **10.113 a.** (Luckiness (L, UL, UC); Competition (C, NC)   **b.** no evidence of interaction or main effects

10.115 b.

Source	df	SS	MS	F
Prompt	4	1185.00	296.25	39.87
Week	5	386.40	77.28	10.40
Error	20	148.60	7.43	
Total	29	1720.00		

c. yes, $F = 39.87$ **d.** $\mu_{Control} < (\mu_{Int-Low}, \mu_{Int-Hi}) < (\mu_{Freq-Low}, \mu_{Freq-Hi})$ **10.117** Thickness: $F = 11.74$, p-value $= .000$, reject H_0: $\mu_{Barn} = \mu_{Cage} = \mu_{Free} = \mu_{Organic}$; Overrun: $F = 31.36$, p-value $= 000$, reject H_0: $\mu_{Barn} = \mu_{Cage} = \mu_{Free} = \mu_{Organic}$; Strength: $F = 1.70$, p-value $= .193$, do not reject H_0: $\mu_{Barn} = \mu_{Cage} = \mu_{Free} = \mu_{Organic}$; thickness and overrun   **10.119 a.** $F = 3.96$; reject H_0 **b.** $\mu_{Sad} < \mu_{Happy}$, $\mu_{Sad} < \mu_{Angry}$

10.121 a.

Source	df	SS	MS	F	p
Diet (D)	1	0.0124	0.0124	0.22	.645
Size (S)	1	8.0679	8.0679	141.18	.000
D × S	1	0.0364	0.0364	0.64	.432
Error	24	1,3715	0.0571		
Total	27	9.4883			

b. D × S interaction: fail to reject H_0; main effect Diet: fail to reject H_0; main effect Size: reject H_0 **10.123 a.** df(Period) = 1df, (Gender) = 1, df, (P × G) = 1, df(Error) = 120, df(Total) = 123   **10.125 a.** 2 × 2 factorial **b.** factors: tent type and location; treatments: (treated, inside), (treated, outside), (untreated, inside), (untreated, outside) **c.** number of mosquito bites **d.** effect of tent type on mean number of bites depends on location   **10.127** yes, $F = 34.12$; System 1 or System 4

Chapter 11

11.7 $\beta_1 = 1/3$, $\beta_0 = 14/3$, $y = 14/3 + (1/3)x$ **11.11** difference between the observed and predicted   **11.13** true   **11.15 b.** $\hat{y} = 7.10 - .78x$ **11.17 c.** $\hat{\beta}_1 = .918$, $\hat{\beta}_0 = .020$ **e.** −1 to 7   **11.19 a.** $y = \beta_0 + \beta_1 x + \varepsilon$ **b.** $\hat{y} = 19.393 - 8.036x$ **c.** y-intercept: when concentration $= 0$, predicted wicking length $= 19.393$ mm; slope: for every 1-unit increase in concentration, wicking length decreases 8.036 mm   **11.21 c.** hoop pine   **11.23 a.** positive   **11.25 a.** $y = \beta_0 + \beta_1 x + \varepsilon$ **b.** $\hat{y} = 250.14 - .629x$ **e.** slope **11.27 a.** $y = \beta_0 + \beta_1 x + \varepsilon$ **b.** positive **c.** $\hat{\beta}_1 = 210.8$: for every additional resonance, frequency increases 210.8; $\hat{\beta}_0 = 1469.4$: no practical interpretation   **11.29 a.** $\hat{y} = 86.0 - .260x$ **b.** yes **c.** positive trend for female students; positive trend for male students **d.** $\hat{y} = 39.3 + .493x$; for every 1-inch increase in height for females, ideal partner's height increases .493 inch **e.** $\hat{y} = 23.3 + .596x$; for every 1-inch increase in height for males, ideal partner's height increases .596 inch **f.** yes   **11.31** yes, $\hat{y} = 5.22 - .114x$; decrease by .114 pound   **11.35 a.** 57.5; 3.194 **b.** 257.5; 6.776 **c.** 9.288; 1.161   **11.37** 11.14: SSE $= 1.22$, $s^2 = .244$, $s = .494$; 11.17: SSE $= 5.134$, $s^2 = 1.03$, $s = 1.01$ **11.39 a.** SSE $= 22.268$, $s^2 = 5.567$, $s = 2.3594$ **b.** $\approx 95\%$ of wicking length values fall within 4.72 mm of their respective predicted values   **11.41 a.** SSE $= 2760$, $s^2 = 307$, and $s = 17.51$ **11.43 a.** 5.36 **b.** 3.42 **c.** reading score   **11.45 a.** $\hat{y} = 23.3 + .596x$; $s = 2.06$ **b.** $\hat{y} = 39.3 + .493x$; $s = 2.32$ **c.** males   **11.47** 0   **11.49** divide the value in half **11.51 a.** 95%: 31 $\pm$ 1.13; 90%: 31 $\pm$ .92 **b.** 95%: 64 $\pm$ 4.28; 90%: 64 $\pm$ 3.53 **c.** 95%: $-$ .84 $\pm$ .67; 90%: $-$ .84 $\pm$ .55 **11.53 b.** $\hat{y} = 2.554 + .246x$ **d.** $t = .627$ **e.** fail to reject H_0 **f.** .246 $\pm$ 1.81   **11.55 a.** negative linear trend **b.** $\hat{y} = 9,658.24 - 171.573x$; for each 1-unit increase in search frequency, the total catch is estimated to decrease by 171.573 kg **c.** H_0: $\beta_1 = 0$, H_a: $\beta_1 < 0$ **d.** .0402/2 $= .0201$ **e.** reject H_0 **11.57 a.** $t = -6.42$, reject H_0 **b.** $-.305 \pm .135$ **11.59** $-.0023 \pm .0019$; 95% confident that change in sweetness index for each 1-unit change in pectin is between $-.0042$ and $-.0004$ **11.61 a.** $y = \beta_0 + \beta_1 x + \varepsilon$ **b.** $\hat{y} = -8.524 + 1.665x$

d. yes, $t = 7.25$ **e.** $1.67 \pm .46$ **11.63 a.** $t = 6.29$, reject H_0 **b.** $.88 \pm .236$ **c.** no evidence that slope differs from 1
11.65 a. $\hat{\beta}_0 = .515, \hat{\beta}_1 = .000021$ **b.** yes **c.** very influential **d.** $\hat{\beta}_0 = .515, \hat{\beta}_1 = .000020$, p-value $= .332/2 = .166$, fail to reject H_0
11.67 true   **11.69 a.** perfect positive linear **b.** perfect negative linear **c.** no linear **d.** strong positive linear **e.** weak negative linear
f. strong negative linear   **11.71 a.** $r = .985, r^2 = .971$ **b.** $r = -.993, r^2 = .987$ **c.** $r = 0, r^2 = 0$ **d.** $r = 0, r^2 = 0$ **11.73** .877
11.75 a. 18% of sample variation in points scored can be explained by the linear model **b.** $-.424$ **11.77 a.** moderate positive linear
relationship; not significantly different from 0 at $\alpha = .05$ **c.** weak negative linear relationship; not significantly different from 0 at
$\alpha = .10$ **11.79 b.** .185   **11.81 a.** moderately strong negative linear relationship between the number of online courses taken and
weekly quiz grade **b.** yes, $t = -4.95$ **11.83 b.** piano: $r^2 = .1998$; bench: $r^2 = .0032$; motorbike: $r^2 = .3832$, armchair: $r^2 = .0864$;
teapot: $r^2 = .9006$ **c.** Reject H_0 for all objects except bench and armchair   **11.85 a.** $H_0{:}\beta_1 = 0, H_a{:}\beta_1 < 0$ **b.** reject H_0 **c.** no; be
careful not to infer a causal relationship   **11.87** $r = .570, r^2 = .325$ **11.89** $E(y)$ represents mean of y for all experimental units with
same x-value   **11.91** true   **11.93 a.** $\hat{y} = 1.375 + .875x$ **c.** 1.5 **d.** .1875 **e.** $3.56 \pm .33$ **f.** 4.88 ± 1.06 **11.95 c.** 4.65 ± 1.12
d. $2.28 \pm .63; -.414 \pm 1.717$ **11.97 a.** Find a prediction interval for y when $x = 10$ **b.** Find a confidence interval for $E(y)$ when
$x = 10$ **11.99** (92.298, 125.104)   **11.101** run 1: 90% confident that for all runs with a pectin value of 220, mean sweetness index will
fall between 5.65 and 5.84   **11.103 a.** (67.16, 76.53); 95% confident that ideal partner's height is between 67.16 and 76.53 in. when a
female's height is 66 in. **b.** (58.37, 66.85); 95% confident that ideal partner's height is between 58.37 and 66.85 in. when a male's
height is 66 in. **c.** males; 66 in. is outside range of male heights in sample   **11.105 a.** (2.955, 4.066) **b.** (1.020, 6.000) **c.** prediction interval; yes   **11.107 a.** Brand A: $3.35 \pm .59$; Brand B: $4.46 \pm .30$ **b.** Brand A: 3.35 ± 2.22; Brand B: 4.46 ± 1.12 **c.** $-.65 \pm 3.61$
11.109 yes; $\hat{y} = 7.77 + .000113x, t = 4.04$, reject $H_0{:}\beta_1 = 0$ **11.111** $E(y) = \beta_0 + \beta_1 x$ **11.113** true   **11.115 b.** $\hat{y} = x; \hat{y} = 3$ **c.** $\hat{y} = x$
d. least squares line has the smallest SSE   **11.117 a.** $y = \beta_0 + \beta_1 x + \varepsilon$; negative **b.** yes **c.** no   **11.119 a.** positive **b.** yes
c. $\hat{y} = -12.62 + .363x$ **e.** slope: for each additional hit per 1,000 at bats, estimate number of games won to increase by .363
f. $t = 1.47$, fail to reject $H_0{:}\beta_1 = 0$ **g.** .1535; ≈ 15 of sample variation in games won is explained by the linear model **h.** no
11.121 a. $y = \beta_0 + \beta_1 x + \varepsilon$ **b.** $\hat{y} = 175.70 - .8195x$ **e.** $t = -3.43$, reject H_0 **11.123 a.** $y = \beta_0 + \beta_1 x + \varepsilon$ **b.** 92% of sample variation
in metal uptake is explained by the linear model   **11.125 a.** $\hat{y} = 560.1 + 63.3x$; for every 1-unit increase in JIF, estimated cost
increases by \$63.30 **b.** \$908.50 **c.** 63.3 ± 232.7 **d.** $\hat{y} = 326.5 + 1.48x$; \$736.66; $1.48 \pm .81$ **e.** $\hat{y} = 338.9 + 197.21x$; \$846.73;
197.21 ± 177.31 **11.127 a.** yes **b.** $\hat{\beta}_0 = -3.05, \hat{\beta}_1 = .108$ **c.** $.t = 4.00$, reject H_0 **d.** $r = .756, r^2 = .572$ **e.** 1.09 **f.** yes
11.129 a. $\hat{\beta}_0 = -13.49, \hat{\beta}_1 = -.0528$ **b.** $-.0528 \pm .0178$; yes **c.** $r^2 = .854$ **d.** (.5987, 1.2653)   **11.131 a.** yes; positive
b. $y = \beta_0 + \beta_1 x + \varepsilon$ **c.** $\hat{\beta}_0 = 20.13, \hat{\beta}_1 = .624$ **11.133 a.** $\hat{y} = 46.4x$ **b.** $\hat{y} = 478.44 + 45.15x$ **d.** no, $t = .91$ **11.135 a.** no; $r^2 = .748$
b. yes, $18(\hat{\beta}_1) = -.98$

Chapter 12

12.1 a. $E(y) = \beta_0 + \beta_1 x_1 + \beta_2 x_2$ **b.** $E(y) = \beta_0 + \beta_1 x_1 + \beta_2 x_2 + \beta_3 x_3 + \beta_4 x_4$ **c.** $E(y) = \beta_0 + \beta_1 x_1 + \beta_2 x_2 + \beta_3 x_3 + \beta_4 x_4 + \beta_5 x_5$
12.5 test the null hypothesis that all the beta parameters (except β_0) are equal to 0   **12.7 a.** $t = 1.45$, do not reject H_0 **b.** $t = 3.21$,
reject H_0 **12.9** df $= n - (k + 1)$ **12.11 a.** yes **b.** yes, $F = 55.2$ **12.13 a.** sum of errors $= 0$; SSE is minimized **b.** For each unit
increase in "betweenness centrality," lead-user rating is estimated to increase by .42, holding all other x's constant **c.** reject H_0
12.15 a. $E(y) = \beta_0 + \beta_1 x_1 + \beta_2 x_2$ **b.** test $H_0{:}\beta_2 = 0$ vs $H_a{:}\beta_2 > 0$ **c.** reject H_0 **12.17 a.** $\hat{y} = 3.70 + .34x_1 + .49x_2 +$
$.72x_3 + 1.14x_4 + 1.51x_5 + .26x_6 - .14x_7 - .10x_8 - .10x_9$ **c.** $t = -1.00$, do not reject H_0 **d.** (1.412, 1.608)   **12.19 a.** $E(y) = \beta_0 +$
$\beta_1 x_1 + \beta_2 x_2 + \beta_3 x_3$ **c.** reject $H_0{:}\beta_1 = \beta_2 = \beta_3 = 0$ **d.** reject H_0 **e.** fail to reject H_0 **f.** fail to reject H_0 **12.21 a.** $E(y) = \beta_0 + \beta_1 x_1 +$
$\beta_2 x_2 + \beta_3 x_3$ **b.** $\hat{y} = -86,868 - 2,218.8x_1 + 1,542.2x_2 - .3496x_3$ **d.** 103.3 **e.** $R^2 = .128, R_a^2 = .120$ **f.** $F = 15.80$, reject H_0
g. possibly not (low R^2 values)   **12.23 a.** $\hat{y} = 14.0107 - 2.1865$ (GENDER) $- .04794$ (SELFESTM) $- .3223$ (BODYSAT)
$+ .4931$ (IMPREAL) **c.** yes, reject $H_0(p\text{-value} \approx 0)$ **d.** $R_a^2 = .485$; 48.5% of sample variation in desire level can be explained by
the model (after accounting for sample size and size of the model) **e.** reject $H_0{:}\beta_3 = 0$ in favor of $H_0{:}\beta_3 < 0$ (p-value $= .013$)
f. (.24, .74)   **12.25 a.** $E(y) = \beta_0 + \beta_1 x_1 + \beta_2 x_2 + \beta_3 x_3 + \beta_4 x_4 + \beta_5 x_5$ **b.** $\hat{y} = 13,614.5 + .089x_1 - 9.20x_2 + 14.39x_3 + .35x_4 - .85x_5$
d. 458.83 **e.** .917 **f.** yes, $F = 147.30$ **12.27 a.** $\hat{y} = 1.81231 + .10875x_1 + .00017x_2$ **c.** (.026, .192) **d.** (.00009, .00025)
e. $\hat{y} = 1.20785 + .06343x_1 + .00056x_2$; (.016, .111); (.00025, .00087)   **12.31 a.** 3.78 **b.** 4.68   **12.33 a.** 95% confident that true mean
desire falls between 13.42 and 14.31 for all females with a self-esteem score $= 24$, a body satisfaction score $= 3$, and a reality
TV impression score $= 4$ **b.** 95% confident that true mean desire falls between 8.79 and 10.89 for all males with a self-esteem
score $= 22$, a body satisfaction score $= 9$, and a reality TV impression score $= 4$ **12.35 a.** 95% confident that true heat rate falls
between 11,599.6 and 13,665.5 kj/kw/hr for an engine with a speed $= 7,500$ rpm, inlet temperature $= 1,000°$C, exhaust
temperature $= 525°$C, cycle pressure ratio $= 13.5$, and air mass flow rate $= 10$kg/s **b.** 95% confident that true mean heat rate falls
between 12,157.9 and 13,107.1 kj/kw/hr for all engines with a speed $= 7,500$ rpm, inlet temperature $= 1,000°$C, exhaust
temperature $= 525°$C, cycle pressure ratio $= 13.5$, and air mass flow rate $= 10$kg/s **c.** yes   **12.37** (0, 207.25)   **12.39** (24.03,
440.64)   **12.41 a.** $E(y) = \beta_0 + \beta_1 x_1 + \beta_2 x_2 + \beta_3 x_1 x_2$ **b.** $E(y) = \beta_0 + \beta_1 x_1 + \beta_2 x_2 + \beta_3 x_3 + \beta_4 x_1 x_2 + \beta_5 x_1 x_3 + \beta_6 x_2 x_3$
12.43 a. .956 **b.** yes; $F = 202.8$ **d.** yes; $t = 2.5$ **12.45 a.** $E(y) = \beta_0 + \beta_1 x_1 + \beta_2 x_2 + \beta_3 x_1 x_2$ **b.** $\beta_1 + 10\beta_3$ **c.** $\beta_1 + 25\beta_3$
12.47 a. $E(y) = \beta_0 + \beta_1 x_1 + \beta_2 x_2 + \beta_3 x_1 x_2$ **c.** $\beta_3 < 0$ **12.49 a.** 99.4% of sample variation in amplitude (y) can be explained by the
model **b.** the relationship between amplitude (y) and cross position (x_2) depends on probe position (x_1) **c.** slope of line for
$x_1 = 3.5$: $- .165$; slope of line for $x_1 = 6.5$: $- .255$ **12.51 a.** $E(y) = \beta_0 + \beta_1 x_1 + \beta_2 x_2 + \beta_3 x_3 + \beta_4 x_4 + \beta_5 x_5 + \beta_6 x_1 x_2$
b. $H_0{:}\beta_4 = 0$ **c.** reject H_0 **d.** yes   **12.53 a.** effect of client credibility on likelihood depends on the level of linguistic delivery style
b. $H_0{:}\beta_1 = \beta_2 = \beta_3 = 0$ **c.** $F = 55.35$, reject H_0 **d.** $H_0{:}\beta_3 = 0$ **e.** $t = 4.01$, reject H_0 **f.** .114 **g.** .978   **12.55 a.** $E(y) = \beta_0 + \beta_1 x_1 +$
$\beta_2 x_2 + \beta_3 x_3 + \beta_4 x_4 + \beta_5 x_5 + \beta_6 x_2 x_5 + \beta_7 x_3 x_5$ **b.** $\hat{y} = 13,646 + .046x_1 - 12.68x_2 + 23.00x_3 - 3.02x_4 + 1.29x_5 + .016x_2 x_5 - .04x_3 x_5$

c. $t = 4.40$, reject H_0 **d.** $t = -3.77$, reject H_0 **12.57 a.** $E(y) = \beta_0 + \beta_1 x + \beta_2 x^2$ **b.** $E(y) = \beta_0 + \beta_1 x_1 + \beta_2 x_2 + \beta_3 x_1 x_2 + \beta_4(x_1)^2 + \beta_5(x_2)^2$ **c.** $E(y) = \beta_0 + \beta_1 x_1 + \beta_2 x_2 + \beta_3 x_3 + \beta_4 x_1 x_2 + \beta_5 x_1 x_3 + \beta_6 x_2 x_3 + \beta_7(x_1)^2 + \beta_8(x_2)^2 + \beta_9(x_3)^2$ **12.59 a.** $t = 3.133$, reject H_0 **b.** $t = 3.133, t > 1.717$, reject H_0 **12.61 b.** moves graph to right or left **c.** controls whether graph opens upward or downward **12.63 b.** 22.6% of sample variation in points scored (y) can be explained by the model **c.** no **d.** $H_0: \beta_2 = 0$ **12.65 b.** first-order model, $\beta_1 > 0$; first-order model, $\beta_1 < 0$; second-order model **12.67 a.** $E(y) = \beta_0 + \beta_1 x_1 + \beta_2 x_2 + \beta_3 x_1 x_2 + \beta_4(x_1)^2 + \beta_5(x_2)^2$ **b.** β_4 and β_5 **12.69 b.** 6.25 **c.** 10.25 **d.** 200 **12.71 a.** $E(y) = \beta_0 + \beta_1 x_1 + \beta_2 x_2 + \beta_3 x_1 x_2 + \beta_4(x_1)^2 + \beta_5(x_2)^2$ **b.** 40.2% of sample variation in satisfaction level (y) can be explained by the model **c.** $F = 6.99$, reject $H_0: \beta_1 = \beta_2 = \beta_3 = \beta_4 = \beta_5 = 0$ **d.** $t = -.60$, do not reject $H_0: \beta_4 = 0$ **e.** $t = -1.83$, reject $H_0: \beta_5 = 0$ **12.73 a.** $\hat{y} = -288 + 1.395x + .0000351x^2, t = .36$, fail to reject H_0 **b.** outlier **c.** yes; $R^2_{adj} = .996$, model statistically useful (p-value = .000), evidence of curvature (p-value = .000) **12.75** $E(y) = \beta_0 + \beta_1 x; x = \{1$ if level 2, 0 if level 1$\}$ **12.77 a.** 10.2, 6.2, 22.2, 12.2 **b.** $H_0: \beta_1 = \beta_2 = \beta_3 = 0$ **12.79 a.** β_0 **b.** $\mu_{SetNet} - \mu_{GillNet}$ **c.** test $H_0: \beta_1 = \beta_2 = 0$ **12.81 a.** Race: $x_1 = \{1$ if black, 0 if white$\}$; Availability: $x_2 = \{1$ if high, 0 if low$\}$; Position: $x_3 = \{1$ if QB, 0 if not$\}$, $x_4 = \{1$ if RB, 0 if not$\}$, $x_5 = \{1$ if WR, 0 if not$\}$, $x_6 = \{1$ if TE, 0 if not$\}$, $x_7 = \{1$ if DL, 0 if not$\}$, $x_8 = \{1$ if LB, 0 if not$\}$, $x_9 = \{1$ if DB, 0 if not$\}$ **b.** $E(y) = \beta_0 + \beta_1 x_1$ **c.** $E(y) = \beta_0 + \beta_1 x_2$ **d.** $E(y) = \beta_0 + \beta_1 x_3 + \beta_2 x_4 + \beta_3 x_5 + \beta_4 x_6 + \beta_5 x_7 + \beta_6 x_8 + \beta_7 x_9$ **12.83 a.** 4; AA, AB, BA, BB **b.** $E(y) = \beta_0 + \beta_1 x_1 + \beta_2 x_2 + \beta_3 x_3$, where $x_1 = \{1$ if AA, 0 if not$\}$, $x_2 = \{1$ if AB, 0 if not$\}$, $x_3 = \{1$ if BA, 0 if not$\}$ **d.** $H_0: \beta_1 = \beta_2 = \beta_3 = 0$ **12.85 a.** $E(y) = \beta_0 + \beta_1 x_1 + \beta_2 x_2$, where $x_1 = \{1$ if major depression only, 0 if not$\}$, $x_2 = \{1$ if personality disorder only, 0 if not$\}$ **b.** $\beta_1 = \beta_2 = 0$ **c.** test $H_0: \beta_1 = 0$ vs. $H_a: \beta_1 < 0$ **12.87** $E(y) = \beta_0 + \beta_1 x_1 + \beta_2 x_2 + \beta_3 x_1 x_2$ **12.89 a.** $E(y) = \beta_0 + \beta_1 x$, where $x = \{1$ if flightless, 0 otherwise$\}$ **b.** $E(y) = \beta_0 + \beta_1 x_1 + \beta_2 x_2 + \beta_3 x_3$, where $x_1 = \{1$ if vertebrates, 0 if not$\}$, $x_2 = \{1$ if vegetables, 0 if not$\}$, $x_3 = \{1$ if invertebrates, 0 if not$\}$ **c.** $E(y) = \beta_0 + \beta_1 x_1 + \beta_2 x_2 + \beta_3 x_3$, where $x_1 = \{1$ if cavity within ground, 0 if not$\}$, $x_2 = \{1$ if trees, 0 if not$\}$, $x_3 = \{1$ if cavity above ground, 0 if not$\}$ **d.** $\hat{y} = 641 + 30,647x$ **e.** $F = 33.05$, reject H_0 **f.** $\hat{y} = 903 + 2,997x_1 + 26,206x_2 - 660x_3$ **g.** $F = 8.43$, reject H_0 **h.** $\hat{y} = 73.732 - 9.132x_1 - 45.01x_2 - 39.51x_3$ **i.** $F = 8.07$, reject H_0 **12.91 a.** $\beta_0 + \beta_2$ **b.** $\beta_0 + \beta_1 + \beta_2 + \beta_3$ **c.** $\beta_1 + \beta_3$ **d.** $\beta_0; \beta_0 + \beta_1; \beta_1$ **f.** evidence of interaction **g.** change in crime rate after murder in Jasper: $\hat{\beta}_2 + \hat{\beta}_3 = -169 + 255 = 86$; change in crime rate after murder in Center: $\hat{\beta}_2 = -169$ **12.93 a.** $E(y) = \beta_0 + \beta_1 x_1 + \beta_2(x_1)^2$ **b.** $E(y) = \beta_0 + \beta_1 x_1 + \beta_2(x_1)^2 + \beta_3 x_2 + \beta_4 x_3 + \beta_5 x_1 x_2 + \beta_6(x_1)^2 x_2 + \beta_7 x_1 x_3 + \beta_8(x_1)^2 x_3$ **c.** $E(y) = \beta_0 + \beta_1 x_1 + \beta_2(x_1)^2 + \beta_3 x_2 + \beta_4 x_3$ **d.** $\beta_5 = \beta_6 = \beta_7 = \beta_8 = 0$ **e.** $\beta_2 = \beta_5 = \beta_6 = \beta_7 = \beta_8 = 0$ **f.** $\beta_3 = \beta_4 = \beta_5 = \beta_6 = \beta_7 = \beta_8 = 0$ **12.95 a.** $E(y) = \beta_0 + \beta_1 x_1; E(y) = (\beta_0 + \beta_2) + \beta_1 x_1; E(y) = (\beta_0 + \beta_3) + \beta_1 x_1$ **b.** $\hat{y} = 44.8 + 2.2x_1; \hat{y} = 54.2 + 2.2x_1; \hat{y} = 60.4 + 2.2x_1$ **12.97 a.** $\hat{y} = 11.779 - 1.972x_1 + .585x_4 - .553x_1 x_4$ **b.** 9.97 **c.** $F = 45.09(p$-value $< .0001)$, reject H_0 **d.** 43.9% of sample variation in level of desire (y) can be explained by the model **e.** $\approx 95\%$ of sampled desire levels fall within 4.70 points of their respective predicted values **f.** $t = -2.00(p$-value $= .0467)$, reject H_0 **g.** .585 **h.** .032 **12.99 a.** $E(y) = \beta_0 + \beta_1 x_1; \beta_1$ **b.** $E(y) = (\beta_0 + \beta_2) + (\beta_1 + \beta_3)x_1; \beta_1 + \beta_3$ **c.** evidence of interaction at $\alpha = .05$ **12.101 a.** $E(y) = \beta_0 + \beta_1 x_1 + \beta_2 x_2 + \beta_3 x_3$, where $x_1 = $ water depth, $x_2 = \{1$ if set net, 0 if not$\}$, $x_3 = \{1$ if pots, 0 if not$\}$ **c.** $E(y) = \beta_0 + \beta_1 x_1 + \beta_2 x_2 + \beta_3 x_3 + \beta_4 x_1 x_2 + \beta_5 x_1 x_3$ **e.** $\beta_1 + \beta_4$ **f.** $\beta_1 + \beta_5$ **g.** β_1 **h.** test $H_0: \beta_4 = \beta_5 = 0$ **12.103 a.** $x_2 = \{1$ if low, 0 if not$\}$, $x_3 = \{1$ if neutral, 0 if not$\}$, $\{$base level = high$\}$ **b.** $E(y) = \beta_0 + \beta_1 x_1 + \beta_2 x_2 + \beta_3 x_3$ **c.** $E(y) = \beta_0 + \beta_1 x_1 + \beta_2 x_2 + \beta_3 x_3 + \beta_4 x_1 x_2 + \beta_5 x_1 x_3$ **d.** part c **12.105 a.** $E(y) = \beta_0 + \beta_1 x_1 + \beta_2 x_2 + \beta_3 x_3$ **b.** $E(y) = \beta_0 + \beta_1 x_1 + \beta_2 x_2 + \beta_3 x_3 + \beta_4 x_1 x_3 + \beta_5 x_2 x_3$ **12.107** (a and b); (a and d); (a and e); (b and c); (b and d); (b and e); (c and e); (d and e) **12.109** model with a small number of β parameters **12.111 a.** 5; 3 **b.** $H_0: \beta_3 = \beta_4 = 0$ **c.** $F = .38$, do not reject H_0 **12.113 a.** complete: 12.103c; reduced: 12.103b **b.** $H_0: \beta_4 = \beta_5 = 0$ **c.** model 12.103c **d.** model 12.103b **12.115 b.** $H_0: \beta_5 = \beta_6 = \beta_7 = \beta_8 = 0$ **c.** yes **d.** reject H_0; complete model better **e.** $E(y) = \beta_0 + \beta_1 x_1 + \beta_2 x_2 + \beta_3 x_3 + \beta_4 x_4 + \beta_5 x_5 + \beta_6 x_6 + \beta_7 x_7 + \beta_8 x_8 + \beta_9 x_5 x_6 + \beta_{10} x_5 x_7 + \beta_{11} x_5 x_8 + \beta_{12} x_6 x_7 + \beta_{13} x_6 x_8 + \beta_{14} x_7 x_8$ **f.** do not reject H_0: no evidence of interaction **12.117 a.** $E(y) = \beta_0 + \beta_1 x_1 + \beta_2 x_2 + \beta_3 x_3 + \beta_4 x_4 + \beta_5 x_5 + \beta_6 x_6 + \beta_7 x_7 + \beta_8 x_8 + \beta_9 x_9 + \beta_{10} x_{10}$ **b.** $H_0: \beta_3 = \beta_4 = \cdots = \beta_{10} = 0$ **c.** at least one of the additional variables is important **e.** (8.12, 19.88) **f.** yes **g.** $E(y) = \beta_0 + \beta_1 x_1 + \beta_2 x_2 + \beta_3 x_3 + \beta_4 x_4 + \beta_5 x_5 + \beta_6 x_6 + \beta_7 x_7 + \beta_8 x_8 + \beta_9 x_9 + \beta_{10} x_{10} + \beta_{11} x_1 x_2 + \beta_{12} x_2 x_3 + \beta_{13} x_2 x_4 + \beta_{14} x_2 x_5 + \beta_{15} x_2 x_6 + \beta_{16} x_2 x_7 + \beta_{17} x_2 x_8 + \beta_{18} x_2 x_9 + \beta_{19} x_2 x_{10}$ **h.** $H_0: \beta_{11} = \beta_{12} = \cdots = \beta_{19} = 0$; nested model F-test **12.119 a.** $H_0: \beta_2 = \beta_3 = \beta_4 = \beta_5 = 0$ **b.** $E(y) = \beta_0 + \beta_1 x_1$ **c.** mean lengths of entangled whales for 3 gear types differ **d.** $H_0: \beta_4 = \beta_5 = 0$ **e.** $E(y) = \beta_0 + \beta_1 x_1 + \beta_2 x_2 + \beta_3 x_3$ **f.** rate of change of whale length with water depth is same for all 3 gear types **12.123 a.** x_2 **b.** yes **c.** fit all models of the form $E(y) = \beta_0 + \beta_1 x_2 + \beta_2 x_j$ **12.125 a.** $E(y) = \beta_0 + \beta_1 x_1 + \beta_2 x_2$, where $x_1 = $ pressure and $x_2 = $ leg length **b.** 77.1% of sample variation in heel depth (y) can be explained by the model **c.** p-value $< .001$, reject H_0 **d.** 15 **e.** very high **12.127 a.** 11 **b.** 10 **c.** 1 **d.** $E(y) = \beta_0 + \beta_1 x_{11} + \beta_2 x_4 + \beta_3 x_2 + \beta_4 x_7 + \beta_5 x_{10} + \beta_6 x_1 + \beta_7 x_9 + \beta_8 x_3$ **12.129 a.** 11 **b.** 10 **c.** model is statistically useful (p-value $= .001$) **d.** large number of t-tests performed; no higher order terms (e.g., interactions) in model **e.** $E(y) = \beta_0 + \beta_1 x_1 + \beta_2 x_2 + \beta_3 x_1 x_2 + \beta_4(x_1)^2 + \beta_5(x_2)^2$ **f.** test $H_0: \beta_4 = \beta_5 = 0$ **12.131** residual that lies more than 3 standard deviations from 0 **12.133** false **12.135** (1) Global F-test significant, but all t-tests insignificant; (2) β estimates with opposite signs from expected; high pairwise correlations among x's **12.137** yes; x_4 is highly correlated with both x_2 and x_5 **12.139** no **12.141 a.** no **b.** yes **12.143** yes **12.145** multicollinearity **12.147 a.** reasonably satisfied **b.** likely violated **c.** 8 outliers **d.** likely violated **e.** no multicollinearity **12.149** no **12.151** $E(y) = \beta_0 + \beta_1 x_1 + \beta_2 x_2 + \beta_3 x_3; x_1 = \{1$ if level 2, 0 if not$\}$; $x_2 = \{1$ if level 3, 0 if not$\}$; $x_3 = \{1$ if level 4, 0 if not$\}$ **12.153 a.** $E(y) = \beta_0 + \beta_1 x_1 + \beta_2 x_2 + \beta_3 x_3$, where $x_1 = $ quantitative, $x_2 = \{1$ if level 2, 0 if not$\}$, $x_3 = \{1$ if level 3, 0 if not$\}$ **b.** $E(y) = \beta_0 + \beta_1 x_1 + \beta_2 x_1^2 + \beta_3 x_2 + \beta_4 x_3 + \beta_5 x_1 x_2 + \beta_6 x_1 x_3 + \beta_7 x_1^2 x_2 + \beta_8 x_1^2 x_3$ **12.155 a.** yes, $F = 24.41$ **b.** $t = -2.01$, reject H_0 **c.** $t = .31$, do not reject H_0 **d.** $t = 2.38$, reject H_0 **12.157** yes; $x_1 = 60, x_2 = .4, x_3 = 900$ are outside range of sample data **12.159** df(Error) $= 0$ **12.161 a.** 31% of sample variation in $\ln(CO_2$ emissions) can be explained by the model **b.** $F = 3.72$, reject H_0 **c.** no **d.** $H_0: \beta_1 = 0$ **e.** $t = 2.52$, reject H_0 **f.** x_4 is highly correlated with x_5 **12.163 a.** $E(y) = \beta_0 + \beta_1 x_1 + \beta_2 x_2 + \beta_3 x_3 + \beta_4 x_4 + \beta_5 x_5$ **b.** model is statistically useful **c.** $E(y) = \beta_0 + \beta_1 x_1 + \beta_2 x_2 + \beta_3 x_3 + \beta_4 x_4 + \beta_5 x_5 + \beta_6 x_6 + \beta_7 x_7$ **12.165 a.** $E(y) = \beta_0 + \beta_1 x + \beta_2 x^2$ **b.** positive **c.** no; $E(y) = \beta_0 + \beta_1 x$ **12.167 a.** $E(y) = \beta_0 + \beta_1 x$, where $x = \{1$ if enriched pond, 0 if natural pond$\}$ **b.** β_0 mean larval density of natural pond $= \mu_{natural}$;

$\beta_1 = \mu_{\text{enriched}} - \mu_{\text{natural}}$ **c.** $H_0: \beta_1 = 0$, $H_a: \beta_1 > 0$ **d.** reject H_0 **12.169 b.** $F = 5.11$, reject H_0 **12.171 a.** $\hat{y} = 12.180 -$
$.0265x_1 - .4578x_2$ **c.** $t = -.50$, do not reject H_0 **d.** $-.4578 \pm .3469$ **e.** $R^2 = .529$, $R_a^2 = .505$; R_{adj}^2 **f.** yes, $F = 21.88$
g. $E(y) = \beta_0 + \beta_1x_1 + \beta_2x_2 + \beta_3x_1x_2$, where $x_2 = \{1 \text{ if plant, 0 if duck chow}\}$ **h.** $\hat{y} = 8.14 - .016x_1 - 10.4x_2 + .095x_1x_2$ **i.** .079
j. $-.016$ **k.** $t = .67$, fail to reject H_0 **12.173 a.** race and foreign status, year GRE taken and years in graduate program
12.175 a. negative **b.** no, $F = 1.60$ **c.** no, $F = 1.61$ **12.177 b.** $\hat{y} = 42.25 - .0114x + .00000061x^2$ **c.** no, $t = 1.66$
12.179 a. $E(y) = \beta_0 + \beta_1x_1 + \beta_2x_6 + \beta_3x_7$, where $x_6 = \{1 \text{ if good, 0 if not}\}$, $x_7 = \{1 \text{ if fair, 0 if not}\}$ **c.** excellent:
$\hat{y} = 188,875 + 15,617x_1$; good: $\hat{y} = 85,829 + 15,617x_1$; fair; $\hat{y} = 36,388 + 15,617x_1$ **e.** yes, $F = 8.43$ **f.** x_1, x_3, and x_5 are highly
correlated **g.** assumptions are satisfied   **12.181 a.** 7 **b.** 6 **c.** 5   **12.183 a.** number of trees-QN(x_1); transect location-QL:
$x_2 = \{1 \text{ if SPF, 0 if not}\}$ and $x_3 = \{1 \text{ if SAF, 0 if not}\}$; land use-QL: $x_4 = \{1 \text{ if pasture, 0 if arable}\}$ **b.** $E(y) = \beta_0 + \beta_1x_1$
c. $E(y) = \beta_0 + \beta_1x_1 + \beta_2x_2 + \beta_3x_3$ **d.** $E(y) = \beta_0 + \beta_1x_1 + \beta_2x_2 + \beta_3x_3 + \beta_4x_4$; β_1 **e.** $E(y) = \beta_0 + \beta_1x_1 + \beta_2x_2 + \beta_3x_3 + \beta_4x_4 +$
$\beta_5x_2x_4 + \beta_6x_3x_4$; no **f.** $E(y) = \beta_0 + \beta_1x_1 + \beta_2x_2 + \beta_3x_3 + \beta_4x_4 + \beta_5x_2x_4 + \beta_6x_3x_4 + \beta_7x_1x_2 + \beta_8x_1x_3 + \beta_9x_1x_4 + \beta_{10}x_1x_2x_4 +$
$\beta_{11}x_1x_3x_4$; LAF/Arable: β_1; LAF/Pasture: $\beta_1 + \beta_9$; SPF/Arable: $\beta_1 + \beta_7$; SPF/Pasture $\beta_1 + \beta_7 + \beta_9 + \beta_{10}$; SPF/Arable: $\beta_1 + \beta_8$;
SAF/Pasture: $\beta_1 + \beta_8 + \beta_9 + \beta_{11}$ **12.185** model includes $x_1 =$ DOT estimate, either $x_2 =$ low bid ratioor $x_3 =$
$\{1 \text{ if fixed, 0 if competitive}\}$, and $x_5 =$ estimated days to complete ; $\hat{y} = 8.14 - .016x_1 - 10.4x_2 + .095x_1x_2$

Chapter 13

13.1 (1) n identical trials, **(2)** k possible outcomes to each trial, **(3)** probabilities of outcomes sum to 1, **(4)** probabilities remain the same
from trial to trial, **(5)** trials are independent   **13.3 a.** 18.3070 **b.** 29.7067 **c.** 23.5418 **d.** 79.4900   **13.5 a.** $\chi^2 > 5.99147$ **b.** $\chi^2 > 7.77944$
c. $\chi^2 > 11.3449$ **13.7 a.** no, $\chi^2 = 3.293$ **c.** $.288 \pm .062$ **13.9 a.** jaw habits; bruxism, clenching, bruxism and clenching, neither **c.**
$H_0: p_1 = p_2 = p_3 = p_4 = .25$, H_a: At least one p_i differs from .25 **d.** 15 **e.** $\chi^2 = 25.73$ **f.** $\chi^2 > 7.81473$ **g.** reject H_0 **h.** (.37, .63)   **13.11**
a. Pottery type; burnished, monochrome, painted, other **b.** $p_1 = p_2 = p_3 = p_4 = .25$ **c.** $H_0: p_1 = p_2 = p_3 = p_4 = .25$
d. $\chi^2 = 436.59$ **e.** p-value ≈ 0, reject H_0 **13.13 a.** $P(GG) = P(BG) = P(GB) = P(BB) = 1/4$ **b.** 10,722 **c.** 187.04 **d.** reject H_0
e. $E_{GG} = 10,205.2$, $E_{BG} = 10,715.6$, $E_{GB} = 10,715.6$, $E_{BB} = 11,251.7$; $\chi^2 = 64.95$; reject H_0 **13.15 a.** Answer 1: 450; Answer 2: 627;
Answer 3: 219; Answer 4: 23 **b.** $H_0: p_1 = p_2 = p_3 = p_4 = .25$ **c.** 329.75 for each category **d.** 634.36 **e.** reject H_0 **f.** $\chi^2 = 39.29$, reject
H_0 **13.17** $\chi^2 = 4.84$, do not reject H_0 **13.19 a.** yes, $\chi^2 = 360.48$ **b.** (.165, .223)   **13.21** column totals for one qualitative variable are
known in advance   **13.23** expected cell counts are all at least 5   **13.25 a.** H_0: Row & Column are independent, H_a: Row & Column
are dependent **b.** $\chi^2 > 9.21034$
c.

	Column 1	Column 2	Column 3
Row 1	14.37	36.79	44.84
Row 2	10.63	27.21	33.16

d. $\chi^2 = 8.71$, fail to reject H_0 **13.27** $\chi^2 = 12.33$, reject H_0 **13.29 a.** .078 **b.** .157 **c.** possibly **d.** H_0: Presence of Dog & Gender are
independent **e.** $\chi^2 = 2.25$, fail to reject H_0 **f.** $\chi^2 = .064$, fail to reject H_0 **13.31 a.** .653 **b.** .555 **c.** .355 **d.** yes **e.** H_0: Authenticity of news
story & Tone are independent **f.** $\chi^2 = 10.427$ (p-value $= .005$), reject H_0 **13.33 a.** masculinity risk (high and low); event (violent and
avoided-violent) **b.** 1,507 newly incarcerated men **c.** 260.8 **d.** 118.2, 776.2, 351.8 **e.** $\chi^2 = 10.1$ **f.** reject H_0 **13.35** $\chi^2 = 23.46$, reject H_0
13.37 yes; $\chi^2 = 3.29$, fail to reject H_0 **13.39** yes, $\chi^2 = 7.38$, p-value $= .025$; no   **13.41 a.** no
b.

SHSS:C	CAHS		
	Low	Medium	High
Low	32	14	2
Medium	11	14	6
High	6	16	29

c. yes **d.** $\chi^2 = 46.71$, reject H_0 **13.43 a.** $\chi^2 = 4.407$, reject H_0 **b.** no **c.** .0438 **d.** .0057; .0003 **e.** p-value $= .0498$, reject H_0 **13.45** false
13.47 a. yes, $\chi^2 = 54.14$ **b.** no **c.** yes
d.

	Col. 1	Col. 2	Col. 3	Totals
Row 1	.400	.222	.091	.200
Row 2	.200	.222	.636	.400
Row 3	.400	.556	.273	.400

13.49 a. Location (downtown, central city, suburban area) **b.** $H_0: p_1 = .40$, $p_2 = .30$, $p_3 = .30$ **c.** 45.2, 33.9, 33.9 **d.** 6.17 **e.** .046; reject H_0

13.51 a.

| | LLD | | |
Genetic Trait	Yes	No	Total
Yes	21	15	36
No	150	306	456
Total	171	321	492

b. $\chi^2 = 9.52$, reject H_0 **13.53** $\chi^2 = 35.41$, reject H_0 **13.55 a.** $\chi^2 = 1.14$, do not reject H_0 **b.** $.179 \pm .119$ **13.57** Internet: $\chi^2 = .512$, do not reject H_0; party:$\chi^2 = 164.76$, reject H_0; military: $\chi^2 = 8.3$, reject H_0; views:$\chi^2 = 174.39$, reject H_0; race: $\chi^2 = 69.18$, reject H_0; income: $\chi^2 = 16.39$, reject H_0; community: $\chi^2 = 17.62$, reject H_0 **13.59 a.** $\chi^2 = 17.16$, reject H_0 **b.** $.376 \pm .103$ **c.** yes **13.61** $\chi^2 = 6.17$, reject H_0 **13.63 a.** no, $\chi^2 = 7.39$ **b.** $(.456, .518)$ **13.65 a.** $\chi^2 = 164.90$, reject H_0 **b.** Gender: $\chi^2 = 6.80$, reject H_0; Education: $\chi^2 = 6.59$, fail to reject H_0; Work: $\chi^2 = 11.69$, fail to reject H_0; Satisfaction: $\chi^2 = 30.42$, reject H_0 **13.67 a.** $\chi^2 = 9.65$ **b.** 11.0705 **c.** no **d.** $.05 < p\text{-value} < .10$ **13.69** $\chi^2 = 2.28$; insufficient evidence to reject the null hypothesis of independence

Chapter 14 (available on CD)

14.1 population not normal **14.3 a.** .063 **b.** .500 **c.** .004 **d.** .151; .1515 **e.** .212; .2119 **14.5** $S = 16$, $p\text{-value} = .115$, do not reject H_0 **14.7 a.** H_0: $\eta = 15$, H_a: $\eta \neq 15$ **b.** data not normal **c.** $S = 14$ **d.** .0309; reject H_0 **14.9 a.** H_0: $\eta = 2.8$, H_a: $\eta > 2.8$ **b.** $S = 21$, $z = 2.01$ **c.** .0222 **d.** do not reject H_0 **14.11 a.** invalid inference **b.** sign test of H_0: $\eta = 95$ vs. H_a: $\eta \neq 95$ **c.** $S = 5$ **d.** .454 **e.** do not reject H_0 **14.13** no; $S = 10$, $p\text{-value} = .3145$ **14.15** $S = 8$, $p\text{-value} = .8204$, do not reject H_0 **14.17** true **14.19 a.** $T_1 \leq 41$, $T_1 \geq 71$ **b.** $T_1 \geq 50$ **c.** $T_1 \leq 43$ **d.** $|z| > 1.96$ **14.21** yes, $z = -2.47$ **14.23 a.** $T_1 = 62.5$, reject H_0 **b.** yes, $T_1 = 62.5$ **14.25 a.** rank sum test **b.** H_0:$D_{Low} = D_{High}$, H_a: $D_{Low} > D_{High}$ **c.** $T_2 \leq 41$ **d.** do not reject H_0 **14.27 a.** H_0:$D_{Bulimic} = D_{Normal}$, H_a: $D_{Bulimic} > D_{Normal}$ **c.** 174.5 **d.** 150.5 **e.** $z > 1.28$ **f.** $z = 1.72$, reject H_0 **14.29** yes, $z = 2.86$ **14.31 a.** rank sum test **b.** H_0: $D_{CMC} = D_{FTF}$, H_a: $D_{CMC} < D_{FTF}$ **c.** $z < -1.28$ **d.** $z = -.21$, fail to reject H_0 **14.33 a.** data not normal **b.** $z = -6.47$, reject H_0 **c.** $z = -.39$, do not reject H_0 **d.** $z = -1.35$, do not reject H_0 **14.35** data must be ranked **14.37 a.** H_0: $D_A = D_B$, H_a: $D_A > D_B$ **b.** $T_- = 3.5$, reject H_0 **14.39 a.** H_0: $D_A = D_B$, H_a: $D_A > D_B$ **b.** $z = 2.5$ **c.** .0062 **14.41 a.** before and after measurements not independent **b.** scores not normal **c.** reject H_0, ichthyotherapy is effective **14.43 a.** $z = -4.638$ **b.** yes, $p\text{-value} \approx 0$ **14.45 a.** H_0: $D_{good} = D_{average}$, H_a: $D_{good} \neq D_{average}$ **d.** $T = 3$ **e.** $T \leq 4$ **f.** reject H_0 **14.47** $T_+ = 3.5$, reject H_0; program was effective **14.49** no, $T_- = 7$ **14.51** yes, $T_- = 50.5$ **14.53** samples size for each distribution is more than 5 **14.55 a.** completely randomized design **b.** H_0: Three probability distributions are identical **c.** $H > 9.21034$ **d.** $H = 14.53$, reject H_0 **14.57 a.** H_0: $D_{1dog} = D_{2dogs} = D_{3dogs}$ **b.** $H > 5.99147$ **c.** do not reject H_0 **14.59 b.** 84 **c.** 145 **d.** 177 **e.** $H = 18.4$ **f.** reject H_0 **g.** $z = -3.36$, reject H_0 **14.61 a.** $H = 5.67$, do not reject H_0 **b.** $z = -2.36$, reject H_0 **14.63** $H = 19.03$, reject H_0 **14.65 a.** normality assumption violated **b.** $H = 13.66$, reject H_0 **14.67** ranking b **14.69 a.** 6 **b.** H_0: The probability distributions for the four treatments are identical **c.** $F_r = 15.2$, reject H_0 **d.** $p\text{-value} < .005$ **14.71** $F_r = 13$, reject H_0 **14.73 a.** $R_1 = 25.5$, $R_2 = 11.0$, $R_3 = 18.5$, $R_4 = 25.0$ **b.** 10.39, yes **c.** .016 **d.** reject H_0 **14.75 a.** $F_r > 9.21034$ **b.** reject H_0 **c.** do not reject H_0 **d.** reject H_0 **14.77** $F_r = 11.10$, reject H_0 at $\alpha = .05$ **14.79** $F_r = 6.78$, fail to reject H_0 **14.81** random sample, continuous variables **14.83 a.** $|r_s| > .648$ **b.** $r_s > .450$ **c.** $r_s < -.432$ **14.85 b.** $r_s = .745$, do not reject H_0 **c.** $.05 < p\text{-value} < .10$ **14.87 b.** $r_s = .9429$ **c.** do not reject H_0 **14.89 c.** .713 **d.** $|r_s| > .425$ **e.** reject H_0 **14.91 b.** perfect positive rank correlation **c.** actual relationship may be nonlinear **14.93 a.** .714 **b.** reject H_0 **14.95 a.** $-.877$ **b.** $-.907$ **c.** reject H_0 **d.** reject H_0 **14.97 b.** invalid inference **c.** .507 **d.** yes **14.99 a.** rank sum test **b.** sign test **c.** Kruskal-Wallis test **d.** Spearman's test **e.** signed rank test **f.** Friedman's test **14.101 a.** no, $r_s = .40$ **b.** yes, $T_- = 1.5$ **14.103** yes, $F_r = 14.9$ **14.105** $S = 8$, $p\text{-value} = .0391$, reject H_0 **14.107** $T_+ = 27$, do not reject H_0 **14.109 a.** $-.733$ **b.** do not reject H_0 **14.111 a.** no, $S = 14$, $p\text{-value} = .058$ **b.** no, $S = 12$, $p\text{-value} = .180$ **c.** $T_- = 50$, do not reject H_0 **d.** $r_s = .774$, reject H_0 **14.113 a.** zinc measurements non-normal **b.** $T_1 = 18$, do not reject H_0 **c.** $T_2 = 32$, do not reject H_0 **14.115** yes, $F_r = 6.35$ **14.117** yes, $H = 7.154$ **14.119 a.** moderate positive association between the two scores **c.** reject H_0 **14.121 a.** data not normal **b.** Kruskal-Wallis **c.** $H = 11.20$ **d.** reject H_0 **14.123** yes, $S = 17$

Index

Photo Credits

Chapter 1
p. 1 Anson0618/Shutterstock; **pp. 2, 9, 14, 17** TheProductGuy/Alamy; **p. 7** Monkey Business Images/Shutterstock; **p. 9** TheProductGuy/Alamy; **p. 10** DIGIcal/iStockphoto; **p. 13** Justin Horrocks/iStockphoto; **p. 16** (top) Jupiterimages/Thinkstock, (bottom) SFC/Shutterstock

Chapter 2
p. 25 Beaucroft/Shutterstock; **pp. 26, 31, 44, 70, 84** Luis Louro/Shutterstock; **p. 29** Elena Elisseeva/iStockphoto; **p. 51** Diane Labombarbe/iStockphoto; **p. 55** Andy Z./Shutterstock; **p. 69** Hywit Dimyadi/Shutterstock; **p. 82** Goodluz/Shutterstock; **p. 88** Wavebreakmedia ltd/Shutterstock

Chapter 3
p. 108 Alexandr Shebanov/Shutterstock; **pp. 109, 119, 130, 144** Simon Askham/iStockphoto; **p. 111** Vladimir Wrangel/Shutterstock; **p. 114** (top) Lorraine Kourafas/Shutterstock, (bottom) iStockphoto; **p. 124** Luminis/iStockphoto; **p. 125** Yvonne Chamberlain/iStockphoto; **p. 136** Sculpies/Shutterstock; **p. 138** Darren Brode/Shutterstock; **p. 151** Matti/Shutterstock; **p. 152** Geotrac/iStockphoto; **p. 156** iStockphoto; **p. 157** P. Wei/iStockphoto; **p. 159** Molotovcoketail/iStockphoto; **p. 166** Micimakin/Shutterstock

Chapter 4
p. 179 Dibrova/Shutterstock; **pp. 180, 204, 214** Melinda Fawver/Shutterstock; **p. 181** Sambrogio/iStockphoto; **p. 184** Vladimir Wrangel/Shutterstock; **p 185** JohnKwan/Shutterstock; **p. 196** Irina Tischenko/Shutterstock; **p. 197** MistikaS/iStockphoto; **p. 209** Ibsky/Shutterstock

Chapter 5
p. 224 Noam Armonn/Shutterstock; **pp. 225, 240, 247** Tracing Tea/Shutterstock; **p. 228** J. lsohio/iStockphoto; **p. 236** Nathan Gutshall-Kresge/iStockphoto; **p. 237** Morgan Lane Studios/iStockphoto; **p. 239** JC559/iStockphoto; **p. 245** Diane Labombarbe/iStockphoto; **p. 254** iStockphoto; **p. 258** Matt Matthews/iStockphoto

Chapter 6
p. 271 Keith Bell/Shutterstock; **pp. 272, 288** Stephanie Horrocks/iStockphoto; **p. 275** JohnKwan/Shutterstock; **p. 287** Hywit Dimyadi/Shutterstock

Chapter 7
p. 298 Michael Shake/Shutterstock; **pp. 299, 315, 324, 331** Dewayne Flowers/Shutterstock; **p. 301** Wavebreakmedia Ltd/Shutterstock; **p. 312** Elena Elisseeva/iStockphoto; **p. 314** Tatniz/Shutterstock; **p. 320** Uyen Le/iStockphoto; **p. 322** Dmitry Naumov/Shutterstock; **p. 329** Dan Thornberg/iStockphoto; **p. 330** Berislav Kovacevic/Shutterstock; **p. 335** DIGIcal/iStockphoto

Chapter 8
p. 349 Fotocrisis/Shutterstock; **pp. 350, 359, 371, 384** Niki Crucillo/Shutterstock; **p. 357** Lisa F. Young/Shutterstock; **p. 358** Russell Gough/iStockphoto; **p. 370** Wavebreakmedia Ltd/Shutterstock; **p. 375** Robert Byron/iStockphoto; **p. 382** iStockphoto; **p. 397** Andrew Johnson/iStockphoto

Chapter 9
p. 409 Robyn Mackenzie/Shutterstock; **pp. 410, 421, 443** Andresr/Shutterstock; **p. 411** Oliver Hoffmann/Shutterstock; **p. 417** Andrzej Tokarski/iStockphoto; **p. 433** iStockphoto; **p. 447** Chas/Shutterstock; **p. 452** Russell Gough/iStockphoto

Chapter 10
p. 474 Bobby Deal/RealDealPhoto/Shutterstock; **pp. 475, 491, 501, 530** Lee Pettet/iStockphoto; **p. 478** B. Hathaway/Shutterstock; **p. 482** Monticello/Shutterstock; **p. 512** Leon Forado/Shutterstock

Chapter 11
p. 549 Fotorich01/Shutterstock; **pp. 550, 559, 574, 584, 592** Sarah Angeltun/Shutterstock; **p. 580** Angelo Gilardelli/Shutterstock

Chapter 12

p. 612 LampLighterSDV/Shutterstock; **pp. 613, 635, 676, 700** Sean Locke/iStockphoto; **p. 617** David Stockman/iStockphoto; **p. 640** Ilya Andriyanov/Shutterstock; **p. 646** LittleMiss/Shutterstock; **p. 650** Didon/Shutterstock; **p. 658** B. Hathaway/Shutterstock; **p. 665** Dmitry Kalinovsky/Shutterstock; **p. 673** Tommounsey/iStockphoto; **p. 683** Kutay Tanir/iStockphoto

Chapter 13

p. 725 Dmitry Yashkin/Shutterstock; **pp. 726, 732, 744** Lisegagne/iStockphoto; **p. 727** Walik/iStockphoto

Chapter 14 (available on CD)

p. 14-1, Nito/Shutterstock; **pp. 14-2, 14-7, 14-14, 14-30, 14-44** Luchschen/Shutterstock